W9-CRB-737

A
GUIDE
TO THE
POLYAMINES

A
GUIDE
TO THE
POLYAMINES

QP
801
P683
C64
1998
CHEM

SEYMOUR S. COHEN

New York   Oxford
OXFORD UNIVERSITY PRESS
1998

Oxford University Press

Oxford   New York

Athens   Auckland   Bangkok   Bogota   Bombay   Buenos Aires
Calcutta   Cape Town   Dar es Salaam   Delhi   Florence   Hong Kong
Istanbul   Karachi   Kuala Lumpur   Madras   Madrid   Melbourne
Mexico City   Nairobi   Paris   Singapore   Taipei   Tokyo   Toronto   Warsaw

and associated companies in
Berlin   Ibadan

Copyright © 1998 by Oxford University Press, Inc.

Published by Oxford University Press, Inc.
198 Madison Avenue, New York, New York 10016

Oxford is a registered trademark of Oxford University Press

All rights reserved. No part of this publication may be reproduced,
stored in a retrieval system, or transmitted, in any form or by any means,
electronic, mechanical, photocopying, recording, or otherwise,
without the prior permission of Oxford University Press.

Library of Congress Cataloging in Publication Data
Cohen, Seymour S. (Seymour Stanley), 1917–
A guide to the polyamines / by Seymour S. Cohen.
p.   cm.
Includes bibliographical references and index.
ISBN 0-19-511064-1
1. Polyamines—Physiological effect.   2. Polyamines—Metabolism.
3. Polyamines—Synthesis.   I. Title.
QP801.P683C64   1998
574.19′24—dc20   96-27561

1 3 5 7 9 8 6 4 2

Printed in the United States of America
on acid free paper

Preface

The subject of the polyamines has grown so large and its study has become so active that it appeared useful to prepare a text on this subject for students and developing research workers. Also, the literature of biomedical science has become so vast that the tasks of maintaining a familiarity even with one's own chosen interest are very difficult and time consuming. For this reason research workers in any biological field who have reason to believe that cations, particularly the organic cations, may warrant their attention, may find it useful to browse in this volume. Many large and serious texts of biochemistry and molecular biology have completely omitted discussion of the polyamines. It is hoped that this text will make it clear that these disciplines cannot afford to neglect or to minimize an area of knowledge relevant to a wide array of biological and medical problems.

In the preparation of my lectures, *Introduction to the Polyamines*, given in 1970 at the Collège de France, examination of the literature had focused initially on the papers and references derived from a few well-established investigators: E. Herbst, A. D. Hershey, H. Mahler, C. W. Tabor, H. Tabor, and H. G. Williams–Ashman. These references then led to the work in Finnish, Italian, and Israeli laboratories and other American centers. These examinations of the literature could not be considered as other than fragmentary. In an era in which nucleotides and nucleates were among the most fashionable metabolites and products, the organic cations had not achieved widespread interest, prominence, or concern.

The next 25 years from 1970 to 1995 mark the period of the growth of an invisible college of polyamine workers. International meetings and symposia enlarged personal contact and familiarity with a much broader literature. Research on the polyamines is currently very active in Japan, Europe, and the Americas. The weekly compilation of new titles, for example, *Current Contents* (*Life Sciences*), contains over 5000 new papers per week. A search of the subject index for titles containing the key words *decarboxylase, ornithine, polyamine, putrescine, spermidine, spermine,* and *transglutaminase*

picks up about 10–15 papers per week, or a polyamine-oriented production of over 500 papers per year. In addition, because biomedical workers may not think polyamines are a central theme of their papers, "key" words relating to these cations may not appear in their titles, nor would many polyamine derivatives such as nor-, homo-, -amide, etc. be picked up. Moreover, papers on Z-DNA or recA proteins, or ethylene or alkaloid production in plants, among many other topics, may involve the use of polyamines or their precursors or products as natural products or as laboratory reagents. Therefore, it became necessary to scan the title pages of the entire issue for relevant papers. Another 10–15 papers per week that contain data relating to these cations are found in this way. Increasingly titles in clinical journals contain the key words, because the clinicians have attempted to relate this "new" parameter to their clinical interest. In the basic sciences, words such as spermidine have almost achieved the status of a cation such as Mg^{2+}, which is included in a reaction on the basis of past reports of empirically determined utility. As a result of this recent improvement in the status of the polyamines, the inclusion of the words putrescine, spermidine, and/or spermine in the titles of basic science papers almost disappeared in the late 1980s, as compared to their more frequent appearance in titles in the 1970s.

Inevitably many papers are not located by this method of search. Of the titles of interest that are found, two-thirds to three-quarters are obtained in the splendid Library of the Marine Biological Laboratory in Woods Hole, Massachusetts. The remainder are sought as reprints. In this way more than 10,000 papers that contain data on the polyamines have been accumulated. To keep the present work at a reasonable size, almost all papers that appeared before 1970 have been omitted from the list of references, although the authors and year of publication of many significant discoveries have been noted in the text.

In addition, two growing fields of polyamine research have not been discussed in a systematic fashion. These relate to the roles of the polyamines in nerve function and in immunology. Other areas that have not been discussed include initial approaches to the relations of the polyamines to the structure and function of individual animal tissues. In any case, this book is based on the accumulation and examination of many thousands of papers on the polyamines, papers sought and obtained by the methods described above. The problems of understanding the almost innumerable communications, and of analyzing the data they contain in an accurate, clear, and concise form within a well-organized whole, are inevitably more serious than that of mere accumulation.

Among those whom I wish to thank for help essential to this undertaking are the many younger colleagues who shared in our own laboratory research on the polyamines. Among those are the postdoctoral fellows who participated in the explorations, discoveries, and interpretations presented in our research papers. The first of these was Aarne Raina of Finland, who brought a refreshing vigor and rigor into the work. Of major importance also are colleagues in whose welcoming laboratories I was able to learn aspects of the biological and biochemical systems with which my own laboratory eventually became concerned. These were Jan Kaper of the U.S. Department of Agriculture in Beltsville, Maryland, Roger Stanier of the Pasteur Institute in Paris, Moshe Shilo of the Hebrew University in Jerusalem, and Masamichi Tsuboi of the University of Tokyo. On a broader scale, I am indebted to the many members and friends of the "invisible college" of creative, contributing, and critical polyamine workers with whom I share one or more weeks each year in our now numerous national and international meetings. Indeed, their enthusiasm, interest, and productivity, as well as those features of their colleagues and students, have convinced me that a text can have an interested readership and a possible use. This text constitutes my own instrument of thanks to this specialized, growing, and cheerful community.

In the years of preparation of this volume, I am particularly grateful for the financial support of the American Cancer Society, which has encouraged the development of the many facets of my efforts since I was appointed as an American Cancer Society Research Professor in 1957. The Marine Biological Laboratory, in which I am a reader in its Library, has been my host in the collection of the papers and writing of the book. I am happy to acknowledge the friendly and reliable secretarial assistance initially of Sandy Hansen and subsequently the interested and knowledgeable editorial help of Dr. Virginia Eckenrode. Barbara F. Atwood has greatly assisted in proofreading this large volume. Finally, I can scarcely begin to indicate the extent of my gratitude to my recently deceased wife, Elaine, who helped me to grow in science and to undertake the many efforts which this growth seemed to warrant. In the end her insistence that I complete this book was determining.

Woods Hole, Massachusetts   S. S. C.
August 11, 1997

Contents

Chapter 9
Bacterial Paths to Spermidine
and Other Polyamines   142

Chapter 10
Microbial Eucaryotic Systems:
Fungi and Slime Molds   165

Chapter 11
Polyamines in the Animal Cell   184

A
GUIDE
TO THE
POLYAMINES

CHAPTER 1

Historical Introduction

> ... *Among the various classes of organic substances, there is perhaps*
> *none of which, from an early period, chemists have so constantly*
> *endeavoured to attain a general conception as the group of compounds*
> *which have received the name of organic bases, all—and they are now*
> *very numerous—being capable of combining, like the metallic oxides, with*
> *acids, and being derived either from vital processes in animals or plants,*
> *or from a variety of artificial reactions conducted in the laboratory.*
>
> A. W. HOFMANN (1850)

This essay by Hofmann, which contained the excerpt quoted above, presents the first clear description of the nature of primary, secondary, and tertiary amines in which hydrogen atoms of ammonia were replaced by organic substituents. Hofmann synthesized such "ammonia-type" derivatives of aniline, a compound he identified in 1843 in tar oil. His subsequent essay in 1850 also described the synthesis of ethylamine, diethylamine, and triethylamine, and pointed to the synthesis of a comparable series of derivatives of arsenic and phosphorus (Hofmann, 1850). In 1851 Hofmann went on to synthesize "nonvolatile" quaternary nitrogen derivatives (e.g., tetraethylammonium iodide) and subsequently the "polyammonias." Hofmann's extensive chemical work was discussed in a Hofmann Memorial Lecture (Armstrong, 1896).

The development of our knowledge of the polyamines is inextricably tied to the history of our understanding of the nature and roles of nitrogen and its compounds. In addition, this saga relates to the early development of organic chemistry in the past two centuries. In short, a discussion of the prehistory of the polyamines intersects with and enlarges upon the story of the development of chemistry and biochemistry in general.

Early History of Chemistry

The early history of chemistry may be subdivided into three major periods: an ancient period preceding the

evolution of alchemy, a middle period in which alchemy began to organize chemical thought, and a recent period in which the chemical experiments of scientists such as J. van Helmont, R. Boyle, and J. Mayow revealed possible generalizable explanations. Van Helmont, for example, attempted to examine physiological processes as chemical phenomena and described digestion as a form of fermentation.

In these early periods the accumulation of knowledge concerning nitrogen-containing substances related to such important practical matters as the disposal of animal excrement and dead bodies, including those of humans, the improvement of agriculture, and the evolution of warfare, in part as a result of the acquisition of metals and eventually of gunpowder. Medical aspects of the human condition obviously participated in this gathering of information, and the satisfaction of intellectual interest and curiosity became a significant human incentive, particularly in the third period. The interplay of practice, partial knowledge, conjecture or theory, and eventually experiment are exemplified in the Time Line (Appendix to this chapter).

Discovery of Spermine Phosphate

In the period of the Scientific Revolution, Antonie van Leeuwenhoek explored the previously invisible world with the aid of a microscope. He examined various fluids, including human semen. In a famous letter to the Royal Society of London in 1678, he reported the existence of motile sperm in semen. He also observed the slow crystallization of a mysterious substance that was later described as the salt, spermine phosphate. In 1791 Louis Nicolas Vauquelin recorded the rediscovery of these crystals. Young Vauquelin's efforts at characterization of the crystals were inadequate; reproducible and satisfying descriptions of the isolation, composition, structure, and synthesis of spermine were not achieved for another 135 years. Dissection of the nature of the first discovered natural polyamine, spermine, thus had to wait almost the entire period of evolution of early modern "classical" and organic chemistry.

The chemistry of these early periods is described in books such as those of Multhauf (1966), Crosland (1962), Ihde (1964), and Partington (1965), which are interesting and useful guides for developing chemists. However, it is evident from the Time Line that compounds of nitrogen were noted very early in the course of human activities. The odors of amines and ammonia were detected in ordure and in the putrefaction of urine and other biological materials. Certain substances, such as nitrites and nitrates, as well as alkaline salts, minimized putrefaction and were useful in the preservation of corpses. According to Multhauf (1966), the Greeks and Romans referred to salts now known to contain nitrogen, such as *sal ammoniac* or NH_4Cl, which volatilized and deposited after the boiling of putrefied urine, and to saltpeter or KNO_3, which crystallized at the surface of putrefying manure piles. The former was useful in cleaning metallic surfaces; in later centuries the latter, which is nonhygroscopic in contrast to $NaNO_3$, was found to be essential in the preparation of gunpowder. Of course the requirement for biological processes or residues in the preparation of KNO_3 relates to the concentration of K^+ in cells and the relative exclusion of Na^+. The nitrates permitted the preparation of nitric acid, which facilitated the distinction between gold and other metals, and eventually nitrates were exploited at many stages in the evolution of chemical knowledge.

As described by Crosland (1962), these poorly characterized substances were given many different names in the many lands in which the substances were used. In any case, in the eighth or ninth century A.D., Arabs are believed to have distilled the subliming salt, *sal ammoniac*, from hair. Originally called a "spirit," the terms salt, sal, or sel began to be adopted for this substance in the 12th century. The presence of saltpeter in gunpowder appeared in the writings of Roger Bacon in the 13th century and its preparation in "saltpeter plantations" in India stemmed from the 15th century. Its import as a major product from India lasted well into the 19th century. Indeed the problem of obtaining adequate supplies of KNO_3 for gunpowder concerned such responsible national agents as Benjamin Franklin, Antoine Lavoisier, David Rittenhouse, Jean Chaptal, and Humphry Davy. Recipes for the production of nitric, sulfuric, and hydrochloric acids were also written in the 16th century. These acids proved to be major tools in the routes to important stages of chemical understanding as in the work of R. Boyle, J. Mayow, and J. Priestley.

Fractionation of Air

The evolution of early modern chemistry in the mid-and late 18th century significantly relates to the study of the formation, analysis, and clarification of the nature of the gases of atmospheric air (i.e., CO_2, O_2, and N_2) and other "elastic fluids." The addition of an acid to magnesium carbonate by Joseph Black in 1754 led to the discovery of "fixed air" or CO_2 that was subsequently found to be present in the atmosphere and was a product of animal respiration and a substrate for plant growth and photosynthesis. In 1766 Henry Cavendish collected H_2 over water and over mercury, and this gaseous element was eventually proven in the next three decades to be a component of water, acids, and ammonia. Techniques of gas collection were developed further in the 1770s by Joseph Priestley in his studies of gases, which included the nitrogen oxides and NH_3, which was produced from HNO_3 and nitrates. In 1774 Priestley demonstrated the evolution of oxygen ("dephlogisticated air") by heating HgO produced from $Hg(NO_3)_2$. He observed that a mixture of water-insoluble O_2 and NO permitted the formation of water-soluble NO_2 and on this basis developed an analysis (now termed eudiometry) of O_2 in atmospheric air and in respired air. The residual gas or nitrogen of atmospheric air was described by Daniel Rutherford in 1772 (Dobbin, 1935; Weeks, 1934). Priestley called nitrogen "phlogisticated air" and C. W. Scheele, who also isolated oxygen and nitrogen, called the latter "spent air."

The reduction of the odors of sewage and the putrefaction of the dead were major problems of the 18th and early 19th century (Corbin, 1986). In addition to the discussions of the efficacy of perfumes, aromatics, and fumigants, an experiment in 1773 by L. B. Guyton de Morveau, a chemist of Dijon, is particularly interesting. The stench of bodies in the church vaults of Dijon compelled the most drastic steps, and his mixture of salt and concentrated H_2SO_4 generated HCl, which caused the very unpleasant odor to completely disappear overnight. His method was not widely adopted. In retrospect, the result appears to suggest the possible conversion of a volatile amine to a nonvolatile amine hydrochloride.

The rediscovery of oxygen and nitrogen ("azote") in 1775 enabled Lavoisier to readdress the theories of "phlogiston," which attempted to explain the mechanisms of calcination and combustion. Rejecting the notion of the loss of phlogiston from metals during calcination in air, he proposed the addition of oxygen to metals and to phosphorus, sulfur, and carbon to form oxides. Concerned also with the problem of the origin of animal heat, Lavoisier then studied the combustions of the gases during animal respiration and detected the consumption of oxygen and the formation of carbon dioxide, without utilization of nitrogen. His analyses of fermentative products, such as glucose and ethanol, introduced oxidative reactions for the estimation of the carbon, hydrogen, and oxygen they contain. When oxygen and hydrogen were sparked electrically to produce water, traces of residual nitrogen in these partially purified gases formed nitrogen oxides and soluble acid, confusing early chemists such as Priestley as to the composition of water. This disconcerting experimental fact temporarily impeded the transition from phlogistic chemistry to modern chemistry.

Early Modern Chemistry

This period, described as the Chemical Revolution, is often discussed primarily in terms of the major achievements in "pneumatic chemistry" of C. W. Scheele, H. Cavendish, J. Priestley, and A. Lavoisier. It was also the product of mineralogical exploration and study in the laboratories of T. Bergman and his collaborators, who assisted Scheele. In 1782 Scheele synthesized the first organic compound of nitrogen, a cyanide, from ammonia, carbon, and alkali; and over many years he isolated and characterized many substances of biological origin.

In the decades from the 1770s to the 1820s, the path of studies on nitrogen, reduced nitrogen (i.e., ammonia), and the various forms of oxidized nitrogen evoked discussions and innovations of nomenclature. In 1787 the joint efforts of the Swedish scientist Bergman and the French scientist Guyton de Morveau in improving nomenclature were brought to fruition by the Parisian group led by Lavoisier. Nitrogen was designated azote in the new nomenclature; despite the protests of J. Chaptal who preferred "nitrogène" as a term more consistent with the nomenclature for the nitrogen oxides or nitric acid, azote found its way into the French literature and is retained even more widely in "azo," "diazo," etc. Oxygen, thought by Lavoisier to designate the acid-forming component of acids, was found to be absent from hydrocyanic and hydrochloric acids.

The existence and availability of the new gases led to measurement of their properties at various temperatures and pressures, and their ordered behavior permitted the evolution of gas laws and concepts of atomic and molecular theory. As noted in the Time Line, the extraordinary achievements of John Dalton, Claude Louis Berthollet, and Joseph Gay–Lussac in the development of theories of atomic composition and molecular structure emerged in part as a result of their studies with the nitrogen oxides and ammonia. Ethylene, now known as an important plant hormone that is synthesized from a major polyamine precursor, was discovered in 1794 and within a decade became a crucial target of Dalton's scrutiny and theories. Its reactions with halogens were observed and its role in the evolution of aliphatic chemistry was considerable.

Origins of Biochemistry

Analyses were developed for combined nitrogen in animal and plant products and led to the discovery and synthesis of organic nitrogen compounds and the detection of isomeric structures containing nitrogen, such as cyanide and the isomeric fulminates. The historic conversion of ammonium cyanate to urea by F. Wohler has been frequently described as a breakthrough in the thinking of the time, establishing biochemical substances and events as appropriate subjects of chemical study. At the turn of the 19th century, the Swedish scientist J. Berzelius began an extensive series of analyses of animal fluids as an approach to improving the diagnostic capabilities of physicians. The French team of A. Fourcroy and his younger collaborator L. N. Vauquelin discovered the high content of combined nitrogen in animal muscle. Exploring its possible source in plants, they described nitrogen-containing protein material and amino acids in plants and isolated characterizable nitrogenous products in animal urine and feces. In 1800 these workers described the kidney as the major organ of nitrogen excretion, a subject elaborated by W. Prout (Brock, 1985), who was among the first to propose nitrogen excretion in urine analysis as a measure of kidney health. The French workers, as well as Prout, understood the importance of isolating pure urea and Prout analyzed crystalline urea with sufficient precision to bolster Dalton's law of atomic proportions. Earlier, in 1815, Prout offered the hypothesis that all the then known elements might be multiples of the weight of hydrogen, the "first matter." In 1824, this chemist–physician discovered the existence of HCl as the predominant acid of gastric juice.

Interest in the improvement of farm yield, as seen in the early publications of Jethro Tull and Arthur Young and the analyses of F. Home and J. Wallerius, led to the studies of H. Davy and J. von Liebig on the applications of chemistry to this important subject. In 1771 Priestley observed the evolution of O_2 by plants and postulated a cycle in which CO_2 enabled the growth of plants (e.g., algae) with the concurrent restitution of O_2 to the atmosphere. In 1779 Jan IngenHousz, a Dutch physician, detected the requirement of light for CO_2 fixation and O_2 evolution and demonstrated the evolution of CO_2 in the dark. Men like Fourcroy, Vauquelin, Wollaston, and Prout examined the nature of the components of organic manures and their relation to the stimulation of plant growth. Whereas excessive ammonia might be toxic, more complex components, such as the uric acid of guano, can better supply essential nitrogen for plant components. The presence of nitrogenous plant components, for example, asparagine, gluten, etc., obviously related to the nitrogenous constituents of herbivores and to animal muscle in general, as well as to their elimination of nitrogen in urine and feces. Fourcroy, Vauquelin, Prout, and others considered the roles of animal organs, such as the kidney, in the elimination of urea; and the work of J. Prevost and J. B. A. Dumas in 1821 revealed that urea must be produced in a site other than the kidney. Chemists such as F. Serturner in 1817, P.-J. Pelletier and J.-P. Caventou in 1818, and J. B. A. Dumas demonstrated the existence of complex "alkaline" nitrogen-containing plant alkaloids such as morphine and strychnine. Physiologists such as F. Magendie revealed the requirement for nitrogenous components in animal diets. In the immediate aftermath of the Napoleonic Wars, the disposition and possible use of gelatin derived from the hooves of dead horses was a significant problem. At this time the preparation of nutritious soups was also a concern of Benjamin Thompson (i.e., Count Rumford), who had taken on the task of organizing the feeding of the poor in Munich.

The existence of nitrogen in animal muscle, even that of herbivores, posed the multiple problems of the nature of nitrogen in plants, the sources of nitrogen for plants, the manner in which a plant substance became animal protein, and the role of excreted urea in the overall management of ingested nitrogen. The nitrogen of asparagine or of the soon to be coined "proteine" stemmed from simpler nitrate, ammonia, or even urea. Could the plant obtain this simpler form of nitrogen from the atmosphere as it did CO_2 or O_2? J. Boussingault showed that, with the exception of legumes, most plants could not use atmospheric nitrogen and instead obtained their nitrogen from nitrate or ammonia in the soil. Plants were clearly able to synthesize complex nitrogen-containing materials. Were these plant proteins simply assimilated by the animal to form comparable structures in the animal? Introducing the study of nitrogen balance, Boussingault showed that the intake of plant nitrogen was approximately equal to the excretion of nitrogen in the cow and other organisms. The demonstrations that animals were capable of degrading plant protein to amino acids and of synthesizing new kinds of proteins required an additional century of biochemical investigation.

Studies on Amines

The chemical work on urea led to an almost concomitant detection of nitrogen in more complex substances, such as uric acid, allantoin, asparagine, alkaloids, and the nature of the structural roles of nitrogen in these substances were appropriate subjects of structural and synthetic chemical investigation. Simple aliphatic amines were synthesized in the late 1840s and the 1850s, and some derivatives of these, such as triethylamine, were discovered as natural products.

The synthesis of methylamine and ethylamine by C. Wurtz in 1849 was of considerable theoretical interest, lending support to Liebig's view (1837) that volatile nitrogenous bases were substituted ammonia compounds. A. W. von Hofmann, a young assistant of Liebig, developed a synthesis of aniline by 1845 and in 1850 demonstrated the production of mono-, di-, and trialkyl amines by the reaction of ammonia with alkyl iodides. By 1851 he had prepared crystalline quaternary salts, thereby posing the question of the valence of nitrogen, a problem clarified only many decades later. In the late 1850s Hofmann undertook the preparation of "polyammonias" and his description of large-scale syntheses of compounds, such as ethylenediamine, appeared in the 1870s. By the end of the 19th century the metallic am-

mines, such as the various cobalt complexes with ethylenediamine, became of great interest, as in A. Werner's clarification of the tetrahedral nature of the nitrogen atom and its valence. In the early 1850s Hofmann's work on substituted derivatives of nitrogen, phosphorus, arsenic, and antimony enabled D. Mendeleyev and L. Meyer to relate these elements closely in their periodic tables. Simultaneously Hofmann's continuing work on the reactions of aniline with nitrous acid contributed to the discovery of diazo compounds. One part of this line of work led to the discovery of phenylhydrazine by Emil Fischer and provided an important reagent for subsequent analyses of monosaccharide structure. Hofmann, who resided in England from 1845 to 1865 and who directed the Royal College of Chemistry in London, trained many chemists who contributed in major ways to the evolution of industrial chemistry, particularly to the development of the dyes in which the exploitation of reactions of nitrogen (e.g., azo compounds) were so important.

From Amines to Chemical Industry

The significant work on the organic chemistry of nitrogen compounds in the last half of the 19th century not only revealed the simplest structures of the amines and polyamines but also had enormous consequences in the development of many branches of chemistry as well as an extensive impact on the industry and structure of nations. The large-scale requirements for nitrogen compounds in agriculture and in warfare were met by the successful studies of F. Haber and C. Bosch on the high pressure catalytic synthesis of ammonia from atmospheric nitrogen and hydrogen. The ammonia was used in the preparation of fertilizers or for the synthesis of explosives after oxidation to nitric acid. In its turn, ethylene, the curious gas discovered in 1794 that is now known as a plant hormone derived from methionine, became not only a model for the development of molecular theory in the hands of Dalton and Gay–Lussac but was also a substrate in the synthesis of alkyl halides, pointing to the evolution of the chemistry of alkyl substances. In the hands of P. Sabatier, addition of hydrogen to ethylene to form ethane led to the discovery of relatively inexpensive metal catalysts for industrial use. The study of such catalysts by K. Ziegler and G. Natta culminated in the polymerization of ethylene and other olefines to form the polymers, which are quantitatively major organic products of chemical industry.

Professionalization of Biochemical Research

In the late 19th century, the increasing competence and discoveries of the chemists began to be increasingly applied to the questions of the nature of proteins and other natural products. These had been posed by the agricultural community, which was concerned with animal nutrition and growth; the physiologists such as F. Magendie and C. Bernard, who had begun to study the interrelations of anatomical structure, organ function, and animal metabolism; and the science-minded physicians who noted pathology and its chemical consequences. The theories and tasks of medicinal chemistry had evolved from the iatrochemical speculations of van Helmont to a somewhat more experimental physiological chemistry of Fourcroy and Vauquelin and the French school of pharmaceutical chemists. Prout's contributions led to the professional biochemists whose colleges, societies, and journals emerged after 1870. Subsequent decades produced the knowledge of the structure and origin of the natural polyamines (Mann, 1954). The crystals of spermine phosphate had been seen by observant chemically minded physicians such as a young J. Charcot in 1853 and A. Boettcher in 1865. The nature of the substance as the phosphate salt of an organic base was described by Schreiner (1878) and the origin of seminal spermine in the prostate was clarified a few years later by P. Fuerbringer in 1881.

Microbiology and Diamines

The late 19th century also saw the development of microbiology and its insights into the nature and control of infectious disease. The applications of the rapidly growing discipline to the fermentative production of beer and wine were soon seen to be relevant to the nature of putrefactions of all kinds, as seen in the odoriferous decays of the animal tissues with which the pathologists were working, the deterioration of urine samples, and the putrefaction of plant and animal extracts. In the late 1880s L. Brieger began his isolation of ptomaines, the compounds associated with corpses, from putrefying animal organs and soon found it possible to isolate basic compounds, such as the diamines and other amines and their salts (i.e., hydrochlorides, picrates, and platinum and gold chloride double salts). Brieger went on to isolate the diamines, including putrescine and cadaverine, from cultures of bacterial pathogens, such as streptococci and the bacilli of tetanus and cholera, and compared these with synthetic samples obtained from A. Ladenburg (Guggenheim, 1951).

The isolation of the bases involved the use of improved precipitants such as phosphotungstic acid; and the methodology of selective precipitation was extended to the isolation of the basic amino acids arginine, lysine, and ornithine. With these amino acids in hand, their chemical and bacterial degradation to the respective decarboxylated amines was soon observed, a result that contributed to the clarification of the structures of the amino acids and the metabolic origins of the amines.

Knowledge of the existence of spermine and some other organic bases as mammalian products and of the diamines as bacterial products initially evoked some confusion. Von Udránszky and Baumann (1889) isolated diamines, with putrescine greatly in excess of cadaverine, from the urine of cystinurics, and they wondered if this meant that the compounds had been formed by intestinal bacteria. Actually the decarboxylation of ornithine to putrescine by a liver extract is mentioned

casually by Dakin (1906) in the first volume of the *Journal of Biological Chemistry*.

Essentiality of Amino Acids and Amines

The problem of the need for specific amino acids was explored early in the 20th century. In the period preceding World War I, some plant proteins were found to be deficient in certain amino acids and animal requirements were demonstrated for lysine and tryptophan. The determination of requirements for arginine and the sulfur-containing amino acids became a more complicated story, and the synthesis of arginine in mammals, such as the rat, was established eventually (Vickery & Schmidt, 1931). The discovery and data on the essentiality of methionine depended on the evolution of microbiology. In its turn demonstrations of nutritional requirements for specific microorganisms resulted in the use of these organisms for the microestimation of a particular essential component, such as an amino acid or various B vitamins. It was not until 1948 that E. J. Herbst and E. E. Snell described the requirement of exogenous putrescine for the growth of the bacterium *Hemophilus parainfluenzae*.

Clarification of the structures of the amino acids and other natural products had become an important activity among organic chemists and physiological chemists in the period before World War I and indeed was essential to further progress prior to the dissection of the structure of proteins, nucleic acids, and other polymers. Knowledge of the structures was also crucial in the development of hypotheses concerning the biosynthetic origin of the compounds. Fairly clear ideas of the biosynthetic origins of the diamines had been evoked in studies of the crude putrefactive systems examined by Ellinger (1900) and other workers. Nevertheless the uncovering of important metabolic details, such as the existence of an ornithine cycle in mammalian liver by H. Krebs and K. Henseleit in 1932, required knowledge of the occurrence (1914) and structure (1930) of citrulline. The discovery of specific amino acid decarboxylases was a relatively late event that was dependent on the enzymology of appropriate bacterial systems in the early 1940s. The existence of enzymes (i.e., amine oxidases) that altered the polyamines was first detected by E. Zeller in the late 1930s.

The growth of microbiology as a discipline had produced pure culture techniques in which the problems of the nature of essential nutrients inevitably arose. As the growth medium for a given organism was simplified, it became clear that an original complex medium contained unknown growth factors. In 1921–1922 the pursuit of one of these systems led to J. Mueller's belated discovery of methionine as essential for the growth of hemolytic streptococci and of some pneumococci. Indeed the sulfur-containing amino acid was an important constituent of many proteins, although frequently in a fairly small amount and absent from some. The structure and synthesis of methionine as a methylthio derivative of α-aminobutyric acid were developed by G. Barger and F. Coyne in 1928. The clarification of the mechanism of methyl donation by methionine required a long period of laboratory experience with nucleosides and nucleotides before the discovery of S-adenosyl-methionine by G. Cantoni in 1953. As we now know, methionine and its adenosyl derivative are precursors in the biosynthesis of triamines and tetramines.

Structure of Spermine and Discovery of Spermidine

The need for a moiety derived from methionine stemmed from the clarification of the structure of spermine; steps in this chemical dissection went back to World War I and the work of Mary Christine Rosenheim (1917). Although in 1878 Schreiner described spermine as a substance present in many mammalian tissues, it was widely believed that the compound was uniquely present in human semen. In 1917 Mary Rosenheim briefly recorded her isolation of spermine from many tissues, including the roe of cod and yeast, of which the latter is described as the most convenient source for its preparation. Her husband, Otto Rosenheim, subsequently described the necessary improvement of Schreiner's method, eliminating the contamination of seminal protein. Spermine was purified by several methods appropriate for low molecular amines, including dialysis or separation of an alkaline solution by butanol or steam distillation (Rosenheim, 1924). Such purified materials that were reactive as amines were easily converted to insoluble crystalline salts such as the phosphate or picrate.

H. Dudley and the Rosenheims (1924) also reported the distribution of spermine in many tissues, noting the fact of the "extraordinarily insoluble phosphate," and commenting also, presumably thinking of Leeuwenhoek and Vauquelin, that "this salt is of the greatest historical importance." Tissues were extracted with hot acetic acid; after treatment with lead acetate and removal of the lead, the extract was precipitated with phosphotungstic acid. Acetone extraction of the phosphotungstates left insoluble spermine salts, which were extracted and precipitated as the crystalline phosphates. The existence in yeast of a novel unidentified soluble base, as well as the presence of putrescine, were also recorded. Modifications of the method of isolation and of various derivatives were presented in detail, as well as properties of the salts and of the free base and several organic derivatives, including a tetrabenzoylspermine. A footnote in the essay indicated that the work had begun in 1910.

In a subsequent study in 1925, spermine was converted by exhaustive methylation to a decamethyl derivative. The compound was a quaternary ammonium derivative, indicating the existence in spermine of two primary amines and two secondary amines. At this point several early reports of novel amines, designated as "neuridine" by Brieger, and "musculamine" and "gerontine" were clarified by Dudley and O. Rosenheim, who identified these as spermine.

Dudley, Rosenheim, and Starling (1926) described the destructive distillation of spermine derivatives, with results suggesting two possible structures of spermine. Spermine hydrochloride yielded pyrrolidine, indicating the presence of 1,4-diaminobutane, while decamethyl-spermine sulfide formed tetramethyl trimethylenediamine, suggesting the addition of aminopropyl moieties on the diaminobutane. They synthesized the possible tetramines, and one structure, in which both amino groups of 1,4-diaminobutane are substituted by aminopropyl moieties, proved to be identical to spermine. A mild oxidation product of spermine was a volatile base with the characteristic odor of semen and was thought to be an N-γ-aminopropylpyrroline. As a result of the work of the Tabors some 30 years later, we now know that the aminopropyl moieties are derived from decarboxylated S-adenosylmethionine.

Dudley and colleagues (1927) described the isolation of spermidine phosphate from ox pancreas and derived a possible structure from data of the molecular weight and elementary analysis of the nitrobenzoyl derivative. The monoaminopropyl derivative of 1,4-diaminobutane was synthesized and its derivatives were demonstrated to be identical with comparable salts and derivatives of isolated spermidine. The name spermidine was coined for this obvious relative of spermine. Spermine and spermidine were the first examples of derivatives of trimethylenediamine among natural products. Their classic work noted that the coexistence of both bases in animal tissues "strongly suggests that they may be metabolically related."

In studies seeking initially to clarify earlier reports of the existence of spermine among bacterial polyamines, F. Wrede and E. Banik (1923) concluded that the identity of the polyamine present in and isolable from *Vibrio cholerae* was in fact the diamine, cadaverine. Wrede and his collaborators then independently isolated and determined the structure of seminal spermine (Wrede, Fauselow, & Strack, 1926, 1927). Crucial to their structural analysis was the isolation of tetra-m-nitrobenzoyl spermine, the determination of a molecular weight of this compound, and subsequent isolation and characterization of degradation products of spermine phosphate (Wrede, 1924). These products included aminopropyl-pyrrolidine and diaminopropane, and Wrede and his collaborators (Wrede et al., 1926, 1927) went on to a synthesis of spermine via a method different from that described by Dudley et al. (1926).

Spermine Content of Human Tissues

Harrison (1931) undertook the determination of the content of spermine in human tissues. Because Dudley and Rosenheim failed to detect the base in bull semen, the compound was assumed to not be essential to fertilization in this mammal. Harrison extracted alcohol-soaked tissues with acetic acid and eventually steam distilled the base from an alkaline solution. The base was recovered initially as the hydrochloride, then as the picrate, and finally as the phosphate, with an estimation of recoveries of about 86% in these steps. Harrison obtained results similar to the analyses of human tissues by Wrede (1924), who used a different less stringent method. Harrison (1933) then extended his studies to analyses of single organs. In contrast to modest variations in the content of spermine in the pancreas and testis, he observed very large variations in the spermine content of the prostate and failed to obtain evidence bearing on the functional significance of the tetramine.

In 1935 U. Puranen studied the properties of the insoluble spermine diflavianate, the 2,4-dinitro-α-naphthol-7-sulfonate. Both the study and the compound were made the bases of a method for the quantitation of spermine in human tissues in 1941 by the Finnish researcher R. Hämäläinen, whose extensive work was not published until 1947. The methods of selective precipitation of amino acids as an approach to the study of protein composition was similarly applied at about this time. All of these elegant chemical manipulations were to be replaced soon after World War II by the discoveries of paper and ion exchange chromatography, methods that markedly altered the activities and pace of biochemical study of the polyamines.

APPENDIX

A Time Line:
Events in Chemical and Biochemical
Prehistory of Polyamines

The items presented in this Time Line were gleaned from many volumes, some of which are given in the References. The dates are approximate in that the various authors may differ slightly, the time at which the work was done inevitably precedes the year of publication, sometimes by several years, and occasionally the work was never published by the experimenter but was discovered among notes found after death. It is hoped that the Time Line will elicit corrections, significant additions, and, more importantly, a sense of the many areas of human activity that have contributed to the field as well as to the slow but increasing pace of developing knowledge.

"What is meant by the discovery of a substance? . . . Mayow and Hales, though they worked with nitric oxide, did not point to it as a gas distinct from ordinary air. Priestley is the discoverer of nitric oxide in the sense that he prepared it, worked with it, discovered many of its properties and recognized it as a distinct substance: He gave it, in token, the name of 'nitrous air'." (Meldrum, 1933).

This table was composed in Woods Hole, Massachusetts, an early center for the import of guano, its preparation as fertilizer and its subsequent shipment around the country.

The author chose to bring the Time Line to the end of World War II or the beginning of the postwar explosion of biomedical science. Despite the growth of metabolic knowledge in the period prior to World War II, as exemplified in the ornithine cycle of H. Krebs, little was known of the origin of the polyamines in 1945, as can be seen in the review of Blaschko.

Approximate Date (A.D.)	Scientist	Event
8th–9th century	Geber (Jabir ibn Hayyan)	Distillation from hair of *sal ammoniac* (NH_4Cl), used in cleaning metal surfaces
9th century	Nai-An Chao	Invention of a gunpowder mixture
13th century	R. Bacon	Noted presence of saltpeter in gunpowder
	R. Lully	Described the distillation of nitric acid
15th century		"Saltpeter plantations" are formed in India
1556	G. Agricola	Demonstrated the extraction of wood ashes in the preparation of saltpeter
1563	B. Palissy	Recommended the use of manure or urine to restore exhausted fields
1609	O. Croll	Wrote symbols for sal ammoniac and saltpeter; latter symbol subsequently used for nitrogen by Dalton in 1810
1620	A. Sala	Synthesized sal ammoniac
1630	J. Rey	Reported the increase in weight of tin and lead on calcination
1648	J. van Helmont	Explored chemical approaches to disease, iatrochemistry; coined "gas" for vapors, including those of fermentations and putrefactions; described solubility of AgCl in ammonia solutions; described *aqua regia*

8

Approximate Date (A.D.)	Scientist	Event
1651	J. Glauber	Described the preparation of saltpeter from dung, and of a complete mineral fertilizer containing saltpeter
1664	N. Le Febure	Distinguished different kinds of *niter*
1666	R. Boyle	Reported on sal ammoniac as a preservative
1669	J. Becher	Proposed existence of "a principle of combustibility" in combustible substances
1673–1674	R. Boyle; J. Mayow	Reported niter supported combustion in vacuo; gunpowder burned in the absence of air
1674	J. Mayow	Described displacement of a more volatile HNO_3 from niter by H_2SO_4; collected nitric oxide; detected nitrogen dioxide
1675	R. Boyle	Distilled ammonium carbonate
1677	F. Bacon	Described the use of dung or urine in hastening seed germination
1678	A. Leeuwenhoek	Observed the crystallization of spermine phosphate in human semen
	J. Kunckel	Isolated phosphorus from urine
1689	J. Glauber	Separated sal ammoniac into "volatile salt of urine" (NH_3) and the "spirit of common salt" (HCl); prepared "secret sal ammoniac" (($NH_4)_2SO_4$)
1692	W. Homberg	Distilled plant extracts in acid and alkali
1715	G. Stahl	Designated Becher's "principle of combustibility" as "phlogiston"
1717	N. Lemery	Found niter in plants and postulated a biological nitrogen cycle
1746	G. Rouelle	Replaced "volatile urinous salts" by salts of ammoniac
1749	P. Macquer	Described vapors as "gases," following van Helmont
1756	L. Spallanzani	Detected digestion in gastric juice in vitro
1757	F. Home	Classified "manures" and discussed biological decay in saltpeter production
1759	A. Margraff	Used microscope and flame tests to distinguish "cubic niter" ($NaNO_3$) and "prismatic niter" (KNO_3)
1761	J. Wallerius	Discussed joint roles of salts and manures in promoting plant growth
1764	P. Macquer	Described saltpeter and sal ammoniac as neutral salts composed of an acid and an alkali
1771	P. Macquer	Distinguished putrefaction from other fermentations
1772	J. Priestley	Discovered nitrogen oxides, i.e., nitric oxide (NO), nitrous oxide (N_2O) and nitrogen dioxide (NO_2); collected NH_3 over mercury
	D. Rutherford	Discovered nitrogen, a gas residual in air after animal respiration and removal of carbon dioxide, and which extinguishes "both flame and life"
	H. Cavendish	Prepared nitrogen after removing oxygen and CO_2 from air
	C. W. Scheele	Discovered "fire air" (oxygen); described the residual air as "foul air"
1773	H. Rouelle	Isolated urea from urine
	L. B. Guyton de Morveau	Eliminated putrefactive odors of corpses with gaseous HCl
	A. Baumé	Used "niter" to describe salts formed with nitric acid
1774	J. Priestley	Discovered "dephlogisticated air" (oxygen) by heating HgO derived from $Hg(NO_3)_2$; detected an acid when electric sparks are passed through air; determined oxygen by loss of volume in reaction with NO and absorption of the product (eudiometry)
1776	C. W. Scheele	Found uric acid in urinary calculi among studies of biological products

Approximate Date (A.D.)	Scientist	Event
1777	A. Lavoisier	Observed O_2 utilization and CO_2 production during respiration, without utilization of N_2
1779	J. Priestley	Discovered nitrososulfuric acid, the absorption product of NO_2 in H_2SO_4
1780	L. B. Guyton de Morveau	Titrated contaminating KCl in saltpeter with $Pb(NO_3)_2$
1781	H. Cavendish	Showed many atmospheres to be nearly constant mixtures of "dephlogisticated air" and "phlogisticated air"
	J. Priestley	Prepared nitrogen dioxide by heating $Pb(NO_3)_2$, and NH_3 by the action of iron on dilute HNO_3
1782–1783	C. W. Scheele	Synthesized hydrocyanic acid; described "murexide" test for uric acid
1784	T. Bergman & L. B. Guyton de Morveau	Proposed "ammoniacum" for base in NH_4 salts
1785	C. Berthollet	Analyzed the elemental composition of NH_3
	A. Lavoisier	Stated the principle of the conservation of mass
	H. Cavendish	Demonstrated the formation of nitrate in mixed gases sparked over alkali
1786	A. Fourcroy	Classified 16 gases, including nitrogen, its oxides, and ammonia; found gaseous nitrogen in the swim bladder of carp
1787	A. Fourcroy	Named "uric acid"; found nitrogen, ammonia, carbonic acid, and prussic acid in uric acid calculi
	A. Lavoisier et al.	Introduced a new chemical nomenclature, designating nitrogen as "azote"
	L. B. Guyton du Morveau et al.	Introduced *nitrates* and *nitrites* as salts derived from nitric and nitrous acids, respectively
1788	A. Fourcroy	Discovered the high content of nitrogen in animal muscle, including that of herbivores
	C. Berthollet	Reported the absence of oxygen in hydrocyanic acid
	A. Lavoisier	Oxidized compounds to analyze C as CO_2 and H as H_2O
1789	A. Fourcroy	Reported nitrogen in many plants, a possible source of nitrogenous foods for herbivores
1790	J. Chaptal	Introduced a more consistent "nitrogène" to replace "azote"
	A. Fourcroy	Discovered the relatively insoluble salt, $MgNH_4PO_4$, later found as calculus
1791	L. N. Vauquelin	Rediscovered crystals of spermine phosphate in human semen; reported some of its properties
1793–1794	J. Chaptal	Organized saltpeter production in France
1794	J. Deiman et al.	Synthesized ethylene, known as "olefiant gas"
1799	A. Fourcroy	Named "urea"
	A. Fourcroy & L. N. Vauquelin	Isolated urea as a crystalline nitrate; reported loss of urea and appearance of ammonia in putrefying urine
1800	A. Fourcroy & L. N. Vauquelin	Described the kidney as the major organ of nitrogen excretion
	L. N. Vauquelin	Discovered allantoin
	H. Davy	Heated ammonium nitrate to produce nitrous oxide ("laughing gas")
1802	Du Pont family	Organized gunpowder production in the United States
1803	J. Priestley	Discovered nitrate and nitrite in snow
	J. Berzelius & W. Hisinger	Decomposed ammonia electrolytically
1803–1804	J. Dalton	Applied atomic theory to derive the law of multiple proportions from analyses of ethylene, ammonia, and nitrogen oxides
1805	A. Fourcroy & L. N. Vauquelin	Found nitrogen in gluten of wheat flour; showed a loss of NH_3 on putrefaction

Approximate Date (A.D.)	Scientist	Event
1806	L. N. Vauquelin & P. Robiquet	Discovered the amide, asparagine, in asparagus
	J. Berzelius	Began analyses of animal fluids to guide medical practice
1807	H. Davy	Showed electrolytically decomposed cyanogen moved like chloride in an electric field
1808	A. Fourcroy & L. N. Vauquelin	Purified urea and described its conversion by heat to ammonium carbonate; heated uric acid to produce urea and ammonium carbonate
1808–1809	J. L. Gay–Lussac & L. Thenard	Improved elementary analysis of oxidized organic substances to determine ratio of C as CO_2, oxygen, nitrogen, and water vapor; found the combining volumes of gases with each other to form gases to be whole numbers, e.g., 1 vol nitrogen and 1 vol oxygen form 2 vol nitric oxide
1810	W. Wollaston	Found uric acid in bird droppings and South American guano
1811	A. Avogadro	Stated relations between volumes of gases and numbers of component molecules
	J. Davy	Prepared urea in a reaction between phosgene ($COCl_2$) and NH_3
1812	H. Davy	Calculated the optimal concentrations of components in a gunpowder mix
1813	J. Berzelius	Summarized knowledge of animal chemistry
1815–1816	W. Prout	Proposed that all atomic weights are multiples of that of hydrogen, the "first matter"
1815	W. Prout	Described the uric acid content of snake excrement and murexide as the ammonium salt of a purine base
1815–1816	J. L. Gay–Lussac	Studied cyanogen and its radical; distinguished oxides of nitrogen
1816	F. Magendie	Reported diets deficient in nitrogenous substances to be inadequate for animal survival
1817–1818	W. Prout	Analyzed urea and discussed its composition in support of Dalton's laws; recommended analyses of fasting urine as a measure of kidney health
1817	F. Serturner	Isolated crystalline morphine from opium
1818	L. Thenard	Prepared H_2O_2 beginning with $Ba(NO_3)_2$
1818–1821	P.-J. Pelletier & J.-P. Caventou	Isolated plant alkaloids, e.g., quinine and caffeine
1821	J. Prevost & J. B. A. Dumas	Discovered urea formation in animals whose kidneys had been removed
1823	P.-J. Pelletier & J. B. A. Dumas	Demonstrated nitrogen as a component of many plant alkaloids
	F. Wohler	Determined the composition of cyanic acid
1824	J. L. Gay–Lussac & J. von Liebig	Determined the composition of the fulminates
	W. Prout	Discovered HCl in gastric juice
1825	T. Thomson	Stated atomic weight of nitrogen as 14
1825–1852	J. von Liebig	Routinized combustion analyses; determined N separately
1827	J. Berzelius	Recognized the identical composition of fulminic and cyanic acids, and defined *isomerism*
1828	F. Wohler	Prepared urea from ammonium cyanate, and established a link between organic and biological chemistry
1830	J. B. A. Dumas	Described urea as the amide of carbonic acid; determined nitrogen in organic compounds as a gas (N_2) after combustion
1832	E. Turner	Gave atomic weight of nitrogen as 14.0
	E. Mitscherlich	Prepared nitrobenzene, a substrate in the subsequent reduction to aniline

Approximate Date (A.D.)	Scientist	Event
1831–1845	C. Sprengel	Prepared treatises on soil chemistry for agriculture; classified fertilizers, including manures
1833	J. Berzelius	Proposed compounds containing —NH_2 group should be called *amides*
1834	J. von Liebig	Proposed quantitative balance studies to determine biochemical transformations
1836	T. Schwann	Discovered pepsin
1837–1838	F. Wohler	Decomposed uric acid to allantoin; defined murexide with von Liebig
	G. Mulder	Coined "proteine"
1838	J. Boussingault	Obtained evidence of fixation of atmospheric nitrogen by legumes
	H. Dyar & J. Henning	Patented the ammonia-soda process, producing NH_4Cl and $NaHCO_3$
1839	J. Boussingault	Studied nitrogen balance in a cow, horse, turtle, and bird
1840	J. von Liebig	Discussed the sources of plant nutrients, the kidney as an organ of nitrogen excretion
1842	J. von Liebig	Published *Animal Chemistry*, with speculations on metabolism; stimulated active discussion and work by physiologists and physicians
1843	C. Gerhardt	Defined acids in terms of displaceable hydrogen
1844	A. Laurent	Detected the optical activity of natural organic bases
1845	A. Hofmann	Synthesized aniline
1846	A. Sobrero	Prepared nitroglycerin
1848	W. Prout	Considered urinary urea to be a measure of nitrogen assimilation
1849	C. Wurtz	Synthesized methylamine and ethylamine from alkyl isocyanates
1850	N. Sinin & A. Hofmann	Treated ammonia with ethyl iodide to form mono-, di-, and triethylamines, named primary, secondary, and tertiary amines, respectively, by C. Gerhardt (1856)
1851	T. von Wertheim & A. Hofmann	Isolated and defined triethylamine contained in herring brine
	F. Genth; E. Fremy; F. Claudet	Independently synthesized cobalt–ammines
1851–1863	A. Hofmann	Prepared quaternary amines and "polyammonias," i.e., diamines and triamines
1854	J. Boussingault	Disproved the fixation of atmospheric nitrogen by many non-leguminous plants
1855	J. Boussingault	Demonstrated the use of nitrogen of KNO_3 for growth of sunflowers
	C. Bernard	Proposed concept of homeostasis and regulation of internal environments of animals
1857	E. Pugh	Showed utilization of soil NH_3 for plant growth
1858–1876	J. Boussingault	Demonstrated nitrification in the soil
1862	A. Butlerov	Proposed tetrahedral carbon
1865	A. Hofmann	Popularized atomic models
	E. Erlenmeyer	Defined ethylene as $H_2C{=}CH_2$
1869–1871	D. Mendeleyev; L. Meyer	Developed periodic tables of elements in eight columns, the fifth of which related N to P, As, Sb, and Bi
1872	V. Meyer	Studied aliphatic nitro compounds, e.g., nitroethane and isomeric ethyl nitrite
	J. van't Hoff	Developed concept of tetrahedral carbon to explain optical activity
1877	E. Hoppe–Seyler	Founded *Zeitschrift für Physiologischen Chemie*, which published many early essays on polyamines

Approximate Date (A.D.)	Scientist	Event
	M. Jaffe	Isolated ornithuric acid, i.e., dibenzoylornithine, from chickens fed benzoic acid
1878	P. Schreiner	Identified spermine phosphate as a phosphate salt of an organic base, and as a source of a semenlike odor
	S. Jørgensen	Began to study metal–ammine complexes
1881	P. Fuerbringer	Found prostatic secretion rich in an organic base, later identified as spermine
1882	A. Hofmann	Converted amides to amines
1882–1900	E. Fischer	Studied the structure and synthesis of purines
1883	E. Schulze; J. Kjehldahl	Converted organic nitrogen to NH_3 and determined the latter
1885–1887	L. Brieger	Isolated and named diamines from putrefying animal organs; isolated diamines from cultures of *Vibrio cholerae*
1886	E. Schulze & E. Steiger	Isolated arginine from aqueous extracts of plant seedlings
	A. Ladenburg	Described cadaverine as pentamethylene diamine
1887	E. Drechsel	Introduced phosphotungstic acid as a precipitant for organic bases
	S. Gabriel	Synthesized primary amines via phthalimido derivatives
1888	A. Ladenburg & J. Abel	Suggested "spermin" for crystals isolated from semen
1889	L. von Udránszky & E. Baumann	Identified putrescine as tetramethylene diamine; described isolation of diamines from urine of cystinurics after benzoylation; considered the problem of the metabolic origin of amines
1889–1890	S. Jørgensen	Prepared complexes of cobalt and ethylenediamine
1889–1891	E. Dreschsel	Isolated lysine as a platinum salt from a casein hydrolysate
1893	A. Werner	Studied cobalt–ethylenediamine coordination complexes and formulated the theories of coordination chemistry and of tetrahedral trivalent nitrogen
1894	T. Curtius	Prepared amines from carboxylic acids via the acid azide
1897	E. Schulze & E. Winterstein	Converted arginine to urea and ornithine with alkali
1898	A. von Poehl	Inhibited bacterial growth with spermine
1899–1902	A. Ellinger	Demonstrated a putrefactive conversion of lysine to cadaverine, and of ornithine to putrescine
1901	E. Fischer	Synthesized ornithine, i.e., 2,5-diaminovaleric acid
1903	K. Birkeland & S. Eyde	Converted atmospheric N_2 to NO and N_2O_4, and then to HNO_3 and $Ca(NO_3)_2$, i.e., commercial "Norwegian saltpeter"
	W. Ostwald & E. Brauer	Converted NH_3 catalytically to HNO_3 for explosives and fertilizers
1904	A. Kossel & H. Dakin	Degraded arginine with arginase to ornithine and urea
1905	F. Haber & C. Bosch	Converted N_2 and H_2 to NH_3 catalytically under pressure
	M. Barberio	Developed a forensic test for spermine in human semen
1906	H. Dakin	Noted the decarboxylation of ornithine to putrescine by liver; studied the chemical decarboxylation of amino acids
1907–1909	S. Sørensen	Developed the formol titration of amino acids; defined pH, studied buffers, and enzyme-pH optima
1909	G. Barger & G. Walpole	Demonstrated the production of amines in putrefaction of mammalian organs
1910	D. Ackermann & Fr. Kutscher	Described putrefactive decarboxylation of arginine, lysine, and glutamic acid
	A. Kossel	Isolated agmatine from hydrolysed herring testes
	K. Yoshimura	Isolated putrescine and cadaverine from putrefied soybean extract
1911	A. Werner	Demonstrated the isomerism of cobalt–ethylenediamine complexes

Approximate Date (A.D.)	Scientist	Event
1914	Y. Koda & S. Odake	Isolated citrulline from watermelon juice
	T. Osborne & L. Mendel	Demonstrated the requirement for lysine in animal nutrition
1917	M. Rosenheim	Discovered spermine, a simple organic base, in many mammalian organs and in yeast
1918	J. Drummond	Studied precipitation of amino acids and amines by phosphotungstic acid
1921	J. Mueller	Described the requirement for methionine in bacterial nutrition
	F. Hopkins	Isolated and characterized glutathione
1923	J. Mueller	Reported the isolation and composition of methionine
	J. Brønsted & J. Lowry	Defined an acid as a proton donor and a base as a proton acceptor
1923–1927	F. Wrede et al.	Probed nature of previously described organic bases; described the isolation, composition, structure, and synthesis of spermine
1924	O. Rosenheim et al.	Described the isolation and composition of spermine phosphate from different sources, and the isolation of free spermine and some derivatives
1925		Found spermine to be a tetramine; revealed some bases of other workers to be spermine; reported distribution of spermine in mammalian tissues
1926		Described structure and synthesis of spermine as a derivative of tetramethylene diamine and trimethylene diamine
1926	J. Sumner	Crystallized urease
1927	H. Dudley et al.	Discovered spermidine and described its constitution and synthesis
1928	G. Barger & F. Coyne	Described the constitution and synthesis of methionine
1929–1930	C. Best	Discovered histaminase, an oxidase sensitive to carbonyl reagents
1930	M. Wada	Isolated citrulline, described its properties, and synthesized the amino acid
1931	G. Harrison	Determined the spermine content of human tissues; proposed prostate as the tissue of origin of seminal spermine; reported trace amounts of spermine in feces
1932	H. Krebs & K. Henseleit	Demonstrated the ornithine cycle in liver
	M. Damodaran et al.	Isolated glutamine from a protein hydrolysate
1938	E. Zeller	Discovered diamine oxidase and spermine oxidation in seminal plasma
1940–1946	E. Gale	Studied specific bacterial amino acid decarboxylases, including those of lysine, arginine, and ornithine
1941	A. Martin & R. Synge	Described partition chromatography of N-acetyl amino acids on silica gels
1944	R. Consden et al.	Developed paper chromatography of amino acids, detected by ninhydrin
1945	H. Blaschko	Stated that the mammalian origins of putrescine, cadaverine, and spermine were unknown.

REFERENCES

Armstrong, H. E. (1896) *Journal of the Chemical Society,* **69,** 637–732.

Brock, W. H. (1985) *From Protyle to Proton. William Prout and the Nature of Matter, 1785–1985.* Bristol, U.K.: Adam Hilger, Limited.

Corbin, A. (1986) *The Foul and the Fragrant.* Cambridge, Mass.: Harvard University Press.

Crosland, M. (1962) *Historical Studies in the Language of Chemistry.* Cambridge, Mass.: Harvard University Press.

Dakin, H. D. (1906) *Journal of Biological Chemistry,* **1,** 171–176.

Dobbin, L. (1935) *Journal of Chemical Education,* **12,** 370–375.

Dudley, H. W., Rosenheim, M. C., & Rosenheim, O. (1924) *Biochemical Journal,* **18,** 1263–1272.

Dudley, H. W., Rosenheim, O., & Starling, W. W. (1927) *Biochemical Journal,* **21,** 97–103.

——— (1926) *Biochemical Journal,* **20,** 1082–1094.

Ellinger, A. (1900) *Hoppe–Seyler's Zeitschrift für Physiologischen Chemie,* **29,** 334–348.

Harrison, G. A. (1931) *Biochemical Journal,* **25,** 1885–1892.

——— (1933) *Biochemical Journal,* **27,** 1152–1156.

Guggenheim, M. (1951) *Die Biogenen Amine.* Basel: S. Karger A. G. Verlag.

Hofmann, A. W. (1850) *Philosophical Transactions of the Royal Society,* **140,** 93–131.

Ihde, A. (1964) *The Development of Modern Chemistry.* New York: Harper & Row.

Mann, T. (1954) *The Biochemistry of Semen.* London: Methuen and Company, Limited.

Meldrum, A. N. (1933) *Journal of the Chemical Society Transactions,* Part 2, 905–915.

Multhauf, R. P. (1966) *The Origins of Chemistry.* London: Oldbourne.

Partington, J. R. (1965) *A Short History of Chemistry,* 3rd ed. New York: Harper & Row.

Rosenheim, M. C. (1917) *Journal of the Physiological Society,* **51,** vi–vii.

Rosenheim, O. (1924) *Biochemical Journal,* **18,** 1253–1262.

Schreiner, P. (1878) *J. Liebig's Annalen der Chemie,* **194,** 68–84.

Von Udránszky, L. & Baumann, E. (1889) *Hoppe–Seyler's Zeitschrift für Physiologischen Chemie,* **13,** 562–594.

Vickery, H. B. & Schmidt, C. L. H. (1931) *Chemical Reviews,* **9,** 169–318.

Weeks, M. E. (1934) *Journal of Chemical Education,* **11,** 101–107.

Wrede, F. (1924) *Hoppe–Seyler's Zeitschrift für Physiologischen Chemie,* **138,** 119–135.

Wrede, F., Fauselow, H., & Strack, E. (1926) *Hoppe–Seyler's Zeitschrift für Physiologischen Chemie,* **161,** 66–73.

——— (1927) *Hoppe–Seyler's Zeitschrift für Physiologischen Chemie,* **163,** 218–228.

CHAPTER 2

Structures and Syntheses of Polyamines

The polyamines discovered and characterized by 1930 comprised five compounds: putrescine, cadaverine, 1,3-diaminopropane, spermidine, and spermine. Putrescine and cadaverine are primary diamines (1,4-diaminobutane and 1,5-diaminopentane, respectively) and were known as products of bacterial metabolism. Spermidine and spermine are a triamine and a tetramine, respectively, that contain primary and secondary amines (R_1R_2NH). As presented in Figure 2.1, spermidine is mono-*N*-3-aminopropyl-1,4-diaminobutane or 1,8-diamino-4-azaoctane, and spermine is bis(*N*-3-aminopropyl)-1,4-diaminobutane or 1,12-diamino-4,9-diazadodecane. Both of these natural amines can be thought of as derivatives of putrescine. The designation of these compounds and the diamines, as well as other novel amines discovered subsequently in biological materials, as "polyamines" is evidently not quite appropriate. The suggestion of "oligoamines" has not achieved wide acceptance. The term polyamine is useful in that it is brief, widely accepted, and defines a small and relatively discrete class of biological substances, that is, low-molecular aliphatic nonprotein nitrogenous bases.

In addition, the monoguanido derivative of putrescine, agmatine, was isolated from hydrolysed animal tissue in 1910 by A. Kossel, who proposed that this was a decarboxylation product of arginine. This is a frequent intermediate in the conversion of arginine to putrescine by routes that will be explored in later chapters.

In 1887 Brieger isolated a base from cultures of *Vibrio cholerae*, whose elementary composition was that of 1,3-diaminopropane. The aminopropyl moiety was recognized in the 1920s to be a portion of spermine, and 1,3-diaminopropane was synthesized and characterized by Wrede et al. in 1926 in their work on 3-aminopropylpyrrolidine. Its identity as a metabolically active polyamine stems from the nutritional work of Herbst in the 1950s and from later studies on amine oxidases (chapter 6 in this volume).

Two additional classes of natural polyamines were discovered more recently, tertiary amines (R_3N) and quaternary amines (R_4N^+OH). In addition, numerous natural polyamines were isolated that contain derivatives of diaminopropane and diaminopentane. Two natural polyamines were found that contain one carbon atom in excess of spermidine and were named *homo*spermidines. A symmetric homospermidine, bis(aminobutyl)amine, is found in some plants and bacteria whereas an asymmetric homospermidine, aminopropylcadaverine, is found in cells relatively rich in cadaverine and depleted of putrescine. Other natural polyamines that contain moieties with one carbon atom less than that of the usual 1,4-diaminobutane are designated as *nor*spermidine or *nor*spermine.

Isolation of Diamines

In early studies the isolation of the diamines was accomplished initially by precipitation of a deproteinized extract with a heavy metal salt, for example, as a double salt of mercuric chloride. The double salts of the diamines were then dissolved in hot water and decomposed with H_2S; the filtrate was dried and treated with absolute ethanol to extract cadaverine hydrochloride, leaving an insoluble residue of water-soluble putrescine hydrochloride.

A reaction of amines to form benzoyl derivatives was exploited in the isolation of the diamines from the urine of cystinuric patients. These were water insoluble, and the derivatives of putrescine and cadaverine were also separable in ethanol.

Diamines as Precursors and Analogs

As the ubiquity and metabolic activities of the polyamines became clearer, the biogenesis and abiogenesis of these bases were explored, with appropriate attention to the synthesis of possible precursors. The derivation of putrescine from L-ornithine and L-arginine and of cadaverine from L-lysine was recognized early, and the various isotopic forms of these amino acids were synthesized and became available. L-(1-^{14}C)-ornithine is

Names	Structure	Shorthand Representation
Putrescine 1,4-diaminobutane	$H_2NCH_2CH_2CH_2CH_2NH_2$	
Cadaverine 1,5-diaminopentane	$H_2NCH_2CH_2CH_2CH_2CH_2NH_2$	
1,3-diaminopropane	$H_2NCH_2CH_2CH_2NH_2$	
Spermidine N-3-aminopropyl-1,4-diaminobutane, or 1,8-diamino-4-azaoctane	$H_2NCH_2CH_2CH_2NHCH_2CH_2CH_2CH_2NH_2$	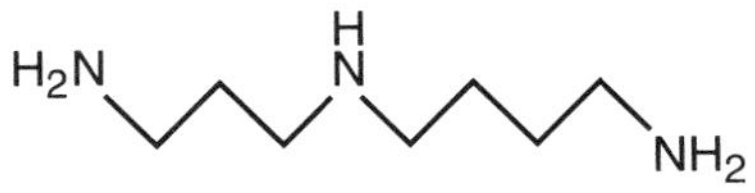
Spermine bis(N-3-aminopropyl)-1,4-diaminobutane, or 1,12-diamino-4,9-diazadecane	$H_2NCH_2CH_2CH_2NHCH_2CH_2CH_2CH_2NHCH_2CH_2CH_2NH_2$	

FIG. 2.1 Early known polyamines.

widely used to follow the production of putrescine via decarboxylation of the amino acid. A considerable thermal decarboxylation to form $^{14}CO_2$ and nonradioactive putrescine was found to proceed in sealed tubes at 100–210°C in the pH 4–11 range. This result was taken to imply the possibility of a prebiotic synthesis of polyamine (Wong et al., 1991).

Variously labeled putrescines containing tritium or deuterium were prepared via reductions of precursors of the carbon chain (Marchand et al., 1988). Olefinic or acetylenic amine hydrochloride precursors were hydrogenated by tritium or deuterium gas (Saljoughian et al., 1988). The use of ^{13}N- or ^{15}N-diamines facilitates the application of nuclear magnetic resonance (NMR), and the preparation of ^{13}N-labeled putrescine was described (Kabalka et al., 1992). ^{15}N-Putrescine and ^{15}N-spermidine were used to explore reactions of bacterial membranes and of tRNA by NMR techniques. These studies will be discussed in later chapters.

Many substituted aliphatic diamines have been prepared as synthetic building blocks or possible inhibitory analogues. A 2-methylcadaverine is advertised as a useful component of polyamides for adhesives, resins, plastics, etc. Three dimethylcadaverines were tested as possible inhibitors of barley infections (Brown et al., 1990). Cadaverines containing ethoxyalkyl groups in position 2 were prepared as models for the study of diamine oxidase (Bertini et al., 1993). Substituted putrescines were prepared by many workers. For example, isomers of 1,4-dimethylputrescine, that is, 2,5-hexanediamine, as well as a series of 2-alkyl putrescines, proved to be inhibitors of ornithine decarboxylase (Garrido et al., 1988).

Isolation and Properties of Spermine

The characterization of spermine from semen by the Rosenheim group (cf. chapter 1 in this volume) required the elimination of proteins and permitted the spontaneous crystallization of a relatively clean phosphate salt of the base. In semen, phosphate is slowly liberated in the hydrolysis of phospholipid fragments. When such semen is precipitated with ethanol, choline and other dialysable substances are removed. The "dry semen" can then be boiled in hot water and dialysed, as a result of which the spermine phosphate appears in a concentrated dialysate. Also, free spermine can be steam distilled from a strong alkaline solution of dry semen and crystallized as the phosphate in the distillate on addition of ammonium phosphate. The base (in alkali) is also extractable into n-butanol and can then be precipitated as a salt. These isolation procedures obviously take advantage of spermine as a relatively small organic amine. The water-insoluble picrate of spermine is the basis of Barberio's test of the presence of human semen in forensic medicine. These and other characteristic salts were then isolated from extracts of many mammalian organs and other biological specimens. The isolation of

spermine from ox pancreas was described in detail by Dudley et al. (1924).

Structure and Synthesis of Spermine

In characterizing the structure of spermine (chapter 1), oxidation was unable to detect a resistant ring structure. The compound did not contain unsaturated bonds and was optically inactive. Unexpectedly the reaction with nitrous acid was complex, but exhaustive methylation yielded a quaternary base, decamethylspermine. Destructive distillation of spermine and of putrescine yielded pyrrolidine, suggesting the presence of a tetramethylenediamine within spermine, whereas similar treatment of the decamethylspermine produced tetramethyltrimethylenediamine. The possible structure of spermine may be described in a useful shorthand as **3,4,3** or **3,3,4**, of which **3** represents the trimethylene moiety and **4** the tetramethylene portion of the structure.

The possible structures of the original spermine then were

$$H_2NCH_2CH_2CH_2$$
$$\mathbf{3}$$
$$NHCH_2CH_2CH_2CH_2NH$$
$$\mathbf{4}$$
$$CH_2CH_2CH_2NH_2,$$
$$\mathbf{3}$$

or

$$H_2NCH_2CH_2CH_2$$
$$\mathbf{3}$$
$$NHCH_2CH_2CH_2NH$$
$$\mathbf{3}$$
$$CH_2CH_2CH_2CH_2NH_2$$
$$\mathbf{4}$$

The structure of spermine was determined by synthesis, in which 1,4-diaminobutane was reacted with an excess of α-phenoxy-γ-bromopropane to yield α,δ-bis(γ-phenoxypropyl)aminobutane. Treatment with HBr replaced the phenoxy group with bromine to form α,δ-bis(γ-bromopropylamino)butane, and in turn the bromine was replaced with an amino group by heating with ethanolic ammonia. After removal of reactants and steam distillation of the amine, an insoluble phosphate was isolated as well as five additional derivatives, which proved identical with those derivatives of natural spermine, the **3,4,3** derivative. *Thermospermine* or **3,3,4** was found more recently in bacteria.

To synthesize spermine, the Wrede group (chapter 1) had chosen a shorter route, the Gabriel synthesis, reacting an excess of the iodopropylphthalimide with 1,4-diaminobutane to produce the diphthalimide of spermine. The **3,4,3** base was liberated by hydrolysis and crystallized as the phosphate. This group compared the tetra-m-nitrobenzoyl derivatives of the natural and synthetic product. In addition they used dimethylputrescine to prepare novel dimethylspermine in which the secondary amines were selectively methylated.

During World War II, possible roles for spermine and spermidine were seen in the program developed for the chemotherapy of malaria. It had been observed that the compounds partially prevented the inhibition of bacterial growth by atabrine and quinine. A large-scale synthesis was devised for the tetramine consisting of the following sequence:

1. hydrogenation of succinonitrile to 1,4-diaminobutane;
2. dicyanoethylation of putrescine to

$$NC(CH_2)_2NH(CH_2)_4NH(CH_2)_2CN;$$

3. hydrogenation of the N,N'-bis-(2-cyanoethyl)-putrescine to spermine.

Spermine was isolated in an overall yield of 51% (Schultz, 1948).

The synthesis of derivatives of spermine aroused greater interest as a result of the recognition of the marked affinity of the tetramine in reactions with nucleates, both RNA and DNA, and certain proteins (see chapters 22 and 23). These resulted in the production of 5-carboxy-substituted lipophilic spermines (Fig. 2.2A), which can coat genetically active DNA and facilitate penetrations through eucaryotic cell membranes (Behr et al., 1989). The possibility of reactions with specific nucleotide regions was exploited with photoactivable derivatives of spermine, which can serve to mark particular fragments of a cut nucleate. Thus, p-diazonium anilides of L-5-carboxyspermine (Fig. 2.2B) were used to cut RNA at specific spermine binding sites in a particular tRNA species (Garcia et al., 1990).

Structure and Synthesis of Spermidine

Spermidine was found initially by Dudley et al. (1927) as a soluble base present in tissue extracts after the removal of spermine phosphate. After the precipitation of the latter in 25% ethanol, 50% ethanol permitted crystallization of spermidine phosphate. The compound, like spermine, was optically inactive, an aliphatic compound resistant to permanganate, and a base whose phosphotungstate salt was insoluble in acetone. It yielded a tri-m-nitrobenzoyl derivative of molecular weight 592. The structure of spermidine was postulated to be a monoaminopropyl derivative of 1,4-diaminobutane, that is, a **3,4** compound. The proposed structure was proven by a synthesis in which a monophenoxypropylamino derivative of putrescine was converted by HBr to the bromopropylamino derivative and then by ammonolysis to spermidine. Various derivatives of the synthetic product were identical with those of the natural base.

An improved synthesis of spermidine involved the monocyanoethylation of γ-aminobutyronitrile ($H_2N(CH_2)_3$-CN) and catalytic hydrogenation to spermidine from the N-(2-cyanoethyl)-γ-aminobutyronitrile (Danzig & Schultz, 1952). A synthesis was also devised beginning with monoacetylputrescine, which was cyanoethylated and

hydrogenated to form monoacetylspermidine. Three nonidentical monoacetyl derivatives of spermidine are possible and are delineated in the following:

$$H_2NCH_2CH_2CH_2NHCH_2CH_2CH_2CH_2NHCOCH_3. \quad (2.1)$$

This compound, described as N^8-acetylspermidine, is the product of the above synthesis. In short, the nitrogen in the primary amine of the residual aminopropyl moiety is described as N^1, a designation equally applicable to spermine. The other isomers include:

$$CH_3CONHCH_2CH_2CH_2NHCH_2CH_2CH_2CH_2NH_2, \quad (2.2)$$

that is, N^1-acetylspermidine, and

$$H_2NCH_2CH_2CH_2N(COCH_3)CH_2CH_2CH_2CH_2NH_2, \quad (2.3)$$

that is, N^4-acetylspermidine.

N^1- and N^8-spermidines are both known as metabolic products. Evidently selective derivatization of the primary amines is required for the synthesis of these naturally occurring compounds. Derivatized or substituted diamines as precursors for subsequent syntheses or the invention of selective reactions for one or another end of the triamine or tetramine were both used.

Syntheses and Reactions of Amines

The examples of isolation, characterization, and synthesis in proof of structure given in the preceding include many of the generalizable properties, reactions, and syntheses available for the amines. As Hofmann observed, the displacement of halogen by excess ammonia from an alkyl halide can produce primary, secondary, and tertiary amines as well as quaternary ammonium salts. The conversion of the latter to a tertiary amine by heat (i.e., the Hofmann elimination) was used industrially in the production of tertiary amines. The process of exhaustive methylation of spermine and degradation of the quaternary salt, decamethylspermine, to produce identifiable fragments in determining the structure of the tetramine was rooted in this familiar set of reactions.

A controlled synthesis of primary amines, the Gabriel synthesis, was used in the syntheses of spermine by Wrede et al. and involved the reaction of phthalimide with an alkyl halide. Phthalic anhydride and ammonia are heated to form phthalimide (a). The acidic hydrogen of the imide permits a conversion to the K+ salt.

The phthalimide anion reacts easily with halide (b) and the substituted imide is released as the amine by hydrolysis (c). The hydrolysis was improved by a reaction with hydrazine. The addition of HCl precipitates insolu-

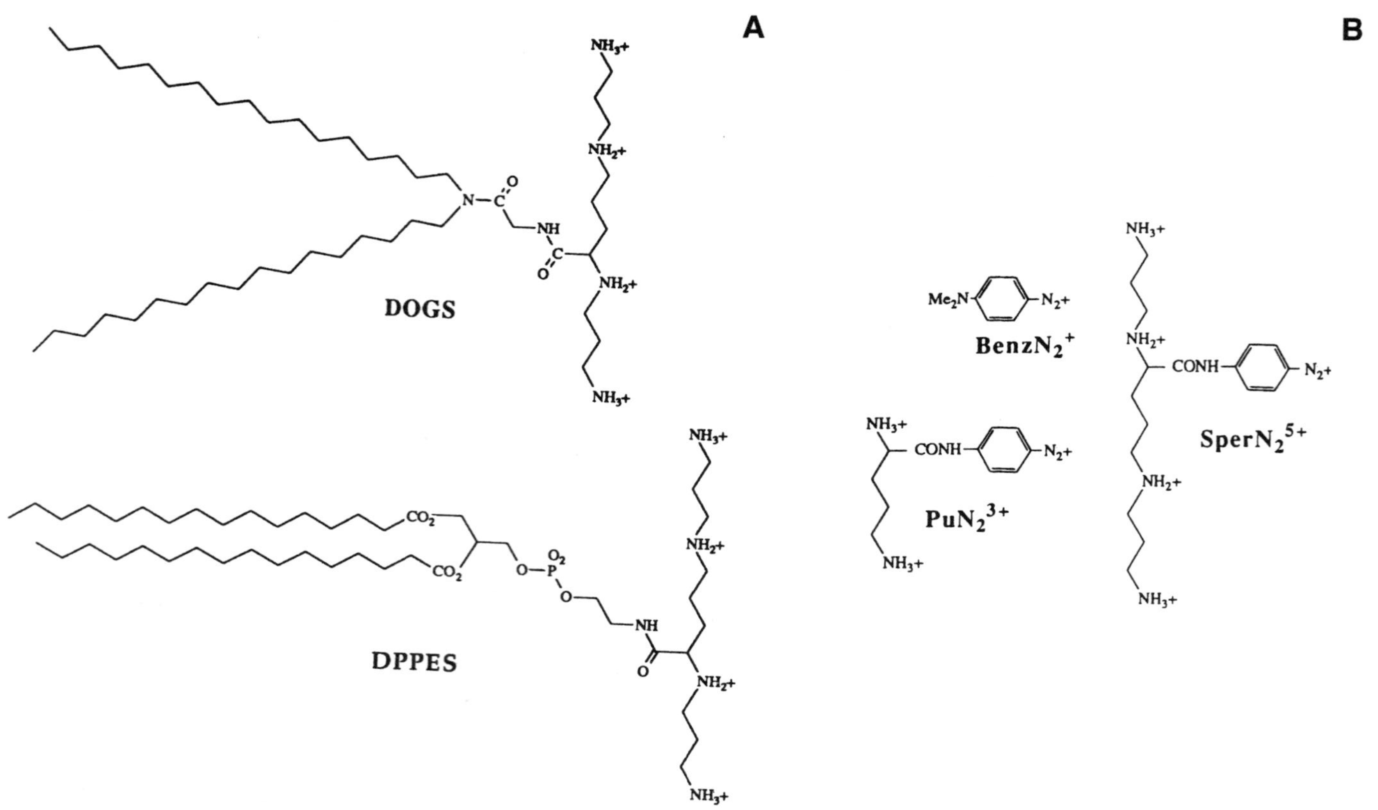

FIG. 2.2 (**A**) Amphiphilic derivatives of spermine facilitating transfection, redrawn from Behr et al. (1989). (**B**) Photochemical probes of polyamine binding sites, redrawn from Garcia et al. (1990). Benz N_2^+ = p-N-dimethylaminobenzenediazonium; Pu N_2^{3+} = p-diazoniumanilide of L-2-carboxyputrescine; Sper N_2^{5+} = p-diazoniumanilide of L-5-carboxyspermine.

ble phthalhydrazide (d) and soluble amine hydrochloride.

(b)

(c)

(d)

In the synthesis of monoacetyl spermidine by Jackson (1956), monoacetyl putrescine was prepared via the formation of γ-phthalimidobutyronitrile. The nitrile was reduced to the amine in acetic anhydride. Monoacetyl putrescine was released by heating with hydrazine.

In the ground-breaking work of Schoenheimer and Ratner in developing isotopic derivatives for metabolic study, ^{15}N amino acids were prepared via the phthalimido synthesis (Schoenheimer & Ratner, 1939). The procedure was exploited more recently in the production of ^{15}N-enriched polyamines (Samejima et al., 1984). Such compounds are of particular value in dissecting the physiological significance of the compounds. ^{15}N compounds enable the application of ^{15}N-NMR and mass spectrometry in the study of the polyamines in intact cells and in their association with large structures and molecules. In the synthesis of Samejima et al. (1984), a ^{15}N-enriched potassium phthalimide was used in the synthesis of ^{15}N-enriched putrescine via reaction with 1,4-dibromobutane. Such labeled putrescine was then employed in the synthesis of spermidine via reaction with ^{15}N-enriched N-(3-bromopropyl)-phthalimide. Both compounds were also used in the synthesis of spermine. In this synthesis, an N,N^{1}-bis(benzyl) putrescine was prepared and reacted with the bromopropylphthalimide. The product was cleaved with hydrazine and the benzyl derivatives were removed by catalytic hydrogenation to form a ^{15}N-enriched spermine (**3,4,3**). The authors noted the use of these reagents to form a wide range of symmetric (**3,3,3**; **4,4,4**) and asymmetric (**3,3,4**; **4,3,4**; **3,4,4**) tetramines. Similar procedures were used in the synthesis of a series of 10 N-pentamines (Niitsu & Samejima, 1986) with three or four methylene chain intervals, many of which were subsequently found in nature.

The discovery of apparent tertiary and quaternary polyamines in thermophilic bacteria posed the problem of the confirmation of identity by comparison with au-

thentic samples. Initially one method for the synthesis of a symmetric compound, *tris*(3-aminopropyl)amine, included an alkylation of ammonia with acrylonitrile and reduction of the nitrile. A more general method was developed involving the formation of bisphthalimido derivatives of a triamine containing an unsubstituted secondary amine and the N-alkylation by another phthalimido derivative to form the tertiary amine. The further addition of a monophthalimidopolymethyleneiodide resulted in the synthesis of the quaternary amine iodide from which the phthalimido substituents were removed with hydrazine and acid. The numerous tertiary and quaternary amines produced by these procedures were described by Niitsu et al. (1992). Bradshaw et al. (1992) reviewed several newer methods for the extension of polyamines by 3-aminopropyl and 4-aminobutyl units. Borane reductants and other derivatives proved to be major acquisitions in recent synthetic methodologies (Carboni et al., 1993).

Preparing for Metabolic Study

The improved synthesis of primary amines by catalytic reduction of nitriles, as in the Danzig and Schultz (1952) and Jackson and Rosenthal (1960) studies, was extended to a synthesis of [^{14}C]-spermidine and [^{14}C]-spermine (Jackson & Rosenthal, 1960). A series of linear aliphatic triamines and tetramines was prepared by a reaction of 1 or 2 equivalents (equiv) of acrylonitrile (CH$_2$=CHCN) with the appropriate diamine. The products were then reduced catalytically to monoaminopropyl or bis(aminopropyl) derivatives and tested against mouse tumors (Israel et al., 1964). Spermine analogs made by the insertion of α,ω diaminopolymethylene derivatives containing eight or nine CH$_2$ groups between 3-aminopropyl groups do have significant antitumor activity (Edwards et al., 1990). Products in which sulfur or oxygen replaced the secondary nitrogens are essentially inactive.

Site-Specific Syntheses

Recognition of the roles of the polyamines in proliferative and other physiological processes and the possibility of preparing inhibitory or modulating analogs evoked an increased interest in improving the syntheses. Furthermore, the many natural products, for example, antibiotics, toxins, siderophores, and normal metabolites, were discovered to be relatively complex conjugates of the polyamines. Coupling reactions with peptides (Shieh et al., 1984) and newer coupling reagents (Morin & Vidal, 1992) were described. Site-specific reactions were required to prepare such structures, and several of these will be described in the following text.

In the preparation of the alkaloid, *maytenine*, a dicinnamoyl spermidine, the 1-hydroxypiperidine ester of *trans*-cinnamic acid was shown to react relatively specifically with the primary amines of the polyamine (Husson et al., 1973).

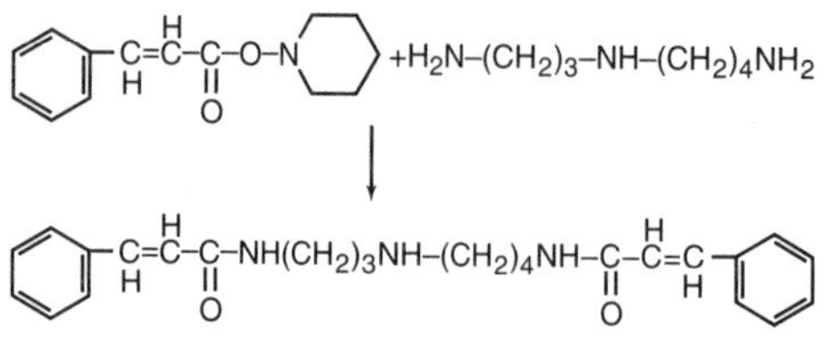

Bis-(*trans*-cinnamoylspermidine)

Acylimidazoles can also effect a symmetrical bis-acylation of the primary amines of spermidine and other linear triamines (Joshua & Scott, 1984).

Extending such reactions, the N^4-benzylspermidine was synthesized that contained the substituted secondary amine. The desired bis-acyl derivatives of the N^1,N^8 primary amines were prepared in the synthesis of a known bacterial siderophore. The benzyl group was removed by catalytic hydrogenation to produce the required N^1,N^8 bis(acyl)spermidines (Bergeron et al., 1980). A novel N^4-acyl derivative of the diacyl spermidine was then made to complete the total synthesis of the bacterial siderophore (Fig. 2.3). The strategy of using eliminable N^4-benzyl substitutions on the secondary amine was developed for the acylation of primary amines and for the construction of homospermidine (4,4) and norspermidine (3,3), as well as spermidine (3,4) (Bergeron et al., 1981).

In another approach to this problem, the reaction of spermidine with 1 equiv formaldehyde selectively produced the six-membered hexahydropyrimidine ring:

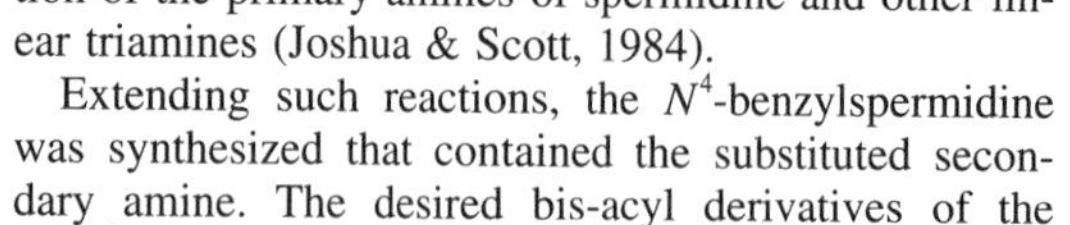

Successive acylation and ring cleavage produced terminal *N*-acyl derivatives such as the bis-(*trans*-cinnamoylspermidine) described in the preceding (Ganem, 1982). Furthermore this N^1,N^4-protected triamine was easily convertible via addition of acrylonitrile and subsequent reduction to spermine. However, because the initial reaction distinguishes three and four carbon moieties, the reaction is not easily adapted to symmetrical triamines but it has been used in the syntheses of many complex natural products (Ganem, 1982). The aminobutyl- and aminopropyl-hexahydropyrimidines compete with spermidine for transport by mouse cells and are active inhibitors of cell growth (Bergeron & Seligsohn, 1986). An effort to explore the applicability of this method to the synthesis of 2-hydroxyputrescine and several natural products, such as the unusual amino acid, hypusine, led to the discovery of oxygen-containing ring structures that served as precursors to the synthesis of the desired natural compounds (Tice & Ganem, 1983a,b).

Several different reagents were reported for a simple one-step protection of the primary amino functions (Almeida et al., 1989; Sosnovsky & Lukszo, 1986). New routes to selectively *N*-alkylated polyamines were also described (Almeida et al., 1989; Barluenga et al., 1993; Golding et al., 1988; Levchine et al., 1994; Nordlander et al., 1984; Ramiandrasoa et al., 1984). Approaches to the problem of selective modification are now seen to require stepwise additions and specific elimination of independent protecting groups interspersed with additions of modifiable carbon chains. In one such exercise, formation of a trisubstituted spermidine was accomplished stepwise by the synthesis of a monobenzyl diamine, substitution of tertiary butoxycarbonyl groups at the other primary amine, addition of acrylonitrile at the benzylated nitrogen, reduction of the nitrile to the amine, and final derivatization of this newly acquired primary amine by a trifluoroacetyl group. Each substituent can be eliminated selectively and each nitrogen can then be resubstituted at will (Bergeron et al., 1984). Modifying the sequence and introducing a new protective agent, an initial disubstituted spermidine was converted into a tetraprotected spermine in which each protecting group can be removed independently and refunctionalized at will (Bergeron & McManis, 1988).

Increased knowledge of the participation of the tetramines (e.g., spermine, etc.) in physiological processes increased activity in the syntheses of analogs of these and other polyamines (Chu et al., 1995; Fiedler & Hesse, 1993). The discovery of the bizarrely structured polyamine-containing toxins of spiders and wasps, which block voltage-sensitive calcium channels, led to the production of a large array of these compounds and variants (Blagbrough & Moya, 1994). Many of the polyamine components of the toxins are terminated with guanido groups.

Guanidine Derivatives

The decarboxylation of arginine can produce agmatine, the substrate in one important route to the biosynthesis of putrescine. Arginase converts arginine to urea and ornithine, the substrate of ornithine decarboxylase, a major source of putrescine in most organisms. Therefore, it is evident that the reactions of this guanidine-containing amino acid are of great importance. In addition to arginine, several guanidine-containing amino acids, such as homoarginine, γ-hydroxyhomoarginine, and canavanine (Fig. 2.4), and some of their metabolic products are known in nature.

Arginine also participates in transamidation reactions leading to the synthesis of many guanidine-containing compounds, for example, creatine, streptomycin, and octopine. In addition polyamine derivatives such as arcaine (1,4-diamidino putrescine), audouine (1,5-diamidino cadaverine), and hirudonine (1,8-diamidinospermidine) (Fig. 2.4) are known and are formed by this route (Audit et al., 1967).

The increased basicity of the guanido group confers an increased affinity for acidic groups, and it may be anticipated that guanidinated compounds can compete effectively with lesser bases such as the primary amines. Several antibiotics that contain such groups are known

N^4–benzylspermidine

$\downarrow$

N^4–benzyl–N^1,N^8–bis(2,3–dihydroxybenzoyl)spermidine

$\downarrow$

N^1,N^8–bis(2,3–dihydroxybenzoyl)spermidine

$\downarrow$

N^1–[N–(2–hydroxybenzoyl)threonyl]–
N^1,N^8–bis(2,3–dihydroxybenzoyl)spermidine

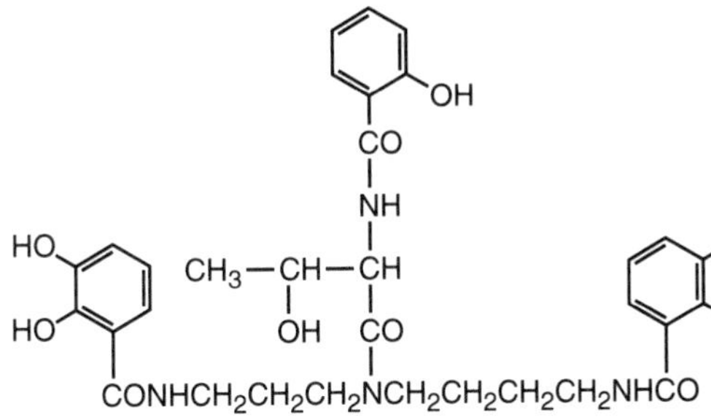

FIG. 2.3   Stepwise synthesis of the siderophore of *Micrococcus denitrificans*, redrawn from Bergeron et al. (1980).

and include bleomycin B_2, which terminates in agmatine as an essential cation, edeine B_1, which terminates in a monoamidinospermidine, and spergualin, which contains a spermidine at one end of the molecule and a guanidine group at the other (Umeda et al., 1987a, b).

The synthesis of amidino derivatives from amines is

of interest, therefore, in the proof of structure of the natural products, in the preparation of analytical standards and in the synthesis of competitive analogs. The conversion of putrescine, cadaverine, or spermidine to salts of guanidine derivatives was easily effected by reaction with the sulfate of *S*-methylisothiourea (Robin &

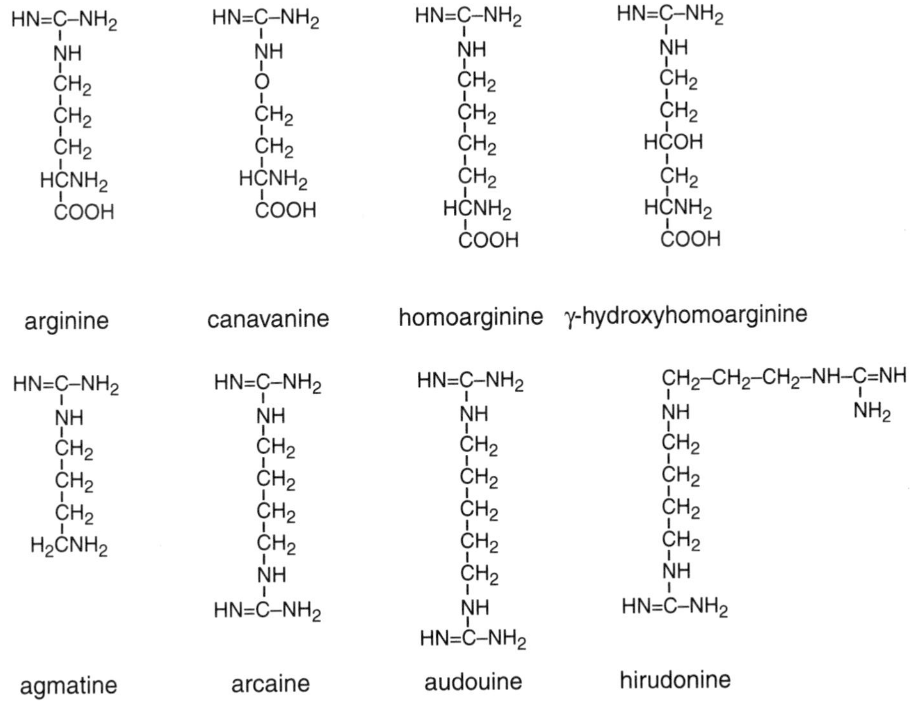

FIG. 2.4   Some guanidine-containing amino acids and polyamines.

van Thoai, 1961). The insoluble sulfate salts are readily recrystallizable from ethanol or water.

The acarnidines, antimicrobial and antiviral compounds isolated from the sponge, were found to terminate in a monoamidino derivative of an aminopropylcadaverine. In the syntheses of acarnidine and its analogs, it was helpful to reduce the basicity of the guanido group by preparing an N-nitro derivative in which the protecting nitro group was removed in the last step of the synthesis (Boukouvales et al., 1983). Such derivatives of the guanidine group of arginine are inhibitory to the enzyme that converts arginine to citrulline and generates nitric oxide.

REFERENCES

Almeida, M. L. S., Grehn, L., & Ragnarsson, U. (1989) *Acta Chemica Scandinavica,* **43,** 990–994.

Audit, C., Viala, B., & Robin, Y. (1967) *Comparative Biochemistry and Physiology,* **22,** 775–785.

Barluenga, J., Kouznetsov, V., Rubio, E., & Tomás, M. (1993) *Tetrahedron Letters,* **34,** 1981–1984.

Behr, J., Demeneix, B., Loeffler, J., & Pérez-Mutul, J. (1989) *Proceedings of the National Academy of Sciences of the United States of America,* **86,** 6982–6986.

Bergeron, R. J., Burton, P. S., McGovern, K. A., & Kline, S. J. (1981) *Synthesis,* **15,** 732–733.

Bergeron, R. J., Garlich, J. R., & Stolowich, N. J. (1984) *Journal of Organic Chemistry,* **49,** 2997–3001.

Bergeron, R. J., McGovern, K. A., Channing, M. A., & Barton, P. S. (1980) *Journal of Organic Chemistry,* **45,** 1589–1592.

Bergeron, R. J. & McManis, J. S. (1988) *Journal of Organic Chemistry,* **53,** 3108–3111.

Bergeron, R. J. & Seligsohn, H. W. (1986) *Bioorganic Chemistry,* **14,** 345–355.

Bertini, V., Lucchesini, F., Pocci, M., DeMunno, A., Picci, N., & Iemma, F. (1993) *Tetrahedron,* **49,** 8423–8432.

Blagbrough, I. S. & Moya, E. (1994) *Tetrahedron Letters,* **35,** 2057–2060.

Boukouvalas, J., Golding, B. T., McCabe, R. W., & Slaich, P. K. (1983) *Angewandte Chemie—International Edition in English,* **22,** 618–619.

Bradshaw, J. S., Krakowiak, K. E., & Izatt, R. M. (1992) *Tetrahedron,* **48,** 4475–4515.

Brown, A. M., Walters, D. R., & Robins, D. J. (1990) *Letters in Applied Microbiology,* **11,** 130–132.

Carboni, B., Benalil, A., & Valtier, M. (1993) *Journal of Organic Chemistry,* **58,** 3736–3741.

Chu, P., Shirahata, A., Samejima, K., Saito, H., & Abe, K. (1995) *Japan Journal of Pharmacology,* **67,** 173–176.

Danzig, M. & Schultz, H. P. (1952) *Journal of the American Chemical Society,* **74,** 1836–1837.

Edwards, M. L., Prakash, N. J., Stemerick, D. M., Sunkara, S. P., Bitonti, A. J., Davis, G. F., Dumont, J. A., & Bey, P. (1990) *Journal of Medicinal Chemistry,* **33,** 1369–1375.

Fiedler, W. J. & Hesse, M. (1993) *Helvetica Chimica Acta,* **76,** 1511–1519.

Ganem, B. (1982) *Accounts of Chemical Research,* **15,** 290–298.

Garcia, A., Giege, R., & Behr, J. (1990) *Nucleic Acids Research,* **18,** 89–95.

Garrido, D. O., Buldain, G., Ojea, M. I., & Frydman, B. (1988) *Journal of Organic Chemistry,* **53,** 403–407.

Golding, B. T., O'Sullivan, M. C., & Smith, L. L. (1988) *Tetrahedron Letters,* **29,** 6651–6654.

Husson, H., Poupat, C., & Potier, P. (1973) *Comptes Rendus de l'Academie des Sciences Paris,* **276C,** 1039–1040.

Israel, M., Rosenfield, J. S., & Modest, E. J. (1964) *Journal of Medicinal Chemistry,* **7,** 710–716.

Jackson, E. L. (1956) *Journal of Organic Chemistry,* **21,** 1374–1375.

Jackson, E. L. & Rosenthal, S. M. (1960) *Journal of Organic Chemistry,* **25,** 1055–1056.

Joshua, A. V. & Scott, J. R. (1984) *Tetrahedron Letters,* **25,** 5725–5728.

Kabalka, G. W., Wang, Z., Green, J. F., & Goodman, M. M. (1992) *Applied Radiation and Isotopes,* **4,** 389–391.

Levchine, I., Rajan, P., Borloo, M., Bollaert, W., & Haemers, A. (1994) *Synthesis,* **1** 37–40.

Marchand, A. P., Satyanarayana, N., McKenny, R. L., & Struck, S. R. (1988) *Journal of Labelled Compounds and Radiopharmaceuticals,* **25,** 971–976.

Morin, C. & Vidal, M. (1992) *Tetrahedron,* **48,** 9277–9282.

Niitsu, M. & Samejima, K. (1986) *Chemical and Pharmaceutical Bulletin,* **34,** 1032–1038.

Niitsu, M., Sano, H., & Samejima, K. (1992) *Chemical and Pharmaceutical Bulletin,* **40,** 2958–2961.

Nordlander, J. E., Payne, M. J., Balk, M. A., Gress, J. L., Harris, F. D., Lane, J. S., Lewe, R. F., Marshall, S. E., Nagy, D., and Rachlin, D. J. (1984) *Journal of Organic Chemistry,* **49,** 133–138.

Ramiandrasoa, F., Milat, M., Kunesch, G., & Chuilon, S. (1989) *Tetrahedron Letters,* **30,** 1365–1368.

Robin, Y. & van Thoai, N. (1961) *Comptes Rendus de L'Academie des Sciences Paris,* **252,** 1224–1226.

Saljoughian, M., Morimoto, H., Williams, P. G., & Rapoport, H. (1988) *Journal of Labelled Compounds and Radiopharmaceuticals,* **25,** 313–328.

Samejima, K., Takeda, Y., Kawase, M., Okada, M., & Kyogoku, Y. (1984) *Chemical and Pharmaceutical Bulletin,* **32,** 3428–3435.

Schoenheimer, R. & Ratner, S. (1939) *Journal of Biological Chemistry,* **127,** 301–313.

Schultz, H. P. (1948) *Journal of the American Chemical Society,* **70,** 2666–2667.

Shieh, H., Campbell, D., & Folkers, K. (1984) *Biochemical and Biophysical Research Communications,* **122,** 21–27.

Sosnovsky, G. & Lukszo, J. (1986) *Zeitschrift für Naturforschung,* **41b,** 122–129.

Tice, C. M. & Ganem B. (1983a) *Journal of Organic Chemistry,* **48,** 5043–5047.

Tice, C. M. & Ganem, B. (1983b) *Journal of Organic Chemistry,* **48,** 5048–5050.

Umeda, Y., Moriguchi, M., Ikai, K., Kuroda, H., Nakamura, T., Fuji, A., Takeuchi, T., & Umezawa, H. (1987b) *Journal of Antibiotics,* **40,** 1316–1324.

Umeda, Y., Moriguchi, M., Kuroda, H., Nakamura, T., Fujii, A., Iinuma, H., Takeuchi, T., & Umazawa, H. (1987a) *Journal of Antibiotics,* **40,** 1303–1315.

Wong, C., Santiago, C., Rodríguez–Paez, L., Ibáñez, M., Baeza, I., & Oró, J. (1991) *Origins of Life and Evolution of the Biosphere,* **21,** 145–156.

CHAPTER 3

Properties and Analysis of Polyamines

The isolation, characterization, and determination of the natural polyamines have employed almost every physical and chemical advance of the last century. These advances and their applications to the amines are sketched briefly in the standard texts of organic chemistry; and as the chemistry of the amines and other nitrogen compounds assumed ever greater importance in industry, medicine, and biology, many specialized volumes (e.g., Tabor and Tabor, 1983) and reviews (e.g., Seiler, 1986) dealing with these topics have become available.

The existence of the tetrahedral ammonium ion was originally postulated by A. Werner who earned the Nobel Prize in 1913 for this contribution and his work on the isomerism of coordinate complexes containing metal-ammines. As a consequence of the theories of G. N. Lewis, I. Langmuir, and N. V. Sidgwick, the structure of the ion is considered to arise from the sharing of an unshared pair of electrons in ammonia with a proton donated by an acid, as given below:

$$\text{H:}\overset{\text{H}}{\underset{\text{H}}{\text{N:}}} + H^+ + X^- \longrightarrow \left[\text{H:}\overset{\text{H}}{\underset{\text{H}}{\text{N:}}}\text{H} \right]^+ + X^-$$

The amines may be considered to be substituted ammonia molecules, as described by Hofmann; that is, the hydrogen atoms may be successively replaced to form the various classes of amines: primary, secondary, and tertiary. As in the case of ammonia, a trisubstituted or tertiary amine can be converted to a tetrasubstituted (quaternary) protonated amine. Thus ammonia can form the ammonia ion or be methylated in organisms to form the naturally occurring tetramethyl ammonium ion, and spermine can be exhaustively methylated to form the quaternary ammonium derivative, decamethylspermine, as in the structural determination of Dudley and Rosenheim (1925). The presence of substituents on the nitrogen, for example, alkyl groups, directs electrons from carbon toward nitrogen, which then share a fourth pair of electrons more readily with a reactant, such as an acid.

The loss of the proton from the protonated amine at an alkaline pH permits the generation of the free amine. Similar reactions of formation of a nitrogenous cation or reversion to the nonprotonated amine as a function of pH are obviously crucial in our isolation and analysis of the natural polyamines. The distinctive names given to the diamines, putrescine and cadaverine, relate to the odors of the free volatile amines, which lack volatility and odor when the compounds exist as their cationic salts. Let us recall Guyton de Morveau and his deodorization of grave sites with gaseous HCl.

In the first clean isolations of spermine (Dudley et al., 1924, Chap. 1), the nonprotonated amine was extracted from an alkaline solution into butanol or was steam distilled, neither of which is possible with a spermine salt. The extraction from tissue of a soluble spermine cation was carried out in somewhat strong acid, for example, trichloroacetic acid, which was also a better protein denaturant (Harrison, 1933). An insoluble picrate was readily convertible to the hydrochloride, a salt insoluble in the organic solvent acetone. The hydrochloride is soluble in water and was crystallized as the insoluble phosphate for purposes of gravimetric analysis. This sequence of steps minimized the contaminants of protein and other substances contained in a precipitate of spermine phosphate formed directly in human seminal fluid or from an extract of prostatic or pancreatic tissue. This sequence of interconversion of free and protonated amine thereby permitted an estimation of spermine in relatively small amounts of some tissues, that is, prostate, testis, pancreas. The gravimetric method of Hämäläinen (1947) similarly required the removal of protein from a tissue extract prior to formation of an insoluble salt at pH 7–8 with flavianic acid. Thus appropriate combinations of pH and solubilities of the amines and their salts in aqueous and organic media were exploited fairly early in the isolation and estimation of the polyamines. Unlike many organic derivatives of spermine, such as the tetrabenzoyl derivative, which has a sharp melting point, the ionic salts of spermine begin to decompose before melting. The resulting temperature range eliminates the melting point as a reliable characteristic of the easily isolated polyamine salts.

The separation of free simpler polyamines from animal and plant tissues and their homogenates is effected by extraction with an acid that precipitates most proteins. However, the problem of estimating these substances in some other situation, for example, as water-soluble excretion products of bacteria cultured in mineral media, required the methods that free the amines from salts and some nitrogenous components like amino acids. One such procedure required a strongly alkaline pH, successive extractions of free amine with tertiary butanol in the presence of added salt, acidification of the solvent before concentration to dryness, and solution of the residual amine salt in water before the actual estimation of the amines. Obviously these tedious manipulations required the introduction of correction factors for the losses sustainable at each step. The determination of recoveries was improved by the use of radioactive polyamines or an unnatural polyamine, such as 1,6-hexanediamine. By the end of the 1960s methods leading to the formation of separable colored or fluorescent products had reduced the amounts of amines needed to form a detectable product separable from the interfering contaminants. The new methods of analysis and their sensitivity tended to minimize the need for the application of the classical methods of separation and purification of the natural polyamines.

Basicity of Amines

Many of the reactivities of the amines depend on their donation of free electron pairs. As defined by G. N. Lewis (1916), the amine, in which the nitrogen has an unshared pair of electrons, acts as a base and can accept a proton in an equilibrium to form a positively charged ion. The amount of the ion formed is a function of the basicity of the amines: the stronger the basicity, the more the amount of charged amine and the larger the equilibrium constant (K_b) in the equilibrium:

$$R_1R_2R_3N + H_2O \rightleftharpoons \left[R_1R_2R_3NH\right]^+ + OH^-$$

$$K_b = \frac{\left[R_1R_2R_3NH^+\right]\left[OH^-\right]}{\left[R_1R_2R_3N\right]}$$

with water in excess.

The equilibrium constant (or basicity constant), K_b, increases as a function of protonation in the displacement of hydroxyl ion from water. The K_b of aliphatic amines falls in the range of 10^{-3}–10^{-4} mol/L; their basicities are on the order of 50–100 times stronger than that of NH_3 ($K_b = 1.8 \times 10^{-5}$ mol/L). Basicity is often reported as a pK_a, the pH at which the ratio of protonated amine and nonprotonated amine is 1 (see Table 3.1).

Substitution of ammonia by an alkyl group markedly increases basicity. It may be supposed that the alkyl

group, which is a weak electron donor, directs these to the nitrogen, thereby making the fourth unshared pair more available for sharing with an acid, as in the reaction

$$R \rightarrow \overset{\overset{\displaystyle H}{|}}{\underset{\underset{\displaystyle H}{|}}{N}}: +H^+ \rightleftharpoons \left[R \rightarrow \overset{\overset{\displaystyle H}{|}}{\underset{\underset{\displaystyle H}{|}}{N}}H\right]^+$$

The pK values of the various amines in the polyamines have been estimated by a variety of methods, and are given in Tables 3.1 and 3.2. In the studies of Palmer and Powell (1974), hydrochlorides were titrated potentiometrically with NaOH in the presence of 0.1 M NaCl at 25°C. The shifts of pH with spermidine and spermine did not reveal clear inflexions, implying overlap in the proton dissociations of several primary and secondary amines.

Thermodynamic data on a series of diamines separated by two to six methylenes, thereby including putrescine and cadaverine, indicated an effect of the length of an intervening aliphatic chain on the amine protonation (Barbucci et al., 1970). Extending this work to spermidine, spermine, and their analogues, Gold and Powell (1976) concluded that the initial protonations are on the primary amines and secondary amines are significantly less basic. Indeed the basicities of the primary amines of spermidine are now thought to be very similar (Kanavarioti et al., 1995).

A quaternary amine, lacking a pair of free electrons, is unable to displace OH⁻ from water and exists in a quaternary ammonium salt solely as a positively charged ion. The formation of such a salt is given below.

$$\overset{\overset{\displaystyle R_1}{|}}{\underset{\underset{\displaystyle R_3}{|}}{R_2-N}} + R_4I \longrightarrow \left[\overset{\overset{\displaystyle R_1}{|}}{\underset{\underset{\displaystyle R_3}{|}}{R_2-N-R_4}}\right]^+ + I^-$$

Whereas uncharged amines can frequently penetrate lipid-containing membranes by diffusion (Guarino & Cohen, 1979), charged amines, including the quaternary ammonium salts, do not cross membranes easily, if at all, in most cellular environments.

Estimation of pK by Nuclear Magnetic Resonance (NMR)

Potentiometric titrations were described as an approach to *macroscopic* protonation constants, whereas the dissection of multiple parallel protonation pathways in a polyamine were termed a study of *microscopic* protonation constants. Since 1980 the latter estimations have been based mainly on NMR chemical shift measurements.

Becker (1993) described the discovery in the 1920s

Table 3.1 pK_a values of polyamines.

Potentiometry determined		
Polyamine	pK_a	
Spermine[a]	10.80, 10.02, 8.85, 7.96	
Spermine[b]	10.94, 10.12, 9.04, 7.97	
Spermidine[b]	10.95, 9.98, 8.56	
Spermidine[c]	10.89, 9.81, 8.34	
Putrescine[d]	10.80, 9.63	
Potentiometry and NMR determined		
Polyamine	pK_a	Method
Spermidine (3,4)	11.16, 10.06, 8.51, —	Potentiometry[e]
	11.56, 10.80, 9.52, —	^{15}N-NMR[e]
	11.88, 10.77, 9.60, —	^{13}C-NMR[e]
	10.96, 9.91, 8.51, —	Potentiometry[f]
	11.02, 10.02, 8.57, —	^{1}H-^{13}C correlated NMR[f]
	10.90, 9.71, 8.25, —	^{13}C-NMR[g]

[a]Adapted from Palmer and Powell (1974) in 0.10 M NaCl.
[b]Adapted from Templeton and Sarkar (1985) in 0.15 M NaCl.
[c]Adapted from Gold and Powell (1976) in 0.10 M NaCl.
[d]Adpated from Barbucci et al. (1970) in 0.50 M NaNO$_3$
[e]Adapted from Takeda et al. (1983) at 0.1 M by potentiometry and 0.8–1.6 M by NMR.
[f]Adapted from Onash et al. (1984) at 0.1 M by potentiometry.
[g]Adapted from Kimberly and Goldstein (1981) at 0.037–0.160 M.

of electron spin and magnetic moment of the electron and the extension of these phenomena to certain nuclei, as well as the later demonstration of the absorption of electromagnetic energy at sharply defined radiofrequencies. An early review of the applications of spectroscopy to amines summarizes the initial contribution of NMR, among other methods, to the structure, identification, and reactivities of the amines and the inception of biochemical application (Eisdorfer et al., 1968).

NMR explores the reactivities of various atomic nuclei, that is, ^{1}H of hydrogen, ^{2}H of deuterium, ^{13}C of carbon, ^{31}P of phosphorus, ^{25}Mg of magnesium, ^{14}N of nitrogen, and ^{15}N of nitrogen. Some of the isotopic elements exist in nature in abundance, permitting the development of a signal, for example, ^{13}C; the abundance of the isotope in the compound under study can be greatly increased through synthesis.

An NMR spectrum displays the absorption of energy by the nuclei at a particular radiofrequency in a magnetic field. A plot of signal frequency versus peak inten-

Table 3.2 Reactions of primary, secondary, and tertiary amines.

Reaction	Primary (RNH$_2$)	Secondary (R$_2$NH)	Tertiary (R$_3$N)
Addition or removal of proton	RNH$_2$ + H$^+$ ⇌ RNH$_3^+$	R$_2$NH + H$^+$ ⇌ R$_2$NH$_2^+$	R$_3$N + H$^+$ ⇌ R$_3$NH$^+$
Nitrous acid	RNH$_2$ $\xrightarrow{\text{HONO}}$ N$_2$↑ +	R$_2$NH $\xrightarrow{\text{HONO}}$ R$_2$N—N═O (nitrosamine)	No reaction
Alkylation	RNH$_2$ $\xrightarrow{\text{RX}}$	R$_2$NH $\xrightarrow{\text{RX}}$	R$_3$N $\xrightarrow{\text{RX}}$ R$_4$N$^+$X$^-$
Formation of amide	RNH$_2$ $\xrightarrow{\text{R'COCl}}$ RNHCOR'	R$_2$NH $\xrightarrow{\text{R'COCl}}$ R$_2$NCOR'	No reaction
	RNH$_2$ $\xrightarrow{\text{R'SOCl}}$ RNHSO$_2$R'	R$_2$NH $\xrightarrow{\text{R'SOCl}}$ R$_2$NSO$_2$R'	No reaction

sity indicates the numbers of nuclei absorbing at that frequency. The environment surrounding the absorbing nucleus affects the magnetic field strength at which resonance is achieved, and neighboring atoms will evoke specific *chemical shifts* in the absorption and expression of resonance. The integrated intensity or area of peaks at the spectral positions characteristic of the chemical shifts are proportional to the numbers of protons involved in the structure being studied.

In an initiating ^{13}C-NMR study of the effects of protonation of the primary and secondary amines of triamines on the various methylenes, Delfini et al. (1980) observed marked chemical shifts of the various carbons as a function of pH. These workers concluded that in spermidine the inner secondary amine is protonated only after the complete protonation of the primary amines. Both primary nitrogens appeared to be protonated simultaneously, despite attachment to aliphatic chains of different length. Similarly using natural abundance ^{13}C-NMR, Kimberly and Goldstein (1981) determined the distribution of variously protonated spermidines as a function of pH (Fig. 3.1). The nonprotonated species, that is, an uncharged spermidine, makes its appearance after pH 9.5.

Takeda et al. (1983) explored both ^{13}C- and ^{15}N- natural-abundance NMR of a new asymmetric bacterial tetramine, thermospermine (3,3,4). Aiken et al. (1983) also compared pK values and protonation sequences for spermidines, spermines, and their analogues that were determined by potentiometric titration and ^{13}C-NMR. This group then analyzed the protonation of spermidine with ^{1}H- and ^{13}C-correlated NMR, on the supposition that it is important to focus on effects at the α-carbons and to minimize the long-range effects seen with ^{13}C and ^{15}N alone (Onasch et al., 1984). These results, given in Table 3.1, agree with those of Kimberly and Goldstein (1981), but differ significantly from those of Delfini et al. (1980) and of Takeda et al. (1983) More recent explorations of these techniques simplified the methods of analyzing protonation sequences in the polyamines (Hague & Moreton, 1994).

Reactions of Amines

In addition to the basicity of the amines, defined by the reversible addition or removal of a proton, aliphatic amines are characterized most easily by their reactions with nitrous acid, the addition of alkyl groups, and their conversion into amides (Table 3.2). The reactivities of the amines depend on the availability of replaceable hydrogen; in effect these permit the classification of an unknown substance as a primary, secondary, or tertiary amine. New isolated products whose nature may be somewhat obscure require additional study by modern methods such as the characterization of a suitable derivative by gas chromatography and of its fragments by mass spectroscopy (GC-MS), as well as other spectroscopic methods such as NMR.

Reaction with nitrous acid

Limited reactivities with H_2O_2 and $KMnO_4$, as well as the formation of various salts, and methyl-, benzoyl-, and phenylureido derivatives, led to the conclusion that spermine was an aliphatic tetramine (Dudley et al., 1926). Nitrous acid is known to react with amino groups, such as the ε-amino group of lysine, and has been used for many years in the estimation of free amino groups in amino acids and proteins. Nevertheless, the standard reaction system did not give unequivocal results with any of the polyamines.

Nitrate is widely used as a food preservative, for example, for inhibiting the growth of toxic *Clostridia* in canned food or sausage. Nitrite is also recognized to be a chemical mutagen, apparently as a result of the formation of nitrosamines with secondary amines (see Table 3.2). Although some but not all dietary nitrosamines are mutagens in bacterial tests (Ohta et al., 1990), there has been some interest in the detected mutagenic effects of polyamines, mainly spermidine, treated with nitrous acid. The significant dietary intake of the polyamines poses an interesting problem for the possible generation of mutagenic and/or carcinogenic nitrosamine in the presence of nitrite in acidic gastric juice. Several mutagenic nitrosamines have been described as products of the nitrosation of spermidine (see Chap. 15). The use of nitrosatable secondary amines has been determined as an approach to the content of possible carcinogens in various foods (Atawodi et al., 1995).

Many mammalian tissues oxidize L-arginine to citrulline, thus liberating physiologically active nitric oxide. Nitric oxide is converted to both nitrate and nitrite and is currently believed to activate the formation of cyclic guanylate in mammalian cells. A pertinent question is whether any portion of the nitrite generated physiologically participates with spermidine or spermine in the formation of endogenous mutagen and/or carcinogen.

Alkylation

Hofmann began the systematic study of the amines by observing the reaction of ammonia with ethyl iodide to form the series of ethylamine, diethylamine, triethylamine, and the quaternary salt of tetraethylamine. These reactions, in which the nitrogen atom displaces the halide anion, were exploited in the methylation of spermine by methyl iodide to form the quaternary ammonium base, decamethyl spermine. This indicated the presence of two primary amines and two secondary amines in spermine.

Methylation has also been used to produce volatile tertiary amines suitable for GC (Giumanini et al., 1976). The polyamines are derivatized by reaction with an excess of formaldehyde in aqueous acid to form a Schiff base. Sodium borohydride then reduces the methylene moieties to methyl groups. The now tertiary diamines, as well as similar derivatives of the higher polyamines, that is, pentamethylspermidine and hexamethylsper-

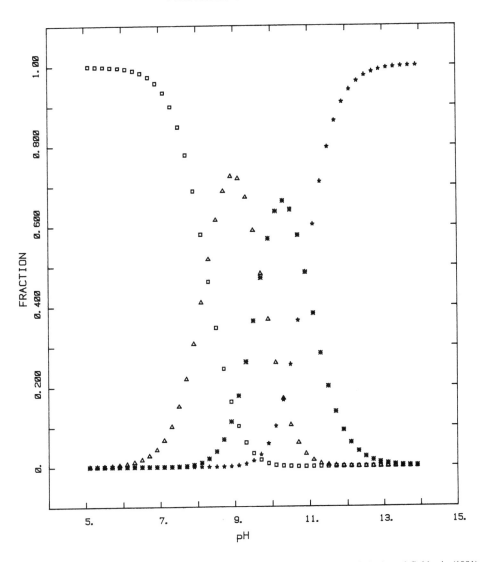

FIG. 3.1 Study by ^{13}C-NMR of species distribution of spermidine; redrawn from Kimberly and Goldstein (1981). (□) Triprotonated; (Δ) diprotonated; (○) monoprotonated; (★) nonprotonated.

mine, are extractable in ether and separate well on many chromatographic columns. It may be mentioned that Schiff bases, comprising polyamines and pyridoxal or pyridoxal phosphate, have been detected in human urine at nanomole levels (Aigner–Held et al., 1979). These compounds are similarly reduced with sodium borohydride to stable derivatives that are suitable for further steps of characterization.

As mentioned earlier, N-alkylated amines have been synthesized as potentially active therapeutic agents. Two methylated homospermidines have been described as components of tumor-inhibitory plant alkaloids. In each of these the primary amines are dimethylated to form tertiary amines, while the secondary nitrogen is acylated with a fatty acid (Kupchan et al., 1969). The structures are

$$(CH_3)_2N(CH_2)_4 \diagdown$$
$$N{-}R_{1 \text{ or } 2}$$
$$(CH_3)_2N(CH_2)_4 \diagup$$

$$R_1 = -\overset{\overset{O}{\|}}{C}-(CH_2)_{14}CH_3 \text{ in Solapalmitine}$$

$$R_2 = -\overset{\overset{O}{\|}}{C}-C{=}C-(CH_2)_{12}CH_3 \text{ in Solapalmitenine}$$

Spermine and spermidine may be described as alkyl derivatives, in which aminopropyl moieties were added to the nitrogen of diaminobutane in reactions similar to that in which the synthesis of amines was described initially. In the first syntheses by Hofmann, the halogen of an alkyl halide was displaced in a nucleophilic substitution by ammonia to form a primary alkyl amine, with halide anion as the leaving group. Thus the aminopropyl moieties of spermidine were first added in a nucleophilic substitution via the reaction of 1,4-diaminobutane and α-phenoxy-γ-bromopropane ($C_6H_5O \cdot CH_2CH_2CH_2Br$) to form $C_6H_5O \cdot CH_2 \cdot CH_2 \cdot CH_2 \cdot NHCH_2CH_2CH_2NH_2$ and bromide ion. A second bis-addition of an aminopropyl occurred in the presence of a second equivalent of alkyl bromide to form the sequence needed for spermine. The biological alkylation of putrescine to form spermidine via an aminopropyl transferase is considerably more complex, involving an unusual nucleophile that possesses a transferable aminopropyl group essential for the biosynthesis of spermidine and spermine.

Finally we note that the displacement of halogen by ammonia or an amine can proceed on an aryl moiety such as benzene. The reaction of primary and secondary amines with 2,4 dinitrofluorobenzene is an example of such a reaction.

The dinitrophenyl group was used initially to label the free amino groups of proteins and peptides and subsequently for the quantitation of free amino acids (Sanger, 1945). This method was applied to the determination of the polyamines (Dubin, 1960; Rosenthal & Tabor, 1956). The spectra of the products of reaction with primary and secondary amines differed significantly, permitting gross estimation at an isoclastic point or estimation of the separated product at each spectral maximum.

Conversion to amides

The addition of groups to form an amide, for example, N-acetylation, as in the formation of N-acetylspermidine, or of N-benzoylation, as in the formation of tetrabenzoylspermine or ornithuric acid (i.e., dibenzoylornithine), are reactions that are biologically and chemically significant acylations.

Acylation

The chemical synthesis of N-acetyl or N-benzoyl derivatives via reaction of the amine with an acid chloride is also describable as a nucleophilic substitution in which the leaving group is the halide anion. Biological acylation is commonly effected in the reaction of the nucleophilic amine with a complex acyl donor, resulting in the transfer of the acyl moiety to the nucleophilic nitrogen and the release of the nonacylated complex donor.

Among the natural acylated polyamines are the N-acetyl polyamines and the more recently recognized compounds such as the agrobactins and kukoamine (Funayama et al., 1995). The immobilizing venoms of wasps and spiders contain compounds described as philanthotoxins, which are tyrosyl derivatives of the primary amine of an unusual natural spermine analogue (4,3,3). Solapalmitine and solapalmitenine are examples of acyl substitution on the secondary nitrogen of homospermidine.

After isolation of protein-free amine salts from a biological sample, the separation of the amines can be attempted and the separate amines estimated via an acylation reaction, or the amine mixture can be acylated first and their derivatives separated and estimated. In either case acylation to form a characteristic volatile, colored, or fluorescent derivative is usually an essential step. Mono N-acetylated polyamines are found so frequently in nature that their identification in assay systems has been well developed. The synthetic volatile trifluoroacetyl and heptafluorobutyryl derivatives of the polyamines have been useful in the GC of the polyamines.

N-Benzoyl derivatives of the polyamines have not been reported in organisms, for example, in birds fed benzoic acid. Benzoyl moieties provide a group that absorbs in ultraviolet light at characteristic wavelengths, and possess properties that facilitate separation in various chromatographic systems. However, several difficulties (to be noted later) have emerged in developing benzoylation as an approach to polyamine estimation.

Sulfonamides are easily formed by treatment of primary and secondary amines with sulfonyl chlorides. Indeed the reactions of benzene sulfonyl chloride in potassium hydroxide with the various classes (given in Table 3.2) constitute a classical approach (the Hinsberg method) to the typing of the compounds. In contrast with the carboxamide, the sulfonamide derived from a primary amine contains an acidic hydrogen, which permits the formation of a soluble potassium salt, or an insoluble product in acid. A sulfonamide derived from a secondary amine lacks the acidic hydrogen and is insoluble in alkali or acid. Finally a tertiary amine tends to be insoluble in alkali and soluble in acid. The sulfonamides hydrolyze more slowly than do carboxamides, and this is useful in the isolation and characterization of fluorescent sulfonamides. The dimethylaminonaphthylsulfonamido compounds, commonly described as dansyl derivatives, are widely used in the separation and estimation of the polyamines (Seiler and Wiechmann, 1967). The reagent dansyl chloride was introduced initially to produce fluorescent proteins (Weber, 1952) in which the dansyl group could be excited by polarized light and the reaction of the group would relate to the behavior of the modified protein.

Dansyl chloride   A dansyl amine

All primary and secondary amino groups of the diamines, spermidine, and spermine and their most common derivatives are labeled quantitatively with this reagent. The fluorescent derivatives are readily separated in thin-layer chromatography and can be scanned on a thin-layer plate or extracted and read in a spectrofluorimeter.

However, the dansyl reagent does react with sugar hydroxyls, nucleic acid bases, and some nucleotides to yield separable and determinable fluorescent products. Some biological systems may be rich in reactive polyols or other compounds and many of these dansyl derivatives are selectively cleaved and eliminated by hydrolysis in heated methanolic KOH (Seiler & Deckardt, 1975).

Reactions of Quantitative Analysis

Ninhydrin

This substance (triketohydrindene hydrate) is well known in the detection and quantitation of amino acids and was adapted to a sensitive estimation of the polyamines (Hammond & Herbst, 1968). Amines are oxidized to form an aldehyde and ammonia and the products of most polyamines combine with a molecule of ninhydrin to form a purple product. Many of the recently discovered polyamines or the newly synthesized polyamine analogues have been shown to react with ninhydrin on thin layer plates (Shirahata et al., 1983). One proposed reaction sequence, extrapolated from that given for most amino acids, is given below.

It is possibly relevant to the proposed mechanism that the colored products produced with 1,3-diaminopropane and norspermidine are red.

The ninhydrin reaction carried out in alkaline solution (0.8 M NaOH) on guanidine compounds, for example, homoagmatine and homoarginine (Ramakrishna & Adiga, 1973a,b), was recently found to give rise to fluorescent products that are separable in reversed-phase liquid chromatography and adaptable for sensitive and specific quantitation at the nanogram level (Boppana & Rhodes, 1990).

Fluorogenic reactions

The reaction of ninhydrin with phenylalanine produced phenylacetaldehyde, and it was observed that this product gave rise to a fluorescent product when a primary amine was added to the system. This reaction became useful in the sensitive detection and assay of serum phenylalanine in phenylketonuria, and the nature of the fluorescent product was subsequently determined to be a pyrrolinone. This sequence of findings led to the design of a fluorogenic reagent, fluorescamine, specific for primary aliphatic and aromatic amines (de Silva & Strojny, 1975; Imai et al., 1974). The following is the chemical reaction of fluorescamine with primary amines (redrawn from de Silva & Stojny, 1975):

Fluorescamine   Ninhydrin
                                                    +

                                         OHC

                     R-NH₂
                  [primary aliphatic
                       or   Phenylacetaldehyde
                  aromatic amine]

"Major fluorophor"
Excit : Max : 275, 390/Em : Max : 480 nm

This reagent was employed to estimate diamines and higher amines separated on phosphocellulose columns (Furuta et al., 1985).

Another sensitive fluorogenic reagent is *o*-phthalaldehyde (OPT), which has proven quite useful in derivatization in liquid chromatographic systems. Although the formation of fluorescent products in reactions of OPT and the polyamines and histamine had been observed earlier, the demonstration of a role for a thiol in a reaction of OPT and amino acids at pH 9.0 (Roth, 1971) led to the present exploitation of the reagent. A fluorescent product is also formed with glutathione in the absence of an added sulfhydryl. The reagent was eventually adapted to the fluorometric analysis of the polyamines and histamine of urine samples separated in a liquid chromatographic system (Perini et al., 1979). The reagent is specific for primary amines and, as with fluorescamine, can detect polyamine in the nanomolar range.

The fluorescent product is a 1-alkylthio-2-alkyl substituted isoindole. If the reaction is carried out with ethanethiol instead of mercaptoethanol, the rearrangement of the initial product (given below) to a substance of lower fluorescence occurs far less rapidly (Simons & Johnson, 1977).

Other Reactions

Chloranil forms a relatively stable complex (1:1) with 1,3-diaminopropane at pH 9 with a relatively high extinction coefficient ($E = 28,000$ mol^{-1}cm^{-1}) at the λ_{max} (350 nm) (Sulaiman et al., 1984). The significance of this reaction is possibly increased by the apparent similarity of the structure of chloranil to the toxic agent, 2,3,7,8-tetrachlorodibenzo-p-dioxin. The latter compound apparently reduces the polyamine content of several tissues in mice. The toxicity of the agent was markedly heightened by the lowering of the polyamine content of the tissues and was reduced by the administration of putrescine (Thomas et al., 1990). The structures of the compounds are given below.

Chloranil   2,3,7,8–Tetrachlorodibenzo–p–dioxine

Oxidative Reactions

Biological oxidation of polyamines is understood to be a most complex set of reactions that involves the possible production of many potentially oxidizing radicals in normal metabolism and in the response to chemical and physical agents.

$$CH_3CHNH_2COOH \xrightarrow[H_2O_2]{Fe^{2+}} CO_2 + NH_3 +$$

$$CH_3CHO \xrightarrow{O} CH_3COOH$$

Several intermediates that may be generated in these reactions are toxic and may be eliminated by enzymatic reactions. Reference will be made in a later chapter to the toxic effects of the widely used herbicide, paraquat, mediated via a polyamine transport system.

On the other hand the polyamines have been described as antioxidants in some systems. For example, hydroxyl radicals (ȮH) generated by photolysis will react with polyamines and other amines below pH 7 to remove a hydrogen atom from a carbon not adjacent to the protonated amine. In alkaline solution, hydrogen is abstracted from the amine nitrogen. The different products will react characteristically with 2-methyl-2-nitrosopropane as a spin trap to form structures identifiable by electron spin resonance (Mossoba et al., 1982).

Separation Techniques

As described by Herbst et al. (1958), paper chromatography of the polyamines had not been actively pursued until relatively late. Hammond and Herbst (1968) performed assays for total growth factor activity for bacteria followed by fractionation to detect one or more biologically active polyamines by paper chromatography and paper electrophoresis. They also detected the position of the amines with ninhydrin and, as noted earlier, subsequently adapted this reaction to the quantitation of the polyamines in thin-layer chromatography.

In the 1950s an automated ion-exchange system was devised for the complete fractionation and detection of the components of a protein hydrolysate by ninhydrin (Spackman et al., 1958). This apparatus, which was a model for the development of commercially available equipment, facilitated the evolution of the modern ion-exchange or reversed-phase separation of biological components, including the polyamines, in high-pressure liquid chromatographs. This automated separation technique (HPLC), combined with one of the several extraordinarily sensitive fluorescent labeling approaches, has become the analytical method of choice in many laboratories. Nevertheless, many other laboratories have continued to use less elaborate, less expensive, and often less cumbersome techniques that are suitable for their needs and resources or have devised useful improvements on even the most recent advanced equipment. Difficulties in the simultaneous estimation of basic amino acids and polyamines in a crude extract have compelled the development of an intermediate separation of neutral and acidic amino acids prior to the use of the HPLC equipment (Gilbert et al., 1991).

The evolution of our progress in detecting and estimating ever smaller amounts of the polyamines was summarized (Table 3.3) by Bachrach (1978). It may be mentioned that the finding of one to two molecules of polyamine per tRNA molecule was entirely dependent on the then recent introduction of the sensitive fluorescent dansyl reagent and scanning instruments for the commercially prepared thin-layer chromatographic plates

Table 3.3 Sensitive analytical methods for polyamine assay.

Method	Details	Sensitivity (pmol)
Thin-layer chromatography	Dansyl derivatives	20–100
Mass spectrometry	Dansyl derivatives	
Automated ion-exchange chromatography	Ninhydrin conventional	1,000–2,000
	o-Phthalaldehyde	100–200
	Fluorescamine	200
Gas–liquid chromatography	Putrescine	100,000
	N-Trifluoroacetyl derivatives	200–600
		100
	Pentafluorobenzyl derivatives, electron capture	1
Gas chromatography–mass spectrometry	N-Trifluoroacetyl derivatives, deuterated internal standard	1
High-pressure liquid chromatography	Ninhydrin	25–100
	Dansyl derivatives	25–50
Radioimmunoassay		1
	Purified antibodies	0.05
Immunonephelometry	Latex beads	1–10

Adapted from Bachrach (1978).

on which the polyamines were separated (Cohen et al., 1969).

Paper Chromatography

In 1957 Hershey used this procedure to study the metabolism of ^{14}C-labeled arginine in bacteriophage-infected cells and detected some unknown basic products of this metabolism of arginine in the isolated virus. Paper chromatography revealed the substances to be putrescine and spermidine, and these components were also found to be present in high concentrations in *Escherichia coli* (Ames et al., 1958). Although 2-dimensional chromatographic systems had been developed for the separation of numerous types of compounds, often this was not needed for the limited number of polyamines sought in many organisms. Nevertheless, Hershey used such a system to separate amino acids from the polyamines.

Paper Electrophoresis

This technique was an obvious extrapolation of the basicity of the free amines, which facilitated a separation at low pH, from many possible contaminants such as amino acids. Figure 3.2 presents a ninhydrin-stained 2-dimensional paper chromatogram and electrophoresis of amines detectable in a deproteinized acid extract of cow manure (van Rheenen, 1962). Putrescine and cadaverine were separable from each other in a phenol–water system but were not clearly separable from lysine and taurine, although these latter acidic molecules were readily separated electrophoretically from the free fully charged diamines. The rapidity of the process was increased in "high voltage" electrophoresis and subsequently this was used in scanning the amines extracted from the

urines of cancer patients (Russell, 1971). Paper electrophoresis provided the first demonstration that many cancer patients excreted more polyamines than do normal and other noncancerous individuals. However, the method is not very sensitive and yielded questionable values for urinary spermine.

Thin-Layer Chromatography

Similarly adaptable to small inexpensive jars and providing even more rapid separations, thin-layer plates can handle more material than can paper and became a

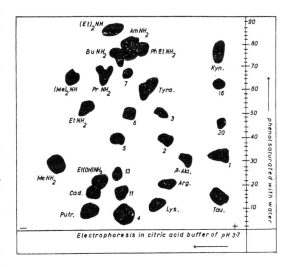

FIG. 3.2 Chromatogram and electrophoretic pattern of a feces extract; redrawn from van Rheenen (1962). Putr., putrescine; Cad., cadaverine; Lys., lysine; Arg., arginine.

separation technique of choice for the organic chemist. Several adsorbents, such as silica gel and cellulose, and complex solvents, such as the carbitol and methyl cellosolve systems, were exploited by Hammond and Herbst (1968) in the 2–5 h separation of the common polyamines. In one of these systems, the separate mobilities of 13 polyamines and five basic amino acids were determined, as well as the colors of the ninhydrin stain.

The application of the dansyl reaction to the thin-layer chromatography of the amines was carefully standardized by Seiler and Wiechmann (1965) and was shown to be capable of the determination of picomole amounts of amine. This was 50–60 times more sensitive than were other methods and was adapted to analysis of the amines, including the N-acetyl polyamines in tissues and biological fluids. Limitations in the separations of the dansyl derivatives of putrescine from ammonia and of spermine from dansyl by-products of the reaction were noted. Fluorescent impurities detected on commercially available precoated plates were eliminated and the quality of the reagent itself was improved. This step was rendered particularly necessary by efforts to use tritium-labeled dansyl chloride and ^{14}C-labeled amines in more precise analyses by double-isotope dilution techniques (Ni, 1986; Paulus & Davis, 1983). The discussion of the dansyl reaction and its application in thin-layer chromatography by Seiler (1983) is a valuable summary of 20 years of experience with the method.

This method has been useful in my laboratory, with the aid of a continuous scanning device, thereby avoiding the tedious scraping and extraction techniques. Eventually sophisticated sensitive double-beam reflectance spectrophotometers became available (Gurkin & Cravitt, 1972) with which the plates could be scanned at wavelengths in the ultraviolet and visible spectral regions. The instruments are capable of integration of area and determination of total absorbing material within the spots, as well as the determination of excitation and emission spectra and visible and ultraviolet absorption. Such an instrument is obviously suitable for estimation of dansyl derivatives as well as for many other aspects of the thin-layer chromatography of the amines. This type of scanning system has been used in various other thin-layer techniques such as the estimation of the fluorescence of urinary polyamines treated with OPT after electrophoresis on cellulose acetate as in Figure 3.3 (Kanda et al., 1985), or measurement of the intensity of ninhydrin-stained tissue polyamines after chromatographic separation on cation-exchange coated papers (Bardócz et al., 1985).

Ion-Exchange Separation and HPLC

Tabor and Rosenthal (1956) estimated the polyamines as dinitrophenyl derivatives. The amines were absorbed on columns of cation-exchange resins and separated by batch elution with acid, or by gradient elution. In the selective and rapid estimation of tightly bound spermine, it is convenient to elute a column with a single

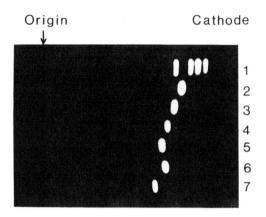

Origin   Cathode

FIG. 3.3 Electrophoretic patterns of polyamines, diamines, and monoamines; redrawn from Kanda et al. (1989). Lane 1, authentic polyamine mixture (putrescine, cadaverine, spermidine, and spermine in this order from the right); lane 2, 1,7-diaminoheptane; lane 3, 1,8-diaminooctane; lane 4, 1,9-diaminononane; lane 5, 1,10-diaminodecane; lane 6, n-propylamine; lane 7, n-butylamine.

concentration of HCl to remove essentially everything but spermine and to obtain this fraction at a higher concentration of acid. It is possible to remove 99% of the excess spermidine from a defined Dowex-H$^+$ column with 3.3 N HCl, while 93% of the spermine remains bound and is eluted with 6 N HCl (Sindhu & Cohen, 1983).

In a recent study upgrading the early analysis depicted in Figure 3.2, a recovery by ion-exchange separation of amines extracted from samples (1 g) of human feces was used in the synthesis of dansyl derivatives, which were then separated by 2-dimensional thin layer chromatography. The many dansyl spots were analyzed by field desorption MS and were found to include a large number of polyamines and their acyl derivatives, including 2-acetoxy derivatives of putrescine and cadaverine (Murray et al., 1993).

In the earliest studies, gradient elutions were unable to provide clean separations of the numerous now-known overlapping compounds of polyamine metabolism. As the systems were automated over a decade, the elution patterns were improved as smaller amounts of the amines could be analyzed (Fig. 3.4). Similarly automated equipment, primarily designed originally for the estimation of amino acids, confirmed the enhanced polyamine content of the urine of many cancer patients. This study noted the elimination of an artifact of an apparently exaggerated spermine content of such urines (Marton et al., 1973). These methods were subsequently adapted to the analyses of amino acids and polyamines in cerebrospinal fluid, gingivalcrevicular fluid, ocular humors, bile, synovial fluids, etc., and many tissues.

Automated detection of separated amines with fluorogenic reagents, for example, OPT, permitted the continuing use of the commercial automated amino acid ana-

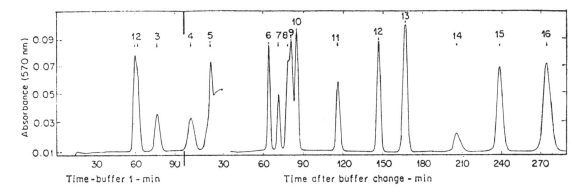

FIG. 3.4 Elution pattern of amines; redrawn from Tabor et al. (1973). A mixture containing 20 nmol of each amine was applied to the column. Buffer and temperature were changed at 95 min.

lyzers with ion-exchange resins. As can be seen in Figure 3.5, the separation and detection of the multitude of biologically significant metabolites in urine in such a system poses problems of the identity of a particular peak. Indeed the sensitivity of the method, that is, a picomole level of determination, requires considerable care in preparation of the buffers, elimination of contaminating compounds, and frequent standardization of the system. Elimination of many of the possible complicating compounds was accomplished by pretreatment of the samples (Whitmore & Slotkin, 1985).

Difficulties with the instability of fluorescamine led to prederivatization and separation of the fluorescamine derivatives on reversed phase columns. These were more rapid than the ion-exchange systems and even more discriminating in some analyses. An early study, involving partition chromatography on an octadecylsi-

lane column with a gradient elution of methanol in 0.1 M borate at pH 8.0, detected picomole levels and differences of a single methylene group among the compounds tested (Samejima, 1974). The fluorescamine had formed stable derivatives of only the primary amines.

The column chromatography of dansyl polyamines was described in 1975 and subsequently adapted to the isolation and characterization of the N-acetyl polyamines in the urine of cancer patients (Abdel-Monem & Ohno, 1977). A routinized HPLC procedure directed to prederivatized dansyl polyamines appeared somewhat later. The derivatives, that is, standards or tissue extracts, were extracted in toluene and separated from dansyl amino acids on a silica gel column by ethyl acetate. The dansyl amines were dissolved in aqueous methanol, applied to a prepacked commercial column, and eluted in a water–methanol gradient. The separation

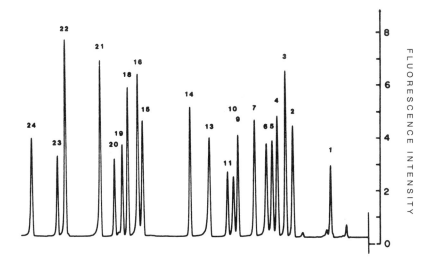

FIG. 3.5 Separation of polyamines and related compounds; redrawn from Seiler and Knödgen (1985). Fluorescence intensity after reaction with *o*-phthaldehyde-2-mercaptoethanol. Amount of each compound is 1 nmol/0.1 mL.

of the dansyl amines from mouse liver is presented in Figure 3.6. Because dansyl chloride reacts with the primary and secondary amines, the fluorescence intensities of equimolar concentrations of the polyamines increases: spermine > spermidine > diamines. Dansyl ammonia and some side products are eluted early, and many of the dansyl biogenic amines (histamine, tyramine, serotonin, etc.) are normally low in tissues and have a low fluorescence quantum yield. The method was modified to include a rapid determination in tissue extracts of γ-aminobutyric acid (GABA), which is estimated as N-dansyl-2-pyrrolidinone (Desiderio et al., 1987).

The system was used to study the blood of cancer

patients, and revealed unexpectedly the accumulation of blood polyamines in erythrocytes (Saeki et al., 1978). Most recently a procedure has been devised to detect and determine the very low polyamine content of sea water (Cann–Moisan et al., 1994). The method of examining prederivatized polyamines by reversed-phase chromatography has achieved broad attention and use and has been applied to the examination of the OPT derivatives also, because this eliminates the apparatus associated with post-column derivatization (however, see Saito et al., 1994 and Watanabe et al., 1993).

The study of the partitioning of ion-pair complexes of amines and acidic organic compounds, such as picric acid and alkyl and aryl sulfonic acids, between immiscible solvents added an additional differentiating characteristic to the amines and their derivatives. It was found that the addition of 1-heptane sulfonic acid to an acetonitrile gradient did in fact facilitate the separation of the dansyl polyamines of hydrolyzed urine (Brown et al., 1979). Nevertheless, Bidlingmeyer (1980) concluded that the success of the ion-pair addition was not well understood. In any case the addition of an acidic ion, such as alkyl sulfonate or dodecyl sulfate, has indeed been useful in the separation of free amino acids or free amines followed by post column derivatization with OPT-2-mercaptoethanol. This procedure and its improved separation (see Fig. 3.6) extended the range of detectable substances potentially present in human urine.

The addition of an ultraviolet absorption detector (fixed at 254 nm) to HPLC systems also enabled the discovery of several nucleosides involved in polyamine metabolism (Wagner et al., 1982, 1984). A system will be demonstrated in Chapter 20 in which these nucleosides are in fact separated from the simpler array of polyamines present in plant cells in a single chromatographic run and the ^{14}C of methionine is found in S-adenosylmethionine and its metabolic derivatives, which include nucleosides and certain polyamines (Greenberg & Cohen, 1985). Thus, desirable analytical features in polyamine research include the separation of certain amino acids and nucleosides, as well as the amines themselves.

In systems in which sensitivity at picomole levels was not a major consideration, the possible use of an ultraviolet detector was expected to simplify the equipment to be used. Several groups attempted to use benzoyl derivatives, and a sensitive procedure suitable for mammalian cells and fluids was described (Wongyai et al., 1989). However, the derivatization with benzoyl chloride does require incubations of the reagents, extraction with an organic solvent (i.e., chloroform or ether), washing of the solvent, and concentration before reversed-phase HPLC. In plant extracts it was found that an early benzoylation method was inadequate for certain metabolites, such as agmatine, the product of the important plant enzyme arginine decarboxylase. One laboratory improved the gradient elution to permit an improved isolation of agmatine and other amines (Slocum et al., 1989). Another compared the benzoyl method with a reversed-phase procedure using the derivatives

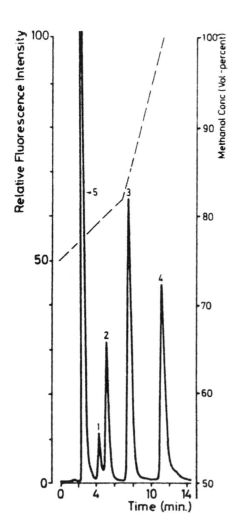

FIG. 3.6 Separation of dansylamine derivatives of a normal mouse liver using an elution program adjusted for rapid determination of spermidine and spermine; redrawn from Seiler et al. (1978). (1) bis-Dns-1,4-diaminobutane (putrescine); (2) bis-Dns-1,6-diaminohexane (internal standard); (3) tris-Dns-spermidine; (4) tetrakis-Dns-spermine; (5) Dns-ammonia.

of OPT-mercaptoethanol. Reportedly some polyamines were markedly underestimated in the benzoyl method (Corbin et al., 1989). These workers prepurified the extracts by an ion-exchange procedure and presented some evidence that plant extracts may contain components that are inhibitory to benzoylation and may be partially removed by ion exchange. Wehr (1995) has described a purification of plant polyamines in an extract with anion exchangers prior to derivatization.

Pressure to simplify, accelerate, and detect a broader range of polyamines and their derivatives at ever smaller levels evoked a continuing stream of trials of new fluorescent reagents. Thus 9-fluorenylmethylchloroformate is reported to serve as a useful precolumn derivatizing agent (Wickström & Betnér, 1991).

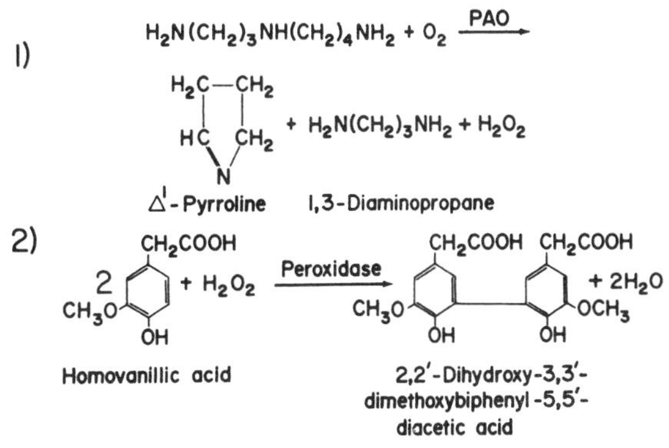

Enzymes and Enzymatic Assays

The enzymatic assay of the polyamines has been a useful addition to the armamentarium of available chemical and physical methods. Following the early discovery of the amine oxidase, histaminase, which oxidatively deaminates histamine to an aldehyde and hydrogen peroxide, many different amine oxidases of various specificities were described. The various amine oxidases will be discussed in greater detail in Chapter 6. The sensitive detection of the various products, particularly that of H_2O_2, facilitated assay by several different methods that were explored. The growing evidence of significant roles of the polyamines in the physiology of the organisms and apparent distortions of their normal concentrations in pathologies posed the problem of the possible performance of numerous analyses in clinical laboratories, as well as in biological research in general. Thus one might be interested in total polyamine excretion or in that of particular amines, and indeed commercial kits have become available to assist and simplify such analyses.

As one example of the direct estimation of a specific polyamine, the polyamine oxidase of oat seedlings is unreactive with putrescine and is useful for the specific assay of spermidine and spermine. The enzyme has been used for the assay of spermidine synthase and permits the assay of 100 samples in 3 h (Sindhu & Cohen, 1983; Suzuki et al., 1981). The reactions to form a fluorescent product proportional to the spermidine present in the range of 1–10 nmol are given in Figure 3.7.

The amine oxidase of soybean seedlings has been reported to generate H_2O_2 from putrescine, spermidine, and spermine. The chemiluminescence generated by luminol in the presence of H_2O_2 and peroxidase has been used to estimate total polyamine in human serum (Fagerström et al., 1984). Differential assays have been described for the various specific and total polyamines in biological fluids (Bachrach & Plesser, 1986).

Because much of the polyamine of urine is present as acetyl derivatives, some methods have employed an initial acid hydrolysis while others have exploited an acyl-polyamine amidohydrolase of bacterial origin (Kubota et al., 1983). Reportedly recoveries of added polyamines

FIG. 3.7 Coupled enzymatic analysis of spermidine; redrawn from Sindhu and Cohen (1983).

in the 93–96% range have been obtained in acid hydrolysis of urine samples. It may be asked if similar recoveries would occur with the urine of a diabetic who excretes significant amounts of carbohydrate, the acidic degradation products of which can react with amines.

The detection, isolation, and estimation of amines as reaction products have been extended to the analysis of seven different aliphatic amino acid decarboxylases, including those for arginine, lysine, and ornithine, to give rise to agmatine, cadaverine, and putrescine, respectively, as shown in Figure 3.8 (Kochhar et al., 1989). In another study, the isolation of the polyamine products has been useful in the measurement of tissue polyamine oxidase (Halline & Brasitus, 1990).

A postseparation column comprising an immobilized plant or microbial polyamine oxidase reacts with the normal range of mammalian polyamines to produce H_2O_2 that is detectable electrochemically (Maruta et al., 1989; Watanabe, 1992). This detection system was used for the urinary polyamines in clinical investigation and was described as rapid, relatively reproducible, and durable. One group of workers used immobilized polyamine oxidases to generate H_2O_2, which would then generate chemiluminescence to be read with a luminescence detector. It was reported that the sensitivity and accuracy of this method in analysis of urine compared favorably with and potentially improved upon a less specific OPT method (Kamei et al., 1989).

It may be noted that a growing pharmacological and therapeutic interest in the metabolism of polyamine analogues may now lead to the study of compounds that are difficult to separate, which may also prove insensitive to the detecting system or require changes in derivatization. For example, tetrafluoroputrescine, of interest in tumor localization, ionizes at a much lower pH than does putrescine and requires a markedly different pH condition for dansylation (Hawi et al., 1988).

Additionally the cellular fate of some potent inhibitors has been followed by HPLC methods. Thus guanidine compounds, including difluoromethyl arginine and hirudonine, which are detectable by 9,10-phenanthrene-

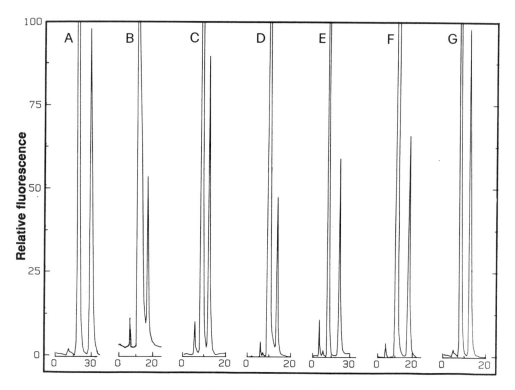

Retention time, min

FIG. 3.8 HPLC separation of substrates and products of aliphatic amino acid decarboxylases; redrawn from Kochar et al. (1989). Of the large peaks, the first is the substrate and the second is the product. (**A**) arginine decarboxylase; (**B**) aspartate α-decarboxylase; (**C**) 2,6-diaminopimelate decarboxylase; (**D**) glutamate decarboxylase; (**E**) histidine decarboxylase; (**F**) lysine decarboxylase; (**G**) ornithine decarboxylase.

quinone, have been studied for their effects on trypanosomes (Hunter & Fairlamb, 1990). Uptake and cellular content of aminooxy analogues of the polyamines have been monitored with OPT after reaction with [2-^{14}C] acetone (Hyvönen et al., 1992). Some enantiomers of chiral derivatives of ornithine, for example, α-difluoromethyl ornithine, can be separated in HPLC systems as L-proline–copper complexes (Wagner et al., 1987).

Gas-Liquid Chromatography (GLC) and GC-MS

Early initial studies in the GLC of the methylamines, and subsequently with pyridines, led to studies of the volatile pyridine alkaloids of tobacco. It was then seen that the polar diamines presented many problems of loss on the columns, tailing patterns of elution, difficulties of improvements, detections, etc. Van den Heuvel et al. (1964) studied many derivatives of putrescine and cadaverine, and pointed to the excellence of results obtainable with trifluoroacetyl and heptafluorobutyryl derivatives. Nevertheless, attempts were made to improve the separation of the free amines by adding alkali to

the various phases, thereby converting the amines to the uncharged compounds; apparently successful initial results were not reproducible. In one study of the GLC of the free amines, the polyamines extracted from turnip yellow mosaic virus were reported to lack spermidine and to contain norspermidine, a polyamine not found in the normal plant. The subject was reexamined by several groups who reported a large amount of spermidine in the virus and who were unable to find norspermidine. The absence of norspermidine in the trifluoroacetylated viral polyamines was proven with the aid of GC-MS (Cohen & Greenberg, 1981).

In the early 1970s Gehrke et al. (1973) demonstrated the applicability of GLC to the analysis of the trifluoroacetyl polyamines of acid-hydrolyzed urine samples. The method was reported to give recoveries of 90, 84, and 56% for putrescine, spermidine, and spermine, respectively. Similar results were obtained by Walle (1973), who also described the presence of a monoacetylspermidine in the urine of a leukemic patient. Walle extended the work to the determination of the mass spectra of several trifluoroacetyl polyamines. In an analysis of a mixture, single volatile components can be drawn into the mass spectrograph. The formulae of the deriva-

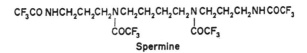

A $CF_3CONHCH_2CH_2CH_2CH_2NHCOCF_3$

putrescine

$CF_3CONHCH_2CH_2CH_2CH_2\,NCH_2CH_2CH_2NHCOCF_3$
$\qquad\qquad\qquad\qquad\quad |$
$\qquad\qquad\qquad\qquad COCF_3$

Spermidine

$CF_3CO\,NHCH_2CH_2CH_2N\,CH_2CH_2CH_2CH_2\,N\,CH_2CH_2CH_2NHCOCF_3$
$\qquad\qquad\qquad\quad |\qquad\qquad\qquad\qquad\quad |$
$\qquad\qquad\qquad COCF_3\qquad\qquad\qquad COCF_3$

Spermine

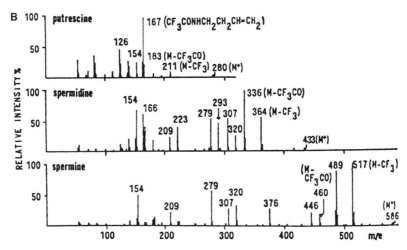

FIG. 3.9   Structures and mass spectra of trifluoroacetylated polyamines; redrawn from Walle (1973). (**A**) Trifluoroacetylated putrescine, spermidine, and spermine; (**B**) normalized mass spectra of trifluoroacetylated putrescine, spermidine, and spermine.

tized compounds and the mass spectral fragmentation patterns are presented in Figure 3.9.

Characteristic molecular ions of mass per charge (m/e) were obtained, as well as identification of the most intense fragment ions. All the molecular ions had also lost CF_3 to yield fragments decreased in m/e by 69. A characteristic ion fragment ($154 = m/e$) is given by trifluoroacetylated di- and polyamines and is given by the structure ($CF_3CONHCH_2CH_2CH_2$). It is possible to record the mass spectra of only those compounds giving rise to such an ion. This procedure, known as single or selected ion monitoring or SIM (Jenden & Cho, 1979), may be applied to a complex mixture of urinary trifluoroacetylated amines and other compounds and reveals in a far more sensitive manner those compounds, that is, the polyamines, that produce such a characteristic fragment and ion.

In the GC-MS study of the triamines of turnip yellow mosaic virus, norspermidine and spermidine gave rise to characteristic molecular ions of 350 and 364, respectively (Fig. 3.10). The viral polyamines only showed a component of 364 m/e (Cohen & Greenberg, 1981).

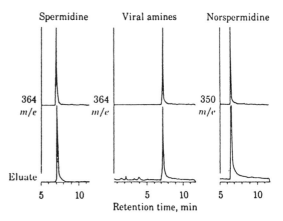

FIG. 3.10   Gas chromatography and mass monitoring of trifluoroacetylated derivatives of triamines and viral di- and triamines; redrawn from Cohen and Greenberg (1981). Derivatives of spermidine and norspermidine were monitored at 364 and 350 m/e, respectively. The viral triamines were similarly monitored; a signal was obtained only at 364 m/e.

The synthesis of ^{15}N-enriched polyamines was mentioned earlier. Although the pentafluoropropionyl derivatives of a mixture of ^{14}N- and ^{15}N-spermidines could not be resolved, the fragmentation patterns of the ^{14}N-spermidine and differently labeled ^{15}N-spermidines were detectably different (Samejima et al., 1985). This permitted the development of a computer program for the analysis of mixtures of the various spermidines. One application established that ^{14}N-putrescine and ^{15}N-putrescine were handled identically by a rat spermidine synthase. Nevertheless $^{15}NH_3$ may be readily distinguished as the pentafluorobenzamides, a method that has proved useful in following the oxidation of putrescine in plasma (Fujihara et al., 1986).

Internal standards to improve the sensitivity, selectivity, and accuracy in analysis of the trifluoroacetyl derivatives at subnanomolar levels were discussed by R. G. Smith et al. (1978). This laboratory prepared the deuterated forms of the natural compounds and used these as internal standards for purposes of close comparison and quantitative assay.

In a study of putrescine and its relation to GABA in regions of the mouse brain, very small samples of brain tissue were extracted and putrescine was converted to the pentafluoroproprionyl derivative for GC-MS by the SIM method. The strongest and most specific ion peak for the putrescine derivative (217 m/e) was monitored, as was that for the peak at 232 m/e, characteristic of the internal standard 1,6-diaminohexane (Noto et al., 1987).

Because urine is so complex and acid hydrolysis can add to its complexity, various cleanup procedures were developed prior to the actual GLC procedure to remove amino acids before derivatization and to expand the analysis to include some polyamine metabolites. One such analysis measures urinary creatinine as well as polyamine as their trifluoroacetyl derivatives and expresses the polyamines as a function of creatinine excretion (Jiang, 1990). Few efforts have been made to estimate airborne polyamines. Muskiet et al. (1995) discussed in detail the numerous cleanup and preparative steps necessary for the GC-MS and SIM approaches to the characterization of the numerous acidic and basic metabolites of the polyamines present in the tissues, cells, and extracellular fluids of rats and humans.

Recently, the GC and GC-MS properties of 27 heptafluorobutyryl derivatives of linear di-, tri-, tetra-, penta-, and hexamines have been described and compared (Niitsu et al., 1993). A selective trifluoroacetylation of primary amino groups of the polyamines has been reported (O'Sullivan & Dalrymple, 1995).

Immunoassay

There have been limited efforts to develop this approach to the quantitation of the polyamines. A major problem has been that of the specificity of the antibody produced. In general a polyamine has been used as a haptene linked to a protein and the complex has been mixed with an adjuvant and injected into experimental animals, for example, rabbits, goats, etc. After an appropriate course of immunization, the presumed antisera were tested in precipitation with free polyamine or the polyamine bound to an unreactive protein or polypeptide, for example, poly-L-glutamic acid. The latter would obviously neutralize the charge of the polyamine, for example, spermine, and minimize the possibility of nonspecific precipitation of an anionic serum protein.

In an early study an antiserum was produced that was capable of forming a characteristic antigen–antibody precipitation curve between the isolated γ-globulin and spermine-polyglutamate (Quash et al., 1973). The precipitate was solubilized by excess antigen; it also fixed the complement. The presence of free polyamine inhibited antibody precipitation, and it was shown that the inhibition was affected to slightly different degrees by the different polyamines but not by lysine. Absorption of different antibodies by the different polyamine complexes indicated that an antiserum prepared against spermine contained antibody relatively specific for spermine and spermidine when antibody against the diamines had been removed.

Immunization against a spermidine–thyroglobulin conjugate produced an antibody essentially specific for spermidine, with a very slight cross-reactivity (2% or less) with spermine, putrescine, and cadaverine and essentially unreactive with ^8N-monoacetyl spermidine, monoacetyl putrescine, lysine, and ornithine (Bartos et al., 1978). Such an antiserum bound tritium-labeled spermidine. Free (^3H)-spermidine was eliminated by absorption on charcoal and the tritium remaining in solution was measured. This procedure was converted to a radioimmunoassay (RIA) by diluting standard ^3H-spermidine with the spermidine of assayable materials. The dilution of the standard ^3H-spermidine by known amounts of spermidine gave rise to a calibration curve, linear over the range of 10^2–10^4 pg. The procedure permitted measurement of as little as 1.5 pmol of spermidine per sample and gave values for nonextracted, unhydrolyzed, and unfractionated human serum in close agreement with those obtained by GC-MS.

Earlier efforts to produce antibody to spermine had resulted in somewhat less specific antispermine antisera that possessed significant cross-reactions with spermidine (Bartos et al., 1975). A similar antispermine antibody cross-reactive with spermidine was described more recently (Fujiwara et al., 1983) and the estimation of the polyamines in serum employed a more complex enzyme immunoassay. Antisera prepared against spermidine conjugates also proved to be nonspecific (Fujiwara & Kitagawa, 1993). A similar result was obtained for monoclonal antibody (Schipper et al., 1991). In addition to the difficulty of producing specific antisera, specific anti-putrescine antibodies have not yet been generated in response to protein-bound putrescine.

Quash and colleagues (1973) pressed the problems of immunoreactivity in somewhat different analytical directions. Coupling an antipolyamine γ-globulin to latex spheres chemically, they studied the deposition of the spheres on cells to detect the presence of the polyamines on cell membranes. Such spheres could also be used in a nephelometric agglutination test to detect the reaction with a polyamine–polyglutamate complex. They might also detect the possible presence of protein-bound polyamines in a biological fluid, for example, serum, or detect the inhibition of coated latex-sphere agglutination by free polyamine (Bonnefoy–Roch & Quash, 1978). The group also explored the reactivities of spheres chemically coupled with polyamines.

Analytical Accuracy

Our problems entail the estimation of the kind and amount of polyamines in synthetic and isolated compounds, in chemical fractions and biological fluids, and in cells, tissues, and organisms. The various syntheses, reactions, and methods of separation are important guides to the strategy to be followed in an analytical procedure. This strategy will include a method of obtaining the maximal amounts of the substances to be analyzed and the minimal amount of interfering contaminant, as well as the selection of a characterizing reaction with a specific, sensitive, and possibly linear response. The general problem was conceived initially in terms applicable to any quantitative microchemical analysis: sensitivity, range of linearity, duplicability, variance, and corrections with respect to recovery of the standard in the preparation of the sample and in the analytical reaction. However, the concern for accuracy and the meaning of the analytical result were heightened by the report that the polyamine content of the urine of cancer patients was frequently higher than that of individuals without cancer (Russell, 1971). The problem soon spread to the estimation of polyamines in the various fluids, cells, and tissues of "normal" and cancer-bearing animals, including humans, and has extended broadly to the general significance of polyamine content in normal and pathology-bearing cells, tissues, and organisms. These considerations then tied the problems of analytical accuracy to the physiological, pharmacological, and pathological roles of the polyamines in medicine and biology generally.

The problems of the relation of polyamine content to clinical significance will be considered briefly in a later chapter in the context of the polyamines and cancer. For the present let us note that the matters relating to the rapid, sensitive, and accurate measurement of these compounds were markedly hastened by the observations of Russell (1971), the efforts to enlarge and confirm her report, and the debate that ensued as to their clinical and predictive value.

REFERENCES

Abdel–Monem, M. M. & Ohno, K. (1977) *Journal of Pharmaceutical Sciences*, **66**, 1089–1094.

Aigner–Held, R., Campbell, R. A., & Daves, G. D. (1979) *Proceedings of the National Academy of Sciences of the United States of America*, **76**, 6652–6655.

Atawodi, S. E., Mende, P., Pfundstein, B., Preussmann, R., & Spiegelhalder, B. (1995) *Food and Chemical Toxicology*, **33**, 625–630.

Bachrach, U. (1978) *Advances in Polyamine Research*, vol. 2, pp. 5–11, R. A. Campbell, D. R. Morris, D. Bartos, G. D. Daves, & F. Bartos, Eds. New York: Raven Press.

Bachrach, U. & Plesser, Y. M. (1986) *Analytical Biochemistry*, **152**, 423–431.

Barbucci, R., Paoletti, P., & Vacca, A. (1970) *Journal of the Chemical Society*, **A**, 2202–2206.

Bardócz, S., Karsai, T., & Elödi, P. (1985) *Chromatographia*, **20**, 23–24.

Bartos, D., Campbell, R. A., Bartos, F., & Grettie, D. P. (1975) *Cancer Research*, **35**, 2056–2060.

Bartos, F., Bartos, D., Dolney, A. M., Grettie, D. P., & Campbell, R. A. (1978) *Research Communications in Chemical Pathology and Pharmacology*, **19**, 295–309.

Becker, E. D. (1993) *Analytical Chemistry*, **65**, 295A-302A.

Bidlingmeyer (1980) *Journal of Chromatographic Science*, **18**, 527–539.

Bonnefoy–Roch, A. & Quash, G. A. (1978) *Advances in Polyamine Research*, vol. 2, pp. 55–63, R. A. Campbell, D. R. Morris, D. Bartos, G. D. Daves, & F. Bartos, Eds. New York: Raven Press.

Boppana, V. K. & Rhodes, G. R. (1990) *Journal of Chromatography*, **506**, 279–288.

Brown, N. D., Sweet, R. B., Kintzios, J. A., Cox, H. D., & Doctor, B. P. (1979) *Journal of Chromatography*, **164**, 35–40.

Cann–Moisan, C., Caroff, J., Hourmant, A., Videau, C., & Rapt, F. (1994) *Journal of Liquid Chromatography*, **17**, 1413-1417.

Cohen, S. S. & Greenberg, M. L. (1981) *Proceedings of the National Academy of Sciences of the United States of America*, **78**, 5470–5474.

Corbin, J. L., Marsh, B. H., & Peters, G. A. (1989) *Plant Physiology*, **90**, 434–439.

Delfini, M., Segre, A. L., Conti, F., Barbucci, R., Barone, V., & Ferruti, P. (1980) *Journal of the Chemical Society— Perkin Transactions II*, 900–903.

Desiderio, M. A., Davalli, P. & Perin, A. (1987) *Journal of Chromatography*, **419**, 285–290.

de Silva, J. A. F. & Strojny, N. (1975) *Analytical Chemistry*, **47**, 714–718.

Dudley, H. W. & Rosenheim. O. (1925) *Biochemical Journal*, **19**, 1032–1033.

Dudley, H. W., Rosenheim, O., & Starling, W. W. (1926) *Biochemical Journal*, **20**, 1082–1094.

Eisdorfer, I. B., Warren, R. J., & Zarembo, J. E. (1968) *Journal of Pharmaceutical Sciences*, **57**, 195–217.

Fagerström, R., Seppänen, P., & Jänne, J. (1984) *Clinica Chimica Acta*, **143**, 45–50.

Fujihara, S., Nakashima, T., & Kurogochi, Y. (1986) *Journal of Chromatography*, **383**, 271–280.

Fujiwara, K., Asada, H., Kitagawa, T., Yamamoto, K., Ito, T., Tsuchiya, R., Sohda, M., Nakamura, N., Hara, K., Tomonaga, Y., Ichimaru, M., & Takahashi, S. (1983) *Journal of Immunological Methods*, **61**, 217–226.

Fujiwara, K. & Kitagawa, T. (1993) *Journal of Biochemistry*, **114**, 708–713.

Funayama, S., Zhang, G., & Nozoe, S. (1995) *Phytochemistry*, **38**, 1529–1531.

Furuta, H., Yamane, T., & Sugiyama, K. (1985) *Journal of Chromatography*, **337**, 103–109.

Gehrke, C. W., Kuo, K. C., Sumwalt, W., & Waalkes, T. P. (1973) *Polyamines in Normal and Neoplastic Growth*, pp. 343–353. D. H. Russell, Ed. New York: Raven Press.

Giumanini, A. G., Chiavari, G., & Scarponi, F. L. (1976) *Analytical Chemistry*, **48**, 484–489.

Gold, M. & Powell, H. K. J. (1976) *Journal of the Chemical Society—Dalton Transactions*, 230–233.

Greenberg, M. L. & Cohen, S. S. (1985) *Plant Physiology*, **78**, 568–575.

Guarino, L. A. & Cohen, S. S. (1979) *Proceedings of the National Academy of Sciences of the United States of America*, **76**, 3184–3188.

Gurkin, M. & Cravitt, S. (1972) *American Laboratory*, **January**.

Hague, D. N. & Moreton, A. D. (1994) *Journal of the Chemical Society—Perkin Transactions II*, 265–270.

Halline, A. G. & Brasitus, T. A. (1990) *Journal of Chromatography*, **533**, 187–194.

Hämäläinen, R. (1947) *Acta Society Medicine Fenn. "Duedecim,"* **23A**, 97–165.

Harrison, G. A. (1933) *Biochemical Journal*, **27**, 1152–1156.

Hawi, A. A., Yip, H., Sullivan, T. S., & Digenis, G. A. (1988) *Analytical Biochemistry*, **172**, 235–240.

Herbst, E., Weaver, R., & Keister, D. (1958) *Archives of Biochemistry*, **75**, 171–177.

Hunter, K. J. & Fairlamb, A. H. (1990) *Analytical Biochemistry*, **190**, 281–285.

Hyvönen, T., Keinänen, T. A., Khomutov, A. R., Khomutov, R. M., & Eloranta, T. O. (1992) *Journal of Chromatography*, **574**, 17–21.

Imai, K., Böhlen, P., Stein, S., & Udenfriend, S. (1974) *Archives of Biochemistry and Biophysics*, **161**, 161–163.

Jenden, D. J. & Cho, A. K. (1979) *Biochemical Pharmacology*, **28**, 705–713.

Jiang, X. (1990) *Biomedical Chromatography*, **4**, 73–77.

Kamei, S., Ohkubo, A., Saito, S., & Takagi, S. (1989) *Analytical Chemistry*, **61**, 1921–1924.

Kanavarioti, A., Baird, E. E., & Smith, P. J. (1995) *Journal of Organic Chemistry*, **60**, 4873–4883.

Kanda, S., Takahashi, M., & Nagase, S. (1989) *Analytical Biochemistry*, **180**, 307–310.

Kimberly, M. M. & Goldstein, J. H. (1981) *Analytical Chemistry*, **53**, 789–793.

Kochhar, S., Mehta, P. K., & Christen, P. (1989) *Analytical Biochemistry*, **179**, 182–185.

Kubota, S., Okada, M., Imahori, K., & Ohsawa, N. (1983) *Cancer Research*, **43**, 2363–2367.

Marton, L. J., Russell, D. H., & Levy, C. C. (1973) *Clinical Chemistry*, **19**, 923–926.

Maruta, K., Teradaira, R., Watanabe, N., Nagatsu, T., Asano, M., Yamamoto, K., Matsumoto, T., Shionoya, Y., & Fujita, K. (1989) *Clinical Chemistry*, **35**, 1694–1696.

Mossoba, M. M., Rosenthal, I., & Riesz, P. (1982) *Canadian Journal of Chemistry*, **60**, 1493–1500.

Murray, K. E., Shaw, K. J., Adams, R. F., & Conway, P. L. (1993) *Gut*, **34**, 489–493.

Muskiet, F. A. J., Dorhout, B., Van den Berg, G. A., & Kessels, J. (1995) *Journal of Chromatography B*, **667**, 189–198.

Ni, Y. (1986) *Journal of Chromatography*, **354**, 534–538.

Niitsu, M., Samejima, K., Matsuzaki, S., & Hamana, K. (1993) *Journal of Chromatography*, **641**, 112–123.

Noto, T., Hasegawa, T., Kamimura, H., Nakao, J., Hashimoto, H., & Nakajima, T. (1987) *Analytical Biochemistry*, **160**, 371–375.

Ohta, T., Inoue, T., & Takitani, S. (1990) *Agricultural and Biological Chemistry*, **54**, 2559–2564.

Onasch, F., Aikens, D., Bunce, S., Schwartz, H., Nairn, D., & Hurwitz, C. (1984) *Biophysical Chemistry*, **19**, 245–253.

O'Sullivan, M. C. & Dalyrmple, D. M. (1995) *Tetrahedron Letters*, **36**, 3451–3452.

Palmer, B. N. & Powell, H. K. J. (1974) *Journal of the Chemical Society—Dalton Transactions*, 2086–2092.

Paulus, T. J. & Davis, R. H. (1983) *Methods in Enzymology*, vol. 94, pp. 36–42, H. Tabor & C. W. Tabor, Eds. New York: Academic Press.

Perini, F., Sadow, J. B., & Hixson, C. V. (1979) *Analytical Biochemistry*, **94**, 431–439.

Quash, G. A., Fresland, L., Delain, E. & Huppert, J. (1973) *Polyamines in Normal and Neoplastic Growth*, pp. 157–165, D. H. Russell, Ed. New York: Raven Press.

Ramakrishna, S. & Adiga, P. R. (1973a) *Journal of Chromatography*, **86**, 214–218.

——— (1973b) *Phytochemistry*, **12**, 2691–2695.

Roth, M. (1971) *Analytical Chemistry*, **43**, 880–882.

Russell, D. (1971) *Nature New Biology*, **233**, 144.

Saeki, Y., Uehara, N., & Shirakawa, S. (1978) *Journal of Chromatography*, **145**, 221–229.

Saito, K., Horie, M., & Nakazawa, H. (1994) *Analytical Chemistry*, **66**, 134–138.

Samejima, K. (1974) *Journal of Chromatography*, **96**, 250–254.

Samejima, K., Furukawa, M. & Haneda, M. (1985) *Analytical Biochemsitry*, **147**, 1–9.

Schipper, R. G., Jonis, J. A., Rutten, R. G. J., Tesser, G. I., & Verhofstad, A. A. J. (1991) *Journal of Immunological Methods*, **136**, 23–30.

Seiler, N. (1983) *Methods in Enzymology*, vol. 94, pp. 3–29, H. Tabor & C. W. Tabor, Eds. New York: Academic Press.

——— (1986) *Journal of Chromatography*, **379**, 157–176.

Seiler, N. & Deckardt, K. (1975) *Journal of Chromatography*, **107**, 227–229.

Seiler, N. & Knödgen, B. (1985) *Journal of Chromatography,* **339**, 45–57.

Seiler, N., Knödgen, B., & Eisenbeiss, F. (1978) *Journal of Chromatography*, **145**, 29–39.

Shirahata, A., Takeda, Y., Kawase, M., & Samejima, K. (1983) *Journal of Chromatography*, **262**, 451–454.

Simons, Jr., S. & Johnson, D. F. (1977) *Analytical Biochemistry*, **82**, 250—254.

Sindhu, R. K. & Cohen, S. S. (1983) *Advances in Polyamine Research*, vol. 4, pp. 371–380, U. Bachrach, A. Kaye, & R. Chayen, Eds. New York: Raven Press.

Slocum, R. D., Flores, H. E., Galston, A. W., & Weinstein, L. H. (1989) *Plant Physiology*, **89**, 512—517.

Smith, R. G., Daves, G. D. Jr., & Grettie, D. P. (1978) *Advances in Polyamine Research*, vol. 2, pp. 23–35, R. A. Campbell, D. Bartos, D. R. Morris, G. D. Daves, Jr., & F. Bartos, Eds. New York: Raven Press.

Sulaiman, S. T., Saleem, L. M. N., & Al-Nuri, I. J. (1984) *Microchemical Journal*, **29**, 228–231.

Suzuki, O., Matsumoto, T., Oya, M., Katsumata, Y., & Samejima, K. (1981) *Analytical Biochemistry*, **115**, 72–77.

Tabor, H. & Tabor, C. W., Eds. (1983) *Methods in Enzymology*, vol. 94. New York: Academic Press.

Tabor, H., Tabor, C. W., & Ineverre, F. (1973) *Analytical Biochemistry*, **55**, 457–467.

Takeda, Y., Samejima, K., Nagano, K., Watanabe, M., Sugeta, H., & Kyogoku, Y. (1983) *European Journal of Biochemistry*, **130**, 383–389.

Templeton, D. M. & Sarkar, B. (1985) *Canadian Journal of Chemistry*, **63**, 3122–3128.

Thomas, T., MacKenzie, S. A., & Gallo, M. A. (1990) *Toxicology Letters*, **53**, 315–325.

van Rheenen, D. L. (1962) *Nature*, **193**, 170–171.

Wagner, J., Danzin, C., & Mamont, P. (1982) *Journal of Chromatography*, **227**, 349–368.

Wagner, J., Danzin, C., Huot–Olivier, S., Claverie, N., & Palfreyman, M. G. (1984) *Journal of Chromatography*, **290**, 247–262.

Wagner, J., Geget, C., Heintzelmann, B., & Wolf, E. (1987) *Analytical Biochemistry*, **164**, 102–116.

Walle, T. (1973) *Polyamines in Normal and Neoplastic Growth*, pp. 355–365, D. H. Russell, Ed. New York: Raven Press.

Watanabe, N. (1992) *Biomedical Chromatography*, **6**, 1–3.

Watanabe, S., Sato, S., Nagase, S., Tomita, M., Saito, T., & Ishizu, H. (1993) *Journal of Liquid Chromatography*, **16**, 619–632.

Wehr, J. B. (1995) *Journal of Chromatography A*, **709**, 241–247.

Whitmore, W. L. & Slotkin, T. A. (1985) *Experientia*, **41**, 1209-1211.

Wickström, K. & Betnér, I. (1991) *Journal of Liquid Chromatography*, **14**, 675–697.

Wongyai, S., Oefner, P. J. & Bonn, G. K. (1989) *Journal of Liquid Chromatography*, **12**, 2249–2261.

CHAPTER 4

Macrocyclic Polyamines—A Digression?

*Self-assembly and self-organization have recently been implemented in
several types of organic and inorganic systems. By clever use of metal
coordination, hydrogen bonding, and donor-acceptor interactions,
researchers have achieved the spontaneous formation of a variety of novel
and intriguing species such as the inorganic double and triple helices
termed helicates, catenanes, threaded entities (rotaxanes), cage
compounds, and so forth.*

J.-M. Lehn (1993)

As described in Chapter 2, much modern interest in synthetic methods in this field stemmed from efforts to prepare natural products, many of which are conjugated or modified polyamines. Some of these are presented below.

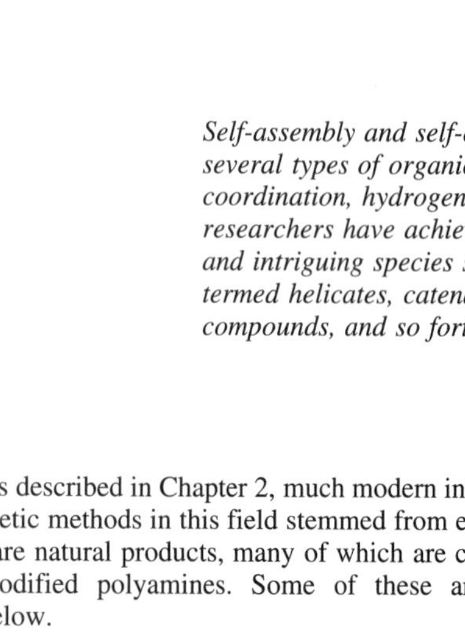

Agrobactin
R = OH

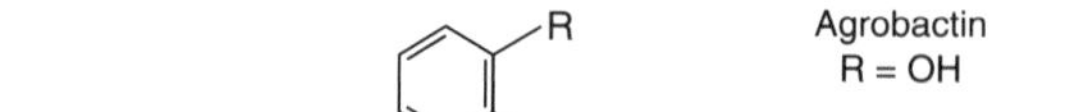

Deferoxamine

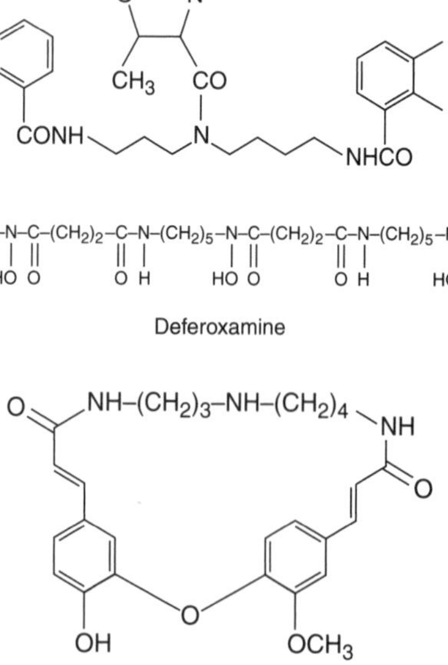

Codonocarpine

The agrobactins are examples of siderochromes (Neilands, 1973; Peterson et al., 1980), which complex FeIII. The siderochromes produced by many bacteria sequester iron through their 2,3-dihydroxybenzoyl substituents, which, in the case of the agrobactins, have been added to spermidine (Chap. 2, Fig. 2.3). Several siderophores derived from *Vibrio* species are built on norspermidines (Okujo et al., 1994; Yamamoto et al., 1993).

Deferoxamine is a *Streptomyces* derived siderochrome that is widely used to treat iron poisoning. The nitrogen of each of the four amides is hydroxylated and these hydroxamates bind the iron most effectively. In this class of hydroxamate siderochromes, the hydroxyamine moiety may be derived from ornithine or lysine. Many natural hydroxamates and their iron-binding properties were reviewed by Neilands (1967). Whereas the catechols are stronger Fe chelators at high pH, the hydroxamates hold the metal more readily at a lower pH. Desferrioxamine B hydrochloride, one of nine produced biosynthetically, was synthesized by Bergeron and Pegram (1988). One recent derivative was shown to produce irreversible damage to developing *Plasmodium falciparum* (Cabantchik, 1995). Other *N*-hydroxycadaverine-containing siderophores, bisucaberin and nannochelin A, have also been synthesized (Bergeron & McManis, 1989; Mulqueen et al., 1993). A novel siderophore, rhizobactin 1021 derived from a *Rhizobium* species, appears to contain the component, α-*N*-hydroxy 1,3-diaminopropane (Persmark et al., 1993). In addition, spermidine-based trihydroxamates, termed *spermexatins*, and spermidine-based mixed (hydroxamate- and catechol-) siderophores (*spermexatols*) are known (Sharma et al., 1989).

Finally codonocarpine is included as an example of a closed ring compound, in which spermidine serves to cross-link the two ends of a diphenolic *trans* cinnamate to form a macrocycle. Studies on the synthetic macrocyclic compounds merged with an even earlier discovery of a class of antibiotics, *ionophores*, many of which proved to be cyclic peptides capable of carrying small cations across lipid barriers (Pressman, 1973). Interest in this combination of chemical and biological properties as well as those of the crown ethers led initially to the design and synthesis of cyclic compounds capable of the specific binding of ions, and eventually to the analysis of molecular interactions pointing to the creation of a supramolecular chemistry (Lehn, 1988). These studies have not only posed the narrow problem of the potentially comparable physiological activities of the natural polyamines, but also the broader problem of the simulation of polymeric interactions leading to specific catalyzed reactions (Cram, 1988).

The natural siderophores, such as the agrobactins, have been modified synthetically to improve their potential therapeutic application of reducing iron overload and of sequestering other toxic Fe-like metals, such as plutonium. In one such effort, the spermine- and spermidine-containing enterobactin analogues containing catecholcarboxamides were alkylated at the terminal nitrogen atoms of the bound polyamine to increase lipophilicity and to improve the distribution of the compound in tissues (Weitl & Raymond, 1981). The attachment of the siderophore, deferoxamine, to a biocompatible polymer such as dextran prolonged its plasma half-life and reduced toxicity without affecting metal chelation (Hallaway *et al.*, 1989).

With bound metal, the siderophores are classifiable as coordinate complexes. In the agrobactins and parabactins, the polyamine chain is clearly the frame on which the metal-binding dihydroxybenzoyl moieties are organized in space. Nevertheless, in other natural compounds the nitrogen atoms themselves participate in the formation of coordinate covalent bonds, as in the binding of iron in hemoglobin or heme, or of magnesium in chlorophyll.

metalloporphyrin

Coordination Compounds and Some Reactions of Copper

The formation of the startlingly blue tetrammine copper II ion $[Cu(NH_3)_4]^{2+}$ was an early discovery, recorded in 1597 by the alchemist Libavius, who mixed an alkaline solution containing ammonium chloride with copper-containing brass. Kauffman (1973, 1981) described the growth of knowledge of these somewhat unexpected metal complexes with ammonia to form *metal-ammines*. In 1893 Alfred Werner discussed the previously unexplained "metal-ammonia salts" within the context of a proposed tetrahedral ammonium ion. His postulates included the concepts of primary valence (oxidation state) and secondary valence or coordination number. Werner's theories were then extended to include the concept that, in the formation of the ammonium ion, an unshared pair of electrons on the nitrogen atom are donated to the proton derived from an acid to form a coordinate bond, for example

The unshared electrons of the nitrogen atoms of various primary, secondary, and tertiary amines are similarly available, after deprotonation, to form coordinate covalent bonds with metal ions. The properties of such coordinate complexes obtained with various polyamines have been studied.

The reaction of copper with proteins was noted in 1833 and that with biuret in 1848. The structure of the latter was formulated in 1907.

As discussed by Sigel and Martin (1982), formation of a stable Cu complex depends on the substitution of the metal for a nitrogen-bound amide hydrogen. In the case of the biuret reaction, which requires alkali, deprotonation occurs above pH 11.5 and the characteristic violet color of the Cu-biuret complex does not form until this high pH is achieved. At neutrality, weak differently colored complexes are formed linking copper to two oxygen atoms in two biuret molecules.

Many studies have been made of metal complexes with amines, amino acids, and peptides. A. Albert (1950) began studies of the binding of copper with α-amino acids following a 1941 publication by J. Bjerrum. Albert demonstrated a stepwise liberation of hydrogen ions as complex formation occurred in the binding of copper and other metals to glycine, and he used the titrations of various amino acids in the estimation of stability constants for their metal complexes. He then developed equations for the estimation of such constants for amino acids (e.g., ornithine, aspartic acid, etc.) containing three ionizing groups (Albert, 1952). Potentiometric titrations and spectrophotometric and electrophoretic analyses of copper complexes defined the formation of the complexes (Dobbie et al., 1955).

In early studies with ethylenediamine(*en*) as a ligand in cobalt complexes, it had been found that ethylenediamine(*en*) can replace two molecules of ammonia to form a stable complex containing a five-membered ring. The greater stability of the former *en* complex is attributed to a "chelate" effect. Indeed, 1,3-diaminopropane can form a six-membered ring, although the complex is less stable (Bertsch et al., 1958). In the chelation of metal with two nitrogen atoms in a given molecule, a five-membered ring is more stable than a six-membered ring and larger coordinate chelates are quite unstable (Hedwig & Powell, 1973). This is the interpretation usually given to the observation that complexes of copper with the ligands 1,3-diaminopropane, 1,4-diaminobutane, and 1,5-diaminopentane cannot be detected in one analytical system that can readily determine complexes with 1,2-diamines (Selig, 1982a,b). However, complex formation was detected with macrocyclic polyamines and even in larger linear polyamines, which may assume a partial ringlike configuration and approach the conformation of macrocyclic polyamines (Palmer & Powell, 1974). Complete cyclization confers a marked increase of several orders of magnitude in the stability of the copper–polyamine complexes.

Copper II complexes of spermidine and spermine were inhibitory to tumor cell lines that were normally stimulated by the free polyamines. The Cu complexes were studied by several spectroscopic methods, and the spermidine complex appears to contain a seven-membered chelate ring fused to another six-membered ring (Anichini et al., 1977). The structure of the Cu complex of spermine, determined by X-ray crystallography, also appears to contain a seven-membered chelate ring.

The binding of copper(II) to spermidine and spermine has been studied (Templeton & Sarkar, 1985, chap. 3) by titration and spectrophotometry. Copper begins to complex with spermidine above pH 6 to form a structure containing a single protonated amine, that is, a (Cu H ligand)$^{3+}$ species. Further deprotonation permits formation of a (Cu ligand)$^{2+}$ species that predominates in the range of pH 7.5–9.5. Additional complexes are generated at higher pH. In the titration of spermine containing two secondary nitrogens, a complex comprising (Cu H$_2$ ligand)$^{4+}$ forms above pH 5. This is converted almost quantitatively to (Cu ligand)$^{2+}$ between pH 7 and 9 with all of the primary and secondary nitrogens linked to copper. The study also determined the absorption maxima of the various species of the copper complexes with spermidine and spermine. These analyses established the preconditions for the isolation and crystallographic analyses of the several complexes of spermidine and spermine. Thus these natural tri- and tetramines do form complexes with copper at physiological pH, although the natural diamines do not, with the possible exception of 1,3-diaminopropane.

As noted earlier in the discussion of copper complexes of biuret, copper may bind to carboxylate oxygen at neutral pH. Some amino acids form very insoluble copper complexes and in one instance the formation of such a complex can be used to quantify the presence of a toxic amino acid, the neurotoxin β-*N*-oxalyl-L-α,β-diaminoproprionic acid (Davis et al., 1990).

Copper is an essential nutrient of many organisms and is known to be a structural element of numerous enzymes, including several amine oxidases. The element is widely distributed in nature and is normally supplied in a mammalian diet. A copper deficiency has been induced in rabbits fed a 2,3,2-tetramine (Hing & Lei, 1991). Copper-binding proteins appear to act in insuring absorption and in minimizing absorption of toxic levels of the metal (Evans, 1973). No role has yet been found or sought for spermidine or spermine in copper homeostasis.

In the severe pathologies of copper accumulation in the liver, central nervous system, and other organs, known as Wilson's disease, it might be asked if the content of spermidine or spermine in the liver is abnormal or if copper is bound to these substances. To deplete copper in Wilson's disease a patient may be treated with the thiolated amino acid, penicillamine, resulting in the urinary excretion of an unusual caged complex (Muetterties, 1982).

Although copper is believed to be essential for the biosynthesis of the bleomycins antibiotics, Fe(II) is the major metallic cation in the functioning antibiotic. Any antibiotic of the series (to be discussed in a later chapter) consists of a complex glycopeptide bound to an intercalatable dithiazole linked via an amide bond to a terminating di- or polybasic amine. The latter appears essential for an initial binding to nucleic acid, and bleomycins terminated with spermidine or spermine are among the most toxic of the series. The bleomycins can serve as a nuclease, nicking DNA and certain RNA molecules. An oxygen-dependent cleavage is effected via a coordinately bound metal ion (Fe^{2+}). The polyamines in the bleomycins, although important in the attachment to the nucleic acids, do not appear to participate in the metal-binding or cleavage reactions.

Copper ions can combine selectively to purine bases, at the N$_7$ of nucleosides (Martin, 1985). Because compounds such as spermidine or spermine that are cationic at physiological pH bind to the phosphates of nucleotides and can also bind to copper, it has been asked if the polyamines participate in the synthesis of ternary compounds in which the polyamine is cross-linked to the purine base through copper. A study of complexes of copper (II) with adenosine-5′-monophosphate and spermine detected the formation of a major binary complex (CuSpe)$^{2+}$ of Cu and spermine (1:1). A ternary complex of composition [Cu(AMP)(Spe)H$_4$]$^{4+}$ was formed only at a 5–10 fold excess of AMP and it was suggested that in this complex copper(II) was bound to the purine whereas tetraprotonated spermine was bound to dianionic phosphate (Antonelli et al., 1988). The authors concluded that spermine did not serve to cross-link from phosphate to copper to purine nitrogen.

Platinum and Palladium Complexes

The antitumor activity of bis(platinum) complexes has led to studies of the synthesis of other complexes containing two platinum-amine units linked by diamine chains of variable length (Qu & Farrell, 1990, 1991). The linkers used were 1,4-butanediamine and 2,5-dimethyl-2,5-hexanediamine, substances intended to improve water solubility and to increase cross-links to opposing strands in target DNA.

Various palladium complexes of putrescine, spermidine, and spermine were also prepared, characterized in some detail, and compared in their effects on DNA and tumor cells (Navarro–Ranninger et al., 1992, 1993). Three different complexes, [PutH$_2$][PdCl$_4$], [Pd$_2$Cl$_4$(Put$_2$)], and [PdCl$_2$][Put]$_2$, proved to be more active than the chemotherapeutic drug, *cis*-diamine dichloroplatinum II (*cis*-DDP). They were also markedly more active than the two spermine complexes [PdCl$_2$(SpeH$_2$)][PdCl$_4$] and [Pd$_2$Cl$_4$(Spe)]. Several spermidine complexes, [Pd$_3$Cl$_6$-(Spd)$_2$], proved to be at least as active as the *cis*-DDP and to affect the conformation of plasmid DNA.

Enter the Macrocyclic Polyamines

In the 1950s the discovery and characterization of the ionophore valinomycin as a cyclic compound affecting the transport of K$^+$, stimulated efforts to isolate and develop other specific substances possessing similar biological and chemical properties. Valinomycin is cyclized by means of alternating amide and ester groups and forms ligand complexes with potassium, rubidium, and cesium far more readily than with sodium or lithium (Pressman, 1973). In 1960 the industrial removal of autoxidants such as copper and vanadium from petroleum and rubber by the formation of multidentate complexes with N,N'-bis(o-hydroxy-benzylidene)-1,2-propanediamine encouraged further study of these agents (Pedersen, 1988). This led to the study of phenolic ligands containing ether groups and resulted in the production of the crown ethers, whose solubility in methanol was increased specifically by the presence of sodium ions. The first crown ether, 2,3,11,12-dibenzo-1,4,7,10,13,16-hexaoxacyclooctadeca-2,11-diene, is presented below:

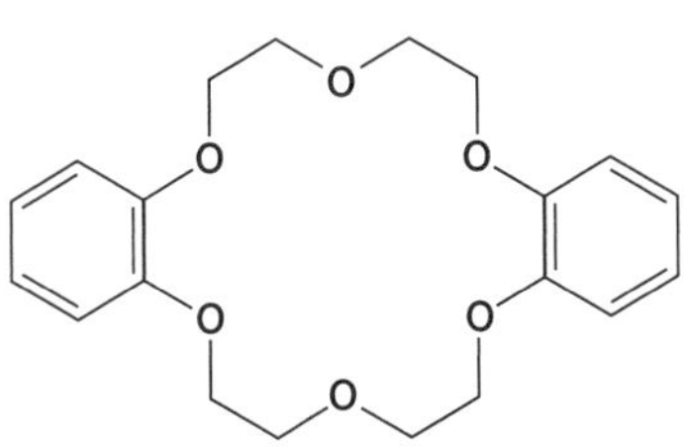

Pedersen was awarded the Nobel Prize in 1987 for his synthesis of the first neutral compound capable of complexing the alkali metal cations.

J. M. Lehn and his collaborators, who noted the studies on chemically reactive valinomycin and the more stable crown ethers, undertook to develop compounds whose internal cavities might form even stronger complexes (Lehn, 1988). In 1968 the macrobicyclic ligands containing linkers of both secondary nitrogen and oxygen ethers possessed the desired properties. His new structures could bind the ammonium ion more strongly than did the crown ethers. The crown ether, 18-crown-6, as an ammonium tosylate has been used in the selective acetylation of a hexahydropyrimidine in the synthesis of N^1-acetyl spermidine (Tice & Ganem, 1983).

Subsequent "cryptands," as they were called, were macrotricyclic and were now tailored to improve the specificity of binding for one or another molecule, which might be a tetrahedral cation RNH$_3^+$, neutral H$_2$O, or an anionic Cl$^-$. Lehn (1988) considered the macrocyclic compounds as receptors and generalized the possibility of developing potentially interacting structures that can both emulate and extend biological function.

D. Cram and his coworkers also generalized consideration of the structures of complexes. These were postulated as arising in the interactions of a host that provided converging binding sites for a guest molecule or ion that might have divergent binding sites within the complex. Cram formulated several types of complexes related to the relative position of guest and host. These include nesting complexes and capsular complexes (Cram, 1988). He noted that in their crystal forms the initial crown ethers or cryptands lacked cavities and that cavities were in fact generated during binding, the energy being used in forming the appropriate cavity. Preorganized ligands were synthesized with cavities required by a particular anticipated guest. Such preorganized macrocyclic ligands were in fact shown to possess markedly increased binding power for defined guests.

A rather early observation was that nitrogen macrocycles had an enhanced selectivity for K$^+$ over Na$^+$. The studies soon led then to the synthesis of macrocyclic polyamines, linked entirely through their nitrogen atoms (Richman & Atkins, 1974). The cyclization involved a reaction between bis-sulfonamide sodium salts and sulfonate ester-leaving groups to prepare substances such as

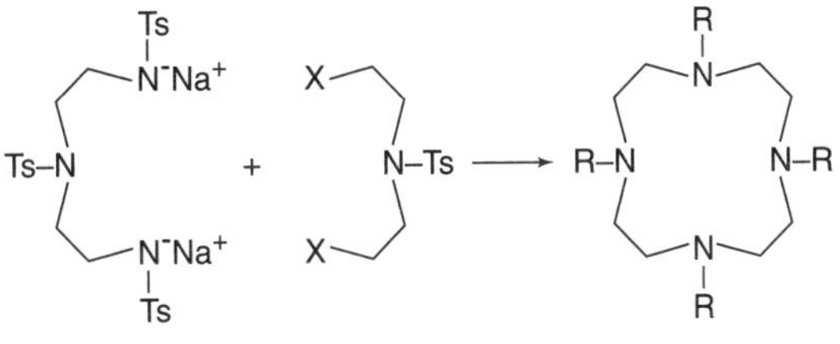

In the initial procedure it was noted that syntheses attempting to incorporate hydrocarbon segments longer than three methylene units markedly decreased yields. Improvements in the Richman–Atkins syntheses and in the isolation of the desired compounds have been described (Searle & Geue, 1984). In addition macrocycles have been coupled directly through their sulfonamide derivatives to form macrobicyclic polyamines (Dietrich et al., 1985). Recently macrocycles have been linked through polymethylene bridges that are then capable of forming sandwiched or encapsulated complexes (Kimura et al., 1990). Kimura and colleagues (1980, 1981) developed syntheses and analytical procedures to identify the macrocyclic polyamines.

The design of macrocyclic ligands to optimize the binding of specific metals was discussed by McMurry et al. (1989). A series of macrocyclic polyamine ligands were made in which the size of the cavity was increased and in which the binding of cations such as CuII was studied in comparison with the linear amines. Whereas the macrocyclic and open chain polyamines of comparable size and composition react at similar rates in their unprotonated forms with metal ions, the protonated macrocycles react more slowly than the linear amines. Among the macrocyclic polyamines, increased size does increase the rate constant of the binding of copper.

Macrocyclic polyamine ligands, possessing five and six nitrogen donor atoms and 15–18 members, were effective in binding metal ions such as those of Cu^{2+} or the bivalent transition metals (Kodama et al., 1980). The penta-aza cyclic compounds of 15 and 16 members bound CuII 10^4–10^5 times more stably than did the related linear structures that lacked a linking di- or trimethylene bridge. This "macrocyclic effect" was also seen for the tetraaza compounds (Kodama & Kimura, 1978).

Hori et al. (1993) recently calculated the relation of hole size to the number of atoms in the rings of cyclic ethers and amines. New syntheses of polyaza compounds that generate differently sized cavities were summarized by Thummel (1991). The structures may comprise components of molecular cages in which otherwise difficult reactions, for example, the synthesis of cyclobutadiene, may be carried out. The recognition of the complementarity of hydroxy- and amino groups as hydrogen-bond donors and acceptors (Ermer & Eling, 1994) introduced the possibility of new elements of a self-assembling supramolecular chemistry (Lehn, 1993).

The macrocyclic polyamines have been found to serve as selective receptor molecules for polyoxyanions such as succinate and citrate (Kimura et al., 1981). These reactivities occurred at physiological pHs, even as they did for phosphates, which often exchange for the carboxylates in biological transport systems. One such protonated macrocycle, 18-azacrown-6, which can bind with polycarboxylates and carbonate as well as various phosphates including nucleotides, will also interact with various catechols and catecholamines (Kimura et al., 1983).

Macrocycles bearing an intraannular phenolic group were found to bind divalent metal ions (e.g., Mg^{2+}) relatively specifically (Kimura et al., 1987). Compounds possessing pyridines as sidearms are Na^+ specific (Tsukube et al., 1991a). These workers developed amide-functionalized macrocyclic polyamines selective for alkali-metals (Tsukube et al., 1991b). Borane-containing macrocycles are amine-selective macrocycles (Reetz et al., 1992).

These findings raised the problem of the possible applications of these activities of the compounds to analytical or medicinal applications or to their relevance to problems of transport and metabolism. As examples of analytical applications we may note the following: a lipophilic macrocyclic pentamine, 15-hexadecyl-1,4,7,10, 13-pentaazacyclohexadecane, serves as a potentiometric sensor for ATP^{4-} when incorporated into a membrane. This sensor operates in the very wide range of 10^{-7}–10^{-2} M and is more sensitive for this purpose than most electrochemical methods (Umezawa et al., 1988). Some spermine-or homospermine-like macrocycles have also been constructed to be optically active, as an approach to tailoring an analyzable complex more readily reactive with ATP (Marecek et al., 1988). Indeed several such macrocycles, containing a hydroxymethyl substituent, proved to recognize specifically the γ-phosphorus of ATP (Prakash et al., 1991).

One potential application envisioned the coupling of a complexing agent to an antibody that can target the radioactive metal complex for diagnostic or therapeutic use. Parker (1990) discussed the construction, suitability, and limitations of a macrocyclic polyamine bound through several linked moieties to a specific antibody to direct [54]Cu to a specific antigenic site as an imaging radioisotope or of yttrium-90 (^{90}Y) for radioimmunotherapy. The structure of a possible bifunctional complexing agent containing antibody at one end is given below:

Most recently cyclic polyamines have been prepared from the polyamines by adding polymethylene sequences linking the nitrogens of putrescine. Adding an aminopropyl moiety to one nitrogen of this cyclo-pu-trescine produced a cyclo-spermidine; adding two such moieties produced a cyclo-derivative of spermine (Brand et al., 1994).

Biological Activities of Macrocyclic Polyamines

Some macrocyclic polyamines have been compared with the natural polyamines in several metabolic systems of protein biosynthesis, in which the natural compounds were stimulatory. A system comprising bacterial ribosomes in a soluble extract was capable of synthesizing polyphenylalanine and was stimulated by linear and macrocyclic polyamines at low Mg^{2+} concentrations. However, the latter compounds increased incorporation of phenylalanine into polypeptide by only 30% of the stimulation obtained with 2 mM spermidine. A similar result was obtained with the incorporation of leucine into globin in a wheat germ cell-free system (Igarashi et al., 1980). These workers also examined the rate of aminoacyl-tRNA formation with partially purified iso-leucyl-tRNA synthetase obtained from rat liver and noted a 40% stimulation by a macrocyclic tetramine compared to that produced by spermine.

Hydrolysis of poly C by RNase A was stimulated similarly by a macrocycle and spermine but the former required significantly higher concentrations. The macrocyclic polyamines inhibited ATPase slightly in contrast to a far more inhibitory spermine and failed to substitute for polyamine in the growth of a polyamine-requiring mutant of *Escherichia coli*, in contrast to the marked activity of spermine. Whereas spermine slightly inhibited lipid peroxidation, several of the macrocycles were very inhibitory to this process; this effect was attributable to chelation of Fe^{2+}. The experiment with the polyamine-deficient bacteria suggests that the macrocyclic compounds are not taken up into the organism, nor are they simple simulants of the natural polyamines in a variety of enzymatic reactions. A lanthanum (Ln^{3+}) complex of a macrocyclic tetramine containing acetamide substitutions on the nitrogen was recently found to be a stable catalyst active in the rapid cleavage of RNA oligomers (Amin et al., 1994).

An examination of the binding of the macrocyclic polyamines with ATP and other phosphoanhydrides revealed some surprising actions. Not only do the protonated macrocycles bind ATP, ADP, and AMP stably, but they enhance the rate of hydrolysis of ATP by several orders of magnitude over a wide pH range. An acyclic hexamine, which does bind ATP, does not increase the rate of hydrolysis. In a hydrolytic system stimulated by a macrocycle, the products are orthophosphate and ADP, which is also hydrolyzed somewhat slowly to AMP. In the cleavage of ATP, the formation of an intermediate phosporamidate was detected and the possible form of an initial "perched" complex and a mechanism of hydrolysis were postulated (Hosseini et al., 1983,

1987). These macrocyclic catalysts are described as "functional mimics" of ATPases (Bencini et al., 1992; Mertes & Mertes, 1990). The hydrolysis of ATP involved an exchange of oxygen at the β-phosphate of ATP and occurred only in the presence of Ca^{++}. Under these conditions, subsequent hydrolysis of ADP was decreased and the phosphorylated macrocycle accumulated. When this reaction mix was adjusted to pH 4.5, pyrophosphate was formed (Bethell et al., 1988). Actually the macrocyclic phosphoramidate was shown to be capable of synthesizing ATP from ADP.

In the reaction of the active mixed macrocyclic poly-amine [24]-N_6O_2 with acetyl phosphate (Hosseini & Lehn, 1987), acetyl phosphate was cleaved to ortho-phosphate; the reaction proceeded to the synthesis of pyrophosphate. The reaction mechanism, analyzed by nuclear magnetic resonance (NMR), indicated a course represented in Figure 4.1.

It is supposed that an initial transfer to the macrocycle forms the phosphoramidate and releases acetate. Some orthophosphate is released eventually and then bound to a phosphoramidate macrocycle, which transfers phosphate within the second complex to form pyrophosphate. Pyrophosphate is released to begin a new catalytic cycle.

In the reactions of formyl phosphate with both linear and macrocyclic polyamines, aminolysis was detected by NMR spectroscopy at pH 7 and involved mainly the cleavage of the formyl (C—O) bond (Jiang et al., 1989). In contrast to the acetyl phosphate cleavage, this reaction led to the formation of *N*-formyl amines. Linear tetramines and pentamines also catalyzed the aminolysis, although the natural tetramines and pentamines have not been tested in this reaction. The results on hydrolysis of formyl phosphate by [24]N_6O_2 have been discussed in the context of ATP hydrolysis, formation of the phosphoramidate-activated macrocycle, and phosphoryl transfer (Fig. 4.2). It had been observed that the macrocycle could activate formate in an ATP-dependent reaction in the presence of Ca^{2+} or Mg^{2+} (Jahansouz et al., 1989). The activation appeared to proceed via the hydrolysis of ATP to generate the macrocyclic phosphoramidate. The supposition is that the latter formed the proposed intermediate formyl phosphate, and this was cleaved on the macrocycle to produce a macrocyclic formamide.

Research has suggested that this set of reactions might mimic the ATP-dependent enzymatic synthesis of N^{10}-formyl tetrahydrofolate and poses the problem of the nature of the active site in N^{10}-formyl tetrahydrofolate synthetase. Because the linear polyamines can enhance aminolysis, we may ask if the natural polyamines participate in these significant biochemical reactions.

The phosphatase activity of the compounds was explored with Zn(II) complexes of triaza- and tetraaza-macrocyclic amines (Koike & Kimura, 1991). One of these, the ZnII complex of 1,5,9-triazacyclododecane, which can be thought of as a cyclic norspermine or thermine (3,3,3), is also active as a carbonic anhydrase

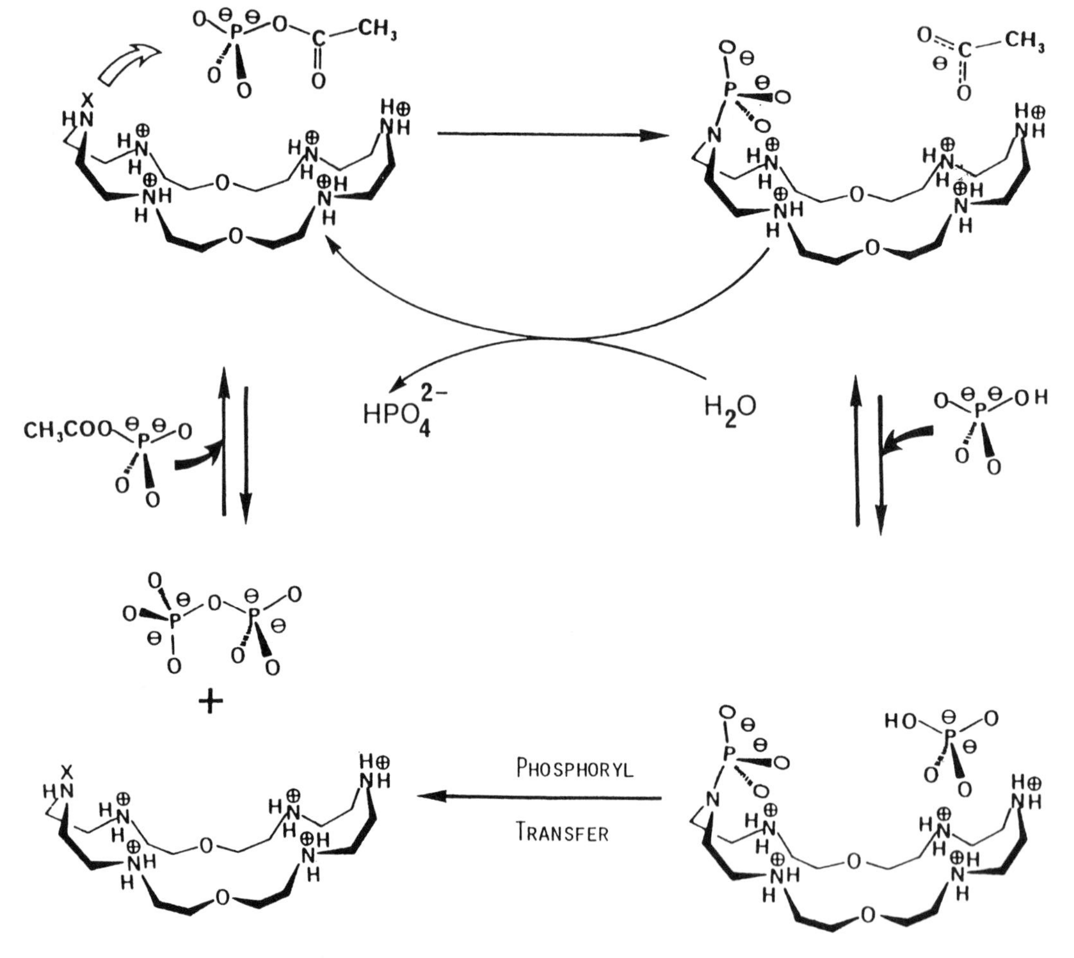

FIG 4.1 Schematic representation of the catalytic cycle for acetyl phosphate hydrolysis and pyrophosphate synthesis by a macrocyclic catalyst; redrawn from Hosseini and Lehn (1987).

in the hydrolysis of carboxylic ester; it proved to be most active in the hydrolysis of a phosphodiester and a phosphotriester. Kimura (1992) discussed these catalytic activities of Zn complexes in some detail, as well as the potential development of additional cyclic analogues of well-known spermidine- and spermine-alkaloids. Some macrocyclic polyamines and their metal complexes inhibit the proliferation of human immunodeficiency virus (HIV) in tissue culture systems and the divalent metal complexes are more active and less toxic than the free macrocycles (Inouye et al., 1994,1995). Also, dimeric macrocyclic polyamines are more active than the monomers.

Summary

The ability of amines to serve as ligands in coordinate linkage to various metals has been pursued. The study of colored metal complexes with ammonia or various nitrogen-containing derivatives has provided analytical tools or theoretical clarifications of such problems as the chemical architecture of bonding atoms, isomerism, and the nature of bonds themselves. The discoveries of metal-binding antibiotics and siderophores led to the detection of the polyamines in these structures and their possible roles as ligands. Study of the nature of effective ligands produced the crown ethers and subsequently the macrocyclic polyamines, whose properties have been compared with those of the natural linear polyamines. For the most part the comparisons have suggested that the linear and cyclic polyamines probably do not share biological roles. However, analyses of the catalytic activities of the macrocyclic polyamines and related activities of the linear polyamines have revealed possible similarities in biochemical functions such as phosphoryl transfer.

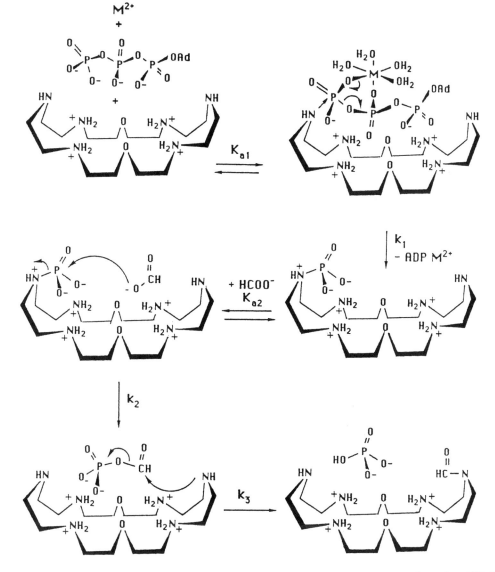

FIG 4.2 Schematic representation of the route for formate activation and *N*-formylation by a macrocyclic catalyst, ATP, divalent cation, and formate; redrawn from Jahansouz et al. (1989).

REFERENCES

Amin, S., Morrow, J. R., Lake, C. H., & Churchill, M. R. (1994) *Angewandte Chemie—International Edition in English*, **33**, 773–775.

Anichini, A., Fabbrizzi, L., Barbucci, R., & Mastroianni, A. (1977) *Journal of the Chemical Society—Dalton Transactions*, 2224–2228.

Antonelli, M. L., Balzamo, S., Carunchio, V., Cernia, E., & Purrello, R. (1988) *Journal of Inorganic Biochemistry*, **32**, 153–161.

Bencini, A., Bianchi, A., García–Espana, E., Scott, E. C., Morales, L., Wang, B., Deffo, T., Takusagawa, F. Mertes, M.

P., Mertes, K. B., & Paoletti, P. (1992) *Bioorganic Chemistry*, **20**, 8–29.

Bergeron, R. J. & McManis, J. S. (1989) *Tetrahedron*, **45**, 4939–4944.

Bergeron, R. J. & Pegram, J. J. (1988) *Journal of Organic Chemistry*, **53**, 3131–3134.

Bethell, R. C., Lowe, G., Hosseini, M. W., & Lehn, J. (1988) *Bioorganic Chemistry*, **16**, 418–428.

Brand, G., Hosseini, M. W., & Ruppert, R. (1994) *Tetrahedron Letters*, **35**, 8609–8612.

Cabantchik, Z. I. (1995) *Parasitology Today*, **11**, 74–78.

Cram, D. J. (1988) *Angewandte Chemie—International Edition in English*, **27**, 1009–1020.

Davis, A. J., Nunn, P. B., O'Brien, P., Pettit, L. D., & Wang, G. (1990) *Journal of Inorganic Biochemistry*, **39**, 209–216.

Dietrich, B., Hosseini, M. W., Lehn, J., & Sessions, R. B. (1985) *Helvetica Chimica Acta*, **68**, 289–299.

Ermer, O. & Eling, A. (1994) *Journal of the Chemical Society —Perkin Transactions II*, 925–944.

Evans, G. W. (1973) *Physiological Reviews*, **53**, 535–570.

Hallaway, P. E., Eaton, J. W., Panter, S. S., & Hedlund, B. E. (1989) *Proceedings of the National Academy of Sciences of the United States of America*, **86**, 10108–10112.

Hedwig, G. R. & Powell, H. K. J. (1973) *Journal of the Chemical Society—Dalton Transactions*, 793–797.

Hing, S. A. O. & Lei, K. Y. (1991) *Biological Trace Element Research*, **28**, 195–211.

Hori, K., Haruna, Y., Kamimura, A., Tsukube, H., & Inoue, T. (1993) *Tetrahedron*, **49**, 3959–3970.

Hosseini, M. W. & Lehn, J. (1987) *Journal of the American Chemical Society*, **109**, 7047–7058.

Hosseini, M. W., Lehn, J., Maggiora, L., Mertes, K. B., & Mertes, M. P. (1987) *Journal of the American Chemical Society*, **109**, 537–544.

Hosseini, M. W., Lehn. J.-M., & Mertes, M. P. (1983) *Helvetica Chimica Acta*, **66**, 2454–2466.

Igarashi, K., Kashiwagi, K., Kakegawa, T., Hirose, S., Yatsunami, T., & Kimura, E. (1980) *Biochimica et Biophysica Acta*, **633**, 457–464.

Inouye, Y., Kanamori, T., Sugiyama, M., Yoshida, T., Koike, T., Shionoya, M., Enomoto, K., Suehiro, K., & Kimura, E. (1995) *Antiviral Chemistry and Chemotherapy*, **6**, 337–344.

Inouye, Y., Kanamori, T., Yoshida, T., Bu, X., Shionoya, M., Koike, T., & Kimura, E. (1994) *Biological & Pharmaceutical Bulletin*, **17**, 243-250.

Jahansouz, H., Jiang, Z., Himes, R. H., Mertes, M. P., & Mertes, K. B. (1989) *Journal of the American Chemical Society*, **111**, 1409–1413.

Jiang, Z., Chalabi, P., Mertes, K. B., Jahansouz, H., Himes, R. H. & Mertes, M. P. (1989) *Bioorganic Chemistry*, **17**, 313–329.

Kauffman, B. (1981) *Inorganic Coordination Compounds*. Philadelphia: Heyden & Son, Incorporated.

Kauffman, G. B. (1973) *Isis*, **64**, 78–95.

Kimura, E. (1992) *Tetrahedron*, **48**, 6175–6217.

Kimura, E., Kimura, Y., Yatsunami, T., Shionoya, M., & Koike, T. (1987) *Journal of the American Chemical Society*, **109**, 6212–6213.

Kimura, E., Kuramoto, Y., Koike, T., Fujioka, H., & Kodama, M. (1990) *Journal of Organic Chemistry*, **55**, 42–46.

Kimura, E., Sakonaka, A., Yatsunami, T., & Kodama, M. (1981) *Journal of the American Chemical Society*, **103**, 3041–3045.

Kimura, E., Watanabe, A., & Kodama, M. (1983) *Journal of the American Chemical Society*, **105**, 2063–2066.

Kimura, E. & Yatsunami, T. (1980) *Chemical & Pharmaceutical Bulletin*, **28**, 994–997.

Kodama, M. & Kimura, E. (1978) *Journal of the Chemical Society—Dalton Transactions*, 1081–1085.

Kodama, M., Kimura, E., & Yamaguchi, S. (1980) *Journal of the Chemical Society—Dalton Transactions,* 2536–2538.

Koike, T. & Kimura, E. (1991) *Journal of the American Chemical Society*, **113**, 8935–8941.

Lehn, J. (1988) *Angewandte Chemie—International Edition in English*, **27**, 89–112.

——— (1993) *Science*, **260**, 1762–1766.

Marecek, J. F., Fischer, P. A., & Burrows, C. J. (1988) *Tetrahedron Letters*, **29**, 6231–6234.

Martin, R. B. (1985) *Accounts of Chemical Research*, **18**, 32–38.

McMurry, T. J., Raymond, K. N., & Smith, P. H. (1989) *Science*, **244**, 938-943.

Mertes, M. P. & Mertes, K. B. (1990) *Accounts of Chemical Research*, **23**, 413–418.

Muetterties, E. L. (1982) *Chemical and Engineering News*, **30**, 28–41.

Mulqueen, G. C., Pattenden, G., & Whiting, D. A. (1993) *Tetrahedron*, **49**, 9137–9142.

Navarro–Ranninger, C., Pérez, J. M., Zamora, F., González, V. M., Masaguer, J. R., & Alonso, C. (1993) *Journal of Inorganic Biochemistry*, **52**, 37–49.

Navarro–Ranninger, C., Zamora, F., Pérez, J. M., López–Solera, I., Martínez–Carrera, S., Masaguer, J. R., & Alonso, C. (1992) *Journal of Inorganic Biochemistry*, **46**, 267–279.

Neilands, J. B. (1973) *Inorganic Biochemistry*, pp. 166–201, G. I. Eichhorn, Ed. New York: Elsevier.

Okujo, N., Saito, M., Yamamoto, S., Yoshida, T., Miyoshi, S., & Shinoda, S. (1994) *BioMetals*, **7**, 109–116.

Palmer, B. N. & Powell, H. K. J. (1974) *Journal of the Chemical Society—Dalton Transactions*, 2089–2092.

Parker, D. (1990) *Chemical Society Reviews*, **19**, 271–291.

Pedersen, C. J. (1988) *Angewandte Chemie—International Edition in English*, **27**, 1021–1027.

Persmark, M., Pittman, P., Buyer, J. S., Schwyn, B., Gill, P. R., & Neilands, J. B. (1993) *Journal of the American Chemical Society*, **115**, 3950–3956.

Peterson, T., Falk, K., Leong, S. A., Klein, M. P., & Neilands, J. B. (1980) *Journal of the American Chemical Society*, **102**, 7715–7718.

Prakash, J. P., Rajamohanan, P., & Ganesh, K. W. (1991) *Journal of the Chemical Society—Perkin Transactions I*, 1273–1278.

Pressman, B. C. (1973) *Inorganic Biochemistry*, pp. 202–225, G. I. Eichhorn, Ed. New York: Elsevier.

Qu, Y. & Farrell, N. (1990) *Journal of Inorganic Biochemistry*, **40**, 255-264.

——— (1991) *Journal of the American Chemical Society*, **113**, 4851–4857.

Reetz, M. T., Niemeyer, C. M., Hermes, M., & Goddard, R. (1992) *Angewandte Chemie*, **31**, 1017–1019.

Richman, J. E. & Atkins, T. J. (1974) *Journal of the American Chemical Society*, **96**, 2268–2270.

Searle, G. H. & Geue, R. J. (1984) *Australian Journal of Chemistry*, **37**, 959–970.

Selig, W. (1982a) *Microchemical Journal*, **27**, 102–111.

——— (1982b) *Microchemical Journal*, **27**, 200–209.

Sharma, S. K., Miller, M. J., & Payne, S. M. (1989) *Journal of Medicinal Chemistry*, **32**, 357–367.

Sigel, H. & Martin, R. B. (1982) *Chemical Reviews*, **82**, 385–426.

Templeton, D. M. & Sarkar, B. (1985) *Canadian Journal of Chemistry*, **63**, 3122–3128.

Thummel, R. P. (1991) *Tetrahedron*, **47**, 6851–6886.

Tice, C. M. & Ganem, B. (1983) *Journal of Organic Chemistry*, **48**, 2106-2108.

Tsukube, H., Adachi, H., & Morosawa, S. (1991a) *Journal of Organic Chemistry*, **56**, 7102–7108.

Tsukube, H., Yamashita, K., Iwachido, T., & Zenki, M. (1991b) *Journal of the Chemical Society—Perkin Transactions I*, 1661–1665.

Umezawa, Y., Kataoka, M., Takami, W., Kimura, E., Koike, T., & Nada, H. (1988) *Analytical Chemistry*, **60**, 2392–2396.

Weitl, F. L. & Raymond, K. N. (1981) *Journal of Organic Chemistry*, **46**, 5234–5237.

Yamamoto, S., Okujo, N., Fujita, Y., Saito, M., Yoshida, T., & Shinoda, S. (1993) *Journal of Biochemistry*, **113**, 538–544.

CHAPTER 5

Free Polyamines and Their Distribution

The distribution of the polyamines relates to practical considerations of our interactions with the biological world, as well as to the roles of the compounds. The interest in the amines of foods, including agricultural products (i.e., oat flakes, cereals, grapes, and wine) has focused largely on the many compounds that possibly affect odor and taste. Occasionally these studies have reported on the presence of the di-, tri-, and tetramines (Table 5.1).

The possible nutritional roles of the amines in growth and development is also beginning to be explored. For example, the study of human milk in the first month of lactation after birth has revealed significant concentrations of nucleosides, nucleotides, and polyamines (in the range of 1–2, 3–8, 3–8 μM putrescine, spermidine, and spermine, respectively). Commercial infant formulae are uniformly low in spermidine and nucleic acid derivatives. In contrast to human milk, early rat milk is somewhat higher in putrescine and spermidine and lower in spermine (Buts et al., 1995).

The presence of fairly large amounts of compounds containing secondary amines in our food stuffs has posed the potentiality of the formation of carcinogenic nitrosamines via a reaction with exogenous and endogenous nitrite (Yamamoto et al., 1982). More recent analyses of the amines have attempted also to define the microbial dangers in our supplies of fish and beef (Ryser et al., 1984; Sayem–El-Daher & Simard, 1985). The latter workers have reported a significant correlation between the amount of cadaverine in ground beef and its content of coliform bacteria.

The early data on the distribution of the polyamines suffered primarily from the narrow scope of the biological material studied and from limitations of the analytical methods. Before 1970 it had appeared that spermine was a unique tetramine, found only in most eucaryotic cells, whereas putrescine and spermidine were ubiquitous in almost all cells. This division implied an acquisition of metabolic function in eucaryotic cells. It suggested further that whereas spermine was perhaps associated with the new biochemical activities of eu-

caryotic cells, putrescine and spermidine fulfilled major functions essential for the growth and multiplication of procaryotic cells.

The studies of Herbst and his collaborators had in fact suggested that the roles of the polyamines in supporting bacterial growth might be more complicated than that presaged by this simple picture. Putrescine, agmatine, spermidine, and spermine would satisfy a nutritional requirement for the growth of *Hemophilus parainfluenzae* (Herbst & Snell, 1948, 1949); but the satisfaction of the growth requirement was also fulfilled by some diamines, such as 1,3-diaminopropane, and triamines, such as norspermidine and homospermidine, neither of which had yet been discovered in biological material (Herbst et al., 1955). Neither ethylenediamine nor cadaverine satisfied the requirement.

Exogenous spermidine and spermine were actively degraded to diaminopropane by several microorganisms. Further, improved micromethods detected this diamine in mammalian tissues and fluids for the first time (Weaver & Herbst, 1958). Additionally, their experiments revealed active catabolic systems, which required a more detailed characterization. Some catabolic systems, particularly the amine oxidases, will be discussed in Chapter 6. Some of these enzymes do convert spermidine to 1,3-diaminopropane, which accumulates in several *Myxamoebae* and *Acanthamoebae*, among other organisms (Fig. 5.1).

In addition, bacteria of the genus *Vibrio*, which synthesizes 1,3-diaminopropane, possess a novel enzyme for the decarboxylation of the unusual amino acid L-2,4-diaminobutyric acid (Nakao et al., 1989). The organism also has a novel pathway for the synthesis of norspermidine, which is found in large amounts in *Vibrio*. Thus, the examination of the compounds found in common organisms revealed new polyamines and new metabolic routes. At present there are many new compounds describable as polyamines, and the simple classification of procaryotic and eucaryotic polyamines given above is not valid.

The distribution of the polyamines in the biological

TABLE 5.1 Polyamine content of some foods.

Food type	Polyamines (nmol/g or mL)		
	Putrescine	Spermidine	Spermine
Meat			
Cod	300–337	7–11	15–32
Trout	21–22	24–35	22–30
Fish sausage	180–185	27–29	43–45
Chicken	32–33	63–65	291–296
Pork, lean	34–35	20–34	149–348
Pork, bacon	45–46	27–41	100–249
Pork, smoked ham	45–49	14–61	199–249
Pork, roasted ham	101–103	41–43	199–299
Meat sausage	157–165	40–44	119–128
Beef, lean, raw	63–67	126–136	152–208
Beef, cooked	22–32	39–47	113–165
Ground beef	100–101	487–503	229–235
Milk products			
Cheddar, matured	7409–7427	1361–1392	115–198
Cheddar, fresh	115–227	557–751	118–194
Full cream milk	1	1–3	1–3
Semiskimmed milk	1–2	2–4	1–2
Cereals			
Whole bread	6–10	147–189	35–45
White bread	17–21	57–59	17–19
Bran, rivita	7–11	28–64	2–4
Pasta, cooked	11–12	48–50	52–64
Rice, cooked	11–15	9–11	40–50
Vegetables			
Cabbage leaves	4–18	22–35	16–18
Onion	62–82	38–56	4–6
Potato, raw	108–112	76–78	14–16
Potato, cooked	229–261	101–109	24–28
Mushroom	1–2	236–279	4–6
Carrot	14–20	53–57	10–14
Cauliflower	35–51	150–192	48–64
Lettuce	37–55	29–57	0
Tomato	106–1386	11–17	0
Cucumber	36–37	10–11	1–3
Radish	2–3	3–4	6–8
Soybean	18–74	229–428	147–170
Red kidney bean	4–5	131–138	113–127
Green beans, cooked	49–61	53–61	23–27
Green peas, cooked	61–67	428–470	166–355
Fruits			
Orange	1081–1579	61–67	0
Orange, canned	307–341	5–7	0
Apple	5–19	15–19	0
Pears	268–275	208–524	40–244
Grapefruit juice	1120	0	0

Adapted from Bardócz et al. (1993).

kingdoms may relate to the problem of the functionality of these amines. Ubiquity and conservation of the amines and their biosyntheses may suggest essentiality or some function conferring an evolutionary advantage. The identity of the amines and their biosynthetic mechanisms in various organisms may be taken as an indication of evolutionary relationship. In fact these simple approaches have been almost overwhelmed by the di-versity of the rapidly accumulating data. For this reason only some of the free polyamines will be included in this discussion, and the discussion of amides of the polyamines, guanidine-containing compounds, and other derivatives will be put off to chapters concerned with the biosynthesis of the compounds and their metabolic fate and function. In general the free polyamines may be extracted readily and essentially completely from cells

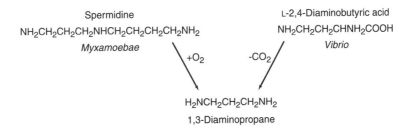

FIG 5.1   Some sources of 1,3-diaminopropane.

and tissues by aqueous acid and can be separated, analyzed and characterized by the methods described in Chapter 3. An apparent exception to this generalization will be noted in Chapter 20, which is on plants.

The compounds described as the natural polyamines are given in Table 5.2. The major expansion of the list occurred after 1970 as a result of the adoption of sensitive analytical methods for the polyamines and their application to broad groups of biological materials.

Odors and Amines

Despite the discoveries and suggestive names given to the diamines found in putrefying systems, and of spermine and spermidine in odoriferous ejaculates, the amines are relatively minor subjects of study in olfaction. This is surprising in view of the overwhelming commercial attention currently paid to human presentation and hygiene. Head-space techniques of trapping emanations, combined with gas chromatography and mass spectrometry, have generally failed to detect volatile underivatized polyamines obtained from human vapors (Ellin et al., 1974; Preti et al., 1988; Sastry et al., 1980). The characteristic odor of semen has been attributed to an oxidation product of spermidine, N-aminopropylpyrroline (Mann, 1954). Several "popular" volumes on human odor have focused on several urinary and salival odorous sterols, suspected to be active as pheromones (Bird & Gower, 1983). Human axillae have been found to exude some contributory straight-chain, branched, and unsaturated fatty acids, for example, (E)-3-methyl-2-hexenoic acid (Zeng et al., 1991). These odors are eliminated by treatment with base. On the other hand, the use of a vaginal collagen sponge as a sperm barrier was found to lead to the production of malodorous propylamine from seminal spermine and spermidine (Eskelson et al., 1978).

Periodontal disease and halitosis are frequently associated. These are correlatable with very high concentrations of putrescine in gingival crevicular fluid (Lamster et al., 1987). Nevertheless, the major components of malodor in such patients have been reported to be volatile sulfur compounds, such as CH_3SH, H_2S, etc. (Yaegaki & Sanada, 1992).

Odor as evidence of bacterial putrefaction, perhaps as advanced as that occurring in animal death, has been studied in animals that bury their dead. The addition

(sprinkling) of 1 g of free diamine on an anesthetized rat resulted in much effort by conspecific rats to bury this treated animal (Pinel et al., 1981). However, the very large amount of diamine used in such experiments raises questions of the biological significance of these results.

A penetrating mouth and body odor (i.e., the fish odor syndrome) may arise from the bacterial degradation of choline and carnitine to trimethylamine. This is oxidized normally to odorless and excretable trimethylamine-N-oxide. The lack of a genetically determined flavin monooxygenase is believed to be the major cause of this relatively rare and unpleasant syndrome (Ayesh et al., 1993). Tetramethylamine has been found in several marine phyla, and organisms that contain this metabolite are quite toxic when ingested by humans.

Several diamines, as well as other amines, have been cold trapped within the inflorescence of very large arum lily species that exude a warm insect-attracting stench (Smith & Meeuse, 1966). However, no diamines have been detected in over 100 studies of volatile plant components since that report (Knudsen et al., 1993). The possibility of a polyamine serving as a behavioral scent seems possible in an aqueous environment. Chemoreceptor cells in several pairs of walking legs of the American lobster, which function in feeding and egg-grooming behavior, do respond, albeit weakly, to putrescine and cadaverine.

Discovery of Hydroxyputrescine and Hydroxyspermidine

The N-acetylation of the polyamines, putrescine, spermidine, and spermine was demonstrated in Escherichia coli (Dubin & Rosenthal, 1960). The formation and excretion of the N-acetyl derivatives were of interest for their possible role in the regulation of the functions of the polyamines, whatever their roles might be in growth and multiplication. The discovery of the presence of the amines in T-even bacteriophages and their association with nucleic acid in these viruses suggested the desirability of testing the essentiality of the amines in virus multiplication by developing host bacteria depleted of their free polyamines, perhaps by acetylation. A strain of Pseudomonas was found that was thought to contain putrescine as the sole polyamine, and it appeared to excrete and occasionally contain acetyl putrescine (Kim,

TABLE 5.2   Natural free polyamines.

Trivial name	Chemical		Structure
	Abbreviation	Name	
Diamines			
Putrescine	4	1,4-Diaminobutane	$H_2N(CH_2)_4NH_2$
Cadaverine	5	1,5-Diaminopentane	$H_2N(CH_2)_5NH_2$
Diaminopropane	3	1,3-Diaminopropane	$H_2N(CH_2)_3NH_2$
2-Hydroxyputrescine	—	S(+) 1,4-Diaminobutane-2-ol	$H_2NCH_2CHOHCH_2CH_2NH_2$
Triamines			
Spermidine	3,4	1,8-Diamino-4-azaoctane	$H_2N(CH_2)_3NH(CH_2)_4NH_2$
sym-Homospermidine	4,4	1,9-Diamino-5-azanonane	$H_2N(CH_2)_4NH(CH_2)_4NH_2$
Aminopropylcadaverine	3,5	1,9-Diamino-4-azanonane	$H_2N(CH_2)_3NH(CH_2)_5NH_2$
Norspermidine (caldine)	3,3	1,7-Diamino-4-azaheptane	$H_2N(CH_2)_3NH(CH_2)_3NH_2$
7-Hydroxyspermidine	—	1,8-Diamino-4-azaoctane-6-ol	$H_2N(CH_2)_3NHCH_2CHOHCH_2CH_2NH_2$
Tetramines			
Spermine	3,4,3	1,12-Diamino-4,9-diazadodecane	$H_2N(CH_2)_3NH(CH_2)_4NH(CH_2)_3NH_2$
Thermine (norspermine)	3,3,3	1,11-Diamino-4,8-diazaundecane	$H_2N(CH_2)_3NH(CH_2)_3NH(CH_2)_3NH_2$
Thermospermine	3,3,4	1,12-Diamino-4,8-diazadodecane	$H_2N(CH_2)_3NH(CH_2)_3NH(CH_2)_4NH_2$
Canavalmine	4,3,4	1,13-Diamino-5,9-diazatridecane	$H_2N(CH_2)_4NH(CH_2)_3NH(CH_2)_4NH_2$
Aminopropylhomospermidine	3,4,4	1,13-Diamino-4,9-diazatridecane	$H_2N(CH_2)_3NH(CH_2)_4NH(CH_2)_4NH_2$
N,N^1-bis(3-Aminopropyl)-cadaverine	3,5,3	1,13-Diamino-4,10-diazatridecane	$H_2N(CH_2)_3NH(CH_2)_5NH(CH_2)_3NH_2$
Aminopentylnorspermidine	5,3,3	1,13-Diamino-6,10-diazatridecane	$H_2N(CH_2)_5NH(CH_2)_3NH(CH_2)_3NH_2$
Aminobutylhomospermidine	4,4,4	1,14-Diamino-5,10-diazatetradecane	$H_2N(CH_2)_4NH(CH_2)_4NH(CH_2)_4NH_2$

N^4-Aminopropylnorspermidine	3(3)3	tris(3-Aminopropyl)ammine	$H_2N(CH_2)_3N(CH_2)_3NH_2$
			                 $\|$
			            $(CH_2)_3NH_2$

N^4-Aminopropylspermidine	3(3)4	N^4-(3-Aminopropyl)-1,8-diamino-4-azaoctane	$H_2N(CH_2)_3N(CH_2)_4NH_2$
			                 $\|$
			            $(CH_2)_3NH_2$

Pentamines			
Caldopentamine	3,3,3,3	1,15-Diamino-4.8.12-tri-azapentadecane	$H_2N(CH_2)_3NH(CH_2)_5NH(CH_2)_3NH(CH_2)_3NH_2$
Homocaldopentamine	3,3,3,4	1,16-Diamino-4,8,12-triazahexadecane	$H_2N(CH_2)_3NH(CH_2)_5NH(CH_2)_3NH(CH_2)_4NH_2$

N^4-bis(Aminopropyl)norspermidine	3(3)(3)3	tetrakis(3-Aminopropyl)ammonium	$H_2N(CH_2)_3\ \ (CH_2)_3NH_2$
			         \\ /
			           N^+
			         / \\
			 $H_2N(CH_2)_3\ \ (CH_2)_3NH_2$

Thermopentamine	3,3,4,3	1,16-Diamino-4,8,13-tri-azahexadecane	$H_2N(CH_2)_3NH(CH_2)_3NH(CH_2)_4NH(CH_2)_3NH_2$

N^4-bis(Aminopropyl)spermidine	3(3)(3)4	N^1-tris(Aminopropyl)-ammoniobutylamine	$H_2N(CH_2)_3\ \ (CH_2)_3NH_2$
			         \\ /
			           N^+
			         / \\
			 $H_2N(CH_2)_3\ \ (CH_2)_4NH_2$

Hexamines			
Caldohexamine	3,3,3,3,3	1,19-Diamino-4,8,12,16-tetraza-ane	$H_2N(CH_2)_3NH(CH_2)_3NH(CH_2)_3NH(CH_2)_3NH-$ $(CH_2)_3NH_2$
Homothermohexamine	3,3,4,3,3	1,20,Diamino-4,8,13,17-tetraza-ane	$H_2N(CH_2)_3NH(CH_2)_3NH(CH_2)_4NH(CH_2)_3NH-$ $(CH_2)_3NH_2$
Thermohexamine	3,3,3,4,3	1,20-Diamino-4,8,12,17-tetraza-ane	$H_2N(CH_2)_3NH(CH_2)_3NH(CH_2)_3NH(CH_2)_4NH-$ $(CH_2)_3NH_2$
Homocaldohexamine	3,3,3,3,4	1,20-Diamino-4,8,12,16-tetraza-ane	$H_2N(CH_2)_3NH(CH_2)_3NH(CH_2)_3NH(CH_2)_3NH-$ $(CH_2)_4NH_2$

1966). The presumed acetyl putrescine was subsequently discovered to be 1,4-diaminobutane-2-ol or (+)-hydroxyputrescine (Rosano & Hurwitz, 1969; Tobari & Tchen, 1971). Further, the studies detected the conversion of ^{14}C-putrescine to labeled hydroxyputrescine. Rosano and Hurwitz (1969) noted that the new amine was found in the ribosomes of the pseudomonad and that its concentration fell with increasing Mg^{2+} content in the medium. Indeed the hydroxyputrescine concentration in

those ribosomes approached that of spermidine in the ribosomes of *E. coli*.

The configuration of the natural hydroxydiamine was established by synthesis of the identical S-(+)-2-hydroxyputrescine·2HCl from the known S-(−)-hydroxysuccinamide (Kullnig et al., 1973). This group then discovered a 7-hydroxyspermidine (originally designated as a 2-hydroxyspermidine) in these organisms (Rosano et al., 1978). The *Pseudomonas* sp. strain Kim, does in fact make spermidine, which is rapidly degraded to putrescine intracellularly, and does excrete detectable amounts of *N*-acetylspermidine (Rosano et al., 1989). Nevertheless, spermidine does not accumulate in *Pseudomonas* sp. strain Kim and is not easily detectable in the growing organism.

The Proteobacteria, which include the purple photosynthetic bacteria and their nonphotosynthetic relatives, do contain 2-hydroxyputrescine uniquely in all members of the beta subclass. These organisms also contain putrescine as a major polyamine and possess very low levels of spermidine (Busse & Auling, 1988). This diagnostic feature has been pursued with two carbon monoxide utilizing organisms, formerly designated as pseudomonads. Because only one of these contained the hydroxypolyamines, it was suggested that the organisms belonged to different subclasses of the Proteobacteria (Hamana & Matsuzaki, 1990b). Similarly an examination of the polyamines of 12 species of sulfur-oxidizing eubacteria, designated as *Thiobacillus*, revealed five different types, only one of which, comprising four species, contained hydroxypolyamines (Hamana & Matsuzaki, 1990a). The presence of hydroxyputrescine is also reported in a species of *Flavobacterium* and *Cytophaga*, groups whose major polyamine is homospermidine (Hamana & Matsuzaki, 1991). A cyclic siderophore, named *alcaligin*, has been isolated from strains of *Alcaligenes*. The macrocyclic compound consists of two molecules of amine-hydroxylated 2-hydroxyputrescine, that is, hydroxamates, linked by two molecules of succinate (Bergeron et al., 1991).

Spermine among Microbes

Although a spermine salt was the first of the polyamines to be seen and separated as a crystalline substance, many of the key biochemical problems posed by its existence are among the least understood. In the studies of spermine isolated between 1880 and 1910, the data on its elementary composition were faulty. Eventually apparently new bases in calf's muscle (musculamine), human brain (neuridine), and dog's liver (gerontine) were shown to be spermine. Its distribution in procaryotic and eucaryotic organisms was not adequately explored in early studies, which followed the simple notion that *E. coli* might serve as the guide to all biochemistry. The apparent inability of *E. coli* and several other bacteria to synthesize spermine evoked the hypothesis that this incapacity might point to an acquisition of function by eucaryotic cells.

At this time an enzymatic mode of biosynthesis of spermine via aminopropyl transfer in eucaryotic organisms is known but our knowledge is incomplete concerning this aminopropyl transferase, that is, its structure, specificity, variability, inheritance, regulation, and physiological role or roles. A variety of reasons relating in part to the role of microbiology in the history of biochemistry—the toxicity of the amine to mammalian cells and organs, the tetravalence of spermine and its tight binding to and precipitation of nucleic acids, and

the slow development of analytical methods for the polyamines—has delayed our inquiries concerning spermine. The presence of a toxin-generating enzyme in many mammalian sera minimized the pursuit of the fate of exogenous spermine in cultured animal cells, and the precipitation by spermine of functional components of cell extracts suggested that this basic substance might well be omitted from a reagent shelf.

However, the interest of "general" microbiologists in organisms living in unusual ecological niches has led eventually to a detailed characterization of many such organisms. Analysis of the sequences of nucleotides in ribosomal RNA resulted in the discovery of a second kingdom of procaryotes, the Archaebacteria, contrasting with the Eubacteria, such as *E. coli*. Both kingdoms include extreme thermophiles, and thermophilic bacteria frequently contain tetramines, including spermine, and even larger polyamines. The evolutionary relationships of the Archaebacteria, Eubacteria and Eucaryota are presented in Figure 5.2. For reasons of habit, as well as the arguments of H. Jannasch (1993), I have chosen to use procaryotes or eucaryotes instead of prokaryotes or eukaryotes.

Polyamines and Thermophilic Microorganisms

After 1970 questions were posed concerning the biochemistry of organisms existing in unusual ecological niches; as discussed in Chapter 3, tools for detecting, isolating, and characterizing novel polyamines became available. For example, it was asked how a thermophilic eubacterium, *Thermus thermophilus*, managed to grow and multiply optimally at temperatures in the range of 65–75°C, temperatures at which protein synthesis did not proceed in the extracts of most microorganisms. It was known that the structures of nucleic acids essential to protein synthesis were markedly altered on elevation of temperature, and that the melting of the nucleic acids was related to base composition and hydrogen bonding within the structure. It had been found that the polyamines reduced the temperature effect and prevented this structural alteration in the order of $Mg^{2+} \leq$ putrescine < spermidine < spermine. Tetramines clearly stabilized double-stranded or looped nucleates. Other effects of the polyamines in stabilizing the protein-synthesizing machinery had also been reported. It was observed that an extract of *T. thermophilus* did not easily support polypeptide formation at 65°C unless spermine in particular was added to the system (Ohno–Iwashita et al., 1975). Unusual polyamines were then sought and found in extracts. Two tetramines were found, one of which was reported to be spermine, and the second was the new 3,3,3 tetramine, that is, norspermine or 1,11-diamino-4,8-diazaundecane. This was named *thermine* by its discoverer (Oshima, 1975). The further study of *T. thermophilus* revealed new pentamines and hexamines as well as branched tertiary and quaternary polyamines.

The structure of thermine was determined by chromatography and comparison of its nuclear magnetic reso-

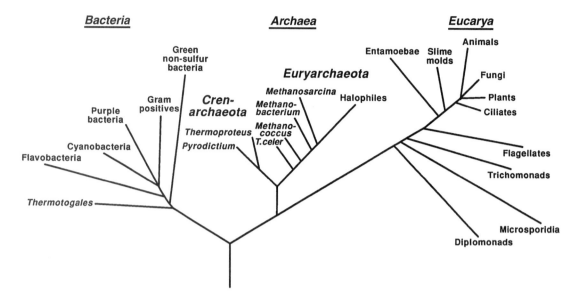

FIG 5.2   Universal phylogenetic tree with three domains; redrawn from Woese (1993).

nance (NMR) spectrum with a synthetic sample. Almost simultaneously, thermine (3,3,3), also occasionally designated as *sym*-nor-spermine, was also found in *Caldariella acidophila*, an extreme thermoacidophilic bacterium, now described as an archaebacterium (DeRosa et al., 1976). This organism was also found to contain *sym*-nor-spermidine or 1,7-diamino-4-aza-heptane, named *caldine* by its discoverers. However, this organism lacked a sperminelike tetramine. In any case, both eubacterial and archaebacterial thermophiles contained tetramines.

Oshima (1979) reexamined the sperminelike polyamine of *T. thermophilus* and discovered it to be a new polyamine, "thermospermine," in which an aminobutyl moiety was added to norspermidine to form the 3,3,4-tetramine, 1,12-diamino-4,8-diazadodecane. The biosynthesis of thermospermine (3,3,4) might also be formulated as the addition of an aminopropyl substituent to the existing aminopropyl of spermidine (3,4). Thermospermine was recently described as a component of wasp and spider toxins, the philanthotoxins. Thermospermine was separable from spermine by ion-exchange chromatography and was distinguishable from true spermine (3,4,3) by its infrared spectrum and its proton and [^{13}C]-NMR spectra. In the infrared spectra, absorption bands characteristic of spermine were not seen in the new tetramine. In the [^{13}C]-NMR spectrum, the new compound was clearly asymmetric in the distribution of its butyl and propyl groups, in contrast to the evidence of the spectrum for spermine. The mass spectra of the two compounds were also different in the distribution of the fragments produced, despite the identity of the mass of the individual fragments.

As discussed by De Rosa et al. (1980), the methods then available were capable of effectively separating tetramines as large as spermine and thermine, differing in a single methylene group. Not only was this realizable by high voltage paper electrophoresis, but gas–liquid chromatography of the trifluoroacetylated derivatives was useful in distinguishing *sym*-nor-spermidine from spermidine, and *sym*-nor-spermine from spermine. These separations further facilitated characterization of these compounds by mass spectrometry, in which it was shown that specific higher fragments of *sym*-nor-spermidine and *sym*-nor-spermine were 14 mass units less than those of spermidine and spermine, respectively. This unequivocally demonstrated the distinction between norspermidine and spermidine by a methylene moiety. This is the converse of the previously mentioned demonstration of the identification of spermidine, rather than norspermidine, in turnip yellow mosaic virus.

The power of NMR spectroscopy is exemplified in the discussion of DeRosa et al. (1980). The symmetry of norspermidine, as compared to the structure of spermidine, was revealed in the number and separateness of the signals developed in the [^{13}C]-NMR spectra presented in Figure 5.3(**A**,**B**). The additional complexity of norspermine and increased symmetry of spermine are revealed in the [^{13}C]-NMR spectra of these compounds. The current power of spectrometry is well-exemplified by the application of mass spectrometry and various NMR techniques, as well as other more familiar spectrophotometry (i.e., visible), ultraviolet, and infrared spectroscopy in the study of newly discovered polyamine-derived alkaloids.

The examination of the chromatographic separations was fine tuned to define the significance of minor bumps in the elution pattern, as well as the identification of less easily elutable components, for even a single

A   NH₂-CH₂-CH₂-CH₂-NH-CH₂-CH₂-CH₂-NH₂

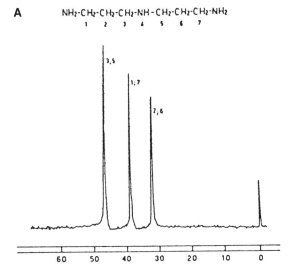

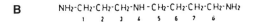

B   NH₂-CH₂-CH₂-CH₂-NH-CH₂-CH₂-CH₂-CH₂-NH₂

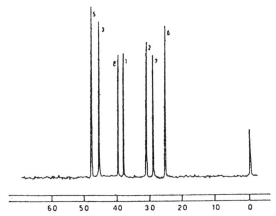

FIG 5.3   [¹³C]-NMR spectra of polyamines; redrawn from De Rosa et al. (1980).

organism, such as *T. thermophilus*. As the search was extended to other thermophiles, many new natural compounds were detected and characterized (Fig. 5.4). The availability of synthetic standards (described in Chap. 2), as well as the spectroscopic methods of modern chemistry noted above, were essential to this obvious and startling progress.

Although the chart (Fig. 5.4) concludes with 1990, it should be noted that new polyamines, linear (e.g., a heptamine) and branched (i.e., tertiary and quaternary), have since been found.

Although putrescine, cadaverine, 1,3-diaminopropane, and spermidine were found frequently in many

procaryotes, Table 5.2 notes the relatively recent discovery of compounds in bacteria and other cells, such as *sym*-homospermidine and *asym*-homospermidine (i.e., aminopropylcadaverine, norspermidine, norspermine, caldopentamine, homocaldopentamine, etc.). The slightly halophilic *Halococcus acetinofaciens* contains components such as aminopentylnorspermidine and *N,N'*-bis(3-aminopropyl)cadaverine (Hamana et al., 1988a). As a greater number of microorganisms were collected, classified, and analyzed for polyamines, it became evident that the polyamines themselves might serve as chemotaxonomic markers. For example, diaminopropane was found to be the major polyamine of the eubacterial genus *Acinetobacter* (Hamana & Matsuzaki, 1992b). This perception was strengthened by the systematic studies on the Proteobacteria, discussed earlier in connection with their occasional content of hydroxypolyamines (Busse & Auling, 1988; Hamana & Matsuzaki, 1993). Progress in the use of the polyamines in bacterial systematics has been summarized in the extensive review by Hamana and Matsuzaki (1992a). For example, it has been possible to distinguish species of *Aeromonas* from *Vibrionaceae*, in that both contain diaminopropane but the former is devoid of norspermidine (Kämpfer et al., 1994). Recent analyses of genera including *Flavobacteria* and *Cytophaga* also make this point (Hamana et al., 1995a).

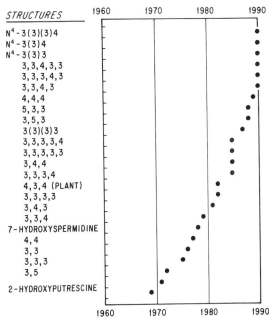

FIG 5.4   Discovery of microbial polyamines.

Polyamines in Microbial Evolution: The Problem of the RNA World

As described in later chapters, double strandedness reduces temperature-induced melting of nucleates; the polyamines, particularly the triamines and tetramines, are effective in organizing and maintaining this type of complementary structure. Actually the stabilization and maintenance of the secondary and tertiary configurations of single-stranded RNA is at least as important as this function of stabilizing DNA in contemporary cells. The observations of RNA splicing and RNA-determined catalysis have encouraged the speculation that RNA preceded DNA in evolution, and the following remarks that place stabilizing polyamines in the most primitive cells may add another parameter to the concept of an RNA world.

Extreme thermophilic eubacteria containing higher polyamines were described in 1975. Some eubacteria live at temperatures even higher than *Thermus* species, for example, *Thermatoga maritima*, and it was asked if the original eubacteria were thermophiles (Achenback–Richter et al., 1987). Some *Thermatoga* species grown on glucose at 78°C contain the long-chain polyamines, caldopentamine and caldohexamine (Zellner & Kneifel, 1993). Thermophilic Gram-positive eubacteria of the genera *Saccharococcus*, which are related to *Bacillus stearothermophilus* (Rainey & Stackebrandt, 1993), have also been found to contain tertiary and quaternary branched tetra- and pentamines. However, the linear caldopentamine and caldohexamine are more efficient in increasing the melting temperatures of DNA. Organisms such as *Saccharococcus thermophilus*, which has a composition of DNA, 16S rRNA, and a cell wall similar to those features of *B. stearothermophilus*, are selected, and concentrate in beet sugar extraction plants that operate at relatively high temperatures (Nystrand, 1984).

Among two groups of Archaebacteria other than the thermophiles (i.e., the halophiles and strictly anaerobic methanogens), the former contain very little polyamine, about 1–10% of that found in *Thermus* species (Hamana & Matsuzaki, 1992a; Kneifel et al., 1986). More recently many extremely halophilic Archaebacteria have been found to contain agmatine as the major polyamine (Hamana et al., 1995b). Four known families of methanogens appeared to possess distinctively different patterns of polyamine content, none of which were proven to contain tetramines or other higher polyamine. One of the four families was very low in polyamine, two were rich in spermidine, and one was high in homospermidine (Scherer & Kneifel, 1983). The survey of Kneifel et al. (1986) describes the types of polyamines, many of which are of the norspermine and spermine types found in thermophilic Archaebacteria. The presence of a tetramine aminobutylhomospermidine (4,4,4) occurs in Japanese volcanic ash soils, which are also relatively high in homospermidine (Fujihara & Harada, 1989b).

The hypotheses that life on earth began in the thermal springs found in volcanic regions, possibly even below oceanic bottoms, has been bolstered by the findings of the sulfur-metabolizing thermoacidophilic Archaebacteria, for example, *Sulfolobus acidocaldarius*, in extreme ecological niches. Nucleate and protein sequences have placed these organisms among the earliest procaryotes (Lake, 1994). Relatively high concentrations of tetramines and higher polyamines in such organisms may point to their possible primitive role in the survival of genetic materials as well as their requirement for "higher" or polyvalent polyamines in protein synthesis (Chap. 22).

Further, the unusual spermidine derivatives deoxyhypusine and hypusine, found uniquely in eucaryotic cells in the initiation factor of protein synthesis, EIF5a, have also been found in thermophilic Archaebacteria (see Chap. 22) and not in any Eubacteria. These findings also imply early roles for the polyamines in the evolution of primitive metabolism and pose the problem of the distribution and structures of genes for these functions, the in vivo roles of these activities and compounds and the characterization of the biosynthetic reactions of the higher polyamines themselves. Thus a systematic study of the biosynthesis of the higher polyamines may reveal very early evolutionary links among the Eubacteria, Archaebacteria, and Eucaryota.

Spermine in Bacteria

The presence of spermine in yeast posed the problem of the microbial distribution of this compound found initially in seminal fluid. The distinction between procaryotic and eucaryotic cells, in which the latter, such as yeast, possessed many "new" biosynthetic capabilities, suggested that spermine might be associated with one of these new activities. The problem of determining the distribution of spermine has been complicated by the nutrients used to grow the organism. If a microorganism is grown in a medium containing spermine, this substance is frequently concentrated selectively within the organism and replaces the normally synthesized amines (Dubin & Rosenthal, 1960). Thus an analysis of a microorganism reported to contain spermine must assure the absence of spermine from the growth medium. Many essays have not been clear on this experimental point. This requirement in both experiment and publication is clearly presented by Hamana (1994) and Hamana and Satake (1995) who described the presence or absence of polyamines in various Gram-positive cocci grown in synthetic media.

It was reported in 1968 that *B. stearothermophilus* contained spermine. Some 10 years later it was stated that the compound might have been incorporated from the medium. Indeed considering the difficulties of an unequivocal separation and identification of spermine from the many tetramines now known in thermophilic organisms, it would have been difficult to be certain at that time that the observed tetramine was in fact spermine. However, in another decade spermine (3,4,3) was found in this organism grown in a synthetic medium devoid of spermine (Hamana et al., 1989a). Spermine was also found in a few obligate moderately thermophilic and thermoacidophilic bacilli grown in such media, but not in many other species of bacillus, unless they were grown in spermine-containing media. A possibly coeluting thermospermine (3,3,4) can be distinguished from coeluting spermine by the products formed with putrescine oxidase (Hamana & Matsuzaki, 1987). This method was used to identify spermine in several thermophilic actinomycetes. Matsuzaki et al. (1992) developed the combined use of high performance liquid chromatography and selected oxidases in identification of tetramines.

The review by Hamana and Matsuzaki (1992a) on the polyamines as taxonomic markers in bacterial systematics records the presence of major amounts of spermine in thermophilic Gram-positive eubacteria, that is, Bacilli, Clostridia, Actinomycetes, as well as its occurrence as a minor component in certain Agrobacteria, Pseudomonads, and Purple Sulfur bacteria (Chromatiaceae). For the most part, the reports would benefit from a stated upper limit of the possible concentration of spermine in the medium used and more rigorous evidence of biosynthesis of the spermine measured, for example, the incorporation of $[^{14}C]$-putrescine into the isolated spermine.

Many studies indicate large groups of microorganisms that lack this tetramine when grown in the apparent absence of the compound. The presence of spermine in bacteria under several conditions poses many problems of biosynthesis and its regulation, as well as the possible cellular function of the amine. Of great interest is a study of the effects of iron depletion and siderophore production on spermine biosynthesis in *Paracoccus* species. These eubacteria, described as Gram-negative facultative chemolithotrophs, contain putrescine and spermidine as major polyamines and may also produce higher polyamine derivatives of diaminopropane and cadaverine (Hamana & Matsuzaki, 1992c). In that study, coeluting spermine and thermospermine in *Paracoccus denitrificans*, which were grown in a synthetic medium, were at trace levels. However, *P. denitrificans* appears to synthesize significant amounts of spermine in iron-deprived cultures, under conditions of derepression of siderophore synthesis and excretion. The nature of the tetramine was established by mass spectrometry of tetradansyl spermine. After iron supplementation, spermine synthesis was nondetectable (Bergeron & Weimar, 1991).

In a study of the polyamines of *Rickettsia prowazeki*, the etiologic agent of epidemic typhus, the organism was found to contain high levels of putrescine, spermidine, and spermine (Speed & Winkler, 1990). This organism is an obligately parasitic Gram-negative eubacterium, but isolated rickettsiae could not be shown to accumulate putrescine or spermidine. The bacteria contained arginine decarboxylase, and $[^{14}C]$-arginine was incorporated into rickettsial polyamine within infected cells. However, the proof of the inference that the rickettsiae can synthesize their polyamines, including spermine, is incomplete.

Studies with Cyanobacteria

Like *E. coli*, the detectable free polyamines of some cyanobacteria included putrescine and spermidine solely and appeared to lack a tetramine (Ramakrishna et al., 1978). Although *E. coli* grown in a synthetic medium contains 10–30 mM putrescine and 1–3 mM spermidine, one of the Cyanobacteria, *Anacystis nidulans*, now known as a *Synechococcus*, contained about 0.15 mM putrescine and 1–2 mM spermidine during growth. Toward the end of growth, *Anacystis* had degraded 90% of its spermidine (Guarino & Cohen, 1979). Unlike *E. coli*, which generates several conjugates, that is, *N*-acetylated polyamines and a glutathionylamide of spermidine, *Anacystis* did not appear to form soluble conjugated derivatives.

Putrescine proved to be toxic to cultures of *Anacystis* during growth at a high pH (>9) produced during photosynthetic assimilation of CO_2. This effect will be discussed in Chapter 20. Exogenous spermidine and spermine were metabolized quite rapidly. The former was converted largely to compounds whose dansyl derivatives were similar to those of diaminopropane and an otherwise undefined tetramine (Ramakrishna et al., 1978).

The presence of spermidine in excess of putrescine and the apparent absence of spermine in *A. nidulans* have been confirmed (Hamana et al., 1983). However, this broader survey of the cyanobacteria also revealed the presence of homospermidine as the most abundant polyamine in the nitrogen-fixing species (seven analyzed), whereas spermidine was the major polyamine in the cyanobacteria incapable of fixing nitrogen (eight analyzed). Several species in both groups were reported to synthesize small amounts of a tetramine.

Hamana et al. (1988b) sought homospermidine in many N_2-fixing eubacteria and eucaryota (i.e., algae and ferns). The compound was a major polyamine in microorganisms such as *Rhizobium* (Smith, 1977), some photosynthetic Rhodopseudomonads, and sulfur-oxidizing Thiobacilli.

Polyamines in Nucleated Cells

Photosynthetic algae

Kneifel (1979) summarized early studies of amines of algae, both eucaryotic and bacterial (i.e., Cyanobacteria). His extensive survey described aliphatic monoamines and complex alkaloids, as well as their polyamines. Spermine, in addition to putrescine and spermidine, is found among certain red algae: unicellular *Porphyridium cruentum* and multicellular *Porphyra tenera* in the phylum Rhodophyta. One strain of *Cya-*

nidium caldarium also possessed this pattern and lacked other tetramines, including thermospermine (Hamana et al., 1990). However, another strain of *Cyanidium* contained diaminopropane, norspermidine, and norspermine in addition to those listed above. Both organisms increased their spermine content when the temperature of cultivation was increased from 20 to 50°C.

Many other eucaryotic algal groups (i.e., Pyrrophyta, Chrysophyta, Chlorophyta, and Euglenaphyta), appeared to lack spermine while being rich in spermidine (Hamana & Matsuzaki, 1985a). Homospermidine was not present in most of the eucaryotic algae. A careful examination of the tetramine isolated from *Euglena gracilis* revealed it to be norspermine (Kneifel et al., 1978).

Fungi

It was reported that unlike the yeasts that contained putrescine, spermidine, and spermine, several filamentous fungi did not contain the latter polyamine. However, reexamination of many of these organisms revealed both the presence of spermine and the ability to incorporate [^{14}C]-putrescine into this tetramine (Hart et al., 1978). If the tetramine is truly characterized to exclude thermospermine, this biosynthetic property may be described as definitive.

Paulus and Davis (1983) devised an elegant analytical method involving the formation of doubly labeled [^{3}H,^{14}C]-dansyl spermine to determine small amounts of spermine in the filamentous *Neurospora crassa* and its polyamine-deficient mutants. Spermine is a minor constituent of the polyamines of *N. crassa*. In studies of the growth of mycelia in chemically defined media, the decrease of Mg^{2+} in the media evoked increases in the concentrations of spermidine and spermine within *N. crassa* and *Aspergillus nidulans* (Stevens & Winther, 1979).

The spermine content of several of these fungi may vary considerably as a function of the growth phase. In the case of the plant pathogenic fungus *Sclerotium rolfsii* (Fig. 5.5), spermine accumulated primarily as mycelial growth concluded and sclerotia were formed and disappeared as active synthesis of putrescine and spermidine began and mature sclerotia germinated (Shapira et al., 1989). Nevertheless it was not proven in this instance that the spermine had been newly synthesized and that the growth medium, which contained agar, was totally devoid of the tetramine.

Myxamoebae

In analyses of several *Myxamoebae* or slime molds grown on bacteria on agar, the organisms were reported to contain small amounts of spermine in addition to spermidine and putrescine. Curiously *Dictyostelium discoideum* contained large amounts of diaminopropane. In a later study *D. discoideum* and *Physarum polycephalum*, grown axenically on completely synthetic media, were found to contain putrescine and spermidine and to

lack spermine (Hamana & Matsuzaki, 1984). They did contain diaminopropane, which may be generated from spermidine in *Dictyostelium*. *Physarum* also contained homospermidine.

Animals

As one ascends an apparent evolutionary ladder of the animal kingdom, it has become evident that some eucaryotic organisms exist either without spermine or at levels of this polyamine undetectable by current analytical procedures. For example, among the protozoans, *Tetrahymena thermophila*, which contains putrescine, spermidine, and cadaverine, lacks spermine when grown on defined media (Eichler, 1989).

Other protozoans contain spermine only during specific stages of their growth cycle. For example, among the Trypanosomatidae, which include the parasitic hemoflagellates, *Leishmania donovani* lacks spermine when grown as a promastigote in a complex medium in vitro. When grown as an amastigote in the hamster, the isolated washed cells incubated with [^{14}C]-ornithine were capable of synthesizing labeled putrescine, spermidine, and spermine (Morrow et al., 1980). In its bloodstream form *Trypanosoma brucei brucei* contained and synthesized putrescine and spermidine. Washed cells did lack spermine and the ability to synthesize the tetramine, although they were able to incorporate exogenous spermine (Bacchi et al., 1979). However, Bacchi (1981) stated that spermine can be found in culture-form trypanomastigotes.

The common white shrimp *Penaeus setiferus* is rich in norspermidine and norspermine but appears to lack spermine (Stillway & Walle, 1977). The inability to detect spermine in another species of *Penaeus* was confirmed and the presence of norspermidine and norspermine was also reported in these and other marine organisms (Zappia et al., 1978). Nevertheless, spermine was found in Porifera, Cnidaria, Ctenophora, Mollusca, Annelida, Arthropoda, Echinodermata, Tunicata, as well as all Vertebrata examined. Low levels of spermine in a few organisms recorded as Porifera, Annelida, Crustacea, Echinoderms, Tunicates, and Reptilia are evidently uninterpretable as evidence of a capability to synthesize this tetramine. On the other hand unusually high concentrations of diaminopropane in *Acanthamoeba* (Poulin et al., 1984), as in *Dictyostelium discoideum* (Mach et al., 1982), are suggestive of a metabolic pathway for the synthesis of this amine, possibly via acetylpolyamine (see Chap. 6) in these amebae, as well as among bacteria. Indeed *Acathamoeba castellanii* was shown to synthesize putrescine and spermidine as well as diaminopropane, and spermidine was found to be a primary source of the diaminopropane (Kim et al., 1987). In this organism spermine was found only after growth in a spermine-containing medium. *Acanthamoeba* is also reported to contain a small amount of acetyl spermine (Zhu et al., 1989).

The relatively high concentrations of spermine found in organisms of the phyla Cnidaria, Ctenophora, Mol-

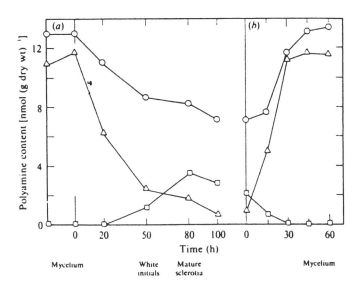

FIG 5.5 Changes in polyamine content during formation and germination of sclerotia of *S. rolfsii* grown in solid medium; redrawn from Shapira et al. (1989). (**a**) Maturation of mycelium and sclerotium formation; (**b**) germination of mature sclerotia. (△), putrescine; (○), spermidine; (□), spermine.

lusca, Arthropoda, Echinodermata, and Tunicata, as well as among most Vertebrata are suggestive of the possession of a biosynthetic capability for this tetramine (Zappia et al., 1978). In several instances, such as cephalopods, some crustacea, and some arthropods, in which spermine is accompanied by significant amounts of norspermine, it may be asked if the two compounds are synthesized by the same enzyme.

In some broad surveys of the animal kingdom, unusual polyamines have been found unexpectedly. Whereas putrescine, spermidine, and spermine were detected easily in most mammalian tissues, homospermidine, canavalmine, and aminopropylhomospermidine were found uniquely in the hamster epididymis. The origins of these compounds in this specialized tissue are quite obscure. The testes of amphibians frequently contain homospermidine, as do many invertebrates.

Many novel tetramines, pentamines, and hexamines have been found in the sea urchin, sea cucumber, sea squirt, and bivalves (Hamana et al., 1991a). Arthropods may contain the three "common" triamines as well as several tetramines and a pentamine (Hamana et al., 1991b). In addition to spermine, several common insects (e.g., cricket and cockroach) contain diaminopropane, norspermidine, norspermine, and a wide range of higher polyamines. The polyamines found in these collected organisms may reflect their diets or the products of symbiotic microorganisms rather than the biosynthetic capabilities of "sterile" cells.

A high level of spermine, crystallizable as spermine phosphate, is found in the spermatophore of a particular tick (Feldman–Muhsam et al., 1980). As noted earlier, the bulk of spermine, which in the human is made in the prostate gland, is a soluble component of the prostatic

secretion in the seminal fluid. When human semen is sedimented, spermine and amine oxidase may be found associated with the spermatozoa. Washing the cells with high concentrations of salts (>0.5 M NaCl or >0.1 M $MgCl_2$) will remove essentially all of the spermine but little of the amine oxidase. Thus spermine may be adventitiously and reversibly associated with the cell surface and the internal structures of the sperm may be devoid of the polyamine (Pulkkinen et al., 1975). Human seminal fluid also contains low quantities of spermidine, putrescine, and diaminopropane.

Several studies have been made of the release of spermine from testicular cells during spermatogenesis. Rat germ (Sertoli) cells, derived from rats of different ages, were found to release putrescine and spermidine but not spermine during culture (Shubhada et al., 1989). Spermine achieved a maximal level in the Sertoli cells of 4–5-week-old rats. During human spermatogenesis, the nuclei of developing spermatids were continually transformed with a progressive loss of spermine content (85%) in the change from primary spermatocytes to elongated spermatids (Quemener et al., 1992). This release of spermine occurred as transcription decreased, histones were replaced by protamines, and the chromatin itself was compacted.

It may be expected that mammalian cell death would also result in the release of the tetramine, which is found in the circulation (Sarhan et al., 1991). Circulating spermine is found mainly in the enucleate red blood cells, where it is also gradually converted to spermidine. This subject, which is of some clinical interest, will be discussed in later chapters.

The possible functions of spermine in human seminal fluid may also be discussed in the context of the virile bull. The bull lacks a functional prostate and possesses a seminal ejaculate devoid of spermine; the ejaculate, however, is rich in a protein, seminal plasmin, inhibitory to many microorganisms (Shivaji, 1984). Nevertheless, spermine is reported to stimulate processes of bovine sperm (Chap. 11).

Putrescine and Homospermidine

In 1917 R. Robinson proposed a key role for ornithine in the biosynthesis of certain plant alkaloids (Robinson, 1955). Study of the biosynthesis of the alkaloids in *Atropa belladonna* and *Datura stramonium* indicated putrescine as an intermediate in the synthesis of hyoscyamine (Cromwell, 1943). The synthesis of putrescine via metabolism of arginine and ornithine in a mixture of washed suspensions of *Streptococcus faecalis* and *E. coli* was described by Gale in 1940.

Agmatine, or decarboxylated arginine, a precursor of putrescine, was recently reported as an agonist at certain mammalian brain receptors (Li et al., 1994; however see Atlas, 1994). Bovine brain was reported to contain 1.5–3.0 nmol agmatine per gram and arginine decarboxylase. The search for agmatine in mammalian tissues has recently become quite active (Kalra et al., 1995; Wang et al., 1995).

As we know, putrescine is a major intermediate in the synthesis of many polyamines. In the mammal, putrescine is metabolized to many derivatives (Fig. 5.6) and these metabolic paths, as well as some of those leading to compounds such as nicotine in tobacco, will be discussed in later chapters. Tetramethylputrescine was first isolated from plants in 1907 and was isolated more recently from *Ruellia* species (Johne et al., 1975). Cinnamic acid amides of the polyamines are well known, but rust-infected wheat have been found to contain amides (i.e., ferulic acid and *p*-coumaric acid) of 2-hydroxyputrescine (Samborski & Rohringer, 1970). The diversity of the plant polyamines and their derivatives has been summarized by several plant biochemists (Guggisberg & Hesse, 1983; Smith, 1971).

The distribution of the symmetrical homospermidine has been linked to that of putrescine because of reports that two molecules of 1,4-diaminobutane are the metabolic precursors of homospermidine. The compound was discovered initially as a backbone component of five isolated alkaloids of the Solanaceae (see solapalmitine and solapalmitenine in Chap. 2) and soon thereafter as a free amine in extracts of the dried leaves of the sandalwood tree, *Santalum album* L. (Kuttan et al., 1971).

Homospermidine was rediscovered in legume root nodules and in the bacterial genus *Rhizobium*, which exists symbiotically within the nodules. The compound was not found in nodule-free roots of the host plant (Smith, 1977). Homospermidine proved to be the major polyamine in all rhizobia tested (Fujihara & Harada, 1989a). Among the Agrobacteria, of the family Rhizobiaceae, a strain of *A. rhizogenes* has been found that contains only putrescine and homospermidine (Hamana et al., 1989b). This would appear to be an organism of choice for the study of the enzymology of synthesis of homospermidine.

A new polyamine, aminobutylhomospermidine, has been detected. In a rapidly growing soybean rhizobial strain, neither spermidine nor spermine were found in cells grown in polyamine-free media. The new tetramine was found only in the strains of the fast-growing Rhizobia.

These results have been extended to many other eubacteria that form nitrogen-fixing nodules or fix nitrogen in nonsymbiotic association (Hamana et al., 1988b). Homospermidine has been found in the N_2-fixing species of the photosynthetic eubacteria, such as *Rhodospirillum* and *Chromatium* (Hamana et al., 1985), as well as in N_2-fixing cyanobacteria (Hamana et al., 1983). Nevertheless several species of nitrogen-fixing *Azotobacter* did not contain detectable homospermidine. Some other invasive Agrobacteria, which invade dicotyledonous plants and are gall-formers, do contain homospermidine and putrescine. When given diaminopropane or cadaverine, these organisms can also generate aminopropyl or aminopentyl derivatives of the various triamines (Hamana et al., 1989b). Many of these organisms also contain spermine and thermospermine.

The finding of unusual polyamines in the seed of the sword bean, *Canavalia gladiata*, may relate to the formation of aminobutyl derivatives by symbiotic bacteria. These seed extracts contained homospermidine and the then new polyamine, canavalmine (4,3,4) (Fujihara et al., 1982). Within the plant normal levels of putrescine, spermidine, and spermine are found in all major organs, such as the root, stem, and leaves, and significant additional levels of homospermidine in active bacteroid-infected root nodules. Canavalmine increased somewhat in the nodule as it aged. The latter increased significantly in the developing seed, which also contained homospermidine. It has been suggested that both unusual polyamines are synthesized in the seed (Fujihara et al., 1986). Small amounts of several additional new pentamines, the aminopropyl and aminobutyl derivatives of canavalmine, (i.e., 3,4,3,4 and 4,4,3,4, respectively) have been found in the sword bean (Matsuzaki et al., 1990). A new polyamine, 4-aminobutyl cadaverine (4,5), has been described in the root nodules of the adzuki bean plant (Fujihara et al., 1995).

The seeds of vetch, *Vicia sativa*, contain a great many new derivatives of homospermidine, including linear and branched pentamine, hexamines and heptamines (Hamana et al., 1991d). In addition to homopentamine (4,4,4,4) and homohexamine (4,4,4,4,4), some eight tertiary branched polyamines were identified: N^5-aminobutylhomospermine [4(4),4,4], N^5-aminobutylhomopentamine [4(4),4,4,4], N^{10}-aminobutylhomopentamine [4,4(4),4,4], N^5,N^{10}-bis(aminobutyl)homospermine [4(4),4(4),4], N^5-aminobutylhomohexamine [4(4),4,4,4,4], N^{10}-aminobutylhomohexamine [4,4(4),4,4,4], N^5,N^{10}-bis(aminobutyl)homopentamine [4(4),4(4),4,4], and N^5,N^{15}-bis(aminobutyl)homopentamine [4(4),4,4(4),4].

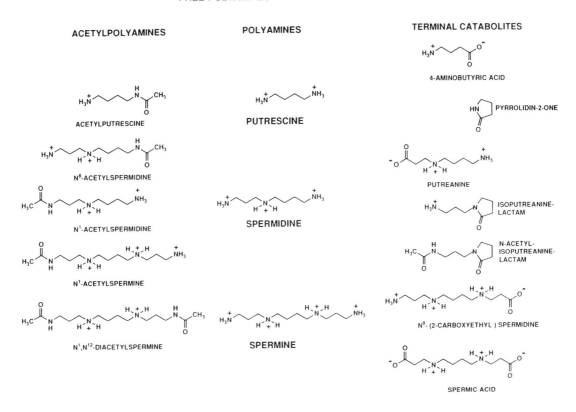

ACETYLPOLYAMINES

ACETYLPUTRESCINE

N^8-ACETYLSPERMIDINE

N^1-ACETYLSPERMIDINE

N^1-ACETYLSPERMINE

N^1,N^{12}-DIACETYLSPERMINE

POLYAMINES

PUTRESCINE

SPERMIDINE

SPERMINE

TERMINAL CATABOLITES

4-AMINOBUTYRIC ACID

PYRROLIDIN-2-ONE

PUTREANINE

ISOPUTREANINE-LACTAM

N-ACETYL-ISOPUTREANINE-LACTAM

N^8-(2-CARBOXYETHYL) SPERMIDINE

SPERMIC ACID

FIG 5.6 Natural derivatives of putrescine; redrawn from Seiler (1986).

In the past decade, homospermidine has been found in many other types of organisms, including some methanogenic archaebacteria. It is a major polyamine among all of the chemoorganotrophic Gram-negative eubacterial members of the genus *Flavibacterium* (Hamana & Matsuzaki, 1991). It was also among the various triamines, including spermidine and norspermidine, produced by the extreme thermophile, *T. thermophilus*, as was homocaldopentamine, 1,16-diamino-4,8,12-triazahexadecane (Oshima & Kawahata, 1983). Homospermidine is a major polyamine in mosses and lichens, which also contain norspermidine and norspermine. The former is also present in most species of ferns, whereas the latter are absent in the ferns (Pteridophyta), gymnosperms (such as the ginkgo), and fungi (Hamana & Matsuzaki, 1985b). In the plant kingdom the distribution of the various triamines can serve as a taxonomic marker among the lower plants.

When homospermidine is added to the media of fungi, which lack this triamine, and some thermophilic bacteria, these organisms may form a new tetramine, namely aminopropylhomospermidine (3,4,4). Because yeasts and thermophilic bacteria are normally capable of generating aminopropyl-containing tetramines, such as spermine in the first group and thermine in the second, the new polyamine may reflect the activity of the normal spermine synthase (Matsuzaki et al., 1987a).

Microorganisms unable to generate spermine are unable to produce aminopropylhomospermidine.

The presence of homospermidine, aminopropylhomospermidine (3,4,4), and canavalmine (4,3,4) has been reported in the Syrian hamster epididymis but not in other tissues of this animal (Matsuzaki et al., 1987b). The compounds may also be present in the epididymis of the Chinese hamster and the guinea pig. This was a first report of tetramines other than spermine in the mammal. The presence of homospermidine in all tissues of the Japanese newt (Hamana & Matsuzaki, 1979) was the first report of this compound among "lower" vertebrates. Although it is evident that the distribution of the polyamines and the appearance of "unusual" compounds such as homospermidine will eventually assist in describing the evolutionary relations of organisms, it is also clear that the task has barely begun.

Adaptation to Salt

E. coli, grown in the laboratory in common liquid media, is very high in putrescine (10–30 mM) in contrast to many other bacteria, whose putrescine content is about 1/100th as high. When *E. coli* is placed in media of high salt content putrescine is excreted, and the bacteria adapt to changes in external osmolarity (Peter et al., 1978). In some organisms certain polyamines, such as spermine, prevent the lysis of osmotically sensitive membranes. Many organisms have used other devices in meeting these needs, of which the synthesis and control of proline content and the production of sulfonium ions and quaternary ammonium salts (Cantoni, 1960) are examples.

Some bacteria live and grow in waters of high salinity and these Halobacteria contain very high concentrations of Na^+ and K^+, some 10–20 times higher than that found in nonhalophiles (Bayley & Morton, 1978). Some of these, the halophilic Archaebacteria, have developed unusual biochemical features, for example, structures of lipids and carotenoids and sequences of nucleic acids and proteins, which separate them in major respects from eubacteria. Nevertheless, the organisms have retained many common biochemical mechanisms (e.g., protein synthesis on ribosomes), which nevertheless operate maximally at 3–4 M K^+. Although extreme halophiles may not contain the usual detectable polyamine (Chen & Martynowicz, 1984), agmatine has been found in many of these organisms (Hamana et al., 1995b; Kamekura et al., 1987).

Hamana (1994) and Hamana and Satake (1995) reported that many Gram-positive cocci grown in polyamine-deficient media are devoid of polyamine. It will be of interest to see if the excess Mg^{2+} found in many Gram-positive organisms serves to regulate polyamine biosynthesis.

Summary and Conclusions

Examination of many different types of microbe, plant, and animal has revealed numerous patterns of polyamine content, in which a previously highly limited range of di-, tri-, and tetramines has been significantly enlarged. The earlier distinction between procaryotic and eucaryotic organisms has broken down in several ways. Spermine and new polyamines have been found in procaryotic cells, particularly in those whose ecological niches differ significantly from that of the mammal. Particularly the exploration of thermophiles, both eubacterial and archaebacterial, has revealed new higher polyamines, some of which are essential in protein synthesis and possibly in the survival of their nucleates. Some halophiles functioning in very high salt lack the normal polyamines altogether but contain agmatine. Many eucaryotic organisms lack spermine and some of these contain unusual polyamines. The new data on the distribution of the many polyamines thus poses many problems of metabolic origin and role, problems that will be seen to bear on the specific biological properties and ecological niches of bacteria, plants, and animals.

REFERENCES

Achenbach–Richter, L., Gupta, R., Stetter, K. O., & Woese, C. R. (1987) *Systematic and Applied Microbiology*, **9**, 34-39.
Atlas, D. (1994) *Science*, **266**, 462–463.
Ayesh, R., Mitchell, S. C., Zhang, A., & Smith, R. L. (1993) *British Medical Journal*, **307**, 655–657.

Bacchi, C. J. (1981) *Journal of Protozoology*, **28**, 20–27.
Bacchi, C., Vergara, C., Garofalo, J., Lipschik, G. Y., & Hutner, S. H. (1979) *Journal of Protozoology*, **26**, 484-488.
Bardócz, S., Grant, G., Brown, D. S., Ralph, A., & Pusztai, A. (1993) *Journal of Nutritional Biochemistry*, **4**, 66–71.
Bayley, S. T. & Morton, R. A. (1978) *CRC Critical Reviews in Microbiology*, **6**, 151–205.
Bergeron, R. J. & Weimar, W. R. (1991) *Journal of Bacteriology*, **173**, 2238–2243.
Bird, S. & Gower, D. B. (1983) *Experientia*, **39**, 790–792.
Busse, J. & Auling, G. (1988) *Systematic and Applied Microbiology*, **11**, 1–8.
Buts, J.-P., De Keyser, N., De Raedemaeker, L., Collette, E., & Sokal, E. M. (1995) *Journal of Pediatric Gastroenterology and Nutrition*, **21**, 44–49.
Chen, K. Y. & Martynowicz, H. (1984) *Biochemical and Biophysical Research Communications*, **124**, 423–429.
De Rosa, M., De Rosa, S., Gambacorta, A., Cartenì–Farina, M., & Zappia, V. (1976) *Biochemical and Biophysical Research Communications*, **69**, 253–261.
De Rosa, M., Gambacorta, A., Cartini–Farina, M., & Zappia, V. (1980) *Polyamines in Biomedical Research*, pp. 255–272, J. M. Gaugas, Ed. New York: Wiley.
Eichler, W. (1989) *Biological Chemistry Hoppe–Seyler*, **370**, 1113–1126.
Ellin, R. I., Farrand, R. L., Oberst, F. R., Crouse, C. L., Billups, N. B., Koon, W. S., Musselman, N. P., & Sidell, F. R. (1974) *Journal of Chromatography*, **100**, 137–152.
Eskelson, C. D., Chvapil, M., Chang, S. Y., & Chvapil, T. (1978) *Biomedical Mass Spectrometry*, **5**, 238–242.
Feldman–Muhsam, B., Bachrach, U., & Ben-Joseph, M. (1980) *Journal of Insect Physiology*, **26**, 407–413.
Fujihara, S., Abe, H., & Yoneyama, T. (1995) *Journal of Biological Chemistry*, **270**, 9932–9938.
Fujihara, S. & Harada, Y. (1989a) *Biochemical and Biophysical Research Communications*, **165**, 659–666.
——— (1989b) *Soil Biology and Biochemistry*, **21**, 449–452.
Fujihara, S., Nakashima, T., & Kurogochi, Y. (1982) *Biochemical and Biophysical Research Communications*, **107**, 403–410.
Fujihara, S., Nakashima, T., Kurogochi, Y., & Yamaguchi, M. (1986) *Plant Physiology*, **82**, 795–800.
Guarino, L. & Cohen, S. S. (1979) *Analytical Biochemistry*, **95**, 73–76.
Guggisberg, A. & Hesse, M. (1983) *The Alkaloids: Chemistry and Pharmacology XXII*, pp. 85–188. A. Brossi, Ed. New York: Academic Press.
Hamana, K. (1994) *Journal of General and Applied Microbiology*, **40**, 181-195.
Hamana, K., Akiba, T., Uchino, F., & Matsuzaki, S. (1989a) *Canadian Journal of Microbiology*, **35**, 450–455.
Hamana, K., Hamana, H., & Itoh, T. (1995b) *Journal of General and Applied Microbiology*, **41**, 153–158.
Hamana, K., Kamekura, M., Onishi, H., Akazawa, T., & Matsuzaki, S. (1985) *Journal of Biochemistry*, **97**, 1653–1658.
Hamana, K. & Matsuzaki, S. (1979) *FEBS Letters*, **99**, 325–328.
——— (1984) *Journal of Biochemistry*, **95**, 1105–1110.
——— (1985a) *Journal of Biochemistry*, **97**, 1311–1315.
——— (1985b) *Journal of Biochemistry*, **97**, 1595–1601.
——— (1987) *FEMS Microbiology Letters*, **41**, 211–215.
——— (1990a) *FEMS Microbiology Letters*, **70**, 347–352.
——— (1990b) *FEMS Microbiology Letters*, **70**, 353–356.
——— (1990c) *Canadian Journal of Microbiology*, **36**, 228–231.
——— (1991) *Canadian Journal of Microbiology*, **37**, 885–888.

———— (1992a) *CRC Critical Reviews in Microbiology*, **18**, 261–283

———— (1992b) *Journal of General and Applied Microbiology*, **38**, 191–194.

———— (1992c) *Journal of General and Applied Microbiology*, **38**, 93–103.

———— (1993) *Canadian Journal of Microbiology*, **39**, 304–310.

Hamana, K., Matsuzaki, S., Niitsu, M., & Samejima, K. (1989b) *FEMS Microbiology Letters*, **65**, 269–274.

Hamana, K., Matsuzaki, S., Niitsu, M., Samejima, K., & Nagashima, H. (1990) *Phytochemistry*, **29**, 377–380.

Hamana, K., Matsuzaki, S., & Sakakibara, M. (1988b) *FEMS Microbiology Letters*, **50**, 11–16.

Hamana, K., Mijagawa, K., & Matsuzaki, S. (1983) *Biochemical and Biophysical Research Communications*, **112**, 606–613.

Hamana, K., Nakagawa, Y., & Yamasato, K. (1995a) *Microbios*, **81**, 135–145.

Hamana, K., Niitsu, M., Samejima, K., & Matsuzaki, S. (1988a) *FEMS Microbiology Letters*, **50**, 79–83.

———— (1991a) *Comparative Biochemistry and Physiology B*, **100**, 59–62.

———— (1991b) *Comparative Biochemistry and Physiology B*, **100**, 399–402.

———— (1991c) *Journal of Biochemistry*, **109**, 444–449.

———— (1991d) *Phytochemistry,* **30**, 3319–3322.

Hamana, K. & Satake, S. (1995) *Journal of General and Applied Microbiology*, **41**, 159–163.

Hamana, K., Suzuki, M., Wakabayashi, T., & Matsuzaki, S. (1989c) *Comparative Biochemistry and Physiology B*, **92**, 691–695.

Hart, D., Winther, M., & Stevens, L. (1978) *FEMS Microbiology Letters*, **3**, 173–175.

Jannasch, H. W. (1993) *American Society of Microbiology News*, **59**, 217-218.

Johne, S., Gröger, D., & Radeglia, R. (1975) *Phytochemistry*, **14**, 2635-2636.

Kalra, S. P., Pearson, E., Sahu, A., & Kalra, P. S. (1995) *Neuroscience Letters*, **194**, 165–168.

Kamekura, M., Hamana, K., & Matsuzaki, S. (1987) *FEMs Microbiology Letters*, **43**, 301–305.

Kämpfer, P., Blaszcyk, K., & Auling, G. (1994) *Canadian Journal of Microbiology*, **40**, 844–850.

Kim, B. G., Sobota, A., Bitonti, A. J., McCann, P. P., & Byers, T. J. (1987) *Journal of Protozoology*, **34**, 278–284.

Kneifel, H. (1979) *Marine Algae in Pharmaceutical Science*, pp. 365–401, H. A. Hoppe, T. Levring & Y. Tanaka, Eds. New York: Walter de Gruyter.

Kneifel, H., Schuber, F., Aleksijevic, A., & Grove, J. (1978) *Biochemical and Biophysical Research Communications*, **85**, 42–46.

Kneifel, H., Stetter, K. O., Andreesen, J. R., Wiegel, J., König, H., & Schoberth, S. M. (1986) *Systematic and Applied Microbiology*, **7**, 241–245.

Knudsen, J. T., Tollsten, L., & Bergström, L. G. (1993) *Phytochemistry*, **33**, 253–280.

Kullnig, R., Rosano, C. L., Coulter, M., & Hurwitz, C. (1973) *Journal of Biological Chemistry*, **248**, 2487–2488.

Kuttan, R., Radhakrishnan, A. N., Spande, T., & Witkop, B. (1971) *Biochemistry*, **10**, 361–365.

Lake, J. A. (1994) *Proceedings of the National Academy of Sciences of the United States of America*, **91**, 1455–1459.

Lamster, I. B., Mandella, R. D., Zove, S. M., & Harper, D. S. (1987) *Archives of Oral Biology*, **32**, 329–333.

Li, G., Regunathan, S., Barrow, D. J., Eshraghi, J., Cooper, R., & Reis, D. J. (1994) *Science*, **263**, 966–970.

Mach, M., Kersten, H., & Kersten, W. (1982) *Biochemical Journal*, **202**, 153–162.

Matsuzaki, S., Hamana, K., Niitsu, M., Samejima, K., & Yamashita, S. (1987a) *FEMS Microbiology Letters*, **48**, 1–4.

Matsuzaki, S., Hamana, K., Okada, M., & Isobe, K. (1992) *Dokkyo Journal of Medical Science*, **19**, 117–124.

Matsuzaki, S., Hamana, K., Okada, M., Niitsu, M., & Samejima, K. (1990) *Phytochemistry*, **29**, 1311–1312.

Matsuzaki, S., Xiao, L., Suzuki, M., Hamana, K., Niitsu, M., & Samejima, K. (1987b) *Biochemistry International*, **15**, 817–822.

Morrow, C. D., Flory, B., & Krassner, S. M. (1980) *Comparative Biochemistry and Physiology B*, **66**, 307–311.

Nakao, H., Ishii, M., Shinoda, S., & Yamamoto, S. (1989) *Journal of General Microbiology*, **135**, 345–351.

Nystrand, R. (1984) *Systematic and Applied Microbiology*, **5**, 204–219.

Ohno–Iwashita, Y., Oshima, T., & Imahori, K. (1975) *Archives of Biochemistry and Biophysics*, **171**, 490–499.

Oshima, T. (1975) *Biochemical and Biophysical Research Communications*, **63**, 1093–1098.

———— (1979) *Journal of Biological Chemistry*, **254**, 8720–8722.

Oshima, T. & Kawahata, S. (1983) *Journal of Biochemistry*, **93**, 1455–1456.

Paulus, T. J. & Davis, R. H. (1983) *Methods in Enzymology*, vol. 94, pp. 36–42, H. Tabor & C. W. Tabor, Eds. New York: Academic Press.

Peter, H. W., Ahlers, J., & Günther, T. (1978) *International Journal of Biochemistry*, **9**, 313–316.

Pinel, J. P., Gorzalka, B. B., & Ladak, F. (1981) *Physiology & Behavior*, **27**, 819–824.

Poulin, R., Larochelle, J., & Nadeau, P. (1984) *Biochemical and Biophysical Research Communications*, **122**, 388–393.

Preti, G., Labows, J. N., Kostele, J. G., Aldinger, S., & Daniele, R. (1988) *Journal of Chromatography*, **432**, 1–11.

Pulkkinen, P., Kanerva, S., Elfving, K., & Jänne, J. (1975) *Journal of Reproduction & Fertility*, **43**, 49–55.

Quemener, V., Blanchard, Y., Lescoat, D., Havouis, R., & Moulinoux, J. P. (1992) *American Physiological Society*, **263**, C1-C5.

Rainey, F. A. & Stackelrandt, E. (1993) *Systematic and Applied Microbiology*, **16**, 224–226.

Ramakrishna, S., Guarino, L., & Cohen, S. S. (1978) *Journal of Bacteriology*, **134**, 744–750.

Rosano, C. L., Braun, C. B., & Hurwitz, C. (1989) *Journal of Bacteriology*, **171**, 1223–1224.

Rosano, C. L., Hurwitz, C., & Bunce, S. C. (1978) *Journal of Bacteriology*, **135**, 805–808.

Ryser, E. T., Taylor, S. L., & Marth, E. H. (1984) *Systematic and Applied Microbiology*, **5**, 545–554.

Samborski, D. J. & Rohringer, R. (1970) *Phytochemistry*, **9**, 1939–1945.

Sarhan, S., Quemener, V., Moulinoux, J., Knödgen, B., & Seiler, N. (1991) *International Journal of Biochemistry*, **23**, 617–626.

Sastry, S. D., Buck, K. T., Janak, J., Dressler, M., & Preti, G. (1980) *Biochemical Applications of Mass Spectrometry*, vol. 1, supplemental, pp. 1085–1129, G. R. Waller & O. C. Dermer, Eds. New York: Wiley–Interscience.

Sayem El Daher, N. & Simard, R. E. (1985) *Journal of Food Protection*, **48**, 54–58.

Schere, P. & Kneifel, H. (1983) *Journal of Bacteriology*, **154**, 1315–1322.

Seiler, N. (1986) *Journal of Chromatography,* **379**, 157–176.

Shapira, R., Altman, A., Henis, Y., & Chet, I. (1989) *Journal of General Microbiology*, **135**, 1361–1367.

Shivaji, S. (1984) *Trends in Biochemical Science*, **March**, 104–107.

Shubhada, S., Lin, S. N., Qian, Z. Y., Steinberger, A., & Tsai, Y. H. (1989) *Journal of Andrology*, **10**, 145–151.

Smith, T. A. (1971) *Biological Reviews*, **46**, 201–241.

—— (1977) *Phytochemistry*, **16**, 278–279.

Speed, R. R. & Winkler, H. H. (1990) *Journal of Bacteriology*, **172**, 5690-5696.

Stevens, L. & Winther, M. D. (1979) *Advances in Microbial Physiology*, **19**, 63–148.

Stillway, L. W. & Walle, T. (1977) *Biochemical and Biophysical Research Communications*, **77**, 1103–1107.

Tobari, J. & Tchen, T. T. (1971) *Journal of Biological Chemistry*, **246**, 1262–1265.

Wang, H., Regunathan, S., Youngson, C., Bramwell, S., & Reis, D. J. (1995) *Neuroscience Letters*, **183**, 17–21.

Woese, C. R. (1993) *The Biochemistry of Archaea*, pp. vii–xxix, M. E. A. Kates, Ed. New York: Elsevier Science Publishers.

Yaegaki, K. & Sanada, K. (1992) *Journal of Periodontal Research*, **27**, 233–238.

Yamamoto, S., Itano, H., Kataoka, H., & Makita, M. (1982) *Journal of Agricultural and Food Chemistry*, **30**, 435–439.

Zappia, V., Porta, R., Cartenì–Farina, M., De Rosa, M., & Gambacorta, A. (1978) *FEBS Letters*, **94**, 161–165.

Zellner, G. & Kneifel, H. (1993) *Archives of Microbiology*, **159**, 472–476.

Zeng, X., Leyden, J. J., Lawley, H. J., Sawano, K., Nohara, I., & Preti, G. (1991) *Journal of Chemical Ecology*, **17**, 1469–1492.

Zhu, C., Cumaraswamy, A., & Henney, H. R. (1989) *Molecular and Cellular Biochemistry*, **90**, 145–153.

CHAPTER 6

Polyamine Oxidases and Dehydrogenases

Diamines and polyamines widely occur in the living kingdom. Diamine oxidases, as potential regulators of the cellular concentration of all these amines, therefore may be assumed to participate in an endless number of biological reactions.

E. A. Zeller (1963)

Studies on the fate of histamine revealed enzymes that oxidized this substance and other amines. The latter were both "polyamines" and "biogenic" physiologically active monoamines. The normal progress of these initial findings ensued: purification of an enzyme, study of its specificity, definition of the reaction it catalyzed, and detection and distribution of similar enzymes. The metabolic roles and fates of the reaction products were determined as were the structure of the enzyme molecule and finally the relation of the detailed structure to the mechanism of the catalytic function. The genetic (DNA) and physiological (RNA) determinants of the enzyme provide data on the structure of the protein and on the regulation of its synthesis.

Histaminase and Discovery of Diamine Oxidase (DAO)

The vasodilator histamine, present in extracts of normal lung, disappeared during autolysis of the tissue (Best, 1929). The enzyme "histaminase" was inhibited by lack of oxygen and by cyanide (Best & McHenry, 1930). The histaminase that catalyzed the oxidative deamination of histamine and aliphatic diamines was inhibited by semicarbazide, a well-known carbonyl reagent. A pig kidney histaminase oxidized putrescine, cadaverine, and spermine, producing ammonia and hydrogen peroxide, and evoked the designation DAO (Zeller, 1938). The reaction appeared to follow the course:

$$RCH_2NH_2 + O_2 + H_2O \rightarrow RCHO + NH_3 + H_2O_2. \quad (1)$$

Bases active as substrates included ethylenediamine, 1,3-diaminopropane, spermidine and several spermidine homologues, and agmatine. Numerous carbonyl reagents were inhibitory at 10^{-5} M. DAO was present in several tissues of many animal and several plant species.

The enzyme was purified from pig kidney in order to eliminate catalase or peroxidase (Tabor, 1951). After a partial purification, equation (1) did describe the DAO reaction. In addition, the imidazole ring of histamine was not cleaved as had been suspected. Although putrescine and cadaverine were probably converted to amine aldehydes, their further oxidation was not catalyzed by these additional enzymes. It seemed likely that the amine aldehydes cyclized rapidly to Δ^1-pyrroline and Δ^1-piperideine, respectively. The former presumed cyclic product was detected by Tabor in a reaction with *o*-aminobenzaldehyde to give a specific colored product (Holmstedt et al., 1961), as in (2) below.

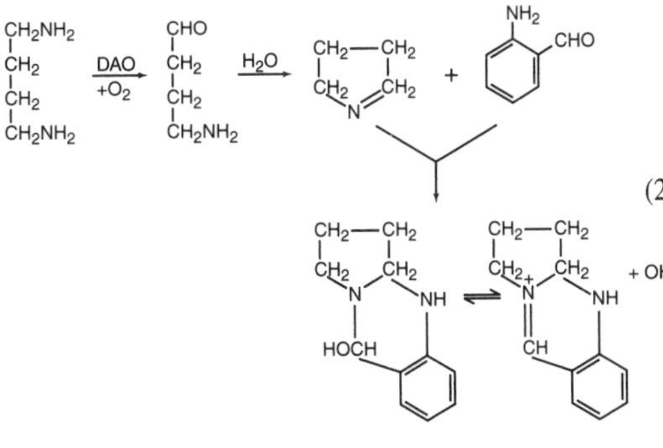

(2)

A DAO was isolated from pea seedlings and reacted similarly with putrescine and cadaverine. The initial product of the reaction with cadaverine is thought to be the amino aldehyde that cyclizes to the Δ^1-piperideine or 2,3,4,5-tetrahydropyridine (Mann & Smithies, 1955). Polymeric products of the latter accumulate, as in (3).

(3)

α-tri-piperideine

reactive intermediate

and

iso-tri-piperideine

Toxicity of Spermine

The ubiquity of spermine in mammalian tissues encouraged searches for a possible function of this amine. Spermine at 1.5 mM levels did not stimulate the respiration of brain slices or homogenates but significantly inhibited oxygen consumption in the presence of glucose, lactate, or pyruvate (Evans et al., 1939).

In the next decade spermidine and spermine were revealed to be growth factors for some bacteria and antibacterial for others under certain conditions. Spermine also induced severe renal damage in mice, rats, and rabbits (Rosenthal et al., 1952) whereas spermidine, putrescine, and cadaverine were far less toxic. The blood and plasma levels of spermine were followed also in rabbits injected with toxic amounts of spermine (30 mg/kg). Interestingly, two-thirds of the spermine in the blood was present in blood cells 1 h after injection. In 5 h, when blood spermine had fallen about 50%, plasma spermine was less than 5 ng/mL. Very little spermidine was present in the blood and less than 10% of the injected spermine appeared as spermidine in the urine. Only 4–20% of injected spermine appeared in the urine whereas injected spermidine was largely metabolized. Injected spermine produced a transitory fall in blood pressure, as well as respiratory difficulty in the mice, of which 80% died when a single intraperitoneal dose of spermine exceeded 0.1 mmol/kg.

This laboratory also purified the amine oxidase from beef plasma; and the products of oxidation of spermine and spermidine were toxic aldehydes, H_2O_2, and NH_3 (Tabor et al., 1954). The enzyme was virtually inactive on putrescine and cadaverine but was quite active on furfurylamine and benzylamine. The rate of production of benzaldehyde from benzylamine, determined spectrophotometrically, was proportional to enzyme concentra-

tion and paralleled the rates of oxidation of spermine at a defined pH. Unfortunately the enzyme described in some laboratories then bears the confusing names of "monoamine oxidase" (MAO) and "benzylamine oxidase." It was not then clear whether spermine or the DAO-oxidation products were the nephrotoxic compounds. However, the beef plasma DAO was used to demonstrate the great toxicity of oxidized spermine to the spermatozoa of many mammalian species, including that of humans. Motility was rapidly arrested in trypanosomes as well, although these organisms were somewhat more resistant than sperm (Tabor & Rosenthal, 1956).

Antibacterial Action of Oxidized Spermine

Tissue extracts contained a substance inhibitory to the growth of tubercle bacilli. The antibacterial agent proved to be spermine, isolated as the crystalline phosphate (Hirsch & Dubos, 1952). This result was surprising in view of reports that spermine had protected non-acid-fast organisms from some other antibacterial agents. Spermine was relatively specific in inhibiting the growth of tubercle bacilli in a system containing bovine serum. Spermidine, but not putrescine and cadaverine, was also active in inhibiting mycobacterial growth (Hirsch, 1953), and a bovine plasma amine oxidase relatively specific for spermine and spermidine was essential to developing the toxic agent. Carbonyl reagents were good inhibitors of the enzyme, now dubbed a "spermine oxidase."

Spermine and spermidine were subsequently found to inhibit the growth of *Staphylococcus aureus*, a Gram-positive organism, under both aerobic and anaerobic conditions (Grossowicz et al., 1955). Sheep and bovine sera reduced this activity of spermine, as did cultivation at 6°C rather than 37°C, implying a direct action of spermine. Indeed, it will be seen that spermine and spermidine do have direct effects on many types of bacterial structures, such as membranes and nucleic acids.

Role of Spermine in Human Prostatic Fluid and Semen

It is commonly observed that women have a higher incidence of urinary infections than men and it has been asked if this effect is due to the high concentration of antibacterial spermine in prostatic secretion and semen. Although spermine is bactericidal to many Gram-positive bacteria and markedly less so for Gram-negative organisms (Rozansky et al., 1954), the prostatic fluid and semen of the dog and human are highly bactericidal to both groups of organisms (Stamey et al., 1968). The antibacterial spectrum of spermine itself did not in fact fill the requirements of the presumed natural defense mechanism (Fair & Wehner, 1971). Nevertheless, the content of oxidizable spermine and spermidine in seminal plasma is thought to be clinically significant (Fernandez et al., 1994) and to suppress formation of antisperm antibodies (Maayan et al., 1995).

Zeller found a DAO in human semen. This is active on putrescine and cadaverine, and at somewhat lower rates on spermine and spermidine, to produce H_2O_2 and possibly toxic aldehydes. A correlation was found among DAO activity, spermine concentration, and decreased motility of spermatozoa. Thus the presence of the enzyme in an aerobic environment and the production of the toxic aldehydes might affect the motility and survival of the postejaculatory sperm. Also, the presence of these products might be lethal to invading bacteria and in fact be responsible for the reported bactericidal substances.

Although spermine and spermidine, and some diamines, stimulated the metabolism of the seminal carbohydrate fructose by rat epididymal sperm, oxidized spermine was extremely inhibitory in the conversion of fructose, glucose, and pyruvate to CO_2 and lactate (Pulkkinen et al., 1978).

A purified fraction of the seminal plasma DAO is also reported to oxidize the ε-amino group of peptidic lysine; that is, it is also a lysyl oxidase (Crabbe & Kavanagh, 1977). This activity was also found in the human placental DAO (Crabbe et al., 1976). The purified placental DAO was also described as a histaminase that was active on putrescine and cadaverine, inactive on benzylamine, and inhibited by aminoguanidine and less so by semicarbazide (Baylin & Margolis, 1975).

Studies of fertilization in the mammal revealed new types of phenomena. In some species newly ejaculated sperm are activated by incubation within the female reproductive tract, and the increased ability of the sperm to release its acrosomal hydrolytic enzymes within the fused sperm–egg complex depends on a prior incorporation of Ca^{2+} into the sperm. Proteolytic and other enzymes derived from the sperm acrosome are believed to be essential to effective fertilization (McRorie & Williams, 1974). The proteinase *acrosin* was isolated from the sperm of the rabbit, boar, and human, as well as an acrosin precursor or zymogen, *proacrosin*, which is autoactivable to acrosin. At ejaculation, at least 99% of the potential acrosin is present as the zymogen and the polyamines slow the activation of proacrosin to the fully active proteinase (Parrish & Polakoski, 1977). Nevertheless, spermine and spermidine markedly stimulate the active proteinase, as does Ca^{2+}, while the diamines stimulate acrosin to a much slighter extent. Furthermore, spermine or Ca^{2+} protect acrosin from autoproteolysis. Spermine does not affect the activation of trypsinogen or the activity and stability of an analogous proteinase, trypsin. Although polyamine appears absent from mature sperm, a polyamine-dependent (i.e., spermidine or spermine) protein kinase was found in bovine epididymal spermatozoa (Atmar et al., 1981). Fertilization and exposure of the protein kinase to polyamine in the egg may be an initiating event.

Toxic Aldehyde Products of Polyamine Oxidation

All of the products of the DAO reaction (aldehyde, H_2O_2, and NH_3) are toxic to cells, but the observations of the toxicity of oxidized spermine and oxidized spermidine were interpreted initially to imply the formation of an aminoaldehyde as the most toxic component of the various reaction products. A wide distribution of the enzyme was demonstrated early, as was the isolation and crystallization of the bovine plasma amine oxidase. In the presence of sheep serum, which contained a DAO, spermine disappeared and this was accompanied by the formation of spermidine, which accumulated and then was degraded as well (Bachrach & Bar-Or, 1960). These reactions are not observed with a purified plasma amine oxidase (Tabor & Tabor, 1964). In subsequent collaborative studies, the product of the oxidation of spermine by a purified bovine plasma enzyme was found to be the dialdehyde, in which the primary amines had been oxidized completely and the secondary amines were intact (Tabor et al., 1964). The dialdehyde was unstable and this product was reduced by $NaBH_4$ to the corresponding stable and isolable dialcohol. O_2 consumption and NH_3 production were also determined in the system and the stoichiometry of the reaction was defined as

$$NH_2(CH_2)_3NH(CH_2)_4NH(CH_2)_3NH_2 + 2O_2 + 2H_2O \rightarrow$$
$$OHC(CH_2)_2NH(CH_2)_4NH(CH_2)_2CHO + 2NH_3 + 2H_2O_2.$$

No evidence of the formation of spermidine was seen in this system. In similar experiments, spermidine was oxidized as follows:

$$NH_2(CH_2)_3NH(CH_2)_4NH_2 + O_2 + H_2O \rightarrow$$
$$OHC(CH_2)_2NH(CH_2)_4NH_2 + NH_3 + H_2O.$$

This aminoaldehyde was also reduced and isolated chromatographically.

The purified spermine/spermidine amine oxidase reacts entirely with the terminal amines and does not cleave at the secondary nitrogen. However, when an incubation mixture intermediate in the oxidation of spermine is heated to 100°C, spermidine is produced. Similarly, when the intermediates generated in the oxidation of spermine or spermidine are heated, putrescine is formed (Tabor et al., 1964). In the first instance, it may be supposed that a significant amount of the monoaldehyde derived from spermine is formed and may be decomposed to the stable spermidine. This may also account for the early result of spermidine production from spermine with sheep serum.

In a study of analogues of spermidine and spermine and their toxicity to the growth of animal cells in the presence of bovine sera (Israel & Modest, 1971), a potentially toxic compound contains a primary amine separated from a secondary amine by three methylene units. Further, the next basic center must be separated from the secondary amine by three or more methylene units. Thus canavalmine, a 4,3,4 tetramine, will not be oxidized to a toxic aldehyde in bovine sera; it is incorporated intact within cultured cells and is inhibitory to cell growth, apparently by competition for essential functions with spermidine and spermine (Fujihara et al., 1984).

Problem of Primary Toxic Agent

The presence of both bovine plasma and spermine in an aerobic culture of some organisms, for example, bacterial or mammalian cells, is inhibitory if not lethal to the growth of the cell. Indeed all of the products of oxidation of spermine are toxic. The serum or plasma of essentially all adult ruminants contains an active DAO, as does an occasional nonruminant, such as the hippopotamus or members of the order *Hyracoidea*, which possess an accessory cecum (Blaschko, 1968). For those finding it necessary to use intact sera in tissue culture media, it is important to use the serum of fetal calves or a medium that contains little of the enzyme.

Studies on inhibition can be conducted with freshly cleaved dialdehyde derived from the protected bishemiacetal of the spermine-derived iminodialdehyde. Nevertheless, it was reported initially by Alarcon (1964, 1970) that oxidation products of spermine and spermidine break down spontaneously to acrolein, a highly toxic compound. Many studies have been conducted to determine if the toxicities detected with the dialdehydes are due to these substances or to a toxic decomposition product, acrolein. Many investigators concluded, however, that the effects of acrolein do not entirely correspond with the toxicities of spermine and spermidine or of their oxidation products.

Nevertheless, it is appropriate to indicate some of the properties of acrolein, which is found as a ubiquitous environmental pollutant, a component of tobacco smoke, and even as a product of the degradation of the antitumor agent, cyclophosphamide (Witz, 1989). Several steps in the degradation of oxidized spermine and of cyclophosphamide (Patel, 1990) are given in Figure 6.1.

Acrolein is reactive with thiols, such as cysteine and glutathione, and indeed glutathione serves as a detoxifying agent in the metabolic fate and urinary excretion of herbicides giving rise to acrolein derivatives (Hackett et al., 1990). Acrolein may form either Schiff bases or Michael addition products with primary amines, as in the reactions

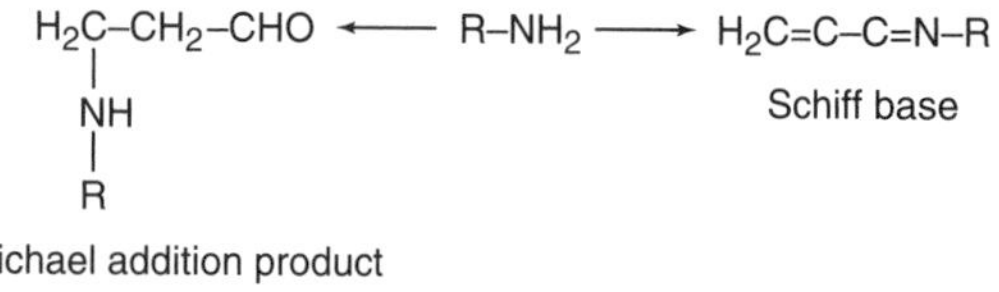

H_2C–CH_2–CHO ⟵ R–NH_2 ⟶ H_2C=C–C=N–R

    |
    NH   Schiff base
    |
    R

Michael addition product

Acrolein also reacts with nucleosides and nucleotides of cytosine, adenine, and guanine (Smith et al., 1989, 1990). A reaction occurs with guanine in DNA to form cyclic $1,N^2$-propanodeoxyguanosine adducts (Chung et al., 1984), as in the following:

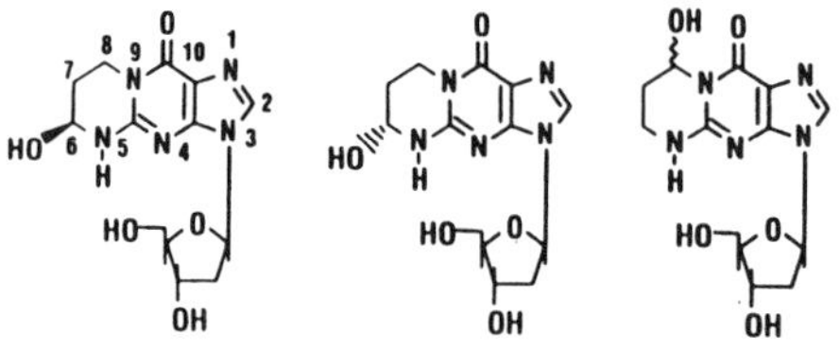

Acrolein also reacts with specific proteins as in the production of formylethyl derivatives of exposed histidine and lysines of bovine carbonic anhydrase II (Tu et al., 1989). The toxicity and carcinogenicity of cyclophosphamide, attributed to acrolein, led to studies of the effect of the toxic aldehyde on biochemical controls of gene expression. Acrolein appears to inhibit DNA methylase and to react with DNA itself; either reaction can result in an inhibition of DNA methylation (Cox et al., 1988).

Aldehyde dehydrogenases capable of oxidizing allyl alcohol to acrolein have been found in many mammalian tissues. Rat liver contains an aldehyde dehydrogenase with a low Michaelis constant (K_m) for acrolein and converts acrolein to a less toxic acrylic acid, as in the sequence

$$CH_2=CHCH_2OH + NAD \overset{I}{\rightleftharpoons} CH_2=CHCHO + NADH$$

$$CH_2=CHCHO + NAD \overset{II}{\rightleftharpoons} CH_2=CHCOOH + NADH$$

Inhibitors active on enzyme II result in severe liver damage in rats treated with allyl alcohol (Rikans, 1987).

In the mammal, the aminopropyl moieties of spermine and spermidine are derived from the α, β, and γ carbons of methionine. Methionine labeled at one of these carbon atoms will give rise to labeled spermidine and spermine, which might be decomposed to this reactive aldehyde. No report has appeared on an attempt to

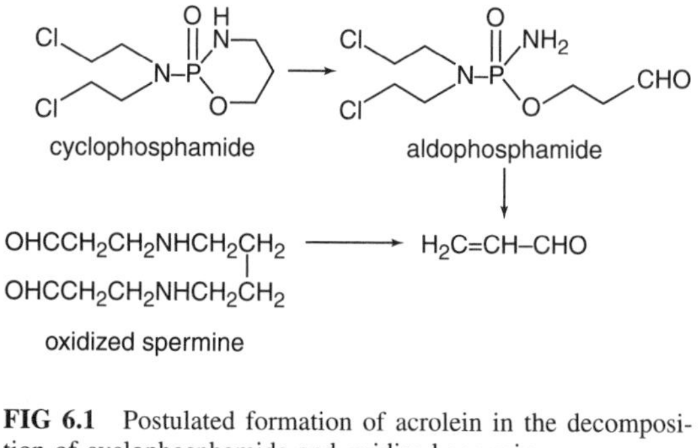

FIG 6.1 Postulated formation of acrolein in the decomposition of cyclophosphamide and oxidized spermine.

determine acrolein production from polyamine in organisms and its possible disposal in urinary or other metabolites.

Alarcon (1968) developed a simple and relatively specific fluorimetric assay for acrolein and other α,β-unsaturated compounds by reaction with *m*-aminophenol to produce fluorescent 7-hydroxyquinolines.

HO— (benzene ring) —NH$_2$ + H$_2$C=CH–CHO ⟶ HO— (quinoline ring) —N

Kimes and Morris (1971a,b) studied the production of acrolein from the purified products of a bovine serum amine oxidase acting on spermine and spermidine. The oxidation of spermine occurred stepwise, with the partial accumulation of a compound with a remaining primary amine before formation of the dialdehyde. On incubation of the final products, dioxidized spermine decomposed at 37°C in phosphate buffer (pH 7), mainly to oxidized spermidine and putrescine, oxidized spermidine to putrescine, and monooxidized spermine to spermidine. Acrolein was found as a major product (60–80% of the predicted amount) in these reactions. In addition some of the spermine dialdehyde was presumably converted via aldol condensation to higher molecular weight materials containing carbonyl groups. The various polyamine aldehydic products markedly inhibited protein and nucleic acid synthesis in *Escherichia coli*. However, these effects occurred with a short lag and this was consistent with, but did not prove, the possibility that a decomposition product, such as acrolein, was responsible for the delay in the observed inhibitions. Nevertheless, a failure to detect acrolein production from oxidized spermine and from analogues of this substance was reported in conditions simulating those described by Alarcon (Israel et al., 1973).

Comparison of the detailed effects of oxidized spermine and of acrolein by several groups revealed differences in the actions of the substances. The most significant of these relate to the systems in which active metabolic transformation is excluded, as in the inhibitory action of oxidized spermine or acrolein on a virus, a nucleic acid, or an enzyme. As shown by Bachrach, oxidized spermine rapidly inactivates the readily penetrable T5 bacteriophage and other T-odd viruses of the *E. coli*–T phage group, but not the more tightly packaged T4 phage. Oxidized spermine can react with a DNA of the virus and an infecting complex does not produce viable phage. Oxidized spermine is about 1000 times more phagocidal on T5 than is acrolein. Both substances inactivate T4 DNA in infecting *E. coli* spheroplasts, but oxidized spermine is still far more active than acrolein in this property (Nishimura et al., 1972).

In studies with the DNA of a virus, the single-stranded circular DNA of ϕx174, oxidized spermine produced twisted, rapidly sedimenting molecules resistant to enzymatic cleavage (Morita et al., 1978). The oxidized polyamine appeared to produce cross-links between DNA strands, as measured in heating of the modified DNA. Both carbonyl groups appeared necessary because an aminomonoaldehyde (oxidized spermidine) did not affect the hyperchromicity of the DNA (Bachrach & Elion, 1967). Numerous effects were also reported on nucleosides, nucleotides, and polynucleotides containing either ribose or deoxyribose; several plant viruses and myxoviruses, which contain RNA, were also inactivated by oxidized spermine. In addition, the products of reactions of oxidized spermine with nucleates were cleaved less rapidly by nucleases. Nevertheless, the difficultly attainable proof that the toxicity of oxidized spermine in metabolizing systems is not due to the liberation of acrolein will require the minimal demonstrations that DNA bases of killed cells are essentially devoid of the types of acrolein adducts described by Nath and Chung (1994), as well as evidence on the relation between adducts formed and cellular survival. It has also been asked if another group of potentially toxic catabolites, the aminoaldehydes, are not the actively inhibitory agents (Smith et al., 1985).

In a recent reexamination of the cytotoxicity of spermine on cultured cells, Chinese hamster ovary (CHO) cells were exposed to spermine and the purified bovine serum amine oxidase separately and together. The separate agents were not cytotoxic. Removal of evolved H$_2$O$_2$ by catalase inhibited early cytotoxicity markedly but not completely (Averill–Bates et al., 1993). Aldehyde dehydrogenase inhibited a late evolution of cytotoxicity and it is relevant that spermine mono- and dialdehyde, spermidine monoaldehyde, and acrolein are all good substrates for the dehydrogenase (Ambroziak & Pietruszko, 1991). A combination of catalase and aldehyde dehydrogenase completely prevented cytotoxicity (Averill–Bates et al., 1994). Thus, more than one of the oxidation products of spermine cause cytotoxicity. Further, human monocytes are sensitive to the NH$_3$ produced from polyamines by the "polyamine oxidase" (PAO) of the cells (Flescher et al., 1991).

New Amine Oxidases and Problems of Nomenclature

Research on the degradation of histamine led to the discovery and partial purification of pig kidney DAO, and subsequently to the purification of beef plasma amine oxidase, and almost simultaneously to the detection of a significantly less specific pea seedling enzyme. All of these proved to be copper-containing proteins. Because not all histaminolytic extracts of tissues can oxidize putrescine and cadaverine, Zeller (1963) suggested that histaminase be reserved for histamine-attacking enzymes uncharacterized for diamine activity.

However, the functions of the polyamines and their oxidation products were less clear than those of many other biologically active amines (e.g., adrenaline, serotonin, tyramine, etc. as well as histamine). The synthesis and metabolic fates of these latter compounds were closer to perceivable phenomena of clinical interest, and the amine oxidases for these compounds, the "biogenic amines," were soon explored on an independent and important track of investigation (Winkler, 1993). Activities were found in many tissues for the oxidation of the biogenic monoamines, and these activities did not extend for the most part to the polyamines. The MAOs were particulates and were present in outer mitochondrial membranes (Greenawalt & Schnaitman, 1970). They were eventually solubilized and demonstrated to be flavoproteins. An irreversible inhibitor, iproniazid (i.e., 1-isonicotinyl-2-isopropylhydrazine) was discovered by Zeller and his colleagues (1952). The inhibitor, which had been used in the treatment of tuberculosis, had been seen to evoke curious behavioral side effects. These directions of study, which might have had a bearing on mental disorders, soon outstripped the development of work on the oxidation of the polyamines.

The introduction of the MAOs was soon confused by the observation that many of the copper enzymes, active on diamines and polyamines, were also active on substrates such as benzylamine. Some early essays on the isolation of the bovine plasma enzyme and its essential copper present this as a MAO (Yamada & Yasunobu, 1962). The more accurate current label, "bovine plasma amine oxidase," is designated by the International Enzyme Commission as EC 1.4.3.6 [amine:oxygen oxidoreductase (deaminating) (copper containing)]. Studies on plasma amine oxidases have exploited the substrate benzylamine in the assay of the enzyme, although benzylamine is not its natural substrate.

Whereas some inhibitors distinguished the copper-containing and other amine oxidases, some "carbonyl reagents" (e.g., phenylhydrazine) inhibited both MAO and the copper-containing amine oxidases. However, the mechanism of the inhibition in the MAO series is not a reaction with a carbonyl group in the enzyme whereas the Cu series proved to contain an essential carbonyl-containing cofactor. Some inhibitory carbonyl reagents such as semicarbazide do distinguish the latter series (Callingham et al., 1991) and identify the semicarbazide-sensitive amine oxidases (SSAO).

Soluble PAOs are now known in mammalian tissues (Höltta, 1977) which have flavin-containing cofactors as do the mitochondrial MAO. The natural substrates of the former in animals are the monoamines and the N-monoacetylated polyamines (i.e., N-acetylspermidine and N-acetylspermine) whereas the "classical" MAO enzymes are inactive on these substrates. The former, the Höltta enzyme, should not be described as a mitochondrial MAO and is currently designated as PAO as a distinction from the copper-containing amine oxidases. PAO is used largely for the soluble non-copper-containing flavoprotein that is active in vertebrates in a pathway degrading the monoacetylated spermine and spermidine. PAOs of comparable mechanism and specificity are also known in fungi (Kobayashi et al., 1983).

Many PAOs or dehydrogenases containing flavins or other non-copper-containing prosthetic groups have been described. For example, a relatively specific spermidine dehydrogenase of *Serratia marcescens* is a hemoprotein containing both flavin adenine dinucleo-

tide (FAD) and protoporphyrin IX (Tabor & Kellogg, 1970). An inducible amine dehydrogenase of *Pseudomonas putida* is a hemoprotein that is apparently devoid of FAD and copper and acts on spermidine, putrescine, and spermine at rates of 22, 7, and 3%, respectively, relative to that of the rate on *n*-butylamine, the best known substrate (Durham & Perry, 1978a,b). In these instances and others it appears best to describe the enzyme in terms of the organism and tissue of origin, as well as in terms of an optimal natural substrate.

Some Characteristics of Classical MAO

Two different forms of mitochondrial MAO, known as MAO-A and MAO-B, are distributed differently in mammalian tissues, which usually contain a mixture of both enzymes. The enzymes, described as amine:oxygen oxidoreductase (deaminating) (flavin-containing), EC 1.4.3.4, differ in substrate preference. MAO-A oxidizes serotonin, whereas MAO-B, present in lymphocytes and platelets, is active on β-phenylalkylamines. The cofactor of these enzymes is FAD, which is covalently bound to cysteine in a pentapeptide common to both enzymes (Fig. 6.2A).

A bacterial trimethylamine dehydrogenase also contains a cysteinyl flavin thioether at the active site. This enzyme as well as MAO are inhibited irreversibly by alkyl and aryl hydrazines. The hydrazines are converted to form diazenes that react with the oxidized form of the flavin. The inactivated adduct is presented in Figure 6.2B. Acetylenic compounds such as pargyline (anti MAO-A and -B), deprenyl (anti MAO-B) and clorgyline (anti MAO-A) are considered to react at the N_5 of the flavin structure (Singer, 1985). Inhibitors of these enzymes are of interest as possible antidepressants and as adjuncts in the treatment of disease, such as Parkinsonism by L-dihydroxyphenylalanine (dopa).

Because the polyamines exist in cells at millimole levels and neurotransmitters and biogenic amines are

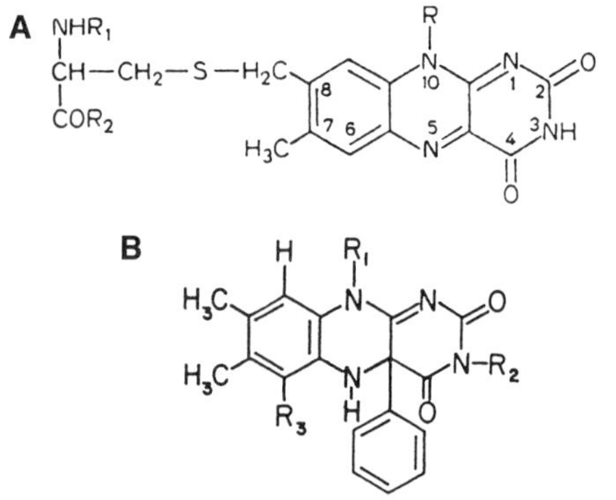

FIG 6.2 Structure of the (**A**) flavin peptide from monoamine oxidase and (**B**) phenyldihydroflavins, products of phenylhydrazine inactivations; redrawn from Singer (1985).

frequently active at two orders of magnitude less, the latter effects must be expressed in a milieu 100 times richer in spermidine. Nevertheless the physiological interactions of the polyamines and monoamines have not been explored in detail. Finally it should not be supposed that the MAO systems are entirely separate from the PAOs. A vertebrate brain MAO, inhibited by pargyline, participates in polyamine metabolism by oxidizing monoacetyl putrescine in a pathway leading to the formation of 4-aminobutyrate (Seiler, 1987; Youdin et al., 1991).

Some Early Copper-Containing Proteins and Enzymes

The presence of copper in biological material was long thought to be accidental. The metal was detected in plant and animal tissues in the early 19th century but essentiality to life processes was demonstrated in the 1920s. Protein containing tightly bound copper, a hemocyanin, was found in octopus blood by L. Fredericq in 1878. The hemocyanins are O_2 carriers and do not appear to possess oxidase functions. The copper in the molecule binds oxygen reversibly to form a blue protein, an oxyhemocyanin.

G. E. Bertrand, who detected a plant phenol oxidase (laccase) in the early 1890s, postulated a catalytic role of metals in oxidations. The demonstrations that the plant polyphenol oxidases were copper proteins in which the copper was essential were carried out by F. Kubowitz (Warburg & Keilin, 1966). An extraordinary competition developed in these laboratories on cytochrome oxidase (a_3), which also proved to contain essential catalytically active copper. A. Szent–Gyorgyi (1931) described an ascorbate oxidase from cabbage leaves, and some 20 years later this somewhat difficultly purifiable protein was shown to be a "blue" copper enzyme. The protein catalyzes a four electron reduction of molecular oxygen (dioxygen) to water with the conversion of the enediol of ascorbate to the dicarbonyl-containing lactone. It has been convenient to compare the properties of the enzyme and its reaction mechanism to that of laccase.

Another Cu enzyme, tyrosinase or polyphenol oxidase, was detected in a fungus by Bertrand in 1895 in his search for organisms containing laccase. The Cu enzyme, tyrosinase, found in many animals and plants will carry out two consecutive reactions, a conversion of a phenol to an o-diphenol and, as a catechol oxidase, the oxidation of the diphenol to the o-diquinone.

The copper protein and enzyme ceruloplasmin was discovered in the course of fractionating human plasma in the late 1940s. It was easily crystallized and was found to contain some 95% of plasma Cu, with some six to seven copper atoms per molecular weight of 134,000. The protein is thought to be a vehicle in Cu transport in mammals. In addition to an apparent role in conversion of Fe^{2+} to Fe^{3+}, prior to incorporation of iron in transferrin, ceruloplasmin is also an oxidase capable of converting molecular oxygen to water in the presence of appropriate substrates such as dopamine and tryptamine.

The obvious possible relation of the valency of Cu, monovalent Cu^{1+} and divalent Cu^{2+}, to the oxidation-reduction function of the copper proteins has led to tools capable of defining this valency within the isolated proteins. Electron paramagnetic resonance (EPR) selectively detects the divalent state of copper (Finazzi Agrò & Rossi, 1992). The copper in blue proteins, such as laccase, ascorbate oxidase, and ceruloplasmin, give signals (type 1 copper) that suggest the presence of Cu^{2+} in a relatively simple chelate. The copper-containing amine oxidases give type 2 signals (i.e., low absorption in the visible range) and a "normal" EPR spectrum, suggesting a more complex structure with coordinating ligands of both nitrogen and oxygen. These signals then have the potential of detecting the reduction of Cu^{2+} to Cu^{1+} or of other structural interactions of the metal.

Copper-Containing Bovine Plasma Amine Oxidase

This most extensively studied amine oxidase, the enzyme of adult ruminant sera, has a marked specificity for spermine and spermidine. Fortunately several enzymes of this group that have now been found in animals, plants, fungi and bacteria, although distinctly different in origin, composition, structure, and specificity, have many common features. In general the enzymes consist of two subunits, each of which contains a copper atom; these can react independently. On isolation, the copper is divalent and in many instances is removable from the protein after reduction to the cuprous state by dithionite and in a second step by dialysis against cyanide. The inactive proteins may be restored to activity by reintroduction of Cu^{2+}.

The enzymes are pink and have an optical absorption with a broad maximum between 470 and 480 nm. This peak disappears on reduction of the enzyme with dithionite or by anaerobic treatment of the enzyme with substrate. The peak reappears on oxidation of the enzyme. The enzyme and cofactor are irreversibly inactivated by a carbonyl reagent such as phenylhydrazine. The purification and crystallization of the bovine plasma oxidase was achieved by Yamada and Yasunobu (1962). The enzyme was assayed spectrophotometrically by following the conversion of benzylamine to benzaldehyde. Simplified purifications of highly active enzyme were described with data on stability, spectral properties, copper content, and specificity. Essentially all of the amine oxidases in vertebrate sera, ruminant (e.g., cow, sheep) and nonruminant (e.g., pig, horse, etc.), are very low in activity on putrescine and cadaverine. However, they are reactive with benzylamine. The recrystallized bovine enzyme is capable of oxidizing the ε-amino group of lysyl peptides (Oda et al., 1981); it is thus also a peptidyl lysyl oxidase.

The importance of working with purified enzymes is highlighted by the complications observed in comparing the amine oxidases of the adult bovine serum and fetal bovine serum. Although the purified adult enzyme is clearly spermine and spermidine specific, with an apparent K_m for these substrates of about 40 µM, the adult sera do have some slight activity on putrescine. Actually the same enzyme, the sole amine oxidase of the adult serum, is active on putrescine with an apparent K_m of 2 mM. A separate fetal enzyme, putrescine oxidase, with an apparent K_m for putrescine of 2.6 µM, has only slight activity on spermine and spermidine and is absent from adult serum (Gahl & Pitot, 1982). The adult serum is apparently capable of the following four reactions:

$$\text{spermine} \rightarrow \text{spermidine} \rightarrow \text{putrescine,}$$

$$N^8\text{-acetylspermidine} \rightarrow N^4\text{-acetylputrescine}$$
$$+ \text{H}_2\text{O}_2 + \text{acrolein,}$$

$$\text{spermidine} \rightarrow \text{putrescine} + \text{H}_2\text{O}_2 + \text{acrolein,}$$

$$\text{spermine} \rightarrow \text{spermidine} + \text{H}_2\text{O}_2 + \text{acrolein.}$$

The first reaction is essentially that described for sheep serum by Bachrach and Bar-Or (1960). The copper contents of the fractionated preparations increased with increasing specific activity and the homogeneous protein contained 1 g atom Cu/70,000 g protein. In preparations that minimized a tendency of the enzyme to aggregate, the protein was found to have a molecular weight of about 170,000 and comprised equivalent subunits linked through a disulfide bond. In addition the enzyme contains two sulfhydryl groups that are not essential for activity. The copper atoms are not bound to the sulfhydryl groups (Knowles & Yadav, 1984).

The pink color of the enzyme (e.g., adult plasma amine oxidase) is discharged anaerobically by a reducing substrate (e.g., benzylamine) in the absence of oxygen. Restoration of the broad 480 nm peak and NH₃ release are effected in oxygen; this clearly divides the overall reaction into two successive reactions (Janes & Klinman, 1991; Rius et al., 1984). Analysis of the anaerobic reductive half-reaction by the latter workers demonstrated 2 mol aldehyde/mol enzyme, indicating a functionally active cofactor at each enzyme subunit. Nevertheless some aromatic hydrazides reacted with only one of these, whereas other hydrazides reacted with both, one reaction occurring much faster than the first (Bossa et al., 1994; Morpurgo et al., 1992).

The reduced enzyme is insensitive to inhibitors such as 2-bromoethylamine, which is considered to be a "suicide" inactivator. This compound reacts as a substrate with the active oxidized enzyme to form an inhibitory bromoacetaldehyde. The latter is itself an effective alkylating agent and eventually jams the site after several turnovers of substrate and product (Neumann et al., 1975). Similar observations have been made with the β-Br-ethylamine in the inhibition of the crystalline amine oxidase isolated from *Aspergillus niger* (Kumagai et al., 1979).

The kinetic analyses and the lack of evolution of ammonia in the first step were consistent with an aminotransferase mechanism, and spectroscopic evidence and sensitivity to carbonyl reagents both suggested the existence of a carbonyl cofactor, possibly like or identical with pyridoxal. Anaerobic reduction of the enzyme by substrate was found by ESR techniques to liberate one sulfhydryl group per subunit. The possibility that electron transfer might proceed from a reduced sulfide ion to Cu^{2+} was detected in some bacterial blue copper proteins, the azurins. However, this amine oxidase sulfhydryl is not readily interactive with the essential copper of the subunit, as studied by EPR (Zeidan et al., 1980). Nevertheless, exogenous sulfide is an irreversible inactivator of the enzyme (Dooley & Coté, 1984).

In early EPR experiments the copper of the enzyme appeared to be in the cupric form in both stages of the overall reaction; it could be reduced to the cuprous state, which was reoxidizable to the cupric state (Yamada & Yosunobu, 1962). Both copper and specific activity were lost in parallel on dialysis in the range of pH 3–5. A dithionite-reduced bleached copper-depleted bovine serum amine oxidase did not restore its 480 nm maximum in oxygen; readdition of copper was essential in this step. Further, a recent EPR study of the oxidative half-reaction indicated that the oxidation of the reduced cofactor to a semiquinone is coupled to Cu(I) formation (Dooley et al., 1991). This sequence would then involve two sequential one-electron oxidations, which were also schematized (Fig. 6.3) by Mure and Klinman (1995).

Nature of Organic Cofactor

The apparent presence of a distinctive spectral absorption in the absence of copper and of an essential moiety inactivated by carbonyl reagents suggested the presence of pyridoxal (free or covalently bound, with or without esterified phosphate) in the enzymes. This compound could not be isolated in any form from the crystalline active enzyme. A few workers have continued to believe in the existence of covalently bound pyridoxal phosphate in a highly purified pig plasma amine oxidase (Buffoni & Cambi, 1990). It may be mentioned that pyridoxal phosphate has been tested as an antidote for many toxicants, including cyanide, dopamine, and spermine. In the latter instance pyridoxal phosphate blocks

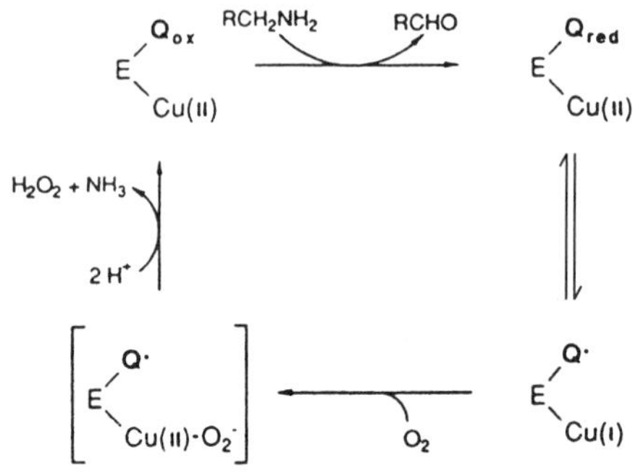

FIG 6.3 Possible catalytic cycle of copper-containing amine oxidases; redrawn from Dooley et al. (1991). Q_{ox}, oxidized quinone; Q·, semiquinone; Q_{red}, two-electron reduced quinone.

spermine-induced renal failure in the rat and markedly improves survival (Keniston et al., 1987).

A comprehensive review of the molecular properties of the amine oxidases, detailing the central position of this unknown cofactor in the state of the field, did appear (Yasunobu et al., 1976). In 1979 tricarboxylated quinoline quinone, with unusual redox properties, was isolated from bacterial cultures (Fig. 6.4A). This substance, known as PQQ, was highly reactive with carbonyl reagents and assays were developed based on this reactivity. PQQ was found to be the dissociable prosthetic group for several bacterial oxidoreductases described as quinoproteins, and the activation of the apoproteins for the enzymes has been the ultimate test for

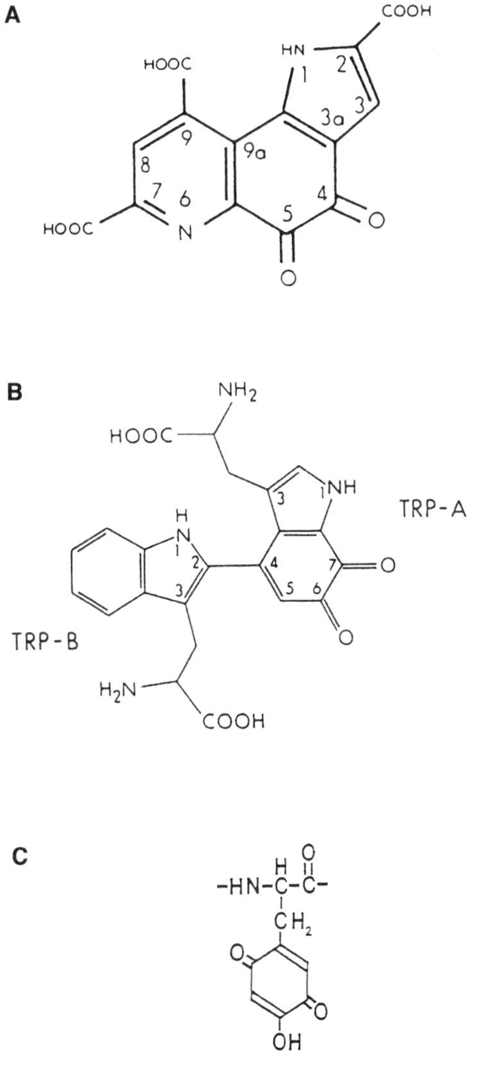

FIG 6.4 Quinones of oxidases: (**A**) pyrroloquinoline quinone (PPQ); (**B**) Tryptophantryptophyl quinone (TTQ); (**C**) 6-hydroxydopa(topa)quinone (TOPA).

this prosthetic group (Duine, 1989; Shinagawa et al., 1989). In 1984 it was reported that the beef plasma amine oxidase, other amine oxidases, as well as various animal, plant, and fungal enzymes were probably PQQ proteins. Indeed the presumption of the presence of PQQ, containing vicinal carbonyl groups, in the human lysyl oxidase led to the study of vicinal diamines as irreversible inhibitors of this enzyme (Gacheru et al., 1989).

Derivatized peptides isolated from the homogeneous beef plasma amine oxidase failed to reveal the presence of derivatized PQQ (Kumazawa et al., 1990b). Similarly the presence of PQQ in plant, animal, and bacterial enzymes could not be confirmed. It was then reported that a cofactor-containing pentapeptide Leu-Asn-x-Asp-Tyr was isolated in high yield from the bovine amine oxidase. The residue x was identified as 6-hydroxydopa or *topa* quinone (TOPA) (Janes et al., 1990; Janes & Klinman, 1995). The structure of three proposed cofactors are presented in Figure 6.4.

The enzyme had been labeled with radioactive phenylhydrazine to produce an adduct absorbing at 350 nm. The labeled enzyme was degraded proteolytically and the peptides were separated with high performance liquid chromatography. A radioactive fraction of appropriate spectral properties was obtained and characterized as a pentapeptide devoid of PQQ. After subtraction of the four characterizable known amino acids, one possibility of x was the phenylhydrazone of TOPA. Peptides and derivatives of TOPA were found to develop spectrophotometric shifts comparable to that of the intact enzyme in the presence of phenylhydrazine (Palcic & Janes, 1995).

Assays previously reported to detect PQQ in mammalian cell extracts are also given positive responses by derivatives of TOPA. Earlier investigators of these assays confirmed this activity for TOPA-containing proteins and certain amine oxidases (e.g., pig kidney diamine oxidase) that were presumed previously to contain PQQ. A more recent analysis of the phenylhydrazine adducts of various amine oxidases by resonance Raman spectroscopy (Brown et al., 1991) extended the presence of a TOPA-like structure to the above as well as the pea seedling enzyme and a bacterial methylamine oxidase previously believed to contain PQQ. In contrast, however, to the finding of TOPA, Chen et al. (1991) found a bound TOPA-like structure in a bacterial methylamine dehydrogenase to be tryptophan tryptophylquinone (TTQ) (Fig. 6.4B) as the presumably active cofactor.

It is broadly accepted at present that the organic redox active cofactor present in the copper-containing amine oxidases of beef serum, yeast, pig kidney, and pea seedlings is a protein-bound amino acid TOPA. Model compounds which attempt to emulate the numerous steps of the reductive and oxidative stages of the cyclic reaction have been prepared and tested (Mure & Klinman, 1995; Wang et al., 1994). Analysis of the amine oxidase gene of a yeast revealed a tyrosine codon at the position assumed by TOPA quinone in an isolated tryptic peptide of the protein (Mu et al., 1992). Thus the formation of TOPA from tyrosine within the protein is considered to be a posttranslational event. The properties of TOPA-containing enzymes have been summarized by Klinman and Mu (1994).

The existence of such enzymes has now been extended to bacteria. MAOs containing copper and TOPA were recently isolated from *E. coli* and were studied crystallographically (Cooper et al., 1992; Roh et al., 1995).

How close is the copper to other essential sites, such as the inactivable carbonyl-containing catalytic center? Covalent attachment of a single molecule of carbonyl reagent was shown to inhibit activity completely and irreversibly. NMR studies with the pig enzyme appear to have demonstrated that the two copper atoms are well separated and that one of these is some distance, slightly less than 10 Å, from the carbonyl substituent (Williams & Falk, 1986). In studies of energy transfer in a hydrazinoacridine-substituted enzyme, the separation of metal and cofactor was found to be on the order of 11 to 15 Å. These results were interpreted to preclude the possibility of direct interaction between Cu^{2+} and the organic cofactor or between Cu^{2+} and substrate (Lamkin et al., 1988), although the conclusion was drawn that the copper is not close to and cannot interact directly with the organic cofactor (Bossa et al., 1994; Greenaway et al., 1991). Studies have led to the formulation of specific mechanisms involving interactions of Cu^{2+} and TOPA (Belleli et al., 1991; Kleutz et al., 1980).

Animal Enzymes: Histaminase and Amiloride-binding Protein

The pursuit of histaminase and the serum amine oxidases led to the determination of suitable and abundant tissue sources, resulting in the detection of the pig kidney histaminase and DAO. Partial purification of this enzyme assisted in clarifying the stoichiometry of the reaction (Tabor, 1954). Although this enzyme does not oxidize benzylamine, the oxidation of a substituted benzylamine (i.e., *p*-dimethylaminomethylbenzylamine) has been useful in its assay (Bardsley et al., 1972). Although the substrate lacks significant absorption at 250 nm, formation of the aldehyde results in an absorption spectrum with a maximum at that wavelength. The reaction is

An improved purification facilitated some physical studies on its size, structure, and reactivity (Kleutz & Schmidt, 1977a,b).

Obviously the addition of an olefinic bond evoked the increased absorption at 250 nm. Thus cinnamyl amine, $\Phi-C{=}C-CNH_2$, with an initial maximum near 250 nm may be used as a substrate for the mitochondrial MAOs A and B. It is readily converted to cinnamyl aldehyde with a maximum at 290 nm (Williams et al., 1988).

The spectrophotometric assay of the oxidation of benzylamine facilitated the purification of the beef plasma enzyme (Tabor et al., 1954) and eventually its crystallization (Yamada & Yasunobu, 1962). The molecular characterization of the enzyme and the detection of essential copper helped to begin the many studies on the structure of the enzyme and its mechanism of action.

The pig plasma amine oxidase was far more active on benzylamine than on spermidine and spermine, differing in this respect from the bovine plasma enzyme and differing also from the pig kidney enzyme, which is inactive on benzylamine. The crystallized pig plasma enzyme was found to contain copper (Buffoni & Blaschko, 1964). Several studies of this enzyme dissected the structure and mechanism of the amine oxidases (Petterson, 1985). Unlike the pig kidney enzyme, which is active on putrescine and cadaverine, this enzyme shows low activity to these diamines and high activity on biogenic amines (e.g., tyramine, dopamine, etc.) in addition to histamine. The stereochemistry of benzylamine oxidation by various active amine oxidases was also studied (Alton et al., 1995).

Despite the seeming irrelevance of this benzylamine oxidase and histaminase to the polyamines, a very active plasma histaminase and DAO was found in women in the third trimester of pregnancy. The enzyme, which is easily purified by binding to and elution from a gel coupled to cadaverine, is unable to oxidize benzylamine (Baylin & Margolis, 1975). A similar enzyme was found in the human placenta, but it is not clear that the plasma and placental enzymes are identical. The placental enzyme is immunologically identical to the DAO found in human neutrophilic cells and in amniotic fluid (Morel et al., 1992).

A copper-containing DAO was also purified from human kidney (Suzuki & Matsumoto, 1987). It is active mainly on the aliphatic diamines, with some additional activity on histamine, spermidine, spermine, and monoacetyl polyamines. Whereas the simple diamines and histamine have apparent K_m values on the order of 0.01–0.03 mM, the apparent K_m values for the monoacetyl diamines, spermidine and spermine, are much higher. Using mM putrescine as substrate, the enzyme is totally inhibited by mM aminoguanidine, but not at all by mM pargyline.

The human placental enzyme and the human and pig kidney DAO have been found to be homologous to a human kidney protein, previously considered to be a component of the amiloride-sensitive Na^+ channel. Amiloride and some derivatives have been used clinically as diuretics and antihypersensitive agents. The binding site (ABP protein) now proves to be a DAO (histaminase) and inhibition by amiloride occurs at the active site of the enzyme (Novotny et al., 1994). The enzyme occurs on kidney cell membranes and may be displaced by heparin, facilitating several purification steps. The cDNA prepared from this protein has been used not only to characterize the structure of the ABP gene and a derived mRNA, but also to detail the gene DNA and the position of the codon for tyrosine destined to become TOPA (Chassande et al., 1994). The active site peptide of the bovine serum amine oxidase proved to be identical to that in the amiloride-binding protein (Mu et al., 1994). The surface site of cellular binding by DAO was characterized by Baenziger et al. (1994).

Peptidyl-Lysyl Amine Oxidases and Copper-Related Cross-Linking

In the early 1960s it was observed that copper-deficient nutritional regimens in pigs evoked rupture in essential blood vessels in the heart and aorta. The plasma amine oxidase had virtually disappeared in the copper-deficient animals (Blaschko et al., 1965). Analysis of the proteins of the deficient aortas revealed an altered elastin content, an increased solubility of this structural element, and an increased lysine content of the soluble protein (Smith et al., 1968). This pointed to the participation of lysine in the cross-linking of elastin to form the insoluble polymer, and a soluble amine oxidase was obtained from the bone of embryonated chicks that converted peptide-bound lysine to α-aminoadipic-δ-semialdehyde (Pinnell & Martin, 1968).

[^{3}H-6]-Lysine was incorporated into chick proteins and was followed by the appearance of tritiated H_2O_2, degradable to ^3H_2O in the presence of catalase. The reaction is given as

$$
\begin{array}{ccc}
NH_3^+ & & \\
| & & \\
CH_2 & \xrightarrow[\;H_2O\;]{O_2} & CHO \\
| & & | \\
(CH_2)_3 & & (CH_2)_3 \\
\end{array}
$$

$$
\underset{R_1-C-NCH-C-NR_2}{\overset{O \quad H \quad O \quad H}{\| \quad | \quad \| \quad |}} \qquad
\underset{R_1-C-N-CH-C-NR_2}{\overset{O \quad H \quad O \quad H}{\| \quad | \quad \| \quad |}}
$$

$$+ NH_3 + H_2O_2 \longrightarrow H_2O + \tfrac{1}{2}O_2$$

Nonenzymatic condensation of the aldehyde and an amine to form a Schiff base or an aldol condensation of two aldehydes are believed to account for the lysine-derived cross-links of elastin and collagen. Some biochemical attempts to reverse the pathology of copper deficiency have been discussed (Gacheru et al., 1993; Tinker et al., 1990). Additional biologically significant cross-linking mechanisms are known in the formation of disulfide bonds or in the more novel tyrosine-derived pulcherosine (Nomura et al., 1990). Peroxidase-catalyzed cross-links to products of amine oxidases in plants will be mentioned later, and the roles of the cross-linking animal transglutaminases will be discussed in Chapter 21.

The solubility of the chick lysyl oxidase was thought

surprising, but a similarly soluble enzyme was found in the aorta of cattle that was more active on benzylamine than on peptidyl lysine (Rucker et al., 1970). In many tissues the association of the enzyme with large and near-insoluble structural proteins led to the use of urea or nonionic detergents in purification. In humans and pigs, skin is a rich source of a solubilizable enzyme that has been localized extracellularly and intracellularly by immunohistochemical studies (Kobayashi et al., 1994).

Bovine and rat lysyl oxidases analyzed by immunological cross-reactivities and cDNA sequences proved to be nearly identical (Trackman et al., 1990). A chick aortic lysyl oxidase was also studied in detail (Wu et al., 1992). In humans a single gene controls the production of multiple mRNA transcripts (Boyd et al., 1995; Svinarich et al., 1992).

The purified bovine aortic enzyme (32 kDa) contains a single tightly bound copper atom per molecule, whose presence is essential for the anaerobic production of aldehyde from p-hydroxybenzylamine (Gacheru et al., 1990). This result has implicated copper in the initial anaerobic phase in the reaction sequence. The human, rat, and mouse lysyl oxidase cDNAs have revealed many conserved similarities among these enzymes, as well as a probable coordination site for Cu^{2+} (Krebs & Krawetz, 1993). The organic cofactor is now thought to be a cross-linked TOPA-lysine (Klinman, 1996).

Some Plant Amine Oxidases

Werle and Pechmann (1949) detected H_2O_2 in amine oxidation in plant extracts, and Kenten and Mann (1952) focused on the possible role of this reaction product in plant metabolism. Studies on the pea seedling enzyme led to a purification of the protein sufficient to provide the first evidence of the color and spectral properties of the enzyme, as well as the presence of an organic cofactor and essential copper (Hill & Mann, 1962; Mann, 1961). Plant copper amine oxidases were recently reviewed by Medda et al. (1995). In the plant kingdom, legumes (Leguminosae) have proven to be excellent sources of the copper-containing amine oxidases, in contrast to cereal groups such as the Graminae, which are rich in the flavin-containing amine oxidases. However, Cu-containing DAO may be isolated from some other plants, such as barley or corn (Suzuki & Hagiwara, 1993).

Seedlings of the pea (Pisum sativum) and several other plants (e.g., Lathyrus sativus, Lens esculenta), have extraordinarily high concentrations of a dimeric diamine oxidase, containing one copper atom per subunit (Rinaldi et al., 1985; Smith, 1985). The amine oxidase present in pea cotyledons is 100 times more concentrated per unit weight in the plant than in many animal tissues like hog kidney. Only a 30–50-fold purification via affinity chromatography has been required to produce a homogeneous enzyme containing a single carbonyl group per mole of enzyme. Such large amounts of the stable enzyme has facilitated studies of protein structure, catalytic mechanism, and substrate specificity

(McGuirl et al., 1994). Another simple and mild procedure has been developed for the similar soybean enzyme (Vianello et al., 1993).

In addition to being very active on putrescine and cadaverine, the pea enzyme is also active on agmatine, 2-hydroxyputrescine, 2- and 3-hydroxycadaverine, histamine, and many aromatic and aliphatic monoamines. The enzyme can also slowly oxidize D- and L-lysine. The presence in crude plant extracts then of oxidase substrates, and H_2O_2-utilizing enzymes, such as peroxidase and catalase, as well as substrates for peroxidase, can result in complex and confusing interactions. One unusual substrate of the pea amine oxidase is monodansylcadaverine, which has often been used in studies of transglutaminase in animal tissues. This reactivity of the pea amine oxidase precludes the use of this reagent in exploring transglutaminase in pea tissues.

Studies of amine oxidases in some other legumes, such as that from *Lathyrus sativus* and *Lens esculenta*, have revealed additional quirks. The former (Suresh et al., 1976) is most active on the normal diamines and only slightly active on monoamines, including benzylamine. However, it shares an intermediate activity on agmatine, spermidine, and spermine with homoagmatine, *sym*-homospermidine, aminopropylcadaverine, as well as *N*-carbamylputrescine and *N*-carbamylcadaverine. The amine oxidase from *L. esculenta* seedlings has proven to be of interest in the study of molecular properties and mechanisms (Rinaldi et al., 1985; Tipping & McPherson, 1995). The lentil enzyme converts spermine to a dialdehyde, as was seen with the bovine plasma amine oxidase (Cogoni et al., 1991). This dimeric enzyme contains an active site at each subunit (Padiglia et al., 1992) and a characteristic TOPA-containing hexapeptide was isolated (Rossi et al., 1992).

Antibodies to the pure enzyme (Angelini et al., 1985) have been used to detect the control of the activity of the enzyme in light and dark development of the plant and to demonstrate actual changes in enzyme protein during development (Federico et al., 1985). The antisera have also detected the presence of the enzyme in the cell walls of the lentil seedling, extending earlier observations of the presence of amine oxidases and of amines in the cell walls of oat leaves (Kaur–Sawhney et al., 1981; Li, 1993) and mung beans. This distribution of reactants and the oxidation of the amines might then produce the hydrogen peroxide essential to the peroxidase-catalyzed development of oxidized and cross-linked polymers in the cell walls of the growing plant. The time course of enzyme increase in response to wounding of chick pea internodes and healing supports this hypothesis (Scalet et al., 1991).

In addition to its presence in the seedlings of legumes, a copper-containing DAO has been isolated from the latex of a Mediterranean shrub, *Euphorbia characias*. The enzyme is active on putrescine and cadaverine and inactive on benzylamine, histamine, spermidine, or spermine. It contains a copper atom for each of its two subunits, like other enzymes of this type, and

is inhibited irreversibly by isoniazid, phenylhydrazine, and semicarbazide.

An *N*-methylputrescine oxidase isolated from tobacco roots is involved in the culturally significant conversion of *N*-methylputrescine to nicotine. Although this enzyme is inhibited by carbonyl reagents, such as phenylhydrazine, it has not yet been rigorously demonstrated if this protein is a copper-containing amine oxidase (Davies et al., 1989).

Stereochemical Specificity of Animal and Plant Enzymes

Despite clear differences of the proteins, substrate specificity, and immunological reactivities of the animal and plant amine oxidases, the common "conserved" features (i.e., Cu content, quinone cofactor, subunit organization, etc.) of the copper-containing amine oxidases may have been expected to extend to the common features in the details of their catalytic activities. The existence of a quinone with the liberation of an aldehyde had been supposed to reflect a common mechanism. However, isotopic exchange and a comparison of the NMR of the substrates and products (e.g., tyramine) revealed that the reactions may proceed stereospecifically or nonspecifically, depending on the specific enzyme used (Coleman et al., 1991) (Table 6.1).

Fungal Copper-containing Oxidases

Particular growth conditions may maximize the production of a desired enzyme. Growth of *Aspergillus niger* on *n*-butylamine as the sole source of nitrogen resulted in high enzyme levels and permitted the isolation and crystallization of a copper-containing amine oxidase. The enzyme was quite active on aliphatic monoamines and much less active on C_4 to C_6 aliphatic diamines. The enzyme was also active on benzylamine, histamine, and agmatine.

Many fungi, grown on agmatine, putrescine, spermidine, or spermine as the sole nitrogen source, contain an agmatine oxidase. One of these was isolated from *Penicillium chrysogenum* and was crystallized (Isobe et al., 1982). The dimeric enzyme also contained two Cu atoms per mole and had apparent K_m values for agma-

TABLE 6.1 Stereochemical studies on copper amine oxidases.

Enzyme source	C-1 Proton abstraction	C-2 Solvent exchange
Pig plasma	Pro-R	Yes
Bovine plasma	Nonstereospecific	Yes
Pea seedling	Nonstereospecific	No
Pig kidney	Pro-S	No

Adapted from Coleman et al. (1991).

tine, histamine, and putrescine of 2.5×10^{-4} M, 4.3×10^{-4} M, and 1.6×10^{-2} M, respectively. Indeed putrescine was inhibitory at concentrations above its K_m value.

The purified methylamine oxidase, from a methylotrophic yeast *Hansenula polymorpha*, proved to contain TOPA quinone and is designated a *quinoprotein* (Cai & Klinman, 1994). The formation of TOPA quinone on the protein appeared to occur posttranslationally after transfection of *S. cerevisiae*, an organism that does not utilize amines as a nitrogen source and does not normally contain Cu amine oxidases.

Some Bacterial Enzymes

Few copper proteins are known among bacteria and it has been suggested that the presumedly anaerobic environment of the early earth in which bacteria evolved contained very low concentrations of cupric ion. The acquisition of photosynthesis and an oxygen-containing environment supposedly led to the evolution of several mechanisms for the metabolism of molecular oxygen, reactions in which copper was particularly useful. Several copper-containing proteins (e.g., azurins and methylamine oxidase) are found in some aerobic organisms. Nevertheless, the isolation and crystallization of a copper amine oxidase containing TOPA from *E. coli* has been described, as noted earlier. Also, some methylotrophic bacteria convert methylamine to formaldehyde and ammonia. A Gram-positive *Arthrobacter* P1 carries out the following reaction:

$$CH_3NH_2 + H_2O + O_2 \rightarrow CH_2O + H_2O_2 + NH_3.$$

This enzyme proved to be a typical copper-containing amine oxidase, bleachable anaerobically by substrate and inhibited by phenylhydrazines. Spermine and putrescine were not substrates. The *Arthrobacter* enzyme has been found to contain a moiety very like TOPA in a phenylhydrazine-reactive peptide (Brown et al., 1991).

Methylamine Dehydrogenases

The ability of aerobic organisms, such as strains of the methylotrophic pseudomonads, to grow on methylamine or other simple aliphatic amines as the sole nitrogen sources, led to the discovery of methylamine dehydrogenases.

The crystallographic analysis of the methylamine dehydrogenase of *Thiobacillus versutus* indicated that PQQ was not the essential cofactor in this presumed "quinoprotein." The reexamination of the methylamine dehydrogenase isolated from *Methylobacterium extorquens* AM1, and the enzymes from *Pseudomonas* and *Thiobacillus* revealed linked semicarbazide-derivatized peptides containing unknown amino acids. These peptides did not contain PQQ or TOPA, and are believed to contain a cofactor derived from cross-linked tryptophan moieties (i.e., tryptophan tryptophylquinone or TTQ; Fig. 6.4) (McIntire et al., 1991).

Amine Oxidation and Polyamine Assay

The classical text of Marjorie Stephenson on bacterial metabolism contains the tables of den Dooren de Jonge, which record the growth of many organisms on organic amines as sole sources of carbon and nitrogen or as a sole source of nitrogen. The use of glucose and the diamine, cadaverine, extended the range of growth potentiality. Such studies have been done with other microbes, such as yeasts, and these results indicate the organisms suitable for further study of amine metabolism.

The discovery of organisms requiring diamine or polyamine as an essential exogenous nutrient provided the basis for a microbial assay based on the amount of growth. As described by Martin et al. (1952), the turbidimetric growth response of cultures of *Neisseria perflava* approached linearity in the range of 0.05 and 0.25 µg putrescine/mL. The ability of the cultures to respond to other polyamines, such as spermidine, induced Herbst's group to turn to separations and colorimetric assays of the separate amines. In 1958 Weaver and Herbst showed that this organism and others, grown in the presence of spermine, produced 1,3-propanediamine, NH_3, H_2O_2, and an aldehyde.

By the end of the 1960s studies on the oxidation of the polyamines had revealed many different reactions of polyamine degradation. The organisms performing these degradations were available in stock cultures, and in several instances the specificity of the reaction for substrate permitted the use of the particular enzyme for a relatively specific assay. The formation of the common products of the enzymatic oxidations, ammonia and hydrogen peroxide, has been made the basis of the many enzymatic assays for particular polyamines. In other instances the nature and quantity of the organic product has been determined.

Organic Products of Copper Amine Oxidases

1. Oxidation by bovine plasma amine oxidase, lentil amine oxidase:

   Spermine →

   $OHCCH_2CH_2NHCH_2CH_2CH_2CH_2NHCH_2CH_2CHO.$

2. Oxidation by pea seedling amine oxidase:

   putrescine → $NH_3 + H_2O_2 + \Delta^1$-pyrroline,

   spermidine → $NH_3 + H_2O_2 +$ 1-(3-aminopropyl)-pyrroline.

3. Oxidation by agmatine oxidase from *Penicillium chrysogenum* (Isobe et al., 1982):

   agmatine → $NH_2(NH)CNH(CH_2)_3CHO.$
   γ-guanidinobutyraldehyde

The enzyme is inducible when the fungus is grown on any natural polyamine. These authors also isolated and crystallized two flavin-containing polyamine oxidases from the fungi *Penicillium chrysogenum* and *Aspergillus terreus* (see below).

Organic Products of Some Plant, Fungal, and Microbial PAOs and Dehydrogenases

1. PAOs of plants (Graminae: maize, barley, oat, wheat, rye)

Spermine $\longrightarrow$ H$_2$NCH$_2$CH$_2$CH$_2$NH$_2$ +

1,3 diaminopropane

H$_2$NCH$_2$CH$_2$CH$_2$NHCH$_2$CH$_2$CH$_2$CHO

8-amino-5-aza-octanal

H$_2$O | spontaneous
+H$^+$ |

1-(3-aminopropyl)pyrrolinium

This substance (Smith, 1983; Smith et al., 1986), which exists as the iminium cation, is converted to a bicyclic compound (Brandänge et al., 1984).

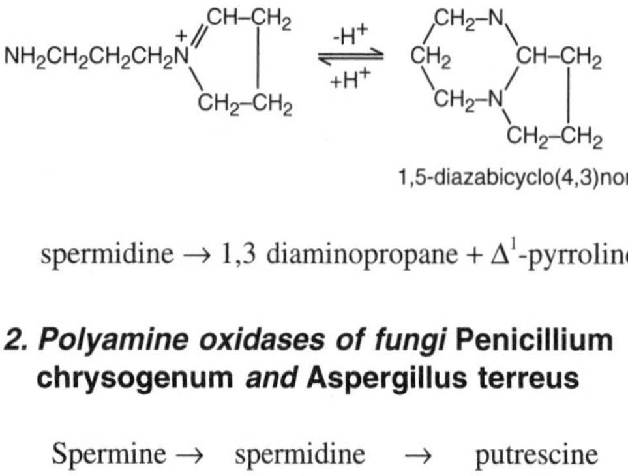

NH$_2$CH$_2$CH$_2$CH$_2$N⁺ ⟨CH–CH$_2$ / CH$_2$–CH$_2$⟩ $\underset{+H^+}{\overset{-H^+}{\rightleftharpoons}}$ ⟨CH$_2$–N / CH$_2$ CH–CH$_2$ / CH$_2$–N / CH$_2$–CH$_2$⟩

1,5-diazabicyclo(4,3)nonane

spermidine $\rightarrow$ 1,3 diaminopropane + Δ^1-pyrroline.

2. Polyamine oxidases of fungi Penicillium chrysogenum and Aspergillus terreus

Spermine $\rightarrow$ spermidine $\rightarrow$ putrescine
        +   +
     NH$_2$(CH$_2$)$_2$CHO   NH$_2$(CH$_2$)$_2$CHO

These fungal flavoprotein enzymes (Isobe et al., 1980a,b, 1981) are relatively specific for spermine and spermidine and provide final products, putrescine and H$_2$O$_2$, that are easily estimated. These enzymes do not oxidize putrescine, an activity provided in differential assay systems by the enzyme putrescine oxidase, obtainable from the bacterium *Micrococcus rubens* (see below).

Many fungi can be grown in media containing spermidine or spermine as the sole source of nitrogen and the formation of these amine oxidases has been studied as a function of exogenous polyamine (Yamada et al.,

1980). Many of these fungal PAOs oxidize acetylpolyamines as does the rat liver enzyme.

3. PAOs of bacteria

a. *Neisseria perflava.*

    Spermine, spermidine $\rightarrow$ 1,3-diaminopropane + unknown product.

The formation of γ-aminobutyraldehyde and a cyclic product was postulated (Weaver & Herbst, 1958).

b. *Mycobacterium smegmatis.*

    Spermine, spermidine $\rightarrow$ 1,3-diaminopropane + γ-aminobutyric acid.

It is supposed that the latter product was generated from δ-aminobutyraldehyde (Bachrach et al., 1960; Zeller et al., 1951).

c. *Pseudomonas* species.

1. Spermidine $\rightarrow$ putrescine + 3-aminoproprionaldehyde (Padmanabhan & Kim, 1965).
2. *Pseudomonas putida* NP (Durham & Perry, 1978a,b)

Growth on benzylamine or other monoamines induced primary amine dehydrogenase of broad specificity. Many alkyl and aryl monoamines were deaminated;

benzylamine $\rightarrow$ benzaldehyde,

whereas diamines including 1,3-diaminopropane and polyamines were oxidized at low rates. This enzyme was purified to homogeneity and was found to contain two molecules of heme per molecule, both of which were bound covalently to one of the two nonequivalent subunits. Flavin could not be detected in the enzyme.

d. *Serratia marcescens.*

    spermidine $\rightarrow$ 1,3-diaminopropane + γ-aminobutyraldehyde.

The latter product cyclizes to Δ^1-pyrroline (Bachrach, 1962). Indeed all amines containing a 4-aminobutylimino group formed Δ^1-pyrroline stoichiometrically, whereas compounds such as spermine that contain 3-aminopropyl imino groups liberate one molecule of 1,3-diaminopropane. This suggests that the reaction has stopped at the formation of the aminopropyliminobutyraldehyde (Okada et al., 1979b). Spermine and putrescine are less than 1% as active as spermidine in forming pyrroline. Norspermidine is oxidized but cannot form pyrroline.

Although a crude extract utilized oxygen, some purified enzymes required additional electron carriers to consume oxygen. Such a system has been used to estimate amines oxidized by a relatively nonspecific bacterial amine dehydrogenase. The *Serratia* enzyme is sufficiently specific for spermidine to be of interest as an assay tool in which the product, Δ^1-pyrroline, is determined as a colored derivative after reaction with o-aminobenzaldehyde (Bachrach & Oser, 1963). A more sensitive assay for Δ^1-pyrroline was obtained in a reaction with ninhydrin in acid medium (Naik et al., 1981). This enzyme proved to contain flavin adenine dinucleotide and heme as an additional electron carrier (Tabor & Kellogg, 1970) but required an additional electron carrier to interact with oxygen (Okada et al., 1979a; Tabor, 1985).

e. *Micrococcus rubens*.

$$\text{Putrescine} \rightarrow \Delta^1\text{-pyrroline} + NH_3 + H_2O_2,$$

$$\text{spermidine} \rightarrow \Delta^1\text{-pyrroline} + 1,3\text{-diaminopropane} +$$
$$NH_3 + H_2O_2.$$

This enzyme proved to be a flavoprotein oxidase, containing FAD, and did not require an additional electron carrier (Okada et al., 1979b). Unlike most mammalian or plant DAOs, the enzyme was markedly inhibited by —SH reagents, such as p-chloromercuribenzoate. Some higher alkylmonoamines and diamines (e.g., dodecylamine) and 1,8-diaminooctane are active inhibitors. Affinity chromatography on amine-Sepharose has been a simple and effective final purification step for both the Micrococcal oxidase and the Serratia dehydrogenase. The oxidase attacks putrescine and spermidine. Unlike the bovine plasma amine oxidase, which attacks the aminopropyl moieties of spermidine and spermine, the Micrococcal oxidase is essentially inactive on spermine (3,4,3) and quite active on canavalmine (4,3,4) and thermospermine (4,3,3). However, it has a low activity on aminobutylhomospermidine (4,4,4).

Matsuzaki et al. (1992) (see Chapter 5) used a micrococcal enzyme to distinguish spermine (3,4,3), thermospermine (3,3,4), and canavalmine (4,3,4). Because the enzyme will distinguish spermidine and spermine, it is possible to create differential enzyme assays with the capability of estimating the total putrescine, spermidine, spermine, and cadaverine in urine samples in which the acetylated polyamines were first hydrolyzed with an acylpolyamine aminohydrolase. Kits containing the essential enzymes are available (Mashige et al., 1988). The purified putrescine oxidase of *M. rubens* has been used to prepare peptides, oligonucleotides, and cDNAs, which are active in *E. coli* for the large-scale production of the oxidase (Ishizuka et al., 1993).

Other types of differential enzymatic assays have been discussed by Kumagai and Yamada (1985). Recently a differential assay was applied to the analysis of prostatic fluids, and the estimation of the individual amines was proposed for the routine assessment of male sexual function. Despite the diversity and apparent utility of the now numerous systems for amine oxidation applicable to the assay of the free polyamines, a large fraction of the urinary amines are present as acetyl derivatives and the situation in other materials (e.g., prostatic fluid) has not been determined. The acylpolyamine hydrolase used by some analysts does not act on acetyl spermine, which might be present. The usual approach of hydrolyzing samples with acid to provide unconjugated amines requires two analyses to determine the conjugation of any polyamine by difference. Of even greater possible significance is the potential loss of polyamine in reacting under acid conditions with carbohydrate in the sample. Obvious examples of potential difficulty might include the sugar-laden urine of a diabetic or a polysaccharide-rich extract of a plant.

f. Coupled bacterial systems. A Gram-positive soil bacterium grown on putrescine as the sole carbon and nitrogen source contained an active putrescine oxidase (Shimizu et al., 1988). The enzymatic products of the reaction were

$$\text{putrescine} + O_2 \rightarrow \gamma\text{-aminobutyraldehyde} + H_2O_2 + NH_3.$$

The specificity of the oxidase was similar to that of the micrococcal enzyme.

A strain of *Pseudomonas putida*, grown on a complex medium, contained a γ-aminobutyraldehyde dehydrogenase, converting

$$\gamma\text{-aminobutyraldehyde} + NAD \rightarrow$$
$$\gamma\text{-aminobutyrate (GABA)} + NADH.$$

H_2O_2 in Assay of Amine Oxidases and Polyamines

The product H_2O_2 was used to assay the presence of the enzyme and the amount of a polyamine. The presence of catalase or peroxidase and metabolites can minimize or eliminate the amount of H_2O_2 to be found. Even amniotic fluid, which is relatively dilute and presumably free of significant concentrations of endogenous peroxidase, can yield aberrant results (Neufeld & Chayen, 1979). In the case of tests of amniotic fluid, the use of a radioactive substrate (i.e., [^{14}C]-putrescine) and the extraction of labeled product (i.e., Δ^1-pyrroline) into toluene was consistent and linear with time, and otherwise satisfactory.

On the other hand, a peroxidatic assay to quantitate H_2O_2 produced from a purified polyamine by use of a specific enzyme assay is much used. Interest in the metabolism of dopamine (3,4-dihydroxyphenylethylamine) led to methods for the determination of homovanillic acid. The oxidation of the nonfluorescent homovanillic acid to a highly fluorescent $2',2'$-dihydroxy-$3,3'$-dimethoxy-$5,5'$diacetic acid is presented in Chapter 3 (Fig. 3.9). This reaction occurs with H_2O_2 in the presence of a horseradish peroxidase, and the fluorescence evoked provides a sensitive and accurate measure of the H_2O_2 generated.

Efforts have also been made to integrate the peroxide-generating oxidases directly as postcolumn reactions into column fractionations. Many of the numerous oxidases have been immobilized on inert solid carriers, for example, glass beads of controlled pore size derivatized with aminopropyl residues. Hydrogen peroxide can be detected and determined by electrochemical oxidation on a platinum electrode.

Animal Flavin-Containing PAO

The discovery of the mitochondrial MAOs (Blaschko, 1963) and the significance of these enzymes led to the discovery of many other systems active in the metabolism of the biogenic amines (Axelrod, 1991). A carryover effect to polyamine metabolism can be seen in the adoption of the peroxidatic measurement of H_2O_2 produced by both types of oxidase (Snyder & Hendly, 1968).

Monoacetyl derivatives of the polyamines were detected as metabolites in *E. coli* (Dubin & Rosenthal, 1960), but several monoacetylated polyamines (i.e., monoacetyl putrescine, monoacetyl cadaverine, and N^1-monoacetylspermidine) were discovered in the human brain and urine almost a decade later (Nakajima et al., 1969; Perry et al., 1967). Indeed N^1- and N^8-monoacetyl spermidine were found at similar levels in the urine of normal human adults, and the amounts of acetyl putrescine and the acetyl spermidines were in great excess (ca. 10X) over the free amines (Abdel–Monem et al., 1978).

The assay of the acetyl and other bound derivatives in biological fluids (e.g., urine, serum, tissue extracts) called attention to the distinction between free and conjugated amines, as well as questions concerning the origin and fate of these compounds (Löser et al., 1988). In one early study of bound amine products, the presumed oxidation of diaminopropane to β-aminoproprionaldehyde by a pig kidney DAO was considered as a possible starting point in exploring intermediates in polyamine metabolism (Quash & Taylor, 1970). Hog kidney does degrade spermidine to this product by a mechanism seen earlier in studies of a *Pseudomonas* species, that is,

$$NH_2(CH_2)_4NH(CH_2)_3NH_2 \rightarrow NH_2(CH_2)_4NH_2 + NH_2(CH_2)_2CHO.$$
spermidine → putrescine + β-aminoproprionaldehyde

The degradation in the rat of some spermine to spermidine and of some spermidine to putrescine had been detected some years earlier (Rosenthal & Tabor, 1956; Simes, 1967). A search for a rat liver enzyme catalyzing the conversion of spermidine to putrescine detected a PAO acting specifically at the secondary amine of spermidine (Hölttä, 1977). The reaction emulated that catalyzed by the pseudomonas enzyme (Padmanabhan & Kim, 1965) and seen also in the hog kidney (Quash & Taylor, 1970). The liver PAO was purified to homogeneity, and was found to have a molecular weight of about 60,000, and to contain FAD as a prosthetic group. Hölttä (1983) described the isolation, assay, and properties of the enzyme. An improved purification method has been published (Gasparyan & Nalbandyan, 1990). The enzyme is localized within the peroxisomes of rat liver cells (Den Munckhof et al., 1995), but children lacking peroxisomes (i.e., with Zellweger's syndrome) have normal excretion patterns of polyamines (Govaerts et al., 1990).

The nature of the products was carefully established, putrescine by its chromatographic properties and utilization as a substrate in the enzymatic formation of spermidine, and unstable 3-aminopropionaldehyde by several chemical tests. In one of these the compound was reduced to the known 3-aminopropanol, and in another the aldehyde was oxidized by acid dichromate to the well-characterized β-alanine (Hölttä, 1977). The activity of the enzyme on free polyamines is greatly stimulated by the addition of aldehydes (e.g., benzaldehyde), an effect that is not understood (Morgan, 1989). The enzyme is inhibited by sulfhydryl reagents and "carbonyl" reagents. The activity of the enzyme in the liver does not change a great deal in response to treatments evoking large changes in polyamine metabolism (i.e., partial hepatectomy, CCl_4 poisoning, thioacetamide, etc.). These results suggested that the activity of the enzyme is not very carefully regulated.

The study of this enzyme has merged with the slowly growing body of data on the *N*-acetyl polyamines. Hölttä (1977) mentioned briefly that N^1-acetyl spermidine inhibited the oxidation of spermine and served as a substrate, and he suggested that acetylation might have a special role in the degradative pathway served by this PAO. An enzyme using acetyl coA to convert putrescine to *N*-acetyl putrescine had been detected in liver and brain nuclei and microsomes (Seiler & Al-Therib, 1974a, b), and this activity was significantly greater in growing cells (Seiler, 1981). Nuclear systems, located in chromatin, were found to acetylate spermidine and spermine as well as putrescine. Both the N^1- and N^8-acetyl spermidines were reaction products; indeed in this system, the N^8 derivative was the major product (Blankenship & Walle, 1978). Two distinct chromatin-associated enzymes also appear to be histone acetylases (Libby, 1978). A cytoplasmic fraction was capable of deacetylating N^8-acetyl spermidine more actively than N^1-acetyl spermidine (Blankenship, 1978).

Diacetyl spermidine and diacetyl spermine have recently been found in normal human urine (Hiramatsu et al., 1995). Among the various substrates for the PAO of rat liver, N^1-acetyl spermidine, N^1-acetyl spermine, and N^1,N^{12}-diacetyl spermine were found to be better than the free polyamines. A specific assay was devised for the oxidase based on the oxidation of N^1,N^{12}-diacetyl spermine to N^1-acetyl spermidine. An assay based on the conversion of N^1-[^{14}C]acetyl spermine to an easily isolated 3-[^{14}C]acetamidopropanal has also been reported (Kumazawa et al., 1990a). Extracts of various rat tissues demonstrated high activities in most tissues. These were not inducible and appear to function mainly in the diminution of high levels of the polyamines in the tissues.

The unusual range of substituted polyamines attacked by the enzyme has led to the discovery of new interesting substrates. For example, *N*-alkyl α,ω-diaminoalkanes (e.g., *N*-benzyl-1,4-diaminobutane) are oxidized, in this instance to benzaldehyde and putrescine (Bolkenius & Seiler, 1989):

Studies of the bis (benzyl) derivatives of spermidine, spermine, and other polyamines have led to compounds active in inhibiting malaria in a mouse model (Bitonti et al., 1990; Edwards et al., 1991).

Nonacetylated Polyamines and Plant Flavin-Containing Oxidases

The amine oxidases of plants were observed to comprise copper-containing and flavin-containing enzymes in distinct plant groups, such as the legumes (Leguminosae) for the former and the cereals (Gramineae) for the latter. This distribution had been observed prior to the more recent analyses of free and conjugated amines. A retrospective examination of the literature suggested the relative absence of N-acetylated polyamines from plants, although these have been discovered recently (Chap. 20). The numerous early reports of pyrrolines and piperidines and their derivatives in plants might have suggested the availability of a spontaneously reactive amine with newly formed aldehyde in contrast to the relative absence of such compounds in animals. The detection and significance of compounds such as the pyrrolidinones in animal tissue were eventually pursued in connection with the study of the origins of putreanine, isoputreanine, and γ-aminobutyric acid (GABA) and other metabolites (see below).

Two types of amine oxidases active on nonacetylated amines have been described in plant tissues: the very active pea seedling enzyme is DAO, which oxidizes primary amino groups and contains essential copper. This enzyme is present at low activity or lacking in the Graminae (e.g., barley). Whereas the DAO of many legumes appears to be associated largely with cell wall structures during stages of elongation (i.e., stiffening and lignification), some members of the Leguminosae (e.g., tubers of *Helianthus tuberosus*) do contain the enzyme in mitochondria, a cytoplasmic compartment. In such plants the enzyme is presumed to regulate intracellular polyamine levels (Scoccianti et al., 1991). The second type is an amine oxidase, which oxidizes spermine and spermidine at the secondary amino group. This was found initially in barley and then purified. In addition to 1,3-diaminopropane, this gave rise to aminopropylpyrroline and pyrroline, respectively. Unlike the pea seedling enzyme, the barley enzyme, a PAO free of copper, is inactive on putrescine (Smith, 1974; Smith & Bickley, 1974).

Significant amounts of 1-(3-aminopropyl)pyrrolinium cation are found in leaves of oat, maize, barley, and wheat seedlings (Smith et al., 1986). This is believed to arise in the oxidation of spermine by PAO and yields

1,3-diaminopropane as well. A similar PAO from the leaves of oat seedlings was also purified to homogeneity (Smith, 1983). Maize also contains an enzyme similar to the barley enzyme; the maize enzyme required FAD and this electron carrier can be coupled to other carriers, like cytochrome c, via p-benzoquinone (Hirasawa & Suzuki, 1975). The maize enzyme was purified to homogeneity and had a molecular weight of about 65,000 and one molecule of FAD per mole of enzyme (Suzuki & Yanagisawa, 1980). Unlike the animal enzyme, this enzyme was not inhibited by sulfhydryl or carbonyl reagents; like the animal enzyme, quinacrine was inhibitory. The oxidase can be solubilized with a very high specific activity by washing with 0.5 M KH_2PO_4. An additional column purification (five-fold) results in a homogeneous enzyme preparation (Federico et al., 1989). The enzyme is solubilized and lost in protoplast preparations; histochemical methods have placed 90% of the activity on seedling cell walls (Augeri et al., 1990; Slocum & Furey, 1991). It is of interest that N^1-acetyl spermine, a good substrate for the animal enzyme, is a noncompetitive inhibitor of the maize and oat enzymes (Federico et al., 1990). A similar FAD-containing PAO of a non-Graminae plant (i.e., water hyacinth) has also been isolated from the insoluble structure of the ground plant (Yanagisawa et al., 1987). In contrast to these polymer-bound FAD PAOs, the barley enzyme appears to be soluble within the cells (Li & McClure, 1989).

Pyrroline Oxidation Products

The oxidation of putrescine by pig kidney DAO is known to produce pyrroline, and it was shown that this product can be oxidized to 2-pyrrolidinone (Lundgren et al., 1980). Rat liver can also convert the pyrrolidinone to 5-hydroxy-2-pyrrolidinone. The overall sequence is effected in rat liver (Lundgren & Fales, 1980; Lundgren & Hankins, 1978), and mouse brain can convert 2-pyrrolidinone (i.e., the lactam of GABA) to the γ-amino acid (Callery et al., 1979), as in the proposed sequence (Fig. 6.5). This route to GABA differs from the major route involving the decarboxylation of glutamic acid.

Mouse brain can convert isotopically labeled putrescine and monoacetyl putrescine to GABA (Seiler & Al-Therib, 1974b). Brain mitochondria oxidize N-acetyl putrescine to N-acetyl γ-aminobutyrate and this reaction is inhibited by pargyline, a known inhibitor of MAO. It is supposed then that the conversion in the brain of putrescine to GABA, which is significantly but not completely inhibited by pargyline, is largely a route in which MAO plays a major role.

Whereas many plants convert glutamate to GABA (Narayan & Nair, 1990), some plants have devised yet another pathway for the synthesis of GABA via conjugated polyamines (Chap. 20). Also, an enzyme from wheat leaves can effect an oxidative decarboxylation of ornithine to 4-aminobutanamide and a hydrolysis to

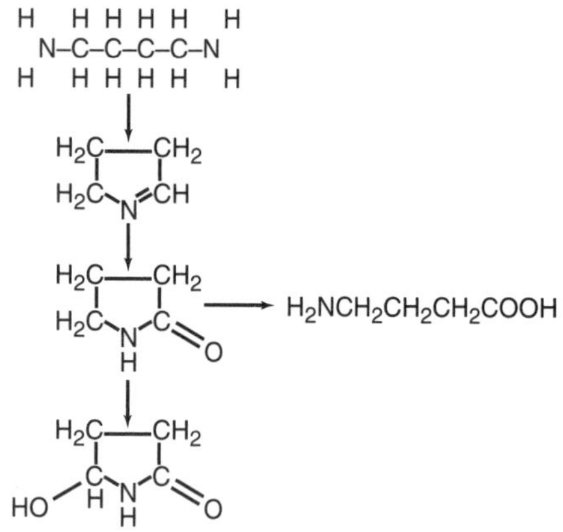

FIG 6.5 Conversion of putrescine in rodent tissues to 2-pyrrolidinone and then to 5-hydroxy-2-pyrrolidinone or γ-aminobutyric acid; redrawn from Lundgren et al. (1980).

GABA (Smith & Marshall, 1988). Reactions such as these, which can convert the carboxyl to CO_2 without forming putrescine, are a source of possible concern in the study of the decarboxylation of ornithine in crude extracts.

Several oxidized metabolites of spermidine (i.e., putreanine and isoputreanine) are found in tissues and in human urine (Fig. 6.6). Putreanine occurs uniquely in mammalian brain (Kakimoto et al., 1969) and is formed in the oxidation of spermidine in tissues to the aminoaldehyde, which is oxidized in turn by an aldehyde dehydrogenase. A newly found metabolite, 2-oxo-1-pyr-

rolidinepropionic acid, has been found in rat urine by Kawase et al. (1994), who postulated pathways for the various detected metabolites. An initial oxidation is sensitive to aminoguanidine and aminoacetonitrile and is therefore suggested to be a function of a copper-dependent amine oxidase (Seiler et al., 1983). The excretion of isoputreanine is decreased, as an effect of aminoguanidine and a resulting increased excretion of spermidine and N^1-acetylspermidine.

Seiler et al. (1982) also showed that N^1-acetylspermidine is the precursor of N-(3-acetamidopropyl) pyrrolidin-2-one, via reactions catalyzed by a flavin-containing PAO and an aldehyde dehydrogenase. Free isoputreanine is formed by hydrolysis. The detection in urine of the acetamido-derivative was described (Van den Berg et al., 1986).

Thus pyrrolines and their derivatives may be the product of any of the amine oxidases, although these appear to be formed as spontaneous cyclizations. These reactions and possible intermediates have begun to be scrutinized more closely. As indicated earlier, Brandänge et al. (1984) proposed that a cyclized aminoaldehyde derivable from the oxidation of spermine formed an N-protonated pyrrolinium structure, which existed in equilibrium with a bicyclic nonane (Zoltewicz & Bloom, 1990).

Inhibitors and Roles of Amine Oxidases

Knowledge of the isolated enzymes and their inhibitors provided an experimental approach to the in vivo control and roles of the amines and enzymes in intact or-

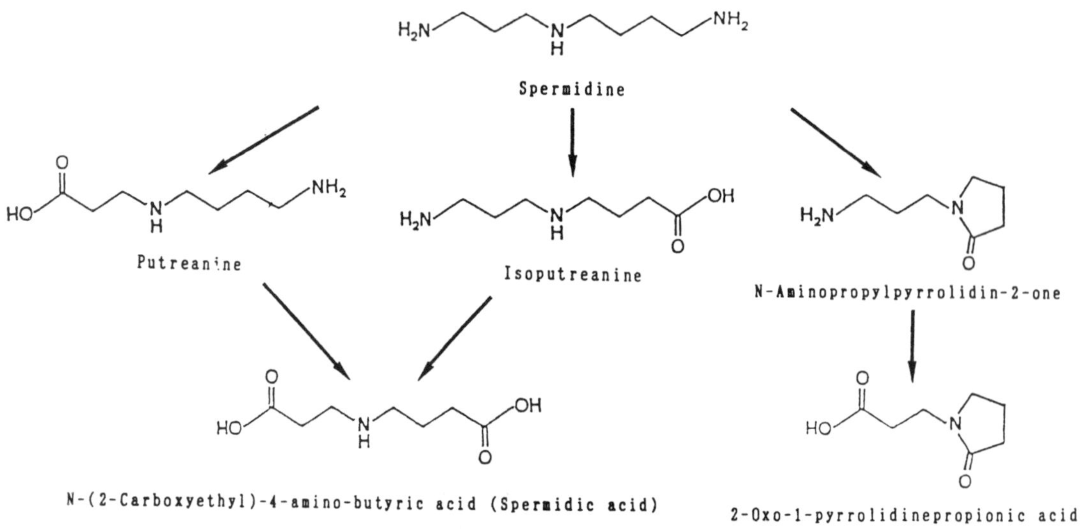

FIG 6.6 Postulated pathways for the oxidation of spermidine; redrawn from Kawase et al. (1994).

ganisms. A simplified classification of major inhibitors and their effects is given in Table 6.2.

Many other inhibitors are known for the three major groups of amine oxidases (Seiler, 1987). The kinetic behavior of the enzymes and their interactions with inhibitors have been discussed by Bardsley (1985). Seiler et al. (1995) recently characterized the amine oxidases of human macrophages by their substrate and inhibitor patterns. The presence of FAD PAO on the cell walls of oat and maize has implicated the oxidation product, H_2O_2, in the peroxidatic fashioning of cell wall modification and cross-linking. In light-grown chick pea seedling, the activities of the copper-containing DAO and peroxidase change in parallel in the various phenomena of growth and wound response (Angelini et al., 1990). These authors concluded that the pattern evoked in development of the Graminae closely resembles that of the Leguminosae. Thus, the two groups produce quite different enzymes, which nevertheless synthesize a common much needed metabolite, H_2O_2, via oxidation of the amines.

The amine oxidases of animals have received more attention and the relatively broad type of generalization noted above concerning the amines, oxidation, H_2O_2, and peroxidase in plant development has not emerged in the animal. There are at least several copper amine oxidases but only one of these, the peptidyl-lysine oxidase, has been of obvious essential function in the development of connective tissue and skin and in the life of the organism. The effects of iproniazid in tuberculosis patients revealed a sympathetic stimulation thought to relate to an inhibition of the degradation of adrenaline and noradrenaline by MAO. Stronger and more specific inhibitors, the *cis*- and *trans*-2-phenylcyclopropylamines, were then discovered (Zeller & Sarkar, 1962):

iproniazid

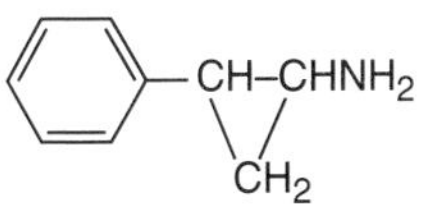

2-phenylcyclopropylamine

Iproniazid is a competitive inhibitor under many conditions whereas the phenylcyclopropylamine is an irreversible inhibitor when used with a presumably natural substrate, tyramine. It was found that the phenylcyclopropylamine was oxidized to the imine, which reacted irreversibly at an essential site of the MAO (Paech et al., 1980). More recently *trans*-2-phenylcyclopropylamine has been found to be an inhibitor and substrate of lysyl oxidase (Shah et al., 1993).

Pargyline (*N*-methyl-*N*-2-propynylbenzylamine),

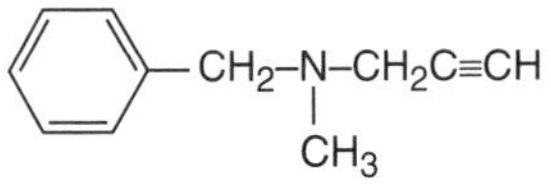

also combines covalently to the active site of the enzyme (Ho, 1972). The powerful and useful α-allenic amines, suicide inhibitors (i.e., 2,3-butadienylamines), have been synthesized and studied as probes of the active site (Smith et al., 1988). Despite a very large assort-

TABLE 6.2 Inhibitors in study of amine oxidase function.

Amine oxidase	Inhibitor	Effect	Suggested role of enzyme
I. Cu amine oxidase	Aminoguanidine	Increase of putrescine and spermidine in tissues and urine (other effects)	Normal terminal oxidation of these polyamines
II. Classical monoamine oxidase	Pargyline	Inhibits conversion of putrescine and *N*-acetyl putrescine to GABA in brain	Initial oxidation of *N*-acetyl putrescine
III. Flavin polyamine oxidase	bis-Butadienyl putrescine	Increases excretion of *N*-acetyl isoputreanine lactam	Participates in degradation of spermidine
		Increases excretion of N^1, N^{12}-diacetyl spermine	Participates in degradation of spermine and in removal of products of dead cells

ment of available MAO inhibitors, the role of an MAO in polyamine metabolism has largely been explored with the aid of pargyline.

Mono- and bis-N-2,3-butadienyl derivatives of putrescine have been synthesized and shown to be irreversible inactivators of the flavin PAO.

Mono-N-2,3-butadienyl putrescine

Bis-N-2,3-butadienyl putrescine

The bis derivative is highly specific in vitro, noninhibitory for the classical MAO or the DAO, and highly potent against mammalian tissue PAO in vitro (Bey et al., 1985).

The inhibitory activity of guanidine derivatives against the DAOs (histaminases), including one of the most potent of these, aminoguanidine, was discovered quite early (Schuler, 1952). This inhibitor was the most active on the DAO of the human placental enzyme (Paolucci et al., 1971). Aminoguanidine is a time-dependent inhibitor of several mammalian DAOs and the activity can be slowly restored by incubation with pyruvate, suggesting the slow trapping of the pyruvoyl Schiff base of the inhibitor (Crabbe et al., 1975). Many aromatic diamidines are reversible inhibitors of the pig kidney DAO (Cubrìa et al., 1993).

Aminoguanidine is active in vivo in reducing the activity of the DAO in rabbit intestine and decreased the survival of animals with occluded mesenteric arteries (Kusche et al., 1973). Putrescine and spermidine increase in the small intestine of treated rodents. It was concluded that the high level of the enzyme in intestinal tissues participates in the elimination of toxic amines in the normal gut, which accumulate in the arterially blocked and enzyme-inhibited intestine. Pregnant rats injected with aminoguanidine excreted 5–10 times more putrescine (and far more cadaverine) than the nontreated controls (Rojansky et al., 1979). A similar result was obtained in the excretion of putrescine, cadaverine, and spermidine in normal aminoguanidine-treated men (Chayen et al., 1985). Thus, the copper amine oxidases (i.e., inhibitable by aminoguanidine) appear to be involved in the normal degradation of putrescine. Much of this diamine occurs in the enzyme-rich small intestine, where biosynthesis and nutritional intake are in considerable excess of minimally essential levels. Seiler (1987) commented on the surprising apparent lack of toxicity of aminoguanidine and many other inhibitors of steps of polyamine metabolism.

Aminoguanidine also markedly inhibited the excretion of the isoputreanine lactam, a result interpreted to signify a role for a Cu DAO in the conversion of spermidine to this excretion product (Seiler, 1987). On the other hand, the treatment of rats with the inhibitor did not increase the excretion of acetylisoputreanine lactam, a normal constituent of human and rat urine. This product increased 12-fold after treatment with the specific inhibitor of the flavin PAO, bis-2,3 butadienylputrescine (Hessels et al., 1990). Thus the flavin-containing system is very much involved in the elimination of spermidine via N^1-acetyl spermidine, which is shunted into another metabolic path when the flavin enzyme is inhibited. This result has been interpreted to suggest that N^1-acetyl spermidine can be oxidized at the aminobutyl moiety by a copper enzyme, or that the flavin enzyme is responsible for the conversion of the lactam to the aminopropyl pyrrolidinone. The inhibitor of the flavin enzyme also causes an increase in the excretion of N^1,N^{12}-diacetyl spermine, signifying the normal metabolism of this product via the flavin enzyme. Under conditions causing gross liver damage, such as treatment with CCl$_4$, the excretion of diacetyl spermine is markedly increased in animals treated with the bis-butadienyl putrescine. This suggests that the diacetyl spermine may be a sensitive marker for cell death.

The conversion of putrescine and acetyl putrescine to GABA in sheep and mouse brain is inhibited by pargyline, the inhibitor of classical MAO (Kremzner, 1973; Seiler & Al-Therib, 1974b). This reaction can be effected by brain mitochondria, which are inhibited by pargyline but not by aminoguanidine. These results appear to indicate that the initial oxidation of N-acetyl putrescine to N-acetylaminо butyraldehyde is effected by the classical mitochondrial MAO of the brain. In rat ovary stimulated by gonadotrophin, an increase of GABA was attributed to a DAO inhibited by aminoguanidine and not by pargyline.

The active role of the intestinal Cu DAO in controlling the level of toxic amines, including the concentration of putrescine, has been noted. The enzyme appears to be a good marker of mucosal integrity and is thought to be an indicator of inflammation in acute appendicitis and of risk in appendectomies (Menningen et al., 1986). The enzyme is synthesized in enterocytes, from which it may be released into plasma by injection of heparin or glycosaminoglycans. Both oxidase and ornithine decarboxylase are found within the mature cells of intestinal mucosa (Baylin et al., 1978). The development of colon adenocarcinomas induced in rats by 1,2-dimethylhydrazine is significantly inhibited by treatment with the inhibitor of the flavin PAO (Halline et al., 1990). These data imply the existence of several oxidases in intestinal tissue participating in polyamine metabolism in the tissue.

An increase of PAO in the serum of pregnant mammals occurs in the last trimester of pregnancy, concomitantly with a great increase in the mass of the fetus and placental tissue (Morgan, 1989). The latter contains a great deal of DAO, similar in many properties to the serum enzyme. In undisturbed pregnancy, the excretion of amines increases some eight- to 17-fold in the third trimester. Aminoguanidine does not substantially change the pattern of increase but markedly increases the total amount of the amines excreted. This implies a role for a copper enzyme in the metabolism and excretion of putrescine in normal and pregnant animals.

In many instances, experiments with inhibitors thought to be quite specific provide "peculiar" results, which may eventually prove to arise from inhibitions at sites different from that originally supposed. For example, aminoguanidine has activities other than that on DAO. It will inhibit protein reactions in cross-linking by reducing sugars (Hayase et al., 1991) and inhibits glycosylation in hyperglycemic sera by binding reactive aldehydes (Picard et al., 1992). The compound is an irreversible inhibitor of catalase (Ou & Wolff, 1993) and is a "selective" inhibitor of the inducible nitric oxide synthase of rat aorta (Joly et al., 1994). Thus results obtained with aminoguanidine used to inhibit copper amine oxidases must be viewed with some caution.

Hydrogen Peroxide as Toxic Product of Amine Oxidases

Many mechanisms have evolved to minimize the toxicity of oxygen and its by-products. The products of the detoxification reactions may have important adaptive or structural roles in defining the oxygen-utilizing organism. Among these may be noted the diverse reactions of oxygen with metabolites which lead to bioluminescence, the biosynthesis of sterols, or the synthesis of tryptophan in eucaryotic organisms. The enzymes of amine oxidation may possess different structural features, such as a cupric site or a FAD, but one end product of the reaction, H_2O_2, is common to both enzyme types.

In addition to H_2O_2 there can be many species of toxic oxygen in the forms of the superoxide radical, O_2^-, singlet oxygen, the hydroxyl radical, $\cdot OH$, some of which may react with metallic ions to produce other oxidizing radicals. Pauling predicted the existence of superoxide anion radical in 1932 and the production and possible toxicity of this radical has been studied extensively (Halliwell, 1984; Halliwell & Gutteridge, 1986). It is thought to be formed as an early step in the four electron reduction of O_2 to water. As described by Fridovich (1978, 1989), superoxide is converted to H_2O_2 and O_2 by superoxide dismutases. In its turn H_2O_2 may be decomposed by catalase, which produces O_2 and water, or removed by peroxidases in the presence of cellular reductants. Evidence was presented to indicate that oxidation of cadaverine by the pig kidney amine oxidase at a pH of 9.2 led to the production of superoxide anion radical, whereas H_2O_2 was the major product at pH 7.4 (Rotilio et al., 1970). In studies on the oxidation and toxicity of mixtures of spermine and bovine serum amine oxidase, superoxide dismutase did not affect the formation of the initial products (Gaugas & Dewey, 1980).

H_2O_2 may be a useful metabolite in the development of cross-linked polymers and lignification in many plants. The gradient of DAO in growing tissues of legumes correlates with lignification in these tissues, implicating H_2O_2 and a peroxidatic step in this process. The observed increase of the oxidase in the dark may evoke a protective sterilizing effect of H_2O_2 by roots against invading pathogenic microorganisms (Federico & Angelini, 1988).

H_2O_2 is often removed by peroxidases. In mitochondria the oxidation of benzylamine by MAO gives rise to peroxide and a subsequent oxidation of glutathione (Werner & Cohen, 1991):

$$monoamine + O_2 + H_2O \rightarrow aldehyde + H_2O_2 + NH_3,$$

$$\begin{array}{c}reduced \\ glutathione\end{array} (2GSH) + H_2O_2 \xrightarrow[peroxidase]{glutathione} GSSG + 2H_2O.$$

Organisms deficient in glutathione or in this peroxidase are sensitive to oxygen; in the case of the mammal, the deficiency produced by an inhibitor of glutathione synthesis leads to mitochondrial degeneration. It has been found that the inhibition of animal cell growth by 2 mM spermine in tissue culture is caused by a marked depletion of glutathione, although the mechanism of this effect is still obscure (Brunton et al., 1990). It has not yet been demonstrated that the detoxification of H_2O_2 produced in polyamine oxidation is a function of glutathione and its peroxidase. Although mono- and bis-glutathionyl spermidines are reactive with H_2O_2 and specific peroxidases, their oxidation by amine oxidases has not yet been described.

Heat shock (43°C) in CHO cells in serum appears to evoke cytotoxicity by both an increased action of the serum Cu amine oxidase and an increased endogenous acetylation of polyamine and oxidation by PAO. Aminoguanidine is protective against the former and the bis-*N*-2,3-butadienyl putrescine is partially active against the latter in cells depleted of glutathione. Thus the inhibitors can modulate the oxidative stress induced by heat shock (Harari et al., 1989).

Under conditions in which H_2O_2 cannot be removed, the cells confronted by newly generated H_2O_2 may die. The study of an early stage of embryonic development (i.e., limb bud development in a mammalian blastocoele) revealed the production of cytotoxic H_2O_2 derived from polyamine oxidation within the limb bud. It has been proposed that polyamine oxidation provides the lethal agent in this system of programmed cell death or apoptosis (Coffino & Poznanski, 1991; Pierce et al., 1991).

Summary

The study of the amine oxidases and their inhibitors has revealed two major paths of polyamine catabolism. In one, a terminal oxidation pathway in which copper enzymes play a major role, the oxidation products of the polyamines are no longer useful metabolites and are excreted. In a second, in which flavin enzymes are most active, primary amines of the polyamines may be acetylated in animals and an N^1-acetyl spermine can be degraded to spermidine, an N^1-acetyl spermidine to putrescine, and an N^1-acetyl putrescine to GABA. This pathway, which emphasized a key role of N-acetylation of the polyamines, has been termed an "interconversion" pathway, which assures a conservation and reutilization of major moieties of the compounds. Finally we note the involvement of other products of the oxidase reactions in the lives of the organisms. Thus, a potentially toxic product may become a useful metabolite, a detoxified reagent, or a useful lethal agent.

REFERENCES

Abdel–Monem, M. M., Ohno, K., Newton, N. E., & Weeks, C. E. (1978) *Advances in Polyamine Research*, vol. 2, pp. 37–49, R. A. Campbell, D. R. Morris, D. Bartos, G. D. Daves, & F. Bartos, Eds. New York: Raven Press.

Alarcon, R. A. (1970) *Archives of Biochemistry and Biophysics*, **137**, 365–372.

Alton, G., Taher, T. H., Beever, R. J., & Palcic, M. M. (1995) *Archives of Biochemistry and Biophysics*, **116**, 353–361.

Ambroziak, W. & Pietruszko, R. (1991) *Journal of Biological Chemistry*, **266**, 13011–13018.

Andersson, A., Henningsson, S., Persson, L., & Rosengren, E. (1978a) *Acta Physiologica Scandinavica*, **102**, 159–166.

Andersson, A., Henningsson, S., & Rosengren, E. (1978b) *Journal of Physiology*, **285**, 311–324.

Angelini, R., Di Lisi, F., & Federico, R. (1985) *Phytochemistry*, **24**, 2511–2513.

Angelini, R., Manes, F., & Federico, R. (1990) *Planta*, **182**, 89–96.

Atmar, V. J., Kuehn, G. D., & Casillas, E. R. (1981) *Journal of Biological Chemistry*, **256**, 8275–8278.

Augeri, M. I., Angelini, R., & Federico, R. (1990) *Journal of Plant Physiology*, **136**, 690–695.

Averill–Bates, D. A., Agostinelli, E., Przybytkowski, E., Mateescu, M. A., & Mondovì, B. (1993) *Archives of Biochemistry and Biophysics*, **300**, 75–79.

Averill–Bates, D. A., Agostinelli, E., Przybytkowski, E., & Mondovì, B. (1994) *Biochemistry and Cell Biology*, **72**, 36–42.

Axelrod, J. (1991) *Current Contents, Citation Classics*, **34**, 10.

Baenziger, N. L., Mack, P., Jong, Y. I., Dalemar, L. R., Pérez, N., Lindberg, C., Wilhelm, B., & Haddock, R. C. (1994) *Journal of Biological Chemistry*, **269**, 14892–14898.

Bardsley, W. G. (1985) *Structure and Function of Amine Oxidases*, pp. 136–152, B. Mondovi, Ed. Boca Raton, Fla: CRC Press.

Bardsley, W. G., Crabbe, M. J. C., Shindler, J. S., & Ashford, J. S. (1972) *Biochemical Journal*, **127**, 875–879.

Baylin, S. B. & Margolis, S. (1975) *Biochimica et Biophysica Acta*, **397**, 294–306.

Baylin, S. B., Stevens, S. A., & Shakir, K. M. M. (1978) *Biochimica et Biophysica Acta*, **541**, 415–419.

Bellelli, A., Agrò, A. F., Floris, G., & Brunori, M. (1991) *Journal of Biological Chemistry*, **266**, 20654–20657.

Bey, P., Bolkenius, F. N., Seiler, N., & Casara, P. (1985) *Journal of Medicinal Chemistry*, **28**, 1–2.

Bitonti, A. J., Dumont, J. A., Bush, T. L., Stemerick, D. M., Edwards, M. L., & McCann, P. P. (1990) *Journal of Biological Chemistry*, **265**, 382–388.

Blankenship, J. (1978) *Archives of Biochemistry and Biophysics*, **189**, 20–27.

Blankenship, J. & Walle, T. (1978) *Advances in Polyamine Research*, vol. 2, pp. 97–110, R. A. Campbell, D. R. Morris, D. Bartos, G. D. Daves, & F. Bartos, Eds. New York: Raven Press.

Bolkenius, F. H. & Seiler, N. (1989) *Biological Chemistry Hoppe–Seyler*, **370**, 525–531.

Bossa, M., Morpurgo, G. O., & Morpurgo, L. (1994) *Biochemistry*, **33**, 4425–4431.

Boyd, C. D., Mariani, T. J., Kim, Y., & Csiszar, C. (1995) *Molecular Biology Reports*, **21**, 95–103.

Brandänge, S., Eriksson, L., & Rodríguez, B. (1984) *Acta Chemica Scandinavica B*, **38**, 526–528.

Brown, D. E., McGuirl, M. A., Dooley, D. M., Janes, S. M., Mu, D., & Klinman, J. P. (1991) *Journal of Biological Chemistry*, **266**, 4049–4051.

Brunton, V. G., Grant, M. H., & Wallace, H. M. (1990) *Biochemical Pharmacology*, **40**, 1893–1900.

Buffoni, F. & Cambi, S. (1990) *Analytical Biochemistry*, **187**, 44–50.

Cai, D. & Klinman, J. P. (1994) *Biochemistry*, **33**, 7647–7653.

Callery, P. S., Stogniew, M., & Geelhaar, L. A. (1979) *Biochemical Mass Spectrometry*, **6**, 23–26.

Callingham, B. A., Holt, A., & Elliott, J. (1991) *Biochemical Society Transactions*, **19**, 228–233.

Chassande, O., Renard, S., Barbry, P., & Lazdunski, M. (1994) *Journal of Biological Chemistry*, **269**, 14484–14489.

Chayen, R., Burke, M., & Goldberg, S. (1985) *Israel Journal of Medical Science*, **21**, 543–545.

Chen, L., Mathews, F. S., Davidson, V. L., Huizinga, E. G., Vellieux, F. M. D., Duine, J. A., & Hol, W. G. J. (1991) *FEBS Letters*, **287**, 163–166.

Chung, F., Young, R., & Hecht, S. S. (1984) *Cancer Research*, **44**, 990–995.

Coffino, P. & Poznanski, A. (1991) *Journal of Cellular Biochemistry*, **45**, 54–58.

Cogoni, A., Padiglia, A., Medda, R., Segni, P., & Floris, G. (1991) *Plant Physiology*, **95**, 477–479.

Coleman, A. A., Scaman, C. H., Kang, Y. J., & Palcic, M. M. (1991) *Journal of Biological Chemistry*, **266**, 6795–6800.

Cooper, R. A., Knowles, P. F., Brown, D. E., McGuirl, M. A., & Dooley D. M. (1992) *Biochemical Journal*, **288**, 337–340.

Cox, R., Goorha, S., & Irving, C. C. (1988) *Carcinogenesis*, **9**, 463–465.

Crabbe, M. J. C., Childs, R. E., & Bardsley, W. G. (1975) *European Journal of Biochemistry*, **60**, 325–333.

Crabbe, M. J. C. & Kavanagh, J. P. (1977) *Biochemical Society Transactions*, **5**, 735–737.

Crabbe, M. J. C., Waight, R. D., Bardsley, W. G., Barker, R. W., Kelly, S. D., & Knowles, P. R. (1976) *Biochemical Journal*, **155**, 679–687.

Cubrìa, J. C., Fouce, R. B., Alvarez–Bujidos, M. L., Negro, A., Ortiz, A. I., & Ordóñez, D. (1993) *Biochemical Pharmacology*, **45**, 1355–1357.

Davies, H. M., Hawkins, D. J., & Smith, L. A. (1989) *Phytochemistry*, **28**, 1573–1578.

Den Munckhof, R. J. M., Denyn, M., Tigchelaar-Gutter, W., Schipper, R. G., Verhofstad, A. A. J., Van Noorden, C. J. F., & Frederiks, W. M. (1995) *Journal of Histochemistry and Cytochemistry*, **43**, 1155–1162.

Dooley, D. M. & Coté, C. E. (1984) *Journal of Biological Chemistry*, **259**, 2923–2926.

Dooley, D. M., McGuirl, M. A., Brown, D. E., Turowski, P.

N., McIntire, W. S., & Knowles, P. F. (1991) *Nature*, **349**, 262–264.

Duine, J. A. (1989) *Antonie van Leeuwenhoek*, **56**, 3–12.

Durham, D. R. & Perry, J. J. (1978a) *Journal of Bacteriology*, **134**, 837–843.

————— (1978b) *Journal of Bacteriology*, **135**, 981–986.

Edwards, M. L., Stemerick, D. M., Bitonti, A. J., Dumont, J. A., McCann, P. P., Bey, P., & Sjoerdsma, A. (1991) *Journal of Medicinal Chemistry*, **34**, 569–574.

Fair, W. R. & Wehner, N. (1971) *Applied Microbiology*, **21**, 6–8.

Federico, R., Alisi, C., & Forlani, F. (1989) *Plytochemistry*, **28**, 45–46.

Federico, R. & Angelini, R. (1988) *Planta*, **173**, 317–321.

Federico, R., Angelini, R., Cesta, A., & Pini, C. (1985) *Plant Physiology*, **79**, 62–64.

Federico, R., Cona, A., Angelini, R., Schininà, M. E., & Giartosio, A. (1990) *Phytochemistry*, **29**, 2411–2414.

Fernandez, C., Sharrard, R. M., Monks, N., & Barrett, C. L. R. (1994) *Clinica Chimica Acta*, **227**, 201–208.

Finazzi Agrò, A. & Rossi, A. (1992) *Biochemical Society Transactions*, **20**, 369–373.

Flescher, E., Fossum, D., & Talal, N. (1991) *Immunology Letters*, **28**, 85–90.

Fridovich, I. (1978) *Science*, **201**, 875–880.

————— (1989) *Journal of Biological Chemistry*, **264**, 7761–7764.

Fujihara, S., Nakashima, T., & Kurogochi, Y. (1984) *Biochimica et Biophysica Acta*, **805**, 277–284.

Gacheru, S., McGee, C., Uriu-Hare, J. Y., Kosonen, T., Packman, S., Tinker, D., Krawetz, S. A., Reiser, K., Keen, C. L., & Rucker, R. B. (1993) *Archives of Biochemistry and Biophysics*, 325–329.

Gacheru, S. N., Trackman, P. C., Calaman, S. D., Greenaway, F. T., & Kagan, H. M. (1989) *Journal of Biological Chemistry*, **264**, 12963–12969.

Gacheru, S. N., Trackman, P. C., Shah, M. A., O'Gara, C. Y., Spacciapoli, P., Greenaway, F. T., & Kagan, H. M. (1990) *Journal of Biological Chemistry*, **265**, 19022–19027.

Gahl, W. A. & Pitot, H. C. (1982) *Biochemical Journal*, **202**, 603–611.

Gasparyan, V. K. & Nalbandyan, R. M. (1990) *Biokhimiya*, **55**, 1632–1637.

Gaugas, J. M. & Dewey, D. L. (1980) *British Journal of Cancer*, **41**, 946–955.

Govaerts, L. C. P., van den Berg, G. A., Theeuwes, A., Muskiet, F. A. J., & Monnens, L. A. H. (1990) *Clinica Chimica Acta*, **192**, 61–68.

Greenawalt, J. W. & Schnaitman, C. (1970) *Journal of Cell Biology*, **46**, 173–179.

Greenaway, F. T., O'Gara, C. Y., Marchena, J. M., Poku, J. W., Urtiaga, J. G., & Zou, Y. (1991) *Archives of Biochemistry and Biophysics*, **285**, 291–296.

Hackett, A. G., Kotyk, J. J., Fujiwara, H., & Logusch, E. W. (1990) *Journal of the American Chemical Society*, **112**, 3669–3671.

Halline, A. G., Dudeja, P. K., Jacoby, R. F., Llor, X., Teng, B., Chowdhury, L. N., Davidson, N. O., & Brasitus, T. A. (1990) *Carcinogenesis*, **11**, 2127–2132.

Halliwell, B. (1984) *Copper Proteins and Copper Enzymes*, vol. II, pp. 63–102, R. Lontie, Ed. Boca Raton, Fla.: CRC Press.

Halliwell, B. & Gutteridge, J. M. C. (1986) *Archives of Biochemistry and Biophysics*, **246**, 501–514.

Harari, P. M., Fuller, D. J. M., & Gerner, E. W. (1989) *Interna-

tional Journal of Radiation Oncology, Biology, Physics*, **16**, 451–457.

Hayase, F., Kim, Y. H., & Kato, H. (1991) *Agricultural and Biological Chemistry*, **55**, 1435–1436.

Hessels, J., Ferwerda, H., Kingma, A. W., & Muskiet, F. A. J. (1990) *Biochemical Journal*, **266**, 843–851.

Hiramatsu, K., Sugimoto, M., Kamei, S., Hoshino, M., Kinoshita, K., Iwasaki, K., & Kawakita, M. (1995) *Journal of Biochemistry*, **117**, 107–112.

Hirasawa, E. & Suzuki, Y. (1975) *Phytochemistry*, **14**, 99–101.

Ho, B. T. (1972) *Journal of Pharmaceutical Sciences*, **61**, 821–837.

Höltttä, E. (1977) *Biochemistry*, **16**, 91–100.

————— (1983) *Methods in Enzymology*, **94**, 306–311.

Ishizuka, H., Horinouchi, S., & Beppu, T. (1993) *Journal of General Microbiology*, **139**, 425–432.

Isobe, K., Tani, Y., & Yamada, H. (1980a) *Agricultural and Biological Chemistry*, **44**, 4651–2658.

————— (1980b) *Agricultural and Biological Chemistry*, **44**, 2749–2751.

————— (1981) *Agricultural and Biological Chemistry*, **45**, 727–733.

————— (1982) *Agricultural and Biological Chemistry*, **46**, 1353–1359.

Israel, M. & Modest, E. J. (1971) *Journal of Medicinal Chemistry*, **14**, 1042–1047.

Israel, M., Zoll, E. C., Muhammad, N. M., & Modest, E. J. (1973) *Journal of Medicinal Chemistry*, **16**, 1–5.

Janes, S. M. & Klinman, J. P. (1991) *Biochemistry*, **30**, 4599–4605.

————— (1995) *Methods in Enzymology*, **258**, 20–34.

Janes, S. M., Mu, D., Wemmer, D., Smith, A. J., Kaur, S., Maltby, D., Burlingame, A. L., & Klinman, J. P. (1990) *Science*, **248**, 981–987.

Joly, G. A., Ayres, M., Chelly, F., & Kilbourn, R. G. (1994) *Biochemical and Biophysical Research Communications*, **199**, 147–154.

Kaur–Sawhney, R., Flores, H. E., & Galston, A. W. (1981) *Plant Physiology*, **68**, 494–498.

Kawase, H., Hasegawa, T., Hashimoto, H., Noto, T., & Nakajima, T. (1994) *Journal of Biochemistry*, **115**, 356–361.

Keniston, R. C., Cabellon, Jr., S., & Yarbrough, K. S. (1987) *Toxicology and Applied Pharmacology*, **88**, 433–441.

Kimes, B. W. & Morris, D. R. (1971a) *Biochimica et Biophysica Acta*, **228**, 223–234.

————— (1971b) *Biochimica et Biophysica Acta*, **228**, 235–244.

Klinman, J. P. (1996) *Journal of Biological Chemistry*, **271**, 27189–27192.

Klinman, J. P. & Mu, D. (1994) *Annual Review of Biochemistry*, **63**, 299–344.

Kluetz, M. D., Adamsons, K., & Flynn, J. E. (1980) *Biochemistry*, **19**, 1617–1620.

Kluetz, M. D. & Schmidt, P. G. (1977a) *Biochemical and Biophysical Research Communications*, **76**, 40–45.

————— (1977b) *Biochemistry*, **16**, 5191–5199.

Knowles, P. F. & Yadav, K. D. S. (1984) *Copper Proteins and Copper Enzymes*, vol. II, pp. 103–129, R. Lontie, Ed. Boca Raton, Fla.: CRC Press.

Kobayashi, H., Ishii, M., Chanoki, M., Yashiro, N., Fushida, H., Fukai, K., Kono, T., Hamada, T., Wakasaki, H., & Ooshima, A. (1994) *British Journal of Dermatology*, **131**, 325–330.

Kobayashi, Y., Higashi, T., Machida, H., Iawsaki, S., & Korikoshi, K. (1983) *Biochimica et Biophysica Acta*, **743**, 431–436.

Krebs, C. J. & Krawetz, S. A. (1993) *Biochimica et Biophysica Acta*, **1202**, 7–12.

Kremzner, L. (1973) *Polyamines in Normal and Neoplastic Growth*, pp. 27–40, D. H. Russell, Ed. New York: Raven Press.

Kumagai, H., Uchida, H., & Yamada, H. (1979) *Journal of Biological Chemistry*, **254**, 10913–10919.

Kumagai, H. & Yamada, H. (1985) *Structure and Functions of Amine Oxidases*, pp. 37–49, B. Mondovi, Ed. Boca Raton, Fla.: CRC Press.

Kumazawa, T., Seno, H., & Suzuki, O. (1990a) *Analytical Biochemistry*, **188**, 105–108.

Kumazawa, T., Seno, H., Urakami, T., & Suzuki, O. (1990b) *Archives of Biochemistry and Biophysics*, **283**, 533–536.

Kusche, J., Richter, H., Schmidt, J., Hesterberg, R., Specht, C., & Lorenz, W. (1973) *Agents and Actions*, **3**, 182–183.

Lamkin, M. S., Williams, T. J., & Falk, M. C. (1988) *Archives of Biochemistry and Biophysics*, **261**, 72–79.

Li, Z. (1993) *Phytochemistry*, **34**, 611–612.

Li, Z. & McClure, J. W. (1989) *Phytochemistry*, **28**, 2255–2259.

Libby, P. R. (1978) *Journal of Biological Chemistry*, **253**, 233–237.

Löser, C., Wunderlich, U., & Fölsch, U. R. (1988) *Journal of Chromatography*, **430**, 249–262.

Lundgren, D. W. & Fales, H. M. (1980) *Journal of Biological Chemistry*, **255**, 4481–4486.

Lundgren, D. W. & Hankins, J. (1978) *Journal of Biological Chemistry*, **253**, 7130–7133.

Lundgren, D. W., Lloyd, H. A., & Hankins, J. (1980) *Biochemical and Biophysical Research Communications*, **97**, 667–672.

Maayan, R., Zukerman, Z., & Shohat, B. (1995) *Archives of Andrology*, **34**, 95–99.

Mashige, F., Tanaka, N., Murakami, T., Shimosaka, H., Kamei, S., & Ohkubo, A. (1988) *Clinical Chemistry*, **34**, 2271–2274.

McGuirl, M. A., McCahon, C. D., McKeown, K. A., & Dooley, D. M. (1994) *Plant Physiology*, **106**, 1205–1211.

McIntire, W. S., Wemmer, D. E., Chistoserdov, A., & Lidstrom, M. E. (1991) *Science*, **252**, 817–824.

McRorie, R. A. & Williams, W. L. (1974) *Annual Review of Biochemistry*, **43**, 777–803.

Medda, R., Padiglia, A., & Floris, G. (1995) *Phytochemistry*, **39**, 1–9.

Menningen, R., Günther, M., Bönninghoff, N., Stahlknecht, C. D., Kubitza, I., & Kusche, J. (1986) *Agents and Actions*, **18**, 38–40.

Morel, F., Surla, A., & Vignais, P. V. (1992) *Biochemical and Biophysical Research Communications*, **187**, 178–186.

Morgan, D. M. L. (1989) *Physiology of Polyamines*, pp. 203–229, U. Bachrach & Y. M. Heimer, Eds. Boca Raton, Fla.: CRC Press.

Morita, J., Nishimura, K., Komano, T., Kumagai, H., & Yamada, H. (1978) *Agricultural and Biological Chemistry*, **42**, 1971–1972.

Morpurgo, L., Agostinelli, E., Mondovi, B., Avigliano, L., Silvestri, R., Stefancich, G., & Artico, M. (1992) *Biochemistry*, **31**, 2615–2621.

Mu, D., Janes, S. M., Smith, A. J., Brown, D. W., Dooley, D. M., & Klinman, J. P. (1992) *Journal of Biological Chemistry*, **267**, 7979–7982.

Mu, D., Medzihradszky, K. F., Adams, G. W., Mayer, P., Hines, W. M., Burlingame, A. L., Smith, A. J., Cai, D., & Klinman, J. P. (1994) *Journal of Biological Chemistry*, **269**, 9926–9932.

Mure, M. & Klinman, J. P. (1995) *Methods in Enzymology*, **258**, 39–52.

Naik, B. I., Goswami, R. G., & Srivastava, S. K. (1981) *Analytical Biochemistry*, **111**, 146–148.

Narayan, V. S. & Nair, P. M. (1990) *Phytochemistry*, **29**, 367–375.

Nath, R. G. & Chung, F. (1994) *Proceedings of the National Academy of Sciences of the United States of America*, **91**, 7491–7495.

Neufeld, E. & Chayen, R. (1979) *Analytical Biochemistry*, **96**, 411–418.

Neumann, R., Hevey, R., & Abeles, R. H. (1975) *Journal of Biological Chemistry*, **250**, 6362–6367.

Nishimura, K., Komano, T., & Yamada, H. (1972) *Biochimica et Biophysica Acta*, **262**, 24–31.

Nomura, K., Suzuki, N., & Matsumoto, S. (1990) *Biochemistry*, **29**, 4525–4534.

Novotny, W. F., Chassande, O., Baker, M., Lazdunski, M., & Barbry, P. (1994) *Journal of Biological Chemistry*, **269**, 9921–9925.

Oda, O., Manabe, T., & Okuyama, T. (1981) *Journal of Biochemistry*, **89**, 1317–1323.

Okada, M., Kawashima, S., & Imahori, K. (1979a) *Journal of Biochemistry*, **85**, 1235–1243.

—— (1979b) *Journal of Biochemistry*, **86**, 97–104.

Ou, P. & Wolff, S. P. (1993) *Biochemical Pharmacology*, **46**, 1139–1144.

Padiglia, A., Medda, R., & Floris, G. (1992) *Biochemistry International*, **28**, 1097–1107.

Padmanabhan, R. & Kim, K. (1965) *Biochemical and Biophysical Research Communications*, **19**, 1–5.

Paech, C., Salach, J. I., & Singer, T. P. (1980) *Journal of Biological Chemistry*, **255**, 2700–2704.

Palcic, M. M. & Janes, S. M. (1995) *Methods in Enzymology*, **258**, 34–38.

Paolucci, F., Cronenberger, L., Plan, R., & Pacheco, H. (1971) *Biochemie*, **53**, 735–749.

Parrish, R. F. & Polakoski, K. L. (1977) *Biology of Reproduction*, **17**, 417–422.

Patel, J. M. (1990) *Pharmcology & Therapeutics*, **47**, 137–146.

Petterson, G. (1985) *Structure and Functions of Amine Oxidases*, pp. 106–120, B. Mondovi, Ed. Boca Raton, Fla.: CRC Press.

Picard, S., Parthasarathy, S., Fruebis, J., & Witztum, J. L. (1992) *Proceedings of the National Academy of Sciences of the United States of America*, **89**, 6876–6880.

Pierce, G. B., Parchment, R. E., & Lewellyn, A. L. (1991) *Differentiation*, **46**, 181–186.

Pulkkinen, P., Puk, K., Koso, P., & Jänne, J. (1978) *Acta Endocrinologica*, **87**, 845–854.

Quash, G. & Taylor, D. R. (1970) *Clinica Chimica Acta*, **30**, 17–23.

Rikaus, L. E. (1987) *Drug Metabolism and Disposition*, **15**, 356–362.

Rinaldi, A., Floris, G., & Giartosio, A. (1985) *Structure and Functions of Amine Oxidases*, pp. 51–62, B. Mondovi, Ed. Boca Raton, Fla.: CRC Press.

Rius, F. X., Knowles, P. F., & Petterson, G. (1984) *Biochemical Journal*, **220**, 767–772.

Roh, J. H., Takenaka, Y., Suzuki, H., Yamamoto, K., & Kumagai, H. (1995) *Biochemical and Biophysical Research Communications*, **212**, 1107–1114.

Rojansky, N., Neufeld, E., & Chayen, R. (1979) *Biochimica et Biophysica Acta*, **586**, 1–9.

Rossi, A., Petruzzelli, R., & Agrò, A. F. (1992) *FEBS Letters*, **301**, 253–257.

Rotilio, G., Calabrese, L., Finazzi–Agrò, A., & Mondovi, B. (1970) *Biochimica et Biophysica Acta*, **198**, 618–620.

Rucker, R. B., Roensch, L. F., Savage, J. E., & O'Dell, B. L. (1970) *Biochemical and Biophysical Research Communications*, **40**, 1391–1397.

Scalet, M., Federico, R., & Angelini, R. (1991) *Journal of Plant Physiology*, **137**, 571–575.

Scoccianti, V., Torrigiani, P., & Bagni, N. (1991) *Journal of Plant Physiology*, **138**, 752–756.

Seiler, N. (1981) *Polyamines in Biology and Medicine*, pp. 127–149, D. R. Morris & L. J. Marton, Eds. New York: Marcel Dekker.

——— (1987) *Inhibition of Polyamine Metabolism*, pp. 49–77, P. P. McCann, A. E. Pegg, & A. Sjoerdsma, Eds. New York: Academic Press.

Seiler, N. & Al-Therib, M. J. (1974a) *Biochimica et Biophysica Acta*, **354**, 206–212.

——— (1974b) *Biochemical Journal*, **144**, 29–35.

Seiler, N., Bolkenius, F. N., Knödgen, B., & Mamont, P. (1980) *Biochimica et Biophysica Acta*, **615**, 480–488.

Seiler, N., Knödgen, B., Bink, G., Sarhan, S., & Bolkenius, F. (1983) *Advances in Polyamine Research*, vol. 4, pp. 135–154, U. Bachrach, A. Kaye, & R. Chayen, Eds. New York: Raven Press.

Seiler, N., Knödgen, B., & Haegele, K. (1982) *Biochemical Journal*, **208**, 189–197.

Seiler, N., Moulinoux, J., Havouis, R., & Toujas, L. (1995) *Biochemistry and Cell Biology*, **73**, 275–281.

Shah, M. A., Trackman, P. C., Gallop, P. M., & Kagan, H. M. (1993) *Journal of Biological Chemistry*, **268**, 11580–11585.

Shimizu, E., Tabata, Y., Hayakawa, R., & Yorifuji, T. (1988) *Agricultural and Biological Chemistry*, **52**, 2865–2871.

Shinagawa, E., Matsushita, K., Adachi, O., & Ameyama, M. (1989) *Agricultural and Biological Chemistry*, **53**, 1823–1828.

Singer, T. P. (1985) *Structure and Functions of Amine Oxidases*, pp. 219–229, B. Mondovi, Ed. Boca Raton, Fla.: CRC Press.

Slocum, R. D. & Furey, III., M. J. (1991) *Planta*, **183**, 443–450.

Smith, C. J., Hussain, J. I., & Allen, J. C. (1985) *Biochemical Society Transactions*, **13**, 326–328.

Smith, R. A., White, R. L., & Krantz, A. (1988) *Journal of Medicinal Chemistry*, **31**, 1558–1566.

Smith, R. A., Williamson, D. S., Cerny, R. L., & Cohen, S. M. (1990) *Cancer Research*, **50**, 3005–3012.

Smith, R. A., Williamson, D. S., & Cohen, S. M. (1989) *Chemical Research in Toxicology*, **2**, 267–271.

Smith, T. A. (1974) *Phytochemistry*, **13**, 1075–1081.

——— (1983) *Methods in Enzymology*, vol. 94, pp. 311–314, H. Tabor & C. W. Tabor, Eds. New York: Academic Press.

——— (1985) *Biochemical Society Transactions*, **13**, 319–322.

Smith, T. A. & Bickley, D. A. (1974) *Phytochemistry*, **13**, 2437–2443.

Smith, T. A., Croker, S. J., & Loeffler, R. S. T. (1986) *Phytochemistry*, **25**, 683–689.

Smith, T. A. & Marshall, J. H. A. (1988) *Phytochemistry*, **27**, 703–710.

Suresh, M. R., Ramakrishna, S., & Adiga, P. R. (1976) *Phytochemistry*, **15**, 483–485.

Suzuki, O. & Matsumoto, T. (1987) *Biogenic Amines*, **4**, 237–245.

Suzuki, Y. & Hagiwara, M. (1993) *Phytochemistry*, **33**, 995–998.

Suzuki, Y. & Yanagisawa, H. (1980) *Plant and Cell Physiology*, **21**, 1085–1094.

Svinarich, D. M., Twomey, T. A., Macauley, S. P., Krebs, C. J., Yang, T. P., & Krawetz, S. A. (1992) *Journal of Biological Chemistry*, **267**, 14382–14387.

Tabor, C. W. (1985) *Structure and Functions of Amine Oxidases*, pp. 45–49, B. Mondovi, Ed. Boca Raton, Fla.: CRC Press.

Tabor, C. W. & Kellogg, P. D. (1970) *Journal of Biological Chemistry*, **245**, 5424–5433.

Tinker, D., Romero–Chapman, N., Reiser, K., Hyde, D., & Rucker, R. (1990) *Archives of Biochemistry and Biophysics*, **278**, 326–332.

Tipping, A. J. & McPherson, M. J. (1995) *Journal of Biological Chemistry*, **270**, 16939–16946.

Trackman, P. C., Pratt, A. M., Wolanski, A., Tang, S., Offner, G. D., Troxler, R. F., & Kagan, H. M. (1990) *Biochemistry*, **29**, 4863–4870.

Tu, C., Wynns, G. C., & Silverman, D. N. (1989) *Journal of Biological Chemistry*, **264**, 12389–12393.

Van den Berg, G. A., Kingma, A. W., Elzinga, H., & Muskiet, F. A. J. (1986) *Journal of Chromatography*, **383**, 251–258.

Vianello, F., Di Paolo, M. L., Stevanato, R., Gasparini, R., & Rigo, A. (1993) *Archives of Biochemistry and Biophysics*, **307**, 35–39.

Wang, F., Bae, J., Jacobson, A. R., Lee, Y., & Sayre, L. M. (1994) *Journal of Organic Chemistry*, **59**, 2409–2417.

Werner, P. & Cohen, G. (1991) *FEBS Letters*, **280**, 44–46.

Williams, C. H., Lawson, J., & Backwell, F. R. C. (1988) *Biochemical Journal*, **256**, 911–915.

Williams, R. J. & Falk, M. C. (1986) *Journal of Biological Chemistry*, **261**, 15949–15954.

Winkler, H. (1993) *Trends in Pharmacological Sciences*, **14**, 264.

Witz, G. (1989) *Free Radical Biology and Medicine*, **7**, 333–349.

Wu, Y., Rich, C. B., Lincecum, J., Trackman, P. C., Kagan, H. M., & Foster, J. A. (1992) *Journal of Biological Chemistry*, **267**, 24199–24206.

Yamada, H., Isobe, K., & Tani, Y. (1980) *Agricultural and Biological Chemistry*, **44**, 2469–2476.

Yanagisawa, H., Kato, A., Hoshiai, S., Kamiya, A., & Torii, N. (1987) *Plant Physiology*, **85**, 906–909.

Yasunobu, K. T., Ishizaki, H., & Minamiura, N. (1976) *Molecular and Cellular Biochemistry*, **13**, 3–29.

Youdin, M. B. H., Harshak, N., Yoshioka, M., Araki, H., Mukai, Y., & Gotto, G. (1991) *Biochemical Society Transactions*, **19**, 224–228.

Zeidan, H., Watanabe, K., Piette, L. H., & Yasunobu, K. T. (1980) *Journal of Biological Chemistry*, **255**, 7621–7626.

Zoltewicz, J. A. & Bloom, L. B. (1990) *Bioorganic Chemistry*, **18**, 85–90.

CHAPTER 7

Bacterial Metabolism and Polyamines

Why Study Bacteria?

The discovery of bacteria (e.g., *Hemophilus parainfluenzae, Neisseria perflava*) that require polyamines for growth asserted an essential role for these compounds in the lives of these organisms. The use of microorganisms to study nutritional requirements (Snell, 1989) had been an active pursuit in the period following World War I, leading to the discovery of methionine and the requirements of bacteria for compounds like nicotinic acid. The discovery of the requirement of NAD for the growth of *H. parainfluenzae* introduced this organism to E. E. Snell and E. J. Herbst (1949), who subsequently discovered the additional requirement for putrescine or another polyamine. Herbst and his later colleagues then demonstrated the ubiquity of "putrescine equivalents" in various microorganisms, although many Gram-positive organisms contained significantly less of these substances than did Gram-negative organisms (Herbst et al., 1958). The metabolic conversion of labeled putrescine to labeled spermidine in *Escherichia coli, Aspergillus nidulans*, and minced rat prostate was first reported in 1956 by the Tabor group, and the degradation of exogenous polyamines by bacteria and fungi was described shortly thereafter (Weaver & Herbst, 1958).

In the mid 1950s the kinds of polyamines and the routes to and from these compounds appeared initially to be limited in bacteria. The organisms themselves were easy to grow under controlled conditions and had been found to be genetically stable and manipulable. At this time investigators had also learned how to obtain enzymes from the bacteria, to isolate their cellular components, and to dissect cellular structure. Of the many kinds of bacteria available, the successes and opportunities offered by *Escherichia coli* in the study of virus multiplication and of mechanisms of inheritance soon spread the useful strains of *E. coli* as major biological tools. In the period of 1950–1970 this microorganism was of major interest in the dissection of the origins and metabolism of putrescine, cadaverine, and spermidine and in the seeking of the possible roles of the amines in

the life of the bacterium. In subsequent decades it became possible to study the structural features of some of the polyamine-determining enzymes of this organism, as well as to generally study the inheritance, biosynthesis, and regulation of the enzymes in *E. coli*. For these reasons our major discussion of polyamines in bacterial physiology will be concerned largely with the data concerning *E. coli*. Nevertheless, an ever-present concern for infectious diseases and their etiological agents compelled studies on the laboratory growth of the latter organisms. Such a pursuit led to the discovery of a role for spermine and spermidine in permitting the cultivation of *Pasteurella tularensis* (Traub et al., 1955).

Several interesting amide conjugates of spermidine were found in *E. coli* in the 1960s, but their functions were not clear. Diaminopropane was also detected in procaryotic and eucaryotic metabolism of exogenous polyamines. 2-Hydroxyputrescine was found in a pseudomonad in 1969, but it was not until the 1970s that many new polyamines and their derivatives, siderophores, were discovered in bacteria. These included thermophilic Eubacteria and Archaebacteria, nitrogen-fixing Cyanobacteria and Vibrio, etc. (i.e., many organisms other than *E. coli*).

A Biological and Biochemical Context of the Bacteriological World

Bacteria are termed procaryotic organisms because their cells are devoid of a structurally limited nucleus, in contrast to nucleated cells, which are called eucaryotic. We note the report of a membrane-bounded nuclear body in one unusual eubacterium (Fuerst & Webb, 1991). Bacteria are very small, about 1–10 μm in length, and are frequently organized as spherical cocci or as cylindrical bacilli. The large exception in this instance lives in the intestinal tract of a surgeonfish (Angert et al., 1993). The bacteria may exist and multiply as single units, growing and dividing to separate units, or continue to multiply in clusters or chains. Clearly the separation of viable single units, capable of multiplication to form colonies on the surface of a solid nutritional medium, is an experimental plus, possibly permitting the extrapolation of data on bacterial mass and composition in a given culture, such as data on cultures of *E. coli*, to be that of single cells. This simple fact tends to minimize experimental studies on many bacteria that multiply within groups or chains. However, many organisms that multiply to form separated cells fail to form colonies readily on a culture medium. The multiplication of bacteria such as the photosynthetic cyanobacterium *Anacystis nidulans*, with a "low efficiency of plating," may be followed by a tedious direct-counting procedure in a calibrated Petroff–Hauser chamber under a microscope.

Heated bacteria were found by H. Gram to absorb a basic dye, which might be fixed with a solution of KI containing I_2. Organisms that retain the basic dye after washing with acetone or alcohol are called Gram positive; cells that lose the dye but can then be counterstained by another dye are termed Gram negative. Thus, the spore-forming hay bacillus, *B. subtilis*, is Gram positive; our convenient model, *E. coli*, is Gram negative.

The wall of a bacterium is a rigid structure; it surrounds and protects a lipid-rich bacterial membrane. The latter is far more responsive to differences in osmotic pressure between the external environment and the internal cell contents. In the absence of the wall, low salt content of a medium may cause a rapid entrance of water and enlargement of the cell, lysing the membrane. The wall may be removed enzymatically to generate protoplasts, which can be protected in media of high osmolarity. Wall-less bacteria (e.g., *Mycoplasma*) exist in nature and may have chemical components in their membrane that are protective against lysis.

The tough bacterial cell wall is usually a biochemically unique tight network of glycopeptides, whose branched structures are frequently characteristic of the organism. In certain cases the cell wall peptidoglycans are cleavable by enzymes, such as lysozyme in the absence of protective ions, or may be prevented from formation by specific antibiotics, such as penicillin. The bared membrane may then be lysed to liberate cytoplasmic constituents. The wall and membrane may be cracked by many different physical techniques to liberate the aqueous constituents into an appropriate diluent.

In the mid 1950s the internal macromolecular constituents of bacteria were found to include the readily sedimentable protein-synthesizing ribosomes, which are rich in characteristic molecules of RNA and contain some polyamines, and a DNA-containing chromosome. A nonsedimentable phase was obtained that is rich in proteins and enzymes, some nucleic acids, and many other smaller molecules including nucleotides and many of the cell polyamines. A description of the participation of the polyamines in the inner life of the bacteria is the major task of this chapter, which introduces the subjects of cell structure, function, growth, and metabolism.

In the late 1950s Herbst and colleagues sought organisms containing polyamine growth factors for *H. parainfluenzae*. Far more Gram-negative species contained relatively large amounts of such growth factors than did the Gram-positive species. In contrast to their Gram-negative species that are relatively rich in putrescine, the predominant growth factor in the Gram-positive organisms was spermidine. It is of interest that the multiplication of Bdellovibrios in the periplasm of a bacterial host at low cell densities is markedly stimulated by putrescine (Varon et al., 1983).

The polyamines are present in only a very few instances as significant structural components of a peptidoglycan polymer of the cell wall. The membrane, in *E. coli* at least, is essentially devoid of polyamine as well as some other common constituents, such as phosphatidyl choline and sphingomyelin. It constitutes an osmotic barrier that relates to some phenomena of polyamine excretion or synthesis and to specific transport systems for the polyamines and for the amino acids essential for polyamine synthesis (i.e., arginine, lysine, methionine, and ornithine). The membrane also contains the components of the electron transport systems that determine much of the energetics of bacterial function, particularly of aerobic metabolism for *E. coli*. In photosynthetic bacteria the photosynthetic apparatus is in many or possibly all instances derived from the membrane. In organisms utilizing light to assist in fixing CO_2, the production of an alkaline environment (pH 10–11) creates new types of polyamine metabolism. At an elevated pH, transport is facilitated for uncharged amines that may then be trapped and accumulated within the cell as the protonated compounds. Such events have been observed for the eubacterial Cyanobacterium, *Anacystis nidulans*.

Of the many thousands of bacterial components, water may be about 70% of the cell weight, and some 2000–3000 proteins and nucleates may approach 20–25% of the remainder. In addition to the chromosomal DNA, small circular genetically active DNA molecules may be present as well. The composition and sequential arrangement of nucleotides of bacterial DNA or genes will determine the composition of bacterial RNA and of the structural and metabolically important proteins. The average base compositions of various bacterial DNAs differ considerably, ranging from 25 to 75% guanine (G) + cytosine (C)/total base among Gram(+) organisms to 50–70% G + C among Gram(−) eubacteria. It is known that their differences in DNA composition are reflected in the amino acid compositions of the homologous proteins (e.g., ribosomal proteins) of bacteria whose genomic GC varies from 25 to 74% (Osawa et al., 1990).

The RNA molecules vary quite considerably in structure, size, and number. Three ribosomal nucleates are found in the 10,000–20,000 or more ribosomal subunits present in a cell and comprise the bulk of the cell RNA. Some 50–70 different soluble tRNA molecules are also present; *E. coli* possesses 45 different species. Molecules of bacterial tRNA vary from 75 to 93 nucleotides (i.e., about 25,000 molecular weight), and there is frequently more than one tRNA for each amino acid in a particular bacterium. Bacterial tRNA may amount to 15% of the total RNA of the cell. A bacterium can synthesize about as many messenger RNA molecules as bacterial proteins. In *E. coli* this fraction is a bare 1–3% of the total nucleic acid. Discontinuous reading frames (i.e., a sequence of exons and introns) are found in eucaryotic DNA and the latter are removed by splicing from the functional mRNA. Introns have not been detected among most of these bacterial mRNAs.

The sequences of some ribosomal RNAs have been useful in determining evolutionary relations. It is now believed that many organisms found in different ecological niches have in fact been separated in their evolution and may have derived from a long-gone ancestral procaryote. Thus, the procaryotic Eubacteria are genetically and biochemically distinguishable from the procaryotic Archaebacteria. The latter fall in turn into two major groups. These groups include the thermophiles, such as *Caldariella* (now known as *Sulfolobus*), in which the novel natural polyamines caldine and thermine were first found. A second group includes bacteria such as the extreme halophiles, some of which lack a common polyamine and contain agmatine, and the anaerobic methanogens, one type of which synthesizes spermine. An inferred phylogeny of Eubacteria includes a branch leading to the Cyanobacteria, of which *A. nidulans* is an

example. A still later evolution led to the appearance of the even more highly evolved *E. coli* (Sogin et al., 1986).

The polyamine content of bacteria can differ quite considerably as a function of the growth stage of the organism. After an overnight cultivation in a defined glucose-containing medium, *E. coli* can contain less than 1% of the spermidine content attained during exponential growth. *Lactobacillus casei* may have no more than 200 molecules of spermidine per cell after overnight growth, and this value can increase to 1.4×10^5 molecules of this amine by the inception of exponential growth (Fig. 7.1). It will fall once again during its growth to the low level characteristic of the beginning of a stationary phase. We wish to understand then if the increase of spermidine, or other polyamines, relates to any particular element of bacterial structure and cell function and the role and essentiality of the polyamine in the life of the organism, as well as the molecular nature of this interaction between polyamine and other cellular components.

It was found in the 1940s that propamidine, quinacrine, and aromatic diamidines inhibited the growth of various bacteria including *E. coli* and that these toxic effects were largely eliminated or minimized by spermidine or spermine. This type of study of growth inhibition and reversal or protection that proved so useful at that time in clarifying the mode of action of many toxic agents (e.g., the sulfa drugs) did not develop with the polyamines for many years.

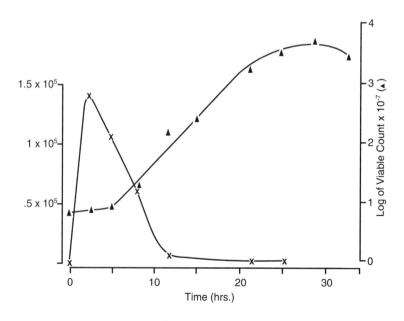

FIG. 7.1 Relationship of (▲) growth of *L. casei* (viable count) and (X) spermidine content per cell; redrawn from Elliott and Michaelson (1969).

Maintaining an Intact Cell

Polyamines and the bacterial envelope

The outer layer of many bacteria contains an anionic lipopolysaccharide comprising sugars, ketodeoxyoctanoic acid, and esterified fatty acids. These complex substances bind polyvalent cations at low ionic strength. Spermine will aggregate the substance from *E. coli* C at relatively high concentrations (50 μM); the polyamine is thought to bind at a low-affinity site (Field et al., 1989).

Filamentous appendages, *fimbriae*, are present on many bacteria and facilitate adherence of the organisms to various types of animal and plant cells. The types of bacterial fimbriae are genetically determined and their association with cells (e.g., in the agglutination of treated erythrocytes) may be used to classify the organism. In one study spermine was found to specifically inhibit the hemagglutination caused by type 3 fimbriae (Gerlach et al., 1989). Indeed the polyamine at millimolar concentrations can dissociate bacterium–erythrocyte complexes.

As described in Chapter 6, the toxicities of spermidine and spermine fell into two groups: organisms affected in the presence of polyamine plus ruminant sera, and those affected by the polyamine directly. Among the latter, spermine proved to be toxic to *Staphylococcus aureus*. On the other hand, the growth of some organisms such as *Pasteurella tularensis* appeared to be enhanced by the addition of spermine (Traub et al., 1955). This organism was protected by spermine in media of low tonicity, implying an effect in preserving membranes. It was shown subsequently that spermine did protect osmotically fragile structures, such as the cell walls of *P. tularensis* and protoplasts of *E. coli* (Mager, 1959).

Although in some instances spermine was inhibitory to enzymes lytic to a cell wall (Brown, 1960), the major effect appeared to be on the stabilization of the membrane (Grossowicz & Ariel, 1963). This result, obtained initially with *Micrococcus lysodeckticus* treated with lysozyme, was extended to the protection of protoplasts of other organisms, to *Mycoplasma*, and even to mitochondria. Spermine was also found to stabilize the membrane-localized enzymes of extreme halophiles, such as an NADH dehydrogenase or menadione reductase.

Both cell wall preparations and protoplasts of *B. subtilis* bound [^{14}C]-spermidine, appearing to indicate binding sites on these structures for the polyamine (Bachrach & Cohen, 1961). Thus the compounds might be both toxic and growth stimulating, depending on the organism, interfering in the first instance with the normal permeability of the wall and membrane or in the second strengthening the membrane against osmotic effects and preventing cellular loss of essential nutrients. In studies on the uptake of lysine, arginine, and ornithine, by amino acid requiring strains of *E. coli*, the normal diamines, putrescine and cadaverine, compete with the amino acid at a surface receptor, block the entrance of the amino acid, and create a deficiency in the supply of the essential requirement (Mandelstam, 1956a,b). Spermine is able to replace protamine in facilitating the transfection of *E. coli* spheroplasts by phage DNAs (Henner et al., 1973). Labeled spermine, taken up by growing *E. coli*, was mainly bound within the cells or on the cell membranes. It was not present in significant amounts on the cell walls and was readily exchangeable with unlabeled spermine, implying the absence of tight binding sites in the walls of this organism (Johnson & Bach, 1968).

Cell envelopes or membranes isolated in a zonal centrifuge from the deoxyribonuclease-treated spheroplasts of *E. coli* contained 92% of the cellular phospholipid and less than 4% of the cellular diaminopimelic acid, a characteristic cell wall component. The purified envelope retained up to 7% of the protein, 3% of the RNA, and 4–10% of the spermidine; this polyamine was exchangeable during isolation. The spermidine was not proven to be a characteristic membrane component and may have been held in the small fraction of residual RNA, which was sufficient to account for all the cation provided by spermidine (Quigley & Cohen, 1969).

In some marine bacteria the polyamine and phospholipid contents are concomitantly a function of the ionic strength of the media in which they are grown (Peter et al., 1978). These results were interpreted to suggest that the polyamines participate in membrane stabilization and phospholipid synthesis in these organisms. In contrast to the minimal association of spermidine with cell membrane in *E. coli*, it was recently found that spermidine complexes of membrane can exist in the ice-nucleating bacterium, *Erwinia uredovora*. This polyamine was found to contribute significantly to the ice-nucleating capacity of the membrane (Kawahara et al., 1993).

One inventory of compounds found in *E. coli* estimated the existence of some 20×10^6 molecules of lipid, 5.6×10^6 molecules of putrescine, and 1.1×10^6 molecules of spermidine per cell. There appears to be some 130 μmol of membrane-bound phospholipid per gram dry weight (Neidhardt, 1987). The phospholipid composition of the membrane of Gram-negative and Gram-positive organisms may be modified significantly by the conditions of growth, and it has been shown that depletion of the polyamine content (65%) of *E. coli* by growth in the presence of an inhibitor of ornithine decarboxylase reduces phospholipid content about 20% (Peter et al., 1979). This and various intermediate effects on phospholipid content are reversed by the addition of putrescine but not spermidine to the medium.

The intracellular penetration and binding of putrescine to *E. coli* was studied by ^{13}C-nuclear magnetic resonance (NMR) (Frydman et al., 1984). Exposure to ^{13}C-putrescine resulted in broadening of line spectra because the compounds were bound by different fractions (Fig. 7.2A, B). The broadening arises from the restriction in tumbling of bound putrescine. Thus Figure 7.2A(A)

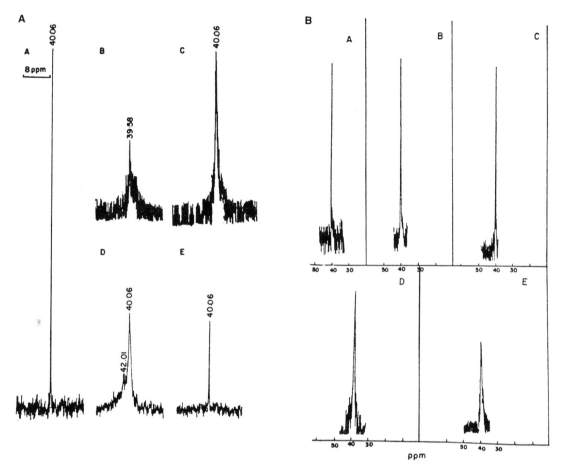

FIG. 7.2 In vivo binding of [1,4-^{13}C] putrescine to macromolecular components of *E. coli* cells; redrawn from Frydman et al. (1984). (**A**) Spectra of [^{13}C] putrescine in: A, buffer; B, addition to cell suspension (6 min); C, addition to cell suspension (30 min); D, centrifuged, washed cells; E, supernatant fluid from D. (**B**) Spectra after interaction for 10 min with: A, DNA; B, tRNA; C, soluble proteins; D, membrane fraction; E, ribosomes.

demonstrates the spectrum of the free [1,4-^{13}C]putres-cine and its change on binding to the cells [Fig. 7.2**A**(D). The spectra of various cellular fractions are presented in Figure 7.2**B**, with that of the noncovalent putrescine–membrane complex in Figure 7.2**B**(D).

Charge, hydrophobicity, and amine effects

Essentially all outer cellular membranes (i.e., the plasma membrane) in procaryotes or eucaryotes, which separate the extracellular medium from the cytoplasm, comprise a hydrophobic lipid bilayer. Zachowski (1993) discussed some details of the movement of lipid and protein within membranes. It is relevant that membrane ghosts of *S. faecalis*, rich in protein and cardiolipin, contain a membrane-bound ATPase and GTPase, which is released from the structure by 1 mM spermidine (Abrams, 1965).

Most bacteria do not contain sterols as a result of the

inability of these organisms to effect reactions utilizing molecular O_2 essential to sterol biosynthesis. Nevertheless, some mycoplasmas may incorporate sterols into their membranes, thereby stabilizing these structures. This nutritional evidence possibly relates to studies on a new polyamine-containing antibiotic squalamine and to data on the contribution of positively charged residues in membrane topology.

A broad spectrum steroidal antibiotic was isolated from tissues of the dogfish shark, *Squalus acanthias*, and was shown to consist of a sulfated bile acid fused to the aminopropyl primary amine of spermidine (Moore et al., 1993). The recently synthesized compound squalamine (Pechulis et al., 1995) is presented below. The antibiotic is not only bactericidal to many Gram-negative and Gram-positive bacteria but is also fungicidal and lytic to some protozoa. It is assumed that the compound is lethal as a result of its action on the cell membrane and its effect on lipid-containing sites. Bile acids containing basic groups possess antibiotic activity, which is enhanced when these modifications occur at the C-3 position of the bile acid.

squalamine

Ornithine and diamines as cell wall and membrane constituents

Ornithine or 2,5-diaminovaleric acid is one amino acid precursor of putrescine, isolated initially as a dibenzoyl derivative (ornithuric acid) in the excreta of chickens fed benzoic acid. In 1889 A. Ellinger described conversion of the amino acid to 1,4-diaminobutane in a putrefying system, presumably by growing bacteria. In 1897 the amino acid and urea were found to be produced in the treatment of arginine by alkali and a few years later (1901) ornithine was synthesized by Emil Fischer. Its role in the formation of urea was established by Krebs and Henseleit (1932), and the compound has been studied mainly for its metabolic role in this activity as well as a precursor of putrescine. It has also been found as an unexpected structural component of some antibiotics: the polypeptide gramicidin S, siderophores, lipids, peptidoglycans, etc. Although no codon has been found for ornithine, the amino acid has been found in a very few proteins (Akers & Dromgoole, 1982). Ornithine-activating and -dependent ATP-pyrophosphate exchange has been detected for the addition of ornithine to the polypeptide gramicidin S in the absence of tRNA, perhaps via ornithine adenylate (Kleinhauf & von Döhren, 1990).

Ornithine was identified in mycobacterial phosphatides in 1955, and since then it has been found in the lipid constituents of many bacteria. In some Gram-positive organisms it was found that ornithine or diaminobutyric acid occasionally replaced the characteristic diamino acids in cell wall peptidoglycans (Cummins, 1965), and this finding has been used to interpret the classification and physiology of the ornithine-containing organisms (Jürgens et al., 1987).

Cadaverine, a metabolite of diaminopimelic acid and of lysine, has been found in the peptidoglycan of two organisms: *Selenomonas ruminantium* and *Veillonella alcalescens*. The former is a strictly anaerobic Gram-negative organism in which the wall peptidoglycan is unexpectedly stable. The isolated peptidoglycan lacks lysine, and cadaverine is linked to a D-glutamic acid residue in a peptide containing diaminopimelic acid (Kamio et al., 1981a, 1981b, 1982). Cadaverine is an essential constituent of the structure during growth because inhibition of lysine decarboxylase by difluormethyllysine leads to cell lysis. This lytic effect is prevent-

able by the presence of exogenous cadaverine during growth (Kamio et al., 1986).

A strain of the anaerobic Gram-negative coccus, genus *Veillonella*, was also found to have a nutritional requirement for cadaverine or putrescine, which was not met by spermidine (Ritchey & Delwiche, 1975). In several species of this genus, putrescine or cadaverine has been found as an essential constituent of the peptidoglycan (Kamio & Nakamura, 1987).

Osmoregulation in bacteria

Bacterial species multiply in media of widely varying osmolarity and have many adaptations that permit growth under these potentially variable conditions. In addition to the frequently protective corsetlike cell wall, the penetrability and retention of solutes is regulated by several types of transport systems for uptake and extrusion that operate on inorganic and organic cations (e.g., K^+, Mg^{2+}, polyamines). In addition the cells can rapidly synthesize several kinds of osmotically effective metabolites such as proline (anionic), glycerol (neutral), and putrescine (cationic). Turgor pressure describes the stress in the bacterial envelope by the difference in osmotic pressure across the membrane. A high concentration of an impermeant solute in the medium causes a loss of cell water and retraction of the membrane from the wall, that is, *plasmolysis*. The minimum osmolality in the medium evoking plasmolysis is taken to define the osmolality within the cell. Such a procedure indicated that exponentially growing *E. coli* have a turgor pressure several times that of cells in stationary phase. A rapid decrease in turgor pressure by increasing medium osmolality is known to cause an arrest of synthesis of protein and nucleic acid until a positive turgor pressure is restored (Epstein & Schultz, 1966). Ingraham (1987) discussed various environmental effects, including that of water activity, on the growth of *E. coli*.

Putrescine and diamines in regulation of cellular ionic balance and pH

Positively charged onium compounds occur abundantly in many higher marine animals and in some plants. More than 100 classes of quaternary ammonium compounds as metabolic products and their distribution in the biosphere were described in some detail (Anthoni et al., 1991). Their roles as sources of chemical energy for purposes of biosynthesis were discussed by G. Cantoni (1960), who also commented on their distributions in organisms overproducing a variety of storable chemical groups (e.g., methyl, pyridinium, etc.).

Cantoni (1960) was led to *S*-adenosylmethionine by an interest in the formation of methylated compounds. The formation of compounds such as tetramethyl ammonium hydroxide relates to the disposal of an excessive production of methyl groups and their excretion in the removal of undesirable base. Plants may also accumulate dimethylpropriothetin, *S*-methylmethionine, and the alkaloids, and also eliminate compounds such as *N*-

methyl nicotinamide, as do vertebrates. The variety of such compounds is relatively limited in bacteria, and in these organisms the concentrations of metabolically active onium compounds such as *S*-adenosylmethionine and di- and triphosphopyridine nucleotides and thiamine are never very high and are closely regulated. These substances do not serve in storage functions in bacteria.

Bacteria may change their polyamine content quite directly in response to ionic challenge. Conditions markedly changing the osmolarity of growth media for plant and animal cells can also evoke large effects in the intracellular content of the polyamines. Both plants and animals produce large amounts of putrescine in response to hypoosmotic media. In higher organisms these responses are considered to represent homeostatic mechanisms for the maintenance of ionic balance. In plants, amine formation and excretion are also suggested to serve as internal compensating mechanisms to maintain a fairly constant intracellular pH. Nevertheless, polyamine contributions to ionic homeostasis in bacteria are quite complex.

The relations of nitrogen metabolism and excretion to the habitat of multicellular organisms in the oceans, estuarine conditions, fresh water, and various water-limited land conditions were discussed by Baldwin (1939). The polyamines participate in the solutions to the problems posed by these environmental changes. Polyamines have been found to be important in the growth of heterotrophic bacteria of the genus *Acidiphilium*, which may be isolated from acidic mineral environments. A mutant derived from one of these strains, *Acidiphilium facilis* 24R, was able to grow at pH 5.5 but was unable to grow at pH 3.0 in the absence of polyamine (Kawahara et al., 1991). One part of the essential polyamine became conjugated, probably within the cell wall.

The presence of a tough, relatively rigid cell wall is a protective device in plants and bacteria. As noted, there are few instances of the structural or metabolic involvement of the polyamines in the formation of the rigid protective corseting systems of Gram-negative and Gram-positive organisms. Many solutions to the development of tolerance to osmotic stress have been found in a wide array of organisms, and some of these (e.g., the production of proline; Dandekar & Uratsu, 1988) are known in *E. coli*. In addition, a possible role for the polyamines in maintaining ionic balance has been posed.

The variation in the polyamine contents of common strains of *E. coli* as a function of the osmolarity of the medium is presented in Figure 7.3. The putrescine content is high in organisms grown at low ionic strength and is much less when these cells are placed in media of higher osmolality; spermidine remains relatively constant during these changes. When the cells at low osmolality are transferred to media of high salt or sucrose content, the putrescine is rapidly released, a result also produced by merely adding NaCl to the growth medium. These treatments did not damage the cell membrane, because neither spermidine nor amino acids were

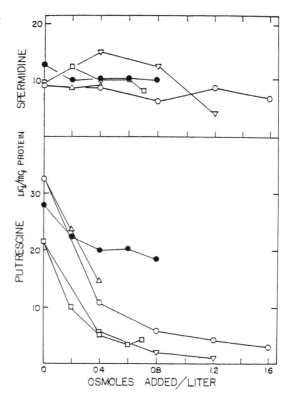

FIG. 7.3 Effect of osmolarity of the medium on polyamine content of *E. coli* B; redrawn from Munro et al. (1972). A fresh culture of *E. coli* grown in low salt nutrient broth was diluted 1/100 with broth containing (○) NaCl, (△) MgCl₂, (▽) KCl, (□) sucrose, or (●) glycerol; the cultures were grown to an optical density of 0.3–0.4.

lost from the cell, nor did a significant loss of water occur in the cells in a high salt medium. Actually, growth at high osmolarity markedly increased the amino acid pool of the cells, which became unusually high in glutamate, lysine, and glutathione (Munro et al., 1972).

In these experiments, K^+ uptake occurred under conditions of putrescine excretion; indeed such excretion was K^+ dependent and evidence was obtained of a putrescine^{2+}/2K^+ antiporter system (Tkachenko & Chudinov, 1994). On the other hand, K^+ uptake could occur at high osmolarity in the absence of putrescine excretion from previously depleted cells.

Günther and Peter (1979) attempted to test the participation of putrescine synthesis and the intracellular content of the diamine in osmoregulation in high and low osmolarity media. Difluoromethyl ornithine, an inhibitor of ornithine decarboxylase, was used to inhibit production of putrescine and spermidine in these media. The presence of the inhibitor in the growth medium sharply reduced spermidine and putrescine content of the bacteria at low osmolarity but had very little effect in media of higher salt. Despite the clear reduction of polyamine content by the inhibitor in media of low osmolarity, the Na^+ and K^+ contents were not affected by the inhibitor in either medium. These results were interpreted to indicate that the polyamines were not involved in the long-term steady-state osmoregulation by the bacteria (Ingraham, 1987).

Cation availability and activation of putrescine biosynthesis

The excretion of putrescine has been decoupled from K^+ uptake in several mutant strains of E. coli (e.g., strain B207 or the derived strain B697), which are unable to retain this ion (Rubenstein et al., 1972). Intracellular K^+ in E. coli normally approaches 200 mM. When placed in a Na^+-containing medium, otherwise adequate in carbon, nitrogen, and salts, intracellular K^+ falls to about 10 mM in these E. coli strains, which become unable to synthesize protein. Nevertheless, the bacteria increase putrescine production about five- to eightfold, which they excrete. The biosynthetic supply of ornithine proved to be limiting in the Na^+ medium, and addition of this amino acid to the medium doubles the already enormous overproduction and excretion of putrescine in K^+-depleted cells. It may be noted that the E. coli mutants used in these studies, E. coli strain B207 and the derivative strain B697, are known to take up K^+ at a greater than normal rate. When mutant strains truly deficient in K^+ transport were tested in their ability to excrete putrescine after a transfer from a low to high osmolarity medium, the rates of putrescine excretion were sharply reduced, suggesting some structural relation of K^+ uptake to elimination of the diamine (Munro & Sauerbier, 1973).

Many strains of E. coli have two routes for the biosynthesis of putrescine. These are:

1. the decarboxylation of ornithine to putrescine by an ornithine decarboxylase (ODC) and
2. the decarboxylation of arginine to agmatine by an arginine decarboxylase and the conversion of agmatine to putrescine and urea via an agmatine ureohydrolase.

The production of urea then becomes a measure of the pathway involving the synthesis of putrescine via the decarboxylation of arginine, and this test demonstrated that the K^+-deficient mutants of E. coli employed only the first route, that of putrescine synthesis via the decarboxylation of ornithine.

Exogenous spermidine in the culture and its introduction into the cells severely inhibits putrescine biosynthesis. This is additional evidence of a feedback mechanism regulating ODC. A similar result is obtained when methionine is fed to the methionine-requiring K^+-depleted strain B207. In this system spermidine is synthesized at a low rate and leads to a significant but less-pronounced inhibition of putrescine biosynthesis.

More than 60 enzymes have been described as activated by K^+ (Suelter, 1970), and these included enzymes for the synthesis of purines and pyrimidines. In studies on nucleate synthesis in depleted cells, the addition of the purine ribonucleosides, adenosine and guanosine, to the Na^+ medium increased RNA synthesis approximately sixfold. Enhanced putrescine biosynthesis was not affected significantly. Spermidine biosynthesis was largely inhibited and accumulation of this polyamine did not occur in either case in K^+-deficiency, with or without enhanced RNA synthesis. In this instance, the accumulation of RNA does not appear to be paralleled by a concomitant accumulation of spermidine.

In media of low osmolality that contain arginine, Vibrio parahaemolyticus, which contains large amounts of norspermidine and is normally low in putrescine and spermidine, also overproduces and excretes putrescine (Yamamoto et al., 1988). NaCl and KCl inhibit arginine decarboxylase, which is the major enzyme in putrescine production in this system; the salts do not inhibit the agmatine ureohydrolase. Incubation in arginine at low osmolarity does not increase the amounts of these enzymes but releases the activity of inactive enzymes. This result is comparable to the release of activity of previously inactive ODC in the K^+-deficient E. coli in K^+-free media, cells that fail to synthesize any protein.

Ingraham (1987) noted the probable role of antiport systems (i.e., Na^+/proton and K^+/proton systems) for controlled acidification and homeostasis of pH in E. coli. The existence of a putrescine^{2+}/K^+ antiport may also be imagined to contribute to homeostasis. It is known that the expression of some genes is affected by external pH, resulting in the removal of acid from protein-rich media, particularly under anaerobic conditions. In one such case, the production of cadaverine at low pH occurs via the action of lysine decarboxylase on lysine. An operon in E. coli DNA contains the gene (cadA) for this enzyme, as well as a gene (cadB) for a lysine-cadaverine exchanger, and the gene (cadC) that encodes a protein required for the transcription of genes A and B. CadC is activated by low pH and a high concentration of lysine and participates thereby in the production of the enzyme and cadaverine, as well as the excretion of the latter (Neely et al., 1994). The E. coli cad operon has also been implicated in the supply of CO_2 via the decarboxylation of lysine (Takayama et al., 1994). Similar effects have been detected with arginine decarboxylase and arginine degradation in arginine-rich media (Slonczewski, 1992).

Early observations on transport of polyamines

Despite the minimal participation of the polyamines themselves as structural elements in the bacterial cell wall and envelope, these structures can react with added amine. Several classes of polyamine derivatives containing lipophilic regions affect bacterial transport, apparently by disruption of the bacterial membrane (Silver et al., 1970). The fatty acid conjugated aliphatic triamines (e.g., solapalmitine) cause the leakage of K^+ ion and of thiomethylgalactoside from the cells. Another class includes natural steroidal diamines that inhibit virus multiplication and damage the bacterial membrane. The effects on permeability are reversed by Mg^{2+}. The fact that the steroidal diamines are also mutagenic (Mahler & Baylor, 1967) implies that at certain concentrations the compounds themselves penetrate the bacterial membrane sufficiently to interact with DNA without great damage to biosynthesis. An in vitro interaction of the steroidal diamines with DNA was demonstrated by Gabbay et al. (1970). Compounds such as the morphine analog, levorphanol, appear to affect polyamine content and the *E. coli* membrane by inhibiting uptake of putrescine, spermidine, and amino acids, and by increasing putrescine efflux (Simon et al., 1970). There is clearly a range of effects to the membrane, from outright damage to a reversible increase of efflux determining the cell's ability to contain the diamine pool.

Studies on the transport of the basic amino acids in *E. coli* revealed several distinct transport systems for these compounds in this organism (Celis, 1984; Oh et al., 1994; Rosen, 1973). The polyamines themselves are also transported and accumulated within cells by specific mechanisms. Putrescine, spermidine, and spermine were found to react with and to be accumulated by *E. coli* by several mechanisms (Tabor & Tabor, 1966). Radioactive amines were taken up by the bacterium and were detached by washing with a salt solution or with nonradioactive amines. In addition an energy-dependent uptake occurred at 37°C but not at 0°C; the amines were released only by disruption of the membrane (Fig. 7.4). Addition of exogenous putrescine to cell-accumulated

putrescine revealed an exchange of the cell-held amine; this exchange did not appear to occur between cell spermidine and spermidine in the medium. The uptake of putrescine was markedly inhibited (40%) by omission of glucose from the medium; spermidine uptake is exceedingly slight in the absence of glucose. Active spermine uptake was relatively slow, metabolism dependent, and nonexchangeable.

An early study of the metabolism of arginine appeared to demonstrate a very rapid and relatively specific conversion of arginine to agmatine and putrescine in passage through the cell membrane and suggested the existence of the arginine decarboxylase in or very near the cell membrane (Wilson & Holden, 1969). Some decades later this conclusion proved to be essentially correct. The enzyme is fairly large, is not a soluble component of the periplasmic space between cell wall and cell membrane, and appears to be located as a fixed outer component of that membrane (Buch & Boyle, 1985).

The demonstrations of the existence of transport systems for the polyamines in *E. coli* and the role of these compounds in osmotic adaptations led to the study of the development of transport systems for putrescine in media of high and low osmolarity. When grown in a medium of high salt content, the bacteria were found to have a transport system for putrescine of high affinity (low Michaelis constant, K_m) and low rates of uptake (low maximum velocity, V_{max}). After growth in media of low osmolarity, the bacteria appeared to have added an additional system of higher K_m and V_{max} (Munro et

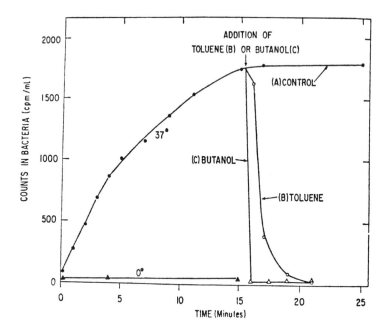

FIG. 7.4   Effect of temperature on the uptake of $[1,4-^{14}C]$ putrescine by *E. coli*; redrawn from Tabor and Tabor (1966). Accumulated counts were released by toluene or 1-butanol.

al., 1974). Mutants were isolated that preserved one or the other system, both of which were partially energy dependent.

These studies encouraged study of the binding proteins for the specific polyamines and for the isolation of mutants, genes, and proteins involved in the bacterial transport mechanisms. These will be discussed later. It will be noted that these systems are quite different from some antiporter systems (Driessen et al., 1989). A system mediated by an agmatine-putrescine antiporter was found in *Enterococcus faecalis* (*Streptococcus faecalis*). This organism can use arginine or agmatine as its sole energy source via a deiminase pathway to produce ornithine or putrescine, respectively. Thus, the organism can take up agmatine and extrude putrescine, generating ATP and liberating CO_2 and NH_3 (Driessen et al., 1988).

Transport and synthesis in an obligate intracellular parasite

The parasitic intracellular Gram-negative bacterium *Rickettsia prowazekii* (the louse-borne etiologic agent of epidemic typhus), when grown in an animal cell deficient in ODC, contains high concentrations of polyamines: 6 mM putrescine, 5 mM spermidine, and 3 mM spermine. Transport systems for these polyamines could not be detected in the organism that appeared to be able to synthesize the compounds from arginine and did contain arginine decarboxylase but lacked detectable ODC (Speed & Winkler, 1990). Although it was claimed that all of the polyamine, including spermine, was synthesized in the bacterium, the presence of each enzyme of polyamine synthesis was not demonstrated.

Inorganic Cations and Polyamines

It is often asked if the polyamines are not simply replacements for a divalent cation, such as Mg^{2+}, that can fulfill the requirements for a given enzyme system. Certain biosynthetic reactions, such as those of protein synthesis or as a requirement for a cation like Mg^{2+}, may be reduced but not eliminated by the addition of spermidine or spermine; joint application of the inorganic and organic cations is frequently found to markedly increase the rate of the reactions. However, in some instances the two cations do function at different sites, as in the organization of 3-dimensional structure in tRNA (Chap. 23). Eventually reference is made to early nutritional observations in which a polyamine is an essential additional component of a medium otherwise rich in presumably essential divalent cations, as in the demonstrations of the growth requirements of *Hemophilus* or *Neisseria*. Obviously there are some essential functions for the polyamines in these organisms, even as there are such roles for divalent cations, of which Mg^{2+} has had the major attention.

Calcium has been found to be essential for growth of some organisms (e.g., *Azotobacter*) and stimulatory for others (e.g., some staphylococci and clostridia) (Norris et al., 1991). Calcium is tightly regulated by several mechanisms to 0.1 μM in *E. coli* (Gangola & Rosen, 1987) and has been shown to be stimulatory to several enzymatic proteins of this organism. The concentration of Ca^{2+} is significant in the cell envelope and cytoplasm (Table 7.1) but still only one-third to one-sixth that of Mg^{2+} in these components (Chang et al., 1986).

According to Table 7.1 the inorganic ions can account for about half of the positive charge necessary to neutralize cell phosphate. Mg^{2+} represents about half of this, at about 200 mmol/kg (dry weight). Mg^{2+} is largely bound to structure and is uniformly distributed within the cell. Therefore, Mg^{2+} is neither concentrated in the DNA of the nucleoid nor absent from this structure. Among the cell-filling structures to which Mg^{2+} is certainly bound are the ribosomes, accounting for the uniformly high concentrations of Mg^{2+} and P. Mg^{2+}-depleted *E. coli* lose ribosomes without loss of viability. Restoration of Mg^{2+} permits resynthesis of rRNA and then the appearance of ribosomal subunits. Several workers detected a specific role for the cation in membrane development in Mg^{2+}-starved cultures (Fiil & Branton, 1969).

The activity of Mg^{2+} as a freely soluble ion within the bacteria is not known. Although most of the Mg^{2+} is sequestered to soluble and insoluble nucleates and metabolically active nucleotides in all cells, methods involving NMR spectroscopy and fluorescent Mg indicators have been developed largely for the measurement of cytosolic free Mg^{2+} concentration in animal cells and tissues (London, 1991; see Günther et al., 1995).

Limited Mg^{2+} in cultures of *Aerobacter aerogenes* reduced bacterial yield and RNA synthesis within the bacteria (Tempest et al., 1965). In bacteria grown at various growth rates at controlled Mg^{2+} concentrations in a chemostat, 4 mol of RNA nucleotide were synthesized per mole of Mg^{2+} present in the culture. K^+ was similarly essential for bacterial growth and multiplication; in K^+-limited cultures of various growth rates the molar stoichiometry of cellular Mg^{2+}, K^+, and P appeared approximately constant at 1:4:8 (Tempest et al., 1966).

TABLE 7.1 Elemental concentration of dividing washed cells of *Escherichia coli* B.

Element or ion	Concentration (mmol/kg, dry weight)	
	Cytoplasm	Cell envelope
Na	146 ± 7.7	117 ± 5.3
Mg	201 ± 12.5	114 ± 8.3
P	1543 ± 55.4	1060 ± 44.4
S	139 ± 11.2	130 ± 9.9
Cl	14.0 ± 1.5	12.0 ± 1.3
K	169 ± 12.8	128 ± 11.3
Ca	32.6 ± 2.2	34.2 ± 2.1

Adapted from Chang et al. (1986).

About 70% of the cell-bound phosphorus in this K^+-limited organism is present as nucleic acid P and about 90% of the total nucleate is RNA, indicating a relation of Mg^{2+} to RNA-P of about 1:5. More elaborate studies have included carbon, nitrogen, and sulfur contributions to cell mass (Dean & Rogers, 1967).

Bacteria permitted to accumulate Mg^{2+} during exponential growth were harvested and transferred to media devoid of Mg^{2+} but otherwise adequate for growth. Although *Aerobacter aerogenes* continued to synthesize RNA, DNA, and proteins for many hours, the ribosomes were degraded and resynthesized to a greater or lesser degree depending on the extent of Mg^{2+} depletion during continuing growth (Kennell & Kotoulas, 1967). The ribosomes of *E. coli* are even more sensitive to Mg^{2+} deficiency than those of *A. aerogenes*. In the *E. coli* studies it was estimated that stationary phase bacteria grown in limiting Mg^{2+} (10^{-5} M) nevertheless possessed a Mg content about half that of cells grown in 10^{-3} M Mg^{2+}. About 15% of the cell Mg^{2+} was present in a soluble nonsedimentable fraction at a concentration of about 4 mM (Lusk et al., 1968). The functional aggregation of 30S and 50S ribosomal subunits to form a 70S ribosome requires at least 3 mM Mg^{2+}. Cells limited in Mg^{2+} increased their K^+ content.

A significant fraction of *E. coli* spermidine is bound to a sedimentable ribosome fraction and spermidine reduces the concentration of Mg^{2+} needed to form 70S ribosomes (Cohen & Lichtenstein, 1960). Beyond demonstration that both cations acted synergistically in the functional association of ribosomal subunits, few workers analyzed the interrelations of Mg^{2+} and polyamine in the development of bacterial ribosomes. One study of ribosomes in *E. coli* grown at various Mg^{2+} concentrations concluded that ribosomes derived from Mg^{2+}-limited cultures contained less Mg^{2+} and more spermidine (Table 7.2). It appeared that a spermidine molecule replaced about 20 Mg^{2+} ions, as the Mg^{2+} content of the ribosome fraction decreased. In subsequent studies with the pseudomonad strain Kim, which contains putrescine and 2-hydroxyputrescine and little or no detectable spermidine, a similar inverse relation was found between the ribosomal content of Mg^{2+} and the polyamines (Rosano & Hurwitz, 1969). The authors pointed to the similarity of 2-hydroxyputrescine to spermidine in its behavior in their associations with bacterial ribosomes. More recently Mg^{2+} was found to regulate the homospermidine content of *Rhizobium fredii* as a complex function of external salinity, K^+ availability, and glutamate content (Fujihara & Yoneyama, 1994).

Specific transport mechanisms are present in *E. coli* for both K^+ and Mg^{2+}. At least 10 genes are involved in the accumulation of K^+ and two separate types of K^+ transport systems were discovered and analyzed (Epstein & Laimins, 1980). Specific transport systems were also described for Mg^{2+} (Nelson & Kennedy, 1972; Silver & Clark, 1971). Genes for three Mg transport systems have now been detected in *E. coli* (Park et al., 1976). Three genes governing such transport have also been cloned from *Salmonella typhimurium* (Snavely et al., 1989).

Polyamines and Isolation of Internal Organelles: Ribosomes

The isolation of enzymes from disrupted bacteria is facilitated by their solubility. An extract of crushed fragmented *E. coli* sedimented at low centrifugal speeds provides a soluble phase rich in nucleates of DNA and RNA. Much of the latter is present in the particulate ribosomes. A typical bacterial ribosome (e.g., of *E. coli*), has a molecular weight of 2.3×10^6 and a sedimentation coefficient, under standard conditions, of 70S. It comprises two unequal subunits (50S and 30S), which associate during protein synthesis. The 50S subunit contains two RNA chains and 35 proteins; the 30S subunit contains a single RNA chain of 8.5×10^5 and 21 proteins.

The presence of nucleates and nucleoproteins complicated the fractionation of soluble proteins. The basic antibiotic streptomycin precipitated the nucleates and was used widely in an initial elimination of the nucleic acids; and indeed a stepwise precipitation of DNA, ribosomes, and other nucleates is possible (Cohen & Lichtenstein, 1960). In an aqueous extract DNA possessed a low sedimentation rate, and the ribosomes often existed as dissociated subunits that were associable (Cohen & Lichtenstein, 1960). A role for Mg^{2+} at 10^{-3}–10^{-2} M in the association of the subunits to form a more easily sedimentable fraction was confirmed. More significantly the addition of 5×10^{-3} M spermidine to 5×10^{-3} M Mg^{2+} markedly increased the association of the 30S and 50S subunits to form 70S particles, and it was suggested that Mg^{2+} and spermidine act together in the or-

TABLE 7.2 Ribosome-bound Mg^{2+}, spermidine, and putrescine in *Escherichia coli* ML35 grown in different concentrations of Mg^{2+}.

Concentration of Mg^{2+} in medium	Mg^{2+} (μmol)	Per mole RNA–ribose	
		Putrescine	Spermidine
4×10^{-5} M	0.090 ± 0.010	0.071 ± 0.066	0.027 ± 0.004
4×10^{-4} M	0.138 ± 0.015	0.059 ± 0.008	0.014 ± 0.002
4×10^{-3} M	0.240 ± 0.025	0.063 ± 0.009	0.007 ± 0.002

Adapted from Hurwitz and Rosano (1967).

ganization of ribosomal structure. The presence of spermidine also affected the sedimentation of DNA. When 5×10^{-3} M spermidine in 5×10^{-2} M Tris (pH 7.5) was directly used to extract alumina-crushed *E. coli*, the soluble extracts were devoid of much of the DNA (Cohen & Barner, 1962).

When *E. coli* is grown in the presence of $2–5 \times 10^{-4}$ M spermine and extracted in the presence of 10^{-3} M Mg^{2+}, the nonnative polyamine is found in ribosomal aggregates (Colbourn et al., 1961). Millimolar spermine can precipitate ribosomes quantitatively (Selman et al., 1965).

In early studies of the role of ribosomes in protein synthesis, the particles isolated from *E. coli* were shown to be split by removal of Mg^{2+} with ethylenediaminetetraacetate. Putrescine and cadaverine were found among the products released from the degraded ribosomes of *E. coli*; spermidine was among the amines present on liver particles (Zillig et al., 1959). These authors estimated that the amines were in sufficient concentration to neutralize at least a third of the RNA phosphorus.

In an examination of the polyamines present in ribosomes of exponentially grown *E. coli*, ribosomes extracted in H_2O or Mg^{2+} and washed with NaCl or Mg^{2+} contained 12–15% of the total polyamine of the extract (Cohen & Lichtenstein, 1960). This was calculated as putrescine, although much of it was spermidine. The polyamine represented some 5–8% of the molar content of the ribosomal RNA phosphorus. The apparent constancy of this polyamine content suggested the existence of relatively tightly bound polyamine. To test the possibility of gratuitous absorption or exchange of polyamine from the extract to the ribosomes, a sedimented ribosome-free extract containing radioactive spermidine and putrescine was prepared from cells grown in radioactive glucose and was used to extract cells grown in nonisotopic medium. The ribosomes derived from this latter extract, which contained radioactive amines, nevertheless proved to yield amines of very low specific radioactivity, for example, putrescine with <1% of the specific activity of the putrescine of the radioactive extracting medium. This experiment was interpreted to indicate that the polyamines are associated with the ribosomes in the cell and are not adsorbed in a significant amount to the ribosomes after formation of the extract. It was supposed that cells grown exponentially in the glucose medium regulated their synthetic patterns rather closely and that this may be extended to their ribosomal polyamine contents.

The exchangeability of polyamines on ribosomes was reinvestigated in experiments in which the ribosomes were most frequently isolated from cells grown in the presence of radioactive spermidine or putrescine (Tabor & Kellogg, 1967). In these studies the polyamine content of the ribosomes was a function of the polyamines present in the media from which they were prepared. Thus, in contrast to the earlier experiments, amines could be taken up from the media or wash fluids onto the ribosomes and displace amines from the ribosomal structure.

To explain the dicordant results, it can be asked if a

regulated cell saturates functional tight-binding sites first and limits the synthesis of polyamine necessary to saturate more readily exchangeable and less functional binding sites. An organism fed polyamine may saturate both types of site, of which one group is the more readily exchangeable. An increase in the net uptake of medium spermidine to spermidine-containing ribosomes was demonstrated in the experiments of Tabor and Kellogg (1967); however, the above hypothesis of two types of binding sites for polyamines within bacterial ribosomes has not been tested systematically.

As noted earlier, cells grown at low Mg^{2+} increased their cellular and ribosomal content of polyamine, and this finding and the problem of the exchangeability of ribosomal cation was pursued by Weiss and Morris (1970). Dialysis of *E. coli* ribosomes against spermidine or putrescine replaced bound Mg^{2+} by the polyamine. Reduction to less than 20% of the bound Mg^{2+} inactivated peptide synthesis by these ribosomes, implying a class of essential Mg^{2+}-binding sites nonreplaceable by spermidine (Weiss et al., 1973). In additional classes of sites at which greater than 20% of the RNA phosphorus is neutralized, Mg^{2+} and spermidine are interchangeable without marked functional disturbance. In this range of neutralization of charge, NH_4^+ and K^+ may replace some of the Mg^{2+}. A comparable effort to eliminate spermidine or other polyamines from ribosomes and to follow the functional activity of the structure and its subunits has not yet been described. Weiss and Morris (1970) established an essential role for some Mg^{2+} and demonstrated the existence of a class of Mg^{2+} easily replaceable by polyamine, creating the conditions for the net addition of spermidine to ribosomes detected by Tabor and Kellogg.

Once essential Mg^{2+} was lost, the ribosomal RNA was more easily cleaved by ribonuclease. Reduction of the activity and damage of ribosomal structure by ribonuclease has been a fairly frequent observation and several groups have noted that the presence of spermidine and a high concentration of Mg^{2+} protect the ribosomal RNA and subunits of *E. coli* and other organisms from such degradation (Erdmann et al., 1968; Igarashi et al., 1982). The inclusion of 8 mM putrescine, 1 mM spermidine, and 5 mM Mg^{2+} into an extracting buffer containing additional Ca^{2+}, NH_4, and K^+ permitted the rapid isolation by gel filtration of a high yield of stable active preparation of 70S ribosomes of *E. coli* (Jelenc, 1980). This preparation also minimized misincorporation of the wrong amino acid in polypeptide synthesis. The isolation of stable 68S monosomes from the chloroplasts of *Euglena gracilis* has also been accomplished by the use of extracting and washing buffers containing Mg^{2+} and spermidine (Schwartzbach et al., 1979).

The conformational change of a subunit after binding a fluorescent probe to the subunit protein has been studied as a function of Mg^{2+}-dependent association of the *E. coli* subunits. Spermidine was shown to markedly decrease the need for Mg^{2+} in the association, and the polyamine appeared to be directly involved in the coupling (Favaudon & Pochon, 1976).

The Pseudomonas species in which putrescine and 2-hydroxyputrescine are present, largely replacing spermidine, also contained up to 15% of the cell polyamine on the ribosome fraction. In the earliest studies with thermophiles, in which *Bacillus stearothermophilus* was grown in a spermine-containing medium of tryptone and yeast extract, it was found that growth at higher temperatures accumulated ribosomes of high spermine content (Stevens & Morrison, 1968). It can be seen in Table 7.3 that the organism grown at 45°C had a higher spermine content than that grown at 65°C. However, the ribosomes of the latter had accumulated far more spermine, and the ribosomes of cells grown at 45°C were markedly less stable at higher temperatures than those of cells grown at 65°C. Thus, the presence of a tetramine confers a selective advantage in the survival of a thermophile.

Stevens and Pascoe (1972) studied the binding of the intercalating dye, ethidium bromide, to double-stranded regions of the ribosomal RNA of *B. stearothermophilus*, as well as competition of ethidium with spermine in such binding. The dye is superficially analogous to spermidine, in that the primary amino groups (albeit of low pK) of the phenanthridinium structure are located three and four carbon atoms from the very basic quaternary nitrogen.

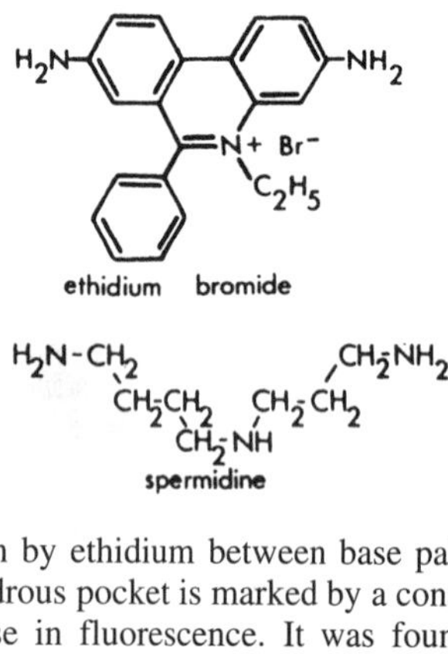

ethidium bromide

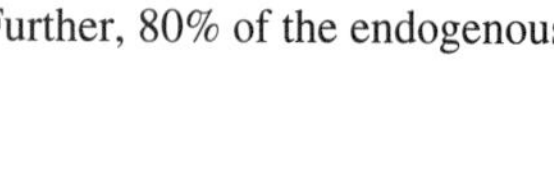

spermidine

Intercalation by ethidium between base pairs in a relatively anhydrous pocket is marked by a considerable 25-fold increase in fluorescence. It was found that spermine competes with ethidium for double-stranded sites in rRNA. Further, 80% of the endogenously bound spermine is displaced from ribosomes by fluorescent ethidium, implying that most of the spermine-binding sites in the ribosome are in double-stranded regions of the RNA. It was demonstrated also that the presumably RNA-bound [14C]-spermine could be cross-linked to a significant number of identifiable ribosomal proteins from both subunits (Kakegawa et al., 1986). Bound [14C]-spermidine, cross-linked with dimethylsuberimidate, was bound to eight different proteins.

The association of polyamine (i.e., spermidine and spermine) with ribosomes maintains relatively compacted structures, in contrast to the effects of decreased Mg^{2+} or the binding of high concentrations of ethidium. Isolated nucleates, RNA and DNA, are also compacted by the polyamines and the highest sedimentation coefficients of the RNA of the 50S ribosomal subunit were obtained by the addition of spermidine (Shatsky & Vasiliev, 1988). The presence of these polyamines appears to impose a higher degree of internal order and external regularity on many nucleates, facilitating the crystallization of these polymeric structures and the subsequent crystallographic analyses. The 50S ribosomal subunits of *B. stearothermophilus* have been crystallized (Yonath et al., 1980) as well as 2-dimensional sheets of ribosomal subunits suitable for electron micrography (Arad et al., 1987). Spermine (1 mM) was used in generating these latter structures. Halophiles lacking polyamines yielded crystals of ribosomal subunits without polyamines among the precipitants. However, in the formation of large single crystals of the 50S ribosomal subunit of *Halobacterium marismortui*, 10 mM spermidine was present in the crystallizing mix.

Despite their frequent use of the polyamines for the development of crystals, Yonath et al. (1982) considers the presence of the polyamine in the crystallizing mix as less influential than the purity of the ribosomal preparation. The formation of 3-dimensional crystals of *E. coli* monosomes was reported to be unaffected by the addition of spermidine (Wittmann et al., 1982). In any case, the current view of the structures of ribosomal subunits and monosomes has incorporated the knowledge derived from the many types of empirically obtained crystalline arrays in some of which polyamine preexists and in others to which polyamine has been added to impose order (Yonath & Wittmann, 1989).

TABLE 7.3 Polyamine concentration in *Bacillus stearothermophilus* grown in presence of spermine.

Growth temp. (°C)	Polyamines in whole cells (μmol/mmol RNA-P)		Polyamines in ribosomes (μmol/mmol RNA-P)	
	Spermidine	Spermine	Spermidine	Spermine
45	14–22	22–29	3.8	6.0
55	12–22	30–31	8.2	12.1
65	4–9	11–23	2.2	21.0

Adapted from Stevens and Morrison (1968).

Nucleoid

The electron microscopy of rapidly frozen thin sections of *E. coli* fixed with OsO_4 revealed the existence of a membrane-free internal cavity filled with ribosomes and fibrous DNA (Hobot et al., 1985). The latter may be localized in ribosome-free space, of which it is the main component, and in which the *nucleoid* is considered to exist as a more or less structured and folded entity (Robinow & Kellenberger, 1994). This structure lacks an enveloping membrane, although it is now believed that in *E. coli* and *B. subtilis* the bacterial chromosome is frequently attached to the membrane. Indeed the addition of 0.02 M spermidine to the cell prior to the fixation with OsO_4 produces bacteria whose thin sections indicate multiple regions of close proximity between DNA and membrane (Driedger, 1970). Apparent artifacts of preparation and fixation (e.g., the existence of mesosomes) are common in electron microscopy, and the significance of the observations of apparent chromosome attachment to cytoplasmic membrane is being explored by more chemical and genetic methods. In any case, the appearance of the nucleoid, whether relatively disperse or centrally spherical, is considered to reflect a real description of the DNA structure in various physiological states.

The existence of an *E. coli* strain that produces "minicells" (i.e., miniature DNA-deficient cells) appeared to provide an opportunity to determine the relative polyamine content of the nucleoid. As seen in Table 7.4, minicells are not deficient in total putrescine or spermidine, nor in the ratio of polyamine N to RNA-phosphate. The comparison does not provide evidence for a preferential binding of polyamine to DNA.

Condensed nucleoids have been isolated from exponentially growing *E. coli* (Stonington & Pettijohn, 1971). The cells were lysed in buffer of high ionic strength (2 M NaCl), and lysis was effected in part with the possibly compacting basic protein, lysozyme. The sedimented DNA complex (ca. 3200S) had a very low viscosity, which was greatly increased by addition of ribonuclease, implying a role for RNA in maintaining a folded structure. The DNA complex contained about 10% protein, stated initially to be largely components of RNA polymerase. Incomplete chains of messenger and ribosomal RNA (i.e., "nascent"), about 10% by weight, were also found associated with the DNA in the complex.

The molecular weight of the particle did not differ greatly from that calculated for the *E. coli* genome, about 2.5×10^9. This result then confirmed the electron microscopic view of the *E. coli* chromosome as a single closed ring of double-stranded DNA, packaged into tiny nuclear bodies of 1 µm diameter, representing only a small fraction of the cell volume. Other strains, grown in synthetic media, gave rise to somewhat more heterodisperse complexes of lower sedimentation rates, and complexes derived from cells beginning a replication cycle had a significantly lower rate of 1300S. It was estimated that a chromosome contained about 50 loops, each of which was about 20 µm long (Worcel & Burgi, 1972).

The isolation of the nucleoids in the presence of a high concentration of spermidine (8 mM in the lysing medium) and 5 mM $MgCl_2$ in the fractionating sucrose density gradient produced stably folded chromosomes. In the presence of *E. coli* DNA polymerase and DNA ligase, such folded structures, maintained in 4 mM spermidine, were excellent templates for semiconservative DNA replication. The added spermidine did not appear to affect the rate or extent of DNA synthesis, but an adequate amount of the amine may have been present in the originally isolated structure (Kornberg et al., 1974).

Extraction of alumina-ground phage-infected *E. coli* with buffer containing 5 mM spermidine eliminated 85% of the DNA and a relatively high percentage of the phage-induced RNA with the easily sedimentable bacterial walls and membranes (Cohen & Barner, 1962). Difficultly characterizable DNA–membrane complexes have been isolated from *E. coli* by various physical methods. The introduction of 5 mM spermidine in the absence of high concentrations of other salts for the isolation of folded nucleoids permitted the demonstration of membrane fragments containing particular identifiable proteins (Drlica et al., 1978; Portalier & Worcel, 1976). A portion of the *E. coli* origin of DNA replication, *oriC*, binds with high affinity to the bacterial membrane (Ogden et al., 1988; Rothfield, 1994). The association and release of DNA and membrane have been postulated to relate to the mechanism of chromosome segregation in dividing *E. coli*. About 100 proteins, which are components of some six different groups, have been detected in spermidine-isolated nucleoids and up to seven of these have been identified as membrane proteins (Moriya & Hori, 1981).

TABLE 7.4   RNA, DNA, and polyamine content of *Escherichia coli* and minicells.

E. coli K12 P678-54	Total content (mg/g, wet cells)		Polyamine/nucleotide (mol)		Polyamine N/RNA-P
	RNA	DNA	Putrescine	Spermidine	
Normal cells	56	5.2	0.06	0.006	0.14
Minicells	52	0.15	0.05	0.017	0.15

Adapted from Michaels and Tchen (1968).

The stabilization of the DNA folds at various temperatures were studied further, as shown in Figure 7.5. The efficacy of 2 mM spermidine in this function is evident; the aggregation of isolated nucleoids at higher concentrations of spermidine was also noted by Flink & Pettijohn (1975). Giorno et al. (1975) examined transcription in spermidine-stabilized nucleoids and observed that such structures facilitated the association of RNA polymerase and chain initiation of RNA on the condensed DNA.

Condensed nucleoids containing folded chromosomal DNA have been isolated from both Gram-negative and Gram-positive organisms, such as *B. subtilis* (Dworsky, 1977). Folding in the nucleoids of the latter, which also contain RNA, is sensitive to ribonuclease. The examination of the latter, lysed in the presence of spermidine, revealed folded nucleoids from *B. subtilis*, which proved to be similar in many respects (protein composition, transcription etc.) to those from *E. coli*.

Activities of Bacterial Chromosome: RecA and Homologous Recombination

The homologous recombination of genetic material in *E. coli* is determined by the protein product of the *recA* gene (Griffith & Harris, 1988). The recA protein can be

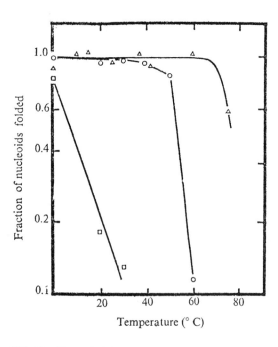

FIG. 7.5 Thermally induced unfolding of bacterial nucleoids in the presence or absence of spermidine; redrawn from Flink and Pettijohn (1975). Isolated nucleoids at similar DNA concentrations were incubated in 0.1 M NaCl and (Δ) 5 mM, (○) 2 mM, or (□) no spermidine. The solutions were heated for 5 min, chilled to 0°C, and sedimented.

crystallized in high yield from a crude cell lysate of *E. coli* by low concentrations of spermidine (Griffith & Shores, 1985). The 38,000 molecular weight protein is multifunctional and promotes pairing and strand exchange between DNA molecules in an ATP-dependent reaction. It also acts as a coprotease and promotes the proteolytic degradation of a *Lex-A* repressor via cleavage of the repressor protein at a specific peptide bond. Homologous recombination promoted by *recA* is believed to involve a triple-stranded DNA intermediate (Shchyolkina et al., 1994). The recA protein, isolated by spermidine precipitation, can bind to ssDNA and dsDNA, reactions essential to the ATPase action of the protein.

RecA protein (260 μg/mL) in 20 mM Tris (pH 7.5) can be maximally precipitated with 2 mM spermine or 3–4 mM spermidine. The microcrystals are generally too small to be useful for X-ray analysis. Precipitation by spermidine is inhibited above 50 mM anion, that is, chloride, sulfate, or phosphate. The preparations have high ssDNA-dependent ATPase activity and are effective in genetic recombination. However, the purified recA protein tends to aggregate and self-assemble into rods of varying length and bundles. Self-assembly is promoted by millimolar Mg^{2+} and indeed by conditions typical of in vitro strand exchange reactions (Brenner et al., 1988). The bundling of the protein may suppress the activity on ssDNA, as well as on inappropriate substrates such as dsDNA and RNA. In these terms, spermidine may be reducing and thereby regulating the active concentration of the occasionally overproduced recA protein (Story et al., 1992).

In crystallization from a crude lysate, the recA protein was contaminated with nucleates and prior chromatographic steps were introduced before precipitating with spermidine. Although spermidine was reported to be removed readily by dialysis, various samples have not been reported to be spermidine-free. Neither the spermidine-binding sites nor the association constants of these sites have been determined. It was observed that a ssDNA-binding protein (SSB) can assist some functions of the recA protein. The binding of SSB to complementary ssDNA strands and the rapid annealing of these strands requires micromolar levels of spermidine (Wei et al., 1992). It is possibly relevant that the recent crystallization of the SSB protein was effected in a mix containing 1 mM spermidine (Thorn et al., 1994).

Proteins other than RNA and DNA polymerases, the recA protein and the SSB, have also been found in the nucleoid. The possible analogies of some of these more easily localized and visualized constituents (e.g., histones) to the chromosomal constituents of eucaryotic cells induced studies of the bacterial proteins and their possible roles in the many complications of strand and chromosome replication and separation. Spermidine is possibly inadequate for such functions. A histonelike protein was described initially in a thermophilic mycoplasma in 1975, and within a decade at least six of these small proteins and their genetic determinants had been demonstrated in *E. coli*. Numerous similar members of this class have been found in eubacteria and archaebacteria. The genes for some of these proteins appear to determine nucleoid condensation (Barry et al., 1992), closure of DNA rings, and the regulation of synthesis of acid-induced basic amino acid decarboxylases (Shi & Bennet, 1994).

Repair of DNA

The problems of DNA repair and DNA replication have intersected at many specific points. For example, the recA protein discovered in the dissection of the functions promoting DNA repair, the so-called SOS functions, is assembled cooperatively onto ssDNA, generated by the unwinding of dsDNA as a result of the binding of the SSB protein. As noted above, the SSB is also active in the annealing of complementary strands of ssDNA, a reaction markedly enhanced by polyamine.

Early schemes for repair reactions foresaw roles of nucleases, DNA polymerases, and DNA ligases. Incision nucleases, which might recognize and eliminate ultraviolet-damaged DNA, containing compounds such as thymine dimers or thymine glycol, were not stimulated markedly by polyamine (Kleppe et al., 1981). *E. coli* DNA polymerase I, which might excise DNA damage and fill in the gap with new DNA, presented a more complex picture in which relatively high concentrations of the polyamines, >2 mM spermidine, were inhibitory. Nevertheless, putrescine and spermidine enhanced the activity with small homopolymers of d(A-T).

Two enzymes detected in T-even phage-infected cells, DNA ligase and polynucleotide kinase, were both increased in activity by polyamines at low concentrations. The bacteriophage-induced polynucleotide kinase adds a 5′-phosphate to a nucleic acid (both DNA and RNA) nicked to produce a 3′-phosphate and a free 5′-hydroxyl. This kinase activity assures that the gap to be closed by DNA ligase contains a 5′-phosphate. The active homogeneous enzyme contains four identical monomers of 33,000 molecular weight and has been studied in detail (Kleppe & Lillehaug, 1979). The enzyme is increased (fourfold) in activity by monovalent cations (0.1–0.15 M). This is also effected maximally by 2 mM spermidine or spermine concentrations that stabilize the oligomeric structure. Anionic polymers (e.g., heparin) are powerful inhibitors of the enzyme.

In the case of the polyamine-stimulated T4-induced polynucleotide ligase, ATP is used as a cofactor in the joining of the oligonucleotides. In the activity of the *E. coli* DNA ligase, the cofactor is NAD and the enzyme is stimulated by NH_4^+ ions but not by polyamines. The DNA ligase isolated from the thermophile, *Thermus thermophilus*, like the T4 DNA ligase, can catalyze duplex joining at base-paired ends of DNA (i.e., blunt-end ligation); this reaction is supported by the unusual amines of the thermophiles (Takahashi & Uchida, 1986). These include caldopentamine, thermine, spermine, and thermospermine, which were optimally active at 0.05, 0.1–0.2, 0.2, and 0.4 mM, respectively. The Thermus ligase will catalyse these reactions with the aid

of polyethylene glycol in the absence of polyamine. This agent of steric exclusion will markedly increase DNA concentration and possibly provoke a reaction of aggregation comparable to that obtained with spermidine in the DNA gyrase system. It may then be asked if this aggregation of substrate, produced in essentially all instances by the polyamines (Osland & Kleppe, 1977), is not a major cause of stimulation effected on the T_4 and Thermus ligases.

DNA Replication

The nucleate polymerases convert single-stranded RNA or DNA to double-stranded replicative forms (RF). The assay of the incorporation of a labeled nucleotide is far more sensitive than the complex reactions involved in recombinational events. The preparation of a single-stranded nucleate from tissue dsDNA is relatively simple, and the discovery of single-stranded DNA viruses (e.g., Φx174 and M13) introduced a relatively simple, genetically defined, and hence chemically definable DNA template for such studies.

Studies in DNA repair called attention to enzymes capable of filling in DNA gaps, polynucleotide kinase, DNA polymerase (i.e., I in *E. coli*), and DNA ligase. In the context of damaged and fragmented nucleates, a possible role for a polyamine, which might hold the pieces together, seemed reasonable. As seen earlier, this hypothesis led to observations suggesting a role for polyamine in the stabilization of the kinase and a cofactor role in certain DNA ligases. The formation of the double-stranded RF of viral DNA was used to test the possibility of an effect of polyamines in this reaction. With enzymes of a crude *E. coli* extract, spermidine (1.7–5.1 mM) was found to stimulate (5–20-fold) the formation of the Φx174 RF (Schekman et al., 1972). However, the experiments also revealed the need for a priming sequence of short RNA chains (Okazaki fragments) and the inability of the DNA polymerase alone to initiate a chain. Several genes and their gene products were required, and it was evident that the replication of a single-stranded template might involve an enzyme complex. Such a complex, PolIII star, was detected and purified; replication of ss Φx174 with this enzyme continued to require spermidine, a primer, and an additional protein (Wickner et al., 1973a). The removal of the RNA primer, and the filling in of the gap in formation of the Φx174 RF, was effected by the exonuclease of the *E. coli* or T4-induced DNA polymerase I and the DNA ligase of either *E. coli* or of T4-infected cells. In either of the latter instances it was convenient to include 3 mM spermidine in the mix (Westergaard et al., 1973).

As the numerous gene products essential to the synthesis of Φx174 RF were separated and analyzed by complementation assay, the spermidine effects became less striking (Wickner et al., 1973b). In the study of DNA synthesis in *B. subtilis*, the formation of an effective ternary complex with DNA polymerase and an RNA-primed DNA template was found to require either Mg^{2+} or spermidine (Low et al., 1974). At that time of

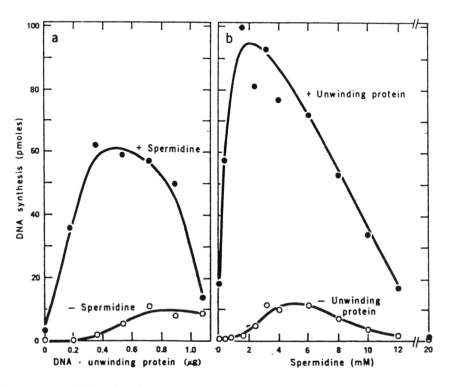

FIG. 7.6   Requirements in DNA synthesis for DNA-unwinding protein and spermidine; redrawn from Schekman et al. (1974).

dissection, spermidine facilitated reconstitution of *E. coli* systems essential for synthesis of Φ*x*174 but was far less active for the synthesis of a related phage (G4) RF (Schekman et al., 1974). Further, the replication of an unprimed Φ*x*174 template required spermidine and the unwinding protein, SSB (Fig. 7.6). Of three different replicative systems for the DNA of these *E. coli* phages, spermidine was found to be important in only one of these, Φ*x*174. Furthermore, the conversion of Φ*x*174 DNA to its RF occurred without any effect of exogenous spermidine (0.5–5.0 mM) in a system composed of five purified DNA-essential gene products plus SSB, two DNA elongation factors, two additional replication factors, and Mg^{2+}, ATP, and deoxynucleoside triphosphates (Wickner & Hurwitz, 1974). In short, it appeared as if the assembly of the total system (Kornberg, 1988) substituted proteins for the functions of the polyamines. Nevertheless, when Geiger and Morris (1980) returned to the unfractionated system, spermidine or various aminopropyldiamine analogues evoked a 20-fold activation of the formation of the RF.

The isolation of nucleoids with folded chromosomes from *E. coli* lysed with lysozyme and detergent in the presence of 10 mM spermidine made available fragments of *E. coli* DNA, which were replicable with added DNA polymerase I and DNA ligase (Kornberg et al., 1974). Although this system was unaffected in rate or extent by added spermidine, the preparation of the

DNA may have contained sufficient polyamine (untested) to have evoked this response in replication.

Synthesis, Accumulation, Regulation, and Elimination in Developing Cultures

Growth phases

Tradition and the relative ease of experimental manipulation tended to condition the form of many polyamine studies. It was convenient initially to use a nutritionally sufficient complex medium in which organisms might be grown in an overnight culture. Organisms might then be classified according to their need for air, and the overnight growth of cultures of *E. coli* was frequently assisted by swirling or bubbling to increase aeration. Nevertheless, the organisms stopped multiplying and the bacteria in this *stationary* phase, if diluted about 50–100-fold in fresh media, could begin multiplication again after a *lag phase* of about an hour. Cells in the stationary phase were markedly smaller than those in the phase of exponential growth, and the reentrance into multiplication required active growth of cell components in the so-called lag phase. Thus, the biological material (i.e., the cells of *E. coli*), indeed of any bacterial species, was quite variable depending on the phase of growth. As noted earlier, the polyamine contents of cells of *E. coli* and *L. casei*, determined in mass culture, varied widely as a function of the stage of growth, markedly increasing their spermidine contents before entering exponential growth. Simplification and definition of the nutrients of the medium did not change this pattern of growth very much; lag phase → exponential growth → stationary phase, but did permit some increased measure of experimental control in *E. coli* grown at optimal aeration and temperature. For example, limitation and exhaustion of the carbon source (e.g., glucose, 1 mg/mL), stopped growth in the exponential phase at 1×10^9 colony-forming bacteria/mL. A culture diluted 1 : 20 in the complete medium possessed a significant low turbidity that after a short lag increased its rate of growth and multiplication for a bit more than three divisions (or 3 h). Because turbidity was directly proportional to bacterial mass, one reference could be made to a mass doubling time of just about 1 h during exponential growth in glucose. Cells were harvested by centrifugation during exponential growth, resuspended in a complete defined medium, and permitted to develop for several hours. Samples were removed for analysis of viability, the polyamines, nucleic acid, etc. These were the preparations and manipulation of an *E. coli* culture in which the cells were examined and analyzed. Essentially all of the cells were in an exponential phase of growth in the range of 1×10^8 to 8×10^8 cells/mL.

In a typical early experiment of this design (Fig. 7.7) it was found that polyauxotrophic *E. coli* strain TAU, incubated in the exponential phase at 3×10^8 bacteria/mL containing 1 mg glucose/mL, grew and exhausted glucose in 2 h at 1.3×10^9; the bacteria remained at this level for an additional hour. In the period of incubation, the contents of putrescine, spermidine, and RNA had increased exponentially for 2 h and remained constant in the third hour when the cells stopped growing (Raina

et al., 1967). Two antibiotics, chloramphenicol and streptomycin, permitted a normal increase of spermidine and a significant increase of RNA for an hour, after which both remained approximately constant for two additional hours. Both antibiotics led to a decrease of intracellular putrescine, under conditions in which chloramphenicol was bacteriostatic and streptomycin was bacteriocidal.

Chloramphenicol and streptomycin inhibit protein synthesis but permit the synthesis of ribosomal RNA. In cultures of 15 TAU in which arginine is withheld and chloramphenicol is added, the RNA and spermidine both accumulate for an hour within the cells. Chloramphenicol is not lethal and its effect may be reversed by washing the bacteria.

Streptomycin does permit the increase of spermidine and ribosomal RNA in cells in the absence of protein synthesis. In this condition, however, there is an initial delayed extrusion of putrescine and a relatively prolonged accumulation of ribosomal RNA and spermidine. In all of the combinations studied, therefore, the spermidine contents of the cells harvested in the exponential phase and manipulated by subsequent controls of polymer synthesis paralleled their accumulation of ribosomal RNA. These results suggested that the spermidine content of *E. coli* might reflect the ribosome content of the cells (Raina et al., 1967).

Exogenous spermidine stimulated the incorporation of uracil into growing *E. coli*. Methionine is a precursor to spermidine in *E. coli*; when this amino acid, not an exogenous requirement for growth of strain 15 TAU, was added to the culture, the RNA and spermidine content of the cells were increased without a change in growth rate (Table 7.5). Tkachenko and colleagues (1989) studied the effects of a transition from aerobic growth to anaerobiosis in batch cultures of *E. coli*. As anaerobiosis developed, there were marked decreases in

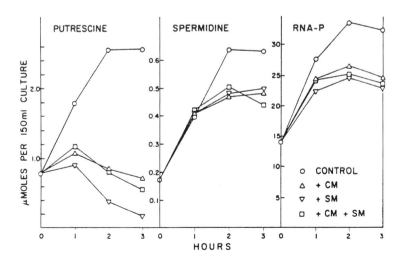

FIG. 7.7 Effect of chloramphenicol and streptomycin on the intracellular accumulation of polyamines and RNA in *E. coli* 15 TAU in a complete medium; redrawn from Raina et al. (1967).

TABLE 7.5 Effect of methionine on accumulation of RNA and polyamines in cultures of *Escherichia coli* 15 TAU.

Complete medium supplemented with	RNA-P	Polyamines (µmoles/150 mL)					
		Cellular				Total: cells + medium	
		Unconjugated		Total			
		Pu	Spd	Pu	Spd	Pu	Spd
None	15.7	0.80	0.25	0.94	0.44	3.40	0.56
Methionine	19.4	0.88	0.47	1.08	0.78	3.47	0.91

Pu, putrescine; Spd, spermidine. Adapted from Raina et al. (1967).

ODC, putrescine, spermidine, and ATP. The effects were reversed by oxygenation. The correlation of ATP and ODC content suggested a close coupling of energy status to polyamine biosynthesis (Tkachenko & Chudinov, 1990). Anaerobiosis also markedly reduced the phospholipid content of the bacterial membranes, raising the question of polyamine intermediation in these effects (Tkachenko et al., 1991).

In batch technique studies on the growth and composition of the unicellular Cyanobacterium *A. nidulans*, a culture was swirled in the light at 28°C in a mineral medium containing CO_2 (as 1% bicarbonate). The pH increased rapidly in the culture as the green cells grew with a generation time of 9 h, and the cell number increased as well as spermidine, chlorophyll, RNA, and protein for approximately 100 h (Fig. 7.8). After achieving maximal density and a pH of almost 11, the cells began to yellow but decreased in number very slowly. They lost more than 90% of their spermidine, which had been in the range of 1–2 mM during growth. Putrescine had been about 10% of the level of spermidine. Other components such as RNA and protein were also lost in the aged culture (Ramakrishna et al., 1978).

Exogenous ^{3}H-spermidine labeled in the aminopropyl moiety was rapidly converted by growing cells to 1,3-diaminopropane, which was also metabolized. Putrescine and cadaverine proved to be quite toxic to the growth of the cells, whereas the triamines and tetraamines were less so (Ramakrishna et al., 1978). When *A. nidulans* is grown for some hours in the presence of [2-^{14}C]-methionine, the cell spermidine is heavily labeled. The cells were harvested and washed and grown to maximum density in the absence of exogenous methionine. The total isotope content of the spermidine fraction remained constant for many hours despite a severalfold decrease of the specific radioactivity of the spermidine. Thus, cell spermidine did not turn over until the rapid postgrowth decay ensued. In the exponential phase of growth the decrease of specific radioactivity to one-half took 8.3 h, compared to a doubling time of 7.5 h in this experiment (Guarino & Cohen, 1979a).

Chemostat and control of rate of growth and cell size

The overnight culture revealed differences in cell size as a function of the phase of growth, and particularly in triamine content from a maximal rate of growth to the stationary phase. It was asked if the limitation of an essential nutrient during exponential growth might control the rate of cell division and cell size. The device that permitted such studies was the *chemostat*, in which the addition of a solution of a limiting essential nutrient to the growth chamber and the removal by siphoning of an equal amount of culture maintained the fed culture at constant volume in a medium of constant composition. Bacteria in such a system establish a constant rate of utilization of the limiting nutrient and a constant growth rate, which might be far less than that established as exponential in the ordinary overnight culture rich in nutrient. In cultures of *Aerobacter aerogenes* in a chemostat, the content of Mg^{2+} related directly to the RNA and ribosome content over a wide range of growth rates and cell size. Under conditions of Mg^{2+} limitation of *E. coli*, the inorganic cation was replaced to some extent by spermidine.

The chemostat has also been used to explore the routes of polyamine biosynthesis at various growth rates imposed by restriction of ornithine in an ornithine-requiring auxotroph of *E. coli* (Tabor & Tabor, 1969). The parental strain grown in a minimal medium directs the products of ornithine and arginine to form putrescine and spermidine at 25–30% of the amount of the protein-bound arginine. In limiting [^{14}C]-ornithine at various growth rates, some 5–18% of the ornithine is still directed to putrescine and spermidine, indicating a still significant role for these organic cations in the life of the organism. [^{14}C]-Arginine may also be used to replace ornithine and is also directed to the synthesis of polyamines.

The addition of exogenous putrescine or spermidine to the cultures markedly reduced the conversion of the amino acids to the cellular polyamines. Cells grown in

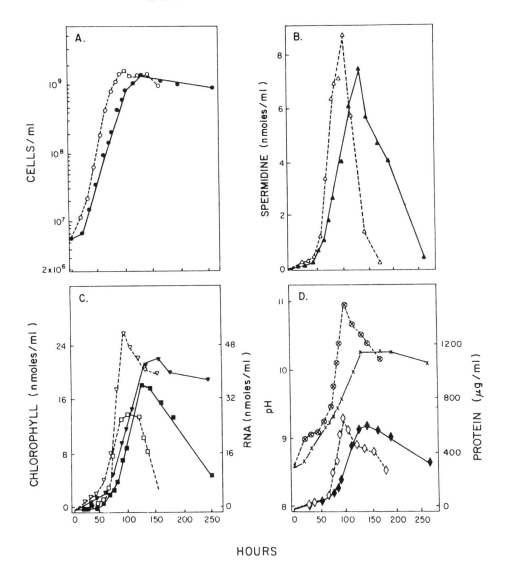

FIG. 7.8 Changes during growth of *Anacystis nidulans* in cell number, RNA, protein, chlorophyll, pH, and spermidine in green (closed symbols) or white (open symbols) light; redrawn from Ramakrishna et al. (1978). (○, ●) cell number; (△, ▲) spermidine; (∇, ▼) RNA; (◊, ◆) protein; (□, ■) chlorophyll; (⊗, x) pH.

the presence of exogenous polyamines were rapidly inhibited in CO_2 production from carboxyl-labeled ornithine. This inhibition was also demonstrable with putrescine or spermidine in cell extracts. Inhibitory effects of the polyamines (ca. 50%) on the decarboxylase in the extracts were detected at 7×10^{-4} M putrescine and $<1.4 \times 10^{-4}$ M spermidine. Because concentrations in *E. coli* strain B were at levels greater than 19 mM putrescine and 6 mM spermidine, much of the cell polyamine appeared associated with structure or these enzymes were partially inhibited or repressed.

Acetylation in batch cultures

When *E. coli*, grown in log phase in minimal medium, was filtered rapidly from the medium, it was found that cellular polyamine had been completely retained within the bacterium. However, the normal biochemical operation of chilling and storage of the samples before separation of cells and medium led to the loss of putrescine to the medium, and the conversion of spermidine to an *N*-acetylspermidine (about 59% conversion at 5°C in 2 h) (Tabor, 1968). Very little *N*-acetylspermidine was present in bacteria separated rapidly at the exponential phase of the growth cycle (Tabor & Dobbs, 1970). The leakage of intracellular putrescine, significant in stored cultures, was of the order of 50%; some acetylation of the diamine occurred as well.

Cultures of *E. coli* did not degrade spermidine or spermine but did produce monoacetylputrescine and equal amounts of the N^1-acetylspermidine and N^8-acetylspermidine isomers (Dubin & Rosenthal, 1960). These workers found monoacetylputrescine, N^1-acetylspermidine, and N^8-acetylspermidine within *E. coli* strain B, without evidence of diacetylputrescine or diacetylspermidine. When spermine was added to the culture, this was taken up and converted to mono- and diacetylspermine, concomitantly with a marked reduction of the internal concentration of putrescine and spermidine. These acetylations were also found in Staphylococcus (Rosenthal & Dubin, 1962).

A spermidine N^1-acetyl transferase, capable of acetylating spermine as well, was eventually found in *E. coli*, but this enzyme scarcely acetylated putrescine or formed N^8-acetylspermidine (Matsui et al., 1982). Various stress responses of *E. coli* (i.e., heat shock, cold shock, ethanol, etc.) evoked the enhanced production of almost equal amounts of N^1- and N^8-monoacetylspermidine in the cultures. However, the level of *N*-acetyl transferase was not affected by these stresses (Carper et al., 1991). Because the enzyme is not induced by stress reactions, the nature of the controls on acetylation within the bacterium remains a mystery. Furthermore, the problem of the mechanism of formation of the N^8-acetylspermidine in *E. coli* has not been clarified.

Biochemical studies of bacteria and media of cultures developed in normal and inhibited growth (Raina & Cohen, 1966) had been pursued with steps of chilling and centrifugation of the cultures and the findings were the cells of *E. coli* strain TAU contained only 20% of the total putrescine and 60% of the total spermidine as conjugated derivatives when harvested during exponential growth in a complete medium. In subsequent studies with strains of *E. coli* (e.g., TAU) in which the precursors for polyamine synthesis were available but the synthesis of negatively charged cellular constituents such as ribosomes had been prevented, the more highly charged spermidine was often excreted as the conjugated amine. Spermidine synthesis is not inhibited in *E. coli*, but acetylation minimizes charge, increases dissociation from anionic cell structure, and facilitates elimination of the potentially toxic amine by excretion as the acetyl spermidine.

In *E. coli* strain TAU, the elimination of the essential amino acid arginine from the medium prevented the intracellular synthesis of ribosomal protein and of ribosomal RNA, the latter being known as the "stringent" response. Within such an organism intracellular spermidine remained essentially constant, but polyamine continued to be excreted. In such a system putrescine was largely unsubstituted, while spermidine was largely conjugated. In these systems it appears that the synthesis of RNA does not control the synthesis of spermidine but

does determine its intracellular accumulation. The toxicity of spermidine accumulation in *E. coli* in the absence of acetylation has been demonstrated in a strain lacking the spermidine N^1-acetyltransferase (Fukuchi et al., 1995).

Studies on Stringent and Relaxed Responses

The acquisition of the polyauxotrophic strains of *E. coli* strain 15 TAU, deficient in the ability to synthesize thymine (T), arginine (A), and uracil (U), permitted studies on the biosynthesis and accumulation of the polyamines in the presence and absence of the synthesis of DNA (T⁻), protein (A⁻), and RNA (U⁻). Despite the artifact introduced in these studies by the "normal" biochemical procedure of chilling to stop the reaction, it was found that conditions of incubation of the bacteria permitting RNA synthesis and accumulation led to a concomitant and parallel accumulation of spermidine (Raina et al., 1967). The putrescine content of the cells doubled in the absence of arginine in the medium or following the exhaustion of glucose in the culture; these increases were also accompanied by a marked extrusion of free putrescine into the medium.

Strain 15 TAU stops rRNA synthesis in the absence of an essential amino acid; it is known as stringent. Many biosynthetic reactions are blocked in the "stringent response," which arrests the synthesis of ribosomal RNA (and many other cell components) when an amino acid deficiency prevents protein synthesis; this arrests an unbalanced accumulation of such RNA. An organism capable of such a response is designated as possessing a *Rel⁺* gene; a mutant strain lacking the gene is described as *relaxed*. The protein determined by the *E. coli Rel⁺* gene has been described as the stringent starvation protein, which was thought to form an inhibitory complex with a portion of the RNA polymerase of the bacterium. The structure and composition of the protein has been deduced from the structure of the isolated gene (Serizawa & Fukuda, 1987).

Under conditions of amino acid starvation, the accumulation of spermidine was markedly greater in the relaxed cells, because those cells also accumulated ribosomal RNA. In addition, relaxed strains synthesized far more spermidine in the absence of protein and RNA synthesis than did the stringent organism (Cohen et al., 1967). This result appeared to relate the phenomenon of stringency to some consequences of polyamine synthesis.

The product of the *RelA* gene also catalyzed the formation of the compound pppGpp in a ribosome-dependent reaction containing unacylated tRNA, ATP, and GTP. The pentaphosphate is thought to be a precursor of a major mediator of the stringent response (Cashel & Rugg, 1987); it is formed in the reaction,

$$\text{ATP} + \text{GTP} \rightarrow \text{pppGpp} + \text{AMP}.$$

The pentaphosphate is converted to guanosine 3′-diphosphate 5′-diphosphate, ppGpp. Fatty acid starvation in *E. coli* also provokes an accumulation of ppGpp (Seyfzadeh et al., 1993).

A complex of ppGpp with the elongation factor Tu active in protein synthesis, that is, EF-Tu·ppGpp, binds to ribosomes and inhibits the formation of ternary complexes essential to protein synthesis comprising aminoacyl tRNA + EF-Tu + GTP (Dix & Thompson, 1986). Although the effect of ppGpp in inhibiting protein synthesis has been accepted (Svitil et al., 1993), the possibility that the nucleotide is also the major inhibitor of the synthesis of ribosomal and mRNA remains under investigation (Hernandez & Bremer, 1993; Sørensen et al., 1994). In many systems the formation and concentration of ppGpp is consistent with the many inhibitory effects observed in the stringent response (Justensen et al., 1986).

Some relations have been recorded linking the polyamines to the accumulation of ppGpp. Igarashi et al. (1983) reported that 2–3 mM spermidine sharply stimulated the formation of ppGpp in a ribosome-containing cell-free system low in Mg^{2+} (Fig. 7.9). Such a system required the stringent factor protein, mRNA, ribosomes, and tRNA. A similar stimulation of ppGpp synthesis by spermidine was demonstrated with a polyamine-requiring mutant with or without amino acid starvation.

In another *E. coli* strain unable to synthesize putrescine, it was found that the organism was relaxed in the absence of polyamine. On addition of putrescine or spermidine, the organism became stringent in its synthesis of RNA and concomitantly synthesized large amounts of ppGpp (Goldemberg, 1984). Thus, both groups appear to have demonstrated a polyamine requirement in intact cells for the formation of the presumed inhibitory nucleotide, ppGpp. Nevertheless, both groups have demonstrated the development of abnormal ribosomes

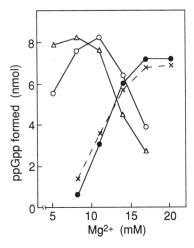

FIG. 7.9 Effects of polyamines on ppGpp formation in a ribosomal system; redrawn from Igarashi et al. (1983). (●) No addition; (○) 2 mM spermidine; (Δ) 3 mM spermidine; (×) 10 mM putrescine.

under conditions of polyamine depletion. Because the polyamines are important in maintaining ribosomal stability and functionality and the synthesis of ppGpp is ribosome dependent, the causal relation between polyamine depletion and ppGpp formation is not immediately evident.

Formation of Glutathionylspermidine

A covalent compound containing spermidine and glutathione was isolated from an acid extract of *E. coli* harvested from a glucose-salts medium (Dubin, 1959). When the cells were grown in the presence of spermine, a spermine derivative of glutathione was obtained. The spermidine–glutathione compound could be produced in extracts of *E. coli* containing ATP and Mg^{2+}. An enzyme with this activity, glutathionyl spermidine synthetase, was purified, and recently to near homogeneity (Bollinger et al., 1995; Tabor & Tabor, 1971). It carries out the reaction,

$$\begin{array}{c} \text{glutathione} \\ + \\ \text{spermidine} \end{array} \xrightarrow[\text{Mg}^{2+}]{\text{ATP}} \text{glutathionylspermidine.}$$

The product has been found to be γ-glutamylcysteinylglycylspermidine, that is,

$$HOOCCHCH_2CH_2CONHCHCONHCH_2CONH(CH_2)_3NH(CH_2)_4NH_2$$

with NH₂ on the first carbon, and CH₂–SH on the cysteinyl residue

γ-glutamyl – cysteinyl – glycyl – N^1–spermidine

The enzyme also hydrolyzes the glutathionylspermidine amide bond. As a result of the exhaustion of glucose, the acidification of an inadequately buffered medium, or anaerobiosis in a slightly acidic medium (pH 5–6), the glutathionyl derivative appears to be formed almost entirely when the culture enters the stationary phase with a high concentration of cells under anaerobic conditions at a pH <7.0. Essentially all of the cell spermidine is derivatized under these conditions (Tabor & Tabor, 1970). On dilution and aeration of the culture in media of pH 7.0, the newly formed glutathionyl derivative is hydrolyzed within the cells to form free spermidine. A similar hydrolysis was found in a bacterial extract. This finding suggests important, if not essential, roles for the sequestration of spermidine in the generation of the stationary phase and for its release during the lag phase.

Nevertheless, glutathionylspermidine was not found in other organisms for many years. Recently it was detected in a species of *Citrobacter* (Hisano et al., 1991), and more significantly as a major and crucial component of the biosynthesis effected by certain protozoan parasites, the Trypanosomatids and Leishmanii. These protozoans contain disubstituted spermidine, bis-glutathionyl-spermidine, termed *trypanothione* (Chap. 18).

E. coli Deficient in Polyamine

Essential or stimulatory effects of polyamine in the growth media were noted in studies on *Hemophilus parainfluenzae* and *Neisseria perflava*. Study of *Pasteurella tularensis* and *Mycoplasma* var. detected stabilization of their membranes by polyamine. Cadaverine was found to be an essential component of the cell wall of *Selenomonas ruminantium*. *Veillonella alcalescens* will lyse in the absence of putrescine or cadaverine, suggesting a possible role of a diamine in synthesis of the cell wall.

The clarification of structures and metabolic pathways of the bacterial cell, that is, ribosomes and nucleoid, began to point to roles for the polyamines in the "model" system, *E. coli*. The genetic and metabolic manipulation of the model to depress or eliminate routes leading to the polyamines now permitted some analysis of the causal relations of polyamine depletion on essential structure or metabolism. The earliest studies of a polyamine-deficient strain of *E. coli* were on a mutant defective in the conversion of arginine to putrescine and inhibited in another route from ornithine to putrescine. The organism grew poorly and developed in "snakelike forms," indicative of an effect on cell division (Hirshfield et al., 1970). The further study of similar polyamine auxotrophs of *E. coli* revealed such phenomena as the partial replacement of putrescine by cadaverine and the latter's conversion to the spermidine analogue, aminopropylcadaverine, transport systems for polyamines, and their precursors, etc. In one study of the effects of polyamine deficiency in this mutant, a slow turnover of *E. coli* phospholipid, entirely localized in the membrane, was markedly stimulated by putrescine or spermidine (Munro & Bell, 1973).

A polyamine deficiency also revealed an early arrest of protein synthesis (Young & Srinivasan, 1972) and significant alterations of the ribosome population in the deficient cells (Algranati et al., 1975). The cells contained fewer 30S subunits and 70S monomers, which were restored after addition of putrescine to the depleted cells. The reduced rate of protein synthesis was ascribed to a poorly efficient initiation step (Algranati & Goldemberg, 1981).

1,3-Diaminopropane stimulated growth rate as well as putrescine did, and norspermidine was as active as spermidine in filling the growth requirements of several *E. coli* mutants (Morris & Jorstad, 1973). A systematic study of the stimulation of growth and protein synthesis by putrescine derivatives revealed partial or considerable decreases of these effects by *N*-methylation and other alkyl derivatives. *N*-Propylputrescine, lacking only one amino group of spermidine, was inactive in enhancing growth (Goldemberg et al., 1983).

In subsequent studies with this group of depleted mutants and derived strains, readdition of polyamine (i.e., putrescine or spermidine) rapidly and preferentially stimulated the synthesis of a specific protein, P1, that had a 62,000 molecular weight (Mitsui et al., 1984a). This was implicated as a periplasmic binding protein in an oligopeptide transport system (Kashiwagi et al., 1990a). This result also indicates an indirect role for polyamines in supplying nutrients during cell growth. Similarly the addition of polyamines to a polyamine-starved organism was found to stimulate the synthesis of several subunits ($\beta\beta'$) of RNA polymerase severalfold (Mitsui et al., 1984b). This result suggests a role in RNA synthesis.

In one depleted mutant of *E. coli*, K101, with an 8.5-h mass doubling time, added polyamine greatly stimulated the early synthesis of the specific essential ribosomal protein, S1 (Kashiwagi et al., 1989). The synthesis of a few other essential proteins was also stimulated, although somewhat later and less actively. Putrescine appeared to stimulate the synthesis of a ribosome protein and its association to depleted ribosomes. The various steps of stimulation of proteins produced, type of RNA evoked, and translation effected varied as a function of the mutant bacterium employed and in various in vitro studies varied as a function of the type of DNA template, RNA polymerase, and source of translating components (Watanabe et al., 1983).

Recent Observations on Polyamine Transport

The experimental need to separate uptake and efflux led to the study of transport in biosynthetically deficient mutants and in membrane vesicles. The uptake of an amine appeared to be unidirectional as a result of the binding of the previously free amines to major bacterial components: cytoplasmic nucleates and membrane phospholipids. Using both a polyamine-deficient mutant and vesicles, thereby eliminating a potential biosynthetic supply, uptake was inhibited by protonophores, for example, carbonyl cyanide *m*-chlorophenylhydrazone (CCCP), and seemed almost entirely dependent on a proton motive force (Kashiwagi et al., 1986). CCCP does not markedly inhibit the internal biosynthesis of osmolytes during hyperosmotic adaptation and growth on glucose as a carbon source. This group also demonstrated the putrescine stimulation of the synthesis of a periplasmic oligopeptide-binding protein in a polyamine-requiring mutant (Kashiwagi et al., 1990a). Induction of this binding protein by putrescine facilitated transport and use of a labeled tripeptide.

The existence of a membrane, a hydrophobic permeability barrier to many hydrophilic solutes, suggests the need for specific transport mechanisms mediated by proteins. Several types of membrane transport systems described earlier were found to be protein based. However, studies with cyanobacteria in the dilute media in which alkalinity was generated by photosynthetic growth, unexpectedly revealed a major role for mere diffusion for the uptake of the polyamines. The rate of uptake of putrescine was a function of the external pH and the concentration of exogenous amine (Guarino & Cohen, 1979b). The former determined the fraction of the nonprotonated amine, as determined by the p*K* values of the amino groups. Thus putrescine with p*K* values of

9.0 and 10.5 is found to be taken up actively above pH 8.5 (Fig. 7.10). A growth medium containing Na_2CO_3 from which CO_2 is removed photosynthetically can rapidly achieve a pH of 10.5. At pH 9.5 (0.15 mM) putrescine is rapidly taken up to a concentration of 80–90 mM and is lethal to >99% of the cells in 40 min. The nonprotonated putrescine is trapped intracellularly as the protonated cation and is retained at the intracellular pH of 7.5–7.6 (Guarino & Cohen, 1979b). In this system the uptake of putrescine was partially inhibited by the inhibition of photosynthesis or by the protonophore CCCP, indicating some dependence of accumulation on energy production. Protein synthesis was cut off first in this system, and it was found that the isolated ribosomal subunits contained covalently bound putrescine and had

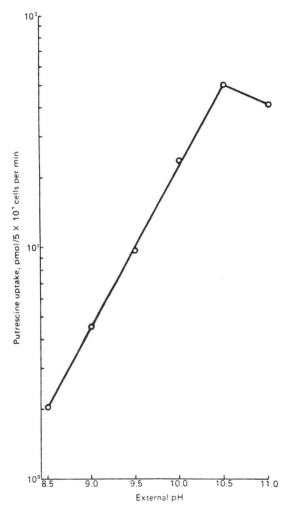

FIG. 7.10 Initial rate of putrescine uptake by *Anacystis nidulans* as a function of external pH; redrawn from Guarino and Cohen (1979b). Exponential phase cells were washed and resuspended in growth media that was adjusted to the indicated pH value.

been irreversibly dissociated. Other cell polymers contained even larger amounts of covalently bound putrescine.

Chen and Cheng (1988) found that putrescine uptake at pH 11.0 in an obligately alkalophilic bacterium, *B. alkalophilus*, is Na^+ dependent and that antibiotics disrupting Na^+/K^+ balance were inhibitory to the uptake. These workers postulated a Na^+/putrescine symport mechanism.

Mutants of *E. coli* deficient in polyamine transport were obtained by Kashiwagi et al. (1990b) from a biosynthetically deficient strain. There were four types: one specific for uptake of spermidine alone, one specific for spermidine and putrescine (II), and two (I and III) specific for putrescine alone. Genes were mapped and isolated for transport systems I–III (Table 7.6). These were placed in variously deficient strains, and the physiology of the transport systems was then examined. In each system, transport was most sensitive to CCCP. In the mutant system II, with an apparently common spermidine and putrescine mechanism, spermidine uptake completely inhibited putrescine transport. Several strong inhibitors for putrescine uptake were detected, including N^1-acetylspermidine and N^1-acetylspermine.

Putrescine transport systems I and III are genetically separable in *E. coli* and III appeared more active than I, which was more strongly inhibited by monovalent cations. The gene for I encoded a protein of 439 amino acids (46,494 calculated kilodaltons), whose sequence appears to be organized into 12 linked transmembrane spanning segments. The I gene, termed *potE* (Kashiwagi et al., 1991), was present in an operon that contained a gene (*speF*) for an inducible ODC; the operon appeared to be repressed normally. The inducible protein determined by *speF* was homologous (65%) to a constitutive ODC but these differed significantly in pH optima. *SpeF* was scarcely expressed at pH 5.2, unlike the constitutive enzyme produced at that pH.

The "gene" (II) for the spermidine/putrescine transport system contained open reading frames (termed *potA, potB, potC,* and *potD*) for four distinct proteins, all of which were necessary for optimal transport (Furuchi et al., 1991). One of the proteins was membrane-associated whereas two others contained transmembrane-spanning segments. A fourth, encoded by *potD*, was a periplasmic polyamine-binding protein, capable of binding [^{14}C]-spermidine or [^{14}C]-putrescine specifically. The far greater affinity of the potD protein for spermidine was demonstrated. The genes *potA* and *potD* were essential for spermidine uptake; spermidine transport in membrane vesicles obtained from cells supplied with *potA, potB,* and *potC* was dependent on the addition of the potD protein (Kashiwagi et al., 1993). Despite great inhibition of spermidine uptake by CCCP, this transport was completely dependent on ATP, thereby also involving this energy source in polyamine transport. It was found that ATP was bound to the potA protein, and that the elimination of the ATP-binding site in the potA protein blocked spermidine uptake (see Kashiwagi et al., 1995).

TABLE 7.6 Polyamine transport systems of *Escherichia coli.*

System	Plasmid designation and gene location	Protein transported	Protein designation	Molecular weight (kDa)	Position and function
I	pPT71, 16 min	Putrescine	potE	46	Transmembrane excretion
II	pPT104, 15 min	Spermidine and putrescine	potA	43	Membrane associated ATP binding
			potB	31	Transmembrane channel forming
			potC	29	Transmembrane channel forming
			potD	39	Periplasmic binding protein
III	pPT79, 19 min	Putrescine	potF	38	Periplasmic binding protein
			potG	45	Membrane-associated ATP binding
			potH	35	Transmembrane channel forming
			potI	31	Transmembrane channel forming

The genetic determinants of the putrescine transport system (III) were located in a distinctive mapping position in the *E. coli* chromosome and were found to contain four open reading frames, *potF, potG, potH,* and *potI,* coding for distinctive proteins. The expression of the four proteins was essential for maximal putrescine transport activity. The potF protein is believed to be a putrescine-specific binding protein, whereas the potG protein has a distinctive nucleotide binding site, as does the potA protein. The potH and potI proteins were similar in their content of transmembrane spanning elements, like those of system II. The various components of system III, F, G, H, and I, possessed some homology to comparably functional components of system II: D, A, B, and C.

The roles of system I and the gene *potE* were explored in the excretion of putrescine. Putrescine was taken up into inside-out membrane vesicles (i.e., excreted) only when the vesicles were prepared from cells carrying system I and *potE* (Kashiwagi et al., 1992). This uptake (excretion) was not affected by CCCP, and it was asked if the system did not serve as an antiporter. Exogenous ornithine or lysine strongly inhibited the incorporation of exogenous putrescine. When the vesicles were preloaded with ornithine, an equal number of molecules of the amino acid was removed from the vesicles as putrescine was taken up. The antiporter system could also exchange ornithine for ornithine and putrescine for putrescine, or operate with right-side out membranes. However, it is not thought to function actively in the bacterial uptake of putrescine. It was suggested that the *potE* product is concerned mainly with the elimination of excessive intracellular levels of the diamine under conditions in which the endogenous diamine inhibits ODC activities. The potE protein does not function with N^1-acetylspermidine, whose excretion from *E. coli* would require yet another system.

Table 7.6 summarizes the recent data on these *E. coli* transport systems. Sequences similar to those of *potA, potB,* and *potC,* for the transport of putrescine and spermidine, were reported recently in the tiny chromosome of *Mycoplasma genitalium.* Sequences for the transport

of the polyamines as well as one for ODC have been reported in the genome of *Haemophilus influenzae.*

Essentiality of Polyamines

From the point of view of Darwinian evolution, it might be supposed that the study of "essentiality" is merely a compulsion created by the history of biochemical research, because we now consider that a capability facilitating successful survival need not be all or none. Even so, an answer is relevant to the potential success of an inhibitor of a polyamine pathway and to the dissection of the roles of the compounds.

Given the *E. coli* auxotrophs unable to synthesize putrescine and spermidine, it was found that the organisms were not killed by depletion and indeed grew quite slowly. A polyauxotrophic strain is known that lacks this biosynthetic equipment and doubles in a polyamine-deficient synthetic medium at a rate of 12% of normal growth (Kashiwagi et al., 1989). A mutant strain lacking detectable putrescine or spermidine grew indefinitely in the absence of exogenous polyamine with a growth rate one-third that in the presence of polyamines (Hafner et al., 1979), but had virtually lost the ability to produce its lysogenic phage, λ. Several other functions were inhibited, for example, ca. 90% inhibition for T4 multiplication, but were otherwise completed. These putrescine- and spermidine-deficient strains still produced small amounts of cadaverine, that is, 1% of the normal wild type concentration of 1,4-diaminobutane. However *E. coli* is a biochemical freak, containing perhaps 100 times the diamine content of many other organisms. In this study, then, the cadaverine content of the polyamine-deficient organisms approaches the normal diamine content of many otherwise "normal" organisms. Earlier it was observed that an *E. coli* strain, deficient in putrescine and spermidine and actively producing cadaverine and aminopropylcadaverine, was still capable of a retarded and low level of virus DNA synthesis (Dion & Cohen, 1972). The cadaverine–aminopropylcadaverine "back-up" system for polyamine deficiency would have to be eliminated to permit the analysis of polyamine essentiality.

Continuing studies of *E. coli* mutants, with additional deficiencies in cadaverine synthesis and streptomycin resistance (Tabor et al., 1980), led to an apparent demonstration of a so-called "absolute" requirement of polyamine for growth (Tabor & Tabor, 1982). Further studies with such mutants indicated a key role of polyamines in the organization of protein synthesis (Chaps. 22 and 23). It is apparent that the study of polyauxotrophic mutants of *E. coli*, which are deficient in polyamine biosynthesis, continues to reveal complexities of polyamine metabolism and possible roles of these compounds in critical structures and functions.

REFERENCES

Akers, H. A. & Dromgoole, E. V. (1982) *Trends in Biochemical Sciences*, **April**, 156–157.

Algranati, I. D., Echandi, G., Goldemberg, S. H., Cunningham–Rundles, S., & Maas, W. K. (1975) *Journal of Bacteriology*, **124**, 1122–1127.

Algranati, I. D. & Goldemberg, S. H. (1981) *Biochemical and Biophysical Research Communications*, **103**, 8–15.

Angert, E. R., Clements, K. D., & Pace, N. R. (1993) *Nature*, **362**, 239–241.

Antoni, V., Christophersen, C., Hougaard, L., & Nielsen, P. H. (1991) *Comparative Biochemistry and Physiology B*, **99**, 1–18.

Arad, T., Piefke, J., Gewitz, H. S., Romberg, B., Glotz, C., Müssig, J., Yonath, A., & Wittmann, H. G. (1987) *Analytical Biochemistry*, **167**, 113–117.

Barry, III, C. E., Hayes, S. F., & Hackstadt, T. (1992) *Science*, **256**, 377–379.

Bollinger, Jr., J. M., Kivon, D. S., Huisman, G. W., Kolter, R., & Walsh, C. T. (1995) *Journal of Biological Chemistry*, **270**, 14031–14041.

Brenner, S. L., Zlotnick, A., & Griffith, J. D. (1988) *Journal of Molecular Biology*, **204**, 959–972.

Buch, J. K. & Boyle, S. M. (1985) *Journal of Bacteriology*, **163**, 522–527.

Carper, S. W., Willis, d. G., Manning, K. A., & Gerner, E. W. (1991) *Journal of Biological Chemistry*, **266**, 12439–12441.

Cashel, M. & Rudd, K. E. (1987) In *Escherichia coli* and *Salmonella typhimurium*: Cellular and Molecular Biology, Vol. 2, pp. 1410–1438, F. C. Neidhardt, Ed. Washington, D.C.: American Society for Microbiology.

Celis, R. T. F. (1984) *European Journal of Biochemistry*, **145**, 403–411.

Chang, C., Shuman, H., & Somlyo, A. P. (1986) *Journal of Bacteriology*, **167**, 935–939.

Chen, K. Y. & Cheng, S. (1988) *Biochemical and Biophysical Research Communications*, **150**, 185–191.

Dandekar, A. M. & Uratsu, S. L. (1988) *Journal of Bacteriology*, **170**, 5943–5945.

Dion, A. S. & Cohen, S. S. (1972) *Journal of Virology*, **9**, 423–430.

Dix, D. B. & Thompson, R. C. (1986) *Proceedings of the National Academy of Sciences of the United States of America*, **83**, 2027–2031.

Driedger, A. A. (1970) *Canadian Journal of Microbiology*, **16**, 881–882.

Driessen, A. J. M., Moelnaar, D., & Konings, W. N. (1989) *Journal of Biological Chemistry*, **264**, 10361–10370.

Driessen, A. J. M., Smid, E. J., & Konings, W. N. (1988) *Journal of Bacteriology*, **170**, 4522–4527.

Drlica, K., Burgi, E., & Worcel, A. (1978) *Journal of Bacteriology*, **134**, 1108–1116.

Dworsky, P. (1977) *FEMS Letters*, **1**, 95–98.

Elliott, B. & Michaelson, I. A. (1969) *Proceedings of the Society for Experimental Biology and Medicine*, **131**, 105–108.

Epstein, W. & Laimins, L. (1980) *Trends in Biochemical Sciences*, **January**, 21–23.

Favaudon, V. & Pochon, F. (1976) *Biochemistry*, **15**, 3903–3911.

Field, A. M., Rowatt, E., & Williams, R. J. P. (1989) *Biochemical Journal*, **263**, 695–702.

Flink, I. & Pettijohn, D. E. (1975) *Nature*, **253**, 62–63.

Frydman, B., Frydman, R. B., De Los Santos, C., Garrido, D. A., Goldemberg, S. H., & Agranati, I. D. (1984) *Biochimica et Biophysica Acta*, **805**, 337–344.

Fuerst, J. A. & Webb, R. I. (1991) *Proceedings of the National Academy of Sciences of the United States of America*, **88**, 8184–8188.

Fujihara, S. & Yoneyama, T. (1994) *Microbiology*, **140**, 1909–1916.

Fukuchi, J., Kashiwagi, K., Yamagishi, M., Ishihama, A., & Igarashi, K. (1995) *Journal of Biological Chemistry*, **270**, 18831–18835.

Furuchi, T., Kashiwagi, K., Kobayashi, H., & Igarashi, K. (1991) *Journal of Biological Chemistry*, **266**, 20928–20933.

Gabbay, E. J., Glasser, R., & Gaffney, B. L. (1970) *Annals of the New York Academy of Sciences*, **171**, 810–826.

Gangola, P. & Rosen, B. P. (1987) *Journal of Biological Chemistry*, **262**, 12570–12574.

Geiger, L. E. & Morris, D. R. (1980) *Biochimica et Biophysica Acta*, **609**, 264–271.

Gerlach, G., Allen, B. L., & Clegg, S. (1989) *Infection and Immunity*, **57**, 219–224.

Giorno, R., Stamato, T., Lydersen, B., & Pettijohn, D. (1975) *Journal of Molecular Biology*, **96**, 217–237.

Goldemberg, S. H. (1984) *Biochemical Journal*, **219**, 205–210.

Goldemberg, S. H., Algranati, I. D., Miret, J. J., Garrido, D. O. A., & Frydman, B. (1983) *Advances in Polyamine Research*, **4**, 233–244.

Griffith, J. D. & Harris, L. D. (1988) *Critical Reviews in Biochemistry*, **23**(Suppl. 1), S43–S86.

Griffith, J. & Shores, C. G. (1985) *Biochemistry*, **24**, 158–162.

Guarino, L. A. & Cohen, S. S. (1979a) *Analytical Biochemistry*, **95**, 73–76.

——— (1979b) *Proceedings of the National Academy of Sciences of the United States of America*, **76**, 3184–3188.

Günther, T., & Peter, H. W. (1979) *FEMS Microbiology Letters*, **5**, 29–31.

Günther, T., Vormann, J., McGuigan, J. A. S., Luthi, D., & Gerber, D. (1995) *Biochemistry and Molecular Biology International*, **36**, 51–57.

Hafner, E. W., Tabor, C. W., & Tabor, H. (1979) *Journal of Biological Chemistry*, **254**, 12419–12426.

Henner, W. D., Kleber, I., & Benzinger, R. (1973) *Journal of Virology*, **12**, 741–747.

Herbst, E. J. & Snell, E. E. (1949) *Journal of Biological Chemistry*, **181**, 47–54.

Hernandez, V. J. & Bremer, H. (1993) *Journal of Biological Chemistry*, **268**, 10851–10862.

Hirshfield, I. N., Rosenfeld, H. J., Leifer, A., & Maas, W. K. (1970) *Journal of Bacteriology*, **101**, 725–730.

Hisano, T., Abe, S., Osamura, N., Song, X., Murata, K., & Kimura, A. (1991) *Journal of General Microbiology*, **137**, 845–850.

Hobot, J. A., Villiger, W., Escaig, J., Maeder, M., Ryter, A., & Kellenberger, E. (1985) *Journal of Bacteriology*, **162**, 960–971.

Hurwitz, C. & Rosano, C. L. (1967) *Journal of Biological Chemistry*, **242**, 3719–3722.

Igarashi, K., Kakegawa, T., & Hirose, S. (1982) *Biochimica et Biophysica Acta*, **697**, 185–192.

Igarashi, K., Mitsui, K., Kubota, M., Shirakuma, M., Ohnishi, R., & Hirose, S. (1983) *Biochimica et Biophysica Acta*, **755**, 326–331.

Ingraham, J. (1987) *Escherichia coli* and *Salmonella typhimurium: Cellular and Molecular Biology*, Vol. 2, pp. 1543–1554, F. C. Neidhardt, Ed. Washington, D.C.: American Society for Microbiology.

Jelenc, P. C. (1980) *Analytical Biochemistry*, **105**, 369–374.

Jürgens, V. J., Meissner, J., Fischer, V., König, W. A., & Weckesser, J. (1987) *Archives of Microbiology*, **148**, 72–76.

Justensen, J., Lund, T., Pedersen, F. S., & Kjelgaard, N. O. (1986) *Biochimie*, **68**, 715–722.

Kakegawa, T., Sato, E., Hirose, S., & Igarashi, K. (1986) *Archives of Biochemistry and Biophysics*, **251**, 412–420.

Kamio, Y., Itoh, Y., & Terawaki, Y. (1981a) *Journal of Bacteriology*, **146**, 49–53.

Kamio, Y., Itoh, Y., Terawaki, Y., & Kusano, T. (1981b) *Journal of Bacteriology*, **145**, 122–128.

Kamio, Y. & Nakamura, K. (1987) *Journal of Bacteriology*, **169**, 2881–2884.

Kamio, Y., Pösö, H., Terawaki, Y., & Paulin, L. (1986) *Journal of Biological Chemistry*, **261**, 6585–6589.

Kamio, Y., Terawaki, Y., & Izaki, K. (1982) *Journal of Biological Chemistry*, **257**, 3326–3333.

Kashiwagi, K., Endo, H., Kobayashi, H., Takio, K., & Igarashi, K. (1995) *Journal of Biological Chemistry*, **270**, 25377–25382.

Kashiwagi, K., Hosokawa, N., Furuchi, T., Kobayashi, H., Sasakawa, C., Yoshikawa, M., & Igarashi, K. (1990a) *Journal of Biological Chemistry*, **265**, 20893–20897.

Kashiwagi, K., Kobayashi, H., & Igarashi, K. (1986) *Journal of Bacteriology*, **165**, 972–977.

Kashiwagi, K., Miyamoto, S., Nukui, E., Kobayashi, H., & Igarashi, K. (1993) *Journal of Biological Chemistry*, **268**, 19358–19363.

Kashiwagi, K., Miyamoto, S., Suzuki, F., Kobayashi, H., & Igarashi, K. (1992) *Proceedings of the National Academy of Sciences of the United States of America*, **89**, 4529–4533.

Kashiwagi, K., Sakai, Y., & Igarashi, K. (1989) *Archives of Biochemistry and Biophysics*, **268**, 379–387.

Kashiwagi, K., Suzuki, T., Suzuki, F., Furuchi, T., Kobayashi, H., & Igarashi, K. (1991) *Journal of Biological Chemistry*, **266**, 20922–20927.

Kashiwagi, K., Yamaguchi, Y., Sakai, Y., Kobayashi, H., & Igarashi, K. (1990b) *Journal of Biological Chemistry*, **265**, 8387–8391.

Kawahara, H., Hayashi, Y., Hamada, R., & Obata, H. (1993) *Bioscience, Biotechnology and Biochemistry*, **57**, 1424–1428.

Kawahara, H., Inagaki, K., Kishimoto, N., Sugio, T., & Tano, T. (1991) *Agricultural and Biological Chemistry*, **55**, 679–685.

Kleinhauf, H. & von Döhren, H. (1990) *European Journal of Biochemistry*, **192**, 1–15.

Kleppe, K. & Lillehaug, J. R. (1979) *Advances in Enzymology*, **48**, 245–275.

Kleppe, K., Osland, A., Rosse, V., Male, R., Lossius, I., Helland, D., Lillehaug, J. R., Raae, A. J., Kleppe, R. K., & Nes, I. F. (1981) *Medical Biology*, **59**, 374–380.

Kornberg, A. (1988) *Journal of Biological Chemistry*, **263**, 1–4.

Kornberg, T., Lockwood, A., & Worcel, A. (1974) *Proceedings of the National Academy of Sciences of the United States of America*, **71**, 3189–3193.

London, R. E. (1991) *Annual Review of Physiology*, **53**, 241–258.

Low, R. A., Rashbaum, S. A., & Cozzarelli, N. R. (1974) *Proceedings of the National Academy of Sciences of the United States of America*, **71**, 2973–2977.

Matsui, I., Kamei, M., Otani, S., Morisawa, S., & Pegg, A. E. (1982) *Biochemical and Biophysical Research Communications*, **106**, 1155–1160.

Michaels, R. & Tchen, T. T. (1968) *Journal of Bacteriology*, **95**, 1966–1967.

Mitsui, K., Igarashi, K., Kakegawa, T., & Hirose, S. (1984a) *Biochemistry*, **23**, 2679–2683.

Mitsui, K., Ohnishi, R., Hirose, S., & Igarashi, K. (1984b) *Biochemical and Biophysical Research Communications*, **123**, 528–534.

Moore, K. S., Wehrli, S., Roder, H., Rogers, M., Forrest, J. N., McCrimmon, D., & Zasloff, M. (1993) *Proceedings of the National Academy of Sciences of the United States of America*, **90**, 1354–1358.

Moriya, T. & Hori, K. (1981) *Biochimica et Biophysica Acta*, **653**, 169–184.

Morris, D. R. & Jorstad, C. M. (1973) *Journal of Bacteriology*, **113**, 271–277.

Munro, G. F. & Bell, C. A. (1973) *Journal of Bacteriology*, **116**, 1479–1481.

Munro, G. F., Bell, C. A., & Lederman, M. (1974) *Journal of Bacteriology*, **118**, 952–963.

Munro, G. F., Hercules, K., Morgan, J., & Sauerbier, W. (1972) *Journal of Biological Chemistry*, **247**, 1271–1280.

Munro, G. F. & Sauerbier, W. (1973) *Journal of Bacteriology*, **116**, 488–490.

Neely, M. N., Dell, C. L., & Olson, E. R. (1994) *Journal of Bacteriology*, **176**, 3278–3285.

Nelson, D. L. & Kennedy, E. P. (1972) *Proceedings of the National Academy of Sciences of the United States of America*, **69**, 1091–1093.

Norris, V., Chen, M., Goldberg, M., Voskuil, J., McGurk, G., & Holland, I. B. (1991) *Molecular Microbiology*, **5**, 775–778.

Ogden, G. B., Pratt, M. J., & Schaechter, M. (1988) *Cell*, **54**, 127–135.

Oh, B.-H., Ames, G. F.-L., & Kim, S.-H. (1994) *Journal of Biological Chemistry*, **269**, 26323–26330.

Osawa, S., Muto, A., Ohama, T., Andachi, Y., Tanaka, R., & Yamao, F. (1990) *Experientia*, **46**, 1097–1106.

Osland, A. & Kleppe, K. (1977) *Nucleic Acids Research*, **4**, 685–695.

Park, M. H., Wong, B. B., & Lusk, J. E. (1976) *Journal of Bacteriology*, **126**, 1096–1103.

Pechulis, A. D., Bellevue, III, F. H., Cioffi, C. L., Trapp, S. G., Fojtik, J. P., MacKitty, A. A., Kinney, W. A., & Frye, L. L. (1995) *Journal of Organic Chemistry*, **60**, 5121–5126.

Peter, H. W., Ahlers, J., & Günther, T. (1978) *International Journal of Biochemistry*, **9**, 313–316.

Peter, H. W., Günther, T., & Seiler, N. (1979) *FEMS Microbiology Letters*, **5**, 389–393.

Portalier, R. & Worcel, A. (1976) *Cell*, **8**, 245–255.

Quigley, J. W. & Cohen, S. S. (1969) *Journal of Biological Chemistry*, **244**, 2450–2458.

Raina, A., Jansen, M., & Cohen, S. S. (1967) *Journal of Bacteriology*, **94**, 1684–1696.

Ramakrishna, S., Guarino, L., & Cohen, S. S. (1978) *Journal of Bacteriology*, **134**, 744–750.

Ritchey, M. B. & Delwiche, E. A. (1975) *Journal of Bacteriology*, **124**, 1213–1219.

Robinow, C. & Kellenberger, E. (1994) *Microbiological Reviews*, **58**, 211–232.

Rosen, B. P. (1973) *Journal of Bacteriology*, **116**, 627–635.

Rothfield, L. I. (1994) *Cell*, **77**, 963–966.

Rubenstein, K. E., Streibel, E., Massey, S., Lapi, L., & Cohen, S. S. (1972) *Journal of Bacteriology*, **112**, 1213–1221.

Schekman, R., Weiner, A., & Kornberg, A. (1974) *Science*, **186**, 987–993.

Schekman, R., Wickner, W., Westergaard, O., Brutlag, D., Geider, K., Bertsch, L. L., & Kornberg, A. (1972) *Proceedings of the National Academy of Sciences of the United States of America*, **69**, 2691–2695.

Schwartzbach, S. D., Freyssinet, G., Schiff, J. A., Hecker, L. I., & Barnett, W. E. (1979) In *Methods in Enzymology*, **59**, 434–437, K. Moldave & L. Grossman, Eds. New York: Academic Press.

Serizawa, H. & Fukuda, R. (1987) *Nucleic Acids Research*, **15**, 1153–1163.

Seyfzadeh, M., Keener, J., & Nomura, M. (1993) *Proceedings of the National Academy of Sciences of the United States of America*, **90**, 11004–11008.

Shatsky, I. N. & Vasiliev, V. D. (1988) *Methods in Enzymology*, **164**, 76–117, H. F. Noller Jr. & K. Moldave, Eds. New York: Academic Press.

Shchyolkina, A. K., Timofeev, E. N., Borisova, O. F., Ilicheva, I. A., Minyat, E. E., Khomyakova, E. B., & Florentiev, V. L. (1994) *FEBS Letters*, **339**, 113–118.

Shi, X. & Bennett, G. N. (1994) *Journal of Bacteriology*, **176**, 6769–6775.

Silver, S. & Clark, D. (1971) *Journal of Biological Chemistry*, **546**, 569–576.

Silver, S., Wendt, L., Bhattacharyya, P., & Beauchamp, R. S. (1970) *Annals of the New York Academy of Science*, **171**, 838–862.

Simon, E. J., Schapira, L., & Wurster, N. (1970) *Molecular Pharmacology*, **6**, 577–587.

Slonczewski, J. L. (1992) *ASM News*, **58**, 140–144.

Snavely, M. D., Florer, J. B., Miller, C. G., & Maguire, M. E. (1989) *Journal of Bacteriology*, **171**, 4761–4766.

Snell, E. E. (1989) *Annual Review of Nutrition*, **9**, 1–19.

Sogin, M. L., Elwood, H. J., & Gunderson, J. H. (1986) *Proceedings of the National Academy of Sciences of the United States of America*, **83**, 1383–1387.

Sørensen, M. A., Jensen, K. F., & Pedersen, S. (1994) *Journal of Molecular Biology*, **236**, 441–454.

Speed, R. R. & Winkler, H. H. (1990) *Journal of Bacteriology*, **172**, 5690–5696.

Stevens, L. & Morrison, M. R. (1968) *Biochemical Journal*, **108**, 633–640.

Stevens, L. & Pascoe, G. (1972) *Biochemical Journal*, **128**, 279–289.

Stonington, O. G. & Pettijohn, D. E. (1971) *Proceedings of the National Academy of Sciences of the United States of America*, **68**, 6–9.

Story, R. M., Weber, I. T., & Steitz, T. Z. (1992) *Nature*, **355**, 318–325.

Suelter, C. H. (1970) *Science*, **168**, 789–795.

Svitil, A. L., Cashel, M., & Zyskind, J. W. (1993) *Journal of Biological Chemistry*, **268**, 2307–2311.

Tabor, C. W. & Dobbs, L. G. (1970) *Journal of Biological Chemistry*, **245**, 2086–2091.

Tabor, C. W. & Tabor, H. (1966) *Journal of Biological Chemistry*, **241**, 3714–3723.

——— (1970) *Biochemical and Biophysical Research Communications*, **41**, 232–237.

Tabor, H., Hafner, E. W., & Tabor, C. W. (1980) *Journal of Bacteriology*, **144**, 952–956.

Tabor, H. & Tabor, C. W. (1971) *Methods in Enzymology*, **17B**, 815–817.

——— (1982) *Proceedings of the National Academy of Sciences of the United States of America*, **79**, 7087–7091.

Takahashi, M. & Uchida, T. (1986) *Journal of Biochemistry*, **100**, 123–131.

Takayama, M., Ohyama, T., Igarashi, K., & Kobayashi, H. (1994) *Molecular Microbiology*, **11**, 913–918.

Thorn, J. M., Carr, P. D., Chase, J. W., Dixon, N. E., & Ollis, D. L. (1994) *Journal of Molecular Biology*, **240**, 396–399.

Tkachenko, A. G. & Chudinov, A. A. (1990) *Microbiology*, **59**, 6–10.

——— (1994) *Current Microbiology*, **28**, 81–83.

Tkachenko, A. G., Chudinov, A. A., & Churilova, N. S. (1989) *Mikrobiologyia*, **58**, 709–715.

Tkachenko, A. G., Rosenblat, G. F., Chudinov, A. A., & Raev, M. B. (1991) *Current Microbiology*, **22**, 151–153.

Varon, M., Fine, M., & Stein, A. (1983) *Archives of Microbiology*, **136**, 128–159.

Watanabe, Y., Igarashi, K., Mitsui, K., & Hirose, S. (1983) *Biochimica et Biophysica Acta*, **740**, 362–368.

Wei, T., Bujalowski, W., & Lohman, T. M. (1992) *Biochemistry*, **31**, 6166–6174.

Weiss, R. L., Kimes, B. W., & Morris, D. R. (1973) *Biochemistry*, **12**, 450–456.

Weiss, R. L. & Morris, D. R. (1970) *Biochimica et Biophysica Acta*, **204**, 502–511.

Westergaard, O., Brutlag, D., & Kornberg, A. (1973) *Journal of Biological Chemistry*, **248**, 1361–1364.

Wickner, S. & Hurwitz, J. (1974) *Proceedings of the National Academy of Sciences of the United States of America*, **71**, 4120–4124.

Wickner, S., Wright, M., & Hurwitz, J. (1973b) *Proceedings of the National Academy of Sciences of the United States of America*, **70**, 1613–1618.

Wickner, W., Schekman, R., Geider, K., & Kornberg, A. (1973a) *Proceedings of the National Academy of Sciences of the United States of America*, **70**, 1764–1767.

Wittmann, H. G., Müssig, J., Piefke, J., Gewitz, H. S., Rheinberger, H. J., & Yonath, A. (1982) *FEBS Letters*, **146**, 217–220.

Worcel, A. & Burgi, E. (1972) *Journal of Molecular Biology*, **71**, 127–147.

Yamamoto, S., Nakao, H., Yamasaki, K., Takashina, K., Suemoto, Y., & Shinoda, S. (1988) *Microbiology and Immunology*, **32**, 675–687.

Yonath, A. E., Müssig, J., Tesche, B., Lorenz, S., Erdmann, V. A., & Wittmann, H. G. (1980) *Biochemistry International*, **1**, 428–435.

Yonath, A., Müssig, J., & Wittmann, H. G. (1982) *Journal of Cellular Biochemistry*, **19**, 145–155.

Yonath, A. & Wittmann, H. G. (1989) *Trends in Biochemical Sciences*, **14**, 329–335.

Young, D. V. & Srinivasan, P. R. (1972) *Journal of Bacteriology*, **112**, 30–39.

Zachowski, A. (1993) *Biochemical Journal*, **294**, 1–14.

CHAPTER 8

Biosynthesis and Metabolism of Diamines in Bacteria

Approaching Metabolism

Much of our current metabolic map of nitrogen-containing compounds was established in the 1950s and 1960s by the analysis of microbial mutants, blocked particularly in the biosynthesis of nitrogen-containing metabolites. In the previous century biochemists applied techniques of chemical isolation and characterization to nutrients, metabolites, and excretion products; constructed hypotheses to tie the appearance of some compounds to the transformation of others; and devised experiments and developed quantitation to test for the postulated reactions in intact organisms and organs. The investigation of putrefying fluids and tissues, microbes, and eventually relatively undecomposed tissue brei established a biochemistry of animals, plants, and microbes and distinguished the outstanding and relatively easily detectable reactions of these organisms: respiration, fermentation, the degradation of proteins and carbohydrates, etc. The introduction of inhibitory analogues, for example, malonate, limited an experimental design to questions involving a relatively few metabolic steps. The necessity of limiting events to permit a more specific interpretation of results developed a reductionist bent, which was soon expressed in the isolation and characterization of enzymatic catalysts for single metabolic reactions.

Isotope experiments with intact organisms were designed to test old hypotheses of metabolic events: fatty acid oxidation, routes of glucose utilization, amino acid degradation, and biosynthesis. Genetically manipulable microorganisms provided an additional important approach to limit biosynthesis in an intact cell. An accumulation or excretion of identifiable metabolic precursor or side product permitted a dissection at an early stage of investigation of a presumed reaction sequence. Genetics had become an alternative or complement to biochemical investigation.

Early Studies on Decarboxylation of Ornithine and Arginine

The biochemistry of the polyamines began about 50 years ago with the study of amino acid decarboxylation by the use of intact bacteria and their extracts, described in the papers of E. Gale (1940, 1946) in the early 1940s. With a given bacterium, often *Escherichia coli*, grown in a mineral medium, supplementation by different amino acids resulted in the production of an amine corresponding to that amino acid; the conversions of histidine to histamine, tyrosine to tyramine, leucine to isoamylamine, and arginine to putrescine were well established (Gale, 1940, 1946). Gale also demonstrated the existence of specific bacterial amino acid decarboxylases, all of which functioned optimally in the pH range of 2.5–6.0. He distinguished an ornithine decarboxylase (ODC) giving rise to putrescine from an arginine decarboxylase (ADC), which produced agmatine in a first step. In addition he demonstrated the existence of a specific lysine decarboxylase (LDC), which formed cadaverine. The enzymes were synthesized most actively toward the end of the growth cycle, mainly in response to the presence of the amino acid. Of 151 strains of *E. coli* over 90% possessed ODC and at least two-thirds possessed ADC and LDC. Some soluble enzymes from the organisms were also purified up to 100-fold, and pH optima for the activities and Michaelis constants (K_m) for the substrates were determined. The enzymes, as well as a transaminase, were demonstrated to require a common codecarboxylase, which was separated from the protein; this proved to be pyridoxal phosphate (PLP; Gunsalus et al., 1945). A brief chronology of the discovery of this coenzyme and its place in the development of microbiological growth factors was summarized by E. Snell (1989).

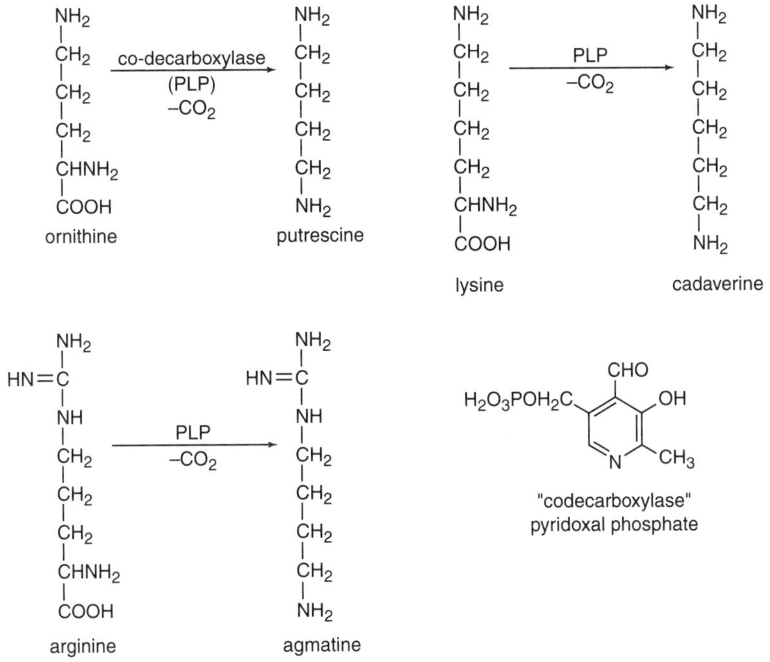

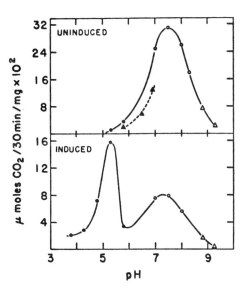

FIG. 8.1 The enzymatic decarboxylations of ornithine, arginine, and lysine, as well as the structure of the "codecarboxylase," pyridoxal phosphate.

Several of the reactions described by Gale are presented in Figure 8.1. These included the isolation of agmatine as the product of the bacterial degradation of arginine (Gale, 1940). He stressed the inducibility of the decarboxylases in acidic growth media and noted the inactivation of "amino acid deaminases" under such conditions.

The demonstration in 1948 that a diamine was essential for bacterial growth, and subsequently that actively growing *E. coli* contained "putrescine equivalents" posed the problem of the mechanism and significance of the biosynthesis of these equivalents during growth. By 1958 ornithine had been shown to be converted to putrescine in intact cells. Further, the enzyme, induced late in growth at acidic pH, did not account for the formation of putrescine during normal or exponential growth. An extract of *E. coli* grown in minimal medium was found to contain an ODC functional at pH 7.5, requiring PLP after dialysis (Morris & Pardee, 1965). Cells grown under the condition of induction possessed two peaks of activity as a function of pH (Fig. 8.2) whereas an uninduced cell (i.e., a cell growing in the absence of ornithine) contained only the enzyme active at pH 7.5. The pH 7.5 enzyme was dubbed "biosynthetic," in contrast to the induced pH 5 enzyme dubbed "catabolic." After purification to homogeneity the inducible degradative ODC (see Table 8.1) has been reported to have a significantly higher pH optimum (6.9) than that (ca. 5.3) shown in Figure 8.2.

FIG. 8.2 The pH dependence of the biosynthetic and catabolic ODC activities; redrawn from Morris and Pardee (1965). An arginine-deficient strain of *E. coli* K-12 was grown in inducing and noninducing conditions. Extracts were prepared and assayed in several buffered solutions at the indicated pH values.

TABLE 8.1　Basic amino acid decarboxylases of *E. coli*.

Enzyme	Optimum pH	Turnover number [pmol/s (μmol pyridoxal phosphate)$^{-1}$]	Substrate K_m (mM)	Cofactors	Subunit (M × 10^{-3})	No. subunits
bODC	8.3	157	5.6	PLP	81	2
dODC	6.9	173	3.6	PLP	80	2
bADC	8.4	18	0.03	PLP, Mg^{2+}	74	4
dADC	5.2	547	0.65	PLP	82	2 or 10
LDC	5.7	1330	1.5	PLP	80	2 or 10

Adapted from Morris and Boeker (1983).

In the absence of arginase in *E. coli*, which lacks urease, ornithine-requiring mutants should be incapable of synthesis of putrescine from arginine unless another as yet undiscovered route existed for such a sequence. Although the *E. coli* conversion of arginine to putrescine had been detected earlier, the decarboxylation of arginine produced agmatine. Morris and Pardee (1966) described agmatine ureohydrolase (AUH), which converted agmatine to putrescine plus urea. The search for AUH in normal growth also revealed a new ADC, which in contrast to the older acid-inducible enzyme was found to have an absolute Mg^{2+} requirement. Thus normal growth involved two biosynthetic pathways to putrescine via arginine and ornithine as biosynthetic precursors, and two catabolic pathways to putrescine via the same precursors in an acidic inducing medium.

The separation and dissection of these pathways exploited knowledge developed on the biosyntheses of ornithine and arginine. Genetic and biochemical analyses revealed a pathway for the formation of ornithine from glutamate. Ornithine can be decarboxylated directly to putrescine or converted to arginine via several steps, to include the addition of carbamyl phosphate to form the intermediate, citrulline. The *E. coli* strain 15 TAU lacks carbamyl phosphate synthetase and growth requires exogenous arginine and a pyrimidine (e.g., uracil) synthesized with carbamyl phosphate. The presence of arginine in the medium also inhibits steps leading to ornithine and hence confines putrescine biosynthesis almost entirely to the arginine pathway (Morris & Koffron, 1969). These relations are presented below.

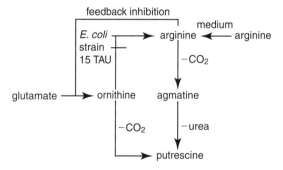

On the other hand *E. coli* strain 607, requiring high K$^+$ concentration in the medium for growth, lacks a func-

tional pathway for converting arginine to putrescine and does not produce urea during growth.

The existence of alternate paths of putrescine biosynthesis posed the problem of choices used by *E. coli* in different growth conditions. Spermidine is not degraded in any case; the spermidine content accounts for only about an eleventh or twelfth of the total amount of putrescine synthesized in the bacterium during growth on minimal medium. Urea is entirely a product of agmatine ureohydrolase and also is not destroyed (Morris & Koffron, 1967). Thus the estimation of the total polyamine (putrescine + spermidine) biosynthesis and of urea production under various conditions of growth of nonauxotrophic strains in a minimal medium supplemented with various amino acids establishes the relative ratio of use of the two pathways. Supplementation with arginine cuts off ornithine biosynthesis and markedly shifts the pathway of putrescine synthesis away from the ornithine pathway to that of ADC and AUH (Morris & Koffron, 1969). Because these shifts in the use of the pathways occurs without major changes in synthesis of putrescine or spermidine, the precise regulation of the totality of polyamine biosynthesis and its maintenance implies significant physiological roles for these compounds.

Assay of ODC (EC 4.1.1.17)

The biochemist is concerned initially with several simple aspects of the reaction under study; that of stoichiometry is primary. Among the many possible parameters of the study, it is asked if an extract carries out several types of reaction with the substrate and if the major product is reacted upon further. These potential aberrancies are seen in a determination of the full course of the reaction and its products. The rates of the reaction as a function of substrates and of the amount of extract added indicate more clearly the range of extract to be added in an assay system, and the absence of linearity in the response to the amount of added extract suggests the possible existence of activators and inhibitors.

The earliest assays of the catabolic decarboxylases (i.e., ODC, LDC, and ADC) were conducted by the assay of CO$_2$ in Warburg respirometers. ODC and LDC rapidly approached and did not exceed 1.0 mol CO$_2$ per mole of substrate (Gale, 1940). The product, putrescine, was not determined. In a study of the crude biosynthetic ODC, using [U-^{14}C]-ornithine as substrate, 1.0 mol CO$_2$

was produced per mole of substrate. In this assay radioactive CO_2 is released by acid and trapped on filter paper moistened with alkali. The paper is then added to scintillation fluid and the radioactivity is counted. The efficiency of counting and reproducibility are of interest; the entire procedure of estimating decarboxylation is reported to be approximately 50% efficient. In this experiment only 0.87 mol putrescine were recovered per mole of substrate metabolized and the question is whether this is due to an overestimation of CO_2 production or to a problem in recovery of amine. If the discrepancy is real, does another route exist in a crude extract to affect the production of the diamine? Among the possibilities it can be imagined that transamination occurs with either ornithine or putrescine or both to yield small amounts of unidentified products.

Amino acid pools tend to be low in Gram-negative organisms and considerable in Gram-positive bacteria. The presence of the bacterium's own amino acids in a crude extract may significantly dilute the labeled amino acid used in the assay, and thereby reduce the apparent recovery of the product. This systematic source of error could be serious in assays of some microorganisms or of crude plant extracts (Sindhu & Cohen, 1983).

As the assay for ODC became a matter of increasing interest, methods were devised to improve speed and the recovery of crucial products. For example CO_2-selective electrodes were devised for the study of decarboxylases; these were shown to be rapid, potentially continuous, specific, and sufficiently precise in the study of a partially purified bacterial LDC (Tonelli et al., 1981). However, in rat liver nuclei the decarboxylation of 1-[^{14}C]-ornithine to yield $^{14}CO_2$ did not produce putrescine, and the exact nature of the reactions has not been established (Grillo & Fossa, 1983). Simple and direct methods were developed for the rapid isolation of radioactive putrescine by cation exchange paper (Djurhuus, 1981) or resins (Heerze et al., 1990).

Initially Gale's discovery of the decarboxylases and amine formation was thought to relate to the Enterobacteriaceae particularly and therefore to the detection of fecal bacterial pollutants. In the period of 1960–1973, the detection of ODC was included among tests of motility and other chemical tests in the classification of *Klebsiella, Enterobacter, Serratia, Escherichia*, and *Proteus*, among others. Amine production in the presence of ornithine was frequently detected by a pH change. Although the excessive production of putrescine by *E. coli* was extensively confirmed, it is not clear whether this is common or distinctive among members of the Enterobacteriaceae. Nevertheless, the tests in clinical laboratories revealed correlations of the apparent presence of ODC in difficultly classifiable, possibly pathogenic organisms (Hickman–Brenner et al., 1987).

On the other hand, a test for the presumably less essential cadaverine and hence an LDC does distinguish many groups of microorganisms. For example, many species of pseudomonads, *Proteus*, and *Citrobacter* appear to lack this enzyme in routine clinical tests, which is found in *Salmonella* and *Escherichia* (Brooker et al.,

1973; Elston, 1971). Furthermore, it can be imagined that a more complete characterization of bacterial amines might provide a better guide to a search for distinctive decarboxylases. For example, species of genera such as *Enterobacter, Citrobacter*, and *Acinetobacter* contain large amounts of diaminopropane (DAP) in addition to the amines found in *E. coli*, putrescine and spermidine. These organisms also proved to contain a decarboxylase for 2,4-diaminobutyric acid (Ikai & Yamamoto, 1994; Yamamoto et al., 1992). Certain polluting organisms, such as members of the Vibrionaceae, also produce DAP and a specific and unique vibrio decarboxylase of 2,4-diaminobutyric acid synthesized DAP in a sequence leading to norspermidine via carboxynorspermidine (Nakao et al., 1989, 1990). That pathway in *Vibrio alginolyticus* is:

$$\begin{array}{c} NH_2 \\ | \\ HOOC-C-CH_2-CH_2NH_2 \quad \text{2,4-Diaminobutyric acid} \\ | \\ H \end{array}$$

$$\downarrow -CO_2$$

$$H_2NCH_2CH_2CH_2NH_2 \quad \text{Diaminopropane}$$

$$\downarrow \text{via carboxynorspermidine}$$

$$H_2NCH_2CH_2CH_2NHCH_2CH_2CH_2NH_2$$

Norspermidine

In some instances, the recognition of amine formation in marine bacteria shifted the problem of identifying polluting bacteria to a search for the detection and prevention of putrefaction or spoilage in marine products. For example, a marine organism, *Shewanella putrefaciens*, formerly described as a pseudomonad, contaminates fish and the degree of the contamination is estimated by the diamines produced. The organism was found to contain both the biosynthetic and biodegradative ODC (Suzuki et al., 1993). Almost all eubacteria, including marine forms, contain ODC. Several halococci contain ODC and a system for converting *N*-carbamylputrescine to putrescine but are unable to metabolize agmatine (Hamana & Matsuzaki, 1991). Extracts of *Streptococcus faecalis* contain an inducible putrescine transcarbamoylase that can reversibly cleave *N*-carbamylputrescine to putrescine and carbamyl phosphate (Roon & Barker, 1972).

Decarboxylases of Basic Amino Acids

The existence of biodegradative and biosynthetic routes to putrescine from ornithine and arginine set the task of determining the molecular nature of the four decarboxylases: degradative ODC (dODC), biosynthetic (bODC), degradative ADC (dADC), and biosynthetic (bADC). Such a program included a definition of the size of the purified proteins and their enzymatic properties: pH optimum, turnover number, primary amino acid sequence, the binding site of PLP, the mode of action of the complex in decarboxylation, etc.

The inducible biodegradative LDC, whose existence and production of cadaverine was established by Gale (1940), was also examined. In the early 1970s, as *E. coli* mutants deficient in putrescine became available, it was noted that putrescine-depleted cells greatly enhanced the synthesis of cadaverine, even in the absence of exogenous lysine, and indeed enabled the synthesis of aminopropylcadaverine (i.e., *N*-3-aminopropyl-1,5-diaminopentane) (Dion & Cohen, 1972). The latter served in part to replace depleted spermidine (i.e., *N*-3-aminopropyl-1,4-diaminobutane) in the bacterial economy. Several *Mycoplasma* strains proved to be relatively high in putrescine and cadaverine. An inhibitor of ODC enhanced the production of cadaverine and aminopropyl cadaverine (Alhonen–Hongisto et al., 1982). The hypothesis of partial replacement suggested the possible existence of a biosynthetic LDC. The existence of two LDCs is supported by some genetic evidence as well as by differences in the activities of LDC in extracts of bacteria grown under inducing and noninducing conditions. For example, the inducible LDC is essentially uninhibited by putrescine and spermidine whereas constitutive LDC is markedly inhibited by these amines (Wertheimer & Leifer, 1983).

The bacterial syntheses of the biodegradative enzymes requires the presence of substrate and an acidic pH. Curiously the induction of the biodegradative ADC has an obligate iron requirement, significantly higher than that required for growth (Melnykovych & Snell, 1958). The activity of the enzyme is not decreased by chelating agents, nor is the activity of extracts of cells grown in limiting iron increased by addition of the metal. In addition to the requirement for iron, production of the enzyme is increased by addition of several specific amino acids.

The purifications of the several decarboxylases of *E. coli* to homogeneity have been described (Applebaum et al., 1977; Boeker & Fischer, 1983; Morris & Boeker, 1983). The crystallization of the *E. coli* bADC led to an undesirable aggregation. The main features of the enzymes are presented in Table 8.1.

Of the preparations presented in Table 8.1, the enzymes were quite specific for their amino acid substrate. It may be mentioned that N^G-monomethyl-L-arginine can be decarboxylated by an *E. coli* enzyme to N^G-methylagmatine (Paik et al., 1981). In studies of eucaryotic ODC, it was found that this enzyme also decarboxylated lysine at about 1% of the rate on ornithine, and that animal cells or yeasts lack an independent LDC.

The biosynthetic ADC is the only member of the group with an Mg^{2+} requirement; this requirement is absolute at pH 8 (Wu & Morris, 1973a). A constitutive ADC from a *Pseudomonas* species has a similar requirement for Mg^{2+} (Rosenfeld & Roberts, 1976). In this organism and in *E. coli* the diamines and spermidine are inhibitory and inhibit the enzyme more significantly at limiting Mg^{2+} (Wu & Morris, 1973b).

The proteins show no immunological cross-reactions, although they are structurally similar. Four form dimers, whereas the biosynthetic ADC acts as a tetramer of subunits that are slightly smaller than the subunits of bODC, dODC, dADC, and LDC. The inactive dimers of dADC and LDC assemble further to form pentamers, which stack as active decamers at a pH below 6. This aggregation is assisted by an increase of Na^+, presence of substrate, etc. The dADC and LDC approach molecular weights of 820,000 and 800,000, respectively.

Boeker (1978) studied the association and dissociation of the *E. coli* dADC. The latter process appears to involve the successive binding of an Na^+ ion and the dissociation of a proton for each dissociation step of a subunit. The dimer of this enzyme is probably inactive; the decamer is the active form. The dADC may control intracellular pH or the intracellular formation of CO_2, functions that become less urgent as the pH increases. Thus, the rate of dissociation from the active decamer to inactive dimer increases with increasing pH.

A similar LDC of about 10^6 molecular weight was isolated from *Bacterium cadaveris* and was crystallized (Soda & Moriguchi, 1969). It was found to have 10 mol nonresolvable PLP per mole of enzyme. If the analogy to the *E. coli* enzyme as decamer holds, this result implies 1 mol PLP/subunit.

LDC is found in chloroplasts and cadaverine is a key intermediate in the synthesis of quinolizidine alkaloids. Most recently efforts to increase alkaloid production have approached the problem via the transformation of LDC-deficient plants by the insertion of the LDC gene of an appropriate bacterium like *Hafnia alvei*, an enterobacterium rich in LDC (Beier et al., 1987; Fecker et al., 1986). The enzyme of *H. alvei* was purified to homogeneity and contained 10 subunits of 80 kDa; each contained one molecule of PLP. These are properties similar to that of the inducible LDC of *E. coli*.

The LDC of *Selenomonas ruminantium*, which contains cadaverine as an essential component of its mucopeptide, was also purified to homogeneity. The enzyme is constitutive, requires PLP, and in its native form is a dimer of two identical subunits of 44,000 molecular weight (Kamio & Terawaki, 1983). The size of this enzyme is quite different from the previously described LDCs.

It has been possible to purify ODC from organisms other than *E. coli*. An inducible enzyme active with PLP, from *Lactobacillus* sp. 30a, formed dimers, which assembled to dodecamers. The electron microscopy of these forms revealed hexamers that had stacked in pairs to form the dodecamer (Guirard & Snell, 1980). A crystallization of the lactobacillus enzyme (Momany & Hackert, 1989) permitted crystallographic analysis that described a unit cell accommodating a dodecamer and the PLP binding sites (Momany et al., 1995). The gene was isolated and sequenced (Hackert et al., 1994).

Guanine Nucleotide Activators and Inhibitors of ODC

Examination of Table 8.1 reveals an expectedly high substrate K_m for ODC; and it was found that certain nucleotides, particularly the GTPs, markedly activate the biosynthetic enzyme (Höltta et al., 1972). In these studies with a 100-fold purified enzyme, an apparent K_m for L-ornithine of about 2 mM was found, whereas millimolar GTP lowered the apparent K_m for ornithine to about 0.2 mM.

A marked inhibition of ODC was found in an organism incubated under stringent conditions, as well as a direct inhibition of the enzyme by ppGpp (Höltta et al., 1974). However, a role of guanine nucleotides in regulating putrescine biosynthesis could not be detected in the ODC-utilizing K^+-dependent organism in which a switch to Na^+ in the medium stimulated putrescine synthesis and ODC activity (Sakai & Cohen, 1976). The cells did not change their GTP or ppGpp contents in this pliable bacterium.

Polyamines as Inhibitors of Decarboxylases

End product inhibition of the biosynthetic ODC by putrescine was recognized to be a significant regulatory control (Morris & Fillingame, 1974). Additionally, spermidine inhibited the enzyme 50% at a concentration in the millimolar range at which this triamine exists in the cell (5 mM spermidine compared to 20 mM putrescine at saturating ornithine). However, the intracellular binding of spermidine to negatively charged nucleates and other ions prevents a clear understanding of the *activity* of "free" polyamines. Feedback inhibitions of the Mg^{2+}-requiring bADC were demonstrated by Wu and Morris (1973a), as shown in Figure 8.3. Higher concentrations of Mg^{2+} also revealed a partial competition between Mg^{2+} and putrescine. The inhibition of bADC and LDC by spermidine reveals one possible method of regulating an early step of polyamine biosynthesis by a late product.

Polyamine generally inhibits the bacterial decarboxylases. The Mg^{2+}-requiring ADC of a pseudomonad is sensitive to agmatine and cadaverine, as well as to putrescine and spermidine in a manner partially competitive with Mg^{2+}. The non-Mg^{2+}-activated specific ADC of *Mycobacterium smegmatis* is inhibited by agmatine, putrescine, spermidine, and spermine (Balasundaram & Tyagi, 1989).

In addition to end product inhibition, the activities of the pathways are also regulated by repression by the polyamines. The experiments on the formation of the polyamines in intact cells grown in a chemostat (Tabor & Tabor, 1969) demonstrated a marked decrease in their biosynthesis from labeled ornithine or arginine on addition of exogenous polyamine. In addition to polyamine inhibition of ODC and ADC in extracts, the Tabors demonstrated the slow but marked decrease of

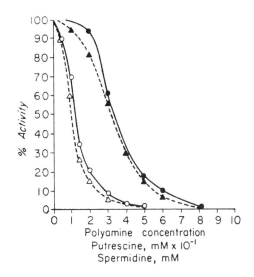

FIG. 8.3 The inhibition of the biosynthetic arginine decarboxylase of *E. coli* by polyamines; redrawn from Wu and Morris (1973a). The polyamine inhibitors tested were (○, ●) putrescine or (△, ▲) spermidine at 1-mM (open symbols) or 2-mM (closed symbols) concentrations of $MgSO_4$. $^{14}CO_2$ derived from [1-^{14}C]-arginine was measured 2 min after addition of the enzyme at 37°C.

ODC and ADC in cells of the cultures after exposure to the polyamine in the media.

In more recent experiments, various polyamine-deficient mutants were fed particular polyamines or their amino acid precursors. Polyamine, which greatly stimulated growth, also markedly decreased the biosynthesis and intracellular concentrations of potentially toxic ornithine (Cataldi & Algranati, 1989).

PLP Binding Sites in Decarboxylases

Despite gross dissimilarities in primary amino acid sequences of the homogeneous enzymes, their PLP binding sites proved to have interesting similarities. The addition of the spectrophotometrically detectable aldehydic coenzyme to the protein formed a Schiff base with the ε-amino group of lysine (Table 8.2). This double bond could be reduced readily and irreversibly by sodium borohydride. The proteolytic degradation of the altered protein enabled the isolation and sequencing of pyridoxyl-peptides (Applebaum et al., 1975; Boeker et al., 1971; Sabo & Fischer, 1974).

The three proteins have a histidine residue on the amino side of the phosphopyridoxyl lysine [(ε-Pxy)Lys] and a hydrophobic sequence on the carboxyl side. The similarities of the sequences of LDC and dADC are evident, whereas the site of dODC is quite different from the others. Differences among various PLP binding sites were discussed by Momany et al. (1995).

A recent isolation and characterization of ADC from

TABLE 8.2 Pyridoxal phosphate binding site of biodegradative decarboxylases.

LDC	Glu-Thr-Glu-Ser-Thr-His-(ε-Pxy)Lys-Leu-Leu-Ala-Ala-Phe
dADC	Ala-Thr-His-Ser-Thr-His-(ε-Pxy)Lys-Leu-Leu-Asn-Ala-Phe
dODC	Val-His-(ε-Pxy)Lys-Glu-Gln-Ala -Gly-Gln

Adapted from Morris and Boeker (1983).

Mycobacterium smegmatis revealed a molecule of 232,000 Da, comprising four identical subunits. Unlike the *E. coli* bADC, the specific mycobacterial enzyme did not have an Mg^{2+} requirement, but it was entirely dependent on PLP. Two types of binding were found; the tight-binding mode is related to activation of the enzyme.

Aminotransferases and Metabolism of Putrescine

Early studies of organisms capable of using an organic amine as the sole source of carbon and nitrogen detected strains of pseudomonads active in these feats of metabolism. Jacoby and Fredericks (1959) described a strain of *Pseudomonas fluorescens* that could use either putrescine, pyrrolidine, or Δ^1-pyrroline in this way. In this organism, putrescine may be converted to Δ^1-pyrroline by an amine oxidase and subsequently to γ-aminobutyric acid (GABA; Chap. 6). GABA was postulated to serve as a precursor to succinic semialdehyde by transamination with α-ketoglutarate initially and then to succinate. A transaminase was subsequently purified from pseudomonads that transfers an amino group to α-ketoglutarate from putrescine and cadaverine, as well as from γ-aminobutyrate, ornithine, and lysine (Padmanabhan & Tchen, 1972; Voellmy & Leisinger, 1976). Thus this organism can metabolize putrescine by a route other than that of an amine oxidase.

The product, GABA, of some marine bacteria, induces the settling of marine invertebrate larvae. Several of these bacterial strains can convert putrescine to GABA in a manner modulated by osmolarity and pH; this is unaffected by aminoguanidine, an inhibitor of a diamine oxidase (Mountfort & Pybus, 1992).

The search for a putrescine-metabolizing enzyme in *E. coli* began in an attempt to deplete the diamine in the organism. At first, putrescine was found to serve as a nitrogen source during growth on glucose, and the reactions, via an initial transamination, were identical to that found in *Pseudomonas fluorescens*:

$$\text{putrescine} + \alpha\text{-ketoglutarate}$$
$$\downarrow$$
$$\gamma\text{-aminobutyraldehyde} + \text{glutamate}.$$

The γ-aminobutyraldehyde was converted successively in extracts to GABA and eventually to succinate (Kim & Tchen, 1962). Kim (1963) also isolated a mutant strain of *E. coli* containing a constitutive putrescine-α-ketoglutarate transaminase and thus obtained an organism capable of growth on putrescine as its sole source of carbon and nitrogen. This bacterium markedly reduced its putrescine content under conditions of several hours of nitrogen starvation, but in this period lost only a part of its considerable spermidine content. More recently an *E. coli* strain capable of growth on GABA and of converting this to growth-sustaining succinate was selected for its ability to grow on putrescine as its sole carbon and nitrogen source. This proved to contain the putrescine-α-ketoglutarate transaminase and a γ-aminobutyraldehyde dehydrogenase necessary to form GABA and eventually to succinate (Large, 1992; Prieto-Santos et al., 1986).

E. coli and *Aerobacter aerogenes* were found to contain high levels of transaminase but to lack diamine oxidase, in contrast to many organisms, such as pseudomonads and mycobacteria, which possess both (Michaels & Kim, 1966). The partially purified amine transaminase of *E. coli* was active with putrescine, cadaverine, and 1,6-hexanediamine but inactive with monoamines, spermidine, and spermine. Treatment with acid ammonium sulfate removed the coenzyme, leaving an apoenzyme reactivable with either PLP or pyridoxamine phosphate.

L-Ornithine, L-arginine, agmatine, and putrescine can serve as the sole nitrogen source for *E. coli* strain K12. These amino acids and agmatine are converted to putrescine, whose further metabolism via transaminases has been described. The isolation of *E. coli* mutants blocked in one or another of these metabolic steps demonstrates the essentiality of these steps in the production of utilizable nitrogen (Shaibe et al., 1985a). Although ADC and ODC are fully constitutive in these studies, AUH and putrescine aminotransferase were controlled by nitrogen availability and repressed by products (catabolites) of glucose metabolism. The repression of AUH was reduced by the presence of cyclic AMP (cAMP; Schaibe et al., 1985b). Some anaerobic bacteria, which use putrescine as their sole carbon and nitrogen source, can convert the diamine further to acetate, butyrate, and molecular hydrogen (Matthies et al., 1989).

PLP-Mediated Reactions

PLP, which is recognized as the coenzyme of amino acid decarboxylases and transaminase, was synthesized by Gunsalus et al. (1945). It participates in many unusual reactions, of which 12 different types have been described (Martell, 1989; Metzler, 1977). The mechanistic diversity and similarities among these systems were discussed by Hayashi (1995). The conversion of S-adenosylmethionine to 1-aminocyclopropane-1-carboxylate on the pathway to plant ethylene (Chap. 20), also proved to be a PLP-determined reaction. Phosphorylated derivatives of vitamin B_6 (pyridoxine) serve as active coenzymes.

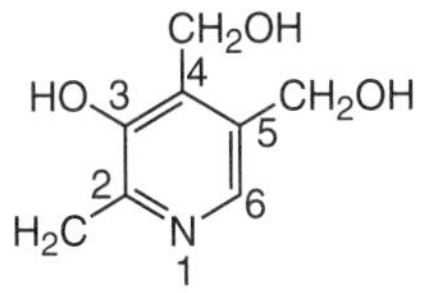

CH$_2$OH
HO 3 4 CH$_2$OH
2 5
H$_2$C 6
N
1

vitamin B_6

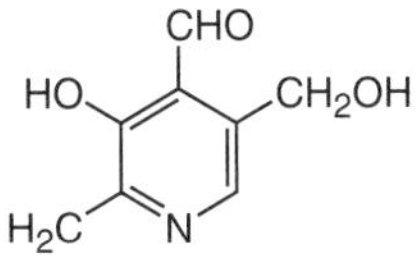

CHO
HO CH$_2$OH
H$_2$C
N

pyridoxal

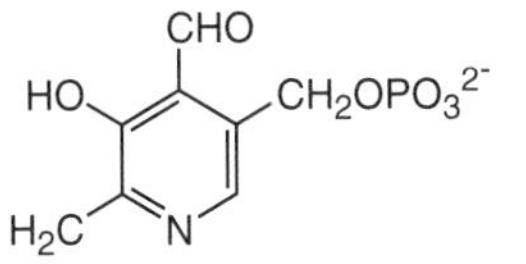

CHO
HO CH$_2$OPO$_3^{2-}$
H$_2$C
N

pyridoxal phosphate

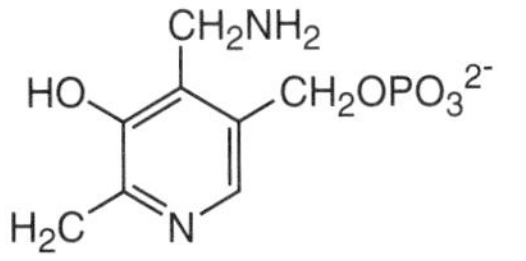

CH$_2$NH$_2$
HO CH$_2$OPO$_3^{2-}$
H$_2$C
N

pyridoxamine phosphate

The presence of the phosphate on the hydroxymethyl group prevents the formation of a fairly stable hemiacetal and permits the carbonyl group of the aldehyde to form a Schiff base with a reactive amine. The reduction of the planar Schiff base with sodium borohydride creates a stable bond, and the absorbing aromatic structure can label a substrate or a portion of the protein enzyme.

As presented below, an amino acid and the coenzyme form a Schiff base on the appropriate enzyme. This may be decarboxylated to an amine and regenerate PLP or transfer an amino group to form pyridoxamine phosphate and a keto acid. Pyridoxamine phosphate can transfer its amine to another keto acid to regenerate PLP.

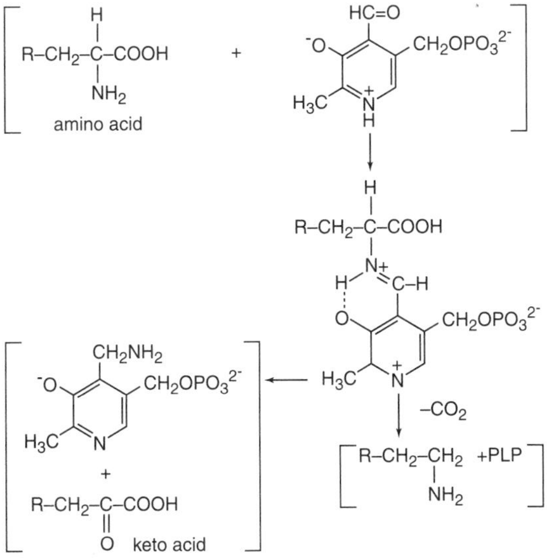

The detailed mechanisms of the reactions and the nature of the quinoid intermediates were discussed by Metzler (1977), Martell (1989), and Hayashi (1995) and were clarified in some measure initially by the ability of the nonphosphorylated pyridoxal to react with amino acids at elevated temperatures in the presence of metal ions.

Stereochemistry of Enzymatic Decarboxylation

The replacement of the carboxyl group by H after decarboxylation creates a new prochiral center in the amine. As discussed above, the loss of CO_2 occurs from the Schiff base formed from PLP and the α-amino group of the amino acid. Rearrangement and protonation (or deuteration or tritiation) produce an amine whose original carbon-bound proton may be compared with the configuration of the carbon-bound proton in the original amino acid. The estimation of configuration turns on the retention of the isotope of one chiral form or loss from the enantiomer after treatment by previously characterized enzymatic reactions, such as those of yeast alcohol dehydrogenase or pea seedling diamine oxidase. Although protonation might have occurred from either side of the planar —C=N— bond, the original hydrogen had been retained in its initial configuration in the case of the decarboxylations of lysine, ornithine, and arginine by either bacterial or plant enzymes (Asada et al., 1984; Robins, 1983). Despite the apparent generality of this result, several instances are known of inversion of configuration at C^α during enzymatic decarboxylation (Gani, 1991).

Synthetic Inhibitors of Bacterial Decarboxylases

Competitive and reversible inhibitors

Compounds were sought that might inhibit the decarboxylases in vitro and in vivo in efforts to minimize the availability of the polyamines. Work in the 1970s provided insights concerning the types of compounds that might be useful in such experiments. The first indications of medically effective inhibitors of polyamine metabolism were obtained with eucaryotic (not procaryotic) systems. Early observations on the inhibition of spermidine synthesis in a mammalian system by a previously known antitumor agent (Williams–Ashman & Schenone, 1972) encouraged the development of other compounds potentially inhibitory to specific steps in the biosynthetic pathways of the bacteria. One of the first of these, α-hydrazinoornithine, was directed to the easily available E. coli ODC and the ODC of rat prostate (Harik & Snyder, 1973).

The inhibitor, α-hydrazinoornithine or 5-amino-2-hydrazinopentanoic acid, was viewed, as is another inhibitory compound, α-methylornithine, as an analogue of ornithine. These are quite potent in assay with the partially purified ODC (i.e., of E. coli and of rat prostate)

H
R–CH$_2$–C–COOH
NH$_2$
amino acid

+

HC=O
$^-$O CH$_2$OPO$_3^{2-}$
H$_3$C N$^+$
H

H
R–CH$_2$–C–COOH
N$^+$
H C–H
O CH$_2$OPO$_3^{2-}$
H$_3$C N$^+$
H
–CO$_2$

R–CH$_2$–CH$_2$ +PLP
NH$_2$

CH$_2$NH$_2$
$^-$O CH$_2$OPO$_3^{2-}$
H$_3$C N
+
R–CH$_2$–C–COOH
O keto acid

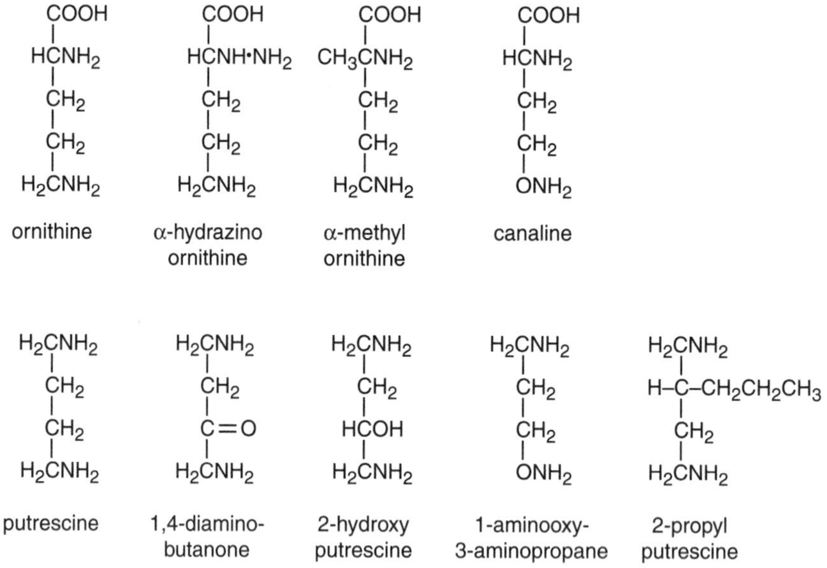

FIG. 8.4 Competitive and other reversible inhibitors of bacterial ODC.

producing 50% inhibition (K_i) at 5×10^{-7} M and 5×10^{-6} M, respectively, with concentrations of PLP at 10^{-5} M and L-ornithine at the K_m values for the respective enzymes. The inhibitor is competitive with ornithine; but its action is also reduced by increase of the concentration of PLP, suggesting an additional interaction with the coenzyme. Although other α-hydrazino derivatives of the amino acids are far less specific for ODC, the α-hydrazinoornithine was somewhat inhibitory for other PLP decarboxylases of mammalian tissues.

Several other competitive inhibitors of ODC, including α-methylornithine, and several analogues of the inhibitory product, putrescine (i.e., 1,4-diaminobutanone and 2-hydroxyputrescine) were found to be active on *E. coli* (Bitonti & McCann, 1987). Many other simple derivatives of putrescine (i.e., *N*-, *N,N′*-, 1-alkyl-, and 2-alkyl-) (Ruiz et al., 1986, 1988) proved to be relatively weak inhibitors of the *E. coli* ODC. The analogue of canaline, 1-aminooxy-3-aminopropane, was found to be by far the most active of these inhibitors with a K_i of 1.0 nM (Paulin, 1986). The structures of these related structures are given in Figure 8.4.

The ε-aminocaproic acid > L-lysine ethyl ester > α-aminopimelic acid were competitive inhibitors of the inducible LDC. The product of the reaction, cadaverine, proved to be the least active of this group and L-ornithine was neither a substrate nor an inhibitor of the enzyme (Sabo et al., 1974).

Canaline, 2-amino-4(aminoxy)butyric acid, the product of ureohydrolysis of canavanine, γ-guanidino-oxy-2-aminobutyric acid, inhibited several PLP enzymes, including mammalian ODC (Rahiala et al., 1971). The inhibitions were reversed by excess PLP, suggesting, as

in the case of α-hydrazinoornithine, the formation of a complex between canaline and the coenzyme. Despite the potency of canaline in inhibiting PLP enzymes, such as ornithine aminotransferase in vitro, it has not been an effective inhibitor in vivo.

Stable reduced PLP-substrate adducts of the dibasic amino acids lysine and ornithine (i.e., the *N*-(5′phosphopyridoxyl) derivatives) were also prepared (Heller et al., 1975). The lysine derivative was active in the inhibition of the *E. coli* LDC, and the ornithine derivative was active on mammalian ODC but was not tested on the bacterial enzyme. The study is of interest as an early example of an effort to simulate the transition state of an intermediate along the polyamine pathway. Indeed, the inhibition was noncompetitive with ornithine and PLP, suggesting a binding to the protein at the site of the Schiff-base intermediate.

Various derivatives of arginine, including canavanine, can be decarboxylated by *E. coli* to produce guanidinoxypropylamine, an analog of agmatine (Makisumi, 1961). Toxic canavanine participates in numerous reactions of polymer biosynthesis in the organism and has proved useful in the study of repression of enzymes of arginine biosynthesis.

Irreversible inhibitors

The development of reagents for exploring the active sites of various enzymes in the 1960s led to haloketones as covalent affinity labeling reagents and inhibitors for several proteases. L-Lysine bromomethyl ketone was shown to be an almost specific irreversible inhibitor of the crude LDC of *Bacterium cadaveris* (Miller & Rod-

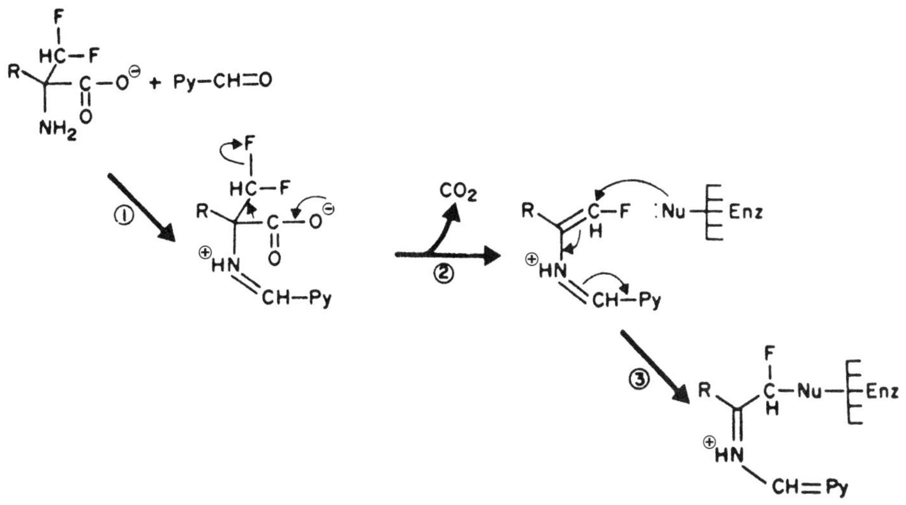

FIG. 8.5 Early postulate of irreversible inactivation of ODC by DFMO; redrawn from Fozard and Koch–Weser (1982). A Schiff base between the inhibitor and coenzyme at the active site loses CO_2 and a fluorine atom. The reactive intermediate alkylates a nucleophilic residue (Nu) at the active site and irreversibly inactivates the enzyme. The reactive moieties are described in Chapter 13.

well, 1971). The mechanism of the inhibition has not been pursued with the LDC of *E. coli*.

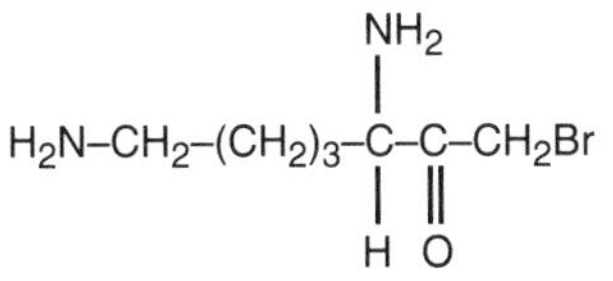

L-lysine bromomethylketone

Interest in the medical potential of ODC inhibitors developed after 1969 when the unusual expansion and decay of this enzyme activity had been demonstrated in actively growing mammalian tissues. However, some competitive inhibitors (e.g., α-hydrazinoornithine and α-methylornithine) were inadequate as inhibitors of polyamine synthesis in vivo and more particularly markedly increased the half-life of ODC in the liver. It was supposed that a persistent irreversible inhibitor might effect a sustained inhibition in tissues. These led initially to the synthesis of ornithine and putrescine analogues that might react specifically with the enzyme to unmask latent reactive groups in the inhibitor, facilitating covalent binding to an essential site of the protein. These were the "suicide inhibitors" (Metcalf et al., 1978).

The mechanism of the powerful inactivating effect by difluoromethyl ornithine (DFMO) on an animal enzyme will be discussed in some detail in Chapter 13. An early formulation is presented in Figure 8.5 (Fozard & Koch–Weser, 1982). Unfortunately DFMO has far less effect on the *E. coli* ODC, and this result has led to continuing searches for compounds affecting this enzyme.

The putrescine analogues, 5-hexyne-1,4-diamine and *trans*-hex-2-en-5-yne-1,4-diamine, are also irreversible inhibitors that take advantage of the reversibility of enzymatic reactions, despite the low affinity to the enzyme of an enzymatic product of decarboxylation. An amine of this type that did prove to be a potent inhibitor of the *E. coli* biosynthetic ODC was α-monofluoromethylputrescine (or 1,4-diamino-5-fluoropentane). The compound was less inhibitory on the ODC derived from *Pseudomonas aeruginosa* (Kallio et al., 1982). Indeed it was not particularly inhibitory to the growth of intact *E. coli*, although it reduced the ODC by about 95%.

In combination with monofluoromethylornithine, difluoromethylarginine might be expected to eliminate almost all of the ODC and ADC; indeed at 1-mM levels it inhibits both of them at >90% in 4-h incubations (Kallio et al., 1982), as shown in Table 8.3.

However, putrescine was reduced only 60–70%, spermidine was not reduced at all, and these irreversible inhibitors were not effective in controlling polyamine metabolism in this organism. Nevertheless, continuing

TABLE 8.3 Enzymes and polyamines of *E. coli* after growth in presence of inhibitors.

Inhibitor	ODC	ADC	Putrescine	Spermidine
			(nmol/10^8 cells)	
None	0.97	0.64	1.58	0.34
MFMP	0.05	0.73	1.63	0.33
DFMA	1.28	0.03	1.45	0.33
MFMP + DFMA	0.06	0.05	0.54	0.46

MFMP, monofluoromethylputrescine; DFMA, difluoromethylarginine. Adapted from Bitonti and McCann (1987).

efforts to prepare new effective compounds resulted in the synthesis of a new type of irreversible inhibitor of *E. coli* ODC, namely α-allenyl putrescine (5,6-heptadiene-1,4-diamine). The α-allenyl amines had been developed initially as inhibitors for the flavin-enzyme monoamine oxidases. Only the (R)-enantiomer, with a K_i of 140 μM, produced time-dependent inactivation of the bacterial enzyme, which could not be reactivated by prolonged dialysis against PLP and an —SH reagent, dithiothreitol (Danzin & Casara, 1984).

Inhibition of *E. coli* bODC and bADC by *E. coli* Proteins

Studies on the regulation of mammalian ODC demonstrated its very low concentration in many tissues. It was also found that an unusual protein inhibitor (dubbed "antizyme"), induced by high concentrations of polyamines, exists in some of these tissues (Canellakis et al., 1978). A protein inhibitor of bacterial ODC was similarly found in a polyamine-deficient strain of *E. coli* grown in the presence of putrescine and spermidine. Its concentration in the bacteria increased with increasing polyamine concentrations in the medium. This ODC inhibitor formed a complex with ODC, dissociable by salt, and the enzyme and inhibitor were then separable on a Sephadex gel. Surprisingly the *E. coli* antizyme similarly inactivated the rat liver ODC (Kyriakidis et al., 1978). The *E. coli* inhibitor was resolved into three protein components: an acidic protein of 49,500 molecular weight and two basic proteins, antizyme 1 and antizyme 2, of 11,000 and 9000 molecular weights, respectively (Kyriakidis et al., 1983). The basic components account for about 90% of the total inhibitory activity.

The gene for the acidic protein, now designated "the ODC antizyme of *E. coli*," was isolated, cloned, and sequenced (Canellakis et al., 1993). Overproduction of the protein in *E. coli* is inhibitory to ODC and ADC. Indeed the ODC of an inactivated ODC–antizyme complex can be reactivated by the addition of ADC; the converse also occurs (Pangiotidis et al., 1994). This technique enabled the demonstration of much inactive ODC and ADC in *E. coli* extracts, and growth in the presence of polyamines increases the fractions of inactive ODC and ADC.

The inhibition of the *E. coli* bODC by the basic proteins, initially dubbed antizymes 1 and 2, is considerable (Heller et al., 1983). These antizymes were found to be inactive on the inducible decarboxylases including LDC, but were inhibitory to bODC and bADC, although more so to the former. The authors suggested that in *E. coli* ornithine and bODC are probably the main source of bacterial putrescine and that the various antizymes do regulate putrescine biosynthesis in this organism. Antizyme 1 and antizyme 2 were found to behave similarly and to have identical sequences with the ribosomal proteins S20/L26 and L43, respectively (Panagiotidis & Canellakis, 1984). Thus, the data suggested that polyamine biosynthesis is also regulated by unintegrated, perhaps excess, elements of ribosome structure.

Many detached ribosomal proteins were found to possess antizyme activity. Actually one protein found on the 50S subunit was more active than the previously detected proteins, which had accounted for only 10% of the total antizyme activity of ribosomal proteins. Although it was observed that intact ribosomes were inactive as antizymes, actually some bacterial extracts contained ribosomes that held all of the antizymic activity of the organism and supernatant fractions were devoid of antizymic activity (Kashiwagi & Igarashi, 1987). These workers proposed that *E. coli* ribosomal proteins should not be termed antizymes because they are not specific nor induced by putrescine.

Huang et al. (1990) undertook to determine if certain genes for ribosomal proteins are selectively transcribed during growth in the presence of polyamines, even as the transcription of genes for ODC and ADC is decreased. A selective stimulation of the production of the ODC-inhibitory ribosomal proteins S20 and L34 was observed in a nonpolyamine requiring strain of *E. coli* grown in the presence of putrescine and spermidine. The studies, although pointing to an interrelationship between the polyamines and potentially ODC-inhibitory ribosomal proteins, have not yet demonstrated a role of these proteins in determining polyamine biosynthesis in vivo. The ODC of *Mycobacterium smegmatis* has been reported to be regulated by an RNA whose concentration is increased by polyamines in the growth medium (Balasundaram & Tyagi, 1988). The problem of the in vivo role of antizyme will be discussed in a later chapter that considers the regulation of mammalian ODC.

Biosynthesis of Ornithine and Arginine

The study of the genetic and regulatory controls of the pathways leading to and from these amino acids has been a center of interest since the 1950s. The pathway in *E. coli* leading from glutamate to ornithine and arginine, that is, details of the metabolic steps and the genetic controls of the biosynthetic enzymes, have been summarized in some detail. Glansdorff (1987) considered these systems and the biosynthesis of the polyamines as a related biosynthetic unit. Many "minor" variations in the biosynthetic pathway have been seen in organisms other than *E. coli*. As in the study of the polyamines themselves, the study of thermophiles has also revealed some "quirks" (Sakanyan et al., 1993). A current puzzle relates to the inability to demonstrate carbamyl phosphate synthetase in several archaebacteria, although ornithine carbamoyltransferase was detected in these organisms (Van De Casteele et al., 1990).

In the dissection of the pathways, a route was found in *E. coli* for the conversion of ornithine to proline (Vogel, 1956). The main route of the latter is a reduction of glutamate to glutamic γ-semialdehyde, which cyclizes spontaneously to Δ^1-pyrroline carboxylic acid. This compound is reduced enzymatically to proline. In *E. coli* exogenous ornithine may also enter the sequence via a transamination at the δ-NH_2 group to form glutamic γ-semialdehyde. An ornithine cyclodeaminase or cyclase, which converts L-ornithine directly to L-proline plus ammonia, is known in some organisms.

Biochemical research demonstrated the role of carbamyl phosphate in the biosynthesis of arginine and the pyrimidines. This is exemplified in the nutritional requirements of *E. coli* strain TAU, blocked in the synthesis of carbamyl phosphate synthetase. Further, putrescine can be produced from either arginine or its precursor, ornithine. How were such intermediates and pathways controlled and balanced? The existence of product inhibition of the decarboxylases and repression or inhibition of the synthesis of these enzymes was demonstrated (Tabor & Tabor, 1969; see Chap. 7). The term "repression" to describe a relative decrease in the rate of synthesis of an enzyme by exposure of cells to a substance had been coined over a decade earlier to describe the inhibition of synthesis of *N*-acetyl-ornithinase by exogenous arginine. It was known that arginine, used as a nutrient, could limit the synthesis of ornithine, and arginine is now known to repress many steps, if not all, in the biosynthesis of the amino acid. Many of these decreases in enzyme synthesis have been attributed to the inhibition of transcription of the genes. Many of the genes do not exist in clusters and the repression response is frequently incompletely coordinated. Repression is now thought to be affected by an arginine repressor protein, determined by the *argR* (repressor) gene, and activated by a corepressor thought to be arginine itself, acting on ARG box operators in the various *arg* genes. In addition, the first step of the sequence (i.e., acetylglutamate synthetase) has been shown to be controlled by end product inhibition and by repression. The inhibition of this enzyme, although strongly effected by L-arginine and somewhat by L-citrulline, is not produced by polyamines. In many other organisms, arginine also serves as an inhibitor of several of the enzymes of the pathway.

The study of various questions concerning the arginine pathway and the origin of putrescine led to the first isolations of mutants largely deficient in this biosynthetic function. In the absence of strains of *E. coli* deficient in the ability to synthesize the diamine or of inhibitors unable to markedly reduce their content (i.e., lacking putrescine-depleted cells), the role of the diamine was difficult to assess. Repressor effects, in which exogenous arginine decreases the formation of the arginine biosynthetic enzymes, are most marked with strain K12. One early isolation of a putrescine-deficient strain developed in the course of studies of arginine repression and many of the mutants were then derived from the K12 strains.

Arginine analogues, such as the toxic plant amino acid canavanine, and homoarginine were tested in repression. Homoarginine inhibited the growth of *E. coli* strain B but not of strain K12. It does not enter bacterial protein. However, canavanine, which enters protein via formation of a canavanine-charged tRNA, was found to repress K12 strains. Canavanine-resistant strains were isolated, all of which had a defective repression mecha-

nism. The levels of arginine-tRNA in repressed or derepressed strains did not indicate that this substance was related to the mechanism of repression.

Bacterial Pathways of Arginine and Ornithine Catabolism

Interest in the origin and metabolism of the polyamines in the entire biological world suggests the desirability of an at least minimal sketch of the variability of the metabolism of the precursors of the polyamines in organisms other than *E. coli*. A comparison of the biosynthetic pathway of arginine and its controls in various bacteria was made by Cunin et al. (1986). This review also examined the complex literature concerning some 40 enzymes involved in the catabolism of this amino acid and of ornithine, including some data on the formation and utilization of putrescine.

Arginase pathway

Arginase converts arginine to ornithine and urea; urease converts urea to NH_3 and CO_2. In contrast to *E. coli*, which lacks these enzymes, many organisms contain one or another or both. For example, *Proteus* strains contain both enzymes as separable entities; the presence of urease is useful in the diagnosis of infections caused by these organisms. Some cyanobacteria contain the two activities, possibly in a single protein (Meyer & Pistorius, 1989). However, these bacteria do not metabolize ornithine further, suggesting that arginine may be metabolized primarily as a source of nitrogen. The cyanobacterium *Nostoc* is capable of a greater use of the derived ornithine (Martel et al., 1993). The origins of the diamines have not been studied in these cyanobacteria.

Arginase has been characterized from many organisms; a thermostable Mn^{2+}-containing enzyme has recently been isolated and characterized from an extreme thermophile (Patchett et al., 1991). Many arginase-containing mesophiles and thermophiles also contain several enzymes of a classical arginase pathway: arginase, ornithine transaminase, and pyrroline-5-carboxylate dehydrogenase.

Arginine-deiminase pathway

In this widely distributed energy-producing system, three enzymes constitute the pathway and are presented in Figure 8.6. Arginine serves as the sole energy source for many organisms, among which are some spirochetes, pseudomonads, and streptococci. The production of NH_4^+ by the system has been suggested to assist in some degree of acid tolerance by streptococci and pseudomonads.

Although the pathway was believed to be the major energy source for nonglycolytic arginine-utilizing *Mycoplasma*, the system is now thought to be an alternate energy source for these organisms. In *Mycoplasma*-infected animal cells in culture, putrescine stimulates animal cell growth and minimizes arginine depletion by the

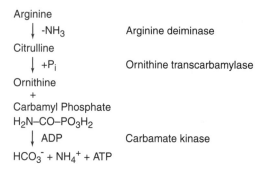

Arginine
  ↓ -NH$_3$ Arginine deiminase
Citrulline
  ↓ +P$_i$ Ornithine transcarbamylase
Ornithine
  +
Carbamyl Phosphate
H$_2$N–CO–PO$_3$H$_2$
  ↓ ADP   Carbamate kinase
HCO$_3^-$ + NH$_4^+$ + ATP

FIG. 8.6 The arginine-deiminase pathway.

bacteria. Although the arginine deiminase and ornithine transcarbamylase of extracts of the infected cells were decreased by growth in the presence of putrescine, these enzymes were not inhibited by the diamine added to the extracts of infected cells (Kamatani et al., 1983). The isolated mycoplasma deiminase may effectively deplete arginine from growth media and has been shown to inhibit the growth of animal cells in tissue culture by this mechanism.

Arginine succinyltransferase pathway and other systems in pseudomonads

The N^2 succinylation of arginine via succinyl coenzyme A (coA) was discovered in many pseudomonads (Cunin et al., 1986). In at least one the degradation of N^2-succinyl arginine to succinyl ornithine and eventually to succinate was the sole source of utilizable carbon. In most strains having this pathway (e.g., *P. aeruginosa*), ornithine was succinylated initially and metabolized similarly to glutamate (Wauven et al., 1988). In some others, ornithine was converted to proline and then metabolized further (Stalon et al., 1987).

    P. aeruginosa also contains the arginine-deiminase pathway and an inducible arginine decarboxylase leading to the formation of agmatine. Agmatine is degraded by an inducible deiminase in a reaction producing N-carbamoylputrescine. A further hydrolysis produces putrescine, which is also degraded to succinate. (See Figure 8.7.)

An extract of *E. coli* is also reported to decarboxylate citrulline to N-carbamoylputrescine (Akamatsu et al., 1978). The formation of this compound also occurs in *Streptococcus faecalis* grown on agmatine as an energy source (Roon & Barker, 1972; Simon & Stalon, 1982). In addition to an agmatine deiminase this organism also contains a putrescine transcarbamoylase that synthesizes N-carbamoylputrescine from putrescine and carbamyl phosphate. These reactions are given below.

$$\text{Citrulline} \underset{E.\ coli}{\overset{-CO_2}{\longrightarrow}} \overset{\text{carbamyl}}{\underset{\text{putrescine}}{}} \overset{\text{putrescine}}{\underset{\text{carbamyl}}{\overset{+}{\underset{\text{phosphate}}{}}}} \overset{}{\underset{S.\ faecalis}{\longleftarrow}}$$

The role and quantitation of the citrulline pathway to carbamyl putrescine and putrescine has not been evaluated in *E. coli*.

    In addition to the three degradative paths noted above, *P. aeroginosa* possesses an arginine dehydrogenase that converts D-arginine selectively to 2-ketoarginine. The organism also contains an arginine racemase, permitting L-arginine to be catabolized via this new route. The keto acid was decarboxylated to 4-guanidinobutyraldehyde and eventually to δ-aminobutyrate and succinate (Jann et al., 1988). The main route of arginine degradation in strains of *P. putida* proved to be the oxidase pathway (Tricot et al., 1991).

Additional mechanisms of ornithine utilization

In clostridia, ornithine is converted by a 4,5-amino mutase containing a vitamin B$_{12}$ coenzyme to the 2,4-diaminopentanoate, which is transformed to alanine and acetyl coA. In arginine-deficient media, some streptococci form N^5-(1-carboxyethyl)-ornithine, related to the N^2-(1-carboxyethyl)-ornithine (octopinic acid) of the octopine family of compounds found in crown gall tissue. The bacterial synthesis of the N^5 derivative from ornithine and pyruvate is effected by an NADP oxidoreductase (Thompson, 1989). An N^6-(carboxyethyl)lysine is known as well, but the functions of these compounds in bacteria are not known. Several inhibitory agents (antibiotics), derived in part from ornithine (e.g., clavulanic acid, acivin, L-N^5-(1-iminoethyl)ornithine), are synthesized by *Streptomyces* species.

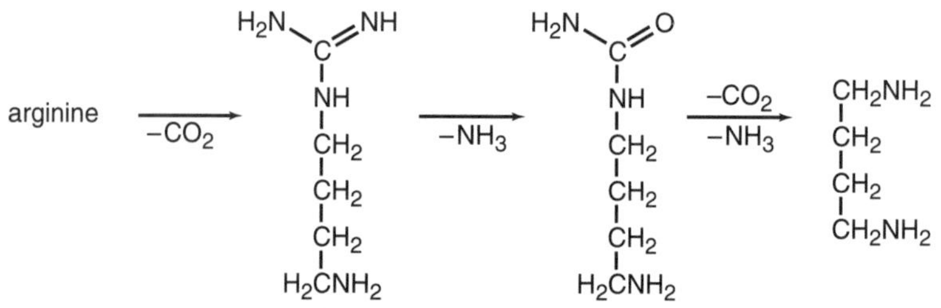

FIG. 8.7 An agmatine-deiminase pathway to putrescine.

Genetic Blocks in Synthesis of Putrescine

Knowledge of the two main routes to the formation of putrescine from arginine or ornithine led to an interest in blocking these syntheses as a possible approach to the functions of the diamine. The first isolated mutants defective in this biosynthesis were conditionally deficient; that is, they contained blocks in arginine degradation that became evident in the presence of arginine, which had almost completely repressed ornithine biosynthesis.

In one group of investigations a search for the corepressor of arginine biosynthesis explored K12 strains resistant to the toxic analogue, canavanine. A canavanine-resistant organism, after treatment with a mutagenizing agent, permitted the selection of a strain that grew quickly on canavanine but slowly in the presence of arginine. The organism proved to have a normal level of arginine decarboxylase but only 4% of the parental content of AUH. Additional mutagenesis and selection from cultures of this treated organism led to further reduction in the content of AUH and to a slower growth in the presence of arginine. This slow growth rate in the presence of arginine was enhanced by putrescine > spermidine > spermine. Growth in the presence of arginine resulted in long snakelike forms; these did not appear in the presence of putrescine (Hirshfield et al., 1970).

Cadaverine had essentially no effect on the growth of these mutants, which had enhanced production of cadaverine and aminopropylcadaverine under conditions of growth in arginine (Dion & Cohen, 1972). The presence of exogenous lysine markedly increased the production of cadaverine and the triamine. This switch to a pathway of lysine degradation thus provided a diamine that increased growth rate somewhat, and introduced a new complexity in the study of arginine-repressed putrescine-deficient organisms.

In another approach a strain low in inducible ADC was mutagenized and plated on ornithine and putrescine to permit the growth of organisms unable to degrade arginine. Individual colonies were grown in the presence of [^{14}C]-*guanidino*-arginine and the media were tested for [^{14}C]-urea. Two of the almost 900 colonies tested were very low in this degradative step. One of the mutants was found to be deficient in ADC and the other was markedly low in AUH. As shown in Table 8.4, the polyamine contents of the parental strain and the mutant organisms were consistent with these enzymatic constitutions. Surprisingly, putrescine-depleted cells (i.e., CJ556), depleted to 1% of the normal level of the diamine, had a very small decrease, ca. 10%, of growth rate. The spermidine content of this strain appeared to be increased by about 50%. The putrescine-deficient strain had a normal morphology and normal levels of protein and RNA content.

Following these inconclusive results on the role of putrescine in *E. coli*, an organism known to contain an unusually large amount of putrescine, efforts were made to introduce more stringent genetic blocks in the pathways of polyamine biosynthesis. The genes controlling ADC and AUH, designated as *speA* and *speB*, respectively, were found in various types of genetic crosses (i.e., mating experiments and phage transductions), to be in close proximity in the *E. coli* chromosome. The genes for the transport of arginine, *argP*, and for the synthesis of *S*-adenosylmethionine (*metK*), a metabolite essential for the synthesis of spermidine, were also within this region of the chromosome (Maas et al., 1970). In contrast the genes for the bADC and inducible ADC were in readily separable sites.

Pursuing this problem, mutants for bODC were sought using a method of localized mutagenesis from strains blocked in AUH (Cunningham–Rundles & Maas, 1975). One such organism, in which both routes of putrescine synthesis were blocked, was found and required putrescine or spermidine for growth in the absence of arginine. The organism grew slowly in the absence of exogenous polyamine at a growth rate of about 25% that of the growth in the presence of putrescine and contained about 3% of the diamine present in the sup-

TABLE 8.4　Enzyme and polyamine levels in putrescine-deficient mutants.

Strain	Growth condition	Enzyme (U/mg protein)			Polyamine (nmol/10^9 cells)		
		ODC	ADC	AUH	Putrescine	Spermidine	Agmatine
Parent							
BR9a	−Arg	24	16	23	22.1	3.8	<0.3
BR9a	+Arg	33	17	25	26.4	4.0	<0.3
ADC deficient							
CJ556	−Arg	33	1	18	25.6[a]	3.7	<0.3
CJ556	+Arg	119	4	16	<0.3	5.5	<0.3
AUH deficient							
CJ866	−Arg	33	21	5	24.8	1.7	4.8
CJ866	+Arg	44	20	3	5.0	2.2	20.2

Adapted from Morris and Jorstad (1970).

[a]The ADC-deficient strain used for polyamine studies was different from that used in the enzyme studies, although the two grew similarly in minimal media.

plemented bacteria. Much of the diamine in the unsupplemented organism was cadaverine. Genetic crosses demonstrated that the gene for ODC (*speC*) was also close to those of *speA* and *speB*.

A mutant containing AUH but devoid of ODC was derived from the putrescine-requiring double mutant. Its slow rate of growth implies a bacterial preference for the decarboxylation of ornithine to putrescine over the ADC pathway, although it can synthesize arginine and degrade it to putrescine internally. It had been shown earlier that exogenous arginine, which markedly increases growth rate, is preferred over internally generated arginine.

Other strains have been isolated with deletions in *speA*, *speB*, and *speC*, as well as that for the gene, *speD*, which determines *S*-adenosylmethionine decarboxylase, a first step in the synthesis of spermidine from putrescine. Indeed a point mutation governing LDC (*cadA*) has been added to minimize synthesis of cadaverine. To obtain mutants unable to decarboxylate particular substrates, a general mass screening method was developed to seek the presence or absence of $^{14}CO_2$ production above a colony grown on a plate containing the labeled substrate (Tabor et al., 1983a).

As in the earlier studies, a triple deletion mutant ($\Delta speA$, $\Delta speB$, $\Delta speC$) grew indefinitely in the absence of exogenous amines at a rate of about a third that of a supplemented organism; the organism lacked "detectable" amine. Such strains became far more or even "completely" polyamine dependent when a gene for streptomycin resistance (*strA*) was introduced and the organism was only partially depleted of amine. The particular gene for streptomycin resistance bearing the *rpsL* mutation affects the S12 ribosomal protein; this result was interpreted to suggest that the mutation increased the polyamine requirement in ribosomal structure and protein biosynthesis (Tabor et al., 1981; Tabor & Tabor, 1983).

Spontaneous revertants of a polyamine-dependent strain were grown with and without added polyamine and were tested for the ability to convert [³H]-ornithine to polyamine and arginine. It was found that such strains did actually synthesize very small amounts of putrescine and spermidine and large amounts of arginine. They were essentially devoid of bODC, dODC, and bADC. The strain did contain dADC. Some very small degree of conversion of agmatine to putrescine occurred in bacterial extracts. It was concluded that putrescine can be synthesized from arginine via the dADC and the subsequent cleavage of agmatine. The diamine content, determined by an isotopic dilution technique, of the deletion mutant grown in ornithine amounted to 8000–8400 molecules of putrescine per cell and an approximately equivalent amount of spermidine (Panagiotidis et al., 1987). Despite this small amount of these two amines, the deletion mutant of *E. coli* contained only 16% less RNA than a normal cell, which contains about 10,000–20,000 ribosomes and 150,000–200,000 molecules of tRNA. It is evident that it is very difficult to exclude diamines completely from *E. coli*.

cAMP and Expression of *SpeA*, *SpeB*, and *SpeC* Genes

The expression of the biosynthetic decarboxylases in *E. coli* is controlled by the polyamines acting in feedback inhibition and repression, by guanine nucleotides that activate or inhibit ODC, by antizyme induced by the polyamines, and by monovalent cations. The control of bacterial growth rate by different carbon sources (e.g., growth rate in glucose > glycerol > acetate) is paralleled by the activities of the bacterial decarboxylases grown in these media. It was surmised that the parallelism related to the production of cAMP and it was shown that exogenous cAMP added to *E. coli* deficient in adenylate cyclase (cya mutants) decreased the production of ODC and ADC. This did not occur in mutants (crp mutants) deficient in the cAMP receptor protein (CRP) (Wright & Boyle, 1982). The phenomenon of repression of ODC by cAMP was explored further in minicells of *E. coli* that lacked the bacterial chromosome and had been transformed by a plasmid containing *speC*. The repression of the synthesis of ODC by cAMP in the minicells or in an extract of these cells required the presence of CRP (Wright et al., 1986). The site of inhibition of transcription was thought to be at a site of binding of a cAMP–CRP complex at a promotor region of the *speC* gene.

Satishchandran and Boyle (1984) noted that the synthesis of AUH could also be repressed by exogenous cAMP, an effect that could be reversed by the inducer, exogenous agmatine. Study of the gene *speB*, which encodes AUH, revealed that agmatine induced the activity of the *speB* promoter (Szumanski & Boyle, 1992). The nature of the inhibitory cAMP effect was not demonstrated.

Cloning and Expression of *SpeA* (bADC), *SpeB* (AUH), and *SpeC* (bODC) Genes

The procedures employed in these studies were described (Boyle et al., 1984; Tabor et al., 1983b). Their details were elaborated in a methodological text on recombinant DNA (Wu, 1979). Portions of *E. coli* DNA were introduced into plasmids used subsequently to transform polyamine-deficient bacteria. Cells containing plasmids with the desired genes were grown to amplify plasmid yield and the plasmid DNA was isolated. Restriction endonucleases fragmented this DNA, portions of which were separated and introduced into other plasmid vectors. Transformation of selected *E. coli* strains resulted in the overproduction of the desired proteins, which were detected by electrophoresis in denaturing polyacrylamide gels. The transformation, carried out in minicells incubated in a radioactive amino acid (^{14}C-lysine), permitted the demonstration that a particular plasmid carried the structural gene for the desired enzyme. The positions of the genes on the circular *E. coli* chromosome were determined by genetic crosses, and with the aid of these data the physical positions of the genes were established with an an array of complementary transducing λ bacteriophages (Satischandran et al., 1990). A comparison of the genetic map and physical map of the three genes is given in Table 8.5.

DNA fragments containing the desired gene for *speA* and *speB* were removed from plasmid DNA and subcloned into new plasmids that were used to transform polyamine-deficient cells. The DNA of one of these subclones that overproduced ADC or AUH was sequenced by a method of sequencing DNA fragments. The determination of the entire sequence was obtained from analysis of overlapping DNA fragments.

In the case of the *speA* gene determining the biosynthetic ADC, a DNA piece of 3236 base pairs was found to contain an open-reading frame (ORF) of 1974 nucleotides. This began with an AUG initiation codon (methionine) at nucleotide 987, and ended with a GAG codon (glutamic acid) at nucleotide 2960. This ORF encoded 658 amino acids comprising a protein of 73,980 Da. The expression of the *speA* gene in minicells led to the appearance of two species of immunoprecipitable ADC: one of 74,000 Da and another of 70,000 Da. The 74,000 species, set by the gene, is considered to be the precursor of the mature 70,000 form. The ADC is thought to have lost a signal sequence for rapid export to the periplasm (Buch & Boyle, 1985). A PLP binding site was found in the protein and this same sequence was found in several different PLP-dependent decarboxylases (Moore & Boyle, 1990).

The enzyme agmatine ureohydrolase was purified some 1600-fold from *E. coli* transformed with a plasmid containing the *speB* gene. Its activity increased significantly during purification, indicating the removal of an inhibitor in the extract. The enzyme was found to have a molecular weight of about 30,000 and after denaturation possessed an apparent weight of about 38,000 (Satischandran & Boyle, 1986). The enzyme was not inhibited by nucleotides or by polyamines; apparently it was not under end product control. The enzyme was inhibited by ornithine and arginine with K_i values approaching 10^{-2} M.

An analysis of *E. coli* DNA containing the *speB* gene involved a sequencing of a 2.97 kilobase pair fragment (Szumanski & Boyle, 1990). One ORF, of four ORFs detected, was found to contain the *speB* gene. Its sequence coded for a 33,409 molecular mass AUH, which

was some 4.6 kDa less than that detected by physical measurement in an electrophoretic gel. The nature of the discrepancy is not clear. The sequence of AUH had three regions of homology to three sequenced arginases of eucaryotic cells, suggesting an evolutionary relationship or common origin between *E. coli* and eucaryotes, or the remote possibility of a functional convergence, reflecting the common task of removing urea hydrolytically from their substrates.

Most recently as the amino acid sequences of nine different PLP-dependent amino acid decarboxylases became available, studies of their alignments and differences revealed the existence of four apparently unrelated groups (Sandmeier et al., 1994). One group contained the procaryotic ODC (constitutive and inducible) and LDC (see below) as well as the bADC. Another contained the distinctively different eucaryotic ODC and the procaryotic bADC and diaminopimelate decarboxylase.

Biosynthesis of Lysine

Although cadaverine in polyamine biosynthesis and metabolism appears to be a relatively minor component in most bacteria, its role in partially replacing putrescine functionally suggests the desirability of filling in the available data on this diamine. The biosynthesis and metabolism of lysine and cadaverine in plants, and more particularly in chloroplasts, strengthens this perspective. Furthermore, lysine stems from a bacterial pathway that also gives rise to methionine, which serves in turn as a precursor of spermidine, aminopropylcadaverine, and spermine.

In the late 1940s a new amino acid, α,ε-diaminopimelic acid (DAPA), was discovered in Corynebacteria. DAPA was present in many procaryotic organisms and was later shown to be an essential component of many bacterial cell wall mucopeptides. Extracts of *E. coli* contained diaminopimelic acid decarboxylase and produced lysine (Dewey et al., 1954). The enzyme, which specifically decarboxylates the mesoisomer of DAPA and yields L-lysine, was purified free of LDC and of a diaminopimelate epimerase, shown to have an absolute requirement for PLP, and was activated by thiols (White & Kelly, 1965). The synthesis of this decarboxylase was repressed by growth of *E. coli* in the presence of lysine, as well as by some other amino acids and pyridoxine.

The biosynthesis of DAPA in *E. coli* begins with a condensation of aspartate semialdehyde and pyruvate to form dihydropicolinate. This initial step is inhibited by lysine. The complex sequence is presented in Figure 8.8; the individual enzymes and their genetic controls were described by Cohen and Saint–Girons (1987). Three different pathways of lysine biosynthesis are in fact known in procaryotes. Succinylation is replaced by acetylation in some *Bacillus* species, and a route to DAPA via a Δ^1-piperideine dehydrogenase exists in some Corynebacteria (Schrumpt et al., 1991). It appears

TABLE 8.5 Locations of genes.

| Gene | Map location | |
	Genetic (min)	Physical (kbp)
speA	63.5	3099–3097
speB	63.5	3097–3096
speC	64.0	3123–3121

Adapted from Satischandran et al. (1990).

kbp, kilobase pairs.

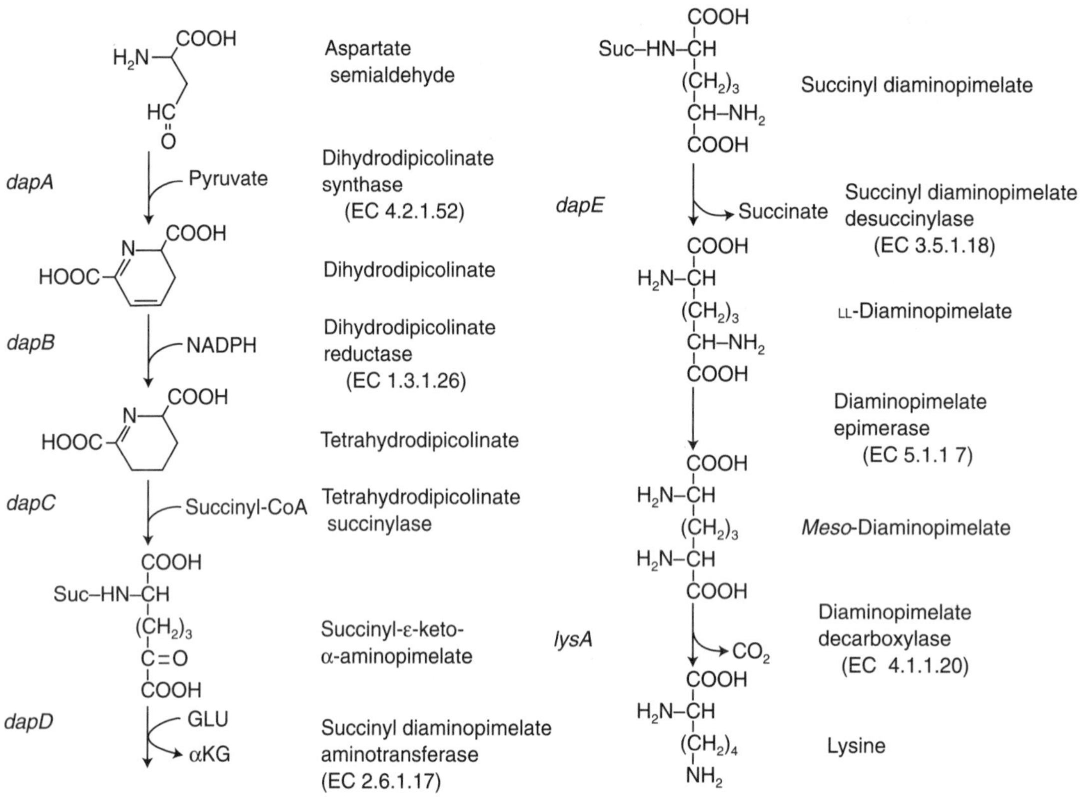

FIG. 8.8 Biosynthetic pathway for diaminopimelate and lysine; redrawn from Cohen and Saint–Girons (1987).

that *E. coli* may bypass a block in the succinylation route via the dehydrogenase but not via acetylation.

The major immediate step to lysine, DAPA decarboxylase, a protein of 200,000 molecular weight, is encoded by the *lysA* gene. The gene, whose sequence has been determined, defines a polypeptide of 46,099 molecular weight, a size suggesting a tetrameric nature of the enzyme. The *lysA* genes of several other organisms, corynebacteria, pseudomonads, etc., have been sequenced and comparisons have revealed many similarities among them. No similarities were found to the bacterial amino acid decarboxylases. However, DAPA decarboxylases are described in many sequences as homologous to mouse ODC (Sandmeier et al., 1994).

LDC and Sources of Cadaverine

LDC, determined in *E. coli* by the *cadA* present in the *cad* operon, is usually produced in dense bacterial cultures in the presence of lysine, low pH, and anaerobiosis. Induction by low pH was found to be independent of the presence of O_2 or the phase of growth. The structure of the *cad* operon and its regulation by low pH have been studied (Meng & Bennett, 1992; Shi and Bennett, 1995; Watson et al., 1992). The regulation of

expression of the *cadA* gene in acidic media appears to be somewhat different from that of the *adi* gene, which determines the similarly acidic induction of the biodegradative ADC. The latter gene was recently sequenced by Stim and Bennett (1993). The LDC of *Vibrio parahaemolyticans* was purified to apparent homogeneity and characterized and compared with that of *E. coli* and a yeast (Yamamoto et al., 1991).

E. coli deficient in ODC and AUH were mutagenized and screened for the inability to produce $^{14}CO_2$ from [U-^{14}C]lysine (Goldemberg, 1980). Extracts of strains low in this activity were tested for LDC, and three were found to have 1–3% of the parental activity. Two forms of this small amount of LDC were detected, one thermolabile and the other thermostable. The latter was considered to be the usual inducible form. The mutant strains produced only traces of cadaverine in the absence of putrescine in the media.

As in the case of arginine biosynthesis and metabolism, the search for resistance to toxic analogues can permit the selection of strains affected in various regulatory properties. The presumed inhibitor of LDC, α-difluoromethyllysine, was synthesized (Bey et al., 1979) and reported to be an active inhibitor of the LDC of *Mycoplasma dispar* (Pösö et al., 1984). The retardation of growth of this organism and the inhibition of the activity of the enzyme were prevented by the presence of lysine.

However, a lysine analogue, thiosine (*S*-aminoethylcysteine), like canavanine, enters protein and inhibits the growth of many microbes in the absence of exogenous lysine. One resistant mutant (*CadR*) had a lowered rate of lysine uptake and synthesized and excreted lysine more readily than the parent strain. It was also derepressed in its content of LDC (Popkin & Maas, 1980).

A polyamine-deficient LDC-overproducing strain, cadR⁻, was used to construct a strain lacking the *cadA* gene, the structural gene for LDC. The *cadR* and *cadA* genes were found to be widely separated and relatively distant from the *speA* and *speB* genes. This strain, lacking both genes, was also reported to form only traces of cadaverine on an amine-deficient medium (Tabor et al., 1980). A study of four of the various putrescine-requiring mutants revealed mass doubling times inversely proportional to their contents of aminopropylcadaverine plus spermidine. It appeared that cadaverine and aminopropylcadaverine were in fact "compensatory polyamines" (Igarashi et al., 1986).

Other Routes in Metabolism of Cadaverine and Lysine

Organisms that produce the siderophore, desferrioxamine B, contain LDC. The siderophore is a trihydroxamate, consisting of three residues of 1-amino-5-hydroxyaminopentane, thought to originate from cadaverine. The compound is linked in a linear structure by two molecules of succinate and is acetylated at one end. Limitation of iron in cultures of *Streptomyces pilosus* leads to the production of the siderophore, for which cadaverine is thought to be a major precursor. LDC production precedes siderophore appearance in the growth medium and mutants blocked in synthesis of the siderophore lack this enzyme. The presence of Fe^{3+} in the medium represses LDC and siderophore production (Schupp et al., 1987).

In many *Streptomyces* species lysine may serve as the sole nitrogen source, and in these organisms the entire pathway will pass via LDC through cadaverine. In a second pathway, contained in some *Streptomyces* penicillin producers, as well as in *Flavibacteria*, an initial transamination from lysine gives rise to Δ^1-piperideine carboxylate.

In pseudomonads grown in the presence of lysine an inducible lysine oxygenase initiates a degradation to 5-aminovalerimide, which is converted to 5-aminovalerate. Lysine is also an activator of oxygenase (Van De Casteele & Hermann, 1972). The subsequent pathway leads to glutarate semialdehyde, glutarate, and eventually to acetyl coA + CO_2.

In *Clostridia*, in which lysine may serve as a source of carbon, nitrogen, and energy, L-lysine is converted from the 2,6 diamino compound to β-L lysine, the 3,6 diamino acid. The enzyme carrying out this reaction, in which the 2-amino group of lysine is interchanged with

a hydrogen at carbon 3, is known as a lysine 2,3-aminomutase. The enzyme contains Fe and PLP, and the reaction requires the participation of *S*-adenosylmethionine.

This chapter closes therefore with reference to an unexpected activity of *S*-adenosylmethionine, whose properties and activities as an important component of polyamine metabolism in procaryotic and eucaryotic cells will be examined in greater detail in subsequent chapters.

REFERENCES

Akamatsu, N., Oguchi, M., Yajima, Y., & Ohno, M. (1978) *Journal of Bacteriology*, **133**, 409–410.
Alhonen–Hongisto, L., Veijalainen, P., Ek-Kommonen, C., & Jänne, J. (1982) *Biochemical Journal*, **202**, 267–270.
Applebaum, D., Sabo, D. L., Fischer, E. H., & Morris, D. R. (1975) *Biochemistry*, **14**, 3675–3681.
Appelbaum, D. M., Dunlap, J. C., & Morris, D. R. (1977) *Biochemistry*, **16**, 1580–1584.
Asada, Y., Tanizawa, K., Nakamura, K., Moriguchi, M., & Soda, K. (1984) *Journal of Biochemistry*, **95**, 277–282.
Balasundaram, D., and Tyagi, A. K. (1988) *Archives of Biochemistry and Biophysics,* **264**, 288–294.
Balasundaram, D. & Tyagi, A. K. (1989) *European Journal of Biochemistry*, **183**, 339–345.
Beier, H., Fecker, L. F., & Berlin, J. (1987) *Zeitschrift für Naturforschung C*, **42**, 1307–1312.
Bey, P., Ververt, J., Dorsselaer, V. V., & Kolb, M. (1979) *Journal of Organic Chemistry*, **44**, 2732–2742.
Bitonti, A. J. & McCann, P. F. (1987) *Inhibition of Polyamine Metabolism*, pp. 259–275, P. P. McCann, A. E. Pegg, & A. Sjoerdsma, Eds. New York: Academic Press.
Boeker, E. A. (1978) *Biochemistry*, **17**, 258–263.
Boeker, E. A. & Fischer, E. H. (1983) *Methods in Enzymology*, **94**, 108–184.
Boeker, E. A., Fischer, E. H., & Snell, E. E. (1971) *Journal of Biological Chemistry*, **246**, 6776–6781.
Boyle, S. M., Markham, G. D., Hafner, E. W., Wright, J. M., Tabor, H., & Tabor, C. W. (1984) *Gene*, **30**, 129–136.
Brooker, D. C., Lund, M. E., & Blazevic, D. J. (1973) *Applied Microbiology*, **26**, 622–623.
Buch, J. K. & Boyle, S. M. (1985) *Journal of Microbiology*, **163**, 522–527.
Canellakis, E. S., Heller, J. S., Kyriakidis, D., & Chen, K. Y. (1978) *Advances in Polyamine Research*, vol. 1, pp. 17–30, R. A. Campbell, D. R. Morris, D. Bartos, G. D. Daves, & F. Bartos, Eds. New York: Raven Press.
Canellakis, E. S., Paterakis, A. A., Huang, S., Panagiotidis, C. A., & Kyriakidis, D. A. (1993) *Proceedings of the National Academy of Sciences of the United States of America*, **90**, 7129–7133.
Cataldi, A. A. & Algranati, I. D. (1989) *Journal of Bacteriology*, **171**, 1998–2002.
Cohen, G. N. & Saint–Girons, I. (1987) *Escherichia coli* and *Salmonella typhimurium*: Cellular and Molecular Biology, vol. I, pp. 429–444, F. C. Neidhardt, Ed. Washington, D.C.: American Society for Microbiology.
Cunin, R., Glansdorff, N., Piérard, A., & Stalon, V. (1986) *Microbiological Reviews*, **50**, 314–352.
Cunningham–Rundles, S. & Maas, W. K. (1975) *Journal of Bacteriology*, **124**, 791–799.
Danzin, C. & Casara, P. (1984) *FEBS Letters*, **174**, 275–278.
Dion, A. S. & Cohen, S. S. (1972) *Journal of Virology*, **9**, 423–430.

Djurhuus, R. (1981) *Analytical Biochemistry*, **113**, 352–355.

Elston, H. R. (1971) *Applied Microbiology*, **22**, 1091–1095.

Fecker, L. F., Beier, H., & Berlin, J. (1986) *Molecular and General Genetics*, **203**, 107–184.

Fozard, J. R. & Koch–Weser, J. (1982) *Trends in Pharmacological Sciences*, **March**, 177–110.

Gale, E. F. (1940) *Biochemical Journal*, **34**, 392–413.

———— (1946) *Advances in Enzymology*, **6**, 1–32.

Gani, D. (1991) *Philosophical Transactions of the Royal Society of London Series B*, **332**, 131–139.

Glansdorff, N. (1987) *Escherichia coli* and *Salmonella typhimurium*: Cellular and Molecular Biology, vol. I, pp. 321–344, F. C. Neidhardt, Ed. Washington, D.C.: American Society for Microbiology.

Goldemberg, S. H. (1980) *Journal of Bacteriology*, **141**, 1428–1431.

Grillo, M. A. & Fossa, T. (1983) *International Journal of Biochemistry*, **15**, 139–141.

Guirard, B. M. & Snell, E. E. (1980) *Journal of Biological Chemistry*, **255**, 5960–5964.

Hackert, M. L., Carroll, D. W., Davidson, L., Kim, S.-O., Momany, C., Vaaler, G. L., & Zhang, L. (1994) *Journal of Bacteriology*, **176**, 7391–7394.

Hamana, K. & Matsuzaki, S. (1990) *Canadian Journal of Microbiology*, **37**, 350–354.

Harik, S. I. & Snyder, S. H. (1973) *Biochimica et Biophysica Acta*, **327**, 501–509.

Hayashi, H. (1995) *Journal of Biochemistry*, **118**, 463–473.

Heerze, L. D., Kang, Y. J., & Palcic, M. M. (1990) *Analytical Biochemistry*, **185**, 201–205.

Heller, J. S., Canellakis, E. S., Bussolotti, D. L., & Coward, J. K. (1975) *Biochimica et Biophysica Acta*, **403**, 197–207.

Heller, J. S., Rostomily, R., Kyriakidis, D. A., & Canellakis, E. S. (1983) *Proceedings of the National Academy of Sciences of the United States of America*, **80**, 5181–5184.

Hickman–Brenner, F. W., MacDonald, K. L., Steigerwalt, A. G., Fanning, G. R., Brenner, D. J., & Farmer, III, J. J. (1987) *Journal of Clinical Microbiology*, **25**, 900–906.

Hirshfield, I. N., Rosenfeld, H. J., Leifer, Z., & Maas, W. K. (1970) *Journal of Bacteriology*, **101**, 725–730.

Hölttä, E., Jänne, J., & Pispa, J. (1972) *Biochemical and Biophysical Research Communications*, **47**, 1165–1171.

———— (1974) *Biochemical and Biophysical Research Communications*, **59**, 1104–1111.

Huang, S., Panagiotidis, C. A., & Canellakis, E. S. (1990) *Proceedings of the National Academy of Sciences of the United States of America*, **87**, 3464–3468.

Igarashi, K., Kashiwagi, K., Hamasaki, H., Miura, A., Kakegawa, T., Hirose, S., & Matsuzaki, S. (1986) *Journal of Bacteriology*, **166**, 128–134.

Ikai, H. & Yamamoto, S. (1994) *FEMS Microbiology Letters*, **124**, 225–228.

Jann, A., Matsumoto, H., & Haas, D. (1988) *Journal of General Microbiology*, **134**, 1043–1053.

Kallio, A., McCann, P. P., & Bey, P. (1982) *Biochemical Journal*, **204**, 771–775.

Kamatani, N., Willis, E. H., McGarrity, G. J., & Carson, D. A. (1983) *Journal of Cellular Physiology*, **114**, 16–29.

Kamio, Y. & Terawaki, Y. (1983) *Journal of Bacteriology*, **153**, 658–664.

Kashiwagi, K. & Igarashi, K. (1987) *Biochimica et Biophysica Acta*, **911**, 180–190.

Kyriakidis, D. A., Heller, J. S., & Canellakis, E. S. (1978) *Proceedings of the National Academy of Sciences of the United States of America*, **75**, 4699–4703.

———— (1983) *Methods in Enzymology*, **94**, 193–199.

Large, P. J. (1992) *FEMS Microbiology Reviews*, **88**, 249–262.

Maas, W. K., Leifer, Z., & Poindexter, J. (1970) *Annals of the New Academy of Sciences*, **171**, 957–967.

Martel, A., Jansson, E., García–Reiner, G., & Lindblad, P. (1993) *Archives of Microbiology*, **159**, 506–511.

Martell, A. E. (1989) *Accounts of Chemical Research*, **22**, 115–124.

Matthies, C., Mayer, F., & Schink, B. (1989) *Archives of Microbiology*, **151**, 498–505.

Meng, S. & Bennett, G. N. (1992) *Journal of Bacteriology*, **174**, 2670–2678.

Metcalf, B. W., Bey, P., Danzin, C., Jung, M. J., Casara, P., & Vevert, J. P. (1978) *Journal of the American Chemical Society*, **100**, 2551–2553.

Metzler, D. E. (1977) *Biochemistry*, pp. 444–461. New York: Academic Press.

Meyer, R. & Pistorius, E. K. (1989) *Biochimica et Biophysica Acta*, **975**, 80–87.

Miller, D. L. & Rodwell, V. W. (1971) *Biochemical and Biophysical Research Communications*, **44**, 1227–1233.

Momany, C., Ernst, S., Ghosh, R., Chang, N., & Hackert, M. L. (1995) *Journal of Molecular Biology*, **252**, 643–655.

Momany, C. & Hackert, M. L. (1989) *Journal of Biological Chemistry*, **264**, 4722–4724.

Moore, R. C. & Boyle, S. M. (1990) *Journal of Bacteriology*, **172**, 4631–4640.

Morris, D. R. & Boeker, E. A. (1983) *Methods in Enzymology*, **94**, 125–134.

Morris, D. R. & Fillingame, R. H. (1974) *Annual Review of Biochemistry*, **43**, 303–325.

Morris, D. R. & Jorstad, C. M. (1970) *Journal of Bacteriology*, **101**, 731–737.

Morris, D. R. & Pardee, A. B. (1965) *Biochemical and Biophysical Research Communications*, **20**, 697–702.

Mountfort, D. O. & Pybus, V. (1992) *FEMS Microbiology Ecology*, **101**, 237–244.

Nakao, H., Ishii, M., Shinoda, S., & Yamamoto, S. (1989) *Journal of General Microbiology*, **135**, 345–351.

Nakao, H., Takeuchi, K., Shinoda, S., & Yamamoto, S. (1990) *FEMS Microbiology Letters*, **70**, 61–66.

Padmanabhan, R. & Tchen, T. T. (1972) *Archives of Biochemistry and Biophysics*, **150**, 531–541.

Paik, W. K., Kim, S., Hutchins, M. G. K., & Swern, D. (1981) *Canadian Journal of Biochemistry*, **59**, 131–136.

Panagiotidis, C. A., Blackburn, S., Low, K. B., & Canellakis, E. S. (1987) *Proceedings of the National Academy of Sciences of the United States of America*, **84**, 4423–4427.

Panagiotidis, C. A. & Canellakis, E. S. (1984) *Journal of Biological Chemistry*, **259**, 15025–15027.

Panagiotidis, C. A., Huang, S., & Canellakis, E. S. (1994) *International Journal of Biochemistry*, **26**, 991–1001.

Patchett, M. L., Daniel, R. M., & Morgan, H. W. (1991) *Biochimica et Biophysica Acta*, **1077**, 291–298.

Paulin, L. (1986) *FEBS Letters*, **202**, 323–326.

Popkin, P. S. & Maas, W. K. (1980) *Journal of Bacteriology*, **141**, 485–492.

Pösö, H., McCann, P. P., Tanskanen, R., Bey, P., & Sjoerdsma, A. (1984) *Biochemical and Biophysical Research Communications*, **125**, 205–210.

Prieto–Santos, M. I., Martin–Checa, J., Balaña–Fouce, R., & Garrido–Pertierra, A. (1986) *Biochimica et Biophysica Acta*, **880**, 242–244.

Rahiala, E., Kekomäki, M., Jänne, J., Raina, A., & Räihä, N. C. R. (1971) *Biochimica et Biophysica Acta*, **227**, 337–343.

Robins, D. J. (1983) *Phytochemistry*, **22**, 1133–1135.

Roon, R. J. & Barker, H. A. (1972) *Journal of Bacteriology*, **109**, 44–50.

Rosenfeld, H. J. & Roberts, J. (1976) *Journal of Bacteriology*, **125**, 601–607.

Ruiz, O., Alonso–Garrido, D. O., Buldain, G., & Frydman, R. B. (1986) *Biochimica et Biophysica Acta*, **873**, 53–61.

Ruiz, O., Buldain, G., Garrido, D. A., & Frydman, R. B. (1988) *Biochimica et Biophysica Acta*, **954**, 114–125.

Sabo, D. L., Boeker, E. A., Byers, B., Waron, H., & Fischer, E. H. (1974) *Biochemistry*, **13**, 662–670.

Sabo, D. L. & Fischer, E. H. (1974) *Biochemistry*, **13**, 670–676.

Sakai, T. T. & Cohen, S. S. (1976) *Proceedings of the National Academy of Sciences of the United States of America*, **75**, 3502–3505.

Sakanyan, V., Charlier, D., Legrain, C., Kochikyan, A., Mett, I., Piérard, A., & Glansdorff, N. (1993) *Journal of General Microbiology*, **139**, 393–402.

Sandmeier, E., Hale, T. I., & Christen, P. (1994) *European Journal of Biochemistry*, **221**, 997–1002.

Satischandran, C. & Boyle, S. M. (1984) *Journal of Bacteriology*, **157**, 552–559.

——— (1986) *Journal of Bacteriology*, **165**, 843–848.

Satischandran, C., Markham, G. D., Moore, R. C., & Boyle, S. M. (1990) *Journal of Bacteriology*, **172**, 4748.

Schrumpf, B., Schwarzer, A., Kalinowski, J., Pühler, A., Eggeling, L., & Sahm, H. (1991) *Journal of Bacteriology*, **173**, 4510–4516.

Schupp, T., Waldmeier, U., & Divers, M. (1987) *FEMS Microbiology Letters*, **42**, 135–139.

Shaibe, E., Metzer, E., & Halpern, Y. S. (1985a) *Journal of Bacteriology*, **163**, 933–937.

——— (1985b) *Journal of Bacteriology*, **163**, 938–942.

Simon, J. & Stalon, V. (1982) *Journal of Bacteriology*, **152**, 676–681.

Sindhu, R. K. & Cohen, S. S. (1983) *Advances in Polyamine Research*, vol. 4, pp. 371–380, U. Bachrach, A. Kaye, & R. Chayen, Eds. New York: Raven Press.

Snell, E. (1989) *Annual Review of Nutrition*, **9**, 1–19.

Stalon, V., Wauven, C. V., Momin, P., & Legrain, C. (1987) *Journal of General Microbiology*, **133**, 2487–2495.

Stim, K. P. & Bennett, G. N. (1993) *Journal of Bacteriology*, **175**, 1221–1234.

Suzuki, S., Kubo, A., & Takama, K. (1993) *Journal of Marine Biotechnology*, **1**, 47–50.

Szumanski, M. B. W. & Boyle, S. M. (1990) *Journal of Bacteriology*, **172**, 538–547.

——— (1992) *Journal of Bacteriology*, **174**, 758–764.

Tabor, C. W., Tabor, H., & Hafner, E. H. (1983a) *Methods in Enzymology*, **94**, 83–91.

Tabor, C. W., Tabor, H., Hafner, E. W., Markham, G. D., & Boyle, S. M. (1983b) *Methods in Enzymology*, **94**, 117–121.

Tabor, H., Hafner, E. W., & Tabor, C. W. (1980) *Journal of Bacteriology*, **144**, 952–956.

Tabor, H. & Tabor, C. W. (1983) *Advances in Polyamine Research*, vol. 4, pp. 455–465, U. Bachrach, A. Kaye, & R. Chayen, Eds. New York: Raven Press.

Tabor, H., Tabor, C. W., Cohn, M. S., & Hafner, E. W. (1981) *Journal of Bacteriology*, **147**, 702–704.

Thompson, J. (1989) *Journal of Biological Chemistry*, **264**, 9592–9601.

Tonelli, D., Budini, R., Gattavecchia, E., & Girotti, S. (1981) *Analytical Biochemistry*, **111**, 189–194.

Tricot, C., Stalon, V., & Legrain, C. (1991) *Journal of General Microbiology*, **137**, 2911–2918.

Van De Casteele, J. & Hermann, M. (1972) *European Journal of Biochemistry*, **31**, 80–85.

Van De Casteele, M., Demarez, M., Legrain, C., Glansdorff, N., & Piérard, A. (1990) *Journal of General Microbiology*, **136**, 1177–1183.

Voellmy, R. & Leisinger, T. (1976) *Journal of Bacteriology*, **128**, 722–729.

Watson, N., Dunyak, D. S., Rosey, E. L., Slonczewskii, J. L., & Olson, E. R. (1992) *Journal of Bacteriology*, **174**, 530–540.

Wauven, C. V., Jann, A., Haas, D., Leisinger, T., and Stalon, V. (1988) *Archives of Microbiology*, **150**, 400–404.

Wertheimer, S. J. & Leifer, Z. (1983) *Biochemical and Biophysical Research Communications*, **114**, 882–888.

Williams–Ashman, H. G. & Schenone, A. (1972) *Biochemical and Biophysical Research Communications*, **46**, 288–295.

Wright, J. M. & Boyle, S. M. (1982) *Molecular and General Genetics*, **186**, 482–487.

Wright, J. M., Satischandran, C., & Boyle, S. M. (1986) *Gene*, **44**, 37–45.

Wu, R. (1979) *Methods in Enzymology*, **68**, R. Wu, Ed., New York: Academic Press.

Wu, W. H. & Morris, D. R. (1973a) *Journal of Biological Chemistry*, **248**, 1687–1695.

——— (1973b) *Journal of Biological Chemistry*, **248**, 1696–1699.

Yamamoto, S., Imamura, T., Kusaba, K., & Shinoda, S. (1991) *Chemical and Pharmaceutical Bulletin*, **39**, 3067–3070.

Yamamoto, S., Tsuzaki, Y., Tougou, K., & Shinoda, S. (1992) *Journal of General Microbiology*, **138**, 1461–1465.

CHAPTER 9

Bacterial Paths to Spermidine and Other Polyamines

Escherichia coli as an Experimental System

Although *E. coli* is atypical in its excessive production of putrescine, it has been a useful organism not only because of the exaggerated synthesis of diamines and its genetic manipulability, but also because it synthesizes large amounts of spermidine as the sole major triamine. Therefore, in many physiological situations confronting this organism we are exploring the limited problem of the roles of only the two native amines, the diamine putrescine, and the triamine spermidine. Exogenous spermine can be incorporated into the cells and may replace or displace spermidine. In some bacteria (Chap. 5) other triamines (e.g., norspermidine and homospermidine) have been found.

As described in earlier chapters, a few putrescine-deficient strains of *E. coli*, with only a very small percentage of their normal putrescine content (6–20 mM), do grow almost normally and this is sometimes attributed to a residual relatively high spermidine content of the putrescine-deprived cells. Actually many other organisms normally possess a relatively low putrescine content (0.1–0.2 mM) and high spermidine content as in *E. coli* (1–3 mM). It has been suggested that in those organisms putrescine serves mainly as a precursor to spermidine. In *E. coli* (Chap. 7) putrescine has been implicated in several additional activities: osmoregulation, association with macromolecular structure, and phospholipid biosynthesis.

Study of the roles of the polyamines called for an effort to block the biosynthesis of spermidine, with the aid of analogues, inhibitors, or genetically modified strains. Some putrescine-deficient strains of *E. coli* did grow significantly less rapidly in defined media than the parent, and such strains grew normally when putrescine or spermidine was added to the media. Aberrant products of metabolism (e.g., accumulated agmatine or cadaverine, or aminopropylcadaverine) are not thought to be inhibitory in these systems; as discussed in Chapter 7, the latter partially compensates for the lack of normal polyamine. In studies with analogues of aminopropylputrescine (i.e., spermidine) in which the putrescine moiety was replaced by diamine containing C_2–C_8 (Morris, 1981), the C_3 analogue (3-aminopropyl diaminopropane or norspermidine) did as well as spermidine in restoring growth to the control rate. The C_5- and C_6-containing aminopropyl derivatives possessed decreasing ability to support growth. The C_7 and C_8 compounds were completely ineffective. These experiments suggested that the polyamine requirement is not simply a matter of appropriate charge (i.e., any trivalent amine), but demands a particular spatial distribution of the protonated amines. Thus the specific structure of spermidine fulfills specific roles in the lives of cells.

Severe polyamine depletion in some diamine-deficient strains was observed to evoke an inhibition of protein synthesis and a decreased ability to divide. The pursuit of these effects demonstrated not only the dependence of ribosome composition and structure on polyamine availability but also provided evidence of the stimulatory activity of spermidine in many of the individual steps of protein synthesis studied in vitro (to be discussed in Chap. 22). The effect of spermidine on the DNA substrate of an enzyme, DNA gyrase, which can control DNA structure, synthesis, and expression, will also be discussed in a later chapter. In addition the availability of the diamines in these mutant strains was found to determine the expression of the stringent response (Goldemberg, 1984, see Chap. 7), perhaps indirectly through the decreased availability of ppGpp (Nastri & Algranati, 1988).

Transformation of *E. coli* by a plasmid containing the ornithine decarboxylase (ODC) gene led to the overproduction of the enzyme and putrescine (Kashiwagi & Igarashi, 1988). The response of the bacteria to this surfeit of ODC and putrescine is presented in Table 9.1. Surprisingly the contents of putrescine and spermidine

TABLE 9.1 Polyamine contents in *E. coli* producing normal and excessive amounts of ODC.

Strain	A_{600} at harvest	Exogenous amino acid	Amount of polyamine (nmol/mg protein)			Amount of diamine in medium (nmol/mL)	
			Putrescine	Cadaverine	Spermidine	Putrescine	Cadaverine
DR112	0.3	0	95.7	2.3	14	3.5	0.31
	0.3	Ornithine (0.1)	104	1.6	11	33.4	0.11
	0.3	Arginine (1.0)	0.6	5.1	5.1	ND	ND
DR112 with p(ODC)	0.3	0	117	2.1	12.3	12.3	0.35
	0.3	Ornithine (0.1)	122	0.9	10.3	361	0.28
	0.3	Arginine (1.0)	36.5	22.8	14.2	0.75	0.03

Adapted from Kashiwagi and Igarashi (1988). ND, not detectable.

within the cells were not markedly different. Both types of cells were limited in ornithine, and if this were provided much more putrescine was made; however, much of the diamine was not retained in the cell but was excreted. Excretion of the diamine appeared to be the major control on putrescine content.

In these systems spermidine, which was not excreted, was maintained at similar concentrations in the bacteria. During growth in the presence of arginine, which repressed synthesis of ornithine and further reduced the availability of this amino acid, putrescine production fell precipitously. Nevertheless, the diamine was funneled into an almost normal concentration of bacterial spermidine. Indeed this relative control of spermidine concentration is maintained significantly in cells containing an excess of a gene determining the first step of spermidine synthesis. In this system regulation of spermidine biosynthesis and content during the exponential growth of *E. coli* appeared to be determined by the internal inhibitory activity of the triamine.

In these studies the harvesting of *E. coli* was accomplished uniformly during the exponential phase. As described earlier, the ribosome and spermidine contents of *E. coli* are known to be markedly higher at this phase than in the lag or stationary phases (Igarashi et al., 1975). The ribosomes of exponential phase cells are capable of a greater rate of polypeptide synthesis than are ribosomes extracted in other growth phases. Spermidine is almost completely sequestered as the monoglutathionyl derivative within the cell during the stationary phase. After dilution in the lag phase this compound is hydrolyzed within the cell to release spermidine. Furthermore, other changes may provoke acetylation and excretion of *N*-acetyl spermidine.

Incubation of *E. coli* and other bacteria in exogenous spermidine does increase the spermidine content of the cells. Indeed in *Bacillus megaterium*, in which spermidine is >95% of all polyamine in all stages of growth, exogenous spermidine could increase spermidine content three- to sixfold without affecting growth, sporulation, or subsequent spore germination. In the latter process, spermidine was not synthesized in the enriched spores until RNA synthesis established a spermidine to RNA ratio comparable to that in a normal unsupplemented log phase. At this point, spermidine synthesis and accumulation began and paralleled synthesis of RNA. This effect possibly indicates a role of unbound spermidine in regulating its own biosynthesis, as has been suggested for *E. coli*. Nevertheless, several apparent exceptions to this generalization were seen in *B. megaterium* (Setlow, 1974). Study of sporulation in *B. subtilis* detected an accumulation of ODC and spermidine in sporulating cells (Yamakawa et al., 1980) and a synthesis of spermidine independent of DNA synthesis during outgrowth of the spores (Ginsberg et al., 1982).

These data enlarged conclusions concerning spermidine in bacteria: a replacement of putrescine for growth in putrescine-deficient *E. coli*, without apparent degradation of the triamine to diamine; a more dominant role of the triamine in the functions of RNA-rich structures, such as ribosomes and protein synthesis; and a careful regulation of synthesis that optimizes accumulation of the triamine during the most rapid phase of cell multiplication. In organisms as different as *E. coli* and *Anacystis nidulans*, the onset of the termination of logarithmic growth is marked by the decline of free spermidine content. In the former case the relatively novel glutathione derivative is formed, of which the spermidine is reutilizable after a hydrolytic reaction. In the latter, free spermidine is actively degraded.

Discovery of Enzymatic Synthesis of Spermidine in *E. coli*

S-Adenosylmethionine (AdoMet) was discovered in studies of the origins of the methyl groups of creatine and *N'*-methylnicotinamide (Cantoni, 1953, 1982). A liver extract, supplemented by ATP and methionine, gave rise to "active methionine" containing adenosine and methionine and capable of transferring a methyl group to guanidoacetic acid. The structure of the compound was defined by Cantoni and his collaborators, who demonstrated its degradation by heat at pH 6.0 to methylthioadenosine (MTA) and homoserine. The structure of MTA was established by Suzuki et al. (1924) and was synthesized in 1951. In the next year the —SCH₃ group was shown to be derived from methionine. These structures and reactions are presented in Figure 9.1.

Methyl transfer from AdoMet leads to the formation of *S*-adenosylhomocysteine (AdoHcy), which is a strong feedback inhibitor of methylation. Enzymatic cleavage of AdoHcy produces adenosine and homocysteine, a reaction that is readily reversible. Curiously this hydrolase has been found to be a major hepatic copper-binding protein (Bethin et al., 1995).

In 1956 and 1957 it was demonstrated that isotopically labeled putrescine provides the four carbon chain for spermidine biosynthesis in *E. coli* and *Azotobacter*

vinelandii. These results were extended to the spermidine made in several fungi, including a putrescine-requiring mutant of a fungus, *Aspergillus nidulans*, and *Saccharomyces cerevisiae*, as well as in rat prostate (Tabor et al., 1958). It was shown also that methionine is the source of the carbon of the aminopropyl moiety of spermidine of *Neurospora crassa* (Greene, 1957).

These studies culminated in the enzymatic synthesis of spermidine in an extract of *E. coli*. In one set of experiments the essential components to produce a labeled spermidine were ¹⁴C-putrescine, ATP, Mg²⁺, and L-methionine. The ATP and methionine were replaceable by adenosyl-L-methionine that was prepared from either *E. coli*, liver enzymes, or yeast. In another set of experiments, label was present in 2-¹⁴C-methionine rather than in putrescine. The incorporation of radioactivity into spermidine occurred if it was present as ¹⁴C at the C-2

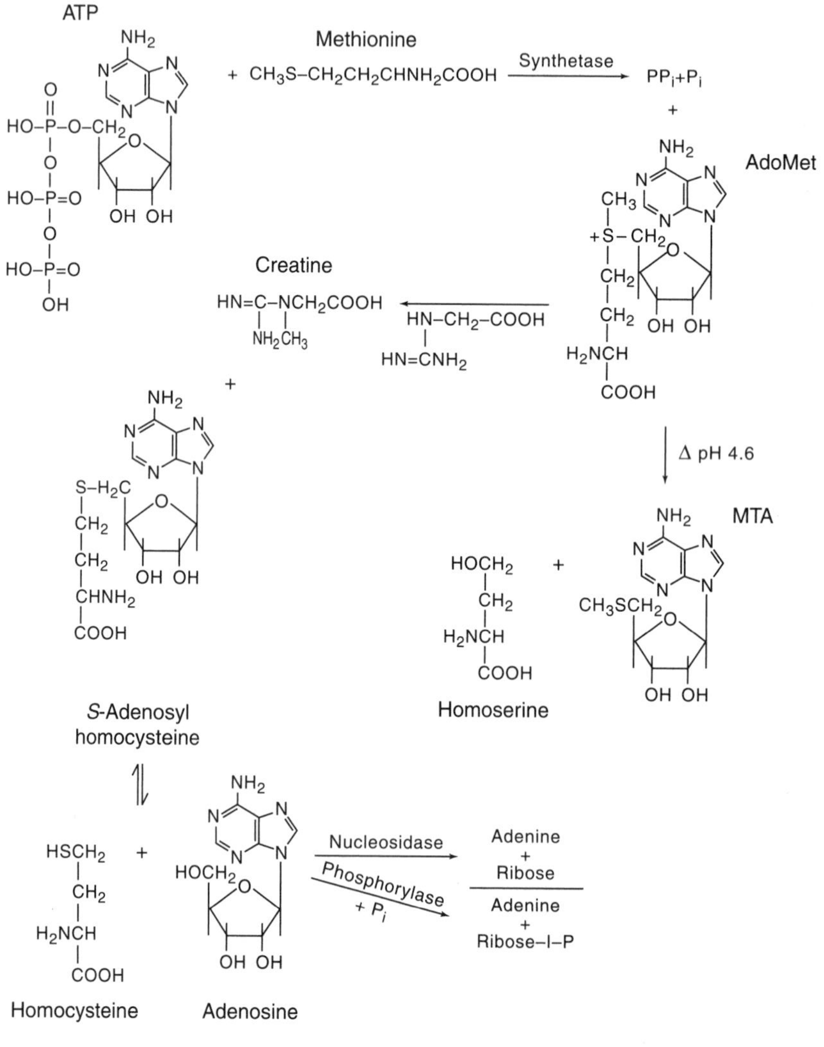

FIG. 9.1   The biosynthesis, structure, and some reactions of *S*-adenosylmethionine.

position of the amino acid, but not if it was originally in the methyl or carboxyl groups, nor in the sulfur of the amino acid. The extract carried out three reactions:

$$ATP + \text{L-methionine} \xrightarrow{\ Mg^{2+}\ } AdoMet, \quad (1)$$

$$AdoMet \rightarrow CO_2 + \text{decarboxylated AdoMet}$$
(i.e., S-adenosyl(5′)-3-methylthiopropylamine), (2)

$$
\begin{array}{ccc}
\text{decarboxylated AdoMet} & & \text{spermidine} \\
+ & \rightarrow & + \\
\text{putrescine} & & \text{5′-methylthioadenosine.}
\end{array}
$$
(3)

About a decade later, it was shown that homogenates of animal tissues will incorporate labeled putrescine into spermidine (Jänne, 1967) and that a similar series of enzymatic reactions can take place in rat prostate (Pegg & Williams–Ashman, 1969). Surprisingly, reaction (2) in the sequence (i.e., decarboxylation of Ado-Met), carried out in the mammalian system, required the presence of putrescine, a property sharply distinguishing some of the eucaryotic AdoMet decarboxylases from the bacterial enzymes. The enzymatic decarboxylation of AdoMet and subsequent propylamine transfer underlined Cantoni's prediction that the three ligands, —CH₃, methylthioadenosine, and 3-aminopropyl, at the sulfonium pole were energetically equivalent.

Expanding Role of AdoMet

In the decade following the discovery of AdoMet, the biochemistry of its synthesis and the mechanism and kinetics of methyl transfer to small molecules were clarified. The methylation of large molecules (e.g., nucleic acids, proteins, and polysaccharides), and the methylation of fatty acids and steroids were then discovered and extended in the 1970s to that of the biogenic amines, such as the catecholamines. In the last group of reactions, the biological import of methyl transfer became particularly evident and the significance of the methylation of proteins (e.g., carboxymethylation in chemotaxis) and of nucleic acids (e.g., of cytosine and other DNA bases in the rejection of foreign DNA) are active subjects of study. Some degradation products of the normal turnover of methylated macromolecules are normal functional metabolites in their own right, for example, ε-trimethyl lysine as a precursor of carnitine. Three different methylated derivatives of arginine, presumably derived from methylated myelin and embryonic muscle appear in human urine:

$$
\begin{array}{c}
NH \\
\parallel \\
CH_3-NH-C-NH-(CH_2)_3-CH(NH_2)COOH
\end{array}
$$

$$
\begin{array}{c}
N(CH_3) \\
\parallel \\
CH_3-NH-C-NH-(CH_2)_3-CH(NH_2)COOH
\end{array}
$$

$$
\begin{array}{c}
NH \\
\parallel \\
(CH_3)_2-N-C-NH-(CH_2)_3-CH(NH_2)COOH
\end{array}
$$

Methylation of tRNA was discovered by Fleissner and Borek (1962) and was soon extended to the finding of methylation of ribosomal RNA, bases, and ribose, and the bases of DNA. Methylases have been described and spermidine can activate some of these reactions.

DNA methylation is a postreplicative event that in procaryotic cells leads to the formation of 6-methyladenine and 5-methylcytosine. Only the latter is known in many eucaryotic cells. Methylation serves as a recognition mechanism for the nucleic acids, particularly for DNA, and was explored initially by Gold et al. (1964). Infection of *E. coli* by a particular T-even bacteriophage (e.g., T2, markedly increased DNA methylation (>100-fold), a reaction later shown to be determined by a phage gene. When infection was carried out by the T3 phage, DNA methylation was prevented whereas that determined by the T3-related T7 phage was essentially unchanged. The T3 effect proved to be due to the production of an enzyme, which specifically cleaved AdoMet (Gefter et al., 1966), and was determined by a gene, the *SAM*ase gene, present only in T3 DNA. SAM was an early abbreviation for AdoMet. The enzyme, SAMase, hydrolyzes AdoMet to MTA and homoserine, as presented in the reaction to heat at pH 4.6 in Figure 9.1. This enzyme, normally absent in *E. coli*, has been found in some other bacteria and yeast.

Infection of *E. coli* with inactivated phage T3, which is unable to multiply as a result of ultraviolet irradiation, leads to destruction of AdoMet and hypomethylation of RNA. The absence of the *SAM*ase gene and the survival of AdoMet and methylation permits a normal multiplication of T3, comparable to that of T7. Prevention of methylation by action of the hydrolase permits the virus to establish a lysogenic infection (Kruger & Schroeder, 1981). In *E. coli* transformed by the gene for the T3 AdoMet hydrolase, the enzyme was expressed and impaired methylase-directed DNA modifications, as well as the synthesis of spermidine from putrescine (Hughes et al, 1987). However, at 1 h after addition of ¹⁴C-putrescine to the hydrolase-containing cells, only a 77% inhibition of spermidine synthesis was found, suggesting a rate of synthesis of AdoMet exceeding that of its hydrolysis.

Properties of AdoMet

A comprehensive early discussion of the chemistry of AdoMet and other biological sulfonium compounds, such as *S*-methylmethionine, was made by Schlenk (1965). The cation of the coenzyme is a molecule of 399.4 molecular weight. The molecular weight of preparations is obviously dependent on the size of the anion neutralizing the sulfonium cation. This point is significant because AdoMet is unstable at neutrality and is best stored in acid. Counterions that reportedly stabilize the coenzyme during storage include the relatively large nonnucleophilic sulfate and tosylate (Matos & Wong, 1987).

The tedious enzymatic synthesis of AdoMet was replaced by an isolation from yeast grown in an excess of methionine (Schlenk & De Palma, 1957). As revealed by ultraviolet microphotography, AdoMet is stored in the yeast vacuoles and may attain high concentrations in yeast in excess of 20 μmol/g (Schlenk, 1965; Shiozaki et al., 1984). The extracted coenzyme may be precipitated by phosphotungstate and can be estimated by ultraviolet spectrophotometry, facilitated by a high molar extinction coefficient. This coefficient was given as 15,200 at pH 1, with a maximum at 256 nm.

In addition to the cleavage of AdoMet to MTA at 100°C at pH 4, incubation at 30°C at a pH above 8.0 splits adenine from the compound. Heating in 0.1 N HCl also cleaves adenine from AdoMet, MTA, and adenosine at similar rates. The ultraviolet absorption spectra of adenine and AdoMet in 0.1 N NaOH permit an estimation of the cleavage of AdoMet to adenine and S-ribosylmethionine at 250 nm (Parks & Schlenk, 1958).

A combination of chromatography and monitoring of ultraviolet-absorbing wavelengths facilitated modern fractionation of acidic extracts and estimation of AdoMet and decarboxylated AdoMet. The details of these procedures for the estimation of adenosyl–sulfur compounds have been described (Pegg & Bennett, 1983; Zappia et al., 1983a,b). Indeed the detection and estimation of these compounds and the polyamines reactive with reagents such as o-phthalaldehyde can be made in the same high pressure liquid chromatography (HPLC) separation (Greenberg & Cohen, 1985; Wagner et al., 1984). Sensitive radioenzymatic methods have also been devised for the estimation of AdoMet in biological fluids. These depend on the enzymatic methylation of tritiated biogenic amines and include an S[methyl-^{14}C]AdoMet as an internal standard (Giulidori & Stramentinoli, 1984; Yu, 1983).

Sulfonium compounds form stereoisomers if, as in the case of AdoMet, different substituents are present on the tetrahedral sulfur, for example,

$$
\begin{array}{ccc}
\text{A} \quad \text{B} & \quad & \text{A} \quad \text{B} \\
\text{S} & & \text{S} \\
\text{Y} \quad \text{Z} & & \text{Z} \quad \text{Y}
\end{array}
$$

The stereoisomerism is in addition to that of the amino acid, which will in almost all discussions of AdoMet refer to L-methionine. The problem arises in the enzymatic synthesis that gives rise predominantly to a configuration dubbed (−), because the sulfonium center contributes a levorotatory component. Chemical synthesis, which most frequently involves a methylation of AdoHcy (SAH) by a compound such as methyl iodide, results in the formation of a diastereomeric mixture, (−) and (+). Enzymatic methylation depletes the (−) form; that is, it preferentially uses this form (Schlenk, 1965). However, reports of the AdoMet isolated from yeast describe a product containing 20–30% of the (+) form (Hoffman, 1986). It is not known if racemization occurs during isolation. Dilution of an enzymatically prepared highly radioactive (−) AdoMet with a partially race-

mized yeast AdoMet, for purposes of assay, will then affect the accuracy of these assays. In any case, small-scale enzymatic preparations of S-(−)-AdoMet have been reported, in addition to controlled chemical syntheses that give higher contents of the (−) epimer (Matos et al., 1987).

Analogues have been prepared to serve as inhibitors or probes of specific reactions. In the study of methylases it has been possible to enzymatically synthesize 8-azidoAdoMet, which can be photochemically attached to a specific functional site on a DNA methylase (Reich & Everett, 1990). The natural product, SAH, is a feedback inhibitor of many methylases, including DNA methylation, and the antibiotic, sinefungin or adenosyl ornithine, acts similarly (Bergerat & Guschlbauer, 1990).

In addition to the role of AdoMet in methylation or in the biosynthesis of spermidine, new unexpected functions of the coenzyme continue to be discovered. For example, some methionine-requiring strains of E. coli are relaxed and synthesize methyl-deficient tRNA, to which AdoMet can transfer its 3-amino-3-carboxylpropyl group to a uridylate (Nishimura et al., 1974). In another instance anaerobic growth of E. coli results in the production of an oxygen-sensitive multicomponent ribonucleotide reductase that requires the participation of AdoMet (Eliasson et al., 1990).

Properties of Decarboxylated AdoMet

The enzymatic decarboxylation of AdoMet produced a difficultly separable dAdoMet. Ion-exchange chromatographic methods were devised for the separation of the substrate and this product (Zappia et al., 1983b). Currently HPLC provides a routine approach to the separation and estimate of these metabolites (Greenberg & Cohen, 1985; Pegg & Bennett, 1983; Zappia et al., 1983a). The decarboxylated product was far more labile to 0.1 M NaOH than AdoMet itself; in addition to adenine the product is 3-methylthiopropylamine. On the other hand the decarboxylated AdoMet was stable to heating at 100°C at pH 4.5, conditions that completely hydrolyzed AdoMet to MTA and homoserine. This differential sensitivity to hydrolysis was exploited in the study of the AdoMet decarboxylase (AdoMetDC; Allen & Klinman, 1981).

The study of aminopropyl transfer required dAdoMet. An initial synthetic effort by Jamieson (1963) involved the condensation of 3-thiopropylamine and 2′,3′-isopropylidene-5′-toluene-p-sulfonyladenosine, removal of protecting groups, and methylation by methyl iodide. Poor yields and noncrystalline products compelled improvements; the addition of silver perchlorate in the methylation step markedly reduced the reaction time and improved the yield of the sulfonium salts. The dAdoMet and various analogues were converted to sulfates, and after trituration with ethanol the oily precipitates were obtained as hygroscopic powders (Samejima et al., 1978). The dAdoMet prepared by this method was characterized by infrared and nuclear magnetic resonance

(NMR) spectra; the latter revealed the existence of sulfonium diastereoisomers. This dAdoMet is active in the enzymatic synthesis of spermidine from putrescine.

It is of interest that dAdoMet is a weak inhibitor of some methylases and also participates in some methylation reactions (Zappia et al., 1969). In the latter instance the product would be decarboxylated SAH (i.e., S-adenosyl-3-thiopropylamine), which has been isolated from the eye of the sea catfish (Ito & Nicol, 1976). Unlike AdoMet, this is toxic in animal tissue cultures, possibly as a result of its subsequent oxidation by serum amine oxidase and the liberation of acrolein (Kawase et al., 1979). In a plant extract in which AdoMetDC was being sought, the addition of putrescine as a possible cofactor led to the production of H_2O_2 from a diamine oxidase; this product was also shown to decarboxylate AdoMet (Suresh & Adiga, 1977). However, the precise mode of origin of S-adenosyl-3-thiopropylamine has not been clarified.

AdoMet Synthetase (Methionine Adenosyltransferase)

Cantoni and Durell (1957) purified the AdoMet synthetase from rabbit liver and defined the stoichiometry of the synthetic reaction; that is,

L-methionine + ATP →

$$S\text{-adenosylmethionine} + PP_i + P_i,$$

where PP_i is pyrophospate. Neither triphosphate nor ADP were free intermediates; the γ phosphate of ATP forms orthophosphate and the α and β phosphates were the source of the pyrophosphate. However, the yeast enzyme formed and bound inorganic tripolyphosphate from ATP in the presence of methionine and proved to be a tripolyphosphatase (Mudd, 1963). The enzyme is relatively specific for L-methionine but is also active on the slightly toxic selenomethionine and quite toxic ethionine, compounds used to affect enzymes of methionine utilization and to screen for genetic alterations relating to this pathway.

The addition of methionine to a growth medium represses the enzymes of methionine biosynthesis; the amino acid also represses formation of AdoMet synthetase. Mutants constitutive for these enzymes are also high in the AdoMet synthetase, suggesting a similar control for this enzyme by the regulatory system for methionine biosynthesis. Such mutants are described as bearing lesions in the regulatory gene, metJ (Cohen & Saint–Girons, 1987). Mutants low in the AdoMet synthetase, isolated as ethionine-resistant, proved to overproduce methionine and to have elevated levels of some of the biosynthetic enzymes (Greene et al., 1970).

AdoMet does not penetrate E. coli and does not cause repression in this organism. On the other hand, AdoMet is utilizable in some organisms, and mutants devoid of the AdoMet synthetase and specifically requiring AdoMet have been isolated. In Corynebacteria AdoMet did repress some of the enzymes of methionine biosynthesis.

The polyamine contents of metJ mutants high in AdoMet synthetase and other mutants low in the synthetase have been compared (Su & Cohen, 1973). As presented in Table 9.2, bacteria containing little enzyme were low in AdoMet and spermidine, in contrast to cells high in the enzyme that were relatively high in AdoMet and spermidine. The former group were high in putrescine and AdoMetDC, consistent with the hypotheses that spermidine regulates the content of putrescine and AdoMetDC.

The structural gene for AdoMet synthetase is designated as metK. The locus of the gene was found to be close to the speA and speB genes on the E. coli chromosome (Maas, 1972), raising the possibility of the existence of an operon comprising these genes. The relative positions of the genes were described by Tabor and Tabor (1984). Although no metK mutant of E. coli isolated in this way is completely devoid of the ability to synthesize AdoMet, one metK mutant was found to contain a temperature-sensitive enzyme, suggesting a relation of the gene to the structure of the protein (Hafner et al., 1977).

AdoMet synthetase was purified to homogeneity from E. coli derepressed for this protein, that is, a metJ mutant transformed by a metK⁺ plasmid (Markham et al., 1980). It has a molecular weight of 180,000 and is composed of four identical subunits. This structure is similar to that of a yeast enzyme but different in some respects from various mammalian AdoMet synthetases (Mitsui et al., 1988; Tabor & Tabor, 1984). The bacterial enzyme is functionally similar to a purified yeast enzyme: catalyzing hydrolysis of a tripolyphosphate; requiring a divalent cation such as Mg^{2+}, Mn^{2+}, or Ca^{2+}; and activated by some monovalent cations such as K^+, NH_4^+, and Na^+ for maximal activity. Selenomethionine is a good substrate, as are several nucleoside triphosphates, such as 3′-deoxy-ATP and 8-bromo-ATP. Nevertheless, the E. coli and yeast enzymes were kinetically dissimilar, with the bacterial reaction beginning with a burst of 1 mole AdoMet per mole of enzyme, followed by a rate of synthesis proportional to enzyme concentration.

The bacterial enzyme was crystallized and studied by X-ray diffraction techniques (Gilliland et al., 1983). The metK gene was cloned (C. W. Tabor et al., 1983b) and sequence data for the protein identified the DNA sequence of the metK gene as the structural gene for the enzyme (Markham et al., 1984). An open reading frame of the gene extends 1152 bases, coding for a protein of 41,941 molecular weight, whereas the value of 43,000–44,000 was obtained in a gel electrophoresis of the enzyme subunit.

Null mutants of metK proved to have exceedingly low contents of AdoMet, that is, 0.5% of the wild type. Growth of two such strains required exogenous methionine. A third strain at the nonpermissive temperature had even lower AdoMet content and required an unknown growth factor that permitted the expression of a new type of AdoMet synthetase (Satishchandran et al., 1990).

Extreme thermophilic archaebacteria, such as Sulfolobus sulfataricus which grow optimally at 87°C, might be supposed to have difficulties in preserving the ther-

TABLE 9.2 Polyamine contents of *E. coli* strains with varying AdoMet contents.

Strains	Relative AdoMet content (%)	Spermidine (μmol/g wet cells)	Putrescine	AdoMet decarboxylase (nmol CO$_2$/30 min/mg protein)
Wild type K12	100	2.7	10.3	47
Low AdoMet				
E4	50	1.3	11.4	62
D7	10	1.1	12.7	60
E40	10	0.8	14.3	59
High AdoMet				
D8	200	4.6	8.8	24
E12	200	4.4	7.5	25
E31	200	6.0	7.2	23

Adapted from Su and Cohen (1973).

molabile AdoMet. Nevertheless, such thermophilic organisms do use AdoMet in methylations and polyamine metabolism and contain the metabolite at a concentration of about 60 nmol/g wet weight similar to the range found in eucaryotic cells (De Rosa et al., 1978). It was found that *Sulfolobus* contained two isoforms of AdoMet synthetase, each of which had an optimum temperature at 90°C and was completely stable at 100°C in the presence of ATP. The kinetic behavior of the enzymes was closer to that of the *E. coli* AdoMet synthetase than to those of other systems (Porcelli et al., 1988). The mechanism of apparent stabilization of AdoMet at the elevated temperature has not yet been clarified.

Analogues of Methionine and Inhibitors of AdoMet Synthetase

In in vivo experiments attempting to explore the effects of analogues on the polyamine pathway, the other diverse roles of methionine in protein synthesis, methylation, etc., must somehow be excluded. Methionine enters protein and is also essential in the initiation of polypeptide synthesis. Selenomethionine sharply represses methionine biosynthesis in *E. coli* but enters protein readily and supports some growth. β-Galactosidase containing selenomethionine is more susceptible than the normal enzyme to heat, urea, and trypsin (Coch & Greene, 1971).

In the initiation of protein synthesis in *E. coli*, methionine is attached to an initiator methionine tRNA and then formylated. The product, fmet-tRNA^{fmet}, initiates protein synthesis; but if the anticodon for this RNA is altered to permit this tRNA to be substituted by other amino acids, these may initiate protein synthesis somewhat inefficiently (Chattapadhyay et al., 1990).

Norleucine, sterically superimposable to methionine, can also replace that amino acid in the growth of *E. coli* but does not appear to repress enzymes of methionine biosynthesis (Kerwar & Weissbach, 1970). In contrast to methionine analogues, such as selenomethionine and ethionine, which can form adenosyl derivatives similar to AdoMet, norleucine provides a tool in which growth may be studied under conditions of severe methionine limitation (Chattopadhyay et al., 1991).

Ethionine permits rapid growth of methionine-requiring *E. coli* at a ratio of ethionine to methionine of <200:1. The adenosylation of ethionine by an *E. coli* extract is about 30% less rapid than that of methionine (Chattopadhyay et al., 1991).

In a systematic study of the structural and conformational analogues of methionine and their activities with the AdoMet synthetase of *E. coli*, yeast, and rat liver, three classes of compounds were made and tested (Fig. 9.2). These included straight carbon chain amino acids, aromatic amino acids, and cyclic amino acids. Among the methionine analogues, methionine methyl ester and *N*-formyl methionine were good substrates of the *E. coli* enzyme; ethionine was a significant but poor substrate. Of the inhibitors, the L-2-amino-4-hexynoic acid, *o*-carbamyl-L-serine, and 1-aminocyclopentanecarboxylic acid (cycloleucine) were among the best, effecting 50% inhibition at about 4 mM in a test system at 37.5 μM L-methionine (Lombardini et al., 1970).

Inhibitory bicyclic compounds have also been synthesized (Sufrin et al., 1979). A spectrophotometric assay for the *E. coli* enzyme coupling the production of pyrophosphate (PP$_i$) to an enzymatic system generating 2 mol NADH per mole of PP$_i$, permitted a more facile screen of newly synthesized substrates and inhibitors (Kim et al., 1992).

Biosynthesis of Methionine

It was of interest to determine if the effects of methionine limitation might be alleviated or spared by the provision of a polyamine. For example, *Myxococcus xanthus*, a bacterium commonly found in nature, aggregates to form a fruiting body and microcysts in nutritionally limiting conditions, and aggregation is inhibited by methionine (Shi & Zusman, 1995). Microcyst formation was shown to be induced by methionine starvation or high concentrations of putrescine in a defined medium and was prevented by the addition of spermidine (Witkin & Rosenberg, 1970). The spermidine content of microcysts is markedly lower than that of growing bacteria.

Methionine supplementation increased RNA synthesis, content, and cell size in the growth of some strains

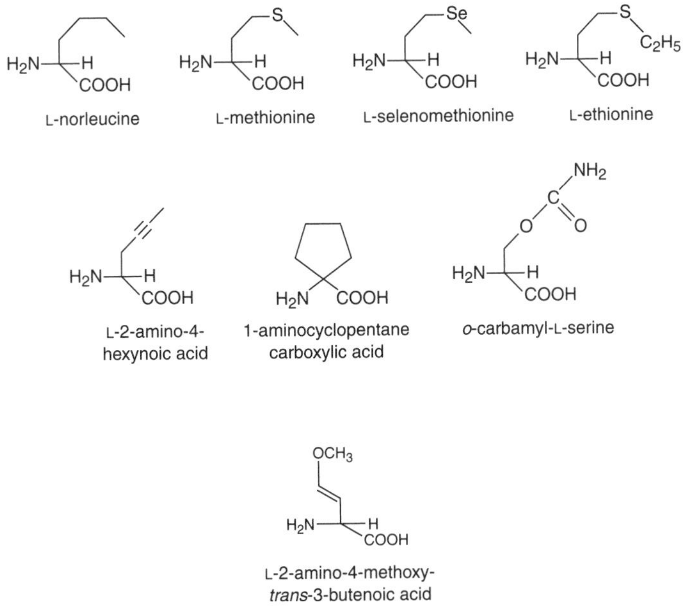

FIG. 9.2   Structures of some methionine analogues.

of *E. coli* in mineral media, suggesting a partial deficiency in the synthesis of the amino acid (Raina et al., 1967; Ron & Davis, 1966). Further exposure to temperatures of 43–45°C leads specifically to methionine starvation and facilitates synchronization of cell division (Lomnitzer & Ron, 1972). Sulfate deprivation blocks cell division selectively in some organisms (Spitznagel, 1961), as did putrescine deficiency in some *E. coli* mutants; whereas spermidine and putrescine were shown to facilitate the synchronization of cell division in other *E. coli* systems (Inouye & Pardee, 1970).

The many possible roles of methionine in a bacterium indicate that the hypothesis of a role for spermidine in accounting for methionine effects must be viewed cautiously. For example, starvation of *E. coli* for Mg^{2+} leads to breakdown of ribosomes, which reassemble following restoration of Mg^{2+} to the medium. Cell recovery involving synthesis of RNA and protein is greatly stimulated by addition of histidine and methionine to the growth medium. Nevertheless, a mixture of histidine and spermidine did not facilitate recovery, implying at least another essential role for methionine in this process. Indeed the presence of spermidine was inhibitory in this system (Cleaves & Cohen, 1970).

Methionine biosynthesis is a branch of the aspartate family of amino acid biosyntheses, members of which include asparagine, homoserine, threonine, diaminopimelic acid, and lysine. The numerous steps in the methionine pathway in *E. coli* and the specific *met* genes controlling biosynthetic enzymes are given in Figure 9.3. Weissbach and Brot (1991) discussed the complex regulation of the process.

Compounds such as α-methylmethionine and methionine are more active than ethionine or norleucine as

feedback inhibitors of homoserine-*o-trans*-succinylase, the enzyme determined by the *met*A gene (Rowbury, 1968). The antibiotic L-2-amino-4-methoxy-*trans*-3-butenoic acid (Fig. 9.2) can also regulate the activity of this enzyme in *E. coli*. An organism carrying multicopy plasmids containing the *met*A gene contains a high level of the homoserine transsuccinylase and overproduces methionine. Such organisms are also high in spermidine and low in putrescine, suggesting that the intracellular level of spermidine depends on the level of available methionine (Michaeli et al., 1983).

Because the various genes are distributed in different regions of the bacterial chromosomes, a coordinate repression by methionine has been assigned to proceed via a single gene, metJ, controlling a coordinate repressor, the *met*J protein. This protein has been isolated and characterized as a dimer whose binding to promoter sequences of all *met* genes except *met*H is enhanced 10-fold by AdoMet. AdoMet is therefore also a corepressor affecting its own biosynthesis.

The terminal reaction of methyl transfer to homocysteine is determined by two enzymes, one of which is a B_{12}-dependent methyltransferase, determined by the *met*H gene or a non-B_{12} methyltransferase, the product of the *met*E gene (Old et al., 1991). These gene products are expressed actively in response to the product of *met*R, a specific DNA-binding protein. *Met*E is also repressed by the *met*J protein and AdoMet. The presence of vitamin B_{12} represses *met*E expression and partially represses *met*F. The donation of formaldehyde from serine to tetrahydrofolate via serine hydroxymethylase, the product of the *gly*A gene, is also activated by the metR protein and homocysteine, the eventual recipient of the methyl group, via the metH or metE products.

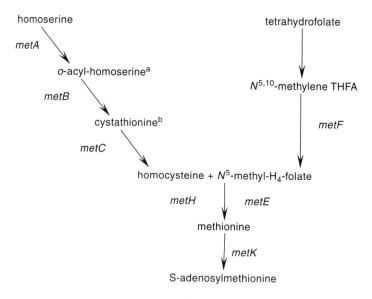

FIG. 9.3 The biosynthesis of methionine and AdoMet in *E. coli.* [a]In enterobacteria (e.g., *E. coli* and *S. typhimurium*) and the cyanobacteria (e.g., *A. nidulans*) the acyl activating group is *succinyl*, in contrast to many other organisms (e.g., *B. subtilis* and *S. cerevisiae*) in which *acetyl* is the activating group. *O*-Phosphoryl homoserine is the major form of activated homoserine in most plants. [b]Cystathionine is synthesized through transsulfuration from L-cysteine.

Studies of the fate of the products of spermidine biosynthesis (i.e., methylthioadenosine) have led to the discovery of still another pathway of methionine biosynthesis, to be discussed below.

Other Reactions of Methionine

Streptomyces species were used to produce various antibiotics in the 1950s and 1960s and the metabolic products of new strains were examined with care. One strain was found to accumulate the decarboxylated product of methionine, 3-methylthiopropylamine, and the organism was found to contain an L-methionine decarboxylase (Hagino & Nakayama, 1968).

3-methylthiopropionamide   L-methionine

3-methylthiopropylamine

The activity of a crude enzyme preparation after dialysis was found to be significantly stimulated by pyridoxal

phosphate (PLP), which also protected the activity during incubation at 50°C.

Some years earlier Mazelis (1962) described a PLP-dependent oxidative decarboxylation of L-methionine and some other amino acids by peroxidase in the presence of Mn^{2+}. With methionine as substrate the product was 3-methylthiopropionamide. A nonenzymatic decarboxylation of AdoMet by PLP and Mn^{2+} or Cu^{2+} was also described (Williams–Ashman et al., 1977; Zappia et al., 1977). A nonenzymatic decarboxylation of AdoMet and *S*-methylmethionine by deoxypyridoxal and Cu^{2+} was also observed (unpublished data), indicating that neither the phosphate ion of PLP nor the adenosyl moiety is essential in the formation of a carboxyl-labile chelate. The decarboxylation of methylmethionine is of interest as a possible step in the synthesis of bleomycin A$_2$ (Chap. 17).

Methionine degradation to the compounds methyl mercaptan (CH_3SH), NH_3, and α-ketoglutarate can be effected by a purified bacterial enzyme known as L-methionine-α-desamino-γ-mercaptomethane-lyase (L-methioninase). The PLP enzyme isolated from *Pseudomonas putida* was used in the microestimation of methionine (Tanaka et al., 1981). A clostridial enzyme can eliminate free methionine from tissue culture media as an approach to the methionine requirement of tumor cells. The production of CH_3SH distinguishes *Proteus* species from *E. coli* in seeking to identify possible causes of urinary tract infections.

The initial product of transamination of methionine, 2-oxo-4-methylthiobutyric acid, is a precursor to the formation of ethylene by many bacteria, including several strains of *E. coli* (Mansouri & Bunch, 1989). The

TABLE 9.3 *S*-Adenosylmethionine decarboxylase of different organisms.

Activation	Bacteria[a] (*E. coli*) Mg^{2+}	Animal[b] (rat prostate) putrescine	Plant[c] (Chinese cabbage) neither
K_m for AdoMet (μM)	60	65	38
K_i for dAdoMet (μM)	20	5	6
K_i for MGBG (μM)	20	0.3	0.6
Mol. wt.	108,000	68,000	35,000
	136,000[d]		

Adapted from
[a]Markham et al. (1982).
[b]Pegg and Jacobs (1983).
[c]Yamanoha and Cohen (1984).
[d]Anton and Kutny (1987a).

production of ethylene in an extract requires Fe^{3+} and NAD(P)H; other products include CO_2 and CH_3SH. However, ethylene, the much-studied plant hormone, is derived from AdoMet by a very different pathway in plants; this will be discussed in Chapter 20.

Isolation and Properties of AdoMetDC

The enzyme was found initially in *E. coli* and was purified to homogeneity (or near homogeneity) from bacteria, yeast, plant, and animal tissues (Tabor & Tabor, 1984). All but the plant enzyme have been shown to contain essential covalently bound pyruvate and do not require PLP for decarboxylation. The plant enzyme was not tested for the presence of pyruvate; it has several differences from the other enzymes of the group (Yamanoha & Cohen, 1985), as shown in Table 9.3.

The bacterial enzyme required Mg^{2+} for activity, in contrast to enzymes from some eucaryotic sources, and was specific for the (−) sulfonium stereoisomer of AdoMet. This steric configuration is retained in the course of the reaction (Allen & Klinman, 1981). These studies, following an enzymatic decarboxylation in tritiated water, used the reactivities of AdoMet and dAdoMet at pH 4 and in alkali to isolate fragments such as the tritiated 3-methylthiopropylamine below.

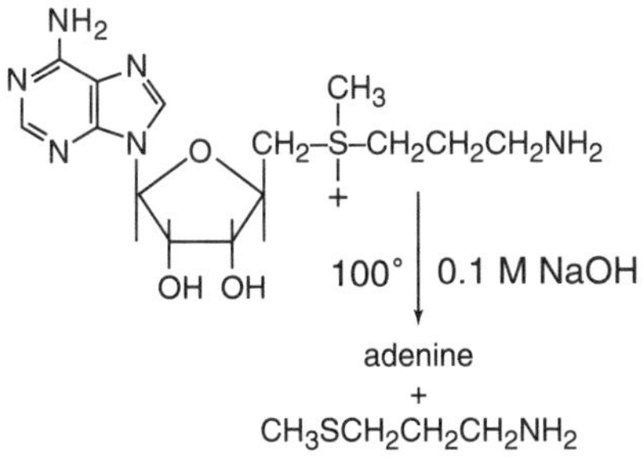

adenine
+
$CH_3SCH_2CH_2CH_2NH_2$

3-methylthiopropylamine

The authors then dissected the distribution of tritium between the protons at C-1 by established enzymatic reactions, beginning with the pea seedling amine oxidase.

The *E. coli* enzyme was purified from an overproducing strain, that is, an organism transformed by a plasmid containing the AdoMetDC gene (*spe*D). The isolation made use of two chromatographic steps, one of which included affinity chromatography (Markham et al., 1982). In this separation a tight-binding relatively specific inhibitor, methylglyoxal-bis(guanylhydrazone) (MGBG) is attached to a sizing polymer (Sepharose) and this reactive material is used to selectively bind the enzyme on the column (Pegg, 1974).

An early purification of AdoMetDC revealed a relatively high molecular weight protein consisting of multiple subunits detectable in acrylamide gel electrophoresis in the presence of the denaturant, sodium dodecyl sulfate (Wickner et al., 1970). The enzyme was inhibited by —SH reagents, such as *p*-hydroxymercuribenzoate, and activity was restored by various thiols. In addition the enzyme was inactivated by carbonyl reagents such as cyanide, phenylhydrazine, or hydroxylamine, but this inhibition was not overcome by PLP. Phenylhydrazine produced an enzyme of altered ultraviolet absorption, with a peak at 323 nm comparable to that of the phenylhydrazone of pyruvic acid and other keto acids. Reduction of the carbonyl by sodium borotritiide inactivated the enzyme and indicated the original presence of about one molecule of pyruvate per mole of enzyme. Hydrolysis of the tritiated enzyme and electrophoresis of the hydrolysate demonstrated the presence of most of the tritium in lactate. It was suggested then that, analogous to the newly discovered pyruvoyl-containing histidine decarboxylase of *Lactobacillus 30a*, the AdoMetDC was a pyruvoyl enzyme.

A later isolation and characterization of the enzyme from the overproducing strain (Markham et al., 1982) described the presence of a pyruvate moiety on each of six subunits of 17,000 molecular weight. It was inhibited competitively at a single binding site by MGBG, although less effectively than the mammalian enzyme.

The bacterial enzyme is similarly inhibited by the enzyme product, dAdoMet, which was thought to have a greater affinity for the enzyme than did the substrate. The reductant sodium cyanoborohydride, which reduces Schiff bases specifically at neutral pH, was demonstrated to be an irreversible inhibitor, thereby demonstrating the role of pyruvate and Schiff base formation in the catalytic center.

A more recent study of the enzyme revealed equimolar amounts of two subunits of 19 and 14 kDa in a protein of 136,000 molecular weight (Anton & Kutny, 1987a). Such new results are attributed to a fine tuning of the physical procedures made possible by newer sizing polymers and newer gadgets. The fixation of [methyl-^{3}H]AdoMet after reduction of the Schiff base revealed four molecules of substrate, suggesting one pyruvate per αβ pair of the subunits. The larger α unit was found to have the pyruvoyl group blocking the amino terminus; reduction in the presence of ammonium acetate converted the pyruvoyl moiety to alanine.

Genetic studies permitted an isolation of an operon containing the gene (speD) for the decarboxylase and that for the spermidine synthase (speE) (C. W. Tabor et al., 1986). The nucleotide sequence of speD showed that it codes for a protein of 30.4 kDa, and subsequent analyses revealed a cleavage at a lysine-serine peptide to form an α subunit of 18 kDa, containing pyruvate, and a β subunit of 12.4 kDa. When a strain containing a speD$^+$ plasmid was given ^{14}C-amino acids under conditions permitting unique expression of the plasmid, the label was found first in a 30-kDa polypeptide and appeared subsequently in the α and β subunits (Tabor & Tabor, 1987). Thus, the decarboxylase is formed initially as a proenzyme in which pyruvate does not exist.

Table 9.3 indicates the marked differences among the AdoMetDCs of various organisms: molecular size, requirements for activation, and sensitivity to some inhibitors. Nevertheless, the remarkable constancy of the pyruvoyl catalytic center is underlined by the pyruvoyl-containing AdoMetDC of the extreme thermophilic archaebacterium Sulfolobus sulfataricus, which has a molecular mass of 32 kDa, is monomeric, and does not require divalent cations or putrescine for activity (Cacciapuoti et al., 1991).

Pyruvoyl Enzymes and Their Precursors

The study of histamine led to the systematic study of amine oxidases, and the exploration of the bacterial origin of histamine from histidine led to the surprising finding several decades ago that certain bacterial histidine decarboxylases contain pyruvate as the catalytically essential prosthetic group (Van Poelje & Snell, 1990). Histidine decarboxylases of animal cells and some bacteria, Morganella morganii, Klebsiella planticola, and Enterobacter aerogenes, require PLP. These PLP-dependent enzymes are inactivated by the α-fluoromethyl amino acid in a manner comparable to the mechanism-based action of such inhibitors on PLP-de-

pendent ODC (Bhattacharjee & Snell, 1990). In common with many PLP enzymes (e.g., E. coli arginine and lysine decarboxylases), these enzymes have the common Ser-X-his-N^ε-lys site in which PLP can form the essential Schiff base (Vaaler & Snell, 1989). Indeed the sequence similarities of procaryotic and eucaryotic pyridoxal-dependent decarboxylases suggested a common origin (Jackson, 1990).

Crystalline preparations of histidine decarboxylase were isolated from Lactobacillus 30a (Rosenthaler et al., 1965). The enzyme was later shown to be decameric and it did not contain and was not dependent on PLP as the cofactor. However, the enzyme was inhibited noncompetitively by cyanide and other carbonyl reagents. The enzyme formed a phenylhydrazone, and a prior reduction with NaBH$_4$ prevented this reaction. Lactate was isolated from the reduced protein. These and other reactions demonstrated the presence of pyruvate at the catalytic center (Riley & Snell, 1968). These workers suggested a role for pyruvate similar to that postulated for PLP. As in the case of AdoMetDC, decarboxylation occurred without inversion at the asymmetric center (Chang & Snell, 1968).

The pyruvoyl group of the enzyme was shown to be derived from the serine of the protein. The enzyme was found to consist of two separable subunits, only one of which contains pyruvate at the N terminus. The pyruvate is easily detectable by assay with lactate dehydrogenase after mild acid hydrolysis. Pyruvoyl residues in proteins generally may be detected following reaction with p-aminobenzoic acid and tritiated sodium cyanoborohydride and hydrolysis to yield tritiated N-(p-carboxyphenyl)alanine (Van Poelje & Snell, 1987). The pyruvate in the enzyme was similarly shown to form a Schiff base with histidine.

The transformation of serine to pyruvate and the presence of N-terminal pyruvate in one of two subunits suggested the possibility of a cleavage of a proenzyme. A large catalytically inactive, immunologically cross-reactive, pyruvate-free protein was found in a decarboxylaseless mutant; on incubation this was converted to an active enzyme that contained two subunits, one of which possessed N-terminal pyruvate derived from an internal serine (Recsei & Snell, 1973). The mechanism of the conversion of the proenzyme to active enzyme proved to involve a serinolysis of the linkage of the pyruvate-destined serine by the hydroxyl of an adjacent serine (Recsei et al., 1983; Van Poelje & Snell, 1990). The structural gene for the Lactobacillus enzyme was cloned and sequenced; it codes for the single precursor polypeptide. A gene for a similar prohistidine decarboxylase from Clostridium perfringens was sequenced and also codes for two contiguous serines, one of which forms the carboxyl terminus of one subunit and the other the N-terminal pyruvate of the other subunit. Analysis of the crystal structure of the Lactobacillus enzyme suggested the possible existence of mechanical strain at the serine–serine cleavage site in the proenzyme (Gallagher et al., 1993).

Thus, the methodology of detecting and dissecting

pyruvoyl enzymes and proenzymes was developed in studies of Gram-positive bacterial histidine decarboxylases. The first pyruvoyl enzyme of the Gram-negative *E. coli* was that of AdoMetDC and was followed by several others, for example, phosphatidyl serine decarboxylase and 4′-phosphopantothenoylcysteine decarboxylase. The proenzymes of some of these were also detected following analyses of their structural genes (Van Poelje & Snell, 1990). These pyruvoyl enzymes also exist as subunit-containing structures derived from a proenzyme. Also, one of the two subunits contains a pyruvate and this is similarly derived from a serine. However, in none of these instances was the contiguous amino acid of the proenzyme a serine. The residue preceding the serine group proved to be quite variable, including lysine and glycine in the *E. coli* enzymes and glutamic acid in eucaryotic AdoMetDC.

Activators and Inactivation of AdoMetDC

As presented in Table 9.3, the enzyme of *E. coli* is activated by Mg^{2+}, which is essential for its activity, in contrast to the mammalian or fungal enzyme that requires putrescine or the plant enzyme that is fully active without either. With the *E. coli* enzyme, Mg^{2+} is essential for the inhibition of the enzyme, as in the case of the dAdoMet analogue in which the 3-aminopropyl moiety is replaced by a 2-(amino-oxy)ethyl moiety.

$$
\begin{array}{cc}
CH_3 & CH_3 \\
| & | \\
+S-Ado & +S-Ado \\
| & | \\
CH_2 & CH_2 \\
| & | \\
CH_2 & CH_2 \\
| & | \\
CH_2 & O \\
| & | \\
NH_2 & NH_2 \\
\text{dAdoMet} & \text{2-(amino-oxy)ethyl}
\end{array}
$$

dAdoMet   2-(amino-oxy)ethyl
(5′deoxyadenosine-5′-yl)
(methyl)sulfonium

This O-substituted hydroxylamine, which is uncharged in the inhibitor, in contrast to the enzymatic product, is a tight-binding irreversible inactivator of the AdoMetDC only in the presence of 1–2 mM $MgCl_2$. This inactivation is decreased by the presence of AdoMet, suggesting a binding of the inhibitor at the active site (Paulin, 1986; Weitkamp et al., 1991). Several other inhibitors also bind more tightly to the *E. coli* enzyme in the presence of Mg^{2+}. In the study of the $NaCNBH_3$ reduction of the Schiff base of the *E. coli* enzyme and AdoMet or dAdoMet, Mg^{2+} was essential to the inactivation by the reductant (Markham et al., 1982). The binding of the substrate can proceed without Mg^{2+}, but the metal ion increases the affinity of the enzyme for AdoMet and is essential to the development of the reac-

tion and sensitivity to $NaCNBH_3$. In other instances it will be noted that some other much studied inhibitors are more active on the putrescine-activated eucaryotic enzyme than on the Mg^{2+}-requiring enzyme.

During the course of the Mg^{2+}-activated reaction with AdoMet, the *E. coli* enzyme is slowly but irreversibly inactivated. This is analogous in part to the inactivation of some PLP-dependent decarboxylases but is complicated by a degradation of the deaminated dAdoMet. It was shown that the inactivation is maximal at 3 mM AdoMet and is half maximal at the Michaelis constant (K_m) for AdoMet. Omission of Mg^{2+} or AdoMet or inhibition by an analogue prevents the inactivation. An enzyme molecule is inactivated after about 6600 turnovers, and a [3,4-^{14}C]methionine-labeled AdoMet was found to be bound to the enzyme, almost entirely to the pyruvoyl-containing subunit. In this inactivated polypeptide, the pyruvate was discovered to have been converted to alanine. Thus, the inactivating transfer of the amino group left a propionaldehyde moiety on the dAdoMet and this was suspected to have hydrolyzed to toxic 2-propenal (i.e., acrolein), which covalently labeled the enzyme at a nucleophilic site (Anton & Kutny, 1987b). This site proved to be a defined cysteine residue that was alkylated by acrolein in the α subunit of the enzyme (Diaz & Anton, 1991).

Development of Inactivating Inhibitors of AdoMetDC

In 1958 a compound known as MGBG or more rigorously as 1,1′-[methylethanediylidene)-dinitrilo]-diguanidine was found to inhibit the growth of L1210 leukemia in mice (Williams–Ashman & Seidenfeld, 1986). In addition to pharmacological studies on the toxicity and possible mode of action of MGBG, a large number of related compounds were then tested for their efficacy in the treatment of various tumors. The inhibitory effect of MGBG on the development of the leukemia was prevented by spermidine (Mihich, 1963). It was subsequently discovered that MGBG was a particularly potent and relatively specific inhibitor of the putrescine-activated AdoMetDCs of animal tissues and yeast (Williams–Ashman & Schenone, 1972).

Although MGBG was far less inhibitory for the *E. coli* enzyme, as can be seen in Table 9.3, it was supposed initially that this related to the activating role of putrescine as compared to Mg^{2+}. Nevertheless, the plant enzyme that requires neither putrescine nor Mg^{2+} for full activity is highly sensitive to MGBG (Yamanoha & Cohen, 1985). Further, an early exploitation of MGBG was its use in affinity chromatography of the various AdoMetDCs, including that of the *E. coli* enzyme (Pegg, 1974). Thus the compound, although more active on the eucaryotic enzymes, does bind adequately to the bacterial enzyme and inhibits many bacteria, as do some newer MGBG analogues (Midorikawa et al., 1991).

A casual comparison of the formulae for MGBG and spermidine suggested that the compounds were analogous. The crystallographic analysis revealed a completely *trans* configuration of the chain of MGBG and its congeners, determined as the sulfates, similar to the contours of an extended spermidine. Some of these structures are indicated in Figure 9.4. Ethylmethylglyoxal-bis(guanylhydrazone) proved to be the most powerful inhibitor of the eucaryotic enzyme, with an apparent inhibition constant (K_i) value of 12 nM; it also inhibited intestinal diamine oxidase (Elo et al., 1986).

Another analogue of MGBG, with an additional amino group on both aminoguanidine moieties, proved to be an irreversible inhibitor of the animal or yeast enzymes and to be quite inactive on the *E. coli* enzyme (Pegg & Conover, 1976). Other antiparasitic drugs, Berenil and Pentamidine, whose guanidine structures are somewhat similar to that of MGBG, also inhibit the rat liver and yeast enzymes and are far less active on the *E. coli* enzyme (Karvonen et al., 1985). However, these substances have been found to be quite toxic in the mammal.

Among the first analogues made of the substrate, AdoMet, or of the reaction product, dAdoMet, the α-methyl derivatives of both of these proved to be irreversible inhibitors of the enzyme and formed reducible Schiff bases with the carbonyl of the pyruvoyl enzyme (Pankaskie & Abdel–Monem, 1980). Early experiences and difficulties with the *S*-adenosyl analogues have been discussed (Kolb et al., 1982). Only one of the latter group, the 5′-(dimethyl sulfonio)-5′-deoxyadenosine, proved to be slightly more inhibitory than dAdoMet with the rat enzyme. A comparison of the effects of many of those newly synthesized compounds on the bacterial and eucaryotic enzymes has been recorded (Pegg & Jacobs, 1983).

Notwithstanding difficulties in the synthesis of effective inhibitory *S*-adenosyl analogues, the recognition of the mechanism of inactivation of the *E. coli* enzyme by transamination to pyruvate, release of MTA, and alkylation of a nucleophilic site by 2-propenal (Anton & Kutny, 1987b) assisted the design of the desired inhibitor. The inhibitor, 5′{[(Z)-4-amino-2-butenyl]methylamino}-5′-deoxyadenosine, in which the methyl sulfonium group was replaced by a methylamino group, proved to be a powerful irreversible inactivator ($K_i = 0.3$

μM) of the *E. coli* enzyme in the characteristic time-dependent manner (Casara et al., 1989).

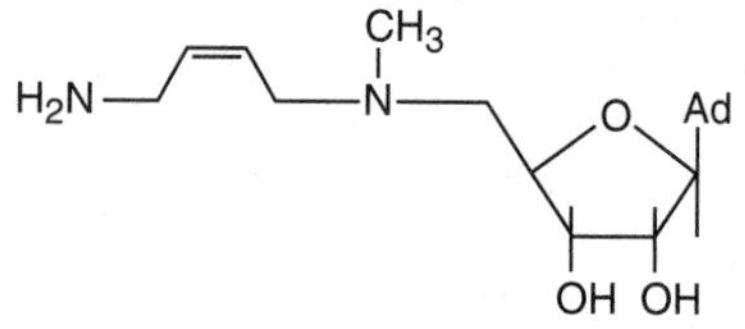

5′-(Methylamino)-5′-deoxyadenosine was released at the rate of inactivation, and Mg^{2+} was required for the inactivation. A mechanism was proposed for the reaction in which a conjugated imine is formed. The proof of this mechanism will require degradation of the inactivated protein and isolation and characterization of an appropriately labeled fragment.

Despite the difficulties in developing suitable methyl sulfonio analogues of AdoMet and dAdoMet, two groups of such irreversible inhibitors of *E. coli* decarboxylase have been reported recently. Replacement of the carboxyl of methionine by a nitrile was a step in producing a series of irreversible inhibitors in which the proximal amine could react irreversibly to form an imine with the terminal pyruvate of the enzyme (Wu et al., 1993). Another series of irreversible inhibitors involved the addition of a cyclopenteneamine to the methylsulfonio group. Of the four diastereoisomeric forms obtained, each was an irreversible inhibitor, although one was most potent (Wu & Woster, 1993).

Bacterial Mutants Deficient in AdoMetDC

E. coli strain K12 was mutagenized with either a chemical mutagen or the bacterial phage Mu (H. Tabor et al., 1983), and organisms were screened for deficiencies in the production of CO_2 from [14]COOH-labeled methionine (C. W. Tabor et al., 1983a). Mu-induced mutagenesis produces deletion mutants most frequently. Strains of *E. coli* deficient in AdoMetDC were isolated and found to also lack spermidine, thereby demonstrating an apparently essential role of the enzyme in the synthesis of spermidine (Tabor et al., 1978). Nevertheless, these organisms grew at 75% of the growth rate of the wild type in a medium supplemented with amino acids, purines, and pyrimidines. The normal growth rate was restored by the presence of spermidine in the medium. Mapping of the position of the gene (*spe*D) in the *E. coli* map showed it to be far from that of the *spe*A, *spe*B, and *spe*C genes, and placed it in an operon with *spe*E, the gene for spermidine synthase (Xie et al., 1989).

Deletion mutants of *E. coli*, lacking the *spe*ED operon (i.e., unable to synthesize spermidine but synthesizing large amounts of putrescine), grow without the triamine at rates 85% of that in the presence of spermidine (Xie et al., 1993). In contrast, similar mutants of fungi, Saccharomyces, or Neurospora have absolute requirements for spermidine.

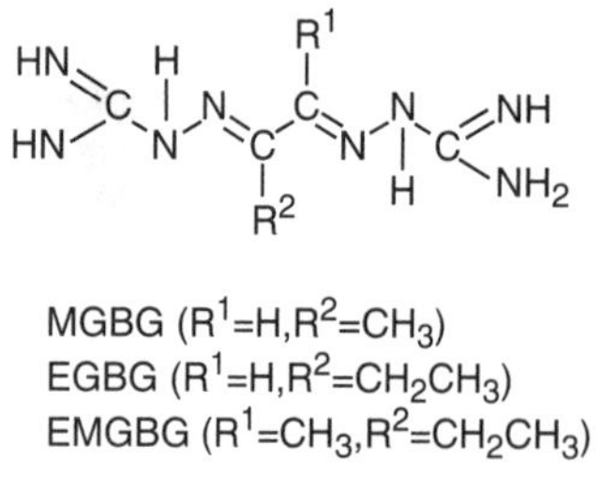

MGBG (R^1=H,R^2=CH$_3$)
EGBG (R^1=H,R^2=CH$_2$CH$_3$)
EMGBG (R^1=CH$_3$,R^2=CH$_2$CH$_3$)

FIG. 9.4   The structural formulae of MGBG, E(ethyl) GBG, and EMGBG; redrawn from Elo et al. (1986).

It was possible to construct polyauxotrophic strains that were apparently unable to synthesize putrescine or spermidine and to test numerous biological activities such as growth rate and the multiplication of numerous viruses (Hafner et al., 1979). The data on virus multiplication will be considered in a later chapter. In general polyamine-deficient strains grew at a rate about one-third that found in the presence of polyamines. Marked increases in the rate of growth were obtained within 1–2 h by 10^{-4} M spermidine > putrescine or spermine > 10^{-3} M cadaverine.

Although Hafner et al. (1979) reported an amine concentration of <0.3% of the wild type level, the paper did not define the upper limit of polyamine content per cell of the polyamine-deficient polyauxotroph. Hafner et al. (1979) noted a small amount of cadaverine in the putrescine-deficient + spermidine-deficient organism. In the course of their examination of control organisms, for example, one mutant deficient in *spe*A, *spe*B, and *spe*C but bearing wild type *spe*D, growth in the presence of spermidine was found to give rise to small amounts of a tetramine, possibly spermine. This observation was placed on firmer ground subsequently by the discovery that the spermidine synthase of *E. coli* is capable of synthesis of an unidentified tetramine.

A DNA fragment containing *spe*D was isolated and cloned and was found to encode not only the proenzyme form of AdoMetDC, but also an adjacent upstream DNA sequence (*spe*E) encoding spermidine synthase (Xie et al., 1989). The RNA for the latter enzyme appeared to be transcribed before that for the decarboxylase. It was found also that an open reading frame of DNA encoding a previously unknown protein of some 115 amino acids was located upstream and adjacent to the *spe*E gene. This region and its promoters are required for the expression of *spe*E and *spe*D.

Growth of the wild type in the presence of exogenous spermidine did not greatly decrease (40–60%) the activities of AdoMetDC and spermidine synthase. This implied that spermidine is not a major regulator in *E. coli* of transcription and translation of these steps of polyamine biosynthesis. However, this result is challenged by experiments of Salmikangas et al. (1989) who used *E. coli* to express the cDNA for rat AdoMetDC and found that exogenous spermidine reduced the bacterial enzyme activity without affecting that of the rat enzyme.

Spermidine Synthase, a Bacterial Propylamine Transferase

The relative unavailability of dAdoMet determined the initially slow pace of work on propylamine transferases. The difficulties encountered by Jamieson were overcome in 1978 by Samejima and others. In the absence of rapid methods of selective purification of the enzyme, such as that of MGBG-affinity chromatography developed in 1974, multiple assays of many fractions were required in the purification scheme. The biochemical preparation of isotopically labeled AdoMet and dAdoMet was possible, but the task was arduous and yields were poor. Nevertheless, the propylamine transferase from *E. coli* was isolated under such conditions (Bowman et al., 1973) in a series of six steps involving salt precipitation, chromatography on three different absorbents, and gel electrophoresis. Fortunately a concentrated enzyme preparation was relatively stable at 4°C. The enzyme was assayed by estimating the production of spermidine by a spectrophotometric method using the triamine-specific dehydrogenase of *Serratia marcescens*, or the more sensitive fluorimetric reaction of spermidine with *o*-phthalaldehyde. In addition, spermidine formation was determined from the incorporation of ^{14}C-putrescine into isolated triamine.

The homogeneous enzyme, purified almost 2000-fold, had a molecular weight of about 73,000 ± 2400 calculated from a study of sedimentation equilibrium; the enzyme comprised two equal subunits. The gene, *spe*E, present upstream in an operon containing *spe*D, was found to possess an open reading frame of 864 base pairs and to code for 288 amino acids (Tabor & Tabor, 1987). The amino acid sequence (30 amino acids) at the N-terminal end of the enzyme agreed completely with that obtained from the nucleotide sequence, although the initial methionine was not present in the purified enzyme (Tabor et al., 1986). Thus, *spe*E is the structural gene for a subunit of the enzyme. The sequence can provide a more accurate estimate of the molecular weight of the enzyme.

The discovery of aminopropylcadaverine in putrescine-deficient *E. coli* compelled a more careful study of the specificity of the newly purified spermidine synthase. It was now found that the enzyme transferred a propylamine moiety from dAdoMet to cadaverine and spermidine at about 10% the rate of transfer to putrescine at pH 8.2. The rates of formation of aminopropylcadaverine and of "spermine" were significantly greater (threefold) at pH 10.0. The reverse reaction did not occur. *N*-Monoacetylputrescine and 1,3-diaminopropane were inactive as acceptors. The discovery of the ability to synthesize "spermine" from spermidine raises many questions of the possible reasons for the absence of spermine in *E. coli*; it will be recalled that mutants lacking *spe*A, *spe*B, and *spe*C did generate small amounts of tetramine from exogenous spermidine (Hafner et al., 1979). This phenomenon also highlights the questions of the nature and relations of the separate eucaryotic spermine synthases to the various spermidine synthases of bacterial and eucaryotic origins.

Pseudomonas acidovorans 29, reported to contain 2-hydroxyputrescine, putrescine, and spermidine, was also found to contain a hydroxylated spermidine. This compound, originally designated as 2-hydroxyspermidine (Rosano et al., 1978), is now described as 7-hydroxyspermidine because the numbering begins on the aminopropyl moiety. Extracts of *E. coli* and *P. acidovorans* 29, capable of converting AdoMet and putrescine to spermidine, also converted AdoMet and 2-hydroxyputrescine to 7-hydroxyspermidine. Both stereoisomers of 2-hydroxyputrescine were equally active in the synthesis in the *E. coli* system.

Mechanism of Spermidine Synthase Reaction

The conversion of $[3,4\text{-}C^{13}]$-methionine to spermidine by *E. coli* revealed the retention of both ^{13}C atoms in the aminopropyl moiety of spermidine without significant dilution (Billington et al., 1980; Golding et al., 1985). This excluded the formation of a protonated azetidine, which then reacted with putrescine, because it was supposed that such an intermediate would be a mix containing C-2 and C-3, and C-3–C-4. The study is the first describing the preparation of ^{13}C-spermidine, a substance that holds the possibility of exploring the cellular distribution of "free" and "bound" spermidine by NMR procedures.

At this point the reaction may be described as involving a nucleophilic attack on the methylene C-3 adjacent to the sulfonium cation of dAdoMet. The enzymatic synthesis of spermidine can then be portrayed as one of two major possibilities (Orr et al., 1988) given in Figure 9.5. In path A the reaction is depicted as a single displacement in which putrescine adds at the carbon bound to the sulfonium ion, displacing methyl thioadenosine and inverting the configuration to the methylene carbon at that site. In path B a reaction of dAdoMet with a nucleophilic site on the enzyme leads to an enzyme

complex containing a propylamine with an inverted hydrogen. In a second reaction of this "ping-pong" mechanism, the original conformation of the complex with putrescine hydrogen is restored. These different possibilities have been tested by several methods.

Despite being a substrate, dAdoMet was found to be a competitive inhibitor whereas putrescine was not inhibitory (Zappia et al., 1980). Deaminated analogues of dAdoMet were inactive as substrates but were competitive with putrescine. Plots of the kinetic data suggested a ping-pong mechanism (path B), and the binding data were interpreted to suggest that the enzyme existed in a "free" and "propylaminated" form.

The mechanism of action of the *E. coli* enzyme was studied by two additional groups, both of whom concluded that path A describes the route of biosynthesis of spermidine. An earlier analysis of the kinetics of the inhibition of the rat enzyme by dAdoMet called into question the interpretation of the reaction as a ping-pong mechanism (Coward et al., 1977).

Golding and Nassereddin (1982) were the first to show inversion of configuration in spermidine at the methylene carbon formerly bound to the sulfonium cation. Using methionine labeled with deuterium of known relative configuration at C-3 and C-4, it was possible to follow the fate of the deuterium in newly synthesized *E. coli* spermidine, now labeled at carbons in the aminopropyl moiety. It had been shown earlier that deuterium is lost from C-2 of AdoMet by the action of AdoMet decarboxylase. Spermidine was converted by acetaldehyde (ethanal) into a hexahydropyrimidine and an

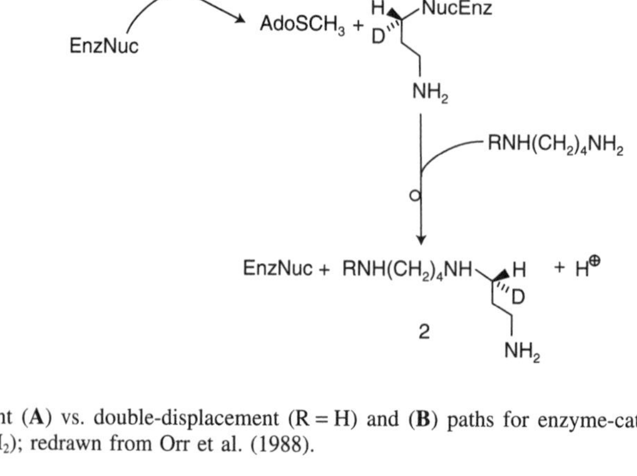

FIG. 9.5 Single-displacement (**A**) vs. double-displacement (R = H) and (**B**) paths for enzyme-catalyzed aminopropyl transfer reactions (R = $CH_2CH_2CH_2NH_2$); redrawn from Orr et al. (1988).

imino hexahydropyrimidine, i.e. the method exploited by Ganem (see Chap. 2). The nuclear magnetic resonance spectrum of the latter compound yielded chemical shifts and coupling constants for the protons of interest and by comparison with model compounds revealed an inversion of configuration of methylene derived from the C-4 of methionine (Golding and Nassereddin, 1985a,b).

Coward and colleagues similarly undertook a study of the configuration of the sulfonium-bound methylene carbon reactive with putrescine. An NMR study of camphanamides of spermidine, containing selectively deuterated carbons adjacent to the substituted secondary nitrogen, clearly distinguished the hydrogens bound to the appropriate methylene (Pontoni et al., 1983). This method was adopted to study the change or retention of configuration (Orr et al., 1988). dAdoMet chirally deuterated at C-3 of the aminopropyl moiety was prepared and used to convert putrescine to spermidine and eventually to the appropriate camphanamide derivative of the secondary nitrogen of spermidine. The restrictive effect of this derivative on the adjacent methylene-bound deuterium was readily discernible and the NMR results demonstrated the inversion of configuration, in agreement with the results of Golding and Nassereddin (1985a,b). Both groups concluded that the reaction proceeds with a single displacement mechanism, similar to that concerning methyl and phosphoryl transfers.

A kinetic study of the purified propylamine transferase of the extreme thermophile, *Sulfolobus solfataricus*, affirmed the view that the reaction carried out by this enzyme is path A (Cacciapuoti et al., 1986). The homogeneous enzyme of 110 kDa was trimeric, comprising identical subunits of 35 kDa and functioning optimally at 90°C. One product, 5′-methylthioadenosine, was an active competitive inhibitor, with a K_i of 3.7 μM. Unlike the *E. coli* enzyme, the *Sulfolobus* enzyme is not affected by thiol reagents.

The pathway of synthesis of *sym*-norspermidine and *sym*-norspermine from AdoMet and dAdoMet was described earlier (De Rosa et al., 1978). In this report, the reactions were deduced from studies of the labeled metabolites isolated from whole cells of *S. sulfobous*, formerly known as *Caldariella acidophila*. In a 1986 study the enzyme was found to be reactive with numerous amine acceptors. In addition to reactivity with putrescine ($K_m = 3.9$ mM), significant activity was also found at relatively high K_m with 1,3-diaminopropane, *sym*-norspermidine, and spermidine. Indeed Table 9.4 reveals that under the conditions of assay, 1,3-diaminopropane is the most active acceptor, followed by putrescine. The products are *sym*-norspermidine and spermidine, respectively. The former is itself an active acceptor and more active in this respect than the somewhat active spermidine. Thus *sym*-norspermine is a tetramine product that is also a detectable acceptor. The presence of an additional methylene in putrescine, spermidine, and spermine render these compounds less ac-

TABLE 9.4 Propylamine transferase activity in presence of various amine acceptors.

Substrates	Activity	
	Enzyme (nmol MeSAdo/min/ mg protein)	Relative (%)
1,3-Diaminopropane	1800	100
1,3-Diaminopropane-2-ol	ND	<0.1
Putrescine	662	36.7
1,5-Diaminopentane	10.8	0.6
1,9-Diaminononane	ND	<0.1
1,10-Diaminodecane	ND	<0.1
sym-Norspermidine	316	17.5
Spermidine	124	6.8
sym-Homospermidine	ND	<0.1
sym-Norspermine	50	2.7
Spermine	ND	<0.1
Caldopentamine	ND	<0.1
Homocaldopentamine	ND	<0.1

Adapted from Cacciapuoti et al. (1986). ND, not detectable.

tive than the trimethylene-containing acceptors with this archaebacterial enzyme. Most recently, the molecular structure of the propylamine transferase of a *Sulfolobus* species was studied as a function of temperature change.

Studies of the formation of the novel aminopropyl-containing polyamines found in the thermophile *Thermus thermophilus* revealed the presence of a heat-stable aminopropyl transferase for which the donor was dAdoMet. Acceptors included spermidine, thermine, caldopentamine, and tris(3-aminopropyl)amine (Hamasaki & Oshima, 1990).

Inihibitors of Spermidine Synthase Reaction

It was reported initially that the propylamine transferase of rat ventral prostate was inhibited competitively by dicyclohexylamine (Hibasami et al., 1980), a substance commercially available from a presumably reliable distributor. The inhibitory properties of the preparation on various bacteria and plants and its reversal by spermidine were reported widely by numerous workers. It was found that the dicyclohexylammonium sulfate salt had been mistaken for the bis-cyclohexylammonium sulfate salt, and that the true nature of the inhibitor was cyclohexylamine (Batchelor et al., 1986).

Cyclohexylamine (10 mM) has little effect on growth and spermidine content of *E. coli* and a partial effect (5 mM) on *Pseudomonas aeruginosa*. The partial efficacy of the mono- and difluoromethyl derivatives of ornithine and arginine in inhibiting bacterial growth suggested the admixture with cyclohexylamine in an inhibitory cocktail. Such a mixture was in fact more efficacious in slowing growth of the bacteria and in reducing their

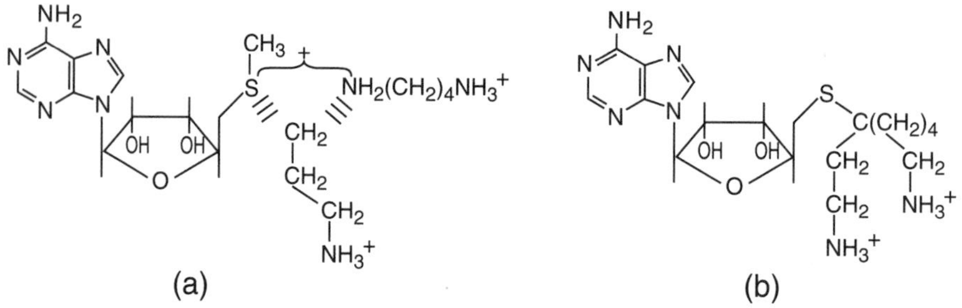

FIG. 9.6 A multisubstrate inhibitor of the spermidine synthase reaction: (**a**) transition state at the catalytic site and (**b**) Ado-DATO.

polyamine content than were the separate components (Bitonti et al., 1982).

At this time also, a new series of aminopropyltransferase inhibitors were designed in the belief that the enzymatic reaction was a single displacement reaction in which the substrate dAdoMet and putrescine were simultaneously at the catalytic site (Tang et al., 1981). Multisubstrate adducts simulating the hypothetical transition state, given in Figure 9.6**a**, could be expected to bind tightly to and be highly specific inhibitors of the reaction.

Surprisingly, the thioether (Fig. 9.6**b**), *S*-adenosyl-1,8-diamino-3-thiooctane (AdoDATO), was a more powerful inhibitor ($IC_{50} = 4 \times 10^{-7}$ M) than the methylsulfonium compound with the rat ventral prostate enzyme (Tang et al., 1980). When the compound AdoDATO was tested on bacterial enzymes of *E. coli*, *P. aeruginosa*, and *S. marcescens*, it was found to have similarly potent inhibitory effects that were greater than those of cyclohexylamine. However, it had little effect on bacterial growth and spermidine content of the cells (Pegg et al., 1983).

The structure of AdoDATO reveals the possibility of stereoisomerism at the adduct carbon attached to the sulfur. The separate isomers were synthesized and one of the stereoisomers was found to be a slightly more potent isomer than the other with the *E. coli* spermidine synthase (Liu & Coward, 1991). We may note that the activities of the compounds in vitro are consistent with the concept of a single displacement reaction and less so with the concept of a ping-pong mechanism.

Bacterial Metabolism of MTA

The synthesis of the polyamines via propylamine transferase reactions leads to the production of MTA (Fig. 9.1). Methionine had been shown to be the source of the methyl group and sulfur of this compound (Schlenk, 1965b). MTA had been found in yeast extracts as early as 1907 and its structure had been established in the mid 1920s. The decomposition of AdoMet by heat at pH 5–7 produced MTA, and a yeast enzyme was described that converted AdoMet to MTA (Mudd, 1959). Homoserine lactone is formed in both instances, and the hydrolysis

of the lactone produces homoserine, which was the isolated product. The reaction sequence is given below:

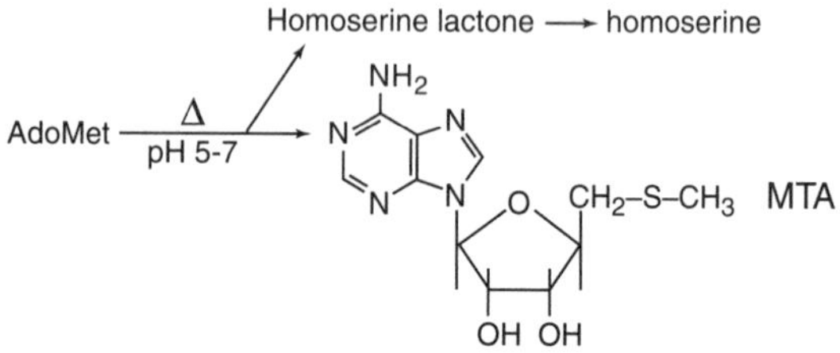

Although such an activity was also found in *Aerobacter (Enterobacter) aerogenes*, as well as a further hydrolysis of MTA to adenine and methylthioribose, the initial cleavage system was much less active in *E. coli* (Shapiro & Mather, 1958).

It had been proposed initially, but not demonstrated, that the biosynthesis of spermidine in *E. coli* produced MTA as an end product. This led eventually to a search in *E. coli* for the compound, which was isolated from this organism and identified (Chu et al., 1968). The isolated nucleoside was characterized by a variety of chemical and physical properties in comparison to the synthesized substance. Tests for the functional groups of MTA are summarized in Table 9.5. No reactions are obtained with ninhydrin for the amino group of adenine, or with sodium *p*-hydroxy-mercuribenzoate for the methylthio group. Electrophoretic and spectrophotometric properties were recorded as a function of pH, consistent with the behavior of the adenine substituent. The mass spectrum of MTA was reported, as well as the behavior of the trimethylsilyl derivatives in chromatography. MTA was found in the cells at a concentration of 0.38 μmol/g dry weight, which is significantly less than the concentration of spermidine found in the cells. MTA was not found in the medium and these results suggested the existence of degradative or other metabolic routes for the compound. Indeed *E. coli* grown on sulfate produced significant amounts of methylthioribose and excreted much of this into the medium (Schroeder et al., 1973).

In the studies on the propylamine transferases of ani-

TABLE 9.5 Functional group tests of methylthioadenosine.

Reagent	Functional group test	Results
Ninhydrin	Aliphatic NH_2	−
2,4-Dinitrofluorobenzene	NH_2	+
Nitrous acid	NH_2	+
Acetic anhydride	NH, SH, OH	+
Sodium p-hydroxymercuribenzoate	SH	−
Iodomethane	RSR′	+
2,4-Dinitrophenylhydrazine	CHO, $\rangle$C=O	−
Periodate	COHCOH, COHCNH$_2$	+
Base	C(=O)OR	−
Acid	Acid-labile linkage	+
Air	Ease of oxidation	+
H_2O_2	Ease of oxidation	+

Adapted from Chu et al. (1968).

mal tissue and bacteria, it was shown that MTA could be degraded by a relatively specific phosphorylase in the former (Pegg & Williams–Ashman, 1969) and by a hydrolytic nucleosidase in *E. coli* (Ferro et al., 1976). The first procaryotic example of an MTA phosphorylase was that reported for *Sulfolobus*, whose thermostable enzyme was purified to homogeneity (Cacciapuoti et al., 1994; Cartenì–Farina et al., 1979). Eubacteria, which contain *S*-adenosylhomocysteine hydrolase, also contain an MTA phosphorylase (Shimizu et al., 1988). In the latter reaction, MTA is degraded to adenine + 5-methyl-thioribose-1-P, whose important subsequent reactions will be discussed below.

With the *E. coli* nucleosidase, MTA is degraded to adenine plus 5-methylthioribose. In this system, the K_m for MTA is 3×10^{-7} M. The enzyme is also active on the 5′-ethyl- and 5′-*n*-propyl-thioadenosines and on AdoHcy and these substrates can also serve effectively as competitor inhibitors for the reaction with MTA. In the cleavage of AdoHcy the products are adenine and ribosyl-homocysteine. The homocysteine moiety of the latter can be incorporated into protein, presumably as methionine (Duerre & Bowden, 1964).

The *E. coli* nucleosidase, now described as the AdoHcy/5′-methyl-thioadenosine nucleosidase, has been purified to homogeneity (Della Ragione et al., 1985). Affinity chromatography on a formycin-derivatized gel proved to be a key step in the purification. Methylthio and thio derivatives of the purine nucleoside analogues, formycin and tubercidin, modified in the imidazole ring of the purine, proved to be powerful inhibitors of the enzyme. It is well known that AdoHcy is a natural powerful inhibitor of methylation by AdoMet, and the concentration of this substance is often carefully regulated in many cells.

In addition to the origin of the methylthio group of MTA in methionine, it was also shown that the methyl-thio group of MTA is recycled to methionine. It was demonstrated in *Enterobacter aerogenes* that the 5-methyl-thioribose cleaved by the nucleosidase from MTA is phosphorylated by a kinase in the presence of ATP to form 5-methylthioribose-1-phosphate, which is in fact an early step in the biosynthesis of methionine (Ferro et al., 1978). The procaryotic system rejoins the eucaryotic system with this second step in forming 5-methylthiori-bose-1-phosphate. Labeled carbon atoms of ribose were then shown to form the four carbon chain of methionine in an extract of *E. aerogenes*, with additional intermediates of labeled 2-keto-4-methylmercaptobutyric acid and 2-hydroxy-4-methylmercaptobutyric acid (Shapiro & Barrett, 1981). The labeling pattern of the various compounds indicated an efficient and direct salvage pathway for the formation of the amino acid.

Work has proceeded in two directions with the occasional human pathogen, *Klebsiella pneumoniae*, a Gram-negative member of the enteric bacilli. The organism contains a nucleosidase for MTA and a kinase for 5-methylthioribose. As adapted from Myers and Abeles (1990) in Figure 9.7, the glycosyl phosphate (3) is converted to the open-chain 2-keto derivative [i.e., a ribulose derivative (4)], which is then dehydrated to the 2,3-diketo-1-phosphate (5). It was shown that the C-1 of methylthioribose is not incorporated into methionine. In an oxidative reaction, the phosphate is removed and the C-1 is eliminated as formate to form the 2′-keto-4-methylthiobutyrate (6). Transamination of the latter from another amino acid provides methionine. Myers et al. (1993) described an enzyme that probably effects the unusual carbon to carbon oxidative cleavage to formate and 2-keto-4-methylthiobutyrate.

The requirement for a methylthioribose kinase in the synthesis of methionine in *K. pneumoniae* suggested that analogues of that sugar might lead to the production of a toxic metabolite of methionine in a pathogenic bacterium. Such a mechanism of incorporation of a sugar analogue would not be present in a kinaseless mammalian host. It was found that sugar analogues, such as 5-trifluoromethylribose, which is a substrate for the kinase, were indeed inhibitory to the growth of *K. pneumoniae*, with an IC_{50} at 40 nM. The analogue was not toxic to animal cells until the compound reached 250–1000 μM levels (Gianotti et al., 1990). Other inhibitors

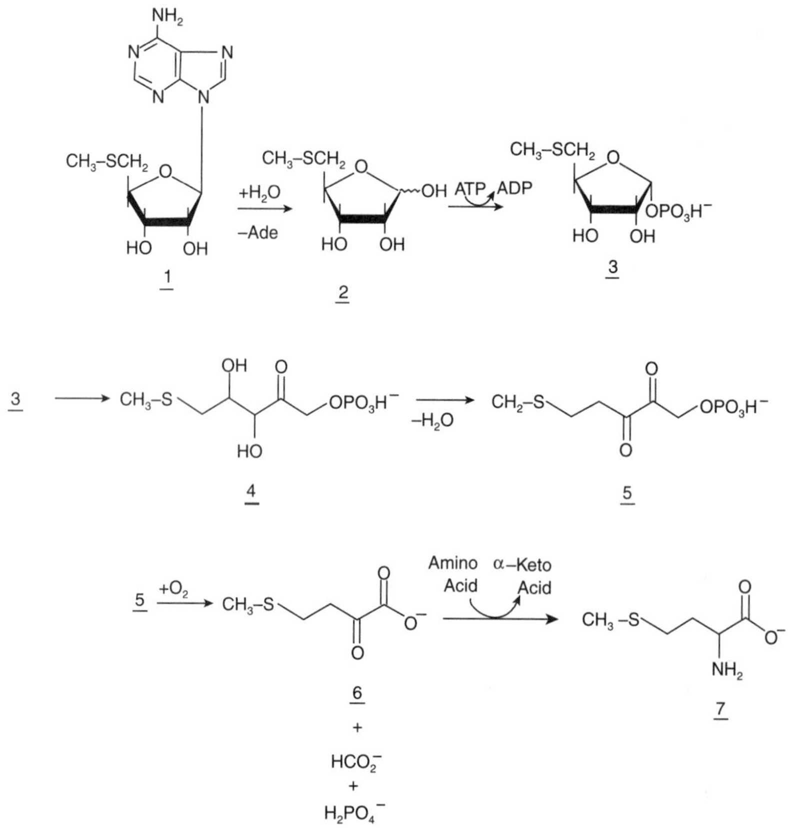

FIG. 9.7 Scheme for the conversion of methylthioadenosine to methionine; redrawn from Myers and Abeles (1990).

of de novo methionine synthesis were found to be synergistic with 5-trifluoromethylribose in inhibiting the growth of the bacterium (Tower et al., 1991). It may be mentioned that xylosyl-methylthioadenosine has been found in some Mediterranean mollusks and appears to arise from MTA by isomerization at the C-3′. The compound is less cytotoxic than MTA (Pani et al., 1991).

Alternative Synthetic Paths of Spermidine and Norspermidine

It has been reported that some organisms that excrete spermidine-containing siderochromes, for example, *Paracoccus denitrificans* and *Rhodopseudomonas spheroides*, do not use methionine in the synthesis of the triamine. Further, the former does not contain AdoMetDC. Both cells are relatively rich in spermidine, containing about 5 times more of the triamine than putrescine. Aspartate and fumarate were good precursors of spermidine, and it was thought that L-aspartic β-semialdehyde, an intermediate in the biosynthesis of several amino acids, might be the donor of the aminopropyl moiety of spermidine. Extracts of the two organisms converted

$[^{14}C]$-putrescine and L-aspartate to spermidine in the presence of ATP, Mg^{2+}, and NADPH and a thiol. The production of a new amino acid believed to be "carboxyspermidine" was demonstrated, and this was thought to be decarboxylated in the presence of PLP. PLP was essential in the reaction. The postulated reactions are presented in Figure 9.8.

The NADPH-dependent enzyme and the decarboxylase are stated to have been separated from one another. The system has also been reported to have been isolated from *A. tumefaciens*, which uses β-aspartic semialdehyde as an aminopropyl donor to spermidine to form spermine and thermospermine (Tait, 1985).

In studies on seedlings of *Lathyrus sativus*, a weak Mg^{2+}-dependent AdoMetDC was present (Suresh & Adiga, 1977), but the activity did not appear sufficient to account for the rate of spermidine formation. Both $[^{14}C]$-aspartate and $[^3H]$-methionine were substrates for spermidine formation after infiltration of these precursors into the plant. Carboxyspermidine was synthesized and used as a standard to characterize the in vitro product of the enzymatic condensation of aspartic-β-semialdehyde and $[^{14}C]$-putrescine. The carboxyspermidine synthase was purified (45-fold) and was able to act more

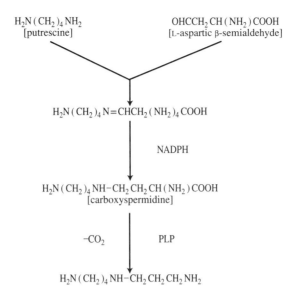

FIG. 9.8   A route for the synthesis of spermidine in *P. denitrificans* and *R. spheroides*; redrawn from Tait (1976).

slowly on cadaverine and 1,3-diaminopropane in place of putrescine. Spermidine was slightly inhibitory. A carboxyspermidine decarboxylase requiring PLP was also isolated and was relatively inert to MGBG; spermidine at 1 mM concentration was significantly inhibitory. The possible existence of a methionine-independent aspartate-dependent spermidine pathway in tobacco cells in culture has also been reported (Lee & Park, 1991).

Vibrio species (e.g., *Vibrio alginolyticus*), which contain norspermidine, have also been reported to synthesize this compound from aspartic acid β-semialdehyde (ASA) and 1,3-diaminopropane (Nakao et al., 1990). It will be recalled that the latter amine is a product of the novel enzyme, L-2,4-diaminobutyric acid decarboxylase. Carboxynorspermidine is formed in the presence of NADPH from diaminopropane and the aldehyde, and this product was decarboxylated by a second homogeneous enzyme, which required PLP for activity. The intermediate, carboxynorspermidine, is thought to be transient and has not yet been isolated from the intact organism. Carboxyspermidine was found to be decarboxylated at <10% that of carboxynorspermidine. The gene for this decarboxylase of *V. alginolyticus* has been isolated and sequenced (Yamamoto et al., 1994).

A carboxynorspermidine synthase was purified to homogeneity from this Vibrio (Nakao et al., 1991). The overall reaction studied was

L-aspartic β-semialdehyde + 1,3-diaminopropane

+ NADPH + H[+] → carboxynorspermidine

+ NADP[+]+ H$_2$O.

The nonenzymatic reaction of ASA and 1,3-diaminopropane to form the Schiff base was largely irreversible, eliminating a potentially inhibitory free ASA. The reduction to form carboxynorspermidine was specific for the L-ASA form of the Schiff base, more reactive with NADPH than with NADH, and essentially irreversible. Putrescine was slightly active in the overall reaction. The activities of all of the enzymes involved in the synthesis of norspermidine, that is, the 2,4-diaminobutyrate decarboxylase, the carboxynorspermidine synthase, and the carboxynorspermidine decarboxylase, were markedly decreased by the growth of the Vibrio in the presence of 5 mM norspermidine. The authors pointed to differences among the various microbial and plant carboxypolyamine synthases.

Although the body of data recorded as yet on this alternative pathway of synthesis of spermidine and norspermidine, based on the formation of a Schiff base with aspartic acid semialdehyde, is far less than that involving AdoMetDC and a propylamine transferase, it has begun to assume a breadth and degree of detail warranting serious attention in dealing with microbes and plants.

Synthesis of Homospermidine

Tait (1979) also described an enzyme system that can form homospermidine. An extract of the photosynthetic *Rhodopseudomonas viridis* produced homospermidine from putrescine in the presence of catalytic amounts of NADH. The organism is reported to be rich in homospermidine but lacks spermidine. It was postulated that putrescine is oxidized to 4-aminobutyraldehyde, which forms a Schiff base with another molecule of putrescine. The Schiff base is reduced by the newly formed NADH to generate homospermidine. A single enzyme of about 73 kDA is reported as effecting these conversions. A similar enzyme was found in the plant, *Lathyrus sativus*.

TABLE 9.6 Comparison of homospermidine synthases.

Property	*A. tartarogenes*	*R. viridis*	*L. sativus*
Native M_r ($\times 10^{-3}$)	102	73	75
Subunit M_r ($\times 10^{-3}$)	52	NR	NR
pI	5.0	NR	NR
K_m for PUT (mM)	0.28	0.2	3
K_m for NAD$^+$ (μM)	18	2.5	NR
Optimal K$^+$ (mM)	50	40	10
Optimal DTT (mM)	1	2	2
Optimal pH	8.7	8.8	8.4
Optimal temperature (°C)	30	37	37
K_i for DAP (μM)	13	2	NR
K_i for NADH (μM)	6	1.5	NR

Adapted from Yamamoto et al. (1993). DTT, dithiothreitol.

Most recently, a homospermidine synthase catalyzing the conversion of putrescine to homospermidine in the presence of NAD$^+$ was isolated from *Acinetobacter tartarogenes* and was purified to homogeneity (Yamamoto et al., 1993). The single protein carries out the following sequence of reactions:

1. putrescine + NAD$^+$ → NH$_3$ + aminobutyraldehyde

    + NADH,

2. putrescine + 4-aminobutyraldehyde → Schiff base,

3. Schiff base + NADH →

    H$_2$N(CH$_2$)$_4$NH(CH$_2$)$_4$NH$_2$ + NAD$^+$.

The properties of the three homospermidine synthases are compared in Table 9.6.

REFERENCES

Allen, R. R. & Klinman, J. P. (1981) *Journal of Biological Chemistry*, **256**, 3233–3239.

Anton, D. L. & Kutny, R. (1987a) *Journal of Biological Chemistry*, **262**, 2817–2822.

——— (1987b) *Biochemistry*, **26**, 6444–6447.

Batchelor, K., Smith, R., & Watson, N. (1986) *Biochemical Journal*, **233**, 307–308.

Bergerat, A. & Guschlbauer, W. (1990) *Nucleic Acids Research*, **18**, 4369–4375.

Bethin, K. E., Petrovic, N., & Ettinger, M. J. (1995) *Journal of Biological Chemistry*, **270**, 20698–20702.

Bhattacharjee, M. K. & Snell, E. E. (1990) *Journal of Biological Chemistry*, **265**, 6664–6668.

Billington, D., Golding, B., & Nassereddin, I. (1980) *Journal of the Chemical Society: Chemical Communications*, 90–91.

Bitonti, A., McCann, P., & Sjoerdsma, A. (1982) *Biochemical Journal*, **208**, 435–441.

Bowerman, W. H., Tabor, C. W., & Tabor, H. (1973) *Journal of Biological Chemistry*, **248**, 2480–2486.

Cacciapuoti, G., Porcelli, M., Bertoldo, C., De Rosa, M., & Zappia, V. (1994) *Journal of Biological Chemistry*, **269**, 24762–24769.

Cacciapuoti, G., Porcelli, M., Cartenì–Farina, M., Gambacorta, A. & Zappia, V. (1986) *European Journal of Biochemistry*, **161**, 263–271.

Cacciapuoti, G., Porcelli, M., De Rosa, M., Gambocorta, A., Bertoldo, C., & Zappia, V. (1991) *European Journal of Biochemistry*, **199**, 395–400.

Cantoni, G. L. (1982) *Biochemistry of S-Adenosylmethionine and Related Compounds*, pp. 3–10, E. Usdin, R. T. Borchardt, & C. R. Creveling, Eds. London: Macmillan Press.

Cartenì–Farina, M., Oliva, A., Romeo, G., Napolitano, G., De Rosa, M., Gambacorta, A., & Zappia, V. (1979) *European Journal of Biochemistry*, **101**, 317–324.

Casara, P., Marchal, P., Wagner, J., & Danzin, C. (1989) *Journal of the American Chemical Society*, **111**, 9111–9113.

Chattapadhyay, R., Pelka, H., & Schulman, L. (1990) *Biochemistry*, **29**, 4263–4268.

Chattopadhyay, M. K., Ghosh, A. K., & Sengupta, S. (1991) *Journal of General Microbiology*, **137**, 685–691.

Chu, T. M., Mallette, M. F., and Mumma, R. O. (1968) *Biochemistry*, **9**, 1399–1406.

Cleaves, G. R. & Cohen, P. S. (1970) *Journal of Bacteriology*, **103**, 697–701.

Coch, E. & Greene R. (1971) *Biochimica et Biophysica Acta*, **230**, 223–236

Cohen, G. N. & Saint–Girons, S. (1987) Escherichia coli and Salmonella typhimurium: *Cellular and Molecular Biology*, pp. 429–444, F. C. Neidhardt, Ed. Washington, D. C.: American Society for Microbiology.

Coward, J., Motola, N., & Moyer, J. (1977) *Journal of Medicinal Chemistry*, **20**, 500–505.

Della Ragione, F., Porcelli, M., Cartenì–Farina, M., Zappia, V., & Pegg, A. E. (1985) *Biochemical Journal*, **232**, 335–341.

De Rosa, M., De Rosa, S., Gambacorta, A., Cartenì–Farina, M., & Zappia, V. (1978) *Biochemical Journal*, **176**, 1–7.

Diaz, E. & Anton, D. L. (1991) *Biochemistry*, **30**, 4078–4081.

Eliasson, R., Fontecave, M., Jornvall, H., Krook, M., Pontes, E., & Reichard, P. (1990) *Proceedings of the National Academy of Sciences of the United States of America*, **87**, 6614–3318.

Elo, H., Mutikainen, I., Alhonen–Hongisto, L., Laine, R., Jänne, J., & Lumme, P. (1986) *Zeitschrift für Naturforschung*, **41C**, 851–855.

Ferro, A. J., Barrett, A., & Shapiro, S. K. (1976) *Biochimica et Biophysica Acta*, **438**, 487–494.

Ferro, A. J., Barrett, A., & Shapiro, S. K. (1978) *Journal of Biological Chemistry*, **253**, 6021–6025.

Gallagher, T., Rozwarski, D. A., Ernst, S. R., & Hackert, M. L. (1993) *Journal of Molecular Biology*, **230**, 516–528.

Gianotti, A. J., Tower, P. A., Sheley, J. H., Conte, P. A., Spiro,

C., Ferro, A. J., Fitchen, J. H., & Riscoe, M. K. (1990) *Journal of Biological Chemistry*, **265**, 831–837.

Gilliland, G. L., Markham, G. D., & Davies, D. R. (1983) *Journal of Biological Chemistry*, **258**, 6963–6964.

Ginsberg, D., Bachrach, V., & Keynan, A. (1982) *FEBS Letters*, **137**, 181–185.

Giulidori, P. & Stramentinoli, G. (1984) *Analytical Biochemistry*, **137**, 217–220.

Golding, B. & Nassereddin, I. (1982) *Journal of the American Chemical Society*, **104**, 5815–5817.

—— (1985a) *Journal of the Chemical Society Perkin Transactions I*, 2011–2015.

—— (1985b) *Journal of the Chemical Society Perkin Transactions I*, 2017–2024.

Golding, B., Nassereddin, I., & Billington, D. (1985) *Journal of the Chemical Society Perkin Transactions I*, 2007–2010.

Greenberg, M. L. & Cohen, S. S. (1985) *Plant Physiology*, **78**, 568–575.

Greene, R. C., Su, C., & Holloway, C. T. (1970) *Biochemical and Biophysical Research Communications*, **38**, 1120–1126.

Hafner, E. W., Tabor, C. W., & Tabor, H. (1977) *Journal of Bacteriology*, **132**, 832–840.

—— (1979) *Journal of Biological Chemistry*, **254**, 12419–12426.

Hamasaki, N. & Oshima, T. (1990) *Abstracts of Symposium on Polyamines*, p. P61. Kyoto, Japan: ISPMMB.

Hibasami, H., Tanaka, M., Nagai J., & Ikeda, T. (1980) *FEBS Letters*, **116**, 116.

Hoffman, J. L. (1986) *Biochemistry*, **25**, 4444–4449.

Hughes, J. A., Brown, L. R., & Ferro, A. J. (1987) *Journal of Bacteriology*, **169**, 3625–3632.

Igarashi, K., Hara, K., Watanabe, Y., Hirose, S., & Takeda, Y. (1975) *Biochemical and Biophysical Research Communications*, **64**, 897–904.

Inouye, M. & Pardee, A. (1970) *Journal of Bacteriology*, **101**, 770–776.

Ito, S. & Nicol, J. A. C. (1976) *Biochemical Journal*, **153**, 567–570.

Jackson, F. R. (1990) *Journal of Molecular Evolution*, **31**, 235–329.

Karvonen, E., Kauppinen, L., Partanen, T., & Pösö, H. (1985) *Biochemical Journal*, **231**, 168–169.

Kashiwagi, K. & Igarashi, K. (1988) *Journal of Bacteriology*, **170**, 3131–3135.

Kawase, M., Samejima, K., Okada, M., & Shibuta, H. (1979) *Journal of Pharmacobio-Dynamics*, **2**, 12–18.

Kerwar, S. S. & Weissbach, H. (1970) *Archives of Biochemistry and Biophysics*, **141**, 525–532.

Kim, H. J., Balcezak, T. J., Nathin, S. J., McMullen, H. F., & Hansen, D. E. (1992) *Analytical Biochemistry*, **207**, 68–72.

Kolb, M., Danzin, C., Barth, J., & Claverie, N. (1982) *Journal of Medical Chemistry*, **25**, 550–556.

Krüger, D. H. & Schroeder, C. (1981) *Microbiological Reviews*, **45**, 9–51.

Lee, S. H. & Park, K. Y. (1991) *Plant and Cell Physiology*, **32**, 523–531.

Liu, C. & Coward, J. (1991) *Journal of Medicinal Chemistry*, **34**, 2094–2101.

Lombardini, J. B., Coulter, A. W., & Talalay, P. (1970) *Molecular Pharmacology*, **6**, 481–499.

Lomnitzer, R. & Ron, E. Z. (1972) *Journal of Bacteriology*, **109**, 1316–1318.

Maas, W. K. (1972) *Molecular and General Genetics*, **119**, 1–9.

Mansouri, S. & Bunch, A. W. (1989) *Journal of General Microbiology*, **135**, 2819–2827.

Markham, G. D., DeParasis, J., & Gatmaitan, J. (1984) *Journal of Biological Chemistry*, **259**, 14505–14507.

Markham, G. D., Hafner, E. W., Tabor, C. W., & Tabor, H. (1980) *Journal of Biological Chemistry*, **255**, 9082–9092.

Markham, G. D., Tabor, C. W., & Tabor, H. (1982) *Journal of Biological Chemistry*, **257**, 12063–12068.

Matos, J. R., Raushel, F. M., & Wong, C. (1987) *Biotechnology and Applied Biochemistry*, **9**, 39–52.

Matos, J. R. & Wong, C. (1987) *Bioorganic Chemistry*, **15**, 71–80.

Michaeli, S., Rozenhak, S., & Ron, E. (1983) *Advances in Polyamine Research*, vol. 4, pp. 519–520, U. Bachrach, A. Kaye, & R. Chayen, Eds. New York: Raven Press.

Midorikawa, Y., Hibasami, H., Gasaluck, P., Yoshimura, H., Masuji, A., Nakashima, K., & Imai, M. (1991) *Journal of Applied Bacteriology*, **70**, 291–293.

Mitsui, K., Teraoka, H., & Tsukada, K. (1988) *Journal of Biological Chemistry*, **263**, 11211–11216.

Morris, D. R. (1981) *Polyamines in Biology and Medicine*, pp. 223–242, D. R. Morris & L. J. Marton, Eds. New York: Marcel Dekker.

Myers, R. W. & Abeles, R. H. (1990) *Journal of Biological Chemistry*, **265**, 16913–16921.

Myers, R. W., Wray, J. W., Fish, S., & Abeles, R. H. (1993) *Journal of Biological Chemistry*, **268**, 24785–24791.

Nakao, H., Shinoda, S., & Yamamoto, S. (1990) *Journal of General Microbiology*, **136**, 1699–1704.

—— (1991) *Journal of General Microbiology*, **137**, 1737–1742.

Nastri, H. G. & Algranati, I. D. (1988) *Biochimica et Biophysica Acta*, **949**, 65–70.

Nishimura, S., Taya, Y., Kuchino, Y., & Ohashi, Z. (1974) *Biochemical and Biophysical Research Communications*, **57**, 702–708.

Old, I. G., Phillips, S. E. V., Stockley, P. G., & Saint Girons, I. (1991) *Progress in Biophysics and Molecular Biology*, **56**, 145–185.

Orr, G., Danz, D., Pontoni, G., Prabhakaran, P., Gould, S., & Coward, J. K. (1988) *Journal of the American Chemical Society*, **110**, 5791–5799.

Pani, A., Marongui, M. E., Obino, P., Gavagnin, M., & La Colla, P. (1991) *Experientia*, **47**, 1228–1229.

Pankaskie, M. & Abdel–Monem, M. M. (1980) *Journal of Medicinal Chemistry*, **23**, 121–127.

Paulin, L. (1986) *FEBS Letters*, **202**, 323–326.

Pegg, A. E. (1974) *Biochemical Journal*, **141**, 581–583.

Pegg, A. E. & Bennett, R. A. (1983) *Methods in Enzymology*, **94**, 69–72.

Pegg, A., Bitonti, A., McCann, P., & Coward, J. (1983) *FEBS Letters*, **155**, 192–196.

Pegg, A. E. & Conover, C. (1976) *Biochemical and Biophysical Research Communications*, **69**, 766–774.

Pegg, A. E. & Jacobs, G. (1983) *Biochemical Journal*, **213**, 495–502.

Pontoni, G., Coward, J., Orr, G., & Gould, S. (1983) *Tetrahedron Letters*, **24**, 151–154.

Porcelli, M., Cacciapuoti, G., Cartenì–Farina, M., & Gambacorta, A. (1988) *European Journal of Biochemistry*, **177**, 273–280.

Recsei, P. A., Huynh, Q. K., & Snell, E. E. (1983) *Proceedings of the National Academy of Sciences of the United States of America*, **80**, 973–977.

Recsei, P. A. & Snell, E. E. (1973) *Biochemistry*, **12**, 365–371.

Reich, N. O. & Everett, E. A. (1990) *Journal of Biological Chemistry*, **265**, 8929–8934.

Rosano, C. L., Hurwitz, C., & Bunce, S. C. (1978) *Journal of Bacteriology*, **135**, 805–808.

Salmikangas, P., Keränen, M., & Pajunen, A. (1989) *FEBS Letters*, **258**, 123–126.

Samejima, K., Nakazawa, Y., & Matsunaga, I. (1978) *Chemical and Pharmaceutical Bulletin*, **26**, 1480–1485.

Satishchandran, C., Taylor, J. C., & Markham, G. D. (1990) *Journal of Bacteriology*, **172**, 4489–4496.

Schroeder, H., Barnes, C. J., Bohinski, R. C., & Mallette, M. F. (1973) *Canadian Journal of Microbiology*, **19**, 1347–1354.

Setlow, P. (1974) *Journal of Bacteriology*, **117**, 1171–1177.

Shapiro, S. K. & Barrett, A. (1981) *Biochemical and Biophysical Research Communications*, **102**, 302–307.

Shi, W. & Zusman, D. R. (1995) *Journal of Bacteriology*, **177**, 5346–5349.

Shimizu, S., Abe, T., & Yamada, H. (1988) *FEMS Microbiology Letters*, **51**, 177–180.

Shiozaki, S., Shimizu, S., & Yamada, H. (1984) *Agricultural and Biological Chemistry*, **48**, 2293–2300.

Su, C. & Cohen, S. S. (1973) *Polyamines in Normal and Neoplastic Growth*, pp. 299–306, D. H. Russell, Ed. New York: Raven Press.

Sufrin, J. R., Coulter, A. W., & Talalay, P. (1979) *Molecular Pharmacology*, **15**, 661–677.

Suresh, R. & Adiga, P. R. (1977) *European Journal of Biochemistry*, **79**, 511–518.

Tabor, C. W. & Tabor, H. (1984) *Advances in Enzymology*, vol. 56, pp. 251–282, A. Meister, Ed. New York: Wiley.

——— (1987) *Journal of Biological Chemistry*, **262**, 16037–16040.

Tabor, C. W., Tabor, H., & Hafner, E. W. (1978) *Journal of Biological Chemistry*, **253**, 3671–3676.

Tabor, C. W., Tabor, H., & Hafner, E. H. (1983a) *Methods in Enzymology*, **94**, 83–91.

Tabor, C. W., Tabor, H., Hafner, E. W., Markham, G. D., & Boyle, S. M. (1983b) *Methods in Enzymology*, **94**, 117–121.

Tabor, C. W., Tabor, H., & Xie, Q. (1986) *Proceedings of the National Academy of Sciences of the United States of America*, **83**, 6040–6044.

Tabor, H., Hafner, E. W., & Tabor, C. W. (1983) *Methods in Enzymology*, **94**, 91–103.

Tait, G. H. (1976) *Biochemical Society Transactions*, **4**, 610–612.

——— (1979) *Biochemical Society Transactions*, **7**, 199–201.

——— (1985) *Biochemical Society Transactions*, **13**, 316–318.

Tanaka, H., Imahara, H., Esaki, N. & Soda, K. (1981) *Analytical Letters*, **14**(B2), 111–118.

Tang, K., Mariuzza, R., & Coward, J. K. (1981) *Journal of Medicinal Chemistry*, **24**, 1277–1284.

Tang, K., Pegg, A. E., & Coward, J. K. (1980) *Biochemical and Biophysical Research Communications*, **96**, 1371–1377.

Tower, P. A., Johnson, L. L., Ferro, A. J., Fitchen, J. H., &

Riscoe, M. K. (1991) *Antimicrobial Agents and Chemotherapy*, **35**, 1557–1561.

Vaaler, G. L. & Snell, E. E. (1989) *Biochemistry*, **28**, 7306–7313.

Van Poelje, P. D. & Snell, E. E. (1987) *Analytical Biochemistry*, **161**, 420–424.

——— (1990) *Annual Review of Biochemistry*, **59**, 29–59.

Wagner, J., Claverie, N., & Danzin, C. (1984) *Analytical Biochemistry*, **140**, 108–116.

Weissbach, H. & Brot, N. (1991) *Molecular Microbiology*, **5**, 1593–1597.

Weitkamp, E., Dixon, H. B., Khomutov, A., & Khomutov, R. (1991) *Biochemical Journal*, **277**, 643–645.

Wickner, R. B., Tabor, C. W., & Tabor, H. (1970) *Journal of Biological Chemistry*, **245**, 2132–2139.

Williams–Ashman, H., Corti, A., & Coppoc, G. L. (1977) *The Biochemistry of Adenosylmethionine*, pp. 493–509, F. Salvatore, E. Borek, V. Zappia, H. Williams–Ashman, & F. Schlenk, Eds. New York: Columbia University Press.

Williams–Ashman, H. G. & Schenone, A. (1972) *Biochemical and Biophysical Research Communications*, **46**, 288–295.

Williams–Ashman, H. G. & Seidenfeld, J. (1986) *Biochemical Pharmacology*, **35**, 1217–1225.

Witkin, S. S. & Rosenberg, E. (1970) *Journal of Bacteriology*, **103**, 641–649.

Wu, Y. Q., Lawrence, T., Guo, J. Q., & Woster, P. M. (1993) *Bioorganic and Medicinal Chemistry Letters*, **3**, 2811–2816.

Wu, Y. Q. & Woster, P. M. (1993) *Bioorganic & Medicinal Chemistry*, **1**, 349–360.

Xie, Q., Tabor, C. W., & Tabor, H. (1989) *Journal of Bacteriology*, **171**, 4457–4465.

Xie, Q., Tabor, C. W., & Tabor, H. (1993) *Gene*, **126**, 115–117.

Yamakawa, T., Taira, H., & Kaneko, I. (1980) *Agricultural and Biological Chemistry*, **44**, 2235–2237.

Yamamoto, S., Nagata, S., & Kusaba, K. (1993) *Journal of Biochemistry*, **114**, 45–49.

Yamamoto, S., Sugahara, T., Tongou, K., & Shinoda, S. (1994) *Microbiology*, **140**, 3117–3124.

Yamanoha, B. & Cohen, S. S. (1985) *Plant Physiology*, **78**, 784–790.

Yu, P. H. (1983) *Methods in Enzymology*, **94**, 66–69.

Zappia, V., Cacciapuoti, G., Pontoni G., & Oliva, A. (1980) *Journal of Biological Chemistry*, **255**, 7276–7280.

Zappia, V., Cartenì–Farina, M., & Galletti, P. (1977) *The Biochemistry of Adenosylmethionine*, pp. 472–492, F. Salvatore, E. Borek, V. Zappia, H. Williams–Ashman, & F. Schlenk, Eds. New York: Columbia University Press.

Zappia, V., Cartenì–Farina, M., Galletti, P., Ragione, F. L., & Cacciapuoti, G. (1983a) *Methods in Enzymology*, **94**, 57–66.

Zappia, V., Galletti, P., Oliva, A., & Porcelli, M. (1983b) *Methods in Enzymology*, **94**, 73–80.

CHAPTER 10

Microbial Eucaryotic Systems:
Fungi and Slime Molds

In exploring and cultivating the fields of nature, the chemists were best provided with the machinery for this cultivation, but the biologists knew best the lay of the land.

F. G. HOPKINS

Fungal Systems

We begin the examination of polyamines in nucleated cells with discussions of certain microbial eucaryotes. These include some fungi such as yeasts and *Neurospora* and the puzzling *Myxomycetes*, the slime molds *Dictyostelium* and *Physarum*. The biologies of these systems are certainly not simple, but for a variety of reasons, historical, technological, and scientific, the organisms have been chosen as experimental systems for the analysis of particular biological and biochemical problems (Fruton, 1972).

The phenomena of decay and putrefaction were undoubtedly noted long before the discovery of culinary arts. The preparation of beer and wine had been described as fermentations that had not passed into noxious phases. To the several kinds of useful fermentation, characterized as the production of gas, alcohol, and acid, was added that of putrefaction. In 1680 the microbial participants of these fermentations, such as the yeasts, had been described by A. Leeuwenhoek as little globules; in 1735 H. Boerhaave described the roles of added yeast in promoting fermentations. J. Black's 1756 discovery of fixed air (CO_2) was soon followed by his demonstration of the evolution of this gas in a typical fermentative process. In the last half of the century, the study of the different gases present in air or generated in combustion culminated in A. Lavoisier's studies on alcoholic fermentation by yeast and with his statement of the law of conservation of mass, derived from his analysis of the conversion of sugar to CO_2 and ethanol.

In 1837 several investigators demonstrated independently that the tiny yeast globules were living reproducing organisms. These findings elicited a major debate on the mechanisms of the reactions carried out by yeasts, a debate in which words such as "catalysis," "protein," etc. were coined. The subject also evoked a debate on the need for intact yeast cells as compared to a possible role for cell-free extracts. By the end of that century some two dozen soluble enzymes had been described, and in 1897 E. Buchner demonstrated that cell-free extracts of yeast were capable of the conversion of sugar to ethanol and CO_2. The preparation of cell-free extracts in the 1920s and 1930s culminated in the isolation of new compounds and enzymatic steps in glycolysis. By the end of the 1930s genetic techniques had been devised to examine steps in the fermentation of sugars by yeasts.

The development of genetics and biochemistry pointed to the advantages of using microbial systems in exploring physiological aspects of heredity and in defining steps of intermediary metabolism. G. Beadle and E. Tatum (1941) introduced the use of an ascomycetous mold, *Neurospora*, for these purposes, as well as the production and preservation of mutants blocked in potentially identifiable biosynthetic reactions. One of the first of the new strains required vitamin B_6 (pyridoxine) for growth and in deficiency was found to be blocked in ornithine metabolism. Studies on ornithine and arginine metabolism in *Neurospora* were among the earliest using deliberately mutagenized organisms, and the methodological advantages led to the extrapolation of mutant techniques to bacterial systems in the mid 1940s.

The exploration of the biology of yeast and *Neurospora*, their life cycles, sexuality, combination, division mechanisms, segregants, etc. not only enlarged the opportunities of genetic analysis but also revealed apparent aberrations, such as the cytoplasmic inheritance of mitochondria, the existence of introns and exons and RNA splicing, the presence of mycoviruses, etc. Simultaneously other fungilike organisms appeared to display even more complex life cycles, opening opportunities of applying the newly enlarged genetic skills to problems of the staging of development. At this point increased active investigation began on slime molds such as *Dictyostelium* and *Physarum*.

In studies of some developmental phenomena like cancer, some aspects of our knowledge of polyamine metabolism in mammalian cells have been explored in more detail than in yeast or the other systems to be described here. Nevertheless, the manipulability of yeast and other microbial systems now compel the exploration of the other eucaryotic genomes, including the DNA of humans in yeast, for example, the isolation and propagation of large pieces of human DNA. The reisolation of these newly propagated yeast artificial chromosomes (YACs) now permits a transformation of yeast spheroplasts and expression of the human DNA within the transformed fungal cells. The pretreatment of YAC DNA with a 1-mM mix of spermidine and spermine, presumably effecting a condensation of the DNA, markedly increases the transformation efficiency with large yeast YACs (Connelly et al., 1991). Polyamine derivatives (i.e., synthetic cationic amphiphiles) have become major aids in transfection and gene transfer in preparation for exercises in gene therapy (Behr, 1994). Thus the methodologies of yeast genetics and the exploitation of the polyamines in the manipulation of DNA have become approaches to the development of our knowledge of the human genome.

Classifying Fungi and Slime Molds

In the 1940s a typical text of botany might have included the true fungi and slime molds, as well as bacteria, in phyla of the subkingdom Thallophyta within the plant kingdom. By the 1970s the fungi had been designated as a kingdom that was separate from those of plants, animals, Protista (containing *Euglena* and protozoa), and Monera (containing bacteria, i.e., eubacteria, mycobacteria, cyanobacteria, and actinomycetes). As described in a text of that time (Ross, 1979), it was necessary to modify the concept used in subdividing the fungal kingdom and to separate the true fungi, division Eumycota, containing *Saccharomyces* and *Neurospora*, from the slime molds, which had been placed in the Protista. By the early 1980s the slime molds were omitted from consideration as fungi by some texts.

In the past decade some evolutionary origins and relationships of proteins and nucleic acids have been clarified as a result of sequence analyses. The dissection of sequence diversity of small subunit ribosomal RNA in various organisms has provided a major tool in studying these evolutionary relationships, which have been pursued by utilizing eucaryotic organisms. Sogin (1991) described the phylogenetic separation of eubacteria and eucaryotic microorganisms revealed by such analyses. He reported that the fungi cluster in one part of his phylogenetic tree is clearly separated from the apparent positions assigned to *Dictyostelium discoideum* and *Physarum polycephalum*. The latter organisms are also separated from each other. Thus the biochemical data on this complex polymer (RNA) has supported the main direction of the past 40 years of biological analysis and its resulting phylogenetic conclusions. It is possibly relevant that the polyamines are probably associated with these conserved RNAs in determining ribosomal structure and thereby may participate in organizing their folding patterns and in selecting for ribosomal stability (Chap. 5).

The interrelations of polyamines, ribosomal structure, and phylogenetic position may also prove to be of some interest in medical practice. A major cause of death in AIDS is known to be a result of a lethal pneumonia caused by *Pneumocystis carinii*. The difficultly cultivable organism was long thought to be a protozoan but, as a result of the analysis of ribosomal RNA sequences, has more recently been described as related to the true fungi. Infections by *P. carinii* in an immunosuppressed rat model were significantly ameliorated by difluoromethylornithine (DFMO), implicating the polyamines in the parasitic process (Sarić & Clarkson, 1994). However, DFMO has not been sufficiently effective in *P. carinii* infections of humans.

Biological Properties of *Saccharomyces Cerevisiae* and *Neurospora Crassa*

The most comprehensive biochemical studies on the polyamines among the fungi have been done on these two species (Tabor & Tabor, 1985). These organisms are considered to be members of the "higher fungi," and more particularly of the subdivision Ascomycotina, separated thereby from that of the Deuteromycotina, or Fungi Imperfecti, *Aspergillus* and *Penicillium*, or the Basidiomycotina, *Ustilago*. Studies of the polyamines have been made with organisms in each of these subdivisions. For example, a putrescine-requiring strain of *Aspergillus nidulans* was described by Sneath (1955). A general review of the roles of the polyamines in fungal development indicates the considerable diversity of the fungi that have been studied (Stevens & Winther, 1979; also see Davis, 1996).

The life cycle of the yeast *S. cerevisiae* is given in Figure 10.1. The organism may frequently be found on fruit in nature. In culture, ascospores give rise to haploid vegetative cells that multiply by budding. A representation of such a cell is given in Figure 10.2 that contains a nucleus entering the bud and the presence of mitochondria in a cytoplasmic layer. This layer surrounds a large vacuole in the parent. The presence of a

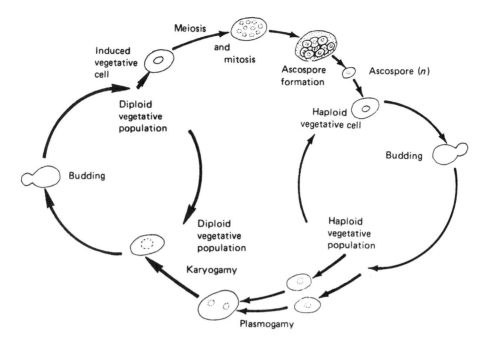

FIG. 10.1 Life cycle of *Saccharomyces cerevisiae*; redrawn from Webster (1970). Haploid and diploid vegetative stages may be present. Karyogamy precedes a multiplying diploid vegetative stage; meiosis may precede a multiplying haploid vegetative stage.

nucleus, in which several chromosomes are contained within a membrane, defines the organism as eucaryotic. An endoplasmic reticulum, with which ribosomes are most frequently associated in many fungi and "higher" organisms, is not given in the figures.

The presence of a large vacuole is seen most frequently in plant cells and infrequently in some specialized animal cells. This feature raises important questions of the possible compartmentation of enzymes and metabolites. The existence of mitochondria and a nucleus also contributes to these problems of distribution within the cell and the division of labor. As discussed by Davis (1986), the solutions to these problems in *Saccharomyces* and *Neurospora* are significantly different in particular features of the biosynthesis and degradation of arginine. Also, the vacuole plays a special role in polyamine metabolism in these fungi.

The organism grows as single cells, unlike the bread mold *Neurospora*, which possesses a multinuclear mycelial stage of multiplication. Both organisms lack chlorophyll and chloroplasts and in nutritional terms may be described as chemoorganoheterotrophs, as is *Escherichia coli*. *N. crassa* is also unable to synthesize essential biotin. The rigid cell walls differ from those of bacteria in containing chitin and insoluble glucans, which may be cleaved enzymatically in hypertonic media to produce easily lysed protoplasts. Methods for the growth of *Saccharomyces* and the preparation of cell extracts for purposes of fractions or protein isolation were described by Jazwinski (1990). The freezing and thawing

of this yeast in the presence of an anionic detergent permeabilizes the cell wall and permits the use of such a cell for many enzyme assays (Miozarri et al., 1978).

The fusion of two haploid cells leads to karyogamy (the fusion of the two nuclei) and a diploid vegetative phase. Meiosis of diploid cells and mitosis precede the formation of haploid ascospores, which are cultivable in a vegetative phase. Haploidy, in which one set of possibly masking chromosomes is avoided, facilitates the analysis of metabolic deficiency and competence. Mutants with blocks in the pathways leading to the biosynthesis of the amino acid precursors of the major yeast polyamines (ornithine, arginine, lysine, and methionine) have been isolated, as well as mutants in the biosynthesis of the polyamines themselves.

Polyamines in Life of *S. Cerevisiae*

Spermine was first detected in yeast by M. C. Rosenheim (1917), who suggested that this was the most convenient source of the amine. Tabor et al. (1958) described the incorporation of $[^{14}C]$-putrescine into spermidine and spermine in cultures of *S. cerevisiae* and a putrescine-requiring mutant of *A. nidulans*. R. Greene (1957) demonstrated the incorporation of $[2\text{-}^{14}C]$-methionine into the spermidine of *N. crassa*. Thus, the various fungi had been used in the earliest demonstrations of the biosynthesis of the polyamines. The enzymology of the process in eucaryotic organisms was accom-

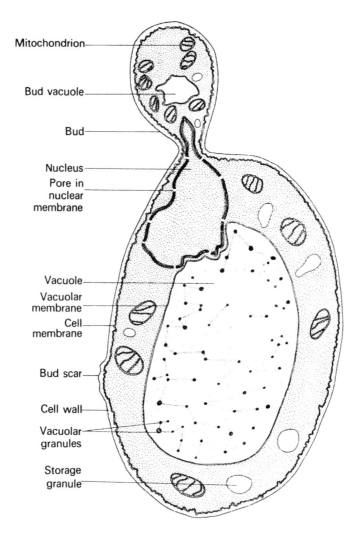

Mitochondrion

Bud vacuole

Bud

Nucleus
Pore in
nuclear
membrane

Vacuole
Vacuolar
membrane
Cell
membrane

Bud scar

Cell wall

Vacuolar
granules

Storage
granule

FIG. 10.2   A budding yeast cell, as seen under an electron microscope; redrawn from Webster (1970).

plished initially with extracts of mammalian tissues, but was soon explored in extracts of *S. cerevisiae* (Jänne et al., 1971).

The accumulation of *S*-adenosylmethionine (AdoMet) in yeasts and particularly in yeast vacuoles had sensitized F. Schlenk and collaborators to recent developments on the life cycle and genetics of yeast, as well as to the significance of polyamine synthesis from AdoMet in the life of the organism. Methionine and adenine were both required for germination of ascospores of *S. cerevisiae*, and the requirement for methionine but not for adenine was satisfied by AdoMet (Choih et al., 1977). Exogenously supplied AdoMet was then shown to stimulate the synthesis of RNA and protein, as well as budding in outgrowth (Brawley & Ferro, 1980).

Several enzymes of AdoMet metabolism increase significantly in the first hour of normal germination in the absence of AdoMet; these include AdoMet synthetase

and AdoMet decarboxylase (AdoMetDC), as well as some methyltransferases. These enzymes increase even more sharply in outgrowth. It had been found earlier that the yeast AdoMetDC is inhibited by methylglyoxal-bis(guanylhydrazone) (MGBG) (Williams–Ashman & Schenone, 1972).

During germination and outgrowth the increase in the spermidine content of the cells paralleled the increase of RNA in the culture (Choih et al., 1978). However, although inhibition by MGBG prevented the increase of spermidine and spermine, there was no effect of this treatment on the accumulation of RNA, DNA, and protein. It was asked, therefore, if the initial and polyamine content of the spores or the additional putrescine accumulated in the presence of MGBG filled a presumed polyamine requirement, or if there was simply not such a requirement at all in these syntheses.

The possible role of putrescine in these events was

pursued with diploid cells (Brawley & Ferro, 1979). Ornithine decarboxylase (ODC) increased and fell rapidly during germination and outgrowth. When the cellular enzyme was inhibited by α-methylornithine, putrescine did not accumulate during germination and growth was significantly slowed after a single division. When MGBG was added to a sporulation medium to deplete spermidine and spermine even further, the diploid cells were unable to develop meiosis and sporulation. It was suggested that a concomitant synthesis of polyamines is not necessary for germination, but that the substances may be essential for meiosis and sporulation.

The finding that α-methylornithine and putrescine depletion affected growth appeared to emphasize a dependence on ODC. In growth of a diploid strain, when stationary phase cultures were diluted into fresh media, the cellular content of ODC increased sharply after the cells began to bud (Kay et al., 1980). Exponentially growing cells had the highest content of enzyme, and ODC content fell as the culture once again approached the stationary phase. However, a high concentration of the enzyme itself was not crucial, because growth occurred in the presence of spermidine or spermine. These supplements almost eliminated the cellular content of ODC.

These results are quite similar to those obtained subsequently with auxotrophic mutants of *Saccharomyces* blocked in different stages of polyamine biosynthesis. The main scheme of polyamine synthesis involving putrescine, spermidine, and spermine proved to be similar to that found in *E. coli* and the mammal. The preparation of polyamine auxotrophs of *S. cerevisiae* was undertaken in laboratories that had seen the power of the approach in similar studies with *E. coli*. It will be recalled that the earliest *E. coli* mutants were obtained by control of ornithine biosynthesis with exogenous arginine.

The pathways of the biosynthesis and metabolism of ornithine and arginine in yeast and *Neurospora* were discussed by Davis (1986). In both organisms the synthesis of ornithine takes place in the fungal mitochondria following the formation of acetylglutamate from glutamate, as given in Figure 10.3. The formation of the early intermediate *N*-acetyl-γ-glutamyl phosphate is feedback inhibited by arginine, theoretically limiting ornithine production. A final $^{\alpha}N$-acetylornithine transfers its acetyl group to glutamate to form ornithine and begins another cycle of metabolism of *N*-acetyl glutamate to ornithine. The existence of this transamination distinguishes this reaction from that in *E. coli* in which $^{\alpha}N$-acetylornithine is hydrolyzed to the free amino acid.

In yeast and *N. crassa* arginine is converted by an arginase to ornithine and urea; the latter is metabolized somewhat differently in the two fungi (Davis, 1986). The presence of arginase and mechanisms for metabolizing urea thereby also distinguish fungal metabolism from that of *E. coli*. The cleavage of arginine by arginase produces more ornithine, and it is obviously desirable to avoid a wasteful cycle of biosynthesis and degradation of arginine. A mode of control of a potentially wasteful cycle was postulated based on the separation of the enzymes of the pathway and certain substrates (Davis, 1986). In *Neurospora*, arginase is cytosolic whereas ornithine transcarbamylase is mitochondrial; this separation is accomplished in yeast by protein–protein interactions in which the cytosolic arginase and ornithine transcarbamylase form a tight 1 : 1 complex in which arginase is a negative allosteric effector for ornithine transcarbamylase (Green et al., 1990). Additionally arginine and ornithine are largely sequestered in the vacuole.

In the construction of ornithine-deficient mutants of *Saccharomyces* it was found that arginase deficiency and arginine excess were insufficient to control the availability of ornithine (Whitney & Morris, 1978). Another mutation for *N*-acetylglutamate kinase was introduced; this led to some dependence of growth rate on ornithine, putrescine, spermidine, and spermine. For two such early mutants, the ornithine-limited cultures had a doubling time 1.5 times that of the supplemented cultures but were apparently devoid of detectable putrescine. Nevertheless, the limited cultures maintained 1–2% and 3–5% of the spermidine and spermine content, respectively, of the supplemented cultures.

Mutagenization of a haploid strain permitted the isolation of strains that grew far less rapidly and whose growth was even more markedly enhanced by polyamines. Some of these were deficient in ODC and some in AdoMetDC. One strain that required spermidine for growth was high in ODC and apparently deficient in propylamine transferase. These types of strains are described in Table 10.1. Their growth rates on polyamine depletion are presented in Figure 10.4; it is evident that *spe*1 and *spe*2 mutants maintain growth at almost normal rates during a 100-fold increase in cell concentration.

When the *spe*1 strain was grown in the presence of spermidine, it was essentially free of putrescine and contained spermidine and some spermine. Growth without spermidine reduced spermidine content to 4% of the supplemented level. When the *spe*2 strain was grown in the presence of spermidine, it contained the three polyamines but putrescine content was 20–25% of the putrescine content of starved cells, indicating a greater activity of ODC in the latter. The spermidine content of starved cells (100-fold growth) fell to 1% of that of spermidine-supplemented cells. Spermidine-limited *spe*2 cells lost viability following 100-fold growth.

In these studies the model yeast system appears to yield cleaner results than the *E. coli* system. Polyamines are required for growth but appear to be present in great excess. There is no evidence for arginine decarboxylase and agmatine ureohydrolase, and the organism depends entirely on ODC for putrescine synthesis. Furthermore, the ODC-deficient strains were unable to produce significant amounts of cadaverine, and therefore lacked any other cadaverine-derived amines that might replace spermidine and spermine.

ODC-deficient (*spe*1) yeast depleted in all polyamines develop ultrastructural abnormalities such as the

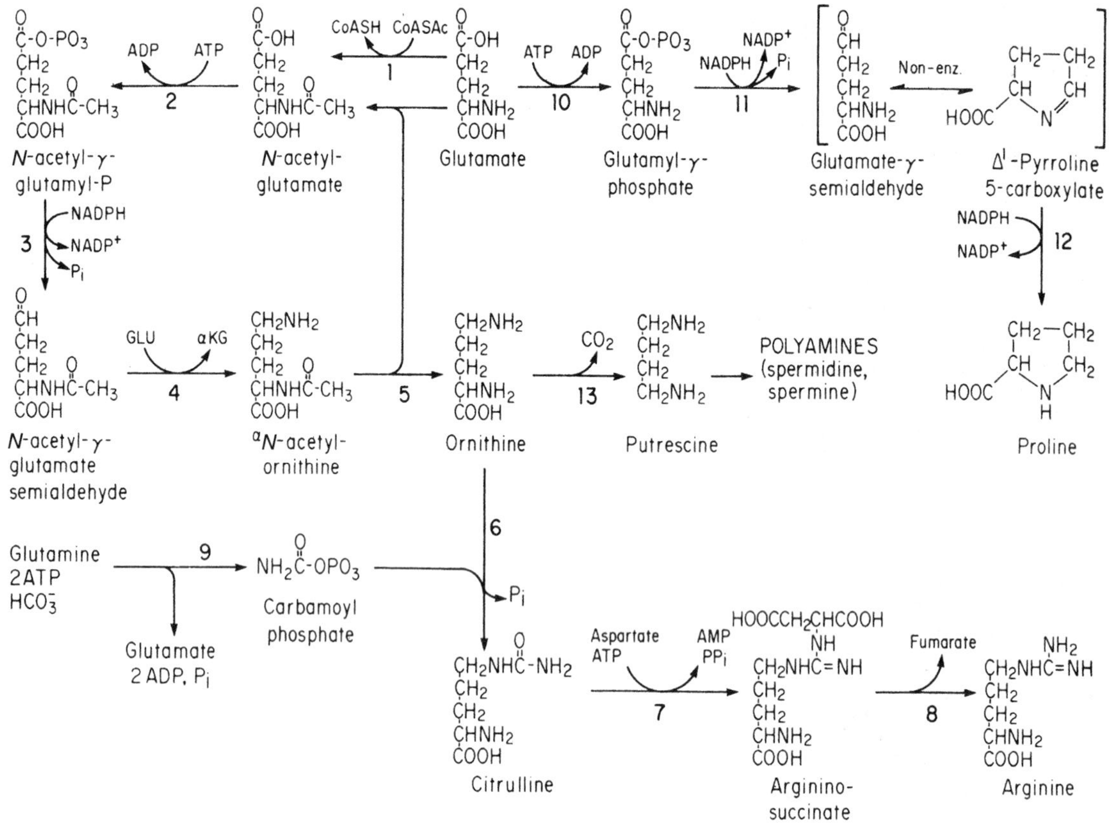

FIG. 10.3 Anabolic reactions leading from glutamate to arginine, proline, and polyamines; redrawn from Davis (1986).

appearance of vacuoles in the cytoplasm and deformation of the nuclei. The budding pattern is altered. More extreme changes appear in a thickened cell wall and irregular cell membrane (Miret et al., 1992). The cell wall is increased in mannan and decreased in glucan. Exogenous spermine thickened yeast cell wall but decreased its content of 1,3-β-glucans, the synthase of which is known to be activated by the tetramine (Poli et al., 1993). In a null ODC spe1 mutant rigorous depletion of the three natural polyamines is not lethal (Schwartz et al., 1995).

Independent studies described several spe2 mutants (Cohn et al., 1978a). These strains were characterized by a mutation in a single chromosomal gene; this was considered to be a structural gene because the AdoMetDC of one of the mutant strains was temperature sensitive. The mutated gene was tightly linked to that for arginosuccinate synthetase (arg1), located in chromosome XV of the 17 present in S. cerevisiae (Hilger & Mortimer, 1980).

Spe2 mutants totally deficient in the enzyme and depleted in spermidine and spermine had doubling times 6 times greater than the parent strain, and the growth rates in the depleted strains were restored by spermidine

or spermine but not by putrescine or cadaverine. The putrescine content of these depleted mutants was 4–5 times greater than that of a wild type strain, indicating once again the control of ODC activity by spermidine.

An absolute requirement for these polyamines was demonstrated in the sporulation of a diploid cross of two spe2 mutants. Spermidine or spermine was essential

TABLE 10.1 Enzyme levels in polyamine mutants of S. cerevisiae.

Strain	Name	ODC[a]	AdoMetDC[a]
M25-4C	Wild type	1.28	5.59
3888	spe1[b]	<0.05	6.53
1009	spe2[b]	2.12	<0.1
4379	spe3[b]	5.98	3.93

Adapted from Whitney and Morris (1978).

[a]Values are nanomoles of CO_2 per milligram of protein per hour.

[b]Whitney and Morris proposed the nomenclature for mutant allelic loci in yeast: spe1 for ODC deficiency, spe2 for deficiency in AdoMetDC, and spe3 for a deficiency in propylamine transferase, presumably spermidine synthase.

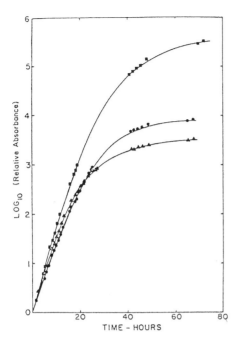

FIG. 10.4 Growth of mutant strains of *S. cerevisiae* in the absence of spermidine; redrawn from Whitney and Morris (1978). Spermidine-grown cultures were filtered and resuspended in fresh spermidine-deficient medium.

for the completion of a sexual cycle (Cohn et al., 1978a). These mutants were also obtained containing a specific "killer" plasmid of double-stranded RNA whose maintenance and multiplication were also totally dependent on the presence of spermidine or spermine in the medium (Cohn et al., 1978b; Wickner, 1989).

Finally a strain totally deficient in ODC, designated *spe*10, did not grow at all in the absence of exogenous polyamine, which restored normal growth (Cohn et al., 1980). Thus, polyamines were found to be absolutely necessary for the growth of yeast. The amount of intracellular spermidine supporting growth of such a strain, although not at maximal growth rate, was of the order of 0.02% of that actually found in wild type strains. The *spe*10 gene is now believed to be identical to the structural gene *spe*1, which has been assigned to chromosome XI (Xie et al., 1990).

Because *spe*2 mutants are unable to sporulate in the presence of putrescine and *spe*1 (*spe*10) mutants grow normally in the presence of spermidine, putrescine in yeast appears to serve mainly as a precursor of spermidine. It will be seen that putrescine is also an essential cofactor for yeast and many eucaryotic AdoMetDCs. Also, a severely depleted ODC-deficient mutant had a low rate of incorporation of [^{14}C]-choline into phosphatidyl choline; this could be restored by incubation in the presence of putrescine, spermidine, or spermine (Hosaka & Yamashita, 1981). The effect appears to relate

primarily to the induction of choline transport by spermidine, which is more active than spermine in this respect. Phosphatidyl choline and other choline derivatives do not occur in *E. coli*.

One ODC-deficient yeast mutant was arrested in growth after a several thousand-fold increase in cell number and depletion in polyamine. The reduced synthesis of the nucleic acids and polyamines was marked by a deterioration of ribosomal structure (Miret & Goldemberg, 1989). These functions were also returned to those of normal growth by a 2-h incubation in putrescine. The restoration of ribosomal association by Mg^{2+} was less effective in the polyamine-depleted preparations.

More recently a *spe*2 mutant totally deficient in the synthesis of spermidine or spermine was found to have an absolute requirement for one of these two compounds for aerobic and anaerobic growth (Balasundaram et al., 1991, 1994). Depleted cells retained vital dyes but were giant in size and deficient in budding; they also had abnormal dispositions of chitinlike and actinlike material. In this last respect they were similar to polyamine-deficient mammalian cells, whose internal actin fiber structures are entirely in disarray (Pohjanpelto et al., 1981). The cells also proved to be defective in mitochondrial function and had suffered damage in the mitochondrial genome.

The arrest of growth is more rapid in an ODC-deficient cell containing AdoMetDC than in one lacking both enzymes. This suggested a toxic effect of decarboxylated AdoMet (dAdoMet) (Balasundaram et al., 1994).

Aminopropyl Transfer in Strains of *S. cerevisiae*

Yeasts appear to lack a specific lysine decarboxylase activity. Indeed L-lysine has been described as a weak inhibitor of the yeast ODC (Tyagi et al., 1981). Although aminopropylcadaverine is not found naturally in *S. cerevisiae*, the ODC of *N. crassa* has a weak lysine decarboxylase activity, like that in animal cells. This permits the appearance of cadaverine and aminopropylcadaverine as a result of the high ODC content of polyamine-deprived *Neurospora* (Paulus et al., 1982). It may be, then, that ODC of yeast has a similar activity and that the conditions for the conversion of lysine to aminopropylcadaverine in yeast have not yet been optimized in growth in most media.

Nevertheless, the "unnatural" triamine is synthesized in various yeasts incubated with cadaverine. Cadaverine and other diamines stimulate the growth of an ODC-less strain of *S. cerevisiae* and new unnatural polyamines are generated, which are also stimulatory (Hamana et al., 1989). This mutant will also respond to homospermidine to produce aminopropylhomospermidine (3,4,4). It is of interest that a strain lacking spermine synthase was unable to add an aminopropyl moiety to homospermidine. Although growth of the ODC mutant is stimulated by spermidine (3,4), homospermidine (4,4), and aminopropylcadaverine (3,5) and their aminopropyl derivatives, it is completely unresponsive to norspermidine (3,3) and norspermine (3,3,3).

The putrescine-deficient organism was also capable of converting the tetramines (3,4,3 and 3,4,4) to the triamines spermidine (3,4), and homospermidine (4,4), respectively. Despite active acetylation of amino acids and the lysines of histones, acetylated polyamines have not been found in yeast. Yeast may contain a polyamine oxidase that acts selectively on the unconjugated amine to remove an aminopropyl moiety.

Enzymes of Polyamine Pathway

ODC

*Spe*2 mutants low in internal spermidine and spermine are derepressed for ODC and have been used in the isolation of the enzyme. Actually growth in the presence of these polyamines (10^{-4} M) reduces enzyme activity to a tenth of the uninhibited level, although these extracts contain the same amount of serologically reactive enzyme protein. ODC was not regulated by the loss of pyridoxal phosphate (PLP) or by incorporation of an inactivating polyamine (Tyagi et al., 1981, 1982). No evidence was found for an inhibitory, possibly regulating protein such as antizyme.

In any case ODC was purified to homogeneity from such an overproducing mutant; the enzyme, stored in dithiothreitol and EDTA, was stable in the cold and had a higher specific activity than that from several other purified eucaryotic ODCs. Ornithine decarboxylation required PLP, which had been lost partially during purification; activity was restored by addition of the coenzyme. However, early studies on the isolation and characterization of ODC produced anomalous results, which have evoked some less direct approaches.

Some of the early puzzles that appeared in the characterization of yeast *spe*1 mutants and yeast ODC led to a renewed analysis of the *spe*1 genes and their isolation and expression in *E. coli*. Expression of constructed plasmids in the bacteria led to production of a yeastlike ODC, that is, possessing similar Michaelis constant (K_m) values for ornithine, insensitivity to guanine nucleotides, and full sensitivity to DFMO (Fonzi & Sypherd, 1985). The plasmid-encoded yeast gene gave rise to several mRNAs, a major one of which is similar in size to the transcript from animal cells, in which ODC had a molecular weight approaching 53,000. The open reading frame of the gene was localized in a DNA fragment and its sequence, containing 1398 nucleotides, was found to encode 466 amino acids with a calculated molecular weight of 52,369 (Fonzi & Sypherd, 1987).

The gene is present in yeast in a single copy, unlike the multicopy gene family of mammalian cells. A reexamination of the regulation of its expression revealed phenomena similar to those of earlier studies. The presence of exogenous spermidine and spermine in the cultures did not appear to affect stages of transcription, amounts of ODC mRNA, the translation of the mRNA,

and the polysome pattern of the cells. Nor did the presence of the polyamines affect the degradation rates of yeast ODC (Fonzi, 1989a). The half-life of ODC was similar to that in mammalian cells.

The puzzle of spermidine control of ODC activity also led to a reexamination of the structure of the enzyme. Molecular sieving (Fig. 10.5) of extracts containing the noninactivating detergent Brij revealed a molecular weight of about 110,000 for the native enzyme, indicating its existence as a dimer of two 53-kDa subunits (Fonzi, 1989b). It is supposed that the dimer is active whereas the ODC monomeric subunit is inactive; and it has been hypothesized that polyamines may repress dimerization, with a loss of monomers by degradation.

AdoMet synthetase

Much work had been done on the yeast synthetase by the mid 1970s: partial purification, stoichiometry of the reaction, and the presence of a tripolyphosphatase. It was then observed that yeast contained two forms of the synthetase; each was approximately 110,000 kDa and comprised two subunits of 55,000 and 60,000, respectively. The isozymes were distinguishable chromatographically, electrophoretically, and in their content of tripolyphosphatase (Chiang & Cantoni, 1977). The enzymes were found to be coded by two independent genes (*SAM*1 and *SAM*2), and in density gradient centrifugation each enzyme was found to exist as a dimer (Cherest & Surdin–Kerjan, 1981). The gene for *SAM*1 was isolated and sequenced; it proved to have a high homology with the structural *met*K gene of *E. coli* (Thomas & Surdin–Kerjan, 1987). *SAM*1 and *SAM*2 are highly homologous and it is believed they are duplicated genes, which are however regulated differently by the presence of exogenous methionine (Thomas et al., 1988). The yeast enzymes have also been shown to be similar in sequence (68% homology) to a rat liver AdoMet synthetase (Horikawa et al., 1989). The development of sequencing techniques for proteins and DNA provided increasing evidence of evolutionary relationships, such as that indicated by the sequences of AdoMet synthetase in procaryotic and eucaryotic organisms.

AdoMetDC

The discovery that this enzyme, isolated from mammalian tissues, was activated by putrescine and not by Mg^{2+}, in contrast to the converse with the *E. coli* enzyme, introduced the concept of comparative biochemistry and phylogenetic relationships to the study of these enzymes of different origins. This avenue was pursued by Williams–Ashman and colleagues and revealed the existence of even a third class of the activity in some plants that was affected by neither Mg^{2+} nor putrescine (Coppoc et al., 1971; Table 9.2). In yeast the activity is quite dependent on the presence of the diamine and this requirement has been used to estimate putrescine with

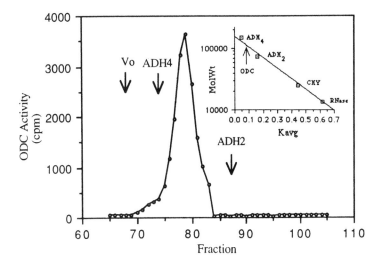

FIG. 10.5 Gel filtration chromatography of native yeast ODC; redrawn from Fonzi (1989b). The elution of the ODC activity was compared with the elution of various molecular weight standards: native yeast alcohol dehydrogenase (ADH), 147,000; alcohol dehydrogenase dimer (ADH₂), 73,000; chymotrypsinogen (CHY), 25,000; RNase A, 13,700.

the aid of the purified yeast enzyme. In early experiments using [2-^{14}C-methionine]AdoMet, it was also possible to demonstrate the synthesis of labeled spermidine and spermine. The experiments thus indicated the existence of a spermidine synthase and an additional activity for the synthesis of spermine.

The yeast enzyme was purified with the aid of affinity chromatography on an MGBG column; AdoMetDC was separated completely from spermidine synthase (Pösö et al., 1975). Putrescine (2.5 mM) activated the decarboxylase best, but cadaverine and 1,3-diaminopropane were about half as active at that concentration and spermidine was inactive. The product of the reaction, dAdoMet, was a powerful inhibitor of the reaction, as was MGBG. Carbonyl reagents, including sodium borohydride, were also strong inhibitors, but PLP did not reverse these inhibitions. Although AdoMet did not protect against NaBH₄, the activating diamines did so to a considerable extent.

The yeast enzyme was purified some 12,000-fold to homogeneity and was found to have a molecular weight of 84,000 by ultracentrifugal sedimentation equilibrium and 88,000 by molecular sieving (Cohn et al., 1977). The spectrum indicated the absence of PLP, and reduction by sodium borotritiide and acid hydrolyses permitted the isolation of the presumed pyruvoyl cofactor as labeled lactate in an estimated 80% recovery. This approached an apparent stoichiometry of 0.9 mol of bound pyruvate per mole of undenatured enzyme subunit, and 1.1 mol of pyruvate per mole of denatured enzyme subunit. The yeast enzyme was the first eucaryotic protein recognized to possess a bound pyruvoyl cofactor.

The later genetic studies of the *E. coli* and mammalian enzymes, as well as the biochemical studies of the origin of the pyruvoyl group in the cleavage of a seryl-containing dipeptide, indicated the probable existence of a proenzyme derived from the primary transcript of the *spe*2 gene. The cleavage of the proenzyme should lead to the existence of separable subunits in the isolated enzyme. The *spe*2 gene was isolated from a bank of *S. cerevisiae* genomic DNA cloned into an appropriate vector. After transformation of an *spe*2 mutant and enrichment for *spe*2⁺ cells by growth in spermidine-deficient media, a plasmid containing the desired DNA insert was isolated from an apparently spermidine-independent large colony. The DNA fragment was used to transform a strain of *E. coli* that lacked AdoMetDC and that was prepared for transcription with a gene for a viral RNA polymerase (Kashiwagi et al., 1990).

The nucleotide sequence for the yeast DNA insert indicated the existence of a 396 amino acid protein with a calculated molecular weight of 46,214. This proenzyme was detected after a brief incorporation (2 min) of [^{35}S]methionine followed by synthesis with unlabeled methionine, that is, in a pulse-chase experiment. Isotope was incorporated into a protein of about 46,000 molecular weight and within 2 min was cleaved to a large carboxy-terminal subunit of 36,000 molecular weight and a small amino-terminal subunit of 10,000 molecular weight, both of which were found in the enzyme isolated from *E. coli*. The cleavage site giving rise to pyruvate was shown to be the linkage between a glutamic acid (position 87) and serine at position 88. Thus the pyruvoyl moiety was at the end of the large subunit opposite to the carboxy-terminal amino acid. The methionine originally at the amino-terminal end of the proenzyme was not found on the small subunit; it had been removed during biosynthesis.

A reductive amination of the pyruvoyl moiety produced alanine. However, even before amination, a high percentage of the large subunits contained N-terminal alanine, indicating a significant inactivation of the enzyme in this isolation system. This result had been seen earlier in a study of substrate inactivation of the *E. coli* AdoMetDC (Anton & Kutny, 1987).

The amino acid sequences of the yeast and *E. coli* enzyme revealed essentially no identity. The yeast and mammalian enzymes do show a small degree of identity. However, the cleavage sites of the proenzymes derived from yeast, human fibroblast, and rat prostate all contain the *glu-ser* sequence and have an identity of seven of eight amino acids. By contrast the cleavage site of the *E. coli* enzyme was *lys-ser* and only one amino acid of seven in addition to *ser* was identical to that of the eucaryotic proenzymes (Kashiwagi et al., 1990).

Yeast AdoMetDC, like the rat enzyme, is strongly inhibited by MGBG. However, the considerable inhibitory activity of MGBG with the Mg^{2+}-, putrescine-independent plant enzyme (Table 9.2) compels a rethinking of the hypothesis that the putrescine-activation site is also that of MGBG inhibition. Newer analogues proved to be even more inhibitory, and the diethylglyoxal-bis-(guanylhydrazone) was the most powerful inhibitor of this group acting on the yeast enzyme, with an inhibition constant (K_i) of approximately 9 nM (Elo et al., 1988).

Inhibitors of ODC have been useful in controlling the growth of several pathogenic fungi (Slocum & Galston, 1987). Although ODC is the sole pathway for the production of putrescine in *S. cerevisiae* and in *N. crassa*, the effectiveness of ODC inhibitors in controlling some phytopathogenic fungi led to studies of the systems present in other fungi. A basidiomycete had been found to contain arginine decarboxylase. The presence of arginase in fungi, which can convert difluoromethylarginine to DFMO, has the potentiality of confusing the identification of the inhibited enzyme. Nevertheless, several phytopathogenic fungi have an apparently biosynthetic ADC (Boyle et al., 1994; Khan & Minocha, 1989). More extensive screens of the various inhibitors of ODC and ADC and of subsequent enzymes of the polyamine pathway are being explored in studies of the inhibition of growth of agriculturally significant phytopathogens (Foster & Walters, 1990; Smith et al., 1990; Walters & Robins, 1994).

García et al. (1991) attempted to detect an effect of MGBG in controlling the mycelial growth and germination of the plant pathogen *Ceratocystis ulmi*. In some systems some inhibition by MGBG was detected but was not reversed adequately by polyamines, suggesting a toxicity resulting from other effects on the fungi. MGBG, tested as a potential antitumor agent, is quite inhibitory to animal cell mitochondria (Pathak et al., 1977). MGBG selectively blocked aerobic growth in *S. cerevisiae*, inhibited the mitochondrial syntheses of cytochromes, and damaged mitochondrial structure (Diala et al., 1980).

It is possibly relevant that spermine (0.01 M) inhibits the growth and respiration of *S. cerevisiae* without af-

fecting rates of fermentation. Spermine-resistant strains form respiration-deficient small colonies, likened to ethidium-induced "petites" with aberrant mitochondria (Sakurada & Matsumara, 1964). Thus toxicity in these instances, produced by MGBG, ethidium, and spermine, may relate to a common mitochondrial site. Genes of *S. cerevisiae* relating to spermine resistance have been isolated and cloned and are being studied (Shiomi et al., 1990).

Propylamine transferases

In yeast the putrescine aminopropyltransferase (spermidine synthase) and spermidine aminopropyltransferase (spermine synthase) are coded by separate genes, *spe*3 and *spe*4, respectively. *Spe*3 mutants lacking spermidine synthase have been isolated and are unable to make spermidine and spermine (Tabor & Tabor, 1985). The cells contain high amounts of AdoMet and dAdoMet, but are low in AdoMetDC. Methods for the mass screening of mutants in polyamine biosynthesis have been described (Tabor et al., 1983). Mutants deficient in ODC and depleted of polyamine were plated; large colonies slowly emerged among the more slowly growing mutants. These were found to contain spermidine and to lack both spermine and spermine synthase (Cohn et al., 1980). The mutants fell into two classes, in one of which *spe*4 was thought to be the structural gene for spermidine aminopropyltransferase, and in the other *spe*40 was thought to be a regulatory gene for this enzyme (Tabor & Tabor, 1985). As indicated earlier a mutant lacking spermine synthase was unable to add aminopropyl moieties to spermidine and homospermidine. Although a single bacterial enzyme can effect syntheses of both triamines and tetramines in *E. coli, Sulfolobus* etc., it will be important to know if eucaryotic microbes have developed a new enzymatic capability for spermine synthesis. Two different enzymes are known in animal cells (Chap. 14).

New Polyamine Functions in Yeast Systems

Hypusine and translation

The unusual amino acid, hypusine, [N^ϵ-(4-amino-2-hydroxybutyl)lysine] was discovered over two decades ago (Shiba et al., 1971). It was found initially in mammalian cells in a single protein serving as the eucaryotic translation initiation factor eIF-5A (formerly 4D) (Cooper et al., 1983). The aminobutyl moiety of spermidine is transferred to the amino group of protein-bound lysine to form protein-bound deoxyhypusine, which is then oxidized in situ to hypusine. The sequence of the reactions is given below; details will be presented in Chapter 22, which discusses the roles of the polyamines in translation.

Spermidine + eIF-5A-lysine →

eIF-5A-deoxyhypusine → eIF-5A-hypusine.

The apparent absence of hypusine in a few members of the procaryotic world eventually led to a broader test of the uniqueness of hypusine in procaryotes and microbial eucaryotes. Deoxyhypusine and hypusine have now been found in many Archaebacteria and in *S. cerevisiae* (Bartig et al., 1990). Among the Archaebacteria the strict anaerobes contain deoxyhypusine and do not hydroxylate it to hypusine. Other aerobically grown organisms contain both products or hypusine alone in their proteins.

The growth rates of *S. cerevisiae* may be altered very considerably in various media; also, transitions in rates of protein synthesis of the organism are defined by development (Rousseau & Halvorson, 1973). The yeast then is a suitable organism to analyze the quantitative relation of its hypusine-containing protein to the spectrum of these rates of protein synthesis. When polyamines were depleted by growth in the presence of difluoromethylornithine and then transferred and grown in the presence of [^{3}H]-spermidine, a single labeled and abundant protein of 18,000 kDa, similar to eIF-5A of mammalian cells, was obtained (Gordon et al., 1987a,b). The amount of the protein varied little under a wide range of growth conditions and rates of protein synthesis. Within eIF-5A, hypusine did not revert to lysine and the half-life of the protein exceeded seven generations in the life of the yeast. Indeed the protein appears to be one of the most stable known.

Studies on the genetics and regulation of anaerobic growth in *S. cerevisiae* suggest functions of the protein other than a role in the initiation of protein synthesis. A gene *ANB1*, essential to anaerobic growth, regulated negatively by oxygen and heme, was found to encode eIF-5A (Mehta et al., 1990). The *ANB1* gene product shows extensive sequence homology to mammalian eIF-5A. Two forms of eIF-5A were detected in spermidine-fed yeast, and these appeared to relate to the two mRNA transcripts found in the cells. One of the latter is expressed anaerobically and the other aerobically, indicating differential and opposite modes of regulation, similar to findings with the two forms of regulation of cytochrome oxidase subunits. Yeast lacking the ANB1 locus grows poorly under anaerobic conditions and has a reduced rate of protein synthesis.

Processing of tRNA precursors

Yeast contains tRNA precursors that possess intervening sequences. These are processed by an endonucleolytic cleavage, followed by ligation. The events are separable and a specific membrane-bound endonuclease has been isolated after solubilization. The activity and accuracy of cleavage of the precursors is markedly increased by spermidine or spermine (Peebles et al., 1983). Mg^{2+} is far less stimulatory and reduces the accuracy of cleavage in the presence of spermidine.

Splicing of mRNA

Constructs of pre-mRNA containing a single intron are spliced in a soluble extract of *S. cerevisiae*. The intermediary intron is converted initially to a circular single-stranded branch, a lariat. The reaction is demonstrated by the linkage of two exons in a specific splice junction whose formation can be assayed (Lin et al., 1985). Splicing requires ATP, monovalent (Na^+) and divalent (Mg^{2+}) cations, and the reaction is optimized by various additions including 1 mM spermidine. The specific role of the polyamine has not been analyzed.

Studies with *N. crassa*

In 1941 Beadle and Tatum described the introduction of this fungus, a salmon-colored pest contaminating bakeries, in the genetic study of inherited metabolic blocks. The life cycle of *Neurospora*, including the discovery of a sexual phase, and the nutritional requirements (inorganic salts, inorganic nitrogen source, carbon source such as glucose, and the vitamin biotin) have been described (Beadle, 1945). The life cycle is given in Figure 10.6. Segmented vegetative hyphae contain many haploid nuclei per segment; these divide mitotically in an asexual phase. Separate mating types unite to form a diploid zygote that undergoes meiotic and mitotic divisions in an ascus to form four pairs of aligned ascospores. Each nucleus is eventually coated in a heavy wall to form a spore. After activation by heat at 60°C for 30 min, each spore can begin a new cycle of mycelial growth. Within the mycelium one may have vegetative fusions of hyphae of the same sex containing two genetically different types of nuclei (i.e., to form a *heterocaryon*) in which mutant haploid nuclei may complement each other to permit normal growth. In initial tests of mutagenized strains, single ascospore-derived strains are grown in minimal and "complete" media to determine the possible blockage of some essential reaction. The germination of isolated conidia is dependent on the presence of a group of iron-transport siderochromes, the major peptide of which contains 3 mol ε-acetylornithine hydroxamate (Horowitz et al., 1976).

Neurospora mutant strains requiring exogenous arginine for growth were found to fall into seven classes (Srb & Horowitz, 1944). As presented in Figure 10.3, *N. crassa* and *S. cerevisiae* possess an ornithine cycle in which biosynthetic ornithine is converted via citrulline and argininosuccinate to arginine. The latter is converted to ornithine by a cytosolic arginase, and the avoidance of a wasteful cycle is effected in *N. crassa* by the separation of key enzymes of a cycle (i.e., carbamylphosphate synthetase within the mitochondrion and arginase in the cytosol). In *S. cerevisiae* both enzymes are cytosolic (Davis, 1986).

Similarly, studies of the biosynthesis of methionine in yeast and *Neurospora* soon pointed to differences in the pathways. The discovery of AdoMet and its high production yield in yeast (Chap. 9) raised the problem of its cellular distribution. Ultraviolet microscopy demonstrated that the vacuoles of yeasts could contain a very high concentration of AdoMet (Schlenk, 1965). This surprising observation led to many studies of the properties and function of the vacuole in fungi. The fun-

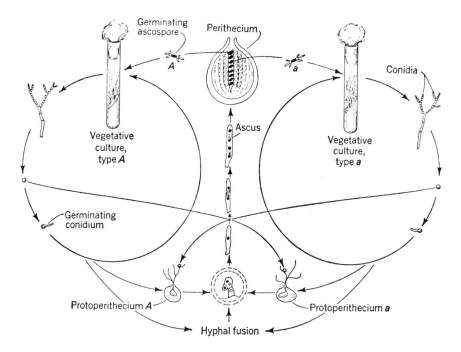

FIG. 10.6 Life cycle of *Neurospora crassa*; redrawn from Hardin (1949).

gal vacuole has been the subject of a recent review, which focuses on its roles in degradative processes and some transport functions (Klionsky et al., 1990). The review also discusses the unsolved problems of vacuolar biogenesis and partitioning in yeast buds, a problem posed in discussing the appearance of preexisting vacuolar AdoMet in the bud vacuoles of multiplying yeast (Suchla & Schlenk, 1960). In addition to a consideration of the mechanisms and functions of vacuole acidification, the review notes the presence of polyphosphates as the only macromolecular anions in the vacuole. These were thought to be involved in this function of the vacuole in holding AdoMet, arginine, and the polyamines.

Exogenous AdoMet can satisfy an adenine requirement, but added methionine inhibits this utilization. These results demonstrated the penetrability of AdoMet in yeast; its degradation, presumably via methyl thioadenosine to utilizable adenine or adenosine; and its removal, presumably by sequestration (Knudsen et al., 1969; Yall et al., 1967). Actually exogenous AdoMet by itself, in some yeasts, did not markedly increase the cellular content of the compound, which was best enriched by exogenous methionine.

A stepwise opening and extraction of yeast cells (*Candida utilis*) indicated the existence of 65–90% of the cell AdoMet in the vacuole (Schlenk et al., 1970). Methods have also been developed for the isolation of vacuoles after osmotic lysis of spheroplasts (Indge, 1968; Nakamura, 1973). Particular strains of *S. cerevisiae* transported exogenous AdoMet to the vacuole; the uptake into the cells was found to be glucose dependent

and quite specific (Nakamura & Schlenk, 1974; Petrotta–Simpson et al., 1975). Isolated vacuoles were also capable of accumulating AdoMet by a glucose-independent transport system, different from that operating in an initial uptake by intact spheroplasts (Schwenke & de Robichon–Szulmajster, 1976). *S*-Adenosylethionine and *S*-adenosylhomocysteine were relatively weak competitive inhibitors in both systems. Preloaded vacuoles were quite stable and did not lose or exchange AdoMet.

It was eventually found that arginine was transportable into isolated vacuoles of *S. cerevisiae* (Boller et al., 1975). During such transport, the amino acid exchanged specifically with the arginine in the organelle and the rate of arginine transport by the vacuole approached that of an entire spheroplast. A similar sequestration of the bulk of arginine of *N. crassa* occurs in a metabolically separated compartment (Subramanian et al., 1973). Thus, the study of methionine metabolism and the production of AdoMet, amino acid transport, and the biosynthesis of ornithine and arginine focused on the vacuole.

In *N. crassa* "exogenous" arginine (added to the medium) was hydrolyzed via arginase at a more rapid rate than internal newly synthesized arginine (Mora et al., 1972). This was interpreted initially as a result of regulation by intermediates of the biosynthetic path but was subsequently reinterpreted as a consequence of differential compartmentation of the parts of the pathway because the bulk of arginine and ornithine was confined to the vacuole (Cybis & Davis, 1975; Subramanian et al., 1973). In these experiments small amounts of labeled metabolites (e.g., [^{14}C]-ornithine) were rapidly incorporated into the cells, and the distribution of label into [^{14}C]-putrescine and other enzymatic products was determined. In the case of the diamine, a very rapid early decarboxylation of [^{14}C]-ornithine to [^{14}C]-putrescine indicated a rapid equilibration of the [^{14}C]-amino acid in the cytosol with about 0.3% of the total ornithine in the cell; that is, new putrescine was made from a pool about 30 times more radioactive than total extractable ornithine (Karlin et al., 1976).

Studies on the uptake and conversion of [^{14}C]-ornithine to [^{14}C]-spermidine revealed that new molecules of putrescine were preferentially converted to new molecules of [^{14}C]-spermidine. The dilution of the newly made spermidine indicated that at least 70% of the cell spermidine did not participate in the metabolic pool. A vacuolar fraction contained about 28% of the cell spermidine, and this suggested that the remainder of the metabolically inactive spermidine is associated fairly tightly with anionic sites in the cell (Paulus et al., 1983). Thus, only a small fraction of cellular polyamine in *N. crassa* and probably many other types of cells can participate in the control and synthesis of the polyamines themselves.

Mutant strains of *N. crassa* that accumulate putrescine have been shown to attempt to store the excess putrescine in the vacuole. When the diamine is indeed excessive putrescine becomes toxic, swelling the vacuoles and flooding the cytosol, and displaces K$^+$ (Davis & Ristow, 1991). Putrescine and other polyamines are not converted to other toxic products by this fungus.

Vacuolar Cations and Anions in Fungi

Studies on vacuoles of yeast revealed a close correlation between the storage of arginine and polyphosphate. Indeed the binding of arginine was related to the concentration of polyphosphate in the vacuoles. Nonbasic amino acids were not retained by the anion. Nevertheless, the cells were able to store and mobilize the polyphosphate and arginine independently (Dürr et al., 1979). Similar results were obtained in a study of these storage substances in *N. crassa*. However, polyphosphate in the cells was not affected by reduction of arginine content by nitrogen starvation or by increasing the cellular arginine. Furthermore, polyphosphate depletion did not affect arginine content or exchange (Cramer et al., 1980).

Vacuoles depleted of polyphosphate were isolated from *N. crassa*, and these were shown to retain basic amino acids. When vacuoles were isolated from normal cells the enclosed polyphosphate (99% of the total phosphate) was found to be half neutralized by spermidine, Mg^{2+}, and Ca^{2+}. The spermidine and Mg^{2+} contents of the vacuoles were estimated to represent 25 and 10%, respectively, of the total cellular contents of these ions. The cationic contents of the vacuoles are given in Table 10.2 (Cramer & Davis, 1984) and contain small amounts of Na$^+$ and K$^+$. Cells starved for phosphate and surfeited with arginine had swollen vacuoles under conditions of osmotic imbalance.

TABLE 10.2 Cations of vacuoles isolated from *Neurospora crassa* grown on minimal medium.

Ions	Concentration in vacuolar extract (μmol/mL)	Charge equivalents (μeq/mL)
Monovalent cations		
Ornithine	18.1	18.1
Arginine	13.6	13.6
Histidine	1.9	1.1[a]
Lysine	1.8	1.8
Na$^+$	1.5	1.5
K$^+$	0.7	0.7
Multivalent cations		
Spermidine	2.5	7.5
Mg^{2+}	2.7	5.4
Ca^{2+}	0.3	0.6
Total monovalents	37.6	36.8
Total multivalents	5.5	13.5
Multivalents + monovalents	43.1	50.3
Total cations (by Dowex titration)		57.2
Total polyphosphate-P	36.7	~36.7

Adapted from Cramer and Davis (1984).

[a]Calculated at intravacuolar pH in vivo, 6.1.

Problem of Concentration of Free Polyamines and Metabolites

Studies of the metabolism of arginine, ornithine, and the polyamines revealed cation sequestration within compartments, for example, vacuoles and mitochondria. These organelles alone could not account for all of the separation of spermidine from the metabolically active pools, and it was supposed that anionic macromolecules such as the nucleic acids presented many additional tight-binding sites. This conclusion concurred with the early finding of spermidine binding tightly to bacterial ribosomes and tRNA. The theoretical problem was posed of defining the concentration of free polyamine within a cell containing much bound polyamine.

Nuclear magnetic resonance (NMR) is a nondestructive methodology that takes advantage of the properties of atoms such as carbon-13, phosphorus-31, and nitrogen-15. Because ^{31}P is the naturally abundant form of phosphorus and compounds rich in ^{31}P abound in cells, the cellular and tissue activities of many different phosphate compounds have been studied (Shulman et al., 1979). These included studies on *E. coli* and *S. cerevisiae* (Ogino et al., 1983). In biological systems of low ^{13}C abundance, the study of particular molecules could be improved by the synthesis of [^{13}C]-containing metabolites (Scott & Baxter, 1981). The sporulation of *S. cerevisiae* incubated in [2-^{13}C]-acetate was followed by NMR spectroscopy and revealed rapid incorporation into glutamate, an early precursor of ornithine and arginine (Dickinson et al., 1983). This approach has also been applied to the study of the behavior of $[1,4\text{-}^{13}C]$-putrescine within *E. coli* (Chap. 7).

Nitrogen metabolism was explored in *N. crassa* with the aid of $[^{15}N]$-NMR (Legerton et al., 1983, 1984). $[^{15}N]$-enriched NH_4Cl is incorporated into all nitrogen compounds. The stability and turnover of the $[^{15}N]$-containing components (e.g., the ureido-*N* of citrulline or the guanidine-*N* of arginine) were determined when the ^{15}N-labeled metabolites were chased by $^{14}NH_4Cl$. Whereas the amide N of glutamine and the ureide N of citrulline turn over rapidly, the $[^{15}N]$ of the guanidine groups of arginine, the $[^{15}N_\epsilon]$ of lysine, and $[^{15}N_\delta]$ of ornithine persist in the sequestrated basic amino acids in the vacuole. Further studies indicated that the ^{15}N spin-lattice relaxation time of the guanidine-*N* of intracellular arginine was markedly decreased, suggesting either an association with a macromolecular component like polyphosphate, or the existence of a high viscosity in the organelle (Kanamori et al., 1982).

In the $[^{13}C]$-NMR study by Frydman and colleagues (1984) who added $[1,4\text{-}^{13}C]$-putrescine (90% enrichment) to *E. coli*, the sharp methylene line of the free diamine was considerably broadened, implying a rapid binding of putrescine. This evidence of bound putrescine was maintained for several hours under aerobic and anaerobic conditions and did not occur with $[^{13}C]$-enriched methionine and glycine. The cell-bound $[^{13}C]$-putrescine was readily extracted by acid or could be exchanged completely by addition of $[^{12}C]$-putrescine to the labeled cell suspension. Binding of putrescine (line-width broadening) was obtained with ribosomes and membrane preparations but not with DNA, tRNA, or soluble protein. No evidence was obtained in these experiments for a conversion of $[^{13}C]$-putrescine to $[^{13}C]$-spermidine, and it was asked if the amount of $[^{13}C]$-spermidine formed was below the limits of detection. These authors also synthesized $[5,8\text{-}^{13}C]$-spermidine and studied the binding of the molecule to *E. coli* and to *E. coli* tRNA (Frydman et al., 1990). The binding of spermidine within the cell (a broadening of signal widths) was observed, as in the study of putrescine binding. The details of the binding to tRNA will be discussed in Chapter 23.

NMR spectroscopy may provide an approach to the problem of the free concentration of the polyamines within cells, including *N. crassa*. However, this concentration is quite low and provides a challenge in the choice of cell, detectable signal-producing isotope, and synthesized metabolite.

Putrescine-Deficient *Neurospora* and Polyamine Pathway

Cytosolic arginase provides a route to ornithine and thence to essential proline via glutamic-γ-semialdehyde (Morgan, 1970). Arginase-deficient mutants of *Neurospora* (*aga*) were selected for an inability to convert arginine to ornithine in strains in which ornithine was essential for proline synthesis. Media high in arginine inhibited growth and this inhibition was relieved by or-

nithine or proline. The inhibition of growth in similar mutant strains was also reversed by putrescine or spermidine (Davis et al., 1970), demonstrating that ornithine is the major source of the diamine, whose major role in *Neurospora* is the formation of the essential growth factor spermidine.

Other putrescine-requiring strains were isolated in the 1970s; one of these was ODC deficient and grew on putrescine, spermidine, or spermine, indicating the decarboxylation of ornithine as the sole route for the generation of the diamine, and the probable role of spermidine as the essential product of putrescine metabolism (McDougall et al., 1977). A role for spermidine in partially substituting for Mg^{2+} in growth and ribosome composition had been detected some years earlier (Viotti et al., 1971), as had been shown in bacteria.

Many workers found small amounts of spermine in filamentous fungi. The ability of the deficient mutants to grow on spermine suggests the existence of degradative oxidases, which might also act on spermidine and produce the small amounts of 1,3-diaminopropane detected in *Neurospora* cultures (Tabor & Tabor, 1985). In any case, the major biosynthetic elements of the polyamine pathway appear uncomplicated accumulating a low level of putrescine (1–2 nmol/mg dry weight of mycelium) during growth and a markedly higher level of spermidine (14–18 nmol/mg) on minimal medium.

However, only small fractions of these amines (10–20%) exist in readily available metabolic pools, unsequestered in the vacuole or bound to macromolecular anions. Arginine supplementation has little effect on polyamine content, whereas accumulation of ornithine leads to a significant increase (50–70%) of mycelial polyamine (Paulus & Davis, 1981). When ornithine is provided to the arginase-deficient putrescine-deficient organism, there is a rapid increase in putrescine; a later decrease in the diamine relates to a marked and prolonged increase in spermidine. This was interpreted to suggest that spermidine concentrations regulated putrescine biosynthesis.

A reexamination of these effects demonstrated that a block of ornithine synthesis in the *arg* mutant sharply increased ODC activity. Spermidine negatively regulated the formation of active ODC. A block of spermidine formation by MGBG or cyclohexylamine evoked a four- to fivefold increase of ODC (Barnett et al., 1988; Davis et al., 1985). Spermidine is therefore described as a specific signal for repression of ODC synthesis.

On the other hand ODC was more rapidly inactivated by increases of putrescine within the mycelium. This was not due to antizyme formation but rather to a 10-fold decrease in half-life that was restored to normal by putrescine or spermidine. The reduction of ODC in ornithine-supplemented cultures was attributed to the rapid production of putrescine (Barnett et al., 1988; Davis et al., 1985).

The puzzles raised by this series of observations led to a renewed effort to obtain mutants in which ODC might be structurally and kinetically altered. A single locus, *spe*1, was found to determine ODC structure and various alleles controlled slightly altered enzymes (Eversole et al., 1985). *Neurospora* mycelia bearing nonsense mutations of the gene grew normally in the presence of spermidine, indicating that ODC had no role other than that of polyamine synthesis (Davis et al., 1987).

The purification of ODC was accomplished from *Neurospora* greatly increased in enzyme activity. ODC proved to be a dimer of 53,000 molecular weight subunits. Inactivation of the enzyme in response to polyamines over a 100-fold range related to the loss of serologically active protein. This is quite different from the result obtained with *S. cerevisiae*, in which inactivation occurred without loss of the ODC protein. However, it was similar to the manner in which the inactivation of ODC is manifested in many animal cells.

Pitkin and Davis (1990) isolated three groups of new mutations in the polyamine pathway in *N. crassa*; the loci and functions have been classified as *spe*1, *spe*2, and *spe*3 as in the yeast system. The three loci are closely linked in the order *spe*2, *spe*1, and *spe*3. *Spe*1 mutants lacking ODC are deficient in all the polyamines, and in unsupplemented media they grow initially with a far longer doubling time than do the other mutants. They also have a smaller requirement for spermidine than do mutants in *spe*2 and *spe*3, which accumulate putrescine. Although this has been taken to suggest that high levels of putrescine can fulfill some functions of spermidine in the latter, a 98% depletion of spermidine does cause all growth to stop.

The ODC gene of this organism has been isolated and sequenced (Williams et al., 1992). Two major control sites of the ODC gene, an activation region and a DNA leader sequence, have been identified (Pitkin et al., 1994). Significant amino acid identity was noted among the ODCs of *Neurospora, Saccharomyces*, and the mouse. The rapid turnover of ODC in *N. crassa* in the presence of polyamines, spermidine particularly, is suggested to relate to the presence of sequences in ODC, designated as PEST sequences, which facilitate rapid proteolysis (see Chap. 13). By contrast to many other eucaryotes, the rate of ODC synthesis appeared to correlate with the abundance of ODC mRNA, that is, to be determined predominantly by the rate of transcription of the gene, with spermidine serving as a main effector of repression.

Although polyamine formation in *Neurospora* appears to be a metabolic "dead-end," spermidine is a precursor to the formation of hypusine on a single molecular species of protein (Yang et al., 1990). In vivo a single protein of 21,000 Da, containing hypusine alone, is produced when [^{3}H]-spermidine is fed to an arginase- and polyamine-deficient strain. An extract of the organism incubated with [^{3}H]-ornithine produced a labeled 21,000 molecular weight protein containing deoxyhypusine and hypusine. The 21,000 molecular weight protein differs from the hypusine-containing protein of yeast and mammalian cells.

Regulation of ODC and Polyamine Pathway

In many studies of the biosynthesis of ornithine, arginine, and methionine in bacteria, major fine controls determining enzymatic activities at various steps were found to be feedback product inhibition and repression. Very early in the study of ODC in mammalian cells, the rapid biosynthesis and disappearance of the enzyme were seen to be methods of regulation of ODC activity, whose mechanisms were different from these other forms of regulation.

In the case of ODC in the fungi, the regulation of the activity of this enzyme does not fall within the general patterns of feedback inhibitory effects on an allosteric system. The effects of spermidine on ODC in *Neurospora* triggers enzyme degradation, as in the mammal, and in yeast appears to relate to the control of inactive monomeric subunits to form an active dimer. Furthermore, a eucaryotic cell that compartments the largest part of its spermidine on macromolecular anions and within organelles possesses a tiny metabolic pool. Small variations in the size of such a pool, effected by many possible changes in the cell and its environment, would be excessively reactive on an allosterically regulated ODC. It is suggested then that the different modes of regulation of the ODC of fungi and of mammalian cells reflects this relation of spermidine to structural and cytological units with which the triamine is associated (Davis et al., 1992).

Systems of excretion of spermidine and spermine as acetyl derivatives were also developed in mammalian cells. Although some yeasts grown on polyamines as their sole nitrogen source develop an *N*-acetyltransferase active on these amines (Haywood & Large, 1985), *Neurospora* and yeast do not acetylate the polyamines. Nevertheless, mechanisms of excretion of saturating spermidine and putrescine have also evolved in these fungi (Davis & Ristow, 1989).

Studies with *Physarum polycephalum*

A brief summary of polyamines in *Aspergillus* has appeared (Tabor & Tabor, 1985), following a more complete discussion of this and other earlier fungal studies (Stevens & Winther, 1979). A more recent review stressed the apparent roles of the polyamines in regulating or determining steps of differentiation found in the *Mucorales*: the appearance of germ tubes or buds, yeast to mycelial transition, sporulation, etc. (Ruiz–Herrera, 1994).

Physarum polycephalum, a true slime mold (Sauer, 1982), is classified as a protist at some phylogenetic distance from the fungi. It is also readily distinguishable from another protist *Dictyostelium discoideum* (Bonner, 1967), which has also proved to possess some interesting features of polyamine metabolism. Both organisms may live as solitary amoebae, move with their pseudopodia, feed via pinocytosis and phagocytosis, and divide by binary fission. However, in its multinucleate phase as a pancakelike moving cell (a plasmodium) or in spherulation in spent media, *Physarum* secretes large amounts of a sulfated polysaccharide, a viscous slime (McCormick et al., 1970); this is not produced by *Dictyostelium*. The fine structures of *Physarum* during changes from a spherule (sclerotium) to plasmodium have been summarized by Anderson (1992) and Burland et al. (1993).

The life cycles of the two organisms are also different. In *Physarum* an amoeba is transformed into a plasmodium (a giant cell) containing many synchronously dividing nuclei. The plasmodium differentiates into a stalked sporangium in which spores are formed. New individual amoebae emerge from the spores. In contrast, roving feeding amoebae of *Dictyostelium* aggregate in response to the signal, cyclic AMP (cAMP), emitted by a central hungry leader. The aggregate produces a cellulose envelope and migrates as a slimeless slug. After the slug stops its migration, the separate cells within elongate to form a stalk and encysted amoebae are moved into the stalk to form a fruiting body. The encysted spores can revive and regenerate the amoebal life cycle. Neither organism can grow on a synthetic medium. *Dictyostelium* feeds by phagocytosis of bacteria, including *E. coli*, and by pinocytosis of soluble components; *Physarum* are grown in shake cultures on a semidefined medium in which hematin has been found to be an essential constituent. Brewer and Rusch (1966) found that the addition of spermine and DNA to nuclei of *Physarum* isolated at different periods of the synchronous mitotic cycle markedly increased DNA synthesis in the organelles. However, the amounts of detectable tetramine in normal spherules or plasmodia were too small to be determined (Mitchell & Rusch, 1973). The growth of plasmodia from resting spherules in a complete medium was accompanied by a rapid increase (four- to fivefold) in putrescine and ODC. The formation of starved spherules paralleled an exaggerated decrease in putrescine with only a 50% drop in protein and spermidine. The latter generally correlated with the protein content and both increased in growth of the plasmodia.

The amounts of putrescine and spermidine in the growing plasmodia were sufficient to neutralize almost 60% of the phosphate present in RNA and DNA. During growth, the ratio of putrescine to spermidine approached 2.5 and suggested a role for the diamine in addition to that of serving as a precursor for spermidine. The initial step in the formation of spermidine was effected by an AdoMetDC that was not stimulated by Mg^{2+} or putrescine. Low concentrations of MGBG also inhibited this putrescine-independent enzyme, as they do a plant enzyme. Both ODC and AdoMetDC were stable cytosolic enzymes and were found to have rapid turnover rates, with half-lives of 14 and 21.5 min, respectively. Short periods of starvation rapidly decreased the amounts of these enzymes. Increases in the amounts of the enzymes were arrested just before and during the synchronous mitoses.

During the isolation of ODC, activities were detected in several different molecular forms; these had different affinities for PLP (Mitchell, 1983). Two alternate states of the apparently interconvertible enzyme were described; the formation of the less catalytically active enzyme could be effected by the addition of spermidine or spermine to cultures (Mitchell et al., 1978). The concentration of active enzyme could be estimated by the binding of isotopically labeled DFMO, which is fixed to the PLP-containing protein only after decarboxylation (Mitchell et al., 1981a). The active A form possessed a much higher affinity for PLP than did the B form and was also shown to have a significantly lower isoelectric point (pH 5.45–5.5) than the B protein (pH 5.65–5.9), despite similar (or identical) molecular weights of the two forms (Mitchell & Wilson, 1983).

In a startling novelty, a protein of 30,000–35,000 molecular weight capable of converting A-ODC to B-ODC was isolated and shown to require spermidine or spermine for its activity; the polyamine was not bound to the newly formed B enzyme (Mitchell et al., 1981b). Norspermidine and 1,3-diaminopropane were almost as active as spermidine in the cell-free system, but putrescine was far less active in the modification of ODC (Mitchell et al., 1982). Nevertheless, exogenously added spermidine or putrescine had far greater effects in reducing ODC activity in the cells than did added spermine or diaminopropane (Mitchell et al., 1978).

The active A-ODC has been isolated as a dimeric protein of 80,000 molecular weight comprising subunits of 43,000 (Barnett & Kazarinoff, 1984). The enzyme was unexpectedly very active, as compared to several other eucaryotic ODCs. It was specific for ornithine, and the activity was not inhibited by a 20-fold excess of lysine or histidine.

The conversion of growing amoebae (*Physarum flavicomum*) to dormant cysts is provoked by nutritional imbalance. Exogenous adenine, leading to an accumulation of AdoMet, blocks the encystment (Henney, 1987). Cysts were very low in polyamine, whereas adenine-treated cells contained almost twice the polyamine content of encysting cells. Putrescine was a moderate inhibitor of encystment but spermidine, spermine, and 1,3-diaminopropane (DAP) were very strong inhibitors with or without adenine. Unexpectedly, cadaverine and cyclohexylamine stimulated encystment. Acetylputrescine was reported in growing cells and acetylspermine, but not free spermine, was found in growing cells and cysts (Zhu & Henney, 1990). It is evident that polyamine metabolism in *Physarum* is very different from that of the fungi.

Dictyostelium discoideum and Polyamines

This organism has been useful in the analysis of controls in the progression of its life cycle. The fate of $[^{35}S]$-methionine was a useful early approach to problems of protein synthesis and inevitably raised the question of the role of AdoMet and its numerous reactions in these events. Indeed methylations of many kinds have been found to occur at the different stages of development, and the regulation of these steps by the production of varying levels of the feedback inhibitor, S-adenosylhomocysteine, has been an active subject of investigation.

Examination of the polyamines of this organism revealed surprisingly high levels of DAP, in addition to putrescine, spermidine, and some spermine (North & Murray, 1980). Surprisingly only two of five members of the Dictyosteliaceae contained DAP, which approached putrescine in concentration (10–20 μmol/mg DNA) in all of the slime molds and was 4–10 times higher than that of spermidine. Cadaverine was not found in members of this group. In contrast to organisms that decarboxylate 2,4-diaminobutyrate to form DAP, *D. discoideum* appears to convert spermidine to diaminopropane, presumably via a polyamine oxidase or spermidine dehydrogenase. Addition of spermidine to the agar + bacterial medium increases the DAP content of the organism and reduces its content of putrescine, suggesting some interchangeability of these diamines. The relatively rapid uptake of [^{14}C]-putrescine, apparently by pinocytosis, is inhibited by the various polyamines, and DAP and cadaverine were effective inhibitors of putrescine uptake at low levels (1 μM) of putrescine (Turner et al., 1979).

Among structural parameters viewed as developmentally significant are the modifications of tRNA, in which the distribution of methylated bases has been seen as an interesting marker. *D. discoideum* has a large number of such methylated bases in its tRNA. It was observed that inhibition of polyamine synthesis by ODC inhibitors, for example, α-methylornithine, markedly reduced the incorporation of methyl groups into tRNA. Spermidine (1–2 mM) and putrescine (10 mM) (but not DAP) maximally and optimally stimulated tRNA methyltransferases in extracts in vitro (Mach et al., 1982). The formation of ribosylthymine and N^2,N^2-dimethylguanosine was particularly affected.

The results on polyamine distribution in *Dictyostelium* and the apparent effects in the control of development eventually led to the study of ODC (North, 1985). The enzyme proved to possess some of the peculiarities of ODC in *Physarum*, high-activity and low-activity forms, generated by osmotic change or treatment of the cells by spermidine. Ammonium salts, as well as salts of sodium and potassium, were inhibitory but the biological significance of these effects is unclear. In any case, the enzyme appeared to increase at an early feeding stage and after aggregation. However, DFMO, which reduced ODC and blocked proliferation, did not affect a later development cycle (Fernández–Pinilla & Pestaña, 1987).

Evidently the pursuit of the polyamines in the slime molds has led to unexpected complexities. In contrast to the fairly large body of work on the fungi, only a few investigators have been drawn into the dissection and characterization of the roles of the polyamines in the developmental sequences of the slime molds.

REFERENCES

Anderson, O. R. (1992) *Journal of Protozoology*, **39**, 213–223.

Anton, D. L. & Kutny, R. (1987) *Biochemistry*, **26**, 6444–6447.

Balasundaram, D., Tabor, C. W., & Tabor, H. (1993) *Proceedings of the National Academy of Sciences of the United States of America*, **90**, 4693–4697.

Balasundaram, D., Xie, Q., Tabor, C. W., & Tabor, H. (1994) *Journal of Bacteriology*, **176**, 6407–6409.

Barnett, G. R. & Kazarinoff, M. N. (1984) *Journal of Biological Chemistry*, **259**, 179–183.

Barnett, G. R., Seyfzadeh, M., & Davis, R. H. (1988) *Journal of Biological Chemistry*, **263**, 10005–10008.

Bartig, D., Schümann, H., & Klink, F. (1990) *Systematic and Applied Microbiology*, **13**, 112–116.

Behr, J. (1994) *Bioconjugate Chemistry*, **5**, 382–389.

Boller, T., Dürr, M., & Wiemken, A. (1975) *European Journal of Biochemistry*, **54**, 81–91.

Boyle, S. M., Szaniszlo, P. J., Nozawa, Y., Jacobson, E. S., & Cole, G. T. (1995) *Journal of Medical and Veterinary Mycology*, **32**(Suppl. 1), 79–89.

Brawley, J. V. & Ferro, A. J. (1979) *Journal of Bacteriology*, **140**, 649–654.

———— (1980) *Journal of Bacteriology*, **142**, 608–614.

Burland, T. G., Solnica–Krezel, L., Bailey, J., Cunningham, D. B., & Dove, W. R. (1993) *Advances in Microbial Physiology*, **35**, 1–69.

Cherest, H. & Surdin–Kerjan, Y. (1981) *Molecular and General Genetics*, **182**, 65–69.

Chiang, P. K. & Cantoni, G. L. (1977) *Journal of Biological Chemistry*, **252**, 4506–4513.

Choih, S., Ferro, A. J., & Shapiro, S. K. (1977) *Journal of Bacteriology*, **131**, 63–68.

———— (1978) *Journal of Bacteriology*, **133**, 424–426.

Cohn, M. S., Tabor, C. W., & Tabor, H. (1977) *Journal of Biological Chemistry*, **252**, 8212–8216.

———— (1978a) *Journal of Bacteriology*, **134**, 208–213.

Cohn, M. S., Tabor, C. W., & Tabor, H. (1980) *Journal of Bacteriology*, **142**, 791–799.

Cohn, M. S., Tabor, C. W., Tabor, H., & Wickner, R. B. (1978b) *Journal of Biological Chemistry*, **253**, 5225–5227.

Connelly, C., McCormick, M. K., Shero, J., & Hieter, P. (1991) *Genomics*, **10**, 10–16.

Cooper, H. L., Park, M. H., Folk, J. E., Safer, B., & Braverman, R. (1983) *Proceedings of the National Academy of Sciences of the United States of America*, **80**, 1854–1857.

Coppoc, G. L., Kallio, P., & Williams–Ashman, H. G. (1971) *International Journal of Biochemistry*, **2**, 673–681.

Cramer, C. L. & Davis, R. H. (1984) *Journal of Biological Chemistry*, **259**, 5152–5157.

Cramer, C. L., Vaughn, L. E., & Davis, R. H. (1980) *Journal of Bacteriology*, **142**, 945–952.

Cybis, J. & Davis, R. H. (1975) *Journal of Bacteriology*, **123**, 196–202.

Davis, R. H. (1986) *Microbiological Reviews*, **50**, 280–313.

Davis, R. H. (1996) *The Mycota III*, pp. 347–356, Brambl & Marzluf, Eds. Berlin: Springer-Verlag.

Davis, R. H., Hynes, L. V., & Eversole–Cire, P. (1987) *Molecular and Cellular Biology*, **7**, 1122–1128.

Davis, R. H., Krasner, G. N., DiGangi, J., & Ristow, J. L. (1985) *Proceedings of the National Academy of Sciences of the United States of America*, **82**, 4105–4109.

Davis, R. H., Lawless, M. B., & Port, L. A. (1970) *Journal of Bacteriology*, **102**, 299–305.

Davis, R. H., Morris, D. R., & Coffino, P. (1992) *Microbiological Reviews*, **56**, 280–290.

Davis, R. H. & Ristow, J. L. (1989) *Archives of Biochemistry and Biophysics*, **271**, 315–322.

—— (1991) *Archives of Biochemistry and Biophysics*, **285**, 306–311.

Diala, E. S., Evans, I. H., & Wilkie, D. (1980) *Journal of General Microbiology*, **119**, 35–40.

Dickinson, J. R., Dawes, I. W., Boyd, A. S. F., & Baxter, R. L. (1983) *Proceedings of the National Academy of Sciences of the United States of America*, **80**, 5847–5851.

Dürr, M., Urech, K., Boller, T., Wiemken, A., Schwencke, J., & Nagy, M. (1979) *Archives of Microbiology*, **121**, 169–175.

Elo, H., Mutikainen, I., Alhonen–Hongisto, L., Laine, R., & Jänne, J. (1988) *Cancer Letters*, **41**, 21–30.

Eversole, P., DiGangi, J., Menees, T., & Davis, R. H. (1985) *Molecular and Cellular Biology*, **5**, 1301–1306.

Fernández–Pinilla, R. & Pestaña, A. (1987) *Revista Espanola de Fisiologia*, **43**, 439–444.

Fonzi, W. A. (1989a) *Journal of Biological Chemistry*, **264**, 18110–18118.

—— (1989b) *Biochemical and Biophysical Research Communications*, **162**, 1409–1416.

Fonzi, W. A. & Sypherd, P. S. (1985) *Molecular and Cellular Biology*, **5**, 161–166.

—— (1987) *Journal of Biological Chemistry*, **262**, 10127–10133.

Foster, S. A. & Walters, D. R. (1990) *Journal of General Microbiology*, **136**, 233–239.

Fruton, J. S. (1972) *Molecules and Life*. New York: Wiley.

Frydman, B., Frydman, R. B., Santos, C. D. L., Garrido, D. A., Goldemberg, S. H., & Agranati, I. D. (1984) *Biochimica et Biophysica Acta*, **805**, 337–344.

Frydman, B., Santos, C. d. L., & Frydman, R. B. (1990) *Journal of Biological Chemistry*, **265**, 20874–20878.

García, J. I., Gómez, A., & Valle, T. (1991) *Journal of Plant Physiology*, **138**, 625–627.

Gordon, E. D., Mora, R., Meredith, S. C., Lee, C., & Lindquist, S. L. (1987a) *Journal of Biological Chemistry*, **262**, 16585–16589.

Gordon, E. D., Mora, R., Meredith, S. C., & Lindquist, S. L. (1987b) *Journal of Biological Chemistry*, **262**, 16590–16595.

Green, S. M., Eisenstein, E., McPhie, P., & Hensley, P. (1990) *Journal of Biological Chemistry*, **265**, 1601–1607.

Hamana, K., Matsuzaki, S., Hosaka, K., & Yamashita, S. (1989) *FEMS Microbiology Letters*, **61**, 231–236.

Hardin, G. (1949) *Biology and Its Human Implications*, New York: W. H. Freeman and Co.

Haywood, G. W. & Large, P. J. (1985) *European Journal of Biochemistry*, **148**, 277–283.

Henney, Jr., H. R. (1987) *Microbios*, **50**, 17–27.

Hilger, F. & Mortimer, R. K. (1980) *Journal of Bacteriology*, **141**, 270–274.

Horikawa, S., Ishikawa, M., Ozasa, H., & Tsukada, K. (1989) *European Journal of Biochemistry*, **184**, 497–501.

Horowitz, N. H., Charlang, G., Horn, G., & Williams, N. P. (1976) *Journal of Bacteriology*, **127**, 135–140.

Hosaka, K. & Yamashita, S. (1981) *European Journal of Biochemistry*, **116**, 1–6.

Jänne, J., Williams–Ashman, H. G., & Schenone, A. (1971) *Biochemical and Biophysical Research Communications*, **43**, 1362–1368.

Jazwinski, S. M. (1990) *Methods in Enzymology*, **182**, 154–174.

Kanamori, K., Legerton, T. L., Weiss, R. L., & Roberts, J. D. (1982) *Biochemistry*, **21**, 4916–4920.

Karlin, J. N., Bowman, B. J., & Davis, R. H. (1976) *Journal of Biological Chemistry*, **251**, 3948–3955.

Kashiwagi, K., Taneja, S. K., Liu, T., Tabor, C. W., & Tabor, H. (1990) *Journal of Biological Chemistry*, **265**, 22321–22328.

Kay, D. G., Singer, R. A., & Johnston, G. C. (1980) *Journal of Bacteriology*, **141**, 1041–1046.

Khan, A. J. & Minocha, S. C. (1989) *Plant and Cell Physiology*, **30**, 655–663.

Klionsky, D. J., Herman, P. K., & Emr, S. D. (1990) *Microbiological Reviews*, **54**, 266–292.

Legerton, T. L., Kanamori, K., Weiss, R. L., & Roberts, J. D. (1983) *Biochemistry*, **22**, 899–903.

—— (1984) *Proceedings of the National Academy of Sciences of the United States of America*, **78**, 1495–1498.

Lin, R., Newman, A. J., Cheng, S., & Abelson, J. (1985) *Journal of Biological Chemistry*, **260**, 14780–14792.

Mach, M., Kersten, H., & Kersten, W. (1982) *Biochemical Journal*, **202**, 153–162.

McCormick, J. J., Blomquist, J. C., & Rusch, H. P. (1970) *Journal of Bacteriology*, **104**, 1110–1114.

McDougall, K. J., Deters, J., & Miskimen, J. (1977) *Antonie van Leeuwenhoek Journal of Microbiology and Serology*, **43**, 143–151.

Mehta, K. D., Leung, D., Lefebvre, L., & Smith, M. (1990) *Journal of Biological Chemistry*, **265**, 8802–8807.

Miozzari, G. F., Niederberger, P., & Hütter, R. (1978) *Analytical Biochemistry*, **90**, 220–233.

Miret, J. J. & Goldemberg, S. H. (1989) *Seventh International Symposium on Yeasts*, pp. 333–337. London: Wiley.

Miret, J. J., Solari, A. J., Barderi, P. A., & Goldemberg, S. H. (1992) *Yeast*, **8**, 1033–1041.

Mitchell, J. L. A. (1983) *Methods in Enzymology*, **94**, 140–146.

Mitchell, J. L. A., Augustine, T. A., & Wilson, J. M. (1981b) *Biochimica et Biophysica Acta*, **657**, 257–267.

Mitchell, J. L. A., Carter, D. D., & Rybski, J. A. (1978) *European Journal of Biochemistry*, **92**, 325–331.

Mitchell, J. L. A., Mitchell, G. K., & Carter, D. D. (1982) *Biochemical Journal*, **205**, 551–557.

Mitchell, J. L. A. & Rusch, H. P. (1973) *Biochimica et Biophysica Acta*, **297**, 503–516.

Mitchell, J. L. A. & Wilson, J. M. (1983) *Biochemical Journal*, **214**, 345–351.

Mitchell, J. L. A., Yingling, R. A., & Mitchell, G. K. (1981a) *FEBS Letters*, **131**, 305–309.

Mora, J., Salceda, R., & Sanchez, S. (1972) *Journal of Bacteriology*, **110**, 870–877.

Morgan, D. H. (1970) *Molecular and General Genetics*, **108**, 291–302.

Nakamura, K. D. (1973) *Preparative Biochemistry*, **3**, 553–561.

Nakamura, K. D. & Schlenk, F. (1974) *Journal of Bacteriology*, **120**, 482–487.

North, M. J. (1985) *Biochemical Society Transactions*, **13**, 313–316.

North, M. J. & Murray, S. (1980) *FEMS Microbiology Letters*, **9**, 271–274.

Ogino, T., den Hollander, J. A., & Shulman, R. G. (1983) *Proceedings of the National Academy of Sciences of the United States of America*, **80**, 5185–5189.

Pathak, S. N., Porter, C. W., & Dave, C. (1977) *Cancer Research*, **37**, 2246–2250.

Paulus, T. J., Cramer, C. L., & Davis, R. H. (1983) *Journal of Biological Chemistry*, **258**, 8608–8612.

Paulus, T. J. & Davis, R. H. (1981) *Journal of Bacteriology*, **145**, 14–20.

Paulus, T. J., Kiyono, P., & Davis, R. H. (1982) *Journal of Bacteriology*, **152**, 291–297.

Peebles, C. L., Gegenheimer, P., & Abelson, J. (1983) *Cell*, **32**, 525–536.

Petrotta–Simpson, T. F., Talmadge, J. E., & Spence, K. D. (1975) *Journal of Bacteriology*, **123**, 516–522.

Pitkin, J. & Davis, R. H. (1990) *Archives of Biochemistry and Biophysics*, **278**, 386–391.

Pitkin, J., Perriere, M., Kanehl, A., Ristow, J. L., & Davis, R. H. (1994) *Archives of Biochemistry and Biophysics*, **315**, 153–160.

Pohjanpelto, P., Virtanen, I., & Höltta, E. (1981) *Nature*, **293**, 475–477.

Poli, F., Romagnoli, C., Dall'Olio, G., & Fasulo, M. P. (1993) *Microbios*, **73**, 261–267.

Pösö, H., Sinervirta, R., & Jänne, J. (1975) *Biochemical Journal*, **151**, 67–73.

Ross, I. K. (1979) *Biology of the Fungi*. New York: McGraw–Hill.

Rousseau, P. & Halvorson, H. O. (1973) *Journal of Bacteriology*, **113**, 1289–1295.

Ruiz–Herrera, J. (1994) *Critical Reviews in Microbiology*, **20**, 143–150.

Sarić, M. & Clarkson, A. B. (1994) *Antimicrobial Agents and Chemotherapy*, **38**, 2545–2552.

Sauer, H. W. (1982) *Developmental Biology of Physarum*. Cambridge, England: Cambridge University Press.

Schlenk, F., Dainko, J. L., & Swihla, G. (1970) *Archives of Biochemistry and Biophysics*, **140**, 228–236.

Schwartz, B., Hittelman, A., Daneshvar, L., Basu, H. S., Marton, L. J., & Feuerstein, B. G. (1995) *Biochemical Journal*, **312**, 83–90.

Schwenke, J. & de Robichon-Szulmajster, H. (1976) *European Journal of Biochemistry*, **65**, 49–60.

Scott, A. I. & Baxter, R. L. (1981) *Annual Review of Biophysics and Bioengineering*, **10**, 151–174.

Shiba, T., Mizote, H., Kaneko, T., Nakajima, T., Kakimoto, Y., & Sano, I. (1971) *Biochimica et Biophysica Acta*, **244**, 521–531.

Shiomi, N., Fukuda, H., Fukuda, Y., Murata, K., & Kimura, A. (1990) *Journal of Fermentation and Bioengineering*, **69**, 63–66.

Shulman, R. G., Brown, T. R., Ugurbil, K., Ogawa, S., Cohen, S. M., & den Hollander, J. A. (1979) *Science*, **205**, 160–166.

Slocum, R. D. & Galston, A. W. (1987) *Inhibition of Polyamine Metabolism*, pp. 305–316, P. P. McCann, A. E. Pegg, & A. Sjoerdsma, Eds. New York: Academic Press.

Smith, T. A., Barker, J. H. A., & Jung, M. (1990) *Journal of General Microbiology*, **136**, 985–992.

Sogin, M. L. (1991) *New Perspectives on Evolution*, pp. 175–188, L. Warren & H. Koprowski, Eds. New York: Wiley–Liss.

Stevens, L. & Winther, M. D. (1979) *Advances in Microbial Physiology*, **19**, 63–148.

Subramanian, K. N., Weiss, R. L., & Davis, R. H. (1973) *Journal of Bacteriology*, **115**, 284–290.

Tabor, C. W. & Tabor, H. (1985) *Microbiological Reviews*, **49**, 81–99.

Tabor, H., Tabor, C. W., & Cohn, M. S. (1983) *Methods in Enzymology*, **94**, 104–108.

Thomas, D., Rothstein, R., Rosenberg, N., & Surdin–Kerjan, Y. (1988) *Molecular and Cellular Biology*, **8**, 5132–5139.

Thomas, D. & Surdin–Kerjan, Y. (1987) *Journal of Biological Chemistry*, **262**, 16704–16709.

Turner, R., North, M. J., & Harwood, J. M. (1979) *Biochemical Journal*, **180**, 119–127.

Tyagi, A. K., Tabor, C. W., & Tabor, H. (1981) *Journal of Biological Chemistry*, **256**, 12156–12163.

——— (1982) *Biochemical and Biophysical Research Communications*, **109**, 533–540.

Viotti, A., Bagni, N., Sturani, E., & Alberghina, F. A. M. (1971) *Biochimica et Biophysica Acta*, **244**, 329–337.

Walters, D. R. & Robins, D. J. (1994) *Biochemical Society Transactions*, **22**, 390S.

Webster, J. (1970) *Introduction to Fungi*. Cambridge, England: Cambridge University Press.

Whitney, P. A. & Morris, D. R. (1978) *Journal of Bacteriology*, **134**, 214–220.

Wickner, R. B. (1989) *FASEB Journal*, **3**, 2257–2265.

Williams, L. J., Barnett, G. R., Ristow, J. L., Pitkin, J., Perriere, M., & Davis, R. H. (1992) *Molecular and Cellular Biology*, **12**, 347–359.

Williams–Ashman, H. G. & Schenone, A. (1972) *Biochemical and Biophysical Research Communications*, **46**, 288–295.

Xie, Q., Tabor, C. W., & Tabor, H. (1990) *Yeast*, **6**, 455–460.

Yang, Y. C., Chen, K. Y., Seyfzaheh, M., & Davis, R. H. (1990) *Biochimica et Biophysica Acta*, **1033**, 133–138.

Zhu, C. & Henney, Jr., H. R. (1990) *Canadian Journal of Microbiology*, **36**, 366–368.

CHAPTER 11

Polyamines in the Animal Cell

Biochemical investigation of the amines has not followed a neat route through biological models of increasing complexity. Despite the disorderly accumulation of data on mammalian systems, some studies have led to surprising and medically promising results, such as the therapeutic efficacy of inhibitors of polyamine metabolism in such pathologies as African trypanosomiasis in humans and certain animal cancers. In current experimental studies animal cell cultures are recognized as an important step in the analysis of the survival and functions of tissue and organisms, and the intact mammal is relegated to later stages of medically oriented investigation.

It will be recalled that the first studied polyamine, spermine, was recognized initially in human seminal fluid and only much later in individual animal tissues and yeast. Spermidine was isolated from ox pancreas as a substance found in the mother liquor obtained after removal of spermine phosphate. Diamines were recognized as microbiological products and somewhat later as products of amino acid degradations in animal tissues. Essential nutritive roles of these compounds were discovered in studies of the growth of some bacteria and fungi. The toxicity of spermine to some microorganisms led to studies of the amine oxidases on the one hand, and pharmacological studies on the other. Studies of the biosynthetic origins and metabolic fates of the compounds were explored initially with cells and extracts of bacteria and fungi. Finally the extension of such studies to animal cells was carried out for the most part in the 1960s by biochemists who exploited interesting tissue systems, such as that of liver regeneration or the hormonally regulated prostate. In the period of 1967–1969, the independent studies of the polyamines in regenerating liver led to the discovery of the startling phenomena of the regulation of mammalian ornithine decarboxylase (ODC), providing problems that are still being explored.

The microbiology of the 1940s and 1950s had produced a powerful methodology for the study of virus multiplication, which was then understood as a phenomenon of individual cells. All the cells of a culture could be infected en masse to establish a timing and progression of synchronized virus-induced cellular pathology. Bacterial genetics and the genetics of eucaryotic microbes had also matured in this period, setting conditions for questions that might be asked in animal cell cultures. After systematic analysis of the growth and properties of animal cells in culture, such studies became a common laboratory procedure by the 1970s (Levintow & Eagle, 1961; Puck, 1961), providing major biological systems for the study of the polyamines. Knowledge of polyamine biosynthesis, metabolism, and function in selected tissues and cell cultures developed independently and in parallel.

Studies with cell cultures have also indicated some limitations of such systems. The nature of tissue culture systems has emphasized cell types that can be subcultured indefinitely and produced from individual cells. These cell types were grown initially in emulation of a bacteriological methodology as colonies from individual cells on solid or semisolid media and then were grown en masse in liquid culture; these were abnormal in significant respects (Hayflick, 1991; Levintow & Eagle, 1961). Differentiation in many animal cells (e.g., nervous tissue) is a terminal process, and such cells are not readily cultivable, if at all. In some systems, continuously cultivable cells had to be grown as confluent monolayers, which became arrested as a result of "contact inhibition" and were then harvested and manipulated as separable aggregates. The frequently used aneuploid human HeLa cells were derived from a cervical tumor (Tijo & Puck, 1958). Fibroblasts taken from mouse tissues proved to harbor various viruses. Thus, much early knowledge in cultured cells was derived from biologically unrepresentative systems that have nevertheless provided useful baselines for comparative studies.

Many apparently quiescent differentiated cells could be stimulated to divide or to initiate a characteristic physiological response after specific hormonal treatment. A fertilized egg, whether of poultry or marine origin, presented a metabolically active mass appropriate for biochemical study. This chapter will discuss results with multiplying cell cultures, both normal and tumorous, and with stimulated differentiated and differentiating systems.

Polyamines as Nutrients in Clonal Growth

Techniques for the clonal growth of cultured mammalian cells were first developed in the mid 1950s. The growth medium contained essential serum factors, which were eventually replaced by purified fractions such as serum albumin and fetuin. Fetuin, which facilitated seeding (attachment) and colony formation by trypsinized cells on the surfaces of plastic Petri dishes, could be replaced by any one of several natural amines (Ham, 1964). Putrescine, spermidine, or spermine were essentially equipotent at micromolar concentrations, as seen in Figure 11.1. Serum albumin could be replaced by linoleic acid in a chemically defined medium (Ham, 1964). Spermidine was shown to be toxic in some media which included serum fractions. Numerous techniques applicable to cloning and clonal growth were developed subsequently (Ham, 1974), and putrescine proved to be stimulatory generally in the growth of human fibroblasts and mammalian epithelial cells (Pohjanpelto, 1973; Roszell et al., 1977; Stoner et al., 1980). The depletion of putrescine in Ehrlich ascites tumor cells and the inhibition of cell proliferation by treatment with difluoromethylornithine (DFMO), the specific irreversible inhibitor of ODC, demonstrated the importance of putrescine and the polyamines in maintaining active cell division (Oredsson et al., 1984). Putrescine was

also found to be greatly stimulatory in animal cells infected by mycoplasma, which deplete the cells of arginine (Kamatani et al., 1983).

Early Data on Tissue Polyamines

The early studies of Harrison and Hämäläinen detected and isolated spermine from many human tissues. High concentrations of the tetramine were found in prostate and pancreas. The more sensitive colorimetric methods and efficient techniques of separation improved the analysis of spermine and spermidine, and in many tissues spermidine was present at a higher concentration than that of spermine (Tabor & Tabor, 1964). Spermine was also found in blood cells, although it was apparently absent in plasma.

Only 20% of the cellular spermine was found in rat liver nuclei (Raina & Teloranta, 1967). The values for spermine and spermidine in rat liver nuclei separated in nonaqueous media proved to be not grossly different (Sarhan & Seiler, 1989). Nevertheless, the small fraction of nuclear volume signifies a markedly higher concentration of these polyamines in the nucleus than in cytoplasm. The concentrations of spermidine and spermine in these nuclei attained levels of 4.5–5.5 mM. Sarhan and Seiler suggested that the ratio of spermidine to spermine is relatively constant through the nondividing liver cell, as demonstrated additionally by a combination of radioactive labeling and electron microscopy of the silver grains deposited in radioactive decay. However, the effects of the cell cycle on the cellular distribution of the polyamines is not clear, despite the cytochemical data summarized by Hougaard (1992).

In his study of polyamine metabolism, Raina (1963) described the great increase in the polyamines in the developing chick embryo and the decrease in their concentration as a function of development. Raina thereby documented the need to include a developmental parameter for polyamine concentration. Embryonic rat tissues were markedly higher in spermidine content than those of aging animals. Aging was generally accompanied by a decrease in the spermidine to spermine ratio, and this was primarily due to a decrease in spermidine concentration (Jänne et al., 1964). In the rat brain both amines decline with age.

In early studies of developing nervous systems the observed distribution of spermidine and spermine in the brain also suggested that spermine might be sequestered in nuclei. This led to a comparison of the polyamine content of enucleated mammalian erythrocytes and the nucleated erythrocytes of other vertebrates (Shimizu et al., 1965). The latter were indeed significantly richer in spermidine and spermine. However, enucleate rabbit erythrocytes were found to be unexpectedly high in spermidine and spermine. The uptake of spermidine and spermine by enucleate erythrocytes unable to synthesize these compounds proved to be diagnostically significant in various pathologies.

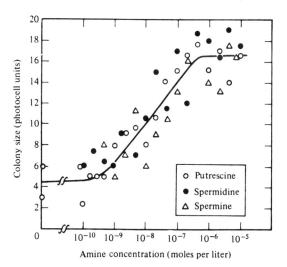

FIG. 11.1 Effects of polyamines on the clonal growth of Chinese hamster cells, strain (CHD-3); adapted from Ham (1964). Incubation time = 12 days. Colony size was determined photometrically.

Data on Development

Raina (1963) demonstrated a 50-fold increase in the spermidine + spermine content of a developing chick embryo over an 18-day period in which the spermidine concentration fell 80%. Inhibition of an active polyamine oxidase in this system evoked abnormally high polyamine levels, concomitantly with increases in RNA and DNA. This suggested a role for the polyamines in some control of the cellular nucleic acids (Caldarera et al., 1965) as well as a possible role in the control of protein synthesis (Moruzzi et al., 1968). Caldarera et al. (1969) subsequently demonstrated the stimulation of RNA and protein synthesis in the chick embryo brain following direct injection of spermidine or spermine into a 10-day embryonated egg. Thus, by the late 1960s results implicating the polyamines in biosynthetic paths of active growth had been established in studies on the embryonated chick.

Investigations began of the roles of the polyamines in embryological development of insects such as *Drosophila* and *Cecropia* that were readily available, easily handled in a laboratory, and well characterized in genetic and other biological parameters. Recent advances in the labeling, microscopy, and microanalytical techniques of the fruitfly were exploited (Herbst & Dion, 1970). *Drosophila melanogaster* was found to contain spermidine and putrescine, of which the former was initially quite high in fertilized eggs and fell to a plateau after fertilization. After hatching, spermidine fell once again and increased just before the inception of pupation. The larvae could take up exogenous polyamine, and this enrichment with spermidine increased the incorporation of ^{3}H-uridine into various classes of RNA (Byus & Herbst, 1976a,b).

The metamorphosis of the giant *Cecropia* silkmoth has been extensively studied. Pupation is characterized by depressed metabolism and a cessation of growth. The steroid hormone, β-ecdysone, reinitiates development with a new synthesis of RNA. After injection of ecdysone there was a 30–40-fold increase in ODC in the fat body after 12 h and in the wing at 18 h. The enzyme then fell and at 36 h the spermidine contents of the fat body and wings had increased (Wyatt et al., 1973). A block of the enzyme with DFMO did not simulate the effect observed with the hormone (Willis, 1981).

Studies on the development of another silk moth, *Bombyx mori*, revealed the presence in many organs of 1,3-diaminopropane (DAP), norspermidine, and norspermine, in addition to putrescine and spermidine. Cadaverine is present in eggs and mucous glands. Furthermore, homospermidine is present in the silk gland and ovary. The apparent absence of the "unusual" amines in the mulberry leaf food implies their synthesis within the organism (Hamana et al., 1984).

Spermidine has been found as a major component during the life cycle of the housefly (Joseph & Baby, 1988). However, several insects (e.g., cricket, cockroach, fruitfly, and midge) have been found to have a bewildering array of minor polyamines, including compounds such as homospermidine, thermospermine, and caldopentamine (Hamana et al., 1989). The central nervous system of the adult cricket is devoid of detectable putrescine and has an unexpectedly high content of spermine, some threefold greater than that of spermidine (Strambi et al., 1993).

Among the early studies it was found that the synthesis of ribosomal RNA in early sea urchin development was stimulated by the addition of spermidine to seawater (Barros & Giudice, 1968). Significant changes occurred in the concentrations of putrescine and spermidine in the early cleavage stages after fertilization (Manen & Russell, 1973a,b). Despite high concentrations of spermine in the gametes, only spermidine increased sharply until gastrulation; after gastrulation the accumulation of ribosomal RNA was paralleled by the accumulation of spermidine and spermine. Increase in the biosynthetic enzymes, that is, ODC and *S*-adenosylmethionine decarboxylase (AdoMetDC), were also detected soon after fertilization. Studies with nudibranchs similarly suggested an important role of the polyamines in development (Manen et al., 1977).

The tissue content of the polyamines in sea stars (Echinoderms) parallels that of the rate of cell proliferation (Asotra et al., 1988). Testes had the highest spermidine to spermine ratio (5.4) and spermatogenesis correlated with high levels of the amines. Indeed, small amounts of exogenous polyamine markedly stimulate mitotic divisions and increase ODC, and the availability of polyamine appears important in the seasonal regulation of spermatogenesis (Sible et al., 1991).

Regeneration in cut Planaria occurs with an early increase of ODC (3–4-fold in 8 h), followed by a slow decline after 3 days. Putrescine increases (2–3-fold) in tissue near the wound in 3 days, followed by slightly lower increases in spermidine and spermine for 5 days. The inhibitors DFMO and methylglyoxal-bis(guanylhydrazone) (MGBG) affect regeneration somewhat differently, with DFMO largely inhibiting proliferation and MGBG particularly reducing the rate of differentiation (Saló & Baguñà, 1989).

Synchronized waves of increase and decrease of the polyamines occur in the first three early cleavages after fertilization of the sea urchin egg (Fig. 11.2). Two peaks of spermidine involving a threefold increase of this substance are generated before the first cleavage and this substance achieves a peak about 10–20 min before a division. Much smaller peaks of spermine are found at about the same times. Small peaks of putrescine are achieved just before those of spermidine. Thus, the changes apparently mirror the expected biosynthetic steps from putrescine to spermine (Kusunoki & Yasumasu, 1976).

In the early manipulable synchronous system of fertilized eggs of a polychaete worm (Emanuelsson & Heby, 1978), spermine is quite high and falls during development. Spermidine does not accumulate at any stage, but putrescine increases to a maximum by the beginning of gastrulation. A block in putrescine synthesis arrests development at gastrulation, whereas a block

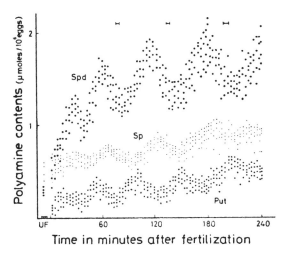

FIG. 11.2 Cyclic alteration in polyamine contents in fertilized sea urchin eggs; adapted from Kusunoki and Yasumasu (1976). Put, putrescine; Spd, spermidine; Sp, spermine. Small horizontal bars show time of egg cleavage.

in synthesis of spermidine and spermine has no effect. Exogenous ^{3}H-ornithine labels the nucleolus at gastrulation, suggesting that the synthesis of putrescine is directly involved in the formation of this nuclear structure (Heby & Emanuelsson, 1978).

Amphibians, taken originally from the wild, have many advantageous features for developmental study. In one early investigation, the polyamine content of embryonic stages of *Xenopus laevis* was determined during various stages of metamorphosis. In the tadpole stages, *Xenopus* increased about 2.5-fold in putrescine and spermidine without a change in the relatively low content of spermine (Russell et al., 1969). These increases were concomitant with increases in ribosomal RNA and protein synthesis, and the microinjection of putrescine into mature *Xenopus* oocytes stimulated the synthesis of major species of ribosomal RNA. During the late stages of oogenesis in *Xenopus*, the level of spermine, hormone responsiveness, and a polyamine-dependent protein kinase appear to be involved (Osborne et al., 1989). Details of the regulation of ODC and the syntheses of the several polyamines in early *Xenopus* development have also been described (Osborne et al., 1991; Rosander et al., 1995). As presented in Figure 11.3, ODC is degraded almost completely at a blastula stage and begins to be actively resynthesized during gastrulation.

Some amphibians contain homospermidine at levels approaching that of spermine; the origins and role of this are quite unclear (Hamana & Matsuzaki, 1979). The relatively rare triamine has been found in regenerating tissues of the Japanese newt (Matsuzaki & Inoue, 1980).

The injection of spermine or spermidine into fertilized amphibian eggs significantly accelerated the formation of cleavagelike furrows in the first division cycle (Grant et al., 1984). Actually unfertilized eggs are

activated to some extent by the addition of polyamines to seawater, with some surface changes and initiation of polymer syntheses similar to effects observed following fertilization (Fujiwara et al., 1983). The microinjection of spermine into *Amoeba proteus* has also been shown to induce cleavage furrows and cytokinesis (Gawlitta et al., 1981). Spermine and spermidine induce the polymerization of actin in vitro; indeed, the macrocyclic polyamines are highly efficient in this activity (Oriol–Audit et al., 1985). In fact spermine organizes microfilaments of actin in the cortices of amphibian eggs (Grant & Oriol–Audit, 1985). This result is consistent with the hypothesis that a polyamine-determined polymerization of actin is involved in the regulation of cell division via the formation of a cortical contractile ring, an interaction that begins with an early increase in polyamine in the fertilized egg prior to the first cleavage (Oriol–Audit, 1991). Exogenous polyamines increase the yield of blastocysts in the in vitro development of fertilized murine oocytes (Muzikova & Clark, 1995).

The concept that advanced stages of embryological development proceed with the genetically programmed death of specific cells has been accompanied by evidence that the production of cytotoxic H_2O_2 resulting from polyamine oxidation is important in early development of mammalian embryos.

Osmotic Effects on Polyamine Metabolism

The changes in external salinity and their mechanisms are major agents of evolution that determine ecological solutions, such as the movement of organisms from the saline oceans to diluted estuaries and land. Stenohaline organisms, which cannot regulate the salt content of their blood, are bound to relatively narrow environmental niches. Euryhaline organisms, such as salmon, can withstand significant changes of external salinity within their life cycle.

The alteration of cellular physiology by volume change has been discussed (Häussinger & Lang, 1991). The manner in which amino acids serve as "volume-regulatory" osmolytes in mammalian cells has recently been reviewed (Law, 1991). The ability to adapt to osmotic change can begin at the cellular level, and at least some responses fall within the area of polyamine metabolism.

Escherichia coli, which survives and grows in a relatively wide range of aqueous media, has a putrescine content inversely related to the osmolarity of the growth medium. When *E. coli* grown in media of low osmolarity and containing unusually high concentrations of putrescine is placed in media of high salt content, there is a rapid extrusion of putrescine to the medium. Putrescine may be reacquired slowly, largely by synthesis, in a low osmolarity medium. This phenomenon was then detected in animal HeLa cells and several types of human fibroblasts cultivated in media of varying salt content (Munro et al., 1975). HeLa cells grown in low NaCl media tended to be rounded; those grown in high salt

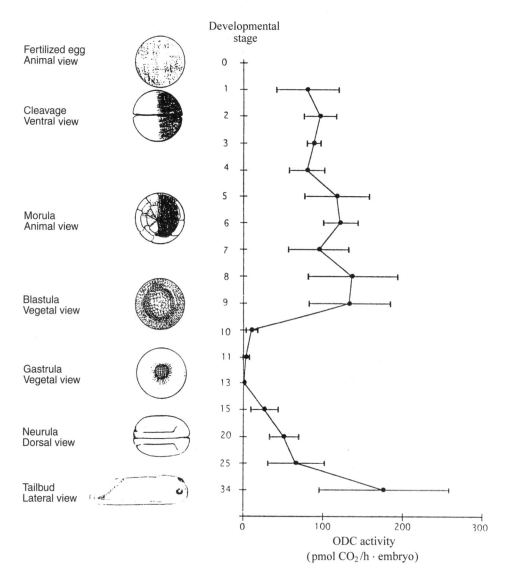

FIG. 11.3 ODC activity during embryonic development of *Xenopus laevis*; adapted from Rosander et al. (1995). Representative developmental stages are shown to the left.

media were polygonal. The former were significantly higher in putrescine and ODC contents. When these cells were placed in media of high NaCl content, the contents of putrescine and spermidine fell without appearance of the compounds in the media. ODC also fell rapidly to undetectable levels without a change in the rate of protein synthesis. When cells of low ODC and putrescine content (i.e., by growth in high osmolarity media) were transferred to low NaCl media, ODC and putrescine content rose quickly (within 30 min).

A biphasic increase of ODC was detected in the early transfer of the mouse mammary gland cultivated in vitro. The first peak, unlike the second, was independent of the hormone content of the medium and proved to be a response to a decrease in osmolarity of the initial incubation medium (Perry & Oka, 1980). There was a 1000-fold stimulation of the enzyme, markedly increasing the putrescine content of the transplant. Hypotonicity also decreased putrescine efflux. Inhibitors of ODC blocked the increase of the enzyme and putrescine in this osmolarity-sensitive system and inhibited a subsequent induction of DNA synthesis by insulin (Inoue & Oka, 1980). Putrescine (but not spermidine, spermine, or cadaverine) reversed this effect, implying a hitherto unsuspected function of putrescine in the control of DNA synthesis in this system.

The powerful induction of ODC by hypoosmotic stress has been studied in various lines of L1210 mouse leukemia cells (Poulin et al., 1991). Hypoosmotic shock (325 → 130 mos mol/kg) slowed cell growth and addition of putrescine to shocked cells increased growth rate, but other polyamines did not. Hypoosmotic conditions markedly reduced ornithine availability and a supply of exogenous ornithine restored putrescine production. An enlarged putrescine pool effected via exogenous ornithine or putrescine appeared essential to the restoration of normal growth when the cells were adapting to low osmolarity media. Thus, a normally high ODC content producing a large putrescine pool permitted a more ready adaptation of cells to media of low osmolarity.

Mutant mouse leukemia cells that overproduced ODC also acquired a very active system for the uptake and accumulation of spermidine (Poulin et al., 1993). This new transport activity was demonstrable in hypoosmotic stress at low spermidine concentrations (1–30 μM) and resulted in toxic concentrations of spermidine in the cells, a property not observed in the normal L1210 cells. The phenomenon of spermidine toxicity has become evident in mutant strains of bacteria or eucaryotic cells that overproduce ODC or have unusually active transport systems for the polyamines. Morris (1991) postulated that the toxicity of the polyamines spermidine and spermine arises in large measure from the displacement of essential Mg^{2+} from native metabolizing structures and has compelled, via selection, the negative regulation of ODC by the polyamines.

Cytochemical Studies

The possibility of polyamine redistribution during isolation procedures led to the search for in situ methods of establishing the intracellular distribution of compounds. We note the apparent identity of results obtained by aqueous and nonaqueous fractionation methods when applied to rat liver. Nevertheless, as discussed earlier, biological systems surfeited with amine might well redistribute this amine during fractionation. Indeed Hougaard (1992) suggested a possible redistribution even in nonaqueous fractionation. In situ histochemistry has provided its own apparent contradictions. A radioautographic experiment on the detection of [³H]-spermidine and [³H]-uridine incubated with the larval salivary glands of *Drosophila melanogaster* provided early evidence that spermidine was relatively concentrated in the nuclei of many cells (Dion & Herbst, 1967). Some concentrations of spermidine (ca. 5×10^{-4} M in the incubation mixture) also stimulated uridine incorporation (presumably in RNA) in the nuclei. The studies on the development of *Drosophila* were extended with the newly devised sensitive dansyl method (Herbst & Dion, 1970).

Fluorescent reagents have proved useful for histochemical detection of the amines. A combination of formaldehyde and fluorescamine are selective for spermidine and spermine. In normal secretory cell systems, the polyamines appear to be concentrated in cytoplasmic structures (Hougaard, 1992; Larsson et al., 1982). Although the amines were found to be associated with

the mitotic chromosomes of HeLa cells and were present in frog erythrocyte nuclei, it was thought that the presence of polyamines in nuclei was an unusual occurrence. This group also used *o*-phthalaldehyde as a specific stain for the amines, confirming their earlier results with the fluorescamine method (Hougaard & Larsson, 1982). Their cytochemical results with secreting cells (i.e., pancreas, prostate) were in agreement with those of their own autoradiographic, immunochemical, and chemical studies. In the pancreas they noted high concentrations of spermidine and spermine, particularly on the secretory granules of insulin-producing β cells. A high level of spermine in the pancreas was found by many workers, and a role in the release of pancreatic enzymes from cell particles was detected for spermine in some early cytochemical studies (Siekevitz & Palade, 1962). However, conclusions on the relative absence of amine in the nuclei have been challenged by Sarhan and Seiler (1989) who suggest that their results may have been due to a quenching of fluorescence or the nonreactivity of the bound amine. The response of polyamine to staining is decreased in the chromatin of interphase nuclei and is increased by chromatin condensation. Furthermore, DNA markedly reduces the reactivity of polyamine with specific antibody, and this reactivity can be released by predigestion with DNase. Thus, amines in some bound states cannot be stained readily by the original procedures that detected the amines largely in the cytoplasm.

Cellular Aging

The spermidine content of rat tissues decreased with increasing age, although spermine content did not (Jänne et al., 1964). However, older cells, or those in the quiescent (G_0) stage of a cell cycle, usually have a decreased spermidine/spermine ratio. Indeed, a linear correlation was found between the growth rate of rat brain tumor cells and the cellular spermidine/spermine ratio (Heby et al., 1975). In a study that possibly confirms this observation, a primary culture of nonproliferating rat hepatocytes (with an initial ratio of spermidine to spermine approaching 0.80) began to lose spermidine rapidly after 4 h, with accumulation of N^1-acetylspermidine and putrescine at this time. By 24 h the spermidine to spermine ratio approached 0.30 and cell growth was quite slow (Wallace, 1989). A ratio of 1.5–2.0 appears to be a good indicator of a rapid rate of cell growth in nontransformed cells. At confluence the ratio approaches 0.25. Scalabrino and Ferioli (1984) reviewed data on changes in the polyamines and their enzymatic complement in the tissues of aging rats.

Attempts have been made to determine polyamine changes in cultures of human diploid fibroblasts that enter senescence (i.e., lose the capacity for self-renewal) after 50 ± 10 doublings, that is, the well-known "Hayflick limit." Hayflick (1991) discussed the history of the concept that diploid somatic cells age and that this biological change, as distinct from "terminal differentiation," can be demonstrated in cultures of nonimmortalized cells. In the first of such studies on human embryonic lung fibroblasts, senescence was marked by a fall in growth rate and by a decrease of the amount of the ODC induced after a transfer to fresh media. Addition of putrescine or an inhibitor of polyamine oxidation did not change the sequence of events, that is, affect the "limit" (Duffy & Kremzner, 1977).

The earlier observation on a decrease of ODC inducibility in aged cells has been confirmed. However, the contents of spermidine and spermine in fibroblasts of different ages were high; these remained unaffected by serum stimulation of aged human diploid fibroblasts. Nevertheless, the latter group was far less able to accumulate putrescine (Chen et al., 1986). The serum-induced cells made similar amounts of ODC mRNA, regardless of the age of the cells and despite the decrease in production of active ODC (Chang & Chen, 1988). The aged cells have a very high rate of conversion of putrescine to amino acids, much greater than in young cells. This effect may be due to observed increases in amine oxidases in the older cells. Conversely, the rate of hypusine formation from spermidine is significantly higher in young cells (Chen & Chang, 1986).

Studies of Suspension Cultures

The microbiological model of growth in suspension culture has the advantage of permitting easy and continuous sampling for purposes of cell count and biochemical estimation, with the additional possibility of plating for determination of viability. When rat hepatoma cells are grown in suspension, the ODC activity falls with increasing cell concentration and is almost undetectable at the stationary phase (Hogan, 1971). Dilution in fresh media results in a rapid reappearance of the enzyme, with a decline after 4 h to a lower steady-state level. These effects are not greatly altered by actinomycin D, but the initial increase is blocked by cycloheximide, suggesting an effect occurring after the formation of mRNA.

McCormick (1978) studied the polyamine synthesis and turnover in suspension cultures of mouse fibroblasts. He prelabeled cells with isotopically labeled polyamines and determined the retention of the isotope and the specific radioactivity of polyamine reisolated from the growing cultures. Rapidly growing cells prelabeled with [^{14}C]-spermine retained the isotopic amine until the cells attained a limit of cell number (i.e., "stationary phase") and then rapidly lost isotope and spermine (Fig. 11.4A). In the same period (Fig. 11.4B) the isotopic spermine of the cells was diluted at a constant rate during growth and in the apparent "stationary phase." In the latter period (40–60 h) the cells continued to synthesize spermine and degraded much of it (50–60%) to extrudable spermine. In this latter period, DNA products were also released, indicating a death of many of the cells. The period in which cell spermine doubled (i.e., in which the specific radioactivity of spermine fell 50%) was 24 h as compared to the cell doubling time of 22 h.

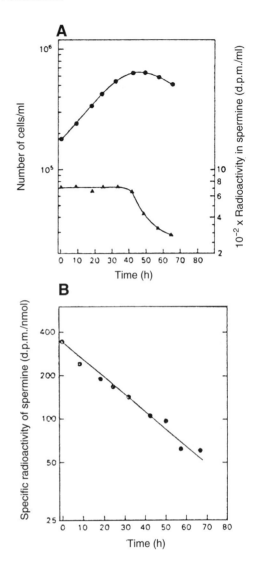

FIG. 11.4 (A) Incorporation of [^{14}C]-spermine in mouse fibroblasts growing in suspension culture; adapted from McCormick (1978). (●) Cell numbers; (△) radioactivity in cell spermine. (B) The rate of decrease in specific radioactivity of cell spermine during exponential growth and the early part of a stationary phase; adapted from McCormick (1978).

In contrast to the results with [^{14}C]-spermine, incorporated isotopic spermidine had a distinct turnover during exponential growth. In this case, 80% of the isotope was lost to spermine and 20% to free spermidine in the medium. The conversion to spermine accounted for all of the synthesis of that polyamine. The doubling of spermidine in the culture, maintaining a constant spermidine content of the cells during exponential growth, took about 20 h. By contrast, putrescine was lost very rapidly from the cells, with a half-life of about 1.5–2 h. In late stages (55–65 h) there was a significant conversion and excretion of putrescine derived from spermidine. Thus, degradation of spermine to spermidine and thence to putrescine only occurred during the stationary phase.

McCormick (1978) concluded that spermine, spermidine, and putrescine comprise essentially all of the detectable polyamine metabolites of these cells; he recorded similar results for HeLa cells that were grown and incubated as monolayers. The rapid turnover of putrescine, in contrast to the slow turnover of spermidine and spermine, was consistent with the frequent observations of rapid uptake and release of the diamine as well as its rapid biosynthesis and regulation of its own synthesis. The simple kinetics of biosynthesis of spermidine and spermine implied a lack of significant compartmentation for these compounds. McCormick further demonstrated a concentration of these amines in the nucleus as compared to the cytoplasm (see Table 11.1) and concluded that the equilibration between these compartments was extremely rapid.

Andersson and Heby (1977) demonstrated that the relations between cell proliferation and polyamine synthesis and content in a growing Ehrlich ascites tumor in mice were similar to observations on these parameters of cells in culture. The growth rate of the tumor was very closely tied to ODC activity, and spermidine and spermine contents were positively correlated with the contents of RNA and DNA. In recent studies it appears that a primary dense culture of tumor cells taken from a tumor-bearing mouse can be stimulated to increase ODC greatly by incubation and perfusion in the presence of 0.5 mM ornithine (Matés et al., 1991; Sanchez–Jiminez et al., 1992). The perfusion technique continuously removes inhibitory products, such as the natural polyamines.

Effects of Inhibitors on Cell Cultures

Ehrlich cells in culture are known to possess the usual sequence of putrescine, spermidine, and spermine and yield easily assayed ODC and AdoMetDC in sonic extracts (Alhonen–Hongisto et al., 1979a,b). These enzymes are inhibited by DFMO and MGBG, respectively. Among the in vivo inhibitors of ODC, 1,3-diamino-2-propanol was found to be particularly effective in eliminating the enzyme from cultured cells. Putrescine only partially reversed the inhibition by the hydroxydiamine. Removal of the latter from the medium by a medium change permitted a complete recovery of the cells.

Intraperitoneal inoculation of mice with cultured cells induced a rapid synthesis of polyamines in the inoculated cells. This synthesis, and that of the enzymes and of DNA, could be inhibited in the animal by injection of ODC inhibitors. Inhibitors of ODC such as α-methylornithine or DFMO produced a rapid depletion of putrescine within 24 h, followed by a slower depletion of spermidine. Spermine content was relatively unaffected; it occasionally increased somewhat. These effects were cytostatic rather than lethal, despite a marked inhibition of cell replication; cell growth and division began again

at normal rates when putrescine and spermidine were restored, even after 8 days of inhibition (Mamont et al., 1978). Ehrlich ascites cells exposed to DMFO can be grown for months in the presence of micromolar levels of cadaverine (Alhonen–Hongisto et al., 1982).

Chick embryo fibroblasts transformed by Rous sarcoma virus (cultures treated with 1–20 mM DFMO) reduced their putrescine content and replaced this diamine by cadaverine, which accumulated mainly in the transformed cells (Bachrach & Shtorch, 1985). Some mycoplasma-free mouse and rat cells normally produce and export significant amounts of cadaverine, as well as putrescine (Hawell et al., 1994). The Ehrlich cells replaced spermidine by N-(3-aminopropyl)cadaverine and spermine by N,N^1-bis(3-aminopropyl)cadaverine. Although the rates of synthesis of nucleic acid were almost normal, growth rate and protein synthesis were inhibited. ODC and AdoMetDC were much higher than in normal cells, as was the rate of MGBG transport. Thus, cadaverine and its derivatives may partially replace the normal amines; but several important functions of the cells, that is, protein synthesis and enzyme regulation, were optimal only with the "normal" compounds.

The ODC content of the cells is regulated by their polyamines, but it was not initially clear whether each normal amine participated or was converted into a single regulating amine, such as spermidine. Ehrlich ascites cells were exposed to nonmetabolizable derivatives of putrescine (1,4-dimethylputrescine) and spermine (5,8-dimethylspermine). Both analogues inhibited synthesis of ODC and increased the relative degradation of preexisting enzyme, implying that all three of the normal polyamines may participate in the regulation of ODC. The spermine analogue (but not that of putrescine) also decreased the activity of AdoMetDC. The analogues accumulated in the cells, while normal putrescine and spermidine content decreased markedly. Only the 5,8-dimethylspermine was inhibitory to cell growth (Holm et al., 1988).

The depletion of polyamines by means of inhibitors was soon observed to evoke several compensating mechanisms in the cell cultures. Many of the inhibitors were found to stabilize ODC and AdoMetDC, leading eventually to a higher than normal level of these activities. In the cultured Ehrlich system, inhibition by DFMO considerably enhanced uptake of the polyamines, with recovery rapidly achieved by exposure to exogenous amine (Alhonen–Hongisto et al., 1980). The uptake and antiproliferative effect of MGBG were also markedly greater when given with DFMO (Seppänen, 1981). Taking advantage of the improved uptake, it was shown that putrescine rapidly reversed the antiproliferative effect of DFMO (Fig. 11.5); DAP or Mg^{2+} did not reverse the effects of DFMO depletion (Oredsson et al., 1984).

The presence of DFMO in ascites cell cultures caused a 1500-fold increase in the content of decarboxylated AdoMet (dAdoMet) (Oredsson et al., 1986). The depletion of putrescine and spermidine eliminated the aminopropyl acceptors for spermidine and spermine synthesis, resulting in the accumulation of this adenine derivative.

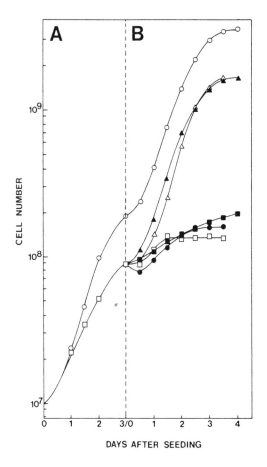

CELL NUMBER

DAYS AFTER SEEDING

FIG. 11.5 Effects of DFMO on the growth of (A) Ehrlich ascites tumor cells and the addition of cations to (B) poly-amine-depleted cells; adapted from Oredsson et al. (1984). (A) The cells are grown in the (□) presence or (○) absence of DFMO. (B) The control cells are reseeded in DFMO-free medium. Polyamine-depleted cells were reseeded in (△) DFMO-free medium or (□) medium containing 5mM DFMO. Cultures of the latter group were supplemented with (▲)putrescine, (●) 1,3-diaminopropane, or (■) Mg^{2+}.

A similar effect occurred in studies on the effects of DFMO on plated mouse fibroblasts and on regenerating rat liver (Pegg et al., 1982). The nucleotide pools of the ascites cells, including those of ADP, ATP, and other ribonucleotides, also increased markedly, suggesting an accumulation as a result of a possible inhibition of the synthesis of the polymeric nucleic acids. More recently it was found that aminoguanidine, an inhibitor of Cu amine oxidases, produces a severalfold increase in the AdoMetDC in L1210 leukemia cells, an effect caused by a stabilization of the enzyme (Stjernborg & Persson, 1993). Early analogues inhibitory to the synthesis of AdoMet reduced pools of that essential metabolite, but did not greatly affect protein or polyamine synthesis. However, one of these analogues did evoke a compensatory increase of the AdoMetDC, pointing once again

to the existence of a problem in the design and development of such "specific" inhibitors.

The supposition that results obtained with tissue cultures (e.g., Ehrlich ascites cells) might be easily extrapolable to the intact tumor-bearing mouse was soon abandoned. The effects of inhibitors on the cultures revealed a series of unexpected complications: enzyme stabilization, stimulation of uptake by polyamine depletion, and cytostatic versus lethal effects of the inhibitors. As additional more specific and powerful inhibitors became available, it was important to explore the effects of the inhibitor on the cell system more thoroughly before turning to cancer chemotherapy. For example, a more potent specific irreversible inhibitor of ODC than DFMO, (2R,5R)-6-heptyne-2,5-diamine, depleted all the polyamines including spermine of a rat hepatoma line (HTC cells) and also killed the cells (Mamont et al., 1984). It appeared that the loss of viable cells related to the limitation of spermine biosynthesis (Gerner & Mamont, 1986). A more specific test of this apparent correlation was subsequently attempted with the aid of inhibitors of spermine synthase.

The bis-ethyl analogues of that polyamine, which were not specific enzyme inhibitors, entered L1210 leukemia cells and displaced normal polyamines. They were found to inhibit cells by a combination of several different mechanisms (Porter et al., 1987). One of these mechanisms was an astonishingly great induction of the spermidine/spermine N^1-acetyl transferase, which led to a depletion of the normal amines. The significance and consequences of this observation will be discussed in Chapter 17. The depletion of normal polyamines by DFMO or a bis-(ethyl) polyamine has been found to produce a small but significant decrease in intracellular pH (Poulin & Pegg, 1995).

Enucleation and Distribution of Amines

When confluent monolayers of mouse fibroblasts are incubated with cytochalasin B, the cell microfilaments are disassembled, and the cells, now incapable of normal division, become multinucleate. Within 30 min of centrifugation in the presence of cytochalasin B, two classes of cellular fragments were obtained. In one of these, nuclei were surrounded by a thin layer of cytoplasm (karyoplasts); the other comprised enucleate cytoplasts. Karyoplasts and cytoplasts are then easily separated by density gradient centrifugation. Karyoplasts can regenerate completely, but only begin making DNA when the cytoplasm is almost completely restored (White et al., 1983).

Polyamine and enzyme analysis of whole cells, karyoplasts and cytoplasts (Table 11.1) revealed a concentration of cell spermidine and spermine to 6.1 and 1.3 mM, respectively, in the nucleated fragment (McCormick, 1977). This compartment, rich in DNA and containing a quarter of the RNA, was almost devoid of ODC. The cytoplasts readily converted ornithine to putrescine but were essentially unable to convert that pu-

TABLE 11.1   Polyamine, ODC, AdoMetDC, and macromolecule content in whole cells, cytoplasts, and karyoplasts.

	Spermidine	Spermine	ODC	AdoMetDC	DNA	RNA	Protein
	(nmol/10^6 particles)		(pmol CO_2/h/10^6 particles)		(μg/10^6 particles)		
Whole cells	5.2 ± 0.58	1.02 ± 0.30	142 ± 25	160 ± 37	11.1 ± 0.9	48 ± 4	312 ± 27
Cytoplasts	2.0 ± 0.23	0.41 ± 0.10	81 ± 15	84 ± 22	0.8 ± 0.2	26 ± 3	144 ± 12
Karyoplasts	2.46 ± 0.46	0.52 ± 0.18	6.7 ± 1.1	50 ± 16	9.7 ± 1.2	11 ± 2	88 ± 8
Recovery (%)	85	91	62	83	94	77	74
Recovered material in (%)							
Cytoplasts	45	44	91	62	3.7	70	62
Karyoplasts	55	56	9	38	96.3	30	38

ODC, ornithine decarboxylase; AdoMetDC, S-adenosyl methionine decarboxylase. Values are mean $\pm$ SEM for six determinations. Adapted from McCormick (1977).

trescine to spermidine, in contrast to whole cells that were very active in the biosynthesis of spermidine from ornithine. A similar result, the inability to convert methionine to spermidine, is also lost in cytoplasts in contrast to whole cells, although about half of the AdoMetDC can be found in the cytoplasts.

Cytochalasin B (10^{-6} M) also inhibits ODC and AdoMetDC (Sunkara et al., 1981a). This causes a rapid fall of the activity of these enzymes in HeLa cells in suspension cultures (within an hour) and a slower but significant decline in putrescine and spermidine (3–6 h). The relatively high concentration of spermine in the nuclei of cultured cells has been found by other aqueous isolation techniques (Mach et al., 1982). The washing of nuclei results in the loss of polyamine, in contrast to the effect of sealing of the nuclei by the cell membrane in karyoplast preparations.

Studies with Monolayers

Many kinds of metabolically active cells capable of cell division will attach to a plastic surface and will multiply in the presence of an appropriate medium to form a confluent monolayer. In general the reseeding and subsequent growth evoke marked increases in the polyamines within 1 or 2 days. In one such experiment with mouse L cells, putrescine, spermidine, and spermine achieved maximal levels 3.3, 2.2, and 1.8 times the initial values, respectively (Gohda et al., 1983). In experiments with cultures of HeLa cells, serum was used to initiate the growth of the quiescent cells. An early production of ODC 2–4 h after addition of serum was followed quickly by production of putrescine and successively by spermidine, which followed the almost simultaneous appearance of AdoMetDC. The rapidity of the rates of polyamine synthesis from labeled ornithine suggested that the amino acid may be a limiting metabolite (Maudsley et al., 1978). An inhibitor of ODC will quickly deplete putrescine, followed by a drop in spermidine, resulting in an inhibition of cell proliferation; one or the other amine will reinstitute proliferation.

The presence of bovine sera in the medium may add an amine oxidase that can convert added spermine or spermidine to toxic materials. Many studies were done with horse sera thought to lack the amine oxidases or with fetal calf sera low in these enzymes. Even in the presence of horse serum, the presence of spermine (50 μg/mL) inhibited the synthesis of protein and RNA during the growth of cancer cells (Goldstein, 1965). Spermine toxicity has been reexamined with baby hamster kidney (BHK) cells grown in monolayers with media containing horse serum. The ID_{50} (dose effective in 50% of subjects) of spermine in a 24-h exposure was about 1 mM and aminoguanidine, an active inhibitor of a bovine serum amine oxidase, prevented an early inhibition of growth (in 12 h) but not a significant later inhibition. If the cells were pretreated with an inhibitor of PAO, which did degrade free spermine, the toxicity of the spermine taken into the cells was even greater. Both effects suggest that high levels of intracellular spermine are toxic in their own right, and this toxicity is unrelated to effects stemming from the oxidation of the amine (Brunton et al., 1991).

BHK cells

Monolayers of BHK cells have been used to study many aspects of polyamine metabolism. Confluent or density-inhibited cultures synthesized little polyamine and took up only small amounts of exogenous putrescine. Free and bound spermidine were continuously excreted, whereas spermine was retained. Addition of fresh serum (which stimulated growth) increased uptake of putrescine, increased biosynthesis of the polyamines, and decreased excretion. Transfer of growing cells to serum-depleted media reversed these processes. Thus, transport, synthesis, and elimination were a function of the growth phase of the cells (Wallace & Keir, 1981).

When the Mg^{2+} content of the medium was increased to 15 mM 30 min prior to the addition of serum, the increase of ODC and the accumulation of cell spermidine were both inhibited somewhat (80 and 27%, respectively) without any effect on the stimulation of cell growth. The spermidine/spermine ratio was significantly reduced in which Mg^{2+} apparently partially substituted for the polyamines (Melvin & Keir, 1979).

The addition of serum to quiescent cultures initiates DNA synthesis and induces an N^1-acetyltransferase concomitantly (Wallace et al., 1985). The induction and synthesis of a spermidine N^1-acetyltransferase in the activation of quiescent bovine lymphocytes was observed by Matsui et al. (1983). They suggested that acetylation serves to reduce spermine content, increases excretion of the oxidation product, N^1-acetylspermidine, and thereby facilitates duplication of freed DNA. Cells exposed to inhibitory nucleosides lose free and bound polyamines when growth is inhibited (Melvin et al., 1978). However, growth-inhibitory drugs such as α-methylornithine and MGBG, which also decrease cellular polyamines, decrease excretion of polyamine (Wallace & Keir, 1986) despite an increase in the activity of an N-acetyltransferase.

The main inducible enzyme is indeed a cytosolic spermidine/spermine N^1-acetyl transferase. However, another enzyme found in the nucleus, the site of much bound spermine, produces an N^8-acetylspermidine and its effect on nuclear spermine is not known. The nuclear enzyme also acetylates MGBG, and Wallace et al. (1988) suggested that other acetyltransferases capable of acetylation of MGBG also exist in tissues. Thus, the control, balance, excretion, and degradation of N-acetyl-polyamines are important in the dissection of the control of growth of BHK cells.

The large number of conditions and compounds evoking an increase in this acetylase suggests the possibility of a common signal, namely the intracellular content of Ca^{2+} for induction of the enzyme (Wallace & Quick, 1994). The transient expression of the enzyme in transfected cells evokes a rapid decrease of triamine and tetramine and increase of putrescine in the cells, reinforcing the hypothesis that this acetylase protects cells against high levels of spermidine and spermine (Vargiu & Persson 1994).

An increase in the activity of the spermidine/spermine N^1-acetyltransferase was detected in the response of cell monolayers to normal polyamines and their analogues (Erwin & Pegg, 1986). Analogues that might affect this N^1-acetyltransferase have been synthesized and a group of seven such analogues, presented in Figure 11.6, was tested as possible substrates and inhibitors of this enzyme and of the spermine synthase. Only compound 5, possessing an unsubstituted aminopropyl moiety and a relatively unhindered secondary amine, served as a significant substrate. None were significant substrates of spermine synthase, although all did inhibit the synthesis of spermine at millimolar levels. Transformed cells (SV-3T3), grown in monolayer cultures in the presence of horse sera and aminoguanidine, were depleted of spermidine by DFMO and were tested with compounds 1–7. All but the specific spermine analogue, compound 7, were stimulatory at 10 μM concentrations and achieved the intracellular concentrations typical of control spermidine. In each case spermine concentrations were markedly lowered. Thus, nonmetabolizable analogues of spermidine and spermine can fulfill the roles of these normal amines in the stimulation and growth of these cultures. Also, the degradation of spermine to spermidine was not an essential step in the re-

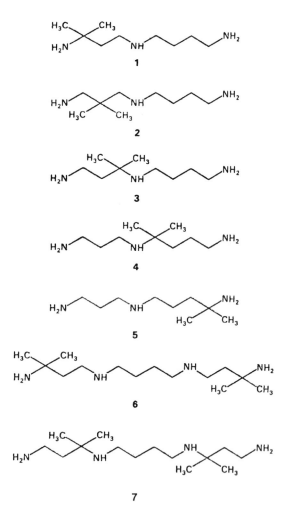

FIG. 11.6 Structure of nonmetabolizable polyamines; adapted from Nagarajan et al. (1988). Labels 1–7 are, respectively: 1,1-dimethylspermidine; 2,2-dimethylspermidine; 3,3-dimethyl-spermidine; 5,5-dimethylspermidine; 8,8-dimethylspermidine; 1,1,12,12-tetramethylspermine; 3,3,10,10-tetramethylspermine.

sumption of growth of spermidine-depleted cells (Nagarajan et al., 1988).

The ability of serum components (e.g., hormones, cAMP) to initiate a cycle of growth of BHK and other cells in monolayers has been used to study the early stages of the transformation of quiescence to active multiplication. Extensive reviews on these effects in mammalian systems have appeared (Scalabrino et al., 1991; Scalabrino & Lorenzini, 1991) and include data obtained with BHK cells. The stimulation of ODC synthesis has been recorded for insulin (Hogan et al., 1974) and several sterol hormones (Lin et al., 1980), as well as a specific effect of parathormone in differentiated chondrocytes but not in the control BHK cells (Takegawa et al., 1980). Stimulatory effects of cAMP have also been discussed. In almost synchronous cultures of

Chinese hamster cells, the increase of cAMP occurs just prior to the increase of ODC and AdoMetDC in the early premitotic phase, G_1, which precedes DNA synthesis (Russell & Stambrook, 1975). Studies with BHK cells similarly suggested a regulation of ODC by cellular cAMP level (Hibasami et al., 1977).

Other cell types

The early stimulation of ODC and other biosynthetic enzymes of the polyamines suggested some relationships of the polyamines to the subsequent synthesis of DNA (Andersson et al., 1978). The inhibition of ODC in synchronized Chinese hamster ovary (CHO) cells by DAP inhibited DNA synthesis but did not inhibit a progression to mitosis. DAP did not affect the concentrations of spermidine and spermine. The conclusion was that putrescine was essential for DNA synthesis but not for mitosis and cell division (Sunkara et al., 1977). Hyperthermia (43°C), which markedly reduced ODC and AdoMetDC in CHO cells, also inhibited DNA synthesis and mitosis until polyamine accumulation occurred (Fuller et al., 1977).

Serum-stimulated CHO cells were depleted of putrescine by α-methylornithine (5 mM) and developed longer doubling times, marked specifically by extension of the premitotic phases of growth. These were markedly reduced to normal intervals by supplementation with putrescine (Harada & Morris, 1981). It may be noted that the inhibitory effects of very high concentrations of this inhibitor (50 mM) applied to rat brain tumor cells cannot be reversed by exogenous putrescine (Seidenfeld & Marton, 1980).

Serum may be replaced by various defined growth factors (i.e., insulin, transferrin, and ferrous sulfate) to permit growth of CHO cells at a rate afforded by serum. However, apparently "normal" growth in the defined medium leads to a markedly lowered putrescine, spermidine, and ODC content of the cells; these polyamines may be restored to the serum-stimulated levels by supplementation with ornithine (Sertich et al., 1986). The AdoMetDC was at normal levels in the defined media, despite wide variations in the spermidine contents of the cells and suggests that intracellular spermidine is not the major regulator of this enzyme.

An arginase-deficient CHO line that was dependent on polyamines for growth permitted the examination of polyamine-starved cells (Pohjanpelto & Knuutila, 1982); such deprivation causes major chromosome aberrations in most mitoses. The staining pattern suggested a prolongation of the cell cycle, as well as the previously observed disappearance of actin filaments and microfilaments.

When monolayers of serum-depleted human fibroblasts were stimulated by fresh serum-containing media, DFMO blocked putrescine and spermidine accumulation severely, but did not inhibit the production of DNA, RNA, and protein. The residual low level of these polyamines may have been adequate for these syntheses. On the other hand, MGBG selectively prevented synthesis of DNA and protein in early periods and partially inhibited RNA synthesis. Exogenous spermidine and spermine permitted DNA synthesis in the presence of MGBG; nevertheless, the data suggest an inhibitory action of MGBG on systems other than those of polyamine synthesis, for example, protein synthesis (Hölttä et al., 1979a).

The relations of limited polyamine availability and the effects of DFMO on DNA synthesis after serum stimulation were explored in an experiment in which murine fibroblasts were depleted of polyamines with DFMO before becoming quiescent. Such depleted cells were not activated by serum, but did so if the serum was supplemented with putrescine. Thus, the polyamine available to quiescent cells did contribute to the commitment to pass from quiescence to DNA replication (Schaefer & Seidenfeld, 1987). Rupniak and Paul (1978) also concluded that the increase in polyamine content does not determine the initial effects of serum stimulation.

The requirement for polyamines in DNA synthesis has also been studied in HeLa cells synchronized by the availability of thymidine. The presence of DFMO during this manipulation essentially eliminated putrescine, all but traces of spermidine, and 80% of the spermine. Resupply of spermidine or spermine with labeled thymidine to DFMO-treated cells restored synchronous DNA synthesis almost completely, whereas the effect of putrescine was less pronounced. Analysis of the polyamine contents of the cells after timed resupplementation suggested that spermidine or spermine are required in preparing the traverse of G_1, the phase prior to the S phase, that is, the period of the synthesis of DNA. Spermidine completely reversed the deficiency of putrescine induced by DFMO (Gallo et al., 1986).

When [^{14}C]-spermidine was incubated with activated CHO cells, a small fraction was present within the cells as the N^1- and N^8-acetylspermidine (Prussak & Russell, 1980). Deacetylation of the N^8 derivative can be inhibited by 7-amino-2-heptanone, but the inhibition did not affect cell growth. The N^8 derivative is normally excreted by CHO cells as a major polyamine product (Hyvönen, 1989). Hyvönen suggested that the formation of the N^8 derivative is entirely a nuclear activity that facilitates the return of a reactivable spermidine to the cytoplasm.

A relatively specific inhibitor of mammalian spermine synthase, S-methyl-5′-methylthioadenosine (AdoS$^+$(CH$_3$)$_2$), was used to explore the role of spermine in the monolayer growth of transformed 3T3 cells. Applied at 10^{-4} M for 2.5 days, the culture content of spermine fell over 80% while spermidine content increased about 65%. The cells were not killed nor was there an observed decrease in cellular growth rate, implying that spermine was not essential for growth (Pegg & Coward, 1985).

When CHO cells are incubated with [^{3}H]-spermidine, the four-carbon moiety of the triamine is linked to lysine in a cell protein of 18 kDa to form hypusine. The reaction has been found in many other cell types, of which human lymphocytes stimulated to grow and divide by phytohemagglutinin (PHA) are a particularly useful test system (Cooper et al., 1982). The formation of hypusine begins actively about 6 h after stimulation and continues to cell division. HTC cells can be depleted in spermidine by DFMO and can be used to study the effects of the concentration of intracellular spermidine on hypusine formation (Gerner & Mamont, 1985). This appears to be maximal at about 7–14 nmol spermidine/mg protein. The turnover of the hypusine-containing protein eIF-5A proved to be very slow, with a half-life approaching that of the doubling time (24 h) of the HTC cells. The rate of synthesis of hypusine in this protein has been shown to be related to the rate of protein synthesis in general and to the growth rate of CHO cells and human diploid fibroblasts (Torrelio et al., 1984).

Some DFMO-resistant HTC cells are ODC overproducers, which can accumulate high toxic levels of exogenous spermidine or spermine by unregulated transport. The experiments point to the existence of a polyamine-regulated transport protein for spermidine and spermine (Mitchell et al., 1992a,b).

Differentiation of mouse neuroblastoma cells

Mouse neuroblastoma cells may be induced by many agents to increase their ODC content. Some of these, such as the induction of ODC by asparagine in minimal (simple salts/glucose) medium, have been surprising (Chen, 1980; Chen & Canellakis, 1977). The effect was ascribed to an intracellular alkalinization (Fong & Law, 1988). In cultured hepatocytes the effect was related to a marked increase in the translation of ODC mRNA; this effect was inhibited by putrescine (Kanamoto et al., 1991). Returning to neuroblastoma cells containing an amplified ODC gene, the increase of activity by asparagine was found to be accompanied by a marked increase of ODC mRNA and relatively stable ODC protein (Chen & Chen, 1992).

The cells have been found to differentiate and lose tumorigenicity as a result of serum deprivation or following the addition of cAMP analogues or DFMO, both of which reduce ODC and polyamine contents. Differentiation is also marked by a steady decrease in the rate of putrescine transport (Chen & Rinehart, 1981). The spermidine content of differentiated cells in extensive neurite webs may also fall to a fifth of that in the undifferentiated cells (Chen et al., 1982). The effects of dibutyryl cAMP or DFMO can be blocked by the exogenous addition of putrescine or spermidine, implicating the latter compounds in differentiation. However, an ODC-overproducing clonal variant, extraordinarily high in polyamine content, was not impelled to differentiate by cAMP or DFMO despite a marked reduction in polyamine content. When this strain was induced to differentiate by serum deprivation, this effect was not prevented by exogenous polyamines (Chen & Chen, 1991). The very high polyamine levels in this mutant were not observed to be toxic.

The labeling of hypusine in an 18-kDa protein by [14C]-putrescine via spermidine is more active in the undifferentiated cells (Chen, 1983; Chen & Liu, 1981). On the other hand, the differentiated cells are more active in the conversion of putrescine to amino acids, as in aging cells. This reaction was selectively inhibited by aminoguanidine.

Activation of Human Lymphocytes

A well-defined plant glycoprotein (PHA) stimulates quiescent lymphocytes to multiply. This system permits the study of the relations of the cell cycle events to polyamine synthesis and accumulations. These events include the synthesis of the protein-synthesizing machinery (i.e., ribosomal RNA and ribosomes) and DNA synthesis. In studies in which purified human lymphocytes were activated by the PHA concanavalin A, AdoMetDC increased before ODC, and two periods of increase could be discerned (Kay & Lindsay, 1973). The first increase appears to relate to an increase in the synthesis of ribosomal RNA, preceding DNA synthesis, which occurs after a lag of about 30 h. AdoMetDC had a rapid turnover, that is, a decrease in the presence of cycloheximide, whereas that of ODC was slower.

The inhibition of spermidine synthesis by high concentrations (1 mM) of MGBG did not markedly impede the synthesis of rRNA for the first 15 h, although protein synthesis was inhibited from the beginning (Kay & Pegg, 1973). The early synthesis and accumulation of polyamines by PHA-activated human lymphocytes was abolished by DFMO and by MGBG, and these were associated with an inhibition of DNA synthesis. The latter could be initiated by exogenous amines. Spermidine also particularly protected against an inhibition of protein synthesis by DFMO or MGBG. These inhibitors and the depletion of spermidine affected RNA synthesis somewhat after some 20 h, but did not affect RNA degradation and turnover in the earlier periods (Hölttä et al., 1979b). The newly synthesized rRNA and ribosomal subunits of the stimulated cultures inhibited by DFMO were found to enter polyribosomal structures less efficiently. This result suggests a role for polyamines in the initiation of protein synthesis (Hölttä & Hovi, 1985).

A number of very early steps had been detected as a result of PHA (or lectin) stimulation. These include an activation of transglutaminase, a Ca^{2+}-requiring cross-linking enzyme, and the activation of a cAMP-dependent protein kinase (see Chap. 21). The sequential activation of the various enzymes is given in Figure 11.7 (Bachrach et al., 1981). These authors place the increase of ODC and RNA synthesis before that of AdoMetDC.

Korpela et al. (1981) demonstrated a later synthesis of spermidine synthase, beginning some 24–48 h after activation by PHA. An increase of polyamine acetylase occurred even later after 48 h, but polyamine oxidase fell in the cells, indicating a minor role, if any, of a catabolic pathway in this system. Nevertheless, the production of small amounts of N-acetyl putrescine and N^1-

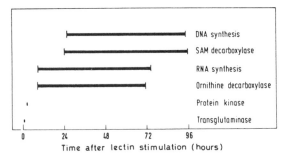

FIG. 11.7 Biochemical events after lymphocyte stimulation; adapted from Bachrach et al. (1981).

acetyl spermidine from $[^{14}C]$-putrescine have been detected in human lymphocytes, with somewhat greater amounts in the PHA-activated cells (Menashe et al., 1980). Inactivated lymphocytes, very low in ODC, may generate putrescine from uptake or degradation of acetylspermidine. The amount of spermidine acetyl-transferase in uncultured resting human lymphocytes has been reported to increase in several diseases, which include psoriasis, chronic lymphocytic leukemia, and acute hepatitis (A and B) (Pezzali et al., 1984).

In rat hepatocytes, which can be induced to synthesize ODC briefly without proliferation, putrescine content increased and then decreased as did spermidine and protein synthesis. DAP hastened and exacerbated the decrease of spermidine and the rate of protein synthesis. This control of the rate of protein synthesis has been ascribed to a lowered content of hypusine in the spermidine-depleted cell (Schulz et al., 1989).

Experiments with Bovine Lymphocytes

Many cells are obtainable from bovine lymph nodes (1 $\times 10^{10}$ small lymphocytes per node), providing material potentially suitable for more extensive biochemical study. After concanavalin A induced lymphocyte transformation, the inception and kinetics of the various syntheses and effects of inhibitors on these events were estimated (Fillingame & Morris, 1973). The addition of 8 μM MGBG permitted sharp cutoffs of the accumulation of spermidine and spermine (Fig. 11.8) and led to a rapid accumulation of putrescine. No effect was detected on the synthesis of RNA or protein. In this synchronized system, the synthesis of RNA occurred prior to the accumulation of spermidine and spermine. It was concluded that net cellular levels of polyamine do not regulate RNA synthesis. Furthermore, the elimination of accumulations of polyamine by MGBG did not lead to a greater rate of degradation of the RNA.

However, low concentrations (2–8 μM) of MGBG did inhibit DNA synthesis (i.e., thymidine incorporation) and the inhibition is reversed by exogenous spermidine. Thus, activated lymphocytes can proceed in the

cell cycle from quiescence (G_0) through active synthesis of RNA and protein (G_1) up to the initiation of DNA synthesis (S phase) without detectable accumulated spermidine and spermine (Fillingame et al., 1975). MGBG does permit an accumulation of putrescine, which enables the synthesis of some DNA. Cutting this off with α-methylornithine increased the inhibition of DNA synthesis, augmenting the polyamine requirement for this function (Morris et al., 1977). Putrescine can therefore replace some need for spermidine and spermine in DNA synthesis. The activity of arginase increases almost fourfold before DNA synthesis and thereby increases the availability of ornithine (Klein & Morris, 1978). DFMO (1 mM), which depressed putrescine and spermidine, also caused a 60% inhibition of DNA synthesis (Seyfried & Morris, 1979).

The activated bovine lymphocyte system has been studied in exploration of a number of other polyamine problems, such as the search for inhibitors of Ado-MetDC that are more specific in this activity than is MGBG. The biphasic increase of this enzyme has led to a study of the regulation of the synthesis of this enzyme with the aid of specific antibody. The initial increase of enzyme was shown to be a function of enzyme synthesis, whereas a secondary increase appeared to be due to enzyme stabilization, that is, a longer half-life (Seyfried et al., 1982).

Activation is also effected by the spermine analogue, N^1, N^{12}-bis(ethyl)spermine, in the bovine lymphocyte system. PHA-induced cells depleted of polyamine by DFMO and MGBG were able to recover the ability to synthesize DNA in the presence of either the analogue or spermine (Igarashi et al., 1990). The relative inability of diamine oxidase to convert the bis-ethyl compound to a toxic metabolite suggests that the analogue may be more useful than spermine itself in treatment of some pathologies overcome by polyamine. The authors have used this lowered toxicity of the analogue to inhibit stress-induced gastric ulceration in mice.

Additional Dissections of Cell Cycle

Efforts to pinpoint the functions of the polyamines in specific stages of the mammalian cell cycle have passed from measurement of their concentration and relation to cell number and apparent growth rate, to the use of synchronized cultures. These have induced studies of:

1. the stationary phase to a period of active growth,
2. the effects of serum and other nutrient or hormonal addition,
3. cytostatic inhibitors of polyamine synthesis to arrest growth and the use of polyamines to reinitiate growth, and
4. polyamine-depleted mutant cells.

In the last group arginase-deficient CHO cells, which require ornithine or polyamines, were shown to lose their internal filamentous structure and to develop chro-

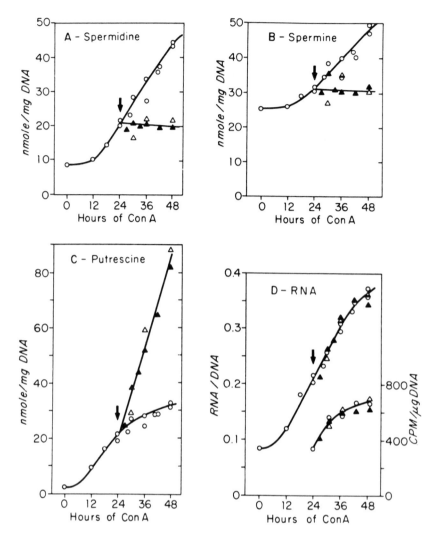

FIG. 11.8 Polyamine contents of concanavaline A stimulated bovine lymphocytes before and after treatment with MGBG; adapted from Fillingame and Morris (1973). MGBG, (▲) 8 μM or (■) 40 μM, is added at 24 h, representing the contents of (A) spermidine, (B) spermine, (C) putrescine, and (D) RNA.

mosome abnormalities under conditions of polyamine depletion. In this condition the depleted cells accumulated in the S phase and the post S-G_2 phase. Exacerbation of the deficiency by incubation in DFMO caused an additional growth suppression and pile up in the S and G_2 phases (Anehus et al., 1984). The disorganization of the cytoskeleton of CHO or HeLa cells by colchicine in a stimulatory medium in itself inhibits the production of ODC (Rumsby & Puck, 1982). The arrest of mitosis by colchicine (colcemide) has been used to develop highly synchronized suspension cultures.

Synchronized cultures of CHO cells and HeLa cells have been used to study the accumulation of polyamines and the incorporation of ^{3}H-uridine in RNA (Goyns, 1981). With both types of cells, spermidine and sper-

mine almost doubled in the cell cycle. However, spermidine increased mainly during the S and G_2 phases, whereas putrescine and spermine increased slightly later. A good correlation was demonstrated only for spermidine and the incorporation of [^{3}H]-uridine, that is, RNA synthesis and content, suggesting an effect of RNA in determining the rate of spermidine accumulation.

A temperature-sensitive mutant of BHK cells, unable to process ribosomal RNA precursor to ribosomal RNA at 39°C, loses ODC selectively at this temperature, perhaps through a reduced rate of synthesis, and is eventually unable to synthesize essential polyamines (Levin & Clark, 1979). There is also a slower partial loss of Ado-MetDC but not of various rRNA methylases. The results have suggested a role of the polyamines in the course of rRNA maturation.

In studies of synchronized HeLa cells grown in suspension culture and subsequently as monolayers, it was found that an accumulation of ornithine in the cells in the G_1 phase rapidly decreased during the S phase. In this system AdoMetDC increased sharply in the G_1 phase, preceding the increase of ODC by approximately 8 h. Diamine oxidase was maximal just before the beginning of the S phase, as was ODC. The latter was at a maximum at the beginning of the S phase and decreased in that 4–5-h period, as did ornithine. Putrescine, spermidine, and spermine were highest during mitosis and putrescine and spermine fell during G_1, to be partially restored early in G_2. The rates of synthesis, determined from the labeling of pools derived from labeled exogenous precursors, were high during mitosis and G_1 and relatively low, although significant, during the S phase (Sunkara et al., 1981b). The results have been interpreted to suggest a high rate of polyamine turnover during mitosis and G_1.

The increases of ODC in newly initiated cell cycles are described as "cell cycle dependent." Some hormones or second messengers may affect ODC as cell cycle independent determinants. Although cAMP can evoke arrest in G_1 in mouse lymphoma cells and cause a decrease of ODC, treatment by dibutyryl cAMP at any stage of the cell cycle of a partially synchronized culture will produce a marked drop of the enzyme over a 6-h period. The regulation of ODC by cAMP has been dubbed "cell cycle independent" (Kaiser et al., 1979).

Changes in Na^+, K^+, Ca^{2+}, and Mg^{2+} during the HeLa cycle have been reported (Morrill & Robbins, 1984). A peak of DNA synthesis in a synchronized culture occurs at a minimum of Mg^{2+} concentration. The effects of K^+ depletion on the inhibition of protein synthesis in tumor cells (Lubin, 1967) and of K^+ excess on the inhibition of DNA synthesis in BHK cells (Orr et al., 1972), as well as the effects of altered osmolarity on the induction of ODC indicate that it has become essential to know the status of the inorganic and organic cations in a given experimental situation.

Reversing Tumorigenicity

In studies on cultured tumor cells the exogenous addition of cAMP changed their structural features (e.g., shape) to that of an apparently nontumorigenic phenotype. The cytoskeleton had been modified in such a way as to modulate the classes of nuclear DNA accessible to a deoxyribonuclease (Puck et al., 1991). The possibility that cancer cells (e.g., mouse neuroblastoma cells) could be induced to become nontumorigenic differentiating cells is obviously of great interest.

One system under active study is that of the Friend erythroleukemia cells, which can form erythroid (hemoglobin-producing) cells when induced by various agents. These virus-transformed murine cells can be induced to develop toward terminal cell division, accumulate globin mRNA, increase their content of enzymes of the heme synthetic pathway, and synthesize several hemoglobins. Several classes of agents that induced this differentiation were discovered, one class of which stimulated ODC. This induction was inhibited by inhibitors of polyamine synthesis, MGBG etc., and the inhibition was reversed by exogenous polyamine (Gazitt & Friend, 1980). In normal rodents DFMO will inhibit the production of peripheral blood elements, such as erythrocytes, and intraperitoneal administration of putrescine will reverse these hematological effects (Luk et al., 1983). Nevertheless, it appeared initially that the target of DFMO, ODC, played a greater role in proliferation than in the differentiation of erythroid precursor cells (Niskanen et al., 1983b). However, cells showing the largest increases in ODC and polyamine levels also exhibited extensive differentiation (Schindler et al., 1985).

Spermidine in particular was subsequently found to be important in the proliferation and differentiation of the murine erythroleukemia cells (Sugiura et al., 1984; Watanabe et al., 1985). The uptake of spermine was shown to be required to induce hemoglobin production in certain cells (Clement et al., 1995). Methylthioadenosine inhibited differentiation in the system without depleting spermidine content; the compound is a known inhibitor of DNA methylation (Shafman et al., 1984). Some inhibitory agents, such as DFMO, which may cause an increase in AdoMet, also lead to hypomethylation of DNA (Papazafiri & Osborne, 1987). However, in some systems DFMO induces terminal differentiation; that is, the precise role of hypomethylation is unclear in these events. The course of putrescine utilization and polyamine levels were found to vary as differentiation progressed (Canellakis et al., 1984). Some changes were similar to those noted earlier in aging, for example, an increased conversion of putrescine to amino acids.

Among compounds stimulating ODC and differentiation of the erythroleukemia cells was a polyamine analogue, hexamethylene bisacetamide (HMBA). The latter, or N,N^1-diacetyl-1,6-diaminohexane, was particularly potent among the bisacetamides, and induced >99% of the cells to differentiate at a 5 mM level (Reuben et al., 1976). It also induced differentiation in vitro in human leukemic cell lines and has been tested as a chemotherapeutic agent in clinical trials. However, although HMBA achieved high plasma concentrations, it was rapidly deacetylated and cleared via the kidney.

Several metabolites were identified from patient urine and a scheme for the metabolism of HMBA was developed (Fig. 11.9). The first metabolite, N-monoacetyl-1,6,-diaminohexane, was far more potent than HMBA itself (Meilhoc et al., 1986; Snyder et al., 1988) and markedly inhibited the levels of ODC, putrescine, and spermidine within the cells. It was not reacetylated to the diacetyl derivative. The monoacetyl compound is thought to be the inducing compound, which is generated during the administration of HMBA. It may be noted that in some cell systems, such as that of Morris rat hepatoma in exponential growth, HMBA and various acetylated polyamines (e.g., N-acetylputrescine, N^1-ace-

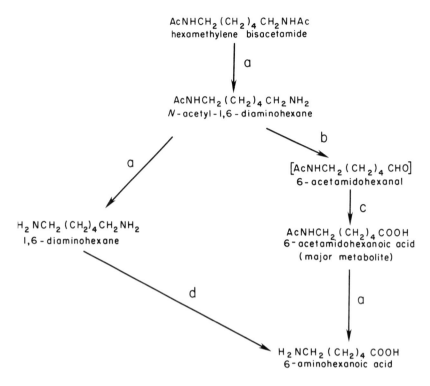

FIG. 11.9   Hexamethylene bisacetamide (HMBA), its known metabolites, and a scheme for their interconversion; adapted from Snyder et al. (1988).

tylspermidine, and N^8-acetylspermidine) are also good inducers of ODC. Nevertheless, diacetyl putrescine and HMBA inhibit the proliferation and immunological functions of B lymphocytes (Canellakis et al., 1989). HMBA also selectively inhibits the proliferation of vascular smooth muscle cells. It appears to act as an inhibitor of ADP-ribosyltransferase, preventing the loss of smooth muscle specific myosin heavy chains (Grainger et al., 1992).

Another metabolite, 6-acetamidohexanoic acid, appeared to enhance differentiation induced by either HMBA or the monoacetyl diaminohexane. It has been reported that the normal nuclear metabolite, N^8-acetylspermidine, is several hundred times more potent as an inducer of differentiation than is HMBA (Snyder et al., 1991). Several bishydroxamic acids also possess this order of activity (Breslow et al., 1991).

Polyamine metabolism appears involved in the mechanism of HMBA action. Whereas DFMO stimulated differentiation in some systems, MGBG added within the first 6 h after application of HMBA inhibited this process, suggesting that polyamine-dependent events were important to early events of commitment to differentiation (Meilhoc et al., 1987). Also, a natural spermine analogue, canavalmine (4,3,4), is an inducer of differentiation (Fujihara et al., 1985). The compound sharply reduces the level of spermidine in the cells, which may be a key to an effect requiring a prolonga-

tion of G_1 (Kiyokawa et al., 1993). Exogenous putrescine or spermidine inhibited induction by canavalmine, whereas spermine did not.

The inhibition of ODC, reduction of the levels of putrescine and eventually of spermidine by DFMO, and the apparent increase of differentiation of mouse and human cancer cells have been noted (Oredsson et al., 1985; Uhl et al., 1986). However, the separation of cell proliferation and differentiation in erythropoietic systems in the intact mouse has not been straightforward (Niskanen et al., 1983a,b; Sharkis et al., 1983). In human promyelocytes (H2-60) it was found that ODC, inhibitable by DFMO, was essential for proliferation but not for cellular differentiation (Luk et al., 1982). Furthermore, the depletion of spermidine blocked the proliferation essential for subsequent differentiation into macrophages and monocytes (Inaba et al., 1986; Kufe et al., 1984). In these studies agents were found that carried polyamine-depleted cells to differentiation. However, inhibition of ODC by DFMO does induce some malignant cell types, such as murine embryonal carcinoma cells, and this is minimized by the addition of putrescine to the medium (Oredsson et al., 1985). Manipulation of polyamine levels by DFMO and/or MGBG in human leukemic patients has not yet succeeded in defining the polyamine contents of leukemic cells at which differentiation might occur (Maddox et al., 1989).

Transformed Cells

The plating of cells and the observation of the colonial growth of primary cultures of cells, some of which proved to be of tumoral origin, revealed significant differences in colony morphologies. These differences related to the immortalization of the cells, their chromosomal complements, the decreased organization of their cytoskeletons, and their requirements for serum as a fresh nutrient. A characteristic colonial morphology was found for tumor cells: this was described as anchorage independent, permitting rounded cells lacking well-developed internal fibrous structures to pile up to a high density. Cells of this type could be isolated from tumor tissue or could be generated from normal cells infected with a tumor virus. Such viruses might be of either an RNA type such as the Rous avian sarcoma virus (RSV) or a DNA type such as simian virus 40 (SV40). The process in which cells become "immortal" is known as "transformation." Transformation may also be carried out by transfecting genetic factors, as well as by carcinogenic agents. Obviously the mechanisms of these processes are of great theoretical and practical interest in the pursuit of the causes and therapy of cancer. Some roles of the polyamines have been detected in these complex processes.

The view that the polyamines might be expected to be involved in transformations stems from observations of the following four types.

1. Many cancer patients excreted increased amounts of the polyamines in urine (Russell, 1971).
2. The growth rates of a series of rat hepatomas were closely correlated with their ODC content (Williams–Ashman et al., 1972).
3. The growth rate of some tumor cells can be correlated with their cellular contents of spermidine, ODC and AdoMetDC (Heby et al., 1975; Woo et al., 1979).
4. Effective therapy of leukemia in mice was shown to depress the abnormally high rates of polyamine synthesis (Heby & Russell, 1973).

The more recent known effects of polyamine-directed agents will be discussed in Chapters 16 and 17.

The transformation of chick embryo fibroblasts by RSV produced an increase in the putrescine content of the transformed cells that was 3–7 times that of normal (Bachrach et al., 1973). When a temperature-sensitive strain of RSV was used, causing transformation only at the permissive temperature and replication at permissive and nonpermissive temperatures, the diamine increased only at the temperature enabling transformation, that is, a change in cellular morphology (Don et al., 1975). An early increase of ODC at the permissive temperature was subsequently related to an increased half-life of the enzyme, that is, a decreased rate of ODC degradation (Bachrach, 1976). The uptake and accumulation of putrescine and spermidine by RSV-transformed cells exceeded these functions of normal cells. The conversion of spermidine to N^1-acetylspermidine and to N-acetyl-

putrescine was also far slower in the normal fibroblasts (Bachrach & Seiler, 1981).

Normal cells appear to have more diamine oxidase than do paired transformed cells (virus infected or tumor derived). Nevertheless, some transformed cells in active growth have higher levels of PAO, an enzyme that may contribute to the higher concentrations of putrescine in such cells (Quash et al., 1979). In contrast to normal cells, in which this enzyme is roughly constant during growth and quiescence, the enzyme falls considerably (fivefold) as the transformed culture approaches confluence.

Mouse fibroblasts (3T3 cells) transformed by the SV40 virus contain high levels of ODC and have been used for the isolation of the enzyme. The early introduction of affinity columns and gel filtration permitted the isolation of an enzyme of high specific activity with accurate estimates of molecular weights of active enzyme and its subunits (Boucek & Lembach, 1977).

The regulation of ODC by putrescine in 3T3 cells has been compared with that of SV40-transformed cells (Bethell & Pegg, 1981). Whereas exogenous putrescine readily blocks the increase of ODC in serum-stimulated normal cells, this inhibition was far less in the transformed cells. Normal cells in a density-inhibited state reduced their rate of putrescine uptake; this increased after serum stimulation. Transformed cells did not change their rates of uptake of the diamine. These phenomena favor an extended polyamine accumulation in transformed cells.

In 1941 I. Berenblum discovered that carcinogenesis in mouse skin is a two-stage process in which tumor development after a single dose of an initiating carcinogen is dependent on application of a tumor promotor. The appearance of ODC is an early response in the carcinogenesis of mouse skin, and high levels of the enzyme are maintained in the epidermal tumors. The question has been asked if the increased level of ODC might be essential for tumor development (Boutwell, 1978).

The induction of ODC by promoters has been studied in hamster embryo cells previously exposed to the carcinogen benzo(a)pyrene. These cells were selected for the ability to produce tumors in hamsters. The promoter, 12-o-tetradecanoyl-phorbol-13-acetate (TPA) was found to produce a far greater induction of ODC (27-fold) in confluent cultures of the transformed cells as compared to the nontumor cells. The putrescine content of the transformed cells was increased fourfold within 4 h of exposure to TPA, attained a further doubling of the diamine by 9 h, and decreased thereafter. Spermidine and spermine levels increased some 30% and slowly declined without increase of DNA synthesis or cell number. Cultures of the transformed cells stimulated by fresh medium produced much more ODC (fourfold) but not AdoMetDC if TPA was added (O'Brien & Diamond, 1977; O'Brien et al., 1980). These investigators did not detect differences in the regulation of ODC production by exogenous putrescine.

TPA stimulated the production of cyclic nucleotides and ODC in uninitiated isolated mouse epidermal cells.

The increase in cyclic nucleotides preceded the appearance of ODC. Increasing the level of such compounds enhanced the TPA-induced stimulation of ODC (Perchellet & Boutwell, 1980), indicating that the cyclic nucleotides play a significant role in the promotion of ODC by TPA. When normal human skin fibroblasts are stimulated by fresh medium in the presence of an agent such as colchicine, which depolymerizes microtubules, ODC induction is almost completely inhibited. When transformed cells are treated similarly, ODC induction remains high and is far less sensitive to colchicine (Rumsby & Puck, 1982).

The working hypothesis of a single control on "immortalization" was abandoned in the 1980s. Tumorigenesis in mouse skin was revealed to involve steps of progression after promotion, and tumorigenesis in humans is believed to be even more complex. In this decade also studies on tumor viruses led to the detection of oncogenes, that is, genes in these viruses and in tumors that were mutated forms of genes present in normal cells. It is thought that mutation of the normal gene can affect physiological targets whose modification may define a step in a multistep tumorigenic or transformation process. Some retroviruses such as RSV contain a single oncogene capable of transforming a normal cell to an "immortal" tumor cell. This gene, labeled *src*, determines the production of a new type of protein kinase, which phosphorylates tyrosine. The availability of poly-

amine is essential to the establishment of a functional *src* gene in a transformable cell (Höltta et al., 1993).

Some DNA tumor viruses contain oncogenes, two or three of which must act in concert to produce a transformation to the full-blown tumor genotype. Studies with two oncogenes of polyoma virus revealed that both oncogenes were required to transform rat embryo fibroblasts (Weinberg, 1989). Similarly at least three genes of Epstein–Barr virus contribute to immortalization (Middleton et al., 1991). Among the oncogenes of tumor cells, neither *ras* nor *myc* alone induce a full transformation of a rat embryo cell that may be carried out by the two oncogenes together. The *ras* oncogene is present in a high proportion of human colon and pancreatic cancers. It has been found that transfection of 3T3 cells by DNA carrying the *ras* oncogene increases the uptake and metabolism of labeled spermidine. In an early phase of growth the transfected cells accumulate polyamines and subsequently excrete acetyl polyamines with a high ratio of N^1-acetylspermidine to N^8-acetylspermidine (Pakala et al., 1988). Such a ratio has been observed in the urine of cancer patients. A comparable excretion of acetyl polyamine did not occur with the uninfected 3T3 cells. The *ras*-transfected cells markedly increased their ODC content; the enzyme was shown to be far more stable in the oncogene-transformed cell (Fig. 11.10) (Höltta et al., 1988).

The increase of ODC was far more pronounced in *ras*

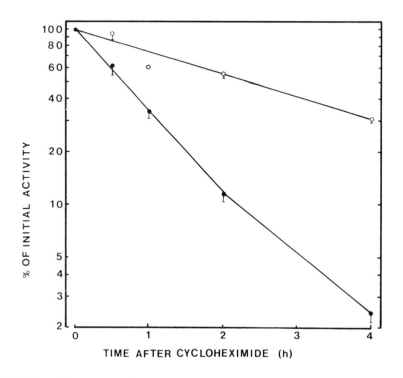

FIG. 11.10   Fall of ODC in (●) normal and (○) *ras* oncogene-transformed cells after exposure to cycloheximide; adapted from Höltta et al. (1988).

transfection of Rat-1 cells than with the *myc* oncogene in these cells, which now had an enhanced polyamine uptake (Chang et al., 1988). Treatment of human colon carcinoma cells with DFMO almost eliminates expression of the *c-myc* protooncogene essential to cell growth. ODC production is increased as a result of the polyamine depletion caused by DFMO. Nevertheless, the fall in *c-myc* expression suggests a role for the polyamines in maintaining that expression (Celano et al., 1988). The production of the mRNA transcribed from the genes for *c-myc* and histone 2A is increased four- to eightfold in isolated control nuclei to which spermidine has been added. Exogenous spermidine has a much lower effect in nuclei from DFMO-treated cells (Celano et al., 1989). The expression of some genes (e.g., *c-myc*) but not others (e.g., ODC, β-actin) is decreased by polyamine depletion, and the results appear to implicate intranuclear polyamines in the regulation of transcription.

Induction of the *ras* oncogene in 3T3 fibroblasts is accompanied by enhanced expression of another oncogene, *c-jun* (Sistonen et al., 1989). It has also been observed that polyamine depletion by DFMO of the normal rat intestinal cell line, IEC-6, decreased expression of several protooncogenes: *c-fos*, *c-myc*, and *c-jun* (Wang et al., 1993). Other workers have described a transactivated expression of the ODC gene by *c-myc* (Bello–Fernandez et al., 1993) or *c-fos* (Wrighton & Busslinger, 1993). In still another system, the inhibition of growth by DFMO inhibited the expression of several immediate early genes, of which *c-fos* and *c-jun* were two. This effect was abolished by a prior addition of putrescine (Schulze–Lohoff et al., 1994). In addition Tabib and Bachrach (1994) reported that the increase in ODC and putrescine in cells activated by *ras* in turn activates two other protooncogenes *c-myc* and *c-fos*; that is, these events can be triggered by putrescine and blocked by DFMO. These results suggest a cascade of oncogene-stimulated events involving the polyamines.

The overexpression of ODC itself in 3T3 cells evokes cellular phenomena very similar to those observed in transformation: a stimulation of cell proliferation, elimination of contact inhibition, anchorage-independent growth, and an induction of tumors in nude mice (Moshier et al., 1993). Similar phenomena of in vitro growth in several 3T3 strains with enhanced production of ODC and AdoMet carboxylase have provided more evidence for the importance of ODC and the polyamines in transformation (Shantz & Pegg, 1994). Höltta et al. (1994) implicated yet another protein, tyrosine kinase, in transformations induced by high levels of cellular ODC.

Summary and Comment

The polyamines are very much involved in the stages of the cell cycle of the animal cell and in its developmental forms, such as the dividing immortal cell or an aging differentiated cell. The transformations between these forms involve significant alterations of cellular activities in polyamine uptake, synthesis, degradation, and excretion. The definitions of these alterations require specific knowledge of the relevant enzymes and their regulation during these changes.

REFERENCES

Alhonen–Hongisto, L., Kallio, A., Pösö, H., & Jänne, J. (1979a) *Acta Chemica Scandinavica*, **B33**, 559–566.
Alhonen–Hongisto, L., Pösö, H., & Jänne, J. (1979b) *Biochimica et Biophysica Acta*, **564**, 473–487.
Alhonen–Hongisto, L., Seppänen, P., Hölttä, E., & Jänne, J. (1982) *Biochemical and Biophysical Research Communications*, **106**, 291–297.
Alhonen–Hongisto, L., Seppänen, P., & Jänne, J. (1980) *Biochemical Journal*, **192**, 941–645.
Andersson, G. & Heby, O. (1977) *Cancer Research*, **37**, 4361–4366.
Andersson, G., Österberg, S., & Heby, O. (1978) *International Journal of Biochemistry*, **9**, 263–267.
Anehus, S., Pohjanpelto, P., Baldetorp, B., Långström, E., & Heby, O. (1984) *Molecular and Cellular Biology*, **4**, 915–922.
Asotra, S., Mladenov, P. V., & Burke, R. D. (1988) *Comparative Biochemistry and Physiology*, **90B**, 885–890.
Bachrach, U. (1976) *Biochemical and Biophysical Research Communications*, **72**, 1008–1013.
Bachrach, U., Don, S., & Wiener, H. W. (1973) *Biochemical and Biophysical Research Communications*, **55**, 1035–1041.
Bachrach, U., Mensche, M., Faber, J., Desser, H., & Seiler, N. (1981) *Advances in Polyamine Research*, vol. 3, pp. 259–274. C. M. Calderara, V. Zappia, & U. Bachrach, Eds. New York: Raven Press.
Bachrach, U. & Seiler, N. (1981) *Cancer Research*, **41**, 1205–1208.
Bachrach, U. & Shtorch, A. (1985) *Cancer Research*, **45**, 215–2164.
Bello-Fernandez, C., Packham, G., & Cleveland, J. L. (1993) *Proceedings of the National Academy of Sciences of the United States of America*, **90**, 7804–7808.
Bethell, D. R. & Pegg, A. E. (1981) *Journal of Cellular Physiology*, **109**, 461–468.
Boucek, R. J. & Lembach, K. J. (1977) *Archives of Biochemistry and Biophysics*, **184**, 408–415.
Boutwell, R. K. (1978) *Carcinogenesis*, vol. 2, *Mechanisms of Tumor Promotion and Cocarcinogenesis*, pp. 49–58. T. J. Slaga, A. Sivak, & R. K. Boutwell, Eds. New York: Raven Press.
Breslow, R., Jursic, B., Yan, Z. F., Friedman, E., Leng, L., Ngo, L., Rifkind, R. A., & Marks, P. A. (1991) *Proceedings of the National Academy of Sciences of the United States of America*, **88**, 5542–5546.
Brunton, V. G., Grant, M. H., & Wallace, H. M. (1991) *Biochemical Journal*, **280**, 193–198.
Byus, C. V. & Herbst, E. J. (1976a) *Biochemical Journal*, **154**, 23–29.
——— (1976b) *Biochemical Journal*, **154**, 31–33.
Canellakis, Z. N., Marsh, L. L., & Bondy, P. K. (1989) *Yale Journal of Biology and Medicine*, **62**, 481–491.
Canellakis, Z. N., Marsh, L. L., Young, P., & Bondy, P. K. (1984) *Cancer Research*, **44**, 3841–3845.
Celano, P., Baylin, S. B., & Casero, Jr., R. A. (1989) *Journal of Biological Chemistry*, **264**, 8922–8927.
Celano, P., Baylin, S. B., Giardiello, F. M., Nelkin, B. D., & Casero, Jr., R. A. (1988) *Journal of Biological Chemistry*, **263**, 5491–5494.

Chang, B. K., Libby, P. R., Bergeron, R. J., & Porter, C. W. (1988) *Biochemical and Biophysical Research Communications*, **157**, 264–270.

Chang, Z. & Chen, K. Y. (1988) *Journal of Biological Chemistry*, **263**, 11431–11435.

Chen, K. Y. (1980) *FEBS Letters*, **119**, 307–311.

——— (1983) *Biochimica et Biophysica Acta*, **756**, 395–402.

Chen, K. Y. & Canellakis, E. S. (1977) *Proceedings of the National Academy of Sciences of the United States of America*, **74**, 6791–6795.

Chen, K. Y. & Chang, Z. (1986) *Journal of Cellular Physiology*, **128**, 27–32.

Chen, K. Y., Chang, Z., & Liu, A. Y. (1986) *Journal of Cellular Physiology*, **129**, 142–146.

Chen, K. Y. & Liu, A. Y. (1981) *FEBS Letters*, **134**, 71–74.

Chen, K. Y., Presepe, V., Parken, N., & Liu, A. Y. (1982) *Journal of Cellular Physiology*, **110**, 285–290.

Chen, K. Y. & Rhinehart, Jr., C. A. (1981) *Biochemical and Biophysical Research Communications*, **101**, 243–249.

Chen, Z. & Chen, K. Y. (1991) *Biochimica et Biophysica Acta*, **1133**, 1–8.

Chen, Z. P. & Chen, K. Y. (1992) *Journal of Biological Chemistry*, **267**, 6946–6951.

Clement, S., Delcros, J.-G., & Feuerstein, B. G. (1995) *Biochemical Journal*, **312**, 933–938.

Cooper, H. L., Park, M. H. & Folk, J. E. (1982) *Cell*, **29**, 791–797.

Dion, A. S. & Herbst, E. J. (1970) *Annals of the New York Academy of Sciences*, **171**, 723–734.

Don, S., Wiener, H., & Bachrach, U. (1975) *Cancer Research*, **35**, 194–198.

Duffy, P. E. & Kremzner, L. T. (1977) *Experimental Cell Research*, **108**, 435–440.

Emanuelsson, H. & Heby, O. (1978) *Proceedings of the National Academy of Sciences of the United States of America*, **75**, 1039–1042.

Erwin, B. G. & Pegg, A. E. (1986) *Biochemical Journal*, **238**, 581–587.

Fillingame, R. & Morris, D. (1973) *Polyamines in Normal and Neoplastic Growth*, pp. 249–260. D. Russell, Ed. New York: Raven Press.

Fillingame, R. H., Jorstad, C. M., & Morris, D. R. (1975) *Proceedings of the National Academy of Sciences of the United States of America*, **75**, 4042–4045.

Fong, W. & Law, C. (1988) *Biochemistry International*, **16**, 279–285.

Fujihara, S., Nakashima, T., & Kurogochi, Y. (1985) *Biochimica et Biophysica Acta*, **846**, 101–108.

Fujiwara, A., Kusunoki, S., Tazawa, E., & Yasumasu, I. (1983) *Development, Growth and Differentiation*, **25**, 445–452.

Fuller, D. J. M., Gerner, E. W., & Russell, D. H. (1977) *Journal of Cellular Physiology*, **93**, 81–88.

Gallo, C. J., Koza, R. A., & Herbst, E. J. (1986) *Biochemical Journal*, **238**, 37–42.

Gawlitta, W., Stockem, W., & Weber, K. (1981) *Cell and Tissue Research*, **215**, 249–261.

Gazitt, Y. & Friend, C. (1980) *Cancer Research*, **40**, 1727–1732.

Gerner, E. W. & Mamont, P. S. (1985) *Recent Progress in Polyamine Research*, pp. 563–574. L. Selmeci, M. E. Brosnan, & N. Seiler, Eds. Budapest: Akademiai Kiado.

——— (1986) *European Journal of Biochemistry*, **156**, 31–35.

Gohda, E., Takigawa, M., Inoue, H., Kato, Y., Daikuhara, Y., & Takeda, Y. (1983) *Journal of Biochemistry*, **94**, 97–106.

Goyns, M. H. (1981) *Experientia*, **37**, 34–35.

Grainger, D. J., Hesketh, T. R., Weissberg, P. L., & Metcalf, J. C. (1992) *Biochemical Journal*, **283**, 403–408.

Grant, N. J., Aimar, C., & Oriol-Audit, C. (1984) *Experimental Cell Research*, **150**, 483–497.

Grant, N. J. & Oriol-Audit, C. (1985) *European Journal of Cell Biology*, **36**, 239–246.

Ham, R. G. (1964) *Biochemical and Biophysical Research Communications*, **14**, 34–38.

——— (1974) *Journal of the National Cancer Institute*, **53**, 1459–1463.

Hamana, K. & Matsuzaki, S. (1979) *FEBS Letters*, **99**, 325–328.

Hamana, K., Matsuzaki, S., & Inoue, K. (1984) *Journal of Biochemistry*, **95**, 1803–1809.

Hamana, K., Suzuki, M., Wakabayashi, T., & Matsuzaki, S. (1989) *Comparative Biochemistry and Physiology*, **92B**, 691–695.

Harada, J. J. & Morris, D. R. (1981) *Molecular and Cellular Biology*, **1**, 594–599.

Häussinger, D. & Lang, F. (1991) *Biochemistry and Cell Biology*, **69**, 1–4.

Hawell, L., Tjandrawinata, R. R., Fukumoto, G. H., & Byus, C. V. (1994) *Journal of Biological Chemistry*, **269**, 7412–7418.

Hayflick, L. (1991) *Mutation Research*, **256**, 69–80.

Heby, O. & Emanuelsson, H. (1978) *Cell and Tissue Research*, **194**, 103–114.

Heby, O., Marton, L. J., Wilson, C. B., & Martínez, H. M. (1975) *Journal of Cellular Physiology*, **86**, 511–522.

Heby, O. & Russell, D. H. (1973) *Cancer Research*, **33**, 159–165.

Herbst, E. J. & Dion, A. S. (1970) *Federation Proceedings*, **29**, 1563–1567.

Hibasami, H., Tanaka, M., Nagai, J., & Ikeda, T. (1977) *Australian Journal of Experimental Biology and Medical Science*, **55**, 379–383.

Hogan, B. L. M. (1971) *Biochemical and Biophysical Research Communications*, **45**, 301–307.

Hogan, B., Shields, R., & Curtis, D. (1974) *Cell*, **2**, 229–233.

Holm, I., Persson, L., Heby, O., & Seiler, N. (1988) *Biochimica et Biophysica Acta*, **972**, 239–248.

Höltttä, E., Auvinen, M., & Andersson, L. C. (1993) *Journal of Cell Biology*, **122**, 903–914.

Höltttä, E., Auvinen, M., Paasinen, A., Kangas, A., & Andersson, L. C. (1994) *Biochemical Society Transactions*, **22**, 853–859.

Höltttä, E. & Hovi, T. (1985) *European Journal of Biochemistry*, **152**, 229–237.

Höltttä, E., Jänne, J., & Hovi, T. (1979b) *Biochemical Journal*, **178**, 109–117.

Höltttä, E., Pohjanpelto, P., & Jänne, J. (1979a) *FEBS Letters*, **97**, 9–14.

Höltttä, E., Sistonen, L., & Alitalo, K. (1988) *Journal of Biological Chemistry*, **263**, 4500–4507.

Hougaard, D. M. (1992) *International Review of Cytology*, **138**, 51–88.

Hougaard, D. M. & Larsson, L. (1982) *Histochemistry*, **76**, 247–259.

Hyvönen, T. (1989) *International Journal of Biochemistry*, **21**, 313–316.

Igarashi, K., Kashiwagi, K., Fukuchi, J., Isobe, Y., Otomo, S., & Shirahata, A. (1990) *Biochemical and Biophysical Research Communications*, **172**, 715–720.

Inaba, M., Otani, S., Matsui–Yuasa, I., Yukioka, K., Nishizawa, Y., Ishimura, E., Morisada, S., Yukioka, M., Morisawa, S., & Morii, H. (1986) *Endocrinology*, **118**, 1849–1855.

Inoue, H. & Oka, T. (1980) *Journal of Biological Chemistry*, **255**, 3308–3312.

Joseph, K. & Baby, T. G. (1988) *Insect Biochemistry*, **18**, 807–810.

Kaiser, N., Bourne, H. R., Insel, P. A., & Coffino, P. (1979) *Journal of Cellular Physiology*, **101**, 369–374.

Kamatani, N., Willis, E. H., McGarrity, G., & Carson, D. A. (1983) *Journal of Cellular Physiology*, **114**, 16–20.

Kanamoto, R., Nishiyama, M., Matsufuji, S., & Hayashi, S. (1991) *Archives of Biochemistry and Biophysics*, **291**, 247–254.

Kay, J. E. & Lindsay, V. J. (1973) *Experimental Cell Research*, **77**, 428–436.

Kay, J. E. & Pegg, A. E. (1973) *FEBS Letters*, **29**, 301–304.

Kiyokawa, H., Richon, V. M., Venta–Pérez, G., Rifkind, R. A., & Marks, P. A. (1993) *Proceedings of the National Academy of Sciences of the United States of America*, **90**, 6746–6750.

Klein, D. & Morris, D. R. (1978) *Biochemical and Biophysical Research Communications*, **81**, 199–204.

Korpela, H., Hölttä, E., Hovi, T., & Jänne, J. (1981) *Biochemical Journal*, **196**, 733–738.

Kufe, D. W., Griffin, J., Mitchell, T., & Shafman, T. (1984) *Cancer Research*, **44**, 4281–4284.

Kusunoki, S. & Yasumasu, I. (1976) *Biochemical and Biophysical Research Communications*, **68**, 881–885.

Larsson, L., Morch–Jorgensen, L., & Hougaard, D. M. (1982) *Histochemistry*, **76**, 159–174.

Law, R. O. (1991) *Comparative Biochemistry and Physiology*, **99A**, 263–277.

Levin, E. G. & Clark, J. L. (1979) *Journal of Cellular Physiology*, **101**, 361–368.

Lin, Y. C., Loring, J. M., & Villee, C. A. (1980) *Biochemical and Biophysical Research Communications*, **95**, 1393–1403.

Luk, G. D., Civin, C. I., Weissman, R. M., & Baylin, S. B. (1982) *Science*, **216**, 75–77.

Luk, G. D., Sharkis, S. J., Abeloff, M. D., McCann, P. P., Sjoerdsma, A., & Baylin, S. B. (1983) *Proceedings of the National Academy of Sciences of the United States of America*, **80**, 5090–5093.

Mach, M., Ebert, P., Popp, R., & Ogilvie, A. (1982) *Biochemical and Biophysical Research Communications*, **104**, 1327–1334.

Maddox, A., Keating, M. J., Freireich, J., & Haddox, M. K. (1989) *Investigational New Drugs*, **7**, 119–129.

Mamont, P. S., Duchesne, M., Grove, J., & Bey, P. (1978) *Biochemical and Biophysical Research Communications*, **81**, 58–66.

Mamont, P. S., Siat, M., Joder–Ohlenbusch, A., Bernhardt, A., & Casara, P. (1984) *European Journal of Biochemistry*, **142**, 457–463.

Manen, C., Hadfield, M. G., & Russell, D. H. (1977) *Developmental Biology*, **57**, 454–459.

Manen, C. & Russell, D. H. (1973a) *Journal of Embryology and Experimental Morphology*, **30**, 243–356.

——— (1973b) *Journal of Embryology and Experimental Morphology*, **29**, 331–345.

Matés, J. M., Sánchez–Jiménez, F., López–Herrera, J., & Núñez de Castro, I. (1991) *Biochemical Pharmacology*, **42**, 1045–1052.

Matsui, I., Otani, S., Kuramoto, A., Morisawa, S., & Pegg, A. E. (1983) *Journal of Biochemistry*, **93**, 961–966.

Matsuzaki, S. & Inoue, S. (1980) *Journal of Experimental Zoology*, **213**, 417–421.

Maudsley, D. V., Leef, J., & King, J. J. (1978) *Advances in Polyamine Research*, vol. 1, pp. 93–100. R. A. Campbell, D. R. Morris, D. Bartos, G. D. Daves, & F. Bartos, Eds. New York: Raven Press.

McCormick, F. (1977) *Journal of Cellular Physiology*, **93**, 285-292.

——— (1978) *Biochemical Journal*, **174**, 427–434.

Meilhoc, E., Moutin, M., & Osborne, H. B. (1986) *Biochemical Journal*, **238**, 701–707.

——— (1987) *Journal of Cellular Physiology*, **131**, 465–471.

Melvin, M. A. L. & Keir, H. M. (1979) *Biochemical Journal*, **178**, 391–395.

Melvin, M. A. L., Melvin, W. T., & Keir, H. M. (1978) *Cancer Research*, **38**, 3055–3058.

Menashe, M., Faber, J., & Bachrach, U. (1980) *Biochemical Journal*, **188**, 263–267.

Middleton, T., Gahn, T. A., Martin, J. M., & Sugden, B. (1991) *Advances in Virus Research*, vol. 40, pp. 19–55. K. Maramorosch, F. A. Murphy, & A. J. Shatkin, Eds. San Diego: Academic Press.

Mitchell, J. L. A., Diveley, Jr., R. R., & Bareyal–Leyser, A. (1992a) *Biochemical and Biophysical Research Communications*, **188**, 81–88.

Mitchell, J. L. A., Diveley, Jr., R. R., Bareyal–Leyser, A., & Mitchell, J. L. (1992b) *Biochimica et Biophysica Acta*, **1136**, 136–142.

Morrill, G. A. & Robbins, E. (1984) *Physiological Chemistry and Physics and Medical NMR*, **16**, 209–219.

Morris, D. R. (1991) *Journal of Cellular Biochemistry*, **46**, 102–105.

Morris, D. R., Jorstad, C. M., & Seyfried, C. E. (1977) *Cancer Research*, **37**, 3169–3172.

Moshier, J. A., Dosescu, J., Skunca, M., & Luk, G. D. (1993) *Cancer Research*, **53**, 2618–2622.

Munro, G. F., Miller, R. A., Bell, C. A., & Verderber, E. L. (1975) *Biochimica et Biophysica Acta*, **411**, 263–281.

Muzikova, E. & Clark, D. A. (1995) *Human Reproduction*, **10**, 1172–1177.

Nagarajan, S., Ganem, B., & Pegg, A. E. (1988) *Biochemical Journal*, **254**, 373–378.

Niskanen, E., Kallio, A., McCann, P. P., & Baker, D. G. (1983a) *Blood*, **61**, 740–743.

Niskanen, E. D., Kallio, A., McCann, P. P. Pou, G., Lyda, S., & Thornhill, A. (1983b) *Cancer Research*, **43**, 1536–1540.

O'Brien, T. G. & Diamond, L. (1977) *Cancer Research*, **37**, 3895–3890.

O'Brien, T. G., Saladik, D., & Diamond, L. (1980) *Biochimica et Biophysica Acta*, **632**, 270–283.

Oredsson, S. M., Anehus, S., & Heby, O. (1984) *Molecular and Cellular Biochemistry*, **64**, 163–172.

Oredsson, S. M., Billgren, M., & Heby, O. (1985) *European Journal of Cell Biology*, **38**, 335–343.

Oredsson, S. M., Kanje, M., Mamont, P. S., Wagner, J., & Heby, O. (1986) *Molecular and Cellular Biochemistry*, **70**, 89–96.

Oriol–Audit, C. (1991) *Les Polyamines*, pp. 97–107. J. Moulinoux & V. Quemener, Eds. Paris: Médecine-Sciences Flammarion.

Oriol–Audit, C., Hosseini, M., & Lehn, J. (1985) *European Journal of Biochemistry*, **151**, 557–559.

Orr, C. W., Yoshikawa–Fukada, M. & Ebert, J. D. (1972) *Proceedings of the National Academy of Sciences of the United States of America*, **69**, 243–247.

Osborne, H. B., Duvall, C., Ghoda, L., Omilli, F., Bassez, T., & Coffino, P. (1991) *European Journal of Biochemistry*, **202**, 575–581.

Osborne, H. B., Mulner–Lorillon, O., Marot, J., & Belle, R. (1989) *Biochemical and Biophysical Research Communications*, **158**, 520–526.

Pakala, R., Kreisel, M., & Bachrach, U. (1988) *Cancer Research*, **48**, 3336–3340.

Papazafiri, P. & Osborne, H. B. (1987) *Biochemical Journal*, **242**, 479–483.

Pegg, A. E. & Coward, J. K. (1985) *Biochemical and Biophysical Research Communications*, **133**, 82–89.

Pegg, A. E., Pösö, H., Shuttleworth, K., & Bennett, R. A. (1982) *Biochemical Journal*, **202**, 519–526.

Perchellet, J. & Boutwell, R. K. (1980) *Cancer Research*, **40**, 2653–2660.

Perry, J. W. & Oka, T. (1980) *Biochimica et Biophysica Acta*, **629**, 24–35.

Pezzali, D. C., De Agostini, M., & Grillo, M. A. (1984) *IRCS Medical Science*, **12**, 750–751.

Pohjanpelto, P. (1973) *Experimental Cell Research*, **80**, 137–142.

Pohjanpelto, P. & Knuutila, S. (1982) *Experimental Cell Research*, **141**, 333–339.

Porter, C. W., McManis, J., Casero, R. A., & Bergeron, R. J. (1987) *Cancer Research*, **47**, 2821–2825.

Poulin, R., Coward, J. K., Lakanen, J. R., & Pegg, A. E. (1993) *Journal of Biological Chemistry*, **268**, 4690–4698.

Poulin, R. & Pegg, A. E. (1995) *Biochemical Journal*, **312**, 749–756.

Poulin, R., Wechter, R. S., & Pegg, A. E. (1991) *Journal of Biological Chemistry*, **266**, 6142–6151.

Prussak, C. E. & Russell, D. H. (1980) *Biochemical and Biophysical Research Communications*, **97**, 1450–1458.

Puck, T. T., Bartholdi, M., Krystosek, A., Johnson, R., & Haag, M. (1991) *Somatic Cell and Molecular Genetics*, **17**, 489–503.

Quash, G., Keolouangkhot, T., Gazzolo, L., Ripoll, H., & Saez, S. (1979) *Biochemical Journal*, **177**, 275–282.

Reuben, R. C., Wife, R. L., Breslow, R., Rifkind, R. A., & Marks, P. A. (1976) *Proceedings of the National Academy of Sciences of the United States of America*, **73**, 862–866.

Rosander, V., Holm, I., Grahn, B., Løvtrup–Rein, H., Mattsson, M.-O., & Heby, O. (1995) *Biochimica et Biophysica Acta*, **1264**, 121–128.

Roszell, J. A., Douglas, C. J., & Irving, C. C. (1977) *Cancer Research*, **37**, 239–243.

Rumsby, G. & Puck, T. T. (1982) *Journal of Cellular Physiology*, **111**, 133–139.

Rupniak, H. T. & Paul, D. (1978) *Journal of Cellular Physiology*, **96**, 261–264.

Russell, D. H. (1971) *Nature (New Biology)*, **233**, 144–145.

Russell, D. H. & Stambrook, P. J. (1975) *Proceedings of the National Academy of Sciences of the United States of America*, **72**, 1482–1486.

Saló, E. & Baguñà, J. (1989) *Journal of Experimental Zoology*, **250**, 150–161.

Sánchez–Jiménez, F., Urdiales, J. L., Matés, J. M., & Núñez de Castro, I. (1992) *Cancer Letters*, **67**, 187–192.

Sarhan, S. & Seiler, N. (1989) *Biological Chemistry Hoppe–Seyler*, **370**, 1279–1284.

Scalabrino, G. & Ferioli, M. E. (1984) *Mechanisms of Ageing and Development*, **26**, 149–164.

Scalabrino, G. & Lorenzini, E. C. (1991) *Molecular and Cellular Endocrinology*, **77**, 37–56.

Scalabrino, G., Lorenzini, E. C., & Ferioli, M. E. (1991) *Molecular and Cellular Endocrinology*, **77**, 1–35.

Schaefer, E. L. & Seidenfeld, J. (1987) *Journal of Cellular Physiology*, **133**, 546–552.

Schindler, J., Kelly, M., & McCann, P. P. (1985) *Journal of Cellular Physiology*, **122**, 1–6.

Schulz, W. Z., Gebhardt, R., & Mecke, D. (1989) *Biological Chemistry Hoppe–Seyler*, **370**, 729–736.

Schulze–Lohoff, E., Fees, H., Zanner, S., Brand, K., & Sterzel, R. B. (1994) *Biochemical Journal*, **298**, 647–653.

Seidenfeld, J. & Marton, L. J. (1980) *Cancer Research*, **40**, 1961–1966.

Seppänen, P. (1981) *Acta Chemica Scandinavica*, **B35**, 731–736.

Sertich, G. J., Glass, J. R., Fuller, D. J. M., & Gerner, E. W. (1986) *Journal of Cellular Physiology*, **127**, 114–120.

Seyfried, C. E. & Morris, D. R. (1979) *Cancer Research*, **39**, 4861–4867.

Seyfried, C. E., Oleinik, O. E., Degen, J. L., Resing, K., & Morris, D. R. (1982) *Biochimica et Biophysica Acta*, **716**, 169–177.

Shafman, T. D., Sherman, M. L., & Kufe, D. W. (1984) *Biochemical and Biophysical Research Communications*, **124**, 172–177.

Shantz, L. M. & Pegg, A. E. (1994) *Cancer Research*, **54**, 2313–2316.

Sharkis, S. J., Luk, G. D., Collector, M. I., McCann, P. P., Baylin, S. B., & Sensenbrenner, L. L. (1983) *Blood*, **61**, 604–607.

Sible, J. C., Marsh, A. G., & Walker, C. W. (1991) *Invertebrate Reproduction and Development*, **19**, 257–264.

Sistonen, L., Hölttä, E., Mäkelä, T. P., Keski–Oja, J., & Alitalo, K. (1989) *EMBO Journal*, **8**, 815–822.

Snyder, S. W., Egorin, M. J., & Callery, P. S. (1991) *Biochemical and Biophysical Research Communications*, **180**, 591–596.

Snyder, S. W., Egorin, M. J., & Geelhaar, L. A. (1988) *Cancer Research*, **48**, 3613–3616.

Stjernborg, L. & Persson, L. (1993) *Biochemical Pharmacology*, **45**, 1174–1176.

Stoner, G. D., Harris, C. C., Myers, G. A., Trump, B. F., & Connor, R. D. (1980) *In Vitro*, **16**, 399–406.

Strambi, C., Raure, P., Renucci, M., Charpin, P., Augier, R., Tirard, A., & Strambi, H. (1993) *Archives of Insect Biochemistry and Physiology*, **24**, 203–217.

Sugiura, M., Shafman, T., & Kufe, D. (1984) *Cancer Research*, **44**, 1440–1444.

Sunkara, P. S., Nishioka, K., & Rao, P. N. (1981a) *European Journal of Cell Biology*, **26**, 154–157.

Sunkara, P. S., Ramakrishna, S., Nishioka, K., & Rao, P. N. (1981b) *Life Sciences*, **28**, 1497–1506.

Sunkara, P. S., Rao, P. N., & Nishioka, K. (1977) *Biochemical and Biophysical Research Communications*, **74**, 1125–1133.

Tabib, A. & Bachrach, U. (1994) *Biochemical and Biophysical Research Communications*, **202**, 720–727.

Takigawa, M., Ishida, H., Takano, T., & Suzuki, F. (1980) *Proceedings of the National Academy of Sciences of the United States of America*, **77**, 1481–1485.

Torrelio, B. M., Paz, M. A., & Gallop, P. M. (1984) *Experimental Cell Research*, **154**, 454–463.

Uhl, L., Kelly, M., & Schindler, J. (1986) *Biochemical and Biophysical Research Communications*, **140**, 66–73.

Vargiu, C. & Persson, L. (1994) *FEBS Letters*, **355**, 163–165.

Wallace, H. M. (1989) *Biochemical Pharmacology*, **38**, 379–381.

Wallace, H. M. & Keir, H. M. (1981) *Biochimica et Biophysica Acta*, **676**, 25–30.

——— (1986) *FEBS Letters*, **194**, 60–63.

Wallace, H. M., Macgowan, S. H., & Keir, H. M. (1985) *Biochemical Society Transactions*, **13**, 329–330.

Wallace, H. M., Nuttall, M. E., & Robinson, F. C. (1988) *Biochemical Journal*, **253**, 223–227.

Wallace, H. M. & Quick, D. M. (1994) *Biochemical Society Transactions*, **22**, 870–875.

Wang, J., McCormack, S. A., Viar, M. J., Wang, H., Tzen, C., Scott, R. E., & Johnson, L. R. (1993) *American Journal of Physiology*, **265**, G331–G338.

Watanabe, T., Shafman, T., & Kufe, D. W. (1985) *Journal of Cellular Physiology*, **122**, 435–440.

Weinberg, R. A. (1989) *Cancer Research*, **49**, 3713–3721.

White, J. D., Bruno, J., & Lucas, J. L. (1983) *Molecular and Cellular Biology*, **3**, 1866–1881.

Williams–Ashman, H. G., Coppoc, G. L., & Weber, G. (1972) *Cancer Research*, **32**, 1924–1932.

Willis, J. H. (1981) *American Zoologist*, **21**, 763–773.

Woo, K. B., Perini, F., Sadow, J., Sullivan, C., & Funkhouser, W. (1979) *Cancer Research*, **39**, 2429–2435.

Wrighton, C. & Busslinger, M. (1993) *Molecular and Cellular Biology*, **13**, 4657–4669.

Wyatt, G. R., Rothaus, K., Lawler, D., & Herbst, E. J. (1973) *Biochimica et Biophysica Acta*, **30**, 482–494.

CHAPTER 12

Pathways of Polyamine Metabolism in Animals

Ultimately we will wish to understand the roles of the polyamines in an intact organism at every stage of its life cycle. Designing meaningful biochemical experiments on intact animals is difficult, and the history of modern biochemistry is replete with efforts to find less complex model systems in which the causal relations of chemical and biological phenomena can be established. Nevertheless, the earliest studies with intact mammals did define some major biochemical problems, for example, the origin of urinary constituents such as urea. This problem led to work on liver and kidney function, nitrogen metabolism, and the metabolic fate of amino acids. Both the polyamines and the existence of new metabolites have been found in excretory products, posing numerous new problems.

Large amounts of somewhat toxic ammonia are excreted by fish and uric acid is excreted by birds and reptiles. Some plants proved to accumulate asparagine. Urea was found in the urine of mammals but also in many amphibians such as frog tadpoles, which excrete ammonia in an aqueous niche and produce urea in a terrestrial niche. The evolution of the pathways to remove ammonia received considerable attention in comparative biochemistry. Some fish living in alkaline lakes have evolved a urea cycle thought to facilitate survival (Randall et al., 1989). Analysis of the proteins and genes of the pathway of urea biosynthesis have revealed many changes in the functions and expression of these elements among eucaryotes and procaryotes (Takiguchi & Mori, 1995).

Discussions of urea synthesis focused on its role in the reduction of the concentration of NH_3 and of ammonia toxicity (Baldwin, 1939; Wright, 1995). In the past decade NH_3 has been discussed in the context of the major problem of pH homeostasis, that is, the problem of minimizing an alkalizing effect of the HCO_3^- generated in protein catabolism (Atkinson, 1992). In these terms, the primary role of urea synthesis in mammals is described as the generation of protons that enable the decomposition of HCO_3^- to CO_2. Atkinson (1992) recognized that two aspects of nitrogen balance exist and can be debated: the toxicity of NH_3 and pH homeostasis maintained by proton liberation. The subject is relevant in as much as a major intermediate in urea formation, ornithine, is also a participant in polyamine biosynthesis.

The ubiquity and the relative lack of biochemical variability of polyamine components in animals, partial and quite incomplete data on some insects and amphibians notwithstanding, defines the conserved nature of the polyamines throughout the eucaryotic world. Some plants have devised alternative routes to the same polyamines (i.e., identical compounds may arise by different paths) requiring possibly novel proteins and novel genetic determinants.

The existence of tissues with specialized functions offers the opportunity of concentrating on key metabolic processes. These include the production of urea in the liver via a cycle of metabolism involving arginine and ornithine, major precursors of the polyamines, or the elaboration of spermine in response to hormonal stimuli in certain mammalian prostate glands. Pursuit of novel constituents, such as the oxidized polyamines in brain and the acetylated polyamines in urine, led to the search for metabolic systems capable of producing these metabolites. Such studies led to the discovery of cycles of polyamine turnover.

The biological features of the tissue cells permitted correlations of polyamine content and metabolism. Mammalian liver largely comprised nondividing (i.e., quiescent) cells that might be induced to multiply by partial hepatectomy. Hormonal development and control of sexual organs such as the prostate is obviously a subject of great interest. The study of polyamines in the nonreproducing cells of an animal brain led to the discovery of a catabolism of the polyamines and the existence of cycles of interconversion. As the metabolic paths were clarified in these tissues, it was necessary to determine the separate reaction steps to explore the enzymology of each reaction and to define its cytological site and regulation. The pursuit of the latter culminated in the study of genetic controls and mechanisms of expression of the various enzymes.

Nitrogen Metabolism, with Special Reference to Arginine and Ornithine

The historical development of this subject was summarized by J. Fruton (1972). Urea and uric acid were isolated early in the 19th century as crystalline substances from animals and were shown to be very rich in nitrogen (Kurzer & Sanderson, 1956). Their excretion in urine removed excess nitrogen derived from food. The urea in the blood was produced by the liver and was removed via the kidney. The routes of production of urea and uric acid from ingested protein was argued until individual amino acids were isolated, some of which, as well as ammonium salts, were shown to be potential sources of urea. Techniques for the perfusion of organs in the 1860s facilitated such studies. "Biuret-free" autolysates of protein could completely replace dietary protein in maintaining nitrogen balance in dogs, opening the path to the discovery of the essentiality of certain amino acids.

The isolation of arginine, and its hydrolytic degradation to urea and ornithine, suggested arginine as a major precursor of urea. Arginase, which liberated urea and ornithine from arginine, was discovered in 1904 and was found to be abundant exclusively in the liver of mammals and other "ureotelic" animals. The path of urea production was clarified in 1932 by Krebs and Henseleit who demonstrated the conversion of ornithine, NH_3, and CO_2 to the ureido amino acid, citrulline, characterized first in 1930 as a plant constituent. Citrul-

line and NH_3 were converted to the guanidino amino acid, arginine, which was degraded by arginase to urea and ornithine.

The demonstration that this ornithine cycle could occur in a cell-free homogenate was reported in 1946. The enzymatic steps of the formation of citrulline required the prior synthesis of the newly discovered carbamoyl phosphate:

$$NH_3 + CO_2 + ATP \rightarrow NH_2CO—OPO_3H_2 + ADP.$$

This reaction of the urea cycle is carried out in the mitochondrial matrix by carbamoyl phosphate synthetase, designated as CPS-I (Morris, 1992). Citrulline formation is also mitochondrial.

$$Ornithine + NH_2CO—OPO_3H_2 \rightarrow H_3PO_4 + citrulline.$$

Carbamoyl phosphate may also be synthesized with glutamine as a nitrogen source. The cytosolic glutamine-dependent enzyme, designated as CPS-II, is largely involved in pyrimidine synthesis and is strongly inhibitable by 0.1–1.0 mM polyamines (Mori & Tatibana, 1975).

In 1954 S. Ratner demonstrated the conversion of citrulline to arginine by the ATP-dependent condensation of the latter with aspartate to form arginosuccinic acid:

citrulline + aspartic acid + ATP →
arginosuccinic acid + ADP + pyrophosphate,

followed by the enzymatic reaction,

arginosuccinate → arginine + fumarate.

Fumarate is a component of the citric acid cycle and is converted to malate and oxaloacetate, which is transaminated to form aspartate. The latter provides the nitrogen for the conversion of the ureido-containing citrulline to the guanidino-containing arginine. The urea cycle is thereby linked to a major energy-producing cycle, the citric acid cycle capable of supplying precursors to some amino acids such as aspartate and glutamate. These linked systems are presented in Figure 12.1.

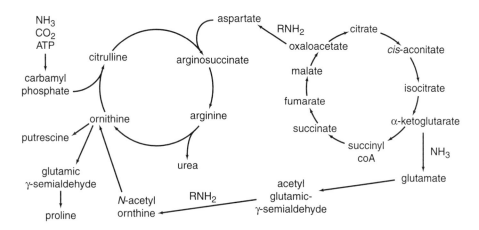

FIG. 12.1   Central roles of the ornithine cycle and the citric acid cycle in providing substrates for the biosynthesis of putrescine.

The array of enzymes, the loci of their reactions, and the genetic determinants for these catalysts have been best characterized for the fungi *Neurospora crassa* and *Saccharomyces cerevisiae* (Davis, 1986). The biosynthesis and additional reactions of ornithine to putrescine and proline in eucaryotic organisms are given in Figure 12.2.

The biosynthesis of ornithine occurs mainly in mitochondria. The amino group of glutamate is acetylated, followed by a phosphorylation and reduction of the acetylated derivative to the *N*-acetyl γ-semialdehyde. Transamination produces the α-*N*-acetylornithine, which can transfer its acetyl moiety to glutamate to form ornithine and regenerate *N*-acetylglutamate. In liver mitochondria, ornithine may enter the urea cycle via the product of the specific mitochondrial CPS-I, which prefers ammonia to glutamine as a substrate. Newly formed citrulline is transferred to the cytosol in which are found the enzymes that generate arginine. Arginase, which generates urea and ornithine, is also cytosolic.

The regulation of CPS-I by varying the concentration of the cofactor *N*-acetylglutamate and the regulation of arginase by Mn^{2+} determine much of the ureagenic activity in vivo (Morris, 1992). The protein precursor of ornithine transcarbamylase is synthesized in the cytoplasm, its transport into mitochondria induced by natural polyamines (Marcote et al., 1994). The precursor is cleaved to the mature enzyme by a matrix protease.

Cytosolic ornithine may be degraded to putrescine via ornithine decarboxylase (ODC) or may be converted to the pyrroline-5-carboxylate and proline. Unmetabolized cytosolic ornithine may also return to the mitochondrial compartment.

Arginine synthesis in the rat was demonstrated almost concurrently with the discovery of the urea cycle (Rose, 1938). Exclusion of arginine from a defined diet led to weight gains of only 70–80% that of control rats receiving the amino acid. Arginine has therefore been described as an essential dietary component whose synthesis in animals can be limiting for normal growth (Narita et al., 1995). The complexities of the pathways involved in its biosynthesis can be expected to affect the availability of arginine and ornithine, as well as the production of the polyamines.

Food restriction in rats leading to deprivation evokes a major decrease in ODC in various organs such as muscle. Rehabilitation via unlimited food facilitates a rapid increase in tissue ODC. Supplementation of a control diet for rats with arginine or arginine plus ornithine produces a major increase of free arginine and ornithine in muscle. With excess arginine in the diet, liver ornithine rises, although free arginine and polyamine do not (Smith et al., 1989). The uricotelic chicken has a high dietary requirement for arginine and responds to dietary ornithine by synthesizing polyamines in the liver. Currently a salt of ornithine and α-ketoglutarate (2:1) is being used to assist in the nutrition of patients affected by severe trauma and malnourishment (Cynober, 1991; Moukarzel et al., 1994). The complex problems of making proline available from ornithine as an approach to the increase of collagen synthesis in wound healing was discussed by Albina et al. (1993).

In the normal development of the rat liver, an enzyme of ornithine metabolism, ornithine-keto acid aminotransferase, which regenerates glutamate, increases some 10- to 20-fold in the liver at birth to that in the adult. This enzyme falls sharply after partial hepatectomy with regrowth of tissue and active biosynthesis of RNA, ODC, and polyamine (Räihä & Kekomäki, 1968). The diversion of ornithine to polyamine synthesis is effected in part by reducing the amount of this enzyme which minimizes the availability of ornithine. Inhibition of the enzyme by 5-fluoromethylornithine (5FMOrn) increases the tissue concentration of ornithine and increases the rates of hepatic urea formation and ornithine decarboxylation (Seiler & Daune–Anglard, 1993). Inhibition of the enzyme in mice by 5FMOrn also prevents the accumulation of NH_3 in the brain after administration of ammonium acetate and protects the mice from ammonia-induced convulsion (Seiler et al., 1993).

A series of rat hepatomas of various growth rates were found to contain elevated levels of ODC and putrescine, compared to the levels in normal rat liver. The amounts of ODC generally paralleled the increasing growth rate (Williams–Ashman et al., 1972) whereas the content of ornithine-carbamoyltransferase, which diverted ornithine to the synthesis of citrulline, was decreased in parallel with the increase in hepatoma growth

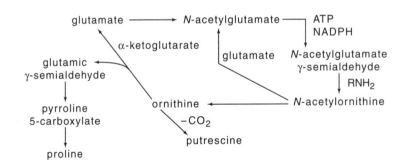

FIG. 12.2   The conversion of glutamate to ornithine and putrescine in eucaryotic organisms.

rate (Weber et al., 1972). The shift to growth and cell division displayed in the tumor was marked by a reduction of a path in which ornithine is used for arginine biosynthesis and an increase in the pathway toward polyamine biosynthesis.

Nitrogen Metabolism, with Special Reference to Methionine

Rose (1938) described methionine as an indispensable amino acid, although the conversion of ingested methionine to cystine in the cystinuric patients posed problems of the quantitative nature of the methionine requirement. This interconversion turned on the formation of homocysteine from S-adenosylmethionine (AdoMet) and the synthesis of cystathionine from serine and homocysteine. A deficiency of cystathionine synthetase and an excess of homocysteine can result in an excretion of homocystine.

The methyl group of methionine can be restored in animal cells by several mechanisms involving direct methyl transfer from various donors or via synthesis of N^5-methyltetrahydrofolate and transfer from this intermediate. The stimulation of the latter methionine synthase by spermidine and spermine has been reported (Kenyon et al., 1995). The bacterial metabolism of 5′-methylthioadenosine as a product of polyamine metabolism revealed the conversion of the ribose carbon chain to methionine, and ultimately to homocysteine. The methylthio group is conserved in the newly synthesized methionine. This conversion was detected in animal cells in tissue cultures and is absent in many strains of cancer cells (see Chap. 17).

The mammalian metabolism of the sulfur amino acids has been summarized (Griffith, 1987). In addition to incorporation into protein, methionine is metabolized primarily to AdoMet, which serves as a methyl donor, and thence from adenosylhomocysteine to the free amino acid. AdoMet can also be decarboxylated, and this product can be used to transfer a 3-aminopropyl moiety to form spermidine and spermine, plus 5-methylthioadenosine. The latter is cleaved to adenine and 1-phospho-5-methylthioribose in the course of the resynthesis of methionine. Methionine may also be oxidized to methionine sulfoxide and reduced enzymatically to the original amino acid.

An excess of methionine in the diet of young male rats is toxic, but the toxicity is not revealed in polyamine or AdoMet metabolism in muscle. AdoMet is elevated in the liver and minor decreases of spermine and RNA are seen in that organ. Polyamine metabolism is not involved in the mechanism of methionine toxicity (Smith et al., 1987).

Discovery of Polyamine Metabolites

The compounds with which we are mainly concerned in this section are presented in Figure 12.3.

In the early 1960s two sets of metabolites of putrescine, spermidine, and spermine were described, mainly as a result of studies with microbes. Amine oxidases led to the definition of various types of products of oxidation, and conjugated amines (e.g., N^1-acetylspermidine) were detected as bacterial products. The then recent discovery of the biogenic amines (e.g., serotonin, dopamine, norepinephrine etc.) in the brain and other tissues led to a search for other amines in nerve tissues and resulted in the detection of several new polyamine metabolites. As indicated in Chapter 3, the rapidity of analysis, sensitivity, reproducibility, and accuracy were studied. By the mid 1980s, when almost a dozen polyamine catabolites had been discovered, chromatographic methods had been developed to separate and characterize a large number of free and derivatized amines and their metabolites (Seiler, 1986).

Polyamines of nerve tissue

Spermine was isolated from the brain in the 1920s and 1930s. In 1959 putrescine and spermidine were isolated from hog brain as the classical gold chloride double salts and were characterized by elementary analyses, infrared spectra, and paper chromatography (Kewitz, 1959). The later discovery of γ-aminobutyric acid (GABA) as an oxidation product of putrescine in the brain proved to be significant in the analysis of the catabolism of the polyamines. Shimizu et al. (1964) concentrated on the relative concentrations of spermidine and spermine in various regions of nerve tissue, spermidine being higher in white myelinated tissue of the human cortex and spermine higher in the so-called "gray matter."

Perry and Schroeder (1963), who were the first to report putrescine and cadaverine in *normal* urine, detected a large number of identifiable and non-identifiable amines in the urine of normal and psychotic children. They also concluded that "mammalian brain contains a considerably wider variety of amines than had been supposed." Among the aliphatic amines that this group identified, in addition to the polyamines, were ethanolamine, piperidine, *n*-butylamine, and β-hydroxypropylamine (Perry et al., 1965). Monoacetylputrescine was also detected in the human brain and was thought not to arise via bacterial acetylation in the bowel (Perry et al., 1967).

Nakajima et al. (1967) wondered about the origin of 1,3-diaminopropane (DAP) found in the bovine brain and reported the presence of a very small amount of a presumed precursor, 2,4-diaminobutyric acid. A new amino acid, putreanine [N-(4-aminobutyl)-3-aminopropionic acid] was then found in the brain and spinal cord of mammals and birds (Kakimoto et al., 1969) and was synthesized and characterized (Shiba et al., 1970). This presumed catabolite of spermidine proved to be present in the human brain also (ca. 20 nmol/g or about 30 µmol/1500-g brain). It is excreted in urine at a rate of about 2 µmol/day (i.e., 5–10% of brain putreanine per day) (Nakajima et al., 1970), implying a significant rate of degradation of brain spermidine.

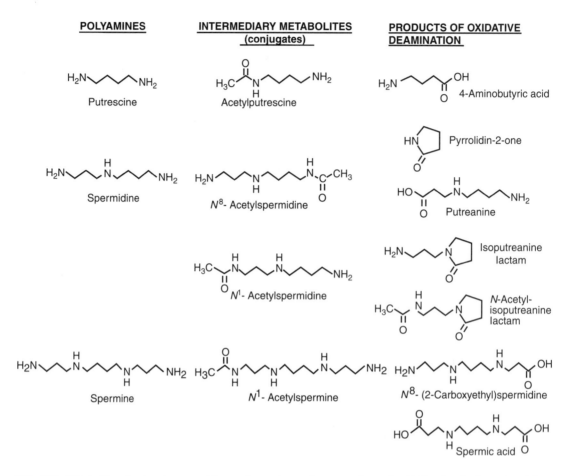

| POLYAMINES | INTERMEDIARY METABOLITES (conjugates) | PRODUCTS OF OXIDATIVE DEAMINATION |

FIG. 12.3 Metabolites of putrescine, spermidine, and spermine, and some products of oxidative deamination; adapted from Seiler (1986).

Putreanine is also found at much lower concentrations in other organs, for example, the liver and lung of the rat and rabbit. When [^{14}C]-spermidine or [^{14}C]-putrescine was injected intraventricularly into a rat brain, polyamines were converted to putreanine. Radioactive putreanine was also found in a rat liver after intraperitoneal injection. The specific activities of the isolated polyamines indicated that spermidine was the immediate precursor. Once formed, putreanine decreased in specific radioactivity in the liver and brain, the rate of decrease in the liver being far greater (Nakajima, 1973). This author noted a sleep-inducing effect of putreanine in cats.

The presence of this presumed oxidation product of spermidine evoked a search for compounds derived from spermine. Spermic acid, N,N^1-bis(2-carboxylethyl)-1,4-diaminobutane, has been isolated from the bovine brain (Imaoka & Matsuoka, 1974). It was also found in the brains of rabbits and rats but not in other organs.

In experiments with labeled polyamines injected intraperitoneally into rats and mice, selected inhibitors were found that blocked the production of putreanine and the intermediary of spermine oxidation, N^8-(2-carboxyethyl)spermidine, in the liver (Seiler et al., 1981b). Liver homogenates were incapable of carrying out this conversion, which required the addition of serum. Aminoguanidine, which inhibits the Cu-containing amine oxidases of serum, blocked the conversion. The inhibitor pargyline, usually thought of as a potent inhibitor of monoamine oxidase (MAO), is also an inhibitor of aldehyde dehydrogenase, but not of the Cu-amine oxidase, and inhibited the formation of the acid in the liver. It was concluded that the polyamines are oxidized first in the circulation, and that the aldehyde is then absorbed and converted to the acid in the liver.

Intraperitoneal injection of putrescine (100–200 mg/kg) in male rats evokes dose-dependent neurotoxicity, that is, shaking and motor disorders. Only about a third of the putrescine is recovered as such; the diamine is not recovered in higher polyamines and appears to be converted to polar metabolites (Camón et al., 1994).

The spermidine content of the rat brain followed that

of histamine in early stages of rat development and at weaning the two compounds were at similar concentrations. Brain concentrations shortly after birth were severalfold greater than that at weaning. However, spermidine concentrations in human fetal and mature brain were similar, although the fetal brain had higher concentrations of putrescine and spermine than did the mature brain (Sturman & Gaull, 1974). The relation of putreanine content to brain weight in the chick and rat was described by Seiler et al. (1985a). These workers demonstrated a close parallelism of the increase of putreanine and brain weight shortly after the hatching of the chick. In the rat, which is born with a significantly larger brain (60% that of the mature animal), putreanine content rises relatively rapidly from a very low value to meet the brain weight curve several months after birth. Many of the developmental relationships are summarized by Shaw (1979). A report that an increase in brain spermidine parallels aggressive behavior was not confirmed by Seiler (1983).

Urinary excretion of polyamines

Putrescine and cadaverine had been found in 1889 in the urine of cystinurics. In the late 1960s similar urines were found to contain both of these diamines, as well as DAP and spermidine (Bremer et al., 1971; Holder & Bremer, 1966). Perry et al. (1967b) detected monoacetylcadaverine and monopropionylcadaverine in the pooled urines of schizophrenic patients and synthesized these compounds as standards.

The researchers in the laboratory of Nakajima isolated N^8-acetylspermidine and N^1-acetylspermidine from urine and found an increase of these derivatives in the urines of cancer patients. This group injected $[^{14}C]$-labeled putrescine, spermidine, and spermine into rats and

systematically fractionated the urinary metabolites (Noto et al., 1978), as described in Figure 12.4.

A nonpolar and acidic fraction (A) contained almost half of the radioactivity derived from putrescine; neither injected spermidine nor spermine contributed isotope to this fraction. Because about half of the injected label of putrescine may be converted to CO_2, a large mixture of products may be expected in this fraction, which was not examined further.

Lundgren and Hankins (1978) described the conversion in the rat liver of putrescine to the nonpolar, neutral 2-pyrrolidone, which is probably contained to some extent in urinary fraction A. The formation of this lactam of GABA poses the problem of its possible origin via: a dehydrogenation of the spontaneously cyclized pyrroline, derived from the deamination product, 4-aminobutyraldehyde; the ring closure of GABA; or the deacetylation of N-acetyl GABA and its subsequent ring closure (Fig. 12.5).

The acid and neutral ampholyte fraction (B) (Fig. 12.4) derived from putrescine proved to contain a significant amount of GABA. The basic ampholyte fraction (C) from the diamine had a small amount of N-monoacetylputrescine whereas the larger polyamine fraction contained 2-hydroxyputrescine, as well as the common polyamine. This was the first report of 2-hydroxyputrescine in urine. These apparent conversions may be summarized as follows:

spermine ← spermidine ← putrescine* → CO_2 + many
metabolites, also 2-hydroxyputrescine, GABA, and
N-monoacetylputrescine

In contrast to the observed degradation of $[^{14}C]$-putrescine, little radioactivity appeared in fraction A from 1,4-diaminobutane-labeled spermidine and spermine. Radioactivity from spermidine was found in fraction B

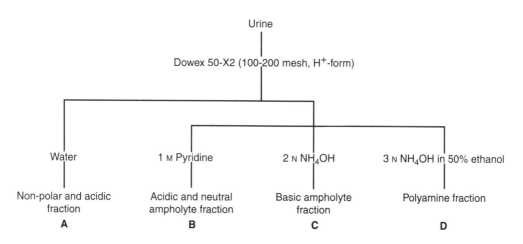

FIG. 12.4 A fractionation of rat urine, adapted to the determination of polyamine-derived metabolites; adapted from Noto et al. (1978).

in three unidentified compounds. Fraction C contained putreanine and N-monoacetylputrescine plus a new amino acid, characterized as N-(3-aminopropyl)-4-aminobutyric acid and named *isoputreanine*. Several unidentified compounds were also present in B. Fraction D contained N^1- and N^8-monoacetyl spermidine, as well as putrescine, spermidine, and spermine. These conversions suggest the following:

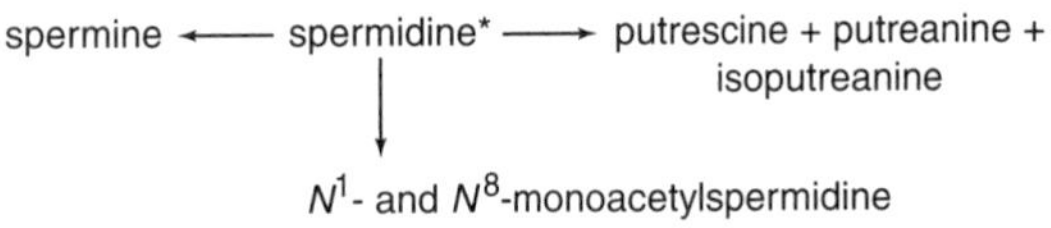

$$spermine \longleftarrow spermidine^* \longrightarrow putrescine + putreanine + isoputreanine$$

$$\downarrow$$

$$N^1\text{- and } N^8\text{-monoacetylspermidine}$$

Finally isotope from radioactive spermine was found in fraction B in spermic acid and several unidentified products. Among the basic ampholytes (fraction C) derived from spermine Noto et al. found putreanine, isoputreanine, and N-monoacetylputrescine. They also described the isolation of the presumed precursor to spermic acid, N-(3-aminopropyl),N^1-(2-carboxyethyl)-1,4-diaminobutane, and its synthesis. The polyamines of fraction D, derived from spermine, included putrescine, spermidine, and spermine. An equivalent amount of isotope present in unidentifiable compounds was also found in this fraction (D). These apparent metabolic events are summarized in Figure 12.6.

The discovery of a ubiquitous new polyamine oxidase (PAO) (Höltta, 1977), which preferred acetyl derivatives of spermidine and spermine to the free amines, introduced a new level of complexity into the initial catabolic schemes, such as those of Noto et al. (1978). This introduction of a degradation of N^1-acetylpolyamine to another N-acetylpolyamine minus an aminopropyl moiety was a key reaction in the development of catabolic cycles of interconversion of the polyamines.

Asatoor (1979) independently discovered isoputreanine in rat urine after injection of [^{14}C]-spermidine. He also synthesized the compound and determined the structure of the isolated and synthetic compounds. The compound was reported to be present, mainly in a conjugated form. However, putrescine and isoputreanine were also suggested to be products of serum oxidases

leading to the formation of aldehydes, which were then oxidized via an aldehyde dehydrogenase to the carboxylic acids.

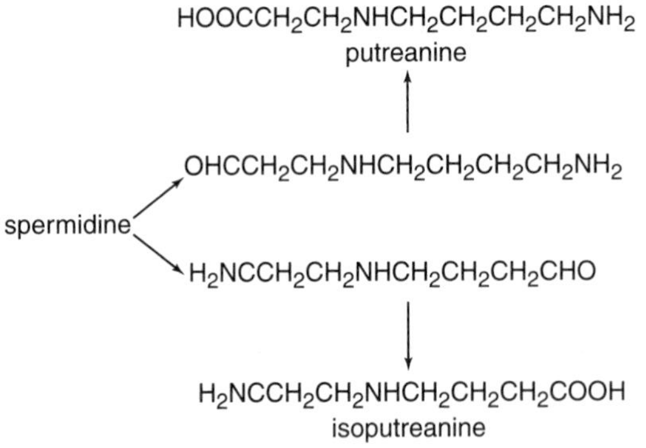

$$HOOCCH_2CH_2NHCH_2CH_2CH_2CH_2NH_2$$
putreanine

$$\uparrow$$

$$OHCCH_2CH_2NHCH_2CH_2CH_2CH_2NH_2$$

spermidine

$$H_2NCCH_2CH_2NHCH_2CH_2CH_2CHO$$

$$\downarrow$$

$$H_2NCCH_2CH_2NHCH_2CH_2CH_2COOH$$
isoputreanine

Seiler and Knödgen (1983) isolated the lactone of isoputreanine, N-(3-aminopropyl)pyrrolidin-2-one, from human and rat urine. On hydrolysis, this gave rise to isoputreanine and represents much of the apparent conjugate described by Asatoor (1979). Treatment of rats with aminoguanidine, which inhibits diamine oxidase (DAO) decreased the excretion of the lactam, thereby excluding PAO as the major enzyme converting spermidine or acetylspermidine to isoputreanine. Because N^1-acetylspermidine is converted to a urinary amino-substituted conjugate that liberates isoputreanine on hydrolysis, Seiler and Knödgen (1983) postulated the existence of N-(3-acetamidopropyl)pyrrolidin-2-one as the simplest conjugate. The existence and excretion of isoputreanine and the 3-acetamido derivative of the lactone in human urine have been demonstrated by Van den Berg et al. (1986) in extensive clinical studies. Using specific inhibitors of the amine oxidases in rats, it also appears that the N-(3-acetamidopropyl)pyrrolidone or N-acetylisoputreanine-γ-lactam is derived via an oxidative deamination of N^1-acetylspermidine by a copper-dependent amine oxidase (Hessels et al., 1991). However, a specific inhibitor of the flavin adenine dinucleotide (FAD) enzyme, PAO, markedly increased excretion

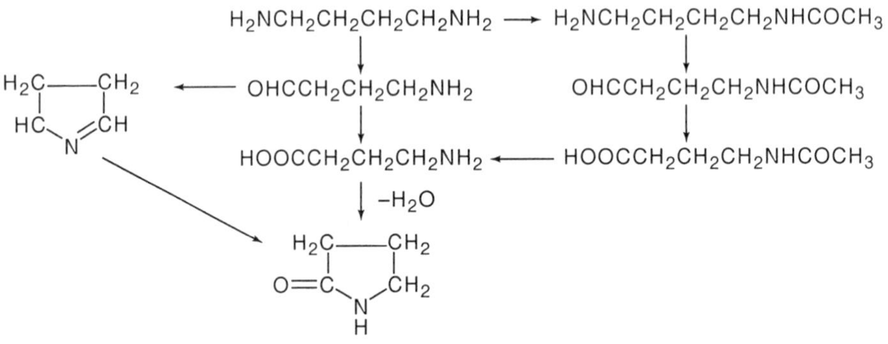

$$H_2NCH_2CH_2CH_2CH_2NH_2 \longrightarrow H_2NCH_2CH_2CH_2CH_2NHCOCH_3$$

$$\downarrow \qquad\qquad\qquad\qquad \downarrow$$

$$OHCCH_2CH_2CH_2NH_2 \qquad OHCCH_2CH_2CH_2NHCOCH_3$$

$$\downarrow \qquad\qquad\qquad\qquad \downarrow$$

$$HOOCCH_2CH_2CH_2NH_2 \longleftarrow HOOCCH_2CH_2CH_2NHCOCH_3$$

$$\downarrow -H_2O$$

FIG. 12.5　The metabolic conversion in rat liver of putrescine to 2-pyrrolidone, and its possible routes.

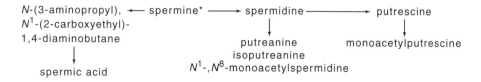

FIG. 12.6   Urinary metabolites of labeled spermine.

of the lactam by increasing the accumulation of N^1-acetylspermidine.

The results of Noto et al. (1978) were largely confirmed and extended by Van den Berg et al. (1984). It was found also that injected DAP was excreted as this form alone, whereas cadaverine was excreted as cadaverine and δ-aminovaleric acid. It was shown at this time that mice injected with DAP had an increased excretion of free spermidine and relatively less spermidine conjugate (Persson & Rosengren, 1984). In addition to the products described by Noto et al. (1978), spermidine appeared as N-(2-carboxyethyl)-4-amino-n-butyric acid, DAP, β-alanine, and putrescine and its metabolites. Injected spermine was excreted with the full array of compounds derived from spermine, spermidine, and putrescine. Van den Berg et al. (1984) drew a network of apparent pathways of excretion products on the supposition that the catabolic cycles operate largely through the acetylated derivatives.

Urinary excretion of polyamines and their metabolites

Concern with the working hypothesis of Russell in 1971, namely that polyamines may serve as markers for tumors and for the therapy of cancer, has produced many investigations of urine, blood, and tissue levels, including an interest in "normal values." These have also included analyses of the polyamines as a function of age, sex, hormonal balance, and obvious pathologies. The compilation presented in Table 12.1 summarizes studies on hydrolysates of human urine. There is considerable individual variation in the urinary excretion of putrescine and spermine in particular, and the excretion of these amines tends to be greater in females than in males (Beninati et al., 1980).

Many metabolic disturbances can produce aberrations in the patterns of polyamine excretion. An increased excretion of free putrescine and monoacetylcadaverine has been found during urinary infection. In normal individuals decontamination of the gastrointestinal tract reduces much of the acetylcadaverine. Therapy for the urinary tract infection reduces the free putrescine levels (Satink et al., 1989). Excretion patterns are even reported to display circadian rhythmicity (Hrushesky et al., 1983). Renal failure may result in a decreased excretion but leads to a rising accumulation in erythrocytes (Campbell et al., 1978; Swendseid et al., 1980). Aminoacidurias, such as cystinuria, are characterized by an increased excretion. Children with the gigantism of Beckwith syndrome have high levels of putrescine excretion and low levels of urinary spermidine and spermine (Barlow, 1980). Hyperargininemia also results in an increased excretion of putrescine (Kato et al., 1987). A lesion to a single tissue (i.e., ultraviolet irradiation of the skin) can evoke increased polyamine excretion for many days (Seiler et al., 1981d). The large number of urinary metabolites and an apparent aberration in the production of one or another may assist in detecting the site, tissue, and metabolic path in which a pathology may have arisen.

Efforts have been made to follow the urinary excretion of polyamines more systematically in various classes of patients. Obese patients maintained on a very low calorie diet have an initially increased excretion of N^1- and N^8-acetylspermidine, relating presumably to a higher rate of catabolism (Uusitupa et al., 1993). Increased rates of excretion of these components are found in surgical and accidental trauma, and in fact relate to the severity of injury, often greater in the latter type of damage. Unexpectedly, support of these patients by parenteral nutrition increased polyamine excretion initially, which presumably related to their increased metabolic activity (Pöyhönen et al., 1993).

Blood polyamines and their conjugates

The result that much of the blood polyamine is sequestered in erythrocytes will be discussed in Chapter 17 describing the status of the polyamines in the diagnosis and treatment of cancer. Although levels of the polyamines in human plasma are low, the levels of plasma putrescine, cadaverine, spermidine, and spermine are in-

TABLE 12.1   Mean urinary excretion of polyamines and their metabolites.

Compound	(μmol)/24 h (10.3 mmol creatinine)
Putrescine	14.8
Spermidine	6.0
Spermine	1.3
Isoputreanine	13.5
Putreanine	1.0
N^8-(2-Carboxyethyl)-spermidine	1.0

The urines of 23 men and 24 women (13–72 years) were hydrolyzed. Putrescine and spermidine were mainly excreted in the monoacetylated form and isoputreanine was as a mixture of the lactam and the N-(3-acetamidopropyl)-pyrrolidine-2-one. Adapted from Seiler (1991).

creased significantly in uremic patients (Takagi et al., 1983). The contents of both free putrescine and free spermidine are given as 0.2 nmol/mL.

When [^{14}C] polyamines are injected into the bloodstream of the rat, the free polyamines disappear quite rapidly from the circulating plasma. Significant amounts of putrescine and spermidine have been found as conjugates in plasma. Conjugation of spermidine was eliminated by hepatectomy (Rosenblum & Russell, 1977). The compounds in plasma were not thought to be acetyl or glutathionyl derivatives.

Human plasma was found to contain covalently bound putrescine on a single polypeptide of 32 amino acids and 4180 molecular weight (Seale et al., 1979). Trace amounts of spermidine were also bound in this material. The amount of putrescine bound to the peptide was stated to be 10–40 times that of unbound putrescine per milliliter of plasma. Peptide-bound conjugated polyamines will be discussed in Chapter 21 as possible products of the action of transglutaminase.

Acetylation and Catabolic Cycles

Seiler and Al-Therib (1974a) described a putrescine N-acetylase in the brain and other organs. The urinary excretion of the polyamines was shown to include significant amounts of the N^1- and N^8-monoacetyl spermidines, substances found in a variety of tissues (Abdel–Monem et al., 1978). An increased excretion of acetyl spermidines with an increased ratio of N^1/N^8-acetylation did not occur in cancer, but an enhanced excretion of the N^1-acetylspermidine was detected in rats after administration of some toxic agents (Seiler et al., 1981a,c). Two enzymes exist in rat tissues, a cytosolic N-acetyltransferase, which probably produced the N^1-acetylspermidine (Matsui & Pegg, 1980), and a nuclear enzyme (Blankenship, 1978), which produced N^8-acetylspermidine. Two nuclear N-acetyltransferases have been found in the rat liver, both of which form a mixture of N^1- (55%) and N^8-acetylspermidine (45%) (Libby, 1983). Both enzymes appear to be histone N-acetyltransferases (Libby, 1978a). The ratios of the N^1- and N^8-acetylspermidines in human urine are similar to those found in these enzymatic reactions.

Matsui and Pegg (1980) suggested that newly acetylated N^1-acetylspermidine was the precursor to putrescine formation in liver treated with carbon tetrachloride. The then recently discovered ubiquitous FAD-containing PAO was found to be capable of degrading N^1,N^{12}-diacetylspermine, and this reaction was suggested as another possible natural route to the synthesis of N^1-acetylspermidine. Indeed, N^1-acetylspermine and N^1-acetylspermidine are also substrates of the oxidase, giving rise to spermidine and putrescine, respectively (Seiler et al., 1981b). The N^8-acetylspermidine is not a substrate of the oxidase but can be converted to free spermidine by a cytosolic N^8-deacetylase (Blankenship, 1978). An acetylspermidine deacetylase that hydrolyzes N^1- and

N^8-acetylspermidine as well as N^1-acetylspermine has been isolated from the rat liver by Libby (1978b).

In his initial study of the oxidation of the free amines by polyamine oxidase, Höltta (1977) demonstrated the reactions,

$$\text{spermidine} + O_2 + H_2O \longrightarrow \text{putrescine} + H_2NCH_2CH_2CHO + H_2O_2$$
$$\text{3-aminoproprionaldehyde}$$
$$\text{spermine} + O_2 + H_2O \longrightarrow \text{spermidine} + H_2NCH_2CH_2CHO + H_2O_2$$

indicating an oxidation of the aminopropyl moiety at the secondary nitrogen of spermidine and spermine.

Soon thereafter in vivo investigations were beginning to demonstrate the N-acetylpolyamines as intermediates in the formation of spermidine and putrescine in reactions that involved the removal of N-acetamidopropanal.

$$N^1\text{-acetylspermidine} + O_2 + H_2O \longrightarrow$$
$$\text{putrescine} + CH_3CONHCH_2CH_2CHO + H_2O_2$$
$$\text{3-acetamidopropanal}$$

The above results implied a normal production of the N-acetyl compounds. The ubiquity of PAO and its greater activity on the N^1-acetylspermidine and N^1-acetylspermine than on the free amines (Seiler et al., 1981b), with a concomitant production of 3-acetamidopropanal, strongly suggested that the acetyl derivatives participated in the catabolism and interconversion of the polyamines.

One test of the hypothesis of acetylation as central in interconversion entailed a reexamination of the polyamine contents of the liver of fasting mice and rats as well as the patterns of urinary excretion of the polyamines. Seiler et al. (1969) had shown earlier that a rigorous correlation existed between the spermidine and RNA contents of the livers of mice subjected to starvation or induced to hypertrophy as a result of various drugs. A change of RNA-P by 15 or 16 μmol was accompanied by a change in spermidine-N of 3 μmol. These early results were confirmed, as seen in Figure 12.7 (Seiler et al., 1981a). Spermine was not much affected in these systems.

Both N^1-acetylspermidine content and the cytosolic acetyl coA:polyamine N^1-acetyltransferase were increased, and N-acetylspermine was detectable in the liver of fasting mice. PAO and the acetylpolyamine deacetylase were decreased during fasting. Early in this process, excretion of spermidine and N^1-acetylspermidine increased, although excretion of putrescine, spermine, and N^8-acetylspermidine decreased. Thus, fasting was a model for excessive polyamine degradation, permitting avoidance of the use of toxic agents such as carbon tetrachloride or thioacetamide.

The discovery of PAO led to the synthesis of specific potent inhibitors, as described in Chapter 6. Interference with the numerous steps of polyamine catabolism involved the use of the inhibitors of the Cu-containing amino oxidases (e.g., aminoguanidine), the inhibitors of the MAO (e.g., pargyline), and the newer potent spe-

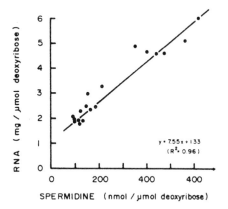

FIG. 12.7   Correlation between RNA and spermidine contents of livers of fasting mice; adapted from Seiler et al. (1981a).

cific inhibitors of PAO. The latter included several industrial products, often designated with a number. These inhibitors were the N^1-methyl,N^2(2,3-butadie-nyl)-1,4-butanediamine (MDL72521) and the N^1,N^2-bis(2,3-butadienyl)-1,4-butanediamine (MDL72527).

Inhibition of PAO in rat hepatoma (HTC) cells in culture by the latter compounds at 10 μM levels was irreversible and led to accumulation of N^1-acetylspermine and N^1-acetylspermidine without affecting the polyamine content and proliferation rate of the cells. A study of the effects of MDL72521 in cultured rodent and human cells similarly indicated that inhibition of PAO had little effect on cell growth or polyamine content in proliferating cells. The inhibitor did block the enzyme used to generate putrescine from N^1-acetylspermidine under conditions of stress, that is, late exponential phase and heat shock (Carper et al., 1991).

With PAO completely inhibited in mice, putrescine and spermidine in the liver fell 30–40% and 10–20%, respectively, while N^1-acetylspermidine and N^1-acetyl-spermine accumulated. These rates of accumulation were considered to relate to the normal rate of degradation of the acetylpolyamines (Bolkenius et al., 1985). When a PAO inhibitor (ML72527) was combined with difluoromethyl ornithine (DFMO), the inhibitor of ODC, the spermidine levels of several nongrowing tissues, the liver, kidney, and brain, declined >50%, more than with either inhibitor alone. Thus, the catabolic pathway had a normal homeostatic role in maintaining spermidine levels when putrescine was in short supply (Bolkenius & Seiler, 1987). It was striking that a complete inhibition of PAO by the inhibitors for many weeks with or without DFMO did not produce severe toxic effects other than some weight loss.

Treatment of mice and rats with ML72527 evoked a marked increase in the concentration of spermine in the blood, which fell when the drug was discontinued. Most of the accumulated spermine was found in erythrocytes, which contain a PAO; this degraded the spermine to

spermidine when inhibition was released. The latter reaction occurred with free spermine in the red cells, and the spermidine released from the erythrocytes was subsequently oxidized by plasma copper oxidases with its products excreted in the urine (Sarhan et al., 1991).

The use of aminoguanidine as an inhibitor of the copper amine oxidases of rats indicated that about 60% of the excreted polyamine products (the terminal putreanine, etc.) stemmed from the Cu-catalyzed oxidation (Seiler et al., 1985b). Even acetylated derivatives are oxidized by copper enzymes and wind up in urine, as given below.

N^1-acetylspermine ⟶ N^1-acetyl-N^8-(2-carboxyethyl)spermidine

N^1-acetylspermidine ⟶ N-acetylisoputreanine lactam

N^8-acetylspermidine ⟶ N-acetylputreanine

Synthesis of GABA from Putrescine

GABA was discovered in the central nervous system by E. Roberts in the late 1940s, and the compound was then shown to be a major inhibitory neurotransmitter. GABA is now known to be present in many branches of the procaryotic and eucaryotic world, and additionally it has been found in all vertebrate tissues. Several GABA conjugates are known in human cerebrospinal fluid, including several GABA-containing peptides (e.g., homo-carnosine, a histidine dipeptide), as well as peptides of lysine and cystathionine (see Miller et al., 1990). Genetic defects in reactions of GABA biosynthesis and degradation have been linked with several neurological disorders (Gibson et al., 1986).

The possibility that GABA might be derived directly or indirectly from putrescine was in fact realized. However, the dissection of the routes and significance of this conversion has proven to be difficult, as well as demonstrating active catabolic reactions in the brain. Indeed the investigations of these catabolic reactions have led to the picture of an elaborate cyclic flow of polyamines in selected tissues, of which the brain has become an important example.

GABA was shown to be formed in the brain largely from L-glutamate via the action of a glutamic acid decarboxylase with pyridoxal phosphate (PLP) as a coenzyme.

$$\begin{array}{ccc}
\text{COOH} & & \text{CO}_2 \\
| & & + \\
\text{HCNH}_2 & \longrightarrow & \text{H}_2\text{CNH}_2 \qquad \text{γ-aminobutyric acid} \\
| & & | \qquad\qquad\quad \text{(GABA)}\\
\text{CH}_2 & & \text{CH}_2 \\
| & & | \\
\text{CH}_2 & & \text{CH}_2 \\
| & & | \\
\text{COOH} & & \text{COOH}
\end{array}$$

Concentrations of GABA at sites corresponding to sites of glutamate decarboxylase have been detected serologically (Seguela et al., 1984). One potent inhibitor of glutamate decarboxylase, allylglycine, decreases brain

GABA. It also markedly increases ODC and decreases AdoMet decarboxylase (AdoMetDC), despite a rise in putrescine concentration (Pajunen et al., 1979), suggesting interaction of GABA metabolism and the regulation of polyamine biosynthesis.

In the brain, GABA and the polyamines, particularly putrescine, are localized at nerve endings and may be isolated in preparations of synaptosomes and their membranes (Seiler & Deckardt, 1976). A high affinity uptake of GABA and ornithine into synaptosomes has been observed, but not of putrescine, spermidine, and acetylated derivatives (Seiler & Deckardt, 1976, 1978), and the polyamines had not been considered to be neurotransmitters. Specific receptors for GABA have been isolated and characterized as components of ligand-gated ion channels (Olsen & Tobin, 1990). At present the polyamine regulation of ligand-gated ion channels is being actively investigated (Rock & Macdonald, 1995).

The existence of GABA in the blood and peripheral tissues has posed the possibility of routes other than glutamate decarboxylase to the compound. The presence of 2-pyrrolidone, that is, the lactam of GABA, as a major product of the metabolism of putrescine by rat liver slices is a demonstration of one such route via an amine oxidase (Lundgren & Hankins, 1978). Because Δ^1-pyrroline, as well as GABA, are found in the incubation media of various tissues, two routes from putrescine to the pyrrolidone become possible: one via the spontaneous cyclization of an aminoaldehyde to the Δ^1-pyrroline and its oxidation to the lactam; another would involve the cyclization of GABA.

The formation of the pyrrolidone was detected in the rat spleen, lung, and liver but not in the kidney, brain, heart, or muscle. The amine oxidases in serum can convert putrescine to the aminoaldehyde, which could be taken up and metabolized further in cells, as demonstrated subsequently in studies on the origin of putreanine (Seiler et al., 1981b).

GABA may be formed from free putrescine outside of the brain and a variety of such peripheral sources, cellular targets, and possible functions of the compound have been discussed (Erdö & Wolff, 1990). Seiler and Al-Therib (1974b) developed the view that, in contrast to these peripheral tissues, the brain largely converts putrescine to N-monoacetylputrescine, which is then oxi-

datively deaminated to N-acetyl GABA; this product is deacetylated to free GABA (Fig. 12.8).

Much attention has been given to GABA transaminase from the point of view of its role as a mitochondrial component in biosynthesis (Choi & Churchich, 1986), as well as its control by various inhibitors. Inhibitors of the GABA transaminase, inasmuch as they block entrance into the tricarboxylic acid cycle via succinic semialdehyde, will inhibit the conversion of putrescine to CO_2 (Seiler & Eichentopf, 1975). Among the inhibitors are active-site-directed irreversible inhibitors such as ethanolamine-O sulfate (Fowler & John, 1972), 4-amino-hex-5-enoic acid (vinyl GABA), 4-amino-hex-5-ynoic acid (acetylenic GABA), and 5-amino-1,3-cyclohexadienyl carboxylic acid (gabaculin). Vinyl GABA inhibited the mammalian brain transaminase specifically (Lippert et al., 1977).

GABA itself appears to act as a feedback regulator of ornithine transaminase in the brain. Specific inhibition of GABA transaminase in the brain by vinyl GABA increased GABA and ornithine levels in the mouse and rat brain (Seiler et al., 1987). It has been suggested that this provides a tight regulatory cycle in the brain in which a metabolic sequence of ornithine to glutamate to GABA in neurons is controlled by the end product. Ornithine does accumulate in GABAergic neurons when the transaminases are inhibited, a result compatible with the view that ornithine may be the precursor for glutamate and GABA, as well as for putrescine (Daune & Seiler, 1988).

A specific irreversible inhibitor of ornithine:2-oxo-acid aminotransferase, 5-fluoromethylornithine, has now been described (Daune et al., 1988). The compound produces near-complete inhibition of the enzyme in all tissues of the rat. The use of this nontoxic inhibitor increases ornithine concentration in many tissues, as well as urinary excretion of the amino acid.

The dissection of the relation of these compounds and pathways by means of inhibitors has proven to be complicated. For example, an inhibitor of ODC, 5-hexyne-1,4-diamine, injected intraperitoneally in rats was in fact oxidized by MAO to acetylenic GABA, which inhibited GABA transaminase and glutamate decarboxylase in the rat brain. This led to a major increase of GABA in the brain. Inhibition of the MAO by pargyline permitted the inhibitor to act specifically on ODC and abolished the effects on GABA metabolism (Danzin et al., 1979).

The appearance of putrescine carbon in GABA in rat liver and brain was detected by Seiler et al. (1971) and was subsequently demonstrated in the fish brain under conditions minimizing the participation of glutamate (Seiler et al., 1973). These results were subsequently confirmed by studies on injected rat brain (Konishi et al., 1977). The fish brain has been found to be inordinately high in putrescine (Seiler & Lamberty, 1973, 1975).

Monoacetyl putrescine was found in the brain somewhat earlier. Putrescine may be monoacetylated with acetyl coA by a nuclear enzyme (Seiler & Al-Therib, 1974a), unlike the acetylation of the diamine noted by Matsui and Pegg (1980). It is not attacked by acetyl-

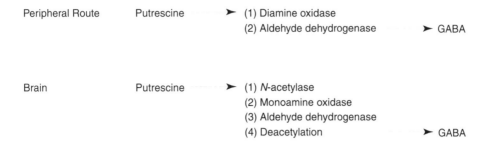

FIG. 12.8   Routes for the conversion of putrescine to GABA.

spermidine deacetylase (Libby, 1978b). Seiler and Al-Therib (1974b) also described the oxidative deamination of monoacetylputrescine to a well-characterized N-acetyl-γ-aminobutyrate by the mitochondrial MAO of the rat brain. This enzymatic action was largely inhibited by pargyline, the MAO inhibitor, and was not affected by aminoguanidine, the inhibitor of DAO. These results pointed to a sequence in the brain in which an acetamidoaldehyde is dehydrogenated to the acetylated GABA, which was somehow transformed to the free acid (Fig. 12.8). In contrast to this result, Tsuji and Nakajima (1978) demonstrated a maximal rate of conversion of putrescine to GABA in the small intestine and described an aminoguanidine-inhibited DAO as a participating enzyme in the sequence. Nevertheless, Seiler and colleagues (1979) showed that a brain homogenate deficient in DAO did not effect the conversion until acetyl coA was added to the system. Thus, the brain system that appeared to form GABA via monoacetylputrescine was suggested to be different from that prevailing in peripheral tissues.

Analytical difficulties for N-acetylputrescine, N-acetamidobutyraldehyde, and N-acetyl GABA appear to have prevented more extensive studies of the acetyl pathway to GABA in the brain. Some DAOs are weakly inhibited by pargyline, as is aldehyde dehydrogenase by propiolaldehyde, a degradation product of pargyline (Fogel, 1986). The existence and purification of a rat brain aldehyde dehydrogenase active on the free γ-aminobutyraldehyde has also been described (Abe et al., 1990). The conversion of ornithine and putrescine to GABA was also demonstrated in mouse neuroblastoma cells. In this system the synthesis of GABA was inhibited by aminoguanidine, implicating DAO (Kremzner et al., 1975). The conversion of putrescine to GABA is most active in stationary phase (Sobue & Nakajima, 1977).

Mice metabolize injected putrescine to a large fraction of expired CO_2, as well as to GABA and polyamines which are found in the kidney and urine (Henningsson & Rosengren, 1976). Aminoguanidine inhibits the metabolism of putrescine to CO_2, once again indicating the participation of DAO in that sequence (Missala & Sourkes, 1980). The role of this enzyme in the formation of GABA from putrescine has been underlined for fetal guinea pig liver (Fogel et al., 1982), rat

pancreas (Caron et al., 1987), and rat adrenal (Caron et al., 1988; Oon et al., 1989).

Because putrescine is also a precursor of spermidine and spermine, the developmental growth of the latter polyamines might be expected to affect the activities of the pathways of putrescine to GABA. There is an early but not immediate increase of the path for conversion of putrescine to GABA in chick embryos and a later increase of two pathways: glutamate to GABA and putrescine to spermidine (Sobue & Nakajima, 1977). However, Seiler and Sarhan (1983) concluded that the route of putrescine to GABA is responsible for only a small fraction of GABA formation in chick embryo brain; this is largely a function of glutamate decarboxylation. Decreasing the putrescine pool by inhibition of ODC did not decrease the rate of GABA accumulation.

Inhibition of GABA turnover in the brain induced ODC and increased the turnover of the polyamines. The latter parameter was estimated by the accumulation of N^1-acetylpolyamines in the presence of an irreversible inhibitor of PAO (Seiler & Bolkenius, 1985). The linear accumulation of N^1-acetylspermidine in the brain led to the conclusion that this compound is removed very slowly via the circulation or the cerebrospinal fluid. The rate of acetylspermidine accumulation then represented the rate of spermidine turnover; this was calculated to have a half-life of 8 days, in contrast to one of 130 days for spermine. Shorter half-lives calculated from the rate at which labeled methionine entered and decreased in these substances were believed to be faulty as a result of the reemergence of labeled methionine from other labeled products such as proteins. Key steps in the overall turnover of the polyamines and GABA are presented in Figure 12.9.

Another effect of inhibition of PAO was a fall in brain putrescine that limited the functioning of AdoMetDC, which is essential for the synthesis of spermidine and spermine. Very early in rat embryo development, before day 10, the low level of this decarboxylase determines the rate at which these polyamines are made. After this time, the turnover of N^1-acetylspermidine provides rate-limiting putrescine for the synthesis. Indeed it appeared that a continuing essential turnover of spermidine occurred until ODC attained a level equal to the need for putrescine (Bolkenius & Seiler, 1986).

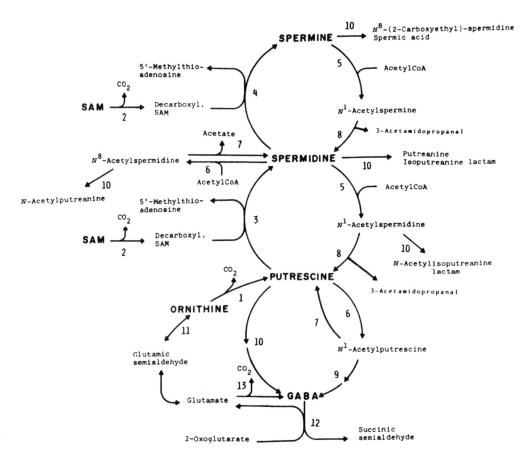

FIG. 12.9 A scheme of synthetic and catabolic reactions of polyamine metabolism in vertebrates; adapted from Seiler and Bolkenius (1985). 1. ODC; 2. AdoMet decarboxylase; 3. spermidine synthase; 4. spermine synthase; 5. acetylcoA : spermidine N^1-acetyltransferase; 6. acetylcoA : spermidine N^8-acetyltransferase; 7. N^8-acetylspermidinedeacetylase; 8. polyamine oxidase (PAO); 9. monoamine oxidase; 10. copper amine oxidases; 11. ornithine : 2-oxoglutarate aminotransferase; 12. 4-aminobutyrate : 2-oxoglutarate aminotransferase; 13. glutamate decarboxylase.

Brief Summary Scheme of Biosynthesis and Catabolism

The studies discussed in this chapter have enlarged the reactions described very early for procaryotes and some eucaryotic microbes. They include major components of terminal oxidized end products (e.g., spermic acid, putreanine), which tend to be excreted. The excretion products included compounds such as the acetylated polyamines, which proved to be metabolic intermediates and were oxidized by newly discovered enzymes to less complex polyamines and fragments destined for excretion. Studies of the oxidation of putrescine indicated routes to the formation of GABA, a compound important in neurological behavior and thought originally to have stemmed entirely from another pathway. The enlarged patterns of metabolism that have given rise to these substances are presented in Figure 12.9, posing the problems of the regulation of the individual reactions and integration of the sequences in growing organisms.

Discovery of Regulation of ODC in Liver

In the early and mid 1960s the cellular content of polyamines appeared to relate to RNA content in the development of the chick and other eucaryotes and in various phenomena of bacterial growth and growth inhibition. The possibility that this correlation also existed in cellular and tissue growth in the mammal and that the polyamines might indeed be determinants in development was then explored. In approaching this set of problems, Raina et al. (1964) tested the effects of the methionine analogue, ethionine. The expected decrease of ADP and ATP as a result of the synthesis of S-adenosylethionine led to an initial decrease in spermidine and a prolonged decrease in spermine. Simes (1967) studied the biosynthesis and turnover of the unconjugated polyamines in all of the tissues, demonstrating spermidine as a precursor of spermine and as a product of degradation of exogenous spermine. The former reaction was blocked by ethionine, whereas the degradation was not. Starvation for 4 days increased the degradative reaction but did not affect the biosynthesis. Young, adult and aging rats were studied with respect to diet, hormonal effects, etc.

In parallel studies J. Jänne (1967) studied the distribution and metabolism of putrescine in rat tissues. He followed its conversion to CO_2, spermidine, and spermine in vivo, and detected an in vitro synthesis of spermidine in a liver homogenate. Jänne determined putrescine and spermidine contents and metabolism in partially hepatectomized animals, as well as the effects of growth hormone.

Regenerating rat liver provides a system in which the rates of synthesis of protein and nucleic acid are greatly increased. The surgical removal of as much as two-thirds of the liver in an otherwise intact animal causes a rapid division and multiplication of cells in the remaining liver in 24 h. The organ is restored completely in about 10 days after resection (Lewan et al., 1977). The system has revealed a reproducible pattern of synthesis and decline of enzymes for the biosynthesis of the polyamines. The activity of RNA polymerase is increased soon after partial hepatectomy, as are the rates of incorporation of nucleic acid components into RNA and DNA. DNA synthesis begins some 16–18 h after partial hepatectomy and is maximal at 24 h, followed by cell division at about 28–30 h.

Major events of polyamine biosynthesis were found long before the onset of DNA synthesis and division. After a day, the spermidine and RNA contents of the regenerating liver were almost twice that of the control liver, remained high for a 3-day period, and then fell to the control levels. However, the spermine content of control and operated livers was similar throughout. The rates of conversion of tritiated putrescine to spermidine doubled in the regenerating liver (Dykstra & Herbst, 1965). The stimulation of spermidine synthesis was detected 4–8 h after operation, long before the observed parallel increase of total spermidine and RNA (Raina et al., 1966). Throughout the experiments, the ratio of polyamine-N to RNA-P was in the range of 0.14–0.16. In the studies of Seiler et al. (1960) of the effect of diet on mouse liver polyamines and RNA (see Fig. 12.7), the linear relation between the contents of spermidine and RNA over a 10-fold range of concentrations indicated that spermidine-N might neutralize some 20% of the RNA-P.

The spermidine content of rat liver was also affected by various hormones (Kostyo, 1966). A hypophysectomized animal contained lowered levels of spermidine than did control animals; administration of growth hormone restored spermidine levels as well as synthesis of RNA and protein. Growth hormone markedly stimulated the synthesis of spermidine from [^{14}C]-methionine in intact rat liver and this related to an early stimulation of synthesis of ODC and putrescine (Jänne et al., 1968; Jänne & Raina 1969a). Both ODC and RNA polymerase increase strikingly some 2–4 h after injection of the hormone and decrease very sharply between 4 and 6 h (Jänne & Raina, 1969b; Russell et al., 1970a). The ef-

fects were prevented by inhibitors of protein synthesis. In studies with adrenalectomized rats, hydrocortisone, insulin, and growth hormone were found to stimulate ODC synthesis (Panko & Kenney, 1971). Cyclic AMP and dexamethasone are also stimulatory (Canellakis & Theoharides, 1976).

A single injection of a weak carcinogen, thioacetamide, markedly stimulates the synthesis of ODC, putrescine, spermidine, and nuclear RNA. These reactions are all blocked by actinomycin D, implicating the intermediary synthesis of messenger RNA and of de novo protein synthesis (Fausto, 1970). Large doses of GABA inhibit the increase of ODC mRNA, ODC, and putrescine in partially hepatectomized rats (Minuk et al., 1991).

Two laboratories (Raina & Jänne, 1968; Russell & Snyder, 1968) began the detailed analysis of the appearance of ODC in regenerating rat liver. Raina and Jänne (1968) began an isolation and characterization of the PLP-requiring ODC of regenerating rat liver and described the early rise and fall of ODC activity in rat liver stimulated by growth hormone. Russell and Snyder (1968) described the rise of ODC specifically within 1–2 h after partial hepatectomy, attaining a maximum at 16 h and declining thereafter. They also reported the extremely rapid turnover of ODC activity. Russell et al. (1970b) underlined the importance of the early accumulation of putrescine for the subsequent elaboration of the polyamines and described the rapid turnover of putrescine ($t_{1/2} = 2$ hours) labeled by exogenous [^{14}C]-ornithine in the regenerating rat liver.

These results presented two major groups of problems. In the first group, the nature of the increase of ODC activity required clarification. In the experiments with growth hormone (Jänne et al., 1968), inhibitors of the synthesis of RNA (e.g., actinomycin D) and protein (e.g., puromycin, cycloheximide) blocked putrescine formation, suggesting that the sharp increase of ODC within 3 h represented the synthesis of new protein, rather than some form of activation (e.g., phosphorylation or proteolytic processing) of a preexisting protein. The increase of ODC after 1 h in regenerating rat liver identified this as one of the earliest events in this system.

The validity of the assay in crude homogenates was challenged. The use of [1-^{14}C]-ornithine to liberate the easily measured $^{14}CO_2$ may be perilous if ornithine is degraded by reactions other than that catalyzed by ODC. For example, the production of α-ketoglutarate via the liberation of ornithine-δ-aminotransferase from damaged mitochondria might result in other sources of $^{14}CO_2$. Inhibitors of the ornithine aminotransferase were added to prevent this contaminating activity (Murphy & Brosnan, 1976). The presence of inhibitors might be demonstrable in the various changes in ODC. These possible complications were eliminated in a careful study of extracts with which [1-^{14}C]-ornithine was also demonstrated to produce equimolar amounts of putrescine and $^{14}CO_2$ (Schrock et al., 1970).

The second major problem is that of the rapid rate of decline of ODC activity. Russell and Snyder (1969) demonstrated that the turnover rate of ODC in the liver, that is, regenerating or normal, was extraordinarily fast. After intraperitoneal injection of cycloheximide or puromycin, which are inhibitors of protein synthesis, the activity of liver ODC fell exponentially with a half-life of 10–12 min (Fig. 12.10). Thus, ODC had the shortest half-life of any then known mammalian enzyme. Did this decline in activity represent an inactivation by addition of an inhibitor or the destruction of the enzyme protein? It was noted that the half-life ($t_{1/2}$) of putrescine in regenerating rat liver was quite rapid ($t_{1/2} = 2$ h) as compared to the turnover of spermidine in normal rat liver ($t_{1/2} = 5$ days), which approached the rate of turnover of ribosomal RNA in that tissue (Snyder & Russell, 1970).

In a more recent study of the relation of the polyamines to the initial liver resection, the putrescine concentrations alone of the remaining hepatic tissue increased in proportion to the extent of hepatectomy. Further, these levels correlated to the restitution of liver mass, protein, and DNA (Minuk et al., 1990), thereby stimulating regeneration as well as serving as precursors to the formation of spermidine.

Hormonal Control of Polyamine Synthesis in Gonadal and Accessory Tissues

H. G. Williams–Ashman had a long-term interest in reproductive physiology and the possible roles of the polyamines. His laboratory initiated studies on the origin of the polyamines of seminal fluid and their synthesis by the rat prostate gland. The spermidine and spermine polyamine content are highest in the ventral portion of the gland, which also contains a cytosolic ODC (Pegg & Williams–Ashman, 1968a). Jänne and Williams–Ashman (1970) subsequently described a great activation of the mammalian enzyme by sulfhydryl compounds, which also stabilize the enzyme at low temperatures. A partially purified ODC was shown to require PLP and to generate approximately equivalent molar quantities of CO_2 and putrescine. Putrescine was a competitive inhibitor of the enzyme ($K_i = 1.2$ mM), as were the less inhibitory spermidine and spermine. Putrescine activation of the decarboxylation of AdoMet by the enzymes present in rat liver or ventral prostate was

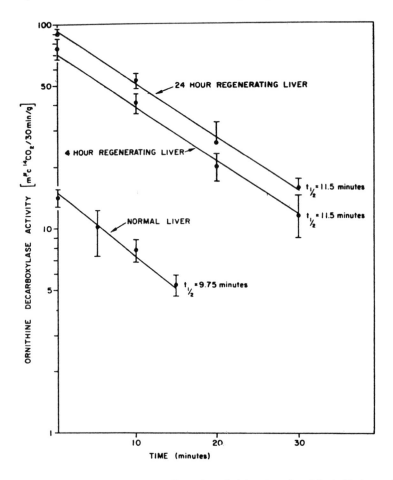

FIG. 12.10 Time course of the decline of ODC in rat liver after administration of cycloheximide intraperitoneally; adapted from Snyder and Russell (1970).

also detected (Pegg & Williams–Ashman, 1968b). The product of the reaction, decarboxylated AdoMet (dAdo-Met), was not a very active inhibitor of the decarboxylase. The extracts were also found to contain spermidine synthase.

The polyamine contents of the ventral prostates fell sharply after castration, as did ODC, AdoMetDC, and spermidine synthase (Pegg et al., 1970). After injection of testosterone propionate to castrated animals, putrescine and spermidine rose strikingly, after an early rapid increase in these enzymes. Spermidine was far more responsive than spermine to castration and hormonal treatment.

Pegg and Williams–Ashman (1970) described activities for the synthesis of spermine and spermidine in the soluble extracts of rat ventral prostate. Spermine synthase transferred an aminopropyl moiety from dAdoMet to spermidine. Thus, the catalytic activities for the biosynthesis of the polyamines had been described in the ventral prostate of the rat. Two of these were clearly under hormonal control by testosterone and related to the secretory functions of the gland, which produced major components of seminal fluid (Williams–Ashman & Lockwood, 1970). Three additional mammalian enzymes, other than ODC, all of which participated in the production of spermine, had been described.

The estrogens (e.g., β-estradiol) had also been shown to increase polyamine and RNA contents in the ovariectomized rat uterus (Moulton & Leonard, 1969). Estradiol stimulated ODC in the oviducts of immature chicks (Cohen et al., 1970). This estrogen also stimulated ODC and AdoMetDCs in rat uteri (Kaye et al., 1971). Luteinizing hormone stimulated the production of this enzyme in rat ovaries (Kobayashi et al., 1971).

Castration was also found to result in a decrease of kidney weight: testosterone produced renal hypertrophy. The kidneys of male mice were found to be unusually high in ODC, resulting in a high rate of putrescine excretion. Indeed mouse kidney subsequently proved to be a major source for the isolation of ODC. Testosterone caused a greater than 1000-fold increase in ODC, as well as a large increase of RNA, in the kidneys of castrated mice. The organ also increased its content of AdoMetDC and polyamines without an increase in DNA (Henningsson et al., 1978).

The androgen induction of ODC in the mouse kidney has been shown to involve increases (8–20-fold) in the level of ODC mRNA (Berger et al., 1984). Androgen will also increase ODC synthesis markedly (400–500-fold) in female mouse kidney, which appears to contain at least two ODC mRNA species (Catterall & Jänne, 1988). However, although renal hypertrophy and induction of ODC occur concomitantly after testosterone administration, inhibition of ODC by DFMO does not prevent kidney enlargement in this or other mouse systems (Tovar et al., 1995).

After the finding of a stimulation of increase of polyamine and RNA in the prostates of testosterone-treated orchiectomized animals, it was shown that the RNA polymerization effected in the prostatic nuclei of orchiec-

tomized rats was significantly stimulated by spermine particularly (Calderera et al., 1968). These workers suggested that the testosterone effects on RNA synthesis may have been mediated primarily by the polyamines.

Following the characterization of AdoMetDC of *Escherichia coli* as a pyruvoyl enzyme, this enzyme was isolated from rat and human prostate and various kinetic parameters were determined (Zappia et al., 1972). The putrescine activation and AdoMet requirements and specificity were examined, as were the pH optima and the activation of the enzyme by putrescine as a function of pH. The rat and human prostatic enzymes have similar patterns of putrescine activation and do not require Mg^{2+}, as does the *E. coli* enzyme. The mechanisms of the different activation effects for enzymes of similar specificities, sites, and patterns of inhibition obviously pose problems concerning the evolution of the structures of these different enzymes.

The startling observation of the extremely rapid rate of ODC decay in rat liver soon led to the study of this parameter for the other biosynthetic enzymes. As seen in Figure 12.11, the disappearance of AdoMetDC in the regenerating liver of rats treated with cycloheximide was significantly slower than that of ODC in that tissue whereas spermidine synthase and spermine synthase were essentially stable for at least 5 h (Raina et al., 1973). The $t_{1/2}$ of ODC and AdoMetDC may vary significantly in the presence of ligands detected initially as inhibitors. Differences are also found with different tissues and the treatments given to a tissue. For example, the $t_{1/2}$ of kidney ODC in androgen-treated animals (140 min) is 4–10-fold greater than that in untreated animals (16 min) (Isomaa et al., 1983).

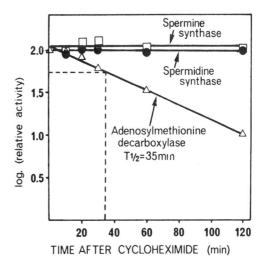

FIG. 12.11 Effect of cycloheximide on activities of AdoMet decarboxylase, spermidine synthase, and spermine synthase in regenerating rat liver; adapted from Raina et al. (1973).

Hypusine, A Posttranslational Product of Spermidine Metabolism

The chapter closes with a discussion of the unexpected discovery of a novel amino acid, hypusine, produced in the metabolism of spermidine and relating the triamine to essential functions of eucaryotic cells. This amino acid is a unique posttranslational derivative of a single lysine contained in a specific eucaryotic protein, eIF-5A. The amino acid was discovered in this protein as a product of the metabolism of labeled putrescine and spermidine when the isotopic label derived from the aminobutyl portion of spermidine was found in a single protein. A comprehensive review of the discovery, distribution, biosynthesis, and possible roles of the amino acid has appeared (Park et al., 1993a,b). The pathway of hypusine biosynthesis is given here.

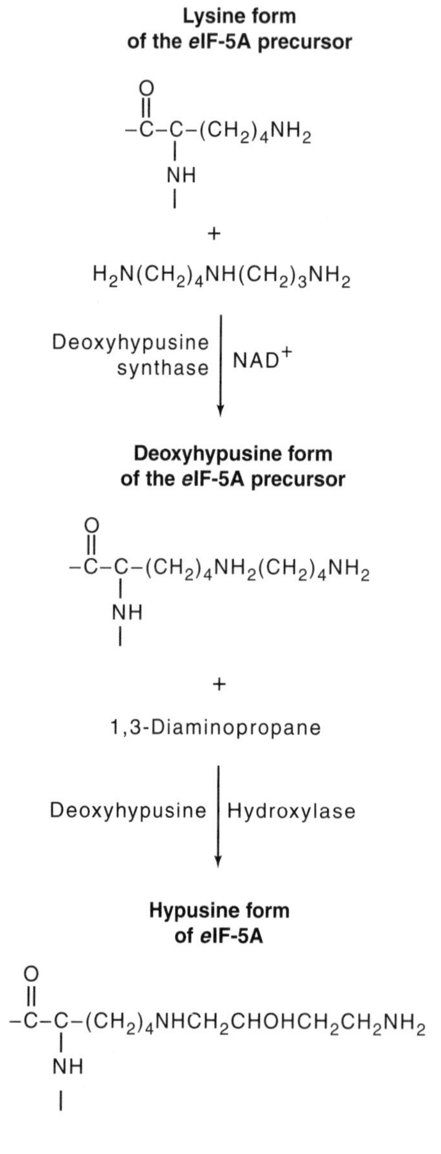

**Lysine form
of the *e*IF-5A precursor**

$$-\overset{\overset{\displaystyle O}{\|}}{C}-\underset{\underset{\displaystyle NH}{|}}{C}-(CH_2)_4NH_2$$

$$+$$

$$H_2N(CH_2)_4NH(CH_2)_3NH_2$$

Deoxyhypusine synthase | NAD$^+$

**Deoxyhypusine form
of the *e*IF-5A precursor**

$$-\overset{\overset{\displaystyle O}{\|}}{C}-\underset{\underset{\displaystyle NH}{|}}{C}-(CH_2)_4NH_2(CH_2)_4NH_2$$

$$+$$

1,3-Diaminopropane

Deoxyhypusine | Hydroxylase

**Hypusine form
of *e*IF-5A**

$$-\overset{\overset{\displaystyle O}{\|}}{C}-\underset{\underset{\displaystyle NH}{|}}{C}-(CH_2)_4NHCH_2CHOHCH_2CH_2NH_2$$

Hypusine was first isolated by ion-exchange chromatography from an acid-soluble extract of bovine brain; its structure was determined by NMR, the analysis of the fragments produced in mass spectrometry, and by chemical degradation. Shiba et al. (1971) also synthesized the compound, which was demonstrated to be N^ε-(4-amino-2-hydroxybutyl)lysine. It was present in human urine (1–3 nmol/mg creatinine) and in essentially all mammalian tissues. A peptide of GABA, namely α-(γ-aminobutyryl)-hypusine, was subsequently isolated from some mammalian brains (Sano et al., 1987). Thus, the amino acid has been found in the free state and in peptides and in protein.

Several groups found that [^{3}H]-putrescine or the [^{3}H] or [^{14}C] of labeled spermidine was incorporated predominantly into a single cellular protein (Chen & Liu, 1981; Park et al., 1981). Park et al., working with mitogen-stimulated human lymphocytes, obtained evidence that indicated that spermidine was an immediate precursor of the 4-amino-2-hydroxybutyl portion of protein-bound hypusine. Chen and Liu, who studied the metabolism of [^{14}C]-putrescine in neuroblastoma cells, found not only the conversion of the diamine into GABA, but also its incorporation into a protein of 18,000 molecular weight. This was subsequently shown to be identical to the ubiquitous eucaryotic translation initiation factor, then called eIF-4D, and currently renamed eIF-5A (Cooper et al., 1983). eIF-5A was shown to contain one molecule of hypusine per molecule of protein.

Fertilized sea urchin eggs rapidly incorporate exogenous spermidine into an acid-insoluble fraction, predominantly to a 30-kDa component lacking hypusine. This proved to be covalently bound spermidine present in two fractions. Hypusine subsequently (after 28 h) appeared in these same fractions (Cannellakis et al., 1989). It was suggested that spermidine is bound as such earlier via the action of transglutaminase. Indeed it was subsequently found that eIF-5A was reactive, presumably via transglutaminase, in a guinea pig liver homogenate to produce γ-glutamyl-ω-hypusine. Some of these structures are shown in the following.

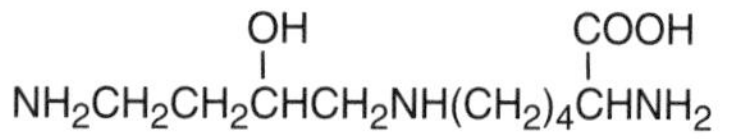

$$\overset{\overset{\displaystyle OH}{|}}{NH_2CH_2CH_2CHCH_2NH(CH_2)_4}\overset{\overset{\displaystyle COOH}{|}}{CHNH_2}$$

hypusine

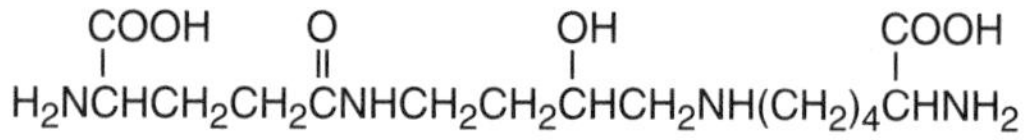

$$\overset{\overset{\displaystyle COOH}{|}}{H_2NCHCH_2CH_2}\overset{\overset{\displaystyle O}{\|}}{CNHCH_2CH_2}\overset{\overset{\displaystyle OH}{|}}{CHCH_2NH(CH_2)_4}\overset{\overset{\displaystyle COOH}{|}}{CHNH_2}$$

γ-glutamyl-ω-hypusine

Some cells may possess a protein somewhat larger than the usual mammalian eIF-5A of 18 kDa. Further, even within mammalian cells, isoforms of that initiation factor also exist with different isoelectric points, in the pI range of 4.7–5.2 (Chen & Dou, 1988). In the earliest studies it was found that the major eIF-5A of 18,000 and pI near 5.1 of Chinese hamster ovary (CHO) cells, rabbit reticulocytes, and human lymphocytes contained

13 or 14 lysines and yielded very similar tryptic peptides. Among these, the single hypusine-containing peptide was essentially or nearly identical in the three species (Park et al., 1984). The mammalian protein proved to contain 154 amino acids of 16.7 kDa, with a blocked amino terminal end and only a single difference between the rabbit and human proteins (Smit–McBride et al., 1989a,b). Exclusion molecular weight analysis, avoiding sodium dodecylsulfate (SDS), revealed that the eIF-5A of human erythrocytes exists normally as a dimer in solution (Chung et al., 1991; Park et al., 1986). However, labeling of chick embryo fibroblasts revealed two eIF-5As of ~20 and ~18 kDa, with comparable activities despite different primary sequences (Wolff et al., 1992).

All eucaryotic cells have been found to contain eIF-5A with hypusine. *N. crassa* contains a 21-kDa protein of low pI, ~3.5 (Yang et al., 1990). The *Neurospora* protein contains 163 amino acids, with considerable similarity to the human and yeast proteins (Tao & Chen, 1994, 1995). The hypusine protein of *Dictyostelium discoideum*, present throughout its life cycle, consists of 169 amino acids amounting to 18.3 kDa (Sandholzer et al., 1989). This protein is encoded by a single gene. Three genes for eIF-5A have been found in tobacco. The polypeptide of one of these showed only a 50–60% identity with other nonplant eIF-5As (Chamot & Kuhlemeier, 1992).

In humans eIF-5A is present at low levels in T lymphocytes and its content in these cells is increased by specific stimuli, including HIV infection (Bevec et al., 1994). A human gene encoding the protein has been isolated and found uninterrupted by introns. Two pseudogenes of the functional protein have also been described (Koettnitz et al., 1994).

In the yeast *S. cerevisiae* eIF-5A is encoded by two genes. Either gene is sufficient for normal growth in aerobic or anaerobic conditions although only one of these is transcribed normally (Magdolen et al., 1994). Each gene determines the formation of a somewhat basic form, which serves as the precursor of a more acidic phosphorylated form (Kang et al., 1993). The elimination of the serine-bound phosphate by phosphatase does not affect the ability of the hypusine-containing eIF-5A to stimulate the synthesis of the first peptide bond in a mammalian assay system. A similar result was seen with a yeast mutant in which the serine site of phosphorylation at the N terminus of the protein had been replaced by alanine (Klier et al., 1993). The N terminus of the phosphorylated protein is a unique doubly modified residue, consisting of an *N*-acetyl, *o*-phosphorylserine.

Hypusine is also present in certain archaebacteria. Although apparently absent in methanococci, hypusine has been found in several aerobic archaebacteria including the thermophile *Sulfolobus acidocaldarius* (Bartig et al., 1992). In affinity chromatography in which specific elution was effected by spermine, the hypusine-containing protein had an apparent mass of 16.8 kDa, but an unusually high pI of 7.8. Unlike most initiation proteins, it

was not ribosome bound but mainly cytosolic. A genomic DNA was found to contain an open reading frame coding for 135 amino acids, pointing to a mass of 15 kDa. The total sequence similarity to other eIF-5A proteins was 33%. Nevertheless, a hypusine-containing nonapeptide was identical to that found in the proteins that had been sequenced. Although it was not possible to test the function of the protein in in vitro polypeptide synthesis, the identities of structure point to analogous function for the *Sulfolobus* protein, as well as an evolutionary link between the archaebacteria and eucaryota. However, the archaebacterial gene does not support growth of yeast deficient in a hypusine gene.

Biosynthesis and Possible Role of Hypusine

The hypusine in protein synthesized in CHO cells has been labeled by growth in the presence of exogenous [4,5-^{3}H]lysine or [terminal methylene-^{3}H]spermidine. The addition of a metal chelator, α,α-dipyridyl, causes the accumulation of the unhydroxylated deoxyhypusine in the cell protein. In washed cells or a lysate, this precursor is converted to protein-bound hypusine (Park et al., 1982). In the initial transfer, the C—H bond at C_1 of the aminobutyl moiety of spermidine is broken. This was thought to suggest the intermediation of an imine (Schiff base) in the transfer to lysine (Park & Folk, 1986). A second C—H bond at C_2 of this moiety is broken during the second step, that of hydroxylation of deoxyhypusine to hypusine.

On readdition of spermidine to HTC or CHO cells depleted of spermidine by treatment with DFMO, the cells synthesize deoxyhypusine and hypusine at a markedly greater rate (10-fold) than do undepleted cells. The rate of hypusine synthesis appeared to be related to the intracellular content of spermidine: a maximal rate was attained at >12 nmol spermidine/mg protein (Gerner et al., 1986; Park et al., 1993a,b). Neither spermine nor N^1-acetylspermidine appeared to serve as precursors. DFMO was shown to permit an accumulation of the lysine form of the protein precursor in mammalian cells; this has been useful in subsequent enzymatic studies.

NAD$^+$ stimulated (150-fold) the labeling of the 18,000-kDa protein by spermidine in cell lysates (Chen & Dou, 1988). The enzymatic activity was many times higher in young than in aged cells in tests with the lysine form of substrate isolated from *Neurospora*. Subsequently this demonstration of the mammalian deoxyhypusine synthase and its NAD$^+$ dependence was achieved with the 18-kDa lysine precursor derived from DFMO-treated neuroblastoma cells (Dou & Chen, 1990). Similar results have been reported with enzymes and substrates from untreated and DFMO-treated CHO cells, including the formation of DAP as a cleavage product derived from spermidine (Park & Wolff, 1988). These workers also demonstrated inhibitory effects of analogues of spermidine and spermine. They purified the enzyme 700-fold from rat testis and determined Michaelis constant values for the substrates. A slow cleavage of spermidine to DAP and Δ^1-pyrroline was observed in the absence of the protein acceptor and ^{3}H from the 5 position of spermidine was found in the NADH and Δ^1-pyrroline products (Wolff et al., 1990).

A rapid sensitive assay has been devised for the deoxyhypusine synthase (Tao et al., 1994) that facilitates the purification of a homogeneous enzyme from *Neurospora* (Tao & Chen, 1994, 1995). One step in the purification of the enzyme has been a selective release by spermidine from a diamine-agarose column.

The lysine and deoxyhypusine precursors of eIF-5A and the lysine precursor were examined as potential stimulators of the formation of methionyl-puromycin, the test reaction for the ability to synthesize the first peptide bond. The lysine precursor of the hypusinated eIF-5A could not stimulate the formation of methionyl-puromycin, a stimulation that was readily obtained with eIF-5A of CHO cells. This result was taken initially to imply an essential role for eIF-5A hypusine in eucaryotic protein synthesis (Park, 1989; Smit-McBride et al., 1989a,b). The deoxyhypusine-containing precursor also activated the synthesis of methionyl-puromycin as well as eIF-5A itself did (Hershey et al., 1990; Park et al., 1991). In the latter study, aminopropylcadaverine was also used to replace spermidine in the formation of a deoxyhypusine analogue, homodeoxyhypusine. Unlike stimulatory deoxyhypusine-eIF-5A, the protein containing the homodeoxyhypusine was not stimulatory in the methionyl-puromycin assay.

The second biosynthetic enzyme, the deoxyhypusine hydroxylase, is present in chick embryo and in all mammalian tissues, of which testis is the most active source. Despite differences of this enzyme from other hydroxylases that have more elaborate cofactor requirements, the activity was inhibited by hydralazine, known to inhibit prolyl hydroxylation in procollagen. The enzyme was inhibited competitively by spermine and thermine and less well by spermidine and norspermidine. A peptide simulating the deoxyhypusine site in the eIF-5A was also inhibitory (Abbruzzese et al., 1989).

Attempting to clarify the roles of the hypusine-containing initiation factor, several groups turned to the study of yeast. The two genes for two eIF-5A proteins found in *S. cerevisiae* shared 90% sequence identity and were 63% identical to the human protein (Wöhl et al., 1993). The disruption of both genes resulted in the loss of cell viability (Schnier et al., 1991). The presence of a plasmid containing an appropriate lysine codon in the sequence permitted survival, whereas a mutant with an arginine codon replacing lysine did not restore growth. Conditions were found to eliminate a destabilized eIF-5A rapidly, within one generation of 2 h; this caused an arrest of cell division despite continued protein synthesis at a reduced rate for over 5 h (Kang & Hershey, 1994). Thus, the function of eIF-5A, apparently nonessential for initiation of protein synthesis, remains obscure.

Nevertheless, the observations that eIF-5A and hypusine are essential for cell growth, cell division, and survival have encouraged studies on inhibitors of the

biosynthetic enzymes. The inhibition of AdoMetDC, and hence of spermidine synthesis, with some newer inhibitors provokes an accompanying cytostasis, suggested to be caused by depletion of hypusine (Byers et al., 1992). Analogues structurally related to spermidine have been found more recently to be excellent inhibitors of deoxyhypusine synthase in vitro and in vivo (Jakus et al., 1993). These include several bis- and mono-guanylated diamines and polyamines, of which N^1-guanyl-1,7-diaminoheptane has proven to be most effective, with a K_i of about 10 nM. This compound at 1–10 μM rapidly inhibits hypusine synthesis in CHO cells without otherwise affecting the cellular levels of polyamines. This inhibitor is taken up by the polyamine transport system and inhibits growth after one generation. The delay in response is a consequence of the very long $t_{1/2}$ of eIF-5A (Park et al., 1993a,b).

The prospect of developing specific inhibitors of synthesis of human hypusine has led to structural studies of the eIF-5A precursor, prepared after transfection of *E. coli* with human cDNA (Joe and Park, 1994). An essential domain of 50 amino acids in the protein corresponded to a highly conserved region of the protein throughout the eucaryotic kingdom.

Summary

In the few years from 1964 to 1971, the study of polyamine biosynthesis in a few choice mammalian model systems capable of rapid growth in response to surgical or hormonal stimuli permitted the demonstration of the existence of the four biosynthetic enzymes. One of these, the catalyst for the synthesis of spermine, was found for the first time. The other three were also found to be synthesized in response to various stimuli, that is, to require protein synthesis, prior to their increase in activity. Unexpectedly, the first two in the biosynthetic sequence, ODC and AdoMetDC, fell precipitously by an unexplained mechanism whereas the third, spermidine synthase, remained at a high level. In some instances enzyme products, for example, putrescine from ODC activity, were inhibitory, although the discovery of how they might do so proved to be surprisingly complex. Putrescine was peculiarly essential in the second and third reactions, as a cofactor for some eucaryotic AdoMetDC, and as a substrate for spermidine synthase, posing the problem of balancing the inhibitory, cofactor, and substrate functions and concentrations of this diamine. The problems of regulation were obviously the next targets for study.

In this period classical biochemical approaches also yielded new information. These involved the identification of natural products in urine and blood and in more stable tissues, of which the nongrowing, nonmultiplying brain was of particular interest. A more refined characterization of fractions detected a series of oxidation products of the polyamines, as well as several acetyl derivatives seen previously in phenomena of bacterial metabolism. The use of isotopically labeled polyamines now provided evidence of a metabolic continuum of these compounds in animal tissues. In the context of the discovery of a new PAO, dissection of the possibilities soon revealed a catabolic cycle for the interconversion of the polyamines, entailing the existence of acetylases and deacetylases.

A large multiplication of the metabolic world of the polyamines had occurred, in which the compounds of interest and their determining enzymes were multiplied manyfold. The regulatory interactions of compounds and enzymes, and more recently their genetic determinants, in a large number of developmental situations, became the foci of work in the 1980s and 1990s. Study of the metabolic fate of spermidine also revealed an unexpected fragmentation of the molecule, one piece of which converted a unique lysine in a special protein, eIF-5A, into a new protein-bound amino acid, deoxyhypusine. This compound was subsequently hydroxylated to a protein-bound product essential to the survival of eucaryotic cells.

REFERENCES

Abbruzzese, A., Park, M. H., Beninati, S., & Folk, J. E. (1989) *Biochimica et Biophysica Acta*, **997**, 248–255.

Abdel–Monem, M. M., Ohno, K., Newton, N. E., & Weeks, C. E. (1978) *Advances in Polyamine Research*, vol. 2, pp. 37–49. R. A. Campbell, D. R. Morris, D. Bartos, G. D. Daves, & F. Bartos, Eds. New York: Raven Press.

Abe, T., Takada, K., Ohkawa, K., & Matsuda, M. (1990) *Biochemical Journal*, **269**, 25–29.

Albina, J. E., Abate, J. A., & Mastrofrancesco, B. (1993) *Journal of Surgical Research*, **55**, 97–102.

Asatoor, A. M. (1979) *Biochimica et Biophysica Acta*, **586**, 55-62.

Atkinson, D. E. (1992) *Physiological Zoology*, **65**, 243–267.

Barlow, G. B. (1980) *Archives of Disease in Childhood*, **55**, 40–42.

Bartig, D., Lemkemeier, K., Frank, J., Lottspeich, F., & Klink, F. (1992) *European Journal of Biochemistry*, **204**, 751–758.

Beninati, S., Piacentini, M., Spinedi, A., & Autuori, F. (1980) *Biomedicine*, **33**, 140–143.

Berger, F. G., Szymanski, P., Read, E., & Watson, G. (1984) *Journal of Biological Chemistry*, **259**, 7941–7976.

Bevec, D., Klier, H., Holter, W., Tschachler, E., Valent, P., Lottspeich, F., Baumruker, T., & Hauber, J. (1994) *Proceedings of the National Academy of Sciences of the United States of America*, **91**, 10829–10833.

Blankenship, J. (1978) *Archives of Biochemistry and Biophysics*, **189**, 20–27.

Bolkenius, F. N., Bey, P., & Seiler, N. (1985) *Biochimica et Biophysica Acta*, **838**, 69–76.

Bolkenius, F. N. & Seiler, N. (1986) *International Journal of Developmental Neuroscience*, **4**, 217–224.

——— (1987) *Biochimica et Biophysica Acta*, **923**, 125–135.

Bremer, H. J., Kohne, E., & Endres, W. (1971) *Clinica Chimica Acta*, **32**, 407–418.

Byers, T. L., Ganem, B., & Pegg, A. E. (1992) *Biochemical Journal*, **287**, 717–724.

Camón, L., De Vera, N., & Martínez, E. (1994) *Neurotoxicology*, **15**, 759–764.

Campbell, R., Talwalkar, Y., Bartos, D., Musgrave, J., Harner, M., Puri, H., Grettie, D., Dolney, A. M., & Laggan, B. (1978) *Advances in Polyamine Research*, vol. 2, pp. 319–343. R. A. Campbell, D. R. Morris, D. Bartos, G. D. Daves, & F. Bartos, Eds. New York: Raven Press.

Canellakis, Z. N., Marsh, L. L., Manabe, Y. C., Infante, A. A., Bondy, P. K., & Scaliser, F. W. (1989) *Biochemistry International*, **19**, 969–976.

Canellakis, Z. N. & Theoharides, T. C. (1976) *Journal of Biological Chemistry*, **251**, 4436–4441.

Caron, P. C., Cote, L. J., & Kremzner, L. T. (1988) *Biochemical Journal*, **251**, 559–562.

Caron, P. C., Kremzner, L. T., & Cote, L. J. (1987) *Neurochemistry International*, **10**, 219–229.

Carper, S. W., Tome, M. E., Fuller, D. J. M., Chen, J., Harari, P. M., & Gerner, E. W. (1991) *Biochemical Journal*, **280**, 289–294.

Catterall, J. F. & Jänne, O. A. (1988) *Modern Cell Biology*, **6**, 1–28.

Chamot, D. & Kuhlemeier, C. (1992) *Nucleic Acids Research*, **20**, 665–669.

Chen, K. Y. & Dou, Q. (1988) *Biochimica et Biophysica Acta*, **971**, 21–28.

Chen, K. Y. & Liu, Y. (1981) *FEBS Letters*, **134**, 71–74.

Choi, S. & Churchich, J. E. (1986) *European Journal of Biochemistry*, **161**, 289–294.

Chung, S. I., Park, M. H., Folk, J. E., & Lewis, M. S. (1991) *Biochimica et Biophysica Acta*, **1076**, 448–451.

Cohen, S., O'Malley, B. W., & Stastny, M. (1970) *Science*, **170**, 336–338.

Cooper, H. L., Park, M. H., Folk, J. E., Safer, B., & Braverman, R. (1983) *Proceedings of the National Academy of Sciences of the United States of America*, **80**, 1854–1857.

Cynober, L. (1991) *Nutrition*, **7**, 313–322.

Danzin, C., Jung, M. J., Seiler, N., & Metcalf, B. W. (1979) *Biochemical Pharmocology*, **28**, 633–639.

Daune, G., Gerhart, F., & Seiler, N. (1988) *Biochemical Journal*, **253**, 481–488.

Daune, G. & Seiler, N. (1988) *Neurochemical Research*, **13**, 69-75.

Davis, R. H. (1986) *Microbiological Reviews*, **50**, 280–313.

Dou, Q. & Chen, K. Y. (1990) *Biochimica et Biophysica Acta*, **1036**, 128–137.

Erdö, S. L. & Wolff, J. R. (1990) *Journal of Neurochemistry*, **54**, 363–372.

Fausto, N. (1970) *Cancer Research*, **30**, 1947–1952.

Fogel, W. A. (1986) *GABAergic Mechanisms in the Mammalian Periphery*, pp. 35–56. S. L. Erdö & N. G. Bowery, Eds. New York: Raven Press.

Fogel, W. A., Bieganski, T., & Maslinski, C. (1982) *Comparative Biochemistry and Physiology*, **73C**, 431–434.

Fowler, L. J. & John R. A. (1972) *Biochemical Journal*, **130**, 569–573.

Fruton, J. (1972) *Molecules and Life*, pp. 424–445. New York: Wiley-Interscience.

Gerner, E. W., Mamont, P. S., Bernhardt, A., & Siat, M. (1986) *Biochemical Journal*, **239**, 379–386.

Gibson, K. M., Nyhan, W. L., & Jaeken, J. (1986) *Bioessays*, **4**, 24–27.

Griffith, O. W. (1987) *Methods in Enzymology*, **143**, 366–376.

Henningsson, S., Persson, L., & Rosengren, E. (1978) *Acta Physiologica Scandinavica*, **102**, 385–393.

Henningsson, S. & Rosengren, E. (1976) *British Journal of Pharmacology*, **58**, 401–406.

Hershey, J. W. B., Smit–McBride, A., & Schnier, J. (1990) *Biochimica et Biophysica Acta*, **1050**, 160–162.

Hessels, J., Kingma, A. W., Sturkenboom, M. C. J. M., Elzinga, H., Van den Berg, G. A., & Muskiet, F. A. J. (1991) *Journal of Chromatography*, **563**, 1–9.

Hölttä, E. (1977) *Biochemistry*, **16**, 91–100.

Hrushesky, W. J. M., Merdink, J., & Abdel–Monem, M. M. (1983) *Cancer Research*, **43**, pp. 3944–3947.

Imaoka, N. & Matsuoka, Y. (1974) *Journal of Neurochemistry*, **22**, 859–860.

Isomaa, V. V., Pajunen, A. E. I., Barden, C. W., & Jänne, O. A. (1983) *Journal of Biological Chemistry*, **258**, 6735–6740.

Jakus, J., Wolff, E. C., Park, M. H., & Folk, J. E. (1993) *Journal of Biological Chemistry*, **268**, 13151–13159.

Jänne, J. & Williams–Ashman, H. G. (1970) *Biochemical Journal*, **119**, 595–597.

Joe, Y. A. & Park, M. H. (1994) *Journal of Biological Chemistry*, **269**, 25916–25921.

Kang, H. A. & Hershey, J. W. B. (1994) *Journal of Biological Chemistry*, **269**, 3934–3940.

Kang, H. A., Schwelberger, H. G., & Hershey, J. W. B. (1993) *Journal of Biological Chemistry*, **268**, 14750–14756.

Kato, T., Sano, M., Mizutani, N., & Hayakawa, C. (1987) *Journal of Inherited Metabolic Disease*, **10**, 391–396.

Kaye, A. M., Icekson, I., & Lindner, H. R. (1971) *Biochimica et Biophysica Acta*, **252**, 150–159.

Kenyon, S. H., Ast, T., Nicolaou, A., & Gibbons, W. A. (1995) *Biochemical Society Transactions*, **23**, 444S.

Klier, H., Wöhl, T., Eckerskorn, C., Magdolen, V. & Lottspeich, F. (1993) *FEBS Letters*, **334**, 360–364.

Kobayashi, Y., Kupelian, J., & Maudsley, D. V. (1971) *Science*, **172**, 379–380.

Koettnitz, K., Kappel, B., Baumruker, T., Hauber, J., & Bevec, D. (1994) *Gene*, **144**, 249–252.

Konishi, H., Nakajima, T., & Sano, I. (1977) *Journal of Biochemistry*, **81**, 355–360.

Kremzner, L. T., Hiller, J. M., & Simon, E. J. (1975) *Journal of Neurochemistry*, **25**, 889–894.

Lewan, L., Yngner, T., & Engelbrecht, C. (1977) *International Journal of Biochemistry*, **8**, 477–487.

Libby, P. R. (1978a) *Journal of Biological Chemistry*, **253**, 233–237.

——— (1978b) *Archives of Biochemistry and Biophysics*, **188**, 360–363.

——— (1983) *Methods in Enzymology*, **94**, 325–328.

Lippert, B., Metcalf, B. W., Jung, M. J., & Casara, P. (1977) *European Journal of Biochemistry*, **74**, 441–445.

Lundgren, D. W. & Hankins, J. (1978) *Journal of Biological Chemistry*, **253**, 7130–7133.

Magdolen, V., Klier, H., Wöhl, T., Klink, F., Hirt, H., Hauber, J., & Lottspeich, F. (1994) *Molecular and General Genetics*, **244**, 646–652.

Marcote, M. J., Corella, D., González–Bosch, C., & Hernández–Yago, J. (1994) *Biochemical Journal*, **300**, 277–280.

Matsui, I. & Pegg, A. E. (1980) *Biochemical and Biophysical Research Communications*, **92**, 1009–1015.

Miller, J. M., Ferraro, T. N., & Hare, T. A. (1990) *Journal of Neurochemistry*, **55**, 769–773.

Minuk, G. Y., Gauthier, T., & Benarroch, A. (1990) *Hepatology*, **12**, 542–546.

Minuk, G. Y., Gauthier, T., Gaharie, A., & Murphy, L. J. (1991) *Hepatology*, **14**, 685–689.

Missala, K. M. & Sourkes, T. L. (1980) *European Journal of Pharmacology*, **64**, 307–311.

Mori, M. & Tatibana, M. (1975) *Biochemical and Biophysical Research Communications*, **67**, 287–293.

Morris, Jr., S. M. (1992) *Annual Review of Nutrition*, **12**, 81–101.

Moukarzel, A., Goulet, O., Salas, J. S., Marti–Henneberg, C., Buchman, A. L., Cynober, L., Rappaport, R., & Ricour, C. (1994) *Americal Journal of Clinical Nutrition*, **60**, 408–413.

Murphy, B. J. & Brosnan, M. E. (1976) *Biochemical Journal*, **157**, 33–39.

Nakajima, T. (1973) *Journal of Neurochemistry*, **20**, 735–742.

Nakajima, T., Matsuoka, Y., & Akazawa, S. (1970) *Biochimica et Biophysica Acta*, **222**, 405–408.

Narita, I., Border, W. A., Ketteler, M., Ruoslahti, E., & Noble, N. A. (1995) *Proceedings of the National Academy of Sciences of the United States of America*, **92**, 4552–4556.

Noto, T., Tanaka, T., & Nakajima, T. (1978) *Journal of Biochemistry*, **83**, 543–552.

Olsen, R. W. & Tobin, A. J. (1990) *FASEB Journal*, **4**, 1469–1480.

Oon, B. B., Scraggs, P. R., & Gillham, B. (1989) *Journal of Endocrinology*, **123**, 227–232.

Pajunen, A. E. I., Hietala, O. A., Baruch–Virransalo, E., & Piha, R. S. (1979) *Journal of Neurochemistry*, **32**, 1401–1408.

Panko, W. B. & Kenney, F. T. (1971) *Biochemical and Biophysical Research Communications*, **43**, 346–350.

Park, M. H. (1989) *Journal of Biological Chemistry*, **264**, 18531–18535.

Park, M. H., Chung, S. I., Cooper, H. L., & Folk, J. E. (1984) *Journal of Biological Chemistry*, **259**, 4563–4565.

Park, M. H., Cooper, H. L., & Folk, J. E. (1981) *Proceedings of the National Academy of Sciences of the United States of America*, **78**, 2869–2873.

Park, M. H., Cooper, H. L., & Folk, J. E. (1982) *Journal of Biological Chemistry*, **257**, 7217–7222.

Park, M. H. & Folk, J. E. (1986) *Journal of Biological Chemistry*, **261**, 14108–14111.

Park, M. H., Liu, T., Neece, S. H., & Swiggard, W. J. (1986) *Journal of Biological Chemistry*, **261**, 14515–14519.

Park, M. H. & Wolff, E. C. (1988) *Journal of Biological Chemistry*, **263**, 15264–15269.

Park, M. H., Wolff, E. C., & Folk, J. E. (1993a) *Trends in Biochemical Sciences*, **18**, 475–479.

——— (1993b) *BioFactors*, **4**, 95–104.

Park, M. H., Wolff, E. C., Smit–McBride, Z., Hershey, J. W. B., & Folk, J. E. (1991) *Journal of Biological Chemistry*, **266**, 7988–7994.

Pegg, A. E., Lockwood, D. H., & Williams–Ashman, H. G. (1970) *Biochemical Journal*, **117**, 17–31.

Pegg, A. E. & Williams–Ashman, H. G. (1970) *Archives of Biochemistry and Biophysics*, **137**, 156–165.

Persson, L. & Rosengren, E. (1984) *Agents and Actions*, **15**, 235–237.

Pöyhönen, M. J., Takala, J. A., Pitkänen, O., Kari, A., Alhava, E., Alakuijala, L. A., & Eloranta, T. O. (1993) *Metabolism*, **42**, 44–51.

Raina, A. & Jänne, J. (1968) *Acta Chemica Scandinavica*, **22**, 2375–2378.

Raina, A., Jänne, J., Hannonen, P., Höltta, E., & Ahonen, J. (1973) *Polyamines in Normal and Neoplastic Growth*, pp. 167–180. D. H. Russell, Ed. New York: Raven Press.

Randall, D. J., Wood, C. M., Perry, S. F., Bergman, H., Maloiy, G. M. O., Mommsen, T. P., & Wright, P. A. (1989) *Nature*, **337**, 165–166.

Rock, D. M. & Macdonald, R. L. (1995) *Annual Review of Pharmacology and Toxicology*, **35**, 463–482.

Rosenblum, M. G. & Russell, D. H. (1977) *Cancer Research*, **37**, 47–51.

Russell, D. H. (1971) *Biochemical Pharmacology*, **20**, 3481–3491.

Russell, D. H., Medina, V. J., & Snyder, S. H. (1970b) *Journal of Biological Chemistry*, **245**, 6732–6738.

Russell, D. & Snyder, S. H. (1968) *Proceedings of the National Academy of Sciences of the United States of America*, **60**, 1420–1427.

Russell, D. H. & Snyder, S. H. (1969) *Molecular Pharmacology*, **5**, 253–262.

Russell, D. H., Snyder, S. H., & Medina, V. J. (1970a) *Endocrinology*, **86**, 1414–1419.

Sandholzer, U., Centea–Intemann, M., Noegel, A. A., & Lottspeich, F. (1989) *FEBS Letters*, **246**, 64–100.

Sano, A., Kotani, K., Ueno, S., & Kakimoto, Y. (1987) *Journal of Neurochemistry*, **48**, 681–683.

Sarhan, S., Quemener, V., Moulinoux, J., Knödgen, B., & Seiler, N. (1991) *International Journal of Biochemistry*, **23**, 617–626.

Satink, H. P. W. M., Hessels, J., Kingma, A. W., Van den Berg, G. A., Muskiet, F. A. J., & Halie, M. R. (1989) *Clinica Chimica Acta*, **179**, 305–314.

Schnier, J., Schwelberger, H. G., Smit–McBride, Z., Kang, H. A., & Hershey, J. W. B. (1991) *Molecular and Cellular Biology*, **11**, 3105–3114.

Schrock, T. R., Oakman, N. J., & Bucher, N. L. R. (1970) *Biochimica et Biophysica Acta*, **204**, 564–577.

Seale, T. W., Chan, W., Shulka, J., & Rennert, O. M. (1979) *Archives of Biochemistry and Biophysics*, **198**, 164–174.

Seguela, P., Geffard, M., Buijs, R. M., & Moal, M. L. (1984) *Proceedings of the National Academy of Sciences of the United States of America*, **81**, 3888–3892.

Seiler, N. (1980) *Physiological Chemistry and Physics*, **12**, 411–429.

—— (1983) *Neurochemistry International*, **5**, 363–364.

—— (1986) *Journal of Chromatography*, **379**, 157–176.

—— (1991) *Les Polyamines*, pp. 221–236. J. Moulinoux & V. Quémener, Eds. Paris: Médecine-Sciences/Flammarion.

Seiler, N. & Al-Therib, M. J. (1974a) *Biochimica et Biophysica Acta*, **354**, 206–212.

—— (1974b) *Biochemical Journal*, **144**, 29–35.

Seiler, N., Al–Therib, M. J., & Kataoka, K. (1973) *Journal of Neurochemistry*, **20**, 699–708.

Seiler, N. & Bolkenius, F. N. (1985) *Neurochemical Research*, **10**, 529–544.

Seiler, N., Bolkenius, F. N., Bey, P., Mamont, P. S., & Danzin, C. (1985a) *Recent Progress in Polyamine Research*, pp. 305–319. L. Selmeci, M. E. Brosnan, & N. Seiler, Eds. Budapest: Akademiai Kiado.

Seiler, N., Bolkenius, F. N., & Knödgen, B. (1985b) *Biochemical Journal*, **225**, 219–226.

Seiler, N., Bolkenius, F. N., Knödgen, B., & Haegele, K. (1980) *Biochimica et Biophysica Acta*, **676**, 1–7.

Seiler, N., Bolkenius, F. N., & Sarhan, S. (1981a) *International Journal of Biochemistry*, **13**, 1205–1214.

Seiler, N. & Daune-Anglard, G. (1993) *Metabolic Brain Disease*, **8**, 151–179.

Seiler, N. & Deckardt, K. (1976) *Neurochemical Research*, **1**, 469–499.

—— (1978) *Advances in Polyamine Research*, vol. 2, pp. 161–167. R. A. Campbell, D. R. Morris, D. Bartos, G. D. Daves, & F. Bartos, Eds. New York: Raven Press.

Seiler, N. & Eichentopf, B. (1975) *Biochemical Journal*, **152**, 201–210.

Seiler, N. & Knödgen, B. (1983) *International Journal of Biochemistry*, **15**, 907–915.

Seiler, N., Knödgen, B., Gittos, M. W., Chen, W. Y., Griesmann, G., & Rennert, O. M. (1981b) *Biochemical Journal*, **200**, 123–132.

Seiler, N., Koch–Weser, J., Knödgen, B., Richards, W., Tardif, C., Bolkenius, F. N., Schecter, P., Tell, G., Mamont, P., & Fozard, J. (1981c) *Advances in Polyamine Research*, vol. 3, pp. 197–211. C. M. Caldera, V. Zappia, & U. Bachrach, Eds. New York: Raven Press.

Seiler, N. & Lamberty, U. (1973) *Journal of Neurochemistry*, **20**, 709–717.

—— (1975) *Journal of Neurochemistry*, **24**, 5–13.

Seiler, N., Richards, W., & Knödgen, B. (1981d) *Archives of Dermatological Research*, **270**, 25–32.

Seiler, N. & Sarhan, S. (1983) *Neurochemistry International*, **5**, 625–633.

Seiler, N., Sarhan, S., & Knödgen, B. (1985a) *International Journal of Developmental Neuroscience*, **3**, 317–322.

Seiler, N., Sarhan, S., Knödgen, B., Nornsperger, J., & Sablone, M. (1993) *Pharmacology and Toxicology*, **72**, 116–123.

Seiler, N., Schmidt–Glenewinkel, T., & Sarhan, S. (1979) *Journal of Biochemistry*, **86**, 277–278.

Seiler, N., Spraggs, H., & Daune, G. (1987) *Neurochemistry International*, **10**, 391–397.

Seiler, N., Wiechmann, M., Fischer, H. A., & Werner, G. (1971) *Brain Research*, **28**, 317–325.

Shaw, G. G. (1979) *Biochemical Pharmacology*, **28**, 1–6.

Shiba, T., Kubota, I., & Kaneko, T. (1970) *Tetrahedron*, **26**, 4307–4311.

Shiba, T., Mizote, H., Kaneko, T., Nakajima, T., Kakimoto, Y., & Sano, I. (1971) *Biochimica et Biophysica Acta*, **244**, 523–531.

Smit–McBride, Z., Dever, T. E., Hershey, J. W. B., & Merrick, W. C. (1989a) *Journal of Biological Chemistry*, **264**, 1578–1583.

Smit–McBride, Z., Schnier, J., Kaufman, R. J., & Hershey, J. W. B. (1989b) *Journal of Biological Chemistry*, **264**, 18527–18530.

Smith, T. K., Hyvönen, T., Pajula, R., & Eloranta, T. O. (1987) *Annals of Nutrition and Metabolism*, **31**, 133–145.

Smith, T. K., Lindqvist, L., Alakuijala, L., & Eloranta, T. (1989) *Annals of Nutrition and Metabolism*, **33**, 143–152.

Snyder, S. H. & Russell, D. H. (1970) *Federation Proceedings*, **29**, 1575–1582.

Sobue, K. & Nakajima, T. (1977) *Journal of Biochemistry*, **82**, 1121–1126.

Sturman, J. A. & Gaull, G. E. (1974) *Pediatric Research*, **8**, 231–237.

Swendseid, M. E., Panaqua, M., & Kopple, J. D. (1980) *Life Sciences*, **26**, 533–539.

Takagi, T., Chung, T. G., & Saito, A. (1983) *Journal of Chromatography*, **272**, 279–285.

Takiguchi, M. & Mori, M. (1995) *Biochemical Journal*, **312**, 649–659.

Tao, Y. & Chen, K. Y. (1995) *Journal of Biological Chemistry*, **270**, 383–386.

—— (1994) *Biochemical Journal*, **302**, 517–525.

Tao, Y., Skrenta, H. M., & Chen, K. Y. (1994) *Analytical Biochemistry*, **221**, 103–108.

Tovar, A., Sánchez–Capelo, A., Cremades, A., & Peñafiel, R. (1995) *Kidney International*, **48**, 731–737.

Tsuji, M. & Nakajima, T. (1978) *Journal of Biochemistry*, **83**, 1407–1412.

Uusitupa, M., Pöyhönen, M., Sarlund, H., Laakso, M., Kari, A., Helenius, T., Alakuijala, L., & Eloranta T. (1993) *Scandinavian Journal of Clinical & Labaratory Investigation*, **53**, 811–819.

Van den Berg, G. A., Elzinga, H., Nagel, G. T., Kingma, A. W., & Muskiet, F. A. J. (1984) *Biochimica et Biophysica Acta*, 175–187.

Van den Berg, G. A., Kingma, A. W., Elzinga, H., & Muskiet, F. A. J. (1986) *Journal of Chromatography*, **383**, 251–258.

Weber, G., Queener, S. F., & Morris, H. P. (1972) *Cancer Research*, **32**, 1933–1940.

Williams–Ashman, H. G., Coppoc, G. L., & Weber, G. (1972) *Cancer Research*, **32**, 1924–1932.

Williams–Ashman, H. G. & Lockwood, D. H. (1970) *Annals of the New York Academy of Sciences*, **171**, 882–894.

Wöhl, T., Klier, H., Ammer, H., Lottspeich, F., & Magdolen, V. (1993) *Molecular and General Genetics*, **241**, 305–311.

Wolff, E. C., Kinzy, T. G., Merrick, W. C., & Park, M. H. (1992) *Journal of Biological Chemistry*, **267**, 6107–6113.

Wolff, E. C., Park, M. H., & Folk, J. E. (1990) *Journal of Biological Chemistry*, **265**, 4793–4799.

Wright, P. A. (1995) *Journal of Experimental Biology*, **198**, 273–281.

Yang, Y. C., Chen, K. Y., Seyfzadeh, M., & Davis, R. H. (1990) *Biochimica et Biophysica Acta*, **1033**, 133–138.

Zappia, V., Cartenì–Farina, M., & Della Pietra, G. (1972) *Biochemical Journal*, **129**, 703–709.

CHAPTER 13

Biosynthetic Enzymes of Animal Tissues and Cells: Part I Properties and Regulation of Ornithine Decarboxylase

The decarboxylation of the dibasic amino acids, ornithine and lysine, by bacterial enzymes is well known. The formation in autolyzing liver of putrescine from ornithine is probably due to bacteria.

H. BLASCHKO (1945)

The decade of the 1960s, encompassing the period of the discovery of the mammalian biosynthetic enzymes and their unexpected responses to growth stimuli, also had important discoveries in the field of gene expression and enzyme regulation. The appearance of new proteins was preceded by the formation of new messenger RNA (transcription). These newly synthesized RNAs were hybridizable to strands of DNA, presumably at the sites of specific genes, and determined the kinds and sequences of amino acids to be assembled as primary chains of protein (translation). The mammalian biosynthetic enzymes, ornithine decarboxylase (ODC) and *S*-adenosylmethionine decarboxylase (AdoMetDC), increased both rapidly and greatly in amount in response to various growth stimuli and, additionally, displayed unusually rapid declines in activity. The enzymatic activities were easily assayed, although *carboxyl*-labeled ornithine became readily available before a similarly labeled AdoMet.

Increases and decreases of particular enzymatic activities are challenges to the biochemist. Is the protein catalyst being synthesized de novo or is a precursor altered? Is the activity proportional to the amount of the unaltered protein enzyme? These questions require the isolation of pure enzyme and the characterization of some specific properties of the protein suitable for assay

in a relatively crude extract, for example, a combination with specific antibody or with a labeled specific irreversible inhibitor of the catalytic protein. Both methods have been used to demonstrate that the rise and fall of several biosynthetic activities have related in most instances to the de novo synthesis and degradation of the protein enzymes.

The task of isolating and characterizing pure animal enzyme was attempted initially with cells at peaks of enzyme activity (e.g., ODC from SV40–3T3 cells) and subsequently from the kidneys of male mice. The development of specific inhibitors, such as difluoromethylornithine (DFMO), led to the detection of drug-resistant mutants, some of which were overproducers of the enzyme. The isolation of the gene for ODC and the transfection of normal cells facilitated an enhanced production of the enzyme. Thus, new cellular sources of a biosynthetic enzyme became available in studies of peaks of normal enzyme production during development, stimuli permitting escape from arrested synthesis, characterization of the genetic determinant, and exploitation of the gene in transformed cells and most recently in transgenic animals. At each stage new information became available concerning the physiological significance of the polyamines and raised the possibility of affecting their production, as well as the polyamine contents and biological behavior of cells and tissues. The synthesis and study of inhibitors of polyamine biosynthesis and interconversion have become major enterprises in the analysis of normal and aberrant physiology and in the attempted control of cancer, parasitic disease, and other pathologies. The discovery of a naturally occurring inhibitory protein, such as ODC *antizyme*, provided a metabolic complication of considerable interest.

Early Studies on ODC

An entire volume has appeared in 1989 on ODC (see Hayashi & Canellakis, 1989), much of it relating to the mammalian enzyme. A historical overview recapitulates the discovery of a bacterial enzyme some 50 years ago, and the existence of biodegradative and biosynthetic forms in *Escherichia coli* about 25 years later. In the mid 1960s the concomitant study of the regenerating liver and androgen-stimulated prostate suggested a possibly central role of the polyamines, leading to the discovery of the mammalian ODC and its unusual increase and decline. These observations and other phenomena of polyamine biosynthesis were then extended to other mammalian systems.

Assay of Tissue ODC

The estimation of radioactive $^{14}CO_2$ generated by ODC from L-[carboxyl-^{14}C]-ornithine, introduced in the early bacterial and mammalian studies in the mid 1960s, proved to be inaccurate with some tissue extracts as a result of two nonspecific processes: nonenzymatic decarboxylation and routes of decarboxylation other than that beginning with ODC. The latter problems were exacerbated by processes disrupting mitochondria, leading to other routes of ornithine metabolism (Gaines et al., 1988). To eliminate these problems in the estimation of ODC some workers adapted the ion-exchange separation and radiometric estimation of [^{3}H]-putrescine from [^{3}H]-ornithine (Weber, 1987).

After initial stages of purification of ODC it is possible to use a nonradioactive spectrophotometric assay. One of these involves the reaction of ornithine and the reaction product with trinitrobenzenesulfonic acid, of which only the colored sulfonic acid derivative of putrescine is soluble in 1-pentanol (Ngo et al., 1987). Another spectrophotometric method applicable to many decarboxylases couples CO_2 production to form oxaloacetate, which in the presence of NADH is reduced to malate and generates NAD with a decreased absorbance at 340 nm.

$$\text{Ornithine} \xrightarrow{\text{ODC}} \text{putrescine} + CO_2 \qquad (1)$$

$$CO_2 + H_2O + \text{phosphoenolpyruvate} \xrightarrow[\text{carboxylase}]{\text{phosphoenolpyruvate}} \text{oxaloacetate} + H_2PO_4^- \qquad (2)$$

$$\text{oxaloacetate} + NADH + H^+ \xrightarrow[\text{dehydrogenase}]{\text{malate}} NAD + \text{malate.} \qquad (3)$$

In any case, the use of [1-^{14}C]-ornithine and the estimation of radioactive CO_2 has remained the most widely used assay. Slotkin and Bartolome (1983) discussed in detail the application of this method of assay to the detection of the ODC response of various types of cells and tissues to neurotransmitters and hormonal peptides. Nonenzymic decarboxylation may be minimized by reduction of the concentration of ornithine and pyridoxal phosphate (PLP) in the presence of added protein. Irreversible enzyme inhibitors, such as DFMO, may be used to establish the level of nonspecific decarboxylation in an assay blank. The presence of high levels of ornithine or putrescine in the extracts may affect the specific radioactivity of the substrate, and as a secondary effect may give end-product competitive inhibition; these effects may require dilution or dialysis of the extracts. Sulfhydryl compounds such as dithiothreitol are essential to maximal ODC activity (Jänne & Williams–Ashman, 1971). Sulfhydryl compounds should also be present during dialysis of the enzyme.

Cellular Sites of ODC Activity

McCormick found 91% of the ODC in the cytoplasts produced by the action of cytochalasin on confluent mouse L cells (Chap. 11). About 9% of the ODC was detected in karyoplasts but the existence of a thin layer of membrane-bound cytoplasm around the nucleus precluded a clear interpretation of this result. The subsequent discovery of the entirely specific and irreversibly bound suicide inhibitor of ODC, DFMO, permitted the use of a fluorescent or radioactive derivative of DFMO for the histochemical localization of the active enzyme. This technique was used to detect the major sites in regions of the developing rat cerebellum and in rat liver (Gilad & Gilad, 1981).

Within the proximal convoluted tubules of the mouse kidney cortex, bound [5-^{14}C]-DFMO was found by autoradiography mainly in the cytoplasm, but 10–15% was present in nucleoli-like structures in the nucleus (Zagon et al., 1984). After treatment with cycloheximide, the labeled protein disappeared rapidly from the cytoplasm and nucleus. Because DFMO binds and is converted to an irreversibly bound form by active ODC alone, the method does not detect inactive or cryptic ODC protein.

The purification of the enzyme enabled the development of a specific precipitating antibody, which could be detected by various histochemical or immunological techniques. Nevertheless, such studies were undertaken even before mammalian enzyme had been purified to homogeneity and yielded apparently specific precipitating antibody (Friedman et al., 1972). Höltta (1975) thereby demonstrated the increase of ODC protein in rat liver during response to growth hormone or in regeneration after partial hepatectomy. In later studies, even more specific monoclonal antibody was used (Matsufuji et al., 1984; Pegg et al., 1984). Methods for the immunocytochemical localization of ODC have been described by Persson et al. (1983).

The distribution of a monospecific antibody within

TABLE 13.1   Properties of purified mammalian ODC.

Source	Final specific activity (μmol/min/mg)	Fold purification	M_r (Subunit)
Mouse kidney	22–70	2,000–10,000	53,000
Mouse cells	—	—	51,150
Rat liver	19–24	350,000–710,000	50,000–55,000 (2)
Murine RAW264 cells	35	ca. 8,000	55,000 (2)
Human liver	—	—	51,150

The number of subunits is given in parentheses. (—)

M_r, relative molecular mass (molecular weight). Adapted from Pegg (1989).

ODC-overproducing Chinese hamster ovary (CHO) cells has been compared to that of labeled DFMO. The antibody, which indicates the presence of ODC protein, confirmed a major presence within the cytoplasm. However, the autoradiographic distribution of bound DFMO indicated that the nucleus contained as much active enzyme as the cytoplasm (Anehus et al., 1984). Despite the similarities of results in adult mouse kidney, that is, the equivalence of immunoreactive ODC and catalytically active enzyme, a great excess of immunoreactive ODC over active enzyme was reported for adult mouse brain, a result largely attributed to the existence of catalytically inactive, immunologically active ODC–antizyme complexes (Laitinen et al., 1985).

A high level of enzyme protein was found in the A-cell type of islets of Langerhans in rat pancreas, whereas other parts of the pancreas lack immunoreactivity with the antibody (Dorn et al., 1986). Immunohistochemical techniques have been used to localize ODC in tissues as diverse as atrial myocytes, phorbol-ester induced epidermal cells, mouse tooth-forming cells, cochlea of the immature rat (Henley et al., 1995), specific areas of developing human brain, and neurons of Alzheimer disease patients (Bernstein & Müller, 1995). An immunofluorescence assay attempted to quantify the protein in situ (Shayovits & Bachrach, 1994).

Studies on the Role and Purification of ODC

The biological evidence suggesting some significant role for ODC in growth led to efforts to purify the enzyme as one approach to understanding this role. The parallelism between the synthesis of the enzyme, polyamine synthesis, and the initiation of the synthesis of ribosomal RNA had suggested the possibility that the ODC protein might be catalytically involved in the latter synthesis. Calf liver ODC was purified some 25,000-fold to apparent homogeneity (Haddox & Russell, 1981), and a preparation of rat liver ODC, purified some 46,000-fold, stimulated the activity of RNA polymerase in several settings (Russell, 1983). Nevertheless, the specific activity of these preparations indicated a purity of no more than 11%. In more recent tests of the hypothesis, inhibitors of ODC induction did not affect induction of RNA polymerase. Antibody to ODC does

not cross-react with nucleolar RNA polymerase. A more highly purified ODC did not stimulate the latter enzyme activity (Urata et al., 1987). ODC and RNA polymerase increase separately in response to various growth stimuli and do not participate in structural interaction affecting function.

Properties of Purified Mammalian ODC

Some of these ODCs are summarized in Table 13.1. The molecular weight of an intact native protein may be determined via molecular sieve separation of a mixture of standards and enzyme. In fact the separated position of the latter may be determined with an enzymatic activity in relatively crude preparations. Gel electrophoresis of a purified protein in a medium containing the detergent sodium dodecyl sulfate (SDS) is used to estimate the size and homogeneity of the subunits in a denatured enzyme. Significant amounts of enzyme protein obtained by more elaborate techniques permit both the sizing of the protein and analysis of its homogeneity in nondenaturing gel electrophoresis, in which the presence of many proteins and their enzymatic activities can be detected by a variety of sensitive tests. Thus, the advent of these methods in the late 1960s and 1970s transformed the strategy of enzyme isolation and the methodologies of protein characterization. Methods involving hydrodynamic properties such as ultracentrifugation, diffusion, and viscosity required far more protein than was initially available in the study of most enzymes. The dearth of homogeneous enzyme protein in many instances has resulted also in a shift to estimation of a primary protein sequence from the nucleotide sequence in the DNA sequence (cDNA) obtained after the isolation and copy of a messenger RNA. Some mammalian ODCs were thereby found to contain 461 amino acids whose sequences could be compared to demonstrate the structural and evolutionary relatedness among them.

The isolation from thioacetamide-induced rat liver by one group required a 350,000-fold purification to obtain an 8% yield of the enzyme (Kameji et al., 1982) and a 71,000-fold purification with a 35% yield by another (Kitani & Fujisawa, 1983). The use of testosterone-induced mouse kidney provided an ODC of specific activity comparable to the most highly purified rat liver enzyme (Persson, 1981a,b), and indeed this value was comparable to that predicted by the binding of [^{14}C]-DFMO to the ODC of a liver extract (Pritchard et al., 1981). It was estimated that the ODC of the thioacetamide-induced rat liver represented no more than 0.00014% of the soluble protein.

The ODC isolated from androgen-stimulated mouse kidney yielded a single component in gel electrophoresis. This proved to have a molecular weight of about 100,000 and consisted of two subunits of relative molecular mass (M_r) 53,000 (Seely et al., 1982a). Michaelis constants (K_m) for L-ornithine and PLP were 75 and 0.3 µM, respectively. The purified protein was quite unstable unless it was stored in the presence of the detergent, 0.02% Brij 35. A minor band capable of binding DFMO was detected by isoelectric focusing, but at this time it was attributed to an artifact in the gel system used.

The purified enzyme bound one molecule of [^{14}C]-DFMO per subunit, affirming data on the DFMO-binding method of estimating the concentration of the enzyme in tissues under various conditions (Seely et al., 1982b). Induced increases or decreases in activity effected by various treatments of rats or mice were paralleled by comparable changes in the irreversible binding of DFMO to a separable protein. Thus, the extent of binding of DFMO measured the existence of proteins containing the catalytic centers of ODC and confirmed the use of androgen-treated mouse kidney as a very good source for purification of the enzyme protein. Nevertheless, in a slightly later study by this group, it was reported that crude kidney homogenates degraded the enzyme from ~55,000 subunit molecular weight to ~53,000 without apparent inactivation (Persson et al., 1984). In later studies, cells overexpressing the enzyme proved to be even better sources for purposes of purification.

Rabbit antibody to the mouse enzyme precipitated and inactivated the protein (Isomaa et al., 1983). The rabbit antibody to apparently homogeneous mouse kidney enzyme is quite specific for the ODC of several mammals, but reacts equivalently with native enzyme, enzyme inactivated by DFMO, or a mouse liver enzyme. A 400-fold increase of the activity in the kidney of testosterone-treated female mice was matched by an increase in immunoreactive protein. Disappearance of the catalytic activity after treatment of the mice with cycloheximide or 1,3-diaminopropane (DAP) led to the disappearance of immunoreactivity. The changes were clearly a function of the increase or decrease of the immunoreactive catalytic protein (Seely & Pegg, 1983a,b). Both active and DFMO-inactivated forms of the protein had identical rates of decay in mice treated with cycloheximide (Seely et al., 1982c). Despite the initial apparent homogeneity of the mouse kidney ODC preparation, 2-dimensional gel electrophoresis revealed charge heterogeneity in the enzyme (Isomaa et al., 1983). The different forms appeared to possess essentially similar molecular weights (Seely et al., 1985).

In more recent studies two sizes of subunits have

been detected in the ODC preparations: a major subunit of 54,000 and a minor subunit of 51,000. In several experiments the minor components were eliminated by the inclusion of specific proteinase inhibitors in the preparation of the extracts (Miyamoto et al., 1989; Nishiyama et al., 1988). Nevertheless, it has also been reported that preparations of the mouse kidney enzyme may contain two species of undegraded ODC of 54,000 and 52,000 molecular weights for the subunits as a function of the mouse strain used as the source. Some mouse strains do contain only the 54,000 subunit species (Kitani & Fujisawa, 1988a,b), and it is believed that protease effects are not the cause of the appearance of the smaller enzyme. Thus, it is thought that the two species of ODC may represent true isozymes of the kidney of particular mouse strains, for which two distinct mRNA molecules have been described. Some workers believe that the rat liver enzyme possesses molecular properties similar to that of the mouse kidney enzyme, while others believe the former is significantly larger by 2.0 kDa.

Charge Heterogeneity and Modification of ODC

Several species of ODC could be detected within the extracts of mammalian tissues and cloned cells by isoelectric focusing in electrophoresis; the species could be separated by anion-exchange chromatography (Mitchell & Mitchell, 1982; Richards et al., 1981). In induced systems, the activities appeared sequentially, suggestive of precursor–product relationships; that is, the change of one charged state to another by some chemical addition to the enzyme. Three mechanisms have been explored for such a change.

It was suggested that transglutaminase, an enzyme that can catalyze the incorporation of polyamines into proteins and thereby increase their basic charge, produces this change. However, in a yeast system in which exogenous polyamine evoked a loss of activity without the loss of ODC protein, the inactive protein did not contain covalently-bound polyamine (Tyagi et al., 1982).

In most preparations the phosphorylation states of the separable enzymes accounts for the observed differences. Of the two enzymatic forms derived from androgen-stimulated mouse kidney, one (20% of the total) could be phosphorylated by a casein kinase whereas the major component could be phosphorylated only after a prior dephosphorylation (Peng & Richards, 1988). Most recently three separate forms of ODC of different isoelectric points, 5.25, 5.15, and 4.95, were isolated by isoelectric focusing gel electrophoresis from murine RAW 264 cells (see Table 13.1). The most basic form was nonphosphorylated and the other more acidic proteins were phosphorylated (Reddy et al., 1994). The addition of one phosphate group will add fewer than 100 Da in molecular weight, an addition undetectable by sizing gels. A high resolution gel isoelectric focusing method has been described that can readily separate the several isoelectric forms of ODC (Reddy et al., 1994).

It was also reported that arginylation of the α-amino end of rat ODC could occur in in vitro systems via a transfer from arginyl-tRNA. An inhibitor of this enzyme significantly increased the half-life of ODC in cultured rat hepatocytes (Kopitz et al., 1990). It was postulated that arginylation is an early step in the degradation and turnover of ODC.

An additional aspect of charge heterogeneity was revealed in a study of ODC preparations fractionated in anion-exchange chromatography in media containing specific salt concentrations, the coenzyme PLP, or the substrate ornithine (Mitchell et al., 1988). In these studies the two major peaks of ODC of rat hepatoma cells contained monomeric subunits of about 55 kDa. Additional detectable forms appeared to be aggregates or mixtures of the combined subunits. Apparently dimers and aggregates can be dissociated into monomeric forms by 0.2 M NaCl. This dissociation is inhibited by PLP, which promotes dimerization, the active state of the predominant cellular ODC. Ornithine also increases dimerization. It was thought that the chromatographic conditions at high enzyme dilution promoted the dissociation of the two enzymes into monomeric subunits, some of which may reassociate as detectable mixed dimers.

That the charged state of ODC is relevant to the physiology of enzyme turnover was suggested but not completely proven. It was observed that the more negatively charged forms may be converted into the less negatively charged forms in cells and in crude cell homogenates at early stages of purification. Exposure of the cells to spermidine, spermine, and inhibitors of their synthesis enhances these changes (Mitchell et al., 1985b). The less negatively charged form was reported to be the more labile form within the cell in the presence of cycloheximide. These workers concluded that the changes of one form to another in these experiments indicates that these forms do not represent isoenzymes but relate to the charge states of a common ODC protein.

In Vitro Activity of ODC

Jänne and Williams–Ashman (1971) reported that mammalian ODC was activated by thiols such as dithiothreitol. Compounds that react with the sulfhydryl groups of the enzyme are inactivators (Fig. 13.1) and in some instances naturally occurring reducing substances, such as the thiol glutathione (GSH), which may attain millimolar levels in cells, can reactivate the enzyme. In the absence of the activating thiol, a partially purified ODC loses activity; this may be reversed by the addition of dithiothreitol or reducing systems containing NADPH or GSH (Flamigni et al., 1988). Oxidized glutathione, that is, GSSG, is inhibitory to the heart enzyme by a mechanism suggested to involve a mixed disulfide comprising ODC-S-S-glutathione (Guarnieri et al., 1982b). Because GSSG is a natural inhibitor, it may be supposed that the oxidizing environment within cells can thereby determine the activity of the enzyme within the

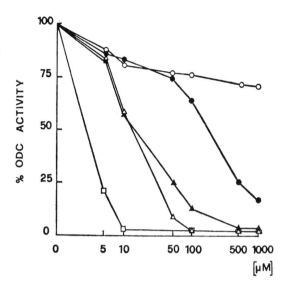

FIG. 13.1   Effect of sulfhydryl reagents on rat heart ODC activity; adapted from Guarnieri et al. (1982a). (○) Diamide; (●) GSSG; (▲) 5,5′-dithiobis-2-nitrobenzoic acid; (□) p-(chloromercuri)-phenylsulfonate.

cell. Hyperoxia and the production of superoxide, which affect the availability of sulfhydryl groups, are also known to inhibit the in vivo activity of ODC.

The activity of ODC-SH groups has been used to devise a thiol-covalent chromatography for the heart enzyme (Guarnieri et al., 1985). Several peaks of the active ODC are displaced from the column by the eluting 10 and 20 mM dithiothreitol, suggesting the existence of several forms of the enzyme with different SH reactivities. The existence of a peak eluted by 5 mM dithiothreitol was evidence for the presence of a mixed disulfide of ODC and glutathione.

The activity of ODC is decreased markedly by increasing salt concentration to 0.25 M NaCl in assay mixtures. This salt alone dissociates active dimers of about 100,000 M_r into inactive monomers of 55,000 M_r, as detected by molecular sieving, that is, gel permeation chromatography. The addition of 0.1 mM L-ornithine partially reverses the dissociation (Solano et al., 1985). Thus, the concentration of ornithine within the cell may determine the activity of intracellular ODC.

$$\text{ODC dimer} \underset{\text{L-ornithine}}{\overset{\text{NaCl}}{\rightleftarrows}} \text{ODC monomer.}$$

Preparations of ODC isolated from livers of thioacetamide-treated rats were stable in the absence of added thiols. However, under this condition the enzyme was rapidly and specifically inactivated by L-ornithine or L-2-methylornithine. The process of inactivation was dependent on the presence of oxygen and was first order, that is, time dependent. Activity was completely recoverable by the addition of dithiothreitol (Danzin & Persson, 1987). It was hypothesized that in the step in which ornithine or 2-methylornithine reacts with PLP, a thiol is unmasked at the active site and is oxidized to form a disulfide in reaction with the many other sulfhydryls in the cysteine-rich enzyme.

Thus, despite the apparent concordance in some tissues of total ODC protein determined serologically and active ODC determined by reactivity with isotopically labeled DFMO, active dimeric ODC may not always be present. Similarly optimizing physiological conditions may not exist in other tissues, and the definition of conditions and tissues in which the equivalence of activity and protein does not exist can be of considerable interest in defining the mechanisms prevailing within the tissue. This also bears on the discovery of antizyme, which is an important determinant of ODC decline.

Rats deficient in vitamin B_6 (pyridoxine) have significantly elevated levels of ODC apoprotein in various tissues (Sampson et al., 1994). In addition to effects of salt, substrate, and essential thiols, the coenzyme PLP appears to modulate various forms of the ODC of unstimulated mouse fibroblasts. Two forms of the enzyme (high K_m and low K_m) have been detected on the basis of association with the coenzyme (Clark & Fuller, 1976). Similar effects had been observed on the ODC of *Physarum polycephalum* (Mitchell & Sedory, 1974). Multiple forms of the *Physarum* enzyme appear in response to cycloheximide (Mitchell et al., 1976).

Examination of the ODC of mouse epidermis and of benign epidermal papillomas revealed the presence of two types of separable enzyme in the latter. One of the papilloma enzymes not only had a lowered affinity (K_m) for L-ornithine and PLP and a greater heat stability, but was also activated specifically by GTP, unlike the normal epidermal enzyme. Two forms of the papilloma enzyme were isolated by gel filtration, of which only a higher molecular weight form was activated by GTP or dGTP (O'Brien et al., 1987). GTP (0.2 mM) markedly increased the affinity of the enzyme for L-ornithine. This effect was apparently irreversible and could not be reversed by dialysis. More recently Kilpeläinen and Hietala (1994) demonstrated a GTP-or ATP-activable form of ODC in all regions of the rat brain. These workers report a greater level of heat stability for this form than for the nonactivable form, which also occurs in the brain and is the exclusive form in the rat kidney.

ODC as Lysine Decarboxylase (LDC)

In contrast to bacterial and plant enzymes, among which ODC and LDC are separable proteins, the ODC of mouse and rat tissues is also a LDC, whose activity is induced, purified, or inactivated concomitantly with that of ODC activity. Crucially the activity on lysine is inactivated by DFMO. The two activities are also found on a single protein in *Neurospora* and yeast.

A constant ratio of the two activities during purification of a rat liver ODC was established by Pegg and McGill (1979). They also showed that the K_m for ornithine (0.09 mM) was a hundredth that for lysine (9 mM). Lysine serves as a weak competitive inhibitor of decarboxylation of ornithine with an inhibition constant (K_i) similar to that of its K_m. The product of lysine decarboxylation is cadaverine; a tissue rich in ODC, such as mouse kidney stimulated by androgen, usually contains some cadaverine derived from endogenous lysine (Persson, 1981a). The increased kidney and urinary levels of cadaverine following treatment with androgen are eliminated by treatment with DFMO.

In some cultured putrescine-deficient cells maximal rates of cell multiplication can be restored by exogenous cadaverine (Alhonen–Hongisto et al., 1987a; Mamont et al., 1978). This diamine is also converted to N-3-aminopropylcadaverine, and slightly to N,N^1-bis(3-aminopropyl)cadaverine. In the synthesis of aminopropylcadaverine by *E. coli* the bacterial enzyme is not an ODC, whereas in *Neurospora*, as well as in the mammalian systems, ODC is the catalyst for the synthesis of the cadaverine.

Mechanism of ODC Reaction

The decarboxylation of L-ornithine, and other L-amino acids, by PLP-dependent decarboxylases occurs with the retention of configuration at the α carbon (Smith et al., 1991). The reaction carried out by the mammalian enzyme was examined with a purified ODC derived from androgen-treated mouse kidney (Orr et al., 1984). In this study deuterated putrescines were prepared by decarboxylation of L-ornithine in D_2O or by decarboxylation of [2-^{2}H]ornithine in H_2O and were compared as bis-camphanamides in proton nuclear magnetic resonance (NMR). A single deuterium atom per molecule was found in each instance. The NMR spectra demonstrated that the insertion of deuterium from D_2O to replace the carboxyl group in the first case differed from that in the second in which the elimination of the carboxyl permitted the interaction of residual deuterium on the α carbon with the camphanamide moiety.

ODC is inhibited by the natural products putrescine and cadaverine and by DAP and some aromatic diamines, such as p-phenylenediamine. The aliphatic diamines, as well as spermidine and spermine, proved to be modest inhibitors at millimolar levels, with putrescine being the most active, 68%. The aromatic diamine, p-phenylenediamine produced a >90% inhibition at the same concentration. The K_i values for these compounds given in Table 13.2 are derived from Lineweaver–Burk plots ($1/v$ vs. $1/s$) of the effects of the inhibitors on the decarboxylation by the rat liver enzyme of ornithine at its K_m (32 M) (Solano et al., 1988).

Other plots of the data appeared to reveal the binding of a second molecule of the natural diamine inhibitors to the enzyme. However, this was not observed for the aromatic diamine, suggesting that the strong interaction of the latter at the active site of ODC precluded the addition of a second molecule. The association of putrescine at the active site supposed the formation of a

TABLE 13.2   Inhibition constants for diamines inhibitory to rat liver ODC.

Inhibitor	K_i (mM)
1,3-Diaminopropane	23.3
Putrescine	0.41
Cadaverine	2.17
p-Phenylenediamine	0.046

Adapted from Solano et al. (1988); K_i, inhibition constant.

Schiff base at PLP and a spanning of a hydrophobic region of the active site with the tetramethylene groups to permit an interaction of the second primary amine with an additional possibly anionic site on the enzyme. Hydrophobic interactions between the side chain of the inhibitor and the enzyme were demonstrated in a study of analogues (Bey et al., 1978). The distance between the two nitrogen atoms (some 6 Å) was important, suggesting that ornithine is fully stretched at the active site. The study with synthetic diamines also demonstrated the potent inhibition by trans-1,4-diaminocyclohexane-1-carboxylic acid, whose K_i of 0.07 mM approached that of p-phenylenediamine.

Regulation of ODC by Polyamines and Discovery of Antizyme

The biosynthesis of amino acids, such as arginine, ornithine, and methionine, is controlled largely by phenomena of feedback inhibition and repression. Pett and Ginsberg (1968) reported that putrescine exerted an inhibition on the conversion of ornithine to putrescine in cultured cells, and this was attributed to a feedback inhibition. ODC activity was considerably inhibited in phytohemagglutinin (PHA)-induced human lymphocytes treated with micromolar levels of putrescine or spermidine (Kay & Lindsay, 1973). As seen in Figure 13.2, the inhibition of the cellular activity was far greater than the in vitro inhibition effected by the polyamines at the same concentrations on the ODCs of extracts.

A similar result was soon obtained with regenerating rat liver after partial hepatectomy or after treatment with growth hormone. A single injection of putrescine caused a rapid and specific decay of ODC (Jänne & Höltta, 1974). Injection of spermidine caused a slower decrease of ODC. Unexpectedly DAP, essentially inactive on the soluble enzyme, evoked a powerful and specific inhibition of the increase of ODC in the regenerating rat liver. Increases of AdoMetDC and spermidine synthase occurred in the inhibited systems, even though synthesis of spermidine was prevented as a result of the lack of newly synthesized putrescine. DAP did not serve as a substrate for spermidine synthase and inhibited DNA synthesis selectively (Kallio et al., 1977).

In studies with various serum-induced mammalian cell cultures, as well as with the liver of rats induced by dexamethasone, it was discovered that putrescine induced the synthesis of a heat-labile, trypsin-and chymotrypsin-sensitive protein(s) that inhibited ODC activity noncompetitively (Fong et al., 1976; Heller et al., 1976). The inhibitors could be detected within an hour after exposure to putrescine and were maximal at 3–4 h. These inhibitors were isolated, had apparent molecular weights of 26,500, and were found to form ODC–inhibitor complexes stable in gel permeation chromatography. The complexes could be dissociated by ammonium sulfate to yield active separable ODC and inhibitor. E. Canellakis and collaborators proposed the name of antizyme to describe this inhibitor.

Some antizyme (perhaps 4–8% of the fully induced

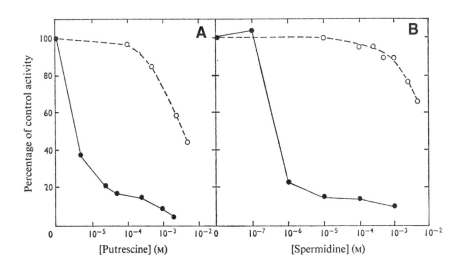

FIG. 13.2   Effects of (a) putrescine or (b) spermidine on ODC activity of phytohemagglutinin-stimulated lymphocytes; adapted from Kay and Lindsay (1973). (●) Extracts of cells incubated 3 h with polyamine; (○) assays of extracts in the presence of polyamine.

level) was detected as normally present in a bound inactive form on particulate fractions of cultured cells or rat liver cells. Much of it was released by 10 mM polyamine. A significant amount appeared to be associated with nuclei (Heller et al., 1977). Furthermore, the induction of antizyme by putrescine varied 100–1000-fold from cell line to cell line, although all of the lines were induced to produce antizyme by high diamine concentrations. The existence of antizyme-type inhibitors induced by DAP, putrescine, or larger diamines was found in various mammalian cells and organs. A nonphysiological inducer of ODC antizyme, 1,3 diamino-2-hydroxypropane, was only slightly less active than DAP (Branca & Herbst, 1980). In the induction of antizyme by putrescine, the exogenous putrescine replaced that which would be synthesized by an active ODC.

Pursuit of Antizyme and Its Role

Initially Seely and Pegg (1983a,b) were unable to detect the formation of ODC–antizyme complexes as intermediates in ODC decay in these systems. Nevertheless, they did obtain antizyme in the DAP-treated liver and suggested that the formation of a complex with antizyme might enhance the degradation of ODC protein. It was recognized that, in addition to ODC–antizyme complexes, an extract might contain other free excess ODC or free excess antizyme that ought to be estimated (Brosnan et al., 1983).

The antizyme for ODC was purified 600,000-fold to homogeneity from the liver cytosol of putrescine-treated rats (Kitani & Fujisawa, 1984, 1985). It was stabilized by a mixture of 2-mercaptoethanol and a Tween detergent. Gel filtration indicated a molecular weight of 22,000, and in SDS-gel electrophoresis the weight appeared to be 19,000, suggesting that antizyme comprised a single polypeptide chain. Studies of the binding of partially purified antizyme to partially purified ODC indicated an equilibrium constant of 1.4×10^{11} M^{-1} in 50 mM sodium phosphate at pH 7.0 and 0°C.

Antizyme, which bound stoichiometrically to ODC and inhibited the enzyme, could be used to explore the possible differences between ODCs of different origins. The crude or purified ODC of mouse or rat origin was inactivated identically by a rat liver antizyme. The enzymes from mouse kidney and rat liver were also labeled to the same extent by isotopic DFMO. Thus, the catalytic centers of these enzymes behaved identically in reacting with both antizyme and DFMO (Marumo et al., 1988).

Another protein factor was discovered, one which inhibits antizyme and reactivates ODC from an ODC–antizyme complex (Fujita et al., 1983). The occasional presence of the antizyme inhibitor in crude extracts diminished the apparent sensitivity of ODC to antizyme. This antizyme inhibitor, although of a size comparable to ODC, was not cross-reactive with antisera to ODC and formed a complex with antizyme, even in the presence of ODC. The antizyme inhibitor was purified some 17,000,000-fold from the liver cytosol of thioacetamide-

treated rats (Kitani & Fujisawa, 1989). It was found to have a molecular weight of 51,000 in a denaturing gel electrophoresis and an apparent weight of 62,000 by molecular sieving, suggesting that the protein consisted of a single polypeptide chain.

This antizyme inhibitor, which was capable of releasing ODC from antizyme complexes, facilitated the estimation of total ODC activity (Fujita et al., 1984). It has been used to demonstrate the presence of a low level of free antizyme and ODC–antizyme complex in the tissue extracts of starved rats. The demonstration that the inhibitor of ODC was "antizyme" included its existence as a heat-labile molecule of about 25,000 molecular weight, time-dependent and stoichiometric inactivation of ODC, formation of a salt–labile complex with the enzyme, and dissociation by antizyme inhibitor. The rapid degradation of antizyme (half-life, $t_{1/2} = 24$ min) and its presence in the liver of starved rats implies its synthesis in organisms even during starvation.

ODC can also be released from a complex by addition of DFMO-inactivated ODC. Large amounts of such a complex were found by this method in extracts of a hepatoma tissue culture (HTC) cell variant containing a relatively stable ODC. The amount of the complex increased as free ODC disappeared in the presence of cycloheximide (Murakami et al., 1985). That the degradation of ODC proceeded via the degradation of the complex was tested by determining the amounts of the complex and the half-lives of the ODC residual at various intervals after treatment with cycloheximide (Murakami & Hayashi, 1985). As shown in Figure 13.3, these workers reported a linear relation between the increasing ratio of complex to total ODC and the inverse of the half-life of ODC activity. Thus, in the cellular sys-

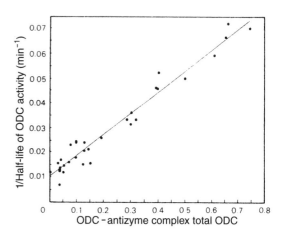

FIG. 13.3 Correlation between reciprocal of half-life of ODC activity and relative amount of ODC–antizyme complex in total ODC activity in HTC cells; adapted from Murakami and Hayashi (1985). Cycloheximide was added to the cells, which were harvested at 20-min intervals for 80 min. Extracts were assayed for ODC and the amount of the ODC–antizyme complex.

tem tested, antizyme participated in the degradation of ODC activity.

This regulatory participation of antizyme has been detected during the cell cycle of Ehrlich ascites tumor cells. Inactive ODC–antizyme complexes are generated mainly in mid-S phase, when antizyme activity is maximal (Linden et al., 1985). In the rat heart the ODC was distributed between equal amounts of free ODC and the complex (Flamigni et al., 1986). In the mouse brain purification of ODC produced a >50-fold increase of the activity of ODC, which thus appeared to exist in large measure in an inactive enzyme–antizyme complex (Laitinen et al., 1986). It should be noted that homogenization of an intact tissue may release unbound antizyme, which may form complexes as artifacts of preparation.

Regulation of Synthesis of Antizyme

Despite some apparent differences among antizymes of different rat tissues (Hu & Brosnan, 1987), the immunological reactivities of the antizymes of many rat and mouse tissues to a series of monoclonal antibodies against rat liver ODC antizyme proved to be similar. This indicated that these proteins of mouse or rat origin shared similar structural features. A very sensitive immunoassay capable of detecting ODC and antizyme was devised (Nishiyama et al., 1989), and this demonstrated that the induction of antizyme after treatment with putrescine represented an increase of antizyme protein. This increase was inhibited by cycloheximide but not by actinomycin D, indicating a control of this increase by posttranscriptional events (Matsufuji et al., 1990a). The antizyme mRNA is thought to be expressed constitutively. A rat gene encoding the ODC antizyme has been isolated and sequenced (Miyazaki et al., 1992).

A cDNA prepared from rat liver mRNA expressed an antizyme in *E. coli*; this had a molecular weight of 23 kDa on SDS-gel electrophoresis. Furthermore, the antizyme mRNAs of all vertebrate organisms studied were all essentially the same size, including those of the chicken and frog, and were hybridized to the rat liver cDNA. Indeed the antizymes of the chicken and frog have been shown to be cross-reactive with antisera to the rat antizyme. Antizyme is clearly a highly conserved protein in all vertebrates (Matsufuji et al., 1990b). The mRNA hybridizable to the antizyme cDNA has not been detected in nonvertebrates nor in yeast, in which antizyme has not been found.

The several antizymes of *E. coli*, some of which are found as ribosomal proteins, have not yet been analyzed in conditions and terms used for the vertebrate antizyme. The antizyme bound to animal cell particulates, which may occasionally amount to a half of the total antizyme, appears to be immunologically similar to that of "true" liver antizyme (Hayashi & Canellakis, 1989).

The isolation of the three major proteins, ODC, antizyme, and antizyme inhibitor, enabled the preparation of specific antisera whose antibody could be used in the formation of specific absorbents for affinity chromatography (Murakami et al., 1989a). These reagents were employed in the systematic separation and estimation of the three components in the various physiological situations, as presented in Table 13.3. Fasting causes an active disappearance of ODC, which can be restored by feeding casein. This is stimulated further by thioacetamide and at least twofold more by β-aminoisobutyric acid (AIB). The other proteins are affected to a far lesser degree. Male mouse kidney has an enormous free ODC content and a significantly large amount of inactive complex (see also Murakami et al., 1988), whereas the mouse brain is essentially devoid of free ODC. In the latter organ, as noted earlier, there is a detectable amount of inactive complex, whose mechanism of release of the free enzyme is not yet understood.

These components have been followed in the thioac-

TABLE 13.3 Activity of antizyme–inhibitor in various tissues.

Tissues	(n)	Free ODC (U/g)	ODC–antizyme complex (U/g)	Antizyme inhibitor (U/g)
Rat liver				
Starved, overnight	14	3.3 ± 2.4	6.0 ± 1.3	2.2 ± 0.6
Fed, 4 h	9	96 ± 10	2.5 ± 1.7	5.9 ± 1.3
Thioacetamide	11	195 ± 94	6.4 ± 3.4	7.9 ± 1.2
Thioacetamide + AIB	151	448 ± 230	8.6 ± 5.0	11.0 ± 3.4
DFMO	15	1.4	3.4	0.9
Rat				
Kidney, fed 4 h	4	295	36.9	0.6
Heart, isoprenaline	49	175	2.2	11.8
Mouse				
Liver	13	0	1.6	1.6
Kidney (male)	13	2570	78.2	1.3
Brain	13	0	3.9	0.5
Calf liver	1	0.3	ND	2.2

ND, not determined. Values are mean ± SD.
Adapted from Murakami et al. (1989a).

etamide-stimulated livers of rats injected with putrescine or cycloheximide. Putrescine evoked a precipitous decline of free ODC and a marked increase of ODC–antizyme complex, without a change in antizyme inhibitor. Cycloheximide, an inhibitor of protein synthesis, evoked a somewhat slower decay of ODC, a slight decrease in the amount of the complex, and a rapid decay of antizyme inhibitor.

Because putrescine is convertible to spermidine and spermine, and the higher amines may be degraded to putrescine, the nature of the polyamine responsible for the induction is not immediately evident. In studies on HTC cells in vitro, it was found that an inhibitor of spermidine synthase, cyclohexylamine, caused a marked increase of ODC, putrescine, and spermine and a 90% decrease of spermidine. ODC was quite stable in this condition and was subsequently rendered unstable by addition of spermidine, suggesting that spermidine controlled the synthesis of a protein involved in the inactivation (Mitchell et al., 1985a). Murakami et al. (1989b) demonstrated that spermidine was far more active than spermine or putrescine in stimulating the formation of ODC–antizyme complexes.

CHO cells maintained in an ornithine-deficient medium, in which intracellular putrescine is low, contain high levels of stable ODC. Exogenous spermidine evokes a rapid decrease of ODC protein after protein synthesis (Glass & Gerner, 1987). In a study of a CHO mutant that overproduced ODC, exogenous polyamine and spermidine in particular similarly caused the disappearance of ODC activity, with a slightly slower decrease of ODC protein (Hölttä & Pohjanpelto, 1986). Pursuing the hypothesis that antizyme is a crucial element in the decay and that spermidine is an inducer of antizyme, which might thereby suppress growth, several new spermidine analogues (e.g., N^l-cyclohexylspermidine) were found to be antizyme inducers (Miyazaki et al., 1990).

The isolation of the antizyme cDNA permitted the transfection of mammalian cells with antizyme cDNA, which then showed decreased levels of cell ODC, polyamine, and cell growth rate (Murakami et al., 1990a,b). Addition of the glucocorticoid, dexamethasone, stimulated production of antizyme mRNA and antizyme in the absence of polyamine and induced rapid decays of ODC activity and ODC protein (Murakami et al., 1994).

It has been shown that the ATP-dependent degradation of ODC in vitro in an extract of CHO cells is in fact markedly stimulated by the addition of purified antizyme (Murakami et al., 1992).

Exploiting Mutations in Polyamine Metabolism

A brief detour will describe the isolation and nature of animal cell mutants possessing one or another aberration in polyamine metabolism. The uses to which such cells have been applied include:

1. studies of the effects of polyamine depletion;
2. production (or overproduction) of key enzymes, such as ODC;
3. isolation of mRNA from an actively producing cell for characterization of the mRNA and the preparation of a hybridizable cDNA; and
4. characterization of transfecting DNA to reveal the primary sequence of a key enzyme and other features of a transcribable DNA.

Roles of Arginine and Arginase

An arginase-deficient CHO cell, defective in the synthesis of ornithine, was isolated and shown to require exogenous polyamine (i.e., putrescine, spermidine, or spermine) for cell growth (Hölttä & Pohjanpelto, 1982). Most sera tested contained arginase and thereby supported growth. With this cell, ODC and AdoMetDC increased during polyamine starvation in serum-free media. Strikingly, polyamine-depleted cells lost their bundles of actin filaments and microtubules (Pohjanpelto et al., 1981) and developed major chromosomal aberrations (Pohjanpelto & Knuutila, 1982).

Selected arginase-deficient cells capable of growth in ornithine-or polyamine-free medium were found to have very large amounts of ODC. The enzyme converted lysine to cadaverine and the latter to the spermidine and spermine analogues N-(3-aminopropyl)cadaverine and N,N^l-bis(3-aminopropyl)cadaverine, which could support the growth of the cells (Pohjanpelto et al., 1985b).

Arginase deficiency, or hyperargininemia, in humans is an inherited disorder of urea cycle metabolism and is accompanied by severe mental retardation. In one such patient the urinary excretion of putrescine was quite high, as were erythrocyte levels of spermidine and spermine. Abnormal amounts of putrescine are produced in response to dietary ornithine, suggesting high levels of tissue ODC.

A mutant strain of the arginase-deficient CHO cell line was isolated and lacked any ODC activity, although it contained a normal amount of the ODC protein, ODC mRNA, and AdoMetDC. This cell line was grown in serum-free media in the presence of 10^{-6} M putrescine, spermidine, or spermine; in the absence of polyamines the cells eventually stop growing. Under conditions of polyamine starvation, putrescine and spermidine disappeared long before the complete loss of spermine. This ODC-deficient mutant did not accumulate cadaverine or its derivatives (Pohjanpelto et al., 1985a). The nature of the mutation that permits the full synthesis of an immunoreactive catalytically inactive protein is unknown.

This totally deficient strain was isolated after mutagenizing an initially arginase-deficient cell by ingestion of tritiated ornithine, according to the method of Steglich and Scheffler (1982). These workers had previously isolated a CHO mutant with only a 3% level of the maximum inducible ODC activity of the parent. These cells of aberrant morphology required 10^{-5} M putrescine or 10^{-7} M spermidine or spermine to maintain a normal growth rate. It was suggested that mutagenesis with tritiated ornithine relates to an association of derived tritiated polyamine with cell DNA. The Steglich–Scheffler mutant of low ODC activity and small amounts of immunologically detectable ODC protein had apparently normal amounts of ODC mRNA. The mutation arose in a substitution of aspartate for glycine at a specific site in the protein (amino acid 381) and an adenine for a guanine 1142 nucleotides from the start of translation of the nucleic acid (Pilz et al., 1990).

Human liver arginase has been produced efficiently in *E. coli* (Ikemoto et al., 1990). The rat liver arginase is a Mn^{2+}-containing protein and catalyzes the hydrolysis of agmatine as well as arginine (Reczkowski & Ash, 1994). N^G-Hydroxy-L-arginine, an intermediate in the biosynthesis of NO, is a potent competitive inhibitor of rat liver arginase (Daghigh et al., 1994). This poses the possibility of a direct relationship between NO production and ornithine availability in cells containing both enzymes.

ODC-Overproducing Mutants

Numerous overproducing mutants have been obtained by selection for resistance in "normal" cell lines, for example, mouse lymphoma cells (McConlogue & Coffino, 1983), mouse myeloma cells (Kahana & Nathans, 1985b), or rat hepatoma HTC cells (Mitchell et al., 1991), in stepwise increasing concentrations of the irreversible ODC inhibitor, DFMO. The overproduction of a drug-targeted protein can occur via an amplification of gene copies (Alhonen–Hongisto et al., 1985a; Hyttinen et al., 1991). There is also a coamplification of certain genes for ODC and other proteins in hydroxyurea resistance.

In one of these resistant mutants ODC represented 15% of the total protein of the cell. The mRNAs isolated from the cell were converted into cDNAs that were incorporated into plasmids. Plasmids containing ODC cDNA were selected via hybridization with ODC mRNA, and the latter was identified by translation to ODC in vitro. The plasmids containing ODC cDNA can be cloned in *E. coli* (Macrae & Coffino, 1987) and the cDNA sequences have been isolated after digestion with appropriate restriction enzymes and identification by hybridization with mRNA. Such cDNA was used to estimate the ODC mRNA in various tissues (McConlogue et al., 1984). The major ODC mRNA of the resistant cells was 2.0 kilobases (kb) long. An mRNA of similar length was present in the mouse kidney and the amount of the latter was increased in the kidneys of mice treated with testosterone. The latter result reflected an increase in transcription.

Kahana and Nathans (1984, 1985a) also isolated and cloned a cDNA for ODC. Treatment of the overproducing cell with putrescine, which evoked a decreased rate of ODC synthesis, production of antizyme, and a fall in ODC resulted in no change in the amount of ODC mRNA. It was concluded that a polyamine, possibly spermidine, regulated the translation of mRNA, as well as the decay of ODC (Kahana & Nathans, 1985b).

The DFMO-resistant mutant HTC cell obtained by Mitchell et al. (1991) increased the cellular content of active ODC some 2000-fold. This cell did not respond to spermidine with a decrease of ODC, which was stable in the presence of cycloheximide. The stable ODC purified from this cell was identical to that of control cells in size, charge, substrate binding kinetics, and sensitivity to DFMO. The mutant thus appeared to be defective in its mechanism of ODC degradation.

A DFMO-resistant HTC cell was studied in the presence and absence of DFMO (Tome et al., 1994). In the presence of DFMO, this cell contains moderate levels of putrescine and grows normally. In the absence of DFMO the cells accumulate high levels of putrescine, are inhibited in growth, and eventually die. In studies with purified ODC from the overproducer, Mitchell et al. (1992) found that an inactivated DFMO complex can release active ODC at low temperatures in the presence of reducing agents. Another DFMO-resistant overproducer, grown in the presence of DFMO, contained an ODC and its mRNA at 60 times the concentration present in the wild-type cells (Kameji et al., 1993). The level of this enzyme could be manipulated by affecting the level of mRNA (transcriptional controls), the efficiency of the mRNA (translational controls), and the stability of the enzyme (posttranslational controls).

Determination of Nucleotide and Amino Acid Sequences

Although the determination of the amino acid sequences in small proteins (e.g., insulin) was an earlier achievement, the determination of sequence in polynucleotides is now less difficult and can provide, as a result of the knowledge of coding equivalence, start and stop codons, etc., a description of amino acid sequence. A cDNA derived from a 2.4-kb ODC mRNA obtained from mouse cells contained 2465 nucleotides (Kahana & Nathans, 1985a). The sequence of polydeoxynucleotide fragments, determined by the dideoxy-chain-termination method, revealed an unusually long 5′-noncoding element, a coding segment of 1383 nucleotides, and a shorter 3′-noncoding segment. The derived protein would consist of 461 amino acid residues and possess a molecular weight of 51,105. This cDNA and the DNA encompassing the ODC-defining sequence hybridized with several fragments of digested cellular DNA. The mouse genome was thus shown to contain several ODC-related genes.

The analysis of another mouse ODC cDNA also found a sequence coding for an ODC of 461 amino acids with a molecular weight of 51,172 (Gupta & Coffino, 1985). Predictions of isoelectric point (pI = 5.1) and the organization of the domains of α helix and β sheets were also presented. These sequences were used to place the site of phosphorylation of mammalian ODC and to analyze the constitution of isolated ODC genes.

The development of methods for the preparation of cDNAs for ODC has permitted comparative studies of the genetic elements of many organisms, that is, from the initial mouse cells to rat kidney (Kranen et al., 1987) to *Xenopus laevis* oocytes (Bassez et al., 1990). Reconstruction of the cDNAs has permitted the manipulation of specific amino acids in the sequence and the determination of the catalytic roles of such amino acids (Viguera et al., 1994). The availability of the cDNAs has facilitated a more ready estimation and isolation of specific mRNAs in various physiological conditions. One interesting application of such a study has described the level of the transcribed ODC mRNA as indicative of mouse embryos capable of developing high mass strains (Gray et al., 1995). The selection of high mass poultry strains containing high levels of ODC and its mRNA has also been reported (Johnson et al., 1995).

Phosphorylation of ODC

Differences in charge among ODCs of a single cell suggested the possibility of phosphorylation of the enzyme. A purified rat heart ODC was shown to be phosphorylated by ATP with a particular casein kinase (type 2) derived from rat liver. The amino acid appeared to be serine-303 in the sequence (303–309) Ser-Asp-Asp-Glu-Asp-Glu-Ser (Meggio et al., 1987).

The serum-precipitated ODC derived from overproducing mouse myeloma cells was found to contain serine phosphate. Purified mouse ODC was also demonstrated to be phosphorylated by casein kinase II at serine-303, almost entirely at a site in a sequence (303–309) identical in the mouse, rat, and human. When a mutant ODC was prepared in which serine-303 was replaced by alanine, a much smaller amount of ^{32}P was incorporated (Rosenberg–Hasson et al., 1991a).

The extent of phosphorylation of ODC by GTP and casein kinase II was increased by spermine, which also increased the maximum velocity (V_{max}) of the kinase reaction. However, this phosphorylated ODC was unchanged in V_{max} and K_m as compared to the original ODC in the conversion of ornithine to putrescine (Tipnis & Haddox, 1990). An epitope in mouse ODC, recognized by a monoclonal antibody, has been observed to be masked by phosphorylation (Donato et al., 1986).

Some mouse cells have been found to contain several forms, one relatively basic nonphosphorylated ODC and two phosphorylated forms. These ODCs were found to contain *o*-phosphoserine and *o*-phosphothreonine, of which the former was present in the largest amount (Worth et al., 1994). The multisite phosphorylation of certain cellular ODCs is now thought to increase enzyme stability and in vivo half-life (Brown et al., 1994).

Isolation and Analyses of ODC Genes of Mouse and Rat

ODC genes are abbreviated as *Odc*. The hybridization of an *Odc* cDNA with fragmented mouse DNA has revealed multiple binding fragments. Many of the latter

are defined as pseudogenes, which are not functional in providing ODC mRNA but have high levels of homology to an active *Odc*. An amplified *Odc*, capable of producing large quantities of the ODC mRNA and enzyme, was used to provide a specific hybridizing fragment from within a transcribable region of the gene; this bound with a single mouse DNA fragment. The structural *Odc* locus is on mouse chromosome 12, whereas many pseudogenes were found on other chromosomes (Cox et al., 1988).

The isolation of a functional mouse *Odc* by molecular cloning was described by Brabant et al. (1988), who reported the coding sequence of the gene as well as the untranslated 5′ leader and the 3′-terminal sites of polyadenylation. These workers noted the classes of regulatory phenomena that might be explored with such isolated genes and stressed the possible significance of the 5′-untranslated leader in the control of ODC expression.

An amplified *Odc* derived from DFMO-resistant mouse myeloma cells contained 12 exons from which a functional ODC mRNA was eventually produced (Katz & Kahana, 1988). Ten of these exons (3–12) code for the ODC protein. Exon 12 also includes the 3′-noncoding region and contains two well-separated polyadenylation signals, as a result of which two species of ODC mRNA are produced. A major species of 2.0 kb and a minor one of 2.4 kb are found in several mouse systems.

The noncoding 5′-region flanking the transcription initiation site is rich in G+C nucleotides, which may form a stable secondary hairpin structure controlling access to the reading frame of the gene (Brabant et al., 1988). This portion of the gene also contained polynucleotide promoters, which bind components of the transcription systems. This 5′-flanking region of the murine *Odc* was also sequenced by Eisenberg and Jänne (1989), who have sought transcriptional control elements that were potentially responsive to estrogens and cyclic AMP (cAMP). Plasmid probes of cDNA fragments have been shown to encompass the full length mRNA derived from the genomic sequence (Katz & Kahana, 1988) and have been used to map the family of approximately 12 mouse ODC-related sequences in numerous inbred strains of mice (Richards–Smith & Elliott, 1992).

A nucleotide sequence of the rat *Odc,* spanning 7.7 kb and comprising 12 exons and 11 introns, was similar to that of the mouse structural *Odc* (Wen et al., 1989). Two mRNAs were detected in rat tissues and two polyadenylation signals (AATAAA) were found in the 3′ region of the gene following the termination codon. Potential regulatory elements (e.g., a cAMP-response element) were also found in the 5′-noncoding flanking region.

Several tests for functionality have been applied to isolated rat *Odc* (van Steeg et al., 1990). Insertion of the true gene into an ODC-deficient hamster cell (i.e., CHO cell) restores ODC activity. Deletion of a portion of a presumed promoter site in the 5′-noncoding region decreases promoter activity. The agreement among the various workers has been striking. A model of the structures of the mouse ODC gene and its two alternative ODC mRNAs is presented in Figure 13.4.

Sequences have also been reported for the *Odc* of the human (van Steeg et al., 1989) and the Chinese hamster (Grens et al., 1989). The coding sequence is conserved among all the mammalian species, whereas the 5′-flanking noncoding regions differ to a greater extent. The ODC of the rat enzyme shows an identity of amino acid sequence of 96% with that of the mouse enzyme and 91% with that of human ODC. Sequence identity falls off to 56 and 33% with that of the enzyme from *Trypanosoma brucei* and yeast, respectively.

Regulatory Function of cAMP

Reference has been made to the presence of a cAMP-responsive element upstream of the mouse ODC gene. A role for cAMP in the induction of ODC has been sought in tissues and cells since the finding of the rapid appearance of ODC in rat liver in regeneration following partial hepatectomy or in response to growth hormone. The role of cAMP as a secondary messenger raised the possibility of an activation via cAMP of a preformed inactive protein. Inducing conditions invoked an even earlier increase in cAMP than in ODC. The induction of cAMP and ODC was seen in quiescent and serum-induced baby hamster kidney (BHK) cells; the induction of the latter was effected by exogenous dibutyryl cAMP (Hogan et al., 1974), and Russell (1981) postulated a causal relation of cAMP in ODC induction.

The sequence analysis of the mouse ODC gene revealed two presumed cAMP-responsive (CRE) sequences in the untranslated 5′-flanking region (1658 nucleotides), as well as sequences for transcription factors potentially sensitive to cAMP in an exon (1) and intron (1) in the gene (summarized by Abrahamsen & Morris, 1991). In a study of the regulation of the murine ODC gene, transfection with DNA plasmids containing deletions of the 5′-flanking region localized a cAMP-responsive element to within the 97 nucleotides proximal to the transcribed gene. In addition, a nuclear protein (70 kDa) was shown to bind to a DNA sequence similar to a CRE in the region (−58 to 37), and this binding was unaffected by phosphorylation by a cAMP-dependent protein kinase (Palvimo et al., 1991). Thus, cAMP does play a role in stimulating transcription from the ODC gene in the mouse, but it is unproven that this function relates to specific phosphorylations. Additional study of a contiguous GC-rich region in the ODC promoter detected the existence of several binding proteins capable of regulating the synthesis of ODC mRNA (Kumar et al., 1995; Li et al., 1994).

Methylation of ODC Gene and Its Expression

The question was asked if the growth-related human gene possessed a different pattern of methylation when present in tumor cells, as compared to normal cells. It was possible to test this with known restriction enzymes that distinguished cytosine sites from 5-methylcytosine sites. One research group found that the *Odc* of L1210 mouse leukemia cells was more methylated than that of Ehrlich ascites tumor cells, and considerably more than that of human normal leucocytes and a human myeloma cell line (Alhonen–Hongisto et al., 1987b). It was subsequently determined that the *Odc* of leucocytes from patients with chronic lymphatic leukemia was distinctly hypomethylated as compared to that of human controls. These observations supported the hypothesis that the level of methylation may affect gene expression (Lipsanen et al., 1988).

In further support of the hypothesis, it was reported that growth inhibition of human myeloma cells and the decrease of ODC expression by the glucocorticoid, dexamethasone, increased methylation of *Odc* whereas resistance to its effects was accompanied by hypomethylation of the ODC gene and an enhancement of accumulation of ODC mRNA (Leinonen et al., 1987). Also, methylation of the cloned *Odc* by a specific methylase decreased an early expression after transfection of CHO cells (Halmekytö et al., 1989). Detailed analyses of the methylation patterns of the ODC genes of control and amplified human myeloma cell lines have been reported (Wahlfors, 1991).

Peña et al. (1993) pointed to the presence of a high GC element in another region of the ODC promoter, which has the potential to bind a protein complex that affects methylations. This binding appears to occur during active cell growth and to regulate the transcription of *Odc*. A recently developed method has pointed to the absence of methylation in all ODC promoters, with extensive methylation of cytosine in protein-coding regions (Myöhänen et al., 1994).

Human *Odc* and Its Effects in Transgenic Mice

Human *Odc* from normal human lymphocytes has been isolated and characterized (Fitzgerald & Flanagan, 1989). Only a single gene was seen, and this possessed a pattern of exons and introns in considerable homology to the mouse and rat pattern. Similar control elements in the 5′ region were found as well. Within the segments of the ODC region, exon segments corresponding to the amino and carboxyl termini of ODC were significantly different in the mammal from those of yeast and the African trypanosome, whereas the central segments of the protein were similar among the different organisms. Additional descriptions of the human *Odc* have included some minor differences in nucleotide sequences.

A full-length ODC-encoding gene derived from amplified human myeloma cells evoked ODC activity when transfected to CHO cells devoid of ODC. The ODC in these cells was induced by serum and decreased rapidly after addition of cycloheximide, putrescine, or spermidine. Briefly then the human *Odc* and human ODC were susceptible to the various forms of regulation of the enzyme found in the CHO cell (Hölttä et

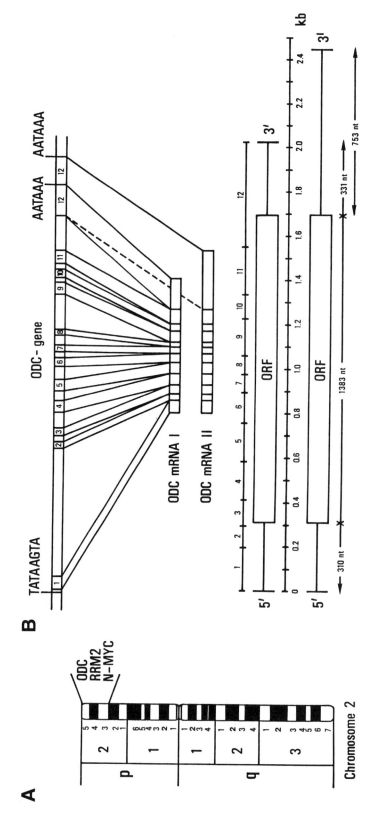

FIG. 13.4   General structures of the mouse ODC gene and its two alternative ODC mRNAs; adapted from Heby and Persson (1990).

al., 1989). A similar isolation of *Odc* was made from a genomic library derived exclusively from human chromosome 2 and was demonstrated to be functional in transfected cells (Hsieh et al., 1990).

It was asked how this human ODC was expressed in intact mammals and what physiological and biochemical effects accompanied this expression. The tests required microinjection of the gene into fertilized oocytes derived from genetically defined mouse strains and transfer of the injected oocytes into the oviducts of pseudopregnant foster females. The DNA of tail samples taken from 3-week-old mouse pups were analyzed for ODC-specific primers by the polymerase chain reaction unresponsive to mouse DNA. Pups derived from zygotes injected with the intact human *Odc* contained the transgene quite infrequently (1 in 117 pups), whereas zygotes injected with a 5'-truncated human *Odc* gave rise to transgenic pups more frequently (1 in 40) (Halmekytö et al., 1991a,b). A human-specific mRNA was detected only in the transgenic pups injected with the whole gene. The other transgenic mice, containing the human gene lacking a 5' promoter, produced mouse-specific mRNA only.

All transgenic mice had high human ODC levels, 20–80 times normal mouse levels, in almost every tissue. Whereas the production of mouse ODC mRNA was stimulated by testosterone in the female kidney, transcription of the human gene was unresponsive to this treatment in this tissue (Halmekytö et al., 1993b). Histological changes were found only in male testis and these animals showed reduced reproductive performance or were infertile. This appeared to relate to the very high ODC and putrescine contents, presumably reducing sperm count and/or producing sperm malformations (Hakovirta et al., 1993). A testicular ODC in the transgenic mouse was found to be more stable than that of a control after treatment with cycloheximide. Polyamine levels, putrescine specifically, were changed most strikingly in testis and brain, and only these tissues had a somewhat increased spermidine to spermine ratio. In these tissues there had been an increase in the biosynthetic enzymes (i.e., AdoMetDC and the spermidine and spermine synthases) whereas enzymes involved in the catabolic paths (i.e., spermidine/spermine acetyl transferase and polyamine oxidase) did not increase. Nevertheless, despite the very high levels, this excess of diamine was not converted actively to triamine and tetramine (Halmekytö et al., 1993a).

High ODC (80 times normal) and putrescine levels in the brain did not result in overt morphological changes in the transgenic mice. However, transgenic mice with very high brain putrescine proved to have impaired performance in spatial learning and memory and were less prone to induced seizures (Halonen et al., 1993). Whereas the diamine could not be detected in the brain of normal mice, it exceeded 60 μmol/g in the brain tissue of transgenic animals. Spermidine and spermine were not particularly different in the five animals examined. An examination of the high putrescine brains by a noninvasive technique, ^{31}P-NMR, revealed a 40% reduction in free Mg^{2+}, suggestive of the interchangeability of Mg^{2+} and a polyamine (Kauppinen et al.,1992). No differences were found in the intracellular pH, phosphorylated metabolites, or in several other groups of nitrogenous metabolites.

Transcription of ODC Gene and Nature of ODC mRNA

The estimation of ODC-specific mRNA requires a cDNA probe, which may be generated by the conversion of an mRNA via reverse transcriptase. The mRNA can be isolated from an overproducing cell, perhaps from active polysomes producing ODC and concentrated by specific antibody to ODC. The ODC mRNA is converted to cDNA that is introduced into plasmids capable of transforming *E. coli* deficient in the enzyme. The plasmid contents of ODC-active cells are amplified and the plasmid DNAs are isolated as the probes that can react with cellular mRNA. The demonstration that such RNA is truly ODC mRNA involves translation into characteristic ODC. Such a procedure, containing many additional specificity-assuring steps, has been devised to study androgen induction of ODC mRNA in the mouse kidney (Kontula et al., 1984). The method detected a 10–20-fold increase in ODC mRNA after 4 days of androgen treatment. Thus, androgens regulate ODC in the mouse kidney by increasing transcription of the ODC gene.

Two molecular sizes of mRNA, 2.15 and 2.7 kb, were found in this system induced by testosterone. Analysis of the RNAs and the cDNA produced from it revealed that the mRNA and cDNA were longer at their 3' end by 429 nucleotides and contained two AATAA signals for poly(A) addition. Both sizes of mRNA were formed in parallel, perhaps coordinately (Hickok et al., 1986). The evolution of the various domains of mammalian ODC mRNA, that is, the 5'-untranslated region, the coding region, and the two domains of the 3'-untranslated region, was discussed by Johannes and Berger (1993). They concluded that the various domains of mouse, rat, hamster, and human mRNA evolved noncoordinately.

An extensive listing of experimental systems in which transcription of the ODC gene leads to increase of mRNA is presented in Table 13.4.

The various systems have displayed many individual differences in their responses. For example, McConlogue et al.'s (1986) study of ODC overproducers revealed several different classes of response, in only one of which did the increase in mRNA account for the increase of enzyme. In another, the rate of translation increased some four- to eightfold. In still others the rate of degradation slowed. In all of these a mere increase in the numbers of ODC genes (i.e., amplification) was insufficient to account for the increase of ODC mRNA.

Induction inhibited by cycloheximide implied a requirement for protein synthesis before transcription occurred (Katz & Kahana, 1987; Pohjanpelto & Hölttä,

TABLE 13.4 Biological systems induced to increase ODC and ODC mRNA.

Biological system	Condition of induction	Reference
1. Transformed hamster fibroblasts	Phorbol ester (TPA)	Gilmour et al. (1985)
2. CHO mutant cells	Arginase deficiency	Pohjanpelto et al. (1985)
3. Mouse L1210 leukemia cells	DFMO resistance and amplification	Alhonen–Hongisto et al. (1985)
4. Mouse S49 cells	DFMO resistance and amplification	McConlogue et al. (1986)
5. Mouse muscle cells	Serum, cycloheximide	Olson & Spizz (1986)
6. 3T3 mouse fibroblasts	Growth factors, TPA	Katz & Kahana (1987)
7. Rat liver	Dexamethasone	Hirvonen et al. (1988)
8. NIH 3T3 cells	Transformation by *ras* oncogene	Höltta et al. (1988)
9. Mouse testis	Development	Alcivar et al. (1989)
10. Mouse T lymphocytes	Interleukin-2	Legraverend et al. (1989)
11. Mouse epidermis	Ultraviolet B radiation	Rosen et al. (1990a)
12. Rat keratinobytes	Ultraviolet B radiation	Rosen et al. (1990b)
13. Human monocytes, mouse	Bacterial lipopolysaccharide	Messina et al. (1990)
14. Rat regenerating liver	Partial hepatectomy	Beyer & Zieve (1990)
15. Intestinal epithelial cells	Cycloheximide	Ginty et al. (1990)
16. CHO mutant cells	Omission of single amino acid	Pohjanpelto & Höltta (1990)
17. Swiss 3T3 fibroblasts	Concanavalin A	Abrahamsen & Morris (1990)
18. LLC-PK$_1$ cells	Hypotonicity	Lundgren (1992)
19. Mouse neuroblastoma cells	Asparagine	Chen & Chen (1992)

1990). Höltta et al. (1988) not only detected regulation of the translation of the mRNA, but also noted situations in which the mRNA itself turned over. In the neuroblastoma cells studied by Chen and Chen (1992), the increase in ODC induced by asparagine was attributed in large part to the stabilization of the ODC mRNA. The mechanism by which ODC mRNA concentration was elevated under conditions of hypotonicity (Lundgren, 1992) remains unclear. Cycloheximide did not affect transcription significantly in that study, suggesting that in this system protein synthesis was not a prerequisite for ODC gene transcription.

The study of the effects of growth factors on myogenesis (Olson & Spizz, 1986) also demonstrated positive and negative signals in determining transcription for ODC. Further, the system of serum-induced growth of intestinal epithelial cells increased its ODC some 20–30-fold without a significant increase of mRNA (Ginty et al., 1990). The downregulation of genes leading to significant decreases in ODC mRNA has also been explored (Barbour et al., 1988; Beyer & Zieve, 1990).

ODC mRNA has been estimated in various phases of the cell cycle in several types of cells induced to divide. Human lymphocytes treated with PHA increase their mRNA in the G_1 phase and this decreases during the S phase (Kaczmarek et al., 1987). In aged human diploid fibroblasts induced by serum, the production of ODC mRNA is similar to that in young cells, although the production of ODC itself is markedly decreased (Chang & Chen, 1988). In Swiss 3T3 cells induced by serum, ODC mRNA does accumulate selectively at first and is expressed throughout the cell cycle (Stimac & Morris, 1987).

Translational Controls of ODC

Studies of this problem may be divided into two groups: experiments dealing mainly with intact cells and those examining in vitro translation. Many systems of ODC synthesis were not found to relate to the accumulation or decline of mRNA and were thought to be controlled by translational or even posttranslational events.

An early dissection by means of inhibitors of a block in ODC synthesis in a temperature-sensitive cell-cycle fibroblast mutant suggested that the block was in the translation of a hypothetical mRNA (Landy–Otsuka & Scheffler, 1978). ODC cDNA was applied by Kahana and Nathans (1985a,b) to the problem of downregulation of ODC in overproducing mouse cells treated with putrescine. A rapid and specific decrease was obtained in the rate of synthesis of ODC, pulse labeled with ^{35}S-methionine, without a decrease of ODC mRNA or the decay rate of ODC. This was ascribed to a negative regulation of the translation of ODC mRNA, possibly through inhibitory effects on the 5' leader. Removal of this leader permitted a 40-fold increase in the in vitro translation of the mRNA.

Inhibitory effects of putrescine and spermidine on ODC activity in Ehrlich ascites cells occurred without changes in the content of mRNA and were attributed mainly to regulation in translation. Although acknowledging the role of ODC turnover in setting the level of ODC activity, study of the incorporation of ^{35}S-methionine into ODC appeared to suggest that the polyamines affected the rate of translation specifically (Persson et al., 1986). Amplification of ODC in such cells is also believed to occur mainly at the translation level (Wallon et al., 1995).

Actinomycin D, an inhibitor of mRNA synthesis, is known to superinduce ODC activity in Ehrlich ascites cells. Examination of the rate of ^{35}S-methionine incorporation into immunoprecipible ODC protein demonstrated an increased rate of translation in this system (Wallon et al., 1990). Hypoosmotic stress, which activates the expression of ODC without a change of mRNA, appears to produce this effect largely by increasing the rate of ODC synthesis and by inhibition of ODC degradation, that is, through several posttranscriptional mechanisms (Poulin & Pegg, 1990).

The inhibitory role of the 5'-leader sequence of the mRNA in determining the rate of translation of mRNA, has been analyzed in greater detail. When the 5'-untranslated sequence is placed before a reporter gene such as that of luciferase, the suppression of translation in a hamster cell was readily demonstrable and was attributed to G-C-rich hairpin structures within this leader (Grens & Scheffler, 1990). This suppression appeared independent of the polyamine levels in the cells. Insulin may facilitate the translation of ODC by activating factors capable of melting the secondary structures of the 5'-leader sequence (Manzella et al., 1991).

In a recent study of the markedly differing levels of ODC found in two murine species, there were many base differences in the 5'-untranslated regions of the mRNAs. The mRNA of the species producing less ODC possessed a changed secondary structure (calculated) in this leader, and it was suggested that this restricted translation (Johannes & Berger, 1992). However, genetic studies with transfecting recombinant DNAs have challenged an inhibitory role for the 5'-leader sequence. An ODC-deficient mouse cell transfected with a recombinant DNA containing only the protein-coding region of the DNA was capable of synthesizing mRNA and ODC (Wetters et al., 1989a). Normal regulation of the ODC mRNA and ODC was found during growth of these cells or after induction of quiescent cells. It was concluded that the 5' leader is not involved in these types of regulation. Further, it was found that the ODC 5'-leader sequence built into another mRNA did not determine polyamine-dependent regulation. This type of regulation did not change the distribution of ribosomes within the affected cells, nor affected the rates of translation (Wetters et al., 1989b). However, Lovkvist et al. (1993) reported that the cellular production of ODC from ODC mRNA devoid of its 5'-nontranslated region does not decrease in response to exogenous spermidine, implying a role for the leader in this response.

The accumulation of putrescine and spermidine in an overproducing mouse cell resistant to DFMO resulted in a 90% fall of the rate of synthesis of ODC, even at short labeling periods, and occurred without change in the steady-state level of mRNA (Stjernborg et al., 1991). Further, the polysome profile indicated a poor translation of the message, whereas the addition of cycloheximide permitted a shift to larger polysomes and greater accessibility to mRNA. A similar shift in polysome composition had been seen earlier in mitogen-activated lymphocytes (White et al., 1987). These results

do support the concept that translational controls operate on both the initiation and elongation of polypeptide chains.

In Vitro Studies of Translation

The analysis has turned therefore to presumably more easily definable systems in which the numerous components of the translation apparatus may be manipulated more readily. The synthesis of manipulable quantities of ODC mRNA from cDNA and then of similarly characterized ODC from this RNA in a reticulocyte lysate has been described as an example of the potential of "in vitro" technology (Glass et al., 1987). The problem of results obtained with isolated components is that of their relevance to events in intact cells. In many instances the final steps of protein synthesis from an ODC mRNA were carried out with quite disparate and only partially defined materials. For example, the frequently employed rabbit reticulocyte lysate is rich in polyamines, which must be removed somehow.

In an early study, ODC mRNA was isolated from the free polysomes of rat liver and translated in a rabbit reticulocyte lysate. Induction of the rat by thioacetamide produced a 100-fold increase of ODC in the liver and an increase of translatable ODC mRNA. However, the increase in mRNA preceded appearance of the enzyme by several hours and was far less (perhaps an eighth) than the increase in enzyme activity and newly synthesized ODC protein (Kameji et al., 1984).

A reticulocyte lysate was used to translate the total RNA from various organs active in the biosynthesis of the polyamines, and the products were analyzed for albumin, ODC, and AdoMetDC after precipitation with specific antisera (Kameji & Pegg, 1987). Gel filtration was used to partially free the reticulocyte lysate of the polyamine (e.g., 0.6 mM spermidine fell to 0.06 mM; 60 μM spermine fell to 13 μM) and the translations were carried out in 1.4 mM Mg^{2+} and various mixes of spermidine and spermine. Significant amounts of the polyamines remain affixed to components of the system. Low concentrations of the added polyamines were slightly stimulatory, whereas higher concentrations were markedly inhibitory. The synthesis of AdoMetDC was particularly inhibited by the polyamines, although that of ODC was also sensitive. Also, it would have been desirable to know the polyamine contents of the specific nucleates being studied in the absence of added exogenous polyamine.

Persson et al. (1988) obtained similar results in a study of the translation of ODC mRNA and total mRNA in gel-filtered reticulocyte lysates. Both of these effects were greater on the translation of ODC mRNA than on that of total cellular mRNA, and suggested that a part of the "feedback" regulation of ODC might arise from a direct effect on translation. In contrast to the effects of spermidine, putrescine did not stimulate or inhibit protein synthesis in this system (Holm et al., 1989).

Lysates of ODC-overproducing cells are capable of

synthesizing ODC with the ODC mRNA present on pre-initiated ribosomes. Low concentrations of added spermidine (0.8–1.0 mM) increase the rate of synthesis significantly. In this system the polysome profile is not affected during this increase of synthesis, which is primarily an increase in the rate of polypeptide extension (Autelli et al., 1990). The above results established spermidine concentrations optimal for ODC synthesis and are similar to those optimizing numerous various steps in cell-free protein synthesis in general.

The 20–40-fold increased rate of translation of ODC mRNA lacking the 5′ leader described by Kahana and Nathans (1985b) was confirmed by Manzella and Blackshear (1990). These workers calculated a secondary stem-loop structure consistent with the composition of this leader and concluded that this was both necessary and sufficient for this inhibition. Van Steeg et al. (1991) similarly revealed the improved efficiency of translation after removing the 5′ leader. In addition these researchers studied the effects of spermidine on translation by intact ODC mRNA and by 5′-leader-truncated ODC mRNA. Stimulation of translation of the intact and truncated mRNAs were achieved at the same spermidine concentrations. This result was interpreted to suggest that the inhibitory secondary structure of the 5′ leader did not require spermidine for its effect to be realized. In contrast, Ito et al. (1990) reported that the stimulatory and inhibitory effects of variously truncated 5′-leader-containing mRNA were markedly increased by spermidine. In their hands these effects were dependent on a limited region of G-C-rich nucleotides upstream from the initiating AUG (Kashiwagi et al., 1991).

Manzella and Blackshear (1992) also demonstrated the existence of a specific protein capable of binding to the 5′ leader. This binding is blocked by mutations in a highly conserved heptanucleotide close to the initiation codon present in the ODC mRNA of many animal species and was not affected by polyamines.

ODC Degradation

The phenomenon of ODC degradation, as well as the rapidity of restoration of the enzyme by increased transcription and translation, are major elements of regulation of the initiation of polyamine synthesis. A degradative mechanism to eliminate ODC itself has replaced the more common regulatory mechanism of feedback inhibition by products of the pathway, which in this case initiate a degradative inhibition related to the formation of antizyme.

The laboratory of S. Hayashi formulated and pursued the hypothesis that the complex of antizyme with ODC is an initiating element in the degradation of the ODC protein. This group demonstrated rigorously that the formation of such complexes does participate in such a proteolytic activity. They also isolated a mammalian antizyme gene, transfected cells with the gene to manipulate their antizyme content, modified the antizyme protein, studied the interaction of antizyme and ODC, and

explored the proteolysis of the ODC–antizyme complex. The existence of antizyme and its participation in ODC degradation in amphibian tissues have also been described (Baby & Hayashi 1991a,b).

The discovery of the unexpectedly short half-life of ODC in mammalian tissues focused attention on the distinctions between short-lived proteins and others. Cellular proteases have included acid-optimal cathepsins that act in endosomal and lysosomal compartments. Many cell proteins are degraded via an ATP-requiring system in which the proteins to be degraded are conjugated covalently to the polypeptide, ubiquitin, and hydrolyzed via a large proteolytic complex, a proteosome (Goldberg & Rock, 1992; Stadtman, 1990). It was asked if the degradation of ODC involves any of these systems.

It was shown fairly early that there was no detectable correlation between the stability of ODC in vitro and its half-life in vivo. However, during early studies with inhibitors of polyamine synthesis on intact cells it was observed that the ODC activity was in fact stimulated and ODC half-life was prolonged in the cells. Thus, ODC was increased after treatment with methylglyoxal-bis(guanylhydrazone) (MGBG) employed as an inhibitor of AdoMetDC (Nikula et al., 1985). Methylglyoxal bis(cyclohexylamidinohydrazone), which inhibits spermidine synthase as well, also stabilized ODC (Hibasami et al., 1988).

Additionally ODC may be stabilized in Friend erythroleukemia cells by the depletion of ATP following an incubation in a medium containing dinitrophenol and 2-deoxyglucose (Flamigni et al., 1989). This occurs even after addition of spermidine, which in normal media increases the subsequent degradative response to cycloheximide. A similar stabilization of the enzyme was obtained by treatment with phenanthroline, a metal-binding agent, evoking the suggestion that degradation involves the participation of both ATP and a metalloprotease (Flamigni et al., 1990). Despite the ATP requirement, ubiquitin was not involved, because degradation in the presence of spermidine proceeded at a nonpermissive temperature in a cell temperature-sensitive for ubiquitin conjugation (Glass & Gerner, 1987).

In studies with rabbit reticulocyte lysate, which contains components of the ubiquitin-dependent proteolytic system, ODC synthesized in vitro was degraded in a lysate by fractions lacking essential elements of the ubiquitin system (Bercovich et al., 1989). The degradation of ODC in the extract of CHO-overproducing cells was then shown to be antizyme dependent, as well as ATP dependent (Murakami et al., 1992). As shown in Figure 13.5, the rate of degradation is maximal when the molecular ratio of antizyme to ODC approaches 1.

Some ODC-overproducing cells, stimulated to degrade ODC by exposure to exogenous spermidine or spermine, were found to become toxic as a result of the overaccumulation of polyamine. It was reasoned that the excess of ODC neutralized newly formed antizyme and thereby permitted an influx of otherwise regulated polyamine. This was tested by blocking such transport by transfection of the cells with an appropriate cDNA

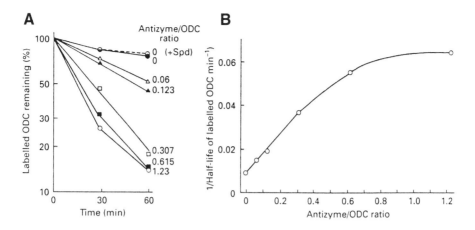

FIG. 13.5   Dose-dependent response of ODC degradation in vitro to antizyme in cell extracts; adapted from Murakami et al. (1992).

coding for antizyme; this resulted in a decrease of polyamine transport, a decay of ODC, and a cellular recovery from a toxic state (Mitchell et al., 1994; Suzuki et al., 1994). These experiments imply that antizyme also participates in polyamine transport.

ODC Structure and Its Stability in Cells

The discovery of rapidly degraded proteins, such as ODC, posed the possibility that such proteins contained specific amino acid sequences sensitive to cellular proteases (Rechsteiner et al., 1987). These were suggested to be sequences rich in proline, glutamate, serine, and threonine and were designated as PEST sequences. Such sequences were found in the carboxyl (C) terminus of ODC, and sequences containing such a C-terminal sequence, or other termini lacking the sequence, were added to dihydrofolate reductase, which lacks a C-terminal PEST sequence. The modified fusion protein, now containing a PEST sequence, was markedly more rapidly degraded in a lysate than was the parent protein or a fusion protein lacking the sequence (Loetscher et al., 1991). Two PEST sequences have been detected in normal mammalian ODC, of which that in the C terminus appears to relate to the stability of the protein. The truncation of 37 residues in the carboxyl terminus does result in a far more stable protein (Ghoda et al., 1989).

As will be discussed in Chapter 18, it was found that the ODC of *Trypanosoma brucei* is stable in the mouse or CHO cells and lacks the specific C terminus that labilizes the enzyme, although the first 376 amino acids, including the amino terminus, of the trypanosome enzyme are very similar in sequence to the mouse enzyme. When a chimeric ODC is made, composed of the amino terminus of the trypanosome ODC and the carboxyl terminus of the mouse enzyme, the chimera was rapidly degraded in CHO cells (Ghoda et al., 1990). Nevertheless, the differences in stability between mammalian

and trypanosome ODC do not reside solely in the structure of the enzyme. The "mouse-labile" mouse enzyme generated in the trypanosome has been found to be completely stable in the latter cell (Bass et al., 1992).

Certain changes in the amino acids of ODC abolished enzyme activity; for example, lysine-169 and histidine-197 were replaced by alanine. Removal of 36 residues (amino acids 426–461) from the C terminus resulted in an active enzyme, whereas this active enzyme was degraded far more slowly in a reticulocyte lysate (Lu et al., 1991). In a similar approach the deletion ODC mutants (minus amino acids 423–461) were found to be stable proteins. Additionally, an ODC mutant lacking amino acids 295–309 was stabilized within cells, although it rapidly degraded in a lysate system (Rosenberg–Hasson et al., 1991a). The latter workers also tested the degradation of an ODC heterodimer comprising a normal subunit and a stable deleted subunit. In a degrading lysate the stable unit remained stable and the sensitive subunit was degraded (Rosenberg–Hasson et al., 1991b).

ODC has been described as linked domains, each of which participate in the various types of degradative regulation (Ghoda et al., 1992). Thus, the carboxy terminus is essential for what is termed "constitutive" degradation. The deletion of five or more carboxy-terminal amino acids eliminates this sensitivity, as does deletion of several other amino acids within a 37 carboxy terminus. This group has shown that trypanosome ODC, which is much like the mouse ODC lacking a carboxy terminus, is stable in cells that have accumulated polyamine; they stated that the truncated mouse ODC is unstable in this condition. Thus, they postulated a structure, that is, a separate domain, outside the carboxy terminus that increases instability in a polyamine-rich cell. Their studies showed that the trypanosome enzyme does not react with antizyme in an in vitro system and that the replacement of 20+ residues in the mouse ODC with the "equivalently placed" residues of the trypansome ODC disrupted antizyme binding and cellular regulation. Also, binding of ODC antibody at the C terminus of the enzyme prevented antizyme-mediated proteolysis. Li and Coffino (1994) concluded that although antizyme binds ODC at its N terminus, this binding evokes a conformational change exposing the C terminus and inactivating the enzyme. The N terminus is essential to subsequent proteolysis. A recent study demonstrated the bacterial synthesis from plasmid cDNAs of separable polypeptide domains capable of spontaneous assembly to form active *T. brucei* ODC (Osterman et al., 1995).

In mutant-overproducing systems, characterized as deficient in ODC turnover, the deficiency may reflect a change in ODC, in antizyme, or in the degrading system. In one such HTC mutant it was found that the ODC itself was altered (Miyazaki et al., 1993). Ichiba et al. (1994) characterized two regions of antizyme essential for binding and another necessary for destabilization.

Activity and Degradation of ODC

The refolding of an inactivated ODC, denatured initially with urea or guanidinium chloride, leads to catalytic reactivation; however, the renatured ODC differs from the native enzyme, for example, in an increase in reactive thiols in the renatured catalytically active protein. Whereas DFMO binds covalently to a specific cysteine in native ODC (Poulin et al., 1992), the renatured ODC is less rapidly inactivated by DFMO, which is no longer bound irreversibly (Tsirka et al., 1993).

The homodimer is the active form of the enzyme, and the position of a single glycine is essential to both dimerization and activity (Tobias et al., 1993). Several other amino acids are essential to activity, a lysine-69 which is important as the PLP-binding site and which can bind DFMO in certain mutant enzymes (Coleman et al., 1993). Nevertheless, replacements of some single amino acids very active in such functions as binding PLP or DFMO, produce partially active ODC capable of generating low levels of catalytic activity. The hybridization in vitro of subunits of catalytically deficient mutants, one of which lacked an essential cysteine-360 and the other essential lysine-69, restored activity to about 25% that of the wild-type enzyme (Tobias & Kahana, 1993); thus, an active site is formed at the interface of two mutant monomers. Similar results were obtained by Coleman et al. (1994) who demonstrated the very rapid exchange occurring among the subunits. The irreversible binding of DFMO to native ODC indicated the existence of two active sites per dimer (Poulin et al., 1992), and the interpretations of active site formation in the heterodimers have been based on the concept of independent activities of each active site in a dimer. It is believed that the binding of antizyme occurs on the monomeric subunits of ODC; the binding appears to be increased under conditions of dimer dissociation (Mitchell & Chen, 1990). The essentiality of various residues in determining catalytic activity, in dimeriza-

tion, and in reactions with antizyme has been summarized by Pegg et al. (1994).

Proteolytic Degradation of ODC

An ODC-overproducing cell stimulated to degrade ODC more rapidly by exogenous spermidine yielded an extract capable of the proteolysis of ODC (Kanamoto et al., 1993). Both phosphorylated and dephosphorylated forms are degraded, the degradation of the latter being slightly slower. Proteolysis in the extract is inhibited by antibody to antizyme. The proteolytic activity of an extract, which required antizyme and ATP, is centered in a particular high molecular weight complex (approximately 700 kDa) termed a 26S proteosome, comprising as many as 25 distinct polypeptides (Murakami et al., 1993; Rivett, 1993). Curiously proteosomes are found in both the nucleus and cytoplasm of animal cells at all phases of the cell cycle. The ODC is degraded specifically to short peptides (Tokunaga et al., 1994). In yeast cells, which apparently lack an antizyme but also degrade ODC, a 26S proteosome complex and additional regulatory subunits have also been implicated in this particular proteolytic reaction (Elias et al., 1995; Mamroud-Kidron & Kahana, 1994).

Inhibition of Mammalian ODC

The surprising correlation of the early rise of ODC preceding many events of cellular and tissue growth implied an important biological role for the polyamines and suggested the desirability of finding inhibitors that might cut off the enzyme activity and the increase of putrescine. The history of biochemical research is intimately bound to the exploitation of inhibitors. The lock and key postulate of P. Ehrlich concerning the nature of an active site and its occupation by a metabolite or an analogue, the exploitation of malonate in the analysis of the citric acid cycle, and the use of penicillin in the study of the bacterial cell wall are examples of the impact of inhibitor theory and practice on the growth of the discipline. Initially synthetic inhibitors of particular enzymes were designed as analogues of reactants or immediate reaction products whose activities served not only to block reactions and reveal function and accumulated precursors, but also sketched the size and possible chemical structures at the enzymatic active site. The discovery of feedback inhibition of an initial enzyme by the end product of a multireaction sequence enlarged the types of analogues to be made and tested. The synthetic inhibitors devised in this early approach were frequently of relatively low affinity for the reversible enzyme and for the most part did not involve covalent binding to the enzyme. As knowledge of protein structure developed, it was possible to produce potential inhibitors that might react irreversibly with the enzyme with the formation of covalent bonds at the active site. This might occur before catalytic processing or afterward as in the case of DFMO. In addition, some antibiotics had been isolated that inactivated enzymes by forming covalent adducts to specific enzymes or by generating noncovalent complexes in which the inhibitor is bound far more tightly than the metabolic substrate. The determination of the type of inhibition produced, reversible or irreversible, is usually made as a result of a study of the kinetics of the enzymatic reaction in the presence of the inhibitor.

The design of inhibitors for ODC has resulted in the development of reversible and irreversible inhibitors, of which the latter, such as DFMO, have proven to be the most useful in the analysis of ODC and polyamine metabolism. Most of the reversible inhibitors (described in Chap. 8) proved not to be tight binding, did not deplete putrescine effectively, and even permitted low levels of replenishment of cellular putrescine. Because the cells most often contain large and excessive amounts of the polyamines, an inhibitor of ODC must deplete both putrescine and the next active metabolite, spermidine, in a major way in a time frame suitable for examination of the effect of the depletion. Normally the weak homeostatic maintenance of cellular polyamine pools is controlled by the balance of diet, transport, synthesis, and the restorative effects of catabolic reactions (Porter et al., 1992); and intervention in the biosynthesis of putrescine (or cadaverine) via an inhibitor of ODC alone is quite demanding on the nature of a useful ODC inhibitor.

An extensive discussion of the inhibition of the decarboxylases of the basic amino acids was prepared by Bey et al. (1987). Inhibitors of ODC that do interfere with polyamine metabolism in cells have included the *reversible* inhibitors (e.g., α-methylornithine described in Fig. 8.4) and the *irreversible* inhibitors (e.g., DFMO and its analogues) and the 6-heptyne-2,5-diamine (RR-MAP). Structures of these compounds are presented in Figure 13.6.

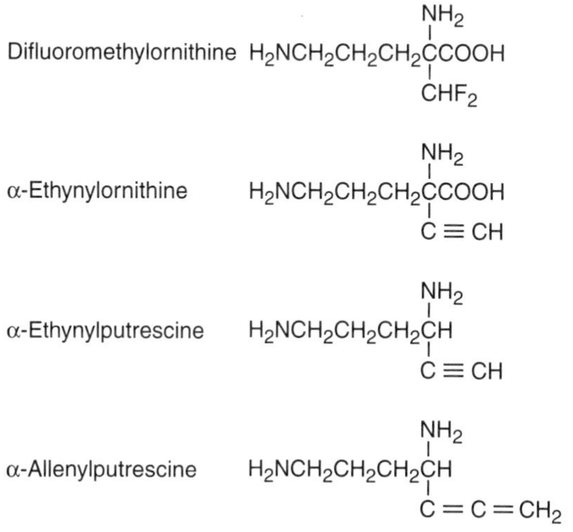

Difluoromethylornithine $\quad$ $H_2NCH_2CH_2CH_2\overset{\displaystyle NH_2}{\underset{\displaystyle CHF_2}{C}}COOH$

α-Ethynylornithine $\quad$ $H_2NCH_2CH_2CH_2\overset{\displaystyle NH_2}{\underset{\displaystyle C\equiv CH}{C}}COOH$

α-Ethynylputrescine $\quad$ $H_2NCH_2CH_2CH_2\overset{\displaystyle NH_2}{\underset{\displaystyle C\equiv CH}{C}}H$

α-Allenylputrescine $\quad$ $H_2NCH_2CH_2CH_2\overset{\displaystyle NH_2}{\underset{\displaystyle C=C=CH_2}{C}}H$

FIG. 13.6 Irreversible enzyme-activated inhibitors of cellular ODC.

Although L-ornithine is quite specific as a substrate for the mammalian enzyme, D-ornithine has some affinity, L-lysine is a poor but significant substrate, and putrescine, spermidine, and spermine are weak competitive inhibitors. The synthesis of α or ε N-substituted ornithines was initiated by Skinner and Johansson (1972). One derivative, α-hydrazino-ornithine (Harik et al., 1974a,b), was found to inhibit the mammalian enzyme in vitro ($K_i = 2$ μM) and within rat hepatoma cells. In the latter case the inhibitor prolonged the apparent half-life of the enzyme and caused the activity of ODC to increase in several tissues. The inhibitor was competitive with ornithine and PLP, and net putrescine synthesis was depressed in HTC cells. However, the compound did not block the synthesis of RNA and DNA (Harik et al., 1974a).

The next in line, α-methylornithine (Abdel–Monem et al., 1974), proved to be a potent reversible inhibitor of rat prostatic ODC and was competitive with ornithine but not with PLP. An ornithine analogue combining the α-methyl and 2-hydrazino structure, 5-amino-2-hydrazino-2-methylpentanoic acid, was an inhibitor whose high activity was equal to that of α-hydrazinoornithine but was also competitive with PLP, which affected the kinetic analysis of the system (Abdel–Monem et al., 1975). A stable adduct of PLP and ornithine, N-(5'phosphopyridoxyl)ornithine, was found to be a weak inhibitor of ODC. The nonphosphorylated N-(4'-pyridoxal)amines can enter liver cells, and can be phosphorylated, oxidized, and cleaved to release PLP and the original amine.

The α-methyl derivative proved to be a far more active inhibitor of ODC than other α-alkyl- or α-benzylornithines. Additionally, it was not a substrate for the enzyme. It was soon found to be active at millimolar levels in the inhibition of putrescine synthesis in HTC cells, with successive decreases in putrescine and spermidine levels. These depletions paralleled the inhibition of DNA synthesis and cell proliferation, functions restored by addition of ornithine or the polyamines, putrescine, spermidine, or spermine (Mamont et al., 1976). The substance also increased the half-life of the enzyme and the ODC activity of the cells, without affecting the content of ODC mRNA (Persson et al., 1985). The high levels of the inhibitor required to act on cells and its stabilizing effect on the enzyme pointed to the need for more effective compounds. Replacement of the carboxyl of ornithine by phosphonic or phosphinic acids did not improve the efficacy of the inhibition.

The inhibitory α-hydrazino amino acids form more stable adducts with the essential cofactor PLP than do the α-amino acids. One ornithine analogue, L-canaline, in which an oxygen replaces the γ-carbon, reacts weakly with the cofactor and is thought to inactivate the enzyme by forming a Schiff base via its α-amino group. However, the putrescine analogue, 1-amino-oxy-3-aminopropane (APA), in which an aminoxy group is isosteric with the aminomethylene of putrescine, is a far more potent inhibitor ($K_i = 3.2$ nM) of mouse kidney ODC (Khomutov et al., 1985). This tight-binding inhib-

252

A GUIDE TO THE POLYAMINES

itor competes exclusively with ornithine for the enzyme and was not found to react with PLP in preincubations or during the enzymatic reaction.

In addition APA proved to be both an active irreversible inhibitor of liver AdoMetDC ($K_i = 50$ μM) and a competitive inhibitor of brain spermidine synthase ($K_i = 2.3$ μM). In the first of these, pyruvate, which is normally present in the enzyme as a covalently bound cofactor, partially prevented inactivation when added to the system. Thus, APA also behaves as a carbonyl reagent capable of reacting with free pyruvate. APA (0.5 mM) inhibited the growth of leukemia cells, and this effect was prevented by 0.1 mM spermidine. The inhibitor reduced spermidine levels markedly and the synthesis of protein and DNA, as well as the proliferation of BHK cells. Nevertheless, APA induced a marked increase (40–50-fold) of ODC and AdoMetDC and the cells were not killed (Hyvönen et al., 1988). Unexpectedly, the inhibitor was not concentrated in the cells, despite the depletion of natural polyamines, and did not compete with the polyamines in tests of polyamine transport (Hyvönen et al., 1990).

Several potent aminoxy derivatives of the polyamines are being studied as antiproliferative agents and microanalytical procedures have been devised for their analysis at picomole and micromole levels (Hyvönen et al., 1992). The 2-hydroxy and 2-fluoro derivatives of the 3-(aminoxy) propanamines have been found to be very powerful inhibitors of cell growth and of ODC with IC_{50} in the nanomolar range (Stanek et al., 1992). These compounds did not inhibit AdoMetDC.

Both putrescine, a weak reversible inhibitor of ODC, and DAP repress the induction of the enzyme activity in various stimulated animal cells. Numerous synthetic α,ω diamines of 3–12 carbon atoms acted similarly to inhibit a serum-mediated increase. However, one diamine, the trans-1,4-diamino-2-butene, proved to be a potent tight-binding competitive inhibitor of the enzyme ($K_i = 2.0$ μM) (Relyea & Rando, 1975), as well as an inhibitor of induction of ODC increase (Kameji et al., 1979).

Variously alkylated derivatives of putrescine have been tested on bacterial and mammalian ODCs. They do not inhibit the bacterial enzyme and inhibit the rat liver ODC poorly. In continuing studies it was found that the three isomers (+, –, and meso) of 1,4-dimethylputrescine, 2,5-hexanediamine, did not inhibit rat liver ODC in vitro. However, all three inhibited ODC in vivo and decreased the putrescine pools of rat liver. The meso isomer induced ODC antizyme and the (–) isomer inhibited synthesis of protein and ODC (Moyano et al., 1990). The three isomers had been found earlier to inhibit mammalian diamine oxidase.

Another putrescine analogue, 2,5-diamino-3-hexyne, containing a triple bond, was a weaker base and was concentrated in the acidic milieu of lysosomes. This uptake evoked the formation of large vacuoles within L1210 cells (Porter et al., 1990b).

Monosubstitution of an amino group of putrescine did not reduce uptake of the diamine by prostatic cancer cells. Alkylation of both nitrogens did reduce putrescine uptake significantly (Heston et al., 1987). Higher alkyl derivatives (N-lauryl, myristyl, etc.) of the diamines (C_3 to C_8) have been reported to be hemolytic and behave as cationic detergents (Miyamoto et al., 1992).

The sole experimentally useful specific competitive inhibitor has been α-methylornithine. APA obviously has many different inhibitory activities. However, after a brief in vivo inhibition, α-methylornithine evoked an increase of the half-life of the enzyme and its accumulation. These results have led to a search for more effective irreversible inhibitors.

It had been found in the late 1960s that acetylenic analogues of a substrate in a dehydrase reaction irreversibly inactivated the enzyme. The investigators postulated an enzymatic rearrangement of the acetylenic derivative to an allene capable of reacting with an essential amino acid in the enzyme. Reactions of this type have been termed suicide inhibitors, or catalytic constant (K_{cat}) inhibitors (Rando, 1974). The designers of similar types of mechanism-based inhibitors in the polyamine field term these "enzyme-activated inhibitors" (Bey et al., 1987). These latter authors indicated kinetic criteria for the definition of such enzymes. These inhibitions were time dependent, were decreased in the presence of a competitive inhibitor or substrate, and were maximal at titratable concentrations of inhibitor. Such inhibitors for the flavin enzymes, such as the monoamine oxidases, and PLP enzymes, such as ODC, have been explored in great detail.

In reaction with an amine-containing substrate with enzyme-bound PLP, a quinonoid intermediate would be formed. It was thought that if the substrate carbon holding the amine were further derivatized with an unsaturated group or a halogenated methyl, these might generate a conjugated imine to alkylate a nearby essential group within the enzyme. Such a postulated mechanism for α-halogeno-methylornithine was presented in Chapter 8 and led to the discovery of the mono- and difluoromethylornithines, of which DFMO has proved to be so useful as a potent irreversible enzyme-activated inhibitor (Metcalf et al., 1978).

The finding that α-monofluoromethyl derivatives of amino acids could be potent irreversible inhibitors of aromatic α amino acids encouraged the synthesis of α-monohalogenomethyl-α amino acids from the easily available α-hydroxymethyl amino acids. This led to the synthesis of various α-halogenomethyl-α amino acids, including α-fluoromethylornithine (Bey et al., 1979), whose inhibitory activities on rat liver ODC have been reported to be quite potent, albeit somewhat variable (Bey et al., 1987). As has been noted, 5-fluoromethylornithine is a potent enzyme-activated irreversible inhibitor of ornithine aminotransferase (Bolkenius et al., 1990).

In any case, DFMO was thought to be a drug candidate of greater interest than the monofluoromethyl derivative and was made more readily available for investigatory studies by the synthetic producer, the Merrell–Dow Company. There is no doubt that this policy

PROPERTIES AND REGULATION OF ODC

253

of distributing the compound to interested investigators markedly increased the pace and discovery in the field of the polyamines.

Pegg (1987) has listed the many uses to which DFMO has been successfully applied:

1. labeling of the enzyme in estimation of the amount of the ODC protein;
2. estimation of the biosynthesis and degradation of the protein;
3. autoradiographic localization of the enzyme within the cell and isolation of the protein in extracts;
4. radioimmunoassay of the enzyme via bound inhibitor;
5. mechanism of the ODC reaction and labeling of the active site;
6. development of resistant cellular clones with amplified genes for gene isolation;
7. exploration of the effects of polyamine depletion on many physiological processes; and
8. chemotherapy of parasitic disease.

DFMO proved to be nonlethal to most cells in relatively short periods (48 h) of inhibition of ODC and depletion of putrescine and a subsequent depletion of spermidine. Most cells are arrested before the S phase when spermidine is depleted. The low level of toxicity of DFMO to intact animals and humans has been explored in studies of the effects of the analogue in the therapy of cancer and parasitic disease. DFMO has also been combined with other inhibitors in an effort to increase therapeutic efficacy (Seiler, 1991).

As indicated in Figure 13.7, a fluorine atom, that is,

a leaving group, is released in the activation of the inhibitor. The release of CO_2 from $[^{14}C]$-DFMO was demonstrated by Pegg et al. (1987), who also studied the binding of the residual carbon (i.e., from admixed $[5-^{14}C]$-DFMO) in the inactivated enzyme. CO_2 release was about 3.3 times more active than the binding to ODC, and this held for several ODCs from different mammalian sources.

In continuing studies of the mechanism of inactivation by DFMO (Poulin et al., 1992), the PLP-binding lysine-69 in mouse ODC was located in a sequence found in all known eucaryotic ODC. The site of adduct formation (for 90% of bound inhibitor) was cysteine-360, which is also in a sequence found in all eucaryotic ODC. As presented in Figure 13.7, the binding of the inhibitor to the enzyme appears to entail release of PLP and permits the formation of the cyclic imine S-((2-(1-pyrroline))methyl)cysteine. This sequence was in accord with spectral changes observed during the inactivation of ODC. Three sets of observations have raised new questions concerning the chemistry of the reactions of DFMO with ODC: the release of DFMO from inactivated ODC and a consequent partial reactivation, the reaction of DFMO with renatured ODC, and the several modes of binding of DFMO to mutant forms of the enzyme.

In addition to the halogenomethyl analogues of ornithine, it was hypothesized that an acetylenic or allenic group at the place of the α-hydrogen atom of ornithine would permit decarboxylation. This would now generate a conjugated imine, as in Figure 13.7, and permit an inactivating alkylation at the active site of the enzyme.

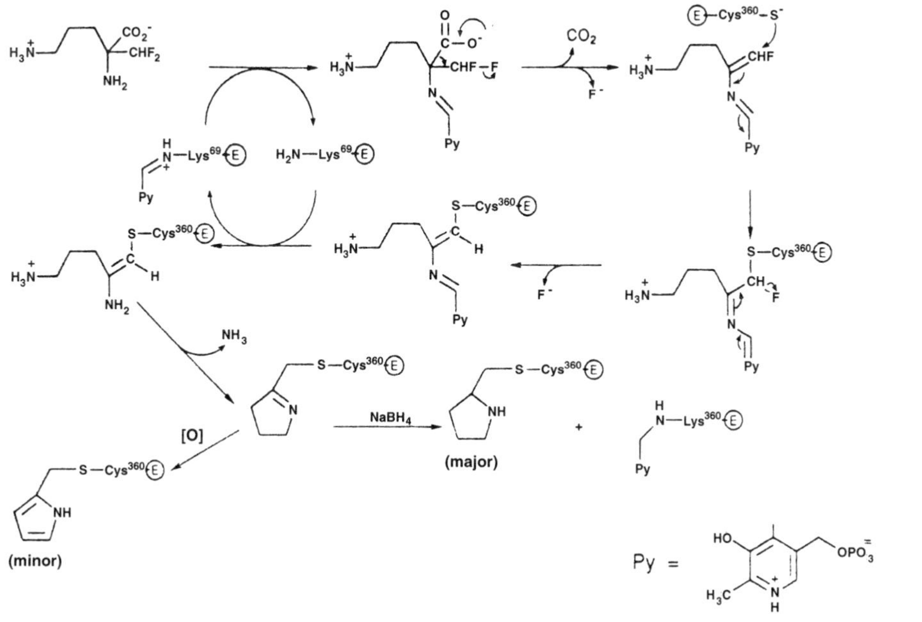

FIG. 13.7 Mechanism proposed for the inactivation of mouse ODC by DFMO; adapted from Poulin et al. (1992).

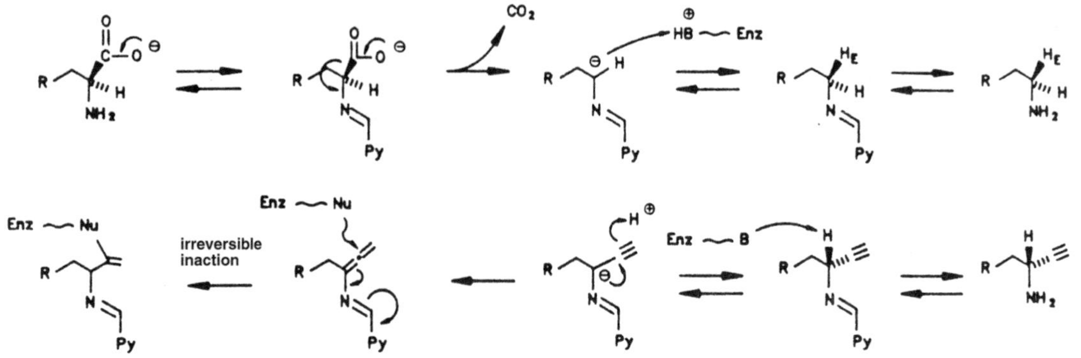

FIG. 13.8 Postulated mechanism for the irreversible inhibition of α-amino acid decarboxylases byproduct analogues; adapted from Bey et al. (1987). In this case, α-ethynylamine is accepted by the decarboxylase active site and forms a Schiff base with the PLP cofactor. Rearrangement occurs to form an electrophilic allenyl imine that alkylates a residue of the enzyme.

The inhibitory activity of α-ethynylornithine has proved to be similar to that of DFMO.

Analogues of the products of decarboxylation of amino acids, the amines, have been prepared (Metcalf et al., 1978); these were excellent inhibitors of the decarboxylases. It is supposed that these amines participate in the reversal of enzyme catalysis and the very active irreversible inhibitor α-ethynylputrescine (5-hexyne-1,4-diamine) is active in only a single enantiomer (Fig. 13.6). The mechanism postulated for the irreversible product analogues and particularly for α-ethynylputrescine is presented in Figure 13.8. This compound effectively reduces ODC in rat prostate, and activates AdoMetDC in vitro as does putrescine (Danzin et al., 1979).

Bey et al. (1987) also summarized data with many putrescine analogues of this type. Many were found to be toxic to the central nervous system (CNS). However, new methyl derivatives (e.g., hept-6-yne-2,5-diamine) also known as S-methylacetylenicputrescine or MAP, are stable to monoamine oxidase and one stereoisomer is a quite potent inhibitor that is 10 times more active than DFMO in vitro and in vivo (Casara et al., 1985). The other potent inhibitory derivative, 2,2-difluoro-5-hexyne-1,4-diamine, is not a substrate of the monoamine oxidase (Kendrick et al., 1989) and is not toxic to the CNS.

The introduction of a double bond between the β and δ carbons in ornithine and putrescine analogues markedly increased the inhibitory potency of α-fluoromethyl derivatives (Bey et al., 1983). One enantiomer of α-allenyl putrescine (5,6-heptadiene,1,4-diamine) has proved to be an active enzyme-activated inhibitor of both the mammalian and *E. coli* ODC (Danzin & Casara, 1984).

Putrescine can also be generated via the degradation of N^1-acetylspermidine by polyamine oxidase. This catabolic reaction can be cut off by the inhibitor of that enzyme N^1,N^4-bis-(2,3-butadienyl)1,4-butanediamine, resulting in an accumulation of N^1-acetylspermidine and a depletion of putrescine (Ientile et al., 1991).

Summary

The early studies demonstrating the rise and fall of ODC in regenerating rat liver have evoked a very large body of research that attempts to dissect the role and structure of the enzyme as well as the numerous genetic and metabolic controls upon it. Many of these problems have proved to be of considerable biological and medical interest. A possible role of the enzyme, over and above that of providing a biosynthetic route to putrescine and cadaverine, has not been demonstrated. Nevertheless, some elements of the regulation of ODC levels, essential to minimizing the toxicity of excessive polyamine production, have demonstrated unique characteristics, such as those involving antizyme and proteolytic degradation. These regulatory features, discussed in some detail by Davis et al. (1992), have bypassed the far more common mechanisms of enzyme regulation, that is, feedback inhibition and repression. In addition, the multiple functions of the novel protein, antizyme, relating to ODC degradation and most recently to polyamine transport, suggest the possibility of multiple functions of other proteins, that is, the possibility of unexpected phenomena. Indeed the formation of antizyme (discussed in Chap. 22) has proven to be a novel aspect of translation. The recent discovery of agmatine and an arginine decarboxylase in the brain and astrocytes is an indication of surprises that may open another bag of experimental worms (Regunathan et al., 1995).

REFERENCES

Abdel–Monem, M. M., Newton, N. E., & Weeks, C. E. (1974) *Journal of Medicinal Chemistry*, **17**, 447.
———— (1975) *Journal of Medicinal Chemistry*, **18**, 645–648.
Abrahamsen, M. S. & Morris, D. R. (1990) *Molecular and Cellular Biology*, **10**, 5525–5528.
———— (1991) *Perspectives on Cellular Regulation*, pp. 107–119. New York: Wiley–Liss.
Alcivar, A. A., Hake, L. E., Mali, P., Kaipia, A., Parvinen,

M., & Hecht, N. B. (1989) *Biology of Reproduction*, **41**, 1133–1142.

Alhonen–Hongisto, L., Hirvonen, A., Sinervirta, R., & Jänne, J. (1987a) *Biochemical Journal*, **247**, 651–655.

Alhonen–Hongisto, L., Kallio, A., Sinervirta, R., Seppänen, P., Kontula, K. K., Jänne, O. A., & Jänne, J. (1985a) *Biochemical and Biophysical Research Communications*, **126**, 734–740.

Alhonen–Hongisto, L., Leinonen, P., Sinervirta, R., Laine, R., Winqvist, R., Alitalo, K., Jänne, O. A., & Jänne, J. (1987b) *Biochemical Journal*, **242**, 205–210.

Anehus, S., Emanuelsson, H., Persson, L., Sundler, F., Scheffler, I. E., & Heby, O. (1984) *European Journal of Cell Biology*, **35**, 264–272.

Autelli, R., Holm, I., Heby, O., & Persson, L. (1990) *FEBS Letters*, **260**, 39–41.

Baby, T. G. & Hayashi, S. (1991a) *Comparative Biochemistry and Physiology B*, **99**, 151–156.

Baby, T. G. & Hayashi, S. (1991b) *Biochimica et Biophysica Acta*, **1092**, 161–164.

Barbour, K. W., Berger, S. H., Berger, F. G., & Thompson, E. A. (1988) *Molecular Endocrinology*, **2**, 78–84.

Bass, K. E., Sommer, J. M., Cheng, Q., & Wang, C. C. (1992) *Journal of Biological Chemistry*, **267**, 11034–11037.

Bassez, T., Paris, J., Omilli, F., Dorel, C., & Osborne, H. B. (1990) *Development*, **110**, 955–962.

Bercovich, Z., Rosenberg–Hasson, Y., Ciechanover, A., & Kahana, C. (1989) *Journal of Biological Chemistry*, **264**, 15949–15952.

Bernstein, H.-G. & Müller, M. (1995) *Neuroscience Letters*, **186**, 123–126.

Bey, P., Danzin, C., & Jung, M. (1987) *Inhibition of Polyamine Metabolism*, pp. 1–31. P. P. McCann, A. E. Pegg, & A. Sjoerdsma, Eds. Orlando, Fla: Academic Press.

Bey, P., Danzin, C., van Dorsselaer, V., Mamont, P., Jung, G. M., & Tardif, C. (1978) *Journal of Medicinal Chemistry*, **21**, 50–55.

Bey, P., Gerhart, F., van Dorsselaer, V., & Danzin, C. (1983) *Journal of Medicinal Chemistry*, **26**, 1551–1556.

Bey, P., Vevert, J., van Dorsselaer, V., & Kolb, M. (1979) *Journal of Organic Chemistry*, **44**, 2732–2742.

Beyer, H. S. & Zieve, L. (1990) *Biochemistry International*, **20**, 761–766.

Blaschko, H. (1945) *Advances in Enzymology*, **5**, 67–85.

Bolkenius, F. N., Knödgen, B., & Seiler, N. (1990) *Biochemical Journal*, **268**, 409–414.

Brabant, M., McConlogue, L., Wetters, T. V. D., & Coffino, P. (1988) *Proceedings of the National Academy of Sciences of the United States of America*, **85**, 2200–2204.

Branca, A. A. & Herbst, E. J. (1980) *Biochemical Journal*, **186**, 925–931.

Brosnan, M. E., Farrell, R., Wilansky, H., & Williamson, D. H. (1983) *Biochemical Journal*, **212**, 149–153.

Brown, P. J., Reddy, S., & Haddox, M. K. (1994) *Biochemical Society Transactions*, **22**, 859–863.

Casara, P., Kanzin, C., Metcalf, B., & Jung, M. (1985) *Journal of the Chemical Society–Perkin Transactions*, **1**, 2201–2207.

Chang, Z. & Chen, K. Y. (1988) *Journal of Biological Chemistry*, **263**, 11431–11435.

Chen, Z. P. & Chen, K. Y. (1992) *Journal of Biological Chemistry*, **267**, 6946–6951.

Clark, J. L. & Fuller, J. L. (1976) *European Journal of Biochemistry*, **67**, 303–314.

Coleman, C. S., Stanley, B. A., & Pegg, A. E. (1993) *Journal of Biological Chemistry*, **288**, 24572–24579.

Coleman, C. S., Stanley, B. A., Viswanath, R., & Pegg, A. E. (1994) *Journal of Biological Chemistry*, **269**, 3155–3158.

Cox, D. R., Trouillot, T., Ashley, P. L., Brabant, M., & Coffino, P. (1988) *Cytogenetics and Cell Genetics*, **48**, 92–94.

Daghigh, F., Fukuto, J. M., & Ash, D. E. (1994) *Biochemical and Biophysical Research Communications*, **202**, 174–180.

Danzin, C. & Casara, P. (1984) *FEBS Letters*, **174**, 275–278.

Danzin, C., Jung, M. J., Metcalf, B. W., Grove, J., & Casara, P. (1979) *Biochemical Pharmacology*, **28**, 627–631.

Danzin, C. & Persson, L. (1987) *European Journal of Biochemistry*, **166**, 45–48.

Davis, R. H., Morris, D. R., & Coffino, P. (1992) *Microbiological Reviews*, **56**, 280–190.

Donato, N. J., Ware, C. R., & Byus, C. V. (1986) *Biochemica et Biophysica Acta*, **884**, 370–382.

Dorn, A., Bernstein, H., Müller, M., Ziegler, M., Järvinen, M., & Pajunen, A. (1986) *European Journal of Cell Biology*, **41**, 127–129.

Eisenberg, L. M. & Jänne, O. A. (1989) *Nucleic Acids Research*, **17**, 2359.

Elias, S., Bercovich, B., Kahana, C., Coffino, P., Fischer, M., Hilt, W., Wolf, D. H., & Ciechanover, A. (1995) *European Journal of Biochemistry*, **229**, 276–283.

Fitzgerald, M. C. & Flanagan, M. A. (1989) *DNA*, **8**, 623–634.

Flamigni, F., Marmiroli, S., Guarnieri, C., & Caldarera, C. M. (1989) *Biochemical and Biophysical Research Communications*, **163**, 1217–1222.

——— (1990) *Biochemical and Biophysical Research Communications*, **172**, 939–944.

Flamigni, F., Steffanelli, C., Guarnieri, C., & Caldarera, C. M. (1986) *Biochimica et Biophysica Acta*, **882**, 377–383.

Fong, W. F., Heller, J. S., & Canellakis, E. S. (1976) *Biochimica et Biophysica Acta*, **428**, 456–465.

Friedman, S. J., Halpern, K. V., & Canellakis, E. S. (1972) *Biochimica et Biophysica Acta*, **261**, 181–187.

Fujita, K., Matsufuji, S., Murakami, Y., & Hayashi, S. (1984) *Biochemical Journal*, **218**, 557–562.

Fujita, K., Murakami, Y., Kameji, T., Matsufuji, S., Utsonomiya, K., Kanamoto, R., & Hayashi, S. (1983) *Advances in Polyamine Research*, vol. 4, pp. 638–691. U. Bachrach, A. Kaye, & R. Chayen, Eds. New York: Raven Press.

Gaines, D. W., Friedman, L., & McCann, P. P. (1988) *Analytical Biochemistry*, **174**, 88–96.

Ghoda, L., Phillips, M. A., Bass, K. E., Wang, C. C., & Coffino, P. (1990) *Journal of Biological Chemistry*, **265**, 11823–11826.

Ghoda, L., Sidney, D., Macrae, M., & Coffino, P. (1992) *Molecular and Cellular Biology*, **12**, 2178–2185.

Ghoda, L., Wetters, T. V. D., Macrae, M., Ascherman, D., & Coffino, P. (1989) *Science*, **243**, 1493–1495.

Gilad, G. M. & Gilad, V. H. (1981) *Journal of Histochemistry and Cytochemistry*, **29**, 687–692.

Gilmour, S. K., Avdalovic, N., Madara, T., & O'Brien, T. G. (1985) *Journal of Biological Chemistry*, **260**, 16439–16444.

Ginty, D. D., Marlowe, M., Pekala, P. H., & Seidel, E. R. (1990) *American Journal of Physiology*, **258**, G454–G460.

Glass, J. R. & Gerner, E. W. (1987) *Journal of Cellular Physiology*, **130**, 133–141.

Glass, J. R., MacKrell, M., Duffy, J. J., & Gerner, E. W. (1987) *Biochemical Journal*, **245**, 127–132.

Goldberg, A. L. & Rock, K. L. (1992) *Nature*, **357**, 375–379.

Gray, A., Tait, A., & Bulfield, G. (1995) *Biochemical Journal*, **308**, 161–166.

Grens, A. & Scheffler, I. E. (1990) *Journal of Biological Chemistry*, **265**, 11810–11816.

Grens, A., Steglich, C., Pilz, R., & Scheffler, I. E. (1989) *Nucleic Acids Research*, **17**, 10497.

Guarnieri, C., Caldarera, C. R., Muscari, C., Flamigni, F., & Caldarera, C. M. (1982b) *Italian Journal of Biochemistry*, **31**, 404–411.

Guarnieri, C., Lugaresi, A., Muscari, C., Flamigni, F., & Caldarera, C. M. (1982a) *Italian Journal of Biochemistry*, **31**, 444–445.

Guarnieri, C., Piazza, G., Rizzuto, S., Flamigni, F., & Caldarera, C. M. (1985) *Italian Journal of Biochemistry*, **34**, 187–190.

Gupta, M. & Coffino, P. (1985) *Journal of Biological Chemistry*, **260**, 2941–2944.

Haddox, M. K. & Russell, D. H. (1981) *Biochemistry*, **20**, 6721-6729.

Hakovirta, H., Keiski, A., Toppari, J., Halmekytö, M., Alhonen, L., Jänne, J., & Parvinen, M. (1993) *Molecular Endocrinology*, **7**, 1430–1436.

Halmekytö, M., Alhonen, L., Alakuijala, L., & Jänne, J. (1993a) *Biochemical Journal*, **291**, 505–508.

Halmekytö, M., Alhonen, L., Wahlfors, J., Sinervirta, R., Jänne, O. A., & Jänne, J. (1991b) *Biochemical and Biophysical Research Communications*, **180**, 262–267.

Halmekytö, M., Hirvonen, A., Wahlfors, J., Alhonen, L., & Jänne, J. (1989) *Biochemical and Biophysical Research Communications*, **162**, 528–534.

Halmekytö, M., Hyttinen, J., Sinervirta, R., Leppanen, P., Jänne, J., & Alhonen, L. (1993b) *Biochemical Journal*, **292**, 927–932.

Halmekytö, M., Hyttinen, J., Sinervirta, R., Utriainen, M., Myöhänen, S., Voipio, H., Wahlfors, J., Syrjänen, S., Syrjänen, K., Alhonen, L., & Jänne, J. (1991a) *Journal of Biological Chemistry*, **266**, 19746–19751.

Halonen, T., Sivenius, J., Miettinen, R., Halmekytö, M., Kauppinen, R., Sinervirta, R., Alakuijala, L., Alhonen, L., MacDonald, E., Jänne, J., & Riekkinen, P. J. (1993) *European Journal of Neuroscience*, **5**, 1233–1239.

Harik, S. I., Hollenberg, M. D., & Snyder, S. H. (1974a) *Molecular Pharmacology*, **10**, 11–17.

——— (1974b) *Nature*, **249**, 250–251.

Hayashi, S. & Canellakis, E. S. (1989) *Ornithine Decarboxylase: Biology, Enzymology, and Molecular Genetics*, pp. 47–58. S. Hayashi, Ed. New York: Pergamon Press.

Heby, O. & Persson, L. (1990) *Trends in Biochemical Science*, **15**, 153–158.

Heller, J. S., Fong, W. F., & Canellakis, E. S. (1976) *Proceedings of the National Academy of Sciences of the United States of America*, **73**, 1858–1862.

Heller, J. S., Kyriakidis, D., Fong, W. F., & Canellakis, E. S. (1977) *European Journal of Biochemistry*, **81**, 545-550.

Henley, C. M., Salzer, T. A., Coker, N. J., Smith, G., & Haddox, M. K. (1995) *Hearing Research*, **84**, 99–111.

Heston, W. D. W., Watanabe, K. A., Pankiewicz, K. W., & Covey, D. F. (1987) *Biochemical Pharmacology*, **36**, 1849–1852.

Hibasami, H., Maekawa, S., Murata, T., & Nakashima, K. (1988) *Biochemical Pharmacology*, **37**, 4117–4120.

Hickok, N. J., Seppänen, P. J., Kontula, K. K., Jänne, P. A., Bardin, C. W., & Jänne, O. A. (1986) *Proceedings of the National Academy of Sciences of the United States of America*, **83**, 594–598.

Hirvonen, A., Immonen, T., Leinonen, P., Alhonen–Hongisto, L., Jänne, O. A., & Jänne, J. (1988) *Biochimica et Biophysica Acta*, **950**.

Hogan, B., Shields, R., & Curtis, D. (1974) *Cell*, **2**, 229–233.

Holm, I., Persson, L., Stjernborg, L., Thorsson, L., & Heby, O. (1989) *Biochemical Journal*, **258**, 343–350.

Hölttä, E. (1975) *Biochimica et Biophysica Acta*, **399**, 420–427.

Hölttä, E., Hirvonen, A., Wahlfors, J., Alhonen, L., Jänne, J., & Kallio, A. (1989) *Gene*, **83**, 125–135.

Hölttä, E. & Pohjanpelto, P. (1982) *Biochimica et Biophysica Acta*, **721**, 321–327.

——— (1986) *Journal of Biological Chemistry*, **261**, 9502–9508.

Hölttä, E., Sistonen, L., & Alitalo, K. (1988) *Journal of Biological Chemistry*, **263**, 4500–4507.

Hsieh, J., Denning, M. F., Heidel, S. M., & Verma, A. K. (1990) *Cancer Research*, **50**, 2239–2244.

Hu, Y. & Brosnan, M. E. (1987) *Archives of Biochemistry and Biophysics*, **254**, 637–641.

Hyttinen, J., Halmekytö, M., Alhonen, L., & Jänne, J. (1991) *Biochemical Journal*, **278**, 871–874.

Hyvönen, T., Alakuijala, L., Andersson, L., Khomutov, A. R., Khomutov, R. M., & Eloranta, T. O. (1988) *Journal of Biological Chemistry*, **263**, 11138–11144.

Hyvönen, T., Keinänen, T. A., Khomutov, A. R., Khomutov, R. M., & Eloranta, T. O. (1992) *Journal of Chromatography*, **574**, 17–21.

Hyvönen, T., Khomutov, A. R., Khomutov, R. M., Lapinjoki, S., & Eloranta, T. A. (1990) *Journal of Biochemistry*, **107**, 817–820.

Ichiba, T., Matsufuji, S., Miyazaki, Y., Murakami, Y., Tanaka, K., Ichihara, A., & Hayashi, S. (1994) *Biochemical and Biophysical Research Communications*, **200**, 1721–1727.

Ientile, R., Fabiano, C., & Trimarchi, G. R. (1991) *Neuroscience Research Communications*, **8**, 191–199.

Ikemoto, M., Tabata, M., Miyake, T., Kono, T., Mori, M., Totani, M., & Murachi, T. (1990) *Biochemical Journal*, **270**, 697–703.

Isomaa, V. V., Pajunen, A. E. I., Bardin, C. W., & Jänne, O. A. (1983) *Journal of Biological Chemistry*, **258**, 6735–6740.

Ito, K., Kashiwagi, K., Watanabe, S., Kameji, T., Hayashi, S., & Igarashi, K. (1990) *Journal of Biological Chemistry*, **265**, 13036–13041.

Jänne, J. & Hölttä, E. (1974) *Biochemical and Biophysical Research Communications*, **61**, 399–406.

Jänne, J. & Williams–Ashman, H. G. (1971) *Journal of Biological Chemistry*, **246**, 1725–1732.

Johannes, G. & Berger, F. G. (1992) *Journal of Biological Chemistry*, **267**, 10108–10115.

Johannes, G. J. & Berger, F. G. (1993) *Journal of Molecular Evolution*, **36**, 555–557.

Johnson, K., Bulfield, G., Tait, A., & Goddard, C. (1995) *Comparative Biochemistry and Physiology*, **110B**, 531–537.

Kaczmarek, L., Calabretta, B., Ferrari, S., & de Riel, J. K. (1987) *Journal of Cellular Physiology*, **132**.

Kahana, C. & Nathans, D. (1984) *Proceedings of the National Academy of Sciences of the United States of America*, **81**, 3645–3649.

——— (1985a) *Proceedings of the National Academy of Sciences of the United States of America*, **82**, 1673–1677.

——— (1985b) *Journal of Biological Chemistry*, **260**, 15390–15393.

Kallio, A., Pösö, H., & Jänne, J. (1977) *Biochimica et Biophysica Acta*, **479**, 345–353.

Kameji, T., Fujita, K., Noguchi, T., Takiguchi, M., Mori, M., Tatibana, M., & Hayashi, S. (1984) *European Journal of Biochemistry*, **144**, 35–39.

Kameji, T., Hayashi, S., Hoshino, K., Kakinuma, Y., & Igarashi, K. (1993) *Biochemical Journal*, **289**, 581–586.

Kameji, T., Murakami, Y., Fujita, K., & Hayashi, S. (1982) *Biochimica et Biophysica Acta*, **717**, 111–117.

Kameji, T., Murakami, Y., & Hayashi, S. (1979) *Journal of Biochemistry*, **86**, 191–197.

Kameji, T. & Pegg, A. E. (1987) *Journal of Biological Chemistry*, **262**, 2427–2430.

Kanamoto, R., Kameji, T., Iwashita, S., Igarashi, K., & Hayashi, S. (1993) *Journal of Biological Chemistry*, **268**, 9393–9399.

Kashiwagi, K., Ito, K., & Igarashi, K. (1991) *Biochemical and Biophysical Research Communications*, **178**, 815–822.

Katz, A. & Kahana, C. (1987) *Molecular and Cellular Biology*, **7**, 2641–2643.

—— (1988) *Journal of Biological Chemistry*, **263**, 7604–7609.

Kauppinen, R. A., Halmekytö, M., Alhonen, L., & Jänne, J. (1992) *Journal of Neurochemistry*, **58**, 831–836.

Kay, J. E. & Lindsay, V. J. (1973) *Biochemical Journal*, **132**, 791–796.

Kendrick, D. A., Danzin, C. & Kolb, M. (1989) *Journal of Medicinal Chemistry*, **32**, 170–173.

Khomutov, R. M., Hyvönen, T., Karvonen, E., Kauppinen, L., Paalanen, T., Paulin, L., Eloranta, T., Pajula, R., Andersson, L. C., & Pösö, H. (1985) *Biochemical and Biophysical Research Communications*, **130**, 596–602.

Kilpeläinen, P. T. & Hietala, O. A. (1994) *Biochemical Journal*, **300**, 577–582.

Kitani, T. & Fujisawa, H. (1983) *Journal of Biological Chemistry*, **258**, 235–239.

—— (1984) *Journal of Biological Chemistry*, **259**, 10036–10040.

—— (1985) *Biochemistry International*, **10**, 435–440.

—— (1988a) *Journal of Biochemistry*, **103**, 547–553.

—— (1988b) *Biochemical and Biophysical Research Communications*, **151**, 450–457.

—— (1989) *Biochimica et Biophysica Acta*, **991**, 44–49.

Kontula, K. K., Torkkeli, T. K., Bardin, C. W., & Jänne, O. A. (1984) *Proceedings of the National Academy of Sciences of the United States of America*, **81**, 731–735.

Kopitz, J., Rist, B., & Bohley, P. (1990) *Biochemical Journal*, **267**, 343–348.

Kranan, H. J. v., Zande, L. v. d., Kreijl, C. F. v., Bisschop, A., & Wieringa, B. (1987) *Gene*, **60**, 145–155.

Kumar, A. P., Mar, P. K., Zhao, B., Montgomery, R. L., Kang, D., & Butler, A. P. (1995) *Journal of Biological Chemistry*, **270**, 4341–4348.

Laitinen, P. H., Hietala, O. A., Pulkka, A. E., & Pajunen, A. E. I. (1986) *Biochemical Journal*, **236**, 613–616.

Laitinen, P. H., Huhtinen, R., Hietala, O. A., & Pajunen, A. E. I. (1985) *Journal of Neurochemistry*, **44**, 1885–1891.

Landy–Otsuka, F. & Scheffler, I. E. (1978) *Proceedings of the National Academy of Sciences of the United States of America*, **75**, 5001–5005.

Legraverend, C., Potter, A., Hölttä, E., Alitalo, K., & Andersson, L. C. (1989) *Experimental Cell Research*, **181**, 273–281.

Leinonen, P., Alhonen–Hongisto, L., Laine, R., Jänne, O. A., & Jänne, J. (1987) *FEBS Letters*, **215**, 68–72.

Li, R., Abrahamsen, M. S., Johnson, R. R., & Morris, D. R. (1994) *Journal of Biological Chemistry*, **269**, 7941–7949.

Li, X. & Coffino, P. (1994) *Molecular and Cellular Biology*, **14**, 87–92.

Linden, M., Anehus, S., Långström, E., Baldetorp, B., & Heby, O. (1985) *Journal of Cellular Physiology*, **125**, 273–276.

Lipsanen, V., Leinonen, P., Alhonen, L., & Jänne, J. (1988) *Blood*, **72**, 2042–2044.

Loetscher, P., Pratt, G., & Rechsteiner, M. (1991) *Journal of Biological Chemistry*, **266**, 11213–11220.

Lövkvist, E., Stjernborg, L., & Persson, L. (1993) *European Journal of Biochemistry*, **215**, 753–759.

Lu, L., Stanley, B. A., & Pegg, A. E. (1991) *Biochemical Journal*, **277**, 671–675.

Lundgren, D. W. (1992) *Journal of Biological Chemistry*, **267**, 6841–6847.

Macrae, M. & Coffino, P. (1987) *Molecular and Cellular Biology*, **7**, 564–567.

Mamont, P. S., Böhlen, P., McCann, P. P., Bey, P., Schuber, F., & Tardif, C. (1976) *Proceedings of the National Academy of Sciences of the United States of America*, **73**, 1626–1630.

Mamont, P. S., Duchesne, M., Joder–Ohlenbusch, A., & Grove, J. (1978) *Enzyme-Activated Irreversible Inhibitors*, pp. 43–54. N. Seiler, M. J. Jung, & J. Koch–Weser, Eds. Amsterdam: Elsevier/North-Holland Biomedical Press.

Mamroud–Kidron, E. & Kahana, C. (1994) *FEBS Letters*, **256**, 162–164.

Manzella, J. M. & Blackshear, P. J. (1990) *Journal of Biological Chemistry*, **265**, 11817–11822.

—— (1992) *Journal of Biological Chemistry*, **267**, 7077–7082.

Manzella, J. M., Rychlik W., Rhoads, R. E., Hershey, J. B., & Blackshear, P. J. (1991) *Journal of Biological Chemistry*, **266**, 2383–2389.

Marumo, M., Matsufuji, S., Murakami, Y., & Hayashi, S. (1988) *Biochemical Journal*, **249**, 907–910.

Matsufuji, S., Fujita, K., Kameji, T., Kanamoto, R., Murakami, Y., & Hayashi, S. (1984) *Journal of Biochemistry*, **96**, 1525–1530.

Matsufuji, S., Kanamoto, R., Murakami, Y., & Hayashi, S. (1990a) *Journal of Biochemistry*, **107**, 87–91.

Matsufuji, S., Miyazaki, Y., Kanamoto, R., Kameji, T., Murakami, Y., Baby, T. G., Fujita, K., Ohno, T., & Hayashi, S. (1990b) *Journal of Biochemistry*, **108**, 365–371.

McConlogue, L. & Coffino, P. (1983) *Journal of Biological Chemistry*, **258**, 12083–12086.

McConlogue, L., Dana, S. L., & Coffino, P. (1986) *Molecular and Cellular Biology*, **6**, 2865–2871.

McConlogue, L., Gupta, M., Wu, L., & Coffino, P. (1984) *Proceedings of the National Academy of Sciences of the United States of America*, **81**, 540–544.

Meggio, F., Flamigni, F., Guarnieri, C., & Pinna, L. A. (1987) *Biochimica et Biophysica Acta*, **929**, 114–116.

Messina, L., Arcidiacono, A., Spampinato, G., Malaguarnera, L., Berton, G., Kaczmarek, L., & Messina, A. (1990) *FEBS Letters*, **268**, 32–34.

Metcalf, B. W., Bey, P., Danzin, C., Jung, M. J., Cagara, P., & Ververt, J. P. (1978) *Journal of the American Chemical Society*, **100**, 2551–2553.

Mitchell, J. L. A., Campbell, H. A., & Carter, D. D. (1976) *FEBS Letters*, **62**, 33–37.

Mitchell, J. L. A. & Chen, H. J. (1990) *Biochimica et Biophysica Acta*, **1037**, 115–121.

Mitchell, J. L. A., Hoff, J. A., & Bareyal–Leyser, A. (1991) *Archives of Biochemistry and Biophysics*, **290**, 143–152.

Mitchell, J. L. A., Judd, G. G., Bareyal–Leyser, A., & Ling, S. Y. (1994) *Biochemical Journal*, **299**, 19–22.

Mitchell, J. L. A., Kurzeja, R. J., Marsh, J. F., & Diveley, R. R. (1992) *Biochemical and Biophysical Research Communications*, **187**, 443–447.

Mitchell, J. L. A., Mahan, D. W., McCann, P., & Qasba, P. (1985a) *Biochimica et Biophysica Acta*, **840**, 309–316.

Mitchell, J. L. A. & Mitchell, G. K. (1982) *Biochemical and Biophysical Research Communications*, **105**, 1189–1197.

Mitchell, J. L. A., Qasba, P., Stofko, R. E., & Franzen, M. A. (1985b) *Biochemical Journal*, **228**, 297–308.

Mitchell, J. L. A., Rynning, M. D., Chen, H. J., & Hicks, M. F. (1988) *Archives of Biochemistry and Biophysics*, **260**, 585–594.

Mitchell, J. L. A. & Sedory, M. J. (1974) *FEBS Letters*, **49**, 120–124.

Miyamoto, E., Murata, Y., Kawashima, S., & Ueda, M. (1992) *Journal of Pharmaceutical Pharmacology*, **44**, 269–271.

Miyamoto, K., Oka, T., Fuji, T., Yamaji, M., Minami, H., Nakabou, Y., & Hagihira, H. (1989) *Journal of Biochemistry*, **106**, 167–171.

Miyazaki, Y., Matsufuji, S., & Hayashi, S. (1992) *Gene*, **113**, 191–197.

Miyazaki, Y., Matsufuji, S., Murakami, Y., & Hayashi, S. (1993) *European Journal of Biochemistry*, **214**, 837–844.

Miyazaki, Y., Matsufuji, S., Murakami, Y., Shirahata, A., Samejima, K., & Hayashi, S. (1990) *International Symposium on Polyamines (Abstracts)*, Kyoto, Japan, p. 136.

Moyano, N., Frydman, J., Buldain, G., Ruiz, O., & Frydman, R. B. (1990) *Journal of Medicinal Chemistry*, **33**, 1969–1974.

Murakami, Y., Fujita, K., Kameji, T., & Hayashi, S. (1985) *Biochemical Journal*, **225**, 689–697.

Murakami, Y. & Hayashi, S. (1985) *Biochemical Journal*, **226**, 893–896.

Murakami, Y., Marumo, M., & Hayashi, S. (1988) *Biochemical Journal*, **254**, 367–372.

Murakami, Y., Matsufuji, S., Miyazaki, Y., & Hayashi, S. (1990a) *International Symposium on Polyamines (Abstracts)*, Kyoto, Japan, pp. 50–51.

——— (1990b) *International Symposium on Polyamines (Abstracts)*, Kyoto, Japan, pp. 196–197.

——— (1994) *Biochemical Journal*, **304**, 183–187.

Murakami, Y., Matsufuji, S., Nishiyama, M., & Hayashi, S. (1989a) *Biochemical Journal*, **259**, 839–845.

Murakami, Y., Matsufuji, S., Tanaka, K., Schihara, A., & Hayashi, S. (1993) *Biochemical Journal*, **295**, 305–308.

Murakami, Y., Nishiyama, M., & Hayashi, S. (1989b) *European Journal of Biochemistry*, **180**, 181–184.

Murakami, Y., Tanaka, K., Matsufuji, S., Miyazaki, Y., & Hayashi, S. (1992) *Biochemical Journal*, **283**, 661–664.

Myöhänen, S., Wahlfors, J., & Jänne, J. (1994) *DNA Sequence*, **5**, 1–8.

Ngo, T. T., Brillhart, K. L., Davis, R. H., Wong, R. C., Bovaird, J. H., Digangi, J. J., Ristow, J. L., Marsh, J. L., Phan, A. P. H., & Lenhoff, H. M. (1987) *Analytical Biochemistry*, **160**, 290–293.

Nikula, P., Alhonen–Hongisto, L., & Jänne, J. (1985) *Biochemical Journal*, **231**, 213–216.

Nishiyama, M., Matsufuji, S., Kanamoto, R., Murakami, Y., & Hayashi, S. (1989) *Journal of Immunoassay*, **10**, 19–35.

Nishiyama, M., Matsufuji, S., Kanamoto, R., Takano, M., Murakami, Y., & Hayashi, S. (1988) *Preparative Biochemistry*, **18**, 227–238.

O'Brien, T. G., Hietala, O., O'Donnell, K., & Holmes, M. (1987) *Proceedings of the National Academy of Sciences of the United States of America*, **84**, 8927–8931.

Olson, E. N. & Spizz, G. (1986) *Molecular and Cellular Biology*, **6**, 2792–2799.

Orr, G. R., Gould, S. J., Pegg, A. E., Seely, J. E., & Coward, J. K. (1984) *Bioorganic Chemistry*, **12**, 252–258.

Osterman, A. L., Lueder, D. V., Quick, M., Myers, D., Canagarajah, B. J., & Phillips, M. A. (1995) *Biochemistry*, **34**, 13431–13436.

Palvimo, J. J., Eisenberg, L. M., & Jänne, O. A. (1991) *Nucleic Acids Research*, **19**, 3921–3927.

Pegg, A. E. (1987) *Inhibition of Polyamine Metabolism*, pp. 107–119. P. P. McCann, A. E. Pegg, & A. Sjoerdsma, Eds. Orlando, Fla: Academic Press.

——— (1989) *Ornithine Decarboxylase: Biology, Enzymology and Molecular Genetics*, pp. 21–34. S. Hayashi, Ed. New York: Pergamon Press.

Pegg, A. E. & McGill, S. (1979) *Biochimica et Biophysica Acta*, **568**, 416–427.

Pegg, A. E., McGovern, K. A., & Wiest, L. (1987) *Biochemical Journal*, **241**, 305–307.

Pegg, A. E., Seely, J. E., Persson, L., Herlyn, M., Ponsell, K., & O'Brien, T. G. (1984) *Biochemical Journal*, **217**, 123–128.

Pegg, A. E., Shantz, L. M., & Coleman, C. S. (1994) *Biochemical Society Transactions*, **22**, 846–852.

Peña, A., Reddy, C. D., Wu, S., Hickock, N. J., Reddy, E. P., Yumet, G., & Soprano, D. R. (1993) *Journal of Biological Chemistry*, **268**, 27277–27285.

Peng, T. & Richards, J. F. (1988) *Biochemical and Biophysical Research Communications*, **153**, 135–141.

Persson, L. (1981a) *Acta Chemica Scandinavica B*, **35**, 451–459.

——— (1981b) *Acta Chemica Scandinavica B*, **35**, 737–738.

Persson, L., Holm. I., & Heby, O. (1986) *FEBS Letters*, **205**, 175–178.

Persson, L., Holm, I., & Heby, O. (1988) *Journal of Biological Chemistry*, **263**, 3528–3533.

Persson, L., Oredsson, S. M., Anehus, S., & Heby, O. (1985) *Biochemical and Biophysical Research Communications*, **131**, 239–245.

Persson, L., Rosengren, E., Sundler, F., & Uddman, R. (1983) *Methods in Enzymology*, **94**, 166–169.

Persson, L., Seely, J. E., & Pegg, A. E. (1984) *Biochemistry*, **23**, 3777–3783.

Pilz, R. B., Steglich, C., & Scheffler, I. E. (1990) *Journal of Biological Chemistry*, **265**, 8880–8886.

Pohjanpelto, P. & Hölttä, E. (1990) *Molecular and Cellular Biology*, **10**, 5814–5821.

Pohjanpelto, P., Hölttä, E., & Jänne, O. A. (1985a) *Molecular and Cellular Biology*, **5**, 1385–1390.

Pohjanpelto, P., Hölttä, E., Jänne, O. A., Knuutila, S., & Alitalo K. (1985b) *Journal of Biological Chemistry*, **260**, 8532–8537.

Pohjanpelto, P. & Knuutila, S. (1982) *Experimental Cell Research*, **141**, 333–339.

Pohjanpelto, P., Virtanen, I., & Hölttä, E. (1981) *Nature*, **293**, 475–477.

Porter, C. W., Regenass, V., & Bergeron, R. J. (1992) *Polyamines in the Gastrointestinal Tract, Falk Symposium 62*, pp. 301–322. R. H. Dowling, V. R. Fölsch, & C. Löser, Eds. Dordrect, The Netherlands: Kluwer Academic.

Porter, C. W., Stanek, J., Black, J., Vaughan, M., Ganis, B., & Pleshkewych, A. (1990b) *Cancer Research*, **50**, 1929–1935.

Poulin, R., Lu, L., Ackermann, B., Bey, P., & Pegg, A. E. (1992) *Journal of Biological Chemistry*, **267**, 150–158.

Poulin, R. & Pegg, A. E. (1990) *Journal of Biological Chemistry*, **265**, 4025–4032.

Pritchard, M. L., Seely, J. E., Pösö, H., Jefferson, L. S., & Pegg, A. E. (1981) *Biochemical and Biophysical Research Communications*, **100**, 1597–1603.

Rando, R. R. (1974) *Science* , **185**, 320–324.

Rechsteiner, M., Rogers, S., & Rote, K. (1987) *Trends in Biochemical Science*, **12**, 390–394.

Reczkowski, R. S. & Ash, D. E. (1994) *Archives of Biochemistry and Biophysics*, **312**, 31–37.

Reddy, S. G., Cochran, B. J., Worth, L. L., Knutson, V. P., & Haddox, M. K. (1994) *Analytical Biochemistry*, **218**, 149–156.

Regunathan, S., Feinstein, D. L., Raasch, W., & Reis, D. J. (1995) *NeuroReport* **6**, 1897–1900.

Relyea, N. & Rando, R. R. (1975) *Biochemical and Biophysical Research Communications*, **67**, 392–402.

Richards, J. F., Lit, K., Fuca, R., & Bourgeault, C. (1981) *Biochemical and Biophysical Research Communications*, **99**, 1461–1467.

Richards–Smith, B. A. & Elliott, R. W. (1992) *Mammalian Genome*, **2**, 215–232.

Rivett, A. J. (1993) *Biochemical Journal*, **291**, 1–10.

Rosen, C. F., Gajic, D., & Drucker, D. J. (1990a) *Cancer Research*, **50**, 2631–2635.

Rosen, C. F., Gajic, D., Jia, Q., & Drucker, D. J. (1990b) *Biochemical Journal*, **270**, 565–568.

Rosenberg–Hasson, Y., Bercovich, Z., & Kahana, C. (1991b) *Biochemical Journal*, **277**, 683–685.

Rosenberg–Hasson, Y., Strumpf, D., & Kahana, C. (1991a) *European Journal of Biochemistry*, **197**, 419–424.

Russell, D. H. (1981) *Polyamines in Biology and Medicine*, pp. 109–125. D. R. Morris & L. J. Marton, Eds. New York: Marcel Dekker.

———— (1983) *Proceedings of the National Academy of Sciences of the United States of America*, **80**, 1318–1321.

Sampson, D. A., Yan X. L., Clarke, S. D., & Harrison, S. C. (1994) *FASEB Abstracts*, **8**, A919.

Seely, J. E. & Pegg, A. E. (1983a) *Journal of Biological Chemistry*, **258**, 2496–2500.

———— (1983b) *Biochemical Journal*, **216**, 701–707.

Seely, J. E., Persson, L., Sertich, G. J., & Pegg, A. E. (1985) *Biochemical Journal*, **226**, 577–586.

Seely, J. E., Pösö, H., & Pegg, A. E. (1982a) *Biochemistry*, **21**, 3394–3399.

———— (1982b) *Biochemical Journal*, **206**, 311–318.

———— (1982c) *Journal of Biological Chemistry*, **257**, 7549–7553.

Seiler, N. (1991) *Progress in Drug Research*, **37**, 107–159.

Shayovits, A. & Bachrach, U. (1994) *Journal of Histochemistry and Cytochemistry*, **42**, 607–611.

Skinner, W. A. & Johansson, J. G. (1972) *Journal of Medicinal Chemistry*, **15**, 427–428.

Slotkin, T. A. & Bartolome, J. (1983) *Methods in Enzymology*, **103**, 590–603.

Smith, D. M., Thomas, N. R., & Gani, D. (1991) *Experientia*, **47**, 1104–1118.

Solano, F., Peñafiel, R., Solano, M. E., & Lozano, J. A. (1985) *FEBS Letters*, **190**, 324–328.

Solano, F., Peñafiel, R., Solano, J. A., & Lozano, J. A. (1988) *International Journal of Biochemistry*, **20**, 463–470.

Stadtman, E. R. (1990) *Biochemistry*, **29**, 6323–6331.

Stanek, J., Frei, J., Mett, H., Schneider, P., & Regenass, U. (1992) *Journal of Medicinal Chemistry*, **35**, 1339–1344.

Steglich, C. & Scheffler, I. E. (1982) *Journal of Biological Chemistry*, **257**, 4603–4609.

Stimac, E. & Morris, D. R. (1987) *Journal of Cellular Physiology*, **133**, 590–594.

Stjernborg, L., Heby, O., Holm, I., & Persson, L. (1991) *Biochimica et Biophysica Acta*, **1090**, 188–194.

Suzuki, T., He, Y., Kashiwagi, K., Murakami, Y., Hayashi, S., & Igarashi, K. (1994) *Proceedings of the National Academy of Sciences of the United States of America*, **91**, 8930–8934.

Tipnis, U. R. & Haddox, M. K. (1990) *Cellular and Molecular Biology*, **36**, 275–289.

Tobias, K. & Kahana, C. (1993) *Biochemistry*, **32**, 5842–5847.

Tobias, K. E., Mamroud–Kidron, E., & Kahana, C. (1993) *European Journal of Biochemistry*, **218**, 245–250.

Tokunaga, F., Goto, T., Koide, T., Murakami, Y., Hayashi, S., Tamura, T., Tanaka, K., & Ichihara, A. (1994) *Journal of Biological Chemistry*, **269**, 17382–17385.

Tome, M. E., Fiser, S. M., & Gerner, E. W. (1994) *Journal of Cellular Physiology*, **158**, 267–244.

Tsirka, S. E., Turck, C. W., & Coffino, P. (1993) *Biochemical Journal*, **293**, 289–295.

Tyagi, A. K., Tabor, H., & Tabor, C. W. (1982) *Biochemical and Biophysical Research Communications*, **109**, 533–540.

Urata, M., Suzuki, N., & Hosoya, T. (1987) *Biochemical Journal*, **241**, 169–174.

van Steeg, H., van Oostrom, C. T. M., Hodemaekers, H. M., Peters, L., & Thomas, A. A. M. (1991) *Biochemical Journal*, **274**, 521–526.

van Steeg, H., van Oostrom, C. T. M., Hodemaekers, H. M., & van Kreyl, C. F. (1990) *Gene*, **93**, 249–256.

van Steeg, H., van Oostrom, C. T. M., Martens, J. W. M., van Kreyl, C. F., Schepens, J., & Wieringa, B. (1989) *Nucleic Acids Research*, **17**, 8855–8856.

Viguera, E., Trelles, O., Urdiales, J. L., Matés, J. M., & Sánchez–Jiménez, F. (1994) *Trends in Biochemical Sciences*, **19**, 318–319.

Wahlfors, J. (1991) *Biochemical Journal*, **279**, 435–440.

Wallon, U. M., Persson, L., & Heby, O. (1990) *FEBS Letters*, **268**, 161–164.

———— (1995) *Molecular and Cellular Biochemistry*, **146**, 39–44.

Weber, L. W. D. (1987) *Experientia*, **43**, 176–178.

Wen, L., Huang, J., & Blackshear, P. J. (1989) *Journal of Biological Chemistry*, **364**, 9016–9021.

Wetters, T. D., Brabant, M., & Coffino, P. (1989a) *Nucleic Acids Research*, **17**, 9843–9860.

Wetters, T. D., Macrae, M., Brabant, M., Sittler, A., & Coffino, P. (1989b) *Molecular and Cellular Biology*, **9**, 5484–5490.

White, M. W., Kameji, T., Pegg, A. E., & Morris, D. R. (1987) *European Journal of Biochemistry*, **170**, 87–92.

Worth, L. L., Cochran, B. J., & Haddox, M. K. (1994) *Cancer Research*, **54**, 3967–3970.

Zagon, I. S., McLaughlin, P. J., Seely, J. E., Hoeksema, G. W., & Pegg, A. E. (1984) *Cell and Tissue Research*, **235**, 371–377.

CHAPTER 14

Mammalian Biosynthetic Enzymes: Part II

S-Adenosylmethionine (AdoMet) was discovered by G. L. Cantoni in 1953 in extracts of mammalian tissues. Within a few years studies with yeast, *Neurospora* and other fungi, and bacteria demonstrated that methionine supplied an aminopropyl moiety for the synthesis of spermidine from putrescine. In 1958 Tabor et al. reported the replacement of methionine by *S*-adenosylmethionine (AdoMet) in spermidine synthesis in *Escherichia coli* extracts. Their formulation of the reaction sequence included the following:

$$\text{L-methionine} \xrightarrow[\text{①}]{\overset{\text{Mg}^{2+} +}{\text{ATP}}} \text{AdoMet} \xrightarrow{\text{②}} CO_2 + \text{dAdoMet}$$

$$\underset{\text{AdoMet}}{\text{decarboxylated}} \xrightarrow[\text{③}]{\text{putrescine}} \underset{\text{methylthioadenosine}}{\text{spermidine} +}$$

The difficulties of labeling and quantifying AdoMet and the polyamines slowed the work on these reactions. Furthermore, tissue culture techniques had not yet been widely adopted. A purification and characterization of the AdoMet decarboxylase (AdoMetDC) of *E. coli* was carried out by R. B. Wickner et al. (1970); these workers also demonstrated that the enzyme contained pyruvate, and not pyridoxal phosphate (PLP), as a prosthetic group (see Chap. 9). Pyruvate was found in 1968 at the active site of a bacterial histidine decarboxylase (Snell, 1977) wherebas in the mammal histidine decarboxylase is a PLP enzyme.

The presence of spermine and spermidine in human seminal fluid had been discussed and their accumulation in particular lobes of the rat prostate was studied by Rhodes and Williams–Ashman (1964). The status of this work and its relation to various facets of reproductive physiology in this period were summarized by Williams–Ashman and Lockwood (1970). Knowledge of the chemistry and biology of AdoMet had also expanded markedly (Shapiro & Schlenk, 1965). The bio-

synthesis of methionine in bacteria and yeast and the accumulation and compartmentation of AdoMet in yeast were discussed in earlier chapters of this book. The mammalian enzyme for the decarboxylation of AdoMet was sought and found in extracts of rat ventral prostate and liver. In contrast to the requirement for Mg^{2+} for the *E. coli* enzyme, the prostatic enzyme was markedly stimulated by putrescine and inhibited by carbonyl reagents (Pegg & Williams–Ashman, 1968). The decarboxylase was soon separated from a spermidine synthase (Jänne & Williams–Ashman, 1971; Jänne et al., 1971).

The active enzymic synthesis of spermine, as well as of spermidine, was demonstrated in rat prostate (Pegg & Williams–Ashman, 1970) and these activities were separated in studies of extracts of rat brain (Raina & Hannonen, 1971). The study of a propylamine transferase, spermidine synthase, in crude extracts was carried out by the activation of the AdoMetDC by putrescine and acceptance of the propylamine by this diamine. Efforts to estimate spermine synthase in crude extracts occasionally used the acceptor spermidine as the activator of the DC. The decarboxylated AdoMet (dAdoMet) was difficult to obtain and infrequently served as the aminopropyl donor. However, AdoMet with an appropriate radioactive label (e.g., containing [2-^{14}C]-methionine) could be prepared.

A number of growth systems found to increase spermidine and RNA were also shown to increase their content of the putrescine-dependent AdoMetDC. A twofold stimulation of the enzyme activity in immature rat uterus occurred after application of estrogen (Kaye et al., 1971). Russell and Lombardini (1971) found a several-fold increase in regenerating liver of young rats (but not in older rats) after 2 days, as well as in developing chick embryos. In contrast to ornithine DC (ODC), the AdoMetDC responded more slowly and subsequently fell less precipitously.

Partial hepatectomy evoked parallel increases of

AdoMetDC, spermidine synthase, and spermine synthase over the initial 50 h of regeneration (Hannonen et al., 1972a). The ratios of increase over the initial liver tissue levels were of the order of 3, 2, and 2. Only spermidine synthase continued to increase slowly, and it fell slowly only after 100 h. The other two fell to the control levels, with a precipitous fall in AdoMetDC between 50 and 100 h. AdoMetDC proved to have a half-life of 35 min compared to 21 min for ODC in the same systems, whereas spermidine synthase was completely stable for 140 min.

A dramatic increase in the AdoMetDC occurred in rat brain beginning some 10 days after birth (Schmidt & Cantoni, 1973). Rat liver, which doubled its nucleic acid and polyamine (spermidine and spermine) contents in a period from 20 to 35 days after birth, increased its content of AdoMetDC very little and slowly (Brosnan et al., 1978). Rat prostate and seminal vesicle in a castrated animal reacted within hours to testosterone with increases of ODC and the AdoMetDC.

In cellular studies concanavalin A-induced lymphocytes were found to increase their AdoMetDC some 20–25-fold, but in the presence of the inhibitor methylglyoxal-bis(guanylhydrazone) (MGBG, 20 μM) the induced lymphocytes continued to increase the enzyme some 2500-fold (Fillingame & Morris, 1973). The short half-life of this enzyme (40 min) had been increased by MGBG to >20 h whereas ODC was not affected by this inhibitor. The inhibitory effects of MGBG on the accumulation of spermidine and spermine are presented in Chapter 11 (Fig. 11.8) whereas the accumulation of putrescine is markedly stimulated. The early detection of MGBG as a potent inhibitor of the enzyme and as a compound that exhibited significant antitumor effects (Mihich, 1963) enticed many workers into studies of the inhibitor. The finding that MGBG could serve effectively in affinity chromatography of the enzyme contributed markedly to its purification. Also, the induction of the enzyme in the presence of MGBG increased the enzyme at least 20-fold in the rat liver as an aid for its subsequent purification.

Methionine and AdoMet

The central role of AdoMetDC is a function of the availability of methionine and the crucial intermediate, AdoMet. Mammalian and other cells have also developed mechanisms for the recapture of critical moieties (e.g., adenine, methylthioribose) produced in the synthesis of the polyamines; indeed this recapturing or recycling of metabolites has become crucial in the maintenance of the function, proliferation, and viability of some growing cells. The significance of this topic was enhanced by observations indicating that many types of tumor cells were selectively unable to synthesize an essential level of utilizable methionine.

Methionine must be made available for the synthesis of spermidine and spermine and for protein synthesis in normal mammalian growth and development. It is converted to AdoMet and participates in numerous AdoMet-dependent transmethylations. The product, adenosylhomocysteine (AdoHcy) releases adenosine (Ado) and homocysteine (Hcy) that is introduced into cystathionine and whose sulfur then becomes the sulfhydryl of cysteine. The shift of sulfur from homocysteine to cysteine requires two PLP enzymes, and a vitamin B_6 deficiency in the rat leads to an increase of homocysteine and concurrent decrease of AdoMet (She et al., 1994). Cysteine is found in many proteins (e.g., ODC), tends to be quite low in the free state in cells (10–100 μM levels), and may be stored in the more stable, less toxic glutathione (Meister, 1991). These reactions are depicted in Figure 14.1; this chapter will be concerned mainly with the reactions of section 1 of the figure, which involves the synthesis of the polyamines and the recycling of methionine. However, the entire network bears on the availability of the amino acid (Finkelstein, 1990).

Although methionine is an essential requirement of some bacteria, most organisms given a utilizable and adequate source of sulfur can synthesize much of the required amino acid. Nevertheless, the routes of biosynthesis, their regulation, the intermediates, and their po-

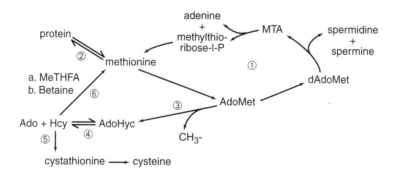

FIG. 14.1 Metabolism relating to the biosynthesis, utilization, degradation, and recycling of methionine.

tential toxicities determine the availability of utilizable amino acid. In an animal, homocysteine can react with serine to form cystathionine, which is cleaved to cysteine. Thus, cystathionine is the key intermediate of transsulfuration.

$$
\begin{array}{ccccc}
& & & & \text{cysteine} \\
& & & & + \\
\text{homocysteine} & \longrightarrow & \text{cystathionine} & \longrightarrow & \alpha\text{–ketobutyrate} \\
+ & & \text{CH}_2\text{–S–CH}_2 & & + \\
\text{serine} & & | \qquad\quad | & & \text{NH}_3 \\
& & \text{CH}_2 \quad \text{HC-COOH} & & \\
& & | \qquad\quad | & & \\
& & \text{H}_2\text{NCH} \quad \text{NH}_2 & & \\
& & | & & \\
& & \text{COOH} & &
\end{array}
$$

In adult rats sulfur proceeds from homocysteine → cysteine. According to Baker (1987) methionine sulfur can be converted quantitatively to cysteine sulfur whereas the reverse reaction does not occur in higher organisms. A diet low in sulfate leads to an oxidation of methionine sulfur, a decrease in tissue AdoMet, as well as an increase in the methionine requirement of the animal. In the rat, dietary cystine can spare about 70% of the requirement for methionine, an effect first detected in 1941. Curiously a dietary supplement of cysteine to a balanced diet increased ODC activity selectively (Acuff & Smith, 1983), an effect which may relate to the —SH requirement of this enzyme and the possible role of cysteine in determining glutathione concentrations. A low sulfate diet fairly high in methionine evokes a deficit in ODC and putrescine in the rat liver but not in spermidine and spermine (Acuff & Smith, 1985). Thus, the rate of degradation of methionine appears to be important in some nutritional conditions (Livesey, 1984). The requirement for methionine is particularly vexing in the development of sheep who are fed on diets containing methionine-poor proteins and must produce a great deal of cystine in their wool (Benevenga et al., 1983).

The estimated daily amino acid requirement of methionine plus cystine of adult humans is about 13 mg/kg. The deficiency of methionine in many plants and plant proteins is well known and produces human nutritional disease (e.g., kwashiorkor) in many underdeveloped countries. Dietary studies have also revealed methionine toxicity, which has often been ascribed to the sequestration of ATP and purine in AdoMet, although this toxicity is not always reversed by adenine. Other products of methionine metabolism can also be toxic, for example, methylthioadenosine (MTA), or a relatively high concentration of AdoHcy.

The development of high-pressure liquid chromatography (HPLC) with quantitative detectors of fluorescence and ultraviolet absorption facilitated a relatively rapid separation and estimation of the major components of methionine metabolism in a tissue extract. Table 14.1 indicates the high levels of spermidine and spermine in particular tissues in contrast to the relatively low levels of methionine, AdoMet, and dAdoMet. Methyl transfer reactions may consume about 95% of the AdoMet formed and in the human polyamine biosynthesis consumes about 2–5% of the AdoMet pool (Griffith, 1987). These figures give some indication of the enormous turnover in that pool.

Studies of the range of metabolites derived from AdoMet in a rat liver extract have provided similar data (Eloranta & Kajander, 1984). A significant nonenzymatic degradation of the compound occurs in the pH 6.9–7.8 range. In a liver homogenate 99% of the AdoMet may be demethylated and the product AdoHcy is hydrolyzed to homocysteine and adenosine. A small amount (1%) is decarboxylated via AdoMetDC and the

TABLE 14.1 Concentrations of analogues of AdoMet (SAM), amino acids, and polyamines in organs of 24-h fasted rats.

	Tissue		
	Pancreas	Prostate	Liver
SAM[a]	32.5 ± 5.5	74.0 ± 8	33.5 ± 5
SAH[a]	7.0 ± 1.5	17.5 ± 5.5	56.0 ± 11.5
MTA[a]	0.9 ± 0.4	4.1 ± 3.9	2.9 ± 0.4
dc-SAM[a]	0.4 ± 0.3	2.1 ± 0.9	≤1.0
Trp[a]	17 ± 6.5	240 ± 75	41 ± 10
Tyr[b]	180 ± 38	930 ± 220	140 ± 30
Met[b]	79 ± 15	775 ± 190	62 ± 9
Put[b]	22.5 ± 11.5	225 ± 50	18 ± 1
Spd[b]	4550 ± 490	8890 ± 1530	1000 ± 130
Spm[b]	660 ± 120	5670 ± 890	1230 ± 110
Mean weight	1.035 ± 0.05	0.355 ± 0.040	8.6 ± 0.3

Adapted from Wagner et al. (1984). SAH, S-Adenosylhomocysteine; dc-SAM, decarboxylated AdoMet; trp, tryptophan; tyr, tyrosine; Met, methionine; Put, putrescine; Spd, spermidine; Spm, spermine. Results are expressed in nmol/g wet wt ± SD and the tissue weights in g ± SD ($n = 5$).

[a]Values obtained by UV detection at 254 nm.

[b]Values obtained by fluorescence detection.

TABLE 14.2 Distribution of enzymes of polyamine synthesis in rat tissues.

	Methionine adenosyl transferase	Ornithine decarboxylase	AdoMet decarboxylase	Spd synthetase	Sp synthetase
Pancreas	16740	7	289	37800	563
Prostate	4230	941	1170	18100	2740
Liver	168300	7	68	3220	260

Adapted from Raina et al. (1976). Values are in pmol/30 min/mg protein.

dAdoMet is used to synthesize spermidine and spermine. Thus, in an extract polyamine synthesis alone represents a minor component of the metabolism of AdoMet. Table 14.2 indicates the large difference in the liver's ability to synthesize AdoMet and the utilization of this metabolite by enzymes of polyamine biosynthesis.

This subject has been carefully reexamined in an in vivo situation. The fate in growing rats of ingested methionine in a casein-based diet has been studied with 2-^{14}C-labeled and/or methyl-^{3}H-labeled L-methionine (Eloranta et al., 1990b). The isotopic amino acid was found in AdoMet in skeletal muscle and both isotopes disappeared similarly, indicating a turnover of the entire methionine molecule within AdoMet. However, in the liver methyl ^{3}H disappeared from AdoMet far faster than ^{14}C, indicating a predominance of transmethylation in this tissue.

Two mechanisms of producing methyl groups for addition to homocysteine to form methionine have been demonstrated. In one of these a moiety at the oxidation level of formaldehyde, derived from serine, is transferred to tetrahydrofolate and reduced in situ to 5-methyltetrahydrofolate. This newly formed methyl group is added to homocysteine with the aid of methionine synthase. Methionine synthase contains cobalamin (vitamin B$_{12}$), which accepts the methyl group from methyltetrahydrofolate to form the methyl donor *methyl cobalamin*. In another path, methyl-rich compounds such as betaine can provide the methyl group to homocysteine via a methyl transferase, as indicated below:

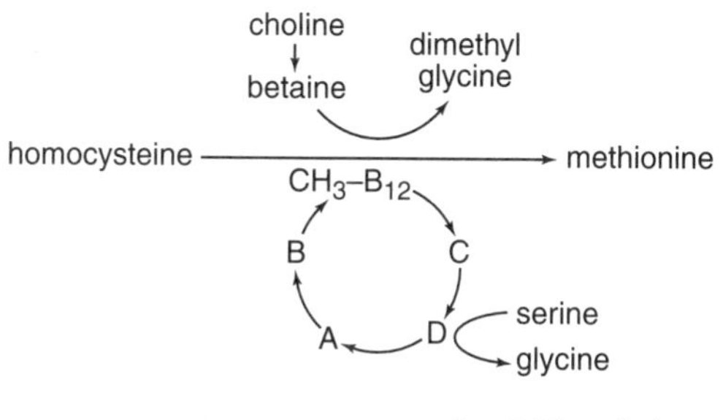

A = 5,10-methylene THFA
B = 5–methylTHFA
C = THFA
D = HCHO

The implications of this schema are that the availability of methionine via biosynthesis may be affected by many factors (Fiskerstrand et al., 1994). Methionine synthase is also stimulated significantly by spermidine and spermine (Kenyon et al., 1996). It is inhibited by nitric oxide, which also decreases levels of all polyamines in human hepatocytes (Anderson et al., 1994).

The amino acid is incorporated largely into AdoMet in a rat liver that contains more than 3 times as much methionine adenosyl transferase as does the kidney, the next most active organ (Finkelstein, 1990). Little increase is found in the content in the liver of methionine, AdoMet, and AdoHyc until dietary methionine is in excess of 1.5%; at 3% these components increase twofold, fourfold, and 20-fold, respectively, with significant increases in transsulfuration and some methyl transferases.

An increase in tissue methionine appears to determine the increase in AdoMet, which in turn is diminished by an increase in the methylating enzyme converting glycine to sarcosine (Eloranta, 1977). Nevertheless, AdoHyc is found to be an inhibitor of certain methylations. Finkelstein (1990) has discussed the regulation of the various components of methionine metabolism and effects of AdoMet and AdoHyc on enzymes of biosynthesis, methyl transfer, and transsulfuration. For example AdoMet inhibits the synthesis of methyl tetrahydrofolate (THFA) and the betaine-homocysteine methyltransferase and can assist in the activation of cystathionine synthase and cystathionase. Nevertheless, the concentration of AdoMet does not appear to be important in the regulation of polyamine metabolism (Eloranta & Raina, 1977). Actually in rats controlled by dietary, hormonal, and other factors, a methionine deficiency did not significantly decrease the AdoMet concentration in the liver and evoked a marked increase of spermidine content (ca. 70%).

An inhibitor of methionine adenosyltransferase depleted AdoMet to undetectable levels, resulting in inhibitions of nucleic acid methylations of 44–87%, two- to fourfold increases of ODC and AdoMetDC and only a minimal decrease of spermidine (25%) and spermine (12%) (Kramer et al., 1987). The antiproliferative effect of inhibitors of this enzyme may be related to the inhibition of specific transmethylations (Porter et al., 1984) and can be exacerbated by a combination with inhibitors of AdoHcy hydrolase, which permits a marked increase in the ratio of the inhibiting AdoHyc to the methyl donor, AdoMet (Kramer et al., 1990). These effects are not reversed by exogenous spermidine, and the cellular spermidine and spermine pools are not greatly depleted.

Toxic agents selenite and ethionine

Selenite is incorporated into the amino acid selenomethionine (SeMet) in fungi and plants and thence into the SeMet analogue of AdoMet. Although SeMet and AdoSeMet are cytotoxic at high concentrations, these products do not block transmethylation or the synthesis of nucleic acids or proteins. AdoSeMet is decarboxylated and this product can participate in the synthesis of spermidine and spermine (Kajander et al., 1990).

Biochemical effects of ethionine have been described by Alix (1982). The amino acid, in which the methyl group of methionine is replaced by an ethyl group, is incorporated into proteins, albeit at a slower rate than methionine, and a few of these proteins may be as active as methionine-containing counterparts. Ethionine is incorporated into S-adenosylethionine (AdoEt) in eucaryotes but not in procaryotes. AdoEt is a poor substrate for methyl transferases and will accumulate in the liver when ethionine is fed; this results in a trapping of purine and ATP. However, AdoEt does participate in the ethylation of some liver nuclear histones and some bases of tRNA (Kuchino et al., 1978). AdoEt is also an inhibitor of DNA synthesis, RNA processing, and some protein methyl transferases. Both AdoEt and AdoSeEt are very poor substrates for AdoMetDC and the former proved to be a very poor intermediate for the synthesis of spermidine (Pegg, 1969). Raina and colleagues had shown earlier that administration of ethionine leads to a decrease in spermidine content in several organisms. Active interest in ethionine has been maintained as a result of the finding that the compound produces cancers in the liver (Hoffman, 1985).

Inhibitors of methionine adenosyltransferase

The importance of methionine and AdoMet soon led to efforts to synthesize inhibitors of synthesis of AdoMet. Studies of the specificity of the mammalian methionine adenosyltransferase revealed that L-2-amino-trans-4-hexenoic acid was an active inhibitor as was the acetylenic L-2-amino-4-hexynoic acid. Unexpectedly 1-aminocyclopentanecarboxylic acid (cycloleucine) was also an efficient inhibitor (Lombardini & Talalay, 1973). In rodents these inhibitors cause marked falls in AdoMet (20–40%), increases in free methionine specifically, and increases of the transferase in the liver. Some rat hepatomas were found to take up cycloleucine rapidly and relatively selectively. The lead of cycloleucine facilitated a systematic exploration of additional analogues of the latter, which revealed an increased inhibitory activity of some bicyclo derivatives (Sufrin et al., 1981). Another effective inhibitor was the L-2-amino-4-

methoxy-cis-but-3-enoic acid (Sufrin et al., 1982); this has been used in various in vivo studies (see Fig. 9.2).

AdoMet synthetase

Studies of the various forms of the enzyme have detected three isozymes, defined by their Michaelis constant (K_m) values as high, intermediate, and low, of which the low K_m type is enhanced in the rapidly growing liver (van Faassen & Berger, 1990). Thus, the differences in the activities and necessarily in the assay conditions for the isozymes creates difficulties for a comparison of the amounts of activity of various cells and tissues (Åkerman et al., 1991). Rat liver activities are largely explained by the presence of two forms differing in molecular weight (MW); the relative amount of the forms varies at various stages of the activity of the liver in development, carcinogenesis, and response to methionine. The two forms, α and β, are the dimer and tetramer of identical subunits and the tetramer may be converted to the dimer by 1.4 M LiBr (Cabrero & Alemany, 1988). The third form (γ) of the enzyme is found in all mammalian tissues, although at low concentrations in the liver. Antibodies to these enzymes have revealed cross-reactivities, as well as relationships to bacterial and yeast enzymes (Kotb et al., 1990).

Deficiency of methionine adenosyltransferase leading to hypermethioninemia has been detected in the livers of some children, and some hereditary deficiencies of the enzyme have been found in rats (Shimizu et al., 1990). However, these defective activities have not been analyzed with immunological tools or those of molecular biology. The cDNA for the liver enzyme has been demonstrated to possess a high degree of sequence similarity (85%) to the cDNA coding for the kidney form, that is, the γ form of the enzyme (Horikawa & Tsukada, 1992). These nucleates indicated a specific molecular weight of 43.7 kDa for the human liver enzyme (Alvarez et al., 1993). The mouse liver enzyme shares a 96% amino acid sequence identity with the human liver enzyme (Sakata et al., 1993).

Transport of AdoMet

AdoMet is not only transported actively in yeast, where it accumulates in vacuoles, but also is taken up by various mammalian cells. This uptake has been demonstrated in the rat liver, whose transport of methionine and MTA occurs at a rate about 3 times that of AdoMet. The apparent K_m for AdoMet transport in the liver (59.5 μM) approaches the concentration of the metabolite in the rat liver and human blood (Zappia et al., 1978). Various labeled carbons of the methionine of injected AdoMet were metabolized actively to CO_2 and appeared in metabolites like creatinine in urine.

The penetrability of this sulfonium compound and its stability in an acid milieu, gastric juice, has permitted its trial as an orally administered agent in several clinical experiments. Orally administered AdoMet has been reported to be an antiinflammatory agent (Stramentinoli,

1987) and effective in the treatment of some symptoms of osteoarthritis (Schumacher, 1987). In conjunction with its significant neuropharmacological actions, it has also been described as an antidepressant (Baldessarini, 1987). Cantoni et al. (1989) proposed a hypothesis on its mode of action.

AdoMetDC

Assay and purification

The putrescine-dependent enzymatic activity and protein has been found almost entirely in the cytosol of various tissues, for example, rat ventral prostate (Gritli–Linde et al., 1995; Pegg & Williams–Ashman, 1968). As described earlier, carboxyl-labeled (^{14}C)AdoMet is decarboxylated to $^{14}CO_2$ and dAdoMet. The reaction with the enzyme from mammalian tissues is markedly stimulated by putrescine; indeed putrescine is essential to the reaction with the enzyme from yeast.

Some workers have reported a significant liberation of $^{14}CO_2$ from carboxyl-labeled AdoMet by membrane fragments of nuclei isolated in the presence of detergent. Two groups have demonstrated the conversion of AdoMet by such membrane fragments to amino acids, to homoserine particularly and then to α-ketobutyric acid, which is decarboxylated. Eloranta and Raina (1978) have attributed the generation of homoserine to a sequence beginning with the demethylation of AdoMet. Wilson et al. (1979) considered the mechanism to begin mainly with a cleavage of AdoMet to homoserine lactone plus 5′-methylthioadenosine. An enzyme for the cleavage of AdoMet in this way is known in bacteria, yeast, and pig liver. These routes are presented below.

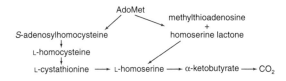

Nonenzymic decarboxylations of AdoMet (described in Chap. 9) result from the production of H_2O_2 [e.g., via diamine oxidases (DAOs)] or the simultaneous presence of Cu^{2+} or Mn^{2+} and compounds such as pyridoxal. Thus, quantification of AdoMetDC requires data on the products of putrescine-dependent reactions and the nature and stoichiometry of the reaction. This is difficult in crude or even partially purified extracts in which the presence of putrescine and spermidine synthase permits the utilization of the initial product dAdoMet in the secondary formation of spermidine. Equivalent production of CO_2 and spermidine was found in such a prostatic enzyme preparation by Pegg and Williams–Ashman (1968).

The existence of dAdoMet as a stoichiometric reaction product could not occur until the DC was separated from the spermidine synthase, permitting the demonstration of the apparently stoichiometric production of CO_2 and a substance chromatographically similar to dAdoMet. The quantitation of dAdoMet is not a simple matter, because a good standard is not easily obtained. dAdoMet synthesized chemically is a racemic compound while the biosynthetic dAdoMet must be separated completely from AdoMet. The chromatographic preparation and estimation of these substances have been described by Zappia et al. (1983) and Pegg and Bennett (1983) in procedures and reactions summarized in Chapter 9.

An extract of rat liver contains a considerable excess of the spermidine synthase up to 50-fold (Hannonen et al., 1972b). This not only accounts for the efficiency of the crude extract in converting AdoMet and putrescine to spermidine, but also establishes the DC as the rate-limiting enzyme for spermidine synthesis in this tissue. This relation is general in eucaryotic organisms. Obviously the activity of the DC can be optimized by adjusting substrate concentrations and activators such as putrescine, but these optimal conditions evoke activities that are always markedly less than the activity of spermidine synthase. On the other hand the apparent activity of spermine synthase was half that of the AdoMetDC and only 1% that of spermidine synthase.

Zappia et al. (1972) studied the relatively stable enzyme of human prostatic tissue and determined kinetic constants, with 4×10^{-5} M and 1.3×10^{-4} M K_m values of AdoMet and putrescine, respectively. Putrescine activation was pH dependent and was minimal at the pH optimum (pH 7.4–7.5) of the enzyme. The enzyme was inactive on many analogues other than AdoMet, including AdoHcy, S-inosylmethionine, and S-methylmethionine.

In the organs of the European sea bass, putrescine does not activate the enzyme at the optimal body temperature for this organism, 25°C, but does stimulate at higher temperatures. A high ODC content and low AdoMetDC activity appears to account for the polyamine distribution of high putrescine and low spermidine and spermine in the organs of the sea bass (Corti et al., 1987).

The mammalian enzyme was inhibited by sulfhydryl and carbonyl reagents, of which the latter were similarly inhibitory to PLP enzymes (Feldman et al., 1972). The discovery of pyruvate and the absence of PLP at the active site of the E. coli enzyme by Wickner et al. (1969) compelled a more careful appraisal of the mammalian enzyme. Using a 1000-fold purified enzyme from rat liver, estimated to have a 68,000 MW, Hannonen (1975) demonstrated an inhibition of the enzyme by reduction with sodium borohydride, implying the presence of a carbonyl group at the active site. This inhibition was largely prevented by a prior incubation with putrescine, suggesting a reactivity of putrescine with this active site. Spermidine and spermine did not provide similar protection. Pyridoxine deficiency in rats or chickens did not result in a decrease of the enzyme (Grillo & Bedino, 1977; Hannonen, 1976). Pegg (1977) isolated lactate after $NaBH_4$ reduction and hydrolysis of the purified enzyme, indicating that PLP was not the coenzyme and that covalently bound pyruvate was, as in the *E. coli* enzyme.

The enzyme was purified to apparent homogeneity from mouse mammary gland and liver using affinity chromatography on an MGBG-bound gel among the steps (Sakai et al., 1979). Various stabilizing agents include putrescine, PLP, and deoxycholate. Gel filtration and sucrose density centrifugation sizing techniques indicated MWs in the range of 68,000–74,000; electrophoresis in denaturing conditions revealed an apparent subunit of 32,000. Stimulation of the purified enzyme was maximal at 5 μM putrescine and Mg^{2+} and PLP were without effect. Millimolar spermidine was slightly stimulatory; spermine was markedly inhibitory with an inhibition constant (K_i) of 5×10^{-4} M and decreased the maximum velocity (V_{max}) of the enzyme. Rabbit antibody inactivated the enzyme and precipitated the protein quantitatively; the liver and mammary gland enzymes appeared identical in all tests.

Thus, the natural amines putrescine and spermine could be significant regulators of the enzyme once formed. In the mouse mammary gland, physiological concentrations of putrescine could be stimulatory. Nevertheless, an increase of this concentration in the tissue by exogenous addition of the diamine or cultivation in hypotonic media reduced the enzyme content whereas an ODC inhibitor, such as α-methylornithine, increased the amount of the enzyme (Sakai et al., 1980). Thus, putrescine appeared to act as a negative regulator.

Large amounts of the enzyme of significantly higher specific activity were also isolated in high yield from the livers of calves treated with MGBG (Seyfried et al., 1982). Rabbit antisera were then used to explore the production of the enzyme in resting or conA-stimulated bovine lymphocytes incubated with [^{3}H]-leucine. The mitogen stimulated an initial phase of enzyme synthesis about 10-fold, as detected by antibody. An apparent second step of enzyme increase was not due to synthesis but was rather an increase in the half-life of the protein from about 80 min at 6 h after mitogenic activation to about 170 min at 24 h.

A detailed method of isolating the enzyme from MGBG-stabilized rat liver and the stability and various properties of a product approaching that of the calf enzyme has been described by Pegg and Pösö (1983). Chapter 20 will show that several plant enzymes (e.g., those of Chinese cabbage and soybean) differ from the several known bacterial and mammalian enzymes and are unaffected by Mg^{2+} or putrescine.

In a study of the activation of a partially purified rat liver AdoMetDC by putrescine and by 2-substituted 1,4-butanediamines, putrescine appeared essentially uncompetitive with AdoMet. The K_m of the latter was 333 μM in the absence of putrescine, whereas the addition of putrescine to achieve maximal activation lowered the K_m of AdoMet to about 50–60 μM. The analogues 2-fluoroputrescine, 2,2-difluoroputrescine, 2-chloroputrescine, 2-methylputrescine, and 2-hydroxyputrescine also activated the enzyme, although much less potently. The analogues did increase the affinity of the enzyme for AdoMet and increased the V_{max} (Dezeure et al., 1989). The results indicated very specific structural requirements for the activation of the enzyme, and the concentrations at which putrescine activates (in excess of 5 μM) supports the hypothesis that the physiological concentration of the diamine can serve to modulate the activity of AdoMetDC in vivo. Similarly the inhibition of the enzyme by spermine is in accord with the concept of a feedback inhibition of the production of dAdoMet.

Later studies of the transcription of the cDNA for the AdoMetDC and the translation of the derived mRNA revealed the existence of a proenzyme whose cleavage to nonidentical subunits generated the essential pyruvoyl moiety. The cleavage reaction itself is stimulated by putrescine.

dAdoMet

The production of this intermediate in the synthesis of spermidine and spermine introduced the possibility of its activity as a substrate or an inhibitor in reactions associated with AdoMet itself. The product of methylation reactions, AdoHcy, is an effective inhibitor of methylation and the ratio of AdoHcy to AdoMet is believed to regulate methylation within cells. dAdoMet was tested for such roles in several methyl transfer reactions: with purified mammalian histamine methyltransferase, mammalian acetylserotonin methyltransferase, and yeast homocysteine methyltransferase (Zappia et al., 1969). It was essentially inactive as a donor and was slightly inhibitory in the first two methyl transfer reactions but served as a fairly good donor in the formation of methionine from homocysteine. The K_m was about 3 times higher for dAdoMet than for AdoMet.

It was found by a sensitive isotope dilution assay with spermidine synthase that dAdoMet in the rat liver (0.8–2.5 μmol/g) was about 2–4% of the concentration of AdoMet in this tissue (Hibasami et al., 1980a,b). Similar results were obtained in other tissues. Treatment of the animal with MGBG did not affect AdoMet concentration in the liver but reduced the concentration of dAdoMet 90% for 8 h. This returned to normal by 24 h. It was suggested that normal concentrations of dAdoMet would not affect methylations.

Nevertheless, dAdoMet [*S*-5′-deoxyadenosyl-(5′)-3-methylthiopropylamine] does inhibit AdoMetDC competitively with AdoMet with a K_i of 6.3 μM. Numerous analogues are also inhibitory, with the *S*-5′-deoxyadenosyl-(5′)-2-methylthioethylamine and the -2-ethylthioethylamine derivatives having even lower K_i values (Yamanoha & Samejima, 1980). The methyl derivative was also an inhibitor of spermidine synthase.

The inhibition by difluoromethylornithine (DFMO) of both ODC and the synthesis of putrescine can lead to a massive increase (500-fold) of dAdoMet in transformed mouse fibroblasts and in regenerating rat liver. The relative absence of putrescine results in a marked decrease in the formation of spermidine and the accumulation of dAdoMet. Depletion of spermidine evokes an increase in the AdoMetDC (Pegg et al., 1982b). Tissues treated with putrescine in the presence of DFMO maintain a low ODC content, increase their content of putrescine and spermidine, and decrease the activity of AdoMetDC.

During a clinical trial of the efficacy of DFMO in the treatment of cancer patients it was found that DFMO increased the urinary excretion of dAdoMet very considerably. The increase was maximal some 4–5 days after treatment and continued for several days after the daily administration was stopped. The excretion of polyamines or their derivatives was not significantly affected, suggesting that the excretion of dAdoMet might be a useful marker of the effectiveness of an ODC inhibitor (Haegele et al., 1987).

Under conditions in which dAdoMet accumulated in cells several hundredfold over the low normal concentration (e.g., inhibition of ODC by DFMO and the relative unavailability of putrescine), a new metabolite of dAdoMet was detected. This proved to be the N-acetyl derivative, which was separated from dAdoMet and was identified by mass spectrometry and comparison with a synthetic sample (Wagner et al., 1985). The dAdoMet was shown to be a substrate for a nuclear N-acetyltransferase in reaction with acetyl coA (Pegg et al., 1986a). It was not a substrate for the spermidine/spermine N^1-acetyltransferase. The reaction with the nuclear enzyme is given below.

$$H_2NCH_2CH_2CH_2-\overset{+}{\underset{\underset{CH_3}{|}}{S}}-Ado \xrightarrow{CH_3CO-ScoA}$$

dAdoMet

$$CH_3\overset{O}{\overset{||}{C}}-NHCH_2CH_2CH_2-\overset{+}{\underset{\underset{CH_3}{|}}{S}}-Ado$$

N–Ac–dAdoMet

The nuclear enzyme acetylated dAdoMet at a rate markedly greater than spermidine. The former substrate is an inhibitor of histone acetylation, and this inhibition by dAdoMet may contribute to the biological effects of ODC inhibition. The effects of DFMO on cells and depletion of polyamines in tissue cultures include a significant inhibition of DNA synthesis (70–80%) and a decrease of active DNA polymerase α (50%) in nuclei isolated from the inhibited and nonproliferating cells (Koza & Herbst, 1992). The inhibited cells may be restored to almost normal function by incubation in media supplemented with 20 µM spermidine.

Physiological regulation of AdoMetDC in animals

Whereas ODC is much higher in the mouse brain (70-fold) at birth than in the adult, the AdoMetDC in this organ is very low at birth and increases 10-fold some 4 weeks thereafter. This increase of enzyme activity occurs with an even greater increase of the essential mRNA. In the adults the polyamines fall to the low values characteristic of nondividing cells.

Weanling rats undernourished for 4 weeks were significantly under the weight (28% for males, 38% for females) for normal animals; however, feeding rapidly restored their weights. Feeding also restored the normal patterns of the biosynthetic enzymes relatively rapidly, as well as the polyamines themselves in deficient organs, despite differences in the behavior of male and female animals (McAnulty & Williams, 1977).

Starving adult female rats for 24 h results in about a 40% decrease in AdoMetDC in the mammary gland, as well as a far greater decrease (>95%) in ODC. A brief refeeding restores both enzymes. Surprisingly the livers of these animals were unaffected by the starvation (Brosnan et al., 1982).

Castration was shown to reduce ODC and AdoMetDC in all male rat accessory sex organs. Subsequent administration of testosterone to the castrated animals evoked a more rapid increase of the latter in these organs. Commenting on the dynamic complexities of the system, Fjösne et al. (1988) concluded that AdoMetDC probably controls the rate of polyamine synthesis at late states of stimulation.

Polyamine concentrations in various tissues are known to control the levels of ODC in the tissue and were explored as an approach to the regulation of the AdoMetDC (Pösö & Pegg, 1981). Administration of spermidine and norspermidine were shown to decrease the activity of the latter enzyme in many rat tissues. Most tissues (but not the heart) responded and several muscles declined more than the liver or kidney, which had taken up more of the exogenous amine. The various tissues also increased their putrescine contents whereas administered putrescine did not affect enzyme activity.

The response to polyamines of rat cells in tissue culture was studied by Mamont and colleagues (1981). Depletion of putrescine and spermidine in the presence of DFMO evoked a marked synthesis of the enzyme, as seen in Figure 14.2. When polyamines were added to these cultures, spermidine produced a rapid decrease of the enzyme whereas putrescine evoked a decrease only after conversion to spermidine. The colon and small intestine of rats administered 1% DFMO in water for up to 15 weeks demonstrated a comparable fall of ODC, putrescine, and spermidine and a great increase of AdoMetDC and dAdoMet in particular intestinal segments (Halline et al., 1989). Alhonen–Hongisto (1980) had also noted a linear inverse correlation between the levels of AdoMetDC in Ehrlich ascites cells and the concentrations of spermidine + spermine.

Physiological regulation in animal cells

The AdoMetDC and ODC are stimulated in the response of bovine lymphocytes to concavalin (conA) (Degen & Morris, 1980). Infection of chick embryo fibroblasts by Rous sarcoma virus also resulted in several-fold higher levels of these enzymes. Increases were found only at the permissive temperature of a tempera-

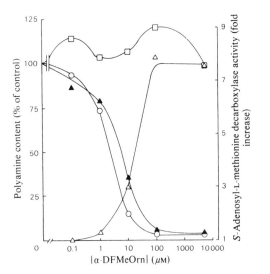

FIG. 14.2 Dose-dependent effects of DMFO (α-DFMeOrn) on AdoMet decarboxylase activity and intracellular polyamine content of rat hepatoma (HTC) cells; adapted from Mamont et al. (1982). (△) S-Adenosylmethionine decarboxylase; (O) putrescine; (▲) spermidine; (□) spermine.

ture-sensitive viral mutant, thereby assuring the transformation-induced specificity of the increase (Bachrach & Wiener, 1980). The half-lives of the AdoMetDC in the normal and virus-transformed fibroblasts treated with cycloheximide were found to be similar.

A comparison of the levels of the various polyamines and the biosynthetic enzymes in the growth cycles of 3T3 mouse fibroblasts and SV40-transformed mouse fibroblasts revealed that transformed cells contained higher levels of putrescine, spermidine, and the biosynthetic enzymes (Bethell et al., 1982). In this system the activity of the AdoMetDC was rate-limiting in determining the levels of spermidine and spermine.

DFMO applied to SV-3T3 transformed fibroblasts produced a 98% drop in putrescine and spermidine and a >90% decrease of growth rate after 2 days without loss of viability. Almost normal growth rates were obtained with various amine analogues, as well as with spermine, in the absence of putrescine and a mere 2% of the normal content of spermidine. The DFMO alone had increased dAdoMet >500 fold, but the addition of the polyamine "analogues" spermine, norspermidine, and homospermidine reduced dAdoMet some 98% (Pegg, 1984). Cells transfected by a cDNA for the enzyme increased production of the enzyme and their content of spermine with decreases in putrescine and spermidine (Manni et al., 1995; Svensson & Persson, 1995).

Depletion of AdoMet by inhibitors of AdoMet synthetase increased ODC and AdoMetDC several-fold. Interestingly, exogenous AdoMet reduced the drug-induced activities (Kramer et al., 1988).

Newer irreversible inhibitors of the AdoMetDC (to be discussed later) also stabilized the enzyme, as had MGBG, without an increase in the specific decarboxylase mRNA (Madhubala et al., 1988). The overall effects, despite an increase in the enzyme protein, included a decrease in cellular spermidine and spermine and a gross increase in ODC. The latter was far more actively decreased to a normal range by exogenous spermidine or spermine than by the high level of internal putrescine. However, the negative control of AdoMetDC by the spermine analogue 6-spermyne was more stringent than that of ODC.

$$H_2NCH_2CH_2CH_2NHCH_2C{\equiv}CCH_2NHCH_2CH_2CH_2NH_2$$

Treatment of L1210 cells with this substance (10 μM) lowered AdoMetDC but not ODC, which increased to a normal level. When the amine oxidases were inhibited by aminoguanidine, neither spermine nor spermyne affected the growth rates of the cells (Porter et al., 1988).

Discovery and processing of enzyme precursor

Specific antibody to AdoMetDC was prepared, and the increased activities found in rat organs after treatment with DFMO and MGBG were shown by radioimmunoassay to be related quantitatively to the amount of enzyme protein (Fig. 14.3). A significant fraction of the increased protein had been stabilized, although the activity disappeared more quickly after cycloheximide than did the immunoreactive protein (Shirahata & Pegg, 1985).

In a second approach, tritium-labeled dAdoMet was added to the cell extract in the presence of an inhibitor of spermidine synthase, and the resulting Schiff base (dAdoMet amine=pyruvoyl carbonyl) was reduced with sodium cyanoborohydride. The enzyme was thereby labeled specifically and stoichiometrically (Shirahata et

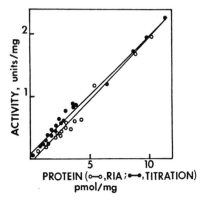

FIG. 14.3 Correlation of AdoMet decarboxylase activity with protein in rat prostate extracts; adapted from Shirahata and Pegg (1985). RIA, radioimmunoassay; titration, a labeling and inactivation of the enzyme protein by [³H]-decarboxylated AdoMet.

al., 1985) and was suitable for radioimmunoassay. When SV-3T3 cells were exposed to exogenous polyamines like spermidine, 90% of the activity and enzyme protein disappeared within 4 h, indicating the existence of a potent degradative system put into play by the polyamine.

The availability of a precipitating antiserum permitted the analysis of the proteins translated in a reticulocyte lysate system in response to the mRNA isolated from rat ventral prostate. Two proteins were precipitated by the antiserum, one of relative molecular mass (M_r) of 32,000 like that of an enzyme subunit and a larger protein of M_r 37,000 (Shirahata & Pegg, 1986). The latter pointed to the existence of an enzyme precursor, the possible origin of the pyruvoyl moiety at the active site, and the possible existence of a bipartite enzyme.

The presence of the precursor in rat prostate was increased by pretreatment with DFMO, which decreased putrescine content in the organ. Pretreatment with MGBG, which also increased the amount of enzyme protein, caused a marked decrease of the precursor. Pegg et al. (1988a) demonstrated that the processing of the precursor was activated by putrescine. Thus, the diamine acts in the formation of the enzyme from its precursor, as well as in its catalytic activity.

Large amounts of the mammalian enzyme were isolated after transfection of Chinese hamster ovary (CHO) cells by a cDNA generated from the mRNA of the prostate of rats treated with DFMO (Pajunen et al., 1988). Clones of cDNA were also isolated from human fibroblasts. Two forms of the decarboxylase mRNA, differing substantially in length, were present in the rat and human tissues. The two forms appear to arise from alternative utilization of two polyadenylation signals in the same gene. These RNAs are unusual in that they contain long 5′-noncoding regions, even as ODC mRNA does. Spermidine depletion, effected by DFMO, evoked a sevenfold increase in the mRNA. Nevertheless, spermidine as well as spermine were active inhibitors of translation of this RNA (Kameji & Pegg, 1987). This was suggested to relate to an action on the 5′-noncoding region as had been supposed for ODC mRNA.

Amino acid sequences were deduced from the cDNAs, and the rat and human enzymes, containing 333 and 334 amino acids, respectively, differed in only 11 of these. After cleavage these gave rise to two subunits of about 32,000 and 6000, respectively, contained within the whole enzyme. Previous studies of the isolated enzyme had failed to detect the smaller subunit.

The human proenzyme (MW 38,000), expressed in a strain of *E. coli* deficient in the enzyme, was shown to be cleaved to form two subunits (MW 7681 and 30,719). The active native enzyme is a cytosolic heterotetramer consisting of two pairs of small and large subunits, an $\alpha_2\beta_2$ protein. The pyruvoyl moiety is at the amino terminus of the large subunit (Stanley et al., 1989). Reduction of the pyruvate to alanine permitted an analysis of the amino terminal end of the large poly-

peptide. Unexpectedly the cleavage site in the sequence proved to be between a glutamate at position 67 and serine-68, which formed the pyruvoyl moiety. The sequence at the cleavage site was found to be

$$66 \quad 67 \quad 68 \quad 69$$
$$\text{—Leu—Ser—}Glu\text{–}Ser\text{—Ser—Met—}$$

Site-directed mutagenesis, which altered serine at positions 66 and 69 to alanine, had little effect whereas such a replacement at position 68 prevented processing and the generation of enzyme activity.

Other replacements tested additional amino acids in the sequence. The insertion of alanine in place of cysteine-82 not only eliminated activity but also markedly diminished the activation of processing by putrescine. A similar essential configuration of cysteine exists in the *E. coli* and *Saccharomyces cerevisiae* enzymes. Replacement of particular glutamates by glutamine (at positions 8 and 11, but not at 15, 61, 67, 247, and 249) resulted in significant effects on processing and enzyme activity (Stanley & Pegg, 1991). These results established significant roles for the smaller subunit in the catalytic activity, as well as for the dicarboxylates in the stimulation of processing by the diamine. Some mutant amino acid replacements abolished processing selectively, confirming the importance of certain glutamate residues in this role, whereas other glutamates proved critical selectively for the catalytic activity (Stanley et al., 1994).

Decreases in putrescine content can have several possible consequences, which may easily confuse the interpretation of experiments affecting this content. For example DFMO leads to a decrease of AdoMetDC in intestinal crypt cells, and this is reversible by small amounts of exogenous putrescine (Wang et al., 1992). In HT$_{29}$ or CHO cells, DFMO evokes a large increase in the enzyme whose product is largely used for the synthesis of spermine (Shantz et al., 1992).

The *S. cerevisiae* enzyme is cleaved from a proenzyme of M_r 46,000 to two associated polypeptide chains of M_r 36,000 and 10,000. Much of the enzyme preparation is inactive and about half of the M_r 36,000 subunits are terminated by alanine instead of pyruvate. Despite a low degree of homology with the mammalian enzyme, the yeast proenzyme is also cleaved between a glutamate and serine (Kashiwagi et al., 1990).

Genetic determinants of AdoMetDC

A rat genomic library has permitted the isolation of sequences hybridizable with cDNAs, including that prepared from the long form of prostatic mRNA of the AdoMetDC. A pseudogene was isolated from such a library and contained mutations incapable of producing an active enzyme (Pulkka et al., 1990). A functional gene was isolated similarly and was demonstrated to consist of eight exons that encode a protein identical to that determined from the rat cDNA (Pulkka et al., 1991). Regulatory elements were detected in the 5′-flanking region or in the first intron. The last long exon included the 3'-noncoding region that spanned the two lengths of mRNA separated by the two polyadenylation signals. Several copies of the gene and pseudogene appear to exist in the rat genome. Two different rat genes have been isolated and sequenced (Pulkka et al., 1993). Although their exon/intron structures appear to be similar, the genes have significantly different 5'-leader regions, suggesting different regulations of the two genes. An intronless mouse gene has also been isolated (Persson et al., 1995).

The chromosomal site for the enzyme in the human was determined by analysis of the DNA of human–mouse somatic cell hybrids. Digested DNA fragments assigned to particular human chromosomes were reacted with a cDNA clone for rat AdoMetDC. DNA sequences for the human enzyme were found on chromosome 6 and on the X chromosome; these differ from the sites of the ODC gene. Unlike the latter, amplification of the AdoMetDC gene was not detected in colorectal neoplasia in which enzyme activity is increased (Radford et al., 1988).

The gene isolated from human chromosome 6 proved to have nine exons in contrast to eight in the rat gene (Maric et al., 1992). Two of the central exons found in the human gene were fused in the rat gene. The human gene contained a complex 5' leader comprising many DNA sites thought to be associated with the binding of transcription factors. The 3' end contained several potential polyadenylation signals.

Molecular regulation of enzyme expression by polyamines and their inhibitors

The considerable increase in the rate of synthesis of the enzyme protein in DFMO-treated Ehrlich ascites tumor cells, concurrent with an increase in the specific mRNA, was markedly reduced by the addition of spermidine (Persson et al., 1989a,b). An increased enzyme synthesis was also observed by White et al. (1990) who measured the ribosomal distribution of the specific mRNA in cells stimulated by DFMO. After addition of DFMO, the increased synthesis was marked by a significant shift of the mRNA to large polysomes, indicating an effect on translation.

When an irreversible aminoxy inhibitor of the enzyme was used with L1210 cells, activity was obliterated despite a 50-fold increase of enzyme protein (Autelli et al., 1991). The increased synthesis and accumulation of the protein in the presence of the inhibitor was almost totally blocked by the additions of either spermidine or spermine to the inhibited cells. Spermidine did not affect the stabilization of the enzyme protein caused by the inhibitor, whereas spermine almost eliminated the stabilization.

When SV-3T3 cells were treated with a specific inhibitor of spermine synthase, S-methyl-5'methylthioadenosine, the decrease in spermine provoked a marked elevation in the activity of AdoMetDC. This was effected via an increase of the specific mRNA, an in-

creased rate of translation, and an increase in the half-life of the enzyme (Pegg et al., 1987).

Analysis of the cDNA encoding the AdoMetDC proenzyme of the mouse revealed a 5'-untranslated region that was highly conserved in many mammals. The presence of spermine determines the formation of a complex in the leader with a cytoplasmic protein. The protein complex can be fixed on the RNA by ultraviolet irradiation (Waris et al., 1992). It was found that the translation of the mRNA for AdoMetDC is stimulated manyfold in a rabbit reticulocyte cell-free system at the low concentrations of spermidine and spermine. The translation is strikingly inhibited at concentrations of spermidine >0.4 mM and spermine >0.06 mM. It appears that the inhibitions are determined by binding to the 5'-untranslated leader sequence, which is very high in GC content (Suzuki et al., 1993b).

Developing this problem, the early sequence of the 5'-transcript leader has been found to contain a curious open reading frame coding for a hexapeptide. The expression of this unit negatively affects translation of the mRNA to form an enzyme in T lymphocytes (Hill & Morris, 1993). When constructs of cDNA altered in the 5'-untranslated leader sequence were transfected into cells, the expression of the enzyme was markedly changed. Deletion of the entire sequence or the extreme 5' end increased enzyme content some 50–60-fold (Shantz et al., 1994). Spermine depletion of cells transfected with an essentially complete cDNA caused a fivefold increase; alteration of the early 5' sequence eliminated this response, indicating a role for spermine in the negative modulation of expression.

MGBG and its toxicity

Despite the differences in cation activation among the enzymes from different biological sources, many of these are quite sensitive to MGBG and other inhibitors (Pegg & Jacobs, 1983). For example, the enzyme from Chinese cabbage (see Chap. 20) is not affected by exogenous Mg^{2+} or putrescine and is nevertheless greatly inhibited by MGBG. Administration of large nontoxic doses of MGBG to rats produced a marked increase (>20-fold) in enzyme activity in the organs, resulting from a striking stabilization of the enzyme (Pegg et al., 1973). The inhibition of this target enzyme led also to a rapid significant increase in ODC and concomitant accumulation of putrescine. MGBG also produces a decrease in DAO in rat tissues (Hölttä et al., 1973).

In kinetic studies with AdoMetDC, MGBG appeared to be competitive with AdoMet with an apparent K_i of <1 μM in saturating concentrations of putrescine and was noncompetitive with putrescine. In studies on the DAO reaction with putrescine as substrate, MGBG was noncompetitive with putrescine and the K_i was about 0.7 μM. However, MGBG does not act simply as a carbonyl reagent and does not inhibit PLP enzymes or react with PLP.

Reexamination of the treatment of L1210 leukemic mice with MGBG revealed a similar array of metabolic disturbances involving changes in the polyamines in the mice. Administration to the mice did produce a depression in AdoMetDC and spermidine only for the first 8–12 h after treatment (Heby & Russell, 1974). Pursuing the myriad organellar reactions it was found that spermidine, spermine, and MGBG affected the structures of isolated nuclei and in binding DNA templates were also somewhat inhibitory at high concentrations to bacterial and mammalian DNA and RNA polymerases (Brown et al., 1975; Nelson et al., 1978). The binding of DNA by antitumor guanylhydrazones has revealed some further details of weak interactions (Dave et al., 1976).

The clinical use of MGBG is limited by host toxicity in proliferating tissues such as intestinal mucosa and bone marrow. Effects on the latter tissue also relate to inhibition of the development of antibody-forming cells (Bennett et al., 1978). Therapeutic doses cause major structural and functional damage to the mitochondria of some affected cells. In L1210 leukemia cells these effects appear as swelling of the organelle, degradation of the internal cristae, and loss of organelle density (Pathak et al., 1977). MGBG also inhibited proliferation without effects on mitochondrial respiration in various types of tumor cells (Loesberg et al., 1991). A study of the inhibition of intestinal mucosa by MGBG revealed a suppression of mitosis in the crypt cells concomitantly with the effects on polyamines and the enzymes seen in the cellular studies, suggesting that the toxicity of MGBG on proliferating tissues is related primarily to the disturbance of polyamine metabolism (Porter et al., 1980).

The promise of MGBG as an antitumor agent compelled a series of new efforts to improve the therapeutic efficacy of a potent inhibitor of AdoMetDC. The aminoguanidine derivative of MGBG (Chap. 9) proved to be an active irreversible inhibitor of the enzyme. It acted in a sequence of binding to the enzyme reversibly and competitively with AdoMet and then inactivated the enzyme irreversibly. Although this substance alone did not block spermidine or DNA synthesis in the livers of partially hepatectomized rats, it did so in conjunction with 1,3-diaminopropane (DAP), the indirect inhibitor of ODC (Wiegand & Pegg, 1978). This result, like that of Morris et al. (1977) with MGBG and α-methylornithine, suggested the combined use of inhibitors to block the accumulation of putrescine and the higher polyamines.

Analyses of MGBG uptake emphasized the relation of efficacy and toxicity to transport. Exposure to MGBG resulted in extraordinary concentrations (500–1000-fold) of the drug in Ehrlich ascites cells (Seppänen et al., 1980). However, micromolar exogenous concentrations of spermidine or spermine blocked uptake and reduced the intracellular MGBG content to noninhibitory levels. Diamines were effective blockers. Thus, effective MGBG levels within the cells were dependent on a spermidine transport system controlled by polyamines of relatively high affinity. Reduction of intracellular spermidine by pretreatment with DFMO enhanced the uptake of polyamines and MGBG in such cells (Alho-

nen–Hongisto et al., 1980). The concept that such a depletion can be used to enhance MGBG effects has been explored in clinical trials.

MGBG-resistant strains of leukemia cells were altered primarily in the uptake of MGBG and spermidine. Resistance to the inhibitor was used in the isolation of mutant cells: these were decreased markedly in the uptake of MGBG and various polyamines (Mandell & Flintoff, 1978). The mutant cells were not found to overproduce AdoMetDC, nor was the enzyme altered in its sensitivity to the inhibitor.

New modifications of MGBG, such as the diethylglyoxal derivatives, possessed new more desirable features. The increased specificity and inhibitory activity of the diethylglyoxal-bis(guanyl hydrazone) ($K_i = \sim 63$ nM) caused a decrease in spermidine and spermine in L1210 cells and an increase in putrescine, permitting growth for only 1–2 days (Svensson et al., 1993). There was a fivefold increase in AdoMetDC and a slight increase in the mRNA for the enzyme (Suzuki et al., 1993a). Some of the new derivatives possessed new unexpected activities. For example, the methylglyoxal-bis(butylamidinohydrazone) (MGBB) simultaneously inhibited ODC, AdoMetDC, and spermidine synthase, but not spermine synthase (Hibasami et al., 1986). Another derivative, methylglyoxal-bis(cyclohexylamidinohydrazone) (MGBC) proved to be a competitive inhibitor of AdoMetDC (with AdoMet) and spermidine synthase (with putrescine). MGBC was similar in K_i to MGBB in inhibition of the first of these and was slightly less active than MGBG. MGBC was significantly more active than MGBB in inhibition of spermidine synthase, and at 20 µM levels could inhibit growth of leukemic cells and depress the spermidine content of the cells.

Conformationally constrained analogues of MGBG have proved to be much more potent and specific than MGBG. One of these amidinohydrazones was 140 times more potent in inhibiting the AdoMetDC (Stanek et al., 1993).

Two earlier compounds of the series appeared to have significant antitumor activity (Regenass et al., 1992).

Irreversible inhibitors

Certain inhibitors of polyamine synthesis (e.g., DFMO) are active inhibitors of some protozoan infections like trypanosomiasis, and this result led to a fresh look at the range of substances previously detected as trypanocides. These included the agents *Berenil* and *Pentamidine*, whose curative effects were found to be blocked by spermidine (Bacchi et al., 1983). These substances were then depicted as MGBG analogues and were found to be irreversible inhibitors of putrescine-activated Ado-MetDC (Karvonen et al., 1985). Intraperitoneal injection of Berenil (50 mg/kg) into rats produced a rapid fall of about 60–70% in the enzyme activity of the liver.

An analogue of putrescine, 1-aminoxy-3-aminopropane was also found to be an irreversible inhibitor ($K_i = 50$ μM) of the rat liver enzyme. This substance also proved to be a potent competitive inhibitor of mouse kidney ODC ($K_i = 3.2$ μM) and bovine brain spermidine synthase ($K_i = 2.3$ μM). At 0.5 mM the aminoxy compound inhibited the growth of HL-60 promyelocytic leukemia cells; this was blocked by spermidine but not by putrescine (Khomutov et al., 1985). The compound blocked proliferation of L1210 cells at 20 μM levels and this was reversed by exogenous spermidine or putrescine. At 100 μM inhibitor, spermidine fell to 1% of the control level and dAdoMet accumulated. However, putrescine was nonexistent and ODC had been essentially eliminated, whereas the AdoMetDC had increased over the control level. In short, 1-aminoxy-3-aminopropane was acting primarily as an ODC inhibitor in a manner similar to the effects of DFMO (Poulin et al., 1989).

The conclusion that this aminoxy derivative was a relatively weak inhibitor of AdoMetDC was supported by Hyvönen and Eloranta (1990) who observed that the substance depleted cell spermidine and not cell spermine under conditions in which the enzyme increased. They concluded that spermidine was the major feedback regulator of this enzyme. The inhibitor is not concentrated in the cells and is not recognized as a polyamine by transport systems; it is also metabolized actively (Hyvönen et al., 1990).

The perceptions of high MGBG toxicity, the improved efficacy and specificity of the ODC-activated irreversible inhibitor DFMO, and the inadequacy of other known inhibitors of AdoMetDC have encouraged efforts to produce new irreversible inhibitors of the enzyme. A compound (see below) that irreversibly inhibited ODC and AdoMetDC (Hibasami et al., 1988),

Difluoro-phenyl-ethyl(4-aminopropylamidinohydrazone)

depressed cell proliferation and intracellular levels of all the polyamines. Unexpectedly it inhibited synthesis of protein more actively than that of the nucleic acids.

The structural analogue of dAdoMet, S-5′-deoxy-5′-adenosyl-methylthioethylhydroxylamine, first prepared by A. R. and R. M. Khomutov, was shown to be an irreversible inhibitor of the bacterial AdoMetDC.

The compound also inhibited the enzyme of L1210 cells irreversibly, producing a marked drop in spermidine and spermine content (Kramer et al., 1989). Nevertheless, the content of enzyme protein was increased 50-fold. Exogenous spermidine restored growth and spermidine content to the cells (but not spermine content) and lowered synthesis of enzyme protein to the control level.

Another group of adenosyl derivatives, containing carbonyl-reactive end groups that might react with the pyruvoyl moiety of the enzyme, was synthesized by Secrist (1987). Among the most active of these, significantly better than MGBG, was one in which nitrogen replaced sulfur, as in

The hydrazino derivative was even more active than the above aminoxy compound. These were irreversible inactivators of the purified enzyme, with 50% inhibition of enzyme activity (I_{50}) values of 400 nM and 70 μM, respectively, and produced hydrazone derivatives of the pyruvoyl moiety. These compounds caused marked decreases in dAdoMet and the cellular content of spermidine and spermine. The inhibition of cell growth was reversed by addition of spermidine. The addition of dAdoMet also partially reversed the inhibition. In combination (100 μM aminoxy derivative and 5 mM DFMO), all polyamine synthesis was blocked (Pegg et al., 1988b). The penetrability of AdoMet analogues and their potency implies a possibly useful position for such compounds in a therapeutic armamentarium.

Finally we note a potent adenosyl derivative that is an irreversible inhibitor after activation by the enzyme. This is the first instance of such an inhibitor for a pyruvoyl enzyme. The compound is 5′-{[(Z)-4-amino-2-butenyl]methylamino}-5′-deoxyadenosine (Casara et al., 1989).

The inactivation is postulated to occur via the protonated form of the 5′-tertiary amine because the enzyme does recognize a sulfonium cation. The pyruvoyl-bound portion of the presumed inhibitor is pictured as a conjugated imine that is cleaved, freeing an aldehydic (deaminated) form of the inhibitor. The pyruvate in the protein is converted in this reaction to an inactivating alanine (Shantz et al., 1992). Pegg and McCann (1992) discussed the inhibition of AdoMetDC via transamination from analogues of AdoMet or dAdoMet; these can form a Schiff base with the pyruvate, and may permit transamination to form alanine at the pyruvoyl carbonyl. The butenyl-containing inhibitor sharply increases the rate of these events on the *E. coli* and rat liver enzyme and is maximally effective in the presence of Mg^{2+} and putrescine, respectively (Danzin et al., 1990). The inactivation of the rat liver enzyme was found to require about 1.5 molecules of inhibitor turned over per enzyme molecule. AdoMetDC of the rat ventral prostate was maximally (93–94%) inhibited 6 h after a single intraperitoneal injection of 100 mg/kg. The level of activity increased thereafter to about 60% of the control value. Seiler et al. (1991) described these effects on the enzymes and on polyamine concentrations in other organs in normal and tumor-bearing animals. A considerable pile-up of putrescine is suggested as serving to maintain a low level of cell function (Byers et al., 1992). The inhibitor increased ODC levels briefly, as well as the level of AdoMetDC protein, which was in fact inactive (Stjernborg et al., 1993). Unexpectedly, the inhibitor was found to cure trypanosomal infections of mice. The compound was taken up selectively by the parasites via their well-developed purine transport system, which does not exist in "normal" mouse cells (Byers et al., 1992).

Propylamine Transferases

The detection of spermidine synthase in mammalian tissues occurred some 10 years after its demonstration in bacteria. *E. coli* extracts were unable to synthesize significant levels of spermine, a substance reported in yeast as well as mammalian tissue, and this posed the new and interesting possibility of an evolutionary acquisition of a new enzyme, spermine synthase, in the eucaryotic organism. The latter enzyme was indeed found and separated from the spermidine synthase. Both enzymes catalyze propylamine transfer from dAdoMet to the acceptors putrescine and spermidine, as in the reactions given below.

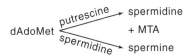

Both reactions give rise to MTA.

As seen in Chapter 5, spermine has now been found in some bacteria, as well as other aminopropylated

polyamines. The degrees of homology among the bacterial enzymes that synthesize the newly discovered bacterial amines and the mammalian spermine synthase are unknown at present.

In many cellular systems in which inhibitors of polyamine synthesis block growth, that block is overcome by the exogenous addition of spermidine. Thus, for the most part spermidine may be thought to be the major functional and essential product of polyamine synthesis. In some mammalian cell systems, however, spermine will also restore growth, perhaps via degradation to spermidine. Possible roles of the tetramine that binds to membrane lipids, several specific proteins, and many nucleates are discussed in Chapters 21–23.

Spermidine appears to regulate early steps of polyamine synthesis, that is, the activities of ODC and AdoMetDC. An increased intracellular concentration of spermidine induces the rapid production of antizyme and the degradation of ODC. In the case of AdoMetDC, an increased concentration of this polyamine arrests or slows the synthesis of the enzyme, also facilitating a decrease via turnover of the enzyme. The enzyme protein is easily stabilized against degradative turnover by combination with active site-directed inhibitors. Spermine also selectively slows the expression of this enzyme.

In every system spermidine synthase is present in excess, and earlier synthetic steps are rate-determining in spermidine production. Thus, the specific inhibition of spermidine synthase would not be expected to be an effective control on polyamine synthesis. However, exogenous spermidine overcomes the effects of inhibitors of the earlier enzymatic steps and thereby compels thought on the regulation of spermidine content. In addition, the work on MGBG has demonstrated that cells have maintained systems for the transport and acquisition of exogenous spermidine, systems that affect the uptake of certain inhibitors and also normally assure an acquisition of spermidine under conditions of potential deficiency.

Assay and purification of propylamine transferases

A procedure involving destruction of a labeled precursor and a retention of isotopic polyamine on a countable ion-exchange paper has been described (Raina et al., 1983). The product of a transferase reaction with methyl-labeled dAdoMet would be a labeled MTA. Details of methods for the isolation and estimation of this product have been given by Raina et al. (1983) and Pegg (1983). Both methods are useful for the assay of synthases for either spermidine or spermine.

In crude extracts the use of labeled substrates must recognize the possible presence of normal unlabeled substrate that will dilute the label in substrate and product. This is particularly worrisome for putrescine that may be present at high concentrations in stressed or growing animal or plant cells (Sindhu & Cohen, 1983a,b). However, the potential error may also be a source of concern for the use of labeled dAdoMet in tissues inhibited to permit the accumulation of this normally sparse metabolite.

A sensitive fluorimetric assay has been devised (Suzuki et al., 1981) in which spermidine (1–10 ng) can be determined by a coupled enzymatic analysis (Chap. 3). Spermidine and spermine are oxidized by the specific oat seedling oxidase to produce H_2O_2 that then oxidizes homovanillic acid in the presence of peroxidase to form a highly fluorescent product. The amine oxidase is inactive on putrescine. It is active on spermine, but the assay was devised for dialyzed extracts of animal tissues essentially devoid of spermidine as a substrate for spermine synthase.

The introduction of an affinity chromatographic adsorbent, S-adenosyl(5′)-3-thiopropylamine linked to Sepharose, has permitted a marked advance in the degree of purification of the enzyme from rat ventral prostate. Salt precipitation, ion-exchange fractionation on DEAE cellulose, and a separation on the affinity gel yielded an enzyme of 90% purity, which was increased to homogeneity by fractionation on Sephacryl S-300. The latter separation revealed a 73,000 MW, comprising two subunits of 37,000 as detected in a denaturing gel. The enzyme was stabilized by dithiothreitol and inhibited by sulfhydryl reagents. The K_m for putrescine was 0.1 mM, and cadaverine and DAP were weak competitive inhibitors with K_i values of 1.9 and 5.2 mM, respectively. The K_m for dAdoMet was 1.1 µM; an analogue was competitively inhibitory with this substrate (Samejima & Yamanoha, 1982). Samejima et al. (1983) also described the isolation of the enzyme from bovine brain. A strong substrate inhibition by dAdoMet was recorded; this is eliminated by increasing the concentration of the substrate putrescine (Raina et al., 1984).

The introduction of spermine coupled to Sepharose resulted in an additional 600-fold purification to apparent homogeneity (i.e., 6000-fold) for the spermine synthase from a bovine brain (Pajula et al., 1979). The affinity of the enzyme for dAdoMet was high ($K_m = 0.6$ µM) and 100-fold less for spermidine ($K_m = 60$ µM), with which putrescine was a weak competitor. Both products, spermine and MTA, were inhibitory, with MTA ($K_i = 0.3$ µM) competing strongly with dAdoMet; spermine is only weakly inhibitory. The enzyme was very sensitive to sulfhydryl reagents and could be restored by incubation in mercaptoethanol or dithiothreitol. Kinetic studies with the brain enzyme have been carried out in the presence of MTA phosphorylase to remove this inhibitory product (Pajula, 1983). dAdoMet is not an inhibitor of spermine synthase.

Specificity of spermidine synthase

Putrescine is the most active aminopropyl acceptor for the eucaryotic spermidine synthase. However, experiments with DFMO in intact cells revealed the formation of aminopropylcadaverine, and in fact cadaverine does serve as a substrate with the purified enzyme. Cadaverine and 1,6-diaminohexane are slightly active, where-

as DAP is inactive (Samejima et al., 1983). 1,4-Diamino-2-butene has also been found to serve as a substrate (Samejima & Nakazawa, 1980).

The unsaturated 2-ene or 2-yne 1,4-butanediamines were poor substrates with the partially purified brain enzyme (Sarhan et al., 1987). The 1-methyl derivative was even less active whereas the 1,4-dimethyl derivative and the 1-hydroxymethyl derivative were completely inactive. On the other hand many derivatives at the 2 position, including the 2-methyl, 2-chloro-, and the 2,2-difluoro derivatives, were quite active. The natural 2-hydroxy-1,4-butanediamine had some activity.

Many of these compounds maintained the survival and growth of DFMO-treated chick embryos. Analogues of spermidine and spermine were found in extracts of these embryos, and the isolated dansyl derivatives were shown to contain the expected derivatized four carbon fragment. All 2-substituted putrescines enable the formation of spermidine and spermine analogues, indicating an activity of the spermidine analogue, with spermine synthase as well as spermidine synthase. Although 1-methylputrescine formed an 8-methylspermidine, this product did not yield an additional spermine analogue.

The 2- and 2,2-fluoroputrescines both form fluorospermidines in vitro as well as in cells depleted by treatment with DFMO (Dezeure et al., 1988). In mice 2-fluoroputrescine will form 6-fluorospermidine and 6-fluorospermine readily, whereas the 2,2-difluoroputrescine is less readily converted to the higher polyamines. The incorporation of these compounds into such polyamines has also been studied in tumor-bearing mice because ^{19}F-containing polyamines might serve as tumor markers detectable by ^{19}F-NMR spectroscopy.

Coward et al. (1977) found that the 7-deaza derivative of dAdoMet will serve as a substrate in the reaction. Samejima and Nakazawa (1980) tested other analogues of dAdoMet and found that substitution of the methyl by ethyl or propyl in the sulfonium cation had little effect on aminopropyl transfer to putrescine. Sulfonium analogues in which the propylamine moiety is replaced by ethylamine or butylamine are inactive as aminoalkyl donors and only slightly active as inhibitors. Only one of the diastereoisomers of dAdoMet is active with the mammalian spermidine synthase.

A derivative of dAdoMet with a methyl group at C_1, containing the amino group in the aminopropyl moiety, was both a substrate for spermidine synthase and an inhibitor of AdoMetDC and spermine synthase (Pankaskie et al., 1981). This structure is given below.

It was inhibitory to the synthesis of DNA and polyamines in conA-stimulated lymphocytes. The addition of putrescine, spermidine, or spermine to the inhibited cultures failed to restore DNA synthesis; and the inability of the polyamines to reverse the inhibitor suggests some site of action other than that affected by the polyamines.

Specificity of spermine synthase

Attempts to determine the role(s) of spermine in cellular function have examined the effects of inhibitors or the effects of spermidine replacements that are almost devoid of activity in the spermine synthase reaction. In one such study, depletion of putrescine and spermidine were carried out in L1210 leukemia cells with DFMO in the presence of the spermidine analogues aminopropylcadaverine, N^4-methylspermidine, N^4-ethylspermidine, or homospermine. Addition of the spermidine analogues, which were not inhibitory in themselves, depleted spermine levels to 15% of the control and growth was then also reduced (Casero et al., 1984). Thus, spermine appears essential for growth in its own right. Although spermine synthase appears specific for spermidine as the aminopropyl acceptor, losing activity with changes in the carbon backbone of the spermidine, in vivo studies with putrescine derivatives have demonstrated the conversion of several of the substituted spermidines to spermine.

A series of mono- or difluorinated spermidines (i.e., derivative of 4-azaoctane-1,8-diamine) were tested as possible substrates with the partially purified rat hepatoma (HTC) cell spermine synthase. The K_m and pK values of their amino groups are presented in Table 14.3. The pK values were determined by potentiometric titration.

It can be seen that 7 substitutions, which markedly decrease the pK of the 8-amino group, also decrease the K_m values of the compound. The K_m of the 7,7-difluoro derivative was quite low and the 7-substituted compounds were inhibitory to the enzyme as a function of pH. Baillon et al. (1988) concluded that protonation of N^1 and N^4 are essential for efficient aminopropyl transfer. The 2-substituted spermidines affect protonation at both N^1 and N^4, whereas 6-substituted spermidines do have lower pKs at N^4.

The cis isomer of the alkene analogues of spermidine, N-(3-aminopropyl)-1,4-diamino-cis-but-2-ene, is a good substrate of the enzyme. The trans isomer and the alkyne analogue were not substrates. The cis alkene analogue of spermine supported the growth of SV-3T3 cells inhibited by DFMO. However, the unsaturated spermidine analogues that could not be converted to spermine analogues also supported the growth of DFMO-inhibited cells. Thus, it is believed that the conversion of the analogues to sperminelike compounds was not essential to growth (Pegg et al., 1991). All of these analogues accumulated within cells.

The 1-methyl spermidine proved to be a poor substrate for spermine synthase and was also acetylated at a very low rate. Nevertheless, this compound reversed the inhibitory effect of DFMO in L1210 cells. Furthermore, the compound was not toxic to cells in the presence of serum amine oxidase, indicating an inability

TABLE 14.3 Properties of spermidine derivatives in spermine synthase reaction.

Amines	Structures and pK Values 1 2 3 4 5 6 7 8	% Fully ionized	K_m (mM)
Spermidine	H₂N ⌇ NH ⌇ NH₂ 9.94 8.40 10.81	90	88
6-Monofluoro-spermidine	H₂N ⌇ NH ⌇F NH₂ 10.40 7.18 F 9.55	37.4	93
7-Monofluoro-spermidine	H₂N ⌇ NH ⌇ F NH₂ 10.28 9.00 7.80	71.0	11
6,6-Difluoro-spermidine	H₂N ⌇ NH ⌇ F F NH₂ 10.34 5.70 9.29	0.20	458
7,7-Difluoro-spermidine	H₂N ⌇ NH ⌇ F F NH₂ 10.25 8.24 6.64	13.20	0.54
2,2-Difluoro-spermidine	H₂N F F NH ⌇ NH₂ 7.30 5.50 10.32	0.6	not a substrate

Adapted from Baillon et al. (1988).

of this enzyme to oxidize the modified aminopropyl moiety.

Comparable spermine analogues, 1-methyl and 1,12-dimethyl, are similarly resistant to acetylation and oxidation. Cells blocked by DFMO were partially restored by the *gem* 1,12-dimethyl spermine despite the absence of metabolism of the compound in the cells.

Serological reactions of propylamine transferases

An apparently homogenous spermidine synthase was isolated from the rat prostate and injected with adjuvant into rabbits to produce an antiserum. The antiserum was used to detect and quantify the enzyme in extracts of rat organs. The quantities of enzyme protein found by this method in normal rat tissues were in accord with the enzyme activities, implying the absence of regulatory factors in the extracts. Enzymes isolated from various mammals (human, rat, mouse, pig, and guinea pig) reacted with the single rabbit antiserum to the rat enzyme with somewhat different intensities, suggesting differences in amino acid sequences. However, precipitations of the different activities were comparable, implying nearly identical conformations (Shirahata et al., 1988b).

Spermidine synthase and spermine synthase were isolated from human tissues and were capable of carrying out only their specific reactions. The former enzyme comprised two subunits of identical molecular weight (35 kDa) whereas the latter comprised two identical subunits (45 kDa). Nevertheless, the spermine synthase revealed two bands by isoelectric focusing. Antisera were prepared separately in mice against both synthases; each antiserum failed to cross-react with the heterologous protein. It was concluded that these enzymes do not have similar antigenic sites and appear to be quite different proteins (Kajander et al., 1989).

The distribution of these enzymes in extracts of human tissues is given in Table 14.4. The distributions suggest sources suitable for enzyme isolation. Although the pancreas has been cited as high in spermine, it appears that this organ is the highest in spermidine synthase. Subsequent studies demonstrated high levels of spermidine synthase in several human malignant lymphoma and leukemia cell lines.

Human gene for spermidine synthase

A peptide was isolated from the bovine enzyme after tryptic hydrolysis and its sequence was used to construct a synthetic deoxyoligonucleotide mixture that could code for the peptide. A human cDNA library constructed on the λ phage was screened by the mixture, and a cDNA insert coding for the enzyme was isolated

TABLE 14.4　Spermidine synthase and spermine synthase activities in human tissues.

Tissue	Spermidine synthase	Spermine synthase
Liver	25 ± 2.7	8 ± 2.4
Spleen	37 ± 6.6	10 ± 3.4
Kidney	21 ± 1.4	77 ± 9.1
Pancreas	203 ± 45	8 ± 1.5
Lung	10 ± 2.7	6 ± 2.1
Full-term placenta	10 ± 1.5	49 ± 6.2
Brain cortex, frontal lobe	108 ± 2.2	38 ± 5.6

Adapted from Kajander et al. (1989). Values are in pmol/min/mg soluble protein. Activities were measured in 100,000 g supernatant fractions after removal of salts with a sieving column. The values are means of 7–8 (autopsy) preparations.

from the phage. A single open reading frame of 906 nucleotides was found that coded for a polypeptide of 302 amino acids with a molecular weight of 33,827 Da (Wahlfors et al., 1990). The 5′-noncoding sequence was high in guanine and cytosine within some 83 nucleotides. The DNA for the human enzyme has been used to estimate the mRNA for the enzyme in various myeloma cell lines.

The human gene was isolated from a genomic library of λ phages constructed with DNA from a human myeloma cell line (Myöhänen et al., 1991). The insert of 10.5 kb contained a full-length human gene of 5818 nucleotides from the cap site of the mRNA to the last polyadenylation site. The gene was sequenced and was found to contain eight exons and seven introns. The gene is reported to be present in human chromosome 1 (Winqvist et al., 1993). Thus, various human genes involved in polyamine synthesis are not grouped at all, with ODC on chromosome 2 and AdoMetDC on chromosomes X and 6.

A human chromosome 1 derived gene was used to produce a transgenic mouse, which contained 2–6 times more spermidine synthase in all mouse tissues tested (Kauppinen et al., 1993). The various polyamine concentrations in each of these tissues were only very slightly changed. No changes were found in the tissues in ODC, AdoMetDC, spermine synthase, and spermidine/spermine acetyltransferase. The level of activity of the DC, determined by the amount of enzyme protein and putrescine, appeared to be rate-limiting in the later biosynthetic sequence. These observations point once again to the existence of rigorous homeostatic mechanisms that prevent the toxic accumulations of spermidine and spermine. The gene of the mouse encodes an enzyme with a 95% similarity to the human enzyme and only 33% similarity to the *E. coli* protein (Myöhänen et al., 1994).

Mitogen-stimulated lymphocytes or medium-stimulated myeloma rapidly increase the mRNA for spermidine synthase but decrease the activity of the enzyme, an effect ascribed to an inhibition of translation by relatively high levels of spermidine (Kauppinen, 1995).

Inhibitors of spermidine synthase

Several groups of inhibitors were found that yielded interesting biological effects. These fall into three categories: nucleosides, amine analogues, and "transition state analogues" or "multisubstrate adduct inhibitors."

The 7-deaza analogue of MTA, a tubercidin derivative, served as a stable competitor of MTA with MTA phosphorylase. And when built into an analogue of dAdoMet, it could serve as both a propylamine donor and a substrate inhibitor (Coward et al., 1977). MTA and the alkylthiotubercidins inhibited spermidine synthase and spermine synthase. MTA was far more potent on the latter. However, the effects on cells were most marked in protein synthesis and these effects appeared unrelated to polyamine depletion (Raina et al., 1982).

Among the nucleosides containing the sulfonium center and thioether centers, most of these were more inhibitory to spermine synthase than to spermidine synthase. However, the sulfoxides and sulfone derivatives of AdoHcy and its decarboxylated product were quite active against the latter (Hibasami et al., 1980a). Among more recently prepared nucleoside analogues active on spermidine synthase is included the 5'-[(3-aminopropyl)amino]-5'-deoxyadenosine, in which nitrogen replaces sulfur at the 5' position.

This substance has an I_{50} of 7 µM, compared to an I_{50} of 17 µM on spermine synthase. It is not very active on AdoMetDC ($I_{50} > 0.2$ mM) (Pegg et al., 1986b).

A specific inhibition of spermidine synthase in cultured cells with 100–200 µM cyclohexylamine leads to an accumulation of the putrescine substrate and a depletion of cellular spermidine that forms spermine (Caruso et al., 1992). The inhibition of proliferation is obtained when cell spermidine falls to 25–50% of its original level; proliferation resumes on addition of exogenous spermidine. Higher concentrations of the inhibitor are lethal to chick embryo fibroblasts. In HTC cells depletion of spermidine in the presence of cyclohexylamine evoked a major increase in the amount and stability of ODC (Mitchell et al., 1985).

Numerous results on the depletion of spermidine by this inhibitor in bacteria, protozoa, plants, and animal cells have been summarized (Pegg & Williams–Ashman, 1987). It may be mentioned that cyclohexylamine is a urinary metabolite of cyclamate, an artificial sweetening agent (Kurebayashi et al., 1989). Although the substance prolonged the survival of tumor-bearing mice, the antitumor effects were manifested in some organs and not in others. Therefore, the promise of the effects of spermidine depletion was pursued with more potent analogues. Two of these were indeed more potent as inhibitors of spermidine synthase competitive with putrescine, N-chlorosulfonyl dicyclohexylamine (Nakashima et al., 1986) and trans-4-methylcyclohexylamine (Shirahata et al., 1988a).

Cyclohexylamine
($k_i = 2 \times 10^{-7}$ M)

N-Chlorosulfonyldicyclohexylamine
($k_i = 1.8 \times 10^{-7}$ M)

trans-4-Methylcyclohexylamine
($k_i = 4 \times 10^{-8}$ M)

The chlorosulfonyl derivative depletes spermidine levels and inhibits the proliferation of human leukemia cells. However, the synthesis of protein but not of the nucleic acids was markedly diminished. In the case of the trans-4-methylcyclohexylamine, spermidine content fell rapidly and precipitously in treated HTC cells but the decrease was compensated by an increase in putrescine with little change in growth rate. An N-(3-aminopropyl) derivative of cyclohexylamine has proved to be a specific and potent inhibitor of spermine synthase and depletes spermine in rats without an effect on growth (Shirahata et al., 1993). There is a concomitant stimulation of the accumulation of spermidine in various tissues, implying a shunt of aminopropyl moieties from dAdoMet to the synthesis of triamine.

The active site of the enzyme is conceived as a large hydrophobic cavity contiguous to a negatively charged site on the protein (Shirahata et al., 1991). A protonated amino group of putrescine would bind to the latter site, while the remainder of the putrescine molecule occupies the hydrophobic cavity. Compounds such as cyclohexylamine compete successfully for the hydrophobic site, as do a series of other alkyl amines such as propylamine, butylamine, and isobutylamine. These studies revealed new aminopropyl acceptors, such as N-monoethylputrescine, cis- and trans-1,4-diaminocyclohexane, and 4-aminomethyl piperidine, which were less active than putrescine.

Propylamine transfer in the spermidine synthase reaction was thought (Tang et al., 1981) to occur in a sequence involving an attack of the nonprotonated amine of putrescine on the bond linking the sulfonium cation of dAdoMet to the propylamine moiety. This structure is presented in Figure 9.7. The transition state analogues S-adenosyl-3-thio-1,8-diaminooctane (AdoDATO) and its methyl sulfonium salt were synthesized to mimic the presumed intermediate of this process and were found to be quite potent and specific in their inhibition of the rat prostate spermidine synthase.

$$Ado-CH_2-S-CH-(CH_2)_5NH_3^+$$
$$\underset{(CH_2)_2NH_3^+}{|}$$

AdoDATO

$$\overset{CH_3}{\underset{}{|}}$$
$$Ado-CH_2-\overset{+}{S}-CH-(CH_2)_5NH_3^+$$
$$\underset{(CH_2)_2NH_3^+}{|}$$

Methyl Sulfonium AdoDATO

The thioether had an I_{50} of 4×10^{-7} M and was almost noninhibitory to spermine synthase and methylthioadenosine phosphorylase. On the other hand the methylsulfonium derivative, like other AdoMet sulfonium derivatives, had significant activity against spermine synthase.

The thioether (AdoDATO) was taken up by mammalian cells and spermidine content fell, with a concomitant increase of putrescine and spermine, the subsequent metabolite of the triamine (Fig. 14.4). Growth rate fell

somewhat (ca. 30%) over a 12-day exposure. AdoMetDC was also increased, as was dAdoMet. It was concluded that the DC was regulated mainly by spermidine. In a combination of AdoDATO and DFMO, all polyamines were markedly reduced. The cells grew very slowly, but this rate was considerably increased by spermidine (Pegg et al., 1982b).

Inhibition of spermidine synthesis by AdoDATO (>95%) in mouse erythroleukemia cells induced to differentiate by dimethyl sulfoxide caused an inhibition of growth rate and hemoglobin production. These effects were reversed by exogenous spermidine and spermine but not by putrescine. Spermidine was more effective than spermine, and the results imply that spermidine is required for induction of differentiation by these cells (Sherman et al., 1986).

Kinetic studies with the *E. coli* spermidine synthase suggested that AdoDATO might compete with dAdoMet and putrescine and might be acting as a structural analogue of both substrates, rather than as a transition state analogue. A mechanism of transfer via a ternary complex has been demonstrated with the spermidine synthase of *E. coli* (see Chap. 9; Orr et al., 1988).

Although the hypothesis that compounds such as AdoDATO resemble the transition state led to the synthesis of potent specific inhibitors and indeed provides an approach to the synthesis of a specific inhibitor of spermine synthase, these substances are now designated more cautiously as "multisubstrate adduct inhibitors." A new class of these inhibitors includes a newly synthesized adenosylspermidine, which has proved to be a significantly more potent inhibitor of spermidine synthase (Lakanen et al., 1995).

Inhibitors of spermine synthase

The product MTA was quite inhibitory of this enzyme (Pajula et al., 1979). Several other nucleoside inhibitors, such as 5'-[(3-aminopropyl)amino]-5'-deoxyadenosine, proved to be inhibitory to spermidine synthase as well. However, the methyl sulfonium derivative of MTA ($AdoS^+(CH_3)_2$) and a diamino octofuranosyl derivative of deoxyadenosine (AdoDap), both given below, proved to be quite specific inhibitors of spermine synthase.

$$\overset{CH_3}{\underset{}{|}}$$
$$Ado-CH_2-\overset{+}{S}-CH_3$$

Methyl Sulfonium MTA
(I_{50} = 8 μM)

$$Ado-CH_2CH(NH_2)CH_2CH_2NH_2$$

AdoDap
(I_{50} = 12 μM)

These were able to reduce the spermine concentration of cells, but there was a compensatory increase of sper-

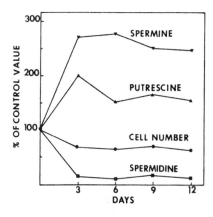

FIG. 14.4 Effect of prolonged exposure to AdoDATO on polyamine content of SV-3T3 cells; adapted from Pegg et al. (1982). Cell number after incubation in AdoDATO was determined by plating cell suspensions and is expressed as a percentage of colonies formed in the presence and absence of AdoDATO.

midine and cell growth was not affected (Pegg et al., 1986b). In combination with inhibitors of spermidine synthase, putrescine and dAdoMet both accumulate. Simultaneous incubation with $AdoS^{+}(CH_3)_2$ and DFMO, in which dAdoMet increases, does not completely deplete spermine. It was suggested that an enlarged dAdoMet pool might limit the competitive effectiveness of the nucleoside inhibitors of the propylamine transferases.

Among the amine analogues, several N-alkylated DAP derivatives proved to be selective inhibitors of spermine synthase (Baillon et al., 1989). N-Butyl-DAP was the most potent of these competitors of spermidine, with a K_i of 12 nM. Interestingly n-butylamine proved to be an active competitive inhibitor of spermidine synthase with a K_i of 0.52 µM. Cells treated with N-butyl-DAP for 3 days contained 25–30% of their spermine content whereas spermidine content increased several-fold. Numerous compounds were found of even greater potency (10–20-fold) than the N-butyl derivatives. One of these was the N-cyclohexyl-DAP. These proved to be at least 100 times more potent than a dimethyl 5′-adenosyl sulfonium salt (Shirahata et al., 1991).

The aminooxy analogues of spermidine, in which carbon was replaced by oxygen in either the propane or butane moieties, proved to be poor substrates of the enzyme and good inhibitors. The latter 1-aminooxy-3-N-[3-aminopropyl]-aminopropane (AP-APA) was the better inhibitor with a K_i of 1.5 µM; neither was active on spermidine synthase. Both compounds also inactivated AdoMetDC and inhibited ODC (Eloranta et al., 1990a).

Finally Coward's work was extended to the design of and synthesis of a multisubstrate adduct inhibitor, S-adenosyl-1,12-diamino-3-thio-9-azadodecane (AdoDATAD) specifically active against spermine synthase (Woster et al., 1989).

AdoDATAD (I_{50} = 20 µM)

The I_{50} is given in a standard assay because of the inhibition by excess dAdoMet. The assay system used contained 5 µM dAdoMet. The 1,12-desamino compounds were essentially inactive as inhibitors. When AdoDATAD was tested in SV-3T3 cells or in L1210 cells, dose-dependent decreases of intracellular spermine were obtained with compensatory increases of spermidine (Pegg et al., 1989). As in studies with AdoDATO, increases in dAdoMet were thought to affect the intracellular inhibition of the enzyme and the effective depletion of spermine. AdoDATAD was cytotoxic in L1210 cells, but this was eliminated by aminoguanidine. An incomplete depletion (80%) of spermine in the SV-3T3 cells permitted a normal growth rate over many (>10) population doublings. This result was possibly due to replacement of a requirement by spermidine. Nevertheless, L1210 cells were inhibited in growth in the presence of AdoDATAD and aminoguanidine, and addition of exogenous spermine did not increase growth. The cells metabolized AdoDATAD to an unknown product similar to AdoDATO.

The recent cloning of a cDNA encoding human spermine synthase may be expected to activate many new studies on the properties and regulation of this enzyme (Korhonen et al., 1995).

Metabolism of MTA

The formation of MTA in the synthesis of spermidine and spermine occurs as a function of propylamine transfer in bacteria, fungi, and other eucaryotic cells (Chap. 9). The generation of MTA also occurs in some other reactions of AdoMet described by Williams–Ashman et al. (1982). However, MTA concentration is low and the nucleoside does not accumulate in mammalian tissue, despite the inability of some deaminases of adenosine or AMP, purine nucleoside phosphorylase, or AdoHcy hydrolase to act upon this product. Nevertheless, MTA is metabolized in these organisms, not only to restore purine requirements, but also to provide methionine in many mammalian cells. An inability to process MTA to form methionine limits growth in a significant number of tumor types.

MTA is inhibitory to many enzymatic reactions (e.g., propylamine transferases), providing another set of reasons why its depletion may be useful to the cells. Product inhibition can also occur in the synthesis of diphthamide from AdoMet, a compound present in the eucaryotic elongation factor EF-2 (Yamanaka et al., 1986). MTA also inhibits adenosine kinase competitively, whereas AdoHcy inhibits this enzyme noncompetitively (Fox et al., 1982).

MTA also inactivates AdoHcy hydrolase irreversibly in the absence of other metabolites (Ferro et al., 1981). This enzyme, which converts demethylated AdoMet to adenosine and homocysteine (reaction 4 in Fig. 14.2) and is important therefore in the resynthesis of ATP and methionine, has increasingly assumed quantitative significance in the performance of these functions in stressed cells, as in some virus infections and some malignancies. AdoHcy is an active inhibitor of transmethylation and methylation steps are essential in the replication of some viruses. Thus, some inhibitors of the reversible AdoHcy hydrolase, which further the accumulation of AdoHcy, have been found to be broad spectrum antiviral agents (Wolfe & Borchardt, 1991).

Analogues of AdoHcy, such as 5′-deoxy-5′-S-isobutylthioadenosine (SIBA), inhibit transmethylations and the conversion of putrescine to the higher polyamines in cultures of chick embryo fibroblasts. The latter inhibitions included the initial uptake of putrescine and the inhibition of spermidine synthase and spermine synthase (Lawrence et al., 1982; Pegg et al., 1981). SIBA also increases the production and content of MTA in intact rat hepatocytes and is thought to compete with MTA for the enzyme believed to be normally responsible for preventing the accumulation of this product, MTA phosphorylase (Schanche et al., 1982).

MTA inhibits the proliferation of murine lymphoid cells and the activity of several membrane receptors (e.g., that for adenosine) (Munshi et al., 1988) and protein kinases (Smith et al., 1989). As summarized by Williams–Ashman et al. (1982), MTA has a large number of pharmacological effects.

The isolation and characterization of MTA in yeast by Schlenk and colleagues and the conversion of labeled MTA (^{3}H in adenine or ^{35}S in the sulfur atom) to AdoMet have been described (Schlenk & Ehninger, 1964). Extracts of liver and prostate were found to cleave MTA in the presence of phosphate to what was thought to be 5-methylthioribose-1-phosphate (Pegg & Williams–Ashman, 1969).

In rat ventral prostate activated by androgen, the biosyntheses of spermidine and spermine lead to an active production of MTA, a feedback inhibitor of the polyamine synthases. Androgen stimulation evokes a concomitant increase of MTA phosphorylase preventing the accumulation of inhibitory MTA (Seidenfeld et al., 1980). The levels of polyamines and MTA also vary inversely in regenerating rat liver, suggesting an increasing degradation of the latter in that system also.

The nucleoside is quite stable but may be cleaved to free adenine by nucleosidases of bacteria or plants or by MTA phosphorylase in procaryotic and eucaryotic cells. There is no evidence of deamination of MTA in eucaryotic cells. It was estimated that 85–97% of the adenine produced by multiplying human cells deficient in adenine phosphoribosyl transferase was derived from MTA generated in the synthesis of spermidine and spermine (Kamatani & Carson, 1981). Inhibitors of polyamine synthesis or of MTA phosphorylase markedly reduced the endogenous production of the free purine.

In a culture of baby hamster kidney (BHK) cells, MTA was cleaved to free adenine that was efficiently incorporated into nucleic acids (Eloranta et al., 1982). All of the sugar moiety was excreted in this cellular system. The influx of MTA into human cells has been reported to be mediated by a nonspecific nucleoside transport system, as well as by passive diffusion increased by the lipophilicity of the methylthio group (Stoeckler & Li, 1987). MTA has been found in the urine of normal subjects (33–447 µg/day) and cancer patients (20–356 µg/day). Despite a frequently increased production of polyamine in cancer patients, a general increase in the level of MTA was not found (Kaneko et al., 1984).

A mollusk has been found to contain the natural D-xylosyl analogue (355 nmol/g) of MTA. MTA is present at a "normal" concentration (2 nmol/g) in the mollusk. Doubly labeled MTA (^{14}C in adenine and methyl) is converted actively, possibly via the 3'-keto derivative, to the xylosyl derivative, which is not cleaved by MTA phosphorylase (Porcelli et al., 1989). The xylosyl ana-

logue does not inhibit this or other enzymes of polyamine synthesis.

D-riboMTA

3'-ketoMTA

D-xyloMTA

Isolation and properties of MTA phosphorylase

The removal and recycling of the moieties of MTA with the aid of this enzyme is an integral aspect of polyamine metabolism. Purified mammalian MTA phosphorylase has been found to have an absolute requirement for inorganic orthophosphate. Thus, MTA is cleaved phosphorylytically to adenine and the α-anomer of 5-methylthioribose-1-phosphate. The reaction is not inhibited by many natural nucleosides (i.e., adenosine, guanosine), nucleotides (i.e., AMP, 3',5'-cyclic AMP), or sugar-1-phosphates except at high concentrations (>2 mM). Also, the enzyme is not inhibited by the natural polyamines.

MTA phosphorylase was found to be relatively high in brain tissue and particularly high in the pig brain (Abruzzese et al., 1986). The activity from the brain of most animals was quite stable in extract at 4°C; that of mouse brain was unexpectedly labile. The discovery that the enzyme was lacking in some human cancers turned attention to the human enzyme, which was purified to homogeneity (30,000-fold) from human placenta (Della Ragione et al., 1986). This protein enzyme was found to have a 98,000 MW and to comprise three identical subunits. It was stabilized to heat considerably by millimolar MTA or by 300 mM potassium phosphate in the optimal pH 7.2–7.6. The enzyme required an —SH reducing agent such as dithiothreitol and was inactivated by thiol-blocking compounds. The equilibrium constant in the direction of phosphorolysis was 1.39×10^{-2} and the K_m values for MTA, inorganic phosphate

(P_i), adenine, and 5-methylthioribose-1-phosphate were 5, 320, 23, and 8 μM, respectively.

An MTA phosphorylase of apparently similar size and subunit composition was isolated from bovine liver and used to prepare antisera. The rabbit antibodies were reactive with both a coupled antigen and the native enzyme; it also reacted, although not as well, with the human enzyme. The antibody was unreactive with the yeast MTA phosphorylase. It was shown with this antiserum that human cancer cells lacking MTA phosphorylase did not contain a cross-reactive enzymatically inactive protein (Della Ragione et al., 1990).

More recently a purification of MTA phosphorylase from human liver (8,000-fold with 25% yield) led to the isolation of a protein of 55,000 MW comprising subunits of 30,000 MW (Toorchen & Mille, 1991). An analysis of the specificity of the human liver enzyme showed that several purines, such as 2-substituted adenines and 1-deazaadenine, are good substrates whereas 3-deazaadenine is an effective inhibitor. In the direction of cleavage, 5'-substituted adenosine, 5'-deoxyadenosine, is a good substrate whereas 5'-deoxy-5'-methylthio-3-deazaadenosine is a good inhibitor (Fig. 14.5).

Many of these effects had been observed earlier in the studies of cytotoxicity of presumed nucleoside analogues of MTA. The toxicity of compounds such as 2-halogeno MTAs and the 5'-ethylthio derivatives arose from the release of toxic products (Savarese et al., 1983). Also, the cleavable 5'-halogenoadenosines were cytotoxic and this toxicity was blocked by an inhibitor of MTA phosphorylase. This result implied that the 5'-halogenated ribose-1-phosphate derivatives also affect the route to methionine (Savarese et al., 1985). One of the best inhibitors of MTA phosphorylase to date has proven to be a 5'-halogenated formycin, in which a cleavable *N*-glycosyl bond is replaced by a stable carbon–carbon bond (Chu et al., 1986). Other 9-deazaadenosines possessing this feature are also potent inhibitors of the enzyme (Pankaskie & Lakin, 1987).

X = Cl, Br, or I

5-Halogeno-formycin A (K_i = ca.10^{-7} M)

Modifications of the 5'-methylthio groups to form the monofluoromethyl and difluoromethyl derivatives of MTA both produced competitive inhibitors of the enzyme (Nishikawa et al., 1987; Sufrin et al., 1989). The difluoromethyl derivative (K_i = 0.48 μM) was not a substrate, inhibited cell growth, and caused the accumula-

tion of MTA in treated cells. However, the toxicity of the compound was partially reversed by exogenous adenine.

From MTA to methionine

The pathway from the product of the phosphorylase, 1-phospho-5-methylthioribose, to methionine (see Chap. 9) was studied in rat liver homogenates (Backlund et al., 1982; Trackman & Abeles, 1981, 1983). The presence of the 1-phosphate was found essential. The oxidation products 2-keto-4-methylthiobutyrate (2,4-KMB) and formate, as well as the 2-hydroxymethyl-thiobutyrate were detected. 2,4-KMB was considered to be the immediate substrate for transamination to methionine and a sensitive assay for the α-keto acid was developed (Toohey, 1983).

The initiation of this pathway in all normal mammalian tissues occurs as a result of the presence of MTA phosphorylase. The gene for the enzyme has been localized to human chromosome 9 by the technique of human somatic cell hybridization (Carrera et al., 1984). This enzyme is absent in some human leukemias, lymphomas, and solid tumors as determined in studies with antibody to the enzyme (Nobori et al., 1991). Some 75% (9 of 12) of the human gliomas examined were completely MTA phosphorylase deficient, that is, unreactive with cross-reactive antibody to bovine liver MTA phosphorylase. This deficiency has been studied in greater detail in two phosphorylase-deficient strains of human fibroblasts, which did not contain inhibitors of the enzyme, an immunoreactive catalytically inactive protein, or an unstable protein.

Although most of the tumor cell lines requiring exogenous methionine for growth have been found to lack MTA phosphorylase, one cell line of colon carcinoma was found that was unable to convert methylthioribose-1-P to methylthioribulose-1-P (Ghoda et al., 1984). It is not known if the lesion in the sugar isomerization existed in the primary tumor. Formate derived from this pathway has been shown to contribute significantly to the pool of single carbon units utilized for purine synthesis in rat bone marrow (Deacon et al., 1990). In any case the biochemical route to the synthesis of methionine from methylthioribose-1-P presented in Chapter 9 is believed to be similar in animal cells and bacteria.

Experiments with methionine-dependent tumor cell lines revealed that tumor cells die in the absence of methionine in the medium; that is, they are unable to form colonies when plated. This apparent loss of viability is also seen in tumor cells inoculated into syngenic animals (Breillout et al., 1990). The nature of the methionine dependency may point to a target for the chemotherapy of such metabolically deficient tumors. For example, when normal human lung fibroblasts and tumor lymphoblastic leukemia cells are grown in mixed culture, depletion of methionine from the medium by means of methioninase (L-methionine-α-deamino-γ-mercaptomethane-lyase) results in growth of the former and

MTA + P$_i$ ⟶ Adenine + 5-methylthioribose-P

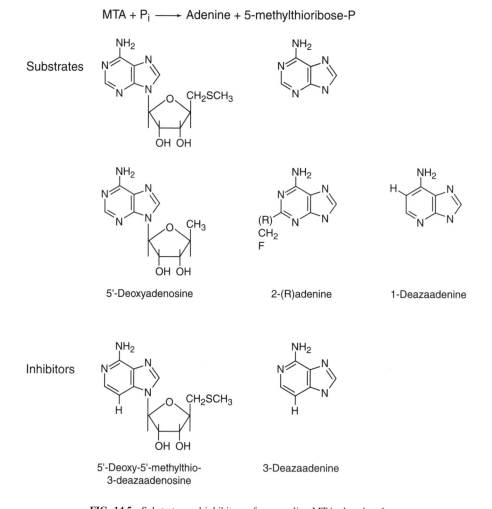

FIG. 14.5 Substrates and inhibitors of mammalian MTA phosphorylase.

a selective and rapid death of the latter (Kreis et al., 1980).

A comparison was made of metabolic activities of some methionine-dependent tumor cell lines and methionine-independent lines, including a revertant from the former (Judde et al., 1989). In this study the methionine-dependent strains contained MTA phosphorylase but had significant increases in transmethylation reactions.

Leukemic cells lacking MTA phosphorylase excrete MTA and the accumulation of this undegraded product increases synthesis of putrescine and suppresses that of spermine and spermidine (Kamatani & Carson, 1980). In such cells MTA itself was found to be extremely antiproliferative (I_{50} = 3–6 μM) in a manner unrelated to the synthesis of nucleic acid bases or of the polyamines (Kubota et al., 1983). MTA was found to induce ODC and AdoMetDC in these cells (Kubota et al., 1985). A combination of DFMO (100 μM) and MTA (50 μM) turned off the synthesis of all the cellular polyamines, although there was a marked accumulation of dAdoMet and its acetyl derivative (Yamanaka et al., 1987). Thus, this combination of compounds acts synergistically and may selectively inhibit the growth of phosphorylase-deficient tumors. However, the latter authors pointed to some potentially toxic effects of MTA.

REFERENCES

Abbruzzese, A., Della Pietra, G., & Porta, R. (1986) *Experientia*, **42**, 820–821.
Acuff, R. & Smith, J. (1983) *Journal of Nutrition*, **113**, 2295–2299.
———— (1985) *Biochemical Archives*, **1**, 167–171.
Åkerman, K., Karkola, K., & Kajander, O. (1991) *Biochimica et Biophysica Acta*, **1097**, 140–144.

Alhonen–Hongisto, L. (1980) *Biochemical Journal*, **190**, 747–754.

Alhonen–Hongisto, L., Seppänen, P., & Jänne, J. (1980) *Biochemical Journal*, **192**, 941–945.

Alix, J. (1982) *Microbiological Reviews*, **46**, 281–295.

Alvarez, T., Carrales, F., Martin–Duce A., & Mato, J. (1993) *Biochemical Journal*, **293**, 481–486.

Anderson, M., Ast, T., Nicolaou, A., & Valko, K. (1994) *Biochemical Society Transactions*, **22**, 2958.

Autelli, R., Stjernborg, T., Khomutov, A., Khomutov, R., & Persson, L. (1991) *European Journal of Biochemistry*, **196**, 551–556.

Bacchi, C., McCann, P., Nathan, H., Hutner, S., & Sjoerdsma, A. (1983) *Advances in Polyamine Research*, **4**, 221–231.

Bachrach, V. & Weiner, H. (1980) *International Journal of Cancer*, **26**, 75–78.

Backlund, Jr., P., Chang, C., & Smith, R. (1982) *Journal of Biological Chemistry*, **257**, 4196–4202.

Baillon, J., Kolb, M., & Mamont, P. (1989) *European Journal of Biochemistry*, **179**, 17–21.

Baillon, J., Mamont, P., Wagner, J., Gerhart, F., & Lux, P. (1988) *European Journal of Biochemistry*, **176**, 237–242.

Baker, D. (1987) *Methods in Enzymology*, **143**, 297–307.

Baldessarini, R. (1987) *American Journal of Medicine*, **83**(Supplement 5A), 95–103.

Benevenga, N., Radcliffe, B., & Egan, R. (1983) *Australian Journal of Biological Science*, **36**, 475–485.

Bennett, J., Ehrke, J., Fadale, P., Dave, C., & Mihich, E. (1978) *Biochemical Pharmacology*, **27**, 1555–1560.

Bethell, D., Hibasami, H., & Pegg, A. (1982) *American Journal of Physiology*, **243**, C262–C269.

Breillout, F., Antoine, E., & Poupon, M. (1990) *Journal of the National Cancer Institute*, **82**, 1628–1632.

Brosnan, M., Ilic, V., & Williamson, D. (1982) *Biochemical Journal*, **202**, 693–698.

Brosnan, M., Symonds, G., Hall, D., & Symonds, D. (1978) *Biochemical Journal*, **174**, 727–732.

Brown, K., Nelson, N., & Brown, D. (1975) *Biochemical Journal*, **15**, 1505–1512.

Byers, T., Casara, P., & Bitonti, A. (1992) *Biochemical Journal*, **283**, 755–758.

Cabrero, C. & Alemany, S. (1988) *Biochimica et Biophysica Acta*, **952**, 277–281.

Cantoni, G., Mudd, S., & Andreoli, V. (1989) *Trends in Neurological Sciences*, **12**, 319–324.

Carrera, C., Eddy, R., Shows, T., & Carson, D. (1984) *Proceedings of the National Academy of Sciences of the United States of America*, **81**, 2665–2668.

Caruso, A., Pellati, A., Bosi, P., Arena, N., & Stabellini, G. (1992) *Cell Biology International Reports*, **16**, 349–358.

Casara, P., Marchal, P., Wagner, J., & Danzin, C. (1989) *Journal of the American Chemistry Society*, **111**, 9111–9113.

Casero, Jr., R., Bergeron, R., & Porter, C. (1984) *Journal of Cellular Physiology*, **121**, 476–482.

Chu, S., Ho, I., Chu, E., Savarese, T., Chen, Z. H., Rowe, E. C., & Chu, M. Y. W. (1986) *Nucleosides and Nucleotides*, **5**, 185–200.

Corti, A., Astancolle, S., Davalli, F., Baciottini, F., Casti, A., & Viviani, R. (1987) *Comparative Biochemistry and Physiology*, **88B**, 475–480.

Coward, J., Motola, N., & Moyer, J. (1977) *Journal of Medicinal Chemistry*, **20**, 500–505.

Danzin, C., Marchal, P., & Casara, P. (1990) *Biochemical Pharmacology*, **40**, 1499–1503.

Dave, C., Ehrke, J., & Mihich, E. (1976) *Chemico-Biological Interactions*, **12**, 183–195.

Deacon, R., Bottiglieri, T., Chanarin, I., Lumb, M., & Perry, J. (1990) *Biochimica et Biophysica Acta*, **1034**, 342–346.

Degen, J. & Morris, D. R. (1980) *Proceedings of the National Academy of the Sciences of the United States of America*, **77**, 3479–3483.

Della Ragione, F., Cartenì–Farina, M., Gragnaniello, V., Schettino, M., & Zappia, V. (1986) *Journal of Biological Chemistry*, **261**, 12324–12329.

Della Ragione, F., Oliva, A., Gragnaniello, V., Russo, G., Palumbo, R., & Zappia, V. (1990) *Journal of Biological Chemistry*, **265**, 6241–6246.

Dezeure, F., Gerhart, F., & Seiler, N. (1989) *International Journal of Biochemistry*, **21**, 889–899.

Dezeure, F., Sarhan, S., & Seiler, N. (1988) *International Journal of Biochemistry*, **20**, 1299–1312.

Eloranta, T. (1977) *Biochemical Journal*, **166**, 521–529.

Eloranta, T. & Kajander, E. (1984) *Biochemical Journal*, **224**, 137–144.

Eloranta, T., Khomutov, A., Khomutov, R., & Hyvönen, T. (1990a) *Journal of Biochemistry*, **108**, 593–598.

Eloranta, T., Martikainen, V., & Smith, T. (1990b) *Proceedings of the Society for Experimental Biology and Medicine*, **194**, 364–371.

Eloranta, T. & Raina, A. (1977) *Biochemical Journal*, **168**, 179–185.

——— (1978) *Biochemical and Biophysical Research Communications*, **84**, 23–30.

Eloranta, T., Tuomi, K., & Raina, A. (1982) *Biochemical Journal*, **204**, 805–807.

Feldman, M., Levy, C., & Russell, D. (1972) *Biochemistry*, **11**, 671–677.

Ferro, A. J., Vandenbark, A., & MacDonald, M. (1981) *Biochemical and Biophysical Research Communications*, **100**, 523–531.

Fillingame, R. & Morris, D. (1973) *Biochemical and Biophysical Research Communications*, **52**, 1020–1025.

Finkelstein, J. (1990) *Journal of Nutritional Biochemistry*, **1**, 228–237.

Fiskerstrand, T., Christensen, B., Tysnes, O., Ueland, P., & Refsum, H. (1994) *Cancer Research*, **54**, 4899–4906.

Fjösne, H., Strand, H., Östensen, M., & Sunde, A. (1988) *The Prostate*, **12**, 309–320.

Fox, I., Palella, T., Thompson, D., & Herring, C. (1982) *Archives of Biochemistry and Biophysics*, **215**, 302–308.

Ghoda, L., Savarese, T., Dexter, D., Parks, R., Trackman, P., & Abeles, R. (1984) *Journal of Biological Chemistry*, **259**, 6715–6719.

Griffith, O. (1987) *Methods in Enzymology*, **143**, 366–376.

Grillo, M. & Bedino, S. (1977) *Italian Journal of Biochemistry*, **26**, 342–346.

Gritli–Linde, A., Holm, I., & Linde, A. (1995) *European Journal of Oral Sciences*, **103**, 133–140.

Haegele, K., Splinter, T., Romijin, J., Schecter, P. J., & Sjoerdsma, A. (1987) *Cancer Research*, **47**, 890–895.

Halline, A., Dudeja, P., & Brasitus, T. (1989) *Biochemical Journal*, **259**, 513–518.

Hannonen, P. (1975) *Acta Chemica Scandinavica*, **B29**, 295–299.

——— (1976) *Acta Chemica Scandinavica*, **B30**, 121–124.

Hannonen, P., Jänne, J., & Raina, A. (1972b) *Biochemical and Biophysical Research Communications*, **46**, 341–348.

Hannonen, P., Raina, A., & Jänne, J. (1972a) *Biochimica et Biophysica Acta*, **273**, 84–90.

Heby, O. & Russell, D. (1974) *Cancer Research*, **34**, 886–892.

Hibasami, H., Borchardt, R., Chen, S., Coward, J., & Pegg, A. (1980a) *Biochemical Journal*, **187**, 419–428.

Hibasami, H., Hoffman, J., & Pegg, A. (1980b) *Journal of Biological Chemistry*, **255**, 6675–6678.

Hibasami, H., Tsukada, T., Maekawa, S., & Nakashima, K. (1986) *Biochemical Pharmacology*, **35**, 2982–2983.

Hibasami, H., Tsukada, T., Maekawa, S., & Nakashima, K. (1988) *Biochemical Pharmacolgy*, **37**, 364–365.

Hill, J. R. & Morris, D. R. (1993) *Journal of Biological Chemistry*, **268**, 726–731.

Hoffman, R. (1985) *Anticancer Research*, **5**, 1–30.

Hölttä, E., Hannonen, P., Pupa, J., & Jänne, J. (1973) *Biochemical Journal*, **136**, 669–676.

Horikawa, S. & Tsukada, K. (1992) *FEBS Letters*, **312**, 37–41.

Hyvönen, T. & Eloranta, T. (1990) *Journal of Biochemistry*, **107**, 339–342.

Hyvönen, T., Khomutov, A., Khomutov, R., Lapinjoki, S., & Eloranta, T. (1990) *Journal of Biochemistry*, **107**, 817–820.

Jänne, J., Schenone, A., & Williams–Ashman, H. (1971) *Biochemical and Biophysical Research Communications*, **42**, 758–764.

Jänne, J. & Williams–Ashman, H. (1971) *Biochemical and Biophysical Research Communications*, **42**, 222–229.

Judde, J., Ellis, M., & Frost, P. (1989) *Cancer Research*, **49**, 4859–4865.

Kajander, E., Harvema, R. J., Kauppinen, T., Ackerman, K. K., Martikainen, H., Pajula, R., & Kärenlampi, S. O. (1990) *Biochemical Journal*, **267**, 767–774.

Kajander, P., Kauppinen, L., Pajula, R., Karkola, K., & Eloranta, T. (1989) *Biochemical Journal*, **259**, 879–886.

Kamatani, N. & Carson, D. (1980) *Cancer Research*, **40**, 4178–4182.

Kamatani, N. & Carson, D. (1981) *Biochimica et Biophysica Acta*, **675**, 344–350.

Kameji, T. & Pegg, A. (1987) *Journal of Biological Chemistry*, **262**, 2427–2430.

Kaneko, K., Fujimori, S., Kamatani, N., & Akaoka, I. (1984) *Biochimica et Biophysica Acta*, **802**, 169–174.

Karvonen, E., Kauppinen, L., Partanen, T., & Pösö, H. (1985) *Biochemical Journal*, **231**, 165–169.

Kashiwagi, K., Taneja, S. K., Liu, T., Tabor, C. W., & Tabor, H. (1990) *Journal of Biological Chemistry*, **265**, 22321–22328.

Kauppinen, L. (1995) *FEBS Letters*, **365**, 61–65.

Kauppinen, L., Myöhänen, S., Halmekytö, M., Alhonen, L., & Jänne, J. (1993) *Biochemical Journal*, **293**, 513–516.

Kaye, A., Icekson, I., & Lindner, H. (1971) *Biochimica et Biophysica Acta*, **252**, 150–159.

Kenyon, S. H., Nicolaou, A., Ast, T., & Gibbons, W. A. (1996) *Biochemical Journal*, **316**, 661–665.

Khomutov, R., Hyvönen, T., Karvonen, E., Kauppinen, L., Paolanen, T., Paulin, L., Eloranta, T., Pajula, L., Andersson, L. C., & Pösö, H. (1985) *Biochemical and Biophysical Research Communications*, **130**, 596–602.

Korhonen, V., Halmekytö, M., Kauppinen, L., Myöhänen, S., Wahlfors, J., Keinänen, T., Hyvönen, T., Alhonen, L., Eloranta, T., & Jänne, J. (1995) *DNA and Cell Biology*, **14**, 841–847.

Kotb, M., Geller, A., Markham, G., Kredich, N., De La Rosa, J., & Beachey, E. (1990) *Biochimica et Biophysica Acta*, **1040**, 137–144.

Koza, T. & Herbst, E. (1992) *Biochemical Journal*, **281**, 87–93.

Kramer, D., Khomutov, R., Bukin, Y., Khomutov, A., & Porter, C. (1989) *Biochemical Journal*, **259**, 325–331.

Kramer, D., Porter, C., Borchardt, R., & Sufrin, J. (1990) *Cancer Research*, **50**, 3838–3842.

Kramer, D., Stanek, J., Diegelman, P., Regenass, U., Schneider, P., & Porter, C. (1995) *Biochemical Pharmacology*, **50**, 1433–1443.

Kramer, D., Sufrin, J. R., & Porter, C. (1987) *Biochemical Journal*, **247**, 259–265.

Kramer, D., Sufrin, R., & Porter, C. (1988) *Biochemical Journal*, **249**, 581–586.

Kreis, W., Baker, A., Ryan, V., & Bertasso, A. (1980) *Cancer Research*, **40**, 634–641.

Kubota, M., Kajander, E., & Carson, D. (1985) *Cancer Research*, **45**, 3567–3572.

Kubota, M., Kamatani, N., & Carson, D. (1983) *Journal of Biological Chemistry*, **258**, 7288–7291.

Kuchino, Y., Sharma, O., & Borek, E. (1978) *Biochemistry*, **17**, 144–147.

Kurebayashi, H., Fukuoka, M., & Tanaka, A. (1989) *Chemical and Pharmaceutical Bulletin*, **37**, 1097–1099.

Lakanen, J. R., Pegg, A. E., & Coward, J. K. (1995) *Journal of Medicinal Chemistry*, **38**, 2714–2727.

Lawrence, F., Bachrach, U., & Robert–Géro, M. (1982) *Biochemical Journal*, **204**, 853–859.

Livesay, G. (1984) *Trends in Biochemical Sciences*, **January**, 27–29.

Loesberg, C., Rooij, H., Romijn, J., & Smets, L. (1991) *Biochemical Pharmacology*, **42**, 793–798.

Lombardini, J. & Talalay, P. (1973) *Molecular Pharmacology*, **9**, 542–560.

Madhubala, R., Secrist, III, J., & Pegg, A. (1988) *Biochemical Journal*, **254**, 45–50.

Mamont, P., Danzin, C., Wagner, J., Siat, M., Joder-Ohlenbusch, A., & Claverie, N. (1982) *European Journal of Biochemistry*, **123**, 499–504.

Mamont, P. S., Joder-Ohlenbusch, A., Nussli, M., & Grove, J. (1981) *Biochemical Journal*, **196**, 411–422.

Mandel, J. & Flintoff, W. (1978) *Journal of Cell Physiology*, 335–344.

Manni, A., Badger, B., Grove, R., Kunselman, S., & Demers, L. (1995) *Cancer Letters*, **95**, 23–28.

Maric, S., Crozat, A., & Jänne, O. (1992) *Journal of Biological Chemistry*, **267**, 18915–18923.

McAnulty, P. & Williams, P. (1977) *Biochemical Journal*, **162**, 109–121.

Meister, A. (1991) *Pharmacology and Therapeutics*, **51**, 155–194.

Mitchell, J., Mahan, D., McCann, P., & Qasba, P. (1985) *Biochimica et Biophysica Acta*, **840**, 309–316.

Morris, D., Jorstad, C., & Segfried, C. E. (1977) *Cancer Research*, **37**, 3169–3172.

Munshi, R., Clanachan, A., & Baer, H. (1988) *Biochemical Pharmacology*, **37**, 2085–2089.

Myöhänen, S., Kauppinen, L., Wahlfors, J., & Alhonen, L. (1991) *DNA and Cell Biology*, **10**, 467–474.

Myöhänen, S., Wahlfors, J., Alhonen, L., & Jänne, J. (1994) *DNA Sequence*, **4**, 343–346.

Nakashima, K., Tsukada, T., Hibasami, H., & Maekawa, S. (1986) *Biochemical and Biophysical Research Communications*, **141**, 718–722.

Nelson, N., Brown, K., Fehlman, B., Stewart, G., & Brown, D. (1978) *Biochimica et Biophysica Acta*, **517**, 429–438.

Nishikawa, S., Ueno, A., Inoue, H., & Takeda, Y. (1987) *Journal of Cellular Physiology*, **133**, 372–376.

Nobori, T., Karras, J., Della Ragione, F., Waltz, T., Chen, P. P., & Carson, D. A. (1991) *Cancer Research*, **51**, 3193–3197.

Pajula, R. (1983) *Biochemical Journal*, **215**, 669–676.

Pajula, R., Raina, A., & Eloranta, T. (1979) *European Journal of Biochemistry*, **101**, 619–626.

Pajunen, A., Crozat, A., Jänne, O., Shalainen, R., Laitinen, P., Stanley, B., Madhubala, R., & Pegg, A. (1988) *Journal of Biological Chemistry*, **263**, 17040–17049.

Pankaskie, M., Abdel–Monem, M., Raina, A., Wang, T., & Foker, J. (1981) *Journal of Medicinal Chemistry*, **24**, 549–553.

Pankaskie, M. & Lakin, D. (1987) *Biochemical Pharmacology*, **36**, 2063–2064.

Pathak, S., Porter, C., & Dave, C. (1977) *Cancer Research*, **37**, 2246–2250.

Pegg, A. (1977) *FEBS Letters*, **84**, 33–36.

——— (1983) *Methods in Enzymology*, **94**, 260–265.

——— (1984) *Biochemical Journal*, **224**, 29–38.

Pegg, A. & Bennett, R. (1983) *Methods in Enzymology*, **94**, 69–72.

Pegg, A., Borchardt, R., & Coward, J. (1981) *Biochemical Journal*, **194**, 79–89.

Pegg, A., Corti, A., & Williams–Ashman, H. (1973) *Biochemical and Biophysical Research Communications*, **52**, 696–701.

Pegg, A., Coward, J., Talekar, R., & Secrist, III, J. (1986b) *Biochemistry*, **25**, 4091–4096.

Pegg, A. & Jacobs, G. (1983) *Biochemical Journal*, **213**, 495–502.

Pegg, A., Jones, D., & Secrist, III, J. (1988b) *Biochemistry*, **27**, 1405–1415.

Pegg, A. & McCann, P. (1992) *Pharmacology and Therapeutics*, **56**, 359–377.

Pegg, A., Nagarajan, S., Haficy, S., & Ganem, B. (1991) *Biochemical Journal*, **274**, 167–171.

Pegg, A. & Pösö, H. (1983) *Methods in Enzymology*, **94**, 234–239.

Pegg, A., Pösö, H., Shuttleworth, K., & Bennett, R. (1982a) *Biochemical Journal*, **202**, 519–526.

Pegg, A., Tang, K., & Coward, J. (1982b) *Biochemistry*, **21**, 5082–5089.

Pegg, A., Wechter, R., Clark, R. S., Wiest, L., & Erwin, B. (1986a) *Biochemistry*, **25**, 379–384.

Pegg, A., Wechter, R., & Pajunen, A. (1987) *Biochemical Journal*, **244**, 49–54.

Pegg, A., Wechter, R., Poulin, R., Woster, P., & Coward, J. (1989) *Biochemistry*, **28**, 8446–8453.

Pegg, A., Wiest, L., & Pajunen, A. (1988a) *Biochemical and Biophysical Research Communications*, **150**, 788–793.

Pegg, A. & Williams–Ashman, H. (1968) *Biochemical and Biophysical Research Communications*, **30**, 76–82.

——— (1970) *Archives of Biochemistry and Biophysics*, **137**, 156–165.

——— (1987) *Inhibition of Polyamine Metabolism*, pp. 33–48, P. McCann, A. Pegg, & A. Sjoersdma, Eds. New York: Academic Press.

Persson, K., Holm, I., & Heby, O. (1995) *Journal of Biological Chemistry*, **270**, 5642–5648.

Persson, L., Khomutov, A., & Khomutov, R. (1989a) *Biochemical Journal*, **257**, 929–931.

Persson, L., Stjernborg, L., Holm, I., & Heby, O. (1989b) *Biochemical and Biophysical Research Communications*, **160**, 1196–1202.

Porcelli, M., Cacciapuoti, G., Cimino, G., Gavagnin, M., Sodano, G., & Zappia, V. (1989) *Biochemical Journal*, **263**, 635–640.

Porter, C., Dworaczyk, D., Ganis, B., & Weiser, M. (1980) *Cancer Research*, **40**, 2330–2325.

Porter, C., McManis, J., Lee, D., & Bergeron, R. (1988) *Biochemical Journal*, **254**, 337–342.

Porter, C., Sufrin, J., & Keith, D. (1984) *Biochemical and Biophysical Research Communications*, **122**, 350–357.

Pösö, H. & Pegg, A. (1981) *Biochemical Journal*, **200**, 629–637.

Poulin, R., Secrist, III, J., & Pegg, A. (1989) *Biochemical Journal*, **263**, 215–221.

Pulkka, A., Keränen, M., Salmela, A. S., Salmikangas, P., Ihalainen, R., & Pajunen, A. (1990) *Gene*, **85**, 193–199.

Pulkka, A., Ihalainen, R., Aatsinki, J., & Pajunen, A. (1991) *FEBS Letters*, **291**, 289–295.

Pulkka, A., Ihalainen, R., Suorsa, A., Riviere, M., Szpirer J., & Pajunen, A. (1993) *Genomics*, **16**, 342–349.

Radford, D., Eddy. R., Haley, L., Henry, W., Pegg, A., Pajunen, H., & Shows, T. B. (1988) *Cytogenetics and Cell Genetics*, **49**, 285–288.

Raina, A., Eloranta, T., & Kajander, O. (1976) *Biochemical Society Transactions*, **4**, 968–971.

Raina, A., Eloranta, T., & Pajula, R. (1983) *Methods in Enzymology*, **94**, 257–260.

Raina, A. & Hannonen, P. (1971) *FEBS Letters*, **16**, 1–4.

Raina, A., Hyvönen, T., Eloranta, T., Voutilainen, M., Samejima, K., & Yamanoha, B. (1984) *Biochemical Journal*, **219**, 991–1000.

Raina, A., Tuomi, K., & Pajula, R. (1982) *Biochemical Journal*, **204**, 697–703.

Regenass, U., Caravatti, G., Mett, H., Stanek, J., Schneider, P., Müller, M., Matter, A., Vertino, P., & Porter, C. (1992) *Cancer Research*, **52**, 4712–4718.

Russell, D. H. & Lombardini, J. (1971) *Biochimica et Biophysica Acta*, **240**, 273–286.

Sakai, T., Hori, C., Kano, K., & Oka, T. (1979) *Biochemistry*, **18**, 5541–5548.

Sakai, T., Perry, J., Hori, C., & Oka, T. (1980) *Biochimica et Biophysica Acta*, **614**, 577–582.

Sakata, S., Shelly, L., Puppert, S., Schutz, G., & Chou, J. (1993) *Journal of Biological Chemistry*, **268**, 13978–13986.

Samejima, K. & Nakazawa, Y. (1980) *Archives of Biochemistry and Biophysics*, **201**, 241–246.

Samejima, K., Raina, H., Yamanoha, B., & Eloranta, T. (1983) *Methods in Enzymology*, **94**, 270–275.

Samejima, K. & Yamanoha, B. (1982) *Archives of Biochemistry and Biophysics*, **216**, 213–222.

Sarhan, S., Dezeure, F., & Seiler, N. (1987) *International Journal of Biochemistry*, **19**, 1037–1047.

Savarese, T., Chu, S., Chu, M., & Parks, Jr., R. (1985) *Biochemical Pharmacology*, **34**, 361–367.

Savarese, T., Dexter, D., Parks, Jr., R., & Montgomery, J. (1983) *Biochemical Pharmacology*, **32**, 1907–1916.

Schanche, T., Schanche, J., & Ueland, P. (1982) *FEBS Letters*, **137**, 196–200.

Schmidt, G. & Cantoni, G. (1973) *Journal of Neurochemistry*, **20**, 1373–1385.

Schumacher, Jr., H. (1987) *American Journal of Medicine*, **83** (Supplement 5a), 1–4.

Secrist, III, J. (1987) *Nucleosides and Nucleotides*, **6**, 73–83.

Seidenfeld, J., Wilson, J., & Williams–Ashman, H. (1980) *Biochemical and Biophysical Research Communications*, **95**, 1861–1868.

Seiler, N., Sarhan, S., Mamont, P., Casara, P., & Danzin, C. (1991) *Life Chemistry Reports*, **9**, 151–162.

Seppänen, P., Alhonen–Hongisto, L., & Jänne, J. (1980) *European Journal of Biochemistry*, **110**, 7–12.

Seyfried, C., Oleinik, O., Degen, L., Resing, K., & Morris, D. (1982) *Biochimica et Biophysica Acta*, **716**, 169–177.

Shantz, L., Holm, I., Jänne, O., & Pegg, A. (1992) *Biochemical Journal*, **288**, 511–518.

Shantz, L., Stanley, B., Secrist, III, J., & Pegg, A. (1992) *Biochemistry*, **31**, 6848–6856.

Shantz, L., Viswanath, R., & Pegg, A. (1994) *Biochemical Journal*, **302**, 765–772.

She, Q.-B., Nagao, I., Hayakawa, T., & Tsuge, H. (1994) *Biochemical and Biophysical Research Communications*, **205**, 1748–1754.

Sherman, M., Shafman, T., Coward, J., & Kufe, D. (1986) *Biochemical Pharmacology*, **35**, 2633–2636.

Shimizu, K., Abe, M., Yokoyama, S., Takahashi, H., Sawada, N., Mori, M., & Tsukada, K. (1990) *Life Sciences*, **46**, 1837–1842.

Shirahata, A., Christman, K., & Pegg, A. (1985) *Biochemistry*, **24**, 4417–4423.

Shirahata, A., Morohoshi, T., & Samejima, K. (1988a) *Chemical and Pharmaceutical Bulletin*, **36**, 3220–3222.

Shirahata, A. & Pegg, A. (1985) *Journal of Biological Chemistry*, **260**, 9583–9588.

——— (1986) *Journal of Biological Chemistry*, **261**, 13833–13837.

Shirahata, A., Takahashi, N., Beppu, T., Hosoda, H., & Samejima, K. (1993) *Biochemical Pharmacology*, **45**, 1897–1903.

Shirahata, A. M., Morohohi, T., Fukai, M., Akatsu, S., & Samejima, K. (1991) *Biochemical Pharmacology*, **41**, 205–212.

Shirahata, A. T., Takeshima, T., & Samejima, K. (1988b) *Journal of Biochemistry*, **104**, 717–721.

Sindhu, R. & Cohen, S. S. (1983a) *Methods in Enzymology*, **94**, 279–286.

——— (1983b) *Advances in Polyamine Research*, pp. 371–380, U. Bachrach, A. Kaye, & R. Chayen, Eds. New York: Raven Press.

Smith, D. S., King, C., Pearson, E., Gittinger, C., & Landreth, G. (1989) *Journal of Neurochemistry*, **53**, 800–806.

Snell, E. (1977) *Trends in Biochemical Science*, June, 131–135.

Stanek, J., Caravatti, G., Frei, J., Furet, P., Mett, H., Schneider, P., & Regenass, U. (1993) *Journal of Medicinal Chemistry*, **36**, 2168–2171.

Stanley, B. & Pegg, A. (1991) *Journal of Biological Chemistry*, **266**, 18502–18506.

Stanley, B., Pegg, A., & Holm, I. (1989) *Journal of Biological Chemistry*, **264**, 21073–21079.

Stanley, B., Shantz, L., & Pegg, A. (1994) *Journal of Biological Chemistry*, **269**, 7901–7907.

Stjernborg, L., Heby, O., Mamont, P., & Persson, L. (1993) *European Journal of Biochemistry*, **214**, 671–676.

Stoeckler, J. & Li, S. (1987) *Journal of Biological Chemistry*, **262**, 9542–9546.

Stramentinoli, G. (1987) *American Journal of Medicine*, **83** (Supplement 5a), 35–42.

Sufrin, J., Dunn, D., & Marshall, G. (1981) *Molecular Pharmacology*, **19**, 307–313.

Sufrin, J., Lombardini, J., & Keith, D. (1982) *Biochemical and Biophysical Research Communications*, **106**, 251–255.

Sufrin, J., Spiess, A., Kramer, D., Libby, P., & Porter, C. (1989) *Journal of Medicinal Chemistry*, **32**, 997–1001.

Suzuki, O., Matsumoto, T., Oya, M., Katsumata, Y., & Samejima, K. (1981) *Analytical Biochemistry*, **115**, 72–77.

Suzuki, T., Kashiwagi, K., & Igarashi, K. (1993b) *Biochemical and Biophysical Research Communications*, **192**, 627–634.

Suzuki, T., Sadakata, Y., Kashiwagi, K., Hoshino, K., Kakinuma, Y., Shirahata, S., & Igarashi, K. (1993a) *European Journal of Biochemistry*, **215**, 247–253.

Svensson, F., Kockum, I., & Persson, L. (1993) *Molecular and Cellular Biochemistry*, **124**, 141–147.

Svensson, R. & Persson, L. (1995) *Biochimica et Biophysica Acta*, **1260**, 21–26.

Tang, K., Mariuzza, M., & Coward, J. (1981) *Journal of Medicinal Chemistry*, **24**, 1277–1284.

Toohey, J. (1983) *Archives of Biochemistry and Biophysics*, **223**, 533–542.

Toorchen, D. & Miller, R. (1991) *Biochemical Pharmacology*, **41**, 2023–2030.

Trackman, P. C. & Abeles, R. (1981) *Biochemical and Biophysical Research Communications*, **103**, 1238–1244.

——— (1983) *Journal of Biological Chemistry*, **258**, 6717–6720.

van Faassen, H. & Berger, R. (1990) *Journal of Biochemical and Biophysical Methods*, **20**, 189–194.

Wagner, J., Claverie, N., & Danzin, C. (1984) *Analytical Biochemistry*, **140**, 108–116.

Wagner, J., Hirth, Y., Piriou, F., Zackett, D., Claverie, N., & Danzin, C. (1985) *Biochemical and Biophysical Research Communications*, **133**, 546–553.

Wahlfors, J., Alhonen, T., Kauppinen, L., Hyvönen, T., Jänne, J., & Eloranta, T. (1990) *DNA and Cell Biology*, **9**, 103–110.

Wang, J., Viar, M., McCormack, S., & Johnson, L. (1992) *American Journal of Physiology*, **263**, G494–G501.

Waris, T., Ihalainen, R., Keränen, M., & Pajunen (1992) *Biochemical and Biophysical Research Communications*, **189**, 424–429.

White, M., Degnin, C., Hill, J., & Morris, D. (1990) *Biochemical Journal*, **268**, 657–660.

Wiegand, L. & Pegg, A. (1978) *Biochimica et Biophysica Acta*, **517**, 169–180.

Williams–Ashman, H. & Lockwood, D. (1970) *Annals of the New York Academy of Science*, **171**, 882–894.

Williams–Ashman, H., Seidenfeld, J., & Galletti, P. (1982) *Biochemical Pharmacology*, **31**, 277–288.

Wilson, J., Corti, A., Hawkins, M., Williams–Ashman, H. G., & Pegg, A. (1979) *Biochemical Journal*, **180**, 515–522.

Winquist, R., Alhonen, L., Grzeschik, K., Jänne, J., & Eloranta, T. (1993) *Cytogenetics and Cell Genetics*, **64**, 64–68.

Wolfe, M. & Borchardt, R. (1991) *Journal of Medicinal Chemistry*, **34**, 1521–1530.

Woster, P., Black, A., Duff, K., Coward, J., & Pegg, A. (1989) *Journal of Medicinal Chemistry*, **32**, 1300–1307.

Yamanaka, H., Kajander, E., & Carson, D. (1986) *Biochimica et Biophysica Acta*, **888**, 157–162.

Yamanaka, H., Kubota, M., & Carson, D. (1987) *Cancer Research*, **47**, 1771–1774.

Yamanoha, B. & Samejima, K. (1980) *Chemical and Pharmaceutical Bulletin*, **28**, 2232–2234.

Zappia, V., Cartenì–Farina, M., & Pietra, G. (1972) *Biochemical Journal*, **129**, 703–709.

Zappia, V., Galletti, P., Oliva, A., & Porcelli, M. (1983) *Methods in Enzymology*, **94**, 73–80.

Zappia, V., Galletti, P., Porcelli, M., Ruggiero, G., & Andreana, A. (1978) *FEBS Letters*, **90**, 331–335.

CHAPTER 15

Thinking About Cancer: Carcinogenesis Initiation

By the end of the 1960s, many workers in the cancer field suspected that human cancer might reflect the presence of human tumor viruses whose isolation, characterization, and further study could reveal more successful approaches to prevention, diagnosis, and therapy. The pursuit of this line of thought, for example, the study of retroviruses, revealed a far greater complexity than had been imagined. Large groups of newly discovered phenomena (e.g., oncogenes), their gene products, and their relations to human tumors are currently subjects of study.

Many new discoveries in the field of the polyamines were made in the cancer research of the past several decades. Newer biological tools have included undying tumor cells, tumor viruses, and viral transformation, as well as choices of animals and tissue in which cancer induction can be dissected into separable stages. The metabolism of the polyamines was shown to be unusually active in almost all of these experimental systems. Cancer-bearing humans possessed exaggerated forms of this metabolism as seen in their excretory products and internal fluids, as well as in the tumors themselves.

Potent mutagens and carcinogens such as the nitrosamines were found to be derived from the interactions of the polyamines and a variety of internal and environmental products, such as the ubiquitous oxides of nitrogen. Further, some antitumoral antibiotics, such as bleomycin, were found to contain polyamine. In this antibiotic polyamine is essential in initiating the action of the antibiotic on the nucleic acids.

Searches were also conducted for inhibitors of polyamine metabolism, substances that might also inhibit tumor development. The place of the polyamines in cancer research will be discussed under three major headings: carcinogenesis, attempts to define the nature and process of cancer, and studies in therapy.

Initiation and Promotion

The development of a cancer occurs in stages, of which the first, *initiation*, is thought to involve some genetic damage to a cellular chromosome. This damage may be effected by radiation, chemicals, or viral infection. Although it is not clear how the hepatocarcinogen ethionine produces this effect in rats, a change in DNA, that is, ethylation of some guanine moieties, was detected in the rat liver DNA (Swann et al., 1971).

A second stage, that of *promotion*, commits the initiated cell to become a cancer cell. Some carcinogens, such as orally administered urethane, require promotion to evoke skin tumors in mice. Other carcinogens, such as the polycyclic hydrocarbons, can induce skin tumors without additional promoters. Tumor promotion, as distinct from initiation, was discovered in the late 1930s and the active promoting compounds contained in croton oil were identified some 30 years later. These proved to be esters of phorbol, and one commonly used phorbol ester is 12-*o*-tetradecanoyl-phorbol-13-acetate or TPA.

In a third stage, *progression*, cancerous cells multiply and develop additional mutational changes that may facilitate survival of cancer cells and tumors. Some of these changes may increase invasiveness (*metastasis*), permit tumor development, spread beyond the possibility of surgical removal, and are frequently life threatening. This chapter will be concerned mainly with some phenomena of initiation.

It has been asked if promotion relates to the reduced polyamine biosynthetic capacity of aged mammals and their cells (Scalabrino & Ferioli, 1984). According to Chen and Chang (1986) (see Chap. 11) aged human diploid fibroblasts have both a low rate of converting putrescine to essential hypusine in the initiation factor (eIF-5A) and a very high rate of converting polyamines to amino acids. However, the effect of age in promotion in mice varies from organ to organ (Pitot, 1978).

Mutagenesis and Alkylation

Inheritable changes in the base composition of DNA can be introduced by many mechanisms. Changes in DNA structure may occur as a result of physical and chemical effects, as in the production of thymine dimers induced by ultraviolet radiation or polynucleotide chain breaks effected by X-radiation. Their repair may introduce base changes. A chemical reaction, such as deamination of an amino group on a purine or pyrimidine base or the oxidation of a methyl group, can activate a specific DNA-base glycosylase to remove the altered base. Alkylation of a DNA base, such as that to the O^6 or N^7 of guanine, can evoke different reactions. A specific enzyme, the O^6-alkylguanine-DNA alkyltransferase, may transfer the alkyl moiety to a cysteine acceptor in its own structure, thereby restoring a DNA which would otherwise be degraded. This reaction may deplete the active enzyme (Pegg & Byers, 1992). The *N*-glycosyl bond of guanine alkylated at N^7 is labile and is easily depurinated. The newly exposed deoxycytidines in single-stranded regions of the DNA can then be detected in individual cells (Frankfurt, 1990). In subsequent replications bases other than guanine may be inserted, as in other reaction sequences in which altered bases are removed from a DNA chain and subsequently replaced.

Free transforming DNA or DNA viruses may be altered directly by treatment with a mutagen, but inheritability most frequently requires the "fixation" of an alteration in the DNA by replication. Treatment of transforming DNA by a *direct-acting* alkylating agent will be revealed by outgrowth of the cell in which the DNA has multiplied. Nevertheless, the term direct acting has also been used to contrast the mutagen from *premutagens*, which are activated by cellular metabolism to become mutagenizing agents. Whereas nitrogen mustard is a direct-acting mutagen, many polycyclic aromatic hydrocarbons must be activated by oxidative reactions.

The premutagen dimethylnitrosamine is toxic and is also a potent hepatocarcinogen in rats. The compound is probably converted to an esterified 1-hydroxy alkylnitrosamine as the actual (or proximate) mutagenic agent (Keefer et al., 1987; Wright, 1980) in the sequence,

$$\begin{array}{c}CH_3\\ \diagdown \\ CH_3 \diagup \end{array}\!\!N-N=O \longrightarrow \begin{array}{c}CH_3\\ \diagdown \\ HOH_2C \diagup \end{array}\!\!N-N=O \longrightarrow \begin{array}{c}CH_3\\ \diagdown \\ AcOH_2C \diagup \end{array}\!\!N-N=O$$

Dimethylnitrosamine was found to produce 7-methylguanine in the DNA of treated rats as the first novel reaction product found between a carcinogen and the genetic material of a target cell (Magee & Barnes, 1967). The problem of the mode of mutagenic and carcinogenic action of the nitrosamines becomes important in view of the mutagenic action of a mixture of nitrous acid and the polyamines, compounds that are both ingested and synthesized by humans.

Detection of Mutagenesis and Antimutagenesis

As an approach to the hypothesis that carcinogenesis is related to mutagenesis, a series of bacterial tests was devised for direct-acting mutagens, using specially constructed mutants of *Salmonella typhimurium* that reverted in well-defined bacterial genes. The subsequent conversion of an apparently nonmutagenic compound to a direct-acting mutagen was determined by incubation in a rat liver fraction rich in cytochrome P-450. The metabolized mix was then tested against these mutant bacteria (McCann et al., 1975). Nitrite has been found to be mutagenic in the Salmonella/microsome test systems (Balimandawa et al., 1994). Base-pair substitutions were induced and frame shift mutations were not obtained.

It was reported that 174 of some 300 chemicals were known to be carcinogenic and of these 158 (i.e., almost 90%) were mutagenic in this assay. Interestingly ethionine was not found to be mutagenic. An improved screen was later applied to 465 known or suspected carcinogens to assess the mutagenicity of toxic agents. Of the 58% found to be carcinogenic, some 77% were shown to be mutagenic in the Salmonella test system (Rinkus & Legator, 1979). For 15 chemical categories with high activity (312 substances) representing electrophiles or chemicals activable to electrophilic species, some 94% of the carcinogens were also mutagens. The correlation between carcinogenicity and mutagenicity was very good in tests in which the bacterial system was fortified by an enzymatic oxidation.

The development of hyperplastic nodules and hepatocellular carcinoma has long been studied in methyl-deficient rats (Rogers, 1993). The accumulation of fat in liver cells initiates reactions that increase mitosis and culminate in cancer. Ethionine contributes to methyl deficiency and the establishment of fatty livers, but it is not known how methyl deficiency enhances carcinogen action.

Polyamines in Tests of Mutagenicity

In the 1950s and 1960s the detection of bacterial mutants was frequently explored in growth patterns in liquid cultures. Such tests were carried out in addition to the appearance of isolable colonies on plates as the progeny of single mutagenized bacteria. It was observed initially that compounds such as spermidine and spermine reduced the emergence in culture of several species of organisms resistant to various antibiotics, such as streptomycin and penicillin (Sevag & Drabble, 1962). This "antimutagenic" effect extended to atabrine and acriflavine derivatives and occurred in systems apparently unrelated to the control of transport.

H. G. Johnson and M. K. Bach (1965) also reported spermine suppression of the mutation rates of emergence of streptomycin resistance in *Escherichia coli* and of the reversion of tryptophan auxotrophy to independence in *Staphylococcus aureus*. These workers extended their observations to the emergence of drug resistance in cultured animal cells. Similar observations have been summarized by Clarke and Shankel (1975). Nestmann (1977) described the antimutagenic effect of spermine in *E. coli* carrying a mutator gene whose increased mutability to phage resistance was dependent on DNA replication. An antimutagenic effect of spermine in the induction of canavanine resistance has been described in yeast. Mutation rates enhanced by spermine have been reported in the emergence of the ability to synthesize adenine.

The antimutagenic effects of spermine were interpreted initially as a stabilization of DNA. The transforming DNA of *Bacillus subtilis* lost its activity rapidly in buffer at 70–75°C, but the addition of 10^{-4} M spermine protected the activity of the DNA up to 90°C. Protection by spermine against shear was also demonstrated for a viral DNA (Kaiser et al., 1963). Systematic studies spelled out the chemical structures of the organic bases necessary to affect the melting of nucleates at high external temperatures (Mahler et al., 1961; Tabor, 1962). This type of stabilization is not known to relate specifically or directly to that of mutagenesis. Mechanisms in which the stabilization of DNA chains by spermine may facilitate antimutagenesis can be imagined, as in the minimization of DNA fragmentation, the lining up of separate damaged DNA chains during replication, or in the maintenance of the continuity of damaged chromosomes during mitotic separation. However, such hypotheses have been difficult to prove. Mutation rates of polyamine-depleted mutant cells have not been studied.

Some substituted diamines, such as the thioxanthenones, are widely used in the treatment of schistosomiasis. Lucanthone or Miracil D inhibits nucleic acid syntheses selectively in HeLa cells and prevents colony formation of the cells.

Lucanthone

Hycanthone

Miracil D is oxidized by a microsomal system to a direct-acting or proximal mutagen thought to be hycanthone, which is a more active antischistosomal agent. The toxicity (e.g., inhibition of RNA and DNA synthesis) and mutagenicity of hycanthone is decreased in *E. coli* by spermine, an effect attributed to a reduction of the incorporation of the drug into the cells. In addition spermine blocks inhibitory actions of lucanthone in subcellular systems (Weinstein et al., 1967).

These inhibitory effects of spermine are also seen with hycanthone in *Salmonella* and it is assumed that the oxidative conversion of lucanthone to the more mutagenic hycanthone is responsible for the increased activity of the latter. Hycanthone causes frame shift mutations that are predominantly of the base-deletion type. The compound is also carcinogenic in in vivo systems and is believed to involve deletions of G/C pairs from G/C rich tracts of DNA (Hartman & Hulbert, 1975).

An implication of the various hypotheses dealing with the conversion of a promutagen (lucanthone) to a proximate mutagen (hycanthone) and with an antimutagenic effect of a polyamine such as spermine is that the latter blocks alkylation of DNA. It was shown that spermine does inhibit the methylation of the DNA in isolated liver chromatin by N-methyl N-nitrosourea. The polyamine inhibited methylation in the nucleate by this agent at the N^7 and O^6 positions of guanine and the N^3 position of adenine, without inhibiting methylation of the free nucleotides (Rajalakshmi et al., 1978). Two other alkylating mutagens, methylmethanesulfonate and dimethylsulfate, were not inhibited by spermine in their rate of formation of 7-methylguanine (Rajalakshmi et al., 1980).

These possibilities of protective action have been pursued with cells in which particular mutagens provoked chromosomal damage, appearing as sister chromatid exchanges. This effect caused by mitomycin C was inhibited by spermidine; that caused by N-methyl-N^1-nitro-

N-nitrosoguanidine was not affected by this amine (Cozzi et al., 1989). However, spermine and spermidine inhibited the formation of these chromatid exchanges after treatment with psoralen and ultraviolet irradiation, which produce adducts and cross-links (Cozzi et al., 1991). Spermine (1 μM) was protective against chromosome damage of human embryo kidney (KB) cells infected by adenovirus 5, but spermidine and putrescine were not (Bellett et al., 1982).

Because spermine and other polyamines are found in the eucaryotic nucleus, it has been asked if depletion of cellular polyamine will facilitate the killing of the cells by various agents. However, methods for the specific depletion of the various amines like spermine have not yet been used. At best, a depletion of cellular (HeLa) polyamine by a combination of difluoromethylornithine (DFMO) and methylglyoxal-bis(guanylhydrazone) (MGBG) for 48 h will markedly reduce putrescine and spermidine content some 90%, but spermine will be reduced only 70%. The DNA of such cells is more readily degraded by deoxyribonucleases than that of untreated cells. Readdition of spermine to the depleted cells does restore some of the resistance to enzymatic cleavage of the DNA (Snyder, 1989).

Oncogenicity of Acrolein

Acrolein is a toxic α,β-unsaturated aldehyde, thought originally to be cleaved from deamination products of spermidine and spermine. It is also formed in the pyrolysis of oxidized fats in various cooking procedures. The compound forms adducts with bases of DNA and has been demonstrated to be mutagenic in assays with bacteria and several types of mammalian cells. The compound is present in cigarette smoke and is liberated from the antitumor agent cyclophosphamide during therapy, both sources of acrolein giving rise to bladder hyperplasia; it has been asked if the compound helps to initiate human bladder cancer. Mice given acrolein orally have decreased weight gain and increased mortality (Parent et al., 1991). Intraperitoneally injected acrolein does initiate a high incidence of formation of bladder papillomas, although somewhat less than did another more potent initiating agent that evoked a higher incidence of bladder carcinomas. The toxicity of the compound has prevented a test of acrolein as a promoter (Cohen et al., 1992).

Chemical Effects of Nitrous Acid

Nitrous acid is an unstable reagent known to convert primary aliphatic amines to intermediate diazonium salts, which decompose quantitatively to nitrogen and a mixture of alcohols and alkenes. The estimation of nitrogen gas was used in the Van Slyke estimation of the amino nitrogen content of amino acids, peptides, and proteins. When applied to nucleic acids, nitrous acid will also deaminate cytosine to uracil, adenine to hypoxanthine, and guanine to xanthine. In the latter reactions, it would be expected to be a direct-acting muta-

gen for DNA. Additionally, nitrous acid might form a nitrosamine with secondary amines such as spermidine and spermine, as in the following reaction:

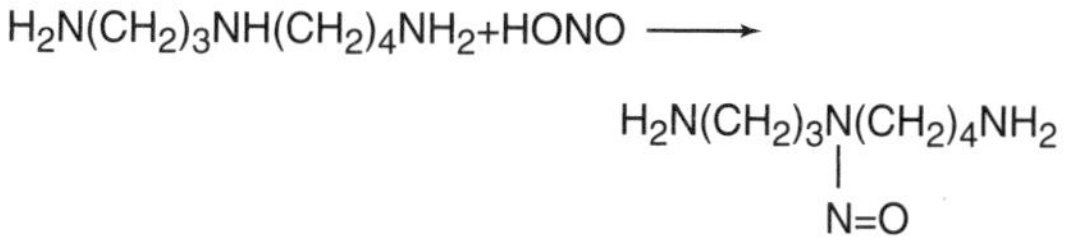

$$H_2N(CH_2)_3NH(CH_2)_4NH_2 + HONO \longrightarrow$$

$$H_2N(CH_2)_3\underset{\underset{N=O}{|}}{N}(CH_2)_4NH_2$$

However, putrescine was shown to be acted upon by nitrous acid to form nitrosopyrrolidine (Warthesen et al., 1975).

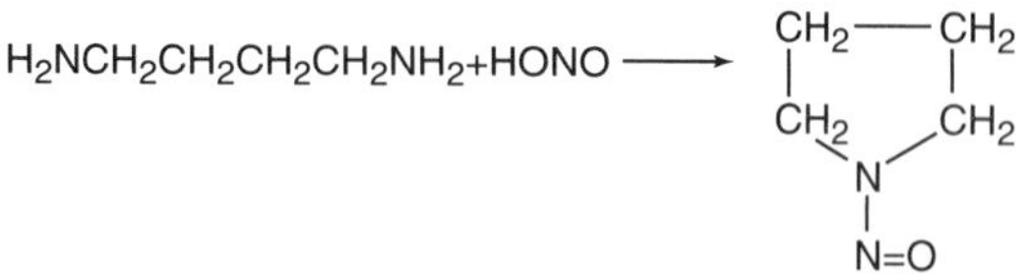

$$H_2NCH_2CH_2CH_2CH_2NH_2 + HONO \longrightarrow$$

Nitrite has been widely used to prevent the growth of pathogenic anaerobic spore formers, such as *Clostridium botulinum*, in various meat products and canned goods. *N*-Nitrosopyrrolidine has been isolated from fried bacon and has been reported to be a carcinogen. Early tests revealed the production of small amounts of the compound by heating several bases, including spermidine and putrescine in the presence of NaNO₂ (Bills et al., 1973). These bases and nitrite are both present to a significant extent in an average Western diet, and it has been estimated that Americans ingest about 2 g of burnt materials daily.

Under the conditions of frying bacon, heating compounds in limited water with nitrite at 160°C, nitrosamines were obtained from putrescine and cadaverine, as well as from ornithine and lysine. When the reactions were carried out in buffer at 100°C, the optimum for the conversion of putrescine to nitrosopyrrolidine was found to be pH 3.8. Putrescine was far more reactive than cadaverine, and ornithine formed significant amounts of nitrosoproline.

The nitrous acid conversion of amino groups in nucleic acid bases to carbonyl groups was the first unequivocal chemical method for altering transforming DNA. This acid is also an interstrand cross-linking agent (Kirchner et al., 1992) affecting guanines on opposite strands to produce the thermostable compound given below.

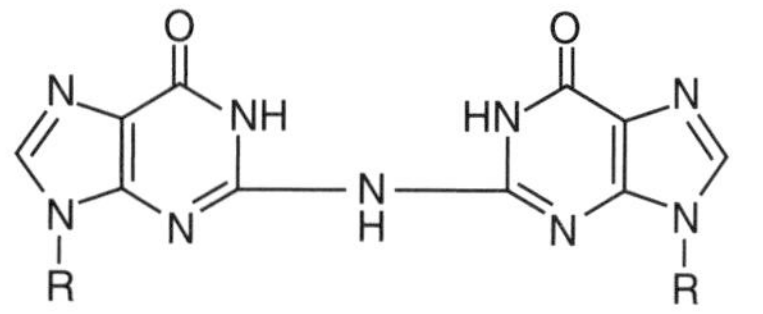

R = 2'-deoxyribosyl

Mutagenicity of Nitrosamines

The mutagenicities of a series of carcinogenic hydroxy-propyl nitrosamines and *N*-nitroso-di-propylamine were compared in the Ames assay after treatment with an oxidizing homogenate containing liver microsomes. The same compounds were compared in an assay for mutagenicity of animal cells to oubain resistance. Liver cells incubated with the mutagen were cocultivated with the indicator Chinese hamster V-79 cells. In contrast to the bacterial assay in which the mutagenic potency of the compounds bore no relation to the known carcinogenic potency of the nitrosamines, the cell-mediated assay revealed a rigorous linear relation to carcinogenicity (Langenbach et al., 1980). Thus, a simple extrapolation of mutagenic data from the bacterial assay to carcinogenicity in the mammal is not possible. A more rigorous test of carcinogenicity requires a cell derived from the mammal of interest, and the carcinogen-exposed cells should subsequently be tested for their cancer-forming activity. In such a recent test a tobacco-specific *N*-nitrosamine, 4-(methylnitrosamine)-1-(3-pyridyl)-1-butanone, was added to growing immortalized nontumorigenic human bronchial epithelial cells, which were then transplanted into athymic nude mice. After some 6 months the exposed cells appeared as invasive adenocarcinomas (Klein–Szanto et al., 1992). This experiment reveals the difficulties of rigorously demonstrating the carcinogenicity of a suspected cancer-forming agent.

Converting Polyamines to Mutagens

Polyamines, nitrates, nitrite, and nitrosamines are present in appreciable amounts in the plant kingdom (Oliveira et al., 1995) and the polyamines at least are a significant continuing component of human and animal diets. The environmental availability of nitrate and nitrite from plant foodstuffs, as well as from the amounts of these salts used to preserve various animal foods, posed the question of the possible production and the potential danger of nitrosamines formed in the interaction of nitrous acid and polyamine. It was found that the actual products of the reaction of spermidine and nitrous acids at pH 3.5 and 80°C included a large number of hydroxyalkyl nitrosamines (Hildrum et al., 1977; Hotchkiss et al., 1977). These were derivatized, separated by gas chromatography, and characterized by mass spectra, NMR, etc., and included substances such as

1. nitrosopyrrolidine,
2. 3-butenyl-(2-propenyl)nitrosamine,
3. 4-hydroxybutyl-(3-hydroxypropyl)-*N*-nitrosamine,
4. 3-hydroxybutyl-(3-hydroxypropyl)-*N*-nitrosamine,
5. 4-hydroxybutyl-(2-hydroxypropyl)-*N*-nitrosamine,
6. 3-hydroxybutyl-(2-hydroxypropyl)-*N*-nitrosamine.

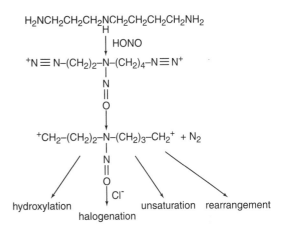

FIG. 15.1   Reactions of nitrous acid with spermidine; adapted from Murray and Correa (1980).

Thus, a single secondary amine, spermidine, can give rise to a range of different nitrosamines, as in the sequence given in Figure 15.1.

Whereas nitrosopyrrolidine is a mutagen only after activation, compounds 2–6 are direct-acting mutagens (Hotchkiss et al., 1979). The direct-acting mutagenicity of several alkyl (ω-hydroxyalkyl) nitrosamines was also reported. The *N*-butyl-*N*-(4-hydroxybutyl)-nitrosamine is a powerful carcinogen that induces bladder cancer in rats and has been used to explore the effect of DFMO in control of polyamine metabolism during carcinogenesis (Uchida et al., 1989). DFMO reduced tumor formation for a significant period.

The direct-acting mutagenicity of the composite products of nitrosation of spermidine had been detected earlier, without identification of these products (Kokatnur et al., 1978). These workers also observed that the production of such mutagens was activated by thiocyanate, a common salivary constituent and a product of cigarette smoke, and was inhibited by ascorbic acid, a common dietary component. These effects are currently interpreted as follows: Protonated nitrous acid

$$H_2O^+\!-\!NO \xrightarrow{\;Y^-\;} YNO$$

combines with an anion Y^- to form $Y{-}NO$, the nitrosating agent capable of reacting with the unprotonated amine substrate. Thiocyanate is such an anion, which forms the catalytic species nitrosyl thiocyanate in saliva (Boyland & Walker, 1974). Nitrosyl iodide, formed from the iodide ion in gastric secretions, is also an accelerating nitrosating agent. Antimutagenic compounds such as thiols, ascorbic acid, and tocopherols reduce nitrous acid to the less active nitric oxide, as in the reactions,

1.

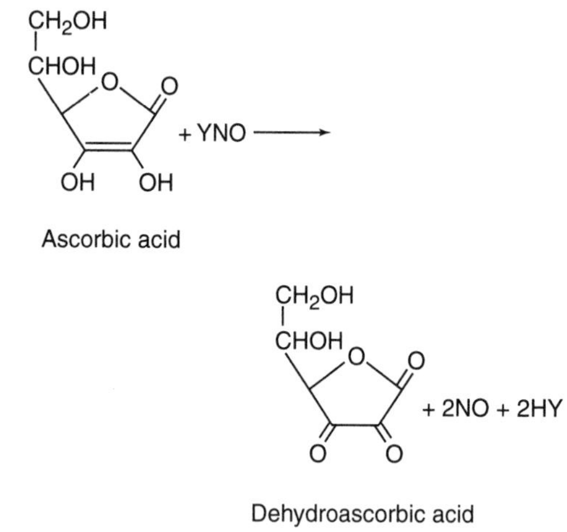

Ascorbic acid

$+ YNO \longrightarrow$

Dehydroascorbic acid

$+ 2NO + 2HY$

2.

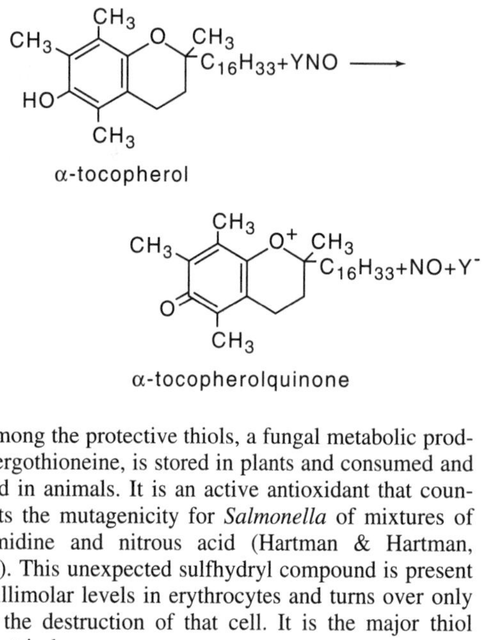

α-tocopherol

$C_{16}H_{33} + YNO \longrightarrow$

α-tocopherolquinone

$C_{16}H_{33} + NO + Y^-$

Among the protective thiols, a fungal metabolic product, ergothioneine, is stored in plants and consumed and stored in animals. It is an active antioxidant that counteracts the mutagenicity for *Salmonella* of mixtures of spermidine and nitrous acid (Hartman & Hartman, 1987). This unexpected sulfhydryl compound is present at millimolar levels in erythrocytes and turns over only with the destruction of that cell. It is the major thiol present in boar semen.

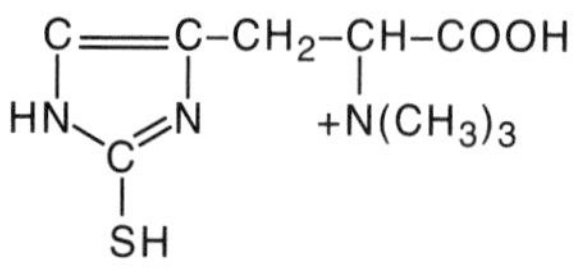

ergothioneine

Thomas et al. (1979) asked if nitrous acid mutagenesis or effects perceived on cells and isolated nucleic acids are not the results of reactions with adherent polyamine. The removal of polyamine from nucleates during isolation procedures requires special attention, for example, fractionation or precipitation of the nucleate in dissociating salt solution. In any case, the studies of Hartman and colleagues demonstrated the increased mutagenicity (base substitutions and not frame shift mutations) as a result of including polyamines in nitrous acid containing systems.

It was evident that carcinogenic nitrosamines might be generated in various cooking techniques. The frying of bacon preserved with added sodium nitrite, or broiling similarly treated meats or fish, can generate many new compounds. Active mutagens are produced from various foods in common cooking procedures. Indeed pyrolysates of spermine and spermidine alone, activated metabolically, were quite mutagenic in the bacterial assay (Ohe, 1982). It has been asked therefore if these mutagens are responsible in some measure for carcinogenesis, as in the production of gastric cancer. Nevertheless, the complexity of the numerous products and the infrequency of the mutagenic events have arrested further direct approaches to the problem, such as an attempted identification of alkylated products of mutagenized DNA.

The Nitrite Problem and Biosynthesis of Nitric Oxide

It has been asked if it is possible or desirable to reduce the dietary intake of nitrate and nitrite as an approach to limiting exposure to nitrosamines. Such exposure is particularly high among smokers of tobacco and workers in industries as diverse as leather tanning, rubber products, and rocket fuel; it constitutes a threat of cancer among these sectors of the population. In many populations, 87% of the nitrate intake is of vegetable origin and unlikely to be modifiable. Vegetables containing ascorbate and polyphenols might inhibit nitrosation. However, almost 75% of the nitrite ingested as such arose from added nitrite in cured meats and other preserved products. The possibility of reducing this intake derived from the curing and preservation of food has been examined (Grossblatt et al., 1982). No simple substitute is known for nitrite in these functions and reductions in the use of nitrite for food preservation were proposed only to the extent that "protection against botulism would not be compromised."

Nitrate is ingested in a normal diet, significantly solubilized in saliva, and much of this is converted to nitrite by oral bacteria (Tenovuo, 1986). In addition it was found that nitrate is produced endogenously, even in germ-free animals, and that the guanidine group of L-arginine was the source of the nitrogen of biosynthesized nitrate (Leaf et al., 1989). Nitrate excretion by humans from NO biosynthesis is of the order of 1–9 mmol/day, perhaps indicating a synthesis of about 200 mmol/day/70-kg individual. The oxidation of NO then leads to a mixture of products, as seen in Figure 15.2.

Despite this common description of the fate of NO, in aqueous solution NO is oxidized primarily to NO_2^- with very little, if any, formation of NO_3^- (Ignarro et

$$2NO + O_2 \longrightarrow 2NO_2$$

$$2NO_2 \rightleftharpoons N_2O_4 \xrightarrow{H_2O} NO_2^- + NO_3^- + 2H^+$$

$$NO + NO_2 \rightleftharpoons N_2O_3 \xrightarrow{H_2O} 2NO_2^- + 2H^+$$

FIG. 15.2 Chemical species arising from nitric oxide in the presence of oxygen; adapted from Leaf et al. (1989).

al., 1993). The conversion to NO_3^- is thought to require additional oxidations, possibly by hemoproteins.

The synthesis of nitric oxide from arginine is being studied actively in many types of cells, of which three, macrophages, vascular endothelial cells, and nerve cells, have been of particular interest. A nitric oxide synthase is induced in macrophages by materials such as bacterial lipopolysaccharide. The enzyme converts arginine to citrulline and NO, which then may participate in the injury of bacteria and tumor cells. Guanidine derivatives of arginine like N^G-methyl-L-arginine inhibit nitric oxide production by the macrophages and block some of the toxic effects of the lipopolysaccharide. Macrophages generating NO increase their transport of arginine and are capable of resynthesizing arginine from citrulline (Wu & Brosnan, 1992). The reaction catalyzed by nitric oxide synthase is presented in Figure 15.3. The highly reactive NO is reported to be fixed by thiols, such as cysteine to form S-nitrosocysteine, or by this amino acid in proteins to form an S-nitrosoprotein adduct, such as that with the serum albumin in human plasma (Stamler et al., 1992). NO is an important messenger serving as a neurotransmitter in the brain and a relaxer of vascular smooth muscle. Compounds such as glycerol trinitrate, used for the therapeutic treatment of angina pain and cardiac anoxia, are believed to release the oxides to this end (Marks, 1992; Salvemini et al., 1992).

The inducible nitric oxide synthase isolated from mouse macrophages has been identified as a cytochrome P-450 type hemoprotein (White & Marletta, 1992). Nitric oxide is known to bind to the iron of certain Fe porphyrin enzymes, such as guanyl cyclase, which is thereby activated. Although many of its physiological effects have been interpreted by this set of effects (Traylor & Sharma, 1992), ribonucleotide reductase, whose activity requires iron and a tyrosyl radical, is also inactivated by nitric oxide (Kwon et al., 1991; Lepoivre et al., 1991). As noted in Chapters 13 and 14, NO inhibits the cobalamin-containing methionine synthase and exposure to NO has been reported to decrease the cellular content of spermidine and spermine. The intermediate, N^G-hydroxyarginine, has been described as an inhibitor of arginase.

Nitric Oxide Adducts and Mutagenesis

The mutagenic effects of gaseous nitrogen oxides in cultured animal cells (i.e., induced mutations, chromosome breaks, etc.) demonstrated some activities of NO_2 and NO (Gorsdorf et al., 1990; Isomura et al., 1984). The problems of the solubility of the gases, their interconversion to each other and to nitrite, the antimutagenic activities of culture media, etc., have prevented a ready interpretation of these observations and solution of the problem of providing this substance in better defined conditions.

Water-soluble compounds have been prepared comprising adducts of selected nucleophiles and nitric oxide, which release the regenerated gas controllably (Maragos et al., 1991). One such new substance is a separable adduct of a nitric oxide dimer bound to spermine at a nitrogen atom (half-life, $t_{1/2} = 39$ min). The structure of the spermine derivative, dubbed "spermine NONOate," has been presented as the zwitterion,

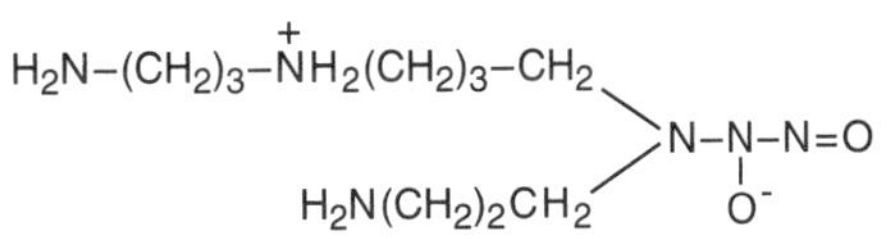

$$H_2N-(CH_2)_3-\overset{+}{N}H_2(CH_2)_3-CH_2$$
$$\diagdown N-N-N=O$$
$$H_2N(CH_2)_2CH_2 \diagup \qquad O^-$$

NO can diazotize primary arylamines in an oxidizing environment and that subsequent hydrolysis of the diazonium ion permits the conversion of cytosine to uracil, adenine to hypoxanthine, and guanine to xanthine.

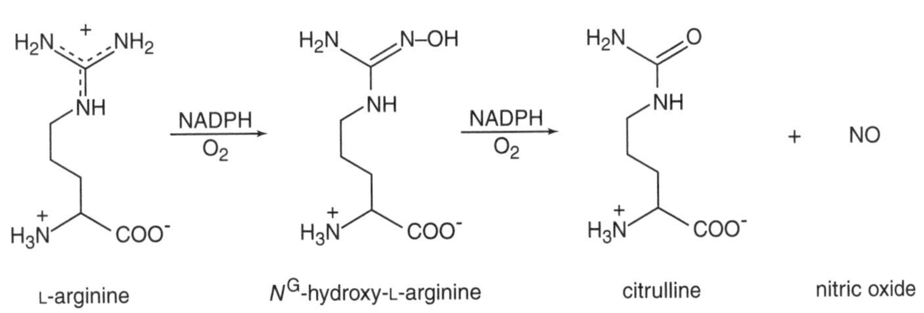

FIG. 15.3 Reaction catalyzed by nitric oxide synthase; adapted from Marletta (1994). N^G-Hydroxy-L-arginine appears to be an intermediate in the reaction.

These reactions were demonstrated with deoxynucleosides and deoxynucleotides presumably giving rise en route to a postulated reactive intermediate (Wink et al., 1991). Nitrite and thiocyanate increased the deamination efficiency of NO and increased the deaminating activity of NO released from NO adducts.

The spermine NONOate was highly active in the direct mutagenization of *Salmonella typhimurium* and almost all of the mutations analyzed (113 of 114) contained C → T transitions, that is, base pairs of G : C in the parent being converted to A : T in the progeny. Such transitions are found frequently in many cancers and hereditary disorders.

The events of the past decade have evoked concerns of the effects of nitrates, nitrites, and the nitrogen oxides. The polyamines have been shown to interact with these effects in several ways, such as the production of more potent mutagenic alkyl nitrosamines or via the preparation of an artifactual mutagenic spermine NONOate. The biogenic source of NO in arginine, a precursor of polyamine, suggests the possible interactions of systems that compete for a common amino acid precursor (Castillo et al., 1995; Soorana & Das, 1995). For example, it would be interesting to know the distribution and regulation of arginase in cells that synthesize both NO and polyamines. One particularly interesting question would be that of the role of the concentration of cellular polyamines on cellular mutagenicity as a function of environmental nitrite or nitric oxide (Morgan, 1994).

REFERENCES

Balimandawa, M., de Meester, C., & Léonard, A. (1994) *Mutation Research Letters*, **321**, 7–11.

Bellett, A., Waldron–Stevens, L., Braithwaite, A., & Cheetham, B. (1982) *Chromosoma*, **84**, 571–583.

Bills, D., Hildrum, K., Scanlan, R., & Libbey, L. (1973) *Journal of Agricultural and Food Chemistry*, **21**, 876–877.

Boyland, E. & Walker, S. (1974) *Nature*, **248**, 601–602.

Castillo, L., Sánchez, M., Vogt, J., Chapman, T. E., De Rojas–Walker, T. C., Tannenbaum, S. R., Ajami, A. M., & Young, V. R. (1995) *American Journal of Physiology*, **268**, E360–E367.

Clarke, C. & Shankel, D. (1975) *Bacteriological Reviews*, **39**, 33–53.

Cohen, S., Garland, E., St. John, M., Okamura, T., & Smith, R. (1992) *Cancer Research*, **52**, 3577–3581.

Cozzi, R., Bona, R., Cundari, E., & Perticone, P. (1989) *Anticancer Research*, **9**, 1129–1132.

Cozzi, R., Perticone, P., Bona, R., & Polani, S. (1991) *Environmental and Molecular Mutagenesis*, **18**, 207–211.

Frankfurt, O. (1990) *Experimental Cell Research*, **191**, 181–185.

Görsdorf, S., Appel, K., Engeholm, C., & Obe, G. (1990) *Carcinogenesis*, **11**, 37–41.

Grossblatt, N. & Committee on Nitrite and Alternative Curing Agents in Food (1982) *Alternatives to the Current Use of Nitrites in Foods*, N. Grossblatt, Ed. Washington, D.C.: National Academy Press.

Hartman, P. & Hulbert, P. (1975) *Journal of Toxicology and Environmental Health*, **1**, 243–270.

Hartman, Z. & Hartman, P. (1987) *Environmental and Molecular Mutagenesis*, **10**, 3–15.

Hildrum, K., Scanlan, R., & Libbey, L. (1977) *Journal of Agricultural and Food Chemistry*, **25**, 252–255.

Hotchkiss, J., Scanlan, R., & Libbey, L. (1977) *Journal of Agricultural and Food Chemistry*, **25**, 1183–1188.

Hotchkiss, J., Scanlan, R., Lijinsky, W., & Andrews, A. (1979) *Mutation Research*, **68**, 195–199.

Ignarro, L., Fukuto, J., Griscavage, J., Rogers, N., & Byrns, R. (1993) *Proceedings of the National Academy of Sciences of the United States of America*, **90**, 8103–8107.

Isomura, K., Chikahira, M., Teranishi, K., & Hamada, K. (1984) *Mutation Research*, **136**, 119–125.

Keefer, L., Anjo, T., Wade, D., Wang, T., & Yang, C. (1987) *Cancer Research*, **47**, 447–452.

Kirchner, J., Sigurdsson, S., & Hopkins, P. (1992) *Journal of the American Chemical Society*, **114**, 4021–4027.

Klein–Szanto, A., Iizasa, T., Momiki, S., García–Palazzo, I., Caamano, J., Metcalf, R., Welsh, J., & Harris, C. (1992) *Proceedings of the National Academy of Sciences of the United States of America*, **89**, 6693–6697.

Kokatnur, M., Murray, M., & Correa, P. (1978) *Proceedings of the Society for Experimental Biology and Medicine*, **158**, 85–88.

Kwon, N., Stuehr, D., & Nathan, C. (1991) *Journal of Experimental Medicine*, **174**, 761–767.

Langenbach, R., Gingell, R., Kuszynski, C., Walker, B., Nagel, D., & Pour, P. (1980) *Cancer Research*, **40**, 3463–3467.

Leaf, C., Weshnok, J., & Tannenbaum, S. (1989) *Biochemical and Biophysical Research Communications*, **163**, 1032–1037.

Lepoivre, M., Fieschi, F., Coves, J., Thelander, J., & Fontecave, M. (1991) *Biochemical and Biophysical Research Communications*, **179**, 442–448.

Magee, P. & Barnes, J. (1967) *Advances in Cancer Research*, **10**, 163–246.

Maragos, C., Morley, D., Wink, D., Dunams, T., Saavedra, J., Hoffman, A., Bove, A., Isaac, L., Hrabie, J., & Keefer, L. (1991) *Journal of Medicinal Chemistry*, **34**, 3242–3247.

McCann, J., Choi, E., Yamasaki, E., & Ames, B. (1975) *Proceedings of the National Academy of Sciences of the United States of America*, **72**, 5135–5139.

Marks, G. (1992) *Journal of Laboratory and Clinical Medicine*, **120**, 826–827.

Marletta, M. (1994) *Journal of Medicinal Chemistry*, **37**, 1899–1907.

Morgan, D. (1994) *Biochemical Society Transcripts*, **22**, 879–883.

Murray, M. & Correa, P. (1980) *Polyamines in Biomedical Research*, pp. 221–235, J. M. Gaugas, Ed. New York: Wiley.

Nestmann, E. (1977) *Molecular & General Genetics*, **152**, 109–110.

Oliveira, C. P., Glória, M. B. A., Barbour, J. F., & Scanlan, R. A. (1995) *Journal of Agricultural and Food Chemistry*, **43**, 967–969.

Ohe, T. (1982) *Mutation Research Letters*, **101**, 175–187.

Parent, R., Caravello, H., & Long, J. (1991) *Journal of the American College of Toxicology*, **10**, 647–659.

Pegg, A. & Byers, T. (1992) *FASEB Journal*, **6**, 2302–2310.

Pitot, H. (1978) *Federation Proceedings*, **37**, 2841–2847.

Rajalakshmi, S., Rao, P., & Sarma, D. (1978) *Biochemistry*, **17**, 4515–4518.

Rajalakshmi, S., Rao, P., & Sarma, D. (1980) *Teratogenesis, Carcinogenesis and Mutagenesis*, **1**, 97–104.

Rinkus, S. & Legator, M. (1979) *Cancer Research*, **39**, 3289–3318.

Rogers, A. (1993) *Journal of Nutritional Biochemistry*, **4**, 666–671.

Salvemini, D., Pistelli, A., Mollace, V., Änggård, E., & Vane, J. (1992) *Biochemical Pharmacology*, **44,** 17–24.

Scalabrino, G. & Ferioli, M. (1984) *Mechanisms of Aging and Development*, **26**, 149–164.

Snyder, R. (1989) *Biochemical Journal*, **260**, 697–704.

Soorana, S. R. & Das, I. (1995) *Biochemical and Biophysical Research Communications*, **212**, 229–234.

Stamler, J., Jaraki, O., Osborne, J., Simon, D., Keaney, J., Vita, J., Singel, D., Valeri, C., & Loscalzo, J. (1992) *Proceedings of the National Academy of Sciences of the United States of America*, **89**, 7674–7677.

Swann, P. F., Pegg, A., Hawks, A., Farber, E., & Magee, P. (1971) *Biochemical Journal*, **123**, 175–181.

Tenovuo, J. (1986) *Oral Pathology*, **15**, 303–307.

Thomas, H., Hartman, P., Mudryj, M., & Brown, D. (1979) *Mutation Research*, **61**, 129–151.

Traylor, T. & Sharma, V. (1992) *Biochemistry*, **31**, 2847–2849.

Uchida, K., Seidenfeld, J., Rademaker, A., & Oyasu, R. (1989) *Cancer Research*, **49**, 5249–5253.

Warthesen, J., Scanlan, R., Bills, D., & Libbey, L. (1975) *Journal of Agricultural and Food Chemistry*, **23**, 898–902.

White, K. & Marletta, M. (1992) *Biochemistry*, **31**, 6627–6631.

Wink, D., Kasprzak, K., Maragos, C., Elespuru, R., Misra, M., Dunams, T., Cebula, T., Koch, W., Andrews, A., Allen, J. S., & Keefer, L. K. (1991) *Science*, **254**, 1001–1003.

Wright, A. (1980) *Mutation Research*, **75**, 215–241.

Wu, G. & Brosnan, J. (1992) *Biochemical Journal*, **281**, 45–48.

CHAPTER 16

Polyamine Metabolism and the Promotion of Tumor Growth

By 1970 the association of an increase of ornithine de-carboxylase (ODC) and putrescine biosynthesis with stimulated cell growth was being explored actively, and it was asked if high concentrations of ODC are charac-teristic of tumors. Although many tumors were high in ODC, a few sarcomas were high in histidine decarbox-ylase and were found to be relatively low in ODC (Sny-der et al., 1970). Nevertheless, in the most actively stud-ied models of chemical and physical carcinogenesis (i.e., mouse skin) and other rodent organs (e.g., liver), ODC did provide an early significant response.

Russell (1971, 1973) noted high levels of urinary ex-cretion of polyamines by many cancer patients and a relatively increased ratio of spermidine to spermine in active hepatomas in mice. William–Ashman et al. (1972) pointed to the parallelism of increased ODC, pu-trescine concentrations, and the increased ratio of ODC to ornithine carbamyltransferase and growth rates in a series of Morris hepatomas.

ODC and Tumor Promotion in Mouse Skin

Some chemicals that stimulate the formation of ODC and polyamine in rodent liver (e.g., thioacetamide, car-bon tetrachloride, phenobarbital) were later found to be carcinogenic or even promoters. The early data on the use of these substances to induce ODC are thus a con-fusing part of the cancer problem, even as are studies with immortal cells in tissue cultures. It seems best to turn to results with somewhat more closely controlled biological systems, such as the staged induction of mouse skin tumors. The complexity of the development and terminal differentiation of skin has introduced some new phenomena, but the availability of genetically de-fined rodent hosts and readily exposed tissue has facili-tated a study of this system.

ODC is very low in normal epidermal cells but very high in tumorous and hyperplastic cells of the epider-mis. As noted in Chapter 15, skin cancers in rodents can be initiated by a single subcarcinogenic dose of a chem-ical such as 7,12-dimethylbenzanthracene (DMBA) and subsequently elicited, initially as papillomas, by promo-tion with croton oil or the phorbol ester O-tetradeca-noyl-phorbol-13-acetate (TPA). In such a sequence, in-direct-acting carcinogens do not induce ODC in the absence of activation by the microsomal enzymes that metabolize various drugs. On the other hand all known tumor promoters do induce ODC. The stimulation of the enzyme is found in tissues other than mouse skin (Weiner & Byus, 1980). A detailed study of promoter-induced ODC in mouse skin demonstrated a 230-fold increase in the activity some 4–5 h after application of TPA or croton oil (O'Brien et al., 1975a). The activity fell to control levels in 12 h. S-Adenosylmethionine de-carboxylase (AdoMetDC) increased to a lesser extent and fell more slowly. The increase of ODC activity was accompanied by an increase of immunoreactive ODC protein, resulting from a transient increase in ODC mRNA and an increase in the rate of synthesis of ODC protein (Gilmour et al., 1987).

Although O'Brien and colleagues (1975b) reported that the induction of ODC occurred after application of promoters exclusively and not with mere irritants or hy-perplastic agents, some inducers of hyperplasia, but not of tumor development, such as the mechanical removal of the uppermost horny layer of mouse skin, also evoke a considerable increase of ODC (Marks et al., 1979). This stimulus, effected by the stripping of an adhesive tape from the epidermis of a hairless rat, has been adopted as a pharmacological model to determine the effects of various agents on the induction of ODC (Bouclier et al., 1987).

Phorbol esters (TPA) induce ODC in in vitro cultures of mouse epidermal cells and will stimulate DNA syn-thesis (Yuspa et al., 1976). A significant level of intracellular Ca^{2+} is required for the induction of ODC by TPA (Verma & Erickson, 1986). TPA induction of ODC occurred without an increase in cAMP; also, induction of cAMP in the epidermal preparations did not induce ODC (Mufson et al., 1977). However, rapid transient increases of cyclic nucleotides heightened by inhibition of the cAMP phosphodiesterase were subsequently observed to be elicited by TPA. Nevertheless, the induction of ODC (150-fold) far outstripped the increase in cyclic nucleotide (six- to 14-fold) (Perchellet & Boutwell, 1981).

The induction of ODC is initiated by an activation of protein kinases, and the analysis of the promoter of the mammalian ODC gene did reveal the presence of two sites that were potentially reactive with proteins responsive to cAMP-dependent protein kinase. The addition of such specific elements in an adrenal carcinoma cell line can increase transcription of the ODC gene (Abrahamsen et al., 1992), further implicating cAMP in the control of ODC inducibility. Kim et al. (1994) described the existence of TPA-responsive sequences in the human ODC gene of HeLa cells, in a region of the gene that bound HeLa nuclear proteins.

ODC change in the in vivo and in vitro systems has been used to screen for tumor promoters and inhibitors of promotion. The polyamine products of the various induced enzymes, which include AdoMetDC, were not sufficient in themselves to support DNA synthesis and cell proliferation (Yuspa et al., 1978). It may be noted that TPA has recently been found to alter cellular morphology and to suppress ODC in human foreskin keratinocytes (see Chap. 21).

Complexities of Promotion and Its Inhibition

The dramatic induction of ODC in mouse skin by TPA has facilitated the tests of various substances of poorly understood function that affect the metabolism of highly active cellular products. For example, aspirin and indomethacin, compounds that interfere with prostaglandin biosynthesis, were found to inhibit induction of ODC by TPA in mouse skin. This inhibition can be counteracted by prostaglandins E_1 and E_2 (Verma et al., 1979; Yamamoto et al., 1992). It has been shown that TPA acting on uninitiated skin to evoke a hyperplasia induces both a lipoxygenase and ODC.

Some oxidized compounds, such as benzoyl peroxide (Binder et al., 1989), 1,4-naphthoquinone, and a 5-hydroxyderivative, also act as promoters and induce ODC (Monks et al., 1990). Plant phenolic compounds that possibly counteract these oxidation products are known to be effective inhibitors of promoter-caused induction of ODC. These substances also proved to be antimutagenic in various test systems and when added to animal diets were found to be protective against several types of tumor induction. In another group of experiments, the hormonally active form of vitamin D_3, $1\alpha25(OH)_2D_3$,

markedly inhibited induction of epidermal ODC effected by TPA (Chida et al., 1984). The induction of epidermal ODC following tape stripping can be inhibited by a vitamin D_3 analogue (Arnold & Van de Kerkhof, 1991).

Retinoic Acid and ODC Induction

Since 1971 vitamin A and some of its analogues have been known to prevent the promotion and development of epithelial cancers. β-Retinoic acid was found to be an excellent inhibitor of the induction of ODC by TPA, although not of the stimulated activity of AdoMetDC. Retinoic acid is currently thought to be a natural morphogen in the development of the chick.

Vitamin A
*trans-*retinol

β-retinoic acid

The compound is not itself an inhibitor of the enzyme and it does not induce the formation of an inhibitor of the ODC. Of some 23 retinoids tested, β-retinoic acid was one of five derivatives that gave 50% inhibition of ODC production at <1 nmol. It also inhibited accumulation of epidermal putrescine and formation of skin papillomas without altering the levels of spermidine and spermine. Subsequent studies suggested a selective inhibition of the synthesis of the ODC protein (Verma, 1985).

A retinol inhibition of ODC formation in cultured Chinese hamster ovary (CHO) cells occurs when the compound is added in the G_1 phase of the cell cycle prior to the increase of the enzyme; it does not affect the G_1-phase increase of cAMP-dependent protein kinase (Haddox et al, 1979). Retinoic acid blocked the synthesis of ODC mRNA in studies with the mouse skin system (Verma, 1988) and in normal human epidermal keratinocytes (Hickok & Uitto, 1992). Several rotenoids of similar potent inhibitory activities and chemotherapeutic promise have recently been found in African plants (Gerhauser et al., 1995).

It was supposed initially that only the cells present in the multiplying underlayer of epidermal cells were the main target of initiation and promotion. In normal development, these cells become terminally differentiated and form piles of cornified cells whose envelope proteins are cross-linked through the action of a membrane-bound Ca^{2+}-requiring tissue transglutaminase (see Chap. 21) (Rice & Green, 1978). The increased concentration of this enzyme can trigger terminal differentiation. TPA enhances transglutaminase activity in mouse epidermal cells; retinoic acid also increases the activity of a cellular transglutaminase but this enzyme is different from that involved in cross-linking cell envelopes (Lichti et al., 1985) and from that induced by TPA (Lichti & Yuspa, 1988). TPA induces ODC in the proliferating basal cells alone (Lichti et al., 1981), and this was inhibited by retinoic acid. The basal cells of developing papillomas, which are high in ODC, are also relatively low in certain normal keratins, implying the existence of an altered pattern of differentiation in the affected cells (Sundberg et al., 1994).

It appears that promotion also affects tumor development via the participation of the several layers, in which TPA both stimulates proliferation of the initiated cells in the basal layer and accelerates terminal differentiation of the keratinocytes (Yuspa et al., 1983). Retinoic acid blocks ODC production in an early stage of the dividing cells and also alters the process of differentiation that relates to the emergence of an epidermal tumor. Within a population of normally promoted cells, subpopulations of keratinocytes of high ODC content have been found to emerge and may be detected by antibody to the enzyme in developing epidermis, papillomas, and tumors (Gilmour et al., 1992). Retinoic acid can affect the emergence of the tumors by altering the multiplication, emergence, and survival of such populations.

Protein Kinases and Control of ODC

Studies in the early 1970s on the reactions stimulated by phorbol esters revealed a rapid early incorporation of $^{32}P_i$ into skin phospholipids, followed by the usual syntheses of the nucleic acids and proteins and the phosphorylation of epidermal histones and other nuclear proteins. The activities of protein kinases raised the problem of their possible roles in the activation of ODC. This may be considered in several forms: whether an inactive protein is somehow activated, as by phosphorylation; whether the induced protein is made de novo, as was subsequently shown; and whether the protein kinases begin a cascade of reactions culminating in the production of active ODC. Mammalian ODC has been isolated in an active nonphosphorylated form, and it can be phosphorylated at several sites by protein kinases with a small change of stability and activity. Our major problem appears to be the dissection of the possible cascade and sequence of reactions from the activation of a protein kinase to the synthesis of the enzyme.

The phorbol esters are bound at the cell membrane by the initiators of reactions activating protein kinase C (PKC). Indeed PKC proved to be the major phorbol ester receptor (Parker et al., 1984). The enzyme can be isolated from several sources, cattle, humans, and rats, and the genes for PKC determine a complex group of homologous protein kinases (Parker et al., 1986). The PKC activity, which is dependent on calcium and phospholipid, transfers phosphate from ATP to serine and threonine in proteins.

The enzyme is activated normally by diacylglycerol in the presence of Ca^{2+} and phosphatidylserine. Diacylglycerol is liberated in the phospholipase C-cleavage of a bound inositol-containing phospholipid. The addition of a diacylglycerol or a phorbol ester to mouse epidermal cells or of phospholipase C to rat epithelial cells induced ODC (Jetten et al., 1985; Sasakawa et al., 1985). ODC can also be induced in guinea pig lymphocytes by a combination of a diacylglycerol and Ca^{2+} (Otani et al., 1985) (Fig. 16.1).

The pleiotropic effects of an activation of a protein kinase by TPA have been revealed in the enhancement of transglutaminase, as well as of ODC. Another finding has been the activation of an Na^+–K^+ pump by TPA in murine leukemic cells; this also appears to relate to an enhancement of spermidine uptake by these cells (Khan et al., 1992).

These treatments increased ODC mRNA in mouse epidermal cells or in skin tumor promotion and suggested an early activation of the PKC functional in ODC gene transcription and tumor promotion. The inhibition by retinoic acid indicated an inhibition in the PKC-mediated accumulation of ODC mRNA (Verma,

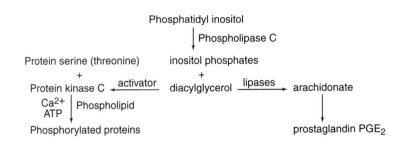

FIG. 16.1 The activation of protein kinase C and the phosphorylation of some proteins.

1988). However, retinoic acid increased PKC activity in mouse melanoma cells (Niles & Loewy, 1989); and quantitative attempts to correlate PKC activity, ODC induction, and DNA synthesis under conditions of promotion by TPA or diacylglycerol were not always successful (Mills & Smart, 1989). A comparative study of ODC induction by TPA or insulin has indicated that induction by the two agents followed different pathways (Butler et al., 1991).

Sphingosine and Inhibition of PKC

The apparent roles of PKC in the governance of growth and cell metabolism, including that of tumor development, led to a search for effective inhibitors of this kinase. The application of a very potent PKC inhibitor, staurosporine, did not inhibit a TPA-induced hyperplasia, ODC induction, and DNA synthesis in mouse skin. The sphingoid bases (e.g., sphingosine), intermediates in the synthesis of ceramides and their derivatives, are compounds present in mammalian epidermis and were found to be potent inhibitors of TPA binding and PKC activity. The PKC activator diacylglycerol overcame the inhibition by sphingosine in various effects on platelets, implicating PKC in signal transduction and suggesting a natural modulating role of sphingosine (Hannun et al., 1987). The early inflammation and hyperplasia caused by application of TPA to mouse skin was inhibited by sphingosine, as was activation of PKC, the induction of ODC mRNA, and synthesis of ODC (Gupta et al., 1988). Sphingosine and other long-chain bases also modulated the transformation of fibroblasts to tumor cells after initiation by radiation and promotion by a phorbol ester (Borek et al., 1991). Sphingosine, one of several naturally occurring sphingoid bases present in animal and plant tissue, may be presented in its Fisher projection formula as

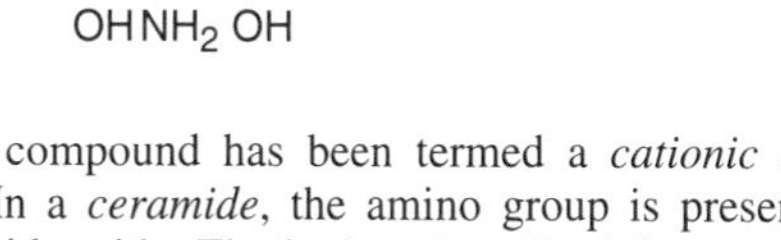

$$H_2C-CH-CH-CH=CH(CH_2)_{12}CH_3$$
$$OHNH_2OH$$

The compound has been termed a *cationic amphiphile*. In a *ceramide*, the amino group is present as a fatty acid amide. The hydroxyl at C_1 of the ceramide is esterified to phosphorylcholine to form sphingomyelin. The presence of the amino and hydroxyl groups at C_2 and C_3, respectively, enable the existence of four stereoisomers, which differ only slightly in inhibitory activities (Merrill et al., 1989). Analysis of porcine epidermis has demonstrated the presence of free sphingosine bases (0.44%) in total lipids, of which some 4–5% is composed of ceramides. These free forms comprise a mixture of C_{16} through C_{20} saturated and unsaturated compounds (Karlsson, 1970; Wertz & Downing, 1989).

Sphingosine is also an activator of casein kinase II, an enzyme also known to be activated by spermidine and spermine (McDonald et al., 1991). It may be asked whether 2-hydroxyputrescine or 7-hydroxyspermidine, compounds substituting for spermidine in some bacterial functions and approaching the structure of sphingosine, can serve as inhibitors or activators of the protein kinases. The possibility of forming intramolecular hydrogen bonds (Fig. 16.2) and its effect in reducing the pK_a of the amino group has been discussed (Merrill et al., 1989). However, its implications in interpreting the activities of hydroxylated polyamines have not been pursued.

Heterogeneity of Tumor ODC

The ODCs of several organisms were found to be heterogeneous in several physical tests that separate the various forms on the basis of charge, for example, electrophoresis, ion exchange, etc. Differences among ODCs were also detectable in stability and activation by various cofactors. The analysis of the ODC of induced mouse epidermal tumors by O'Brien et al. (1989) revealed the presence of two separable ODCs (Fig. 16.3) of identical subunit molecular weights. The molecular weights of the proteins differed, and the higher molecular weight proteins were markedly stimulated by GTP. GTP did not stimulate the apparently single species of ODC obtained from normal skin. The new tumor enzyme also differed from the normal enzyme in the Michaelis constant (K_m) of the substrate, ornithine, and cofactor, pyridoxal phosphate (PLP), heat inactivability, and electrophoretic mobility. No evidence was obtained of a bacterial contamination in the preparations.

A GTP-stimulated ODC was induced in carcinogen-initiated epidermal tumors, promoted by either TPA and a different promoter, chrysarobin, but was not induced by the promoter alone, even though chrysarobin alone did induce ODC in the epidermis (O'Brien et al., 1988). The new ODC appeared in nine of 13 papillomas pro-

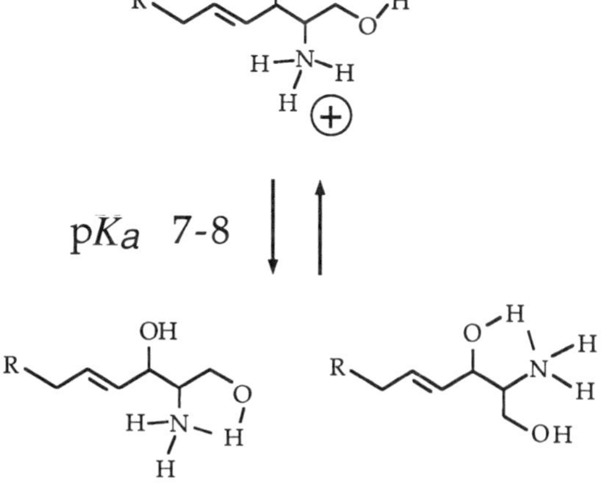

FIG. 16.2 Intramolecular hydrogen bonding in sphingosine; adapted from Merrill et al. (1989).

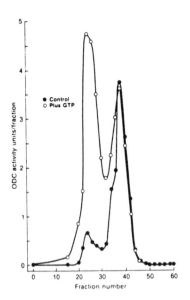

FIG. 16.3 Gel filtration of papilloma ODC; adapted from O'Brien et al. (1989). Fractions containing ODC activity were assayed in the (○) presence or (●) absence of 0.1 mM GTP.

duced in the two-step carcinogenesis. Several human epidermal squamous cell carcinomas (four of seven studied) were found to contain a GTP-activable ODC among a heterogeneous group of ODCs varying in heat stability, kinetic constants, and molecular weights (Hietala et al., 1988). A high constitutive level of ODC was detected in some of the human tumors by reaction with rabbit antibody to mouse kidney ODC.

Two peaks of ODC, separable by DEAE–Sepharose (charge) chromatography were found in extracts of "normal" mouse and human colonic mucosa (Sumiyoshi et al., 1991). The second of these, amounting to 25–30% of the total, could be converted by dephosphorylation to an ODC comparable to the first peak. Both the carcinogen-induced mouse tumors and the human colon tumors contained relatively increased levels of the second peak, and it was asked if the enhanced phosphorylation had caused an escape of this ODC from the normal regulatory controls. This study did not explore the GTP activation reported for mouse and human epidermal tumors by O'Brien et al. (1988, 1989).

The possibility that the GTP-activated enzyme is indeed new and characteristic of the tumor turns on more complete descriptions of the presumably new tumor protein and the presence or absence of the gene of the new tumor protein in the tumor cell. The demonstration of such a gene would obviously be important in the definition of genetic distinction among these tumors and normal tissues. It may also be asked if the apparent GTP activation results from a GTP-induced dissociation of antizyme from an ODC–antizyme complex. The recent report (Kilpeläinen & Hietala, 1994) of a GTP-activated ODC in the mouse brain, a tissue known to be rich in

ODC–antizyme complexes, is consistent with this possibility.

Essentiality of ODC to Tumor Formation in Mouse Skin

The protocol of two-stage carcinogenesis has been used to examine:

a. the levels of activity of ODC in initiated, promoted, and papillomatous mouse tissue;
b. the nature of the ODC proteins found at the different stages;
c. tumor development in the skin of transgenic mice with extraordinarily high levels of human ODC; and
d. the effects of inhibitors of ODC.

A single TPA promotion can induce a rapid transient 500- to 1000-fold increase of the enzyme (Koza et al., 1991). Papillomas proved to be constitutively high in ODC, some 400-fold higher than the basal level. As described above, extracts of most of the papillomas were activable by GTP.

Arginase and ornithine are present in the skin at high levels; the former, which presumably assures the provision of essential ornithine, despite the absence of the remaining enzymes of the urea cycle, is not increased in early phases of ODC induction. Papillomas, however, contain a 20-fold increase in arginase and indeed ornithine content is also markedly higher. Thus, these shifts in the regulation of the polyamine pathway in emerging tumor tissue might affect the balance between proliferation and differentiation.

Various types of overproducing mouse cells have been described (problem c). L1210 leukemia cells resistant to difluoromethylornithine (DFMO) and high in ODC multiplied to form colonies far more readily in soft agar than did the parental line; that is, they appeared to have a growth advantage (Polvinen et al., 1988). Transgenic mice overexpressing the human ODC gene have also been developed. The basal epidermal ODC proved to be 20-fold higher than that of nontransgenic littermates (Halmekytö et al., 1992). A single application of TPA evoked a far greater transient ODC response in the transgenic animal. However, development of papillomas in both groups did require the initiation step of the application of dimethylbenzanthracene, followed by TPA. The transgenic animals then developed almost twice the number of papillomas found on the nontransgenic animals. The high level of ODC in the former had enhanced promotion and also conferred a growth advantage to the tumor cells in vivo. A reexamination of transgenic mice with an overexpression of ODC revealed an increased frequency of spontaneous skin tumors (Megosh et al., 1995). The transfection of mouse 3T3 cells with the DNA encoding human ODC evoked various manifestations of cell transformation. The consistency of this demonstration (see Chap. 17) suggested the designation of the ODC gene as a protooncogene (Auvinen et al., 1992, 1995). Nevertheless, many workers report that ODC overexpression in itself is not sufficient to be tumorigenic (Clifford et al., 1995; Alhonen et al., 1995).

The specific inhibition of ODC (e.g., with DFMO) and its effect in arresting or slowing tumor growth (problem d) will be considered in Chapter 17.

Of Mice, Rats, and Humans

The sequence of initiation and promotion, with a concurrent induction of ODC and production of polyamines stimulating DNA synthesis and growth, is believed to constitute a general pattern of carcinogenesis, which progresses through genetic change and selection to full-blown cancer. These phenomena are found in the development of epidermal tumors in hairless mice induced by long wave ultraviolet light (UVB) (Verma et al., 1979). The system of polyamine and DNA synthesis presented in Figure 16.4 (Seiler & Knödgen, 1979) was

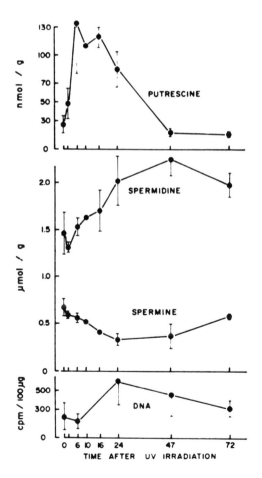

FIG. 16.4   The effect of 10-min exposure to UV light of epidermal polyamines of hairless mice, and on the incorporation of [³H]-thymidine into epidermal DNA; adapted from Seiler and Knödgen (1979).

devised for the screening of drugs potentially useful against psoriasis in humans. The development of tumors induced by UV irradiation in this system is inhibited by retinoic acid, which also inhibits induction of ODC (Connor et al., 1983). Psoralen, which photosensitizes nucleic acids to near UV light and which forms monoadducts and cross-links, blocks synthesis of ODC and DNA in CHO cells (Heimer & Riklis, 1979).

Expression of the ODC gene (i.e., production of ODC mRNA), was increased significantly (10-fold) in one study on the effects of UV irradiation (Fukuda et al., 1990). The expression of ODC in this system has also been used as a measure of potential skin photodamage (Hillebrand et al., 1990a). The levels of basal epidermal ODC and total tumor area in UV-irradiated hairless mice are roughly linearly related (Hillebrand et al., 1990b).

Multistage Carcinogenesis in Rodent Liver

A multistage concept of tumor development was developed early in the 1970s to explain carcinogen-initiated promotion by phenobarbital in the rat liver (Pitot & Sirica, 1992). Although major similarities exist in early stages of tumor development in the skin and liver, the question has been asked if early neoplastic nodules are similar to later tumors. In rat liver foci initiated with diethylnitrosamines, which are high in ODC and DNA synthesis, it was possible to find differences in the early and late expression of the ODC gene. The former was rich in two small ODC mRNAs, 1.8 and 2.1 kb, whereas a major component of mRNA of 2.6 kb appeared actively in the later lesions (Pascale et al., 1993).

Ethionine and the Hypomethylation Problem

Ethionine is a weak hepatocarcinogen in rats and an initiating mutagenesis has been attributed possibly to the formation of 7-ethylguanine (Swann et al., 1971; see Chap. 15). However, it has been asked if a disturbance of methylation patterns may not be mutagenic in itself. Possibly the most significant genetic effect of ethionine relates to the inhibition of DNA methylation. Ethionine is converted to S-adenosylethionine, and this is a good competitor with AdoMet for a DNA methyltransferase. The presence of the analogue results in patterns of DNA hypomethylation, patterns that are also found widely in tumors, including those of humans. Although human neoplastic cells and progression stages of colon cancer may have markedly increased levels (several 100-fold) of DNA methyltransferase mRNA (El-Deiry et al., 1991), it is not evident how this overexpression of the gene relates to the observed hypomethylation.

Hsieh and Verma (1989) found increases (~15-fold) of ODC and proportional increases of ODC mRNA in TPA-treated human bladder cancer cells. The analysis of the ODC gene isolated from such TPA-induced cells by tests with methylation-sensitive restriction endonucleaves did not reveal differences in the methylation patterns between the TPA-induced and noninduced cells. When 5-azacytidine was used in the culture, this did provoke hypomethylation of the ODC gene but did not increase transcription of the gene. Nevertheless, these experiments related only to the effects of promotion on an already initiated cancer cell and did not address the differences between the original bladder cell and the cancer cell that evolved from it. The ODC genes of different human cell lines have large differences in their methylation patterns, and the phenomenon of hypomethylation is most distinct in ODC genes obtained from lymphatic leukemia cells.

ODC in Human Cancers

Scalabrino and colleagues (1980) compared the levels of ODC and AdoMetDC in human normal epidermis, basal cell epitheliomas, and the more rapidly growing squamous cell carcinomas. The levels of ODC in the three systems were in the approximate ratios of $1:9:50$, respectively. With normal ODC taken as 1, the levels of AdoMetDC had ratios of $5:10:40$, respectively. As in the early study of Williams–Ashman et al. (1972) on ODC in Morris hepatomas, the levels of the biosynthetic decarboxylases correlate roughly with the growth rates of the tissues. These results suggested that the levels of ODC can be used as biochemical markers in the progression of the disease, and it has been asked if this presumed marker relates to the nature and condition of the ODC gene.

The autosomal dominant disorder, termed *familial colonic polyposis*, is characterized by the formation of multiple adenomatous polyps that subsequently develop into cancers. It was reported initially that the levels of ODC appeared to increase sharply with the shift to malignancy (Luk & Baylin, 1984). Most subsequent studies of the ODC contents of various types of polyps confirmed these findings. In a more recent study, the contents of ODC and tyrosine protein kinase were compared in benign hyperplastic polyps and malignant adenomatous polyps (Colarian et al., 1991). Whereas both enzymes were elevated in both types of polyps, the development of malignancy significantly increased the content of ODC but not that of the protein tyrosine kinase. Nevertheless, an increase of ODC is characteristic of proliferation and hyperplasia and not of malignancy alone (LaMuraglia et al., 1986; Porter et al., 1987). Further, rectal mucosal ODC was shown not to be a marker of risk for colorectal neoplasia (Sandler et al., 1992). Finally more recent genetic data on familial colonic polyposis related that condition to a mutant gene other than that determining ODC.

Various laboratories have isolated and studied the structure of a human gene for ODC and the expression of the mRNA and the ODC protein (Moshier et al., 1990; Radford et al., 1990). Two such genes have been isolated and localized on chromosomes 2 and 7. Increases in ODC and mRNA were detected in these studies of colorectal neoplasia, but no evidence was found for ODC gene amplification. In another study, colonocytes isolated from cancerous areas were much higher in ODC and polyamine content than were normal colonocytes. Nevertheless, no significant differences were found in their contents of ODC mRNA (Elitsur et al., 1992). However, a study of the ODC gene and its expression in four human breast cancer lines did report higher ODC gene dosage (four- to 12-fold higher) in three of these and increased levels of ODC mRNA in two of the four lines (Thomas et al., 1991). The overexpression of ODC in one human breast cancer cell line was thought to confer a marginal growth advantage to those cells (Manni et al., 1995).

Moshier and colleagues (1992) studied the expression of the human ODC gene in human colon carcinomas cells. It appears, as in the studies of Manzella & Blackshear (1992) (see Chap. 13), that several promoter sites of the gene react with nuclear proteins. These bound sites were detected by the development of resistance to nucleases, that is, DNA band shift and footprint assays. A comparison of the binding patterns to these sites in the cancer cells versus normal fibroblasts suggested that this is more complex in the cancer cells and may relate to their constitutive expression.

In studies of patients with advanced head and neck cancers, the poorly differentiated squamous cell carcinomas were several-fold higher in ODC than were the more differentiated cancers (Westin et al., 1991). Among these patients high ODC activity indicated a poor short-term survival (1 year). Several groups also described a relation between the ODC content of brain tumors and malignancy (Ernestus et al., 1992; Scalabrino et al., 1982). Among astrocytomas, the contents of ODC and putrescine increased with malignancy. However, the point has been made that low levels of ODC do not prove benignity.

Other Enzymes of Polyamine Metabolism as Possible Cancer Markers

The studies on experimental carcinogenesis revealed some unexpected metabolic changes. Despite increases in ODC, changes in transglutaminase did not occur in lymphocytes in cases of lymphocytic leukemia (Vanella et al., 1986), nor was the latter a marker in various classes of brain tumors (Röhn et al., 1992). Actually transglutaminase is active, as are polyamines, in the cross-linking patterns of proteins that take place in various types of skin and mucosal repair (see Chap. 21).

On the other hand several types of human tumors have high levels of diamine oxidase (DAO, histaminase). This was detected in the sera of patients with medullary thyroid carcinoma and small cell lung carcinoma. The tumor DAO was immunologically identical to the immunologically distinctive normal human tissue enzyme and does not appear to be a product of the derepression of the fetal genome (Baylin, 1977). Although the activity has been suggested as a marker for human cancer, the enzyme is not maintained at a high constitutive level in cultures of small (oat) cell lung cancer (Baylin et al., 1980).

The increase of ODC and polyamines in multiplying cancer cells raises the problem of the possible catabolism of spermidine and spermine. The initial steps of this catabolic recycling involves the spermine/spermidine *N*-acetyl transferase (SAT) that is high in tumors, is readily induced like ODC to very high levels, and like the latter enzyme has an unusually short half-life. Thus, a 30-fold increase of the enzyme was found in aggressive prostatic tumors.

On the other hand, the loss of methylthioadenosine (MTA) phosphorylase, which is present in all normal mammalian cells, is a frequent occurrence in human cancer patients. In one study, one of eight leukemia patients and three of 10 solid tumor bearing patients lacked the activity (Fitchen et al., 1986). In another study the urinary excretion of MTA was significantly higher in cancer patients than in normal patients and was reduced by effective antitumor treatment (Kaneko et al., 1990). The inability to complete the cycle of MTA–methionine interconversion creates a growth-retarding deficiency of methionine in such cells. Although the measurement of circulating MTA is not thought to be a very useful diagnostic marker, it may prove to be useful in testing the efficacy of chemotherapy in MTA phosphorylase-deficient patients.

Accumulation of Polyamine in Tumor Cells

The observations of high levels of ODC and polyamine in multiplying and rapidly growing cells (e.g., regrowth after partial hepatectomy) led to studies of tumor cells, as in the growth of a leukemic L1210 tumor in mice (Russell & Levy, 1971). Some 4 days after inoculation with ascites cells the growing tumor had increased its ODC content some 10-fold, as contrasted to normal liver, and AdoMetDC had increased about sixfold. By 6 days ODC had fallen to the near-control level whereas AdoMetDC remained high. At 10 days, putrescine and spermidine content per gram of tissue had doubled in the tumor. Spermine content did not increase, and compounds that slowed tumor growth inhibited the increase of the polyamines.

In tumor growth in mice inoculated by Ehrlich ascites cells, an initially high putrescine content fell after 8 days and spermidine and spermine contents rose significantly in the tumor cells (Andersson & Heby, 1972). Startlingly, the spermidine content of the nontumorous livers of tumor-bearing mice fell initially and then increased markedly (Fig 16.5A, B). Although the RNA contents of the tumor cells and liver were similar, the DNA contents of the tumor cells were about twice that of the liver cells. At 10 days the combined spermidine + spermine nitrogen in the tumor could neutralize about 22% of total cellular nucleic acid phosphorus. The ratio of spermidine to spermine in the tumor was considerably higher than that in normal liver; the ratio of these components in the liver of the tumor-bearing animal

was even higher, approaching the ratio found in regenerating liver.

The increase in the ODC and polyamine in the normal tissues of tumor-bearing animals has been confirmed for some systems. A nondialyzable heat-labile constituent in various tumor homogenates was reported to produce this liver-enhancing effect (Noguchi et al., 1976). A homogeneous glycoprotein of about 70 kDa was isolated from the ascites fluids of an Ehrlich ascites tumor and this can induce ODC (Imamura et al., 1991). This protein, active at 0.5 µg per mouse, stimulated the production in the liver of ODC mRNA and ODC manyfold, but also modified several other types of metabolism.

Heby (1977) analyzed the distribution of the polyamines and some biosynthetic enzymes in rat brain tumor cells. Putrescine concentrations were similar in cytoplasm and nucleus whereas ODC was mainly cytoplasmic, as reported by McCormick for HeLa cells (see Table 11.1). However, the concentrations of spermidine and spermine in nuclei were 8.9- and 5.3-fold higher, respectively, and AdoMetDC had a 7.4-fold higher specific activity in the nucleus than in the cytoplasm.

Increases in all of the polyamines were found in adenocarcinomas of the human thyroid, although a change in the ratio of spermidine to spermine was not seen. The levels of ODC and AdoMetDC were quite high. In one group of medullary carcinomas, DAO (histaminase) was also very high and this enzyme was proposed as a marker for medullary thyroid carcinoma. Some histamine was detected in several adenocarcinomas of the thyroid.

In contrast to these systems, hyperplastic prostate tissue had a markedly increased spermine content whereas spermidine was elevated only slightly and putrescine actually decreased 40–50% (Dunzendorfer & Russell, 1978). These workers reported that renal carcinomas had only slight increases in putrescine and spermidine and modest decreases in spermine content. Another group also reported that renal cell carcinomas increased their spermidine concentrations slightly and they did not find significant changes in other polyamines (Matsuda et al., 1978).

Among patients with lymphomas (both Hodgkin's disease and non-Hodgkin's), urinary polyamines (putrescine and spermidine) were found to increase with progression and severity of the disease. Successful treatment and regression of the lymphoma produced a decrease of these levels of urinary polyamine. It is thought that polyamine excretion reflects the evolution of the disease and may be used to monitor the response to treatment.

Fate of Exogenous Putrescine and Spermidine

When radioactive putrescine was injected subcutaneously into healthy mice and mice inoculated with Ehrlich ascites cells, most of the radioactivity was found quite early in the blood and ascites fluid of the tumor-

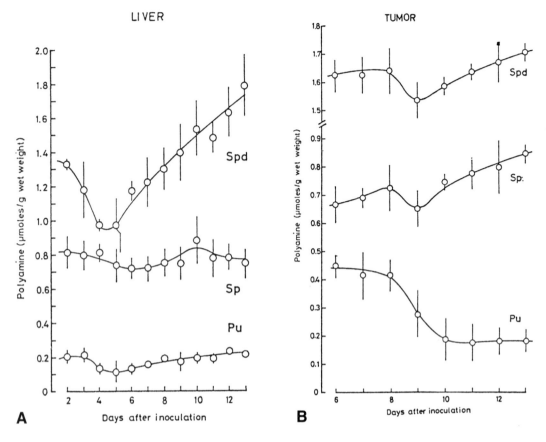

FIG. 16.5 (**A**) Polyamine concentration in the liver of tumor-bearing mice during tumor growth. (**B**) Polyamine concentration in Ehrlich ascites tumor cells during tumor growth. Adapted from Andersson and Heby (1972). Spd, spermidine; Sp, spermine; Pu, putrescine.

bearing mice. Many chromatographically separable metabolites were found (Kallistratos & Pfau, 1965). When [^{14}C]-putrescine and [^{14}C]-spermidine were infused intravenously into humans, both normal volunteers and cancer patients, it was found that intravenously injected polyamine disappeared from plasma in both groups very quickly with half-lives of about 0.5 min (Rosenblum et al., 1978). After 4–5 min almost all of the isotope was present in a conjugated form. A maximum of about 4% of the [^{14}C] from the putrescine appeared in the urine by about 25 h. A maximum of about 65% of the [^{14}C] derived from spermidine was excreted in 50 h. In the four groups, more than 90% of the radiolabel in the urine was in a conjugated form.

Intravenous injection of [^{3}H]-putrescine in the cat produced a similarly rapid clearance of >90% of the tritium from the plasma, with relatively slow accumulations in various organs (Seidenfeld et al., 1979). In 5 h significant amounts of putrescine were detected in the heart, liver, kidney, spleen, lung, and ovary and a lesser conversion to spermidine was found in the liver and spleen. Very little putrescine was found in the brain. Very little newly formed (tritiated) spermine was formed

in any tissue. It was concluded that metabolism of the putrescine took place in the peripheral tissues rather than in the blood.

Injected unlabeled spermidine was not easily detectable in plasma and frequently proved to be relatively toxic in rats. Spermidine appeared to be removed largely by the lung.

Because normal brain tissue proved to be low in free putrescine and was readily detected in brain tumors, it was supposed that an appropriately labeled putrescine might be detected in a brain tumor. Injected circulating [^{3}H]-putrescine in the normal rat is cleared rapidly and appears in a number of tissues. In 2 h [^{3}H] in the brain is about 5% that in the liver and blood and about 1.5% that in skeletal muscle. Injection of [^{14}C]-putrescine into rats with a transplanted gliosarcoma permitted the ready detection of a tumor in the brain by autoradiography of the [^{14}C] polyamine very soon after administration of the putrescine (Volkow et al., 1983). These workers followed the in vivo metabolic conversion of putrescine into spermidine, spermine, and other metabolites in the tumor, and detected a markedly lower level of accumulation and metabolism in the normal brain tissue.

The possible use of $[^{14}C]$-putrescine as a tracer for detection by positron emission tomography (PET) has been explored in the imaging of human tumors. However, studies with $[^{11}C]$-putrescine in human patients followed by PET revealed that the uptake of the isotope in human brain tumors was not sufficiently specific to be useful (Warnick et al., 1988). Furthermore, $[^{11}C]$-putrescine uptake and PET were not found to be a useful approach in determining the growth of human prostate adenocarcinoma (Wang et al., 1994).

The possibility of detecting and localizing tumors as a result of preferential accumulation and metabolism has also been tested with 2,2-difluoroputrescine. This fluoro derivative is metabolized to a 6,6-difluorospermidine and 6,6-difluorospermine, which can be detected by ^{19}F-NMR spectroscopy at levels of 1 nmol/g. Injection of the fluoroputrescine permitted the detection of the fluorinated polyamines in tumors, liver, and urine in 10–20 min, when 2–5% of the natural polyamines had been replaced by the fluoro analogues. Fluorospermidine and fluorospermine appeared preferentially incorporated into tumors. It seemed that such analogues might be useful in diagnosis and localization of tumors particularly active in polyamine metabolism (see Chap. 14).

Can Exogenous Ornithine and Arginine Regulate Tumorigenesis?

When it became clear that ODC was elevated in growing tumors it was asked if growing rodent tumors could be detected in intact animals after the injection of 1-$[^{14}C]$-ornithine. The release of respiratory $^{14}CO_2$ was significantly greater in rats bearing transplanted or carcinogen-induced tumors (Buffkin et al., 1978). When D,L-[5-^{14}C]-ornithine was injected intravenously into mammary carcinoma bearing mice, over 40% of the ^{14}C was found in 60 min in the polyamine fraction of the tumor tissue and only 5% was found in the liver polyamines (Ishiwata et al., 1988).

When exogenous ornithine or arginine was provided to hepatoma cells the relatively low internal pools of these amino acids increased some four- to eightfold; the cells then responded to the tumor promoter TPA by a marked increase in the production of putrescine (Wu & Byus, 1984). The effects of diet restriction were studied in the mouse during two-stage (DMBA + TPA) epidermal carcinogenesis (Gonzalez & Byus, 1991). An arginine-free diet in control animals decreased epidermal levels of arginine and ornithine by 40%. There was a similar decline in the numbers of tumors in the arginine-deprived tumor-bearing animals, but tumor incidence was restored by addition of ornithine to the diet. Arginine-deprived animals overproduced ODC in the epidermis in response to TPA but accumulated only 55% of the putrescine. Addition of ornithine led to a low level of induction of ODC (25% of control) but resulted in an increased production of putrescine (170%). It was clear that dietary arginine can be limiting in tumors and that this limitation can be largely minimized by provi-

sion of ornithine. Nevertheless, the level of putrescine made available was more crucial to the development and number of tumors than the absolute level of ODC produced.

In an Ehrlich ascites tumor model, arginine was the only amino acid undetectable in the cells (Márquez et al., 1989a). However, shortly after transplantation of tumor cells, the host liver lost both arginine and newly synthesized ornithine; the latter appeared in ascitic fluid (Márquez et al., 1989b). Thus, arginine may be limiting in cancer cells and is rapidly converted to ornithine and then via ODC to putrescine, a component that affects tumor development. If in an intact animal the tumor is low in these components, it can cause a transfer from host tissue, liver in these instances, to the tumor.

These effects, which eventually debilitate the host, are not found in all species of mice. Indeed the transfer from host to tumor has been studied thoroughly only in the Ehrlich ascites system (Medina et al., 1992). Further, some laboratories have described a reduction of tumorigenesis by arginine supplementation in some animal tumor models; it has been suggested that these latter effects result from a stimulation of the immune system. Thus, several factors, including those of the diet and withdrawal of essential metabolites from host tissues, can contribute to tumor development and growth.

Detecting Cancer Markers

The presence of relatively high concentrations of polyamines was reported in the urines of cancer patients (Russell, 1971). Some 50 normal volunteers and 35 nontumor patients had mean 24-h excretion values of 2.7–2.9 mg putrescine, 3.1–4.7 mg spermidine, and 3.4–3.5 mg spermine (Russell et al., 1971). Many cancer patients with solid tumors or leukemias were found to have excretion patterns of polyamine several-fold higher. The latter levels of excretion decreased toward normal levels in patients in remission. It was suggested that these excretion levels might serve in the early detection of cancer and in monitoring the development of the disease.

Elevated urinary polyamines were also found in most cases of lung cancer, colorectal cancer, squamous cell carcinoma, etc. (Sanford et al., 1975). However, a significant number of noncancerous individuals excreted higher than normal levels of polyamine, for example, pregnant women or individuals with infectious or inflammatory disease (Dreyfuss et al., 1975). Also, fewer than 15% of patients with breast carcinoma appeared to have abnormal polyamine excretion patterns. This was subsequently increased to 39–50% in preoperative patients (Tormey et al., 1980). In any case, the specificity of the higher levels of excretion was inadequate to permit the use of this urinary excretion pattern to serve in the detection of cancer. However, various early workers noted that temporary remission of the disease following surgery, chemotherapy, or radiation, evoked early changes in the urinary pattern followed by a decrease in the excretion of polyamines. This excretion climbed again early in a relapse.

Reviews of the status of these marker studies pointed to a lack of specificity in an increased urinary excretion of polyamine (Cohen, 1977). Thus, in patients with advanced colon carcinoma, increases of putrescine were noted in only 31%, spermidine in 39%, and spermine in 66%. About 25–30% failed to show an increase in any polyamine. Nevertheless, in selected disease (e.g., Burkitt's lymphoma) urinary polyamine levels correlated well with clinical disease. In general, chemotherapy that killed cancer cells temporarily increased excretion of polyamines, and polyamine levels might be used to monitor the quality and dose of the therapeutic mode. Relapses were indicated by increases of polyamines in urine or cerebrospinal fluid (CSF) (Russell, 1983). Good predictive correlations of putrescine and spermidine excretion in urine were obtained in patients with active large metastatic cancers, but were much less so in tumors of the breast, female genital tract, and prostate (Horn et al., 1984). A recent discussion of biological and NMR markers for cancer (Czuba & Smith, 1991) has not been sanguine on the potential of the polyamines in this role. Nevertheless, Bachrach (1992) has posed the question of their potential more directly in the context of staging, localization, and detection of metastasis, the analysis of the efficacy of therapy, and the detection of recurrence.

Among the compounds detected in urine by means of the newer techniques were the Schiff bases of various polyamines and PLP. The urinary compounds were stabilized by reduction, dephosphorylated in acid, and fractionated on a silica-gel column. Volatile trifluoroacetyl derivatives were characterized by gas chromatography-mass spectroscopy (GC-MS) and compared with standards prepared by reactions of PLP with 1,3-diaminopropane (DAP), putrescine, cadaverine, spermidine, and spermine. Derivatives of all of these were isolated from human urine (Aigner–Held et al., 1979). An eventual adoption of polyamines as tumor markers required more rapid and simplified analyses that led to the development of various types of enzymatic assays, for example, that of Yodfat et al. (1988) in which H_2O_2 liberated by oxidases is estimated by a measurement of luminescence.

Another by-product of this study was the discovery of the increased production and urinary excretion of N^I-acetylspermidine. The detection and initial proof of structure of such products required acid hydrolysis and the evolution of sophisticated methods to find other metabolic products, such as those stemming from the catabolism of the polyamines. GC-MS of the urine samples revealed that this pathway is much more active in neoplastic disease and complements the results indicating increased levels of acetylation of the polyamines in cancer (Van Den Berg et al., 1985). Eventually tests sensitive at nanomole levels did detect putrescine and spermidine at a slightly elevated level in sera of cancer patients (Nishioka & Romsdahl, 1974). These were much lower than the concentrations in urine, and these results led to the exploration of a more sensitive but less

specific radioimmunoassay (Bartos et al., 1975). The appearance of polyamines in selected body fluids was believed to relate to the release of these constituents from dying or dead cells. This hypothesis was developed from observations of the increased polyamine in the blood plasma of patients with *polycythemia vera* (Desser et al., 1975) and in the bone marrow plasma of cancer patients subjected to chemotherapy (Nishioka et al., 1980). One experiment to test the hypothesis correlated the release of a DNA fragment and polyamine into the ascites fluid in which Ehrlich ascites tumor cells were dying (Andersson et al., 1978). Nevertheless, the proof of the hypothesis has been difficult to establish in other tumor systems (Driessen et al., 1989).

In any case a high percentage of the polyamines is transported in blood by the erythrocytes. Takami and Nishioka (1980) were among the first to demonstrate that the blood cells of cancer patients carried the largest proportion of blood polyamines and that the increase in the polyamine of the cellular compartment was far more readily detected than that in plasma. Such an increase in total cellular polyamine was detectable in 77% of the cancer patients in contrast to the 38% found in plasma polyamine.

Erythrocytic Polyamines

In a study of the age distribution of the polyamines in human whole blood, initially high concentrations of spermidine and spermine increased in the hours shortly after birth and then fell to lower levels in infancy and childhood. The levels in this growth period were two- to threefold higher than that in the adult (Casti et al., 1982). In normal healthy adults spermidine plus spermine contents of the red blood cells (RBCs) were of the order of 10–20 nmol/10^{10} erythrocytes. Values of 7–30 nmol/10^{10} RBCs of normal healthy adults have been reported more recently. The latter values were markedly increased in cell populations containing reticulocytes, which are rich in nucleic acids. Enucleate red cells cannot synthesize polyamines.

Three- to fivefold higher contents were found in the erythrocytes of patients with sickle cell anemia (Chun et al., 1976). This was attributed to a binding by a large fraction of the spectrin of the stromal protein (Chun et al., 1977; Natta & Kremzner, 1988). Normal erythrocytes do adsorb some but not the large amounts of polyamine that exist intracellularly in sickled cells (Tadolini et al., 1986). Although plasma transglutaminase can insert polyamine into erythrocyte membranes (see Chap. 21), the compounds are largely transported into the cells, which can carry these substances to the various tissues, including cancers (Khan et al., 1991; Moulinoux et al., 1984b).

Increased spermidine and spermine levels were detected in the erythrocytes of patients with many types of cancer. Significant levels of putrescine were not found in "normal" patients (Gerbaut, 1991) and this component was found to be only slightly elevated (two- to threefold) in the red cells of tumor-bearing animals, including human cancer patients (Grossie et al., 1986). The very low putrescine contents of erythrocytes of women are scarcely elevated in pregnancy or at delivery; a similar finding was noted for the RBCs of newborn children (Hiramatsu et al., 1981). On the other hand, newborn mice had relatively high RBC polyamine levels, including those for putrescine (Quemener et al., 1991). All of these decreased sharply during subsequent growth and the putrescine level was undetectable by the sixth week after birth.

A small temporary increase of putrescine in erythrocytes was found during liver regeneration of the partially hepatectomized rat (Moulinoux et al., 1987). However, the many-fold larger contents of spermidine (10- to 60-fold) and spermine (10-fold) led most workers to concentrate on the latter polyamines. Indeed the spermidine contents in red cells, unlike that of putrescine, correlated well with that of liver spermidine and liver cell proliferation. Moulinoux and colleagues (1989a) demonstrated that the erythrocytes of tumor-bearing mice have a several-fold more rapid uptake of spermidine in vitro than do the cells of normal mice. The increased levels of spermidine found in the cells were formed primarily from variously labeled intermediates in the tumor and released as spermidine in the blood to be taken up by the erythrocytes (Moulinoux et al., 1989b).

Uehara and colleagues (1980a) showed that spermidine and spermine levels of peripheral erythrocytes increased linearly for 10 days after inoculation of Ehrlich ascites carcinoma cells, which showed exponential growth in this period. When tumor cell growth was suppressed, polyamine levels fell. Extrapolating these results to human lymphoma patients, spermidine and spermine levels in the red cells were found to mirror the progression in the various malignant lymphoma patients. Similar results were obtained with many patients of advanced cancer (Uehara et al., 1980b,c).

A study of children with acute lymphoblastic anemia revealed the possibility of classifying the various subpopulations with an eye to setting therapy for a particular subgroup (Quemener et al., 1993). Pursuing the possibility of relating the RBC values to specific types of cancer, these workers observed high levels in prostatic adenocarcinoma patients but found that these values constituted an index of cell proliferation and did not correlate with the usual prostatic tumor markers, phosphatase and specific antigen (Cipolla et al., 1994). Among a mix of patients in remission and in progression, only the RBC polyamines correlated with progression. The spermine values in particular discriminated progressive from nonprogressive patients with this disease with high specificity (95%) and provided a guide to management (Cipolla et al., 1994).

Analysis of CSF and Brain Tumors

To concentrate on brain tumors and the possible significance of polyamines appearing in the CSF, the sensitive ion-exchange column chromatography developed for the amino acids was adapted to the analysis of small amounts of this difficult to obtain fluid (Marton et al., 1974). Elevations of polyamines were found in the CSF of untreated central nervous system (CNS) tumors and successful therapy elicited decreases to the normal range of CSF polyamine levels (Marton et al., 1976). Serial analysis also revealed an inverse relation of polyamine level to clinical status, with increases preceding tumor regrowth. Putrescine, which is at a very low concentration in CNS tissue, is found readily in the CSF and tends to exceed spermidine in this fluid: 182 ± 54 pmol putrescine/mL vs. 150 ± 48 for spermidine.

Among the brain tumor patients analyzed, the elevated polyamine concentrations found in the CSF of individuals with medulloblastoma correlated almost completely with their clinical status. Putrescine concentrations were usually significantly higher in this disease than in others (Marton et al., 1979). A similarly high putrescine level was recorded for astrocytomas and was not found in so-called "benign" tumors (Harik & Sutton, 1979). Although the content of the diamine was proposed as a biochemical marker for malignancy in brain tumors, a correlation was not found with some glioblastomas and astrocytomas (Fulton et al., 1980). The selective correlation of putrescine to medulloblastoma has been attributed to the relative proximity of this tumor to CSF pathways and the rapid rate of diffusion of putrescine from the tumor to the CSF (Pierangeli et al., 1981). Some children with neurological disorders, hydrocephalus, encephaloceles, etc., also have increased levels of CSF putrescine. It has been suggested that the most useful tumor marker is an increasing polyamine level in an individual child.

Adult patients with glioblastomas reportedly have elevated levels of erythrocyte spermidine. This result was noted in all postoperative recurrences, and a significant percentage (30%) of such elevated RBC levels was seen 1–6 months before clinically detectable disease was observed (Chatel et al., 1987).

The analysis of polyamines and their metabolites in the CSF has been extended by combining the GC of derived N-heptafluorobutyryl methyl esters and MS to these CSF fractions. With this technique studies have been reported of the isotopic dilution of standards comprising deuterated DAP, putrescine, cadaverine, spermidine, spermine, putreanine, and isoputreanine. This permitted comparison of the contents of these substances in normal CSF as compared to that of a patient with an astrocytoma (van den Berg et al., 1987). The patient with high initial CSF putrescine was operated on to remove much of the tumor and was treated with a chemotherapeutic agent. Estimation of the fate of isoputreanine, putreanine, and DAP were included as relating to activities of terminal catabolism. Although putrescine fell and remained relatively low, a release of spermidine and spermine was detected shortly after chemotherapy and did not increase markedly. Putreanine, the product of spermidine catabolism, did increase significantly in the course of treatment, and this result was obtained with a large group of similarly studied patients. It will be recalled that the content of putreanine in the brain is five- to 10-fold higher than in other organs.

Putrescine and Tumorigenesis

The demonstrations of the importance of ODC and the accumulations of polyamines in rapidly growing cells, including tumors, posed the possibility of a key role of the ODC product, putrescine. Putrescine appeared to stimulate DNA synthesis in the epidermis of the hairless mouse (Gange & Dequoy, 1980). Obviously the hypothesis of a special role of the diamine is complicated by the subsequent sequential production of spermidine and spermine and the observations that the ratio of spermidine to spermine was maximal during maximal rates of DNA synthesis in TPA-promoted mouse skin (Astrup & Paulsen, 1981).

Injected putrescine decreases the activity of ODC markedly in mouse-grown Ehrlich ascites carcinoma cells (Kallio et al., 1977). This result was also obtained with other exogenous diamines, DAP, cadaverine, and the unnatural 1,6-diaminohexane. In the controlled two-stage system of carcinogen-initiated, TPA-promoted mouse skin tumorigenesis, exogenous putrescine inhibited the induction of epidermal ODC and papillomas (Weekes et al., 1980). These workers did not find an effect on AdoMetDC and duplicated the inhibitions with 1,7-diaminoheptane, spermidine, and spermine. They did not detect the appearance of antizyme, and the mechanism of these inhibitions has not been clarified. More recently exogenous putrescine was shown to inhibit the formation of colon tumors, with concomitant decreases of ODC and DNA synthesis in the colon mucosa during induction by azoxymethane (Tatsuta et al., 1991).

Despite these inhibitory effects of diamines, including putrescine, in tumorigenesis, the requirement for ODC and this polyamine intermediary in tumorigenesis was demonstrated with the aid of DFMO. When this inhibitor of ODC was applied directly to mouse skin in a TPA-promoted system, the formation of papillomas was inhibited 50%. When administered in drinking water, the induction of ODC and the accumulation of putrescine inhibited tumorigenesis by 90% (Takigawa et al., 1983). The invasive capacity of cancer cells has been tested by following their penetration through a monolayer of endothelial cells. DFMO treatment decreased penetrability, and this inhibition was reversed by putrescine but not by spermidine or spermine (Ashida et al., 1992).

Catabolic Source of Putrescine

As discussed in Chapter 12, the presence of acetylated polyamine was detected in urine in the late 1970s. Three of these acetylated compounds, N-monoacetylputrescine, N^1-acetylspermidine, and N^8-acetylspermidine, are found in normal human urine in amounts approaching the amount of total conjugated putrescine and spermidine (Table 16.1).

The urinary excretion of N^1-acetylspermidine is increased in many cancer patients. Persson and Rosengren (1989) reported that human mammary cancers contain much larger amounts of N^1-acetylspermidine and N^1-acetylspermine than do normal breast tissue. They also reported much higher levels (20-fold) of spermidine/spermine N^1-acetyl transferase (SSAT) in the tumors and also asked if the presence of the conjugates can serve as a marker of neoplastic growth. It would be important to know if high concentrations of the acetyl derivatives are found in breast milk. The possible role of polyamines in the evolution of breast cancer is heightened by the report that estradiol-mediated control of the division of breast cancer cells by the cyclins may be inhibited by polyamine depletion (Thomas & Thomas, 1994).

However, in rats the polyamine patterns of urines of normal and tumor-bearing rats were substantially identical, with the exception of a small but significant increase in N^1-acetylspermidine in the cancer-ridden animal (Seiler et al., 1985). The latter excretion was thought too small to serve as a marker, although a considerable excretion of the conjugate did occur at late stages of tumor growth (Seiler et al., 1981). In any case increased practice in the analysis of human urine has demonstrated the consistent excretion of mainly acetylated polyamine by normal children and adults and some upward levels of these conjugates in the urinary contents of leukemic patients. Highly variable responses in these levels were found during chemotherapy (Mach et al., 1983).

McCormick (1978a,b) detected a conversion of spermine to spermidine within growing and virus-infected mouse cells, and gross leakage of these components from the infected mouse cells. These studies preceded the discovery of polyamine oxidase (PAO) and the development of the essential analytical methods for the acetyl derivatives, which were subsequently employed in studies of the polyamine N-acetyl transferase of baby hamster kidney (BHK) cells and excretion of polyamine conjugates by these cells (Wallace et al., 1985). The discovery of the mammalian polyamine oxidase by Hölttä in 1977, a flavoprotein enzyme that preferentially oxidizes N^1-acetylspermidine to putrescine, revealed a route other than that of ODC by which putrescine can be formed. Indeed N^1-monoacetylspermine is also oxidized by this enzyme to spermidine, which in another round of acetylation and oxidation will also produce putrescine, that is,

$$\text{spermine} \xrightarrow{\text{SSAT}} N^1\text{-acetylspermine}$$
$$\downarrow \text{PAO}$$
$$\underset{\text{H}}{CH_3CONCH_2CH_2CHO} + \text{spermidine}$$
$$\downarrow \text{SSAT}$$
$$\text{3-acetamidopropanal} + \text{putrescine} \xleftarrow{\text{PAO}} N^1\text{-acetylspermidine}$$

TABLE 16.1 Urinary polyamine excretion by healthy human volunteers and rats.

Compound	Human Male	Human Female	Rat
Putrescine	Trace	Trace	2.4 ± 0.3
Monoacetylputrescine	16.6 ± 3.2	14.7 ± 4.0	Trace
N^1-Acetylspermidine	5.7 ± 1.3	4.1 ± 1.1	0.24 ± 0.03
N^8-Acetylspermidine	4.8 ± 0.8	3.8 ± 1.4	0.11 ± 0.01
Spermidine			
Free	0.5 ± 0.4	0.4 ± 0.2	0.30 ± 0.07
Total	12.6 ± 3.0	10.7 ± 3.0	0.95 ± 0.14
Spermine			
Free	2.1 ± 1.7	0.3 ± 0.3	Trace
Total	2.1 ± 1.4	0.7 ± 0.3	Trace

Data (μmol/24 h) were taken from N. Seiler (1987).

The oxidase is specific for the three carbon arm, and therefore does not oxidize N^8-acetylspermidine. N^1,N^{12}-Diacetylspermine is oxidized by PAO to produce N^1-acetylspermidine, which is oxidized in a second step. Indeed the oxidation of the diacetyl spermine has been used as an assay for the oxidase (Seiler et al., 1980).

The existence of this catabolic path of putrescine formation, as well as a path of polyamine degradation and a cycle of polyamine biosynthesis and degradation, has posed numerous problems. These include the estimation of the possible mechanisms of depletion of putrescine in normal and tumor-bearing animals, the relative amounts of putrescine derived from ornithine or N^1-acetylspermidine, the number, identity, and regulation of polyamine N-acetyl transferases, etc. All of these have been germane to the problems of the chemotherapy of specific tumors.

When thioacetamide was used to enhance polyamine metabolism in the rat liver, a significant amount of putrescine and monoacetyl putrescine accumulated rapidly in the liver without any change in levels of spermidine and spermine. The urine was found to contain increased (twofold) amounts of putrescine and N^1-acetylspermidine. N^8-Acetylspermidine and spermine excretion remained approximately constant. These results were interpreted to indicate that thioacetamide induced spermidine acetyltransferase as early as it did ODC, and that the formation of N^1-acetylspermidine preceded that of putrescine (Seiler et al., 1980). The early induction of spermidine N^1-acetyltransferase by thioacetamide was demonstrated by Matsui and Pegg. Their experiments also showed that the half-life ($t_{1/2}$) of the enzyme (40 min) was quite brief, only slightly longer than ODC (27 min) in these animals (Matsui & Pegg, 1981).

Seiler (1981) reported that the turnover of spermidine and spermine in various tissues was slow but quite regular and reproducible. Thus, the dilution of liver spermidine prelabeled with [^{14}C]-putrescine and [2-^{14}C]-methionine occurred at similar rates with $t_{1/2}$ of 12–14 days. However, brain spermidine labeled with [^{14}C]-putrescine was diluted with a $t_{1/2}$ of 42 days, whereas that

labeled in the aminopropyl moiety was diluted much more rapidly (Fig. 16.6). The relative retention of the putrescine moiety of spermidine implied a recycling of some 4 or 5 times. In more recent studies spermidine turnover in the rat brain was studied by following the accumulation of N^1-acetylspermidine after intracerebroventricular injection of the PAO inhibitor, N^1,N^{11}-bis-allenylputrescine (Ientile et al., 1990).

Seiler et al. (1985) also dissected the problem of urinary excretion in the rat and of relative roles of the biosynthetic path via ODC with the inhibitor DFMO, and of the catabolic path via SAT and PAO in the presence of the amine oxidase inhibitor, aminoguanidine (AG). The latter inhibitor increased putrescine excretion in normal or leukemic animals very considerably. This data indicated that about 65% of the diamine had been metabolized via oxidation. DFMO decreased polyamine excretion by almost 30%, implying that ODC had contributed this fraction of the total "putrescine equivalents." The levels of cadaverine were unresponsive to DFMO and indicated that most if not all of this diamine was of bacterial origin and should be omitted in balance studies. A comparable study has not been carried out in humans.

Exploitation of the inhibitors has detected some new capabilities in the regrowth of liver in the partially hepatectomized rat. Interestingly, whereas early administration of DAP inhibited ODC induction in this system, perhaps as an effect of antizyme production, hepatic spermidine N^1-acetyltransferase activity was enhanced. DNA synthesis was also high. However, the latter was prevented by the administration of the PAO inhibitor, quinacrine (Sato & Fujiwara, 1988). Thus, the early requirement for putrescine in DNA synthesis that could be made available by ODC might also be provided by the catabolic route of spermidine acetylation and oxidation.

PAO is very low or undetectable in the sera of healthy subjects or cancer patients (Romano & Bonelli, 1988). However, PAO is almost always in excess in various tissues, the liver, brain, etc. The levels of SSAT

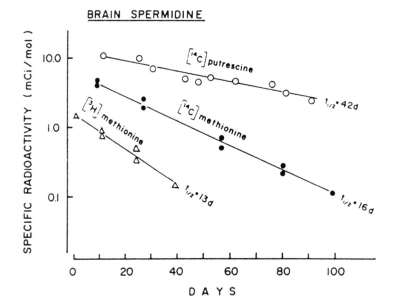

FIG. 16.6 Specific radioactivity of spermidine in the mouse brain after pretreatment with [1,4-^{14}C]-putrescine, [2-^{14}C]-DL-methionine, and [2-^{3}H]L-methionine, respectively; adapted from Seiler (1981). Abscissa: time after the last dose of the radioactive precursor.

appear to be rate limiting for the interconversion pathway. The rapid synthesis and decay of this activity in various tissues has been noted, and these inductions by many chemical and biological stimuli have been tabulated (Seiler, 1987). PAO inhibitors have been used to demonstrate that the increase of rat pancreatic putrescine by the carcinogen azaserine was mainly due to induction of SSAT in the catabolic pathway (Löser et al., 1995).

In many inducible systems it has been difficult to determine whether ODC or the acetyltransferase is induced first (Matsui & Pegg, 1982a), a matter that becomes intriguing in terms of the report that both N-monoacetyl spermidines are themselves inducers of ODC (Cannelakis, 1981). In isoprenaline-stimulated rat salivary glands, marked increases of SSAT and N^1-acetylspermidine occur without any increase of ODC (Ekstrom et al., 1989). Thus, in this system the inductions of N^1-SSAT and ODC appear to be quite independent.

Induction of SSAT in the rat liver was effected by dialkylnitrosamines, without a change in the level of the excess PAO (Matsui & Pegg, 1982b). The tissue rapidly accumulated a small amount of N^1-acetylspermidine and a large amount of putrescine. Spermine was also decreased in a catabolic sequence. Phorbol esters also stimulate SSAT in phytohemagglutinin-stimulated bovine lymphocytes, and a resulting decrease in spermine appeared to inhibit DNA synthesis (Matsui–Yuasa et al., 1985). This inhibition was reversed by the addition of spermine specifically. Injection of spermidine itself into rats induces the SSAT and the accumulation of putrescine in various organs (Matsui et al., 1982). The enzyme

is obviously another means of regulating and removing excess spermidine and spermine.

SSAT: The Inducible N^1-Acetyl Transferase

As therapeutic studies developed to deplete the polyamines, particularly putrescine, with promising inhibitors such as DFMO, it became evident that the supply of putrescine via the catabolic interconversion pathway must also be controlled. This compelled the study of the enzymes involved and their genetic and biosynthetic regulation. In fact the induction of SSAT in tumor tissues proved to be a major clue and approach to the therapy of several tumors.

In the 1970s it became clear that the cytoplasmic enzyme or enzymes, which included SSAT, were different from the nuclear acetyltransferase(s). The latter also acetylated putrescine via acetyl-coA, and also acetylated spermidine in the N^8 position. N^8-Acetylspermidine is also reported to be synthesized in the cytoplasm, but inducible SSAT is stated to be incapable of acetylating at N^8; hence, an enzyme with this specificity, other than SSAT, may be sought in the cytoplasm. As will be seen, the nuclear enzymes also act upon histones.

The unique inducibility of the soluble cytoplasmic N^1-SSAT accounts for both the increased production of N^1-acetylspermidine and the increased ratio of N^1- to N^8-acetylspermidine found in the urine of cancer patients and in their tumors. The inducible enzyme (SSAT) was purified to homogeneity from the liver of rats treated with the toxic inducer, CCl$_4$ (Abdel–Monem & Merdink, 1981; Della Ragione & Pegg, 1982). In the latter study the enzyme had an apparent molecular weight (MW) of 115,000 and comprised two identical subunits. In addition to the formation of N^1-acetylspermidine from spermidine and acetyl-coA, the enzyme acted upon the aminopropyl arms of norspermidine and spermine. Indeed a major step of purification was accomplished by affinity chromatography on Sepharose attached to *sym*-norspermidine, whose high affinity to the enzyme (Table 16.2) is reflected in the K_m values and reaction rates of the various polyamines (Della Ragione & Pegg, 1983). The K_m for acetyl-coA (1.5 μM) is lower than that for norspermidine (9 μM), and free coA is a potent inhibitor. Compounds such as butane-2,3-dione or phenylglyoxal, which are reactive with arginine residues, inactivated the enzyme and this inactivation was prevented by acetyl-coA (Della Ragione et al., 1983). An apparent inactivation of crude enzyme by phosphatase has been suggested to be a result of the dephosphorylation of the 3′-phosphate of coA.

A specific precipitating antibody to the pure rat liver SSAT has been used to study the inducibility of the enzyme. A detergent treatment stabilized the enzyme and permitted the isolation of a protein of 65,000 MW whose subunits under denaturing conditions had an MW of about 18,000. In induction by CCl₄, spermidine, thioacetamide, and methylglyoxal-bis(guanylhydrazone) (MGBG), induction of SSAT activity was accompanied by great increases of precipitable protein (Persson & Pegg, 1984). MGBG has been found to be a competitive inhibitor of SSAT, but it evoked a 700-fold increase in the enzyme as a result of increasing its half-life from a few minutes to many hours (Pegg et al., 1985). Surprisingly the analogous methylglyoxal-bis(butylamidinohy-

drazone), which inhibited AdoMetDC but is a slightly weaker inhibitor of SSAT than MGBG, did not "induce" this enzyme. Despite the inhibition of SSAT by MGBG, the great increase of the enzyme produced an increase of liver acetylspermidine and putrescine. MGBG has been used to increase the SSAT of BHK cells for purposes of isolation (Nuttall & Wallace, 1991). This enzyme was reported not to cross-react with the antibody to the liver SSAT. A human breast cancer cell line is also induced by MGBG to increase its SSAT activity. This induction was potentiated by intracellular Ca^{2+} (Quick & Wallace, 1993).

The pursuit of the duodenal SSAT induced in vitamin D-deficient chicks by injection of 12,25-dihydroxyvitamin D₃, has been most instructive. Nanogram amounts of the vitamin induce SSAT very rapidly in this and other tissues, and these are a tenth of the amount of the vitamin necessary to induce ODC (Shinki et al., 1985). The induction occurs without an increase in the ability to synthesize N^8-acetylspermidine.

Putrescine is made in the chick duodenum by both ODC and the SSAT-PAO route, but after SSAT induction the latter route is predominant (Shinki et al., 1986). Indeed DFMO is ineffective in reducing the putrescine increase in the tissue after induction by the vitamin. An effective inhibitor of PAO permits a marked increase of N^1-acetylspermidine in the vitamin-induced chick duodenum (Shinki et al., 1989). The deficiency of putrescine thereby produced results in an inhibition of Ca^{2+} absorption and a shortening of villus length (Shinki et al., 1991). These effects are reversed by the supplementation of putrescine. The induced SSAT has been purified to homogeneity and despite its lability was found to have an MW of 36,000 with subunits of 18,000, even

TABLE 16.2 Substrate specificity of rat liver spermidine/spermine N^1-acetyltransferase.

Compound	Structure	Apparent K_m (μM)	Maximal rate of reaction (%)
Spermidine	H₂N[CH₂]₃NH[CH₂]₄NH₂	130	100
N^1-Acetylspermidine	CH₃CO · NH[CH₂]₃NH[CH₂]₄NH₂	Inactive	<1
Spermine	H₂N[CH₂]₃NH[CH₂]₄NH[CH₂]₃NH₂	34	24
N^1-Acetylspermine	H₂N[CH₂]₃NH[CH₂]₄NH[CH₂]₃NHCOCH₃	51	16
sym-Norspermidine	H₂N[CH₂]₃NH[CH₂]₃NH₂	9	140
sym-Norspermine	H₂N[CH₂]₃NH[CH₂]₃NH[CH₂]₃NH₂	10	160
3,3-Diamino-N-methyldipropylamine	H₂N[CH₂]₃N[CH₂]₃NH₂ (CH₃ on N)	ND	31
sym-Homospermidine	H₂N[CH₂]₄NH[CH₂]₄NH₂	Inactive	<1
N-(3-Aminopropyl)cadaverine	H₂N[CH₂]₃NH[CH₂]₅NH₂	ND	30
1,3-Diaminopropane	H₂N[CH₂]₃NH₂	460	7.6
1,4-Diaminobutane (putrescine)	H₂N[CH₂]₄NH₂	Inactive	<1
1,5-Diaminopentane (cadaverine)	N₂N[CH₂]₅NH₂	Inactive	<1
N^8-Acetylspermidine	H₂N[CH₂]₃NH[CH₂]₄NHCOCH₃	Inactive	<1
N^1-Acetylputrescine	H₂N[CH₂]₄NHCOCH₃	Inactive	<1

Adapted from Della Ragione and Pegg (1983). The reaction rates are expressed as percentages of that found with spermidine as substrate. K_m values were not determined (ND) for some compounds.

as reported by Persson and Pegg (1984). The kinetic properties of the enzyme were similar to those of the rat liver enzyme (Shinki & Suda, 1989).

In human cancer (melanoma) cells, exogenous spermidine or spermine, which increases intracellular pools, induces significant increases in the SSAT mRNA and SSAT. Conversely, decreases in the pools provoked by inhibitors reduce the SSAT mRNA and the enzyme. Changes in spermine were more effective than those in spermidine (Shappell et al., 1993).

Nuclear Polyamine and Histone N-Acetylases

Seiler and Al-Therib (1974) described an acetylation of putrescine by fractions of rat tissues. The nuclear fraction was most active, although the acetylation of the diamine had a high K_m of 3 mM. Spermidine did not inhibit the reaction competitively, suggesting the existence of a separate putrescine acetylase. Extending the studies of nuclear fractions of rat tissues to the acetylation of spermidine and spermine, Blankenship and Walle (1977) found greater activities and a lower K_m for spermidine and they isolated monoacetyl derivatives of the three polyamines from incubation mixtures. Libby (1978) separated two nuclear N-acetyltransferases from calf liver, each of which transferred acetate from acetyl-coA to either histone or spermidine. In each case both activities appeared to reside in the same molecule. The two purified enzymes differed slightly in MW, sensitivity to heat, and acceptances of the different histones as substrates. Both enzymes appeared to prefer histone to polyamine, and spermidine, spermine, and norspermidine to diamines. Libby found that both N^1- and N^8-spermidine were synthesized by both enzymes.

The existence of two separable transferases in rat liver was affirmed, and their specificities for the various histones were determined (Libby, 1980). The enzymes acetylated spermidine and spermine and were apparently inactive on norspermidine. However, a high rate of nonenzymatic acetylation by acetyl-coA created difficulties in defining low rates of enzymatic activity. The significance of the 11–19% rates of acetylation of the various diamines was thereby obscured.

Additional studies of other associations and interrelations of the histone and polyamine acetyltransferases have exploited unusual multisubstrate analogues of coA. Thus, the multisubstrate analogue formed by joining coA and spermidine through acetic acid was a strong inhibitor ($K_i < 10^{-8}$ M) of the acetylation of spermidine and histones by a histone acetylase of calf thymus (Cullis et al., 1982).

$$\begin{array}{c} O \\ \parallel \\ RN-C-CH_2-ScoA \\ | \\ H \end{array}$$

The coA derivatives of the various polyamines were tested with cytosolic SSAT and with the nuclear acetyl-transferases (Erwin et al., 1984). The norspermidine analogue was by far the most potent inhibitor of the cytosolic SSAT, reflecting the high affinity of norspermidine for this enzyme. An antiserum to this enzyme did not react with the nuclear activities. In reaction with the nuclear activities, the various coA-containing analogues were similarly inhibitory, reducing the activity by 40–50% at low concentrations (up to 20 μM) and then developing a more gradual increase of inhibition requiring much higher concentrations. These results were interpreted to indicate the existence of several enzymes of significantly different sensitivities to the inhibitors. Roblot et al. (1994) recently described regioselective syntheses of acetyl-coA adducts to the N^1 or N^8 positions of spermidine.

The products of spermidine acetylation by the nuclear extracts were about 85% N^8-acetylspermidine and about 15% N^1-acetylspermidine. The formation of both products was inhibited similarly by the norspermidine-containing analogue. In continuing work on the spermidine N^8-acetyltransferase, the activity was extracted from the nuclear structure by detergent at high ionic strength. An apparently homogenous fraction was obtained that had activities on spermidine and histone of 4.6 : 1. The protein was a dimer of 68,000 M_r subunits (Desiderio et al., 1991). However, the addition of serum to quiescent fibroblasts evoked an increase of histone acetyltransferase but not of spermidine N^8-acetyltransferase (Desiderio et al., 1993).

Deacetylation of N^8-Acetylspermidine

A deacetylase of the N^8-acetyl conjugate was described by Blankenship (1978) and Libby (1978). With this substrate, the former found the activity in the cytosolic fraction of rat liver, the organ of highest activity. By contrast, a histone deacetylase is mainly a nuclear activity, at least in calf thymus. Libby partially purified a deacetylating activity, using N^1-acetylspermidine as substrate, and reported a deacetylation of both the N^1- and N^8-acetylspermidines. However, Marchant et al. (1986) described a deacetylase active exclusively on the N^8 conjugate and presented evidence to suggest that PAO participated in the apparent deacetylation with N^1-acetylspermidine. Seiler (1987) stated that the N^8-acetylspermidine deacetylase is specific for 4-acetamidobutyl compounds and will also hydrolyze N-monoacetyl putrescine but not N-monoacetyl DAP.

Several inhibitors have been described for the N^8-deacetylase. One of these is an analogue of monoacetyl putrescine, 7-amino-2-heptanone, which selectively causes an increase of concentration of the N^8-acetylspermidine in cells without affecting growth rate (Desiderio et al., 1992). Subsequently, derivatives of this compound, such as the aminopropyl derivative 7-[N-(3-aminopropyl)amino]heptan-2-ol (APAH), the analogue of N^8-acetylspermidine, proved to be very potent, with a K_i of 0.18 μM, some 60-fold less than the K_m of the substrate (Dredar et al., 1989). Analysis of the inhibitory effectiveness of the various analogues has permitted a rough sketch of the active site of the deacetylase (Fig. 16.7). When APAH was injected into mice, N^8-acetylspermidine accumulated almost immediately, without any effect on histone acetylation (Marchant et al., 1989). Finally, observations of the inhibition of the enzyme by thiol reagents and metal chelators suggested that even more potent inhibitors could be made. One of these, with additional chelating potentialities, has an apparent K_i of 1 nM (Huang et al., 1992). The enzyme is thought to be a Zn^{2+} enzyme. The structures of the most potent inhibitors are

$H_2N(CH_2)_5COCH_3$
7-amino-2-heptanone

$H_2N(CH_2)_3NH(CH_2)_5COCH_3$
7-[N-(3-aminopropyl)amino]heptane-2-one

$H_2N(CH_2)_3NH(CH_2)_5CONHOH$
6-[(3-aminopropyl)amino]-N-hydroxylhexanamide

1,6-Diacetylhexanediamine (HMBA or hexamethylene bisacetamide) was mentioned earlier. The compound, which induces differentiation in many cancer cell lines, is deacetylated to a monoacetyl 1,6-diaminohexane or the free diamine, each of which can be oxidized to the 6-acetamidohexanoic acid or the 6-aminohexanoic acid, respectively. The nature of the active deacetylases has not yet been clarified. The mono-N-acetyl-1,6-diaminohexane is reported to be the more active inducer. Whereas HMBA was found to decrease the transcription of the ODC gene in murine erythroleukemic cells, the N-monoacetyl-1,6-diamine was reported to decrease the half-life of the enzyme in a posttranscriptional event (Papazafiri & Osborne, 1989). Thus, the various acetyl derivatives appear to act at different stages of establishing ODC activity.

SSAT Induction by Polyamine Analogues

The discoveries of catabolic pathways and their physiological roles enlarged ideas concerning possible inhibi-

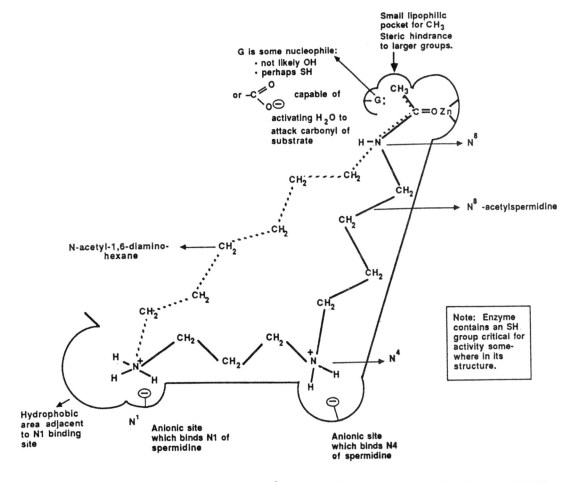

FIG. 16.7 Schematic summary of the active site of the N^8-acetylspermidine deacetylase; adapted from Dredar et al. (1989).

tors of these paths. The unexpected effects of HMBA and its derivatives seemed quite relevant to such a program. New analogues included the bisethyl derivatives of the primary amines (Bergeron et al., 1988). Some of these inhibited proliferation and killed cells resistant to DFMO (Casero et al., 1987). Unexpectedly, these were found to be very good inducers of stable SSAT protein in various mammalian cells. Although N^1,N^8-bis(ethyl)-spermidine and N^1,N^{12}-bis(ethyl)spermine were among the most potent inducers of SSAT in L1210 cells (Libby et al., 1989a), the greatest inducing activity was obtained with bis(ethyl)norspermine (BENSPM) (Libby et al., 1989b). In addition the compounds, particularly bis(ethyl)homospermine (BEHSPM), suppressed ODC and AdoMetDC, depleted polyamine pools, and inhibited cell growth. The structures are presented in Figure 16.8.

In CHO cells BENSPM induced a 600-fold increase in SSAT, demonstrated to arise from an increase in the stabilized enzyme protein, as well as by an increase in mRNA. Bis(ethyl)spermine (BESPM) has been shown to protect SSAT from specific proteases (Coleman et al., 1995). Increase of SSAT also facilitated a subsequent degradation of N^1-acetylspermidine and N^1-acetylspermine and excretion of putrescine. In this system, the content and excretion of N^8-acetylspermidine was unchanged (Pegg et al., 1990).

The bisethyl compounds were markedly cytotoxic (at 10 M levels) to one cell line, derived from an undifferentiated large-cell lung carcinoma resistant to DFMO. ODC fell very sharply and almost all intracellular polyamines were eliminated (Casero et al., 1989a). Extending these results, a 1700-fold increase in SSAT (Fig. 16.9) was demonstrated with BESPM, which killed the cells within several days. This compound effected an extrusion of 100% of the putrescine, 96% of cell spermidine, and 75% of cell spermine. The molar level of

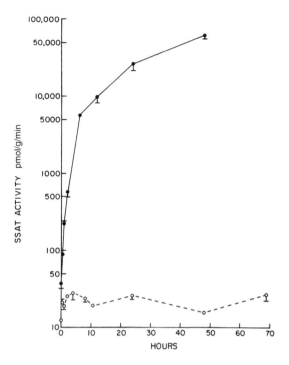

FIG. 16.9 Effects of increasing exposure time to bis(ethyl)-spermine (10 μM) on the induction of spermidine/spermine N^1-acetyltransferase (SSAT) in human cancer cell lines; (—) sensitive and (---) relatively resistant; adapted from Casero et al. (1989b).

intracellular BESPM and residual cell polyamines approached that of the untreated cellular polyamine (Casero et al., 1989b). Similar results were obtained with a human melanoma cell line (Porter et al., 1991). In this line BENSPM proved to be more cytotoxic and 6 times more active in inducing SSAT than was BESPM (some 200-fold). As in the study of Casero et al. (1989a), a cell line was found that was both relatively resistant to these analogues and relatively noninducible for SSAT. These results suggested that high levels of induced SSAT might be found concurrently with active chemotherapeutic agents.

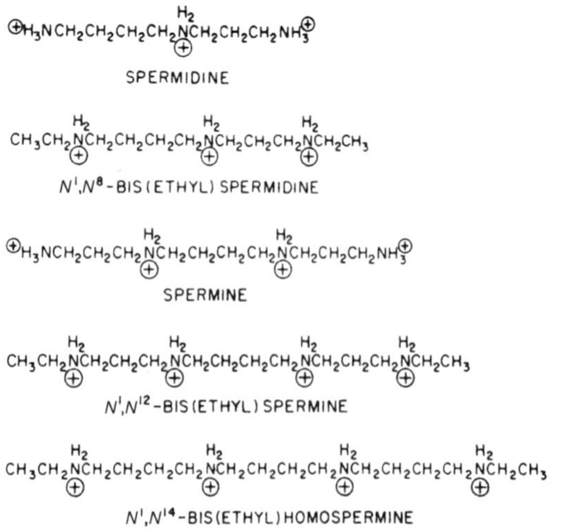

SSAT Protein and Gene

These high levels of induction with BESPM facilitated the isolation of SSAT to homogeneity from the human cell line. The subunit size (ca. 20 kDa) proved to be similar to that of the rat liver enzyme. N-Terminal sequencing of an SSAT fragment yielded a partial (seven amino acid) protein sequence. Isolated mRNA was translated in a cell-free system and produced SSAT (Casero et al., 1990). The crude mRNA was converted to a cDNA library from which the SSAT cDNA was identified by hybridization with a polynucleotide oligomer constructed to code for the previously isolated seven amino acid peptide. In its turn the DNA coded

$$\overset{\oplus}{H_3}NCH_2CH_2CH_2\overset{H_2}{\underset{\oplus}{N}}CH_2CH_2CH_2N\overset{\oplus}{H_3}$$

SPERMIDINE

$$CH_3CH_2\overset{H_2}{\underset{\oplus}{N}}CH_2CH_2CH_2\overset{H_2}{\underset{\oplus}{N}}CH_2CH_2CH_2\overset{H_2}{\underset{\oplus}{N}}CH_2CH_3$$

N^1,N^8-BIS(ETHYL) SPERMIDINE

$$\overset{\oplus}{H_3}NCH_2CH_2CH_2\overset{H_2}{\underset{\oplus}{N}}CH_2CH_2CH_2CH_2\overset{H_2}{\underset{\oplus}{N}}CH_2CH_2CH_2N\overset{\oplus}{H_3}$$

SPERMINE

$$CH_3CH_2\overset{H_2}{\underset{\oplus}{N}}CH_2CH_2CH_2\overset{H_2}{\underset{\oplus}{N}}CH_2CH_2CH_2CH_2\overset{H_2}{\underset{\oplus}{N}}CH_2CH_2CH_2\overset{H_2}{\underset{\oplus}{N}}CH_2CH_3$$

N^1,N^{12}-BIS(ETHYL) SPERMINE

$$CH_3CH_2\overset{H_2}{\underset{\oplus}{N}}CH_2CH_2CH_2\overset{H_2}{\underset{\oplus}{N}}CH_2CH_2CH_2CH_2\overset{H_2}{\underset{\oplus}{N}}CH_2CH_2CH_2CH_2\overset{H_2}{\underset{\oplus}{N}}CH_2CH_3$$

N^1,N^{14}-BIS(ETHYL) HOMOSPERMINE

FIG. 16.8 Structures of the charged polyamines and the bis-(ethyl)polyamine analogues.

for a 20,000 MW protein and hybridized with a cellular mRNA (Casero et al., 1991).

The cDNA that coded for the human SSAT possessed an open reading frame that coded for a protein of 171 amino acids, with a predicted MW of 20,023. In vitro translation of the RNA produced from this cDNA resulted in an active enzyme. The cDNA also hybridized with a large RNA transcript, induced in large amount by BESPM; this work suggested the absence of amplification of the SSAT gene (Casero et al., 1991). A full-length cDNA has been developed from the newly isolated natural mRNA (Xiao et al., 1991). The availability of this cDNA has helped to initiate a localization and analysis of the human SSAT gene, whose 5'- and 3'-regulatory sites are approximately equal in base content to that in the open reading frame (Xiao et al., 1992). The gene, located in the human X chromosome, appears to comprise six exons and five introns. No homology has been found between the sequences of this and any other type of genetic sequence. The gene and cDNA coding for the mouse SSAT have been isolated and characterized by Fogel–Petrovic et al. (1993).

Human SSAT (i.e., the acetyl transferase) has also been isolated from human melanoma cells induced by BENSPM. It had been bound on immobilized spermine and other matrices and eluted initially by spermine and subsequently by coenzyme A. Molecular weights of the enzyme and of subunits were found to be 80,000 and 20,300, respectively. Michaelis constants were norspermidine < norspermine < spermine $\ll N^1$-acetylspermine < spermidine. Inhibition constants for competitive inhibitors of spermidine acetylation included BENSPM < BESPM $\ll$ BEHSPM. Induction of SSAT by BESPM in the human melanoma cells is accompanied by a major increase in SSAT mRNA. This transcriptional control is also effected by spermine and represents one of the few clear evidences of such control by a natural polyamine or its analogues.

SSAT is currently considered to be the rate-limiting enzyme of catabolism and polyamine elimination in many mammalian cells. It is also a major component of the catabolic route to putrescine in normal and tumor cells. The induction of SSAT and ODC has been found in early stages of two types of carcinogenesis in the epithelia of rat bladders. Matsui–Yuasa et al. (1992) proposed that SSAT be considered a biochemical marker for epithelial proliferation in this tissue. The current view of catabolism and this enzyme as possible targets for cancer chemotherapy is a relatively recent advance in the field that previously focused on the biosynthetic path. This advance in this area arose from the development of simple analogues that proved to possess the unusual properties of both inducing SSAT and displacing previously sequestered polyamine (see Chap. 17). Conversion of previously sequestered spermidine to N^1-acetylspermidine reduced the net charge and affinity of the polyamine and prepared it for excretion. The charged bisethyl derivatives can enter the cell and neutralize the anionic sites for the previously residing natural polyamines. Thus, the analysis of the mechanisms of polyamine supply and depletion in normal and tumor tissue merged almost immediately with the experimental study of cancer chemotherapy.

REFERENCES

Abdel–Monem, M. & Merdink, J. (1981) *Life Sciences*, **28**, 2017–2023.
Abrahamsen, M., Li, R., Goetz, W., & Morris, D. (1992) *Journal of Biological Chemistry*, **267**, 18866–18873.
Aigner–Held, R., Campbell, R., & Daves, Jr., G. (1979) *Proceedings of the National Academy of Sciences of the United States of America*, **76**, 6652–6655.
Alhonen, L., Halmekytö, M., Kosma, V., Wahlfors, J., Kauppinen, R., & Jänne, J. (1995) *International Journal of Cancer*, **63**, 402–404.
Andersson, G., Bengtsson, G., Albinsson, A., Rosén, S., & Heby, O. (1978) *Cancer Research*, **38**, 3938–3943.
Andersson, G. & Heby, O. (1972) *Journal of the National Cancer Institute*, **48**, 165–172.
Arnold, W. & Van de Kerkhof, P. (1991) *British Journal of Dermatology*, **125**, 6–8.
Ashida, Y., Kido, J., Kinoshita, F., Nishino, M., Shinkai, K., Akedo, H., & Inoue, H. (1992) *Cancer Research*, **52**, 5313–5316.
Astrup, E. & Paulsen, J. (1981) *Carcinogenesis*, **2**, 545–551.
Auvinen, M., Paasinen, A., Anderson, L., & Hölttä, E. (1992) *Nature*, **360**, 355–358.
Auvinen, M., Paasinen-Sohns, A., Hirai, H., Andersson, L., & Hölttä, E. (1995) *Molecular and Cellular Biology*, **15**, 6513–6525.
Bachrach, U. (1992) *Progress in Drug Research*, **39**, 9–33.
Bartos, D., Campbell, R., Bartos, F., & Grettie, D. (1975) *Cancer Research*, **35**, 2056–2060.
Baylin, S. (1977) *Proceedings of the National Academy of Sciences of the United States of America*, **74**, 883–887.
Baylin, S., Abeloff, M., Goodwin, G., Carney, D., & Gazdar, A. (1980) *Cancer Research*, **40**, 1990–1994.
Bergeron, R., Neims, A., McManis, J., Hawthorne, T., Vinson, J., Bortell, R., & Ingeno, M. (1988) *Journal of Medicinal Chemistry*, **31**, 1183–1190.
Binder, R., Volpenhein, M., & Motz, A. (1989) *Carcinogenesis*, **10**, 2351–2357.
Blankenship, J. (1978) *Archives of Biochemistry and Biophysics*, **189**, 20–27.
Blankenship, J. & Walle, T. (1977) *Archives of Biochemistry and Biophysics*, **179**, 235–242.
Borek, C., Ong, A., Stevens, V., Wang, E., & Merrill, Jr., A. (1991) *Proceedings of the National Academy of Sciences of the United States of America*, **88**, 1953–1957.
Bouclier, M., Jomard, A., Kail, N., Shroot, B., & Hensby, C. (1987) *Laboratory Animals*, **21**, 233–240.
Buffkin, D., Webber, M., Davidson, W., Bassist, L., & Verma, R. (1978) *Cancer Research*, **38**, 3225–3229.
Butler, H., Cohn, W., Mar, P., & Montgomery, R. (1991) *Journal of Cellular Physiology*, **147**, 256–264.
Canellakis, Z. (1981) *Biochemical and Biophysical Research Communications*, **100**, 929–933.
Casero, Jr., R., Celano, P., Ervin, S., Applegren, M., Wiest, L., & Pegg, A. (1991) *Journal of Biological Chemistry*, **266**, 810–814.
Casero, Jr., R., Celano, P., Ervin, S., Porter, C., Bergeron, R., & Libby, P. (1989b) *Cancer Research*, **49**, 3829–3833.
Casero, Jr., R., Celano, P., Ervin, S., Wiest, L., & Pegg, A. (1990) *Biochemical Journal*, **270**, 615–620.

Casero, Jr., R., Ervin, S., Celano, P., Baylin, S., & Bergeron, R. (1989a) *Cancer Research*, **49**, 639–643.

Casero, Jr., R., Go, B., Theiss, H., Smith, J., Baylin, S., & Luk, G. (1987) *Cancer Research*, **47**, 3964–3967.

Casti, A., Orlandini, G., Reali, N., Bacciottini, F., Vanelli, M., & Bernasconi, S. (1982) *Journal of Endocrinological Investigation*, **5**, 263–266.

Chatel, M., Darcel, F., Quemener, V., Hercouet, H., & Moulinoux, J. (1987) *Anticancer Research*, **7**, 33–38.

Chida, K., Hashiba, H., Suda, T., & Kuroki, T. (1984) *Cancer Research*, **44**, 1387–1391.

Chun, P., Rennert, O., Saffen, E., & Taylor, W. (1976) *Biochemical and Biophysical Research Communications*, **69**, 1095–1101.

Chun, P., Saffen, E., DiTore, R., Rennert, O., & Weinstein, N. (1977) *Biophysical Chemistry*, **6**, 321–335.

Cipolla, B., Guille, F., Moulinoux, J., Bansard, J., Roth, S., Staerman, F., Corbel, L., Quemener, V., & Lobel, B. (1994) *Journal of Urology*, **151**, 629–633.

Clifford, A., Morgan, D., Yuspa, S. H., Soler, A. P., & Gilmour, S. (1995) *Cancer Research*, **55**, 1680–1686.

Cohen, S. (1977) *Cancer Research*, **37**, 939–942.

Colarian, J., Arlow, F., Calzada, R., Luk, G., & Majumdar, A. (1991) *Gastroenterology*, **100**, 1528–1532.

Coleman, C. S., Huang, H., & Pegg, A. E. (1995) *Biochemistry*, **34**, 13423–13430.

Connor, M., Lowe, N., Breeding, J., & Chalet, M. (1983) *Cancer Research*, **43**, 171–174.

Cullis, P., Wolfenden, R., Cousens, L., & Alberts, B. (1982) *Journal of Biological Chemistry*, **257**, 12165–12169.

Czuba, M. & Smith, I. (1991) *Pharmacology & Therapeutics*, **50**, 147–190.

Della Ragione, F., Erwin, B., & Pegg, A. (1983) *Biochemical Journal*, **213**, 707–712.

Della Ragione, F. & Pegg, A. (1982) *Biochemistry*, **24**, 6152–6158.

———— (1983) *Biochemical Journal*, **213**, 701–706.

Desiderio, M., Bernhardt, A., & Mamont, P. (1991) *Life Chemistry Reports*, **9**, 57–63.

Desiderio, M., Mattel, S., Biondi, G., & Colombo, M. (1993) *Biochemical Journal*, **293**, 475–479.

Desiderio, M., Weibel, M., & Mamont, P. (1992) *Experimental Cell Research*, **202**, 501–506.

Desser, H., Höcker, P., Weiser, M., & Böhnel, J. (1975) *Clinica Chimica Acta*, **63**, 243–247.

Dredar, S., Blankenship, J., Marchaut, P., Manneh, V., & Fries, D. (1989) *Journal of Medicinal Chemistry*, **32**, 984–989.

Dreyfuss, F., Chayen, R., Dreyfuss, G., Dvir, R., & Ratan, J. (1975) *Israel Journal of Medical Sciences*, **11**, 785–795.

Driessen, O., Hamelink, R., Hermans, J., Franken, H., Emonds, A., & Davelaar, J. (1989) *Methods and Findings in Experimental Clinical Pharmacology*, **11**, 11–16.

Dunzendorfer, V. & Russell, D. (1978) *Cancer Research*, **38**, 2321–2324.

Ekstrom, J., Mansson, B., Nilsson, B., Rosengren, E., & Tobin, G. (1989) *Acta Physiologica Scandinavica*, **135**, 249–254.

El-Deiry, W., Nelkin, B., Celano, P., Yen, R., Falco, J., Hamilton, S., & Baylin, S. (1991) *Proceedings of the National Academy of Sciences of the United States of America*, **88**, 3470–3474.

Elitsur, Y., Moshier, J., Murthy, R., Barbish, A., & Luk, G. (1992) *Life Sciences*, **50**, 1417–1424.

Ernestus, R., Röhn, G., Schröder, J., Klug, N., Hossmann, K., & Paschen, W. (1992) *Acta Histochemica*, **42**, 159–164.

Erwin, B., Persson, L., & Pegg, A. (1984) *Biochemistry*, **23**, 4250–4255.

Fitchen, J., Riscoe, M., Dana, B., Lawrence, H., & Ferro, A. (1986) *Cancer Research*, **46**, 5409–5412.

Fogel–Petrovic, M., Kramer, D., Ganis, B., Casero, Jr., R., & Porter, C. (1993) *Biochimica et Biophysica Acta*, **1216**, 255–264.

Fukuda, M., Kono, T., Ishii, M., Mizuno, N., Tahara, H., Yoshida, H., Matsui–Yuasa, I., Otani, S., & Hamada, T. (1990) *Archives of Dermatological Research*, **282**, 487–489.

Fulton, D., Levin, V., Lubich, W., Wilson, C., & Marton, L. (1980) *Cancer Research*, **40**, 3293–3296.

Gange, R. & Dequoy, P. (1980) *British Journal of Dermatology*, **103**, 27–32.

Gerbaut, L. (1991) *Clinical Chemistry*, **37**, 2117–2120.

Gerhauser, C., Mar, W., Lee, S. K., Suh, N., Luo, Y., Kosmeder, J., Luyengi, L., Fong, H. S., Kinghorn, A. D., Moriarty, R. M., Mehta, R. G., Constantinol, A., Moon, R. C., & Pezzuto, J. M. (1995) *Nature Medicine*, **1**, 260–266.

Gilmour, S., Robertson, F., Megosh, L., O'Connell, S., Mitchell, J., & O'Brien, T. (1992) *Carcinogenesis*, **13**, 51–56.

Gilmour, S., Verma, A., Madara, T., & O'Brien, T. (1987) *Cancer Research*, **47**, 1221–1225.

Gonzalez, G. & Byus, C. (1991) *Cancer Research*, **51**, 2932–2939.

Grossie, Jr., V., Nishioka, K., Ota, D., & Martin, Sr., R. (1986) *Cancer Research*, **46**, 3464–3468.

Gupta, A., Fisher, G., Elder, J., Nickoloff, B., & Voorhees, J. (1988) *Journal of Investigative Dermatology*, **91**, 486–491.

Haddox, M., Scott, K., & Russell, D. (1979) *Cancer Research*, **39**, 4930–4938.

Halmekytö, M., Syrjänen, K., Jänne, J., & Alhonen, L. (1982) *Biochemical and Biophysical Research Communications*, **187**, 493–497.

Hannun, Y., Greenberg, C., & Bell, R. (1987) *Journal of Biological Chemistry*, **262**, 13620–13626.

Harik, S. & Sutton, C. (1979) *Cancer Research*, **39**, 5010–5015.

Heby, O. (1977) *Proceedings of the American Association of Cancer Research*, **18**, 79.

Heimer, Y. & Riklis, E. (1979) *Biochemical Journal*, **183**, 179–180.

Hickok, N. & Uitto, J. (1992) *Journal of Investigative Dermatology*, **98**, 327–332.

Hietala, O., Dzubow, L., Dlugosz, A., Pyle, J., Jenney, F., Gilmour, S., & O'Brien, T. (1988) *Cancer Research*, **48**, 1252–1257.

Hillebrand, G., Winslow, M., Benzinger, M., Heitmeyer, D., & Bissett, D. (1990a) *Cancer Research*, **50**, 1580–1584.

Hillebrand, G., Winslow, M., Heitmeyer, D., & Bissett, D. (1990b) *Journal of the Society of Cosmetic Chemists*, **41**, 187–195.

Hiramatsu, Y., Eguchi, K., Yonezawa, M., Hayase, R., & Sekiba, K. (1981) *Biology of the Neonate*, **40**, 136–144.

Horn, Y., Beal, S., Walach, N., Lubich, W., Spigel, L., & Marton, L. (1984) *Cancer Research*, **44**, 4675–4678.

Hsieh, J. & Verma, A. (1989) *Cancer Research*, **49**, 4251–4257.

Huang, T., Dredar, S., Manneh, V., Blankenship, J., & Fries, D. (1992) *Journal of Medicinal Chemistry*, **35**, 2414–2418.

Ientile, R., Fabiano, C., & Trimarchi, G. (1990) *Italian Journal of Biochemistry*, **39**, 261A–262A.

Imamura, K., Wang, Z., Murayama–Oda, K., Kim, H., Tsuji, T., & Tanaka, T. (1991) *Japanese Journal of Cancer Research*, **82**, 315–324.

Ishiwata, K., Abe, Y., Matsuzawa, T., & Ido, T. (1988) *Nuclear Medical Biology*, **15**, 119–122.

Jetten, A., Ganong, B., Vandenbark, G., Shirley, J., & Bell, R.

(1985) *Proceedings of the National Academy of Sciences of the United States of America*, **82**, 1941–1945.

Kallio, A., Pösö, H., Guha, S., & Jänne, J. (1977) *Biochemical Journal*, **166**, 89–94.

Kaneko, K., Fujimori, S., Kanbayashi, T., Miyazawa, Y., Kumakawa, T., Fujii, H., Miwa, S., Kamatani, N., & Akaoka, I. (1990) *International Journal of Cancer*, **45**, 8–11.

Karlsson, K. (1970) *Chemistry and Physics of Lipids*, **5**, 6–43.

Khan, N., Quemener, V., & Moulinoux, J. (1991) *Cell Biology International Reports*, **15**, 9–24.

—— (1992) *Experimental Cell Research*, **199**, 378–382.

Kilpeläinen, P. & Hietala, O. (1994) *Biochemical Journal*, **300**, 577–582.

Kim, Y., Pan, H., & Verma, A. (1994) *Molecular Carcinogenesis*, **10**, 169–179.

Koza, R., Megosh, L., Palmieri, M., & O'Brien, T. (1991) *Carcinogenesis*, **12**, 1619–1625.

LaMuraglia, G., Lacaine, F., & Malt, R. (1986) *Annals of Surgery*, **204**, 89–93.

Libby, P. (1978) *Journal of Biological Chemistry*, **253**, 233–237.

—— (1980) *Archives of Biochemistry and Biophysics*, **203**, 384–389.

Libby, P., Bergeron, R., & Porter, C. (1989a) *Biochemical Pharmacology*, **38**, 1435–1442.

Libby, P., Henderson, M., Bergeron, R., & Porter, C. (1989b) *Cancer Research*, **49**, 6226–6231.

Lichti, U., Patterson, E., Hennings, H., & Yuspa, S. (1981) *Journal of Cellular Physiology*, **107**, 261–270.

Lichti, U., Ben, T., & Yuspa, S. (1985) *Journal of Biological Chemistry*, **260**, 1422–4126.

Lichti, U. & Yuspa, S. (1988) *Cancer Research*, **48**, 74–81.

Löser, C., Stüber, E., & Fölsch, U. R. (1995) *Pancreas*, **10**, 44–52.

Luk, G. & Baylin, S. (1984) *New England Journal of Medicine*, **311**, 80–83.

Mach, M., Schneider, V., & Kersten, W. (1983) *Recent Results in Cancer Research*, **84**, 413–420.

Manni, A., Wechter, R., Grove, R., Wei, L., Martel, J., & Demers, L. (1995) *Breast Cancer Research and Treatment*, **34**, 45–53.

Marchant, P., Dredar, S., Manneh, V., Alshabanah, O., Matthews, H., Fries, D., & Blankenship, J. (1989) *Archives of Biochemistry and Biophysics*, **273**, 128–136.

Marchant, P., Manneh, V., & Blankenship, J. (1986) *Biochimica et Biophysica Acta*, **881**, 297–299.

Marks, F., Bertsch, S., & Fürstenberger, G. (1979) *Cancer Research*, **39**, 4183–4188.

Márquez, J., Matés, J., Quesada, A., Medina, M., Núñez de Castro, I., & Sánchez–Jiménez, F. (1989b) *Life Sciences*, **45**, 1877–1884.

Márquez, J., Sánchez–Jiménez, F., Medina, M., Quesada, A., & Núñez de Castro, I. (1989a) *Archives of Biochemistry and Biophysics*, **268**, 667–675.

Marton, L., Edwards, M., Levin, V., Lubich, W., & Wilson, C. (1979) *Cancer Research*, **39**, 993–997.

Marton, L., Heby, O., Levin, V., Lubich, W., Crafts, D., & Wilson, C. (1976) *Cancer Research*, **36**, 973–977.

Marton, L., Heby, O., Wilson, C., & Lee, P. (1974) *FEBS Letters*, **46**, 305–307.

Matsuda, M., Osafune, M., Kotake, T., Sonoda, T., Sobue, K., & Nakajima, T. (1978) *Clinica Chimica Acta*, **87**, 93–99.

Matsui, I. & Pegg, A. (1981) *Biochimica et Biophysica Acta*, **675**, 373–378.

—— (1982a) *FEBS Letters*, **139**, 205–208.

—— (1982b) *Cancer Research*, **42**, 2990–2995.

Matsui, I., Pösö, H., & Pegg, A. (1982) *Biochimica et Biophysica Acta*, **719**, 199–207.

Matsui–Yuasa, I., Otani, S., & Morisawa, S. (1985) *Journal of Biochemistry*, **98**, 1591–1596.

Matsui–Yuasa, I., Otani, S., Yano, Y., Takada, N., Shibata, M., & Fukushima, S. (1992) *Japanese Journal of Cancer Research*, **83**, 1037–1040.

McCormick, F. (1978a) *Biochemical Journal*, **174**, 427–435.

—— (1978b) *Virology*, **91**, 496–503.

McDonald, O., Hannun, Y., Reynolds, C., & Sahyoun, N. (1991) *Journal of Biological Chemistry*, **266**, 21773–21776.

Medina, M., Márquez, J., & Núñez de Castro, I. (1992) *Biochemical Medicine and Metabolic Biology*, **48**, 1–7.

Megosh, L., Gilmour, S. K., Rosson, D., Soler, A. P., Blessing, M., Sawicki, J. A., & O'Brien, T. G. (1995) *Cancer Research*, **55**, 4205–4209.

Merrill, Jr., A., Nimkar, S., Menaldino, D., Hannun, Y., Loomis, C., Bell, R., Tyagi, S., Lambeth, J., Stevens, V., Hunter, R., & Liotta, D. C. (1989) *Biochemistry*, **28**, 3138–3145.

Mills, K. & Smart, R. (1989) *Carcinogenesis*, **10**, 833–838.

Monks, T., Walker, S., Flynn, L., Conti, C., & DiGiovanni, J. (1990) *Carcinogenesis*, **11**, 1795–1801.

Moshier, J., Gilbert, J., Skunca, M., Dosescu, J., Almodovar, K., & Luk, G. (1990) *Journal of Biological Chemistry*, **265**, 4884–4892.

Moshier, J., Osborne, D., Skunca, M., Dosescu, J., Gilbert, J., Fitzgerald, M., Polidori, G., Wagner, R., Degen, S. J. F., Luk, G. D., & Flanagan, M. A. (1992) *Nucleic Acids Research*, **20**, 2581–2590.

Moulinoux, J., LeCalve, M., Quemener, V., & Quash, G. (1984b) *Biochimie*, **66**, 385–393.

Moulinoux, J., Quemener, V., & Chambon, Y. (1987) *European Journal of Cancer and Clinical Oncology*, **23**, 237–244.

Moulinoux, J., Quemener, V., Khan, N., Delcros, J., & Havouis, R. (1989a) *Anticancer Research*, **9**, 1057–1062.

Moulinoux, J., Quemener, V., Khan, N., Havouis, R., & Martin, C. (1989b) *Anticancer Research*, **9**, 1063–1068.

Moulinoux, J., Quemener, V., LeCalve, M., Chatel, M., & Darcel, F. (1984a) *Journal of Neurooncology*, **2**, 153–158.

Mufson, R., Astrup, E., Simsiman, R., & Boutwell, R. (1977) *Proceedings of the National Academy of Sciences of the United States of America*, **74**, 657–661.

Natta, C. & Kremzner, L. (1988) *Age*, **11**, 11–14.

Niles, R. & Loewy, B. (1989) *Cancer Research*, **49**, 4483–4487.

Nishioka, K., Ezaki, K., & Hart, J. (1980) *Clinica Chimica Acta*, **107**, 59–66.

Nishioka, K. & Romsdahl, M. (1974) *Clinica Chimica Acta*, **57**, 155–161.

Noguchi, T., Kashiwagi, A. & Tanaka, T. (1976) *Cancer Research*, **36**, 4015–4022.

Nuttall, M. & Wallace, H. (1991) *Biochemical Pharmacology*, **41**, 1090–1092.

O'Brien, T., Dzubow, L., Dlugosz, A., Gilmour, S., O'Donnell, K., & Hietala, O. (1989) *Skin Carcinogenesis: Mechanisms and Human Relevance*, pp. 213–231, T. E. A. Slaga, Ed. New York: Alan R. Liss.

O'Brien, T., O'Donnell, K., Kruszewski, F., & DiGiovanni, J. (1988) *Carcinogenesis*, **9**, 2081–2085.

O'Brien, T., Simsiman, R., & Boutwell, R. (1975a) *Cancer Research*, **35**, 1662–1670.

—— (1975b) *Cancer Research*, **35**, 2426–2433.

Otani, S., Matsui, I., Kuramoto, A., & Morisawa, S. (1985) *European Journal of Biochemisty*, **147**, 27–31.

Papazafiri, P. & Osborne, H. (1989) *European Journal of Biochemistry*, **178**, 789–793.

Parker, P., Coussens, L., Totty, N., Rhee, L., Young, S., Chen, E., Stabel, S., Waterfield, M., & Ullrich, A. (1986) *Science*, **233**, 853–859.

Parker, P., Stabel, S., & Waterfield, M. (1984) *EMBO Journal*, **3**, 953–959.

Pascale, R., Simile, M., Gaspa, L., Daino, L., Seddaiu, M., Pinna, G., Carta, M., Zolo, P., & Feo, F. (1993) *Carcinogenesis*, **14**, 1077–1080.

Pegg, A., Erwin, B., & Persson, L. (1985) *Biochimica et Biophysica Acta*, **842**, 111–118.

Pegg, A., Pakala, R., & Bergeron, R. (1990) *Biochemical Journal*, **267**, 331–338.

Perchellet, J. P. & Boutwell, R. (1981) *Cancer Research*, **41**, 3918–3926.

Persson, L. & Pegg, A. (1984) *Journal of Biological Chemistry*, **259**, 12364–12367.

Persson, L. & Rosengren, E. (1989) *Cancer Letters*, **45**, 83–86.

Pierangelli, E., Levin, V., Seidenfeld, J., & Marton, L. (1981) *European Journal of Cancer*, **17**, 143–147.

Pitol, H. & Sirica, A. (1992) *Current Contents*, **35**, 9.

Polvinen, K., Sinervirta, R., Alhonen, L., & Jänne, J. (1988) *Biochemical and Biophysical Research Communications*, **155**, 373–378.

Porter, C., Ganis, B., Libby, P., & Bergeron, R. (1991) *Cancer Research*, **51**, 3715–3720.

Porter, C., Herrera–Ornelas, L., Pera, P., Petrelli, N., & Mittelman, A. (1987) *Cancer*, **60**, 1275–1281.

Quemener, V., Bansard, J. P., Bouet, F., Bergeron, C., Le Moine, P., Martin, L., Le Gall, E., Kerbaol, M., & Moulinoux, J. P. (1993) *Cancer Journal*, **6**, 208–212.

Quemener, V., Martin, C., Havouis, R., & Moulinoux, J. (1991) *Invivo*, **5**, 91–94.

Quick, D. & Wallace, H. (1993) *Biochemical Pharmacology*, **46**, 969–974.

Radford, D., Nakai, H., Eddy, R., Haley, L., Byers, M., Henry, W., Lawrence, D., Porter, C., & Shows, T. (1990) *Cancer Research*, **50**, 6146–6153.

Rice, R. & Green, H. (1978) *Journal of Cell Biology*, **76**, 705–711.

Roblot, G., Wylde, R., Martin, A., & Parello, J. (1994) *Tetrahedron*, **49**, 6381–6398.

Röhn, G., Ernestus, R., Schröder, R., Hossmann, K., Klug, N., & Paschen, W. (1992) *Acta Histochemica*, **42**, 155–158.

Romano, M. & Bonelli, P. (1988) *Tumori*, **74**, 397–399.

Rosenblum, M., Durie, B., Salmon, S., & Russell, D. (1978) *Cancer Research*, **38**, 3161–3163.

Russell, D. (1971) *Nature New Biology*, **233**, 144–145.

——— (1973) *Polyamines in Normal and Neoplastic Growth*. New York: Raven Press.

——— (1983) *Clinical Reviews of Clinical Laboratory Science*, **18**, 261–311.

Russell, D. & Levy, C. (1971) *Cancer Research*, **31**, 248–251.

Russell, D., Levy, C., Schimpff, S., & Hawk, I. (1971) *Cancer Research*, **31**, 1555–1558.

Sandler, R., Ulshen, M., Lyles, C., McAuliffe, C., & Fuller, C. (1992) *Digestive Diseases and Sciences*, **37**, 1718–1724.

Sanford, E., Drago, J., Röhner, T., Kessler, G., Sheehan, L., & Lipton, A. (1975) *Journal of Urology*, **113**, 218–221.

Sasakawa, N., Ishii, K., & Yamamoto, S., & Kato, R. (1985) *Biochemical and Biophysical Research Communications*, **128**, 913–920.

Sato, N. & Fujiwara, K. (1988) *Journal of Biochemistry*, **104**, 98–101.

Scalabrino, G., Modena, D., Ferioli, M., Puerari, M., & Luccarelli, G. (1982) *Journal of the National Cancer Institute*, **68**, 751–754.

Scalabrino, G., Pigatto, P., Ferioli, M., Modena, D., Puerari, M., & Caru, A. (1980) *Journal of Investigative Dermatology*, **74**, 122–124.

Seidenfeld, J., Levin, V., Devor, W., & Marton, L. (1979) *European Journal of Cancer*, **15**, 1319–1327.

Seiler, N. (1981) *Polyamines in Biology and Medicine*, pp. 169–180, D. R. Morris & L. J. Marton, Eds. New York: Marcel Dekker.

——— (1987) *Canadian Journal of Physiology and Pharmacology*, **65**, 2024–2035.

Seiler, N. & Al-Therib, M. (1974) *Biochimica et Biophysica Acta*, **354**, 206–212.

Seiler, N., Bolkenius, F., & Knödgen, B. (1980) *Biochimica et Biophysica Acta*, **633**, 181–190.

Seiler, N., Graham, A., & Bartholeyns, J. (1981) *Cancer Research*, **41**, 152–153.

Seiler, N. & Knödgen, B. (1979) *Biochemical Medicine*, **21**, 168–181.

Seiler, N., Knödgen, B., & Bartholeyns, J. (1985) *Anticancer Research*, **5**, 371–378.

Shappell, N., Fogel-Petrovic, M., & Porter, C. (1993) *FEBS Letters*, **321**, 179–183.

Shinki, T., Kadofuku, T., Sato, T., & Suda, T. (1986) *Journal of Biological Chemistry*, **26**, 11712–11716.

Shinki, T. & Suda, T. (1989) *European Journal of Biochemistry*, **183**, 285–290.

Shinki, T., Takahashi, N., Kadofuku, T., Sato, T., & Suda, T. (1985) *Journal of Biological Chemistry*, **260**, 2185–2190.

Shinki, T., Tanaka, H., Kadofuku, T., Sato, T., & Suda, T. (1989) *Gastroenterology*, **96**, 1494–1501.

Shinki, T., Tanaka, H., Takito, J., Yamaguchi, A., Nakamura, Y., Yoshiki, S., & Suda, T. (1991) *Gastroenterology*, **100**, 113–122.

Snyder, S., Krenz, D., Medina, V., & Russell, D. (1970) *Annals of the New York Academy of Science*, **171**, 749–771.

Sumiyoshi, H., Baer, A., & Wargovich, M. (1991) *Cancer Research*, **51**, 2069–2072.

Sundberg, J., Erikson, A., Roop, D. R., & Binder, R. (1994) *Cancer Research*, **54**, 1344–1351.

Tadolini, B., Hakim, G., Orlandini, G., & Casti, A. (1986) *Biochemical and Biophysical Research Communications*, **134**, 1365–1371.

Takami, H. & Nishioka, K. (1980) *British Journal of Cancer*, **41**, 751–756.

Takigawa, M., Verma, A., Simsiman, R., & Boutwell, R. (1983) *Cancer Research*, **43**, 3732–3738.

Tatsuta, M., Iishi, H., Baba, M., Ichii, M., Nakaizumi, A., Uehara, H., & Taniguchi, H. (1991) *International Journal of Cancer*, **47**, 738–741.

Thomas, T., Kiang, D., Jänne, D., & Thomas, T. (1991) *Breast Cancer Research and Treatment*, **19**, 257–267.

Thomas, T. & Thomas, T. (1994) *Cancer Research*, **54**, 1077–1084.

Tormey, D., Waalkes, T., Kuo, K., & Gehrke, C. (1980) *Cancer*, **46**, 741–747.

Uehara, N., Kita, K., Shirakawa, S., Uchino, H., & Saeki, Y. (1980b) *Gann*, **71**, 393–397.

Uehara, N., Shirakawa, S., Uchino, H., Saeki, Y., & Nozaki, M. (1980a) *Life Sciences*, **26**, 461–467.

Uehara, N., Shirakawa, S., Uchino, H., & Saeki, Y. (1980c) *Cancer*, **45**, 108–111.

Van Den Berg, G., Nagel, G., & Muskiet, F. (1985) *Journal of Chromatography*, **339**, 223–231.

van den Berg, G., Schaaf, J., Nagel, G., Teelken, A., & Muskiet, F. (1987) *Chemica Chimica Acta*, **165**, 147–154.

Vanella, A., Campisi, A., Guglielmo, P., Cacciola, Jr., E., Cunsolo, F., Geremia, E., Tiriolo, P., Pappalardo, P., & Crisafi, G. (1986) *Acta Haematologica*, **76**, 33–36.

Verma, A. (1985) *Biochimica et Biophysica Acta*, **846**, 109–119.

——— (1988) *Cancer Research*, **48**, 2168–2173.

Verma, A. & Erickson, D. (1986) *Archives of Biochemistry and Biophysics*, **247**, 272–279.

Verma, A., Lowe, N., & Boutwell, R. (1979) *Cancer Research*, **39**, 1035–1040.

Volkow, N., Goldman, S., Flamm, E., Cravioto, H., Wolf, A., & Brodie, J. (1983) *Science*, **221**, 673–675.

Wallace, H., Macgowan, S., & Keir, H. (1985) *Biochemical Society Transactions*, **13**, 1985–1986.

Wang, G., Volkow, N., Wolf, A., Madajewicz, S., Fowler, J., Schlyer, D., & MacGregor, R. (1994) *Nuclear Medicine and Biology*, **21**, 77–82.

Warnick, R., Pietronigro, D., McBride, D., & Flamm, E. (1988) *Neurosurgery*, **23**, 464–469.

Weekes, R., Verma, A., & Boutwell, R. (1980) *Cancer Research*, **40**, 4013–4018.

Weiner, R. & Byus, C. (1980) *Biochemical and Biophysical Research Communications*, **97**, 1575–1581.

Wertz, P. & Downing, D. (1989) *Biochimica et Biophysica Acta*, **1002**, 213–217.

Westin, T., Edström, S., Lundholm, K., & Gustafsson, B. (1991) *American Journal of Surgery*, **162**, 288–293.

Williams–Ashman, H., Coppoc, G., & Weber, G. (1972) *Cancer Research*, **32**, 1924–1932.

Wu, V. & Byus, C. (1984) *Biochimica et Biophysica Acta*, **804**, 89–99.

Xiao, L., Celano, P., Mank, A., Griffin, C., Jaba, E., Hawkins, A., & Casero, Jr., R. (1992) *Biochemical and Biophysical Research Communications*, **187**, 1493–1502.

Xiao, L., Celano, P., Mank, A., Pegg, A., & Casero, Jr., R. (1991) *Biochemical and Biophysical Research Communications*, **179**, 407–415.

Yamamoto, S., Jiang, H., Otsuka, C., & Kato, R. (1992) *Carcinogenesis*, **13**, 905–906.

Yodfat, Y., Weiser, M., Kreisel, M., & Bachrach, U. (1988) *Clinica Chimica Acta*, **176**, 107–114.

Yuspa, S., Ben, T., & Lichti, U. (1983) *Cancer Research*, **43**, 5707–5712.

Yuspa, S., Lichti, U., Ben, T., Patterson, E., Hennings, H., Slaga, T., Colburn, N., & Kelsey, W. (1976) *Nature*, **262**, 402–404.

Yuspa, S., Lichti, U., Hennings, H., Ben, T., Patterson, E., & Slaga, T. (1978) *Carcinogenesis*, pp. 245–255, T. Slaga, A. Swak, & R. Boutwell, Eds. New York: Raven Press.

CHAPTER 17

Surviving Cancers:
Controlling the Disease

Evolution of Therapeutic Modalities

The demonstrations of the increases of polyamines and ornithine decarboxylase (ODC) in growth and proliferation compelled speculation on the relevance of these phenomena to problems of cancer and other cellular pathologies. The somewhat inconsistent evidence of heightened urinary excretion of polyamines among cancer patients and of effects during drug-induced remissions posed the general question of effects of various pathologies and toxic agents on polyamine excretion. The discovery of the effects of methylglyoxal-bis(guanylhydrazone) (MGBG) (Mihich, 1963) and anticancer antibiotics containing polyamines, the bleomycins, encouraged some additional interest. Synthesis of the various inhibitors of ODC and the availability of difluoromethylornithine (DFMO) activated studies in tumorigenic systems. Although polyamine depletion was not easily attained, the discovery of the catabolic systems explained some of the sources of new complicating polyamines. Meanwhile, many other approaches to cancer therapy, for example, the exploitation of apoptosis and hyperthermia, were found to have polyamine components. The newer possibilities of treating cancers by recourse to the many synthesized inhibitors of polyamine metabolism have been discussed by Marton and Pegg (1995).

Monitoring Chemotherapy

The spleens of leukemic mice treated with methotrexate, arabinosylcytosine (Russell, 1972), or 5-azacytidine (Heby & Russell, 1973) were found to have a marked decrease in the levels of the polyamine biosynthetic enzymes and in polyamine accumulation. Treatment by 5-fluorouracil of rats with hepatomas rapidly increased the spermidine content of their sera and concomitantly reduced the spermidine content of the liver (Russell et al.,

1974). Mice bearing Ehrlich carcinoma or S180 sarcoma also responded to cyclophosphamide, 5-fluorouracil, or 6-mercaptopurine with enhanced polyamine excretion, and did so to a greater extent than did tumor-free animals (Osswald et al., 1986). Increased polyamine excretion of cancer patients appears to be a common response to successful chemotherapy or radiation therapy. Partial or complete responses were paralleled by a greater than twofold rise in urinary spermidine within 24–48 h after initiation of therapy. This was interpreted as the result of tumor cell death.

Nutritionally fortified patients increase polyamine excretion to an even greater extent (Pöyhönen et al., 1992). In the latter study a comparison to other types of patients (e.g., multiple trauma) revealed that polyamine excretion reflected the level of metabolism in general and was not specific for any particular disease.

In studies of cultured cancer cells Mackarel and Wallace (1992) described the effect of 5-fluorouracil on polyamine excretion by a human colonic carcinoma cell line. At a level of growth inhibition of 50%, intracellular polyamine fell 60%. Acetylated polyamines were the major excretion product (>60% total) of the control cells, as well as of the treated cells. Studies on the effects of aminoglycoside antibiotics revealed that gentamycin tied up pyridoxal phosphate and initially inhibited ODC, but led to large increases in ODC and spermidine/spermine N^1-acetyltransferase (SSAT) in the cells. Indeed the cells accumulated high levels of acetylspermidine and acetylspermine (Kaloyanides & Ramsammy, 1993). These effects might appear then under conditions of treatment of infections with such drugs.

Anticancer Antibiotics:
The Bleomycins

Many antibiotics proved to be toxic to animal cells and some of these (e.g., actinomycin D, 5-azacytidine) affect tumor growth in animals relatively selectively. The important antimicrobial and antitumor agents, kanamycin and phleomycin, respectively, were discovered in 1956 in a study of water-soluble basic antibiotics (Umezawa, 1983). Phleomycin was found to have a strong renal toxicity, and other antitumor antibiotics were sought that lacked this toxicity. Bleomycins were then isolated from cultures of *Streptomyces verticillium* and were found to include a large family of glycopeptide antibiotics possessing a wide spectrum of antitumor and antimicrobial activities. The structure (Fig. 17.1) containing L-threonine and four new amino acids was linked to a chromophoric carboxydithiazole, combined in turn to any one of a dozen essential bases. The mix of phleomycins differed from bleomycins only in the hydrogenation of one thiazole ring.

The terminal cation of bleomycin A_2 might be the product of the decarboxylation of methylmethionine, a sulfonium-containing amino acid found mainly in plants; however, the mechanism of its biosynthesis is not known. A total synthesis of bleomycin A_2 has been reported (Hecht et al., 1979).

Bleomycin shrinks and even cures some human tumors (e.g., squamous cell carcinomas, Hodgkin's lymphoma) and is used in combination with some other antitumor agents in the treatment of testicular tumors. New types of bleomycins have been synthesized to overcome pulmonary toxicity.

The removal of the terminal cation (agmatine) of bleomycin B_2 has been effected with a mold enzyme, acylagmatine hydrolase (Umezawa et al., 1973), resulting in the production of an inactive bleomycinic acid. This result asserts the importance of the terminal cation

and provides a substrate for the chemical synthesis of artificial bleomycins, such as peplomycin, BAPP, and A5033 (Fig. 17.1). Bleomycinic acid has been coupled chemically to new specific amines to form the semisynthetic bleomycins (Takita et al., 1973). Also, the addition of the various bleomycin amines to fermentation media has increased the rate of production of a specific bleomycin containing an exogenous amine. The addition of either spermidine or spermine to the fermentation markedly increased the synthesis of bleomycin A_5, containing spermidine as the terminal amine, and completely suppressed the formation of other bleomycins (Fujii et al., 1974). In A_5 the amine linked to the carboxydithiazole is that of the aminopropyl moiety and A_5 is essentially inert to many amine oxidases.

Bleomycin commercially available in the United States is a mix of numerous bleomycins, of which A_2 and B_2 are major components; it has been used for most clinical trials in the West. The predominant component found in the clinical mixture in use in Russia and China is the spermidine-containing bleomycin A_5. This is produced readily and was reported to be more active than A_2 and B_2 in some tumor systems and to have less pulmonary toxicity than the commercially available bleomycins. Also the spermidine moiety of A_5 can be linked to an oligonucleotide to effect site-specific cleavage of a DNA target (Sergeyev et al., 1995).

The biological properties of the congeners are not identical. Bleomycin A_5 is far more lethal than A_2 or B_2 in *Escherichia coli* or in mouse fibroblasts (L cells) (Cohen & I, 1976; Lapi & Cohen, 1977). Hirudonine, the diguanido analogue of spermidine, specifically protected the animal cells against A_5. This effect has

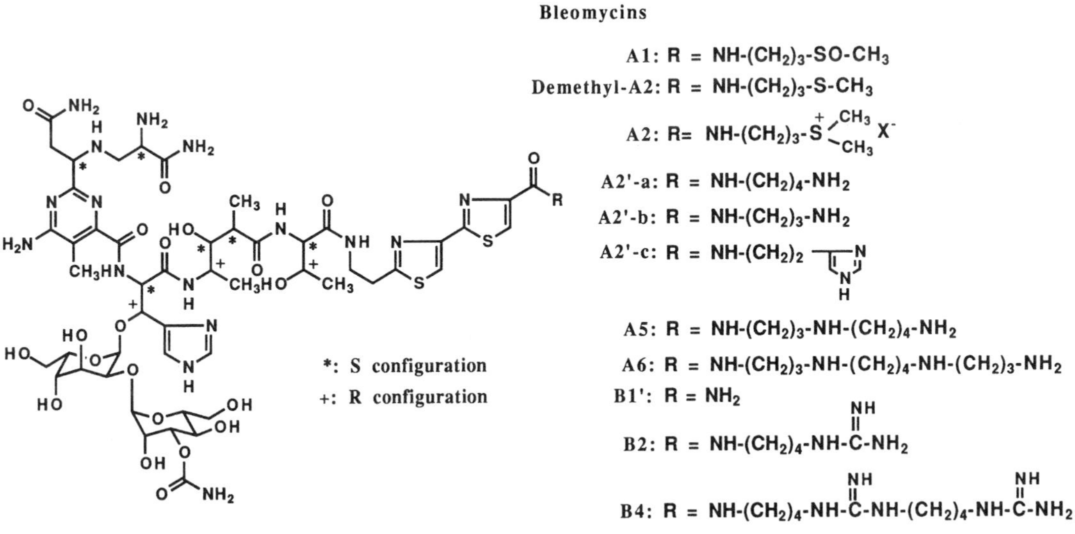

FIG. 17.1 The postulated structure for the bleomycin-Fe(II)-O_2 complex. The R derivatives have been adapted from Umezawa (1983).

been ascribed to effects on the entrance of A_5 into these cells.

$$NH \quad\quad\quad\quad NH$$
$$\| \quad\quad\quad\quad\quad\quad\quad \|$$
$$H_2N-C-NH(CH_2)_4NH(CH_2)_3NH-C-NH_2$$

Hirudonine

Further, although A_2 and B_2 appeared similarly toxic in *E. coli* and fibroblasts, a comparison of the effects of A_2 and B_2 on strains of *Saccharomyces cerevisiae* that were unable to repair DNA lesions produced by the antibiotics revealed that B_2 introduced breaks in DNA at about twice the rate effected by A_2. Nevertheless, approximately equivalent numbers of breaks occurred in equivalent fractions of the surviving cells (Moore, 1990). Thus, the two compounds have significantly different rates of reaction with DNA when offered to cells. However, these differences may relate to different rates of penetration and concentration in cells or to differences in the rates of reaction with DNA. Bleomycin also cleaves cell wall polymers in yeast (Lim et al., 1995).

Bleomycins fragment DNA in cells and inhibit both DNA and RNA synthesis. The antibiotic cleaves vaccinia virus DNA and presumably this inhibits the growth of vaccinia virus; it is also reported to cure the epithelial papillomas (warts) caused by the DNA-containing human papilloma virus. Initially the cleavage of the nucleic acid was thought to be DNA specific but it appears that an activated bleomycin, Fe^{2+}-bleomycin A_2, can cleave certain tRNA precursors oxidatively in specific single-stranded regions of the structure (Magliozzo et al., 1989). Some viral mRNA was also cleaved fairly specifically, suggesting that some RNAs may be a therapeutically significant target for bleomycins (Carter et al., 1990). Morgan and Hecht (1994) described the cleavage of both strands of a DNA–RNA heteroduplex by Fe^{2+}-bleomycin A_2.

In contrast to the extensive fragmentation of isolated DNA, the fragmentation of chromatin is far more limited. The reaction of bleomycin with isolated nuclei produced cleavage of nucleosome linker DNA, and the now familiar "nucleosome ladder" comprising nucleosomal units was demonstrable by gel electrophoresis (Kuo & Hsu, 1978). Bleomycin also released certain nonhistone proteins from chromatin and cleaved nucleosomal linkers not occupied by H_1 histone (Hamana & Kawada, 1989). This selective reaction was blocked by 1 mM spermidine or spermine. Bleomycin and DFMO are markedly synergistic in the cure of trypanosomiasis in mice, implying that the depletion of spermidine may stimulate the bleomycin killing of trypanosomes (Bacchi et al., 1982).

Activated bleomycin is now considered to be a ferric peroxide complex which cleaves DNA oxidatively. (Sam et al., 1994). The structure, presented in Figure 17.1, indicates the role of the amino group of a β-amino-alanine pyrimidine substituent in the formation of the Fe^{2+} chelate. Bleomycin A_2 can be inactivated by enzymatic acetylation of this amino group (Sugiyama et al., 1994). The pK of the amino group turns on the near-neighboring carboxamide, which may be cleaved by a bleomycin hydrolase. The hydrolase, which is responsible in many tissues for the relative inactivation of the antibiotic, releases 1 mol NH_3/1 mol bleomycin to form a desamidobleomycin (Sebti & Lazo, 1988). The absence of the enzyme in lung tissue correlates well with the apparent lung toxicity of this tissue. The resistance of some tumors like Burkitt's lymphoma has been attributed to a rapid metabolism of A_2, in part as a result of the activity of the hydrolase.

The toxicity of bleomycins in lung tissue is manifested in its early stages as alveolar cell hyperplasia and interstitial fibrosis (Lazo et al., 1990). The types of terminal amines affect the incidence of interstitial fibrosis caused by the antibiotic somewhat, but not to a great extent. The semisynthetic peplomycin appears to have a lowered pulmonary toxicity but has some other unfavorable toxicities. Although the toxicities, both fibrosis and metaplasia in mice, were reported to be essentially similar for an antibiotic and its terminal amine (Raisfeld, 1981), it has not been demonstrated that the amine or an active fragment of the antibiotic has been released from the antibiotic. Administration of DFMO failed to protect against bleomycin-induced fibrosis.

Studies with the superhelical SV40 DNA permitted the demonstration of single scissions and subsequently of double-strand breaks. Such dangerous breaks are also introduced exogenously by ionizing radiation and some chemicals, including bleomycin, neocarcinostatin, and toxicants that release or activate endonucleases. Therapeutic compounds like chlorambucil that cross-link double-stranded DNA become far more efficacious (10^4-fold) in this reactivity when conjugated in amide linkage to spermidine (Cohen et al., 1992; Holley et al., 1992).

Modeling studies with A_2 have postulated an electrostatic association of the terminal cation with a phosphate between thymine and adenine nucleotides in a single chain of DNA. This may assist the intercalation of one planar thiazole ring between the bases to place the active oxidizing center of the antibiotic at the cleavable DNA site (Murakami et al., 1976). Fluorescence spectroscopy demonstrated the quenching of bithiazole fluorescence in the binding of A_2 and of the cationic dithiazole amide by thymus DNA. Studies with proton magnetic resonance have shown that the dithiazole and the dimethylsulfonium groups are bound most tightly in this structure. Later studies have revealed details of intercalation of cationic dithiazole amides related to A_2 in poly(dA-dT). Evidence was obtained for an initial ionic interaction to insert the dithiazole system (Sakai et al., 1982). This group also performed a crystallographic analysis of an analogue of the DNA-binding portion of bleomycin A_2. The model derived from the study was in accord with the interpretation of the NMR data: a partial intercalation of the second thiazole ring occurs between the base pairs of a double-stranded deoxydinucleoside phosphate d(CpG) with the sulfonium cation interacting with a backbone phosphate (Kuroda et al., 1982). Addition of various fragments of A_2, dithiazole amines, to DNA did not cause scission of DNA, although the interaction did produce the unwinding of DNA characteristic of intercalation (Fisher et al., 1985). New conjugates containing polyamines linked to an intercalative compound have been found to increase the DNA degradative activity of bleomycin A_2 (Strekowski et al., 1988). The free polyamine also has this effect. Apparently a distorted DNA helix is a better target for bleomycin than is the native form, and polyamine conjugates can distort the structure to this end.

The bleomycins have proven to be complex and to comprise two functional entities, one of which contains a terminal cation or polyamine linked to a partially intercalative planar structure. The association of this with DNA permitted the other functional entity, an oxidizing Fe^{2+} peptide chelate, to cleave DNA at preferential sites. Other antibacterial antibiotics containing polyamines include the edeines, the glysperins, and the glycocinnamoyl spermidines derived from various bacteria. An additional antitumor activity, isolated from *Bacillus laterosporus*, resides in spergualin in which spermidine is bound in amide linkage to 7-guanido-3-hydroxyheptanoic acid. The structures of these compounds have been given by Umezawa (1983).

Provoking Cell Death: Effects of Hyperthermia

Microbial studies of nutritional deficiency, antibiotic action, and radiation damage adopted the principle that a cell was dead when it could no longer multiply, that is, form a viewable microbial colony. Several simple principles emerged:

1. nutritional deficiency was usually cytostatic, rather than cytocidal;
2. a very few nutritional deficiencies, such as thymine deficiency, which creates blocks specifically in DNA synthesis, were cytocidal (thymineless death) and this could be imposed by some toxic agents (e.g., 5-fluorouracil);
3. radiation damage from ultraviolet (UV) irradiation was most lethal at wavelengths absorbed by nucleic acids;
4. some agents that selectively inhibited the biosynthesis of the bacterial cell wall (e.g., penicillin) provoked the lysis of cell membranes.

Because DNA and lipid-rich cell membranes were possible targets for the killing of eucaryotic cells, this was followed by determining the ability of cultured cells to form colonies.

Hyperthermia is synergistic with some types of radiation in killing cells. Polyamine depletion is itself inhibitory to DNA repair in irradiated cells and contributes to the increased radiosensitivity and thermal inhibition of repair (Snyder & Lachmann, 1989). Hyperthermia (43°C) also enhanced the bleomycin-induced kill of

Chinese hamster ovary (CHO) cells. Thus, hyperthermia has been studied mainly as a therapeutic modality in conjunction with other treatments.

The early observations of the stabilization of bacterial membranes in protoplasts by polyamines were extended by Ray and Brock (1971) to the protection of protoplasts against thermal lysis. The shearing of ghosts of erythrocytes to deformable fragments (spherical vesicles) was found to be inhibited by polyamines, notably spermine, which stabilized the membrane skeleton (Ballas et al., 1983). More recently cationic substances (e.g., 2 mM spermidine) were reported to displace proteins and phospholipids from the kidney or brain membranes (Fukushima, 1990). Also, treatment by spermine at 43°C was shown to deplete glutathione in the cells (Russo et al., 1985).

Unexpectedly polyamines enhanced cell kill in CHO cells held at higher than body temperatures (42°C). Spermine also enhanced the lethal effects of the synergistic combination of hyperthermia and γ radiation (Ben-Hur & Riklis, 1979a). Protein synthesis is completely shut off at 42°C, whereas synthesis of RNA and DNA are inhibited more slowly (Ben-Hur & Riklis, 1979b). These inhibitory effects are not increased by spermine, which enhances cell kill at 42°C. At this temperature, ODC decays with a half-life ($t_{1/2}$) of 12 min and disappears completely from the treated cell. Transfer to 37°C permits a linear synthesis of protein and recovery of ODC. However, the presence of spermine at 37°C significantly delayed the recovery of ODC. This may be a useful method of eliminating ODC and facilitating polyamine depletion.

In addition, Gerner et al. (1980) demonstrated a leakage of polyamine from the cell, as well as an uptake of exogenous polyamine, and observed the surprising effect of enhancing cell damage at very low concentrations (10^{-5} M) of exogenous polyamine as compared to the much higher intracellular concentrations (10^{-3} M). The membrane was a major site of heat damage and spermine reactivity. It was also shown that diamines and MGBG might enhance cell kill, and indeed 1,8-diaminooctane was equivalent to spermidine in this sensitization.

Polyamine depletion produced by DFMO evoked a delayed but enhanced hypersensitivity of CHO cells to hyperthermia (43°C). This appeared after an 8-h exposure to DFMO and was partially eliminated by providing putrescine specifically (Fuller & Gerner, 1982, 1987). Gerner et al. (1988) did find significant differences in the plating efficiency of normal and DFMO-depleted CHO cells after X-radiation. It was evident that polyamine deficiency did reduce repair processes and recoveries in repair were restored by exogenous putrescine. However, the DFMO-induced hypersensitization to heat is not readily extrapolated to all cell models, for example, HeLa cells and some tumor models (Lopez Ballester et al., 1991; Roizin–Towle et al., 1990).

Fuller et al. (1990) observed a rapid induction of SSAT in heated CHO cells. DFMO blocked the induction that was restorable by addition of putrescine or spermidine to the cultures. This group went on to discover two mechanisms of increased H_2O_2 production from exogenous and endogenous polyamines induced by hyperthermia (Harari et al., 1989a,b). The first, apparently involving amine oxidases, produced an increased kill, prevented to a significant extent by aminoguanidine (AG). The second degradative sequence, involving SSAT and polyamine oxidase (PAO), led to the accumulation of N^1-acetylspermidine at 42°C in the presence of a PAO inhibitor. In this system, glutathione depletion, produced by an inhibitor of the biosynthesis of this metabolite, markedly increased cell kill at 43°C for 90 min. The presence of a PAO inhibitor increased the survival of these heated cells, perhaps by decreasing the metabolites formed from N-acetylspermidine. The control of newly generated H_2O_2 is obviously relevant to cell survival.

Harari et al. (1990) attempted to extrapolate the data on the sensitization of hyperthermic cell kill by polyamine depletion to the treatment of metastatic melanoma. Heat-inducible polyamine catabolism and heat sensitization by DFMO were observed in human melanoma cultures. The possibility of such a trial with human subjects has been encouraged by improvements in the technology of inducing hyperthermia.

Phenomena of Apoptosis

The term apoptosis was introduced in the early 1970s to describe the conversion of once normal cells (hepatocytes) to small round cytoplasmic masses frequently containing bits of condensed chromatin. "Necrosis" with membrane disruption and lysis did not seem applicable. The detection of apoptotic cell death merged with the view that a less random, even genetically determined, cell death was essential to the evolution of animal embryos and to the modeling of specific tissues (Gershenson & Rotello, 1992). Comparison of cell kill in the range of 42–47°C for 30 min revealed that 43–44°C enhanced an internucleosomal cleavage of DNA, a pattern thought to be characteristic of *apoptosis* (Harmon et al., 1990). At 46 and 47°C there was virtually no DNA degradation, despite cell death as a function of cell necrosis and presumably membrane disruption.

The initial biochemical studies of dying cells, for example, in the involution of androgen-deprived tissue or in the effect of glucocorticoid on lymphocytes, revealed the development of nucleosomal ladders, comparable to that obtained in killing by bleomycin (Tounekti et al., 1995). Apoptosis induced by dexamethasone in mouse thymocytes does not lead to the fragmentation of mitochondrial DNA; it is a specifically nuclear event (Murgia et al., 1992). Significantly, cell necrosis does evoke degradation of mitochondrial DNA (Tepper & Studzinski, 1993). The internucleosomal cleavage is believed to be due to a Ca^{2+},Mg^{2+}-dependent endonuclease optimally active below pH 6.8, which evidently produces breaks in the internucleosomal DNA linker, as does bleomycin. The endonuclease is thought to be activated in some instances by the uptake of Ca^{2+}, and this activa-

tion can be inhibited by spermine (Brüne et al., 1991). An overaccumulation of spermidine and its analogues can induce apoptosis and DNA fragmentation (Poulin et al., 1995). As summarized by Sen and D'Incalci (1992), many antitumor agents (e.g., araC, cisplatin, etoposide, methotrexate, and vincristin) have also been reported to induce this type of cell damage.

Although thymineless death was discovered in bacterial systems, it is also provoked in animal cells deficient in thymidine or treated with 5-fluoro-2′-deoxyuridine, an inhibitor of thymidylate synthase. DNA fragmentation and the formation of a nucleosomal ladder does occur in the latter system (Kyprianou & Isaacs, 1989). Approaches to the therapy of prostatic cancer have pointed to the use of agents that might activate the Ca^{2+}-dependent endonuclease, such as a Ca^{2+} ionophore or 5-fluoro-2-deoxyuridine (Kyprianou et al., 1991). Also, the incubation of human mammary adenocarcinoma cells with a tumor necrosis factor (TNF) results in early DNA fragmentation preceded by increases in intracellular free Ca^{2+} (Bellomo et al., 1992).

Lethality has been correlated with the formation of oligonucleosome-sized fragments in cell cultures such as those of thymocytes induced with adenosine (Kizaki et al., 1990). In this system, fragmentation of DNA requires an increase of cAMP and the synthesis of RNA and protein. The endonuclease and DNA degradation in thymocytes may be activated by glucocorticoid hormones and Ca^{2+} ionophores, but both of these activations can be prevented by the addition of spermine (Brüne et al., 1991). MGBG was also found to cause DNA fragmentation in the thymocyte system. Permeabilized liver nuclei incubated with Ca^{2+} and Mg^{2+} also degrade DNA but this is also blocked by spermine, which modifies chromatin organization. Polyamine depletion is known to increase the digestibility of nuclear DNA by a micrococcal nuclease (Basu et al., 1992). However, few systems have been identified in which the depletion of polyamine has facilitated nucleosomal degradation.

The introduction of breaks in DNA results in the increased synthesis of poly(ADP-ribose), the formation of which has been associated with DNA repair. Poly(ADP-ribose) polymerase, which uses NAD^+ as a substrate, is activated by DNA strand breaks. The inhibition of this synthesis by 3-aminobenzamide or 3-methoxybenzamide significantly potentiates cell kill by bleomycin (Huet & Laval, 1985; Kato et al., 1988). The presence of the poly(ADP-ribose) polymerase at high concentration is correlated with resistance of HeLa cells to bleomycin (Urade et al., 1989).

The increase of poly(ADP-ribose) has been found in apoptotic cell death provoked by antitumor agents (Marks & Fox, 1991). Cell death is therefore preceded by a decrease of nuclear NAD^+ and ATP. Because the cell nucleus generates much of its ATP by glycolysis, it appears that ADP ribosylation and loss of ATP may also contribute to cell death (Mizumoto & Farber, 1995). Recent data have implicated the formation of the polymer in the moving of histone molecules, that is, histone shuttling, essential to the unfolding of chromatin in DNA excision repair (Realini & Althaus, 1992; Thibeault et al., 1992). The inactivation of the poly(ADP-ribose) polymerase in malignant cells by 3-nitrosobenzamide was reported to provoke the activation of the Ca^{2+},Mg^{2+}-dependent endonuclease and the lethal degradation of leukemia cell DNA (Rice et al., 1992). Moreover, elimination of a protease that can degrade the poly (ADP-ribose polymerase) protects cells from apoptosis (Martin & Green, 1995; Nicholson et al., 1995).

The lethality of induced macrophages to tumor cells (Farias–Eisner et al., 1994) and invasive bacteria and fungi (Schmidt & Walter, 1994) has been attributed to the induction of nitric oxide (NO) synthase, as well as to the release of superoxide anion (O_2^-) and H_2O_2. The NO synthases function with tetrahydrobiopterin as a cofactor (Nathan & Xie, 1994) and a group of pteridines is substantially increased in the urine of cancer patients (Fujita et al., 1987). These workers have reported that the levels of urinary pteridines and polyamines were increased and served to mark almost 100% of all cancer patients. Because NO is derived from arginine in NO-producing cells, this biosynthesis could be interactive with that of polyamine formation. For example, the intermediate in NO synthesis, N^G-hydroxy-L-arginine, is an inhibitor of arginases and might restrict the availability of ornithine (Boucher et al., 1994). On the other hand, DFMO and MGBG block the activation of macrophages (Kaczmarek et al., 1992), and TNF has been found to stimulate ODC in various cells (Donato et al., 1991). Whether the use of the inhibitors of polyamine synthesis in cancer chemotherapy does reduce the normal tumoricidal functions of the animal will be a matter of some interest.

Endotoxin, a lipopolysaccharide derived from *E. coli*, is known to evoke necrosis of tumors and to activate macrophages to produce NO. In studies of the S180 sarcoma in mice injected with endotoxin, it was found that prior to overt necrosis, putrescine increased rapidly in the tumor and ATP fell 99% whereas spermidine fell less severely (Roth et al., 1981).

It would be of great interest to know the change of Ca^{2+} and DNA within the tumor cells. It is entirely possible that the lethality now attributed mainly to the action of an endonuclease also involves the toxic activities of ADP ribosylation, which reduces NAD and ATP concentrations. It may also arise from an inability to handle respiratory stresses resulting from the production of NO and other oxidizing reactants, such as superoxide and H_2O_2, with an additional depletion of glutathione and essential sulfhydryl groups.

Manipulation of polyamine levels in the cells can be toxic. Depletion of leukemia cells by DFMO produces significant increases in several nucleotides of the cellular pools as well as arresting cells in the G_1/S transition (Kremmer et al., 1988). Other inhibitors of biosynthesis like MGBG inhibit efflux of free polyamine and increase the excretion of acetyl polyamine (Wallace & Keir, 1986). An excessive accumulation of spermidine and spermine in various cells is characterized by a marked fall of Mg^{2+} and ATP and swelling of mitochondria prior to cell death (He et al., 1993). The proteins of ATP-depleted cells also have an enhanced sensitivity to thermal aggregation (Nguyen & Bensaude, 1994). ODC-overproducing cells that accumulate putrescine display excessive apoptotic death (Tobias & Kahana, 1995). Thus, the possibility of provoking lethal cellular reactions by affecting polyamine content is real.

In contrast to the action of spermine in inhibiting certain apoptotic stimuli (e.g., glucocorticoids, Ca^{2+} fluxes, and MGBG), the oxidation of polyamine has been implicated in a form of programmed cell death detected in developing mouse embryos. In this system, cells derived from an implant of embryonal carcinoma were selectively killed by the H_2O_2 produced in blastocoele fluid via the action of an amine oxidase and released polyamine (Pierce et al., 1990). The lethal action was inhibited either by aminoguanidine (AG) or catalase and has proved to be quite common in the cultures of cells derived from developing mouse limb buds, tissues actively engaged in apoptotic death and tissue modeling (Parchment, 1993). Thus, the phenomenon of amine oxidase-provoked toxicity, inhibitable by AG, has been found to be a normal event in the development of the embryo.

Anticancer Drugs: Blocking Biosynthesis of Putrescine

Early efforts to inhibit ODC and to deplete intracellular polyamines attempted to exploit several competitive ornithine analogues, for example, α-hydrazinoornithine and α-methylornithine. Short-term experiments with cultured cells demonstrated some depletion of putrescine and inhibition of growth rate, as well as a reversibility of these cytostatic effects by exogenous diamine. The conclusion was reached that irreversible inhibitors such as some unsaturated putrescine analogues and the mono- and difluoromethylornithines, might be more effective, that is, analogues whose activation by ODC would result in the formation of a covalent adduct of an inactive enzyme. Time-dependent irreversible inactivations of mammalian ODC by both sets of analogues and their specificities were reported in 1978 (Metcalf et al., 1978).

A pharmaceutical company's decisions on the path of development of an agent for human disease may be extraordinarily complex and require demonstrations on activities of the agent in a wide assortment of biological systems and many levels of biological complexity, from cells to animals to the human species. The costs in scaling up the synthesis and trials of the agent and the stability, specificity, dosage, and toxicity of the agent are all involved, requiring many tests before it is possible to examine the human response. The organization and performance of trials for toxicity, pharmacokinetics of drug distribution and excretion, and finally of therapy are among the most difficult and expensive. A compound such as DFMO may prove to be not quite adequate by itself for most human cancer, but might prove useful in an originally unanticipated biological setting, such as an advanced stage of sleeping sickness (African trypanosomiasis; Chap. 18). Few pharmaceutical companies have the resources to subject their newly discovered active inhibitor to the widest biological screens. In this instance the company that synthesized DFMO and discovered its activity on ODC offered the compound to many academic investigators, some of whom then discovered significant activities in parasitic disease.

Additional work led to the definition of additional potent irreversible inhibitors of ODC and tumor cells, such as 6-heptyne-2,5-diamine (Danzin et al., 1983; Mamont et al., 1984) and the allenyl-putrescine (Casara et al., 1984). The inhibitor 6-heptyne-2,5-diamine, which is 30 times more potent than DFMO, decreased the cellular concentration of putrescine in rat hepatoma cells to a lower level than did DFMO and also decreased the rate of growth more severely. Spermine concentrations fell as well, unlike many results with DFMO, and cloning efficiency, the ability of the cells to form a colony on a plate, fell 88 and 99.9% in 4 and 6 days, respectively. This drug has had striking antitumoral effects in rodent systems. In studies (Pera et al., 1986) with leukemic cell lines the heptyne diamine was found to be more effective than the ornithine analogues in depleting polyamine and inhibiting cell growth. Esters of some of the ornithine analogues, described as *pro-drugs*, were significantly more effective in tumor-bearing mice and were of higher potency and even longer duration of action (Claverie & Mamont, 1989).

DFMO at millimolar levels was shown to deplete polyamines and to slow growth in several lines of cultured human carcinoma cells (Heby, 1986). However, the polyamine contents of various tumor cell lines differed and in some lines were occasionally in sufficient excess of the minimum level essential for cell growth to withstand a partial inhibition established by DFMO (Seidenfeld et al., 1986). An increase in the G_1-phase fraction prior to entrance into a period of DNA synthesis (S phase) was found by two groups (Fredlund et al., 1994). However, there have been conflicting reports on the phase of the cell cycle in which cells treated with DFMO accumulate. For example, in a more recent study, polyamine depletion of a mouse tumor led to the accumulation of cells in the G_2M phase; in contrast the tumor cells in a starved animal accumulate in the G_0G_1 phase (Westin et al., 1991b).

The earliest tests of DFMO in normal and tumor-bearing leukemic mice demonstrated a major reduction of ODC in spleens of the leukemic mice, and a stimulation of S-adenosylmethionine decarboxylase (AdoMet-DC). The putrescine and spermidine contents of the organ fell, and the survival of the tumor-bearing animals was extended from 7.6 to 9.3 days. The ODC of spleens of the normal mice was unaffected, although putrescine and spermidine fell somewhat (Prakash et al., 1978).

DFMO-induced decreases of ODC, putrescine, and spermidine in rat prostate were most precipitous. DFMO did reduce the growth of prostatic tissue in growing animals and shrank prostatic tissue and nucleate content in adult animals. It was noted that Ado-

MetDC was increased sevenfold and produced a major 450-fold increase in decarboxylated AdoMet (dAdo-Met) (Danzin et al., 1982). The urinary excretion of this compound in humans was found to be a good biochemical indicator of the selective effectiveness of the inhibition of ODC by DFMO (Haegele et al., 1987). The growth of prostatic tumor cells in rats was dramatically suppressed by DFMO, in contrast to a much lower inhibition by α-methylornithine (Heston et al., 1982).

In contrast to these promising results, DFMO had no effect on the growth of transplanted Wilms' tumor in rats, although inhibitions of ODC and polyamine synthesis were found. At the same time transplanted renal carcinoma was inhibited with respect to growth and metastasis (Kingsnorth et al., 1983). As we now know, different tumors often vary in their response to these agents. Implants in mice of a human small-cell lung carcinoma cell were completely eliminated if DFMO was begun at low tumor burden, although DFMO was less successful if treatment was started later (Luk et al., 1983). It is significant that this line of human cells is killed by polyamine depletion.

Some lines of melanoma cells are also killed in culture by DFMO, and animals bearing this tumor are also protected to a considerable extent by a DFMO-induced polyamine depletion that also restores some melanin production as an indicator of renewed differentiation (Sunkara et al., 1985). As discussed earlier, low levels of ODC and polyamine may stimulate differentiation.

It has been possible to demonstrate the development of resistance of both mouse and human tumor cell lines to DFMO (Hirvonen et al., 1989). This may occur by several different mechanisms, two of which involve the overproduction of ODC from increased transcription or as a result of amplification of ODC genes. Other enzymes of polyamine metabolism may also affect resistance. In one human line, resistance was based on increased arginase activity. In two such lines, ornithine levels were 5–20 times higher than those in the parental line. Among the mouse lines, increases were also seen in the biosynthetic enzymes and in the catabolic systems, that is, PAO and SSAT.

In one study of DFMO-resistant L1210 leukemic cells, the growth of the cells was nevertheless inhibited by an inhibitor of AdoMetDC, which evoked the expected pattern of decreasing spermidine, spermine, and 5′-methylthioadenosine and increasing putrescine (Pegg et al., 1988). Thus, an inhibitor active in a biosynthetic step subsequent to that affected by DFMO can also be active in controlling tumor cell growth.

It was realized very soon that the cytostasis affected by polyamine depletion, although reducing fears of great toxicity, nevertheless permitted restoration of tumor growth when administration of DFMO was stopped. For this reason it appeared important to combine DFMO treatment with another agent that would irreversibly damage the tumor cells. When rat brain tumor cells grown in culture were depleted by means of DFMO, they were thereby sensitized to an alkylating agent (Hunter et al., 1987). In another test, arabinosyl cytosine (araC), which enters DNA in the S phase and kills cells, did appear to synergize DFMO effects on a murine leukemia. In this system polyamine depletion accumulated tumor cells in the araC-sensitive S phase (Prakash & Sunkara, 1983).

The toxicity of DFMO was also studied in normal and tumor-bearing rats (Ota et al., 1986). In rats, platelet suppression was the major dose-limiting side effect of continuous DFMO infusion. Antitumor activity was found with a dose that increased the content of erythrocyte spermidine but was below that producing platelet suppression.

The early results presented above on the inhibition of animal tumors led to tests of DFMO efficacy on a wide assortment of other tumor systems. In 1991 significant antiproliferative effects were recorded for growth of pancreatic adenocarcinomas in hamsters, fibrosarcomas, prostatic adenocarcinomas, glioblastomas in rats, and mammary carcinomas and glioblastomas in mice. DFMO has also been reported to decrease metastasis of various tumor cells in the mouse lung (Sunkara et al., 1982; Sunkara & Rosénberger, 1987). One possibly relevant finding is the inhibition of tumor-induced angiogenesis that provides an oxygen supply and host metabolites for tumor growth (Jasnis et al., 1994; Takigawa et al., 1990).

As noted earlier, DFMO blocked carcinogenesis in two-stage epidermal carcinogenesis in mice. This was extended to a test of the efficacy of DFMO in inhibiting carcinogenesis after N-methyl-N-nitrosourea treatment of a heterotopically transplanted rat urinary bladder. It was found that the instillation of urine containing DFMO into the bladders inhibited carcinogenesis and the increase of polyamines in the urothelial cells (Homma et al., 1985). An effort to determine if the presumably "low toxicity" DFMO can be used to prevent the development of cancers among certain groups of high risk humans is currently in a preparatory stage and will be discussed.

It is known that starvation diets will reduce the rate of tumor growth in rats. Restoration of a complete diet then permits tumor growth to resume. However, when DFMO was added to the more nutritive restorative diet in the malnourished tumor-bearing animals, tumor growth was considerably inhibited. Many of the deficient nontumorous organs were brought back to more normal weights (Nishioka et al., 1988).

In the first human trials reported in 1986, a small number of patients with metastatic melanoma had complete responses of the elimination of a tumor mass, and about a third of the patients were stabilized for a few months. Toxicity, particularly that of ototoxicity, was significant (Abeloff et al., 1986; Meyskens et al., 1986). In general, hematologic toxicities were mild and reversible on discontinuance of the drug. Intravenous infusion reduced the various toxic manifestations and achieved higher plasma levels than did chronic oral dosing. The somewhat disappointing results suggested the desirability of testing various combinations of other drugs with DFMO. However, a sustained trial on two classes of

brain tumor patients demonstrated that oral administration of DFMO alone did have significant antitumor activity in 45% of 44 anaplastic glioma patients (Levin et al., 1992).

In 1989 a trial was reported on the toxic effects of the ODC inhibitor 6-heptyne-2,5-diamine in humans with advanced cancer (Cornbleet et al., 1989). The urinary excretion of dAdoMet was found to be a conveniently accessible marker of ODC inhibition. Some toxicity related to antiproliferative activity was noted.

The incomplete depletion of putrescine by DFMO led to efforts to cut off putrescine production via the catabolic pathway and an inhibitor of PAO was used in conjunction with the ODC inhibitor (Claverie et al., 1987). It was also important to minimize polyamines in the diet and to prevent the bacterial synthesis by addition of sterilizing antibiotics like neomycin (Hessels et al., 1991; Seiler et al., 1990). In such a test on human glioblastoma xenografted on nude mice, spermidine concentrations were lowered and the spermine level increased in all tissues as well as in erythrocytes. However, putrescine was further decreased in the brain and in the tumor whose growth was almost totally inhibited (Moulinoux et al., 1991).

Comparable results were obtained with a transplanted prostatic carcinoma in rats. The same combination was similarly effective in prolonging the survival of rats with a cortical implant of rat glioblastoma cells (Sarhan et al., 1991). Additional retardation of tumor growth was obtained by further restriction of food intake (Sarhan et al., 1992). However, interruption of polyamine deprivation reversed all of these effects. Thus, control of both the biosynthetic and catabolic paths, as well as diet and bacterial contamination, were not in themselves adequate to the problem of eliminating a tumor. This result was also obtained in attempting to cure L1210 leukemia mice (Ask et al., 1992).

Study of the effects of combinations of DFMO and other antitumor drugs were concentrated on S-phase specific drugs when initial results suggested that polyamine depletion led to arrest of mitosis in G_1 before the S phase. Combination of DFMO with araC or cyclophosphamide enhanced cell kill. Cisplatin appeared additive with DFMO in this regard. Combined treatment with interferons also had this type of effect. Pretreatment with DFMO was also reported to increase subsequent effectiveness of other antitumor drugs in tests with various tumor cell lines. Quemener et al. (1992a,b) studied the effects of DFMO, decontamination of the gastrointestinal tract, and an inhibitor of PAO in the treatment of Lewis lung carcinoma in mice, that is, polyamine deprivation combined with a variety of cytotoxic antitumor drugs. Polyamine deprivation alone reduced tumor volume but did not cure the animals. Additional chemotherapy significantly enhanced the effect of depletion on tumor growth and on survival without apparently exacerbating toxicity (Fig. 17.2).

A particularly intensive effort was made in a possible use of combinations of DFMO and more toxic drugs for the potential therapy of brain tumors. These are known

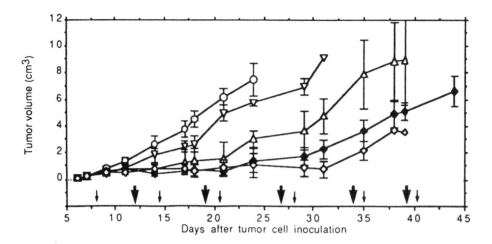

FIG. 17.2 Tumor volume after polychemotherapy with polyamine deprivation; adapted from Quemener et al. (1992b). (O) Controls; (Δ) DR antitumor drugs; (∇) DC-PCD 5/7 discontinuous polyamine deprivation; (◆) DR/DC-PCD 5/7 alternated administration; (◊) DR+ DC-PDC-PDC 5/6 concomitant administration.

to be increased in ODC and putrescine (Ernestus et al., 1993). It was asked initially if polyamine depletion, effected by DFMO, would sensitize rat brain tumor cells to DNA reagents, such as 1,3-bis(2-chloroethyl)-1-nitrosourea (BCNU). It was hypothesized that the reactivity of a DNA stripped of its presumed polyamines with BCNU would be increased. An increased killing was indeed obtained and reported (Hung et al., 1981). Exogenous putrescine prevented this lethal effect. The results in this system, despite some difficulties in the cytological studies, have been confirmed (Wolff et al., 1993).

Radiation therapy has been the most effective treatment of brain tumors, but often provokes edematous change, with a diffuse spread of putrescine beyond the tumor. The volume of edema and degree of breakdown of the blood–brain barrier in the dog brain can be decreased by DFMO before, during, and after irradiation (Fike et al., 1994). In cell lines derived from head and neck carcinomas, a slowing of cell proliferation by DFMO appeared to improve the effectiveness of radiotherapy (Petereit et al., 1994). A replacement in tumor cell lines of natural polyamines by N^1,N^{14}-bis(ethyl)homospermine facilitated an increase of X-radiation induced kill in one cell line of glioblastoma but not in another (Chen et al., 1994).

Polyamine depletion also increases DNA strand breaks in L1210 leukemia cells (Cavanaugh et al., 1984). Increased effects of DFMO depletion to limit normal DNA behavior after drug treatment have been detected in further studies of L-phenylalanine mustard and BCNU in human tumor cell lines. However, sensitization of cisplatin treatment required the prior administration of cisplatin before DFMO (Chang et al., 1987; Marton, 1987). In a study of the combined use of DFMO and X-radiation on a rat glioma model, the combination was at least additive in increasing the significant survival of the animal due to each type of treatment. The

survival periods were inversely related to the tissue levels of putrescine and N^1-acetylspermidine (Tsukahara et al., 1992).

Mention has been made of the use of nuclear ^{3}H-rich polyamine to induce a mutational change in ODC. This lethality of the uptake of [^{3}H]-putrescine has also been studied in brain tumor (gliosarcoma) cells. The cells were permitted to incorporate [^{3}H]-putrescine in DFMO-inhibited cells in a prolonged incubation, and a greater degree of kill was obtained than that found with DFMO alone (Redgate et al., 1993).

The depletion of putrescine and spermidine in animal cells by DFMO is relatively nontoxic, although growth slowing over several normal generation times. Some relatively specific antitumor effects have been noted, and the possibility that a polyamine depletion in rapidly multiplying cells may be sufficiently specific to control a growing tumor without severe damage to the host has encouraged efforts to develop an inhibitor of tumor growth even more effective than DFMO. Newer anti-ODC inhibitors include analogues of 3-aminooxy-1-propanamine (Mett et al., 1993), new halogenated derivatives of ornithine (Trujillo–Ferrara et al., 1992), and antisense oligodeoxynucleotides that would block biosynthesis of the enzyme (Madhubala & Pegg, 1992).

MGBG as Sole or Combined Antitumor Agent

The knowledge that MGBG was itself an antitumor agent, whose action was in part preventable by a concurrent administration of spermidine (1963), and was also a potent inhibitor of the mammalian AdoMetDC (1972), encouraged renewed tests of the compound in combination with other compounds affecting the biosynthesis of the polyamines. Extensive earlier studies on MGBG and other similar derivatives were reviewed by Mihich (1975), who also discussed the relations of the transport of spermidine and MGBG and some effects of the latter on mitochondrial function. The inhibition of AdoMetDC by MGBG in activated lymphocytes resulted in the inhibition of synthesis of spermidine and spermine, the accumulation of putrescine, and the inhibition of cell proliferation (Hölttä et al., 1981). Supplementation with spermidine and spermine, but not putrescine, restored cell proliferation.

However, MGBG inhibited the synthesis of DNA and RNA and damaged mitochondria severely. The swollen organelles slowly lost the ability to produce CO_2 from [2-^{14}C]-pyruvate (Pleshkewych et al., 1983). A DFMO effect on mitochondria was far less severe and was reversed by addition of putrescine to the cells (Oredsson et al., 1983). The actions of MGBG in inhibiting polyamine synthesis and in damaging mitochondrial structure proved to be independent effects. Mitochondrial functions such as pyruvate utilization were inhibited prior to changes in polyamine pools, which in themselves did not affect mitochondrial ultrastructure. Eventually it was found that MGBG inhibited the oxidation of long chain fatty acids in mitochondria. Such oxidation was carnitine-dependent, and the administration of carnitine could prevent both this inhibition and mitochondrial damage at relatively low MGBG concentrations (Nikula et al., 1985).

Quantitation of the drug in plasma and cells of leukemic children has guided the evolution of dosage schedules (Seppänen et al., 1980). MGBG as a sole therapeutic agent had limited effectiveness in treating a wide range of human tumors and was tested in combination with DFMO. A surprising result emerged: a prior depletion by DFMO of the polyamine pool in cultured cells caused a markedly enhanced uptake of MGBG to very high concentrations. The latter effects were strikingly antiproliferative and indeed lethal to many cells. Ehrlich ascites cells were affected in this way in mice, but MGBG did not increase much in other organs (Seppänen et al., 1981a). Tumor cells containing 6–9 mM MGBG were inhibited initially in protein synthesis and disintegrated after 24 h. Although the uptake of spermidine in depleted cells also occurred at an increased rate, this transport stopped when the intracellular spermidine pool reached a normal level. However, MGBG transport did not decrease comparably but continued until cell death (Seppänen et al., 1981b).

These exciting results on the use of a combination were explored in various animal models. For example, cures were obtained with the combination in treatment of rat prostate cancer (Herr et al., 1984). However, the combination was not found to be successful in several trials in adult leukemic patients.

An initial depletion of polyamine caused by DFMO in tumor-bearing mice was eliminated by MGBG as a result of the total inhibition of diamine oxidase (DAO) under conditions in which exogenous polyamines are taken into the cells more rapidly (Kallio & Jänne, 1983). Thus, in the mice with Ehrlich ascites tumor treated with the drug combination the transfer of intesti-

nally generated and conserved cadaverine and putrescine to the tumor was readily demonstrable. As discussed by Jänne et al. (1991), the undesirable effect of MGBG and its congeners on DAO compelled a search for more specific inhibitors of the AdoMetDC.

New MGBG-Like Analogues and Activities

Replacement of the bis-guanidine groups in MGBG by bis(3-aminopropyl), bis(dibutyldiamine), or bis(cyclopentylamidino) groups produced inhibitors of ODC and spermidine synthase, as well as of AdoMetDC. All of these inhibited the growth of cancer cells, and methylglyoxal-bis(butylamidinohydrazone) also blocked induction of ODC by a phorbol ester, an activity not inhibited by MGBG. Carcinogenesis in a two-stage mouse skin system was also blocked by the latter compound (Hibasami et al., 1988). The methylglyoxal-bis(cyclopentylamidinohydrazone), which competitively inhibited AdoMetDC, spermidine synthase, and spermine synthase, inhibited the growth of human leukemia cells and prolonged the survival of leukemic mice. The compound depressed the levels of spermidine and spermine in these cells and affected synthesis of protein but not of the nucleic acids. Additionally the stabilizing effects of this inhibitor on AdoMetDC and the induction of ODC were far less pronounced than the effects of MGBG (Hibasami et al., 1989).

One new highly active compound derived from MGBG, 4-amidinoindane-1-one-2'-amidinohydrazone (Stanek et al., 1993, see Chap. 14), with 50% inhibitory activity of AdoMetDC at 5 nM, has been found to inhibit many human and mouse tumor cell lines in the range of 0.3–3 μM (Regenass et al., 1994). The inhibitor sharply reduced spermidine and spermine levels (ca. 90%) and increased putrescine content in the cells. Intriguingly the compound has shown potent antitumor activity in several model animal systems.

As described in Chapter 14, the search for irreversible inhibitors of AdoMetDC had led to the discovery of the potent enzyme-activated inhibitor, 5'-{(4-amino-2-butenyl)methylamino}-5'-deoxyadenosine (AbeAdo). This compound dramatically reduced spermidine and spermine concentration in L1210 cells, although putrescine was markedly elevated. This substance has been found to have a gradual, but still major effect on growth and on protein synthesis particularly, in which the depletion of spermidine is thought to cause both a lack of hypusine in an essential initiating factor and a diminution of polyamine transport (Byers et al., 1994 a,b). An unsaturated spermidine analogue, which can serve as a substrate in hypusine synthesis, reverses the cytostasis produced by the inhibitor. Thus, the selective effect on protein synthesis seen with the methylglyoxal-bis(cyclopentylamidinohydrazone) may have also related to the depletion of a hypusine-containing initiating factor. These effects may require a protection of spermidine utilization for the synthesis of hypusine. In studies of the inhibition of deoxyhypusine synthase by guanyl diamines, it has been found that the inhibitory compounds, of which guanyl-1,7-diaminoheptane is highly antiproliferative, are cytostatic and their effects are reversible (Park et al., 1994).

Inhibition of Synthesis of Spermidine and Spermine

The study of mutants (Chap. 13) that overexpress ODC and accumulate unusually high concentrations of spermidine and spermine revealed somewhat belatedly the phenomenon of polyamine toxicity. These studies also demonstrated a role of spermidine in regulating antizyme production, and through the latter's inhibition of ODC and polyamine transport, a set of mechanisms of regulating spermidine content.

Numerous putrescine analogues and derivatives, as well as spermidine and spermine analogues have been examined as inhibitors of cells and specific reactions, and some have been found to have significant promise in achieving a selective therapy for tumor cells. Some putrescine analogues with substituents at the C_2 position can be converted to spermidine derivatives. The presence of fluorine at C_2 permits the ready and selective detection of one product, 6,6-difluorospermidine, by ^{19}F-NMR as a tumor marker (Hull et al., 1988).

Some putrescine derivatives, such as the alkylating agent monoaziridinyl-putrescine, proved to be quite toxic to human prostatic cancer cells. The compound competed with putrescine for cell uptake, and its toxicity was reversed by the diamine. Prior depletion with DFMO increased uptake and toxicity (Heston et al., 1987). Aziridinyl derivatives of polyamine-linked cyclphosphazenes proved to be quite active against tumors in mice (Darcel et al., 1990). The aziridinyl derivatives of the diamines were all irreversible inhibitors of DAO (Conner et al., 1992).

Spermidine N^1- and N^8-aziridinyl derivatives have been synthesized, of which the N^1 derivative is slightly more cytotoxic to mouse and human leukemia cells. These effects were potentiated by pretreatment with DFMO. Cytotoxicity was prevented by coincubation with spermidine, but inhibited cells were not restored by spermidine. The analogues inhibited uptake of spermidine. Thus, these aziridinyl derivatives were handled primarily as polyamines (Yuan et al., 1994). As described earlier, a spermidine-linked nitrogen mustard (chlorambucil) has been found to be more cytotoxic than chlorambucil itself. Other agents containing spermidine-linked ornithinamide derivatives of chlorambucil have also proved to inhibit melanoma cells in culture (Stark et al., 1992), but did not take advantage of the spermidine transport system.

Among the inhibitors of spermidine synthase, cyclohexylamine was tested rather early against Ehrlich ascites carcinoma and leukemia in mice and was found to suppress tumor growth and increase survival in both systems (Ito et al., 1982). Additionally, when DFMO and cyclohexylamine were used in combination on a transformed mouse fibroblastic line, the cells assumed the phenotype of the untransformed line. This effect was reversed by exogenous polyamine (Boyd et al., 1987). Methylthiopropylamine has also been shown to be a potent specific inhibitor of spermidine synthase and to inhibit the growth of human leukemic cells, in which the spermidine content was depressed. In this system protein synthesis was inhibited without inhibition of nucleic acid synthesis, pointing once again to the possible production of a hypusine deficiency (Hibasami et al., 1987).

Spermidine synthase is usually in great excess in most cells, including tumors, and spermidine can be depleted readily by inhibitors (e.g., DFMO) of earlier biosynthetic steps. Selectivity in the inhibition of spermidine synthesis has been found to be mainly cytostatic and inadequate for therapeutic ends. In conjunction with DFMO, the tight-binding S-adenosyl-3-thio-1,8-diaminooctane (AdoDATO) has also been used to decrease spermine levels which are refractory to DFMO alone. AdoDATO alone increased spermine levels (Holm et al., 1989) and also inhibited the growth and erythroid differentiation of murine erythroleukemia cells, effects reversed by spermidine (Sherman et al., 1986). The natural spermine analogue, canavalmine (4,3,4), also inhibited the accumulation of spermidine and spermine and the growth of murine erythroleukemia cells (Fujihara et al., 1984).

Both the multisubstrate adduct inhibitor, S-adenosyl-1,12-diamino-3-thio-9-azadodecane (AdoDATAD), and the N-(3-aminopropyl) cyclohexylamine inhibit spermine synthase. Among newer compounds it was found that the aminoxy analogue of spermidine, 1-aminooxy-3-N-[3-aminopropyl]-aminopropane, is also a good competitive inhibitor of spermine synthase with an inhibition constant (K_i) of 1.5 mM (Eloranta et al., 1990). However, the compound also inactivates AdoMetDC and inhibits ODC. On the other hand the compound N-(n-butyl)-1,3-diaminopropane is quite specific in inhibiting spermine synthesis in various tumor cells. However, it also evokes a parallel increase in spermidine and the growth of cells is not inhibited by this inhibitor in the short term. Nevertheless, after 72–144 h growth is slowed and can be restored by spermine specifically (Pegg & Coward, 1993). N-Cyclohexyl-1,3-diaminopropane similarly depletes spermine and inhibits the proliferation of human breast cancer cells (Huber & Poulin, 1995). In combination with DFMO, the inhibition of growth was significant (He et al., 1995).

Effects with Methionine and Its Analogues

Compounds affecting the conversion of methionine to AdoMet, cycloleucine or 1-aminocyclopentane carboxylic acid evoked a 7–20% positive response in human sarcomas but were quite toxic (Aust et al., 1970). The discovery that methionine synthase was a cobalamin enzyme inhibitable by nitrous oxide permitted a trial of a sequential inhibition of the synthesis of methionine and AdoMet on leukemic rats. An additional effect on the production of dAdoMet was tested with MGBG, which proved to be severely toxic. A combination of the three compounds was described as additive compared to the individual effects, but was also stated to be irrelevant to tumor therapy. The combination of nitrous oxide and cycloleucine did inhibit spermine synthesis.

Some methionine analogues, which inhibit AdoMet synthetase (e.g., cycloleucine), are known to deplete spermidine and spermine. However, even more active inhibitors, such as L-2-amino-4-methoxy-cis-but-3-enoic acid, do not deplete these polyamines. Nevertheless, the compounds enhance MGBG effects on L1210 cells. The inhibition of mitogen-stimulated mononuclear cells by methotrexate, a potent inhibitor of dihydrofolate reductase, was reversed by spermidine (Nesher & Moore, 1990). Thus, the enzyme that supplies tetrahydrofolate in the synthesis of methionine may have indeed limited the sequence essential for this arm of polyamine synthesis.

A two-stage system of hepatocarcinogenesis involving diethylnitrosamine and 2-acetylaminofluorene was observed to evoke a fall in the AdoMet content of the liver. Administration of AdoMet intramuscularly partially restored AdoMet in the liver and appeared to promote growth of uninitiated liver cells (Feo et al., 1985). It was found that administration of AdoMet also inhibited ODC activity, caused methylthioadenosine (MTA) accumulation in the liver, and prevented visible nodule development (Garcea et al., 1987). These workers then undertook a long-term chemoprevention of carcinogenesis by AdoMet in this system. They described a decrease of basophilic nodules and a marked reduction of carcinomas in AdoMet-treated animals as compared to the control animals (Pascale et al., 1992). In their hands a 6-month treatment with AdoMet caused an inability of initiated preneoplastic cells to evolve to cancer for a period equal to two-thirds of the life span. Thus, some important facet of AdoMet metabolism is involved in shifting the mutagenized cell to a controlled growth away from cancer or to deletion of the initiated cells by cell death.

Chapter 16 noted the inability of many cancer cell lines to carry out the resynthesis of methionine from MTA. The exacerbation of methionine deficiency in human patients has occasionally been attempted in efforts to control the disease. More recently methods have been developed to identify methionine dependence in primary tumors (Guo et al., 1993a; Nobori et al., 1993) and to study growth in such tumors under conditions of methionine deprivation (Guo et al., 1993b).

Analogues Depleting and Replacing Normal Amines

An early analogue of spermine, containing a central nonyl moiety replacing that of the butyl moiety in spermidine and spermine, was found to inhibit development of a mouse leukemia (Israel et al., 1973). More recently a similar spermine analogue N,N^1-bis-[3-(ethylamino)-propyl]-1-7-heptane diamine has been found to be active against L1210 mouse leukemia. Its activity is protected by an inhibitor of PAO (Prakash et al., 1990) and this combination has been reported to cure 100% of the tumor-bearing animals tested. Two bisethyl (BE) spermine analogues (BE-373 and BE-383) at 10 μM levels have been shown to be lethal (>99% kill/7 days) to a human brain tumor cell line (Basu et al., 1993a).

BE compounds are oxidized by PAO. The active compound BE-373 was cytocidal to the cells in culture and additionally the once cured animals became immune to subsequent challenge by these tumor cells (Bowlin et al., 1991). This BE compound was also found to be selectively concentrated in the tumor cells. In later studies, a BE derivative was found to be an extraordinarily antiproliferative agent as a result of oxidation by PAO to a free amine analogue (Bitonti et al., 1989).

Numerous efforts were made earlier to develop spermidine and spermine analogues that might produce the highly encouraging result just described. Simple analogues of spermidine were tested in combination with DFMO (Casero et al., 1984) and were shown to replace spermidine, to deplete spermine, and to inhibit growth. These were substances such as aminopropylcadaverine, N^4-methylspermidine, N^4-ethylspermidine, and homospermidine. However, the effects of these inhibitors could be readily reversed by exogenous polyamine. Two synthetic triamines, 1,3,6-triaminohexane and 1,4,7-triaminoheptane, restored growth temporarily in DFMO-inhibited cells. This occurred prior to the accumulation of dAdoMet in the DFMO cells. As in the system using spermidine analogues, extensive growth inhibition correlated with significant depletion of spermine (McGovern et al., 1986).

Exogenous norspermidine replaced spermidine in the tumor cells of several mouse tumor models and increased survival of the tumor-bearing mice. Exogenous spermidine antagonized this effect. Although norspermidine alone reduced spermine content somewhat, this did not happen in conjunction with DFMO. The system behaved as if norspermidine made more cell spermidine available for spermine synthesis. In the presence of DFMO, norspermidine reduced the ODC level some 99.5% and prevented (as did spermidine) the major increase of AdoMetDC (Prakash et al., 1988).

N^1,N^8-Bis(ethyl)spermidine (BES), like DFMO, eliminated ODC and depleted putrescine and spermidine in L1210 cells; but unlike the ODC inhibitor, it also decreased spermine pools and did not increase AdoMet or its decarboxylase or stimulate uptake of exogenous polyamine (Porter et al., 1986, 1987). Synthesis of protein was affected more sharply than that of the nucleic acids.

Going to the next step, Bergeron and colleagues (1989) prepared the BE and other alkyl derivatives of spermine and its homologues. The N^1,N^{12}-bis(ethyl)-spermine (BESPM) is quite active against tumor cells in culture and markedly increased survival of mice with L1210 leukemia. Indeed many of the mice proved to be long-term survivors. BESPM proved also to inhibit the growth of hormone-responsive and hormone-resistant human breast cancer cell lines. The effect went hand in hand with an induction of SSAT (Davidson et al., 1993).

BESPM and bis(ethyl)norspermine (BENSPM) produced a very rapid loss of spermidine and spermine from the cells as a result of the extraordinary induction of SSAT (Pegg et al., 1989). BENSPM was 10-fold as active as BESPM in inducing the enzyme; bis(ethyl)homospermine was almost inactive in this role. Although spermine was eliminated more slowly, this was attributed to the conversion of acetylated spermine to spermidine, which was itself excreted as N^1-acetylspermidine and eventually also as putrescine. The excretions can be attributed both to the activation of the catabolic pathway and to the replacement of the cellular polyamines by the BE analogues.

In the lethal activity of BESPM in mouse carcinoma cells, loss of ability of the cells to form colonies was not reversed by spermidine or spermine; polyamine uptake was greatly inhibited, despite the enhanced SSAT activity; and the inhibition of cell growth was correlated with the swelling of mitochondria and a severe decrease of ATP content (Fukuchi et al., 1992). These workers suggested that Ca^{2+} accumulated in mitochondria is released as a lethal event into the cytoplasm by this treatment.

The N^1,N^{14}-bisethylhomospermine (BEHSPM) was more cytotoxic to L1210 cells in a 96-h test than were the BESPM and BENSPM (Bergeron et al., 1989). Lethal effects were more pronounced after 48 h. Despite the low level of activity of the catabolic path, the BEHSPM slowly and more completely displaced the natural cell polyamines. Nevertheless, the total cationic charge associated with the natural polyamines was conserved despite the different impacts of the various analogues. The translational regulation of ODC and AdoMetDC by spermine and BESPM were found to be essentially identical in in vivo and in vitro systems (Porter et al., 1990).

The efficacies of the various BE spermine analogues have been tested against human tumor lines. In a series of melanoma lines the growth sensitivities to BESPM paralleled the inducibility of SSAT. Fogel–Petrovic et al. (1993) studied the production of SSAT mRNA as a result of BESPM treatment. The production of the enzyme itself can be detected in cells by immunohistochemical staining (Casero et al., 1994). However, SSAT can be greatly induced by dimethylspermine without affecting cell growth (Yang et al., 1995).

Xenografts with a human melanoma in mice responded most rapidly and at the lowest dose to BEHSPM. The most effective and best tolerated of the homologues was BENSPM, which evoked an initial tumor regression, suppression of tumor growth, and prolonged inhibition of the cancer (Bernacki et al., 1992; Porter et al., 1993). This substance has also been found to be the best of the series in a test of the treatment of human pancreatic adenocarcinoma xenografts in mice (Chang et al., 1992a, b). It was also markedly antiproliferative

against human bladder cancer cell lines (Chang et al., 1993).

Thus, two lines of spermine analogues are showing promise as antitumor agents. Another line of investigation, involving the bis-alkylated aminopropyl or aminobutyl derivatives of natural diamines, is providing considerable data on their biological and biochemical effects. A particular example of the latter class, BE-4-4-4-4, which has a high affinity for DNA, was strongly lethal to seven human brain tumor cell lines, from which natural polyamines had been largely eliminated and replaced by the analogue (Basu et al., 1993b). These substances led to the displacement of the "normal" cellular polyamines by the analogues. Treatments with the BE analogues also result in an inhibition of protein synthesis and decrease in ATP. Mitochondria are particularly affected; they swell and lose DNA, as well as organellar protein synthesis (He et al., 1994; Igarashi et al., 1995).

The surprising antiproliferative activities of the BE polyamines compelled the synthesis of new polyamine analogues varying in chain length, terminal alkyl group size, and the symmetry of the methylene backbone (Bergeron et al., 1994). Within compounds of defined length, symmetry does not appear very important. However, some unsymmetrically substituted polyamine analogues possessed relatively specific antitumor properties (Casero et al., 1995). In another study substituents have been added to the polyamines to alter charge distribution (Basu et al., 1994; Bergeron et al., 1995). The latter property does have significant effects on growth inhibition.

Approaching Chemotherapy of Gastrointestinal Cancer

The ubiquity and damaging effects of cancer of the gastrointestinal tract have compelled interest in the possible roles of the polyamines in the evolution of the pathological process, as well as in its possible control by affecting these compounds. The study of the normal physiology of the gastrointestinal tract and in the roles of the polyamines in its normal growth and development is described in a large volume of literature and encompasses the polyamine links between the normal and pathological processes (Dowling et al., 1992). McCormack and Johnson (1991) reviewed the role of the polyamines in gastrointestinal mucosal growth.

ODC has been implicated in many aspects of mucosal proliferation and the enzyme has been discussed as a possible tumor marker in the evolution of polyps to adenomas in the large bowel. However, difficulties in accepting this as diagnostically convincing have been analyzed and discussed (Desai et al., 1992). Indeed human familial polyposis has been linked genetically to early mutations in the APC gene (Powell et al., 1992). ODC and high levels of ODC mRNA have been suggested as possible markers for human esophageal cancer (Yoshida et al., 1992), and this hypothesis will also come under clinical scrutiny.

For the most part, however, the discussion of genetic factors (i.e., protooncogenes, oncogenes, suppressor genes, etc.) active in the development or prevention of cancer has not included a discussion of the possible relevance of the data on the polyamines. Similarly, with few exceptions, the data on the induction and activity of the polyamines have not been developed in genetically analyzable systems. This type of linkage between biochemistry and genetics has been examined in cellular systems by Bachrach (1992) and Auvinen et al. (1992). Thus, when 3T3 mouse cells were transfected with plasmid DNA containing the human ODC gene, it was possible to demonstrate the increased expression of the gene, transformation of the cells, and inhibition of both of these by antisense RNA. It has been proposed that the ODC gene is a protooncogene central for the regulation of cell growth and transformation. The formulation of the problem in such terms is certain to begin a more serious interaction of those interested in the origins of cancer and those concerned with its biochemical roots and expression.

Efforts at depletion of gastrointestinal cancers with DFMO, despite yielding suggestive improvement, were found to present experimental difficulties. An increase in studies of ODC in normal and pathological human colorectal mucosa also called attention to some analytical problems (e.g., the nature of the buffer, amount of tissue sample, etc.), and improvements in the methods have been proposed (Garewal et al., 1992). In any case, numerous activities of the normal organs of the tract and their relation to the polyamines have been examined. The endogenous synthesis of arginine in the intestine is important in the growth of the rat, while exogenous compounds such as putrescine and spermine have been shown to stimulate maturation of the intestine in young animals. Both human and rat milk contain significant amounts of the polyamines (Pollack et al., 1992) and the microbial flora also contribute to the supply of these compounds that stimulate mucosa growth. In another context, ingested spermine and spermidine have been reported to inhibit gastric emptying and the secretion of acid in the stomach. Inhibition of polyamine biosynthesis by DFMO inhibits glucose uptake by brush-border membrane vesicles (Johnson et al., 1995).

The marked stimulation of ODC and DNA synthesis in the rat stomach mucosa is evoked by a single oral dose of concentrated NaCl solution. In this model simulating reactions of stress, the peak of ODC at 6 h is preceded by a large peak of ODC mRNA (Ishibashi, 1992). It is known that NaCl enhances two-stage carcinogenesis in rat stomach mucosa. Stressful conditions also evoke increases in ODC and loss of villi in the rat duodenal mucosa. These changes may be prevented by DFMO, which also prevents repair if given after stress. Thus, this and other types of repair also involve increases in ODC and increased polyamine (Fujimoto et al., 1991; Wang & Johnson, 1991).

The ratio of spermidine to spermine in the normal mucosa of the entire length of the rat jejunum and intestine is fairly high, although it falls markedly from cecum to colon. The high ratio parallels the ability of the intestine to respond to various stimuli in which proliferation is an adaptive response (Hosomi et al., 1986). The hypertrophic growth response of the rat small intestine to a plant lectin appeared to involve increased uptake of luminal polyamine, rather than a markedly increased polyamine biosynthesis in situ (Bardócz et al., 1990). More recently it has been observed that the presence of a prostatic tumor in the rat evokes an increased rate of accumulation of putrescine in the brush-border membrane of the small intestine (Brachet et al., 1994, 1995).

In many systems the distinction between an increased uptake and increased biosynthesis can be made with DFMO, which does not abolish the luminal uptake noted above. Because intestinal mucosa is known to be quite sensitive to toxic agents, the response to DFMO was tested (Luk et al., 1980; Yarrington et al., 1983). The inhibition of ODC by DFMO suppressed mucosal maturation and recovery from stress. It also decreased the synthesis of mucosal DNA, RNA, and protein; these polymer syntheses in the inhibited mucosae were restored by treating with luminal polyamine (Wang et al., 1991). Thus, luminal polyamine can contribute to the polyamine requirements of the intestinal tract, as in mucosal repair after stress (Wang & Johnson, 1992a). Nevertheless, increased biosynthesis of polyamine is a major adaptive response in systems involving intestinal obstruction, lactation, and intestinal maturation, as well as pancreatic hyperplasia.

Analysis of polyamine levels in tissues, serum, and urine revealed significant elevations of putrescine in some benign tissues and fluids, as well as in the tumor in four different types of gastrointestinal cancer: esophageal, gastric, pancreatic, and colorectal carcinoma. The urinary excretion of N^1-acetylspermidine, although high in cancer patients and an indicator of an active catabolic path, was similarly unspecific, although the compound was readily detectable only in cancer tissue (Löser et al., 1992). The analyses of the polyamines seemed of greatest interest to workers wishing to follow events after the surgical removal of the cancer, as a possible guide to cure or relapse.

The regulation of gastrointestinal mucosal growth is a function of the hormone gastrin, and this role is insensitive to the inhibition of ODC by DFMO. Gastrin does not induce ODC in the colonic mucosa, but it does induce SSAT and the synthesis of the polyamines essential for mucosal growth (Seidel & Snyder, 1989). This enzyme, and the levels of N^1-acetylspermidine, are also increased slowly after induction of colonic cancer in rats by injection of 1,2-dimethylhydrazine (Halline et al., 1989a,b). These researchers also found that DFMO reduced the concentration of N^1-acetylspermidine and the number of colonic tumors produced. Thus, both the biosynthetic and catabolic pathways are known to be increased during the formation of colonic tumors. Additionally, inhibition of DAO with aminoguanidine was found to promote the formation of large bowel tumors (Kusche et al., 1992), and it might be asked if similar inhibitors (e.g., MGBG) would be contraindicated.

The depletion by DFMO of polyamines from cells of a tumor-bearing tract has been found to be inadequate to the minimal problem of control of the growth of the cancer, much less its cure. Polyamine in the gut may become available from several routes:

1. biosynthesis,
2. the degradative path,
3. transport from normal host tissues,
4. diet, or
5. bacterial synthesis.

A depletion of polyamine in the normal organs (e.g., liver, spleen, kidney) of tumor-bearing animals has been recorded. Although the mechanism of this depletion is presumably a function of erythrocyte transport, there have been few efforts to control or replete the polyamines of such tissues.

Studies on the use of purified diets and oral antibiotics as decontaminating agents have attempted to reveal the possible effects of these routes. It was found that bacterial synthesis in the human gastrointestinal tract is a major source of urinary and tissue cadaverine. Indeed urinary tract infections can produce an increase of urinary putrescine (Satink et al., 1989). Leukemic mice on purified diets with or without variously decontaminating antibiotics were given labeled putrescine and were followed after treatment with DFMO. Only totally decontaminated mice were strongly inhibited by DFMO in the growth of the tumor. The addition of labeled putrescine to the purified diets affected only the incorporation of label into the cellular polyamines after DFMO depletion. It was concluded that the bacterial supply of polyamines was a greater problem in developing DFMO-induced depletion than was dietary putrescine (Hessels et al., 1989).

The control of biosynthesis, catabolic renewal, and control of gastrointestinal dietary and microbial sources of polyamine has been studied in several experimental tumor models: Lewis lung carcinoma, brain tumors, and prostatic carcinoma (Quemener et al., 1992a,b; Seiler et al., 1990). The improvement in the inhibition of growth of the Lewis lung carcinoma by these combinations in the administration of DFMO is presented in Figure 17.3. The effects of the addition of BENSPM or another displacing and replacing analogue in such a system have not yet been reported. Because the N-alkylated polyamine analogues turn off polyamine transport, in contrast to DFMO which stimulates it, it is possible that the addition of one of these substances to the therapeutic regimen may minimize the parasitism of host polyamine by the tumor, as well as increase induction of SSAT and further reduce tumor spermidine and spermine (Chang et al., 1992a, b).

Polyamines in Chemoprevention of Cancer

DFMO has been found to block several types of induced carcinogenesis in animals. In the many studies it was found that normal levels of the polyamines were frequently in great excess of the levels necessary to sustain cell function and survival and very considerable depletions were necessary to develop cytotoxicity. The question has been posed, therefore, if a reduced polyamine level, perhaps effected through administration of the relatively nontoxic highly specific DFMO, can prevent the development of cancer in high-risk individuals. Such trials, carried out at negligible toxicity levels in humans, might include attempts to prevent prostatic carcinoma (Kadmon, 1992) or some forms of bladder cancer (Loprinzi & Messing, 1992). The determination of the lowest daily oral dose of DFMO capable of inhibit-

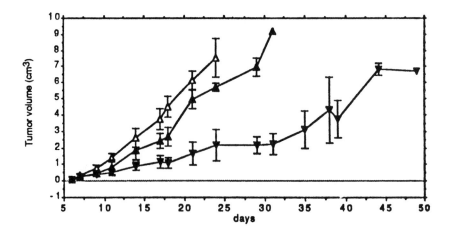

FIG. 17.3 Effects of (▼) continuous polyamine deprivation vs. (▲) discontinuous polyamine deprivation on the growth of Lewis lung carcinoma in grafted mice; adapted from Quemener et al. (1992a). Sources of exogenous polyamine were eliminated by administration of a polyamine-free diet, decontamination of the gastrointestinal tract, and prevention of polyamine reutilization by an inhibitor of polyamine oxidase.

ing ODC induced by TPA in human skin has been made in some cancer-free humans, as well as cancer-bearing patients (Love et al., 1993). Another study determining the levels of DFMO necessary to reduce polyamine levels in human colon polyps established a similar daily oral dose (Meyskens et al., 1994). One or more clinic trials of chemoprevention has been projected (Boone & Wattenberg, 1994). It will be understood that many chemopreventive agents are known, and the possibility of designing a human diet to minimize the threat of cancer is thought to be a desirable outcome of chemoprevention trials.

REFERENCES

Abeloff, M. D., Rosén, S. T., Luk, G. D., Baylin, S. B., Zeitzman, M., & Sjoerdsma, A. (1986) *Cancer Treatment Reports*, **70**, 843–845.

Ask, A., Persson, L., & Heby, O. (1992) *Cancer Letters*, **66**, 29–34.

Aust, J. B., Andrews, N. C., Schroeder, J. M., & Lawton, R. L. (1970) *Cancer Chemotherapy Reports*, **54**, 237–241.

Auvinen, M., Paasinen, A., Andersson, L. C., & Hölttä, E. (1992) *Nature*, **360**, 355–358.

Bacchi, C. J., Nathan, H. C., Hutner, S. J., McCann, P. P., & Sjoerdsma, A. (1982) *Biochemical Pharmacology*, **31**, 2833–2836.

Bachrach, U. (1992) *International Symposium on Polyamines in Cancer*, Houston, p. 28.

Ballas, S. K., Mohandas, N., Marton, L. J., & Shohet, S. B. (1983) *Proceedings of the National Academy of Sciences of the United States of America*, **80**, 1942–1946.

Bardocz, S., Brown, D. S., Grant, G., & Pusztai, A. (1990) *Biochimica et Biophysica Acta*, **1034**, 46–52.

Basu, H. S., Marton, L. J., Pellarin, M., Deen, D. F., McManis, J. S., Liu, C. Z., Bergeront, R. J., & Feuerstein, B. G. (1994) *Cancer Research*, **54**, 6210–6214.

Basu, H. S., Pellarin, M., Feuerstein, B. G., Deen, D. F., & Marton, L. J. (1993a) *Anticancer Research*, **13**, 1525–1532.

Basu, H. S., Pellarin, M., Feuerstein, B. G., Shirahata, A., Samejima, K., Deen, D. F., & Marton, L. J. (1993b) *Cancer Research*, **53**, 3948–3955.

Basu, H. S., Sturkenboom, M. C. J. M., Delcros, J., Csokan, P. P., Szollosi, J., Feuerstein, B. G., & Marton, L. J. (1992) *Biochemical Journal*, **282**, 723–727.

Bellomo, G., Perotti, M., Taddei, F., Mirabelli, F., Finardi, G., Nicotera, P., & Orrenius, S. (1992) *Cancer Research*, **52**, 1342–1346.

Ben-Hur, E. & Riklis, E. (1979a) *Radiation Research*, **78**, 321–328.

——— (1979b) *Cancer Biochemistry Biophysics*, **4**, 25–31.

Bergeron, R. J., Hawthorne, T. R., Vinson, J. R. T., Beck, Jr., D. E., & Ingeno, M. J. (1989) *Cancer Research*, **49**, 2959–2964.

Bergeron, R. J., McManis, J. S., Liu, C. Z., Feng, Y., Weimar, W. R., Luchetta, G. R., Wu, Q., Ortiz–Ocasio, J., Vinson, J. R., Kramer, D., & Porter, C. W. (1994) *Journal of Medicinal Chemistry*, **37**, 3464–3476.

Bergeron, R. J., McManis, J. S., Weimar, W. R., Schreier, K. M., Gao, F., Wu, Q., Ortez–Ocasio, J., Luchetta, G. R., Porter, C., & Vinson, J. R. T. (1995) *Journal of Medicinal Chemistry*, **38**, 2278–2285.

Bernacki, R. J., Bergeron, R. J., & Porter, C. W. (1992) *Cancer Research*, **52**, 2424–2430.

Bitonti, A. J., Bush, T. L., & McCann, P. P. (1989) *Biochemical Journal*, **257**, 769–774.

Boone, C. W. & Wattenberg, L. W. (1994) *Cancer Research*, **54**, 3315–3318.

Boucher, J.-L., Custot, J., Vadon, S., Delaforge, M., Lepoivre, M., Tenu, J.-P., Yapo, A., & Mansuy, D. (1994) *Biochemical and Biophysical Research Communications*, **203**, 1614–1621.

Bowlin, T. L., Prakash, H. J., Edwards, M. L., & Sjoerdsma, A. (1991) *Cancer Research*, **51**, 62–66.

Boyd, D., Bialoski, S., & Brattain, M. G. (1987) *Cancer Research*, **47**, 4099–4104.

Brachet, P., Debbabi, H., & Tomé, D. (1995) *American Journal of Physiology*, **269**, G754–G762.

Brachet, P., Quemener, V., Havouis, R., Tomé, D., & Moulinoux, J.-P. (1994) *Biochimica et Biophysica Acta*, **1227**, 161–170.

Brüne, B., Hartzell, P., Nicotera, P., & Orrenius, S. (1991) *Experimental Cell Research*, **195**, 323–329.

Byers, T. L., Lakanen, J. R., Coward, J. K., & Pegg, A. E. (1994b) *Biochemical Journal*, **303**, 363–368.

Byers, T. L., Wechter, R. S., Hu, R. H., & Pegg, A. E. (1994a) *Biochemical Journal*, **303**, 89–96.

Carter, B. J., DeVroom, E., Long, E. C., van der Marel, G. A., van Boom, J. H., & Hecht, S. M. (1990) *Proceedings of the National Academy of Sciences of the United States of America*, **87**, 9373–9377.

Casara, P., Jund, K., & Bey, P. (1984) *Tetrahedron Letters*, **25**, 1891–1894.

Casero, Jr., R. A., Bergeron, R. J., & Porter, C. W. (1984) *Journal of Cellular Physiology*, **121**, 476–482.

Casero, Jr., R. A., Gabrielson, E. W., & Pegg, A. E. (1994) *Cancer Research*, **54**, 3955–3958.

Casero, Jr., R., Mank, A. R., Saab, N. H., Wu, R., Dyer, W. J., & Woster, P. M. (1995) *Cancer Chemotherapy and Pharmacology*, **36**, 69–74.

Cavanaugh, Jr., P. F., Pavelic, Z. P., & Porter, C. W. (1984) *Cancer Research*, **44**, 3856–3861.

Chang, B. K., Bergeron, R. J., Porter, C. W., & Liang, Y. (1992a) *Cancer Chemotherapy and Pharmacology*, **30**, 179–182.

Chang, B. K., Bergeron, R. J., Porter, C. W., Vinson, J. R. T., Liang, Y., & Libby, P. R. (1992b) *Cancer Chemotherapy and Pharmacology*, **30**, 183–188.

Chang, B. K., Gutman, R., & Chou, T. C. (1987) *Cancer Research*, **47**, 2247–2250.

Chang, B. K., Liang, Y., Miller, D. W., Bergeron, R. J., Porter, C. W., & Wang, G. (1993) *Journal of Urology*, **150**, 1293–1297.

Chen, C. Z., Hu, L. J., Bergeron, R. J., Marton, L. J., & Deen, D. F. (1994) *International Journal of Radiation Oncology Biology Physics*, **29**, 1041–1047.

Claverie, N. & Mamont, P. (1989) *Cancer Research*, **49**, 4466–4471.

Claverie, N., Wagner, J., Knödgen, B., & Seiler, N. (1987) *Anticancer Research*, **7**, 765–772.

Cohen, G. M., Cullis, P. M., Hartley, J. A., Mather, A., Symons, M. C. R., & Wheelhouse, R. T. (1992) *Journal of Chemical Society Chemical Communications*, 298–300.

Cohen, S. S. & I, J. (1976) *Cancer Research*, **36**, 2768–2774.

Conner, J. W., Yuan, Z. M., & Callery, P. S. (1992) *Biochemical Pharmacology*, **44**, 1229–1252.

Cornbleet, M. A., Kingsnorth, A., Tell, G. P., Haegele, K. D., Joder–Ohlenbusch, A., & Smyth, J. F. (1989) *Cancer Chemotherapy and Pharmacology*, **23**, 348–352.

Danzin, C., Casara, P., Claverie, N., Metcalf, B. W., & Jung,

M. J. (1983) *Biochemical and Biophysical Research Communications*, **116**, 237–243.

Danzin, C., Claverie, N., Wagner, J., Grove, J., & Koch-Weser, J. (1982) *Biochemical Journal*, **202**, 175–181.

Darcel, F., Chatel, M., Gautris, P., & Labarre, J. F. (1990) *Anticancer Research*, **10**, 1563–1570.

Davidson, N. E., Mank, A. R., Prestigiacomo, L. J., Bergeron, R. J., & Casero, Jr., R. A. (1993) *Cancer Research*, **53**, 2071–2075.

Desai, T. K., Parikh, N., Bronstein, J. C., Luk, G. D., & Bull, A. W. (1992) *Gastroenterology*, **103**, 1562–1567.

Donato, N. J., Rotbein, J., & Rosénblum, M. G. (1991) *Journal of Cellular Biochemistry*, **46**, 69–77.

Dowling, R. H., Fölsch, V. R., & Löser, C. (1992) *Polyamines in the Gastrointestinal Tract*. Boston: Kluwer Academic Publishers.

Eloranta, T. O., Khomutov, A. R., Khomutov, R. M., & Hyvönen, T. (1990) *Journal of Biochemistry*, **108**, 593–598.

Ernestus, R., Röhn, G., Hossmann, K., & Passchen, W. (1993) *Journal of Neurochemistry*, **60**, 417–422.

Farias-Eisner, R., Sherman, M. P., Aeberhard, E., & Chaudhuri, G. (1994) *Proceedings of the National Academy of Sciences of the United States of America*, **91**, 9407–9411.

Feo, F., Garcea, R., Daino, L., Pascale, R., Pirisi, L., Frassetto, S., & Ruggiu, M. E. (1985) *Carcinogenesis*, **6**, 1713–1720.

Fike, J. R., Gobbel, G. T., Marton, L. J., & Seilhan, T. M. (1994) *Radiation Research*, **138**, 99–106.

Fisher, L. M., Kuroda, R., & Sakai, T. T. (1985) *Biochemistry*, **24**, 3199–3207.

Fogel-Petrovic, M., Shappell, N. W., Bergeron, R. J., & Porter, C. W. (1993) *Journal of Biological Chemistry*, **268**, 19118–19125.

Fredlund, J. O., Johansson, M., Baldetorp, B., & Oredsson, S. M. (1994) *Cell Proliferation*, **27**, 243–256.

Fujihara, S., Nakashima, T., & Kurogochi, Y. (1984) *Biochimica et Biophysica Acta*, **505**, 277–284.

Fujii, A., Takita, T., Shimada, N., & Umezawa, H. (1974) *Journal of Antibiotics*, **27**, 73–77.

Fujimoto, K., Granger, D. M., Price, V. H., & Tso, P. (1991) *American Journal of Physiology*, **261**, G523–G529.

Fujita, K., Shinpo, K., Kimura, E., Sugimoto T., Matsuura, S., Sawada, M., Yamaguchi, T., & Nagatsu, T. (1987) *Biogenic Amines*, **4**, 33–43.

Fukuchi, J., Kashiwagi, K., Kusama-Eguchi, K., Terao, K., Shirahata, A., & Igarashi, K. (1992) *European Journal of Biochemstry*, **209**, 689–696.

Fukushima, Y. (1990) *Biomedical Research*, **11**, 345–352.

Fuller, D. J. M. & Gerner, E. W. (1982) *Cancer Research*, **42**, 5046–5049.

——— (1987) *Cancer Research*, **47**, 816–820.

Garcea, R., Pascale, R., Daino, L., Frassetto, S., Cozzolino, P., Ruggiu, M. E., Vannini, M. G., Gaspa, L., & Feo, F. (1987) *Carcinogenesis*, **8**, 653–658.

Garewal, H. S., Sloan, D., Sampliner, R. E., & Fennerty, B. (1992) *International Journal of Cancer*, **52**, 355–358.

Gerner, E. W., Holmes, D. K., Stickney, D. G., Noterman, J. A., & Fuller, D. J. M. (1980) *Cancer Research*, **40**, 432–438.

Gerner, E. W., Tomé, M. E., Fry, S. E., & Bowden, G. T. (1988) *Cancer Research*, **48**, 4881–4885.

Gerchenson, L. E. & Rotello, R. J. (1992) *FASEB Journal*, **6**, 2450–2455.

Guo, H., Herrera, H., Groce, A., & Hoffman, R. M. (1993a) *Cancer Research*, **53**, 2479–2483.

Guo, H., Lishko, V. K., Herrera, H., Groce, A., Kubota, T., & Hoffman, R. M. (1993b) *Cancer Research*, **53**, 5676–5679.

Haegele, K. D., Splinter, T. A. W., Romijn, J. C., Schechter, P. J., & Sjoersdma, A. (1987) *Cancer Research*, **47**, 890–895.

Halline, A. G., Dudeja, P. K., & Brasitus, T. A. (1989a) *Cancer Research*, **49**, 633–638.

Halline, A. G., Dudeja, P. K., & Brasitus, T. A. (1989b) *Biochimica et Biophysica Acta*, **992**, 106–114.

Hamana, K. & Kawada, K. (1989) *Biochemistry International*, **18**, 971–979.

Harari, P. M., Fuller, D. J. M., Carper, S. W., Croghan, M. K., Meyskens, F. L., Shimm, D. S., & Gerner, E. W. (1990) *International Journal of Radiation Oncology Biology Physics*, **19**, 89–96.

Harari, P., Fuller, D. J. M., & Gerner, E. W. (1989a) *International Journal of Radiation Oncology Biology Physics*, **16**, 451–457.

Harari, P. M., Tomé, M. E., Fuller, D. J. M., Carper, S. W., & Gerner, E. W. (1989b) *Biochemical Journal*, **260**, 487–490.

Harmon, B. V., Corder, A. M., Collins, R. J., Gobé, G. C., Allen, J., Allan, D. J., & Kerr, J. F. R. (1990) *International Journal of Radiation Biology*, **58**, 845–858.

He, Y., Kashiwagi, K., Fukiuchi, J., Terao, K., Shirahata, A., & Igarashi, K. (1993) *European Journal of Biochemistry*, **217**, 89–96.

He, Y., Shimogori, T., Kashiwagi, K., Shirahata, A., & Igarashi, K. (1995) *Journal of Biochemistry*, **117**, 824–829.

He, Y., Suzuki, T., Kashiwagi, K., Kusama-Eguchi, K., Shirahata, A., & Igarashi, K. (1994) *European Journal of Biochemistry*, **221**, 391–398.

Heby, O. (1986) *Advances in Enzyme Regulation*, pp. 103–124, G. Weber, Ed. New York: Pergamon Press.

Heby, O. & Russell, D. H. (1973) *Cancer Research*, **33**, 159–165.

Hecht, S. M., Burlett, D. J., Mushika, Y., Kuroda, Y., & Levin, M. D. (1979) *Bleomycin: Chemical, Biochemical and Biological Aspects*, pp. 48–62, S. M. Hecht, Ed. New York: Springer–Verlag.

Herr, H. W., Kleinert, E. L., Relyea, N. M., & Whitmore, W. F. (1984) *Cancer*, **53**, 1294–1298.

Hessels, J., Kingma, A. W., Ferwerda, H., Keij, J., van den Berg, G. A., & Muskiet, F. A. J. (1989) *International Journal of Cancer*, **43**, 1155–1164.

Hessels, J., Kingma, A. W., Muskiet, F. A. J., Sarhan, S., & Seiler, N. (1991) *International Journal of Cancer*, **48**, 697–703.

Heston, W. D. W., Kadmon, D., Lazan, D. W., & Fair, W. R. (1982) *The Prostate*, **3**, 383–389.

Heston, W. D. W., Yang, C.-R., Pliner, L., Russo, P., & Covey, D. F. (1987) *Cancer Research*, **47**, 3627–3631.

Hibasami, H., Mackawa, S., Murata, T., & Nakashima, K. (1989) *Cancer Research*, **49**, 2065–2068.

Hibasami, H., Sakurai, M., Mackawa, S., & Nakashima, K. (1987) *Anticancer Research*, **7**, 1213–1216.

Hibasami, H., Tsukada, T., Mackawa, S., Sakurai, M., & Nakashima, K. (1988) *Carcinogenesis*, **9**, 199–202.

Hirvonen, A., Eloranta, T., Hyvönen, T., Alhonen, L., & Jänne, J. (1989) *Biochemical Journal*, **258**, 709–713.

Holley, J. L., Mather, A., Wheelhouse, R. T., Cullis, P. M., Hartley, J. A., Bingham, J. P., & Cohen, G. M. (1992) *Cancer Research*, **52**, 4190–4195.

Holm, I., Persson, L., Pegg, A. E., & Heby, O. (1989) *Biochemical Journal*, **261**, 205–210.

Hölttä, E., Korpela, H., and Hovi, T. (1981) *Biochimica et Biophysica Acta*, **677**, 90–102.

Homma, Y., Ozono, S., Numata, I., Seidenfeld, J., & Oyasu, R. (1985) *Cancer Research*, **45**, 648–652.

Hosomi, M., Smith, S. M., Murphy, G. M., & Dowling, R. H. (1986) *Journal of Chromatography*, 375, 267–275.

Huber, M. & Poulin, R. (1995) *Cancer Research*, 55, 934–943.

Huet, J. & Laval, F. (1985) *Cancer Research*, 45, 987–991.

Hull, W. E., Kunz, W., Port, R. E., & Seiler, N. (1988) *NMR in Biomedicine*, 1, 11–19.

Hung, D. T., Deen, D. F., Seidenfeld, J., & Marton, L. J. (1981) *Cancer Research*, 41, 2783–2785.

Hunter, K. J., Deen, D. F., & Marton, L. J. (1987) *Cancer Research*, 47, 5270–5273.

Igarashi, K., Koga, K., He, Y., Shimogori, T., Ekimota, H., Kashiwagi, K., & Shirahata, A. (1995) *Cancer Research*, 55, 2615–2619.

Ishibashi, Y. (1992) *Biomedical Research*, 13, 185–190.

Israel, M., Zoll, E. C., Muhammad, N., & Modest, E. J. (1973) *Journal of Medicinal Chemistry*, 16, 1–5.

Ito, H., Hibasami, H., Shimura, K., Nagai, J., & Hidaka, H. (1982) *Cancer Letters*, 15, 229–235.

Jänne, J., Alhonen, L., & Leinonen, P. (1991) *Annals of Medicine*, 23, 241–259.

Jasnis, M. A., Klein, S., Monte, M., Davel, L., de Lustig, E. S., & Algranati, I. D. (1994) *Cancer Letters*, 79, 39–43.

Johnson, L. R., Brockway, P. D., Madsen, K., Hardin, J. A., & Gall, D. G. (1995) *American Journal of Physiology*, 268, G416–G423.

Kaczmarek, L., Kaminska, B., Messina, L., Spampinato, G., Arcidiacono, A., Malaguarnera, L., & Messina, A. (1992) *Cancer Research*, 52, 1891–1894.

Kadmon, D. (1992) *Journal of Cellular Biochemistry Supplement*, 16H, 122–127.

Kallio, A. & Jänne, J. (1983) *Biochemical Journal*, 212, 895–898.

Kaloyanides, G. J. & Ramsammy, L. S. (1993) *Kidney, Proteins and Drugs: An Update*, pp. 199–205, C. Bianchi, V. Bocci, F. A. Carone, & R. Rabkin, Eds. Basel: Karger.

Kato, T., Suzumura, Y., & Fukushima, M. (1988) *Anticancer Research*, 8, 239–244.

Kingsnorth, A. N., McCann, P. P., Diekema, K. A., Ross, J. S., & Malt, R. A. (1983) *Cancer Research*, 43, 4031–4034.

Kizaki, H., Suzuki, K., Tadakuma, T., & Ishimura, Y. (1990) *Journal of Biological Chemistry*, 265, 5280–5284.

Kremmer, T., Pályi, I., Holczinger, L., Lörincz, I., Boldizsár, M., Paulik, E., & Pokorny, E. (1988) *Experimental Cell Research*, 56, 131–137.

Kuroda, R., Neidle, S., Riordan, J. M., & Sakai, T. T. (1982) *Nucleic Acids Research*, 10, 4753–4763.

Kuo, M. T. & Hsu, T. C. (1978) *Nature*, 271, 83–84.

Kusche, J., Horn, A., & Mennigen, R. (1992) *Polyamines in the Gastrointestinal Tract*, pp. 347–359, R. H. Dowling, U. R. Fölsch, & C. Löser, Eds. Boston: Kluwer Academic.

Kyprianou, N. & Isaacs, J. T. (1989) *Biochemical and Biophysical Research Communications*, 165, 73–81.

Kyprianou, N., Martikainen, P., Davis, L., English, H. F., & Isaacs, J. T. (1991) *Cancer Surveys*, 11, 265–277.

Lapi, L. & Cohen, S. S. (1977) *Cancer Research*, 37, 1384–1388.

Lazo, J. S., Hoyt, D. G., Sebti, S. M., & Pitt, B. R. (1990) *Pharmacology & Therapeutics*, 47, 347–358.

Levin, V. A., Prados, M. D., Alfred Yung, W. K., Gleason, M. J., Ictech, S., & Malec, M. (1992) *Journal of the National Cancer Institute*, 84, 1432–1437.

Lim, S. T., Jue, C. K., Moore, C. W., & Lipke, P. N. (1995) *Journal of Bacteriology*, 177, 3534–3539.

Lopez–Ballester, L. A., Peñafiel, R., Cremades, A., Valcarcel, M. M., Solano, F., & Lozano, J. A. (1991) *Anticancer Research*, 11, 691–696.

Loprinzi, C. L. & Messing, E. M. (1992) *Journal of Cellular Biochemistry Supplement*, 161, 153–155.

Löser, C., Fölsch, V. R., Dittman, K. V., & Paprotny, C. (1992) *Polyamines in the Gastrointestinal Tract*, pp. 155–169, R. H. Dowling, U. R. Fölsch, & C. Löser, Eds. Boston: Kluwer Academic.

Love, R. R., Carbone, P. P., Verma, A. K., Gilmore, D., Carey, P., Tutsch, K. D., Pomplun, M., & Wilding, G. (1993) *Journal of the National Cancer Institute*, 85, 732–737.

Luk, G. D., Abeloff, M. D., Griffin, C. A., & Baylin, S. B. (1983) *Cancer Research*, 43, 4239–4243.

Luk, G. D., Marton, L. J., & Baylin, S. B. (1980) *Science*, 210, 195–198.

Mackarel, A. J. & Wallace, H. M. (1992) *Biochemical Society Transactions*, 21, 508.

Madhubala, R. & Pegg, A. E. (1992) *Molecular and Cellular Biochemistry*, 118, 191–195.

Magliozzo, R. S., Peisach, J., & Ciriolo, M. R. (1989) *Molecular Pharmacology*, 35, 428–432.

Mamont, P. S., Siat, M., Joder–Ohlenbusch, A., Bernhardt, A., & Casara, P. (1984) *European Journal of Biochemistry*, 142, 457–463.

Marks, D. I. & Fox, R. M. (1991) *Biochemical Pharmacology*, 42, 1859–1867.

Martin, S. J. & Green, D. R. (1995) *Cell*, 82, 349–352.

Marton, L. J. (1987) *Pharmacology and Therapeutics*, 32, 183–190.

Marton, L. J. & Pegg, A. E. (1995) *Annual Review of Pharmacology and Toxicology*, 35, 55–91.

McCormack, S. A. & Johnson, L. R. (1991) *American Journal of Physiology*, 260, G795–G806.

McGovern, K. A., Clark, R. S., & Pegg, A. E. (1986) *Journal of Cellular Physiology*, 127, 311–316.

Metcalf, B. W., Bey, P., Danzin, C., Jung, M., Casara, P., & Vevert, J. P. (1978) *Journal of the American Chemical Society*, 100, 2551–2553.

Mett, H., Stanek, J., Lopez–Ballester, J. A., Jänne, J., Alhonen, L., Sinervirta, R., Frei, J., & Regenass, V. (1993) *Cancer Chemotherapy and Pharmacology*, 32, 39–45.

Meyskens, F. L., Emerson, S. S., Pelot, D., Meshkinpour, H., Shassetz, L. R., Einspahr, J., Alberts, D. S., & Gerner, E. W. (1994) *Journal of the National Cancer Institute*, 86, 1122–1130.

Meyskens, Jr., F. J., Kingsley, E. M., Glattke, T., Loescher, L., & Booth, A. (1986) *Investigational New Drugs*, 4, 257–262.

Mihich, E. (1975) *Handbook of Experimental Pharmacology*, pp. 766–788. New York: Springer–Verlag.

Mizumoto, K. & Farber, J. L. (1995) *Archives of Biochemistry and Biophysics*, 319, 512–518.

Moore, C. W. (1990) *Biochemistry*, 29, 1342–1347.

Morgan, M. A. & Hecht, S. M. (1994) *Biochemistry*, 33, 10286–10293.

Moulinoux, J.-P., Darcel, F., Quemener, V., Havouis, R., & Seiler, N. (1991) *Anticancer Research*, 11, 175–180.

Murakami, H., Mori, H., & Taira, S. (1976) *Journal of Theoretical Biology*, 59, 1–23.

Murgia, M., Pizzo, P., Sandoná, D., Zanovello, P., Rizzuto, R., & Di Virgilio, F. (1992) *Journal of Biological Chemistry*, 267, 10939–10941.

Nathan, C. & Xie, Q. (1994) *Journal of Biological Chemistry*, 269, 13725–13728.

Nesher, G. & Moore, T. L. (1990) *Arthritis and Rheumatism*, 33, 954–959.

Nguyen, V. T. & Bensaude, O. (1994) *European Journal of Biochemistry*, **220**, 239–246.

Nicholson, D. W., Ali, A., Thornberry, N. A., Vaillancourt, J. P., Ding, C. K., Gallant, M., Gareau, Y., Griffin, P. R., Labelle, M., Lazebnik, Y. A., Munday, N. A., Raju, S. M., Smulson, M. E., Yamin, T., Yu, V. L., & Miller, D. K. (1995) *Nature*, **379**, 37–43.

Nikula, P., Ruohola, H., Alhonen–Hongisto, L., & Jänne, J. (1985) *Biochemical Journal*, **228**, 513–516.

Nishioka, K., Grossie, Jr., V. B., Ajani, J. A., Patenia, D., Chang, T., & Ota, D. M. (1988) *International Journal of Cancer*, **42**, 744–747.

Nobori, T., Szinai, I., Amox, D., Parker, B., Olopade, O. I., Buchhagen, D. L., & Carson, D. A. (1993) *Cancer Research*, **53**, 1098–1101.

Oredsson, S., Friend, D. S., & Marton, L. J. (1983) *Proceedings of the National Academy of Sciences of the United States of America*, **80**, 780–784.

Osswald, H., Herrmann, R., Jones, G. R. N., Kitta, D., & Kunz, W. (1986) *Journal of Cancer Research and Clinical Oncology*, **111**, 141–148.

Ota, D. M., Grossie, Jr., V. B., Ajani, J. A., Stephens, L. C., & Nishioka, K. (1986) *International Journal of Cancer*, **38**, 245–249.

Parchment, R. E. (1993) *International Journal of Developmental Biology*, **37**, 75–83.

Park, M. H., Wolff, E. C., Lee, Y. B., & Folk, J. E. (1994) *Journal of Biological Chemistry*, **269**, 27827–27832.

Pascale, R. M., Marras, V., Simile, M. M., Daino, L., Pinna, G., Bennati, S., Carta, M., Seddaiu, M. A., Massarelli, G., & Feo, F. (1992) *Cancer Research*, **52**, 4979–4986.

Pegg, A. E. & Coward, J. K. (1993) *Biochemical Pharmacology*, **46**, 717–724.

Pegg, A. E., Secrist, III, J. A., & Madhubala, R. (1988) *Cancer Research*, **48**, 2678–2682.

Pegg, A. E., Wechter, R., Pakala, R., & Bergeron, R. J. (1989) *Journal of Biological Chemistry*, **264**, 11744–11749.

Pera, P. J., Kramer, D. L., Sufrin, J. R., & Porter, C. W. (1986) *Cancer Research*, **46**, 1148–1154.

Petereit, D. G., Harari, P. M., Coutreras, L., Pickart, M. A., Verma, A. K., Gerner, E. W., & Kinsella, T. J. (1994) *International Journal of Radiation Oncology Biology Physics*, **28**, 891–898.

Pierce, G. B., Gramzinski, R. A., & Parchment, R. E. (1990) *Philosophical Transactions of the Royal Society of London Series B–Biological Sciences*, **327**, 67–74.

Pleshkewych, A., Maurer, T. C., & Porter, C. W. (1983) *Cancer Research*, **43**, 646–652.

Pollack, P. F., Koldovskỳ, O., & Nishioka, K. (1992) *American Journal of Clinical Nutrition*, **56**, 371–375.

Porter, C. W., Berger, F. G., Pegg, A. E., Ganis, B., & Bergeron, R. J. (1987) *Biochemical Journal*, **242**, 433–440.

Porter, C. W., Bernacki, R. J., Miller, J., & Bergeron, R. J. (1993) *Cancer Research*, **53**, 581–586.

Porter, C. W., Ganis, B., Vinson, T., Marton, L. J., Kramer, D. L., & Bergeron, R. J. (1986) *Cancer Research*, **46**, 6279–6285.

Porter, C. W., Pegg, A. E., Ganis, B., Madhabala, R., & Bergeron, R. J. (1990) *Biochemical Journal*, **268**, 207–212.

Poulin, R., Pelletier, G., & Pegg, A. E. (1995) *Biochemical Journal*, **311**, 723–727.

Powell, S. M., Zilz, N., Beazer–Barclay, Y., Bryan, T. M., Hamilton, S. R., Thibodeau, S. N., Vogelstein, B., & Kinzler, K. W. (1992) *Nature*, **359**, 235–237.

Pöyhönen, M. J., Takala, J. H., Pitkänen, O., Kari, A., Alhava, E., Alakuijala, L. A., & Eloranta, T. O. (1992) *Journal of Parenteral and Enteral Nutrition*, **16**, 226–231.

Prakash, N. J., Bowlin, T. L., Davis, G. F., Sunkara, P. S., & Sjoersdma, A. (1988) *Anticancer Research*, **8**, 563–568.

Prakash, N. J., Bowlin, T. L., Edwards, M. L., Sunkara, P. S., & Sjoersdma, A. (1990) *Anticancer Research*, **10**, 1281–1288.

Prakash, N. J., Schechter, P. J., Grove, J., & Koch–Weser, J. (1978) *Cancer Research*, **38**, 3059–3082.

Prakash, N. J. & Sunkara, P. S. (1983) *Cancer Research*, **43**, 3192–3196.

Quemener, V., Moulinoux, J.-P., Bergeron, C., Darcel, F., Cipolla, B., Denais, A., Havouis, R., Martin, C., & Seiler, N. (1992a) *Polyamines in the Gastrointestinal Tract*, pp. 375–383, R. H. Dowling, U. R. Fölsch, & C. Löser, Eds. Boston: Kluwer Academic.

Quemener, V., Moulinoux, J.-P., Havouis, R., & Seiler, N. (1992) *Anticancer Research*, **12**, 1447–1454.

Raisfeld, I. H. (1981) *Toxicology and Applied Pharmacology*, **57**, 355–366.

Ray, R. H. & Brock, T. D. (1971) *Journal of General and Applied Microbiology*, **66**, 133–135.

Realini, C. A. & Althans, F. R. (1992) *Journal of Biological Chemistry*, **267**, 18858–18865.

Redgate, E. S., Grudziak, A., Floyd, K. L., Deutsch, M., & Boggs, S. S. (1993) *International Journal of Radiation Oncology Biology Physics*, **25**, 639–646.

Regenass, U., Mett, H., Stanek, J., Mueller, M., Kramer, D., & Porter, C. W. (1994) *Cancer Research*, **54**, 3210–3217.

Rice, W. G., Hillyer, C. D., Harten, B., Schaeffer, C. A., Dorminy, M., Lackey, III, D. A., Kirsten, E., Mendeleyev, J., Buki, K. G., Hakam, A., & Kun, E. (1992) *Proceedings of the National Academy of Sciences of the United States of America*, **89**, 7703–7707.

Roizin–Towle, L., Yarlett, N., Pirro, J. P., & Delohery, T. M. (1990) *Radiation Research*, **124**, S80–S87.

Roth, H., Kitta, D., Jones, G. R. N., Osswald, H., & Kunz, W. (1981) *Cancer*, **48**, 945–950.

Russell, D. H. (1972) *Cancer Research*, **32**, 2459–2462.

Russell, D. H., Looney, W. B., Kovacs, C. J., Hopkins, H. A., Marton, L. J., LeGendre, S. M., & Morris, H. P. (1974) *Cancer Research*, **34**, 2382–2385.

Russo, A., Mitchell, J. B. M., DeGraff, W., Friedman, N., & Gamson, J. (1985) *Cancer Research*, **45**, 4910–4914.

Sakai, T. T., Riordan, J. M., & Glickson, J. D. (1982) *Biochemistry*, **21**, 805–816.

Sam, J. W., Tang, X.-J., & Peisach, J. (1994) *Journal of the American Chemical Society*, **116**, 5250–5256.

Sarhan, S., Knödgen, B., & Seiler, N. (1992) *Anticancer Research*, **12**, 457–466.

Sarhan, S., Weibel, M., & Seiler, N. (1991) *Anticancer Research*, **11**, 987–992.

Satink, H. P. W., Hessels, J., Kingma, A. W., van den Berg, G. A., Muskiet, F. A. J., & Halie, M. R. (1989) *Clinica Chimica Acta*, **179**, 305–314.

Schmidt, H. H. H. W. & Walter, U. (1994) *Cell*, **78**, 919–925.

Sebti, S. M. & Lazo, J. S. (1988) *Pharmacology and Therapeutics*, **38**, 321–329.

Seidel, E. R. & Snyder, R. (1989) *American Journal of Physiology*, **256**, G16–G21.

Seidenfeld, J., Block, A. L., Komar, K. A., & Naujokas, M. F. (1986) *Cancer Research*, **46**, 47–53.

Seiler, N., Sarhan, S., Grauffel, C., Jones, R., Knödgen, B., & Moulinoux, J.-P. (1990) *Cancer Research*, **50**, 5077–5083.

Sen, S. & D'Incalci, M. (1992) *FEBS Letters*, **307**, 122–127.

Seppänen, P., Alhonen–Hongisto, L., & Jänne, J. (1981a) *Biochimica et Biophysica Acta*, **674**, 169–177.

———— (1981b) *European Journal of Biochemistry*, **118**, 571–576.

Seppänen, P., Alhonen–Hongisto, L., Siimes, M., & Jänne, J. (1980) *International Journal of Cancer*, **26**, 571–576.

Sergeyev, D., Godovikova, T., & Zarytova, V. (1995) *Nucleic Acids Research*, **23**, 4400–4406.

Sherman, M. L., Shafman, T. D., Coward, J. K., & Kufe, D. W. (1986) *Biochemical Pharmacology*, **35**, 2633–2636.

Snyder, R. D. & Lachmann, P. J. (1989) *Radiation Research*, **120**, 121–128.

Stark, P. A., Thrall, B. D., Meadows, G. G., & Abdel–Monem, M. M. (1992) *Journal of Medicinal Chemistry*, **35**, 4264–4269.

Strekowski, L., Mokrosz, M., Mokrosz, J. L., Strekowska, A., Allison, S. A., & Wilson, W. D. (1988) *Anti-Cancer Drug Design*, **3**, 79–89.

Sugiyama, M., Kumagai, T., Shionoya, M., Kimura, E., & Davies, J. E. (1994) *FEMS Microbiology Letters*, **121**, 81–86.

Sunkara, P. S., Chang, C. C., Prakash, H. J., & Lachmann, P. J. (1985) *Cancer Research*, **45**, 4067–4070.

Sunkara, P. S., Prakash, N. J., & Rosénberger, A. L. (1982) *FEBS Letters*, **150**, 397–399.

Sunkara, P. S. & Rosénberger, A. L. (1987) *Cancer Research*, **47**, 933–935.

Takigawa, M., Enomoto, M., Nishida, Y., Pan, H., Kinoshita, A., & Suzuki, F. (1990) *Cancer Research*, **50**, 4131–4138.

Takita, T., Fujii, A., Fukuoka, T., & Umezawa, H. (1973) *Journal of Antibiotics*, **26**, 252–254.

Tepper, C. G. & Studzinski, G. P. (1993) *Journal of Cellular Biochemistry*, **52**, 352–361.

Thibeault, L., Hengartner, M., Laguenx, J., Poirier, G. G., & Müller, S. (1992) *Biochimica et Biophysica Acta*, **1121**, 317–324.

Tobias, K. E. & Kahana, C. (1995) *Cell Growth and Differentiation*, **6**, 1279–1285.

Tounekti, O., Belehradek, Jr., J., & Mir, L. M. (1995) *Experimental Cell Research*, **217**, 506–516.

Trujillo–Ferrara, J., Koizumi, G., Muñoz, O., Joseph–Nathan, P., & Yañez, R. (1992) *Cancer Letters*, **67**, 193–197.

Tsukahara, T., Tamura, M., Yamazaki, H., Kurihara, H., & Matsuzaki, S. (1992) *Journal of Cancer Research and Clinical Oncology*, **118**, 117–175.

Umezawa, H. (1983) *Advances in Polyamine Research*, pp. 1–15, U. Bachrach, A. Kaye, & R. Chayen, Eds. New York: Raven Press.

Umezawa, H., Takahashi, Y., Fujii, A., Saino, T., Shirai, T., & Takita, T. (1973) *Journal of Antibiotics*, **26**, 117–119.

Urade, M., Sugi, M., Mima, T., Ogura, T., & Matsuya, T. (1989) *Japan Journal of Cancer Research*, **80**, 464–468.

Wallace, H. M. & Keir, H. M. (1986) *FEBS Letters*, **194**, 60–63.

Wang, J.-W., McCormack, S. A., Viar, M. J., & Johnson, L. R. (1991) *American Journal of Physiology*, **261**, G504–G511.

Wang, J.-Y. & Johnson, L. R. (1991) *Gastroenterology*, **100**, 333–343.

Wang, J.-Y. & Johnson, L. R. (1992a) *Gastroenterology*, **102**, 1109–1117.

———— (1992b) *American Journal of Physiology*, **262**, G818–G825.

Westin, T., Gustafsson, G., Hellander, K., Reinholdtsen, L., Tibell, L., Lundholm, K., & Edström, S. (1991b) *Cytometry*, **12**, 628–635.

Wolff, S., Feeney, L., & Afzal, V. (1993) *Mutation Research*, **289**, 107–114.

Yang, J., Xiao, L., Berkey, K. A., Tamez, P. A., Coward, J. K., & Casero, Jr., R. A. (1995) *Journal of Cellular Physiology*, **135**, 71–76.

Yarrington, J. T., Sprinkle, D. J., Loudy, D. E., Diekema, K. A., McCann, P. P., & Gibson, J. P. (1983) *Experimental and Molecular Pathology*, **39**, 300–316.

Yoshida, M., Hayashi, H., Taira, M., & Isono, K. (1992) *Cancer Research*, **52**, 6671–6675.

Yuan, Z.-M., Egorin, M. J., Rosén, D. M., Simon, M. A., & Callery, P. S. (1994) *Cancer Research*, **54**, 742–748.

CHAPTER 18

Mostly Protozoans:
Some Third World Parasites

A Peek at Helminths

Two major groups of organisms, protozoans and helminths, are important infectious agents seriously affecting about 2 billion people on the face of the earth. Many of these produce widespread disease in the so-called Third World, where poverty has tended to minimize scientific knowledge and medical control. Nevertheless, knowledge of these organisms has developed and in recent years has extended to curious and provocative information concerning the polyamines. The protozoans are unicellular organisms in contrast to the multicellular helminths, and the development of nutritional and physiological data for the former has been more easily adapted to laboratory investigation. The multiplication of the protozoa within their host (helminths do not multiply in this way) has placed study of the protozoa within the tradition of a rapidly developing general microbiology and the Western study of infectious disease. The more complete and promising data have emerged in studies of parasitic protozoa, the Trypanosomatids, including the etiologic agents of the African trypanosomiases, Chagas disease, and Leishmanial disease.

Although the helminthic parasites studied to date contain spermidine and spermine, they lack putrescine or contain relatively low levels of the diamine (Hamana et al., 1995). The soluble extracts of many were apparently devoid of ornithine and arginine decarboxylases (ODC, ADC) or contained very low levels of activity. Difluoromethylornithine (DFMO) is not inhibitory. S-Adenosylmethionine decarboxylase (AdoMetDC) was also very low or negligible, while spermidine and spermine were taken up from the medium quite rapidly. It has been suggested that the studied parasitic helminths lack an active biosynthetic path for the polyamines and are dependent on uptake (Sharma et al., 1991; Tekwani et al., 1995). However, the cDNA for ODC has been cloned from a free-living nematode (Besser et al., 1995).

There have been reports that membrane-associated ODC can be found in both a free-living and a parasitic nematode. This membrane-bound activity had low Michaelis constant (K_m) values for ornithine (2.7 μM) and relatively low inhibition constant (K_i) values for competitive arginine and agmatine, 4.0 and 10 μM, respectively. The K_i for putrescine and spermidine was on the order of 55 μM, suggesting roles as possible feedback inhibitors. Although DFMO was essentially noninhibitory, monofluoromethylornithine was quite inhibitory irreversibly, with a K_i of 15 μM (Schaeffer & Donatelli, 1990). It appears that a new type of ODC may be present in these helminths.

A flavoprotein polyamine oxidase (PAO) has been described in the parasitic nematode *Ascaris suum* (Müller & Walter, 1992). This enzyme is active on spermidine, spermine, and the norhomologues, and inactive with acetylated derivatives of these compounds, as well as with putrescine and cadaverine. Spermine and spermidine are converted to spermidine and putrescine, respectively, as well as H_2O_2 and 3-aminopropanal. This enzyme is inhibited by Berenil in filarial worms; Berenil is lethal to the worms.

Ascaris suum was found to contain soluble putrescine N-acetyltransferase that was active on various diamines and inactive on spermidine and spermine (Wittich & Walter, 1989). A liver fluke, *Fasciola hepatica*, contained a soluble N-acetyltransferase active on putrescine, spermidine, and spermine and a wide range of biogenic amines (i.e., tyramine, tryptamine, histamine, etc.) (Aisien & Walter, 1993). Bispolyamine analogues and Berenil have been described as inhibitors for this system.

These systems and functions are beginning to be explored actively in the human and animal filariae, for example, *Onchocerca volvulus*, which causes river blindness in tropical West Africa. This organism contains putrescine, spermidine, and spermine in the ratios of approximately $1:13:20–50$; it takes up these compounds from a culture medium and acetylates each. ODC and ADC were not detected but a low level of AdoMetDC was found. It was thought that a catabolic pathway was operative through the N-acetyl derivatives. The worms have an active efflux of spermidine, putrescine, and N-acetylputrescine (Wittich et al., 1987; Wittich & Walter, 1990). The inhibition of transport of the polyamines by a bis(benzyl)polyamine in the filarial worm *Brugia pahangi* led to the death of the worm in an in vitro system (Müller et al., 1991). It is evident that the helminths have a plethora of new variants of the systems essential to polyamine metabolism and the inhibition of these systems may affect the survival of the organisms.

Approaching Trypanosomes

Early nutritional studies on protozoans revealed a requirement for protoporphyrin by a parasitic trypanosome of the mosquito and this was related to respiratory function by A. Lwoff. This important linkage of nutrition and the biochemical mechanism encouraged the further study of the parasitic *Trypanosoma* and *Leishmania* among the *Kinetoplastida* (Lwoff, 1951; Hutner et al., 1979).

Two major advances relating to the polyamines have occurred in the therapy of African sleeping sickness, or trypanosomiasis. The first, beginning at the end of the 1970s, demonstrated the startling inhibition, even cure, of the disease by the use of DFMO. The second, beginning in the mid-1980s, relates to the discovery of new spermidine derivatives in trypanosomes and related protozoa and the study of the biosynthesis and metabolic roles of these compounds. This classical set of biochemical tasks uncovered specific essential metabolic functions, the enzymology of which has provided potential targets for chemotherapy.

Fairlamb (1989) has discussed metabolic studies of trypanosomes, the Plasmodia of malaria, Trichomonads, Entamoebae, etc., comparing the limited paths available to the parasites and the more extensive routes used by their hosts. A few novel pathways occur among *Trypanosoma brucei* subspecies and they have developed elaborate mechanisms for carbohydrate metabolism and energy production. Fairlamb also noted the unexpected dependence of these trypanosomes on polyamine metabolism and its role in surviving oxidative stress. Despite our concentration on these advances, it is relevant to present some of the curiosities of polyamine metabolism in the broad spectrum of the free-living as well as parasitic protozoa.

Studies on Tetrahymena

The organism was cultured initially at 27–28°C in a complex medium containing ingestible microparticles. Several sequential heat shocks of an amicronucleate strain of *T. pyriformis* at 34°C could be used to synchronize DNA synthesis and fission, and changes had been found in the contents of the inorganic cations of the synchronized organisms. The cells also contained putrescine and spermidine (22 and 7 nmol, respectively, per 10^6 cells) and a small amount of spermine. The concentrations in the cells were 5 and 8 times higher, respectively, than those in the commonly used complex medium, which contained putrescine, spermidine, and spermine. Studies on Tetrahymena remained dependent on growth of the organism in a complex medium until recently (Wheatley et al., 1994).

Heat shock appeared to synchronize putrescine intake and concentration in the cells (Holm & Emanuelsson, 1971; Holm & Heby, 1971). Exogenous $[^3H]$-labeled putrescine was found only in free putrescine within the cells and not in other amines. Intact 80S ribosomes characteristic of eucaryotic organisms were isolated readily from lysates of this strain in the absence of added Mg^{2+}. This effect is attributed to the high level of bound polyamine, including spermidine, in which the molar ratio of polyamine N to RNA-P was 0.24 (Weller et al., 1968). The presence of a soluble putrescine-insensitive AdoMetDC has been reported (Pösö et al., 1975).

Analysis of crude extracts of *T. thermophila* revealed the presence of an ODC stimulated by pyridoxal phosphate. An arginine decarboxylase and agmatine ureohydrolase could not be detected. Labeled arginine led to the formation of labeled ornithine (Yao et al., 1985). The ODC content of the cells increased prior to division and fell precipitously during the stationary phase. In a system in which a fresh medium was used to synchronize a final activation and decay of ODC, the disappearance of the enzyme had a half-life ($t_{1/2}$) of 15 min and was much more rapid than the half-life of the mRNA and the decline of protein synthesis in the cells (Eichler, 1990). The extensively purified enzyme was a monomer of 68,000 relative molecular mass (M_r) and existed in two forms with slightly differing isoelectric points (Yao & Fong, 1987). No evidence was obtained for an inhibitory "antizyme"-like molecule in the extracts (Koguchi et al., 1996). The presence of polyamines in the media markedly decreased the level of ODC in the cell extracts. ODC is inactivated by a phosphoprotein phosphatase and reactivated by a protein kinase that is not dependent on cAMP (Lougovoi & Kyriakidis, 1989). Eichler (1989c) also isolated the ODC of *T. thermophila* and described its time-dependent inactivation by DFMO. The sensitivity of the enzyme to DFMO was lower than that of any other eucaryotic ODC.

In a complex medium the organism preferentially removes free and peptide-bound arginine. Citrulline and ornithine do not support growth in such an exhausted medium whereas protein-bound arginine reinitiates normal rates of cell multiplication (Eichler, 1989a). In a synthetic medium the organism is dependent on the presence of arginine to maximize the production of ODC. Arginine is selectively removed from among the amino acids of the synthetic medium and can be replaced by protein-bound arginine to stimulate ODC production. This is then released as free arginine. An arginine-degrading complex contains arginine deiminase and citrulline hydrolase, of which the former is rate limiting for the production of ornithine and polyamine. The cells grown in the synthetic medium contain spermidine but are devoid of spermine. Concentrations of putrescine of 1.5 mM are found in arginine-starved cells. Eichler (1989a) has suggested that this pattern is consistent with that expected of an organism that normally lives in fresh water and eats for the most part by phagocytosis with an oral apparatus.

The activity of arginine deiminase is inhibited by relatively high cellular concentrations of putrescine and spermidine (K_i values of 2.8 and 4.3 mM, respectively). Thus, polyamine biosynthesis is regulated at the level of the supply of ornithine and not by the activity of ODC (Eichler, 1989b). Many eucaryotic cells regulate ornithine supply by controls on arginase.

Studies on Acanthamoeba

Three types of human infections are recognized as being caused by free-living amoebae. In meningoencephalitis the etiologic agent is *Naegleria fowleri*, which may be inhibited in culture by sinefungin, an inhibitor of methylase (Ferrante et al., 1987). A granulomatous meningoencephalitis and another disease, *amebic keratitis*, are ascribed to species of *Acanthamoeba*. The etiologic agent of the encephalitis, *A. culbertsoni*, is inhibited in culture by sinefungin and methyglyoxal-bis(guanylhydrazone) (MGBG), albeit at high concentrations of the latter. DFMO is essentially ineffective against the organism. The AdoMetDC of *A. culbertsoni* is inhibited by MGBG, but the inhibition of the organism is not reversed by exogenous polyamine (Kishore et al., 1990). The enzyme of *A. castellani* has been purified extensively and characterized as comprising two α and β subunits (Hugo & Byers, 1993). The α subunit contains an N-terminal pyruvoyl group and possesses a primary sequence homologous to that of yeast and mammalian enzymes. The decarboxylase is stimulated by putrescine, poorly inhibited by MGBG, and most strongly inhibited by Berenil ($K_i = 0.17$ μM). The activity becomes undetectable during encystment.

The *Acanthamoebae* contain low levels of putrescine and spermidine and relatively large amounts of 1,3-diaminopropane (DAP): 40 nmol DAP and 4 nmol spermidine per 10^6 cells. MGBG causes the cells to decrease their DAP content more rapidly than their spermidine content. Most recently *A. culbertsoni* was found to contain a PAO that is relatively specific in oxidizing N^8-acetylspermidine, producing DAP as an end product of catabolism (Shukla et al., 1992). The free polyamines were not substrates but did stimulate the activity of the oxidase slightly. No comparable system is known in mammalian cells, although inhibitors of the mammalian PAO were also inhibitory to the amebal enzyme.

Studies on the *Acanthamoebae* increased with their discovery as etiologic agents in amebic keratitis, a disease associated with the use of inadequately cleansed contact lenses. Proliferation of *A. castellani* can be inhibited by several specific inhibitors of ODC in complex or defined media and this inhibition can be reversed by addition of putrescine or spermidine (Kim et al., 1987a). Analysis of the intracellular polyamines of *A. castellani* has demonstrated high levels of DAP, low levels of spermidine, and very low levels of norspermidine and putrescine. Spermine was not found in organisms grown in a chemically defined medium. The cells contained ODC and arginase and lacked arginine decarboxylase. Cell extracts did not convert spermidine to DAP, although the cells could effect this reaction. Kim et al. (1987b) have postulated a scheme of polyamine biosynthesis. It will be recalled that the main route of biosynthesis of diaminopropane in *Vibrio* species is via the decarboxylation of 2,4-diaminobutyric acid. It would be of interest to know if this reaction exists in the *Acanthamoebae*.

Treatment of infection requires both the inhibition of proliferation and a cytocidal action to prevent encystment of the amebae. Encystment can be induced by inhibitors of polyamine metabolism only in the presence of additional factors, Ca^{2+} or Mg^{2+}, and in the absence of an encystment-enhancing factor. An inhibition of DNA synthesis alone does not appear to induce encystment (Byers et al., 1991b). Polyaminopropyl biguanide is effective in killing cysts of the *Acanthamoebae* (Burger et al., 1994).

How Polyamines Entered Research on Trypanosomes

The class Mastigophora (flagellated protozoans) is a large and heterogeneous group of organisms, and includes the *Trypanosoma*, a genus found in the blood of vertebrates and the gut of some insects, and the *Leishmania*, a genus found in mammalian phagocytes, as well as in the gut of an insect transmitter. The trypanosomatine flagellates, the Trypanosomatidae, constitute a diverse family characterized, as are other members of the order Kinetoplastida, by the presence of a large cytoplasmic element containing stainable DNA (i.e., the kinetoplast) at the base of the single flagellum (Vickerman, 1976). The kinetoplast is thought to be part of the mitochondrial apparatus concerned with respiratory adaptations. The life cycles of these parasites alternating between insect vector and mammalian host show major changes and differences in cellular structure and biochemistry. Comparative studies of the ribosomal RNAs of trypanosomal species and of Leishmaniae have revealed some structural differences from genus to genus as well as major similarities, such as extensive crosshybridization of the RNA with heterologous DNA (Villalba et al., 1985).

A complex group of African trypanosomes described as *T. brucei* are among the many interesting species of the *Trypanosoma*. Subgroups of interest include organisms infecting cattle and horses, such as *T. brucei brucei* (*T. b. brucei*), and those infecting humans, *T. b. gambiense* and *T. b. rhodesiense*. The latter are the etiologic agents of West African and East African sleeping sickness, respectively. The endemic insect vectors, tsetse flies, are species of Glossina.

The requirement of trypanosomes for hemin, including the iron it contains, focused attention on terminal oxidation, which was found to be unusual in many respects and to vary according to the development stage of the organism. For example, the African bloodstream trypanosomes (e.g., *T. brucei*) found in mammalian blood were uninhibited by cyanide, in contrast to the etiologic agent of Chagas disease, *T. cruzi*, which is completely inhibited by this anion. In insects and in culture, however, the African trypanosomes do develop mitochondrial cytochromes and a respiration inhibitable by cyanide (Priest & Hajduk, 1994).

Active glycolysis is organized within a specialized organelle, the glycosome. The triose phosphates usually derived from glycolysis have an unusual fate of funneling into the production of L-glycerophosphate and systems were found to shuttle NAD$^+$ and to oxidize the triose phosphate. Despite the absence of porphyrin-based enzymes, the bloodstream forms of *T. brucei* have an unusual oxidizing system for L-glycero-3-phosphate. The glycerophosphate oxidase is a flavoprotein system containing flavin adenine dinucleotide (FAD), Fe, and thiols. It is a mitochondrial enzyme and is inhibited by *m*-chlorobenzhydroxamic acid. This unusual oxidase, functioning without pyridine nucleotides or cytochrome, is also believed to be active in cultured forms. Although many flavoprotein-catalyzed oxidations give rise to hydrogen peroxide, H_2O_2, this reaction appears to involve a second peroxidatic reaction in which an oxidation of glycerophosphate with newly generated H_2O_2 also yields dihydroxyacetone phosphate and H_2O (Bowman & Flynn, 1976):

$$\text{L-glycerol-3-phosphate} + O_2 \rightarrow$$
$$\text{dihydroxyacetone phosphate} + H_2O_2,$$
$$\text{L-glycerol-3-phosphate} + H_2O_2 \rightarrow$$
$$\text{dihydroxyacetone phosphate} + 2\ H_2O.$$

Compounds that inhibit this oxidase are trypanocidal (Grady et al., 1986). As noted above, glycerol-3-phosphate is generated from the triose phosphates produced in glycolysis. A key enzyme involved in this reaction is an active microbody-located NAD$^+$-dependent glycerophosphate dehydrogenase (Bowman & Flynn, 1976). An interest in this reaction led one active group to isolate and study this enzyme, and they discovered that the requirement for Mg^{2+} in the reaction could be replaced by spermidine or spermine (Bacchi et al., 1974). This observation began a still continuing series of investigations on the roles of the polyamines in the multiplication and survival of these protozoa.

Bacchi and colleagues (1977) began their work on a trypanosome that is parasitic on insects but not on humans, *Crithidia fasciculata*. Many of the known trypanocides are cationic; the hypothesis that these might be affecting presumed crucial roles of the polyamines required analyses of the compounds in the cells and some data on their metabolism. The group studied some of the parasites of insects, *Crithidia*, etc., and mammals, *T. brucei*; the latter were in bloodstream forms of the rat, grown in a system adaptable to tests of chemotherapeutic efficacy. All of these contained putrescine and spermidine but lacked spermine (Bacchi et al., 1977). *T. brucei*, grown in the rat, proved to have a relatively high content of spermidine and a high ratio of spermidine: putrescine.

The partially purified α-glycerophosphate dehydrogenase of the various trypanosomes proved to be markedly activated by 10 mM Mg^{2+} or millimolar spermidine or spermine, but not by putrescine. Several cationic trypanocides like ethidium, which is active against *T. congolense* in cattle, activated the enzyme several-fold at low concentration and inhibited at higher concentrations (Bacchi et al., 1978). These trypanocides have been known to displace spermidine from certain nucleic acids. These observations supported the idea that cellular spermidine may be activating this important enzyme in the trypanosome and that the reduction of cellular spermidine might be chemotherapeutically effective.

Bacchi and his collaborators (1981) asked if spermidine and spermine would abolish the curative effects of several of the cationic trypanocides in *T. brucei* infections of the mouse. They included compounds active against both *Babesia* strains (spiroplasms) and some trypanosomes lethal to mice. Some of the compounds, whose cures were in fact reversed by the polyamines (e.g., amicarbalide and imidocarb), are presented in Figure 18.1. Pentamidine is thought to not affect polyamine metabolism in pentamidine-sensitive trypanosomes (Berger et al., 1993), although this compound and Berenil have been described as inhibitors of trypanosomal AdoMetDC (Bitonti et al., 1986b). In 1978 Berenil remained an acceptable drug for cattle trypanosomiasis because it had a low toxicity and was effectively therapeutic. It was active against many strains resistant to other drugs, such as antricide and ethidium, which had been introduced in the 1950s. Moreover, the latter compounds were damaging at the injection site and by 1965 had been withdrawn in many areas.

The existence of the polyamines in *Trypanosoma* launched a successful search for their known biosynthetic enzymes and tests of previously recorded inhibitors (Chang et al., 1978). MGBG was found to be a selective inhibitor of *T. brucei* as well as of the trypanosomal putrescine-activated AdoMetDC. However, neither spermidine nor spermine reversed the antitrypanosomal action of the inhibitor, which was actually not curative of the infection in mice or rats.

Polyamine metabolism was studied as a possible approach to the chemotherapy of trypanosomiasis in rodents (Bacchi et al., 1979). The strain of *T. brucei* was passed serially in rats and blood was taken from animals after 3–4 day infections. The protozoa were separated by ion-exchange chromatography, and washed cells were incubated with labeled substrate. [^{14}C]-Ornithine or [^{14}C]-arginine were incorporated into putrescine and spermidine, but not into spermine. MGBG (50 μM) inhibited incorporation into spermidine by 40–60% and

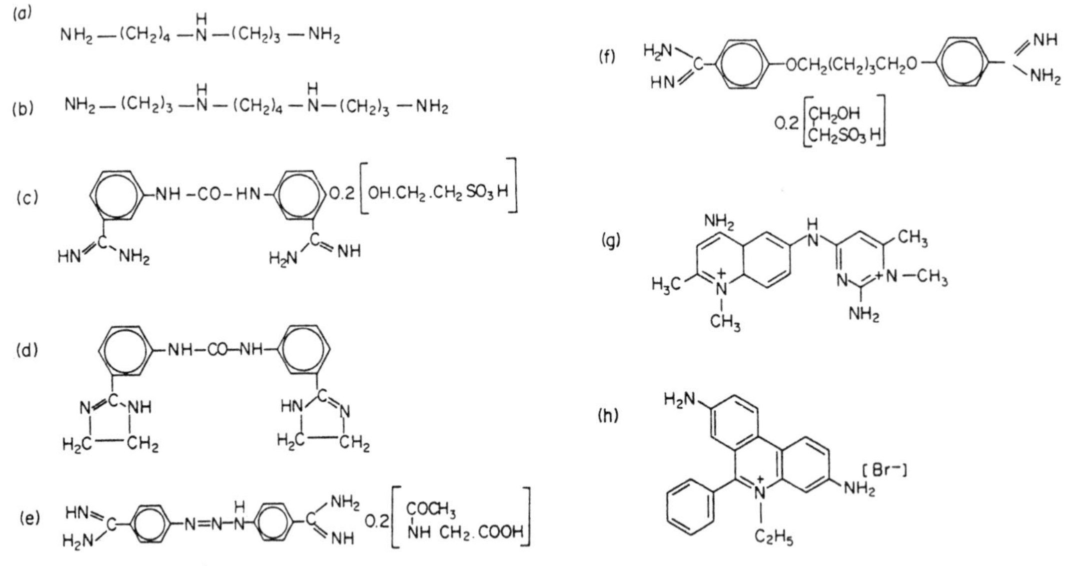

FIG. 18.1   Structures of (**a**) spermidine, (**b**) spermine, (**c**) amicarbalide, (**d**) imidocarb, (**e**) Berenil, (**f**) pentamidine, (**g**) Antrycide, and (**h**) ethidium bromide; adapted from Bacchi et al. (1981).

putrescine accumulated. Conversion of arginine to putrescine was inhibited some 40% by α-methylornithine. Label from [^{14}C]-methionine appeared in spermidine and this was increased by addition of ornithine to the medium. Lysine was not converted to cadaverine in the system. [^{3}H]-Ornithine was rapidly converted to putrescine, suggestive of the presence of an ODC. Although spermine was not synthesized, this polyamine was taken up rapidly, as was spermidine, to levels about 5 times that of putrescine. In later studies of *T. b. brucei* no evidence could be obtained for the presence of agmatine, cadaverine, *N*-acetylspermidine, spermine, or *N*-acetylspermine (Fairlamb et al., 1987).

DFMO became available in 1978 and by 1980 the cure of *T. brucei* infections in mice by orally administered DFMO was readily demonstrable (Bacchi et al., 1980). The infected mice were also cured by tiny amounts of injected Berenil, about 1/1000 of the weight of DFMO used. This datum sets the problem for the practical use of DFMO. Can DFMO be used in a lesser amount or will it be necessary to find another effective, relatively nontoxic, more potent, and less costly inhibitor of polyamine metabolism?

DFMO similarly blocked the infection of mice by the human parasite, *T. b. rhodesiense*. However, it appeared that cures of the infection or long-term survival of the mice required even higher doses of the drug (McCann et al., 1981a). The curative effects of DFMO were also demonstrated in the chicken intestinal coccidia parasite, *Eimeria tenella*. This effect was completely reversed by administration of putrescine. Some malarial parasites were partially inhibited in their development. DFMO was ineffective against *T. cruzi* and *Leishmania donovani*.

Relating Curative Effects of DFMO to Effects on Polyamine Metabolism

T. b. brucei obtained from infections treated with DFMO for 12–15 h were very low in active ODC; they were completely lacking in putrescine, contained about one-third to one-fourth of the control level of spermidine, and had taken in small but detectable amounts of spermine. As described for mammalian cells, DFMO-treated trypanosomes markedly increased their rates of polyamine uptake (McCann et al., 1981b).

The restoration of trypanosomal proliferation in infected mice treated with DFMO was more rapid and complete with spermine and spermidine than with putrescine. Cadaverine and DAP did not affect cures at all (Nathan et al., 1981a). Thus, spermidine appears to be the crucially depleted polyamine. As will be seen, the requirement for spermidine in the trypanosomes is evinced not only in the stimulatory effect on a key shuttle in carbohydrate and lipid metabolism, the NAD-dependent α-glycerophosphate dehydrogenase, and on enzymes of DNA metabolism, but also on its structural role in the crucial coenzyme modulating oxidative stress, trypanothione. This has resulted in a search for a more suitable inhibitor of AdoMetDC, as well as immediate effects to exploit drugs that might serve synergistically with DFMO in the control of protozoan infections.

Nevertheless, a significant reduction (>50%) in the respiratory rate of *T. b. brucei* from DFMO-treated animals is restored to control levels by exogenous putrescine in the cells or mitochondrial preparations but not by exogenous spermidine. Thus putrescine appears to have an activity in addition to that attributable to spermidine (Giffin et al., 1986a).

Synergy of DFMO and Bleomycin

A mixture of bleomycins proved to be effective in increasing the survival of *T. b. brucei* infected mice, and this protection was markedly inhibited by spermidine and spermine. The reversal by spermidine also occurred with the individual bleomycins A_2 and B_2 (Nathan et al., 1981b).

When blenoxane (largely A_2 and B_2) was administered with DFMO, it was possible to cure the animals with a mix comprising 25% of the curative DFMO level and 10% of the curative bleomycin level (Bacchi et al., 1982). These experiments were subsequently extended and cures similarly demonstrated in a mouse model of African trypanosomiasis, which affects the central nervous system (Clarkson et al., 1983). The cures obtained in the initial model were prevented by concomitant doses of polyamine. It was suggested that the depletion of polyamine by DFMO served to expose the cellular DNA to bleomycin.

It was asked if more common low toxicity components might serve in place of bleomycin. Suramin, which was used in early stage disease, was also found to act synergistically with DFMO in the late stage of CNS infection. Suramin is also effective in this way with another highly active inhibitor of ODC, which is accumulated more rapidly than DFMO (Bacchi et al., 1987a). The only drug available earlier for treatment of the CNS stage of African sleeping sickness was a toxic organic arsenical (Mel B). The observation that a combination of suramin and an inhibitor of polyamine synthesis acts against this stage offers a possible alternative to this toxic compound.

From Mice to Humans

Table 18.1 summarizes the cytological and biochemical effects of DFMO on *T. b. brucei* after exposing infected rats to the compound (Bacchi et al., 1983a; Giffin et al., 1986b). Effects on division have been attributed to interference with the organization of actin and microfilaments (McCann et al., 1983).

The ultimate elimination of the infecting organism after treatment with DFMO alone may be the result of the host's immune response to the trypanosomal surface antigen (de Gee et al., 1983). Polyamine depletion effected by DFMO causes the transformation of long slender multiplying forms to short stumpy nonmultiplying forms containing enlarged mitochondria (de Gee et al., 1984). This change goes hand in hand with a decrease in the specific protein synthesis related to the production of the variant-specific glycoprotein that enables an evasion of the host's immune response (Bitonti et al., 1988).

The ODC of the protozoan is not more sensitive to DFMO than is the host enzyme. The apparent inhibition constants (K_i's) of several inhibitors of the trypanosomal ODC, including that of DFMO, have proved to be equal or somewhat higher than those obtained with a rat liver ODC (Bitonti et al., 1985). The simple hypothesis of a selective toxicity of DFMO to the inhibited enzyme of the parasite does not explain the selective elimination of the protozoan from the infected animal. The findings that DFMO prevents the development of antibody-resistant protozoan surface antigens, reduces macromolecular synthesis, and blocks cell division, are partial additions to the explanation. It will be seen that differences in the structure of the host and protozoan genes, which determine the stability and renewal of ODC, contribute in a major way to the explanation of the selective sensitivity of the parasite to DFMO.

Table 18.1 Effects of DFMO on metabolism and morphology of *T. b. brucei*.

Morphological changes	Production of multiple nuclei and kinetoplasts, and enlarged organisms (up to 40 μm long)
ODC activity	>90% inhibition of ODC activity in cells treated for 12 h
Polyamine levels	Putrescine was >90% depleted within 12 h posttreatment; spermidine levels decreased ~60% at 12 h and 70% after 36 h
Polyamine uptake	Putrescine, spermidine, and spermine increased two- to fourfold in 12-h treated cells
Polyamine synthesis by intact treated cells	Putrescine synthesis from [³H]-ornithine decreased >95% in 12-h treated cells; spermidine synthesis from [³H]-putrescine + methionine increased three- to fourfold in treated cells
dAdoMet levels	Increased >1000-fold in 12-h treated trypanosomes
Synthesis of macromolecules	100% inhibition of DNA synthesis and 50–80% inhibition of RNA synthesis resulting after 36 h; protein synthesis increased up to fourfold in 36-h treated cells

Summarized by McCann et al. (1983).

It was important to determine the range of trypanosomes on which DFMO would act. DFMO was obviously effective on *T. b. brucei* infections in mice. In 1982 two humans in a late stage of infection with *T. b. gambiense*, the agent of West African sleeping sickness, were cured with large oral doses of DFMO (400 mg/kg/day) for 44 days. No relapses occurred after 1.5–2 years. This encouraging result led to the successful use of intravenous administration and by 1985 113 cases had been treated, with a high percentage of complete recovery (Schecter & Sjoerdsma, 1986). The treatment was successfully extended to patients with late stage *T. b. gambiense* sleeping sickness refractory to melarsopol. In 1990 the U.S. Food and Drug Administration approved the use of DFMO for human patients and by 1992 clinical trials in more than 700 patients demonstrated effectiveness in over 90% of patients with the gambiense form. The currently recommended regimen is 400 mg/kg/day for 14 days at an estimated cost of $200 per patient. Side effects (i.e., hearing loss, diarrhea, anemia) are reversible on withdrawal of the drug. Currently the case load of gambiense sleeping sickness in Zaire is on the order of 150,000.

DFMO is ineffective with many clinical isolates of *T. b. rhodesiense*, the trypanosomal parasite endemic in Central and East Africa. Of 16 independent isolates injected into mice, seven were partially or totally refractory to DFMO and many were also resistant to other trypanocides. It was evident that other novel agents or appropriate combinations would be necessary to handle the East African disease.

Methionine Metabolism as Target in Trypanosomes

The effects of DFMO can be reversed by exogenous spermidine, indicating a crucial involvement of methionine in later stages of polyamine biosynthesis. MGBG was partially effective but noncurative. Other bisguanyl hydrazones do cure *T. b. brucei* infections in mice but this effect was not reversed by spermidine (Bacchi et al., 1983b). The organism does synthesize spermidine and spermidine synthase has been purified. It is inhibited by cyclohexylamine ($K_i = 3$–15 μM) and by *S*-adenosyl-1,8-diamine-3-octane ($K_i = 25$ μM), but administration of the former class to infected mice did not increase survival (Bitonti et al., 1984). Many structural analogues of MGBG like Berenil (Fig 18.2) were found to be effective inhibitors of the putrescine-stimulated eucaryotic AdoMetDC. Bitonti et al. (1986b) turned to the AdoMetDC of *T. b. brucei* and demonstrated the inhibition of the enzyme by MGBG, pentamidine, and Berenil. The latter blocked the enzyme in vivo and did cure infections in mice; also, this effect was prevented by exogenous spermidine. However, Berenil was known to inhibit kinetoplast DNA synthesis and it could not be concluded that the effects of Berenil were entirely a result of the inhibition of polyamine biosynthesis.

In the next few years attention focused on the metabolism of AdoMet in the trypanosome, which are unable to synthesize purines de novo and rely on salvage pathways. Many unnatural purine nucleoside analogues were metabolized and incorporated into nucleic acids. Some purine analogues, such as sinefungin, formycin B, and 9-deazainosine (Fig. 18.2), suppress *T. b. brucei* infections; the last of these is curative of CNS infections when used in combination with DFMO. This nucleoside is not phosphorylated by mouse cells but is converted to the nucleotide by the parasites. The targets for the purine nucleosides listed in Figure 18.2 include a number of enzymes determining reactions in the biosynthesis of methionine, the control of methylations, and the synthesis of potentially distorted nucleates.

Trypanosomal enzymes metabolizing methionine to AdoMet and beyond were abundantly present in extracts. These included the AdoMet synthetase, various methyltransferases, and an adenosyl-L-homocysteine hydrolase (AdoHcy hydrolase). AdoHcy is a well-known inhibitor of methyltransferases and the ratio of AdoMet to AdoHcy (the methylation index) is one measure of active methylation. As shown in Table 18.2, *T. b. brucei* isolated from rats treated with DFMO contained a major increase in AdoMet, and an even more startling increase in decarboxylated AdoMet (dAdoMet) (Yarlett & Bacchi, 1988a). The result was surprising in that the inhibition of animal cells by DFMO produced a major increase in the latter but was without significant effect on the accumulation of AdoMet itself. This result, in which no increase was noted in AdoHcy, clearly increased the methylation index and posed the problem of a possible damaging increase in some type of methylation. There was a sixfold increase in one class of protein methylases, which transfers methyl groups to carboxyl groups of several proteins, including a group of arginine-rich histones. This methylase was quite sensitive to sinefungin ($K_i = 1.6$ μM), a competitive inhibitor of the AdoMet binding site (Yarlett et al., 1991).

Two different purine nucleoside-cleaving activities were detected in *T. b. brucei*, of which one was a ribonucleoside hydrolase and the other a phosphorylase capable of the phosphorolytic cleavage of methylthioadenosine (MTA) and adenosine. The latter had a broader range than the mammalian enzyme, suggesting the possibility of generating purine analogues and analogues of methylthioribose-1-phosphate (Ghoda et al., 1988). A series of 5′-alkyl-substituted analogues of MTA, such as 5′-deoxy-5′-(hydroxyethyl)thioadenosine (HETA) and its 5′-halogeno derivatives, was cleaved by the *T. b. brucei* enzyme in phosphate-dependent reactions. These nucleosides were inhibitory to growth in vitro at micromolar levels, and these inhibitions were reversed by exogenous methionine or 2-keto-4-methylthiobutyric acid, the intermediate of methionine biosynthesis. The initial compound, HETA, effected high rates of cures in infected mice, implying that the resulting cleavage product, the sugar phosphate analogue, had a significant potential as a specific chemotherapeutic agent (Bacchi et al., 1991).

The new specific inhibitors of AdoMetDC (see compound 5 in Fig. 18.2) proved to protect *T. b. brucei*-

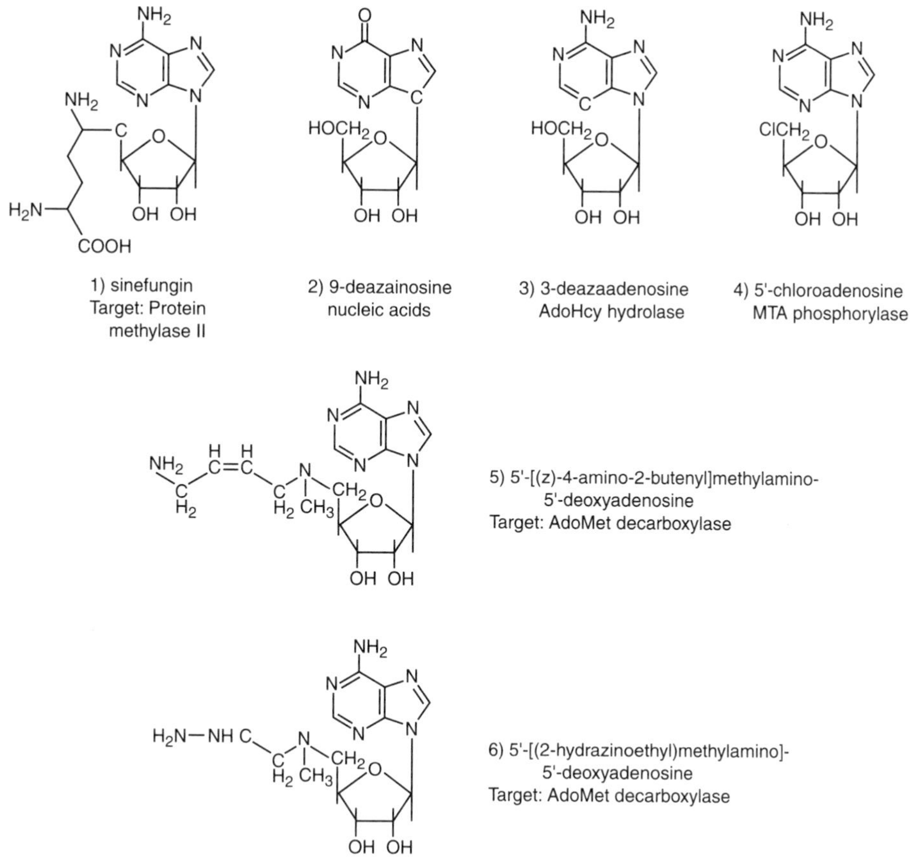

1) sinefungin
Target: Protein
methylase II

2) 9-deazainosine
nucleic acids

3) 3-deazaadenosine
AdoHcy hydrolase

4) 5'-chloroadenosine
MTA phosphorylase

5) 5'-[(z)-4-amino-2-butenyl]methylamino-
5'-deoxyadenosine
Target: AdoMet decarboxylase

6) 5'-[(2-hydrazinoethyl)methylamino]-
5'-deoxyadenosine
Target: AdoMet decarboxylase

FIG. 18.2   Some inhibitory purine nucleoside analogues.

infected rats. The inhibition of the trypanosomal enzyme was rapid, time dependent, and irreversible. Isolated organisms accumulated putrescine and decreased their content of free spermidine. The compound also cured infections of *T. b. rhodesiense* under conditions of continuous administration (Bitonti et al., 1990). Combinations with DFMO were even more effective than a single agent in any of the African trypanosomiases (Bacchi et al., 1992). Nevertheless, a single adenosyl aminoxy derivative has been described both as an effective inhibitor of AdoMetDC and a parasite-specific trypanocidal agent (Guo et al., 1995).

In continuing studies with this compound in *T. b. brucei* infections it was observed that the AdoMet content of the protozoans increased far more (>10-fold) than did the host cells, a result obtained with DFMO as well. Less therapeutic compounds did not increase these AdoMet levels, and it was suggested that the antitrypanosomal effects of the drug were related to the major selective changes in AdoMet in the protozoa (Byers et al., 1991a). Another series of active site-directed inhibitors of AdoMetDC was also found to contain highly potent representatives, of which a hydrazino compound (6 in Fig. 18.2), was the most active (Tekwani et al.,

Table 18.2   Effects on *T. b. brucei* cells of DFMO on AdoMet and its metabolites and methylation index (MI) (AdoMet/AdoHcy).

Time (h)	AdoMet	AdoHcy	dAdoMet	MI
12 control	0.74 (0.67–0.81)	0.11 (0.10–0.12)	0.17 (0.12–0.22)	6.5
12	7.94 (6.07–9.81)	0.13 (0.11–0.15)	15.3 (13.3–17.3)	63
36	35.9 (32.5–39.3)	0.32 (0.27–0.37)	66.9 (63.2–70.6)	114

Adapted from Yarlett and Bacchi (1988).

1992b). Thus, purine nucleosides analogues tailored to the requirements of the parasitic AdoMetDC have provided new tools in the blockage of the infection.

The decarboxylase itself has been purified and, although putrescine activated, has proved to be quite distinct from the mammalian counterpart (Tekwani et al., 1992). It is activated less well by cadaverine and DAP; spermidine and spermine interfere with the activation. The trypanosomal enzyme is not precipitated by antiserum to the human enzyme and binds to MGBG-Sepharose less well. Indeed the subunit structure also appears somewhat different. The hydrazino inhibitor described above completely blocks the binding of labeled AdoMet to the active site of the enzyme.

The pile-up of AdoMet and the increase of protein methylase II in inhibited cultures led finally to the study of the *T. b. brucei* AdoMet synthetase. This enzyme was found to exist in at least two isoforms, the major form of which had a long half-life and was relatively insensitive to high concentrations of the product, AdoMet. This distinguishes this enzyme from the mammalian enzyme (Yarlett et al., 1993a). Furthermore, the protozoal synthetase is increased (doubled) during DFMO treatment. The partitioning of AdoMet between the polyamine pathway and transmethylation reactions is obviously a key problem in any cell, and the lack of careful regulation of the synthetase and of AdoMet production and further reaction in these parasites may prove to be one parameter critical to the survival of organisms exposed to DFMO.

It is of interest that some filarial parasites are devoid of ODC and contain a putrescine-dependent AdoMetDC that is sensitive to the competitive inhibitors MGBG and Berenil (Rathaur et al., 1988). These agents also inhibit spermidine and spermine uptake (Tekwani et al., 1995).

AdoMet and Resistance of Trypanosomes to DFMO

Clinical isolates of *T. b. rhodesiense* that were unexposed to DFMO and proved refractory to DFMO, accumulated that inhibitor somewhat more slowly than did sensitive strains; but this did not appear crucial to sensitivity. More significantly, resistant strains had markedly lower contents of AdoMet synthetase and AdoMetDC; also, "sensitive" strains of *T. b. rhodesiense* accumulated much lower levels of AdoMet pools than did the sensitive *T. b. brucei* strains (Bacchi et al., 1993). These authors attribute toxicity in DFMO-sensitive strains to hypermethylation and have recorded a marked increase of methyl groups into protein in DFMO-treated bloodstream trypanosomes.

In addition, newly synthesized adenosyl analogues have been tested and found active on both *T. b. brucei* and *T. b. rhodesiense* (Guo et al., 1993). Combinations of DFMO and other drugs (e.g., suramin, melarsen oxide) also proved to effect cures in laboratory model infections with the latter organism (Bacchi et al., 1994). It will be noted that the arsenicals are believed to be active in the reaction with the unique trypanosomal cofactor, trypanothione.

Metabolic Stability of Trypanosomal ODC

The causative factors controlling large rapid increases and decreases of ODC in mammalian cells have come under close scrutiny in the past decades (Chap. 13). It was found that unusual systems of regulation existed, including the production of ODC antizyme, specific proteolytic degradation of bound enzyme, and polyamine-dependent mechanisms of transcription and translation of ODC and antizyme. The very short half-life of mammalian ODC became slowly interpretable in terms of these unusual regulatory phenomena. The regulation of ODC in the trypanosomes proved to be quite different.

When *T. b. brucei* is grown in a tissue culture medium to a stationary phase and diluted in fresh medium, ODC is induced in a biphasic curve. A maximum appears at 24 h and a high plateau at 48–72 h, after which the cells enter decline. ODC protein is formed, and this synthesis is inhibited by actinomycin D or cycloheximide. Exogenous putrescine or spermidine also suppress ODC synthesis, although the organisms were more sensitive to the former. Cycloheximide did not cause a marked decline in presynthesized ODC, which had a half-life of >6 h. A similarly prolonged half-life was obtained in mouse-grown parasites. After a single treatment of infected rats with DFMO, the recovery of the enzyme was slow and incomplete after 48 h (Bacchi et al., 1989). Bacchi and colleagues concluded that ODC is neither degraded readily nor rapidly replaced in the trypanosome, unlike the mammalian cell.

Comparison of the structures of the mammalian and trypanosomal ODC and their genes revealed major differences that relate to their regulation. The protozoal enzyme has a molecular weight (MW) of 90,000 and subunit weight of 45,000 (Phillips et al., 1987, 1988). The K_i values for ornithine and DFMO were similar, as were these parameters for the mouse ODC. In these analyses the K_i for DFMO was somewhat lower in the mammalian ODC. An ODC gene was isolated from the parasite, assuming homology to the mouse gene, and this DNA gave rise in transcription and translation to the 45,000 MW subunit protein. Using an appropriate DNA vector, a strain of *Escherichia coli* was transformed and yielded the apparently identical *T. b. brucei* ODC. Greatly improved yields of the enzyme were obtained by adding a highly active promoter to the DNA vector (Kuntz et al., 1991).

Two major differences were found between the protozoan and mouse ODC, which had a significant homology (61.5%) in their amino acid sequences. The trypanosomal enzyme lacked a 36 amino acid C-terminal peptide, whose composition contained presumed PEST sequences believed to relate to the rapid turnover of the mammalian enzyme. As discussed in Chapter 13, the loss of residues from this terminal peptide in mammalian ODC resulted in a marked reduction in its proteolytic degradation. When the *T. brucei* and mouse ODC genes were placed in ODC-deficient Chinese hamster ovary (CHO) cells, the *T. brucei* protein was stable and the latter was labile, affirming the differences in stability as a property of the ODC proteins themselves. An ODC containing the trypanosomal N-terminus and the mouse carboxy-terminus was labile in CHO cells, pointing to the latter domain as the determinant of this phase of regulation. However, mouse ODC is stable in transformed *T. brucei* and this may be due to the absence of antizyme and differences in the proteolytic mechanisms of mouse cells and trypanosomes (Hua et al., 1995). The DFMO inactivation of the ODC of *T. b. brucei* led to depletion of putrescine and spermidine, and the lack of spermidine affected energy production and polymer biosynthesis. The lack of putrescine permitted the accumulation of dAdoMet and unexpectedly of AdoMet itself. The inability of the inhibited organism to restore its essential polyamines turned on the inability of ODC to restart and produce adequate supplies of polyamine (Wang, 1991). A study of a nonpathogenic insect trypanosomatid, *Leptomonas seymouri*, indicated the existence of a more sensitive system of regulation in this parasite, more like that in mammalian cells (Hannan et al., 1984).

Trypanothione: Novel Derivative of Glutathione (GSH)

In addition to various cationic trypanocides, some weakly successful treatments of trypanosomal disease have involved reagents like arsenicals that are capable of reaction with thiol or —SH groups in compounds such as reduced GSH (L-γ-glutamyl-L-cysteinyl-glycine). Indeed inhibition of GSH biosynthesis and its depletion by administration of buthionine sulfoximine in *T. b. brucei* infected mice prolonged survival of the animals. In a study of GSH in trypanosomatids, "GSH" reductases were found in extracts of *T. b. brucei*, *T. cruzi, Leishmania mexicana*, and *Crithidia fasciculata*; these enzymes oxidize NADPH to NADP⁺ in the presence of oxidized GSH (GSSG). However, they required a new isolable intermediate substrate that could only be found in trypanosomatids. A low molecular substrate was isolated from the insect trypanosomatid, *C. fasciculata* (Fairlamb & Cerami, 1985), and the product of the initial reductase reaction contained sulfhydryl groups. The reaction was described as

$$R{-}S{-}S{-}R + 2NADPH \rightarrow 2RSH + 2NADP^{+}.$$

The oxidation of NADPH did not occur in the absence of the substrate contained in the extract. The unknown reduced thiol, RSH, could be reacted nonenzymatically with various oxidized thiols, including GSSG, to produce GSH and the oxidized R—S—S—R. The initial reaction was thus designated a disulfide reductase.

The nature of the new substrate (Fairlamb et al., 1985) was determined to be that of a bis-glutathionyl

derivative of spermidine (Fig. 18.3). Acid hydrolysis liberated cysteic acid, glycine, glutamate, and spermidine in the ratios of 1.0, 1.0, 1.0, and 0.5, respectively. This compound, named *trypanothione*, has not been found in *E. coli*, which does accumulate a monoglutathionylspermidine in the stationary and anaerobic phase of growth (Smith et al., 1995; Tabor & Tabor, 1975). The new (bis) compound proved to be the major low molecular thiol in trypanosomes, and it was suspected that the elimination of the unique trypanothione in the trypanosomes would be deleterious to the survival of the parasites. That thiol metabolism is crucial to the lives of the parasites is still widely held (Krauth–Siegel & Schoneck, 1995).

Bloodstream organisms proved to contain 10–15% of their total spermidine in a bound form and far more trypanothione than monoglutathionylspermidine. The parasites contained significant amounts of free GSH. The levels of spermidine and trypanothione both fell sharply immediately after treatment with DFMO to a level of about 30% in 48 h (Fairlamb et al., 1987). Such treatment causes respiration to decrease 50% in 24-h treatment, DNA and RNA syntheses are blocked, and protein synthesis and mitochondrial biogenesis are increased. The pattern of reactions, including those dependent on trypanothione and thereby affected by DFMO, are presented in Figure 18.4.

The route of synthesis of the new substrate requires a synthesis of GSH. Both an N^{1}- and N^{8}-monoglutathionylspermidine are produced as intermediates, and these

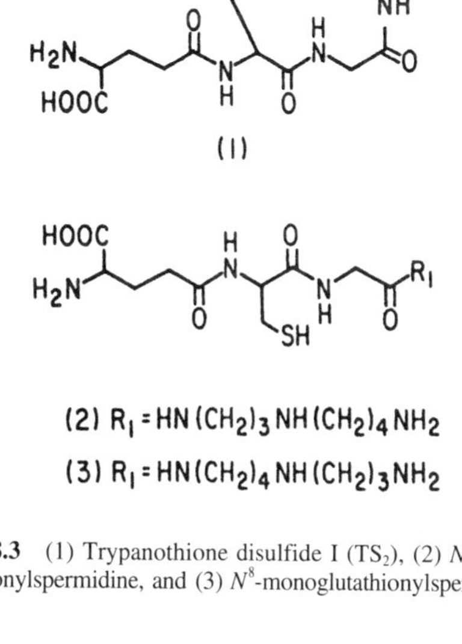

FIG. 18.3 (1) Trypanothione disulfide I (TS₂), (2) N^{1}-monoglutathionylspermidine, and (3) N^{8}-monoglutathionylspermidine.

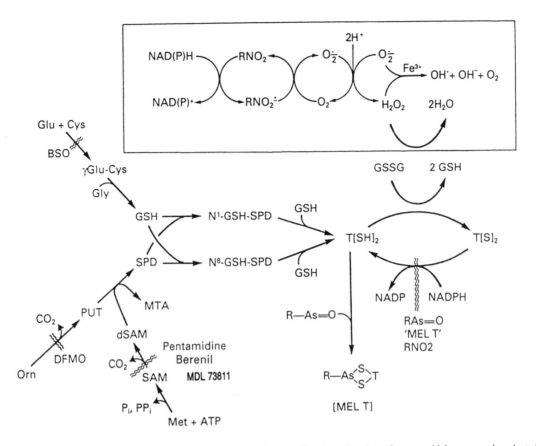

FIG. 18.4 Metabolism and functions of trypanothione, showing possible sites of action of trypanocidal compounds; adapted from Fairlamb and Cerami (1992). BSO, buthionine sulfoximine; Mel T, see Figure 18.6.

Table 18.3 Levels of glutathione, glutathionylspermidine, and dihydrotrypanothione in various trypanosomatids.

Organism	Total GSH[a] (nmol/10⁸ cells)	% GSH recovered as		
		GSH	GSH-spd	T[SH]₂
T. brucei				
Bloodstream forms	5.6	24	5	71
Procyclic forms	11.4	27	4	69
C. fasciculata				
Exponential phase	51.1	12	20	68
Stationary phase	38.7	10	68	22
L. braziliensis guyanensis				
promastigotes	13.7	14	4	82
L. donovani promastigotes	15.6	14	4	82
T. cruzi epimastigotes	9.5	19	2	79

Adapted from Fairlamb and Cerami (1992).
[a] Total GSH = [GSH] + [GSH-spd] + 2[T(SH)₂].

are subsequently converted to the (bis) derivative (Fairlamb et al., 1986).

The concentrations of GSH and its spermidine derivatives in various trypanosomes are presented in Table 18.3. It will be noted that the parasitic forms are particularly high in dihydrotrypanothione. New methods for the qualitative and quantitative analysis of the trypanosomal thiols have been reported recently (Krauth–Siegel et al., 1995; Steenkamp, 1993). The yield of trypanothione from *C. fasciculata* was some 50 mg/100 g (wet weight) of cells.

The trypanothione synthetase was purified some 14,500-fold to apparent homogeneity from the insect trypanosomatid, *C. fasciculata* (Henderson et al., 1990). The enzyme was assayed by estimating the ability to reduce glutathionylspermidine disulfide in the presence of NADPH. This protein catalyzed the synthesis of both the N^1-monoglutathionyl derivatives and trypanothione in the presence of ATP, Mg^{2+}, GSH, and spermidine at a pH optimum of 7.5–7.8. It also converted the N^8-monoglutathionylspermidine to trypanothione with a significantly lower K_m than that for the N^1 derivative. Several substrates such as spermine, which are unnatural to the parasites, can compete with spermidine. In a somewhat different purification, the enzyme appeared to lose its ability to convert monoglutathionylspermidine to trypanothione. The presence of another enzyme, which adds the second glutathione moiety, has been reported.

Trypanothione in Life of *C. fasciculata*

The ability to follow this insect trypanosomatid through cycles of growth on defined media in shake cultures facilitated an analysis of the levels of relevant metabolites. As presented in Figure 18.5, cells in the early phase of exponential growth had quadrupled their content of trypanothione, and this fell sharply as the organism entered the stationary phase. This compound largely appeared in this phase as the monoglutathionyl spermidine, as indeed had been seen in the growth of *E. coli* (Shim & Fairlamb, 1988).

The maintenance of an intracellular reducing activity and its role in protection against toxic oxidations will be discussed in later sections. Trypanothione has been found to be more active than GSH in the formation of S-conjugated xenobiotics (Moutiez et al., 1994), and it appears also that exposure of *Crithidia* to Cd^{2+} induces a 10-fold increase of the compound, which can detoxify this ion (Fairlamb & Cerami, 1992). These authors have also pointed to several functions employing GSH as the coenzyme in nontrypanosomatids. The determination of which type of GSH is active in these reactions in *Crithidia*, for example, has not been explored.

Trypanothione in Chemotherapy of African Trypanosomiasis

The requirement for spermidine in the synthesis of trypanothione posed the question of the effect of DFMO

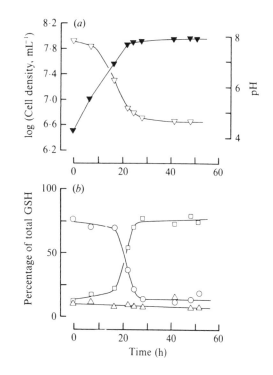

FIG. 18.5 Distribution of glutathione and glutathionylspermidine conjugates during growth of *C. fasciculata*; adapted from Shim and Fairlamb (1988). (**a**) (▼) Cell density and (∇) pH of the medium; (**b**) percentage distribution as (△) free GSH, (○) T(SH)$_2$, and (□) GSH-spermidine.

on trypanothione concentrations and the relevance of those concentrations to sensitivity or resistance to the drug. Resistant strains of *T. b. brucei* were selected by passage in vitro in increasing amounts of DFMO. The resistant strain contained an ODC with an amount and reactivity to DFMO that was similar to that in the unexposed sensitive strain. However, after exposure for 48 h to DFMO, the sensitive strain had half its original concentration of trypanothione whereas this level was not reduced in the resistant organism. The latter proved to contain high protective levels of ornithine, some 80-fold greater than that of the control (Bellofatto et al., 1987).

It was asked if the antitrypanosomal activity of arsenical drugs may arise from their reaction with the sulfhydryls of trypanothione. The rapid formation of a stable macrocyclic adduct (Fig. 18.6) was demonstrated with the arsenical, melarsen oxide, and the coenzyme, in which the arsenic atom links both cysteine thiols of the two molecules of GSH (Fairlamb et al., 1989). Thiols present in free GSH and in free cysteine reacted to form thioarsenite complexes much less readily. When *T. b. brucei* is incubated with melarsen oxide or the 2,3-dimercaptopropanol adduct of melarsen oxide, the trypanothione present in an extract is tied up largely in the compound presented in Figure 18.6. This new arsenic-

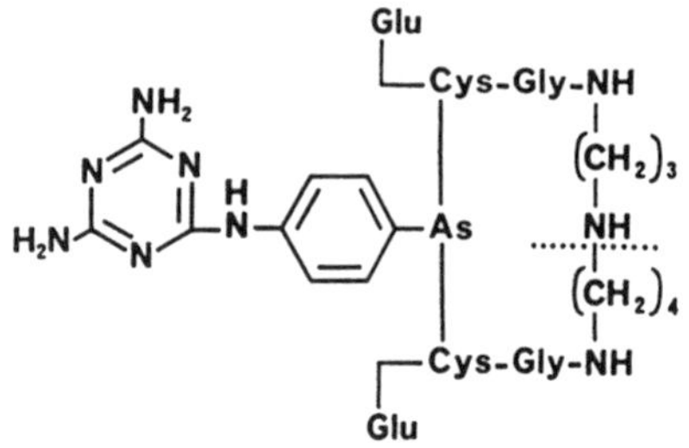

(1)

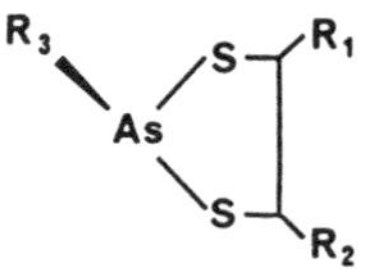

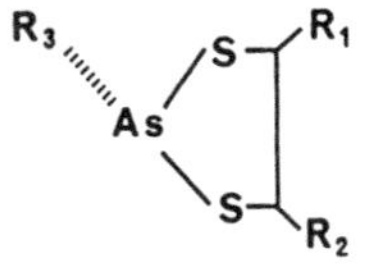

(2)   (3)

FIG. 18.6 Structure of the melarsen oxide-T(SH)$_2$ derivative (Mel T); adapted from Fairlamb et al. (1989). Structures (2) and (3) illustrate the two diastereoisomeric forms arising from the tetrahedral geometry of the arsenic atom. R = melaminophenyl moiety.

containing compound is a potent inhibitor ($K_i = 9$ μM) of trypanothione reductase, the enzyme apparently crucial to the regulation of active thiol in the parasite. Unexpectedly this reductase was not found in the mitochondrial compartment of *T. b. brucei* (Smith et al., 1991).

Melarsen-resistant strains (i.e., resistant to melarsen oxide and melarsoprol) did not contain levels of trypanothione or GSH different from those of sensitive strains. The resistant organisms did contain a somewhat higher level of dihydrolipoamide dehydrogenase. The dithiol derivatives of the cofactor, dihydrolipoic acid and dihydrolipoamide, were both capable of forming six-membered dithioarsenite rings with melarsen oxide, which were hundreds of times more stable than the trypanothione complex (Fairlamb et al., 1992a,b). Melarsen oxide-resistant *T. b. brucei* appears to lack a particular transport system for adenosine and is unable to take up the trypanocide (Carter & Fairlamb, 1993).

Trypanothione Reductase

DFMO is virtually ineffective in the treatment of *T. cruzi*, the etiologic agent of the widespread and important Chagas disease. Because trypanothione is present in all trypanosomatids and is absent from other eucaryotic cells, it was asked if the enzyme crucial to the reducing function of this metabolite should not be a major target for chemotherapeutic drugs. In the late 1970s it had become clear that it was possible to define a target protein in sufficient detail to permit the rational design of an enzymatic inhibitor (Cohen, 1979; Kuntz, 1992).

The choice of trypanothione metabolism as a potential chemotherapeutic target is based on the knowledge of the existence of quite different essential chemical compounds in the parasite and host (Schirmer et al., 1995). In contrast, the work with DFMO has provided an example of therapeutic effects via the different controls on the regulation of ODC in the parasite and host and not on the structure of the active site in the two enzymes.

The enzymology of trypanothione was begun with the isolation and characterization of trypanothione reductase from *C. fasciculata* (Shames et al., 1986) and from *T. cruzi* (Krauth–Siegel et al., 1987). The homogeneous crithidial enzyme was a flavoprotein comprising two identical subunits of *M* 53,800 MW each. After reaction with oxidized disulfide containing trypanothione and NADPH, the enzyme increased its sulfhydryl content by two —SH groups, implying the existence of a dithiol (cystine) at the active site.

Reactions in which reduced trypanothione is reoxidized by rapid disulfide exchange with intracellular disulfides (e.g., glutathione, cystine) have been represented as

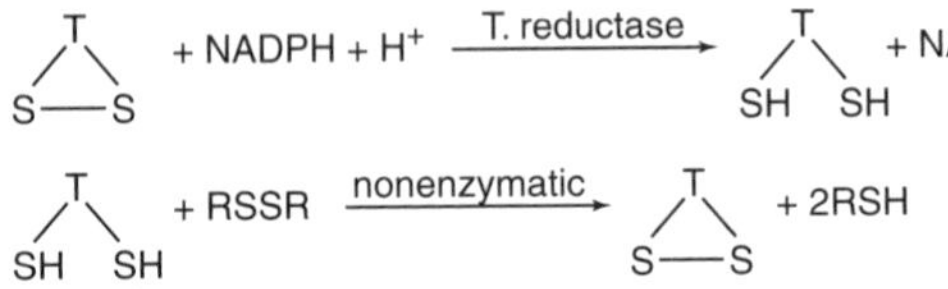

More recently, a new synthetic substrate for this reductase, *N,N*^{1}-bis(benzyloxycarbonyl)-2-cysteinyl-glycine-3-dimethylaminopropylamide, has been prepared to obviate the need for the difficultly obtainable trypanothione (El-Waer et al., 1991). The natural substrate is about 5 times better as a substrate than the synthetic material in reactions with the crithidial and *T. cruzi* enzymes.

The crithidial enzyme contained some 20 amino acids more than the human erythrocytic GSH reductase (Fairlamb & Cerami, 1992). Peptides thought to be present at the active site of the crithidial enzyme were isolated and sequenced to prepare oligonucleotide probes for the isolation of hybridizable trypanosomal DNA (Shames et al., 1988). A DNA fragment containing the entire reductase gene was obtained from *T. congolense*, an organism causing the disease *nagana* in cattle. This has been sequenced and its amino acid composition (492 residues) indicated a subunit MW of 53,443, very similar to that found for the crithidial enzyme. Homologies to the GSH reductases of *E. coli* (450 amino acids) and human erythrocytes (478 amino acids) were on the order of 50%, in which the N-terminal and active site sequences were best conserved. Homology to the crithidial enzyme was considerably higher, on the order of 88%.

The genes for the trypanothione reductase have been isolated from *C. fasciculata* and *T. b. brucei*. The sequences have been compared with each other and those obtained for an enlarged family of flavoprotein disulfide oxidoreductases, which include not only the GSH reductases, but also the dihydrolipoamide dehydrogenases and mercury reductases (Aboagye–Kwarteng et al., 1992). The open-reading frames for the *C. fasciculata* and *T. brucei* sequences code for proteins of 491 and 492 amino acids, respectively. The sequences describing the separate major binding domains for the cofactors FAD and NADP have been defined. The conserved codon usage of the redox active site for several flavoprotein oxidoreductases has also permitted the construction of an oligonucleotide probe enabling the isolation of a putative reductase gene from a leishmanial species (Keithly, 1989). The isolation of the gene from *T. cruzi* by similar methods has enabled the overproduction of the protein in *E. coli* as the beginning of a program of screening for trypanocidal agents for Chagas disease (Sullivan & Walsh, 1991).

Crystallographic analysis has been carried out mainly with well-ordered radiation-stable crystals of the yellow crithidial enzyme (Hunter et al., 1992). The crithidial enzyme was found to share about 40% sequence identity with the human GSH reductase. Indeed the two enzymes were almost, but not quite, identical at the active site (Kuriyan et al., 1991). The human GSH reductase contained arginine residues at the GSH binding site and these were replaced by neutral or acidic residues in the crithidial enzyme. The substrate, trypanothione, had added positive charges conferred by a spermidine that formed amide groups with the glycyl carboxyls of GSH. The arginine-guanido groups at Arg 37 and Arg 347 of GSH reductase, which interact with the glycine carboxyls of GSH, are replaced in the trypanosomal reductase by tryptophan at 20 and by alanine at 342, respectively. Other small variations of the structure have been described (Hunter et al., 1992).

Detailed knowledge of the structure of the crithidial enzyme–substrate complex (Bailey et al., 1993) has permitted the design of substrate analogues. For example, the apparently minimal role of the γ-glutamyl carboxyl or amino groups has suggested a range of replacements, which have been prepared and found to be active as substrates (El-Waer et al., 1993).

It was asked if the structures of these reductases posed the question of whether these would be sufficient to account for the substrate specificities of the enzymes. Mutational changes at crucial sites permitted the conversion of a trypanothione reductase of *T. congolense* to a protein that had acquired GSH reductase activity (Sullivan et al., 1991). Extending this feat, a site-specific mutated human GSH reductase was produced with a 700-fold preference for trypanothione over GSH (Bradley et al., 1991). A similar approach resulted in a change of the *E. coli* GSH reductase, which was no longer reactive with GSH but reduced trypanothione with about 10% of the activity of the natural trypanothione reductase (Henderson et al., 1991).

Inhibitors of Trypanothione Reductase

Efforts were made to understand the mechanism of known inhibitors and to find other compounds expected to inhibit via the trypanothione mechanism. A flavoprotein reduction may operate via a one-electron transfer to a quinone or nitrofuran, and this was found to be an activity of trypanothione reductase. The reduced compounds are readily reoxidized by molecular oxygen but accumulate in anaerobic conditions, then irreversibly inactivate the enzyme and are trypanocidal. The naphthoquinone and nitrofurazone presented below are examples of newly discovered trypanocides that function in this way (Henderson et al., 1988).

Some of these compounds proved to be more active inhibitors for trypanothione reductase than for GSH reductase (Jockers–Scherubl et al., 1989). On the other hand, nifurtimox, a compound of this series that has been used to treat Chagas disease, is a stronger inhibitor of GSH reductase than of trypanothione reductase. Despite this apparent lack of specificity, compounds such as nifurtimox, which may recycle aerobically from the inhibitory reduced form to the oxidized form, are thought to overburden the parasites' ability to handle the various oxidizing moieties and to produce oxidant stress. For example, reduced trypanothione, generated by trypanothione reductase, is used in the removal of peroxides by trypanothione peroxidase.

The concept that a "redox cycling" will augment oxidative stress has suggested the possibility of supplementing the arsenicals that deplete trypanothione with compounds like nifurtimox or chinifur. Such combinations proved highly efficacious in the treatment of mice infected by *T. b. brucei* (Jennings, 1991a,b).

Trypanothione Peroxidase

Many of the parasitic trypanosomes appear to lack catalase, which can decompose H_2O_2. On the other hand some of the insect trypanosomatids do contain good levels of this enzyme (Wertlieb & Guttman, 1963). A common route for the removal of H_2O_2 in many aerobic cells is that of GSH peroxidase, an enzyme absent or very weak in parasitic trypanosomatids (Boveris et al., 1980). Nevertheless, various trypanosomatids (e.g., *T. brucei, T. cruzi*, and leishmanial strains) do consume H_2O_2 at significant rates (Penketh et al., 1987), and a trypanothione-specific peroxidase was demonstrated in extracts of *C. fasciculata* and *T. brucei*. Systems in which GSH and GSH reductase replaced trypanothione and its reductase were inactive, indicating the absence of a GSH peroxidase in the organisms (Henderson et al., 1987). Nevertheless, it has been reported that *T. cruzi* does lack a trypanothione peroxidase and that trypanothione is involved directly in a nonenzymatic elimination of H_2O_2. In many aerobic cells, seleno-cysteine is an essential constituent of the GSH peroxidase. Trypanosomes grown in the presence of radioactive selenite incorporated [^{25}Se] into the proteins but fractionated proteins containing the isotopic Se were not found to possess trypanothione peroxidase activity. Most recently spermine and spermidine derivatives N^1,N^{12}-bis(dihydrocaffeoyl) spermine or -spermidine (see Chap. 20) have been found to inhibit the crithidial trypanothione reductase without affecting the human GSH reductase (Ponasik et al., 1995).

Functions of GSH

GSH is known to be the major nonprotein thiol in most cells and may attain concentrations as high as 10 mM. The compound is both a reducing agent and a nucleophilic reactant, as a function of its free —SH group. The former role is exemplified by enzymatic reactions such as those with the reductase and the peroxidase, and the latter in the formation of GSH conjugates catalyzed by a GSH transferase (Baillie & Slatter, 1991). The depletion of GSH in animals may be achieved by inhibitors of synthesis of GSH, such as buthionine sulfoximine (Meister, 1991), and intracellular GSH may be restored by GSH methyl ester. A lethal GSH deficiency in newborns or in guinea pigs may be counteracted by high levels of ascorbate. Similarly scurvy, the result of ascorbate deficiency, may be delayed by a more readily penetrable GSH methyl ester (Meister, 1994).

GSH is required in *E. coli* in the operation of an alternative and independent ribonucleoside diphosphate reductase involving the enzyme glutaredoxin. The enzyme is inhibited by GSSG (Holmgren, 1989) and it appears that GSH may function in DNA synthesis. The conversion of GSH to the disulfide, GSSG, or its converse has been implicated in enzymatic reactions catalyzed by peroxidases, transhydrogenases, and reductases. Nonenzymatic reactions effecting similar transformations include free radical scavenging and disulfide reduction by thiol/disulfide exchange.

The GSH peroxidase, which was found to contain selenium, is thought to be the enzyme most concerned with the elimination of H_2O_2 in erythrocytes. Several other selenium-containing proteins are present in human cell lines. One of these, distinct from the enzyme discovered earlier, has been demonstrated to be a phospho-

lipid hydroperoxide-GSH peroxidase that is selective in the reduction of membrane hydroperoxides (Maiorino et al., 1991). It may be noted that the polyamines themselves are reported to scavenge activated oxygen and reduce levels of lipid peroxidation (Drolet et al., 1986; Ogata et al., 1992).

In animal cells there appear to be two major sites of GSH metabolism, cytosol and mitochondria. The synthesis of GSH is entirely cytosolic and the thiol is transported rapidly to the mitochondria, from which efflux is relatively slow. Liver mitochondria, which lack catalase and contain about 10–15% of the low molecular thiol, rely on GSH peroxidase to remove hydroperoxides. It has been estimated that 2–5% of mitochondrial O_2 consumption generates H_2O_2. As described by Meister (1991), depletion of GSH may be expected to lead to cytotoxicity, detectable in gross mitochondrial damage, as well as in cell necrosis. Lipid peroxidation may be detected in many membranous structures, including microsomes. The interplay between the effects of thiol depletion in various cellular sites and Ca^{2+} storage and regulation has been discussed by Reed (1990).

Compounds that inhibit hepatic GSH in the rat induce several enzymes, including heme oxygenase, AdoMetDC and ODC. The putrescine contents of the liver are markedly higher. Surprisingly, these inductions are prevented by intraperitoneal injection of GSH (Oguro et al., 1990).

Aspects of Oxidative Stress

Reactive oxygen species derived from a stepwise single-electron reduction of dioxygen are formed in the sequence;

$$O_2 \longrightarrow O_2^{\overline{\cdot}} \longrightarrow H_2O_2 \longrightarrow HO^{\cdot} + H_2O$$

Superoxide dismutase catalyzes the reaction,

$$2O_2^{\overline{\cdot}} + 2H^+ \longrightarrow H_2O_2 + O_2$$

and catalase decomposes H_2O_2 to $O_2 + 2H_2O$. In *E. coli*, superoxide, $O_2^{\overline{\cdot}}$, is produced largely by autoxidation of components of the respiratory electron-transport chain and is held by the intracellular dismutase to an accumulation of 2×10^{-10} M. Dismutase-deficient organisms grown aerobically may show severe growth retardation (Imlay & Fridovich, 1991). The existence of many different superoxide dismutases and their functions in minimizing the toxicity and mutagenicity of superoxide have been summarized recently (Scandalios, 1993).

H_2O_2 generated by substrate oxidations in the presence of Fe^{2+} at metal-binding sites of protein can effect the oxidation and degradation of these polymers (Stadtman, 1992). Among the protein modifications have been found various types of cross-linking of oxidized amino acid residues, many of which are physiologically useful.

The oxidation of lysine in elastin by the Cu-dependent lysyl oxidase was described in Chapter 6. Elastin is toughened by the subsequent formation of cross-links with oxidized residues. A protective fertilization envelope is formed around the sea urchin egg as a result of an oxidative cross-linking reaction during early development (Shapiro, 1991).

Free radicals like superoxide have been demonstrated to be mutagenic (Moody & Hassan, 1982), and many types of oxidation products have been found in the damaged nucleic acids. Nevertheless, it is not entirely clear which forms of reactive oxygen are responsible for the mutagenic change. This is of interest in terms of the hypothesis that the primary function of the polyamines is to protect DNA and RNA from singlet oxygen, the lowest energy-excited state of molecular oxygen (Khan et al., 1992a,b). It was shown that 10 mM spermine or spermidine protects a supercoiled DNA molecule from conversion to an open-chain form by treatment with a presumed source of singlet oxygen. However, a bacterial test of the mutagenicity of singlet oxygen, produced by the illumination of a carefully characterized photosensitizer, has indicated that bacteria are killed but not mutagenized in this system; it was concluded that singlet oxygen has no detectable genetic effects on the bacterial chromosome (Dahl et al., 1988). It is not clear then that the report of genetic effect via this type of damaging treatment is truly specific in providing singlet oxygen and in limiting the formation of other reactive oxygen species (Piette, 1991). Recently spermine has been reported to inhibit superoxide production in human neutrophils by suppressing activation of an NADPH oxidase (Ogata et al. 1996).

Recent studies of the roles of nitric oxide have revealed an array of additional complicating possibilities of oxidizing damage. The biological synthesis of the free radical nitric oxide, which contains an unpaired electron, has provided an active reactant with superoxide to form the stable toxic peroxynitrite anion,

$$O_2^- + NO\cdot \longrightarrow ONOO^- \quad \text{(peroxynitrite anion)}$$

Peroxynitrite can oxidize both nonprotein and thiols and indeed is 1000-fold more reactive with sulfhydryls than is H_2O_2. Radi et al. (1991) suggested that the formation of this oxidant has significant potentiality as a cytotoxic agent. The anion is produced in activated macrophages in amounts accounting for much of the NO produced, and its bactericidal properties have been described (Ischiropoulos et al., 1992a; Zhu et al., 1992). It is also highly lethal to *T. cruzi* (Denicola et al., 1993); nevertheless, *T. brucei* appears to be protected from NO in in vivo conditions (Mabbott et al., 1994).

On protonation, peroxynitrous acid decays rapidly to form the reactive oxidants, hydroxyl radical (HO·) and nitrogen dioxide (NO₂·); these can recombine slowly to form nitric acid (HNO_3). Superoxide dismutase is helpful in preventing the initial formation of peroxynitrite

and the numerous components of toxicity and detoxification. Additionally, peroxynitrite can act to nitrate tyrosine residues within various dismutases. In the presence of a dismutase and Fe^{3+}, phenolics, including tyrosine in other proteins, can be nitrated (Beckman et al., 1992; Ischiropoulos et al., 1992b). It is conceivable then that NO can tie up the essential iron of ribonucleotide reductase or inactivate the essential tyrosyl moiety of that enzyme by nitration. The interrelations of reactive O_2 species in specific pathological systems are being actively studied (Oury et al., 1992).

Exploitation of Oxidative Stress

It was suggested that the sensitivity of trypanosomal parasites to stimulated macrophages might arise from the induced generation of activated forms of oxygen. The role of the nitrogen oxides in the microbiocidal action of activated macrophages on some parasitic organisms was discussed by James and Hibbs (1990). However, some of these inhibitory effects in trypanosome proliferation are quite complex (Vincendeau et al., 1992).

Although lacking catalase or GSH peroxidase, *T. brucei*, *T. cruzi*, and *L. tropica* do possess an Fe-containing superoxide dismutase (Le Trant et al., 1983). It is supposed that the soluble sulfhydryl groups, mainly of GSH in most cells or of trypanothione in trypanosomes, are major players in the elimination of the potentially toxic reactive oxygen species. Compounds that increase the generation of these toxic products (e.g., Nifurtimox and gentian violet) overcome the ability of the trypanosome to detoxify. For example, gentian violet is a cationic dye added to blood banks in areas endemic for Chagas disease; visible light causes the formation of a reduced free radical whose autooxidation generates superoxide anion and H_2O_2 and kills viable parasitic forms (Docampo et al., 1988). The distinctive and active single mitochondrion of *T. cruzi* is a major target of the toxicity of the dye.

The results with Nifurtimox suggested that the effects on energy metabolism are sufficiently different from those on polyamine metabolism to encourage the use of combinations of DFMO and an inhibitor of energy metabolism. Jennings (1991b) reported the promising use of DFMO and known therapeutic antimonials, which are thought to be similar to arsenicals in their action in affecting redox systems.

Efforts to cultivate trypanosomes and other protozoan parasites in vitro have demonstrated the toxic effects of spermidine or spermine to cultures containing a serum amine oxidase generating H_2O_2, ammonia, and aminoaldehydes or dialdehydes. *Plasmodium falciparum* grown in human erythrocytes proved to be quite sensitive to purified aminoaldehydes and dialdehydes, rather than the H_2O_2 and NH_3, derived from spermidine and spermine by bovine serum amine oxidase (Morgan et al., 1986). Antileishmanial effects in mouse infections have been demonstrated by 4-aminobutyraldehyde and

1-(3-aminopropyl)-4-aminobutyraldehyde (Cona et al., 1991).

Approaching Chagas Disease and Leishmaniases

An initial study indicated the relative inefficacy of DFMO in a leishmanial infection. Nevertheless, some *Leishmania* were sensitive to anti-ODC drugs whose effects somehow had to be reinforced (Kaur et al., 1986). The increasing appearance of resistance to the pentavalent antimonials, the usual therapeutic agent for the widespread leishmanial disease kala azar, has compelled an extended search for improved medicinals.

In Chagas disease *T. cruzi*, which is transmitted by infectious feces from the blood-sucking triatomid insect vector, is the etiological agent and the cause of a major epidemic in South America and Latin America (16–18 million cases). Although macrophages may be the most heavily parasitized cells, muscle cells of a mammalian host are frequently heavily colonized and the heart muscle in the infected human is severely weakened (Morris et al., 1991). The striking degeneration of mitochondria in infected muscle raises the question of a possible relation to the unusual GSH metabolism of the protozoan.

Gutteridge and Rogerson (1979) discussed the basic life cycle of *T. cruzi* in which the mitochondrial substance and its metabolic capabilities are transformed in the shifts from an intracellular ellipsoidal amastigote, with abbreviated flagellum, to the elongated flagellated form in the mammalian bloodstream or insect vector. Although DFMO depletion of polyamine in *T. b. brucei* evokes a shift of the long slender bloodstream organism to the short stumpy form, which characteristically contains specific mitochondrial respiratory enzymes, a similar effect has not been found for *T. cruzi* as a result of treatment by DFMO. The unresponsiveness of *T. cruzi* to DFMO is explained by the finding that these epimastigotes scavenge diamines but cannot synthesize them. Fed putrescine is converted to spermidine, trypanothione, and spermine whereas ingested cadaverine forms aminopropylcadaverine, bis(aminopropyl)cadaverine, and the new homologue, homotrypanothione (Hunter et al., 1994).

Curiously, difluoromethylarginine (DFMA) inhibits invasion and growth of *T. cruzi* in mammalian cells and this effect is reversed by agmatine (Schwarcz de Tarlovsky et al., 1993; Yakubu et al., 1992). In *C. fasciculata*, at least, DFMA is converted to DFMO, and the reversal of inhibition by putrescine but not by agmatine suggests a primary effect of DFMO on ODC in the cells (Hunter et al., 1991). *T. cruzi* has an unusually rapid rate of uptake of putrescine, some 50-fold greater than that of some *Leishmania* (González et al., 1992). In any case, several polyamine analogues (e.g., N,N^1-bisbenzyl- and N,N^1-bisthiophene-) were found to limit the ability of *T. cruzi* to invade and multiply in rat heart myoblasts (Majumder & Kierszenbaum, 1993a,b). A similar effect was obtained with a specific irreversible inhibitor of AdoMetDC (Yakubu et al., 1993), thereby asserting an important role for the polyamines in the *T. cruzi* life cycle.

The analyses of ribosomal RNAs from various species of *Trypanosoma* and *Leishmania* have indicated close evolutionary relationships. The metabolic patterns of *T. cruzi* differ from *T. brucei* in several respects, although the organisms are similar in their synthesis of pyrimidines or salvage of purines (Fairlamb, 1989). Taking advantage of the latter property, allopurinol is used to block purine metabolism selectively in the parasite and is now under clinical trial in Chagas disease. Various similarities and differences in pathways leading to energy production among the trypanosomes, and particularly in forms adopted by *T. cruzi* in stages of its life cycles, have also been discussed by Gutteridge and Rogerson (1979).

T. cruzi and the *Leishmania* contain trypanothione and the panoply of intermediates and enzymes required for its biosynthesis and fulfillment of its roles in providing a redox balance and removal of peroxide. Various nitrofurans and quinonoid compounds can be reduced selectively by trypanothione disulfide reductase to form irreversible inactivators of the enzyme under anaerobic conditions (Henderson et al., 1988). Unexpectedly, strains of *T. cruzi* and *L. donovani* that overexpress a functional trypanothione reductase remain as sensitive as the parent strains to drugs like Nifurtimox that are thought to induce oxidative damage (Kelly et al., 1993). Thus, the enzyme does not appear to be rate limiting in the removal of H_2O_2. Although a trivalent antimonial does react with the purified enzyme from *L. donovani*, it appears that this complex is readily reversible (Cunningham & Fairlamb, 1995).

The isolation and characterization of the gene for the reductase from *T. cruzi* and the overproduction of the enzyme in *E. coli* have improved the yields of the protein, the development of crystals, and the analysis of the enzyme structure in the search for improved inhibitors and chemotherapeutic agents (Krauth–Siegel et al., 1993; O'Sullivan & Zhou, 1995; Zhang et al., 1993). Although Nifurtimox was found to be a stronger inhibitor of GSH reductase than of the trypanosomal reductase, some newly synthesized nitrofurans appear to be better inhibitors of the latter (Jockers–Scherubl et al., 1989). A nitrofuran named chinifur has been described as a selective inhibitor of trypanothione reductase (Cenas et al., 1994).

Leishmania species are members of the trypanosomatid genus responsible for the various forms of human leishmaniasis, producing about 12 million new cases a year in many parts of the world. A lethal visceral leishmaniasis, known as kala azar, as well as a cutaneous leishmaniasis, known as "oriental sore," are rampant in Asia and Africa. A cutaneous leishmaniasis, similarly transmitted by a sandfly, is well known in the Americas. There are many species well adapted to many mammals in regions of the world in which this insect vector exists (Lainson, 1983). The flagellated promastigote exists in the sandfly vector and is cultivated easily in vitro. Amastigotes, found in the infected macrophages of the mammalian host, have been isolated and cultivated earlier in peritoneal macrophages. A heat-induced conversion of promastigotes to amastigotes has recently been obtained, as has the in vitro cultivation of the latter (Al-Bashir et al., 1992).

Putrescine and spermidine were found in promastigotes grown on a complex liquid medium. Further [^{14}C]-putrescine was taken up into the cells and converted to spermidine (Bachrach et al., 1979a). Traces of spermine were reported, but this polyamine has not been found in organisms grown on carefully defined media (Kaur et al., 1986). The in vitro growth of the organisms was blocked by compounds such as ethidium and MGBG, which inhibited the uptake of putrescine and its conversion to spermidine. Concomitantly these drugs inhibited the synthesis of RNA and DNA at higher concentrations (Bachrach et al., 1979b). The growth rates were roughly linear functions of the ratio of putrescine to spermidine at different stages of growth of several leishmanial strains (Schnur et al., 1979).

The problem of oriental sore led to attempts to correlate the polyamine metabolism of skin with the presence of ODC and AdoMetDC. Bachrach et al. (1981) demonstrated the presence of ODC in promastigotes, as well as in amastigotes grown in mouse macrophages, and showed marked increases in putrescine and spermidine, as well as in ODC in infected mouse skin. The levels of the enzyme and the polyamines in the skin were suggested as possible markers to test the efficacy of chemotherapeutic agents.

DFMO and α-methylornithine were both cytostatic in the growth of leishmanial promastigotes in defined media (Kaur et al., 1986). The former depleted putrescine from the organisms as did Berenil, which was found to inhibit putrescine uptake noncompetitively. The uptake of putrescine in DFMO-treated *Leishmania* proved to be quite complex and to be activated by putrescine depletion, as had been observed even in mammalian cells (Balana–Fouce et al., 1991). *L. mexicana* promastigotes were not inhibited by DFMO in growth on rich media that supplied polyamine, such as putrescine or spermidine (González et al., 1991). Several *Leishmania* were also able to incorporate and concentrate exogenous AdoMet (1800-fold) (Avila & Polegre, 1993).

The discovery of leishmanial ODC posed the problem of its similarities to that of the *T. b. brucei* and the potential exploitability of its regulation, even as described for African sleeping sickness. Extracts of *L. mexicana* contained an active true ODC and were apparently devoid of arginine decarboxylase. This activity did not decay rapidly in the presence of cycloheximide and was comparable in this respect to the stable ODC of *T. b. brucei* (Mukhopadhyay & Madhubala, 1995b; Sánchez et al., 1989). Sánchez et al. (1995) also described the isolation and some unusual properties of the ODC and its apoenzyme. Nevertheless, despite the activity of DFMO on leishmanial ODC, clones have been found that were almost refractory to the inhibitor (Carrera et al., 1994).

Very high ODC levels (about 15-fold normal wild-type cells) were found in a strain of *L. donovani* that was highly resistant to DFMO; it produced high levels of intracellular putrescine but not of spermidine. This strain did not accumulate ornithine at a rate higher than that of the sensitive strain (Coons et al., 1990). Growth in the absence of DFMO resulted in a reversion of enzyme and putrescine levels to wild-type levels. Further selection of the resistant strain in the presence of DFMO resulted in an amplification of the ODC gene, which was cloned to reveal an open reading frame for a protein of 707 amino acids. The leishmanial ODC has an extra NH$_2$-terminal extension of 200 amino acids and lacks the lability-determining COOH-terminating sequence of the mammalian protein (Hanson et al., 1992). Monomers of the dimeric leishmanial enzyme do not dissociate readily and do not cross-reactivate *T. brucei* or mouse monomers, unlike the cross-reaction of the latter monomers (Osterman et al., 1994).

The search for replacements, even temporary ones, of the failing antimony derivatives, has resulted in two types of advance. The bis(benzyl)polyamine analogue, N,N^1-bis{3-[(phenylmethyl)amino]propyl}-1,7-diaminoheptane (discussed in Chap. 17), eliminates *L. donovani* amastigotes from infected macrophages and greatly suppresses the disease in infected animals (Baumann et al., 1990, 1991). Bis(benzyl) polyamine analogues also proved to block growth in vitro of *L. donovani* promastigotes (Mukhopadhyay & Madhubala, 1995a). The compounds are active in the absence of inhibitors of PAO. The compound is also effective against malaria produced by *Plasmodium falciparum* (Bitonti et al., 1989).

The complexity of the intermediary metabolism of *Leishmania* has been discussed by Blum (1993). Arginine particularly stimulates the respiration of *L. donovani* and is oxidized initially in extracts to γ-guanidinobutyramide + CO$_2$ (Bera, 1987). The former product is deaminated to γ-guanidinobutyrate, which is then converted to urea and γ-aminobutyric acid. It is of interest that this alternate route of arginine metabolism had only been seen in some fungi.

Kinetoplast DNA and DNA Topoisomerases

The kinetoplasts of trypanosomes contain a unique DNA that is on the order of 6–8% of the total DNA of *T. brucei* and 20–25% of the DNA of *T. cruzi*. The kinetoplast DNA (kDNA) molecules are circular duplexes and are interlocked (catenated), comprising mainly minicircles and a much smaller number of maxicircles. The latter are considered to be the equivalent of mitochondrial DNA, while the former have been implicated in the formation of the RNA sequences necessary for editing maxicircle transcripts. The problem of the mode of replication of these DNA tangles has heightened emphasis on the DNA topoisomerases, which can concatenate DNA circles or decatenate the interlocked circles. It was found fairly early that an enzyme, called *DNA gyrase*, or a *topoisomerase* derived from *E. coli*, which introduced superhelical turns into closed double-stranded DNA, was markedly stimulated by spermidine (Gellert et al., 1976). This enzyme was inhibited by a number of antibiotics, including novobiocin and nalidixic acid.

The "supercoiling" of closed circular strands involves the nicking of one strand of a duplex DNA, passing the one strand through the chain break around the other strand, and then closing the nick. Thus, many topoisomerases are described as "nicking–closing" enzymes. Such a mechanism and enzyme are essential for the replication of a closed circular DNA like mitochondrial DNA and have been found in animal mitochondria, as well as in nuclei (Fairfield et al., 1979). The mitochondrial enzyme is far more sensitive to ethidium and Berenil than is the nuclear enzyme. Castora et al. (1983) subsequently found that the replication of mitochondrial DNA was selectively inhibited by inhibitors of the bacterial topoisomerase.

The separation of the intertwined double strands of DNA is essential to replication, repair, recombination, and transcription. Many of these functions require ATP-dependent enzymes that can introduce negative superhelical turns into circular double-stranded DNA or relax supercoiled molecules. The formation of catenanes or interlocked duplex DNA rings by the *E. coli* DNA gyrase was described by Kreuzer and Cozzarelli (1980) who found requirements for ATP, Mg^{2+}, and spermidine. In this special function, the mechanism was thought to involve a double-strand break so that an intact DNA could pass through the gap. Type II topoisomerases have been distinguished from the type I enzyme by the ability of the former to break and reseal both strands of a DNA molecule; a type I enzyme can only nick and close a single strand. Catenation and decatenation is clearly a function of type II topoisomerases; DNA gyrase is a type II topoisomerase. The requirement for spermidine in catenation has been addressed by Krasnow and Cozzarelli (1982) who concluded that the stimulatory effect of the polyvalent cation, spermidine, is a result of the compaction of the DNA, whose aggregates then remain accessible to enzymatic action (Chap. 19).

Decatenation of kDNA by a micrococcal enzyme was demonstrated by Kayser et al. (1982) who then described a topoisomerase in *T. cruzi* that can catenate the circular virus SV40 DNA in the presence of ATP (Riou et al., 1982). Another purified ATP-independent enzyme (type I) of *T. cruzi* was activated by Mg^{2+} and spermidine and relaxed a supercoiled DNA (Riou et al., 1983). The catenation of SV40 DNA also required both Mg^{2+} and spermidine and the activities of the *T. cruzi* enzymes were sensitive to compounds that reacted with DNA, such as ethidium, chloroquine, and the ellipticines (Douc–Rasy et al., 1984).

This group also isolated an ATP-independent catenating enzyme, dubbed a topoisomerase II, from *T. cruzi* (Douc–Rasy et al., 1988). This activity required Mg^{2+}, was activated by spermidine, and appeared more sensitive to a number of antitopoisomerase inhibitors than its

mammalian counterpart. The last result suggested that this enzyme might be an antitrypanosomal target. An enzyme with these properties has also been isolated from *L. donovani* (Chakraborty & Majumder, 1991). A type I DNA topoisomerase has also been isolated from the nuclei of this *Leishmania* (Chakraborty et al., 1993).

Antitrypanosomal drugs such as Berenil or ethidium promote linearization of minicircles and lead to the formation of dyskinetoplastic trypanosomes (Shapiro & Englund, 1990). Suramin, whose antitrypanosomal action has been mysterious for many years, has recently been found to be quite inhibitory to a type II topoisomerase of yeast and animal cells (Bojanowski et al., 1992).

Forays into Malarial Disease

It is estimated that about 600 million people are infected by malaria-causing plasmodia and that malaria leads to the death of about a million children yearly. A time line of our understanding of this mosquito-borne epidemic has been presented by Brown and Nossal (1986), who pointed to the discoveries of the parasites in both human blood and the mosquito stomach in the 1890s. The numerous complexities of the life cycle in two hosts, the mosquito and humans, are presented in Figure 18.7.

Therapeutic chloroquine was developed in 1934, to be followed in a few years by the use of DDT as an insecticide. Widespread resistance to chloroquine of the highly pathogenic *P. falciparum* began to appear in the 1960s and the problem of chloroquine-resistant *P. falciparum* threatens to overwhelm the many populations harboring the mosquito and the parasite. *P. falciparum* contains an ODC that is found to be less active when infected erythrocytes are incubated in chloroquine (Königk & Putfarken, 1983). McCann et al. (1981a) described the effect of DFMO in inhibiting a single schizogonic cycle, reporting an apparent arrest at the trophozoite stage. The parasite was then unable to incorporate labeled hypoxanthine into nucleic acid and to divide mitotically. Gillet et al. (1983) found that DFMO inhibited the multiplication of *P. berghei* in the mosquito during a sporogonic cycle. Although essentially all ODC inhibitors will inhibit erythrocytic schizogony, only DFMO had any effect at all on that parameter in a mouse model of *P. berghei*. However, DFMO did not increase survival of the infected mice (Bitonti et al., 1987).

The inhibitory concentrations of DFMO in all of these studies, including that necessary to reduce polyamine content on the *P. falciparum*-infected human erythrocytes, appeared quite high at about 5–6 mM (Whaun & Brown, 1985). Addition of putrescine to DFMO-inhibited trophozoites permitted a resumption of growth of a now synchronized population (Assaraf et al., 1987a,b). DFMO was cytostatic in this organism in infected erythrocytes over a 2-day period, but a longer exposure killed the intracellular parasite concomitant with a massive accumulation of pigment. DFMO also blocked the synthesis of certain specific parasite pro-

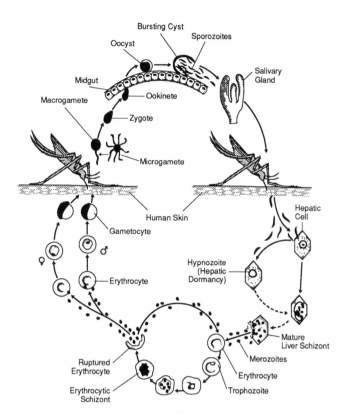

FIG. 18.7 Life cycle of the malaria parasite.

teins. This group purified the ODC of *P. falciparum* and, after labeling with radioactive DFMO, discovered it to be some 49 kDa, significantly smaller than the mammalian enzyme (Assaraf et al., 1988). The *P. falciparum* ODC may be sufficiently different from the host enzyme to permit the design of specific selective inhibitors.

Putrescine-dependent AdoMetDC has been found in *P. falciparum*; it is inhibited by MGBG ($K_i = 0.46 \mu M$) in competition with AdoMet (Rathaur & Walter, 1987). The crude enzyme in parasite extracts is inhibited irreversibly by 5′-{[(Z)-4-amino-2 butenyl]methylamino}-5′-deoxyadenosine, as indeed were the bacterial and trypanosomal enzymes. The compound blocks growth of the parasite in infected erythrocytes, and the inhibition is ameliorated by spermidine but not by putrescine. It also inhibits chloroquine-resistant and -sensitive strains (Wright et al., 1991).

Analogues of 5-methylthioribose, which block methionine biosynthesis in this and other protozoa, are cytocidal to *P. falciparum*. Thus, ethylthioribose enters mammalian cells and is not significantly toxic at 5 mM levels; however, the compound blocks growth of *P. falciparum* in erythrocytes with an IC_{50} of 80 μM (Sufrin et al., 1995). The control of methionine metabolism and levels of AdoMet and AdoHcy appears to be important to the survival of the parasite but may not involve the

polyamines directly. Potent inhibitors of AdoHcy hydrolase, such as the carbocyclic nucleoside antibiotic, neplanocin A, not only form new analogues of AdoMet, but also increase the concentrations of AdoMet and AdoHcy in infected red cells. The growth of the parasite is inhibited in such cells. However, although these cells were somewhat decreased in putrescine content, they were not decreased in spermidine and spermine.

The uptake of the bisbenzyl compounds occurs via a system quite distinct from the polyamine transport system (Byers et al., 1990). When applied to *P. falciparum* infected red cells, the bisbenzyl analogues were accumulated far better than the free amines and were much more inhibitory to the growth of the parasites. In combination with DFMO, these analogues cured mice infected with *P. berghei* (Bitonti et al., 1989). Nucleic acid synthesis of RNA and DNA was inhibited earlier and to a greater extent than protein synthesis. However, the effects of the bisbenzyl derivatives on growth were not reversed by exogenous polyamine and it was suggested that the antimalarial effects are not primarily due to an inhibition of polyamine biosynthesis.

Topoisomerases I and II were purified from *P. berghei* (Riou et al., 1986). Topoisomerase I was described as ATP independent. Mg^{2+} is not required in the relaxation of supercoiled DNA (SV40) but stimulated the reaction about 10-fold, as did spermidine. A DNA polymerase of *P. falciparum* was stimulated about fivefold by spermidine and spermine, and Bachrach and Abu–Elheiga (1990) suggested that the inhibition of DNA synthesis by DFMO in infected red cells was due to a depletion of the polyamines.

Polyamines and Vaginitis

Students of feminine hygiene are aware of the high incidence of vaginitis (180 million new cases per year) and have attempted to determine if the presence of the amines correlate with any particular etiologic agent. In one such study of the contents of vaginal swabs of a large cohort of college women, diamines (i.e., putrescine and cadaverine) were readily detected in 88% of women with nonspecific vaginitis or vaginitis due to *Trichomonas vaginalis*. The diamines were not found in similar swabs of 90% of women without such disease. Metronidazole, a compound used widely for anaerobic organism infections, eliminated diamines from the vaginal smears concurrently with the decline of the disease. Chen et al. (1982) proposed that the presence of the diamines, detected as dansyl derivatives on a thin-layer chromatography plate, be used diagnostically.

The parasitic flagellate *T. vaginalis*, grown in vitro in a complex medium, generates large amounts of putrescine and excretes this into the medium (White et al., 1983). The amount of the diamine and the ratio of putrescine to spermidine is in a range exemplified by *E. coli*, that is, it is quite different from the protozoa discussed in this chapter. Whereas the putrescine is derived from ornithine by the now familiar activity of ODC, the ornithine is produced from arginine via the arginine deiminase pathway (Linstead & Cranshaw, 1983). The organism lacks arginase, urease, and citrulline hydrolase and the arginine deiminase pathway involves a step that releases NH_3 and produces citrulline, which is degraded to ornithine by a catabolic ornithine carbamoyltransferase.

$$\text{arginine} + H_2O \longrightarrow \text{citrulline} + NH_3$$
$$\text{citrulline} + P_i \longrightarrow \text{ornithine} + \text{carbamyl phosphate}$$

$$\text{putrescine} \qquad ATP + H_2CO_3 + NH_3$$

Thus, the generation of putrescine from ornithine is tied to an energy-producing mechanism and possibly replaces the aerobic production of ATP, largely in the early steps of arginine metabolism. Various trypanosomatids isolated from parasitized plants have been found to lack arginase and to contain arginine deiminase, citrulline hydrolase, and ornithine carbamoyltransferase (Camargo et al., 1987).

T. vaginalis was subsequently found to have a higher content of ODC than some other trichomonads and to contain little ADC (North et al., 1986). The ODC was inhibited by DFMO; but this inhibitor, even at high concentrations, has little or no effect on the axenic growth rate. Nevertheless, treated cells had markedly lower lev-

els of ODC and some interference with ornithine decarboxylation could be demonstrated. The enzyme differed from that of *T. b. brucei* in molecular weight and pH optimum, was relatively nonregulated by amines, and possessed a long half-life (Yarlett et al., 1992, 1993b).

Although DFMO protected mammalian cells in culture from the cytocidal action of the protozoan and slowed the course of infection in mice, it did not cure intravaginal infections (Bremner et al., 1987).

When grown on a semidefined medium containing 10% horse serum, labeled putrescine, spermidine, and spermine were taken up but were not interconverted. Yarlett and Bacchi (1994) presented evidence for the existence of an intracellular putrescine antiport with extracellular spermine. AdoMetDC, spermidine synthase, and spermine synthase could not be detected, and inhibition by DFMO did not lead to the accumulation of dAdoMet (Yarlett & Bacchi, 1988b). Despite the presence of enzymes essential to AdoMet synthesis and methyl transfer, it appears that *T. vaginalis* is deficient in the synthesis of any but the diamines. The roles of spermidine, if any, in this organism will be of interest.

Concluding Remarks

Table 18.4 presents a summary of data on the polyamine contents of many of the protozoan parasites. The diversity of the patterns and their underlying enzymatic compositions has been described and underline the need for placing these data within the life-styles of the organisms. The apparent needs for polyamines by each of the parasites has been indicated by the responses to various inhibitors.

Knowledge of the polyamine requirements derives from considerations of both intellectual curiosity and clinical interest. The studies on the polyamines have achieved two extraordinary successes with great promise for the control of trypanosomal disease. In the first,

Table 18.4 Polyamine content of parasites.

Parasite	put	spd	spm	put/spd
Trichomonas vaginalis	38	3.5	1.3	10.9
Entamoeba histolytica	92	2.6	0.03	35.5
Giardia lamblia	10	9.6	0.8	1.0
Trypanosoma brucei	4	24.5	0	0.17
Trypanosoma rhodesiense	4	21.0	1.2	0.18
Leishmania donovani				
Promastigotes	35	37	0	0.9
Amastigotes	2	18.8	3.5	0.1
Trypanosoma cruzi	+	+	+	
Plasmodium falciparum[a]	0	33.0	8.0	0.3
Onchocerca volvulus	0.3	34.0	168.0	0.01

Put, putrescine; spd, spermidine; spm, spermine. Values are in nmol/mg protein. Summarized by Yarlett (1988).
[a]Values are in pmol/10⁶ parasitized erythrocytes.

an inhibitory agent directed against a mammalian ODC proved to have unexpectedly damaging effects that were selective on trypanosomes multiplying within a parasitized host. In a second, the surprising finding that spermidine is an essential component of an essential coenzyme has provided a major demonstration of the potential unity of the academic and clinical enterprises. The almost unique presence of trypanothione in the Trypanosomatidae and the specific enzymes that process it provide a key example of biochemical diversity. This example has established the possibility of controlling a parasite specifically by the exploitation of the chemical differences between the protein target within the parasite and similar but not identical host proteins.

REFERENCES

Aboagye–Kwarteng, T., Smith, K., & Fairlamb, A. H. (1992) *Molecular Microbiology*, **6**, 3089–3099.

Aisien, S. O. & Walter, R. D. (1993) *Biochemical Journal*, **291**, 733–737.

Al-Bashir, N. T., Rassam, M. B., & Al-Rawi, B. M. (1992) *Annals of Tropical Medicine and Parasitology*, **86**, 487–502.

Assaraf, Y. G., Abu–Elheiga, L., Spira, D. T., Desser, H., & Bachrach, U. (1987b) *Biochemical Journal*, **242**, 221–226.

Assaraf, Y. G., Golenser, J., Spira, D. T., Messer, G., & Bachrach, U. (1987a) *Parasitology Research*, **73**, 313–318.

Assaraf, Y. G., Kahana, C., Spira, D. T., & Bachrach, U. (1988) *Experimental Parasitology*, **67**, 20–30.

Avila, J. L. & Polegre, M. A. (1993) *Molecular and Biochemical Parasitology*, **58**, 123–134.

Bacchi, C. J., Berens, R. L., Nathan, H. C., Klein, R. S., Elegbe, I. A., Rao, K. V. B., McCann, P. P., & Marr, J. J. (1987b) *Antimicrobial Agents and Chemotherapy*, **31**, 1406–1413.

Bacchi, C. J., Garofalo, J., Ciminelli, M., Rattendi, D., Goldberg, B., McCann, P. P., & Yarlett, N. (1993) *Biochemical Pharmacology*, **46**, 471–481.

Bacchi, C. J., Garofalo, J., Mockenhaupt, D., McCann, P. P., Diekema, K. A., Pegg, A. E., Nathan, H. C., Mullaney, E. A., Chunosoff, L., Sjoerdsma, A., & Hutner, S. H. (1983a) *Molecular Biochemical Parasitology*, **7**, 209–225.

Bacchi, C. J., Garofalo, J., Santana, A., Hannan, J. C., Bitonti, A. J., & McCann, P. P. (1989) *Experimental Parasitology*, **68**, 392–402.

Bacchi, C. J., Hutner, S. H., Lambros, C., & Lipschik, G. Y. (1978) *Advances in Polyamine Research*, pp. 129–138, R. A. Campbell, D. R. Morris, D. Bartos, G. D. Daves, & F. Bartos, Eds. New York: Raven Press.

Bacchi, C. J., Lipschik, G., & Nathan, H. C. (1977) *Journal of Bacteriology*, **131**, 657–661.

Bacchi, C. J., Marcus, S. L., Lambros, C., Goldberg, B., Messina, L., & Hutner, S. H. (1974) *Biochemical and Biophysical Research Communications*, **58**, 778–786.

Bacchi, C. J., McCann, P. P., Nathan, H. C., Hutner, S. H., & Sjoerdsma, A. (1983b) *Advances in Polyamine Research*, pp. 221–231, U. Bachrach, A. Kaye, & R. Chayen, Eds. New York: Raven Press.

Bacchi, C. J., Nathan, H. C., Clarkson, Jr., A. B., Bienen, E. J., Bitonti, A. J., McCann, P. P., & Sjoerdsma, A. (1987a) *American Journal of Tropical Medicine and Hygiene*, **36**, 46–52.

Bacchi, C. J., Nathan, H. C., Hutner, S. H., Duch, D. S., &

Nichol, C. A. (1981) *Biochemical Pharmacology*, **30**, 883–886.

Bacchi, C. J., Nathan, H. C., Hutner, S. H., McCann, P. P., & Sjoerdsma, A. (1980) *Science*, **210**, 332–334.

Bacchi, D. J., Nathan, H. C., Hutner, S. H., McCann, P. P., & Sjoerdsma, A. (1982) *Biochemical Pharmacology*, **31**, 2833–2836.

Bacchi, C. H., Nathan, H. C., Yarlett, N., Goldberg, B., McCann, P. P., Bitonti, A. J., & Sjoerdsma A. (1992) *Antimicrobial Agents and Chemotherapy*, **36**, 2736–2740.

Bacchi, C. J., Nathan, H. C., Yarlett, N., Goldberg, B., McCann, P. P., Sjoerdsma, A., Sarić, M., & Clarkson, Jr., A. B. (1994) *Antimicrobial Agents and Chemotherapy*, **38**, 563–569.

Bacchi, C. J., Sufrin, J. R., Nathan, H. C., Spiess, A. J., Hannan, T., Garofalo, J., Alecia, K., Katz, L., & Yarlett, N. (1991) *Antimicrobial Agents and Chemotherapy*, **35**, 1315–1320.

Bacchi, C. J., Vergara, C., Garofalo, J., Lipschik, G. Y., & Hutner, S. H. (1979) *Journal of Protozoology*, **26**, 484–488.

Bachrach, U. & Abu–Elheiga, L. (1990) *European Journal of Biochemistry*, **191**, 633–637.

Bachrach, U., Abu–Elheiga, L., Talmi, M., Schnur, L. F., El-On, J., & Greenblatt, C. L. (1981) *Medical Biology*, **59**, 441–447.

Bachrach, U., Brem, S., Wertman, S. B., Schnur, L. F., & Greenblatt, C. L. (1979a) *Experimental Parasitology*, **48**, 457–463.

Bachrach, U., Brem, S., Wertman, S. B., Schnur, L. F., & Greenblatt, C. L. (1979b) *Experimental Parasitology*, **48**, 464–470.

Bailey, S., Smith, K., Fairlamb, A. H., & Hunter, W. M. (1993) *European Journal of Biochemistry*, **213**, 67–75.

Baillie, T. A. & Slatter, J. G. (1991) *Accounts of Chemical Research*, **24**, 264–270.

Balaña–Fouce, R., Escribano, M. I., & Alunda, M. (1991) *Molecular and Cellular Biochemistry*, **107**, 127–133.

Baumann, R. J., Hanson, W. L., McCann, P. P., Sjoerdsma, A., & Bitonti, A. J. (1990) *Antimicrobial Agents and Chemotherapy*, **34**, 722–727.

Baumann, R. J., McCann, P. P., & Bitonti, A. J. (1991) *Antimicrobial Agents and Chemotherapy*, **35**, 1403–1407.

Beckman, J. S., Ischiropoulos, H., Zhu, L., van der Woerd, M., Smith, C., Chen, J., Harrison, J., Martin, J. C., & Tsai, M. (1992) *Archives of Biochemistry and Biophysics*, **298**, 438–445.

Bellofatto, V., Fairlamb, A. H., Henderson, G. B., & Cross, G. A. M. (1987) *Molecular and Biochemical Parasitology*, **25**, 227–238.

Bera, T. (1987) *Molecular and Biochemical Parasitology*, **23**, 183–192.

Berger, B. J., Carter, N. S., & Fairlamb, A. H. (1993) *Acta Tropica*, **54**, 215–224.

Besser, H. V., Niemann, G., Domdey, B., & Walter, R. D. (1995) *Biochemical Journal*, **308**, 635–340.

Bitonti, A. J., Bacchi, C. J., McCann, P. P., & Sjoerdsma, A. (1985) *Biochemical Pharmacology*, **34**, 1773–1777.

———— (1986a) *Biochemical Pharamacology*, **35**, 351–354.

Bitonti, A. J., Byers, T. L., Bush, T. L., Casara, P. J., Bacchi, C. J., Clarkson, Jr., A. B., McCann, P. P., & Sjoerdsma, A. (1990) *Antimicrobial Agents and Chemotherapy*, **34**, 1485–1490.

Bitonti, A. J., Cross–Doersen, D. E., & McCann, P. P. (1988) *Biochemical Journal*, **250**, 295–298.

Bitonti, A. J., Dumont, J. A., Bush, T. L., Edwards, M. L., Stemerick, D. M., McCann, P. P., & Sjoerdsma, A. (1989)

Proceedings of the National Academy of Sciences of the United States of America, **86**, 651–655.

Bitonti, A. J., Dumont, A., & McCann, P. P. (1986b) *Biochemical Journal*, **237**, 687–689.

Bitonti, A. J., Kelly, S. E., & McCann, P. P. (1984) *Molecular and Biochemical Parasitology*, **13**, 21–28.

Bitonti, A. J., McCann, P. P., & Sjoerdsma, A. (1987) *Experimental Parasitology*, **64**, 237–243.

Blum, J. J. (1993) *Parasitology Today*, **9**, 118–122.

Bojanowski, K., Lelievre, S., Markovits, J., Couprie, J., Jacquemin–Sablon, A., & Larsen, A. K. (1992) *Proceedings of the National Academy of Sciences of the United States of America*, **89**, 3025–3029.

Boveris, A., Sies, H., Martino, E. E., Docampo, R., Turrens, J. F., & Stoppani, A. D. M. (1980) *Biochemical Journal*, **188**, 643–648.

Bowman, I. B. R. & Flynn, I. W. (1976) *Biology of the Kinetoplastida*, pp. 436–476, W. H. R. Lumsden & D. A. Evans, Eds. New York: Academic Press.

Bradley, M., Bücheler, U. S., & Walsh, C. T. (1991) *Biochemistry*, **30**, 6124–6127.

Bremner, A. F., Coombs, G. H., & North, M. J. (1987) *Journal of Antimicrobial Chemotherapy*, **20**, 405–411.

Brown, G. V. & Nossal, G. J. V. (1986) *Perspectives in Biology and Medicine*, **30**, 65–76.

Burger, R. M., Franco, R. J., & Drlica, K. (1994) *Antimicrobial Agents and Chemotherapy*, **38**, 886–888.

Byers, T. L., Bitonti, A. J., & McCann, P. P. (1990) *Biochemical Journal*, **269**, 35–40.

Byers, T. L., Bush, T. L., McCann, P. P., & Bitonti, A. J. (1991a) *Biochemical Journal*, **274**, 527–533.

Byers, T. J., Kim, B. G., King, L. E., & Hugo, E. R. (1991b) *Reviews of Infectious Diseases*, **13**, S378–S384.

Camargo, E. P., Silva, S., Roitman, I., de Souza, W., Jankevicius, J. V., & Dollet, M. (1987) *Journal of Protozoology*, **34**, 439–441.

Carrera, L., Balaña–Fouce, R., & Alunda, J. M. (1994) *Parasitology Research*, **80**, 203–207.

Carter, N. S. & Fairlamb, A. H. (1993) *Nature*, **361**, 173–175.

Castora, F. J., Vissering, F. F., & Simpson, M. V. (1983) *Biochimica et Biophysica Acta*, **740**, 417–427.

Cenas, N., Bironaite, D., Dickancaite, E., Anusevicius, Z., Sarlauskas, J., & Blanchard, J. S. (1994) *Biochemical and Biophysical Research Communications*, **204**, 224–229.

Chakraborty, A. K., Gupta, A., & Majumder, H. K. (1993) *Indian Journal of Biochemistry and Biophysics*, **30**, 257–263.

Chakraborty, A. K. & Majumder, H. K. (1991) *Biochemical and Biophysical Research Communications*, **18**, 279–285.

Chang, K., Steiger, R. F., Dave, C., & Cheng, Y. (1978) *Journal of Protozoology*, **25**, 145–149.

Chen, K. C. S., Amsel, R., Eschenbach, D. A., & Holmes, K. K. (1982) *Journal of Infectious Diseases*, **145**, 337–345.

Clarkson, Jr., A. B., Bacchi, C. J., Mellow, G. H., Nathan, H. C., McCann, P. P., & Sjoerdsma, A. (1983) *Proceedings of the National Academy of Sciences of the United States of America*, **80**, 5729–5733.

Cohen, S. S. (1979) *Science*, **205**, 964–971.

Cona, A., Federico, R., Gramiccia, M., Orsini, S., & Gradoni, L. (1991) *Biotechnology and Applied Biochemistry*, **14**, 54–59.

Coons, T., Hanson, S., Bitonti, A. J., McCann, P. P., & Ullman, B. (1990) *Molecular and Biochemical Parasitology*, **39**, 77–90.

Cunningham, M. L. & Fairlamb, A. H. (1995) *European Journal of Biochemistry*, **230**, 460–468.

Dahl, T. A., Midden, W. R., & Hartman, P. E. (1988) *Mutation Research*, **201**, 127–136.

de Gee, A. L. W., Carstens, P. H. B., McCann, P. P., & Mansfield, J. M. (1984) *Tissue and Cell*, **16**, 731–738.

de Gee, A. L. W., McCann, P. P., & Mansfield, J. M. (1983) *Journal of Parasitology*, **69**, 818–822.

Denicola, A., Rubbo, H., Rodriguez, D., & Radi, R. (1993) *Archives of Biochemistry and Biophysics*, **304**, 279–286.

Docampo, R., Moreno, S. N. J., Gadelha, F. R., De Souza, W., & Cruz, F. S. (1988) *Biomedical and Environmental Sciences*, **1**, 406–413.

Douc–Rasy, S., Kayser, A., & Riou, G. F. (1984) *EMBO Journal*, **3**, 11–16.

Douc–Rasy, S., Riou, J., Ahomadegbe, J., & Riou, G. (1988) *Biology of the Cell*, **64**, 145–156.

Drolet, G., Dumbroff, E. B., Legge, R. L., & Thompson, J. E. (1986) *Phytochemistry*, **25**, 367–371.

Eichler, W. (1989a) *Biological Chemistry Hoppe–Seyler*, **370**, 1113–1126.

——— (1989b) *Biological Chemistry Hoppe–Seyler*, **370**, 1127–1131.

——— (1989c) *Journal of Protozoology*, **36**, 577–582.

——— (1990) *Journal of Protozoology*, **37**, 273–277.

El-Waer, A., Douglas, K. T., Smith, K., & Fairlamb, A. H. (1991) *Analytical Biochemistry*, **198**, 212–216.

El-Waer, A. F., Smith, K., McKie, J. H., Benson, T., Fairlamb, A. H., & Douglas, K. T. (1993) *Biochimica et Biophysica Acta*, **1203**, 93–98.

Fairfield, F. R., Bauer, W. R., & Simpson, M. (1979) *Journal of Biological Chemistry*, **254**, 9352–9354.

Fairlamb, A. H. (1989) *Parasitology*, **99S**, 93–112.

Fairlamb, A. H., Blackburn, P., Ulrich, P., Chait, B. T., & Cerami, A. (1985) *Science*, **227**, 1485–1487.

Fairlamb, A. H., Carter, N. S., Cunningham, M., & Smith, K. (1992a) *Molecular and Biochemical Parasitology*, **53**, 213–222.

Fairlamb, A. H. & Cerami, A. (1985) *Molecular and Biochemical Parasitology*, **14**, 187–198.

——— (1992) *Annual Review of Microbiology*, **46**, 695–729.

Fairlamb, A. H., Henderson, G. B., Bacchi, C. J., & Cerami, A. (1987) *Molecular and Biochemical Parasitology*, **24**, 184–191.

Fairlamb, A. H., Henderson, G. B., & Cerami, A. (1986) *Molecular and Biochemical Parasitology*, **21**, 247–257.

——— (1989) *Proceedings of the National Academy of Sciences of the United States of America*, **86**, 2607–2611.

Fairlamb, A. H., Smith, K., & Hunter, K. J. (1992b) *Molecular and Biochemical Parasitology*, **53**, 223–232.

Ferrante, A., Abell, T. J., Robinson, B., & Lederer, E. (1987) *FEMS Microbiology Letters*, **40**, 67–70.

Gellert, M., Mizuuchi, K., O'Dea, M. H., & Nash, H. A. (1976) *Proceedings of the National Academy of Sciences of the United States of America*, **73**, 3872–3876.

Ghoda, L. Y., Savarese, T. M., Northup, C. H., Parks, Jr., R. E., Garofalo, J., Katz, L., Ellenbogen, B. B., & Bacchi, C. J. (1988) *Molecular and Biochemical Parasitology*, **27**, 109–118.

Giffin, B. F., McCann, P. P., & Bacchi, C. J. (1986a) *Molecular and Biochemical Parasitology*, **20**, 165–171.

Giffin, B. F., McCann, P. P., Bitonti, A. J., & Bacchi, C. J. (1986b) *Journal of Protozoology*, **33**, 238–243.

Gillet, J. M., Charlier, J., Boné, G., & Mulamba, P. L. (1983) *Experimental Parasitology*, **56**, 190–193.

González, N. S., Ceriani, C., & Algranati, I. D. (1992) *Biochemical and Biophysical Research Communications*, **188**, 120–128.

González, N. S., Sánchez, C. P., Sferco, L., & Algranati, I. D. (1991) *Biochemical and Biophysical Research Communications*, **180**, 797–804.

Grady, R. W., Bienen, E. J., & Clarkson, Jr., A. B. (1986) *Molecular and Biochemical Parasitology*, **21**, 55–63.

Guo, J., Wu, Y. Q., Rattendi, D., Bacchi, C. J., & Woster, D. M. (1995) *Journal of Medicinal Chemistry*, **38**, 1770–1777.

Guo, J. Q., Wu, Y. Q., Farmer, W. L., Douglas, K. A., Woster, P. M., Garofalo, J., & Bacchi, C. J. (1993) *Bioorganic and Medicinal Chemistry Letters*, **3**, 147–152.

Gutteridge, W. E. & Rogerson, G. W. (1979) *Biology of Kinetoplastida*, pp. 619–652, W. H. R. Lumsden & D. A. Evans, Eds. New York: Academic Press.

Hamana, K., Hamana, H., & Shinozawa, T. (1995) *Comparative Biochemistry and Physiology*, **1211B**, 91–97.

Hannan, J. C., Bacchi, C. J., & McCann, P. P. (1984) *Molecular and Biochemical Parasitology*, **12**, 117–124.

Hanson, S., Adelman, J., & Ullman, B. (1992) *Journal of Biological Chemistry*, **267**, 2350–2359.

Henderson, G. B., Fairlamb, A. H., & Cerami, A. (1987) *Molecular and Biochemical Parasitology*, **24**, 39–45.

Henderson, G. B., Murgolo, N. J., Kuriyan, J., Osapay, K., Kominos, D., Berry, A., Scrutton, N. S., Hinchliffe, N. W., Perham, R. N., & Cerami, A. (1991) *Proceedings of the National Academy of Sciences of the United States of America*, **88**, 8769–8773.

Henderson, G. B., Ulrich, P., Fairlamb, A. H., Rosénberg, I., Pereira, M., Sela, M., & Cerami, A. (1988) *Proceedings of the National Academy of Sciences of the United States of America*, **85**, 5374–5378.

Henderson, G. B., Yamaguchi, M., Novoa, L., Fairlamb, A. H., & Cerami, A. (1990) *Biochemistry*, **29**, 3924–3929.

Holm, B. & Emanuelsson, H. (1971) *Zeitschrift für Zelforschung*, **115**, 593–598.

Holm, B. & Heby, O. (1971) *Zeitschrift für Naturforschung*, **26B**, 604–606.

Holmgren, A. (1989) *Journal of Biological Chemistry*, **264**, 13963–13966.

Hua, S., Li, X., Coffino, Y., & Wang, C. C. (1995) *Journal of Biological Chemistry*, **270**, 10264–10271.

Hugo, E. R. & Byers, T. J. (1993) *Biochemical Journal*, **295**, 203–209.

Hunter, K. J., Le Quesne, S. A., & Fairlamb, A. H. (1994) *European Journal of Biochemistry*, **226**, 1019–1027.

Hunter, K. J., Strobos, C. A. M., & Fairlamb, A. H. (1991) *Molecular and Biochemical Parasitology*, **46**, 35–44.

Hunter, W. H., Bailey, S., Habash, J., Harrop, S. J., Helliwell, J. R., Aboagye-Kwarteng, T., Smith, K., & Fairlamb, A. (1992) *Journal of Molecular Biochemistry*, **227**, 322–333.

Hutner, S. H., Bacchi, C. J., & Baker, H. (1979) *Biology of the Kinetoplastida*, pp. 654–691, W. H. R. Lumsden & D. A. Evans, Eds. New York: Academic Press.

Imlay, J. A. & Fridovich, I. (1991) *Journal of Biological Chemistry*, **166**, 6957–6985.

Ischiropoulos, H., Zhu, L., & Beckman, J. S. (1992a) *Archives of Biochemistry and Biophysics*, **298**, 446–456.

Ischiropoulos, H., Zhu, L., Chen, J., Tsai, M., Martin, J. C., Smith, C. D., & Beckman, J. S. (1992b) *Archives of Biochemistry and Biophysics*, **298**, 431–437.

James, S. L. & Hibbs, Jr., J. B. (1990) *Parasitology Today*, **6**, 303–305.

Jennings, F. W. (1991a) *Tropical Medicine and Parasitology*, **42**, 139–142.

Jennings, F. W. (1991b) *Tropical Medicine and Parasitology*, **42**, 135–138.

Jockers–Scherübl, M. C., Schirmer, R. H., & Kranth–Siegel, R. L. (1989) *European Journal of Biochemistry*, **180**, 267–272.

Kaur, K., Emmett, K., McCann, P. P., Sjoerdsma, A., & Ullman, B. (1986) *Journal of Protozoology*, **33**, 518–521.

Kayser, A., Douc–Rasy, S., & Riou, G. (1982) *Biochimie*, **64**, 285–288.

Keithly, J. S. (1989) *Journal of Protozoology*, **36**, 498–501.

Kelly, J. M., Taylor, M. C., Smith, K., & Fairlamb, A. H. (1993) *European Journal of Biochemistry*, **218**, 29–37.

Khan, A. U., Massio, P. D., Medeiros, M. H. G., & Wilson, T. (1992b) *Proceedings of the National Academy of Sciences of the United States of America*, **89**, 11428–11430.

Khan, A. U., Mei, Y., & Wilson, T. (1992a) *Proceedings of the National Academy of Sciences of the United States of America*, **89**, 11426–11427.

Kim, B. G., McCann, P. P., & Byers, T. J. (1987a) *Journal of Protozoology*, **34**, 264–266.

Kim, B. G., Sobota, A., Bitonti, A. J., McCann, P. P., & Byers, T. J. (1987b) *Journal of Protozoology*, **34**, 278–284.

Kishore, P., Gupta, S., Srivastava, D. K., & Shukla, O. P. (1990) *Indian Journal of Experimental Biology*, **28**, 1174–1179.

Koguchi, K., Murakami, Y., & Hayashi, S. (1996) *Comparative Biochemistry and Physiology*, **113B**, 157–162.

Königk, E. & Putfarken, B. (1983) *Tropical Medicine and Parasitology*, **34**, 1–3.

Krasnow, M. A. & Cozzarelli, N. R. (1982) *Journal of Biological Chemistry*, **257**, 2687–2693.

Krauth–Siegel, R. L., Enders, B., Henderson, G. H., Fairlamb, A. H., & Schirmer, R. H. (1987) *European Journal of Biochemistry*, **164**, 123–128.

Krauth–Siegel, R. L., Jacoby, E. M., & Schirmer, R. H. (1995) *Methods in Enzymology*, **251**, 287–294.

Krauth–Siegel, R. L. & Schoneck, R. (1995) *FASEB Journal*, **9**, 1138–1146.

Krauth–Siegel, R. L., Sticherling, C., Jöst, I., Walsh, C. T., Pai, E. F., Kabsch, W., & Lantwin, C. B. (1993) *FEBS Letters*, **317**, 105–108.

Kreuzer, K. N. & Cozzarelli, N. R. (1980) *Cell*, **20**, 245–254.

Kuntz, D. A., Phillips, M. A., Moore, T. D. E., Craig, III, S. P., Bass, K. E., & Wang, C. C. (1991) *Molecular and Biochemical Parasitology*, **55**, 95–104.

Kuntz, I. D. (1992) *Science*, **257**, 1078–1082.

Kuriyan, J., Kong, X., Krishna, T. S. R., Sweet, R. M., Murgolo, N. J., Field, H., Cerami, A., & Henderson, G. B. (1991) *Proceedings of the National Academy of Sciences of the United States of America*, **88**, 8764–8768.

Lainson, R. (1983) *Transactions of the Royal Society of Tropical Medicine and Hygiene*, **77**, 569–596.

Le Trant, N., Meshnick, S. R., Kitchener, K., Eaton, H. W., & Cerami, A. (1983) *Journal of Biological Chemistry*, **258**, 125–130.

Linstead, D. & Cranshaw, M. A. (1983) *Molecular and Biochemical Parasitology*, **8**, 241–252.

Lougovoi, C. P. & Kyriakidis, D. A. (1989) *Biochimica et Biophysica Acta*, **996**, 70–75.

Lwoff, M. (1951) *The Biochemistry and Physiology of Protozoa*, pp. 129–176, A. Lwoff, Ed. New York: Academic Press.

Mabbott, N. A., Sutherland, I. A., & Sternberg, J. M. (1994) *Parasitology Research*, **80**, 687–690.

Maiorino, M., Chu, F. F., Ursini, F., Davies, K. J. A., Doroshow, J. H., & Esworthy, R. S. (1991) *Journal of Biological Chemistry*, **266**, 7728–7732.

Majumder, S. & Kierszenbaum, F. (1993a) *Molecular and Biochemical Parasitology*, **60**, 231–240.

———— (1993b) *Antimicrobial Agents and Chemotherapy*, **37**, 2235–2238.

McCann, P. P., Bacchi, C. J., Clarkson, A. B., Seed, J. R., Nathan, H. C., Amole, B. O., Hutner, S. H., & Sjoerdsma, A. (1981b) *Medical Biology*, **59**, 434–441.

McCann, P. P., Bacchi, C. J., Hanson, W. L., Cain, G. D., Nathan, H. C., Hutner, S. H., & Sjoerdsma, A. (1981a) *Advances in Polyamine Research*, pp. 97–110, C. M. Caldarera, V. Zappia, & U. Bachrach, Eds. New York: Raven Press.

McCann, P. P., Bacchi, C. J., Nathan, H. C., & Sjoerdsma, H. (1983) *Mechanisms of Drug Action*, pp. 159–173, T. P. Senger & R. N. Ondarza, Eds. New York: Academic Press.

Meister, A. (1991) *Pharmacology and Therapeutics*, **51**, 155–194.

———— (1994) *Journal of Biological Chemistry*, **269**, 9397–9400.

Moody, C. S. & Hassan, H. M. (1982) *Proceedings of the National Academy of Sciences of the United States of America*, **79**, 2855–2859.

Morgan, D. M. L., Bachrach, U., Assaraf, Y. G., Harari, E., & Golenser, J. (1986) *Biochemical Journal*, **236**, 97–101.

Morris, S. A., Tanowitz, H. B., Bilezikian, J. P., & Wittner, M. (1991) *Parasitology Today*, **7**, 82–86.

Moutiez, M., Meziane–Cherif, D., Aumercier, M., Sergheraert, C., & Tartar, A. (1994) *Chemical and Pharmaceutical Bulletin*, **42**, 2641–2644.

Mukhopadhyay, R. & Madhubala, R. (1995a) *Experimental Parasitology*, **81**, 39–46.

Mukhopadhyay, R. & Madhubala, R. (1995b) *International Journal of Biochemistry and Cell Biology*, **27**, 947–952.

Müller, S., Lüchow, A., McCann, P. P., & Walter, R. D. (1991) *Parasitology Research*, **77**, 612–615.

Müller, S. & Walter, R. D. (1992) *Biochemical Journal*, **283**, 75–80.

Nathan, H. C., Bacchi, C. J., Hutner, S. H., Rescigno, D., McCann, P. P., & Sjoerdsma, A. (1981a) *Biochemical Pharmacology*, **30**, 3010–3013.

Nathan, H. C., Bacchi, C. J., Sakai, T. T., Rescigno, D., Stumpf, D., & Hutner, S. H. (1981b) *Transactions of the Royal Society of Tropical Medicine and Hygiene*, **75**, 394–398.

North, M. J., Lockwood, B. C., Bremner, A. F., & Coombs, G. H. (1986) *Molecular and Biochemical Parasitology*, **19**, 241–249.

Ogata, K., Nishimoto, N., Uhlinger, D. J., Igarashi, K., Takeshita, M., & Tamura, M. (1996) *Biochemical Journal*, **313**, 549–554.

Ogata, K., Tamura, M., & Takeshita, M. (1992) *Biochemical and Biophysical Research Communications*, **182**, 20–26.

Oguro, T., Yoshida, T., Numazawa, S., & Kuroiwa, Y. (1990) *Journal of Pharmacobio-Dynamics*, **13**, 628–636.

Osterman, A., Grishin, N. V., Kinch, L. N., & Phillips, M. A. (1994) *Biochemistry*, **33**, 13662–13667.

O'Sullivan, M. C. & Zhou, Q. (1995) *Bioorganic and Medicinal Chemistry Letters*, **5**, 1957–1960.

Oury, T. D., Ho, Y., Piantadosi, C. A., & Crapo, J. D. (1992) *Proceedings of the National Academy of Sciences of the United States of America*, **89**, 9715–9719.

Penketh, P. G., Kennedy, W. P. K., Patton, C. L., & Sartorelli, A. C. (1987) *FEBS Letters*, **221**, 427–431.

Phillips, M. A., Coffino, P., & Wang, C. C. (1987) *Journal of Biological Chemistry*, **262**, 8721–8727.

Phillips, M. A., Coffino, P., & Wang, C. C. (1988) *Journal of Biological Chemistry*, **263**, 17933–17941.

Piette, J. (1991) *Journal of Photochemistry and Photobiology B–Biology*, **11**, 241–260.

Ponasik, J. A., Strickland, C., Faerman, C., Savvides, S., Karplus, P. A., & Ganem, B. (1995) *Biochemical Journal*, **311**, 371–375.

Pösö, H., Sinervirta, R., Himberg, J. J., & Jänne, J. (1975) *Acta Chemica Scandinavica*, **B29**, 932–936.

Priest, J. W. & Hajduk, S. L. (1994) *Journal of Bioenergetics and Biomembranes*, **26**, 179–191.

Radi, R., Beckman, J. S., Bush, K. M., & Freeman, B. A. (1991) *Journal of Biological Chemistry*, **266**, 4244–4250.

Rathaur, S. & Walter, R. D. (1987) *Experimental Parasitology*, **63**, 227–232.

Rathaur, S., Wittich, R. M., & Walter, R. D. (1988) *Experimental Parasitology*, **65**, 277–281.

Reed, D. J. (1990) *Annual Review of Pharmacology and Toxicology*, **30**, 603–631.

Riou, G., Gabillot, M., Douc–Rasy, S., & Kayser, A. (1982) *Comptes Rendus de l'Academie des Sciences, Paris*, **294**, 439–442.

Riou, G. F., Gabillot, M., Douc–Rasy, S., Kayser, A., & Barrois, M. (1983) *European Journal of Biochemistry*, **134**, 479–484.

Riou, J., Gabillot, M., Philippe, M., Schrevel, J., & Riou, G. (1986) *Biochemistry*, **25**, 1471–1479.

Sánchez, C. P., González, N. S., & Algranati, I. D. (1989) *Biochemical and Biophysical Research Communications*, **161**, 754–761.

Sánchez, C. P., Sidrauski, C., Freire, S. M., González, N. S., & Algranati, I. D. (1995) *Biochemical and Biophysical Research Communications*, **212**, 396–403.

Scandalios, J. G. (1993) *Plant Physiology*, **101**, 7–12.

Schaeffer, J. M. & Donatelli, M. R. (1990) *Biochemical Journal*, **270**, 599–604.

Schechter, P. J. & Sjoerdsma, A. (1986) *Parasitology Today*, **2**, 223–224.

Schirmer, R. H., Müller, J. G., & Krauth–Siegel, R. L. (1995) *Angewandte Chemie International Edition English*, **34**, 141–154.

Schnur, L. F., Bachrach, U., Greenblatt, C. L., & Joseph, M. B. (1979) *FEBS Letters*, **106**, 202–206.

Schwarcz de Tarlovsky, M. N., Hernandez, S. M., Bedoya, A. M., Lammel, E. M., & Isola, E. L. D. (1993) *Biochemistry and Molecular Biology International*, **30**, 547–558.

Shames, S. L., Fairlamb, A. H., Cerami, A., & Walsh, C. T. (1986) *Biochemistry*, **25**, 3519–3526.

Shames, S. L., Kimmel, B. E., Peoples, O. P., Agabian, N., & Walsh, C. T. (1988) *Biochemistry*, **27**, 5014–5019.

Shapiro, B. M. (1991) *Science*, **252**, 533–536.

Shapiro, T. A. & Englund, P. T. (1990) *Proceedings of the National Academy of Sciences of the United States of America*, **87**, 950–954.

Sharma, V., Tekwani, B. L., Saxena, J. K., Gupta, S., Katiyar, J. C., Chatterjee, R. K., Ghatak, S., & Shukla, O. P. (1991) *Experimental Parasitology*, **72**, 15–23.

Shim, H. & Fairlamb, A. H. (1988) *Journal of General Microbiology*, **134**, 807–817.

Shukla, O. P., Müller, S., & Walter, R. D. (1992) *Molecular and Biochemical Parasitology*, **51**, 91–98.

Smith, K., Borges, A., Ariyanayagam, M. R., & Fairlamb, A. H. (1995) *Biochemical Journal*, **312**, 465–469.

Smith, K., Opperdoes, F. R., & Fairlamb, A. H. (1991) *Molecular and Biochemical Parasitology*, **48**, 109–112.

Stadtman, E. R. (1992) *Science*, **257**, 1220–1224.

Steenkamp, D. J. (1993) *Biochemical Journal*, **292**, 295–301.

Sufrin, J. R., Meshnick, S. R., Spiess, A. J., Garofalo–Hannan, J., Pan, X.-Q., & Bacchi, C. J. (1995) *Antimicrobial Agents and Chemotherapy*, **39**, 2511–2515.

Sullivan, F. X., Sobolov, S. B., Bradley, M., & Walsh, C. T. (1991) *Biochemistry*, **30**, 2761–2767.

Sullivan, F. X. & Walsh, C. T. (1991) *Molecular and Biochemical Parasitology*, **44**, 145–148.

Tabor, H. & Tabor, C. W. (1975) *Journal of Biological Chemistry*, **250**, 2648–2654.

Tekwani, B. L., Bacchi, C. J., & Pegg, A. E. (1992a) *Molecular and Cellular Biochemistry*, **117**, 53–61.

Tekwani, B. L., Bacchi, C. J., Secrist, III, J. H., & Pegg, A. E. (1992b) *Biochemical Pharmacology*, **44**, 905–911.

Tekwani, B. L., Mishra, M., & Chatterjee, R. K. (1995) *International Journal of Biochemistry and Cell Biology*, **27**, 851–855.

Tsirka, S. A., Sklaviadis, T. K., & Kyriakidis, D. A. (1986) *Biochimica et Biophysica Acta*, **884**, 482–489.

Vickerman, K. (1976) *Biology of the Kinetoplastida*, p. 34, W. H. R. Lumsden & D. A. Evans, Eds. New York: Academic Press.

Villalba, E., Dorta, B., & Ramirez, J. L. (1985) *Journal of Protozoology*, **32**, 49–53.

Vincendeau, P., Daulouede, S., Veyret, B., Darde, M. L., Bouteille, B., & Lemesre, J. L. (1992) *Experimental Parasitology*, **75**, 353–360.

Wang, C. C. (1991) *Journal of Cellular Biochemistry*, **45**, 49–53.

Weller, D. L., Raina, A., & Johnstone, D. B. (1968) *Biochimica et Biophysica Acta*, **157**, 558–565.

Whaun, J. M. & Brown, N. D. (1985) *Journal of Pharmacology and Experimental Therapeutics*, **233**, 507–511.

Wheatley, D. N., Rasmussen, L., & Tiedtke, A. (1994) *BioEssays*, **1616**, 367–372.

White, E., Hart, J., & Sanderson, B. E. (1983) *Molecular and Biochemical Parasitology*, **9**, 309–318.

Wittich, R., Killian, H., & Walter, R. D. (1987) *Molecular and Biochemical Parasitology*, **24**, 155–162.

Wittich, R. M. & Walter, R. D. (1989) *Biochemical Journal*, **260**, 265–269.

——— (1990) *Molecular and Biochemical Parasitology*, **38**, 13–18.

Wright, P. S., Byers, T. L., Cross–Doersen, D. E., McCann, P. P., & Bitonti, A. J. (1991) *Biochemical Pharmacology*, **41**, 1713–1718.

Yakubu, M. A., Basso, B., & Kierszenbaum, F. (1992) *Journal of Parasitology*, **78**, 414–419.

Yakubo, M. A., Majumder, S., & Kierszenbaum, F. (1993) *Journal of Parasitology*, **79**, 525–532.

Yao, K. & Fong, W. (1987) *International Journal of Biochemistry*, **19**, 545–550.

Yao, K., Fong, W., & Ng, S. F. (1985) *Comparative Biochemistry and Physiology*, **80B**, 827–829.

Yarlett, N. (1988) *Parasitology Today*, **4**, 357–360.

Yarlett, N. & Bacchi, C. J. (1988a) *Molecular and Biochemical Parasitology*, **27**, 1–10.

——— (1988b) *Molecular and Biochemical Parasitology*, **31**, 1–10.

——— (1994) *Biochemical Society Transactions*, **22**, 875–878.

Yarlett, N., Garofalo, J., Goldberg, B., Ciminelli, M. A., Rugiero, V., Sufrin, J. R., & Bacchi, C. J. (1993a) *Biochimica et Biophysica Acta*, **1181**, 68–76.

Yarlett, N., Goldberg, B., Moharrami, M. A., & Bacchi, C. J. (1992) *Biochemical Pharmacology*, **44**, 243–250.

——— (1993b) *Biochemical Journal*, **293**, 487–493.

Yarlett, N., Quamina, A., & Bacchi, C. J. (1991) *Journal of General Microbiology*, **137**, 717–724.

Zhang, Y., Bailey, S., Naismith, J. H., Bond, C. S., Habash, J., McLaughlin, P., Papiz, M. Z., Borges, A., Cunningham, M., Fairlamb, A. H., & Hunter, W. N. (1993) *Journal of Molecular Biology*, **233**, 1217–1220.

Zhu, L., Gunn, C., & Beckman, J. S. (1992) *Archives of Biochemistry and Biophysics*, **298**, 452–457.

CHAPTER 19

Viruses

Historical Introduction

The purification of tobacco mosaic virus (TMV), vaccinia virus, and a bacteriophage in the 1930s opened a period of physical and chemical investigation. Viruses that damaged cells were recognized most readily, and the formation of discolored or necrotized tissue or lysed cells (i.e., plaques) was used as a quantitative assay for this biological activity. Viruses were indeed tiny enough to be "filterable" but large enough to be observed directly by the early electron microscopes. Many plant viruses were crystallizable in 2- and 3-dimensional arrays, of which the individual particles were sufficiently homogeneous to fulfill the criteria of "molecularity." This stage of study also demonstrated the presence of protein and of only a single nucleic acid, RNA or DNA, in these infectious agents and the RNA of TMV was shown to be much larger than a tetranucleotide (Cohen & Stanley, 1942). The newly isolated viruses were relatively limited in their biochemical composition and for the most part lacked metabolically active enzymes. These deficiencies defined the metabolic dependence of a virus on its host. This level of knowledge facilitated the preparation of antiviral vaccines. However, an obligate parasitism and metabolic dependence could and did obscure the classification of some etiological agents. Thus, the Rickettsiae were thought of initially as viruses, rather than as metabolically dependent Gram-negative bacteria.

The antibacterial agents were ineffectual in viral disease and the treatment of the latter became a challenge, pointing to ignorance of the nature of virus multiplication. In the early 1940s the biology of the multiplication of several DNA-containing bacteriophages was dissected and a methodology was devised to demonstrate various stages of an infectious cycle: adsorption, virus multiplication (latent period), and lysis. The study of virus multiplication was emphasized as a cellular discipline, which had thereby become amenable to biochemical study. However, at this time the sole system that could be examined easily in biochemical terms was that of the infection by the bacteriophage of the nonpathogenic bacterium, *Escherichia coli*. The simple conditions of mixing to promote adsorption of one or many virus particles, dilution to stop adsorption while multiplication continued, mixing with uninfected cells, and plating to facilitate the search for necrotizing or lysing centers were readily adaptable to the examination of physical and chemical effects in a single cycle of virus multiplication. The changes of substance in multiply-infected bacteria were similarly determinable.

The biochemical study of this system soon demonstrated properties relating to energy production and the syntheses of viral and host nucleic acids and proteins. The relatively new biochemical techniques involving the exploitation of isotopes, new chromatographic methods, ultraviolet spectrophotometry, etc., provided an enlarged vision of the multiplication of the T-even bacteriophages, T2, T4, and T6. These bacteriophages proved to be extraordinarily lethal to their host bacteria. Infection turned off many host functions, including the synthesis of bacterial DNA, ribosomes, and many other components.

However, intact infectious virus particles could not be isolated from within an infected bacterium early in infection. It was suggested by Northrop in 1951 that viral DNA was transferred to the bacterium by the syringelike virus. Hershey and Chase (1952) found that in infection the viral coat was left on the exterior of the bacterium and that the viral chromosome, which included the DNA of the phage as a major component, was transferred to the cellular interior. Efforts to define the viral chromosome and the nature of transferred viral components led to the discovery of the polyamines in viruses (Hershey, 1957). The acid-soluble cationic components of the virus (substance A) were derived from labeled arginine and were not essential to the genetic role of the transferred viral chromosome. The components of substance A were subsequently demonstrated to be putrescine and spermidine (Ames et al., 1958).

In 1952, T-even viral DNA had been found to contain a new pyrimidine, and the analysis of its origin revealed that viruses did not merely redirect the metabolic machinery of the host. Viral insertion of its own DNA into the host introduced new viral genes, many of which determined the synthesis of new, often virus-essential enzymes and other proteins (Cohen, 1968). The T-even phage-infected cell, containing a metabolic apparatus determined originally by the bacterial DNA, was now directed after infection by the viral genome and had also acquired new metabolic capabilities. Some years later it became clear that another phage, grown in a Pseudomonad, contained a DNA in which putrescine was attached covalently to a viral pyrimidine. The biosynthesis of the new putrescinyl pyrimidine required new biochemistry. Further, the existence of this uniquely polyamine-modified DNA posed the problem of the function of this modification. In this context the discovery of the presence of several "minor" components derived from [^{14}C]-arginine in the T-even phage and in phage-infected cells posed a series of biochemical and genetic problems.

1. Are polyamines synthesized in virus-infected cells by bacterial enzymes or by virus-induced systems?
2. Is their synthesis tied to the synthesis of other viral components, such as viral nucleates?
3. Are the polyamines essential to or do they participate in viral multiplication?

A simple hypothesis concerning the role of the polyamines is that the compounds merely participate nonspecifically in the neutralization of viral nucleates and are entrapped with these viral components in the packaged viral protein shells. Tests of this hypothesis were made by depletion of the cellular polyamines with the aid of inhibitors or in polyamine-requiring mutant cells. Because polyamine depletion often severely affects the synthesis of the nucleic acids or proteins, the inhibition of viral synthesis by such depletion does not clearly address the question of a specific role of the polyamines in virus multiplication. Certain mutant strains of *E. coli* may be markedly depleted to 1% of their normal unusually high content of putrescine, but this may not be sufficient to demonstrate a requirement. Indeed under certain circumstances *E. coli* can be stimulated with polyamines derived from the normally reserved cadaverine or by scavenged spermine.

The discovery of the polyamines in the bacteriophages also raised the question of their possible presence in other viruses, all of which contain DNA or RNA. The amines are in fact found in some viruses and are absent or almost absent in others. Some of the viruses in which they are found infect bacteria, animal cells, and plant cells, all of which contain and synthesize polyamines normally. Thus, some models selected to permit the examination of two interacting genetic systems and the survival and development of the viral genome, may be too complex to answer certain apparently simple biochemical questions.

Bacterial Viruses

Polyamines of T-even bacteriophages

These viruses contain a single large molecule of a linear double-stranded DNA, encoding about 200 genes, within a head structure comprising proteins. The growing information on the T-even phages activated study of other phages active on *E. coli* including many other viruses with double-stranded DNA (e.g., T5, T3, and T7) and the lysogenic λ, single-stranded circular DNA (e.g., φX174) and RNA (e.g., R17, MS2). The finding of polyamines in the T-even group led to searches among the others like λ and R17 and tests of the dependence of some of these phages on the polyamines. The adsorption of T3 to *E. coli* was inhibited by spermine (Reiter, 1963) and the dialdehyde of spermine inactivated the DNA of T5, which could nevertheless inject this nonfunctional genetic component (Bachrach & Leibovici, 1966). A phage infectious at 65°C to an extreme thermophilic bacterium was stabilized by broth or 1 mM spermidine (Sakaki & Oshima, 1975). Spermidine and spermine had a variety of effects on phages of various bacterial hosts (Dasgupta & Chakravorty, 1978; Verma, 1989). Thus, the panoply of *E. coli* phages has been a major focus for the study of polyamine effects.

A packaged DNA is transferred through the adsorbing tail appendage from the virus to the host bacterium. Infection with these viruses causes an elimination of many biosynthetic host reactions. Nevertheless, polyamines are made or incorporated in uninfected and infected cells and may be found in viral progeny. The "normal" T-even virus particles are osmotically impermeable, but osmotically permeable O mutants can be obtained that readily exchange their polyamine. Although polyamine-depleted O mutants are active in virus multiplication, they have not been tested on depleted cells. Nor has their efficiency of infection been compared with that of their highly efficient parents: the various steps of adsorption, DNA transfer, and DNA activation of the O mutants in depleted and undepleted cells.

Injected virus DNA causes the sequential synthesis of various specific enzymes and viral structural proteins in infected bacteria (Cohen, 1968; Guttman & Kutter, 1983). Some of these newly synthesized enzymes are essential to an enlarged DNA synthesis, deoxyribonucleotides, including the novel pyrimidine nucleotide, 5-hydroxymethyldeoxycytidylate, found almost uniquely in these phages. Another virus-induced enzyme was that of thymidylate synthetase, which supplements the bacterial enzyme and helps to increase DNA synthesis in infection. Thus, a preexisting bacterial compound may be made via a new enzyme protein determined by the virus. The synthesis of a common bacterial polyamine during virus multiplication need not imply host determination by host systems exclusively.

It is possible that *E. coli*-based systems are peculiarly unsuitable to explore certain polyamine problems during virus multiplication. As discussed in Chapter 7, *E. coli* has an unusual polyamine metabolism characterized by a very active synthesis and accumulation of putrescine, which is often excreted, and by a sequestering of spermidine as *N*-acetyl or monoglutathionyl derivatives in certain types of stress. *N*-Acetylspermidine is also excreted. Nevertheless, some of the enzymes or nucleates active in viral functions are stimulated markedly in in vitro systems by polyamines. It is of interest to determine if the polyamines participate in multiplication of these viruses and if their participation is assured, as is the availability of thymidylate, by a viral genetic determinant.

Hershey (1957) asked if the virus particles contained acid-soluble components in addition to viral DNA and discovered two groups of such materials. One of these, labeled by [^{14}C]-lysine, was a polypeptide comprising about 1% of the total carbon of T2; a second, the low molecular dialyzable substance A, is equal to about 1.5% of the viral carbon and was derived specifically from ^{14}C-arginine. Substance A did not yield arginine or other structures on acid hydrolysis. Although isotopically labeled substance A is transferred unchanged with DNA from the parental to progeny phage, it is not known if A is transferred in toto to early progeny or if it dissociates and equilibrates with an A pool that is depleted during phage multiplication. In any case, A can be separated from viral DNA and protein by dialysis and was shown to contain two arginine-labeled components. Substance A can be made in the presence of chloramphenicol, which blocks protein synthesis and virus multiplication, or can be assimilated from the medium and competes with substance A transferred from the virus in an infection. Substance A was thought to be a normal bacterial constituent, because it is assimilated into uninfected bacteria as an acid-soluble constituent that appears in the phage after infection. Hershey noted that substance A could partially neutralize the negative charge in viral DNA and provided one carbon atom for two atoms of DNA phosphorus.

Ames et al. (1958) identified the two components of substance A as putrescine and spermidine and noted the nonexchangeability of soluble exogenous putrescine with that in the virus particle. The only phages that contained spermine had been grown in organisms in media containing this amine. Ames and Dubin (1960) attempted a complete analysis of the cation content of the T4 phage and reported that detectable cations accounted for about 82–86% of phage phosphate (Table 19.1). Of this total, the polyamine contents of T4 phage incubated in Mg^{2+} or Ca^{2+} amounted to 325 and 377 meq per equivalent P, respectively.

Bachrach et al. (1975) detected small amounts of polyamine in T5, which contains a large double-stranded linear DNA, and in φX174, with a small single-stranded circular DNA. The levels of spermidine, putrescine, ornithine decarboxylase (ODC), and DNA all increase in bacteria infected with these phages.

The T4 phage used was not freely permeable to divalent cations, in contrast to permeable O mutants resis-

Table 19.1 Cation content of T4 phage.

Cation	Phage incubated in	
	MgCl$_2$	CaCl$_2$
Putrescine^{2+}	250	290
Spermidine^{3+}	75	87
Ca^{2+}	<2	96
Mg^{2+}	340	360
Na$^+$	90	30
K$^+$	60	<30
Total	815	863

Adapted from Ames and Dubin (1960). Values are in meq/eq P.

tant to osmotic shock. A washed O mutant was very low in polyamine and could bind spermidine when incubated in the amine. Other permeable phages, such as T3, or the P22 phage grown on *Salmonella typhimurium*, were also low in amine. Some other viruses of plant and animal origin were almost devoid of amine. It was shown also that when bacteria incorporate spermine during growth and are then infected, the amine composition of the phage mirrors the amine composition of the bacteria at the time of infection, including spermine and its acetyl derivatives. The growth of *E. coli* in a tryptone broth with low aeration resulted in an extensive replacement of putrescine by cadaverine derived from lysine. Progeny T4 grown in *E. coli* in such a medium under relatively anaerobic conditions were very high in cadaverine and low in putrescine (Astrachan & Miller, 1973).

When virus DNA was isolated from a T2 or T4 phage by the phenol method carried out at low ionic strength, over 90% of the polyamine remained in the aqueous DNA-containing phase. Ethanol may reprecipitate the polyamine with the nucleate. This widely used preparatory method does not necessarily remove the polyamines that may have been added to activate the enzymes used to prepare polynucleotides.

Synthesis of T-even viral polymers and polyamine biosynthesis

The view had been adopted that the polyamines were not "essential" components of the chromosomes of T2, T4, or T6, relating "essentiality" exclusively to the transfer of genetic information. It was asked if the synthesis of polyamines in infected cells related to the synthesis of specific viral constituents. The study of the T-even bacteriophages and their multiplication has permitted a breakdown of various phases of infection. This dissection, applicable to other double-stranded DNA phages, may be outlined as follows:

1. adsorption and penetration of viral DNA;
2. synthesis of early mRNA and early proteins;
3. synthesis of viral proteins and viral DNA and the turnoff of early syntheses;

4. packaging of DNA into the viral head;
5. assembly of heads, tails, and intact virus; and
6. lysis.

The stimulatory or inhibitory effects of the polyamines have been placed in one or another stage of this development, but it has been difficult to establish the modes of action of these cations.

The availability of mutants of *E. coli* unable to synthesize compounds essential to the synthesis of protein, RNA, and DNA were studied very early, as well as the responses of these strains in virus infection (Cohen & Raina, 1967). It was found that cultures of a stringent *E. coli* strain 15TAU depleted of arginine continued to synthesize large amounts of putrescine, which was almost entirely excreted into the medium, and to make essentially normal amounts of spermidine also, much of which was also excreted. In this stringent system the elimination of a component essential for protein synthesis also blocked the accumulation of ribosomal RNA but did not deter the production of polyamine. Thymine deficiency prevented DNA synthesis and permitted some synthesis of RNA and protein for an hour only. After that time, putrescine synthesis and excretion continued whereas spermidine accumulated in the cells and medium for more than an additional hour. It appeared that the synthesis of the bacterial polyamine was not inhibited by a block in the synthesis of protein or the nucleates. However, the accumulation of spermidine within the bacteria paralleled the accumulation of cell RNA.

Because a T-even phage infection prevents the accumulation of ribosomal RNA, this phage infection provides an excellent model to determine the specific role of such RNA synthesis for polyamine accumulation. T-even phage-infected *E. coli* actively synthesize putrescine and spermidine at rates not markedly different from that of the uninfected cell. Figure 19.1 presents similar high rates of synthesis of these amines in normal and aborted infections, in which the latter is marked by the virtual absence of DNA synthesis (Dion & Cohen, 1972a). In the T4 amber mutant N122 aborted system, polyamine synthesis continues and the rates of polyamine syntheses are not affected by the inability to synthesize and accumulate either RNA or DNA.

Figure 19.1 does reveal a brief lag (ca. 10 min) in the synthesis of the polyamines. This may relate to a brief interval of leakage and membrane repair detected in studies of cation fluxes after infection (Silver et al., 1968). Buller and Astrachan (1968) suggested a role of spermidine in maintaining the cell envelope.

Kuhn and Kellenberger (1985) studied the effects of various phage infections on cation concentrations within the bacteria. T4 and λ phages caused transient reductions (5 min) in K^+, Mg^{2+}, putrescine, and spermidine concentrations. Neither T3 nor T7 had any appreciable effects of this type. The results also indicated that T4, T3, and T7 can multiply over a considerable range of internal ion concentrations.

In addition to the mutants, such as N122, that block

DNA synthesis almost completely, there are DD or DNA-delayed mutants, such as the T4 amber mutant 116, in which the delay in synthesis of DNA is significantly reduced by Mg^{2+} (Guttman & Bagley, 1969) or by spermidine (Dion & Cohen, 1971). The mutant T4 genes of the DD viral strains have been found to relate to subunits of a T4-induced topoisomerase, which may replace or supplement topoisomerase in the initiation of DNA function at the host membrane (Stetler et al., 1979). Thus, the polyvalent cations may be depleted as a result of certain leakage and lack of repair membrane defects, and appear to participate in the initiation of DNA synthesis, perhaps at the membrane. An early doubling of ODC activity was found in the first 6 min of T2 infection, followed by a decline to the original rate (Bachrach & Ben-Joseph, 1971).

Synthesis of viral polymers in polyamine-depleted *E. coli* mutants

One of the first polyamine-requiring mutants of *E. coli* was relatively complex and blocked in growth by the addition of arginine (Maas et al., 1970). Arginine inhibits the conversion of glutamate to ornithine. This organism contained an arginine decarboxylase but was unable to cleave the product agmatine to putrescine. Inhibited cells grow but are unable to divide; they become filamentous. In the presence of inhibitory arginine, either putrescine or spermidine permitted growth and divisions. Lysine also permitted a significant rate of growth and division; this was found to be converted extensively to cadaverine and the then new natural spermidine analogue, aminopropylcadaverine (Dion & Cohen, 1972b). The mutant cells depleted of putrescine and spermidine were infected with T4 in the arginine-containing medium in the presence or absence of putrescine or spermidine. The inhibited cells began a very slow rate of DNA synthesis quite late, after 60 min, whereas infected cells to which putrescine or spermidine had been added began DNA synthesis at 15 min. The rate of synthesis in the supplemented cells approached that of uninhibited cells. These early experiments indicated a marked stimulation of virus multiplication by polyamine in a polyamine-depleted mutant bacterium.

Experiments with bacterial mutants deficient in specific enzymes of polyamine synthesis were extended. Strains of *E. coli* were constructed with deletion mutations in speA (arginine decarboxylase), speB (agmatine ureohydrolase), speC (ODC), and speD [*S*-adenosylmethionine decarboxylase (AdoMetDC)] (Hafner et al., 1979). These organisms, depleted in putrescine and spermidine, grew in the absence of exogenous polyamine at a doubling time one-third that in supplemented media. However, the mutant strains did synthesize small amounts of cadaverine; the amount of this substance increased in the presence of lysine. Such cells supported a slow rate of multiplication of T4 and T7 (Fig. 19.2); the levels and rate of growth were markedly increased by the presence of exogenous polyamines. The data would be clearer if comparisons were presented of the

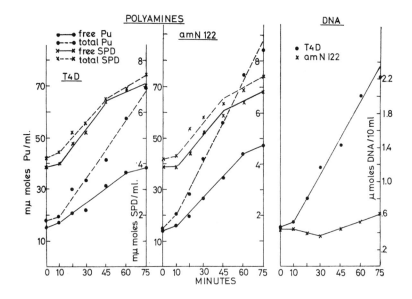

FIG. 19.1 Polyamine levels and DNA synthesis in *E. coli* infected with T4D or T4amN122; adapted from Dion and Cohen (1972a).

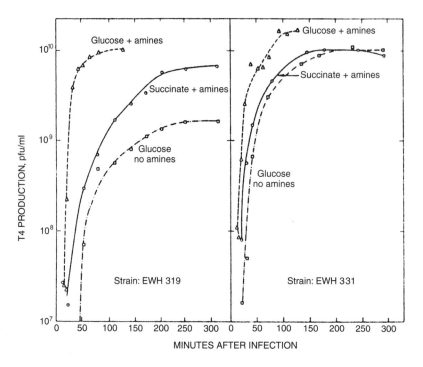

FIG. 19.2 Production of bacteriophage T4 in absence of detectable putrescine and spermidine; adapted from Hafner et al. (1979). *E. coli* strains EWH319 and EWH331 are mutants deficient in genes determining biosynthesis of these compounds and were infected by phage grown in one of these bacteria in the absence of exogenous amines.

one-step growth curves of virus production and of the syntheses of DNA in the infected cells. The T4 virus particles isolated after multiplication in "amine-free" organisms and media contained <1% of the putrescine and spermidine found in normal virus. However, the activities of such particles, adsorbability, transferability of DNA, etc., in normal and depleted cells were not reported. The induced multiplication of lysogenic DNA-containing λ phage was much more severely limited in these polyamine-depleted strains.

An additional deficiency was then imposed on these organisms, that of the inability to synthesize cadaverine from lysine (Tabor et al., 1980). This new depleted strain grew in the absence of added amine at one-tenth the rate of that in the presence of putrescine and spermidine. It was reported to serve as a host for T4 and T7, albeit with even greater reductions of phage yield.

The further development of these mutants via the isolation of a streptomycin-resistant strain resulted in a new organism with an absolute requirement for polyamines for growth (Tabor et al., 1981). The new mutation was believed to affect translation in a ribosomal complex and polyamines were essential to this function. This line of thought led to experiments in which the growth of *amber* mutants of T7 in an *E. coli* suppressor of amber mutations was shown to be dependent on polyamines for the optimal translation of the amber codon (Tabor & Tabor, 1982). The studies on phage multiplication in polyamine deficiency had turned up new phenomena relating the cellular amines to protein synthesis. The role of polyamines in determining the structure and function of suppressor tRNA is discussed in Chapter 23.

The experiments on T-even phage production in polyamine-depleted cells have begun to look like the half-full, half-empty glass. One group might say that the polyamines are relatively important in T-even phage multiplication, another that the compounds are relatively unimportant. The problem has become that of defining "importance." The experimental determination of essentiality in growth or no-growth situations has tended to confer biochemical importance to essential functions and has tended to neglect the significance of biological selection dependent on minor or major growth advantage.

Synthesis of phage mRNA

It is customary to consider the events after the transfer of a virus to its host in terms of the sequence

gene transcription → translation of mRNA →
                                viral-induced or viral proteins.

Viral infection adds the problem of determining if these activities are functions of host enzymes or of viral-induced functions. An early sequential production of mRNA molecules indicates that the host enzyme (the DNA-dependent RNA polymerase) is involved in the earliest production of mRNA in a T-even and T5 phage

infection. Some modifications are made in this complex enzyme shortly before DNA replication, as well as its inception (Rabussay, 1983). Some spermidine effects on the synthesis of RNA by the *E. coli* enzyme will be discussed in Chapter 22.

In contrast, it appears that phage λ uses the unchanged host RNA polymerases throughout infection whereas some phages like T3 and T7 induce new RNA polymerases, and some animal DNA viruses like vaccinia virus carry their own DNA-dependent RNA polymerase.

Phages T3 and T7 provide the classical intermediate, in which the earliest genes to be transcribed use the host enzyme and then induce the formation (via transcription and translation) of a DNA-dependent RNA polymerase that has the major role in transcribing the late T3 or T7 genes that determine the viral structural proteins. It has been observed that a polypeptide synthesis directed by a viral RNA in a cell-free extract derived from *E. coli* was markedly stimulated by spermidine or spermine at low Mg^{2+} concentrations (Takeda, 1969). Extending this, it was found that the synthesis in protein-synthesizing in vitro systems of various virus-induced enzymes, such as these RNA polymerases, were markedly stimulated by spermidine (Fuchs & Fuchs, 1971; Herrlich et al., 1971). The optimal mixes of spermidine and Mg^{2+} in the syntheses of the T3 and T7 enzymes were also determined (Fuchs, 1976). The DNA-dependent T7 RNA polymerase and 2–4 mM spermidine are widely used in experimental transcriptions.

Early mRNA and hydrolysis of AdoMet in T3 infection

An unusual enzyme is produced in T3 infections (Gefter et al., 1966). T3 DNA lacks methyl groups, which probably derive from AdoMet. An extract of T3-infected cells contains an enzyme that destroys AdoMet in the reaction AdoMet → methylthioadenosine + homoserine. The activity of this early enzyme is maximal in 8 min. The mRNA for the AdoMet cleavage enzyme (AdoMet hydrolase) is synthesized early and can be synthesized from viral DNA by either the host or viral RNA polymerase. The elimination of AdoMet in T3 infections permits a partial protection of phage DNA from restriction enzymes (Studier & Movva, 1976). Prevention of methylation appears to inhibit lysis for long periods (Krueger et al., 1975). In cells infected by ultraviolet-irradiated T3, which do synthesize the AdoMet hydrolase but do not lyse, superinfection by T4 permits a continuing synthesis of early T4-induced enzymes, possibly suggesting a role for AdoMet in the normal arrest of synthesis of early T4 proteins. The T3 AdoMet hydrolase occurs both as a free monomer of 17,000 molecular weight (MW) and in a complex with a host protein that does not affect the activity of the associated enzyme (Spoerel & Herrlich, 1979). The gene has been isolated and its sequence has been determined (Hughes et al., 1987).

Splicing and processing of viral RNA

An intron exists within the *tdr* gene of T4 and the extra RNA within the T4-derived mRNA must be removed and the residual RNA spliced prior to the expression of this thymidylate synthase. Each of the transcription products of the few T4 genes that contain introns have similar intron secondary structures; these develop guanosine-mediated autocatalytic splicing reactions. Autocatalytic self-splicing RNAs, known as ribozymes, activated by guanosine or guanine nucleotides, are classified at present in two groups. A third group of splicing reactions relates to the production of individual tRNA molecules (Cech, 1983). In a typical demonstration of a group I ribozyme in the chloroplast ribosomal RNA of a unicellular alga, *Chlamydomonas reinhardtii*, the sequential cleavage and splicing reactions are carried out in 8 mM $MgCl_2$ and 2 mM spermidine (Dürrenberger & Rochaix, 1991). In the in vitro construction and reactions of ribozymes, millimolar levels of spermidine or spermine are often added to maximize the reactions, although virtually nothing is known of the mechanism by which the polyamine is stimulatory in these systems. Spermine increases the rates of ribozyme cleavage reactions run at low concentrations of divalent cations (i.e., Mg^{2+}) and it was suggested that spermine promotes the proper folding of the ribozyme (Dahm & Uhlenbeck, 1991).

Optimization of both the cleavage reaction and the self-ligation reactions of a hairpin form of a ribozyme, present in a satellite RNA of a plant virus, were shown to require a mixture of Mg^{2+} and a polyamine (spermidine or spermine) at millimolar levels (Sekiguchi et al., 1991). It was suggested that these ions were not only general countercations in the hairpin but also bound specifically to certain regions of the ribozyme. More recently, the ionic requirements for RNA binding, cleavage, and ligation by a ribozyme isolated from a plant virus RNA have been defined more carefully. Spermidine was found to induce a slow self-cleavage reaction in the absence of added metallic cations. Both Mn^{2+} or Mg^{2+} markedly increase the activity (>100-fold) of a spermidine-stimulated reaction (Chowrira et al., 1993). It was suggested that spermidine stabilizes a ribozyme–substrate complex whereas Mn^{2+} is probably active at another site.

T4 DNA also contains genes for numerous new tRNA species. In contrast to *E. coli* genes in which tRNA genes are interspersed with genes for ribosomal RNA, T4 genes exist for eight known tRNAs and are found in a single cluster in the genome (Schmidt & Apirion, 1983). The transcription and processing of these tRNAs require cutting, trimming, chain extension, and some base modification. Many of these steps, which are catalyzed by host enzymes, have also been found to be stimulated by spermidine. One of these is catalyzed by a ribozyme contained in bacterial ribonuclease P, an RNA + protein complex, whose RNA alone will cleave nucleotides from a pre-tRNA transcript in the presence of Mg^{2+} plus spermine or spermidine. The RNA will cleave tRNA-precursor molecules in 60 mM Mg^{2+} or in the presence of 10 mM Mg^{2+} plus 1 mM spermidine (Guerrier–Takada et al., 1983). The presence of spermidine in the growth medium of *E. coli* also increases the transcription of the genes for ribonuclease P and its activity in bacterial extracts (Panagiotidis et al., 1992).

The splicing of tRNA precursors in yeast is effected by separable endonuclease and ligase activities (Peebles et al., 1983). The former, isolable from membranes, cleaves precursor tRNAs at two sites and the extent and accuracy of the cleavage is greatly increased by spermidine. The ligation requires ATP and Mg^{2+}.

Two T4-induced enzymes, T4 polynucleotide kinase and T4 RNA ligase, have been thought to function in RNA processing (Snyder, 1983) but have also been implicated in DNA replication. The T4-encoded enzyme, polynucleotide kinase, catalyzes the transfer of the γ-phosphate of ATP to a 5′-hydroxyl at the end of RNA or DNA. This reaction is stimulated about five-to-six-fold on single-stranded substrates by spermidine or spermine over a range of Mg^{2+} concentrations (Lillehaug & Kleppe, 1975). The enzyme is not activated greatly by spermine in reactions with double-stranded DNA (Lillehaug et al., 1976). In the DNA systems, nicks, strandedness, and the existence of protruding 3′-OH determine the effects of polyamines and salt.

The T4-induced RNA ligase in the presence of ATP can catalyze the joining of the 3′-OH end of an oligoribonucleotide to the 5′-phosphoryl group of the polynucleotide to form a circular RNA. The enzyme will also link a nicked tRNA or tie together two separate oligonucleotides. The stable monomeric enzyme has been purified to homogeneity (Last & Anderson, 1976). Low concentrations of diamines and spermidine (1 mM) were modest activators of the joining of the ends of a double-stranded DNA substrate by the T4-induced enzyme (Raae et al., 1975); spermine was quite inhibitory (Raae & Kleppe, 1975). In reactions of this T4 RNA ligase with single-stranded oligodeoxyribonucleotides as acceptors with single nucleotide donors, the favored concentration was 8 mM spermine but 2 mM spermine was used in joining two oligodeoxyribonucleotides (Brennan et al., 1983).

Almost every step in in vitro RNA synthesis and processing appears to be stimulated by polyamines. Nevertheless, there is no unequivocal evidence that these effects operate in a T4-infected bacterium. Further, we remind ourselves that spermine is not a normal constituent of *E. coli*. The roles of spermidine in in vitro splicing reactions have barely been examined, and the two tasks of defining the molecular role of the polyamine in in vitro reactions and its biological role in vivo have only been noted.

DNA phages and topoisomerases

The helical structure of duplex DNA added topological considerations to the problem of DNA replication. The normal existence of coils and supercoils of DNA strands hinders their separation and the formation of replication forks. These topological problems are more urgent when the DNA is circular as in mitochondrial DNA, some DNAs of colicins, and some viruses. In 1971 a protein was isolated from *E. coli* that relaxes negatively supercoiled DNA by cleaving a single-stranded segment and permitting the other strand to pass through the break. A rejoining then occurs at the break site. This ω protein and similarly acting enzymes that introduce a reversible single-strand break in DNA are categorized as type I topoisomerases. Some enzymes may also decatenate plasmid dimers. Whereas relaxation of supercoiled DNA by the *E. coli* type I topoisomerase is inhibited in the presence of spermidine, knotting of single-stranded circular DNA is stimulated by the polyamine (Srivenugopal & Morris, 1985). Topoisomerases of eucaryotic cells are significantly different from the procaryotic enzymes and have few sequence similarities to the bacterial enzymes (Sutcliffe et al., 1989). The relaxing effect of the eucaryotic enzyme is stimulated by spermidine.

Type II topoisomerases introduce reversible double-strand breaks in double-stranded DNA and most frequently require Mg^{2+} and ATP. The binding of ATP is inhibited by the antibiotic novobiocin. In the absence of ATP, negatively supercoiled DNA is relaxed by the type II enzyme, which binds to DNA at the cleavage site and is fixed at the site by inhibitory quinolones like nalidixic acid. Discovery of a bacterial type II topoisomerase, known as DNA *gyrase*, stemmed from in vitro experiments on the insertion of the prophage λ DNA into bacterial DNA. An in vitro system extracted from λ-infected *E. coli* effected this reaction in the presence of ATP, Mg^{2+}, and spermidine (Nash, 1975). The activity was markedly stimulated by spermidine; however, the polyamine was not essential and could be partly replaced by larger amounts of Mg^{2+} (Gellert et al., 1976). Morris and Lockshon (1981) postulated that the depletion of spermidine in *E. coli* is expressed by effects on replication and transcription, mainly via a slowing of DNA gyrase.

The DNA-delay mutants of T4 were found to be largely corrected by the exogenous addition of spermidine (Dion & Cohen, 1971). Analysis of a series of these DD mutants revealed defects in a series of three complementary genes in the virus governing the formation of a novel multicomponent DNA topoisomerase II that was insensitive to novobiocin and a quinolone (Stetler et al., 1979). This topoisomerase, known as T4 topoisomerase, is a new enzyme determined by T4 genes. The deficiency of one of the protein components evoked a low incidence of replicative forks in the viral DNA. Nevertheless, despite a lack of the T4 DNA topoisomerase the synthesis of some viral DNA was permitted in a cell in which the host DNA gyrase was still functional (McCarthy, 1979). Thus, this virus-induced enzyme facilitates viral DNA replication by supplementing the host enzyme.

The T4 topoisomerase proved to be similar to some eucaryotic type II topoisomerases that also catalyze ATP-dependent DNA relaxation but cannot introduce

negative superhelical turns into circular duplex DNA. The latter enzyme of the mammal proved to be sensitive to the antitumor agent, *m*-AMSA [4′-(9 acridinylamino)-methanesulfon-*m*-anisidide] and this compound was also inhibitory to the T4 topoisomerase, forming a defined inactivated complex of the drug with the enzyme and DNA (Rowe et al., 1984). Genetic studies with T4 have also demonstrated that the T4 topoisomerase is the target of *m*-AMSA in virus-infected *E. coli* (Huff et al., 1989). Kreuzer (1989) discussed the manner in which a virus-induced function, the T4 topoisomerase, can be selectively eliminated (with *m*-AMSA) in a virus infection. The fact that the T4 topoisomerase does not entirely determine viral multiplication should not minimize the significance of this new and selective achievement in viral chemotherapy.

It is now thought that the presence of both type I and type II topoisomerases in most, if not all, cells relates to a careful control on the level and direction of coiling of DNA involved in replication, transcription, etc. Catenated and knotted DNA appears to accumulate in cells deficient in topoisomerase activity. The rapid formation of T4 DNA in knots and catenates during viral synthesis can be expected to hinder DNA replication and it was suggested that the T4 topoisomerase may play a role in the deknotting and decatenation of the condensed network of the newly produced T4 DNA (Cozzarelli, 1980; Kreuzer & Huang, 1983).

Spermidine was found to stimulate catenation and knotting of a circular colicin DNA nicked in a single strand by a bacterial type I topoisomerase (Brown & Cozzarelli, 1981). Krasnow and Cozarelli (1982) then demonstrated that spermidine and other tervalent cations can compact the DNA into an aggregate whose very high concentration in the presence of the enzyme favors catenation. The enormous aggregates bind both gyrase and the dispersed rings of DNA, which are freely exchanged from one aggregate to another. The dynamic aggregates were readily accessible to enzymes and are believed to participate in the recombination of phage λ. As described in Figure 19.3, the formation of catenates from the relaxed circular DNA in the presence of gyrase and spermidine occurred when the polyamine had neutralized 90% of the DNA charge. Concentrations of spermidine > 3 mM reduced the level of catenates. A similar result was obtained with spermidine and type I enzyme.

It has been difficult to prove that the deficiency in topoisomerase that restricts the initiation of DNA replication is a function of a deficiency of stimulation by spermidine. Other agents are also involved in steps of DNA replication. A DNA-binding protein without topoisomerase activity, the T4 gene 32 product, was shown to bind cooperatively to single-stranded DNA, and to facilitate denaturation and renaturation of many types of DNA (Alberts & Frey, 1970). This may provide single-stranded templates for the T4 DNA polymerase. On the other hand, in the conversion of a single-stranded circular DNA (M13) to a duplex replicative form, a DNA-unwinding protein of *E. coli* is partially replaceable by

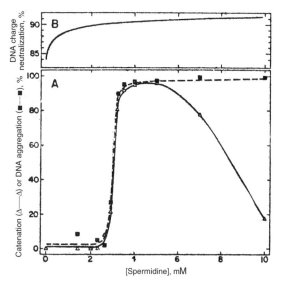

FIG. 19.3   Effect of spermidine on DNA aggregation and on catenation of relaxed circular ColE1 [³H]DNA by *E. coli* DNA gyrase; adapted from Krasnow and Cozzarelli (1982). (**A**) After 30 min, catenation was measured by agarose gel electrophoresis and aggregation by the depletion of labeled DNA after sedimentation. (**B**) Neutralized DNA phosphate charge was calculated.

spermidine (Geider & Kornberg, 1974). In a study of the *E. coli* protein that binds to single-stranded DNA and facilitates the formation of double-stranded DNA, it was shown (Fig. 19.4) that the DNA reassociation was stimulated 5000-fold at pH 7 by 2 mM spermidine (Christiansen & Baldwin, 1977). Thus, spermidine (and spermine) also appear active in the execution of essential steps in DNA replication over and above compaction.

A spermidine-containing antibiotic, *Cinodine*, a glycocinnamoylspermidine (Chap. 20), binds tightly to DNA and inhibits DNA synthesis in *E. coli* and T7-infected bacteria. Cinodine proved to be an active specific inhibitor of DNA gyrase (Greenstein et al., 1981; Osburne et al., 1990).

Compaction of DNA

The head and tail bacteriophages, which include both the virulent T-even phages and the lysogenic λ phage, revealed several significant requirements of the polyamines in λ multiplication. Some of these were mentioned earlier in noting results with depleted *E. coli* mutants. The DNAs of various phages are readily obtainable and are genetically and chemically defined but differ in size, T4 DNA > T7 DNA, serving as DNA standards more suitable for experimental work than are isolated fragments of eucaryotic DNA like thymus DNA.

In studies of the effects of neutral polymers (e.g.,

polyethylene oxide) and salts on the sedimentation of T4 and T7 DNA, certain conditions of polymer concentration and ionic strength caused the condensation of the extended random coil of DNA to produce particles that form a much more dense concentrated phase (Jordan et al., 1972; Lerman, 1971). This phase change was interpreted as resulting entirely from excluded volume interactions (Lerman, 1973) and indeed is very similar to the packing of asymmetric viruses like TMV into 2-dimensional "crystals" by neutral or charged polymers (Cohen, 1942). The T4 DNA used by Lerman was isolated by phenol extraction and may well have contained a high percentage of neutralizing polyamine cations.

However, the condensation of viral DNA and its packaging involved a 10,000-fold reduction of the effective ellipsoid volume of T7 DNA. This could in fact be carried out by low concentrations of spermidine (5 × 10⁻⁵ M) at very dilute concentrations of DNA at low ionic strength (Gosule & Schellman, 1976). Electron microscopy revealed a drastic change in the size and shape of the T7 DNA with the formation of a coiled doughnut-shaped torus. Spermine was also active in effecting this transition, indeed at lower concentrations of the cation than that of spermidine.

Polyamine-induced aggregation of essentially all double-stranded DNA was demonstrated by Osland and Kleppe (1977). The single-stranded circular viral DNA, φX174, was also aggregated quite readily. The effects of different polyamines on the aggregation of T7 DNA are presented in Figure 19.5. In contrast to the much greater effectiveness of spermine over spermidine, it

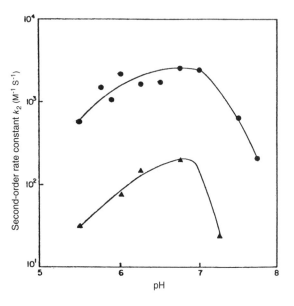

FIG. 19.4   Dependence on spermidine concentration and pH of the rate of the DNA reassociation reaction catalyzed by the *E. coli* DNA binding protein; adapted from Christiansen and Baldwin (1977). (▲) 0.75 mM spermidine; (●) 2 mM spermidine.

can be seen that putrescine is inactive in aggregating T7 DNA whereas cadaverine does condense this DNA, albeit at a high concentration.

The intriguing appearance of polyamine-aggregated DNA as doughnut-shaped toroidal forms, possibly similar to that in a phage head, led to studies of the possible mechanisms of these structural transitions. A distinction was made between condensation and aggregation. Nucleates were considered as polyelectrolytes whose extended structure arises from the repulsive interactions of phosphate groups (Manning, 1978). Cations, both simple and organic, relieve this electrostatic stress and permit transitions of the extended structures to conformations of lower charge density. The complete condensation of DNA in low salt may occur under conditions of very considerable charge neutralization (88–90%). The effects of high ionic strength (0.5–1.0 M) in causing the dissociation of many neutral complexes (nucleohistones, etc.) were also described. Manning (1980) subsequently considered the problem of binding a neutralized DNA into a doughnut-shaped torus. The discussion assumes the existence of a persistence length that relates to the radius of curvature of DNA. The inner radii of the viral DNA doughnuts were seen to be no smaller than 150 Å, and this is taken to define the maximal bend of a DNA molecule, toward which the DNA molecule moves spontaneously on neutralization of its repulsive negative changes.

Wilson and Bloomfield (1979) examined the light-scattering effects of the polyamines and Mg^{2+} on the collapse of T7 DNA. Their work was conducted at low DNA concentrations to attempt to differentiate monomolecular collapse from an aggregation that developed more slowly. Condensation by putrescine and Mg^{2+} did occur in 50% methanol. The counterion condensation theory predicted collapse at an 88% neutralization of DNA charge and these workers saw collapse in mixtures of spermidine and Mg^{2+} at 89–90% neutralization. The absence of changes in the circular dichroism spectrum of the DNA implied an absence of structural change of the DNA. These results differ from the change of state of DNA observed earlier by Lerman (1973). The existence of shells of DNA in the toroids was established by nuclease digestion (Marx & Reynolds, 1982) and by transmission electron microscopy of freeze-fractured toroids (Marx & Ruben, 1983).

Because DNA collapse is produced by charge neutralization, other cations were tested. An inert trivalent metal ion complex, $Co^{3+}(NH_3)_6$, among several active cations, also produced toroidal condensation of T7 or λ DNA and approached the theoretical and experimental results described above (Widom & Baldwin, 1980). The size of collapsed particles is quite insensitive to the size of the DNA. Widom and Baldwin (1980) also appeared to confirm the predictions of Manning (1980) concerning the lack of effects of polyamines at high ionic strength and concluded in general that the electrostatic mechanism of condensation proposed by Manning was correct. Restated, "toroids are the natural physical state of stably bent DNA." However, evidence of anomalies among the diamine competitors for spermidine-binding sites and precollapse changes prior to the collapse transition points suggested that the ion point charge model of compaction should be modified to include some crossbridging of DNA by the potentially dissociable cations between the shells of the toroid (Marx, 1987).

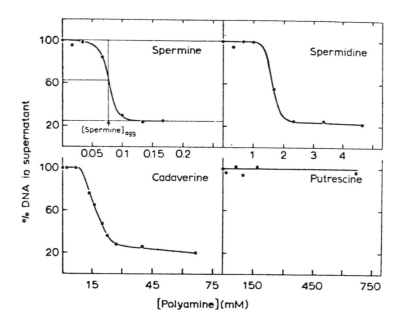

FIG. 19.5 Influence of different polyamines on aggregation of [³H]T7 DNA (15 μM); adapted from Osland and Kleppe (1977).

Head-full hypothesis

The initial studies on DNA collapse as a function of counterion (spermidine) concentrations suggested that the condensed DNA is subsequently coated by a protein capsule. However, genetic studies on phage development revealed the existence of many genes essential to the structure of phage heads and to the development of a DNA-filled head. Phage provided intermediate stages of the packaging of heads and the hypotheses developed to explain the encapsulation of spermidine-compacted DNA did not explain the natural phenomena of T4 multiplication (Murialdo & Becker, 1978). The T4 head is formed initially on the bacterial membrane as a core-containing immature particle in which a protein scaffold is processed and eliminated by proteolysis. The released prohead particle contains an opening into which fully replicated DNA can enter and fill up the cavity. The assembly of head proteins to form the empty proheads of double-stranded DNA viruses has been summarized by Hendrix and Garcea (1994).

The entrance of an extended DNA and its compaction within capsid proteins was thought to be energetically unfavorable. It had been estimated that the major balance of energetic factors governing packaging was that between the free energy of polyelectrolyte repulsion (2.1×10^5 kcal/mol virus, 20°C) and DNA–polyamine interaction (-2.1×10^5 kcal/mol virus) (Riemer & Bloomfield, 1978). Also, the absence or loss of polyamine might be compensated in part by binding of the DNA to protein subunits. It is not yet known if newly synthesized extended DNA is neutralized by polyamine. This may in fact relate to the relatively low concentration of "free" (i.e., unbound) spermidine in a bacterium, in which RNA and ATP hold most of this cation, and to a relatively low association constant of spermidine and DNA (see Chap. 22).

A packaging mechanism that can operate in vitro was discovered by Black (1981). Processed proheads are efficient acceptors of mature T4 DNA and require ATP, Mg^{2+}, and the contents of the extracts of specific mutant-infected cells. The in vitro system was not stimulated by added spermidine. In short the energy of translocation is probably derived from an ATPase, which is localized in a major capsid protein (Black, 1989). This author devised the scheme of Figure 19.6 for prohead DNA packaging.

Arguing that *E. coli* "entirely lacking polyamine allow phage formation" and that a defined T3 packaging system does not require spermidine, Black (1989) concluded that "polyamines are peripheral to the energetics. . . ." However, the data on the rates of synthesis of phage components in polyamine-deficient bacteria are incomplete. Further, the ATP-requiring T3 system (Hamada et al., 1986b) is significantly stimulated to a third of the maximal value by spermidine. These authors have not demonstrated the absence or near absence of polyamine in their enzymatically active system.

If an extended newly synthesized T4 DNA lacks polyamine, how does a completed T4 particle contain polyamine? It may be asked if these neutralizing counterions are present as ATP salts and are freed to bind the newly inserted head DNA when the ratcheting ATPase cleaves that salt. Although significant amounts of ATP and ADP are present within the head, the status of these nucleotides as possible head components is not clear (Kozloff, 1983).

As presented in Figure 19.6, DNA that cannot be crammed into the head is trimmed at the portal. The alteration of head size by amino acid analogues indicated that head size determines the amount of DNA found in T-even phages. Arginine analogues, notably canavanine, produce large heads whereas putrescine and cadaverine induce small-head formation (Couse et al., 1970). These agents affect the structural proteins of the viral head. When exposed to canavanine, some monster phage particles, "lollipops," are formed and contain very large pieces of DNA that are several times the length of normal genomes (Cummings & Bolin, 1975). Canavanine also blocks the appearance of a viral gene-dependent proteolytic enzyme essential for head morphogenesis (Bolin & Cummings, 1975a). Canavanine also inhibits the synthesis of the polyamines putrescine and spermidine in T4-infected cells. Exogenous polyamine did not increase DNA synthesis, suggesting that the latter inhibition was not a result of polyamine depletion. Although the polyamines did not appear causally related to lollipop formation (Bolin & Cummings, 1975b), diamines do provoke an aberration of head size and other phage structures. The neutralization of phage DNA within the particles can be effected interchangeably with either putrescine or spermidine.

Packaging DNAs of T3 and T7

These isolated phages lack internal polyamines and are penetrable by ethidium dye, which can bind to the phage DNA. The dye binds less and more slowly to the DNA within the phage head than to free DNA. These results have been attributed either to hindered binding or a sieving of ethidium through the packaged DNA (Griess et al., 1985, 1986).

The in vitro system of packaging T3 DNA into T3 proheads, described by Hamada et al. (1986b), detected phage mutant strains lacking a gene for a minor head protein (gp8) located at the head–tail junction. The purification of this protein, leading to the crystallization of a head complex containing gp8, has involved the inclusion of spermidine at almost every step, although its presence is unexplained (Nakasu et al., 1985). The packaging mechanism of phage T3 and T7 requires proheads and two additional purified noncapsid virus-induced proteins, gp18 and gp19. The purification of gp18 was found to require spermidine-containing buffers (Hamada et al., 1986a). The packaging system, compressed by steric exclusion in polyvinyl alcohol, was active in the presence of ATP and Mg^{2+} and was inhibited by an ATP analogue. Commenting on the stimulatory but less than absolute requirement for spermidine, Hamada et al. pointed to Hafner et al. (1979) who had

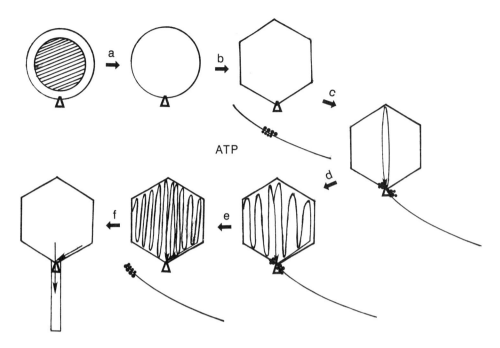

FIG. 19.6 A general scheme for prohead DNA packaging; adapted from Black (1989).

described low phage yields and prolonged latent periods in T7 infections depleted of polyamine.

Continuing with spermidine-containing buffers, Shibata et al. (1987a,b) found an intermediate stage in which, under the impetus of ATP, the prohead is filled with DNA. In a second ATP-requiring step, this intermediate is converted further to a mature viral head. Although these constructed systems will cleave ATP and package T3 or T7 DNA, they do cleave ATP on addition of some initially nonpackageable, heterologous DNA, the T4, λ, and plasmid DNA (Morita et al., 1993). An in vitro system was isolated, beginning with T3 or T7 DNA and proheads, and used to demonstrate that transcriptional specificity of the DNA was essential initially to provide packaging specificity (i.e., *pac* signals for these DNAs). Hashimoto and Fujisawa (1992) included spermidine in this coupled transcription-packaging system. DNA packaging signals were discussed recently by Fujisawa and Hearing (1994).

Packaging DNA in λ bacteriophage

The λ DNA completely fills the head of a λ phage. Deletion mutants are packed somewhat more loosely (Earnshaw & Harrison, 1977). A review of DNA packaging (Earnshaw & Casjens, 1980) argued for a single basic mechanism for all the double-stranded DNA phages. This supposed the universal formation of DNA-accepting proheads, produced around scaffolding proteins, that are eventually cleaved and eliminated. DNA is inserted into various phage proheads, probably by a

ratcheting mechanism, and the heads are enlarged in this step. It has been asked if the volume of the head change in itself is sufficient to suck in all the DNA (i.e., the spaghetti model for DNA packaging), but calculations have suggested that the enlargement of the head alone is inadequate for this task.

Although many similarities of the process in various phages support the concept of a single mechanism, the studies of the individual phages have yielded unique information for each. Some double-stranded DNA phages, such as the lipid-containing PM2 phages, have scarcely been studied from this point of view. Additionally, the study of the lysogenic λ phage has revealed several new roles for the polyamines. The packaging of recombinant phage genomes in λ heads has become a major technique in the steps of directed protein biosynthesis in *E. coli* (Hohn & Murray, 1977).

The isolation of λ mutants deficient in morphogenesis permitted the isolation of λ heads with which it became possible to examine the condensation of DNA within the head. The stability of such filled heads could be tested by the addition and joining of λ tails and the measurement of infectious phage. Filled heads were isolated from some cultures and empty precursor heads from others. Filled heads in lysates leaked DNA on dilution and incubation in salt solutions and extracts containing tails. However, filled heads were stabilized considerably by putrescine, which was more effective than other diamines, spermidine, and spermine. Mg^{2+} and Ca^{2+} hastened expulsion of DNA but this was prevented significantly by 1 mM putrescine. Indeed the addition of the diamine inhibited injection of DNA from the intact phage into a bacterium (Harrison & Bode, 1975).

Hohn and Hohn (1974) prepared head precursors lacking DNA (petit λ) and demonstrated the packing of DNA within the head in extracts. The absence of spermidine and the presence of an ATP analogue each reduced head filling by 90%. Similar results were obtained by Kaiser et al. (1975). A partition of these events into early and later events confirmed the stimulatory roles of spermidine and ATP in early reactions involving the formation of a complex of concatameric λ DNA and the product of the λ gene A (Becker et al., 1977a,b). Although the requirement for ATP appears to be absolute, Syvanen and Lin (1978) reported that spermidine was only stimulatory (some threefold). However, the requirement for putrescine appears to be absolute, possibly at a later maturation step.

The precise cleaving of the linear λ DNA to form a duplex capable of complementary annealing of single-stranded terminal chains is an enzymatic reaction (i.e., the ter or terminase reaction that is catalyzed by the A protein) and is markedly activated by spermidine and ATP (Becker & Gold, 1978). The analysis of DNA termination routinely included spermidine and ATP in the cocktail of the terminase reaction (Feiss & Becker, 1983; Feiss & Widner, 1982).

In the multiplication of λ phage, a linear DNA is injected into the cell and this chromosome is cyclized via the interaction of its complementary cohesive ends. The replication of the DNA circle produces concatemers that can be nicked at the cohesive ends by the phage enzyme, terminase. Although the enzyme functions in vitro at the opening to the phage head, it can produce nicks in vitro. This reaction, involving the initial formation of a terminase-concatameric DNA complex, requires Mg^{2+}, ATP, and salt; specificity is optimal in 2–5 mM Mg^{2+}, 1.5–2 mM ATP, and 5 mM spermidine (Rubinchik et al., 1994a). The presence of both ATP and spermidine facilitates the binding of terminase to the nicking site of the DNA (Higgins & Becker, 1995). The linearized DNA, nicked at the opening to the phage head by terminase, is presumably ratcheted into the head with the aid of the DNA-dependent hydrolysis of ATP to ADP and inorganic phosphorus. Prohead:terminase:DNA ternary complexes have been isolated and are further convertible to an intact phage. The formation of the prohead ternary complex also requires Mg^{2+}, a polyamine, and ATP as well as some host proteins.

Recently the multifunctional λ terminase holoenzyme has been isolated and shown to comprise two protein subunits, (gp)A and (gp)Nu1 (Parris et al., 1994). In addition to its roles in packaging λ DNA, terminase is an endonuclease, an ATPase, and a DNA helicase, which can melt DNA cohesive end sites. These functions are also effected by the A protein, which alone does not package DNA into a prohead (Rubinchik et al., 1994b).

DNA injection

The presence of polyamine in impermeable phages has been demonstrated but it is quite unclear how it entered the phage. Although spermidine-neutralized DNA, lacking negative repulsive charges, tends to coil up, as in a torus, it is evident that an extended DNA structure penetrates and subsequently fills up the phage head. Can an extended spermidine-neutralized DNA be constructed and used to penetrate and fill a phage head or is spermidine provided at or immediately after penetration of an unneutralized DNA?

It has been calculated that the transfer of DNA from virus to host, DNA injection, may be effected by the internal pressure within phage heads. The internal osmotic pressure has been estimated to be in the range of 70–100 atmospheres (Zarybnicky & Horacek, 1968), and these workers pointed to the dissociation of polyamines from DNA as a major component of the ionic strength establishing the internal osmotic pressure. As summarized by Zarybnicky (1969), reactions uncorking and unplugging the tail do permit ejection of the DNA after adsorption on heat-killed E. coli or isolated cell walls. Injection requires an ionic medium equal to 0.02 M NaCl, and higher concentrations (0.1 M NaCl) are needed to dissociate the polyamine–DNA complex more effectively. Removal of the polyamines from phage heads by treatment with polymeric anions like polyglucose sulfate was shown to prevent injection (Mora et al., 1962).

The existence of viable permeable mutants lacking polyamine that inject their DNA into cells appears to minimize the possible validity of a theory of DNA injection depending on the presence of the polyamines. Indeed the addition of putrescine to λ E. coli infections inhibits injection of λ DNA. However, the presence of putrescine does not delay DNA injection in a permeable T4 O mutant. The high efficiency of some phages (T4, λ) in carrying out a multiplication cycle suggests that the injected DNA may make immediate contact with a host factor capable of facilitating integration of the viral DNA into the host machinery initiating multiplication. E. coli integration proteins reactive with end sites of λ DNA have been described and it may be asked if such attachments do not participate in a pulling reaction essential to injection.

Putrescinyl thymine in Pseudomonas phage

The study of bacteria other than E. coli has resulted in the discovery of new polyamines and new metabolic pathways. The study of the bacteriophages of Pseudomonas acidovorans introduced a new type of viral DNA. The host bacterium itself is notable for its production of 2-hydroxyputrescine and 7-hydroxyspermidine. A phage that lysed P. acidovorans 29 under conditions of aerobic growth in a complex medium was isolated, dubbed φW-14, and found to contain a DNA of unusual buoyant density and melting temperature (Kropinski et al., 1973). The very high melting point was characteristic of an unusual DNA, such as that neutralized by polyamines. These results suggested the presence of an unusual base, which was indeed isolated and characterized as 5-(4-aminobutylaminomethyl)uracil or putrescinylthymine.

It was suggested initially that the new compound might aid in DNA packaging by neutralizing negative charge.

After nucleate hydrolysis, the base composition of ϕW-14 DNA was found to be 56% GC and 22% A. The remaining base, paired with adenine, contained 10.3 mol % thymine and 11.9 % putrescinyl-thymine. Thus, about half of the thymine in the phage DNA possessed this unusual substitution. Some free putrescine, 2-hydroxyputrescine, and spermidine were present in the impermeable phage and could neutralize about 15% of the DNA phosphate. However, putrescinylthymine can neutralize an additional fraction of the phosphate.

The new base was positively charged in acidic media and neutral in alkaline media and reacted with a ninhydrin reagent as a primary amine. Spectral properties (ultraviolet absorption, infrared, NMR) were consistent with the structure presented above, and these features were also obtained with a newly synthesized sample that was indistinguishable chromatographically from the natural product. Hydrolysates of the phage DNA did not contain 5-hydroxymethyluracil, which had been found in some other phages. Acid hydrolysis of a mixture of free putrescine and 5-hydroxymethyluracil did not produce the new basic pyrimidine derivative.

The putrescine moiety in the phage DNA could be labeled by addition of isotopic ornithine in the medium but not by arginine. Incorporation of ornithine into the DNA began with a somewhat delayed synthesis of the DNA. Synthesis of the various free polyamines was also increased at that time (Quail et al., 1976). Putrescine uptake into the bacterium, which lacks a system of active transport, occurs by passive diffusion at high pH, pH > 9.5. On return to neutrality, the putrescine accumulated in the cells can be incorporated into 2-hydroxyputrescine and spermidine in the cells and into putrescinylthymine in phage DNA (Bruce & Warren, 1983).

Study of the deoxynucleotide pools revealed that thymidine triphosphate (dTTP) slowly disappeared from the cells and a new pyrimidine deoxyribonucleotide triphosphate appeared. The latter was identified as 5-hydroxymethyldeoxyuridine triphosphate, which entered newly replicated phage DNA (Neuhard et al., 1980). This new nucleotide had been synthesized by a new hydroxymethylase from deoxyuridylate (dUMP), formaldehyde, and tetrahydrofolate. It was concluded that the formation of putrescinylthymine and thymine occurred at the polynucleotide level in reactions of the 5-hydroxymethyldeoxyuridylate incorporated in the DNA (Maltman et al., 1980). Warren and colleagues then isolated

amber mutants of phage, one of which produced an unusual DNA in a nonpermissive host. This DNA lacked putrescinylthymine but contained a new base, in which the hydroxymethyl group of 5-hydroxymethyluracil was pyrophosphorylated. An extract of a bacterium infected with a normal ϕW-14 contained an enzyme that reacted putrescine with this pyrophosphorylated DNA to produce putrescinylthymine. A pyrophosphoryl group had been displaced by the amino nitrogen of putrescine to form a carbon–nitrogen bond (Maltman et al., 1981).

Other amber mutants were found that evoked the synthesis of DNA deficient in putrescinylthymine. The DNAs made by some of these contained thymine and hydroxymethyluracil, others had higher thymine contents and less putrescinylthymine. Many mutants were unable to synthesize DNA at all in nonpermissive hosts. Genetic analysis of the mutants revealed three viral genes involved in DNA synthesis and two more in DNA modification (Miller et al., 1982).

Features of ϕW-14 DNA

The sequence of purine-putrescinylthymine-purine was found more than twice as frequently as purine-thymine-purine nucleotide sequences (Warren, 1981). This result implied a sequence specificity in the enzymes of DNA modification.

Both the primary and secondary amino groups of putrescinylthymine in the phage DNA are reactive with acetic anhydride and the N-acetylated DNA does have a lowered melting temperature, although the buoyant density was unchanged. Only the putrescinylthymine residues were thus modified. Both amino groups appear to neutralize the negative charge and repulsive forces in the DNA (Gerhard & Warren, 1982).

Despite a DNA only slightly smaller in length than that of T4, 60 μm compared to 62 μm, the head of ϕW-14 has a significantly smaller volume, about 72% of that of T4. It was concluded that the DNA of the former was packed far more compactly, and indeed that the tight packing was dependent on the presence of putrescinylthymine (Scraba et al., 1983). The presence of putrescinylthymine in ϕW-14 DNA does modify the in vitro condensation of the DNA by cations such as spermidine and hexaamine cobalt(III) (Benbasat, 1984). As measured by light scattering, about a third as much added spermine was effective in condensing ϕW-14 DNA as was needed for T4 DNA.

RNA bacteriophages

Four groups of these small *E. coli* phages have been recognized on the basis of their serological reactivities, their enzymes for replication (i.e., an RNA replicase), as well as the physical properties of their RNA and proteins. Their different RNAs have also been useful in studies of translation and the properties of translation systems. In one of the earliest of these, Takeda (1969) described the effect of spermidine in decreasing the optimal Mg^{2+} concentration in a cell-free polypeptide synthesis directed by the MS2 phage RNA.

The MS2 phage is a member of the *E. coli* group I phages, which include MS2, R17, fr, and f2 (Witherell et al., 1991). MS2 contains a single-stranded RNA of 3569 nucleotides, 180 units of coat protein (14,000 MW) organized in an icosahedral shell, and a single maturation protein, the A protein. The latter complexes with viral RNA for transfer to the bacterium via the central channel of an F pilus. The coat proteins of MS2 and R17 appear to be identical. The MS2 phage has been crystallized (Min Jou et al., 1979), and a 3-dimensional structure of the icosahedral virus has been determined by X-ray crystallography (Valegard et al., 1990). The protein shell lacks divalent metal cations and it is assumed by analogy to R17, which contains many molecules of spermidine, that MS2 RNA is neutralized largely by these cations (Valegard et al., 1986).

The presence of spermidine and small amounts of putrescine had been found in R17 and its RNA (Fukuma & Cohen, 1975b). The host bacteria and virus had been grown in a glucose–salts medium supplemented by four essential amino acids and thiamine. When isolated from a lysate at low ionic strength, up to 930 molecules of spermidine were found per virion (Table 19.2). Dialysis against solutions of increasing K^+ and Mg^{2+} content reduced the spermidine content to 100 molecules per virion at 1 M KCl + 10 mM $MgCl_2$. In a later study, the virus and another bacterial host were grown in a brain–heart infusion broth. The virus then contained relatively large amounts of cadaverine and spermine, as well as putrescine and spermidine (Sheppard et al., 1980). R17 RNA isolated by the phenol method after viral disruption by sodium lauryl sulfate contained 70–90 molecules of spermidine per viral RNA, all of which could be eliminated by dialysis against 2 M KCl + 10 mM $MgCl_2$ (Table 19.3). The presence of the spermidine on the RNA did not affect the rate or incorporation of [^{14}C]-valine in translation of the RNA in an in vitro translation system activated by 12 mM Mg^{2+} and 0.1 mM spermidine.

The multiplication of the related phage, f2, is much retarded in a putrescine auxotroph of *E. coli* depleted of putrescine and is markedly stimulated by concurrent

Table 19.2 Spermidine release from phage R17 with increasing ionic strength and $MgCl_2$ concentration.

	Spermidine/virion (mol/mol)		
KCl Concn (M)	0 mM $MgCl_2$	4 mM $MgCl_2$	10 mM $MgCl_2$
0.01	930	750	470
0.05	650	560	220
0.15	430	400	170
0.50	250	210	130
1.00	160	150	100

Adapted from Fukuma and Cohen (1975b).

Table 19.3 Spermidine release from phage R17–RNA with increasing ionic strength and $MgCl_2$ concentration.

	Spermidine/RNA (mol/mol)		
KCl Concn (M)	0 mM $MgCl_2$	4 mM $MgCl_2$	10 mM $MgCl_2$
0.01	80	54	57
0.15	76	44	18
0.50	54	19	<5
2.00	10	<5	<5

Adapted from Fukuma and Cohen (1975b).

addition of the diamine, which shortens the latent period and increases burst size (Young & Srinivasan, 1974). Putrescine-supplemented cells contain more of the f2 replicase and have increased rates of viral RNA and protein synthesis. The f2 phage, unlike T4, does not shut down the synthesis of bacterial RNA and protein.

The stimulation of translation of phage RNA and other mRNA by spermidine has also been demonstrated in mammalian cell-free systems, and indeed the presence of the polyamine altered the proportions of various protein products (Atkins et al., 1975). Phage RNA (e.g., MS2 RNA) directs the synthesis of three kinds of protein: the RNA replicase, the A protein, and the coat protein. The syntheses of the replicase and the A protein were stimulated 8-fold and 4.5-fold, respectively, by addition of 1 mM spermidine whereas that of the coat protein was unchanged (Watanabe et al., 1981). As will be discussed in Chapter 22, Igarashi and colleagues have estimated free and bound polyamine in *E. coli*. They have concluded that the concentration of free spermidine and putrescine at the ionic conditions prevailing in *E. coli* (10 mM Mg^{2+} and 150 mM K^+ at pH 7.5) permits a concentration of amines that evokes selective stimulation of the MS2 replicase (Miyamoto et al., 1993).

McMurry and Algranati (1986) examined effects on the fidelity of translation in polyamine auxotrophs of *E. coli* infected by MS2. The coat protein gene of the virus does not code for histidine and the rate of incorporation of this amino acid in the viral protein was unaffected by polyamine supply. However, the supply of polyamine to a depleted infected cell did decrease the rate of substitution of lysine for asparagine.

Among the group I phages, three subgroups have been recognized on the basis of their inhibition of host synthesis. The f2 phages do not markedly affect synthesis of bacterial RNA and relatively small yields of viral particles (5–10% of total new RNA) are obtained. R17 increases the yield of viral RNA to about 30%. The R23 phage, like T4, turns off the synthesis of bacterial RNA almost completely (Watanabe et al., 1968).

In stringent and relaxed strains of arginine-requiring *E. coli*, arginine depletion sharply reduced the accumulation of spermidine and RNA in the stringent cells. R17 infection of these arginine-depleted cells markedly increased the accumulation of both spermidine and RNA

(Fukuma & Cohen, 1973). The newly synthesized RNA labeled by [^{14}C]-uracil included the ribosomal 16S and 23S RNAs and viral RNA. Thus, R17 infection of a stringent cell lacking arginine relaxed the controls on stringency and permitted the synthesis of ribosomal RNA (Fukuma & Cohen, 1975a). As in earlier studies of relaxed cells, spermidine accumulated in the cells in which RNA had also increased.

In the transfer of the viral RNA to the bacterium, the RNA is present in an isolable 1:1 complex with the A protein, which adsorbs to and enters the F pili. The complex, as isolated from phages, is infectious and the low level of infectivity can be increased 20-fold (Fig. 19.7) by incubation with 3×10^{-4} M spermidine for 20–30 min at 37°C (Leipold, 1977). In spermidine the complex develops a more compact structure, which sediments more rapidly in a sucrose gradient. It was suggested that a more compact structure with stabilized double-stranded regions and loops may penetrate the pilus more readily and may be more resistant to cellular nucleases.

Jacobson et al. (1985) explored the structure of MS2

RNA by electron microscopy of samples incubated with spermidine. Two alternate conformations of the RNA were found, and reproducibly spaced hairpins (150 nucleotides) and large open loops (500–700 nucleotides) were located in the structure. Knowledge of the nucleotide sequence permitted a correlation of loop frequency with the possibility of guanine–cytosine pairing, and these calculations facilitated an interpretation of the possible location of viral genes in one of the conformations. The authors concluded that in the presence of spermidine much of the RNA (at least 70%) is double stranded, thereby enabling the identification of helices with long-range contacts in the viral RNA. The data of Leipold (1977) and Jacobson et al. (1985) suggest that the polyamines also function in compaction and the development of specific secondary and tertiary structure within R17, MS2, etc. Fox et al. (1994) discussed the assembly of the icosahedral R17.

Animal DNA Viruses

Vaccinia virus

The introduction in the mid-1950s of tissue culture methods for the study of the animal viruses permitted both the preparation of large amounts of these viruses and studies of viral structure and multiplication. One of the earliest viruses to be analyzed was that of a pox virus, the vaccinia virus. The virus proved to be quite large and complex, containing DNA and some chemical compounds usually associated with metabolically active enzymes. The isolated virions from infected HeLa cells contained an outer layer of lipoprotein and a core. The core contained the genomic DNA, polypeptides released by acid extraction, and the polyamines spermidine and spermine (Lanzer & Holowczak, 1975). The analyses suggested some loss of spermidine in the isolation of the cores.

The polyamines of the virions could be labeled by [^{3}H]-ornithine added to infected cells. Although viral infection imposes a rapid inhibition of synthesis of host proteins, labeled ornithine added to the infected cells 4–6 h postinfection was converted to labeled viral polyamine. Hodgson and Williamson (1975) noted that ODC with a different Michaelis constant (K_m) for ornithine was increased in virus-infected HeLa. It was suggested that the ODC found in infection was virus-specified. Methylglyoxal-bis(guanylhydrazone) (MGBG) partially inhibited viral multiplication and the formation of DNA-containing cytoplasmic inclusions, and this effect was reversed by spermidine (Williamson, 1976). Difluoromethylornithine (DFMO) was quite inhibitory to viral multiplication (Tyms et al., 1983).

This DNA virus was found to be localized and to multiply uniquely in the cytoplasm of infected cells, although separation from the nuclear apparatus involved viral independence from host systems of transcription and replication. Many steps of viral multiplication can occur in enucleated cells.

As described by Obert et al. (1980), arginine is re-

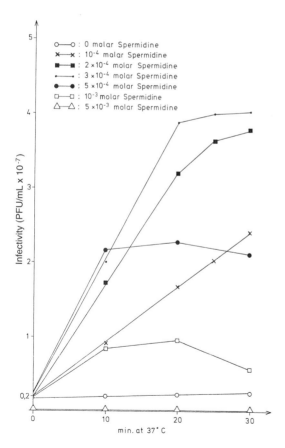

FIG. 19.7 Kinetics of infectivity of the MS2 RNA-A protein complex on *E. coli* containing F pili during incubation with spermidine at 37°C; adapted from Leipold (1977).

quired for the synthesis of late mRNA and viral multiplication in KB cells. Arginine could be replaced by ornithine and the polyamines to which arginine is converted. Arginine was also essential as such for the production of arginine-containing proteins necessary for viral maturation and the production of infectious virus.

Reviews on vaccinial infections have emphasized the regulation of transcription (Moss et al., 1991). Transcription of viral DNA is carried out by a virion-associated DNA-dependent RNA polymerase. Indeed the extruded mRNA contains a poly(A) "cap" and methylated bases. Thus, the capping and methyltransferases and poly(A) polymerase are also considered to be virus-determined enzymes, as are several other proteins like topoisomerase found in the viral cores. A DNA-dependent RNA polymerase was also found in the viral core, which could produce mRNA similar to that in infected cells. The purified enzyme purified from the virus was Mn^{2+} dependent, in contrast to the core enzyme which was Mg^{2+} dependent and almost inactive with Mn^{2+}. Moussatche (1985) demonstrated that the presence of spermidine or spermine stimulated the synthesis of viral RNA several-fold by core enzyme or the viral enzyme in the presence of either Mg^{2+} or Mn^{2+}.

Herpes virus type I (HSV1)

In contrast to vaccinia virus, the DNA-containing HSV1 multiplies in the nuclei of infected cells. The herpes viruses are commonly grouped as a function of the high GC (62–67%) content of these viruses. However, members of the common grouping may be quite different; for example, cytomegalovirus (HCMV) has effects on polyamine metabolism quite different from those produced by HSV1.

HSV1 is found as empty A-capsids and as fully packaged C-capsids, within which the DNA is a dense ball. The protein contents and compositions of the A- and C-capsids are apparently identical. These results have suggested the initial formation of a DNA-free precursor into which the DNA is subsequently drawn. In the formation of C-capsids, a major core protein appears to be expelled when the DNA is pulled in.

Gibson and Roizman (1971) isolated enveloped viral particles from HSV1-infected human carcinoid cells and demonstrated the presence of spermidine and spermine in the molar ratio of spermidine to spermine of 1.6 ± 0.2. Removal of an outer shell by treatment with a nonionic detergent and urea left a nucleocapsid core containing the viral DNA. Much of the spermidine was lost in this process; that is, the triamine appeared to be associated with the outer viral shell. The residual spermine could neutralize about 40–50% of the viral DNA-P. These results were confirmed in the analysis of HSV1 derived from HeLa cells (Cohen & McCormick, 1979) and suggests a role for spermine in the organization of the inner nucleocapsid. Incomplete nuclear herpes viral particles (i.e., with DNA-filled capsids) acquire an additional membrane (i.e., the second shell) in leaving the nucleus. A pattern of packaging similar to that seen in

DNA phages and addition of a nuclear membrane has been suggested in the assembly of avian infectious laryngotracheitis virus, an economically significant member of the herpes virus group (Guo & Trottier, 1994).

Gibson and Roizman (1971) showed that the polyamine of the virus could be labeled by exogenous ornithine only if the amino acid were incorporated before infection. The implication that the conversion of putrescine to higher polyamine is inhibited after infection was subsequently borne out (McCormick & Newton, 1975). McCormick (1978) then showed that synthesis of spermidine and spermine stopped soon after infection. The analyses were complicated by the phenomena of polyamine leakage from the strains of host cells used. Also, infection inhibited leakage in the HeLa cells and in baby hamster kidney (BHK) cells (Wallace & Keir, 1981).

Although exogenous polyamine did not stimulate viral yield significantly in earlier studies, depletion of polyamine by DFMO did reduce the production of infectious HSV1 (Fig. 19.8). This result was thought to arise from effects on DNA synthesis (Pohjanpelto et al., 1988). Neither DFMO nor MGBG blocked multiplication of HSV1 and HSV2 in undepleted cells (McCormick & Newton, 1975; Tyms et al., 1979). However, replication of the human CMV (HCMV) was markedly inhibited by these agents, indicative of a dependence of that virus on de novo biosynthesis of the polyamines.

In studies of the synthesis of HSV1 DNA in the nuclei isolated from infected cells, the organelles were found to be essentially devoid of polyamine. These nuclei did synthesize DNA but these relatively short polynucleotides had been cleaved, unlike a more normal HSV1 DNA made in a crude lysate of infected cells. When polyamines and ATP were added to the nuclei, the combination markedly improved the quality of the HSV1 DNA made in the nuclei (Francke, 1978). The deoxyribonucleases induced by HSV1 and HSV2 were isolated and found to degrade viral DNA (Hoffmann & Cheng, 1978), with an absolute requirement for either Mg^{2+} or Mn^{2+}; however, the enzymes are inhibited 80–90% by spermine and spermidine. Thus, the polyamines appear to minimize DNA degradation in infected nuclei.

A HSV1-induced DNA polymerase has been purified from infected BHK cells (Wallace et al., 1980). The activity of the enzyme was enhanced threefold by spermidine and fourfold by spermine; putrescine had little effect.

HCMV

This virus, a major pathogen, is treated in texts as a member of the herpes virus group (Griffiths & Grundy, 1987). In contrast to HSV1, which codes many new virus-induced enzymes, HCMV is far more dependent on the macromolecular syntheses of its host cell. Replication is characterized by the formation of intranuclear inclusions. Like HSV1, HCMV does induce a virus-specific DNA polymerase, but unlike HSV1, does not inhibit the synthesis of many host cell syntheses. Star-

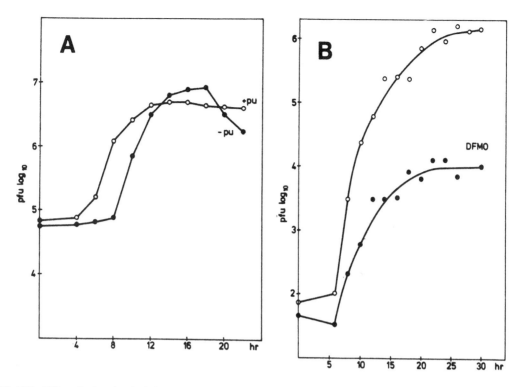

FIG. 19.8 Effect of polyamine depletion on growth of (**A**) Sindbis virus in P22 cells and (**B**) *Herpes simplex* virus in Verocells; adapted from Pohjanpelto et al. (1988). (**A**) The cells grown in the presence or absence of putrescine. (**B**) The cells cultured in the presence or absence of DFMO.

tlingly, unlike HSV1, HCMV induces an ODC activity (Fig. 19.9), which is a function of the multiplicity of infection and the activity of the viral preparation. The virus-induced activity, which was not inhibited by polyamines (Isom, 1979), appeared relatively late, at 10 h. The formation of ODC was tied to DNA synthesis, possibly as a "late" enzyme. The ODC activity evoked by infection was inhibited >99% by DFMO, as was the ODC induced by serum in arrested cells. However, in the experiments of Isom and Pegg (1979), an inhibitory concentration of DFMO (5 mM) or of α-methylornithine did not affect the production of the virus or the stimulation of DNA synthesis during infection. This result appears to challenge the concept of a role for new polyamines during viral multiplication.

A rapid uptake of putrescine by infected cells and a stimulated incorporation of labeled putrescine into spermidine and spermine were demonstrated in infection (Tyms & Williamson, 1980). These events paralleled the relatively late formation of viral DNA. An early accumulation of newly synthesized spermidine and spermine in infected cells and a marked inhibition of viral growth and intranuclear inclusions by MGBG and DFMO (10 mM) were then found (Tyms & Williamson, 1982). Depletion of cellular polyamine by pretreatment of cells with DFMO also inhibited multiplication of the virus, although the depleted infected cells did synthesize

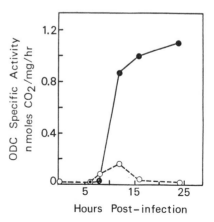

FIG. 19.9 Stimulation of ODC activity in human embryo (Flow 5000) cells infected by human cytomegalovirus; adapted from Isom (1979). (○) Mock infecting fluid or (●) virus was added to low serum-arrested confluent cells.

viral DNA and some empty capsids (Gibson et al., 1984). It was suggested that the polyamines were involved in a late event in replication, such as DNA packaging. Nevertheless, a specific inhibition of DNA synthesis did not affect the syntheses of spermidine and spermine, which were not diminished (Tyms et al., 1987).

The synthesis of spermidine and spermine in HCMV-infected cells has evoked a study of AdoMetDC. Beginning some 5 h after infection, the activity of this enzyme increased some five-to-sevenfold in another 10 h, kinetics similar to that seen for ODC (Fig. 19.9). The putrescine-activated enzyme was isolated from infected cells and, although somewhat similar to the host cell enzyme, had a slightly lower affinity for AdoMet and a much greater stability in NaCl (0.8 M) (White et al., 1994).

Polyamine contents of other DNA viruses

The finding of polyamines in HSV1 in 1971 provoked a search for these compounds in other purified animal viruses, such as the human adenovirus type 5 (Shortridge & Stevens, 1973). The virions were reported to contain enough spermidine and spermine to neutralize about 10% of the virus DNA. The DNA of the virus had been thought to be neutralized largely (60%) by arginine-rich acid-soluble polypeptides. The packaging of DNA within adenoviruses is thought to be similar to that found in some DNA phages (Hendrix & Garcea, 1994).

Incorporation of [^{14}C]-ornithine or [^{14}C]-putrescine into KB cells infected by adenovirus type 5 was used to determine if the polyamines were synthesized after infection (Pett & Ginsberg, 1975). Indeed infected cells stopped converting ornithine to putrescine after 8–12 h and the conversion of labeled putrescine to spermidine was similarly inhibited. The viral particles had smaller levels of polyamine than that reported earlier; a maximum of 3.6% of the viral DNA could be neutralized. It was asked if the polyamine detected was not adherent contaminant derived from the host tissue. Infection of rodent cells by adenovirus type 5 caused chromosome aberrations, and these were minimized by addition of spermine to infected cultures (Bellett et al., 1982).

Some insect DNA viruses have been reported to contain polyamine. Some nuclear polyhedrosis viruses contained high molecular weight double-stranded DNA (Kelly, 1977). One of these, isolated from the cotton leafworm, *Spodoptera littoralis*, contained putrescine and spermidine in amounts sufficient to neutralize 16% of the DNA phosphorus (Elliott & Kelly, 1977). Purification of the virus involved alkaline treatment of polyhedra, suggesting that the polyamines were not adventitious contaminants.

Two single-stranded DNA insect viruses, termed densonucleosis viruses (DNV), were found to contain putrescine, spermidine, and spermine. DNV1 was isolated from wax moth larvae, and the serologically similar DNV2 was obtained from butterfly larvae (Kelly et al.,

1977). The sedimentations revealed virus and top components lacking both DNA and polyamines. The DNV1 and -2 contained almost identical amounts of each of the three polyamines, sufficient to neutralize 26% of the DNV DNA (Kelly & Elliott, 1977). A DNV isolated from the silkworm was found to have a different polyamine composition, despite the presence of a DNA of similar size to DNV1 and -2.

Animal RNA Viruses

Many differences can be found among the animal DNA viruses in their effects on the polyamines. Viruses containing double-stranded DNA and multiplying in nuclei may or may not contain significant amounts of polyamine, may shut off polyamine biosynthesis, or enlarge this synthesis. Viruses with single-stranded DNA can contain significant amounts of these organic cations, and indeed the amounts found in pathogenic insect viruses may be of interest in problems of their control. The data on the polyamines of animal RNA viruses are similarly diverse.

The lethal epidemic of influenza after World War I evoked an early interest in the influenza virus. An acceptable purification of the virus for a "killed" vaccine took many years and the complex virus proved to possess an irregular form and a multipartite genome, containing negative strand RNA and its own unique RNA-dependent RNA polymerase. Spermine, spermidine, and putrescine were found in a myxovirus amd a paramyxovirus (i.e., the PR8 strain of influenza virus and the SP strain of Newcastle disease virus) in amounts capable of neutralizing 30–40% of the viral nucleic acid phosphate (Bachrach et al., 1974).

Data on picornaviruses

The much less complex poliovirus was found to be essentially devoid of polyamine (Ames & Dubin, 1960). In any case, the easily isolable RNAs of viruses of this group were found to be infectious and the infectivities of encephalomyocarditis virus (Moscarello, 1965) and Mengo virus (Cobet & Akers, 1968) were found to be increased by relatively high concentrations of putrescine. Later physical study of these viral RNAs revealed differences of secondary structure among them, properties accentuated by the polyamines. The encephalomyocarditis virus, with an RNA of 7800 nucleotides, was found to have about 200 molecules of putrescine, 100 molecules of spermidine, and 40 molecules of spermine, capable of neutralizing about 11% of the viral genome (Sheppard et al., 1980).

The *picorna* (small RNA) viruses contain several important representatives of the etiologic agents of disease. One of three polio virus strains, type I virions, had been crystallized in 1956 and subjected to crystallographic analysis a few years later. Crystals of the human rhinovirus (HRV) 14 have been found to be similar in structure and crystal packing to those of type I poliovirus, one of the enterovirus group (Erickson et al., 1983).

Extended crystallographic analyses of HRV2, poliovirus, and Mengo virus attempted to delineate immunogenic sites and other functional sites potentially inactivated by molecularly modeled drugs (Blaas et al., 1987; Rossmann et al., 1987).

HRV14 was found to incorporate large amounts of polyamines from polyamine-labeled cells, which did not confer significant levels of polyamine to poliovirus (Shay Fout et al., 1984). HeLa cells were blocked in putrescine synthesis by DFMO (3 mM) and were simultaneously grown in the presence of labeled [^{14}C]-putrescine (10 mM) until the pool sizes and the specific activities of putrescine, spermidine, and spermine were essentially identical. These polyamine-labeled cells were infected with HRV14 or poliovirus in the presence of DFMO and in the absence of exogenous putrescine, resulting in the virus drawing labeled polyamine from the cell pool. The polyamine contents of the two purified viruses are presented in Table 19.4 and show that in the case of HRV14, these could neutralize 27% of the RNA phosphate whereas only 1.6% of the RNA could be so neutralized in poliovirus. No effort has yet been made to localize polyamine within the HRV14 structure. Two minor notes may be added to the puzzle of the absence of polyamine in poliovirus. The "live" oral Sabin vaccine is heat labile and is occasionally inactivated during storage and transport. Of possible practical interest is the fact that the heat stability of the purified type I virus is greatly increased by L- or D-lysine, cadaverine, or spermidine (Dorval et al., 1990).

In the cell-free de novo synthesis of poliovirus from infectious virus RNA in an extract of uninfected HeLa cells (Molla et al., 1991), the synthesis of viral polypeptides was accomplished in a mix containing 1 mM spermidine and 4.4 mM Mg^{2+}. Also, the translation of viral RNA to viral polypeptides was inhibited by 4 μM edeine, a polypeptide antibiotic containing spermidine.

It is known that the RNA of poliovirus melts (i.e., reveals an increase in ultraviolet absorption at 260 μm) at a lower temperature and to a significantly lesser degree than do many other viral RNAs. This is attributed to a smaller proportion of double-stranded regions of RNA, and this may account for the small amount of polyamine in the structure.

Multiplication of Semliki Forest virus (SFV)

This virus is classified as an alphavirus, within the large togavirus group. It is not pathogenic for humans but does produce an encephalitis in mice. SFV contains a single-stranded RNA of 4.0–4.5 × 10^6 Da, which is infectious in tissue culture cells as such. The virus does not carry an RNA polymerase and the RNA may serve as an mRNA for the synthesis of various virus-induced enzymes and structural proteins. Replication involves the formation of a complementary RNA strand that serves as a template for the synthesis of both a viral RNA of 42S, and a smaller (subgenomic) RNA of 26S that codes for four viral structural proteins (Tuomi, 1984).

Purified SFV, grown in BHK cells, was found to contain small amounts of spermidine and spermine, sufficient to neutralize about 3% of the RNA phosphate (Raina et al., 1981). Infection produced a rapid decrease of ODC and AdoMetDC in infected cells. The multiplication of SFV was strongly inhibited in DFMO-depleted cells and this effect was reversed by polyamines. These results implied a dependence on preexisting polyamines before infection (Tuomi et al., 1982). It was also found that SFV-induced RNA polymerase was decreased by depletion of polyamines and this was partly reversible by polyamines, which did not activate the polymerase. These results were interpreted as a requirement for polyamines during normal protein synthesis.

Nevertheless, one major aberration appeared during viral multiplication in depleted cells. The synthesis of intact viral RNA (42S) was markedly inhibited, although the production of the smaller 26S RNA was scarcely reduced. This disproportion was reversed by the addition of spermidine to the infected cells and was interpreted to suggest that the synthesis of the full genomic viral RNA was controlled by polyamine, perhaps in affecting the RNA template at a possible read-through site (Tuomi, 1983).

Other animal RNA viruses

Mention has been made of single-stranded viruses, such as influenza virus or vesicular stomatitis virus, that pos-

Table 19.4 Polyamine content of human rhinovirus 14 and poliovirus.

Sample	Polyamine	Fraction of dpm	Molecules/ virion	Total polyamine charge	Phosphate residues neutralized
HRV-14	Putrescine	0.1	64	2000 ± 30	26.6%
	Spermidine	0.64	404		
	Spermine	0.26	164		
Poliovirus	Putrescine	0.77	42	120 ± 10	1.6%
	Spermidine	0.19	10		
	Spermine	0.04	2		

Determined by Shay Fout et al. (1984). dpm, disintegrations per minute.

sess negative-stranded RNA and are compelled to supply and carry their replication machinery in their virions. Double-stranded RNA viruses, the Birnaviridae, also contain RNA-dependent RNA polymerases (Wickner, 1993). The L-A virus of yeast is such a virus, which encodes a lethal protein toxin. Cohn et al. (1978) described the polyamine requirement of yeast strains carrying a double-stranded RNA plasmid that encodes the "killer" toxin.

The retroviruses, which carry RNA and a reverse transcriptase to synthesize an essential intracellular DNA genome, were investigated initially in the transformation of infected cells. The multiplication of Rous sarcoma virus in chick cells required exogenous arginine (Kotler et al., 1972), and temperature-sensitive transformation markedly increased ODC and the putrescine content of the cells (Don & Bachrach, 1975). When these increases were prevented by DFMO, the transformed state was maintained (Hölttä et al., 1981). Infected cells, maintained at a high temperature to prevent transformation and partially depleted of polyamine by DFMO, nevertheless became transformed at the permissive temperature. Morphological transformation, at least, was not dependent on the normally high concentration of the polyamines.

Spermidine or spermine, in addition to Mg^{2+}, stimulates a retroviral DNA-directed DNA polymerase more than 10-fold (Marcus et al., 1981), an effect relatively specific for mammalian retroviruses. It may be noted that spermine enhances the reverse transcriptase of avian myeloblastosis virus in the presence of a divalent cation (Aoyama, 1989).

The polyamines are beginning to be examined in the pathogenesis of AIDS, a retroviral disease. Lymphocytes from infected humans appear to possess elevated contents of spermidine and spermine. Increased levels of spermidine acetyltransferase have also been reported (Colombatto et al., 1989). Some clinical cases of a virulent infection by *Pneumocystis carinii* in AIDS patients have been reported to be ameliorated by DFMO (Golden et al., 1984).

Plant Viruses

In 1934 the etiologic agent of the tobacco mosaic disease, TMV, was isolated and crystallized in 2-dimensional arrays by W. M. Stanley. Crystalline viral preparations, which had been found to be largely protein, were shown by F. C. Bawden and N. Pirie in 1936 to contain about 5% RNA, and by J. D. Bernal and I. Fankuchen to consist of long thin aligned rods of precisely measurable dimensions. These studies demonstrated the relatively simple nucleoprotein character of this virus. The size and shape of TMV became standards for electron microscopy, as well as proof of the validity of the applicability of hydrodynamic theory to polymer characterization. In the mid-1950s TMV provided the first example of an infectious RNA.

The nature of growth and the spread of infection in intact plants prevented the development of a physiology of plant virus multiplication. In 1967 Takebe discovered that healthy or infected plant cells denuded enzymatically of their rigid protective outer coats by pectinases could be isolated and maintained as suspensions of separate infected or infectable protoplasts. This procedure belatedly introduced the equivalence of bacterial or animal cell culture into plant virology.

About 700 different plant viruses have been described, about 94% of which are RNA viruses (Zaitlin & Hull, 1987). Among the RNA viruses, some 86% are positive stranded. Of the 6% that contain DNA, only a third contain double-stranded DNA. Some RNA viruses under active study possess multicomponent RNA genomes within several nucleoprotein particulate components (Kaper, 1975). For example, cowpea mosaic virus comprises two virions, each with different proteins and positive stranded RNAs, both of which are essential for multiplication in plants. One of these RNAs can cause its replication in cowpea protoplasts whereas the other RNA is completely dependent on the first for its replication.

The separate RNAs have been translated in cell-free wheat germ extract in which 0.4 mM spermidine serves as a significant stimulant for amino acid incorporation in the presence of 3 mM Mg^{2+}. The amine increases both the rate of translation and the size of the products (Davies et al., 1977).

Another multicomponent (4) RNA virus is that of cowpea chlorotic mottle virus, an isolated *bromo* virus, that has been dissociated into separable protein and RNA units. When the dissociation is effected at high ionic strength, reassembly at low ionic strength occurs in 5×10^{-3} Mg^{2+} or spermidine (Hiebert & Bancroft, 1969). Although this isolated virus does not contain polyamine (Nickerson & Lane, 1977), Mg^{2+} or spermidine have been shown to produce significant conformational changes of the viral RNA (Dickerson & Trim, 1978). Subsequent analysis of the virus by laser Raman spectroscopy demonstrated that the bases of the RNA are highly ordered; that is, the genomic bases are largely paired and stacked (Verduin et al., 1984).

Many of the RNA viruses can be obtained in relatively large amounts and crystallize readily. Many of the latter are icosahedral particles and indeed, like the rod-shaped TMV, comprise a shell of identical protein subunits surrounding a core of RNA. X-ray analyses of crystals of these viruses present major opportunities for understanding the packaging of RNA and the relations of protein and RNA within the particles. Nevertheless, only a few of these viruses have been studied for their polyamine content, for example, TMV and turnip yellow mosaic virus (TYMV), and X-ray crystallography has not yet demonstrated the structural relation of the polyamine within the crystallized virions. Studies on polyamine salts of crystallized polynucleotides (Chap. 23) have placed this question on the research agenda. In vitro effects of the polyamines in the expression of the RNA genomes and in their packaging within virions have been described.

TMV

This is probably the most extensively studied virus, with a limited constitution of 95% of a single protein present in 2130 subunits of 17.5 kDa and 5% of a single RNA molecule of 2×10^6 Da. The possible existence of a viral subgenomic TMV RNA will be discussed in a later section. The protein subunits are organized in a helix and the single-stranded RNA occupies a narrow groove in the helix with an association of three nucleotides per protein subunit. The disassembly of the two components and the complex steps of their reassembly have been described in some detail (Butler, 1984; Hirth & Richards, 1981). Assembly of the protein subunits begins *within* the RNA and proceeds bidirectionally with both 5' and 3' ends of the RNA dangling from the growing virion.

One set of properties of TMV is that of ion binding of the virus and protein. Ca^{2+} specifically stabilizes the viral particle and the helical aggregate of the coat protein. Loring et al. (1962) found Ca^{2+} cations on TMV, and the existence of two different calcium ion-binding sites per protein subunit was determined by Gallagher and Lauffer (1983). TMV RNA isolated by the phenol extraction procedure and dialyzed against a chelating agent contained <0.1 calcium ions per nucleotide. Such an RNA bound one Ca^{2+} ion per two nucleotide phosphates at low ionic strength (Gastfriend & Lauffer, 1983), and this was markedly reduced in 0.01 M KCl. It has been asked if a Ca^{2+} binding site on the protein subunit bridging to an RNA nucleotide may serve in forming an RNA-protein nucleation site in assembly. Nevertheless, the extended RNA chain is considered by most workers to be held in place in salt linkage largely by three arginine residues in each of the protein subunits. Aliphatic amino acids may provide hydrophobic surfaces on which the nucleotide bases may stack. Additionally, specific aspartate residues may hydrogen bond to the hydroxyl of the ribose.

Extending their work on the polyamines of phage, Ames and Dubin (1960) reported that TMV contained very small amounts of polyamine, capable of neutralizing <1% of the RNA phosphate. Johnson and Markham (1962) asked if this might play a role in protecting an end of the exposed TMV rod and thereby inhibiting aggregation of the rods. Bawden and Pirie (1972) noted that the presence of spermine in a plant extract reduced the infectivity of viral RNA in the extract. Furthermore, spermine prevented the extraction of the RNA of the plant extract into water in the phenol–water extraction system first used to demonstrate the infectivity of an intact viral RNA.

It was discovered, initially in work with TYMV (Pinck et al., 1970), that the 3' end of many plant virus RNAs may behave as a specific tRNA. Almost all strains of TMV can be charged with histidine specifically (Haenni et al., 1982). Nevertheless, the viral RNAs containing such a 3' end are generally uncharged and can accept specific nucleotides and the specific amino acid after isolation. When charged they serve poorly as donors in the synthesis of polypeptides. These reactions are also obtained with a subgenomic 3' end of viral RNA and can be carried out with otherwise quite specific *E. coli* enzymes. The distribution of plant and animal RNA viruses carrying such structures have been summarized (Haenni et al., 1982). Despite the considerable number of studies on the tRNA-like structures in the RNA viruses, their significance is not known. Because many bacterial and animal cell tRNAs contain discrete amounts of spermidine or other polyamines and these stimulate various tRNA reactions, it may be asked if polyamines are present at the dangling 3' end of assembling virus and may be found at the end of intact virus.

An interesting experiment on TMV assembly bears on this question. The isolated single-stranded RNA molecule in aqueous solution develops many double-stranded regions that bind the dye ethidium to form fluorescent complexes. The assembly of protein subunits on this now fluorescent molecule straightens the RNA and displaces the ethidium (Favre et al., 1972). Ethidium is a competitor of spermidine for binding to tRNA. It would be of interest to see if ethidium can bind to an end of a TMV virion and also if spermidine or spermine is released from complexes of TMV RNA in the presence of TMV protein.

TYMV

The icosahedral RNA virus (a *Tymovirus*), grown commonly on Chinese cabbage (*Brassica chinensis*) for experimental purposes, was first described by R. Markham and K. M. Smith in 1949. Kaper (1975) described its structure in detail. Extracts of infected leaves also contain easily isolated empty viral protein (capsid) shells, and these may be cocrystallizable with intact virions. Infection creates an unusual cytopathology marked by vesiculations and clumping of chloroplasts in the cytoplasm and the presence of empty capsids in the nucleus. The virus is transmitted in nature by insects with biting mouth parts or experimentally by abrasion of leaves with infectious juice to produce a systemic infection. The easy isolation of infected protoplasts in high yield (50%) from plants or the infection of newly isolated protoplasts has permitted the analysis of TYMV multiplication at a cellular level.

Infected leaves of Chinese cabbage markedly increased their content of RNA, the increase being largely that found in the virus (Matthews, 1958). The ease of isolating large amounts of crystalline virus and the possibility of comparing intact virions to their empty protein shells encouraged many structural studies. Some crystalline viral preparations were found to contain fragmented RNA. An apparently intact RNA of ca. 2×10^6 Da was isolable by the phenol–water extraction system and was subsequently shown to be infectious (Haselkorn, 1962). The RNA was fragmented by heat at 60°C and its hyperchromicity revealed a high order of secondary structure in the RNA, both in the virion and in solution. Molecular weights of the large RNA isolated by the phenol method have been estimated by sedimentation after reaction with formaldehyde and by polyacrylamide electrophoresis and yielded average molecular weights of 1.91×10^6 Da (Kaper, 1975). The virion contained 33% RNA, which possessed an unusually high content of cytosine nucleotides (38% of the total). The addition of 5 mM spermidine to a tymovirus RNA increased the hypochromicity of the RNA. The polyamine apparently fulfills the requirement for charge neutralization and compaction of the RNA.

The fractionation of RNA in sucrose gradients after heat denaturation of the virus revealed several classes of RNA, one of which was comparable to the unheated infectious particle of 1.9×10^6. Another was that of a noninfectious RNA of 0.25×10^6. The in vitro translation of the former large infectious RNAs failed to yield significant amounts of coat protein whereas that of the small RNA, which includes the 3′ end, was translated readily to form coat protein. Pleij et al. (1976) concluded that infectious RNA is "closed" and must be replicated to produce an "open" subgenomic 3′ coat protein cistron. The formation of a subgenomic RNA of TMV specific for coat protein has also been reported (Hunter et al., 1976). It was subsequently shown that a TYMV RNA sequence early in the 5′ end of the molecule can base pair with the 3′ end, possibly preventing translation of this portion of the RNA (Briand et al., 1978). The nucleotide sequence of the subgenomic RNA coding for coat protein was found to have a base composition identical to that of the whole RNA (Guilley & Briand, 1978).

Johnson and Markham (1962) reported the presence of an unusual polyamine in TYMV, bis(3-aminopropyl)amine or norspermidine, and the absence of spermine. The specific characterization was based largely on the unreliable gas chromatography of the unsubstituted amines. The authors were unable to detect a triamine in uninfected Chinese cabbage. The possibility of formation of a novel compound and possibly new biosynthetic enzymes in infection had been heightened by the earlier discovery of new virus-induced enzymes in phage systems. Reinvestigation by Beer and Kosuge (1970), largely by thin layer electrophoresis and by isolation of crystalline picrates, revealed the presence of spermidine and spermine in TYMV in amounts of the former sufficient to neutralize about 20% of the RNA. Norspermidine could not be found. The concentration of spermidine in the virus was in excess of 10-fold that of spermine. These findings were confirmed by Nickerson and Lane (1977) who used the thin-layer chromatography of dansyl polyamines to detect these compounds in TYMV and the multigenomic cowpea mosaic virus. Four other purified plant viruses, Bromoviruses, were found to be devoid of detectable polyamine.

Gas chromatography and mass spectrometry of N-acylated derivatives have established unequivocally the structure of the spermidinelike component (see Fig. 3.10). The acylated major amine of TYMV yielded mass ions characteristic of spermidine and not of nor-

spermidine. The latter could not be detected in the analyses (Cohen & Greenberg, 1981).

In the latter study, crystallization of virus in 40% saturated $MgSO_4$ or 7.5% heparin or dialysis did not decrease viral spermidine. The content of this polyamine varied from preparation to preparation in the range of 200–400 molecules of spermidine per virion. The spermine contents of these virus preparations were some 10–20% of the spermidine content. Both of these polyamines were obviously tightly encased in an impermeable protein shell. It was found, however, that TYMV could adsorb spermine relatively selectively and an exogenous spermine contaminant was 95% removable by sedimentation at pH 4.8 in the presence of 0.5 M NaCl or 0.06 M $MgCl_2$, a step recommended in the purification of the virus. RNA-free protein shells did not contain significant amounts of spermidine, which appears associated mainly, if not entirely, with the nucleate.

It has been possible to detect polyamine within TYMV by proton (^{1}H) nuclear magnetic resonance (^{1}H-NMR) (Virudachalam et al., 1983b). Peaks in the spectra of the viral preparation (Fig. 19.10) are obtained at the positions assigned to the protons of the polyamines in solution.

TYMV is an icosahedral particle that is roughly spherical (ca. 150 Å radius) with a mass of 5.5×10^6 Da of which 3.6×10^6 is contained in the 180 identical protein subunits of 20,000 Da (Kaper, 1975). These comprise the protein shell containing the RNA and polyamines. The crystallographic studies of Klug et al. (1966) and electron microscopy of Finch and Klug (1966) clarified the intricate arrangement of RNA and areas of concentration within the shell and virion, as shown in Figure 19.11. This picture set the stage for chemical studies of the interactions of protein subunits to create a stable capsid structure. This structure was in fact rendered less stable in intact virions by the expansive potentialities of the enclosed RNA. This indicated the existence of RNA–protein interactions and emphasized the relative lack of neutralizing basic charge in the amino acids of tymo virus protein subunits (Suryanarayana et al., 1988). A role for the polyamines has therefore been inferred in the important function of neutralizing the RNA, and thus stabilizing the virus.

When the infectious RNA and empty capsids obtained by the urea method of disruption were dialyzed together against pH 4.2 acetate buffer in the presence of 5 mM spermidine or 5 mM Mg^{2+}, some intact infectious "virions" were obtained. However, these dissociated again at neutral pH (Jonard et al., 1972). It was found that only poly(C) was taken up relatively stably by the capsids (Briand et al., 1975). Conversion of many of the cytosine residues (ca. one-sixth) of the TYMV RNA to uracil by treatment with bisulfite eliminated the ability of the RNA to react with the capsids in the presence of spermidine or Mg^{2+} (Bouley et al., 1975). It was concluded that these essential interacting cytosine residues were in single-stranded segments of an otherwise highly ordered RNA, that is, possessing a hypochromicity sug-

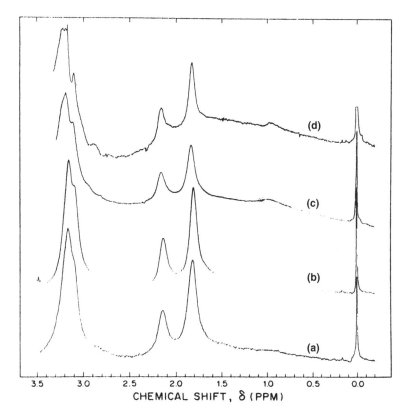

FIG. 19.10 ^{1}H-NMR spectra at 22°C of turnip yellow mosaic virus at (**A**) pH 5.5 and (**B**) pH 8.5 and spermidine-treated belladonna mottle virus at (**C**) pH 5.5 and (**D**) pH 8.5; adapted from Virudachalam et al. (1983b). The peaks at 1.8, 2.1, and 3.2 ppm were assigned to the protons of spermidine and spermine in solution.

gesting >60% base pairing (Jonard et al., 1976). Cross-linking reactions with bisulfite produce a cytosine-lysine couple. It has been concluded that essentially all protein subunits (80–90%) had reacted with RNA at a minimum of three sites (Ehresmann et al., 1980). These phenomena appear relevant to the structure of numerous RNA viruses, such as the positive stranded Mengo virus, encephalomyocarditis virus, and the foot and mouth

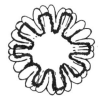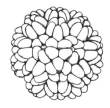

FIG. 19.11 Schematic drawing of turnip yellow mosaic virus (TYMV); adapted from Kaper (1975). Left: Gross RNA distribution within the arrangement of protein subunits in the capsid of the virus. Right: Geometric arrangement of TYMV protein subunits.

disease viruses, which also contain poly(C) tracts in their RNA.

Exposure of TYMV to alkaline conditions at high ionic strength releases a compacted RNA (38S instead of the more normal 28S at pH 6.0). Exposure to pH 10.5 in 1 M KCl at 30°C compacts the RNA within the virion (Pleij et al., 1977). Similar results were obtained with isolated RNA, implying a reorganization of the RNA chain. It was suggested that this might involve the dissociation and rearrangement of the bound polyamines because the effect appears at a pH close to the various pK values of spermidine amino groups. A similar result was obtained at lower pH (7) with the related labile tymovirus, *belladonna mottle virus*. Addition of cations or polyamines stabilized the virus considerably, preventing loss of RNA until the alkalinity was significantly increased to about pH 11 (Virudachalam et al., 1983a).

The release of RNA from the virus by freezing and thawing was shown to result from the formation of a hole in the enclosing protein shell. The hole was estimated as the result of dissociation of five to nine coat protein subunits (Katouzian–Safadi et al., 1983).

Matthews (1974) detected the penetration of various types of TYMV particles by CsCl, permitting their separation by density gradient ultracentrifugation. Some of the minor encapsulated components contain subgenomic RNA alone, whose polyamine content is not known. In the NMR study of the comovirus, cowpea mosaic virus, NMR observation showed that cesium ions can displace spermidine at high pH (Virudachalam et al., 1985). Obviously NMR can become a powerful tool in the study of the tymoviruses. NMR is capable of detecting RNA association with protein, the presence of polyamine in virions, and the displacement of the polyamines by cations such as cesium$^+$.

Another useful tool has been that of circular dichroism (CD) that can indicate the degree of secondary structure in RNA and protein as well as interactions of these components. Conformational changes of the extracted TYMV RNA were detected as a function of ionic strength and as a function of added spermidine (Katouzian–Safadi et al., 1985).

Polyamine metabolism in TYMV-infected plant cells

Structural and genomic similarities of TYMV to animal viruses of the Sindbis group and particularly to SFV have been described (Morch et al., 1988). In recent years much of the genomic analysis has followed the construction of cDNA clones, which permitted the nucleotide sequencing of the virus genomic RNA (Keese et al., 1989; Morch et al., 1988). A full-length cDNA is infectious and initiates viral production. Such DNA is being used in studies of mutagenesis (Weiland & Dreher, 1989).

The positive-stranded infecting monopartite RNA is decapsidated and translated in the plant cell to generate two large overlapping proteins. The largest, derived from 87% of the TYMV genome, is a precursor to both the second and several other products derived by proteolytic maturation of the initial polyprotein translation product (Morch et al., 1989). One of these is found in the RNA replicase (Mouches et al., 1984); another appears to be involved in viral spread (Bozarth et al., 1992).

Much of the dissection of genomic expression has resulted from studies of in vitro translation of various isolated RNA species. Translation in wheat germ extracts, for example, which revealed the selective subgenomic mRNA for the coat protein (Klein et al., 1976) is optimal in the presence of 2.75 mM Mg^{2+} and 0.60–0.70 mM spermidine. Added spermine has been shown to enhance coat protein formation in a wheat-germ extract, despite the presence of some polyamine in the components of the translation system (Benicourt & Haenni, 1976); the addition of spermidine to wheat-germ systems was adopted in subsequent studies (Morch et al., 1982). A read-through translation of viral RNA past termination codons is augmented by excess Mg^{2+} or by polyamine. The RNA replicase generates full-length negative strands and produces full-length double-stranded RNA, whose subsequent transcription forms the subgenomic RNA containing the coat protein sequence and the tRNA at its 3' end (Gargouri et al., 1989). A conserved region (a "tymobox") common to the tymoviruses may serve as a promoter for the synthesis of the subgenomic RNA (Ding et al., 1990). The subgenomic RNA is translated to form coat protein that encapsidates genomic RNA and some less complete RNA intermediates. The detailed roles of the polyamines in these sequences are not known, although they have been implicated in numerous phenomena of viral structure, RNA release, conformation, and encapsidation found in many RNA viruses.

Although techniques of mechanical inoculation have demonstrated a high infectivity of TYMV in cabbage leaves, much recent work has involved the infection of protoplasts. Cationic polymers like polyornithine have facilitated this, but more recently TYMV RNA has been introduced (1 μg/mL) into leaf protoplasts (10^4–10^5/mL) by means of electroporation; the production of viral RNA and proteins was observed (Boyer et al., 1993). The infection of *Brassica* leaf protoplasts by TYMV (1 μg/mL) at pH 5.4–5.8 in the presence of poly-L-ornithine (1 μg/mL) permitted a >90% infection in 24 h. Serologically reactive stainable TYMV protein appeared in the cells at 12 h with the production of 1–2 × 10^6 TYMV particles per cell in 48 h (Renaudin et al., 1975). The infected cells contained characteristic aggregates of chloroplasts. Subsequently Sugimura and Matthews (1981) detected the incorporation of [^{35}S]-methionine into virions and empty shells at approximately equal rates but recorded an earlier appearance of the amino acid into the empty capsids. The appearance of the virus-coded replicase subunit was detected as early as 4 h after TYMV infection of the protoplasts, although replicase activity was found somewhat later (Candresse et al., 1987).

An early study of polyamine synthesis in the leaves of similarly infected and normal plants demonstrated several-fold increases of the three major polyamines over a 2.5-week period (Table 19.5) (Torget et al., 1979). A high proportion of the spermidine and spermine of the plant was found in the virus. In the biological systems also, leaf disks incorporated [2-^{14}C]-methionine into the spermidine, spermine, and AdoMet at similar rates.

Experiments on polyamine metabolism have been done with mechanically infected young leaves on growing plants, in which a very high percentage of subsequently isolated protoplasts was shown to be infected and to produce virus in a few days after inoculation (Balint & Cohen, 1985a). Although such a system cannot be easily used to determine early stages of multiplication, at 7 days after inoculation the isolated protoplasts contain 1–2 × 10^6 virions per isolated protoplast and continue to make virus for at least 48 h. Obviously this system could have been used to study the biochemistry of viral multiplication soon after the virus was first discovered by Markham and Smith in 1949.

Cohen et al. (1981) described a rapid incorporation of

Table 19.5 Polyamine content of healthy and TYMV-infected Chinese cabbage.

Leaves (days)	Pu	Spd	Spm	AdoMet	mg Chlorophyll/g	mg TYMV/g
Healthy						
6	36	250	63	11	1.0	0.0
11	26	270	31	5.6	0.91	0.0
14	23	86	23	7.0	1.1	0.0
18	21	77	42	6.0	1.0	0.0
Infected						
6	41	400	98	14	0.95	≤0.15
11	66	480	110	8.3	0.92	1.5
14	120	490	71	9.1	0.97	0.8
18	200	370	67	10	0.71	0.9

Data from Torget et al. (1979). Polyamine values are in nmol/g.

radioactivity from [2-^{14}C]-methionine into the protein, spermidine, and spermine of normal and infected protoplasts. The spermidine content of protoplasts from 7-day cabbage infections was about twice that of protoplasts from healthy plants. High levels of spermidine synthase were also found in these healthy and infected cells. Balint and Cohen (1985a) described the incorporation of [2-^{14}C]-methionine in infected protoplasts into the protein, spermidine, and spermine of the virus, finding a specific radioactivity of the polyamines in the virus about twice that in the whole cell. This implied a selective concentration of newly synthesized polyamine in the virus, but may only reflect a high level of stored, inactive polyamine in the large vacuole of the plant cells, as in yeast. Exogenous [^{14}C]-spermidine was also readily incorporated into newly formed virus; it was not detectable in empty capsids. Label from the tetramethylene portion of spermidine was found at very high specific radioactivity in viral spermidine and in viral spermine.

Cyclohexylamine, a specific inhibitor of the cabbage spermidine synthase, was used at 1 mM to affect spermidine synthesis from [2-^{14}C]-methionine in the infected protoplast. Incorporation of label into cell spermidine was decreased about 64% whereas that into spermine was increased significantly and exceeded that into spermidine. Incorporation of label into viral spermidine fell about 60% whereas that into spermine almost doubled. In additional experiments in which healthy protoplasts were infected in vitro, cyclohexylamine was used once again to inhibit spermidine synthesis. In this system the inhibitor reduced spermidine synthesis about 85% and stimulated spermine synthesis over 20%. However, the labeling of protein with [^{14}C]-methionine and of RNA with [^{3}H]-uridine were not significantly affected. The production of intact viral particles was reduced <10% but the polyamine composition of the virions had changed significantly (Table 19.6). The spermidine content of the resulting TYMV virions decreased about a third and the spermine content was increased to about a third of the total polyamine (Balint & Cohen, 1985b). Nevertheless the polyamine charge per virion in the spermine-enriched TYMV ap-

proached that found in control TYMV. This result suggests, but does not prove, that the maintenance of polyamine charge is essential to the production of intact TYMV virions. It would be important to repeat this early experiment with a combination of inhibitors of both spermidine synthase and spermine synthase.

The studies described above have definitely eliminated the presence of norspermidine in TYMV and confirmed the presence of spermidine and spermine in the virus. These polyamines were entirely associated with viral RNA. The synthesis of spermidine is achieved by a cytosolic spermidine synthase, and inhibition of this enzyme shunts decarboxylated AdoMet to the synthesis of viral spermine, presumably in the cytosol. In a study of this inhibition, viral spermidine and viral spermine appeared to be interchangable in viral biosynthesis and assembly.

Concluding Remarks

Despite a major reliance on host metabolism and synthetic apparatus, the biochemical diversity among the known viruses is enormous and the synthetic solutions adopted by these minimal organisms can also be quite different. The study of the presence of the polyamines

Table 19.6 Effects of cyclohexylamine (CHA) on polyamine content in TYMV.

	Control		CHA	
	28 h	38 h	28 h	38 h
Spermidine (molecules/virion)	296	305	206	196
Spermine (molecules/virion)	64	63	97	126
Polyamine charge (per virion)[a]	1144	1167	1006	1092

Data from Balint and Cohen (1985b).
[a]Polyamine charge refers to the total positive charge contributed by spermidine (3 per molecule) and spermine (4 per molecule) in the virus.

for any one virus may turn on the permeability of the particle. The apparent absence of spermidine in T3 and T7 phages may reflect an ease of exchange and mask roles for the triamine eventually discovered in packaging DNA, the synthesis of viral mRNA by a virus-induced RNA polymerase, and the synthesis of viral proteins by a complex host apparatus. The biosynthesis of spermidine in T3 may be arrested by the destruction of AdoMet, and the fate of host spermidine and its distribution in an infected cell presents a difficult problem. With infections by some other DNA viruses, such as HCMV, it appears that polyamine synthesis is expanded after infection. Among some RNA viruses, newly synthesized polyamines may help to organize an RNA core, a role otherwise left in some other viruses to the basic charge on capsid proteins.

If one is interested in viral polyamines it is evident that the choice of virus and the nature of the function to be examined deserves some thought. If the focus of attention is the nucleate it would be important to recall that polyamines are known to organize the secondary and tertiary structures of these polymers, and also thereby affect the biosynthesis of viral proteins and nucleates. In these terms knowledge of the roles of the polyamines among the viruses has barely begun. Almost two centuries ago it was recognized that neutral salts contained both cations and anions. The neglect of the roles of the cations in nucleate organization and function is seen in the above discussions on the packaging, compaction, metabolism, and biosynthesis of the viral nucleates.

REFERENCES

Alberts, B. M. & Frey, L. (1970) Nature, 227, 1313–1318.
Ames, B. N. & Dubin, D. T. (1960) Journal of Biological Chemistry, 235, 769–775.
Ames, B. N., Dubin, D. T., & Rosenthal, S. M. (1958) Science, 127, 814–816.
Aoyama, H. (1989) Biochemistry International, 19, 67–76.
Astrachan, L. & Miller, J. F. (1973) Journal of Virology, 11, 792–798.
Atkins, J. F., Lewis, J. B., Anderson, C. W., & Gesteland, R. F. (1975) Journal of Biological Chemistry, 250, 5688–5695.
Bachrach, U. & Ben-Joseph, M. (1971) FEBS Letters, 15, 75–77.
Bachrach, U., Don, S., & Wiener, H. (1974) Journal of General Virology, 22, 451–454.
Bachrach, U., Fischer, R. F., & Klein, I. (1975) Journal of General Virology, 26, 287–294.
Balint, R. & Cohen, S. S. (1985a) Virology, 144, 181–193.
——— (1985b) Virology, 144, 194–203.
Bawden, F. & Pirie, N. W. (1972) Proceedings of the Royal Society of London Series B–Biological Sciences, 182, 319–329.
Becker, A. & Gold, M. (1978) Proceedings of the National Academy of Sciences of the United States of America, 75, 4199–4203.
Becker, A., Marko, M., & Gold, M. (1977b) Virology, 78, 291–305.
Becker, A., Murialdo, H., & Gold, M. (1977a) Virology, 78, 277–290.

Beer, S. V. & Kosuge, T. (1970) Virology, 40, 930–938.
Bellett, A. J. D., Waldron–Stevens, L. K., Braithwaite, A. W., & Cheetham, B. F. (1982) Chromosomes, 85, 571–583.
Benbasat, J. A. (1984) Biochemistry, 23, 3609–3619.
Benicourt, C. & Haenni, A. L. (1976) Journal of Virology, 20, 196–202.
Blaas, D., Kuechler, E., Vriend, G., Arnold, E., Luo, M., & Rossmann, M. G. (1987) Proteins, 2, 263–272.
Black, L. W. (1981) Virology, 113, 336–344.
——— (1989) Annual Review of Microbiology, 43, 267–292.
Bolin, R. W. & Cummings, D. J. (1975a) Journal of Virology, 16, 1273–1281.
——— (1975b) Journal of Virology, 15, 232–237.
Bouley, J. P., Briand, J. P., Jonard, G., Witz, J., & Hirth, L. (1975) Virology, 63, 312–319.
Boyer, J. C., Zaccomer, B., & Haenni, A. (1993) Journal of General Virology, 74, 1911–1917.
Bozarth, C. S., Weiland, J. J., & Dreher, T. W. (1992) Virology, 187, 124–130.
Brennan, C. A., Manthey, A. E., & Gumport, R. K. (1983) Methods in Enzymology, 100, 38–52.
Briand, J., Keith, G., & Guilley, H. (1978) Proceedings of the National Academy of Sciences of the United States of America, 75, 3168–3172.
Briand, J. P., Bouley, J. P., Jonard, G., Witz, J., & Hirth, L. (1975) Virology, 63, 304–311.
Brown, P. O. & Cozzarelli, N. R. (1981) Proceedings of the National Academy of Sciences of the United States of America, 78, 843–847.
Bruce, D. L. & Warren, R. A. J. (1983) Canadian Journal of Microbiology, 29, 827–829.
Butler, P. J. G. (1984) Journal of General Virology, 65, 253–279.
Candresse, T., Batisti, M., Renaudin, J., Mouches, C., & Bové, J. M. (1987) Annals of the Institute Pasteur/Virology, 138, 217–227.
Cech, T. R. (1983) Cell, 34, 713–716.
Chowrira, B. M., Berzal–Herranz, A., & Burke, J. M. (1993) Biochemistry, 32, 1088–1095.
Christiansen, C. & Baldwin, R. L. (1977) Journal of Molecular Biology, 115, 441–454.
Cohen, S. S. (1968) Virus-Induced Enzymes. New York: Columbia University Press.
Cohen, S. S., Balint, R., & Sindhu, R. K. (1981) Plant Physiology, 68, 1150–1155.
Cohen, S. S. & Greenberg, M. L. (1981) Proceedings of the National Academy of Sciences of the United States of America, 78, 5470–5474.
Cohen, S. S. & McCormick, F. P. (1979) Advances of Virus Research, vol. 24, pp. 331–387, M. A. Lauffer, F. B. Bang, K. Maramorasch, & K. M. Smith, Eds. New York: Academic Press.
Cohn, M. S., Tabor, C. W., Tabor, H., & Wickner, R. B. (1978) Journal of Biological Chemistry, 253, 5225–5227.
Colombatto, S., DeAgostini, M., Corsi, D., & Sinicco, A. (1989) Biological Chemistry Hoppe–Seyler, 370, 745.
Couse, N. L., Cummings, D. J., Chapman, V. A., & DeLong, S. S. (1970) Virology, 42, 590–602.
Cozzarelli, N. R. (1980) Cell, 22, 327–328.
Cummings, D. J. & Bolin, R. W. (1975) Bacteriological Review, 40, 314–359.
Dahm, S. C. & Uhlenbeck, O. C. (1991) Biochemistry, 30, 9464–9469.
Dasgupta, B. & Chakravorty, M. (1978) Journal of Virology, 28, 736–742.

Davies, J. W., Aalbers, A. M. J., Stuik, E. J., & van Kammen, A. (1977) *FEBS Letters*, **77**, 265–269.

Dickerson, P. E. & Trim, A. R. (1978) *Nucleic Acids Research*, **5**, 987–998.

Ding, S., Howe, J., Keese, P., Mackenzie, A., Meek, D., Osorio–Keese, M., Skotnicki, M., Srifah, P., Torronen, M., & Gibbs, A. (1990) *Nucleic Acids Research*, **18**, 1181–1187.

Dion, A. S. & Cohen, S. S. (1971) *Journal of Virology*, **8**, 925–927.

Dion, A. S. & Cohen, S. S. (1972a) *Journal of Virology*, **9**, 419–422.

——— (1972b) *Journal of Virology*, **9**, 423–430.

Don, S. & Bachrach, U. (1975) *Cancer Research,* **35**, 3618–3622.

Dorval, B. L., Chow, M., & Klibanov, A. M. (1990) *Biotechnology and Bioengineering*, **35**, 1051–1054.

Dürrenberger, F. & Rochaix, J. D. (1991) *EMBO Journal*, **10**, 3495–3501.

Earnshaw, W. C. & Casjens, S. R. (1980) *Cell*, **21**, 319–331.

Earnshaw, W. C. & Harrison, S. C. (1977) *Nature*, **268**, 598–602.

Ehresmann, B., Briand, J. P., Reinbolt, J., & Witz, J. (1980) *European Journal of Biochemistry*, **108**, 123–129.

Elliott, R. M. & Kelly, D. C. (1977) *Virology*, **76**, 472–474.

Erickson, J. W., Frankenberger, E. A., Rossmann, M. G., Shay Fout, G., Medappa, K. C., & Rueckert, R. R. (1983) *Proceedings of the National Academy of Sciences of the United States of America*, **80**, 931–934.

Favre, A., Guilley, H., & Hirth, L. (1972) *FEBS Letters*, **26**, 15–19.

Feiss, M. & Becker, A. (1983) *Lambda II*, pp. 305–330, R. W. Hendrix, J. W. Roberts, F. W. Stahl, & R. A. Weisberg, Eds. Cold Spring Harbor, N.Y.: Cold Spring Harbor Laboratory.

Feiss, M. & Widner, W. (1982) *Proceedings of the National Academy of Sciences of the United States of America*, **79**, 3498–3502.

Fox, J. M., Johnson, J. E., & Young, M. J. (1994) *Seminars in Virology*, **5**, 51–60.

Francke, B. (1978) *Biochemistry*, **17**, 5494–5499.

Fuchs, E. (1976) *European Journal of Biochemistry*, **63**, 15–22.

Fuchs, E. & Fuchs, C. M. (1971) *FEBS Letters*, **19**, 159–162.

Fujisawa, H. & Hearing, P. (1994) *Seminars in Virology*, **5**, 5–13.

Fukuma, I. & Cohen, S. S. (1973) *Journal of Virology*, **12**, 1259–1264.

——— (1975a) *Journal of Virology*, **15**, 1176–1181.

——— (1975b) *Journal of Virology*, **16**, 222–227.

Gallagher, W. H. & Lauffer, M. A. (1983) *Journal of Molecular Biology*, **170**, 905–919.

Gargouri, R., Joshi, R. L., Bol, J. F., Astier–Manifacier, S., & Haenni, A. L. (1989) *Virology*, **171**, 386–393.

Gastfriend, H. H. & Lauffer, M. A. (1983) *Journal of Molecular Biology*, **170**, 931–937.

Geider, K. & Kornberg, A. (1974) *Journal of Biological Chemistry*, **249**, 3999–4005.

Gellert, M., Mizuuchi, K., O'Dea, M. H., & Nash, H. A. (1976) *Proceedings of the National Academy of Sciences of the United States of America*, **73**, 3872–3876.

Gerhard, B. & Warren, R. A. J. (1982) *Biochemistry*, **21**, 5458–5462.

Gibson, W. & Roizman, B. (1971) *Proceedings of the National Academy of Sciences of the United States of America*, **68**, 2818–2821.

Gibson, W., van Breeman, R., Fields, A., LaFemina, R., & Irmiere, A. (1984) *Journal of Virology*, **50**, 145–154.

Golden, J. A., Sjoerdsma, A., & Santi, D. (1984) *Western Journal of Medicine*, **141**, 613–623.

Gosule, L. C. & Schellman, J. A. (1976) *Nature*, **259**, 333–335.

Greenstein, M., Speth, J. L., & Maiese, W. M. (1981) *Antimicrobial Agents and Chemotherapy*, **20**, 425–432.

Griess, G. A., Serwer, P., & Horowitz, P. M. (1985) *Biopolymers*, **24**, 1635–1646.

Griess, G. A., Serwer, P., Kaushal, V., & Horowitz, P. M. (1986) *Biopolymers*, **25**, 1345–1357.

Griffiths, P. D. & Grundy, J. E. (1987) *Biochemical Journal*, **241**, 313–324.

Guerrier–Takada, C., Gardiner, K., Marsh, T., Pace, N., & Altman, S. (1983) *Cell*, **35**, 849–857.

Guilley, H. & Briand, J. (1978) *Cell*, **15**, 113–122.

Guo, P. & Trottier, M. (1994) *Seminars in Virology*, **5**, 27–37.

Guttman, B. & Kutter, E. M. (1983) *Bacteriophage*, pp. 8–10, C. K. Mathews, E. M. Kutter, G. Mosig, & P. B. Berget, Trans. Washington, D.C.: American Society of Microbiology.

Haenni, A., Joshi, S., & Chapeville, F. (1982) *Progress in Nucleic Acid Research and Molecular Biology* **27**, 85–104.

Hafner, E. W., Tabor, C. W., & Tabor, H. (1979) *Journal of Biological Chemistry*, **254**, 12419–12426.

Hamada, K., Fujisawa, H., & Minagawa, T. (1986a) *Virology*, **151**, 110–118.

——— (1986b) *Virology*, **151**, 119–123.

Harrison, D. P. & Bode, V. C. (1975) *Journal of Molecular Biology*, **96**, 461–470.

Hashimoto, C. & Fujisawa, H. (1992) *Virology*, **191**, 246–250.

Hendrix, R. W. & Garcea, R. L. (1994) *Seminars in Virology*, **5**, 15–26.

Herrlich, P., Scherzinger, E., & Schweiger, M. (1971) *Molecular and General Genetics*, **114**, 31–34.

Hershey, A. D. (1957) *Virology*, **4**, 237–264.

Higgins, R. R. & Becker, A. (1995) *Journal of Molecular Biology*, **252**, 31–46.

Hirth, L. & Richards, K. E. (1981) *Advances in Virus Research*, vol. 26, pp. 145–199. New York: Academic Press.

Hodgson, J. & Williamson, J. D. (1975) *Biochemical and Biophysical Research Communications*, **63**, 308–312.

Hoffmann, P. J. & Cheng, Y. (1978) *Journal of Biological Chemistry*, **253**, 3357–3562.

Hohn, B. & Hohn, T. (1974) *Proceedings of the National Academy of Sciences of the United States of America*, **71**, 2372–2376.

Hohn, B. & Murray, K. (1977) *Proceedings of the National Academy of Sciences of the United States of America*, **74**, 3259–3263.

Hölttä, E., Vartio, R., Jänne, J., Vaheri, A., & Hovi, T. (1981) *Biochimica et Biophysica Acta*, **677**, 1–6.

Huff, A. C., Leatherwood, J. K., & Kreuzer, K. N. (1989) *Proceedings of the National Academy of Sciences of the United States of America*, **86**, 1307–1311.

Hughes, J. A., Brown, L. R., & Ferro, A. J. (1987) *Nucleic Acids Research*, **15**, 717–729.

Hunter, T. R., Hunt, T., Knowland, J., & Zimmern, D. (1976) *Nature*, **260**, 759–764.

Isom, H. C. (1979) *Journal of General Virology*, **42**, 265–278.

Isom, H. C. & Pegg, A. E. (1979) *Biochimica et Biophysica Acta*, **564**, 402–413.

Jacobson, A. B., Kumar, H., & Zuker, M. (1985) *Journal of Molecular Biology*, **181**, 517–531.

Jonard, G., Briand, J. P., Bouley, J. P., Witz, J., & Hirth, L. (1976) *Philosophical Transactions of the Royal Society of London Series B–Biological Sciences*, **276**, 123–129.

Jonard, G., Witz, J., & Hirth, L. (1972) *Journal of Molecular Biology*, **66**, 165–169.

Jordan, C. F., Lerman, L. S., & Venable, Jr., J. H. (1972) *Nature New Biology*, **236**, 67–70.

Kaiser, D., Syvanen, M., & Masuda, T. (1975) *Journal of Molecular Biology*, **91**, 175–186.

Kaper, J. M. (1975) *The Chemical Bases of Virus Structure, Dissociation and Reassembly*. Amsterdam: North-Holland.

Katouzian–Safadi, M., Berthet–Colminas, C., Witz, J., & Kruse, J. (1983) *European Journal of Biochemistry*, **137**, 47–55.

Katouzian–Safadi, M., Charlier, M., & Maurizot, J. C. (1985) *Biochimie*, **67**, 1007–1013.

Keese, P., Mackenzie, A., & Gibbs, A. (1989) *Virology*, **172**, 536–546.

Kelly, D. C. (1977) *Virology*, **76**, 468–471.

Kelly, D. C., Barwise, A. H., & Walker, I. O. (1977) *Journal of Virology*, **21**, 396–407.

Kelly, D. C. & Elliott, R. M. (1977) *Journal of Virology*, **21**, 408–410.

Klein, C., Fritsch, C., Briand, J. P., Richards, K. E., Jonard, G., & Hirth, L. (1976) *Nucleic Acids Research*, **3**, 3043–3061.

Kotler, M., Weinberg, E., Haspel, O., & Becker, Y. (1972) *Journal of Virology*, **10**, 439–446.

Kozloff, L. M. (1983) *Bacteriophage T4*, pp. 25–31, C. K. Matthews, E. M. Kutter, G. Mosig, & P. B. Berget, Eds. Washington, D.C.: American Society of Microbiology.

Krasnow, M. A. & Cozzarelli, N. R. (1982) *Journal of Biological Chemistry*, **257**, 2687–2693.

Kreuzer, K. N. (1989) *Pharmaceutical Therapeutics*, **43**, 377–395.

Kreuzer, K. N. & Huang, W. M. (1983) *Bacteriophage T4*, pp. 90–96, C. K. Matthews, E. M. Kutter, G. Mosig, & P. B. Berget, Eds. Washington, D.C.: American Society for Microbiology.

Kropinski, A. M. B., Bose, R. J., & Warren, R. A. J. (1973) *Biochimie*, **12**, 151–157.

Krueger, D. H., Presber, W., Hansen, S., & Rosenthal, H. A. (1975) *Journal of Virology*, **16**, 453–455.

Kuhn, A. & Kellenberger, E. (1985) *Journal of Bacteriology*, **163**, 906–912.

Lanzer, W. & Holowczak, J. A. (1975) *Journal of Virology*, **16**, 1254–1264.

Last, J. A. & Anderson, W. F. (1976) *Archives of Biochemistry and Biophysics*, **174**, 167–176.

Leipold, B. (1977) *Journal of Virology*, **21**, 445–450.

Lerman, L. S. (1971) *Proceedings of the National Academy of Sciences of the United States of America*, **68**, 1886–1890.

——— (1973) *Physico-Chemical Properties of Nucleic Acids*, vol. 3, pp. 59–76, J. Duchesne, Ed. New York: Academic Press.

Lillehaug, J. R. & Kleppe, K. (1975) *Biochemistry*, **15**, 1225–1229.

Lillehaug, J. R., Kleppe, R. K., & Kleppe, K. (1976) *Biochemistry*, **15**, 1858–1865.

Maas, W. K., Leiger, Z., & Poindexter, J. (1970) *Annals of the New York Academy of Science*, **171**, 957–967.

Maltman, K. L., Neuhard, J., Lewis, H. A., & Warren, R. A. J. (1980) *Journal of Virology*, **34**, 354–359.

Maltman, K. L., Neuhard, J., & Warren, R. A. J. (1981) *Biochemistry*, **20**, 3586–3591.

Manning, G. S. (1978) *Quarterly Reviews of Biophysics*, **11**, 179–246.

——— (1980) *Biopolymers*, **19**, 37–59.

Marcus, S. L., Smith, S. W., & Bacchi, C. J. (1981) *Journal of Biological Chemistry*, **256**, 3460–3464.

Marx, K. A. (1987) *NATO Advanced Studies Institute*, volume E133, *Structures of Dynamics of Biopolymers*, pp. 137–168, C. Nicolini, Ed. Boston: M. Nyhoff.

Marx, K. A. & Reynolds, T. C. (1982) *Proceedings of the National Academy of Sciences of the United States of America*, **79**, 6484–6488.

Marx, K. A. & Ruben, G. C. (1983) *Nucleic Acids Research*, **11**, 1839–1854.

Matthews, R. E. F. (1974) *Virology*, **60**, 54–64.

McCarthy, D. (1979) *Journal of Molecular Biology*, **127**, 265–283.

McCormick, F. (1978) *Virology*, **91**, 496–503.

McCormick, F. & Newton, A. A. (1975) *Journal of General Virology*, **27**, 25–33.

McMurry, L. M. & Algranati, I. D. (1986) *European Journal of Biochemistry*, **155**, 383–390.

Miller, P. B., Maltman, K. L., & Warren, R. A. J. (1982) *Journal of Virology*, **43**, 67–72.

Min Jou, W., Raeymaekers, A., & Fiers, W. (1979) *European Journal of Biochemistry*, **102**, 589–594.

Miyamoto, S., Kashiwagi, K., Ito, K., Watanabe, S., & Igarashi, K. (1993) *Archives of Biochemistry and Biophysics*, **300**, 63–68.

Molla, A., Paul, A. V., & Wimmer, E. (1991) *Science*, **254**, 1647–1651.

Morch, M., Boyer, J., & Haenni, A. (1988) *Nucleic Acids Research*, **16**, 6157–6173.

Morch, M., Drugeon, G., Szafranski, P., & Haenni, A. L. (1989) *Journal of Virology*, **63**, 5153–5158.

Morch, M., Zagorski, W., & Haenni, A. L. (1982) *European Journal of Biochemistry*, **107**, 259–265.

Morita, M., Tasaka, M., & Fujisawa, H. (1993) *Virology*, **193**, 748–752.

Morris, D. R. & Lockshon, D. (1981) *Advances in Polyamine Research*, Vol. 3, pp. 299–307, C. M. Calderera, V. Zappia, & U. Bachrach, Eds. New York: Raven Press.

Moss, B., Ahn, B., Amegadzie, B., Gershon, P. D., & Keck, J. G. (1991) *Journal of Biological Chemistry*, **266**, 1355–1368.

Mouches, C., Candresse, T., & Bove, J. M. (1984) *Virology*, **134**, 78–90.

Moussatché, N. (1985) *Biochimica et Biophysica Acta*, **826**, 113–120.

Murialdo, H. & Becker, A. (1978) *Microbiological Review*, **42**, 529–576.

Nakasu, S., Fujisawa, H., & Minagawa, T. (1985) *Virology*, **143**, 422–434.

Nash, H. A. (1975) *Proceedings of the National Academy of Sciences of the United States of America*, **72**, 1072–1076.

Neuhard, J., Maltman, K. L., & Warren, R. A. J. (1980) *Journal of Virology*, **34**, 347–353.

Nickerson, K. W. & Lane, L. C. (1977) *Virology*, **81**, 455–459.

Obert, G., Tripier, F., Nonnenmacher, H., & Kirn, A. (1980) *Annals of Virology (Institute Pasteur)*, **131E**, 13–24.

Osburne, M. S., Maiese, W. M., & Greenstein, M. (1990) *Antimicrobial Agents and Chemotherapy*, **34**, 1450–1452.

Osland, A. & Kleppe, K. (1977) *Nucleic Acids Research*, **4**, 685–694.

Panagiotidis, C. A., Drainas, D., & Huang, S. (1992) *International Journal of Biochemistry*, **24**, 1625–1631.

Parris, W., Rubinchik, S., Yang, Y., & Gold, M. (1994) *Journal of Biological Chemistry*, **269**, 13564–13574.

Peebles, C. L., Gegenheimer, P., & Abelson, J. (1983) *Cell*, **32**, 525–536.

Pett, D. M. & Ginsberg, H. S. (1975) *Journal of Virology*, **15**, 1289–1292.

Pinck, M., Yot, P., Chapeville, F., & Duranton, H. M. (1970) *Nature*, **226**, 954–956.

Pleij, C. W. A., Eecen, H. G., Bosch, L., & Mandel, M. (1977) *Virology*, **76**, 781–786.

Pleij, C. W. A., Neeleman, A., Van Vloten–Doting, L., &

Bosch, L. (1976) *Proceedings of the National Academy of Sciences of the United States of America*, **73**, 4437–4441.

Pohjanpelto, P., Sekki, A., Hukkanen, V., & von Bonsdorff, C. (1988) *Life Sciences*, **42**, 2011–2018.

Quail, A., Karrer, E., & Warren, R. A. J. (1976) *Journal of General Virology*, **33**, 135–138.

Raae, A. J. & Kleppe, K. (1975) *Biochemistry*, **17**, 2939–2942.

Raae, A. J., Kleppe, R. K., & Kleppe, K. (1975) *European Journal of Biochemistry*, **60**, 437–443.

Rabussay, D. (1983) *Bacteriophage T4*, pp. 167–173, C. K. Mathews, E. M. Kutter, G. Mosig, & P. B. Berget, Eds. Washington, D.C.: American Society of Microbiology.

Raina, A., Tuomi, K., & Mäntyjärvi, R. (1981) *Medicinal Biology*, **59**, 428–432.

Renaudin, J., Bové, J. M., Otsuki, Y., & Takebe, I. (1975) *Molecular and General Genetics*, **141**, 59–68.

Riemer, S. C. & Bloomfield, V. A. (1978) *Biopolymers*, **17**, 785–794.

Rossmann, M. G., Arnold, E., Griffith, J. P., Kamer, G., Luo, M., Smith, T. J., Vriend, G., Ruekert, R. R., Sherry, B., McKinlay, M. A., Diana, G., & Otto, M. (1987) *Trends in Biochemical Science*, **12**, 313–318.

Rowe, T. C., Tewey, K. M., & Liu, L. F. (1984) *Journal of Biological Chemistry*, **259**, 9177–9181.

Rubinchik, S., Parris, W., & Gold, M. (1994a) *Journal of Biological Chemistry*, **269**, 13575–13585.

——— (1994b) *Journal of Biological Chemistry*, **269**, 13586–13593.

Sakaki, Y. & Oshima, T. (1975) *Journal of Virology*, **15**, 1449–1453.

Schmidt, F. J. & Apirion, D. (1983) *Bacteriophage T4*, pp. 208–217, C. K. Matthews, E. M. Kutter, G. Mosig, & P. B. Berget, Eds. Washington, D.C.: American Society of Microbiology.

Scraba, D. G., Bradley, R. D., Leyritz–Wills, M., & Warren, R. A. J. (1983) *Virology*, **124**, 152–160.

Sekiguchi, A., Komatsu, Y., Koizumi, M., & Ohtsuka, E. (1991) *Nucleic Acids Research*, **19**, 6833–6838.

Shay Fout, G., Medappa, K. C., Mapoles, J. E., & Rueckert, R. R. (1984) *Journal of Biological Chemistry*, **259**, 3639–3643.

Sheppard, S. L., Burness, A. T. H., & Boyle, S. M. (1980) *Journal of Virology*, **34**, 266–267.

Shibata, H., Fujisawa, H., & Minagawa, T. (1987a) *Journal of Molecular Biology*, **196**, 845–851.

——— (1987b) *Virology*, **159**, 250–258.

Shortridge, K. F. & Stevens, L. (1973) *Microbes*, **7**, 61–68.

Snyder, L. (1983) *Bacteriophage T4*, pp. 351–355, C. K. Matthews, E. M. Kutter, G. Mosig, & P. B. Berget, Eds. Washington, D.C.: American Society of Microbiology.

Spoerel, N. & Herrlich, P. (1979) *European Journal of Biochemistry*, **95**, 227–233.

Srivenugopal K. S. & Morris, D. R. (1985) *Biochemistry*, **24**, 4766–4771.

Stetler, G. L., King, G. J., & Huang, W. M. (1979) *Proceedings of the National Academy of Sciences of the United States of America*, **76**, 3737–3741.

Studier, F. W. & Movva, N. R. (1976) *Journal of Virology*, **19**, 136–145.

Sugimura, Y. & Matthews, R. E. F. (1981) *Virology*, **112**, 70–80.

Suryanarayana, S., Jacob, A. N., & Savithri, H. S. (1988) *Indian Journal of Biochemistry and Biophysics*, **25**, 580–584.

Sutcliffe, J. A., Gootz, T. D., & Barrett, J. F. (1989) *Antimicrobial Agents and Chemotherapy*, **33**, 2027–2033.

Syvanen, M. & Lin, J. (1978) *Journal of Molecular Biology*, **126**, 333–346.

Tabor, H., Hafner, E. W., & Tabor, C. W. (1980) *Journal of Bacteriology*, **144**, 952–956.

Tabor, H. & Tabor, C. W. (1982) *Proceedings of the National Academy of Sciences of the United States of America*, **79**, 7087–7091.

Tabor, H., Tabor, C. W., Cohn, M. S., & Hafner, E. W. (1981) *Journal of Bacteriology*, **147**, 702–704.

Torget, R., Lapi, L., & Cohen, S. S. (1979) *Biochemical and Biophysical Research Communications*, **87**, 1132–1139.

Tuomi, K. (1983) *Medical Biology*, **61**, 199–202.

——— (1984) Ph.D. Dissertation [Trans.], University of Kuopio, Finland.

Tuomi, K., Raina, A., & Mäntyjärvi, R. (1982) *Biochemical Journal*, **206**, 113–119.

Tyms, A. S., Davies, J. M., Clarke, J. R., & Jeffries, D. J. (1987) *Journal of General Virology*, **68**, 1563–1573.

Tyms, A. S., Rawal, B. K., Naim, H. M., & Williamson, J. D. (1983) *Advances in Polyamine Research*, vol. 4, pp. 507–517, U. Bachrach, A. Kaye, & R. Chayen, Eds. New York: Raven Press.

Tyms, A. S., Scamans, E., & Williamson, J. D. (1979) *Biochemical and Biophysical Research Communications*, **86**, 312–318.

Tyms, A. S. & Williamson, J. D. (1980) *Journal of General Virology*, **48**, 183–181.

——— (1982) *Nature*, **297**, 690–691.

Valegård, K., Liljas, L., Fridborg, K., & Unge, T. (1990) *Nature*, **345**, 36–40.

Valegård, K., Unge, T., Montelius, I., Strandberg, B., & Fiers, W. (1986) *Journal of Molecular Biology*, **190**, 587–591.

Verduin, B. J. M., Prescott, B., & Thomas, Jr., G. J. (1984) *Biochemistry*, **24**, 4301–4308.

Verma, M. (1989) *Archives of Biochemistry and Biophysics*, **270**, 77–83.

Virudachalam, R., Harrington, M., Johnson, J. E., & Markley, J. L. (1985) *Virology*, **141**, 43–50.

Virudachalam, R., Sitaraman, K., Heuss, K. L., Argos, P., & Markley, J. L. (1983b) *Virology*, **130**, 360–371.

Virudachalam, R., Sitaraman, K., Heuss, K. L., Markley, J. L., & Argos, P. (1983a) *Virology*, **130**, 351–359.

Wallace, H. M., Baybutt, H. N., Pearson, C. K., & Keir, H. M. (1980) *Journal of General Virology*, **49**, 397–400.

Wallace, H. M. & Keir, H. M. (1981) *Journal of General Virology*, **56**, 251–258.

Warren, R. A. (1981) *Current Microbiology*, **6**, 185–188.

Watanabe, Y., Igarashi, K., & Hirose, S. (1981) *Biochimica et Biophysica Acta*, **656**, 134–139.

Weiland, J. J. & Dreher, T. W. (1989) *Nucleic Acids Research*, **17**, 4675–4687.

White, E. L., Arnett, G., Secrist, III, J. A., & Ihannon, W. M. (1994) *Virus Research*, **31**, 255–263.

Wickner, R. B. (1993) *Journal of Biological Chemistry*, **268**, 3797–3800.

Widom, J. & Baldwin, R. L. (1980) *Journal of Molecular Biology*, **144**, 431–453.

Williamson, J. D. (1976) *Biochemical and Biophysical Research Communications*, **73**, 120–126.

Wilson, R. W. & Bloomfield, V. A. (1979) *Biochemistry*, **18**, 2192–2196.

Witherell, G. W., Gott, J. M., & Uhlenbeck, O. C. (1991) *Progress in Nucleic Acid Research and Molecular Biology*, vol. 40, pp. 185–220. New York: Academic Press.

Young, D. V. & Srinivasan, P. R. (1974) *Journal of Bacteriology*, **117**, 1280–1288.

Zaitlin, M. & Hull, R. (1987) *Annual Review of Plant Physiology*, **38**, 291–315.

CHAPTER 20

Plant Metabolism

T. A. Smith (1971a) reviewed the distribution and metabolism of the amines in plants, noting the early extensive volume on these compounds by Guggenheim (1951). The latter had recorded the numerous discoveries of putrescine in plants and other organisms, beginning in 1888. In early modern times, Cromwell (1943) had demonstrated a role for the diamines in the biosynthesis of the plant alkaloid hyoscyamine, and numerous "secondary metabolites," including alkaloids containing derivatives of putrescine, cadaverine, spermidine, and spermine, were described subsequently.

Smith had studied the accumulation of putrescine in K^+-deficient barley, an observation first made by Richards and Coleman in 1952 (Smith, 1971b). An exploration of the enzymology of the biosynthesis of putrescine led into many areas of polyamine metabolism. Systematic studies had been begun on the assay of amino acids and amines like putrescine in plants, and the detection and estimation of an array of ureido and guanido derivatives like agmatine (Steward et al., 1960). The presence, biosynthesis, and possible physiological significance of urea, ureides, and guanidines in plants were reviewed by Reinbothe and Mothes (1962). Compounds such as tetramethylputrescine were rediscovered in some plant roots (Johne et al., 1975). Tetramethylhomospermidine, or solamine, had been isolated from roots and characterized in a study of plant alkaloids (Evans et al., 1972).

N. Bagni began the investigations on the relations of the polyamines to growth and development in plants. The polyamines were shown to stimulate cell division in dormant tubers of Jerusalem artichoke in in vitro culture (Bagni, 1966). A significantly greater increase of spermidine than spermine was found in the early development of wheat, and both compounds were synthesized in developed leaves (Bagni et al., 1967). Bagni subsequently correlated changes in polyamine content in bean development with parameters such as the increases in protein and RNA. Thus, numerous questions of the relations of the biosynthesis and metabolism of the poly-

amines to the increase of cellular polymers were being formulated in plant systems.

The plant polyamines had been known to the organic chemists primarily as building blocks in the synthesis of some complex plant products. Currently the biochemical developments have advanced from a focus on the secondary metabolites to: studies relating polyamine content to fundamental biological events; efforts to place the polyamines within plant structures; studies of their biosynthesis, metabolism, and regulation as a function of plant development, growth, and senescence; and their functions and changes as a result of imposed environmental change. The studies on plants have revealed several unique features relating to the special properties of these organisms. Despite the numerous similar biological and biochemical mechanisms among eucaryotes and eucaryotic cells, plants are distinguished from animals by their sessile existence and photosynthetic capabilities, as well as by their particular mode of reproduction via somatically derived gametes. Their form and composition, including that of the polyamines, is a function of age, geographic locale, and ecological niche. They respond to changes within that niche, for example, sunlight or darkness and soil condition, largely by physiological change. Severe stress such as soil deficiency, osmolarity, or abrasion may evoke hyperproduction of putrescine or a synthesis of S-adenosylmethionine (AdoMet) as a precursor in the synthesis of ethylene.

Nitrogenous components, if not supplied in fertilizer, are largely derived via the reduction of nitrate or the evolution of a symbiosis with nitrogen-fixing bacteria. Glutamine, arginine, and asparagine are major components of the amino compounds produced in the roots and transported in the xylem sap, with citrulline and proline as prominent solutes in some root and stem exudates (Sellstedt & Atkins, 1991). Plants produce an extraordinary number of amino acids over and above those found commonly in proteins or as animal metabolites. Some individual substances of this type are found in relatively narrow taxonomic groups. Some of these, such as the arginine analogue, canavanine or 2-amino-4-(guanidinooxy)butyric acid, can inhibit insect predation (Rosenthal, 1990). Legumes often produce a homoglutathione in which glycine is replaced by alanine. Plants of the family Poaceae have been found to produce a new homologue of glutathione, in which serine replaces glycine.

The photosynthetic function is centered within leaf cells in chloroplasts whose biosynthetic mechanisms are mainly responsible for production of cadaverine. The chloroplasts are found in a cytoplasmic layer surrounding a large cell vacuole, whose role in the extensive retention of AdoMet and polyamines may be comparable to that in yeast. Results pointing to that conclusion were mentioned in the previous chapter in the comments on virus-infected protoplasts.

Active cell division in a plant is concentrated in the meristem found in the root and shoot tips (Walbot, 1985). In a shoot, this peripheral tissue may develop to form a stem and leaves or a terminating flower and seed. The biochemistries of these developing tissues appear quite different, and indeed the polyamine metabolism of the developing flower in many species entails new compounds and mechanisms. The extraordinary metabolic potentialities of meristem reflect the totipotentiality of individual plant cells, which can develop from leaf protoplasts into complete plants. Meristematic cloning is a major activity of some agricultural practice, which can give rise to certain types of disease-free plants.

A growth in stems and leaves, with the production of protective cellular walls and the evolution of rigidly structured channels for transport of nitrogenous and other metabolites from roots to leaves and carbon compounds from leaves to other parts of the plant, obviously imposes other demands upon the organism.

Assay of Plant Polyamines

Paper chromatography detected an array of volatile compounds such as putrescine, cadaverine, 1,3-diaminopropane (DAP), and even a trace of *sym*-homospermidine produced in the toasting of oat flakes (Hrdlicka & Janicek, 1964). Significant amounts of cadaverine, putrescine, spermidine, and spermine were found in perchloric acid extracts of cereals (Moruzzi & Calderera, 1964). Thin-layer chromatography (TLC) of the dansyl amines was used to estimate these compounds in the photosynthetic Cyanobacteria and some eucaryotic algae in the 1970s, and various high-pressure liquid chromatography (HPLC) methods were introduced in the 1980s. The simultaneous analysis of the free amines and the ultraviolet-absorbing AdoMet and decarboxylated AdoMet (dAdoMet) at picomole levels was developed for extracts of metabolizing protoplasts (Greenberg & Cohen, 1985).

The various methods are described by M. A. Smith (1991). However, plant tissues may also be rich in amino acids and sugars. The concentration of crude extracts alone results in combinations of amines and sugars and severe losses of these compounds (Dawson &

Mopper, 1978). The addition of glycerol to the crude extracts reduces these losses. Alkaline hydrolysis should be used to eliminate numerous contaminating dansylated compounds like alcohols in the chromatography. Agmatine is often found in extracts. Because it contains a primary amine, the compound will react with the amine reagents and appear in the chromatographic analyses.

Major fractions of both acid-soluble and -insoluble bound polyamines in plants introduce a new set of components, which might be explored by acid hydrolysis of the bound polyamine. This procedure involves treatment with 6 N HCl at 110°C in an N_2 atmosphere for 18 h in a sealed tube and can produce very significant losses of the freed amines (Dinella et al., 1992a). The Maillard or browning reactions between amines and sugars is well known, and the problem of freeing covalently bound polyamine from crude fractions has not been solved satisfactorily. Cohen et al. (1982) reported >60% recoveries of radioactive dansylamine by hydrolysis after fixation of radioactive putrescine in chloroplasts. Of this recovered radioactivity, only 59–75% was dansyl putrescine.

Chromatographic methods for the isolation and separation of phenolic acid amides, found in some meristematic tissue, have been described and will be discussed in a later section. In several HPLC assays of free polyamine in plant tissue, the extracts were prederivatized with the dansyl, o-phthalaldehyde, or benzoyl chloride reagents. Walter and Geuns (1987) introduced a postderivatization step that was crucial in the analysis of lichens and algae isolated from the lichens. A very significant percentage (17%) of the total lichen putrescine was conjugated in both acid-soluble and acid-insoluble fractions (Escribano & Legaz, 1988). A more amine-specific fluorogenic reagent, 9-fluoromethyl chloroformate, enables a rapid derivatization and HPLC procedure that is useful without a prechromatographic purification step (Bartók et al., 1992).

The separation and estimation of benzoylated amines from plant extracts was introduced by Flores and Galston (1982). The separated benzoyl amides were identified unequivocally by mass spectrometry (Roberts et al., 1985). However, a considerable underestimation of the polyamines was obtained in many plants by the benzoyl procedure (Corbin et al., 1989). Reactions of benzoyl chloride to form benzoic acid, benzoic anhydride, and methyl benzoate affect the yields and separations of the benzoyl amides. Difficulties and improvements in the use of benzoyl chloride have been reported (Hauschild, 1993; Kotzabasis et al., 1993a). Nevertheless, precolumn derivatization with benzoyl chloride appears to be useful in a rapid estimation of the free and monoacetylated amines found in the less complex extracts of animal cells (Taibi & Schiavo, 1993).

Data on Plant (and Food) Content of Polyamines

The polyamines of the plants are not trace substances. Depletion of animal polyamine requires the control of food and polyamine intake in prepared diets in animal experimentation (e.g., the maturation and responses of intestinal structures), as well as in the use of inhibitors in human patients. In more recent studies, investigation of the active uptake of polyamines in the gut and the overall patterns of biosynthesis, metabolism, and excretion of the compounds in intact animals has revealed that food is a major source of polyamines in mammals, including humans (see Bardócz et al., 1993, in Chap. 5, Table 5.1). Indeed, in Britain the average daily diet provides enough polyamine for the metabolic needs of most adults (Bardócz et al., 1995). Spermine levels tend to be lower in plant products than in those of animals and many plant products like fruit juices can be very high in putrescine. An early example of the phenolic acid amides, feruloylputrescine, was found in grapefruit juice, which is also very high in free putrescine. Various amines, including products like pyrrolidine and 3-(methylthio)propylamine, are found in grapes and wines. Histamine and the polyamines themselves (DAP, putrescine, cadaverine, spermidine, and spermine) have been found in grapes and derived wines and the concentrations of these substances have been studied during the formation of wine. During spoilage both putrescine and spermidine content may double from an acceptable maximum of 65 and 85 µmol/L, respectively (Desser et al., 1981).

Some odoriferous primary amines are produced actively in the flowering of species of arum lilies and participate in the attraction of insect pollinators. Putrescine, cadaverine, agmatine, and histamine have been identified among the many volatile compounds collected over the warm odor-producing spadix. An aqueous extract of these plant organs is rich in free amino acids, some of which are presumably decarboxylated to these amines (Smith & Meeuse, 1966).

Plant Polyamines

As summarized by Smith (1982), putrescine, spermidine, spermine, and agmatine are ubiquitous in higher plants. A number of additional compounds are found in particular groups in relatively high concentrations (Table 20.1). In 1995 the first evidence was obtained for the existence of N-acetylpolyamines in plants in the chloroplasts (Del Duca et al., 1995). A cyclic regeneration of spermidine and putrescine does not appear to involve an active catabolism of N-acetyl derivatives.

Smith and Best (1977) estimated free putrescine, spermidine, and spermine in the shoots, roots, stem, and leaves of barley seedlings grown in the dark or light. Little difference was detected in the first 3 days after germination. Slow increases (up to 12 days) in the polyamines of plants grown in the light were noted compared to those in the dark (Table 20.2). In general, leaves are higher in spermidine and spermine than roots and fruits. Indeed spermine is frequently undetectable in roots.

Hamana et al. (1992) reported on the distribution of free polyamines in the seeds of 27 different legumes.

Twenty-five of these contained homospermidine. Unexpectedly, the seeds of *Pisum sativum* also contained small amounts of thermospermine (3,3,4), aminopropylhomospermidine (3,4,4), canavalmine (4,3,4), homospermine (4,4,4), aminopropylcanavalmine (3,4,3,4), and aminobutylcanavalmine (4,3,4,4), in addition to putrescine (4), spermidine (3,4), and spermine (3,4,3). The tertiary amine N^4-methylthermospermine is also present in leguminous seeds.

Cucumber DAP (Flayeh et al., 1984) can be the product of an amine oxidase acting on spermidine and appears to serve as a precursor also of β-pyrazol-1-yl alanine in that plant. Aminopropylamino alcohols have also been isolated from certain leguminous seeds.

Homoagmatine and cadaverine were found to increase to high concentrations in developing seedlings of *Lathyrus sativus* (Ramakrishna & Adiga, 1974, 1975). In this legume, the common polyamines and polymers, the proteins and nucleic acids, increased in the seedling shoots as the food reserves in the cotyledons decreased. This was the first instance in which cadaverine appeared to be linked to growth and development. Both lysine and homoagmatine contributed to cadaverine production, with the path from lysine several-fold more active. The transamidination activities of *Lathyrus* facilitated the production of arcaine, the diguanidino derivative of putrescine (Srivenogopal & Adiga, 1980a). Others produced some methylguanidino amines, and one of these, N^G-methylagmatine, was thought to be a product of the decarboxylation of N^G-methylarginine (Matsuzaki et al., 1990).

Cadaverine is present in significant amount in all Leguminosae and serves as a precursor to the quinolizidine alkaloids. In germinating seedlings, the diamine increases initially in the shoots and subsequently declines whereas the dark roots may continue to accumulate cadaverine to a concentration 70-fold greater than that in the leaf (Smith & Wilshire, 1975). Unlike putrescine, cadaverine is not affected by various soil supplementations, Ca^{2+}, NH_4^+, or by K^+ deficiency. The diamine and the mono- and bis-aminopropyl derivatives have been described as products of lysine decarboxylation in leaves and cotyledons of *Vicia faber*. Inhibition of synthesis of the bis-aminopropyl derivative by methylglyoxal-bis(guanylhydrazone) (MGBG) and cyclohexylamine in the plant as in various fungi has suggested dAdoMet as the source of the aminopropyl moieties.

sym-Homospermidine was first isolated from sandalwood leaves (Kuttan et al., 1971). As noted in Table 20.1, tetramethylhomospermidine, or solamine, was discovered in the Solanaceae and had been found in plants in the late 1960s as a component of animal tumor inhibitors. Homospermidine is almost as effective as spermidine in supporting the growth of *Hemophilus parainfluenzae*.

T. A. Smith et al. (1986) showed that oxidation of spermine by the oxidase of leaves of barley and maize produces DAP and 1-(3-aminopropyl)pyrroline. Spermidine and norspermidine were oxidized somewhat more slowly and putrescine was inactive. The reaction with

Table 20.1 Free polyamines found in particular plant groups.

Amine	Structure	Source	Reference
1,3-Diaminopropane	$H_2NCH_2CH_2CH_2NH_2$	Graminae	Smith (1970)
Cadaverine	$H_2NCH_2CH_2CH_2CH_2CH_2NH_2$	Leguminosae (*Lathyrus*)	Ramakrishna & Adiga (1975)
sym-Homospermidine	$H_2N(CH_2)_4NH(CH_2)_4NH_2$	*Santalum*	Kuttan et al. (1971)
Aminopropylpyrroline	$H_2N(CH_2)_3N$	Graminae	Smith (1970)
Δ^1-Pyrroline		Leguminosae	Smith et al. (1986)
N-Carbamylputrescine	$H_2N(CH_2)_4NHCNH_2$ $\parallel$ O	*Sesamum*	Crocomo & Basso (1974)
Homoagmatine	$H_2N(CH_2)_5NHCNH_2$ $\parallel$ NH	*Lathyrus*	Ramakrishna & Adiga (1974)
Arcaine	$H_2NCNH(CH_2)_4NHCNH_2$ $\parallel$ NH $\parallel$ NH	*Lathyrus*	Srivenugopal & Adiga (1980)
Tetramethylputrescine	$(CH_3)_2N(CH_2)_4N(CH_3)_2$	*Ruellia*	Johne et al. (1975)
Tetramethylhomospermidine	$[(CH_3)_2N(CH_2)_4]_2NH$	Solanaceae	Evans & Somanabandhu (1980)
Canavalmine	$H_2N(CH_2)_4NH(CH_2)_3NH(CH_2)_4NH_2$	Leguminosae	Hamana et al. (1984) (in Chap. 5)
N^G-Methylagmatine	NH $H \parallel$ $CH_3N—C—N(CH_2)_4NH_2$	Leguminosae	Matsuzaki et al. (1990)
N^4-Methylthermospermine	$H_2N(CH_2)_3N(CH_2)_3NH(CH_2)_4NH_2$ $\vert$ CH_3	Leguminosae	Hamana et al. (1992)

Adapted from Smith (1982).

Table 20.2 Polyamine distributions within barley seedlings.

Age (days)	Tissue	nmol/g Fresh weight		
		Putrescine	Spermidine	Spermine
12, dark	Leaves			
	Upper	842	231	54
	Lower	338	218	65
	Stem			
	Upper	199	218	28
	Lower	209	189	9
	Root	648	117	<5
12 (16-h day)	Leaves			
	Upper	2980	158	81
	Middle	807	202	131
	Lower	505	320	96
	Stem	707	627	113
	Root	1700	217	<5

Adapted from Smith and Best (1977).

spermidine generates aminopropylpyrroline. These workers also noted the high content of diamine oxidase (DAO) in legumes. This enzyme converts putrescine to 4-aminobutyraldehyde, which cyclizes spontaneously to 1-pyrroline. The polyamine oxidase (PAO) from the Graminae that generates DAP from spermidine also produces 4-aminobutyraldehyde and pyrroline as cleavage products.

The study of potassium-deficient plants demonstrated the accumulation of putrescine, agmatine, and *N*-carbamylputrescine (Fig. 20.1). The subsequent analysis of the origins of the latter also revealed an accumulation of arginine, citrulline, and ornithine in K$^+$-deficient leaves of *Sesamum indicum* (Crocomo & Basso, 1974). Several enzymatic routes may exist in the conversion of arginine to putrescine in higher plants.

Photosynthetic Bacteria and Algae

It had been supposed that an inability to synthesize spermine was characteristic of procaryotic cells. In the 1980s spermine and even larger polyamines were found

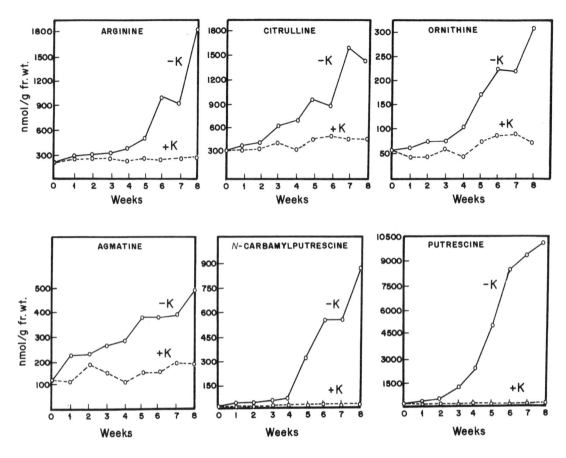

FIG. 20.1 Amino acids and amines in *Sesamum* leaves from plants grown in complete (+K) and potassium-deficient (−K) nutrient solution; adapted from Crocomo and Basso (1974).

in some previously unexamined bacteria grown in media lacking spermine. On the other hand, numerous examples of eucaryotic microorganisms like trypanosomes also lack spermine in particular growth phases. The unicellular alga, *Scenedesmus obliquus*, also does not contain spermine throughout its growth and is inhibited in chlorophyll synthesis by depletion of putrescine. This eucaryotic cell in synchronous culture develops a potentially interesting pattern during the cell cycle of accumulation of peaks of bound and free putrescine and spermidine (Kotzabasis & Senger, 1994).

Various cyanobacteria, for example, the unicellular *Anacystis nidulans*, were found to contain low concentrations of putrescine and 1–2 mM levels of spermidine; they did not appear to contain spermine (Ramakrishna et al., 1978). During the cultivation of *Anacystis* in a mineral medium spermidine was made and conserved without turnover during exponential growth and was rapidly degraded when growth ceased (Guarino & Cohen, 1979a). Exogenous labeled spermidine and spermine were both taken into the cells during growth and the labeled spermidine was rapidly converted to DAP,

presumably by an oxidase. Exogenous putrescine was quite toxic during photosynthetic growth, a result later correlated with the high pH 10–11 generated in the mineral medium during CO_2 fixation. These pHs exceed the pKs of the amino groups of putrescine, permitting the entrance of nonprotonated putrescine and the trapping of cationic putrescine at the lower pH within the cell (Guarino & Cohen, 1979b).

Two eucaryotic unicellular green algae, *Chlorella* and *Scenedesmus*, contain various amines including putrescine and spermidine, but lack cadaverine and spermine. The cells contain slightly more polyamine if grown on nitrate or ammonia salts than if grown on urea. In tests of the growth of various algae and cyanobacteria in the presence of organic nitrogen sources, it was found that most could use urea and nitrate or ammonium ion but that only a few like *Chlorella* and *Scenedesmus* could generate utilizable nitrogen for growth from putrescine. None of the Cyanobacteria tested could do so, although several members of both groups can generate nitrogen for growth from arginine, ornithine, and some other amino acids (Neilson & Larsson, 1980).

Both ornithine and arginine decarboxylases (ODC and ADC) have been detected in *Chlorella vulgaris* grown asynchronously in light (Cohen et al., 1983). Both enzymes require pyridoxal phosphate (PLP) and a sulfhydryl reagent, such as dithiothreitol. Whereas ADC is similarly active in both the exponential and stationary phases, ODC increased 10-fold in the exponential phase and decreased sharply as cell division slowed. Both enzymes, ODC particularly, were feedback inhibited by putrescine. In synchronous cultures, ODC and polyamine content peaked just prior to division, which could be blocked by difluoromethylornithine (DFMO) (Cohen et al., 1984).

In the growth of *Chlorella emersonii* synchronized in a 16:8 h light:dark cycle in which cell mass increased in the light and cell division occurred in the dark, spermidine fell in the dark whereas norspermidine increased comparably. This suggests an oxidative conversion of spermidine to DAP, the precursor of norspermidine (Maiss et al., 1982). Further, the curves of increase of norspermidine and RNA were parallel in both light and dark, and the content of this triamine was 25 ± 5 µg/mg RNA in all conditions tested (Kordy & Soeder, 1984).

The complex alga, *Euglena gracilis*, which can be grown under heterotrophic, photoheterotrophic, or autotrophic conditions, contains DAP, norspermidine, and norspermine, in addition to putrescine and spermidine (Aleksijevic et al., 1979). Small amounts of spermine, homospermidine, and carbamylputrescine, were also detected in a single modified ion-exchange analysis (Adlakha & Villaneuva, 1980). Adlakha et al. (1980) reported peaks of putrescine, spermidine, and particularly norspermine at the end of a division. Norspermine is the major polyamine in synchronously dividing cells, and the content of spermidine is possibly limited by its turnover to form DAP, a major precursor of norspermine (Fig. 20.2) (Villaneuva et al., 1980a,b). The accumulation of spermidine and norspermine in stationary cells does not in itself provoke division.

In a lichen, a symbiotic arrangement in a thallus of fungus and alga, L-arginine may be converted to urea and ornithine by arginase. The lichen, *Evernia prunastri*, which contains two arginases (Vicente & Legaz, 1985) also contains an ADC that is inhibitable by putrescine, urea, and ornithine. Thallus extracts also contain an agmatine amidinohydrolase (agmatine ureohydrolase) that is activated by arginine, ornithine, and putrescine and inhibited by urea (Vincente & Legaz, 1982). Over 90% of this enzyme is contained in the alga. The extracts contain an agmatine iminohydrolase that is slowly activated in the thallus by urea and produces *N*-carbamylputrescine. This enzyme appears restricted to the fungus (Legaz et al., 1983).

Polyamine Distribution in Tissues of Higher Plants

Bagni and Serafini–Fracassini (1974) have reported on the spermidine and spermine contents of the subcellular fractions of etiolated epicotyls of *Pisum sativum*. Of these compounds 70–80% are in the soluble fraction, which include the vacuolar amines. Chloroplasts of spinach leaves contain about 1–2% of the putrescine, spermidine, and spermine of the cells. The mitochondria of *Helianthus tuberosus* tuber slices also contain small amounts of the cellular putrescine, spermidine, and spermine at 0.5, 2, and 1%, respectively (Torrigiani et al., 1986).

Pea epicotyls were somewhat lower in their ratio of free spermidine to spermine than were spinach leaves, 2.1 versus 4.8, although both tissues were similar in their spermidine contents at 153–156 mµmol/g fresh weight. Analysis of various RNA fractions, tRNA and rRNA, detected spermidine in both of these nucleate components (see Chap. 23).

Variability in polyamine content from species to species is considerable, as well as that in individual parts. Stem saps, the xylem and phloem exudates, can be rich in putrescine and spermidine, indicating an active translocation of these substances (Friedman et al., 1986). The compounds are synthesized in both roots and leaves.

Felix and Harr (1987) have determined the levels of free polyamines in the acid extracts of seeds and seedlings of 30 plants from 13 families of monocotyledonous and dicotyledonous plants. Some rapidly growing tissues, such as those of some Cruciferae, may be quite low in the polyamines. Others may be high in putrescine or spermidine or both. Cotyledons are frequently high in these components, which are thought to be synthesized in these tissues. These plant parts do not usually contain covalently bound acid-soluble polyamines, but may contain acid-insoluble bound amines.

High putrescine and spermidine contents of corn (*Zea mays*) kernels correlated with good survivability in storage. In the populations studied, the higher contents were inherited as dominants in the various crosses (Lozano et al., 1989).

Estimates were made in growing plants of free soluble polyamines in the cells and in cell wall fractions, as well as polyamines released by acid hydrolysis of the cell walls of segments of mung bean hypocotyls (Goldberg & Perdrizet, 1984). In the younger segments, soluble polyamines (i.e., putrescine and spermidine) were the major components. Spermidine was a major component bound to the walls and tended to increase with increasing age of the tissue. Indeed most of the polyamines of the oldest cells were associated with the cell wall and may have been present as a salt of the uronic acid component. In the mung bean hypocotyl system, polyamine content was greatest in meristematic and growing cells and least in fully expanded senescent cells, which possess large vacuoles.

Protoplasts derived from mature carrot roots accumulated labeled putrescine and spermidine very rapidly and some of these compounds were also found in the vacuoles (Pistocchi et al., 1988). Some 28 and 42% of the protoplast contents of spermidine and putrescine, respectively, were located in the vacuoles.

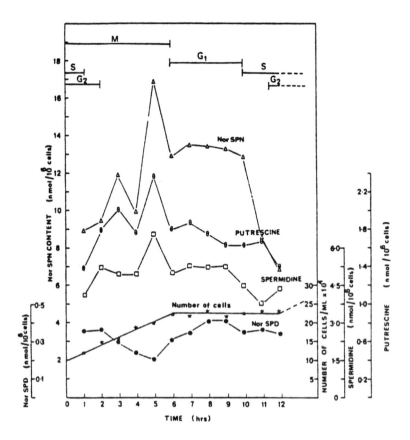

FIG. 20.2 Changes in the concentration of major polyamines during the different phases of the cell cycle in synchronous cultures of *Euglena gracilis* Z; adapted from Villanueva et al. (1980b). NorSPN (*sym*-norspermine) and NorSPD (*sym*-norspermidine).

Putrescine Biosynthesis and Mineral Deficiency

Cromwell (1943) pointed to a possible role of arginine and putrescine as a precursor of the alkaloid hyoscyamine that is produced in *Atropa belladonna*. R. Robinson (1955) was interested in the comparative chemistry of the biosynthesis of tropinone and suggested a role of γ-aminobutyraldehyde in the process, and Cromwell (1943) showed that putrescine is indeed converted to this intermediate by a plant oxidase. This was in fact the first report of an amine oxidase in plants. Cromwell also isolated putrescine from two alkaloid-producing plants. Tetramethylputrescine had been isolated as early as 1907 from *Hyoscyamus muticus*.

Extracts of K^+-deficient plants were abnormally high in ninhydrin-reactive basic substances. One substance isolated from K^+-deficient barley and other plants was putrescine (Richards & Coleman, 1952). Indeed, the diamine fed to cut leaves of high-potassium plants was toxic, evoking symptoms like growth inhibition similar to those of potassium deficiency (Sung et al., 1994). Normal plants did not contain high levels of putrescine, nor were the symptoms characteristic of the high putres-

cine concentrations obtained by feeding ornithine (Coleman & Richards, 1956). In all species studied, barley, wheat, flax, clover, etc., potassium deficiency increased the content of arginine (Fig. 20.1), suggesting that this amino acid, rather than ornithine, was the precursor of putrescine. Indeed, putrescine and agmatine were both found in extracts of K^+-deficient leaves and this suggested the possible existence and change in arginine decarboxylase (Smith & Richards, 1962). T. Smith (1963) demonstrated major increases of ADC in extracts of leaves of deficient barley by determining the rate of agmatine produced from arginine. He partially purified the enzyme, estimated pH optima and the Michaelis constant (K_m), and demonstrated the specificity of the isolated enzyme, but was unable to dissociate PLP from the plant ADC.

Potassium deficiency provoked a rapid loss of plant protein and an increase of amino and amide nitrogen, with the accumulation of six amino acids, including arginine and lysine. The effect is universal among higher plants (Flores, 1991). Low levels of available K^+ evoke decreases in bean leaf cell size and leaf thickness and a degradation of chloroplasts and cell walls. In wheat, high levels of K^+ (15 mM or higher) depress putrescine synthesis and content and stimulate the conversion of the diamine to spermidine (Reggiani et al., 1993). Exogenous putrescine, spermidine, and spermine are inhibitory to K^+ uptake into maize roots (De Agazio et al., 1988). Numerous alkyl guanidines are inhibitors of K^+ transport in isolated barley roots and it may be asked if an accumulation of agmatine thereby exacerbates K^+ deficiency.

In dying tissue putrescine may account for 20% of the total plant nitrogen and it has been suggested that putrescine concentrations might serve as a measure of the ionic status of the plant. Thus, an optimal growth medium would assure a low putrescine content of the plant. Other mineral deficiencies also provoke aberrant accumulations of agmatine and putrescine, and in some plants other than barley increases of spermidine and spermine are found as well (Flores, 1991). A relative increase of phosphate over potassium increased putrescine accumulation (Hackett et al., 1965). Similar results of an increase of arginine, agmatine, and putrescine were obtained by feeding acid to barley seedlings in the presence of an adequate K^+ supply (Smith & Sinclair, 1967). It has been suggested that amine accumulation provides a mechanism of regulating pH and ionic balance within the plant.

No increase could be detected in the soluble ODC of extracts of K^+-deficient oat seedlings (Young & Galston, 1984). When the rate of production of $^{14}CO_2$ from [1-^{14}C]-arginine was estimated, large increases of ADC were detected in the leaves and shoots of the seedlings. However, no mention was made in these studies of the possible dilution of the [1-^{14}C]-arginine by tissue arginine present in the extract, and ADC based on yields of $^{14}CO_2$ may have been significantly underestimated.

N-Carbamylputrescine was detected in the leaves of barley that was fed agmatine and this in turn was converted to putrescine in plant extracts (Smith & Garroway, 1964). K^+-deficient *Sesamum indicum* also accumulated N-carbamylputrescine, which may be derived from either agmatine or via a decarboxylation of citrulline. Citrulline as an intermediate may be postulated as resulting in the synthesis from ornithine or as the less well-known intermediate of arginine degradation via arginine dihydrolase. However, the existence of citrulline decarboxylase in many plants is not well established. Agmatine iminohydrolase, which converts agmatine to N-carbamylputrescine, has been detected in many plants (e.g., maize and cabbage) and this activity was not observed to increase in K^+-deficient plants. Arcaine(1,4-diguanidinobutane) is an inhibitor of this enzyme. A third enzyme, which converts N-carbamylputrescine to putrescine, does increase in deficiency (Smith, 1971b).

Thus, this route has been postulated as the route of putrescine formation in K^+-deficient plants. However, enzyme (2) has not been found in barley.

A Path of Nitrogen Assimilation

Nitrogen is taken up from the soil largely as nitrate and is reduced to NH_3 via nitrate reductase and nitrite reductase. The growth of symbiotic nodulated soybeans is limited primarily by N availability. Nitrate reductase occurs in all organisms capable of utilizing nitrate and the activity of this enzyme regulates the nitrate available for amino acid biosynthesis. The enzyme is inducible by exogenous nitrate and this response is inhibited by polyamines in excised radish cotyledons. The effect is greater with spermine (1 mM) than with the other polyamines. The polyamines do not inhibit the enzyme or the uptake of nitrate (Srivastava & Mungre, 1986). Spermine is essentially absent from barley roots (Table 20.2).

Exogenous NH_4^+ increases the synthesis of free polyamine, particularly putrescine (Le Redulier & Goas, 1979). Free and total putrescine, as well as spermidine, accumulate both in the roots of peas supplied with NH_4^+ and in plants with root nodules (Tonin et al., 1991). The content of the various polyamines found in plants can be manipulated by modification of nitrogen sources, as well as by the use of the inhibitors active on specific pathways (Altman & Levin, 1993). When putrescine itself is given to soybean in the presence of NH_4^+ or NO_3^-, the diamine is rapidly converted to γ-aminobutyric acid (GABA), succinate, and malate, and very little is used for the synthesis of spermidine.

The enzymes essential for the assimilation of nitrite (i.e., nitrite reductase, glutamine synthetase, and glutamine-oxoglutarate aminotransferase) are localized in chloroplasts (Miflin & Lea, 1976). Light generates the reductants and ATP essential for the synthesis of glutamine and glutamate; that is,

arginine $\xrightarrow[①]{-CO_2}$ agmatine $\xrightarrow[②]{-NH_3}$

N-carbamylputrescine $\xrightarrow[③]{}$ putrescine + NH_3 + CO_2

$$\begin{array}{c} COOH \\ | \\ CH_2 \\ | \\ CH_2 \\ | \\ CHNH_2 \\ | \\ COOH \end{array} + NH_3 + ATP \xrightarrow{Mg^{2+}} \begin{array}{c} CONH_2 \\ | \\ CH_2 \\ | \\ CH_2 \\ | \\ CHNH_2 \\ | \\ COOH \end{array}$$

$$\begin{array}{c} CONH_2 \\ | \\ CH_2 \\ | \\ CH_2 \\ | \\ CHNH_2 \\ | \\ COOH \end{array} + \begin{array}{c} COOH \\ | \\ CH_2 \\ | \\ CH_2 \\ | \\ C=O \\ | \\ COOH \end{array} \xrightarrow{2H^+} 2\begin{array}{c} COOH \\ | \\ CH_2 \\ | \\ CH_2 \\ | \\ CHNH_2 \\ | \\ COOH \end{array}$$

Excessive ammonia concentrations are toxic and nitrogen may be accumulated in amides such as asparagine. Glutamine synthesized from NH_4^+ is rapidly converted to ornithine via a pathway of acetyl derivatives.

Shargool et al. (1988) described the pathways and compartmentation of ornithine and arginine biosynthesis from glutamate in plant cells (see Chap. 10). Glutamate is initially converted by acetyl-coA to N-acetylglutamate. This is then phosphorylated to N-acetyl-γ-glutamyl phosphate and to N-acetylornithine in several steps. The crucial kinase is inhibited by arginine and is a key in the regulation of the pathway. Ornithine is generated when the acetyl group of N-acetylornithine is transferred to glutamate. Thus, there are two routes to the formation of N-acetylglutamate but the route via the ornithine acetyl transferase reaction is the major route in the cyclic formation of N-acetylglutamate. Whereas the kinase is a cytoplasmic enzyme, the acetyl transferase is a plastid enzyme. Thus, N-acetylglutamate formed in the plastid must enter the cytoplasm to enable the formation of N-acetyl-γ-glutamylphosphate, which must return to the plastid for subsequent conversions.

The conversion of [^{14}C]-ornithine to [^{14}C]-proline or to [^{14}C]-citrulline and arginine were demonstrated in normal and potassium-deficient barley (Coleman & Hegarty, 1957). Ornithine transcarbamylase (OTC) carries out the reaction,

ornithine + carbamyl phosphate $\rightarrow$ citrulline + P_i.

Glutamine is the immediate source of the NH_3 that enters carbamyl phosphate. Both carbamyl phosphate synthetase and OTC are found within plastids.

Isozymes of the pea and other plant OTC appear to be significantly different from the animal enzyme (Slocum et al., 1990). The amino acid composition of the pea enzyme is markedly different in arginine and leucine content from the mammalian enzyme, and in arginine content from yeast and bacterial enzymes (Slocum & Richardson, 1991). Plant carbamyl phosphate synthetase is stimulated by ornithine, indicating that a major step to arginine production is regulated by this precursor, which does not usually accumulate in the plant. Citrulline is converted to arginine via two steps involving arginosuccinate synthetase and arginosuccinase or arginosuccinate lyase. Both of these enzymes appear to be cytoplasmic. In soybean cell suspension cultures, high intracellular concentrations of arginine inhibit the former and enhance the latter.

Under conditions of phosphate deficiency, citrus (lemon) or squash cultivars accumulate arginine in leaves as a result of increased activity of the arginine biosynthetic pathway. Phosphate-deficient plants accumulate NH_3 initially, followed by a 10-fold increase of arginine biosynthesis (Rabe & Lovatt, 1986). These authors have suggested that essentially all mineral nutrient deficiencies, with the exception of molybdenum essential for nitrate reductase, produce an accumulation of NH_3 and its attempted removal via de novo arginine biosynthesis.

Putrescine and Osmotic Stress

Problems of drought and increasing salinity in soils led to observations of accumulations of putrescine and cadaverine (Flores, 1991). The presence of NaCl in excess of 50 mM in media evoked accumulations of diamine in beans, cotton, peas, etc., although these tended to be less than the changes in putrescine in K^+ deficiency. Some salt-stressed plants may not respond with increases in the free or bound polyamines (Priebe & Jäger, 1978). Thus, salt-shocked barley accumulates proline and other plants may effect osmoregulation with increases of glycinebetaine. Nevertheless, salt stress leading to excessive accumulations of putrescine, as well as of proline, is well known in the growth of salt-sensitive rice. Salt-tolerant cultivars of rice do not have high endogenous levels of putrescine but convert this to increased levels of spermidine and spermine (Krishnamurthy & Bhagwat, 1989).

Exaggerated accumulations of putrescine (i.e., 5–10 times normal concentrations of the diamine) in 2–3 h and relatively minor amounts of spermidine were found during production of oat leaf protoplasts. The protoplasts increase their ADC activity, presumably by de novo synthesis (Flores & Galston, 1984a,b). This accumulation is restricted to cereals and does not occur among dicot genera (e.g., tobacco, etc.), in which leaf protoplasts can regenerate to entire plants. In protoplasts of the latter group, putrescine content decreases and spermidine and spermine increase (Tiburcio et al., 1986a). When cereal leaves are pretreated with difluoromethylarginine (DFMA), the generation of putrescine in protoplasts is markedly reduced and spermidine and spermine are sharply increased, with these cells remaining viable for longer periods (Tiburcio et al., 1986b).

Detached leaves or peeled leaf sections may be floated on media of different osmolalities and their changes studied. The various effects in protoplasts, leaves, and other organs have been summarized by Kaur–Sawhney and Galston (1991). In early studies the deterioration in light and dark of a detached oat leaf in media of physiological osmolality was described as "senescence." The process was characterized by successive increases in ribonuclease activity, protease activity, and a decrease of chlorophyll (bleaching). The loss of chlorophyll in detached oat leaves is preceded by an increase of protease, and this increase in both acidic and neutral proteases and subsequent decrease of pigment are maximally inhibited by spermidine and spermine. These changes were inhibited by 1–10 mM polyamines, with effects of spermine > spermidine > diamines. The polyamines inhibited the fall in RNA and the rise in nonprotein nitrogen in light and dark, and inhibited the bleaching in the dark. The polyamines were dubbed "antisenescent," although they did promote chlorophyll bleaching in the light.

Although oat leaf protoplasts will incorporate uridine and leucine into RNA and protein in culture, they do not incorporate thymidine into DNA. However, the addition of polyamines to such cultures does permit the synthesis of DNA and the development of some mitoses (Kaur–Sawhney et al., 1980). Moreover, guazatine, an inhibitor of the oxidation of spermidine and spermine to DAP and H_2O_2, has been found to prevent chlorophyll loss in osmotically stressed sections of peeled oat leaves (Capell et al., 1993). The provoking agent of "senescence" may prove to have been osmotic stress and an oxidative product such as H_2O_2. However, detached oat leaves, which accumulate DAP in the dark, are inhibited by exogenous DAP in various effects, such as the increase of protease and chlorophyll loss. It is of interest that the production of ethylene in such leaves, which thereby simulates the seasonal event most commonly associated with senescence, is also inhibited by exogenous DAP (93% by 1 mM amine). In summary, then, it appears that in senescing cereal leaves and protoplasts, a supply of exogenous "normal" spermidine and spermine arrests or inhibits the deteriorative processes. In contrast, the protoplasts of regenerative plants increase their content of free and bound spermidine and spermine as they go on to cell division and subsequent organogenesis.

Light Controls

Many developmental processes, germination, growth, greening, and flower formation, are governed by light and several types of photoreceptors are known. One of these, a red/far-red receptor designated "phytochrome," was discovered in the early 1950s (Rüdiger & Thümmler, 1991). Plant meristems have the highest levels of phytochrome. Treatment with red light decreases ADC activity in subapical epicotyl internodes of dark grown pea seedlings, but rapidly increases ADC several-fold in buds. Far-red light reverses these effects on ADC in the two tissues (Dai & Galston, 1981). The red response in apical buds of the pea plants appears to stimulate polyamine synthesis (Goren et al., 1982).

In some plants night/day cycles increase the synthesis of malic acid and release this anion from vacuole to cytoplasm in processes dubbed *Crassulacean acid metabolism*. These acidifications of cytoplasm are balanced, in part, by similar cyclical increases in putrescine, spermidine, and DAP (Morel et al., 1980). The lights-on part of the cycle has been found to markedly stimulate AdoMet decarboxylase (AdoMetDC) in many plants (Kamachi & Hirasawa, 1995).

Other Environmental Effects

Oxygen depletion was found to evoke putrescine accumulation in rice and wheat seedlings (Reggiani & Bertani, 1989). This deficiency also evoked significant accumulations of bound putrescine, spermidine, and spermine. These results were subsequently extended to six species of Gramineae, in which the production of putrescine paralleled tolerance to anaerobic conditions

(Reggiani et al., 1990). Indeed exogenous putrescine increased survival in oxygen-deficit stress.

Chilling during storage also increased putrescine concentrations in various fruits, as well as in bean, wheat, and alfalfa leaves subjected to a vapor saturation deficit. In citrus leaves putrescine did not accumulate, but both spermidine and proline concentrations were observed to increase. A parallelism was found between increases in spermidine and citrus cold hardiness (Kushad & Yelenosky, 1987). It was suggested that the protection by spermidine and spermine relates to the strengthening of membranes and prevention of electrolyte leakage (Kramer & Wang, 1990; Songstad et al., 1990).

Changes in polyamine levels have been sought in current inquiries on the relation of acid rain or other pollution to forest decline. Seasonal patterns of free amino acids have been studied in spruce forests. Although arginine could not be correlated with damage, glutamine, asparagine, and proline were observed to increase in damaged trees. Free putrescine was found in all tissues of growing trees and was the major polyamine of Scots pine needles (Sarjala & Savonen, 1994). The diamine increased rapidly and was particularly high (5–10-fold increase) in the needles of damaged trees as their potassium content fell (Lauchert & Wild, 1995). A combination of ozone and acid mist also increases free and conjugated putrescine in Norway spruce.

A single ozone treatment of tobacco increased the levels of free and bound putrescine in the plant, following a transient rise in ADC. Among the putrescine conjugates was detected the phenolic derivative, monocaffeoyl putrescine, whose structure permits a more efficient scavenging of ozone-derived oxyradicals (Langebartels et al., 1991). Toxic cation, Cd^{2+}, and chromium III or VI also evoke large increases in free and soluble conjugated putrescine, particularly in oat and bean leaves, and an increase in ADC (Weinstein et al., 1986).

Radical Scavenging

Plant systems that convert newly synthesized putrescine to higher polyamines and accumulate the latter amines appear to be more tolerant of environmental stress. These effects have often been attributed to the strengthening of cellular membranes and to the minimization of lipid peroxidation within the membranes. At high concentration (10–50 mM) the amines have been shown to reduce the accumulation of enzymatically generated superoxide and reactive hydroxyl radicals (Drolet et al., 1986).

Tobacco leaf injury by ozone and its prevention by root application of the polyamines were examined more closely by Bors et al. (1989). The rates of reactions of the polyamines with hydroxyl ($^{\bullet}OH$), sulfite ($SO_3^{\bullet-}$), and superoxide/peroxyl ($O_2^{\bullet-}HO_2^{\bullet}$) were relatively slow but the reactions of the natural conjugates of putrescine with the hydroxycinnamates coumarate, caffeate, and ferulate, were far more active (100–1000-fold) in radical scavenging. These results supported more likely roles for polyamine conjugates, which were in fact increased by root application of the free polyamines.

Toxicity of Paraquat

The fact that polyamines are absorbed in roots and translocated to other parts of plants relates to the mode of action of the herbicide paraquat (1,1'-dimethyl-4,4'-bipyridylium) and the toxicity of the herbicide in humans and other mammals. The compound can be degraded by soil bacteria.

CH₃–N⁺ ... N⁺–CH₃   Paraquat

Uptake of the compound in plants and the action of light leads to a very active production of superoxide radical and hydrogen peroxide, substances normally eliminated by chloroplast enzymes involving ascorbate peroxidase, dehydroascorbate reductase, and glutathione reductase. Depletion of iron by the iron chelator desferrioxamine (Chap. 4), increases levels of glutathione reductase in paraquat-treated peas and minimizes lipid peroxidation and protein loss (Zer et al., 1994). Resistant bacteria, e.g., strains of *Escherichia coli*, have been found to overproduce superoxide dismutase or to have reduced rates of paraquat uptake. Putrescine and cadaverine were shown to be competitive inhibitors of a major component of paraquat uptake in the roots of maize seedlings, indicating that paraquat uptake is a function of a diamine uptake system. Spermine proved to be a noncompetitive inhibitor of this uptake system (Hart et al., 1992).

The most characteristic feature of paraquat poisoning in mammals is damage to the lungs. The lung appears to possess a single system for the uptake of diamines and polyamines, and despite a 10-fold higher K_m for paraquat, the maximum velocity (V_{max}) for uptake of this compound approaches that of the natural diamines and polyamines. Uptake by the rat lung system and competition with putrescine uptake relates to the distance between the charged nitrogens, and agmatine proved to be a very effective inhibitor (Smith, 1985). At the cellular level, the mutual competition of paraquat and putrescine were readily demonstrable in rat alveolar type II cells (Chen et al., 1992).

Metabolism of Arginine

In studies of the fate of variously labeled arginines injected into excised cotyledons of soybeans, high percentages of the amino acid appeared first as arginine in protein, although a small amount was metabolized in an arginase-urease sequence. Ornithine-derived glutamate was metabolized to CO_2 (Micallef & Shelp, 1989). Arginase has been found in many plants and the conversion of arginine to ornithine by this Mn^{2+}-activated enzyme has also been studied in the sprouting of tuber

tissues of the Jerusalem artichoke (Wright et al., 1981). Unlike the feedback inhibition of the animal enzyme by urea and products of ornithine catabolism, the artichoke enzyme is inhibited competitively by ornithine and citrulline.

In cell cultures of sugar cane, exogenous arginine is stimulatory to growth. Arginine is converted largely into several amino acids that enter proteins as well as into a nonmetabolized *N*-carbamylputrescine and some guanidine (Maretzki et al., 1969). When [U-¹⁴C]-arginine is fed to tobacco leaf disks, putrescine is formed, and added [¹²C]-agmatine reduces the label greatly in the diamine. Added ornithine was less effective and it was concluded that the main route of putrescine formation in tobacco was through arginine-derived agmatine and not through ornithine (Yoshida, 1969).

ADC

This enzyme has been a major focus in studies of the path of putrescine biosynthesis in plants. Isotopic methods applied to crude extracts have often failed to correct for the presence of unlabeled arginine. Furthermore, as described by Birecka et al. (1985a), extracts high in arginase may also liberate CO_2 from the ornithine derived from labeled arginine. In addition, in many plant extracts the CO_2 derived from ornithine may not arise via ODC. The addition of aminoguanidine was often needed to suppress the oxidative degradation of agmatine and putrescine (Birecka et al., 1985b).

The introduction of DFMA permitted the use of this presumably specific inhibitor to explore the importance of ADC in various developmental roles. In the somatic embryogenesis of the carrot, this inhibitor blocked development and its action was prevented by putrescine whereas DFMO did not significantly affect development (Feirer et al., 1984). However, plants rich in arginase, such as tobacco, can hydrolyze DFMA to DFMO (Slocum et al., 1988). Newer agmatine analogues such as α-monofluoromethylagmatine, a powerful inhibitor of ADC and not a substrate for arginase (Bitonti et al., 1987), may be more useful in analysis of plant pathways.

The oxidation of newly produced agmatine to H_2O_2 by a DAO of pea seedlings permits the assay of the formation of this product via a peroxidatic oxidation of a spectrophotometrically detectable dye such as guaicol. Smith (1983) described the procedure and the purification of ADC from K^+-deficient oat seedlings. The enzyme is activated by added PLP, is specific for L-arginine, and is inhibited 50 and 60% by 1 mM L-canavanine and D-arginine, respectively. ADC was relatively unstable in leaves stored at −15°C but was quite stable when purified.

Somatic embryogenesis of cultured carrot cells results in an early doubling of ADC content, with concomitant RNA and protein synthesis. After addition of cycloheximide, the ADC declined with the short half-life of 30 min (Montague et al., 1979). Low molecular weight inhibitors were detected in the extracts. In studies of the inhibition of cultured tobacco cells by cyclo-heximide it was found that ADC was lost from the cells with a half-life of about 6 h, under conditions in which the half-life of ODC was about 2 h (Hiatt et al., 1986).

The ADC of young oat leaves was isolated to permit the determination of an N-terminal amino acid sequence (Bell & Malmberg, 1990). This permitted the preparation of oligoribonucleotides that hybridized with an ADC cDNA. The initial ADC isolation was established by the formation of an irreversibly bound DFMA–ADC complex. The isolated and sequenced cDNA revealed an open reading frame encoding a protein of 66 kDa, much larger than the actual ADC of 24 kDa encoded in the C-terminal region. Comparison with the amino acid sequence of the *E. coli* ADC revealed several regions very similar to those of the plant enzyme. The C-terminal region bound DFMA, suggesting this region as essential to the active site. The enzyme is reported to be localized in oat chloroplasts (Borrell et al., 1995).

Polypeptides were produced in the expression of the cDNA in *E. coli* and antisera to these detected the 24-kDa polypeptide (ADC) and another 34-kDa protein in the plant extracts. Antisera were developed to the N-terminal and C-terminal regions of the ADC, as well as a monoclonal antibody that precipitated the ADC activity. Analysis of the oat extracts revealed the apparent existence of a 66-kDa precursor polypeptide from which various smaller polypeptides, as well as the 24-kDa enzyme were formed (Malmberg et al., 1992). More recent studies have revealed that the 66-kDa proenzyme is cleaved via a processing enzyme distinct from ADC itself (Malmberg & Cellino, 1994). In this property ADC is different from the pyruvoyl-containing decarboxylases that are self-processing. A cDNA of ADC derived from young *Pisum* fruit can be expressed in yeast to yield an ADC of 79 kDa, with many sequence similarities to the ADC of tomatoes and oats (Pérez–Amador et al., 1995).

Plant ODC

Although the gross level of ODC in leaves is usually much lower than ADC (Birecka et al., 1985b), this difference was not seen in multiplying plant cells. The decarboxylases had been estimated most frequently in readily available tissues, such as leaves in which cells are enlarging but not multiplying. In fact ODC was quite active in tobacco cells multiplying in culture or in tomato ovaries before and after pollination (Heimer et al., 1979) (Fig. 20.3A,B). These systems were used for a partial purification and characterization of the ODC (Heimer & Mizrahi, 1982). The enzyme, which is specific for L-ornithine, required PLP and a thiol and may exist as a dimer of 107,000 Da. Unlike the animal enzyme, the ODC was inhibited >80% by 0.16 mM putrescine.

The enzyme isolated from mung bean seedlings is inhibited by DFMO (Altman et al., 1982). Throughout the various organs of an 8-day-old seedling, the ratios of ADC to ODC are in the range of 2.8–4.7, suggesting a significant role for ODC in the plant, at least at this stage.

Another study of ODC from barley seedlings (Koromilas & Kyriakidis, 1988) described a chromatin-bound monomeric enzyme of 55,000 relative molecular mass (M_r) that was sensitive to DFMO and the natural polyamines. These workers reported on the existence of two plant "antizymes" of relatively low molecular weight (MW), 16,000 and 9000, and pointed to the existence of an inactive ODC–antizyme complex.

The apparently simultaneous existence of ADC and ODC has posed the problem of their possibly different roles and compartmentation: ADC for processes of cell enlargement and ODC for active plant cell multiplication. In tobacco cell cultures, low levels of DFMO eliminate ODC and block cell division but not cell growth and enlargement (Berlin & Forche, 1981). The sorting out of the two enzymatic functions in intact plants has often been confused by a lack of specificity of the inhibitors, such as conversion of DFMA to DFMO by arginase, and the broad range of some inhibitors like 1-aminoxy-3-aminopropane, which inhibits ODC, ADC, and AdoMetDC. Nevertheless, some plants (e.g., wheat shoots) appear to possess 25-fold more ADC than ODC and the selective inhibition of the former by DFMA and the inability to inhibit ODC by DFMA were readily demonstrable (Christ et al., 1989).

As described in Chapter 10, some fungi like *Saccharomyces cerevisiae* are devoid of ADC and contain ODC as a sole route of putrescine formation. The isolated yeast gene for ODC has been integrated into tobacco, and in the transgenic roots a twofold enhanced ODC enabled cultures to accumulate putrescine and the secondary product nicotine (Hamill et al., 1990). Similarly, a mouse ODC cDNA has been integrated into tobacco and has been shown to be expressed in the transgenic plants (DeScenzo & Minocha, 1993).

Curing Plants of Fungal Infections

Rajam et al. (1985) noted selectively inhibitory effects of DFMO on four phytopathogenic fungal infections. The bean rust fungus that damages bean plants (*Phaseolus vulgaris*) was selectively inhibited by DFMO, which protected bean seedlings inoculated with ureidospores of the fungus (Rajam et al., 1986). DFMA did not confer protection. Weinstein et al. (1987) also demonstrated the efficacy of relatively low concentrations of DFMO in controlling three different fungal infections of wheat.

DFMO proved to be selectively inhibitory to mycelial growth and sporulation by *Helminthosporidium maydis*, the pathogen of southern corn leaf blight. Nevertheless, inhibition of some fungi by DFMO is not always demonstrable at low concentrations. More potent inhibitors (25-fold) of fungal ODC than DFMO have been found (e.g., monofluoromethyldehydroornithine methyl ester) for particular fungi (Boyle et al., 1988), but the cost of these agents is an obvious deterrent of field trials. Moreover, it has been reported that several phytopa-

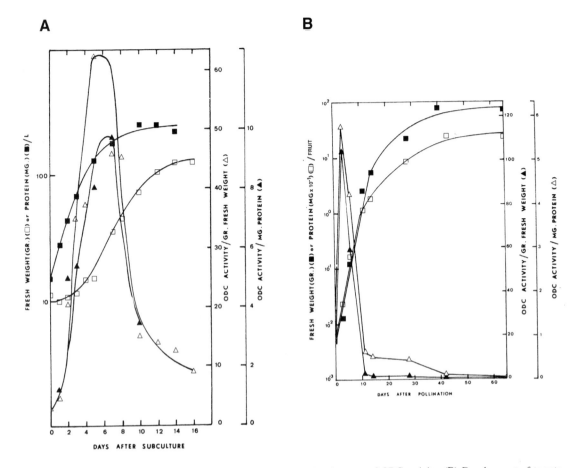

FIG. 20.3 (A) Growth of tobacco cells in suspension culture and development of ODC activity. (B) Development of tomato fruits and ODC activity in the ovaries before pollination and in the developing fruits. Adapted from Heimer et al. (1979).

thogenic fungi do contain ADC inhibitable by DFMA, and little ODC (Khan & Minocha, 1989).

Recently several putrescine analogues have been shown to control five important fungal crop pathogens. The inhibition of fungal growth was found to be accompanied by marked increases of putrescine and spermine and decreases in spermidine in the fungi (Havis et al., 1994a,b).

Path from Agmatine to Putrescine

Agmatine and N-carbamylputrescine have both been found in barley leaf extracts, as well as an enzyme for the conversion of N-carbamylputrescine to putrescine, N-carbamylputrescine amidohydrolase. However, an enzyme for the conversion of agmatine to N-carbamylputrescine, agmatine iminohydrolase, could not be detected in this plant. Nevertheless, the sequence of

arginine → agmatine →
$$N\text{-carbamylputrescine} \rightarrow \text{putrescine}$$

has been postulated as a major route in plants.

Agmatine iminohydrolase, an enzyme that produces NH_3 and N-carbamylputrescine from agmatine, was found in grape, maize, and in sunflower seedlings but not in extracts of many other plants (Mendum & Adams, 1994; Smith, 1969). Although K+-deficient maize accumulated putrescine, the iminohydrolase activity was not increased. Agmatine-fed plants contained N-carbamylputrescine but this product was not found in plants grown without this compound, or indeed grown in the presence of citrulline. An improved separation of agmatine, putrescine, and N-carbamylputrescine set the stage for studies of the enzyme in crude plant extracts.

Highly purified preparations of agmatine iminohydrolase have been isolated from corn, rice, and soybeans but these have been reported to have significantly different properties. The homogeneous corn enzyme was described as a dimer of identical subunits of 43 kDa with an isoelectric point of 4.7 (Yanagisawa & Suzuki, 1981). It was inhibited by arcaine and by the sulfhydryl reagent *p*-hydroxymercuribenzoate. The enzyme of rice seedlings, purified 700-fold, is also a dimer, but about twice the molecular weight of the corn enzyme (Chaudhuri & Ghosh, 1985). A homogeneous enzyme from soybean cytosol appears to be monomeric with a MW of 70,000 and an isoelectric point of 7.5 (Park & Cho, 1991).

Both agmatine and the carbamyl derivative were shown to be used in the formation of nicotine in tobacco roots (Yoshida & Mitake, 1966). A specific enzyme to convert *N*-carbamylputrescine to putrescine, an *N*-carbamylputrescine amidohydrolase, was detected in leaves of barley and many other plants and was increased greatly in K^+-deficient plants (Smith, 1965). The maize group (Yanagisawa & Suzuki, 1982) purified the cytosolic activity 70-fold and estimated an MW of 125,000 Da. The enzyme was sensitive to —SH reagents and was inhibited by Cu^{2+} and Zn^{2+}. The reactions carried out by this purified corn enzyme are given below.

agmatine   *N*-carbamylputrescine   putrescine

$$\underset{\text{agmatine}}{H_2NC-NH(CH_2)_4NH_2} \longrightarrow \underset{\text{N-carbamylputrescine}}{H_2NC-NH(CH_2)_4NH_2} \longrightarrow \underset{\text{putrescine}}{H_2N(CH_2)_4NH_2}$$

with NH_3 and $NH_3 + CO_2$ released.

The carbamylation of putrescine to *N*-carbamylputrescine in the presence of carbamyl phosphate and the diamine by extracts of pea leaves was described by Kleczkowski and Wielgat (1968). These workers also described the linkage of this activity with that of ornithine transcarbamylation to form citrulline. The existence of such synthetic reactions has made it difficult to interpret results obtained with crude extracts.

The clarification of these reaction mechanisms was possibly begun by the discovery of putrescine synthase, an unstable multifunctional enzyme of 55,000 Da (Srivenugopal & Adiga, 1981). This enzyme, with the activities of agmatine iminohydrolase, putrescine transcarbamylase, ornithine transcarbamylase, and carbamate kinase, was found initially in seedlings of *Lathyrus sativus* and was purified from this source. Reactions suggesting a broader distribution of this enzyme have been found in peas, cucumbers, and maize.

In an early purification of agmatine iminohydrolase, the incorporation of inorganic phosphate in the assay mixture resulted in the appearance of putrescine. The *N*-carbamylputrescine amidohydrolase was found to carry out a phosphorylytic reaction that produced carbamyl phosphate, which facilitated the production of citrulline from ornithine. The activities required were all found to be present on the homogeneous enzyme (Fig. 20.4). These authors proposed a cyclic utilization of agmatine that produced putrescine at one end and resynthesized arginine via a transfer of the carbamyl moiety from en-

zyme-bound *N*-carbamylputrescine to newly synthesized ornithine.

Guanidino Compounds and Transamidination

The formation of guanidinated products such as creatine, glycocyamine, taurocyamine, and lombricine in animal tissues and of compounds such as streptomycin by microbes is attributed to arginine : H_2N—R amidinotransferases. Arginine itself was first isolated from plant material by E. D. Schulze and E. Steiger in 1886 and it is present as the free amino acid in relatively high concentration in plant extracts. The application of paper chromatography to plant extracts and specific tests for guanidino compounds and guanidinoxy compounds such as canavanine soon provided evidence of the existence of many free compounds of these types (Bell, 1965). Canavanine proved to be restricted to a particular subfamily of Leguminosae. Early tests revealed several additional guanidinated amino acids in *Lathyrus* species of which homoarginine, γ-hydroxyhomoarginine, and lathyrine are examples.

$$\underset{\text{homoarginine}}{H_2NC(NH)NH \cdot CH_2CH_2CH_2CH_2CH(NH_2)COOH}$$

$$\underset{\text{γ–hydroxyhomoarginine}}{H_2NC(NH)NH \cdot CH_2CH_2CH(OH)CH_2CH(NH_2)COOH}$$

lathyrine

A transamidinase has been partially purified from seedlings of *Lathyrus sativus* (Srivenugopal & Adiga, 1980d). The enzyme has a wide substrate specificity; donors such as arginine, homoarginine, and canavanine transfer amidino groups reversibly to lysine, putrescine, agmatine, and cadaverine. Such reactions can account for the formation of homoarginine, homoagmatine, and arcaine found in *L. sativus* (Ramakrishna & Adiga, 1974). Nevertheless, although arcaine from putrescine or agmatine, audouine from cadaverine, and hirudonine from spermidine are known in animals, audouine and hirudonine have not been reported as plant products.

A modified ornithine-arginine pathway for the biosynthesis of canavanine has been proposed by Rosenthal (1990). The ornithine transcarbamylase of jackbean leaves apparently mediates the formation of *o*-ureidohomoserine from canaline and carbamyl phosphate. In the presence of a jackbean extract and aspartate, ureidohomoserine was converted to canavaninosuccinate, which was then cleaved to canavanine and fumarate. Plants containing canavanine also contain arginase, which can hydrolyze canavanine to urea and canaline. The latter

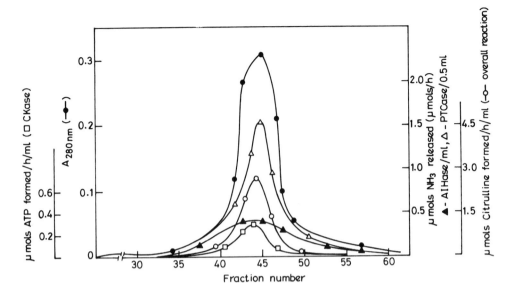

FIG. 20.4 Gel filtrate of activities of putrescine synthase on Sephadex G-200; adapted from Srivenugopal and Adiga (1981). AIHase, agmatine iminohydrolase; PTCase, putrescine transcarbamylase; Ckase, carbamate kinase.

can be reduced to homoserine plus NH₃ by a recently discovered canaline reductase (NADPH : L-canaline oxidoreductase) present in the jackbean (Rosenthal, 1992). Seedlings of the sword bean, *Canavalia gladiata*, contain γ-guanidinoxypropylamine, as well as the triamine canavalmine (Hamana & Matsuzaki, 1985).

Biosynthetic Origins of Plant Methionine

The biosynthesis and metabolism of cysteine and methionine in plants have been described by Giovanelli et al. (1980). The conversion of methylthioadenosine (MTA) to methionine and the pathway to ethylene were added subsequently in an overview (Giovanelli, 1987). Cysteine and methionine are the major end products of sulfate assimilation in plants and these amino acids comprise >99% of the protein sulfur in the ratio of about 66 : 100, respectively. About 10% of total plant sulfur is found in a lipid fraction, glutathione, AdoMet, and several other metabolites.

Within higher plants, enzymes of cystathionine synthesis and cleavage have been isolated from chloroplasts (Droux et al., 1995; Ravanel et al., 1995). Homocysteine, the immediate precursor to methionine, is methylated via a plant methyltransferase using the triglutamyl derivative of N^5-methyltetrahydrofolic acid. In this sequence the methyl carbon is derived from serine at the level of formaldehyde to form N^5,N^{10}-methylenetetrahydrofolate and is reduced by an appropriate NADH-requiring reductase to the N^5-methyl tetrahydrofolate.

A major additional route of methionine formation involves the conversion of MTA derived from AdoMet to methionine. When *Lemna* was grown in the presence of [^{35}S,U-^{14}C]-methionine, it was found that protein methionine contained a lowered ratio of ^{14}C/^{35}S; that is, ^{35}S was conserved relative to the ^{14}C. Data on spermidine synthesis and the regeneration of methionine via MTA, in which the carbon of the amino acid would be derived from unlabeled ribose, indicated a possible mechanism for the ^{14}C dilution in methionine. In addition, the result was also attributed to a rapid turnover of the methyl carbon of methionine via transmethylation and degradation of the methylated compound (Giovanelli et al., 1981). The conversion of MTA to methionine accounted quantitatively for the ^{14}C dilution whereas MTA purine was found in ADP and ATP (Giovanelli et al., 1983). Both exogenous methionine and internally generated methionine reduced the biosynthesis of methionine feeding into AdoMet and other metabolites, indicating that newly available methionine controlled its de novo synthesis (Giovanelli et al., 1985).

AdoMet was detected as a precursor in the formation of ethylene in a process that generates MTA and the fate of the latter product was sought. Murr and Yang (1975) demonstrated recycling and the conversion of methyl-labeled MTA to methionine in apple tissue (Fig. 20.5). Both the methylthio group and the ribose moiety of MTA were equally incorporated into the amino acid, with a modification of the ribose to form the 2-aminobutyrate portion (Yung et al., 1982).

The cleavage of MTA to 5-methylthioribose is effected by a plant nucleosidase, as in some bacterial systems (Kushad et al., 1983), and the sugar is then phosphorylated to the metabolizable 1-phosphate by a specific 5-methylthioribose kinase (Guranowski, 1983). The levels of these enzymes in the developing and rip-

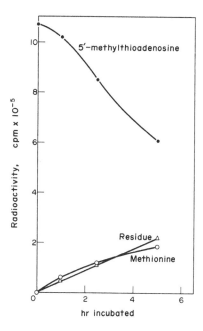

FIG. 20.5 Time course for conversion of [methyl-^{14}C]-5′-methylthioadenosine to radioactive methionine in plugs (1 × 2 cm) of apple tissue at 20°C; adapted from Murr and Yang (1975).

ening tomato has been described, as well as partial inhibitors for the nucleosidase, 5′-chloroformycin, and the kinase, isobutylthioribose (Kushad et al., 1985). The formation of the latter intermediates to methionine, comparable to those in bacterial and animal systems, was described by Miyazaki and Yang (1987a). The inhibition of this formation of methionine in avocado extracts by several inhibitors of the nucleosidase and methylthioribose kinase (e.g., 5′chlororibose and 5′ethylthioribose) was also studied as an approach to the control of ethylene production (Miyazaki & Yang, 1987b). It was concluded that in some plants like ripening apples the conversion of MTA to methionine is so considerable that control of this conversion by inhibitors would not be expected to limit ethylene production.

An oxidation product of MTA, 5′-deoxy-5′methylsulfinyladenosine, has been isolated from a mushroom and has been found to inhibit platelet aggregation and lower blood pressure in the cat (Kawagishi et al., 1993).

A methionine decarboxylase has been purified extensively from membranes of a male fern, *Dryopteris filixmas* (Stevenson et al., 1990). The homodimeric enzyme, containing 57,000 Da subunits, requires PLP for both activity and stability. The normal reaction product is 3-(methylthio)-1-aminopropane. In the absence of added PLP, the enzyme converts methionine abortively to 3-(methylthio)propionaldehyde and residual PLP is converted to pyridoxamine-5′-phosphate. The study of configuration at C_2 by the *N*-camphanamide-NMR procedure, described in Chapter 9, revealed the retention of configuration demonstrated for other amino acid decarboxylations.

Adomet and Its Biosynthesis

The formation of AdoMet in plant extracts (barley) was demonstrated by Mudd in 1960. Slices of pea cotyledons were shown to incorporate labeled methionine rapidly into AdoMet, which was then found to have a rapid turnover. The sulfur and methyl carbon were observed to turn over at different rates (Dodd & Cossins, 1969). A methionine adenosyltransferase was partially purified from pea seedlings and was shown to be dependent on both divalent (Mg^{2+} or Mn^{2+}) and monovalent (K^+ or NH_4^+) cations (Aarnes, 1977). Early purification steps markedly increased the total activity of the enzyme. The products of the reaction with methionine and ATP, tripolyphosphate and AdoMet, were inhibitory.

Three isoforms of the enzyme have been isolated from green pepper fruits (Dogbo & Camara, 1986) and from germinated wheat embryos (Mathur et al., 1991). In the latter case, one isozyme was purified 1530-fold and was characterized as a dimer of 174,000 Da. Triphosphatase activity paralleled the increase in synthetase activity during purification. The enzyme could be labeled by $[^{35}SO_4]^{2-}$ during growth of the embryos. A 2–2.5-fold stimulation of AdoMet synthetase and a similar increase of AdoMet has been observed after stimulation of epicotyls of *Pisum sativum* (dwarf pea) by gibberellic acid. Mathur et al. (1991) noted the considerable diversity of molecular weight subunit structure and forms of the enzyme in bacteria, animals, and plants.

There appear to be two genes for the enzyme in a haploid genome of *Arabidopsis thaliana* and one of these has been cloned and sequenced (Peleman et al., 1989). The polypeptide sequence of the derived enzyme appears to be extensively homologous to the *E. coli* enzyme (49%) and a yeast synthetase (57%).

Origins of Azetidine-2-Carboxylic Acid

Azetidine-2-carboxylic acid was found initially in the Liliaceae in 1955 and was detected subsequently in some legumes. Many plant siderophores that facilitate iron uptake contain the azetidine nucleus (Neilands, 1993). The amino acid replaces proline in protein in some bacteria as a result of formation of the iminoacylated-tRNA^{pro}-amino acid, and its accumulation in protein can be toxic. The aminoacyl-tRNA synthetases of plants containing the free amino acid do not react to form the tRNA derivative. It was suggested to be derived from AdoMet (Leete, 1964), as given below.

$$
\begin{array}{ccc}
 & \text{adenosine} & \text{adenosine} \quad CH_2\text{–}CH_2 \\
 & \quad | & \quad | \qquad | \\
\text{AdoMet} & CH_3\text{–}S\text{–}CH_2\text{–}CH_2 \longrightarrow CH_3\text{–}S \;+\; NH\text{–}CH\text{–}COOH \\
 & \overset{+}{} \; | & \\
 & H_2N\text{–}CHCOOH & MTA \quad \text{azetidine-2-carboxylic} \\
 & & \text{acid}
\end{array}
$$

The four carbon chain of methionine was used in the biosynthesis (Su & Levenberg, 1967) and indeed the amino acid was found to be the most efficient precursor (Leete et al., 1986).

S-Methylcysteine and *S*-Methylmethionine (SMM)

Although the *S*-methylcysteine found in plants like kidney bean seeds and some fungi has been synthesized by yeast enzymes from methylmercaptan and serine, radish leaves may methylate cysteine with methionine methyl (Thompson & Gering, 1966). The formation of SMM, another sulfonium compound that occurs in plants like jackbeans, cabbage, and duckweed, via an AdoMet: methionine methyl transferase has been demonstrated in extracts of at least 20 plants including wheat germ (Allamong & Abrahamson, 1973) and *Lemna* (Mudd & Datko, 1990). The latter group studied the formation of SMM from labeled AdoMet in cultures of *Lemna* and in cells of carrots and soybeans, as well as its degradation in chase experiments in unlabeled media. Extracts were also able to transfer a methyl group from SMM to homocysteine to form two molecules of methionine.

$$
\begin{array}{c}
CH_3 \\
| \\
CH_3\text{–}S\text{–}CH_2CH_2CHNH_2COOH + HS\text{–}CH_2CH_2CHNH_2COOH \\
\overset{+}{} \qquad\qquad\qquad\qquad \downarrow \\
2\ CH_3S\text{–}CH_2CH_2CHNH_2COOH
\end{array}
$$

The functions of SMM are not clearly established. In addition to a possible role in storage of methionine, plant enzymes are known to cleave SMM to dimethylmercaptan and homoserine. SMM may be decarboxylated nonenzymatically in the presence of PLP and a metal ion. A decarboxylated product of SMM is the dimethylsulfonium propylamine found in bleomycin A$_2$. The existence of an enzyme in *Euglena gracilis* that is capable of *S*-methylating methionine moieties of cytochrome c was described by Farooqui et al. (1985).

Methylation Reactions and Fate of *S*-Adenosylhomocysteine (AdoHcy)

Methylation is almost exclusively a major function of AdoMet, which is then converted to AdoHcy. Among the numerous reactions of methyl donation, the formation of quaternary ammonium compounds such as choline and glycine betaine and tertiary sulfonium compounds such as β-dimethylsulfoniopropionate are well known (Rhodes & Hanson, 1993). Many studies have been carried out on the biosynthesis of the free methylated purines theobromine and caffeine in tea and coffee plants (Mazzafera et al., 1994). Additionally, methyl groups are incorporated into tRNA in the plant to form mostly 1-methyladenine in this nucleic acid.

The formation of methylated bases and sugars in plant RNA is also well known and the role of DNA methylation in plant development and differentiation is under active study. Treatment of several plants with 5-azacytidine, which blocks the process and decreases the level of 5-methylcytosine in DNA, has been found to induce early flowering (Burn et al., 1993). However, DNA hypomethylation mutants of *Arabidopsis thaliana* have been obtained that do not exhibit unusual morphological phenotypes (Vongs et al., 1993).

The recovery of methionine from the methylation product AdoHcy involves the adenosylhomocysteinase, which catalyzes the reversible reaction,

$$
\begin{array}{c}
S\text{-adenosyl-}\text{L-homocysteine} + H_2O \rightleftharpoons \\
\text{adenosine} + \text{L-homocysteine}
\end{array}
$$

The enzyme was purified to homogeneity from seeds of *Lupinus luteus* after an initial extraction at low ionic strength (Guranowski & Jakubowski, 1987). The enzyme is a homodimer of 110,000 MW, is optimally active at pH 8.5–9, and requires —SH groups. Although relatively nonspecific for adenosine and its analogues, only cysteine would substitute for homocysteine. Various adenosine analogues are excellent competitive inhibitors of adenosine in the synthetic reaction. The deamination of adenosine shifts the reaction to the formation of homocysteine, which may then be converted to methionine.

Pathways for Synthesis of Spermidine and Spermine

One route found in all plants so far examined is that common to animals and many bacteria, the decarboxylation of AdoMet followed by the transfer of the residual aminopropyl moiety to putrescine to form spermidine, and the transfer of another similar moiety from an additional dAdoMet to spermidine to form spermine. In several plants the initial step is carried out by an AdoMetDC that differs from the enzyme of animals, fungi, and many bacteria in not requiring additional putrescine or Mg^{2+} for activity.

This property was found for two extensively purified enzymes, that from corn seedlings (Suzuki & Hirasawa, 1980) and Chinese cabbage (Yamanoha & Cohen, 1985). The corn seedling was observed to increase in activity 25–50-fold soon after germination and to fall about a week later. The corn preparation was sensitive to sulfhydryl reagents and carbonyl reagents and possessed a molecular weight much lower than that of animal, fungal, or bacterial enzymes. Like other AdoMet-DCs, the enzyme was inhibited by MGBG. The enzyme from Chinese cabbage was purified some 1500-fold with the aid of affinity chromatography on MGBG-Sepharose. After removal of low molecular weight inhibitors by gel filtration, the enzyme was assayed in extracts of normal and virus-infected (high spermidine) plants and activities (twofold higher in extracts of infected cells) were found sufficient to account for all of the spermidine found in growing plants, normal or virus infected. The soluble near-homogeneous cabbage enzyme had an MW of 35,000 in gel filtration, and a similar molecular weight was subsequently found for the overexpressed AdoMetDC produced in tobacco suspension cultures grown in the presence of MGBG (Hiatt et al., 1986). Both MGBG and the reaction product, dAdoMet, were inhibitory with inhibition constants (K_i) of 0.6 and 6 μM, respectively. A novel technique demonstrated that the molecular ratio of the yields of dAdoMet and CO_2 from AdoMet with the cabbage enzyme closely approaches unity (Yamanoha & Cohen, 1985).

The cDNAs for several plant AdoMetDCs have been isolated and characterized and possess low levels of "homology" to the *E. coli* and mammalian enzymes (Bolle et al., 1995; Schröder & Schröder, 1995). Human AdoMetDC can be expressed in tobacco after transfer of the cDNA for the human gene to the plant. Transgenic plants were found to have a two- to fourfold increase of the activity, as well as a two- to threefold increase in spermidine content (Noh & Minocha, 1994).

The study of turnip yellow mosaic virus, a virus containing spermidine, focused attention on the spermidine synthase of the host, Chinese cabbage, and on the affected chloroplasts of the leaves (Chap. 19). This enzyme was present in an over 20-fold excess of the AdoMetDC, and in initial studies an activity of the enzyme (about 10% of the total) was detected accompanying a crude chloroplast fraction. After a purification of intact chloroplasts by density gradient centrifugation in silica sols, it was found that these organelles were essentially devoid of spermidine synthase. The purified organelles nevertheless contained all of the diaminopimelate decarboxylase of the protoplasts. The spermidine synthase is considered to be largely, if not entirely, a cytosolic enzyme (Sindhu & Cohen, 1984b).

Cabbage extracts were found to contain both spermidine synthase and a spermine synthase. In early studies the former was partially purified (150-fold) (Sindhu & Cohen, 1984a). The enzyme was inhibited at micromolar levels by cyclohexylamine and *S*-adenosyl-3-thio-1,8-diaminooctane. Studies on the inhibition of spermidine synthase in barley callus by cyclohexylamine produced results quite similar to those described earlier on Chinese cabbage protoplasts. An inhibition of the

enzyme inhibited formation of spermidine, increased accumulation of dAdoMet, and stimulated spermine synthesis as a result of the availability of the required aminopropyl donor (Katoh & Hasegawa, 1989).

A more extensive purification (1900-fold) of the Chinese cabbage spermidine synthase was effected with the aid of a Sepharose coupled to *S*-adenosyl(5′)-3-thiopropylamine hydrogen sulfate (Yamanoha & Cohen, 1985). This reagent of high affinity for this enzyme had been developed previously (Samejima & Yamanoha, 1982). The purified plant enzyme was found in double peaks in gel filtration and the more slowly eluting peak possessed an MW of 81,000. A partially purified (115-fold) spermidine synthase of maize seedlings was reported to have an MW of about 43,000 (Hirasawa & Suzuki, 1983).

Decarboxylation may occur as a result of the activity of H_2O_2. Indeed, in studies of *Lathyrus sativus* or "grass pea," seedlings containing a high concentration of DAO produced sufficient H_2O_2 in the presence of putrescine to simulate an AdoMetDC. This artifactual result obtained with crude extracts was eliminated by removal of the diamine or the oxidase, as well as by removal of H_2O_2 by catalase (Suresh & Adiga, 1977). After removal of DAO, a Mg^{2+}-dependent AdoMetDC was detected and partially purified, separating this activity from a spermidine synthase. This "classical" activity did not appear adequate to generate the higher polyamines, both spermidine and spermine, contained in the plant and Srivenugopal and Adiga (1980a) sought and found activities detected earlier in *Rhodopseudomonas viridis* by G. H. Tait (see Chap. 7). These involve the condensation of aspartic β-semialdehyde with putrescine to form a Schiff base, subsequently reduced to yield a "carboxyspermidine," which is in turn decarboxylated to spermidine. These workers reported evidence to suggest that the aminopropyl moiety of spermidine might be derived preferentially from aspartate. The authors described the synthesis and isolation of enzymatically generated carboxyspermidine and partially purified the NADPH-requiring carboxyspermidine synthase (putrescine-aspartic β-semialdehyde Schiff base reductase). A feeble SH-dependent carboxyspermidine decarboxylase activity was detected in crude extracts.

A Schiff base intermediate has also been postulated in the synthesis of *sym*-homospermidine found in *Santalum album*. The polyamine might be formed via an initial reaction of putrescine and putrescine-derived γ-aminobutyraldehyde (Kuttan & Radhakkrishnan, 1972). A reductase requiring NAD^+ has been detected in *Lathyrus sativus*, which converts putrescine to homospermidine + NH_3 in the absence of DAO (Srivenugopal & Adiga, 1980c). Earlier, G. Tait observed a similar sequence in *Rhodopseudomonas viridis*, and supposed that NAD^+ oxidized putrescine to the aldehyde + NH_3 and that the resulting NADH reduced the subsequent Schiff base. The activity in *Lathyrus* exists in a plant not known to contain homospermidine. These "nonclassical" reactions obviously warrant extensive study.

Polyamines in Plant Cells: Dormancy Break and Embryogenesis

In 1966–1967, N. Bagni demonstrated the stimulation by polyamines of cell division and tissue growth of explants of initially dormant apparently homogeneous parenchyma of *Helianthus tuberosus*. The explants of *Helianthus* tubers absorbed spermine and spermidine from the medium and appeared to degrade spermine to components similar to spermidine and putrescine (Bagni et al., 1978). Compounds such as the common herbicide, 2,4-dichlorophenoxyacetic acid (2,4-D), which evoked a "break" in dormancy eliciting cell proliferation, rapidly stimulated an early increase in polyamine content, with a corresponding fall in arginine and glutamine (Bagni et al., 1980). Serafini–Fracassini et al. (1980) narrowed the burst of polyamine synthesis to the G_1 phase in the first cell cycle that accompanies the marked increase in RNA and precedes the inception of incorporation of [^{3}H]-thymidine into DNA. Increases of ODC, ADC, and AdoMetDC in the G_1 and S phases of the cell cycle and decreases during cell division were subsequently demonstrated (Torrigiani et al., 1987).

Growing explants of *H. tuberosus* produced significant levels of spermidine and DAP: the latter provides evidence of the degradation of polyamine via the action of an amine oxidase (Phillips et al., 1987). These workers consider that the polyamines are primary regulators of xylogenesis and not of cell division. In discussing these results and their interpretation, Serafini–Fracassini (1991) surveyed the data relating polyamine content, synthesis, and transformation to cell cycle phenomena. A gene markedly affecting tuberization was shown to be homologous to the human gene for AdoMetDC (Taylor et al., 1992).

In another system, totipotent carrot cells maintained in culture can be induced to form embryos in an appropriate medium and the initiation of this process is accompanied by an increase in putrescine and spermidine, with a relatively rapid incorporation of [^{14}C]-arginine into putrescine (Montague et al., 1978). Globular embryos were produced in 2,4-D plus arginine, and developed synchronously in the absence of exogenous arginine (Bradley et al., 1984). Mutant cell lines failing to form embryos were incapable of the increased polyamine synthesis characteristic of embryogenesis (Fienberg et al., 1984). High levels of auxin (indole-3-acetic acid) inhibited the elevation of several polyamine-biosynthetic enzymes. Embryogenesis and polyamine increase were also inhibited specifically by DFMA (Feirer et al., 1984; Robie & Minocha, 1989), and this effect was reversed by an exogenous polyamine (i.e., putrescine, spermidine, or spermine). The subject was reviewed by Minocha and Minocha (1995), whose insertion of a mouse ODC cDNA into carrot cells was found to promote embryogenesis (Bastola & Minocha, 1995).

Nevertheless, nondifferentiating carrot cells, despite DFMA inhibition of ADC, putrescine, and spermidine

at 98, 95, and 50%, respectively, continued to increase in weight with cell expansion (Fallon & Phillips, 1988). In similar nondifferentiated cultures, cell growth was inhibited by MGBG and this effect was almost completely reversed by spermidine or spermine (Kurosaki et al., 1992). The latter authors pointed to an "essential role" of the polyamines in the increase in cell number of these cells. MGBG was also found to inhibit embryogenesis, an effect reversed by spermidine but not by spermine (Minocha et al., 1991). Embryogenesis in carrot cultures is associated with a marked increase in the concentration of AdoMet.

Undifferentiated cells of various plant species grown in suspension cultures have been studied to attempt to correlate content of the polyamines with other cell parameters. In Paul's Scarlet rose cells an increase of RNA correlated with an increase in spermine, preceding the rise in DNA (Smith et al., 1978). In clustered cells of *Catharanthus roseus*, inhibitors of ADC and ODC arrested cell division in the G_1 phase whereas an inhibitor of spermidine synthase did not affect the distribution of the cells in phases of the cycle (Maki et al., 1991). The arrest of division by starvation of cell cultures of *Medicago varia* yielded similar results in which the contents of free putrescine and spermidine were observed to plummet at the beginning of a stationary phase (Pfosser et al., 1992).

The growth of suspension cultures of *Oryza sativa* (rice) depended on the use of a "conditioned" synthetic medium and the essential "conditioning" element can be replaced by spermidine (Manoharan & Gnanam, 1992). When elevated putrescine levels in the cells were reduced, growth inhibition in suspension cultures was reversed (Shih & Kao, 1996) and morphogenesis in callus cultures was improved (Bajaj & Rajam, 1996).

The analysis of the regulation of polyamine synthesis and of the biosynthetic enzymes was undertaken in cultures of tobacco cells derived from callus (Hiatt et al., 1986). Growth of the cells in the presence of MGBG resulted in a 20-fold increase of AdoMetDC, which was useful for isolation and characterization of the enzyme. In this plant cell system MGBG did not increase synthesis but prevented enzyme degradation. MGBG also evoked a rapid loss of ADC without apparent effects on ODC. Exogenous polyamines did not lead to loss of ODC but addition of spermidine and spermine produced a rapid loss of both ADC and AdoMetDC.

DFMO was a lethal inhibitor at all stages of growth, implying a major role of ODC in plant cell multiplication. By contrast, DFMA was lethal only at the initiation of cell growth from callus, a result related to a marked increase of ADC, as in a reaction to ionic stress. These workers explored the effects of acidic stress on tobacco cell lines selected for resistance to DFMA (Hiatt & Malmberg, 1988). Such cells overproduced both ADC and free putrescine and survived acidic stress lethal to wild-type cultures.

The cultivation of plant tissue in various media resulted almost exclusively in the formation of large cell clusters in a callus. Exogenous stimulants of growth and division, such as 2,4-D, kinetin, and abscisic acid, were found to enable some callus cultures to form developing embryos. Early stages in the growth of undifferentiated calli of *Hevea brasiliensis* (rubber), *Mangifera indica* (mango), and *Picea abies* (spruce) were marked by the high level of putrescine (El-Hadrami et al., 1989; Litz & Schaffer, 1987; Santanen & Simola, 1992). Inhibitors of polyamine synthesis (e.g., DFMA, MGBG), at concentrations that did not affect callogenesis, did inhibit somatic embryogenesis (El-Hadrami & D'Auzac, 1992). When calli of *Picea abies* were transferred to maturation media, it was found that the original callus had a high level of free and conjugated putrescine. These fell after some days, as did soluble conjugated spermidine. Subsequent development was marked by an increased level of free and acid-insoluble spermidine as multicellular globular embryos were formed (Santanen & Simola, 1992). The oxidations of putrescine and spermidine were active and a function of the developmental phases of the tissues (Santanen & Simola, 1994).

Similar results have been observed in the embryogenesis of *Oryza sativa* (rice). High levels of free and acid-insoluble putrescine were found as a result of callus induction. Free putrescine, but not the bound diamine, fell during incubation in a maturation medium. DFMA was an inhibitor of embryogenesis, implying a rapid utilization of the free putrescine required for embryogenesis in this system (Koetje et al., 1993).

Asparagus officinalis (asparagus) in in vitro culture contains high levels of free arginine and significant amounts of free putrescine and spermidine. In some genotypes, exogenous polyamines promote the initiation of buds and the formation of crowns (Fiala et al., 1991). In studies of *Vitis vinifera* (grape) early embryos were high in putrescine and the ratio of putrescine to spermidine; these properties were retained in globular and later somatic embryos (Faure et al., 1991). In the more organized zygotic embryos putrescine to spermidine ratios approached one, and a high polyamine content of the somatic embryos was thought to contribute to the disorganization of these forms.

Protoplasts

Initial regeneration of complete plants from single cells was successful in efforts with the Solanaceae: the potato, tomato, and tobacco. Subsequently some successes were obtained with some legumes, but the cereals could not be developed in this way. Newer efforts with cereal plants, corn, rice, etc., have focused on the manipulation of transformation of protoplasts. To produce genetically transformed plants, some of which, such as those of cereal crops like maize, are difficult subjects for gene insertion, it has been possible to prepare protoplasts and to insert DNA by a high voltage electrical pulse (electroporation) (Rhodes et al., 1988). The cells are subsequently grown to mature plants and incorporate the desired inheritable character. Transgenic rice plants may

similarly be prepared from transformed protoplasts matured via the several stages of somatic embryogenesis (Davey et al., 1991).

In such systems the addition of a cell wall to the denuded protoplast membrane, division, and colony formation are preconditions for the formation of viable embryos. In a study of the development of the alder *Alnus glutinosa* from protoplasts, the latter formed new cell walls quite slowly and only a few percent of the cells went on to divide and form colonies. This process was stimulated by ornithine and putrescine whereas spermine and high concentrations of spermidine were inhibitory. Transfer of the colonies to agar media of low osmolarity permitted active callus formation (Huhtinen et al., 1982).

Protoplasts, which lack a cell wall, are inherently osmotically labile and indeed the process of stabilizing in media of high osmolarity induces putrescine accumulation, possibly useful in subsequent development. In many plants DAOs are associated largely with cell walls (Chap. 6), and the elimination of these structures in protoplast preparation both minimizes metabolic complexity and provides a readily soluble relatively purified oxidase. Easily disrupted protoplasts have been used to isolate organelles, such as chloroplasts or the chromatin of tobacco cells. A 2–3 h preparation of cells, available in large amounts suitable for many types of inexpensive biochemical study, warrants the attention of teaching and research laboratories.

Stages of Plant Development

Despite the many experiments suggesting roles of the polyamines in processes of root development, floral initiation and development, and fruit production, it has been difficult to ascribe polyamine functions to specific molecular processes in intact plants (Evans & Malmberg, 1989).

This difficulty may be exemplified by a study of pollen germination (Bagni et al., 1981). Stored apple pollen was rehydrated and formed tubes during growth for 2 h in a germination medium in the dark. The process was cyanide sensitive and evidently required aerobic respiration. An initial decrease of total RNA, protein, and polyamine during rehydration was arrested during germination and early and subsequent syntheses were detected by incorporation of labeled precursors. The incorporation of uridine demonstrated the syntheses of various types of RNA. Incorporation of arginine demonstrated its appearance in protein and the various polyamines. Synthesis of spermidine occurred early during rehydration and less actively during germination. Synthesis of putrescine was active throughout the 2-h incubation. Thus, in this system putrescine and spermidine were formed during periods of both transcription and translation; polyamine syntheses were correlated in time to these processes but their causal relations are unclear. Details of diamine formation by ADC and ODC have been analyzed during pollen germination in tobacco (Chibi et al., 1994).

Seed germination

Numerous studies of seed germination have described changes in polyamine content and plant polymers. Early increases of polyamines, RNA, and proteins were observed in the germination of seeds of mung beans, peas, corn, beans, etc. (Villanueva et al., 1978). These workers recorded major increases of one or another polyamine: spermine in *Phaseolus vulgaris* and spermidine in *Zea mays*.

Spermidine and cadaverine were both found to increase in seeds of *Pisum sativum*. In subsequent studies of the germination of *Glycine max* (soybean) seeds, the polyamines (>97%) were found to be free. Cadaverine, spermidine, and spermine were found in the embryonic plants in ratios of about 2 : 1 : 0.25, while putrescine and agmatine were very much lower in amount in the axes of the emerging embryo (Lin et al., 1984). The very considerable amount and continuing production of cadaverine indicated a high level of activity by lysine decarboxylase (LDC), which was demonstrated (Lin, 1984). At later stages of soybean development, cadaverine was found only in hypocotyls and roots, with spermidine as the most abundant polyamine in buds and leaves (Caffaro et al., 1993). The possible presence of cadaverine-containing analogues of spermidine and spermine was not addressed.

In the germination of corn seed, increase occurs mainly in spermidine content, which was suggested to relate to events of early germination (Anguillesi et al., 1982). In the seedlings the distributions of polyamine, ADC, and ODC were followed in coleoptiles whose growth is entirely due to cell elongation and in roots in which division and DNA synthesis occurs at the tip or apical meristem (Dumortier et al., 1983). In the coleoptile, the tip was low in putrescine and spermidine, but the former increased toward the base. ODC was 10–20-fold higher than ADC in regions of the coleoptile. In the root, the tip was markedly higher in free spermidine and spermine than were other regions, and much lower in putrescine content and ODC. A distinction could apparently be made in this system between the polyamines and enzymes of dividing and enlarging tissues.

Tomato seeds contain only small amounts of putrescine and spermidine and increase these levels considerably, in the hypocotyl particularly, on germination. Significant decreases of polyamine content and growth reduction were observed in tests of MGBG. A lack of response to other inhibitors was attributed mainly to their inability to penetrate the seeds (Felix & Harr, 1989).

Rice seeds increase their polyamine and RNA contents initially after the inception of germination (Mukhopadhyay et al., 1983). An inability to germinate in nonviable embryos appeared to be a function of an unduly increased spermine content of the seed. Nonviable seeds were also higher in DAO and PAO and lower in ADC.

Dormant potato tubers contain the usual polyamines

and the biosynthetic enzymes. The initiation of sprouting in apical buds is accompanied by marked increases in ODC selectively and in AdoMetDC, as well as in the polyamines (Kaur–Sawhney et al., 1982). Node explants were tested in the formation of tubers and this was inhibited selectively by DFMO (Protacio & Flores, 1992). Thus, ODC appears to be of major importance in the early development of this plant system. Taylor et al. (1992, 1993) studied genetic controls on tuberization and detected genes related to the production of several ribosomal proteins and a protein with a sequence similarity to mammalian AdoMetDC. The gene for this potato enzyme has recently been characterized (Arif et al., 1994). Increases in the compounds and enzymes were found early in tuberization, followed by decreases in these components as the tuber enlarged.

In the germination of barley seeds (*Hordeum vulgare*) activated by gibberellic acid, ODC was markedly increased. Although the activity in the cytosol increased continually, an ODC activity associated with chromatin changed greatly after about 90 h. Panagiotidis et al. (1982) report that ODC in germinating barley is mainly a nuclear enzyme tightly bound to chromatin.

Growth of seedlings

The polyamines have been followed in the extension and organogenesis of seedlings, as in the formation of roots, buds, and leaves, and in the fate of seed remnants, such as the cotyledons. The internodes of dwarf pea seedlings was found to have a significant increase in ADC specifically after spraying with gibberellic acid. DFMA inhibited ADC and internode growth (Kaur–Sawhney et al., 1986).

The addition of norspermidine to young cultured corn seedlings was found to inhibit plant growth (Massé et al., 1985). Norspermidine entered the cells, produced severe effects on elongation and root production, and appeared to displace spermidine and spermidine from the tissues.

Formation and growth of roots

Nucleates and proteins increased initially, as did polyamines, in newly emerging roots of germinating *Phaseolus mungo*. In later growth, the concentration of these components declines (Chatterjee et al., 1983). Spermine, but not spermidine, increased the formation of adventitious roots on stem cuttings of *Phaseolus aureus* (mung bean), whereas MGBG inhibited the process (Jarvis et al., 1983). High levels of ADC and ODC and of the major polyamines were observed in all of the most actively growing *P. vulgaris* tissues, including root apices, and a study of root apices revealed a particularly high level of ADC. The inhibition of the enzyme by canavanine reduced root growth and this inhibition was reversed by the addition of putrescine (Palavan–Unsal, 1987). A study of polyamine biosynthesis from labeled amino acids in mung bean cuttings demonstrated a correlation between the formation of putrescine and spermidine and the appearance initially of root primordia and subsequently of root growth (Friedman et al., 1985).

Similar changes were observed in the growth of various types of seedling roots of peas, tomatoes, corn, etc. Putrescine content increased with elongation and spermidine and spermine were highest in the apices, decreasing with cell elongation (Shen & Galston, 1985). Thin layers of the stem of tobacco in cultures formed roots de novo and were inhibited in this process by a combination of DFMA and DFMO, or by MGBG. Cyclohexylamine, which depleted the tissue of free and bound spermidine, did not inhibit root formation very actively, indeed less so than MGBG. Whereas the latter depleted both putrescine and spermidine contents, the former permitted an accumulation of putrescine (Torrigiani et al., 1993).

Dicotyledonous plants infected by *Agrobacterium rhizogenes*, which carries a bacterial plasmid, dubbed T-DNA, are transformed by the plasmid and are induced to form highly branched "hairy" roots at the site of infection. Such transformed plants (tobacco, etc.) regenerated from transformed roots were shown to have an altered nitrogen metabolism revealed in their content of free amino acids and polyamines, as well as some shifts in the content of ODC, ADC, and DAO (Mengoli et al., 1992a,b). After treatment with DFMO, decreases were found of free and bound polyamine (Ben-Hayyim et al., 1994). These parameters were explored, particularly in transformed hairy root cultures of *Hyoscyamus muticus*, in which bound putrescine and spermidine (released by acid hydrolysis) were of the order of 95 and 80%, respectively, of the totals (Biondi et al., 1993). In the latter system MGBG and cyclohexylamine markedly increased the levels of bound putrescine. The possible significance of increases in bound polyamine will be discussed.

Floral Initiation and Polyamine Conjugates

The capability of meristematic tissue to develop to form flowers and fruits introduces an array of novel chemicals and biochemical processes, which are most often absent in leaves. A large group of the new compounds includes the hydroxycinnamic acid amides of the polyamines. Relatively recent reviews are those of Evans and Malmberg (1989), Flores and Martin–Tanguy (1991), and Kakkar and Rai (1993).

The biology of the process of flowering may be focused on apical meristems, and the physiology of photoperiodic control of initiation has led to the demonstration of phytochrome as the receptor pigment. However, the events subsequent to initiation have been obscure. Excised shoot apical meristem will grow and develop in culture on a simple medium containing salts, sucrose, thiamin, and myoinositol if indole acetic acid is present. As summarized by Sussex (1989), mutant meristems with floral aberrations have been described.

Studies on the genetic controls of flowering in species of *Nicotiana*, including tobacco, have been particularly active in exploring their relations to polyamine metabolism. Tobacco contains unusually high levels of conjugated polyamines. Malmberg et al. (1985) described the relative activities of ADC, ODC, and AdoMetDC in tobacco tissues, demonstrating the high activity of the last two in flowers (Fig. 20.6). Cell lines resistant to specific inhibitors of polyamine synthesis were isolated and regenerated to form whole plants. MGBG-resistant lines were all characterized by aberrant flower formation, and other polyamine synthesis variants were also found to display flower abnormalities. Two MGBG-resistant lines contained MGBG-resistant AdoMetDC (Malmberg & Rose, 1987). Others were elevated in the polyamines and/or their conjugates. In any case, many of the polyamine mutants were unable to flower and could not be analyzed in classical genetic crosses.

The study of genotypes of *Petunia hybrida* of altered floral morphology revealed some abnormalities in polyamine content and enzyme content, but these could not always be correlated with the floral aberrations (Gerats et al., 1988). In all of these *Petunia* lines, the levels of polyamine conjugates were very low and this parameter was not distinguishable in wild-type and mutant plants.

On the other hand, some developmental studies linked the polyamines to flowering. In thin-layer tissue cultures of *Nicotiana tabacum*, exogenous spermidine causes floral initiation. The concentration of the polyamine is higher in floral buds than in other parts of the plant and the formation of these buds is inhibited by cyclohexylamine. The developmental choice to form vegetative buds or floral buds is affected by the various polyamine inhibitors (Tiburcio et al., 1988). After initiation with kinetin (N^6-furfurylaminopurine), spermidine greatly increased the emergence of floral buds (Kaur–Sawhney et al., 1990). Spermidine has been implicated in a later stage of the flowering of cultured *Spirodela* (de Cantu & Kandeler, 1989). However, in such systems the major role for spermidine may be expressed subsequent to initiation; and the various effects observed in the culture systems may have been obtained in tissue already committed to flowering.

In intact growing tobacco plants, flowering will occur at 20°C but not at 32°C. Water-soluble amides accumulate in the inflorescences and only in the last initiated leaves. Sterile reproductive organs do not contain the amides. Plants cultivated at 30°C develop low levels of the amides in the shoots and remain low at this temperature, which is incompatible with flowering (Martin–Tanguy et al., 1987). These workers conclude that the appearance of basic amides, even at 30–32°C, indicates that the tobacco is ripening to flower. They also reported that hydroxycinnamic acid amides are found in most plant reproductive tissues. Some sterile mutant plants (tomato) have abnormally high levels of the free polyamines.

In leaf explants of *N. tabacum* cultured in a basal medium containing hormones, benzyladenine permits bud formation while 2,4-D causes callus formation and

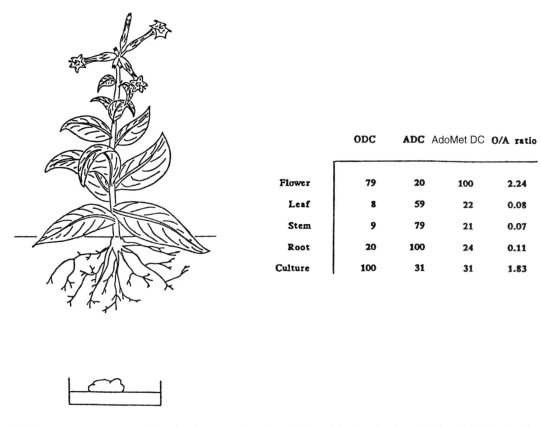

	ODC	ADC	AdoMet DC	O/A ratio
Flower	79	20	100	2.24
Leaf	8	59	22	0.08
Stem	9	79	21	0.07
Root	20	100	24	0.11
Culture	100	31	31	1.83

FIG. 20.6 Developmental regulation of ornithine decarboxylase (ODC), arginine decarboxylase (ADC), and AdoMet decarboxylase (AdoMetDC); adapted from Malmberg et al. (1985). Specific activities are given in percentage of a maximal specific activity. The O/A ratio is the ratio of the specific activities of ODC to ADC.

proliferation. Calli (i.e., proliferating cultures) synthesize higher levels of hydroxycinnamoyl putrescine than do the basal or budding cultures (Martin–Tanguy et al., 1988). The use of the inhibitors DFMA or DFMO in such systems suggested that ADC regulates the levels of free amines whereas ODC is more significant in the accumulation of various putrescine conjugates (Burtin et al., 1989).

A tobacco floral mutant was selected for resistance to MGBG, and this dwarf plant possessed stigmatoid anthers. The mutant plant was significantly low in free spermidine and essentially devoid of conjugated putrescine and spermidine (Trull et al., 1992). Conjugated putrescine and spermidine in the wild type were 60 and 70%, respectively, of the free polyamines.

Floral induction in tobacco is delayed by transformation with the root-inducing plasmid derived from *Agrobacterium rhizogenes*. The delay correlates with a reduction in polyamine accumulation and a delay in the appearance of conjugated polyamines (Martin–Tanguy et al., 1990, 1991). The phenotype of such plants is similar to that of plants treated with DFMO. Although MGBG and cyclohexylamine caused reproductive abnormalities, these were not identical to the DFMO ef-fects (Burtin et al., 1991). A gene in the plasmid (*rol A* or root locus A) that increases polyamine in transformed roots, confers male sterility and abnormal flower growth and appears to block the accumulation of particular conjugates, as well as the cinnamoyl or feruloyl type (see below). The effects of transformation were not reversed by exogenous free polyamine and the search is on for the nature of the protein determined by the *rol A* gene (Sun et al., 1991). Induction of flowers by light in the shoot meristem in *Sinapis alba* involves at least 8% of the 720 detectable proteins in the meristem (Cremer et al., 1992).

Free and Bound Polyamines

The hydroxycinnamoyl amides (Fig. 20.7) are extracted from plant tissues in aqueous acid or with nonpolar solutions, the former solubilizing basic compounds like monocoumaroyl putrescine that contain an ionizable amine. Compounds such as aromatic amides in which all of the amines are substituted (e.g., coumaroyl tyramine), are insoluble in aqueous acid but soluble in solvents such as ethyl acetate. It is important to distinguish

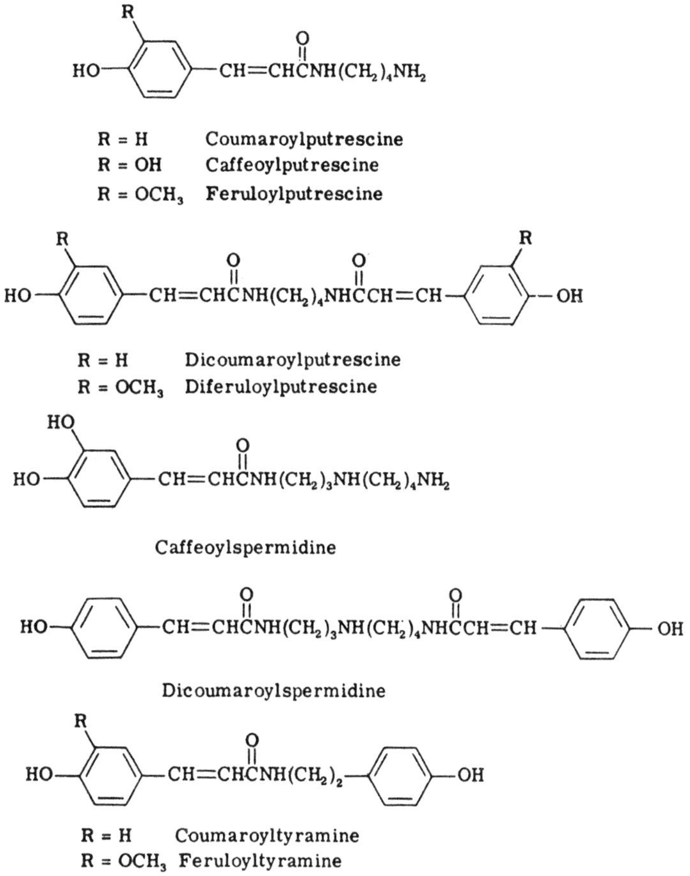

FIG. 20.7 Aromatic amides found in various parts of *Nicotiana* (tobacco) plants; adapted from T. A. Smith (1981).

these from a large class of initially acid-insoluble poly-amines liberated from an acid precipitate of plant tissues by hydrolysis in concentrated acid. As noted earlier, the estimation of polyamine contained in the acid-insoluble fraction is largely uncontrolled and the fractionation of this material has rarely been attempted. A few experiments bear on this problem.

Cohen et al. (1982) reported that preparations of chloroplasts derived from protoplasts of Chinese cabbage converted [^{14}C]-polyamine into an acid-precipitable form (Fig. 20.8) and that this process was stimulated by light, suggesting an intrinsically chloroplast function. After hydrolysis, recovery of the labeled amine as dansyl-labeled amine was on the order of 60–80% in control experiments, suggesting little conversion of the amine to other isotopic products such as amino acids. Unexpectedly, about half of the labeled amine was extractable as free amine from the acid-insoluble precipitate by 80% acetone or 95% ethanol. Nevertheless, half of the acid-insoluble bound amine was not extractable by such solvents.

In recent reports putrescine, spermidine, and spermine were found to be released in particular by hydroly-sis of isolated thylakoid membranes and isolated components of the photosynthetic apparatus of spinach chloroplasts (Kotzabasis et al., 1993b). Del Duca et al. (1994) detected the covalent addition of putrescine and spermidine to specific chloroplast proteins of the Jerusalem artichoke.

In subsequent studies of the transport and effects of exogenous polyamines in plant cells and on the effects of inhibitors such as cyclohexylamine, it was found that these polyamines were also released from acid precipitates by extended acid hydrolysis. An enzyme activity capable of binding polyamines to plant polymers or to casein was detected in the extracts of sprouts from tubers of *Helianthus tuberosus* (Serafini–Fracassini et al., 1988). The bound polyamine was not dissociated by acid or salt nor exchanged with free polyamine, and bound labeled polyamine migrated with various plant proteins in sodium dodecylsulfate gels.

The apical meristem of pea seedlings contains an enzyme that incorporates putrescine into *N,N*-dimethylcasein. The incorporation of putrescine is inhibited by high concentrations of other polyamines. In the uptake of labeled spermidine by floral buds of tobacco ex-

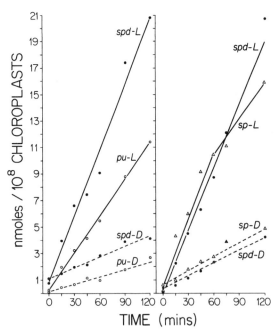

FIG. 20.8 The acid-insoluble associations of polyamines with chloroplasts in light (L) and dark (D); adapted from Cohen et al. (1982). (Left) uptake of putrescine (pu) and spermidine (spd); (right) uptake of spermidine and spermine (sp).

plants, some is bound to acid-precipitable protein. Hydrolysis revealed that some 80% remained as spermidine, although some label (14%) was found in putrescine. Much of the protein contained bound spermidine and lacked hypusine (Apelbaum et al., 1988).

An enzyme described as a transglutaminase has been found in the floral meristem of alfalfa. The enzyme introduces putrescine into ribulose 1,5-biphosphate carboxylase/oxygenase (Margosiak et al., 1990). In an essential characterization of the product of such an enzyme, an extract of silver beet (*Beta vulgaris*) was shown to incorporate putrescine into a glutaminyl-isopeptide bond (Signorini et al., 1991). Despite advances pointing to mechanisms for the formation of acid-insoluble covalently bound polyamine in plant cells (see Chap. 21), the finding of sequestered acid-insoluble noncovalently bound polyamine in chloroplasts of *Brassica* and in parenchyma cells of induced *Helianthus* tubers (Dinnella et al., 1992a,b) raises the problem of the systematic analysis of these compounds.

The existence of structurally bound phenolic acid is well known. The cell walls of the Graminae, the grasses and cereals, contain covalently attached ferulic acid, caffeic acid, and *p*-coumaric acid and their presence in a wall is easily revealed in a microscopic section by their fluorescence after treatment with ammonia. Liberation by hydrolysis with alkali often reveals the presence of cis isomers of the initially *trans*-cinnamic acid derivatives. These products are believed to arise via

photoisomerization in situ. Cis and trans isomers of some soluble cinnamoyl amines like feruloyl tyramine and of cinnamoyl histamine have also been described.

Hydroxycinnamoyl Amides

Several threads of research over four decades have led to increased attention to these polyamine conjugates (Smith et al., 1983). Feruloyl putrescine (Fig. 20.7) was first detected in 1949 and was subsequently described as present in the leaves and fruit of grapefruit and orange (Wheaton & Stewart, 1965). A series of hydroxycinnamoyl putrescines, *p*-coumaroyl- and caffeoyl- as well as feruloyl-, were isolated and characterized from tobacco callus cultures (Mizusaki et al., 1971b). *N*-Feruloyl glycine was found in human urine a decade earlier and a purified plant amidohydrolase that hydrolyzes this and other hydroxycinnamoyl amino acids was described (Martens et al., 1988).

Coumaroyl agmatine was first isolated as an antifungal component present in barley shoots (Stoessl, 1965), and its structural relations to some plant alkaloids and some cell wall constituents like hydroxycinnamoyl esters were noted. The free compound and the more antifungal trans dimers, hordatine A and hordatine B (Fig. 20.9), were also found in barley and were increased in content in the plant as a result of potassium deficiency (Smith & Best, 1978). Coumaroyl and feruloyl derivatives of 2-hydroxyputrescine were detected in wheat. Bird and Smith (1984) subsequently found coumaroyl-coA synthetase and an agmatine coumaroyl transferase in barley seedlings. The oxidative conversion of coumaroyl agmatine to hordatine A in a barley extract by H_2O_2 was described by Negrel and Smith (1984b).

In a third line of investigation, several basic water-soluble antibiotics were discovered in the fermentations of some strains of *Nocardia* (Ellestad et al., 1978). The antibacterial compounds proved to be sugar derivatives of coumaroyl spermidine linked through the phenolic *p*-hydroxyl of the coumaroyl moiety. One of these antibiotics, termed *cinodine*, binds tightly to DNA and is an inhibitor of a DNA gyrase in vitro. The group of compounds is also of interest because hydroxycinnamoyl polyamines had not been found earlier as bacterial products.

A most interesting approach to the amides arose from the observations that the flowers and seeds of virus-infected plants seldom contained transmissible viruses (Martin, 1985). It was surmised that various viruses were eliminated from the meristematic and other tissues by antiviral compounds. It had been observed that the arrest of virus multiplication occurred concomitantly with the accumulation of phenolic compounds, and some of these products in tobacco infected by tobacco mosaic virus included mono- and diferuloyl putrescine (Martin–Tanguy et al., 1973). Feruloyl putrescine, as well as coumaroyl putrescine and caffeoyl putrescine, had been found in substantial amounts in callus tissue cultures of *N. tabacum* (Mizusaki et al., 1971b). Virus-

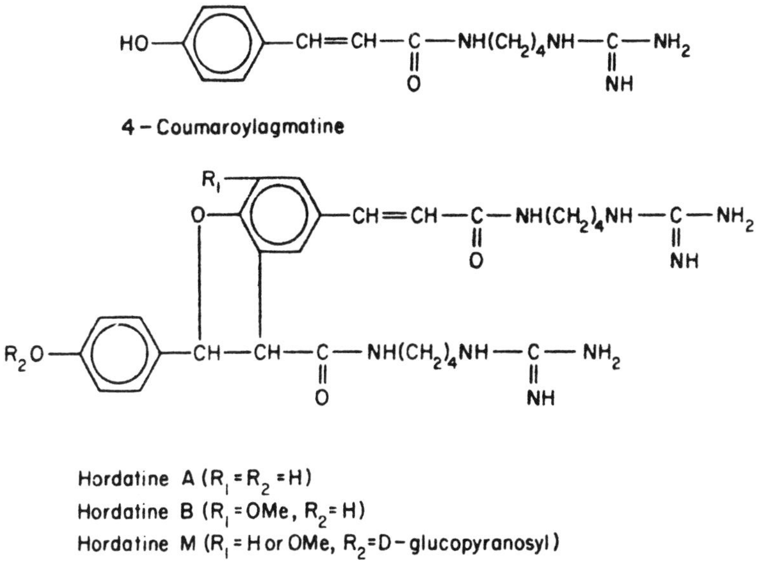

4 – Coumaroylagmatine

Hordatine A (R₁ = R₂ = H)
Hordatine B (R₁ = OMe, R₂ = H)
Hordatine M (R₁ = H or OMe, R₂ = D – glucopyranosyl)

FIG. 20.9 Structure of coumaroylagmatine and the hordatines; adapted from Bird and Smith (1984).

infected strains of particularly susceptible *N. tabacum* produced necrotic lesions, rather than systemic infections, and the surrounding cells were rich in many mono- and dihydroxycinnamoyl derivatives of putrescine. Further, the application of these compounds to disks of the infected leaves blocked the appearance of viral lesions, suggesting that the formation of these substances limited infection and that their lack of formation permitted systemic infection. The hypothesis was specifically formulated that the "aromatic amides" were the inhibitors in the structures of the flower (Martin–Tanguy et al., 1976).

The formation of necrotic lesions and the production of the amides occurred at 20°C but not at 30–32°C, temperatures at which infection was systemic. Free amine levels were not markedly affected at 20°C, but it was found that at that temperature ODC but not ADC increased some 20-fold (Negrel et al., 1984). Similar increases were observed at 20°C in the enzymes synthesizing the phenolic acids and in tyrosine decarboxylase. The synthesis of various *o*-methyl transferases active in the synthesis of phenolic acids during virus infection has also been described (Collendavelloo et al., 1983).

A feruloyl-coA tyramine transferase was also greatly increased after tobacco mosaic virus (TMV) infection at 20°C, although feruloyl tyramine itself was not detected in the healthy or infected leaves of the plant (Negrel & Martin, 1984). The amide appears to leave the cells and bind covalently to lignin components of the cell wall (Martin–Tanguy et al., 1987). *Cis-* and *trans-*feruloyl tyramines and other amides (i.e., derivatives of histamine and phenethylamine) have been detected in several plants of interest in herbal medicine, for example, marijuana. Several 4-hydroxycinnamides have been synthe-

sized and are reported to be specific active inhibitors of the protein tyrosine kinases associated with some oncogenes (Shiraishi et al., 1989).

Useful ion-exchange and HPLC separations of the many amides have been developed (Ponchet et al., 1982) and these data on retention times, as well as retardation factor values in paper chromatography and the distinguishing ultraviolet absorption maxima of several phenolic components, have been summarized (Ibrahim & Barron, 1989). A propos of the focus on the mono- and disubstituted polyamines, it can be noted that tricoumaroyl spermidine has been found in the flowers of the Rosaceae (Strack et al., 1990) and *Aphelandra* (Werner et al., 1995).

The interest in floral initiation resulted in analyses of male and female flower organs. The amides in stamens are largely the basic water-soluble monoamides and are absent in anthers from male sterile plants. The ovaries are stated to be rich in neutral water-insoluble amides (Flores & Martin–Tanguy, 1991). However, the distributions may be more complex. Thus, the pollen of the hazelnut, *Corylus avellana*, was found initially to contain two types of feruloyl spermidines that were nonextractable with water and soluble in 80% methanol. One of these was a bisferuloyl spermidine and the other a mixed caffeoyl-feruloyl-spermidine (Meurer et al., 1986). Pollen from other trees yielded water-soluble N^1,N^5-dicoumaroyl spermidine and N^5,N^{10}-diferuloyl spermidine (Meurer et al., 1988). Quercus pollen contains tricoumaroyl spermidine (Bokern et al., 1995).

In normal and sterile Chrysanthemum, apical meristem contains conjugated polyamines before the appearance of the flower structure. The normal plants accumulated both putrescine and spermidine derivatives, in contrast to the male sterile plants in which only putrescine conjugates were found (Aribaud & Martin–Tanguy, 1994). The latter plants also failed to produce free or bound aromatic amines, such as 2,4-dimethoxyphenylethylamine conjugates that accumulated during floral development.

The presence of hydroxycinnamic amides in many plants during tuberization or flower development has been taken to suggest a possible causal role of the compounds in these processes. In *N. tabacum*, caffeoylputrescine accumulates in pistils, whereas it is found in stamens in some other *Nicotiana* (Leubner–Metzger & Amrhein, 1993). Nevertheless, the hydroxycinnamoyl putrescines do not accumulate in any organ of the potato at any stage of development. This result has been interpreted to exclude a general causal role for the putrescine-containing compounds in the tuberization or flowering of Solanaceae species. Kakkar and Rai (1993) reported that the accumulation of phenolamides is species specific and therefore unlikely to be related generally to flowering.

Biosynthesis and Metabolism of Hydroxycinnamoyl Amides

A summary of the biosynthetic path from phenylalanine and tyrosine to the phenolic acids central to the biosynthesis of the amides is presented in Figure 20.10. In bacteria and plants, phenylalanine and tyrosine are

products of the shikimic pathway; this is a major target of herbicides such as glyphosate. Glyphosate [*N*-(phosphonomethyl)glycine] is a powerful inhibitor of the bacterial synthase that produces the intermediate, 5-*enol*-pyruvylshikimate-3-phosphate. However, in the presence of glyphosate, plant cell cultures not only accumulate shikimic acid but are also inhibited in later stages of the synthesis of certain phenolic acids like caffeic acid, suggesting another level of metabolic control triggered by the inhibitor (Ishikura & Takeshima, 1984).

The sequence of reactions of Figure 20.10 has been presented as reactions of the free cinnamic acids themselves. Although these reactions are known, certain reactions occur with the coA thioesters. The action of phenylalanine ammonia lyase (PAL) produces *trans*-cinnamic acid from phenylalanine. Although tyrosine ammonia lyase, which generates *p*-coumaric acid, has been described, this enzyme is absent from some plants that contain PAL alone or has been separated from PAL. Cinnamate 4-hydroxylases have been described, which produce *p*-OH coumarate.

Monophenols may be hydroxylated by polyphenol oxidase, an enzyme believed to exist exclusively in the plastids. The separation of the frequently latent enzyme from its presumed substrate, *p*-coumaric acid, has raised questions concerning the enzyme involved in the formation of caffeic acid (Vaughn & Duke, 1984). In any case, many derivatives of caffeic acid are widely distributed in various dicotyledonous plants, permitting an

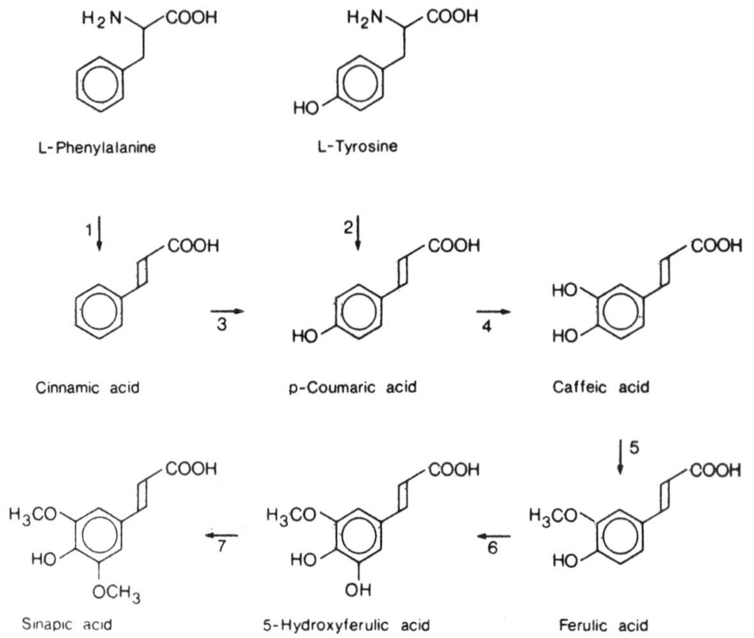

FIG. 20.10   The phenylalanine–cinnamate pathway. (1) phenylalanine ammonia-lyase; (2) tyrosine ammonia-lyase; (3) cinnamate 4-hydroxylase; (4) *p*-coumarate-3 hydroxylase; (5, 7) catechol-*o*-methyltransferase; (6) ferulate 5-hydroxylase.

effort to develop a chemotaxonomy based on the various esters and other compounds present (Mølgaard & Ravn, 1988).

As noted earlier, three *o*-diphenol-*o*-methyltransferases have been detected in tobacco leaves infected by TMV and these differ somewhat in the relative rates of methylation of different *o*-diphenols. Two AdoMet methyltransferases active on caffeic acid, a phenolic acid containing a 3-hydroxyl group, have been isolated from alfalfa cell cultures. The enzymatic methylation of caffeic acid produced ferulic acid (Edwards & Dixon, 1991). An AdoMet *o*-methyltransferase active on both caffeic acid and 5-hydroxyferulic acid produces ferulic acid and sinapic acid, respectively. This homogeneous enzyme has been isolated from the developing xylem of aspen (Bugos et al., 1992).

The hydroxylation of ferulic acid to produce 5-hydroxyferulic acid is carried out by another cytochrome P_{450}-dependent enzyme (Grand, 1984). The enzyme has been isolated from a poplar, *Populus x euramericana*, that is rich in lignin monomers containing sinapic acid. Both 5-hydroxyferulic acid and sinapic acid, in addition to larger amounts of the coumaric and ferulic acids, have been found in the cell walls of corn and barley.

Sinapic acid is found in large amounts in cruciferous species, such as species of *Brassica*, and the compound is present largely in seeds as a choline ester, sinapine (3,5-dimethoxy-4-hydroxycinnamoyl choline). The choline is liberated on germination and serves in the synthesis of membrane lipids. Choline esters of ferulic acid and other phenolic acids have been found in the Cruciferae (Bouchéreau et al., 1991). Sinapine is made by a sinapine synthase isolated from seeds of *Brassica napus* (rapeseed), an enzyme that is a 1-*o*-sinapoylglucose : choline sinapoyltransferase (Vogt et al., 1993). The coA ligase found in *Forsythia*, active on *p*-coumaric acid, caffeic acid, and ferulic acid, is inactive on sinapic acid (Zenk, 1978).

Hydroxycinnamoyl amides have not been described rigorously in *Brassica*. The existence of sinapoyl derivatives of putrescine, spermidine, and spermine in a *Lilium* sp. has been reported briefly, as well as their absence in many other flowering plants (Martin–Tanguy et al., 1978). However, these authors record the apparent existence of a monosinapoyl spermine in *Brassica oleracea*. A tyramine : sinapoyl coA transferase that synthesizes sinapoyl tyramine has been found in the roots of wheat and barley seedlings (Louis & Negrel, 1991).

The various cinnamic acids presented in Figure 20.10 are metabolized in specific plants to a very large number of natural products: alcohols, esters, styrenes, acetophenones, benzoates, flavonoids, coumarins, etc. Harborne (1980) stated that several thousand of such compounds have now been described. Estimates of the distribution of phenolic compounds in immature apple flesh revealed that at least 97% of the free compounds are contained in vacuoles at a concentration of 0.1 M (Yamaki, 1984).

The initial formation of a coA thioester is formed by a coA ligase. The enzyme, which is believed to participate in lignification via the supply of activated phenolic acids like ferulic acid, is reported to be regulated by light in the stems of plantlets. Several isozymes of the 4-coumarate : coA ligase are known in soybeans, and three classes of cDNA have been isolated and analyzed (Uhlmann & Ebel, 1993). Curiously, firefly luciferase has a significant homology in protein sequence to a plant 4-coumarate : coA ligase (Schröder, 1989). The hydroxycinnamoyl-coA derivatives can be degraded in the coA moiety in crude plant extracts by various phosphatases: 3'-nucleotidase, pyrophosphatase, and phosphatase (Negrel & Smith, 1984a).

With the appropriate coA transferases, the coA derivatives of the phenolic acids react with the amines to form the amides, as in the formation of coumaroyl agmatine. The coA ligases found in various plant tissues and the properties of the various cinnamoyl-coA derivatives prepared enzymatically and by chemical syntheses have been summarized by Zenk (1978). This worker also described the roles of cinnamoyl-coA derivatives in the formation of numerous derived natural products.

For example, Kneusel et al. (1989) described a hydroxylase from parsley cells that converts *trans-p*-coumaroyl-coA to *trans*-caffeoyl-coA. These workers also reported on the existence of an AdoMet : *trans*-caffeoyl-coA 3-*O*-methyl transferase that will produce feruloyl-coA (Pakusch et al., 1989).

Metabolic Controls on Synthesis of Phenolic Acid Amides

Interest in the so-called "secondary products" of plants for medicine, foods, cosmetics, etc., have led to studies on the increase of these substances in cultures of plant cells. Spermidine was found to increase the yield of phenolic compounds in Paul's Scarlet rose cultures of increasing age by extending the life span of the cells (Muhitch & Fletcher, 1985). A mix of an elicitor, calcium, and spermine increased PAL and soluble phenolics in suspension cultures of soybean cells (Shetty et al., 1989). A microbial PAL was explored unsuccessfully as a possible medicinal to reduce the dietary intake of phenylalanine in humans potentially afflicted with phenylketonuria.

High levels of PAL and phenolic compounds have been found in plant cell lines (tobacco and carrot) selected for resistance to *p*-fluorophenylalanine. Inhibition of the enzyme in the resistant cells by treatment with α-aminooxy-β-phenylpropionic acid reduced accumulation of phenols by 70% and led to major accumulations (17-fold) of phenylalanine (Berlin & Vollmer, 1979). Bolwell (1992) presented evidence that the degradation of PAL in the French bean is preceded by phosphorylation, which has been suggested to regulate the turnover of the enzyme. PAL has been purified from various plant sources like soybean cell cultures (Havir, 1981) and leaves of *Phaseolus vulgaris* (da Cunha, 1988). The mechanism of the reaction and properties of the enzyme have been described by Hanson and Havir (1981).

PAL activity varies with both the stage of development and various forms of stress. A family of genes encode the enzyme in *Phaseolus vulgaris*. These appear to be expressed differently in the various tissues at different stages of development (Liang et al., 1989). Cinnamic acid above 10^{-4} M can inhibit mRNA production differently from each of the genes and Mavandad et al. (1990) suggested that this compound, or a derivative, regulates gene transcription controlling phenylpropanoid biosynthesis.

The largest number of studies have been done on tobacco cells, which are quite rich in phenolic acid amides. Berlin (1981) obtained cell lines resistant to *p*-fluorophenylalanine, which produced high and low levels of the caffeoyl and feruloyl putrescines. Indeed, one strain (TX4) accumulated 10 times more of these compounds (>10% of the dry weight) than the wild-type strain. Both ornithine and arginine were decarboxylated by ODC and ADC, respectively, to produce the diamine found in the amides. Radioactive phenylalanine (10^{-2} M) appeared in the amides without growth inhibition.

Berlin et al. (1982) developed methods for the quantitation of the separate putrescine-containing amides and demonstrated increased enzymatic activities in the TX4 cell line (Fig. 20.11).

Glyphosate, which inhibited cell growth and evoked an enormous increase in shikimic acid, blocked the accumulation of the amides (Berlin & Witte, 1981). This was not extensively reversed by the exogenous addition of phenylalanine. The accumulation of the amides in a phosphate-free medium was inversely correlated with the phosphate concentration within the cells (Knobloch et al., 1981). Sulfur starvation resulted in a large increase in cellular arginine with some shifts in the concentrations of feruloyl and caffeoyl putrescines (Klapheck, 1983).

According to Martin et al. (1985), the hydroxycinnamoyl putrescines promote the cell multiplication of leaf disks as well as callus formation in vitro, but inhibit bud formation. In in vitro leaf explants, the levels of the amides correlated with growth and bud formation; differentiation occurred when the levels of the amides fell. K^+ deficiency caused a decrease in amide levels, although free putrescine content increased markedly (Klinguer et al., 1986). A basal medium that did not increase cell division or differentiation of the foliar explants sustained low levels of amines and amides. The latter syntheses were increased by 2,4-D as was cell multiplication; as noted above, a fall in the levels of the compounds was followed by differentiation (Martin–Tanguy et al., 1988). These workers suggested that differentiation involves a shift in the metabolism of putrescine and conjugated putrescine.

The formation of roots from foliar explants in media containing indole acetic acid was accompanied by an increase in the amides in the explants before appearance of the roots, or transiently in the roots. These results were taken to support the hypothesis that the hydroxy-

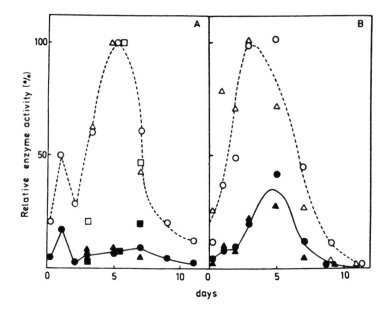

FIG. 20.11 Comparison of enzyme activities involved in the biosynthesis of cinnamoyl putrescines in tobacco cell lines; adapted from Berlin et al. (1982). TX1 (filled symbols) is a wild-type culture; TX4 (open symbols) is a culture of cells resistant to *p*-fluorophenylalanine. (**A**) Values for (○) phenylalanine ammonia-lyase, (△) cinnamate 4-hydroxylase, and (□) *p*-coumarate-coA ligase. (**B**) Data for (△) ODC and (○) ADC.

cinnamoyl putrescines were also significant in root formation (Burtin et al., 1990). On the other hand, stem explants may be inhibited in the increase of feruloyl putrescine and caffeoyl putrescine by inhibitors of PAL without any effect on the increase of dry weight and the formation of cortical callus and floral buds (Wyss–Benz et al., 1990). These workers concluded that these amides do not trigger growth and differentiation in the explants.

Putrescine and spermidine hydroxycinnamoyl transferases have been isolated from extracts of tobacco callus and cell suspensions. Crude soluble enzyme preparations have revealed activities for a full range of phenolic acid-coA thioesters, as well as for putrescine, cadaverine, agmatine, and spermidine (Negrel, 1989). This worker separated a less active spermidine hydroxycinnamoyl transferase, mostly active with p-coumaric acid-coA, from a putrescine hydroxycinnamoyl transferase, mostly active with caffeic acid-coA. The former is relatively specific for spermidine, but the latter is active with putrescine and less so with cadaverine and DAP (Negrel et al., 1991). Putrescine is capable of forming amides with caffeoyl-coA > cinnamoyl-coA > feruloyl-coA > sinapoyl-coA ≫ p-coumaroyl-coA (Negrel et al., 1992).

Some tobacco cells appeared to oxidize a bound form of caffeoyl putrescine to generate utilizable nitrogen (Flores & Martin–Tanguy, 1991). This suggested the formation of caffeoyl GABA, which has been isolated (Balint et al., 1987). The compound has a distribution within tobacco plants similar to that of caffeoyl putrescine that is high in floral tissue and absent from tissues such as the lower leaves. The subsequent fate of bound GABA has not been described.

In a study of the incorporation of amines or specific amino acids into callus cell lines of corn, it was found that [^{14}C]-putrescine was rapidly (<1 min) incorporated into an acid-soluble conjugate. In that time frame putrescine derived from arginine (but not from ornithine) was similarly conjugated despite a much higher level of ODC in the cells than of ADC (Hiatt, 1989). The experiments suggested the existence of a compartment in which a sequestered arginine-derived putrescine was conjugated.

Fruit Development

Although this process usually begins with pollination, it can be initiated by plant growth regulators such as the giberellins or auxins, which start cell division (parthenocarpy). In tomato flowers, which can be induced to begin to fruit by either pollination or auxin, the ovaries increase markedly in weight and in their activity of ODC (Mizrahi & Heimer, 1982). In the fruit pericarp of the tomato, labeled putrescine and spermidine are metabolized to a variety of products, spermidine to putrescine and putrescine to GABA and other products. These reactions indicate the presence of PAO. This tissue is quite low in conjugated polyamines; bound putrescine is significantly higher in the seed (Rastogi & Davies,

1989, 1991). Putrescine has been reported to increase the shelf life for ripe tomatoes as a result of inhibiting late stages of ripening (Dibble et al., 1988).

Slocum and Galston (1985a) followed the polyamines and biosynthetic enzymes in tobacco ovary development in plants regenerated from calli. The highest concentrations of acid-insoluble and acid-soluble conjugated spermidine were found at peaks of DNA content signifying the most active division. The covalently bound putrescines had much lower peaks. At this stage also, ODC activity increased to about 85% of the total decarboxylase, ADC + ODC, and the inhibition of ODC by DFMO selectively inhibited ovarian growth as well as spermidine content (Slocum & Galston, 1985b). The high level of polyamine conjugates was suggested to serve as storage forms of polyamine, of which the putrescine conjugate was hydrolyzed and, in the presence of DFMO was maintained at a more normal level of the diamine. On the other hand, DFMO did not cause a significant decrease in conjugated spermidine, which might release its spermidine in the germinating seed.

It was reported that the use of polyamine sprays of 10^{-5}–10^{-6} M putrescine, spermidine, or spermine on open self-pollinated flowers of apples increased the formation of fruit on the trees. There was an increase in the early stage of fruit growth rate or cell division (Costa & Bagni, 1983). Increases were seen in both ADC and ODC in the developing fruit, of which ADC was the most active. Although high in free putrescine and spermine at bloom and in early stages of fruiting, the apple was relatively low in the soluble conjugated polyamines and free polyamines fell greatly in later stages. Thus, the synthesis of the polyamines appears to relate largely to the initial phenomena associated with fruit set, that is, fertilization, hormone stimulation, ovary growth, and early cell division. A reduction of growth in stimulated unpollinated pea ovaries is accompanied by an increase in the level of spermine, leading to the suggestion that this polyamine is involved in the control of ovary senescence (Carbonell & Navarro, 1989).

Later stages of fruit development (i.e., ripening, abscission, and fruit drop) involve metabolic events related to the conversion of AdoMet to the plant hormone ethylene, a phenomenon that may reflect the existence of an important metabolic branch point determining a choice between growth or senescence.

Ethylene: Multifunctional Plant Hormone

D. Neljubow reported in 1901 that a component of illuminating gas, more particularly ethylene, affected the growth of pea seedlings. In 1910 it was observed that oranges produced a gas that hastened the ripening of bananas; a quarter of a century later, the ripening gas emanating from apples was shown to be ethylene (Gane, 1934). The gas was trapped in bromine, and the product reacted with aniline to form the crystalline N,N′-diphenylethylenediamine, that is,

$$H_2C{=}CH_2 + Br_2 \longrightarrow BrCH_2{-}CH_2Br \xrightarrow{\phi NH_2} \phi NHCH_2{-}CH_2NH\phi$$

Ethylene was first synthesized in the 1790s and was one of the first molecules whose atomic composition was determined by John Dalton. In the 19th and 20th centuries, its reactivity enabled the use of this alkene for the production of organic compounds generally, and in 1987 in the United States the production of ethylene for chemical synthesis like plastics was on the order of 17.5 million tons. Ethylene is considered to be the most important building block of the organic chemical industry in reactions developed with various sophisticated catalytic agents. The obvious importance of its role in plant growth and food production and in the economics of agriculture in fruit harvesting, ripening, and marketing, has evoked a large number of studies on the biosynthesis of ethylene, the regulation of its synthesis, and the mechanism(s) of its action in plants (Haard, 1984). The seasonal changes of fall (i.e., chlorophyll degradation and leaf drop) are in part initiated by ethylene.

In addition to fruit ripening and plant senescence, the actions of the gas in plants include such biological manifestations as the induction of roots and flowers, hypertrophy, etc. Among ethylene-regulated genes may be mentioned those for endochitinase in the bean, one for a proteinase inhibitor in the tomato, a series of plant defense genes in carrot roots, as well as a gene related to ethylene binding in *Arabidopsis*. Ethylene sensitivity has been found to be transduced in some systems via protein phosphorylation and to be effected in others via an increased metabolism and elimination of an active plant hormone such as indole-3-acetic acid (Sagee et al., 1990). Processes such as abscission and fruit drop in citrus are accelerated by ethylene. The hormone increases cellulase and polygalacturonase in the abscission zone and its effects are inhibited by auxins that decrease these enzymes and delay abscission (Huberman & Goren, 1979). The complexities of leaf senescence, involving degradation of chloroplast components (Trebitsh et al., 1993; Woolhouse, 1984), intersect with the problem of the functions of the polyamines, inasmuch as these inhibit the deterioration of detached leaves.

The production of ethylene in plants and the accompanying frequently disastrous effects on plant growth are stimulated by various types of stress, such as wounding, fungal infection, drought, water logging, chilling, and many types of air pollutants (Wang et al., 1990). In one avenue of laboratory research it was found that an excess production of ethylene by a preparation of tobacco leaf protoplasts is a marker for poor survival of the cells and hence for the future of the ongoing experiment.

The major route of ethylene production in higher plants from AdoMet was first described by Adams and Yang (1979). Prior to this discovery routes of biosynthesis had been detected in microbes and other systems and some of these are actually functional, as in the stimulation of ethylene synthesis in evergreen needles stressed by acid mist. One clarified effect of ethylene in animal tissues may be relevant to the breadth of its effects in plants. The gas was used as a surgical anesthetic for some years, until it was observed to have a highly toxic effect on the mammalian liver. The green livers so induced proved to contain an ethylene-inactivated cytochrome P_{450}, whose structure was analysed by Ortiz de Montellano and Mico (1980). These authors suggested that the destruction of cytochrome P_{450} by alkylation may have broader physiological effects, and indeed the formation of a hydroxyethyl derivative of the heme of this enzyme may be one of the possible models for the numerous ethylene effects observed in plants.

Lieberman (1979) surveyed the mechanisms of ethylene production detected prior to the discovery of the role of AdoMet in its biogenesis. Homogenized fruits such as apples or tomatoes, which produce ethylene actively when intact, were unable to synthesize the gas. Many types of fungi, of which *Penicillium digitatum* or *Saccharomyces cerevisiae* are two examples, were found to produce the gas (Arshad & Frankenberger, 1992). L-Methionine could serve as a precursor for ethylene synthesis in yeast and a product of methionine metabolism, 2-keto-4-methylthiobutyric acid (KMBA), has been found to be oxidized to ethylene by others. A homogeneous enzyme has been isolated from *P. digitatum* that can convert α-ketoglutarate to ethylene in the presence of arginine, FeII, dithiothreitol, and O_2 (Fukuda et al., 1989).

There are at least two types of precursors of ethylene in microorganisms including bacteria, that is, methionine, which also serves for *E. coli*, and α-ketoglutarate, a precursor in *Pseudomonas syringae* (Fukuda et al., 1993). A homogeneous enzyme capable of degrading KMBA to ethylene has been obtained from *Cryptococcus albidus*; this reaction proceeds in the presence of NADH, a chelate form of FeIII and O_2 (Ogawa et al., 1990). The classification of many ethylene-producing bacteria also reveals the existence of mechanisms apparently based on other precursors (Nagahama et al., 1992). The existence of microbial producers of ethylene points to the potentiality of disturbances of plant growth due to microbial infections arising from soil or other contamination, as surveyed by Arshad and Frankenberger (1992).

The artifactual production of ethylene has been found in tissue homogenates of plants or animals. A peroxidase found in cauliflower was reported to oxidize KMBA and methional to ethylene and a similar system was found in rat liver extracts. Lieberman (1979) also described a series of model systems involving precursors such as peroxidized linolenate. The latter gives rise to a mixture of hydrocarbons, in contrast to plant systems in which methionine is the precursor and ethylene is the sole product. CuII metal ions will markedly stimulate ethylene production by aquatic plants like *Spirodela*, and inhibitors of lipid peroxidation block this stimulated synthesis. Also, some of the model systems were explored to analyze their relevance to the newly described AdoMet-derived precursor, 1-aminocyclopropane-1-carboxylic acid (ACC). ACC was converted to ethylene in at least two systems (Gardner & Newton, 1987; Osswald et al., 1989). These groups asked if their results had any physiological significance. In any case, the copper-stimulated synthesis of ethylene by the aquatic *Spirodela* does not appear to involve ACC (Mattoo et al., 1986). Yang and Adams (1980) surveyed the extremely numerous known olefin-forming reactions, as well as various methionine-based ethylene-forming reactions, as a preliminary to a discussion of the discovery of the conversion of AdoMet to the hormone.

Biosynthesis of Ethylene and Its Intermediates

It was shown by Adams and Yang (1979), and independently by Lürssen et al. (1979), that ethylene is derived from AdoMet via the formation of the intermediate ACC. In biological terms, ethylene is a hormone that activates plant senescence, among other activities, whereas the polyamines relate largely to phenomena of growth in the synthesis of nucleic acids and proteins. Both groups of substances are derived from AdoMet and the formation and distribution of this substrate might participate in the regulation of these apparent biological antipodes.

In dissecting the pathway it was found that methionine is the major source of ethylene in apple tissue and that carbons 3 and 4 of the former are converted to the alkene. Acting on the hypothesis that AdoMet is the initial intermediate, it was shown that ethylene production in the apple is accompanied by the accumulation of methylthioribose and some of its precursor, MTA (Adams & Yang, 1977). The conversion of methionine into all three products was inhibited by a single inhibitor, 2-amino-4-(2′-aminoethoxy)*trans*-3-butenoic acid (AVG). The reactions proceeded readily in air; but in a nitrogen atmosphere the apple tissue accumulated the ethylene precursor, ACC, which could be converted to ethylene in air. AVG, a known inhibitor of PLP-requiring enzymes, did not inhibit the aerobic conversion of ACC to ethylene. Exogenous ACC was rapidly converted to ethylene in air (Yang & Adams, 1980). Excised senescing petals of *Tradescantia* increased their content of methionine, which was subsequently converted to ACC and then to ethylene (Suttle & Kende, 1980).

Many assays of ACC have been based on the conversion of ACC to ethylene by hypochlorite in the presence of Hg^{2+}. Ethylene was estimated by gas chromatography and was recovered in a yield of 73–80% (Lizada & Yang, 1979); radioactivity was counted after absorption on mercuric perchlorate (Greenberg & Cohen, 1985). Some amines and phenols can reduce ethylene yield from ACC in the hypochlorite reaction (Nieder et al., 1986; Sitrit et al., 1988).

Production of ACC has been studied in excised senescing flowers. Wounded, growing, or ripening tissues were found to generate an ACC synthase as the rate-

controlling enzyme of ethylene synthesis (Kende & Boller, 1981; Yu & Yang, 1980). The synthase was shown to be synthesized de novo after wounding (Acaster & Kende, 1983) and was found to have a short half-life in wounded tissue. Other stresses such as ozone or 2,4-D also induced an AVG-inhibited ACC synthase. The uptake of ACC into leaf disks was inhibited at a transport site by many amino acids with nonpolar side chains: methionine, tyrosine, alanine, leucine, etc. (Lürssen, 1981).

The increased production of ethylene in some virus-infected plants had been detected by Ross and Williamson (1951) and eventually this was shown to occur in infections of TMV via the metabolism of methionine to ACC (De Laat & Van Loon, 1981, 1982). The conversion is particularly active in the necrotic lesions in the localized infections by the virus, suggesting the participation of ethylene in cell death. These sites are surrounded in nonnecrotic areas by accumulations of phenolamides as well as salicylic acid (i.e., 2-hydroxybenzoic acid), a substance produced actively in TMV infections. Both salicylic acid and acetyl salicylic acid have been reported to inhibit ethylene production (Leslie & Romani, 1988). Also salicylic acid and polyamines have both been reported to inhibit formation of ACC synthase mRNA (Li et al., 1992).

A reaction sequence for the biosynthesis of ethylene from methionine is presented in Figure 20.12. ACC was first detected in fruit by Burroughs and by Vahatalo and Virtanen in 1957. Several cyclopropane amino acids are known in nature and the chemistry of the various plant products has been surveyed by Stammer (1990). The synthesis of ACC from AdoMet by ACC synthase results in the production of MTA, which may be converted to methionine, conserving the methylthio group. PLP, which forms a Schiff base with the α-amino group of methionine, is known to participate in reactions catalyzing γ-eliminations, and in this instance the elimination of MTA leads to the formation of a cyclopropane ring. AdoMet has two diastereomers with respect to its sulfonium center and only one of these, the (-)-AdoMet, serves as a substrate of ACC synthase (Khani–Oskouee et al., 1984). However, both isomers inactivate the synthase via a linking of the 2-aminobutyric acid portion of AdoMet (Satoh & Yang, 1989). It has been estimated that an inactivating reaction with (-)-AdoMet as substrate occurs about once in every 30,000 reactions leading to the formation of ACC.

An ACC deaminase, which operates with PLP, has been isolated from a Pseudomonad. ACC is broken down to α-ketobutyrate and NH_3; that is,

ACC synthase and the deaminase are the only PLP enzymes known to determine the synthesis or degradation,

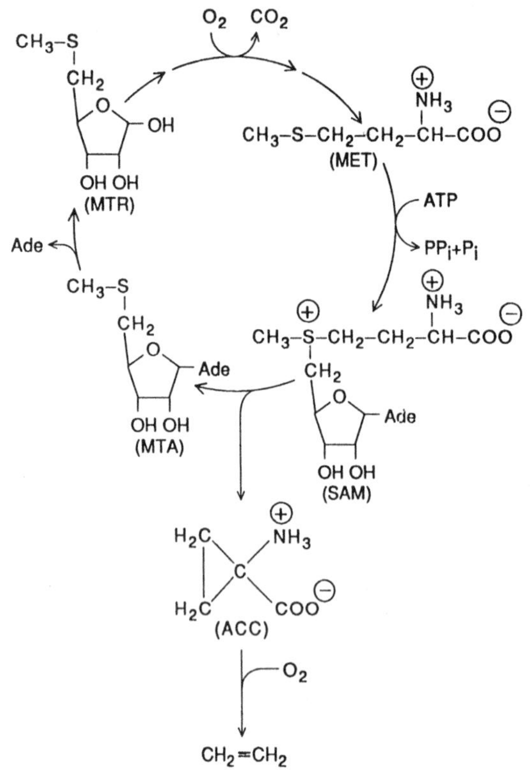

FIG. 20.12 Methionine cycle in relation to the biosynthesis of ethylene in apple tissue; adapted from Yung et al. (1982).

respectively, of the cyclopropane ring. A reaction mechanism for ring opening has been postulated (Walsh et al., 1981) and this has facilitated the design of irreversible or slowly dissociating inhibitors of the deaminase such as β-chloro-D-alanine and 1-aminocyclopropanephosphonate. The enzyme was increased in slices of squash, preceding ACC and ethylene formation. Similar results were obtained in bean leaf disks transferred from 5 to 25°C or in ripening apples. An inhibition by ethylene itself of the de novo synthesis of the enzyme and of the activity of the synthase was detected in a study of citrus peel (Riov & Yang, 1982). The inhibition of the synthase by AVG or AgNO$_3$ both inhibited ethylene production and the abscission of bean explants (Kushad & Poovaiah, 1984).

ACC decreased in stressed (wilted) wheat leaves more rapidly than ethylene was generated and the ACC was isolated as a malonylamino conjugate (MACC) (Hoffman et al., 1982). This derivative was soon found in many other plants and at first did not appear to regenerate utilizable ACC. However, a slow conversion of MACC to ethylene has been demonstrated in various plant tissues (Jiao et al., 1986) and MACC-induced senescence has been observed in preclimacteric carnation petals (Hanley et al., 1989). Ethylene increases conjuga-

tion of ACC in many plant tissues (Liu et al., 1985; Philosoph–Hadas et al., 1985). The malonyltransferase of mung bean seedlings has been purified to homogeneity and is a monomer of approximately 55 kDa (Guo et al., 1992). A somewhat different isoform has been isolated from tomatoes (Martin & Saftner, 1995) and mung beans (Benichou et al., 1995).

In studies with protoplasts, MACC was synthesized in the cytosol and actively transported into the vacuole, a site of largely irreversible sequestration (Bouzayen et al., 1989). The vacuolar uptake of ACC itself has been found to be stimulated by MgATP, as was MACC (Saftner & Martin, 1993). Release of MACC from the vacuole required a vacuolar pH lower than pH 5.5 (Pedreño et al., 1991).

The availability of ACC and the increase of its synthase are among the most important sites of control of the production of ethylene. In the ripening of avocado fruit, ACC and ACC synthase increase sharply during the climacteric rise of the increase of ethylene (Sitrit et al., 1986). Needles derived from acid-mist damaged spruce trees are higher in ACC, MACC, and ACC synthase than those of undamaged trees (Yang et al., 1994). An ethylene-overproducing mutant of the tomato, characterized by thickened tissues and compact growth, is relatively higher in ACC production in early stages of ethylene synthesis than in later stages (Fujino et al., 1988). The ACC synthase is induced by relative anoxia in deep-water rice (Cohen & Kende, 1987) and by abscisic acid in citrus and leaf tissues (Riov et al., 1990). An inhibitor of ethylene action, 2,5-norbornadiene, causes a marked decrease in both ACC synthase and the ethylene-forming enzyme, ACC oxidase, suggesting an autocatalytic action of ethylene in stimulating its synthesis and effects (Peiser, 1989). Whereas indole acetic acid has been found to enhance ACC synthase activity, it also increases the activity of the oxidase and production of ethylene, which also increases the activity of the oxidase (Peck & Kende, 1995).

Numerous workers have purified ACC synthase to apparent homogeneity. The enzyme has been reported as dimers of various sizes from different plant tissues: subunits of the enzyme from wound-induced tomatoes (50 kDa), LiCl-induced tomatoes (45 kDa), ripening tomatoes (47 kDa), hormone-induced cucumbers (46 kDa), mung beans (65 kDa), apples (48 kDa). Several isoforms were detected in tomato extracts by isoelectric focusing, and one of these that was reactive with a monoclonal antibody contained subunits of 67 kDa (Mehta et al., 1988). A wound-inducible tomato gene of ACC synthase was altered to eliminate the carboxyterminal region of the normally dimeric protein (Li & Mattoo, 1994). This was expressed in E. coli as a hyperactive monomeric truncated enzyme of 52 kDa. The apple gene has also been expressed in E. coli to produce a dimer as a major species (White et al., 1994).

In vitro translation products of ACC synthase mRNA were larger than the in vivo enzyme, indicating post-translational processing (Nakajima et al., 1988). This also suggested that such processing was diminished in the mung bean system (Tsai et al., 1991) and possibly very efficient in the low MW systems. In most studies the identity of the isolated activity was established by both the formation of ACC and the binding of AdoMet in inactivation of the enzyme (Yip et al., 1991). A lysine-binding site of the PLP-AdoMet reactant has been found in trypsin digests of inactivated ACC synthases (Yip et al., 1990). The turnover of the labeled enzyme in wounded ripening tomatoes was shown to be rapid, with a half-life of 58 min (Kim & Yang, 1992). The turnover of the synthase in tomato cells has been reported to be regulated by protein phosphorylation and dephosphorylation (Spanu et al., 1994).

The possible existence of several genes for the enzyme, as well as posttranslational modifications was suggested by White and Kende (1990). The isolation of zucchini mRNA for ACC synthase permitted the synthesis of a cDNA that detected the transcription of the ACC synthase gene(s) by various inducers (Sato & Theologis, 1989). The ACC synthase cDNA could be expressed in *E. coli* and was also used to detect two genes within zucchini nuclear DNA (Huang et al., 1991). At least six divergent genes for the synthase are currently recognized in the tomato (Lincoln et al., 1993).

Transcripts of mRNA were isolated from variously induced tomatoes, mung bean hypocotyls, and apple shoots and were used to isolate cDNA. The various cDNAs were used for the detection of the appropriate genes and for the transcription of these genes, as in fruit ripening (Van der Straeten et al., 1990). At least three genes for ACC synthase exist in etiolated mung beans (Botella et al., 1992). Five different synthases were detected in the more easily manipulated *Arabidopsis thaliana* (Liang et al., 1992). The biological activity of one of these was observed in growing and flowering plants, but not under conditions of wounding or auxin stimulation, although the gene was activated by ethylene exposure (Rodrigues–Pousada et al., 1993; Van der Straeten et al., 1992).

Thus, ACC accumulation is a major regulatory step in ethylene production, in many instances as a result of a stimulated synthesis of one or more ACC synthases. Application of ACC alone to the base of cut olive shoots evokes a burst of ethylene from leaves.

A production of ethylene from ACC by plant mitochondria (Vinkler & Apelbaum, 1985) suggests a role for lipoxygenase in ethylene synthesis. However, the production of 1-butene in plant epicotyls and leaves from four stereoisomers of the ACC analogue 1-amino-2-ethylcyclopropane-1-carboxylic acid occurs from only one of these whereas lipoxygenase acts on all four (Wang & Yang, 1987). Such a test excludes some apparently active extracts as the in vivo ethylene-forming enzyme (EFE) and is now a routine test of purported enzymes for similarity to the in vivo stereospecific EFE.

An additional test of the extraction of an in vivo EFE was based on the apparent stoichiometry of ACC degradation. In the vetch (*Vicia sativa*), [^{14}C]-labeled ACC was converted to equal amounts of ethylene, CO_2, and HCN (Peiser et al., 1984), via a route postulated to be

The release of the nitrogen as HCN was also reported by Pirrung (1985). $^{14}CO_2$ was recovered from the labeled carboxyl group.

It was believed initially that the ACC oxidase (EFE) was localized on membranes (Kende, 1993). The ACC oxidase has also been reported as localized on the cell wall in tomatoes and apples (Rombaldi et al., 1994). Studies of the sites of isolable oxidase has indicated an intraprotoplast locus, as well as dependence of the activity on iron (Bouzayen et al., 1991). The key to the isolation of the enzyme developed from a study of the mRNAs produced in both wounding and fruit ripening in tomatos and the structure of the complementary cDNAs (Bouzayen et al., 1992; Hamilton et al., 1990; Holdsworth et al., 1987). The amino acid sequences of the polypeptide (35 kDa) deduced from the cDNA were homologous in significant measure to that of a difficultly isolatable flavone-3-hydroxylase. A soluble ACC oxidase was obtained from a melon (*Cucumis melo*) when O_2 was excluded during the preparation and ascorbic acid and Fe^{2+} were included in the aerobic reaction, conditions comparable to those used for the flavone hydroxylase (Ververidis & John, 1991). The monomeric enzyme (41 kDa) used ascorbate as a cosubstrate and was inhibited completely by SH reagents and partially by 1 mM salicylate (Smith et al., 1992). The protein enzyme was activated by CO_2 and inhibited by the structural analogue, carbonyl sulfide (COS) (Smith & John, 1993). The enzyme isolated from the pear is stated to have an absolute requirement for CO_2, as well as for Fe^{2+} and ascorbate (Vioque & Castellano, 1994). A similar stereospecific oxidase with low K_m for ACC, inhibitable by α-aminoisobutyrate, was solubilized from apple tissue under anaerobic conditions (Dong et al., 1992). Gel filtration revealed a slightly lower molecular weight, 39 kDa, and the catalyzed reaction was shown to be

$$ACC + ascorbate + O_2 \rightarrow C_2H_2 + HCN + CO_2 + dehydroascorbate + 2H_2O.$$

Removing iron from apple extracts by chelation with 1,10-phenanthroline stabilized the enzyme and facilitated purification to homogeneity. Immunolocalization of the tomato oxidase indicated a major presence in cytoplasm and absence from vacuolar or membrane structures (Reinhardt et al., 1994). A successful expression of the tomato oxidase has been achieved in *E. coli* (Zhang et al., 1995).

Great efforts are being expended to determine the details of ethylene production and its controls (Balagué et al., 1993), as well as the mechanisms of its many effects in the hope of solving various economic problems of postharvest physiology, such as deterioration of flowers or overripening of fruit. Transgenic tomato plants carrying the bacterial gene for ACC deaminase have considerably reduced ethylene production and delayed fruit ripening. The genetic and biochemical dissection of the ethylene response pathway is also under active investigation (Kieber & Ecker, 1993).

Possible Reciprocal Relations of Polyamines and Ethylene

The contents of the polyamines and their biosynthetic enzymes increase during many growth processes in plants whereas, for the most part, ethylene production increases as tissues begin to degrade. Does competition for AdoMet exist between ethylene and compounds such as spermidine and spermine? Such a hypothesis was formulated after the initial clarification of the key role of AdoMet in ethylene biosynthesis (Cohen et al., 1981). Transgenic tomato plants expressing the T3 AdoMet hydrolase in the ripening fruit do have a much reduced ability to produce ethylene (Good et al., 1994).

The observations that polyamine retarded the senescence of detached leaves and inhibited enzymes such as proteinases and nucleases were extended to show that the polyamines also inhibited production of ethylene in auxin-mediated soybean hypocotyls and in senescing flower petals (Suttle, 1981). Similar results were obtained with several types of apple and tobacco tissues (Apelbaum et al., 1981). Polyamine concentrations do decrease in senescing oat leaves and exogenous spermidine appeared to inhibit ACC synthase in that system (Fuhrer et al., 1982). Also millimolar polyamines inhibited the ACC synthase of winter squash (Hyodo & Tanaka, 1986). However, the polyamines also inhibited syntheses of proteins and nucleic acids in apple fruit tissues (Apelbaum et al., 1982). In cut carnations putrescine and ethylene did accumulate in parallel without a change in spermidine and spermine (Roberts et al., 1984). A similar result has been seen in the ripening of cherimoya fruit (Escribano & Merodio, 1994). Also, no correlation was detected in growing wheat plants between polyamine content of any tissue and the onset of senescence (Peeters et al., 1993).

Ethylene administration has been observed to decrease enzymes of polyamine biosynthesis. Treated pea seedlings were found to lose about 90% of their ADC in about 10 h (Apelbaum et al., 1985) and about 80% in 18 h (Ickeson et al., 1985). However, cultured cells of N. tabacum increase their levels of spermidine and spermine, as well as ODC and AdoMetDC, in response to treatment with ethylene (Park & Lee, 1994). In one test of competition for AdoMet, a study of spermidine biosynthesis from labeled methionine in aged orange peel, the application of AVG to inhibit ACC synthesis was found to increase label sharply in AdoMet and in spermidine (Even–Chen et al., 1982). Similarly, an inhibition of ethylene production in carrot cells during embryogenesis did increase labeling from methionine in spermidine and spermine. Exogenous ethylene inhibited embryo formation and this was reversed by exogenous spermidine (Roustan et al., 1994).

In another test of the hypothesis, the fate of labeled methionine was compared in healthy and virus-infected Chinese cabbage leaf protoplasts (Greenberg & Cohen, 1985). In the uninfected cells, similar levels of isotope were incorporated into both ACC and spermidine whereas the infected cells tripled incorporation into ACC and halved incorporation into spermidine. Addition of cyclohexylamine, the inhibitor of spermidine synthase, reduced synthesis of spermidine but did not affect the incorporation of methionine into AdoMet. As noted earlier (Chap. 19), the infected cells inhibited by cyclohexylamine incorporated more aminopropyl moieties from the accumulated AdoMet into newly synthesized spermine. However, the level of incorporation into ACC was unaffected by the inhibition of spermidine biosynthesis in this system. More recently cyclohexylamine inhibition of synthesis of spermidine and spermine in germinating chick peas was found to increase the ethylene pathway and products such as ACC and ethylene (Gallardo et al., 1995). Conversely, inhibition of the ethylene pathway in this system by norbornadiene, results in an accumulation of free and bound polyamine (de Rueda et al., 1994).

The hypothesis was formulated for Chinese cabbage, as shown in Figure 20.13. However, the metabolism of AdoMet into polyamine and ACC may be only a small fraction of the total metabolism of AdoMet, which will include numerous reactions of methylation. Thus, the possibility of demonstrating alternative pathways of this compound into polyamine and ACC may be masked by its diversion into other routes that have not been examined quantitatively in these systems. The hypothesis of a simple competition for only two routes of metabolism of AdoMet will require more complete data on the total use of AdoMet in carefully selected plant systems.

Studies on barley callus have revealed inverse relations between production of ethylene and polyamines (Katoh et al., 1987). Similar results on the inverse contents of these compounds have been seen in various systems of storage, fruit development, flower development, etc. However, Cohen and Kende (1986) have described ethylene stimulation of enzymes of polyamine biosynthesis and of polyamine content in deep water rice internodes, that is, the intercalary meristem, following submergence of the plant. Similar results have been seen in rice coleoptiles (Lee & Chu, 1992). Thus, the results have varied from one biological system to another, and a single explanation of the results has not emerged.

Polyamine-Derived Alkaloids

The alkaloids were defined in 1908 by H. Euler as N-heterocyclic basic metabolites. Many biochemical texts have adopted the term, an editorially useful catch-all, to describe miscellaneous groups of several thousand

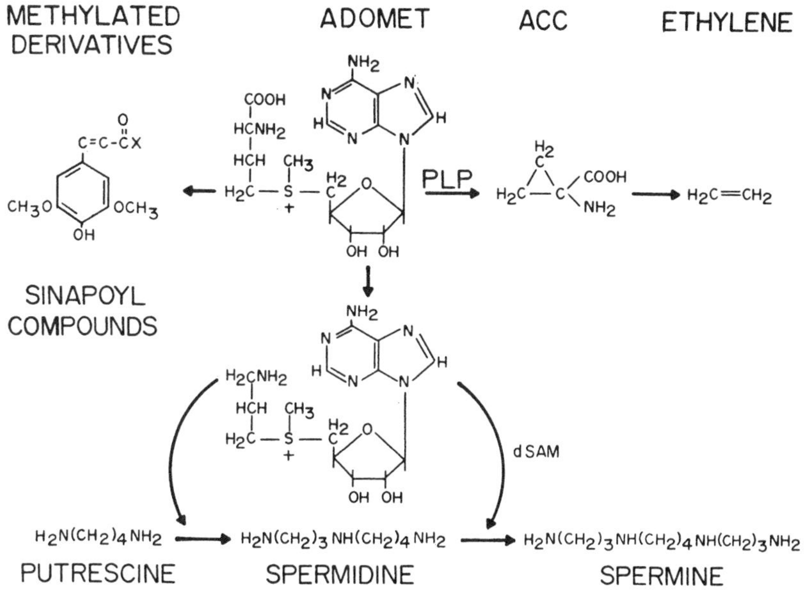

METHYLATED
DERIVATIVES

ADOMET   ACC   ETHYLENE

SINAPOYL
COMPOUNDS

PUTRESCINE   SPERMIDINE   SPERMINE

FIG. 20.13 Central role of *S*-adenosylmethionine (AdoMet) in the formation of the polyamines, ethylene, and methylated compounds; adapted from Cohen et al. (1981).

nitrogen-containing compounds. Many of these are derived in plants from the aromatic amino acids. In numerous instances, the spermidine and spermine alkaloids (see below) are heterocyclic, because the phenolic acid amides of these polyamines have been rearranged to form macrocyclic derivatives. In one of these, the spermidine alkaloid, codonocarpine, is macrocyclic because of the formation of an ether between the feruloyl substituents of bisferuloyl spermidine (Fig. 20.14). In a compound of the pleurostylin series, carbons derived from *trans*-cinnamoyl substituents have formed bonds to nitrogens and carbons of the polyamines, suggesting roles of the cinnamoyl spermidines in the biogenesis of these "alkaloids." It is of interest that N^1,N^8-dibenzoyl spermidine has been isolated from the genus *Cassia* and has been acceptably designated as a "spermidine alkaloid" (Alemayehu et al., 1988).

Among the spermine alkaloids, the construction of the macrocyclic structure may be based on similar enlarging reactions of the phenolic acid amides of spermine. Thus, it can be seen below that verbacine is probably derived from a bis-*trans*-cinnamoyl spermine. Other even more complex spermine alkaloids, such as the aphelandrines, ephedradines, orantines, and schweinines, appear to be derived from mixed phenolic and benzoylated amides to form multicyclic compounds.

An extensive summary of the occurrence and chemistry of the polyamine-derived alkaloids has been published (Guggisberg & Hesse, 1983). Many additions to this catalogue become available each year, besides the many structures, sesquiterpenes, macrocycles, etc., whose biosynthetic origins are thought to be unrelated to the polyamines or their biosynthetic intermediates. In addition to the compounds just noted that are related to or

contain spermidine, spermine, or their derivatives or analogues, compounds derived from putrescine, such as nicotine and the tropane and pyrrolizidine alkaloids, and from cadaverine, such as the quinolizidine alkaloids, will be described.

Methionine, in addition to providing short carbon chains as in the synthesis of azetidines or some phytosiderophores like avenic acid, is also the major donor of methyl groups in many substituted nitrogenous bases, of which solapalmitine is an example. The pursuit of physiological function revealed new substances such as the complex siderophores or the phenolic acid amides, whose antiviral activities were subsequently detected. New vistas in biotechnology have attempted to maximize the availability of physiologically active products (Kutchan, 1995). In other, initially more academic, taxonomic analyses, as in the "mere" search for the full complement of amines in the legumes, many new plant polyamines have been revealed. The studies of Hamana and Matsuzaki (see Chap. 5) in microbial taxonomy based on polyamine identity and distribution became of major interest only after the breadth of their studies permitted some generalization. Similarly, the known distribution of the alkaloids among the higher plants has begun to provide useful evidence of metabolic, taxonomic, and evolutionary relationships among the higher plants (Hegnauer, 1988).

Nicotine and Other Putrescine-Derived Alkaloids

Leete (1980) described alkaloids derived from ornithine, lysine, and nicotinic acid. More recent metabolic studies demonstrated the route of putrescine into the pyrrolidine

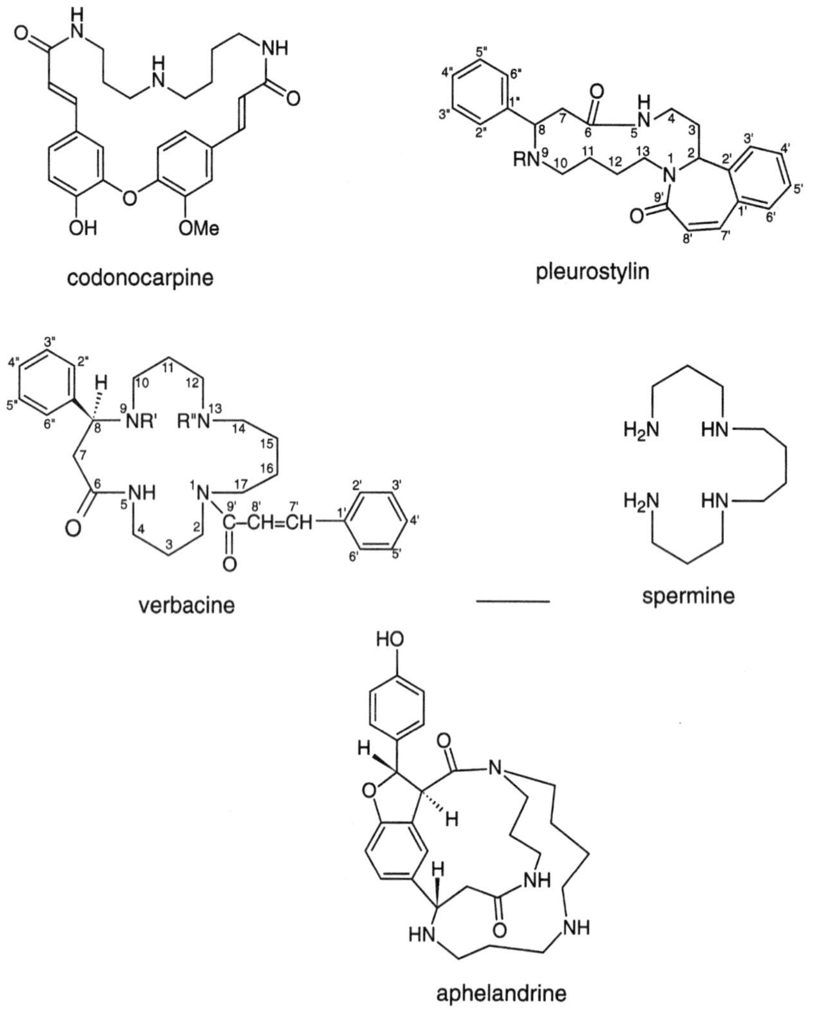

FIG. 20.14 Some macrocyclic alkaloids containing spermidine or spermine as components.

ring of nicotine and into the tropane alkaloids and were summarized by Flores and Martin–Tanguy (1991). Unexpectedly, the synthesis of nicotine was shown in 1941 and 1942 to be effected in tobacco roots rather than in the plant leaves, and this site of synthesis was subsequently extended to the tropane alkaloids. In 1943 Cromwell reported a stimulatory role of putrescine in the synthesis of the tropane alkaloid, hyoscyamine.

Investigations in the 1950s revealed the conversion of glutamate and ornithine in tobacco to a labeled pyrrolidine ring in nicotine. The labeling pattern suggested the intermediary formation of free putrescine and monomethylputrescine; the latter was incorporated into the pyrrolidine ring. The pathways, including the incorpora-

tion of the pyrrolidine into hygrine, are presented in Figure 20.15.

Mizusaki et al. (1971a) described a putrescine N-methyltransferase in the roots of tobacco plants. The substrates included putrescine and AdoMet, and cadaverine and DAP were totally inactive. An N-methylputrescine oxidase was found in tobacco roots, which converted N-methylputrescine to N-methylbutanal, a known precursor in the biosynthesis of nicotine (Mizusaki et al., 1972). [Methyl-^{14}C]-N-methylputrescine was synthesized recently (Secor et al., 1993).

Arginine and ADC appeared to be the more likely system for the provision of substrate (Tiburcio & Galston, 1986). Arginine was the best substrate, DFMA was more inhibitory than DFMO, and ADC levels in the tis-

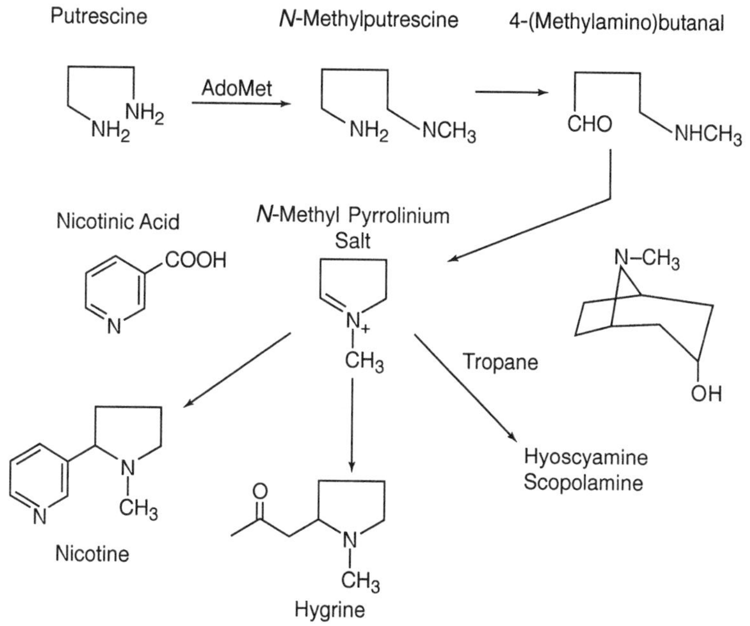

FIG. 20.15 Putrescine in the biosynthesis of nicotine, tropane alkaloids, and hygrine.

sues varied more directly with nicotine production. Nevertheless, the overexpression of a yeast ODC gene in the roots of *N. rustica* does permit an enhanced accumulation of nicotine (Hamill et al., 1990).

Of two genes regulating nicotine levels in tobacco, one determines the production of putrescine *N*-methyltransferase. The cDNA of this gene was found to be significantly homologous to the spermidine synthase of the mammal and of *E. coli*, suggesting the latter enzyme as the evolutionary precursor of the first step of nicotine biosynthesis (Hibi et al., 1994).

The regulation of the nicotine pathway has been studied in tobacco callus, which reproduces the activities found in tobacco roots (Feth et al., 1985a). The formation of the ultraviolet-absorbing *N*-methyl-Δ'-pyrrolinium chloride by methylputrescine oxidase was estimated by HPLC (Feth et al., 1985b). A considerable induction of the methyl transferase and somewhat less of the oxidase occurred at low rates of growth as in auxin-deprived plants (Feth et al., 1986). The extract appeared less specific in its oxidase activity than in its methyl transferase. A purified tobacco oxidase was also active on putrescine. A homogeneous oxidase from tobacco did not cross-react with a pea-seedling DAO (McLauchlan et al., 1993).

Enzymes have been prepared from the roots of plants such as *Datura stramonium*, *Atropa* species, and *Hyoscyanus albus* that produce the tropane alkaloids via the methylpyrrolinium cation. In cultured roots of *H. albus*, which lacked arginase, ADC was twice as active as ODC and putrescine *N*-methyl transferase was quite

high (Hashimoto et al., 1989a). The roots could synthesize spermidine and spermine but did accumulate *N*-methylputrescine. These syntheses, as well as those of hyoscyamine and scopolamine, were inhibited by DFMO, implying a route to putrescine by both ADC and ODC (Hashimoto et al., 1989b). In root cultures of *Datura*, the ADC pathway to hyoscyamine was affirmed in studies on the positive utilization of labeled arginine and agmatine and in the selective inhibition of arginine incorporation by DFMA (Walton et al., 1990). Studies on the complex feedback regulation of the pathway to these alkaloids, particularly by tropine, have been reported (Robins et al., 1991). In *Atropa*, potassium deficiency, which increases putrescine synthesis, increases synthesis of the tropane alkaloids (Khan & Harborne, 1991).

The putrescine *N*-methyltransferase of *H. albus* has been purified and is markedly inhibited competitively in vitro and in vivo by *n*-butylamine ($K_i = 11 \ \mu M$). This compound stimulates the formation of putrescine conjugates (Hibi et al., 1992). The *Datura* enzyme is active on some derivatives and analogues of putrescine, but not on cadaverine (Walton et al., 1994).

In some organisms the pyrroline derived from putrescine may be shunted into other paths. For example, an alkaloid comprising pyrroline bound to a pyrrolidine has been isolated from a *Lilium* sp. In the cyanobacterial toxin, anatoxin-α, the piperidine ring of a tropane alkaloid is replaced by an azacycloheptene ring, in which putrescine is a structural component (Gallon et al., 1990).

Another route for putrescine is that of the pyrrolizidine alkaloids, which are formed from a presumed dimeric intermediate.

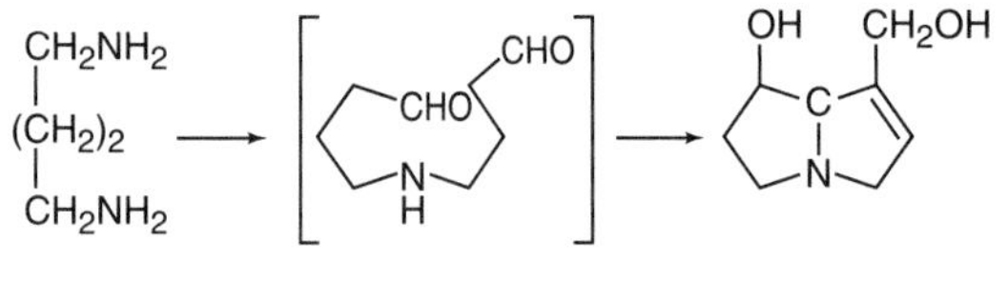

Some 200 known compounds of this series are found in many plants like the genera *Senecio, Crotalaria*, etc., and are frequently acutely toxic, endangering grazing and other animals that ingest these natural products. Modified retronecines substituted at one or another of the hydroxyls are known and found in "medicinal" herbs like comfrey, whose ingestion is considered to be hazardous (Vollmer et al., 1987). The polyhydroxylated pyrrolizidine, australine, is an inhibitor of glycoprotein processing and is a good inhibitor of α-glucosidase amyloglucosidase (Tropea et al., 1989). Monocrotaline, which evokes pulmonary hypertension and chronic lung injury in humans and animals, is thought to increase synthesis of polyamine and DNA and the injury is prevented in rats by a restriction of caloric intake or by administration of DFMO (Hacker, 1993). This pyrrolizidine is also sequestered by moths that feed on *Crotalaria*, and both protects the moths from predatory birds and spiders and serves as a precursor to an insect pheromone (Eisner & Meinwald, 1995).

The complex biosynthetic studies were reviewed by Robins (1989). Studies with inhibitors in various plants indicated that putrescine may be derived entirely from arginine via ADC, or from ornithine via ODC (Birecka et al. 1988). Spermidine and spermine proved to be far better precursors than putrescine in the synthesis of retrorsine in *Senecio isatideus* (Robins & Sweeney, 1979). Homospermidine also served as a precursor of retrorsine; it is synthesized from ornithine in *S. isatideus* and intact [^{13}C]-labeled homospermidine was incorporated into the retronecine structure. Robins (1989) discussed possible biosynthetic schemes for the conversion of homospermidine to retronecine and retrorsine. The oxidation of one primary amino group in homospermidine by a pea seedling DAO resulted in the formation of an *N*-substituted pyrrolinium ion, which appeared to cyclize nonenzymatically to a formylpyrrolizidine. Homospermidine has been found in other plants producing pyrrolizidine alkaloids (Birecka et al., 1984).

The pyrrolizidine alkaloids of seeds of *Crotalaria scassellati* are oxidized to polar N oxides at the juncture of the pyrrolizidine ring. The N oxides are translocated and stored in cell vacuoles and subsequently metabolized in this form in this plant (Toppel et al., 1988). Nitrogen oxidation and some detoxication of toxic pyrrolizidines can occur by cytochrome P_{450} enzymes in the mammalian liver.

Cadaverine-Derived Alkaloids

The administration of cadaverine to the rat leads to the excretion of δ-aminopentanoic acid. In plants, many alkaloids contain the piperidine nucleus, and several of these like anabasine were shown to be derived from [1,5-^{14}C]-cadaverine (Leete, 1958; Leistner & Spenser, 1973) in the following postulated sequence:

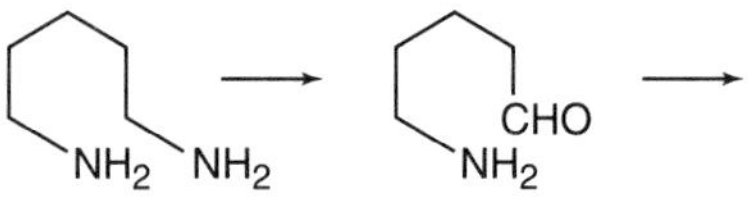

cadaverine   5-aminopentanal

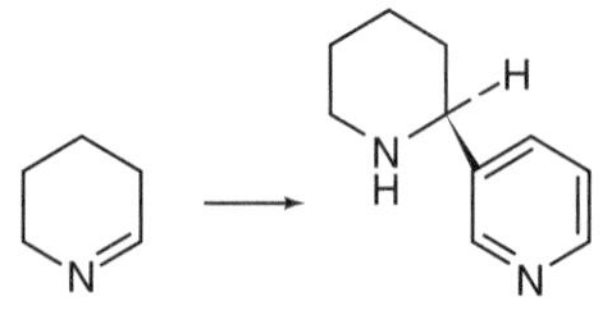

Δ^1-piperideine   anabasine

Δ^1-Piperideine [6-^{14}C] is an active precursor of anabasine in *Nicotiana glauca*. Root cultures of *N. tabacum* that expresses a bacterial gene for LDC synthesize increased amounts of cadaverine and anabasine (Fecker et al., 1993).

The quinolizidine alkaloids are represented by structures such as

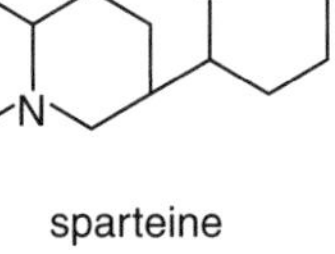

lupinine

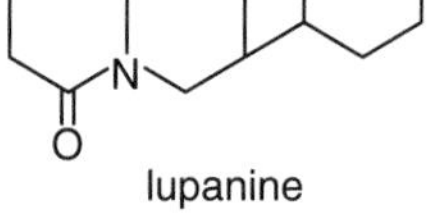

sparteine

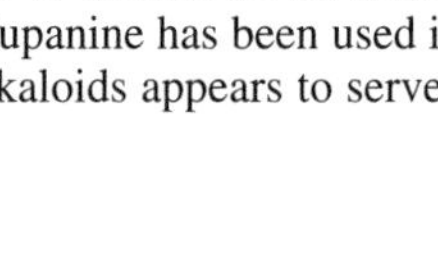

lupanine

These are also products of lysine and cadaverine (Wink, 1987). Sparteine is known as an antiarrhythmic compound whereas lupanine has been used in folk medicine. This group of alkaloids appears to serve defensive functions in plants and some insects, and cell culture techniques have been used to produce the compounds for further study. The quinolizidines are synthesized more actively in chloroplast-containing cells than in callus cultures.

In pursuing this lead it was shown that in the synthesis of lupanine by *Lupinus polyphyllus*, the enzymes LDC and 17-oxosparteine synthase were localized in leaf chloroplasts (Hartmann et al., 1980; Wink & Hartmann, 1982). In the synthesis of several other quinolizidines in in vitro grown *Heimia salicifolia*, the LDC correlated with the chlorophyll content of cell cultures (Pelosi et al., 1986). The localization of these reactions in chloroplasts underlined the earlier demonstration of the formation of lysine via the diaminopimelate pathway found solely in leaf chloroplasts (Mazelis et al., 1976; Wallsgrove & Mazelis, 1980). Indeed, the presence of diaminopimelate decarboxylase in isolated chloroplasts and its absence in other fractions has been used to assess separation procedures.

LDC is found in all tissues in *Nicotiana glauca*, and indeed the enzyme is highest in roots in which anabasine is mainly synthesized. Cadaverine is transported rapidly from the roots to leaves in this plant (Bagni et al., 1986). Cadaverine has also been reported to be essential to normal rooting of germinating seedlings of the soybean (*Glycine max*) (Gamarnik & Frydman, 1991). It would be of interest to know the site of LDC in roots.

REFERENCES

Aarnes, H. (1977) *Plant Science Letters*, **10**, 381–390.

Acastar, M. A. & Kende, H. (1983) *Plant Physiology*, **72**, 139–145.

Adams, D. O. & Yang, S. F. (1977) *Plant Physiology*, **60**, 892–896.

——— (1979) *Proceedings of the National Academy of Sciences of the United States of America*, **76**, 170–174.

Adlakha, R. C. & Villanueva, V. R. (1980) *Journal of Chromatography*, **187**, 442–446.

Adlakha, R. C., Villanueva, V. R., Calvayrac, R., & Edmunds, Jr., L. N. (1980) *Archives of Biochemistry and Biophysics*, **201**, 660–668.

Aleksijevic, A., Grove, J., & Schuber, F. (1979) *Biochimica et Biophysica Acta*, **565**, 199–207.

Alemayehu, G., Abegaz, B., Snatzke, G., & Dudeck, H. (1988) *Phytochemistry*, **27**, 3255–3258.

Allamong, B. D. & Abrahamson, L. (1973) *Analytical Biochemistry*, **53**, 343–349.

Altman, A., Friedman, R., & Levin, N. (1982) *Plant Physiology*, **69**, 876–879.

Altman, A. & Levin, N. (1993) *Physiologia Plantarum*, **89**, 653–658.

Anguillesi, M. C., Grilli, I., & Floris, C. (1982) *Journal of Experimental Botany*, **33**, 1014–1020.

Apelbaum, A., Burgoon, A. C., Anderson, J. D., Lieberman, M., Ben-Arie, R., & Mattoo, A. K. (1981) *Plant Physiology*, **68**, 453–456.

Apelbaum, A., Canellakis, Z. M., Applewhite, P. B., Kaur-Sawhney, R., & Galston, A. W. (1988) *Plant Physiology*, **88**, 996–998.

Apelbaum, A., Goldlust, A., & Icekson, I. (1985) *Plant Physiology*, **79**, 635–640.

Apelbaum, A., Icekson, I., Burgoon, A. C., & Lieberman, M. (1982) *Plant Physiology*, **70**, 1221–1223.

Aribaud, M. & Martin–Tanguy, J. (1994) *Phytochemistry*, **37**, 927–932.

Arif, S. A. M., Taylor, M. A., George, L. A., Butler, A. R., Burch, L. R., Davies, H. V., Stark, M. J. R., & Kumar, H. (1994) *Plant Molecular Biology*, **26**, 327–338.

Arshad, M. & Frankenberger, W. T. (1992) *Advances in Microbial Ecology*, pp. 69–111, K. C. Marshall, Ed. New York: Plenum Press.

Bagni, N., Adamo, P., Serafini-Fracassini, D. & Villanueva, V. R. (1981) *Plant Physiology*, **68**, 727–730.

Bagni, N., Calzoni, G. L., & Speranza, A. (1978) *New Phytology*, **80**, 317–323.

Bagni, N., Creus, J., & Pistocchi, R. (1986) *Journal of Plant Physiology*, **125**, 9–15.

Bagni, N., Malucelli, B., & Torrigiano, P. (1981) *Physiologia Plantarum*, **49**, 341–345.

Bagni, N. & Serafini–Fracassini, D. (1974) *Plant Growth Substances 1973*. Tokyo: Hirokawa Publishing Company.

Bajaj, S. & Rajam, M. V. (1996) *Plant Physiology*, **112**, 1343–1348.

Balagué, C., Watson, C. F., Turner, A. J., Rouge, P., Picton, S., Pech, J., & Grierson, D. (1993) *European Journal of Biochemistry*, **212**, 27–34.

Balint, R., Cooper, G., Staebell, M., & Filner, P. (1987) *Journal of Biological Chemistry*, **262**, 11026–11031.

Bardócz, S., Duguid, T. J., Brown, D. S., Grant, G., Pusztai, A., White, A., & Ralph, A. (1995) *British Journal of Nutrition*, **73**, 819–828.

Bartók, T., Börcsök, G., & Sági, F. (1992) *Journal of Liquid Chromatography*, **15**, 777–790.

Bastola, D. R. & Minocha, S. C. (1995) *Plant Physiology*, **109**, 63–71.

Bell, E. & Malmberg, R. L. (1990) *Molecular and General Genetics*, **224**, 431–436.

Ben-Hayyim, G., Damon, J., Martin-Tanguy, J., & Tepfer, D. (1994) *FEBS Letters*, **342**, 145–148.

Benichou, M., Martínez–Reina, G., Romojaro, F., Pech, J.-C., & Latché, A. (1995) *Physiologia Plantarum*, **94**, 629–634.

Berlin, J. (1981) *Phytochemistry*, **20**, 53–55.

Berlin, J. & Forche, E. (1981) *Zeitschrift für Pflanzenphysiologie*, **101**, 277–282.

Berlin, J., Knobloch, K., Hofle, G., & Witte, L. (1982) *Journal of Natural Products*, **45**, 83–87.

Berlin, J. & Vollmer, B. (1979) *Zeitschrift für Naturforschung C-A*, **34C**, 770–775.

Berlin, J. & Witte, L. (1981) *Zeitschrift für Naturforschung C-A*, **36C**, 210–214.

Biondi, S., Mengoli, M., Mott, D., & Bagni, N. (1993) *Plant Physiology and Biochemistry*, **31**, 51–58.

Bird, C. R. & Smith, T. A. (1984) *Annals of Botany*, **53**, 483–488.

Birecka, H., Birecki, M., Cohen, E. J., Bitonti, A. J., & McCann, P. P. (1988) *Plant Physiology*, **86**, 224–230.

Birecka, H., Bitonti, A. J., & McCann, P. P. (1985a) *Plant Physiology*, **79**, 509–514.

——— (1985b) *Plant Physiology*, **79**, 515–519.

Birecka, H., DiNolfo, T. E., Martin, W. B., & Frohlich, M. W. (1984) *Phytochemistry*, **23**, 991–997.

Bitonti, A. J., Casara, P. J., McCann, P. P., & Bey, P. (1987) *Biochemical Journal*, **242**, 69–74.

Bokern, M., Witte, L., Wray, V., Nimitz, M., & Meurer–Grimes, B. (1995) *Phytochemistry*, **39**, 1371–1375.

Bolle, C., Herrmann, R. G., & Oelmüller, R. (1995) *Plant Physiology*, **107**, 1461–1462.

Bolwell, G. P. (1992) *Phytochemistry*, **31**, 4081–4086.

Borrell, A., Culiañez–Macià, F. A., Altabella, T., Besford, R. T., Flores, D., & Tiburcio, A. F. (1995) *Plant Physiology*, **109**, 771–776.

Bors, W., Langebartels, C., Michel, C., & Sandermann, Jr., H. (1989) *Phytochemistry*, **28**, 1589–1595.

Botella, J. R., Schlagnhanfer, C. D., Arteca, R. N., & Phillips, A. T. (1992) *Plant Molecular Biology*, **18**, 793–797.

Bouchereau, A., Hamelin, J., Lamour, I., Renard, M., & Larher, F. (1991) *Phytochemistry*, **30**, 1873–1881.

Bouzayen, M., Felix, G., Latché, A., Pech, J., & Boller, T. (1991) *Planta*, **184**, 244–247.

Bouzayen, M., Hamilton, A., Picton, S., Barton, S., & Grierson, D. (1992) *Biochemical Society Transactions*, **20**, 76–79.

Bouzayen, M., Latché, A., Pech, J., & Marigo, G. (1989) *Plant Physiology*, **91**, 1317–1322.

Boyle, S. M., Sriranganathan, N., & Cordes, D. (1988) *Journal of Medical and Veterinary Mycology*, **26**, 227–235.

Bradley, P. M., El-Fiki, F., & Giles, K. L. (1984) *Plant Science Letters*, **34**, 397–401.

Bugos, R. C., Chiang, V. L. C., & Campbell, W. H. (1992) *Phytochemistry*, **31**, 1495–1498.

Burn, J. E., Bagnall, D. J., Metzger, J. D., Dennis, E. S., & Peacock, W. J. (1993) *Proceedings of the National Academy of Sciences of the United States of America*, **90**, 287–291.

Burtin, D., Martin–Tanguy, J., Paynot, M., Carré, M., & Rossin, N. (1990) *Plant Physiology*, **93**, 1398–1404.

Burtin, D., Martin–Tanguy, J., Paynot, M., & Rossin, N. (1989) *Plant Physiology*, **89**, 104–110.

Burtin, D., Martin–Tanguy, J., & Tepfer, D. (1991) *Plant Physiology*, **95**, 461–468.

Caffaro, S., Scaramagli, S., Antognoni, F., & Bagni, N. (1993) *Journal of Plant Physiology*, **141**, 563–568.

Capell, T., Lourdes Campos, J., & Tiburcio, A. F. (1993) *Phytochemistry*, **32**, 785–788.

Carbonell, J. & Navarro, J. L. (1989) *Planta*, **178**, 482–487.

Chatterjee, S., Choudhuri, M. M., & Ghosh, B. (1983) *Phytochemistry*, **22**, 1553–1556.

Chaudhuri, M. M. & Ghosh, B. (1985) *Phytochemistry*, **24**, 2433–2435.

Chen, N., Bowles, M. R., & Pond, S. M. (1992) *Biochemical Pharmacology*, **44**, 1029–1036.

Chibi, F., Matilla, A. J., Angosto, T., & Garrido, D. (1994) *Physiologia Plantarum*, **92**, 61–68.

Christ, M., Felix, H., & Harr, J. (1989) *Zeitschrift für Naturforschung C-A*, **44C**, 49–54.

Cohen, E., Arad, S., Heimer, Y. M., & Mizrahi, Y. (1983) *Plant and Cell Physiology*, **24**, 1003–1010.

Cohen, E., Arad, S., Heimer, Y. H., & Mizrahi, Y. (1984) *Plant Physiology*, **74**, 385–388.

Cohen, E. & Kende, H. (1986) *Planta*, **169**, 498–504.

——— (1987) *Plant Physiology*, **84**, 282–286.

Cohen, S. S., Balint, R., Sindhu, R. K., & Marcu, D. (1981) *Medical Biology*, **59**, 394–402.

Cohen, S. S., Marcu, D. E., & Balint, R. F. (1982) *FEBS Letters*, **141**, 93–97.

Collendavelloo, J., Legrand, M., & Fritig, B. (1983) *Plant Physiology*, **73**, 550–554.

Corbin, J. L., Marsh, B. H., & Peters, G. A. (1989) *Plant Physiology*, **90**, 434–439.

Costa, G. & Bagni, M. (1983) *Horticultural Science*, **18**, 59–61.

Cremer, F., Van de Walle, C., & Bernier, G. (1992) *Plant Cell Physiology*, **33**, 1199–1207.

Crocomo, O. J. & Basso, L. C. (1974) *Phytochemistry*, **13**, 2659–2665.

Cromwell, B. T. (1943) *Biochemical Journal*, **37**, 722–726.

da Cunha, A. (1988) *European Journal of Biochemistry*, **178**, 243–248.

Dai, Y. & Galston, A. W. (1981) *Plant Physiology*, **67**, 266–269.

Davey, M. R., Kothari, S. L., Zhang, H., Rech, E. L., Cocking, E. C., & Lynch, P. T. (1991) *Journal of Experimental Botany*, **42**, 1159–1169.

Dawson, R. & Mopper, K. (1978) *Analysis of Biochemistry*, **84**, 186–190.

De Agazio, M., Giardina, M. C., & Grego, S. (1988) *Plant Physiology*, **87**, 176–178.

de Cantu, L. B. & Kandeler, R. (1989) *Plant Cell Physiology*, **30**, 155–158.

De Laat, A. M. & Van Loon, L. C. (1981) *Plant Physiology*, **68**, 256–260.

——— (1982) *Plant Physiology*, **69**, 240–245.

Del Duca, S., Beninati, S., & Serafini–Fracassini, D. (1995) *Biochemical Journal*, **305**, 233–237.

Del Duca, S., Tidu, V., Bassi, R., Esposito, C., & Serafini–Fracassini, D. (1994) *Planta*, **193**, 283–289.

de Rueda, P. M., Gallardo, M., Sánchez–Calle, I. M., & Matilla, A. J. (1994) *Plant Science*, **97**, 31–37.

DeScenzo, R. A. & Minocha, S. C. (1993) *Plant Molecular Biology*, **22**, 113–127.

Desser, H., Bandion, F., & Kläring, W. (1981) *Mitteilungen Klosterneuberg*, **31**, 231–237.

Dibble, A. R. G., Davies, P. J., & Mutschler, M. A. (1988) *Plant Physiology*, **86**, 338–340.

Dinnella, C., Crues, J. A., D'Orazi, D., Encuentra, A., Gavalda, E. G., & Serafini–Fracassini, D. (1992a) *Phytochemical Analysis*, **3**, 110–116.

Dinnella, C., Serafini–Fracassini, D., Grandi, B., & Del Duca, S. (1992b) *Plant Physiology and Biochemistry*, **30**, 531–539.

Dogbo, O. & Camara, B. (1986) *Comptes Rendus de l'Academie des Sciences Serie III–Sciences de la Vie–Life Sciences*, **303**, 93–96.

Dong, J. G., Fernández–Maculet, J. C., & Yang, S. F. (1992) *Proceedings of the National Academy of Sciences of the United States of America*, **89**, 9789–9793.

Drolet, G., Dumbroff, E. B., Legge, R. L., & Thompson, J. E. (1986) *Phytochemistry*, **25**, 367–371.

Droux, M., Ravanel, S., & Douce, R. (1995) *Archives of Biochemistry and Biophysics*, **316**, 585–595.

Dumortier, F. M., Flores, H. E., Shekhawat, N. S., & Galston, A. W. (1983) *Plant Physiology*, **72**, 915–918.

Edwards, R. & Dixon, R. A. (1991) *Archives of Biochemistry and Biophysics*, **287**, 372–379.

Eisner, T. & Meinwald, J. (1995) *Proceedings of the National Academy of Sciences of the United States of America*, **92**, 50–55.

El Hadrami, I. & D'Auzac, J. (1992) *Journal of Plant Physiology*, **140**, 33–36.

El Hadrami, I., Michaux–Ferrière, N., Carron, M., & d'Auzac, J. (1989) *Comptes Rendus de l'Academie des Sciences, Paris*, **308**, 205–211.

Ellestad, G. A., Cosulich, D. B., Broschard, R. W., Martin, J. H., Kunstmann, M. P., Morton, G. O., Lancaster, J. E., Fulmor, W., & Lovell, F. M. (1978) *Journal of the American Chemistry Society*, **100**, 2515–2524.

Escribano, M. I. & Legaz, M. E. (1988) *Plant Physiology*, **87**, 519–522.

Escribano, M. I. & Merodio, C. (1994) *Journal of Plant Physiology*, **143**, 207–212.

Evans, P. T. & Malmberg, R. L. (1989) *Annual Review of Plant Physiology and Plant Molecular Biology*, **40**, 235–269.

Evans, W. C., Ghani, A., & Wooley, V. A. (1972) *Journal of the Chemical Society–Perkin Transactions I*, 2017–2019.

Evans, W. C. & Somanabandhu, A. (1980) *Phytochemistry*, **19**, 2351–2356.

Even–Chen, Z., Mattoo, A. K., & Goren, R. (1982) *Plant Physiology*, **69**, 385–388.

Fallon, K. M. & Phillips, R. (1988) *Plant Physiology*, **88**, 224–227.

Farooqui, J. Z., Tuck, M., & Paik, W. K. (1985) *Journal of Biological Chemistry*, **260**, 537–545.

Faure, O., Mengoli, M., Nougarede, A., & Bagni, N. (1991) *Journal of Plant Physiology*, **138**, 545–549.

Fecker, L. F., Rügenhagen, C., & Berlin, J. (1993) *Plant Molecular Biology*, **23**, 11–21.

Feirer, R. P., Mignon, G., & Litvay, J. D. (1984) *Science*, **223**, 1433–1435.

Felix, H. & Harr, J. (1987) *Physiologia Plantarum*, **71**, 245–250.

——— (1989) *Zeitschrift für Naturforschung C-A*, **44C**, 55–58.

Feth, F., Arfmann, H., Wray, V., & Wagner, K. G. (1985a) *Phytochemistry*, **24**, 921–923.

Feth, F., Wagner, R., & Wagner, K. G. (1986) *Planta*, **168**, 402–407.

Feth, F., Wray, V., & Wagner, K. G. (1985b) *Phytochemistry*, **24**, 1653–1655.

Fiala, V., Querou, Y., Georges, D., & Dore, C. (1991) *Journal of Plant Physiology*, **138**, 172–175.

Fienberg, A. A., Choi, J. H., Lubich, W. P., & Sung, Z. R. (1984) *Planta*, **162**, 532–539.

Flayeh, K. A. M., Najafi, S. I., Al-Delymi, A. M., & Hajar, M. A. (1984) *Phytochemistry*, **23**, 989–990.

Flores, H. E. (1991) *Biochemistry and Physiology of Polyamines in Plants*, pp. 213–228, R. D. Slocum & H. E. Flores, Eds. Boca Raton, Fla.: CRC Press.

Flores, H. E. & Galston, A. W. (1982) *Plant Physiology*, **69**, 701–706.

——— (1984a) *Plant Physiology*, **75**, 102–109.

——— (1984b) *Plant Physiology*, **75**, 110–113.

Flores, H. E. & Martin–Tanguy, J. (1991) *Biochemistry and Physiology of Polyamines in Plants*, pp. 58–76, R. D. Slocum & H. E. Flores, Eds. Boca Raton, Fla.: CRC Press.

Friedman, R., Altman, A., & Bachrach, U. (1985) *Plant Physiology*, **79**, 80–83.

Friedman, R., Levin, N., & Altman, A. (1986) *Plant Physiology*, **82**, 1154–1157.

Fuhrer, J., Kaur–Sawhney, R., Shih, L., & Galston, A. W. (1982) *Plant Physiology*, **70**, 1597–1600.

Fujino, D. W., Burger, D. W., Yang, S. F., & Bradford, K. J. (1988) *Plant Physiology*, **88**, 774–779.

Fukuda, H., Kitajima, H., Fujii, T., Tazaki, M., & Ogawa, T. (1989) *FEMS Microbiology Letters*, **59**, 1–6.

Fukuda, H., Ogawa, T., & Tanase, S. (1993) *Advances in Microbial Physiology*, **35**, 275–307.

Gállardo, M., de Rueda, P. M., Matilla, A. J., & Sánchez–Calle, I. M. (1995) *Plant Science*, **104**, 169–175.

Gallon, J. R., Chit, K. N., & Brown, E. G. (1990) *Phytochemistry*, **29**, 1107–1111.

Gamarnik, A. & Frydman, R. B. (1991) *Plant Physiology*, **97**, 778–785.

Gardner, H. W. & Newton, J. W. (1987) *Phytochemistry*, **26**, 621–626.

Gerats, A. G. M., Kaye, C., Collins, C., & Malmberg, R. L. (1988) *Plant Physiology*, **86**, 390–393.

Giovanelli, J. (1987) *Methods in Enzymology*, **143**, 419–426.

Giovanelli, J., Datko, A. H., Mudd, S. H., & Thompson, G. A. (1983) *Plant Physiology*, **71**, 319–326.

Giovanelli, J., Mudd, S. H., & Datko, A. H. (1980) *The Biochemistry of Plants*, pp. 453–505, B. J. Miflin, Ed. New York: Academic Press.

——— (1981) *Biochemical and Biophysical Research Communications*, **100**, 831–839.

——— (1985) *Plant Physiology*, **77**, 450–455.

Goldberg, R. & Perdrizet, E. (1984) *Plants*, **161**, 531–535.

Good, X., Kellogg, J. A., Wagoner, W., Langhoff, D., Matsumura, W., & Bestwick, R. K. (1994) *Plant Molecular Biology*, **26**, 781–790.

Goren, R., Palavan, N., Flores, H., & Galston, A. W. (1982) *Plant and Cell Physiology*, **23**, 19–26.

Grand, C. (1984) *FEBS Letters*, **169**, 7–11.

Greenberg, M. & Cohen, S. S. (1985) *Plant Physiology*, **78**, 568–575.

Guarino, L. A. & Cohen, S. S. (1979a) *Analytical Biochemistry*, **95**, 73–76.

——— (1979b) *Proceedings of the National Academy of Sciences of the United States of America*, **76**, 3184–3188.

Guggisberg, A. & Hesse, M. (1983) *The Alkaloids*, pp. 85–188, A. Brossi, Ed. New York: Academic Press.

Guo, L., Arteca, R. N., Phillips, A. T., & Lui, Y. (1992) *Plant Physiology*, **100**, 2041–2045.

Guranowski, A. (1983) *Plant Physiology*, **71**, 932–935.

Guranowski, A. & Jakubowski, H. (1987) *Methods in Enzymology*, **143**, 430–434.

Haard, N. F. (1984) *Journal of Chemical Education*, **61**, 277–283.

Hacker, A. D. (1993) *Biochemical Pharmacology*, **45**, 2475–2481.

Hamana, K. & Matsuzaki, S. (1985) *Biochemical and Biophysical Research Communications*, **129**, 46–51.

Hamana, K., Matsuzaki, S., Niitsu, M., & Samejima, K. (1992) *Canadian Journal of Botany*, **70**, 1984–1990.

Hamill, J. D., Robins, R. J., Parr, A. J., Evans, D. M., Furze, J. M., & Rhodes, M. J. C. (1990) *Plant Molecular Biology*, **15**, 27–38.

Hamilton, A. J., Lycett, G. W., & Grierson, D. (1990) *Nature*, **346**, 284–287.

Hanley, K. M., Meir, S., & Bramlage, W. J. (1989) *Plant Physiology*, **91**, 1126–1130.

Hanson, K. R. & Havir, E. A. (1981) *The Biochemistry of Plants*, pp. 577–597, P. K. Stumpf & E. E. Conn, Eds. New York: Academic Press.

Harborne, J. B. (1980) *Encyclopedia of Plant Physiology, Secondary Plant Products*, pp. 329–402, A. Pirson, M. H. Zimmer, E. A. Bell, & B. V. Chalwood, Eds. New York: Springer–Verlag.

Hart, J. J., DiTomáso, J. M., Linscott, D. L., & Kochian, L. V. (1992) *Plant Physiology*, **99**, 1400–1405.

Hartmann, T., Schoofs, G., & Wink, M. (1980) *FEBS Letters*, **115**, 35–38.

Hashimoto, T., Yukimune, Y., & Yamada, Y. (1989a) *Planta*, **178**, 123–130.

——— (1989b) *Planta*, **178**, 131–137.

Hauschild, M. Z. (1993) *Journal of Chromatography*, **630**, 397–401.

Havir, E. A. (1981) *Archives of Biochemistry and Biophysics*, **211**, 556–563.

Havis, N. D., Walters, D. R., Foster, S. A., Martin, W. P.,

Cook, F. M., & Robins, D. J. (1994a) *Pesticide Science*, **41**, 61–69.

Havis, N. D., Walters, D. R., Martin, W. P., Cook, F. M., & Robins, D. J. (1994b) *Pesticide Science*, **41**, 71–76.

Hegnauer, R. (1988) *Phytochemistry*, **27**, 2423–2427.

Heimer, Y. M. & Mizrahi, Y. (1982) *Biochemical Journal*, **201**, 373–376.

Heimer, Y. M., Mizrahi, Y., & Bachrach, V. (1979) *FEBS Letters*, **104**, 146–148.

Hiatt, A. (1989) *Plant Physiology*, **90**, 1378–1381.

Hiatt, A. & Malmberg, R. L. (1988) *Plant Physiology*, **86**, 441–446.

Hiatt, A. C., McIndoo, J., & Malmberg, R. L. (1986) *Journal of Biological Chemistry*, **261**, 1293–1298.

Hibi, N., Fujita, T., Hatano, M., Hashimoto, T., & Yamada, Y. (1992) *Plant Physiology*, **100**, 826–835.

Hibi, N., Higashiguchi, S., Hashimoto, T., & Yamada, Y. (1994) *Plant Cell*, **6**, 723–735.

Hirasawa, E. & Suzuki, Y. (1983) *Phytochemistry*, **22**, 103–106.

Hoffman, N. E., Yang, S. F., & McKeon, T. (1982) *Biochemical and Biophysical Research Communications*, **104**, 765–770.

Holdsworth, M. J., Bird, C. R., Ray, J., Schuch, W., & Grierson, D. (1987) *Nucleic Acids Research*, **15**, 731–739.

Huang, P., Parks, J. E., Rottmann, W. H., & Theologis, A. (1991) *Proceedings of the National Academy of Sciences of the United States of America*, **88**, 7021–7025.

Huberman, M. & Goren, R. (1979) *Physiologia Plantarum*, **45**, 189–196.

Huhtinen, O., Honkanen, J., & Simola, L. K. (1982) *Plant Science Letters*, **28**, 3–9.

Hyodo, H. & Tanaka, K. (1986) *Plant and Cell Physiology*, **27**, 391–398.

Ibrahim, R. & Barron, D. (1989) *Methods in Plant Biochemistry*, vol. 1, pp. 75–111. New York: Academic Press.

Icekson, I., Goldlust, A., & Apelbaum, A. (1987) *Plant Physiology*, **84**, 972–974.

Ishikura, N. & Takeshima, Y. (1984) *Plant and Cell Physiology*, **25**, 185–189.

Jarvis, B. C., Shannon, P. R. M., & Yasmin, S. (1983) *Plant and Cell Physiology*, **24**, 677–683.

Jiao, X., Philosoph–Hadas, S., Su, L., & Yang, S. F. (1986) *Plant Physiology*, **81**, 637–641.

Johne, S., Gröger, D., & Radeglia, R. (1975) *Phytochemistry*, **14**, 2635–2636.

Kakkar, R. K. & Rai, V. K. (1993) *Phytochemistry*, **33**, 1281–1288.

Kamachi, S. & Hirasawa, E. (1995) *Plant and Cell Physiology*, **36**, 915–917.

Katoh, Y. & Hasegawa, T. (1989) *Agricultural and Biological Chemistry*, **53**, 1485–1491.

Katoh, Y., Hasegawa, T., Suzuki, T., & Fujii, T. (1987) *Agricultural and Biological Chemistry*, **51**, 2457–2463.

Kaur–Sawhney, R., Dai, Y., & Galston, A. W. (1986) *Plant and Cell Physiology*, **27**, 253–260.

Kaur–Sawhney, R., Flores, H. E., & Galston, A. W. (1980) *Plant Physiology*, **65**, 368–371.

Kaur–Sawhney, R. & Galston, A. W. (1991) *Biochemistry and Physiology of Polyamines in Plants*, pp. 201–211, R. D. Slocum & H. E. Flores, Eds. Boca Raton, Fla.: CRC Press.

Kaur–Sawhney, R., Kandpal, G., McGonigle, B., & Galston, A. W. (1990) *Planta*, **181**, 212–215.

Kaur–Sawhney, R., Shih, L., & Galston, A. W. (1982) *Plant Physiology*, **69**, 411–415.

Kawagishi, H., Fukuhara, F., Sazuka, M., Kawashima, A., Mitsubori, T., & Tomita, T. (1993) *Phytochemistry*, **32**, 239–241.

Kende, H. (1993) *Annual Review of Plant Physiology and Plant Molecular Biology*, **44**, 283–307.

Kende, H. & Boller, T. (1981) *Planta*, **151**, 476–481.

Khan, A. J. & Minocha, S. C. (1989) *Life Sciences*, **44**, 1215–1222.

Khan, M. B. & Harborne, J. B. (1991) *Phytochemistry*, **30**, 3559–3563.

Khani–Oskouee, S., Jones, J. P., & Woodard, R. W. (1984) *Biochemical and Biophysical Research Communications*, **121**, 181–187.

Kieber, J. J. & Ecker, J. R. (1993) *Trends in Genetics*, **9**, 356–362.

Kim, W. T. & Yang, S. F. (1992) *Plant Physiology*, **100**, 1126–1131.

Klapheck, S. (1983) *Zeitschrift für Pflanzenphysiologie*, **112**, 275–279.

Klinguer, S., Martin–Tanguy, J., & Martin, C. (1986) *Plant Physiology*, **82**, 561–565.

Kneusel, R. E., Matern, V., & Nicolay, K. (1989) *Archives of Biochemistry and Biophysics*, **269**, 455–462.

Knobloch, K. H., Beutnagel, G., & Berlin, J. (1981) *Planta*, **153**, 582–585.

Koetje, D. S., Kononowicz, H., & Hodges, T. K. (1993) *Journal of Plant Physiology*, **141**, 215–221.

Kordy, E. & Soeder, C. J. (1984) *Journal of Plant Physiology*, **117**, 17–28.

Koromilas, A. E. & Kyriakidis, D. A. (1988) *Phytochemistry*, **27**, 989–992.

Kotzabasis, K., Christakis–Hampsas, M. D., & Roubelakis–Angelakis, K. A. (1993a) *Analytical Biochemistry*, **214**, 484–489.

Kotzabasis, K., Fotinou, C., Roubelakis–Angelakis, K. A., & Ghanotakis, D. (1993b) *Photosynthesis Research*, **38**, 83–88.

Kotzabasis, K. & Senger, H. (1994) *Zeitschrift für Naturforschung C-A*, **49C**, 181–185.

Kramer, G. F. & Wang, C. Y. (1990) *Journal of Plant Physiology*, **136**, 115–119.

Krishnamurthy, R. & Bhagwat, K. A. (1989) *Plant Physiology*, **91**, 500–504.

Kurosaki, F., Matsushita, M., & Nishi, A. (1992) *Phytochemistry*, **31**, 3889–3892.

Kushad, M. M. & Poovaiah, B. W. (1984) *Plant Physiology*, **76**, 293–296.

Kushad, M. M., Richardson, D. G., & Ferro, A. J. (1983) *Plant Physiology*, **73**, 257–261.

——— (1985) *Plant Physiology*, **79**, 525–529.

Kushad, M. M. & Yelenosky, G. (1987) *Plant Physiology*, **84**, 692–695.

Kutchan, T. M. (1995) *Plant Cell*, **7**, 1059–1070.

Kuttan, R. & Radhakrishnan, A. N. (1972) *Biochemical Journal*, **127**, 61–67.

Kuttan, R., Radhakrishnan, A. N., Spande, T., & Witkop, B. (1971) *Biochemistry*, **10**, 361–365.

Langebartels, C., Kerner, K., Léonardi, S., Schraudner, M., Trost, M., Heller, W., & Sandermann, Jr., H. (1991) *Plant Physiology*, **95**, 882–889.

Lauchert, V. & Wild, A. (1995) *Journal of Plant Physiology*, **147**, 267–269.

Lee, T. & Chu, C. (1992) *Plant Physiology*, **100**, 238–245.

Leete, E. (1980) *Encyclopedia of Plant Physiology, Secondary Plant Products*, pp. 65–91, A. Pirson, M. H. Zimmerman, E. A. Bell, & B. V. Charlwood, Eds. New York: Springer–Verlag.

Leete, E., Louters, L. L., & Prakash Rao, H. S. (1986) *Phytochemistry*, **25**, 2753–2758.

Legaz, M. E., Iglesias, A., & Vicente, C. (1983) *Zeitschrift für Pflanzenphysiologie*, **110**, 53–59.

Leistner, E. & Spenser, I. D. (1973) *Journal of the American Chemistry Society*, **95**, 4715–4725.

Le Rudulier, D. & Goas, G. (1979) *Comptes Rendus de l'Academie des Sciences, Paris*, **288**, 1387–1390.

Leslie, C. A. & Romani, R. J. (1988) *Plant Physiology*, **88**, 833–837.

Leubner–Metzger, G. & Amrhein, N. (1993) *Phytochemistry*, **32**, 551–556.

Li, N. & Matoo, A. K. (1994) *Journal of Biological Chemistry*, **269**, 6908–6917.

Li, N., Parsons, B. L., Liu, D., & Mattoo, A. K. (1992) *Plant Molecular Biology*, **18**, 477–487.

Liang, X., Abel, S., Keller, J. A., Shen, N. F., & Theologis, A. (1992) *Proceedings of the National Academy of Sciences of the United States of America*, **89**, 11046–11050.

Liang, X., Dron, M., Cramer, C. L., Dixon, R. A., & Lamb, C. J. (1989) *Journal of Biological Chemistry*, **264**, 14486–14492.

Lieberman, M. (1979) *Annual Review of Plant Physiology*, **30**, 533–591.

Lin, P. P. C. (1984) *Plant Physiology*, **76**, 372–380.

Lin, P. P. C., Egli, D. B., Li, G. M., & Meckel, L. (1984) *Plant Physiology*, **76**, 366–371.

Lincoln, J. E., Campbell, A. D., Oetiker, J., Rottmann, W. H., Oeller, P. W., Shen, N. F., & Theologis, A. (1993) *Journal of Biological Chemistry*, **268**, 19422–19430.

Litz, R. E., & Schaffer, B. (1987) *Journal of Plant Physiology*, **128**, 251–258.

Liu, Y., Su, L., & Yang, S. F. (1985) *Plant Physiology*, **77**, 891–895.

Lizada, M. C. C. & Yang, S. F. (1979) *Analysis of Biochemistry*, **100**, 140–145.

Louis, V. & Negrel, J. (1991) *Phytochemistry*, **30**, 2519–2522.

Lozano, J. L., Wettlaufer, S. H., & Leopold, A. C. (1989) *Journal of Experimental Botany*, **40**, 1337–1340.

Lürssen, K. (1981) *Plant Science Letters*, **20**, 365–370.

Lürssen, K., Naumann, K., & Schröder, R. (1979) *Zeitschrift für Pflanzenphysiologie*, **92**, 285–294.

Maiss, B., Kordy, E., Kneifel, H., & Soeder, C. J. (1982) *Zeitschrift für Pflanzenphysiologie*, **106**, 213–221.

Maki, H., Ando, S., Kodama, H., & Komamine, A. (1991) *Plant Physiology*, **96**, 1006–1013.

Malmberg, R. L. & Cellino, M. L. (1994) *Journal of Biological Chemistry*, **269**, 2703–2706.

Malmberg, R. L., McIndoo, J., Hiatt, A. H., & Lowe, B. A. (1985) *Cold Spring Harbor Symposia*, **40**, 475–482.

Malmberg, R. L. & Rose, D. J. (1987) *Molecular and General Genetics*, **207**, 9–14.

Malmberg, R. L., Smith, K. E., Bell, E., & Cellino, M. L. (1992) *Plant Physiology*, **100**, 146–152.

Manoharan, K. & Gnanam, A. (1992) *Plant and Cell Physiology*, **33**, 1243–1246.

Margosiak, S. A., Dharma, A., Bruce–Carver, M. R., González, A. P., Louie, D., & Kuehn, G. D. (1990) *Plant Physiology*, **92**, 88–96.

Martens, M., Cottenie–Ruysschaert, M., Hamselaer, R., De Cooman, L., Van de Casteele, K., & Van Sumere, C. (1988) *Phytochemistry*, **27**, 2465–2475.

Martin, C. (1985) *Endeavour*, **9**, 81–86.

Martin, C., Kunesch, G., Martin–Tanguy, J., Negrel J., Paynot, M., & Carré, M. (1985) *Plant Cell Reports*, **4**, 158–160.

Martin, M. N. & Saftner, R. A. (1995) *Plant Physiology*, **108**, 1241–1249.

Martin–Tanguy, J., Cabanne, F., Perdrizet, E., & Martin, C. (1978) *Phytochemistry*, **17**, 1927–1928.

Martin–Tanguy, J., Martin, C., & Gallet, M. (1973) *Comptes Rendus de l'Academie Sciences Paris*, **276D**, 1433–1435.

Martin–Tanguy, J., Martin, C., Gallet, M., & Vernoy, R. (1976) *Comptes Rendus de l'Academie Sciences Paris*, **282D**, 2231–2234.

Martin–Tanguy, J., Martin, C., Paynot, M., & Rossin, N. (1988) *Plant Physiology*, **88**, 600–604.

Martin–Tanguy, J., Negrel, J., Paynot, M., & Martin, C. (1987) *Plant Molecular Biology*, pp. 253–263, D. von Wettstein & N. Chua, Eds. New York: Plenum.

Martin–Tanguy, J., Tepfer, D., & Burtin, D. (1991) *Plant Science*, **80**, 131–144.

Martin–Tanguy, J., Tepfer, D., Paynot, M., Burtin, D., Heisler, L., & Martin, C. (1990) *Plant Physiology*, **92**, 912–918.

Massé, J., Laberche, J., & Jeanty, G. (1985) *Comptes Rendus de l'Academie des Sciences, Paris*, **301**, 27–32.

Mathur, M., Saluja, D., & Sachar, R. C. (1991) *Biochimica et Biophysica Acta*, **1078**, 161–170.

Matsuzaki, S., Hamana, K., & Isobe, K. (1990) *Phytochemistry*, **29**, 1313–1315.

Mattoo, A. K., Baker, J. E., & Moline, H. E. (1986) *Journal of Plant Physiology*, **123**, 193–202.

Mavandad, M., Edwards, R., Liang, X., Lamb, C. J., & Dixon, R. A. (1990) *Plant Physiology*, **94**, 671–680.

Mazelis, M., Miflin, B. J., & Pratt, H. M. (1976) *FEBS Letters*, **64**, 197–200.

Mazzafera, P., Wingsle, G., Olsson, O., & Sandberg, G. (1994) *Phytochemistry*, **37**, 1577–1584.

McLauchlan, W. R., McKee, R. A., & Evans, D. M. (1993) *Planta*, **191**, 440–445.

Mehta, A. M., Jordan, R. L., Anderson, J. D., & Mattoo, A. K. (1988) *Proceedings of the National Academy of Sciences of the United States of America*, **85**, 8810–8814.

Mendum, M. L. & Adams, D. O. (1994) *Plant Physiology*, **105**(Supplement), 129.

Mengoli, M., Chriqui, D., & Bagni, N. (1992a) *Journal of Plant Physiology*, **139**, 697–702.

——— (1992b) *Journal of Plant Physiology*, **140**, 153–155.

Meurer, B., Wray, V., Grotjahn, L., Wiermann, R., & Strack, D. (1986) *Phytochemistry*, **25**, 433–435.

Meurer, B., Wray, V., Wiermann, R., & Strack, D. (1988) *Phytochemistry*, **27**, 839–843.

Micallef, B. J. & Shelp, B. J. (1989) *Plant Physiology*, **91**, 170–174.

Miflin, B. J. & Lea, P. J. (1976) *Phytochemistry*, **15**, 873–885.

Minocha, S. C. & Minocha, S. (1995) *Biotechnology in Agriculture and Forestry*, **30**, 53–70.

Minocha, S. C., Papa, N. S., Khan, A. J., & Samuelsen, A. I. (1991) *Plant and Cell Physiology*, **32**, 395–402.

Miyazaki, J. H. & Yang, S. F. (1987a) *Plant Physiology*, **84**, 277–281.

——— (1987b) *Phytochemistry*, **26**, 2655–2660.

Mizrahi, Y. & Heimer, Y. M. (1982) *Physiologia Plantarum*, **54**, 367–368.

Mizusaki, S., Tanabe, Y., Noguchi, M., & Tamaki, E. (1971b) *Phytochemistry*, **10**, 1347–1350.

——— (1971a) *Plant and Cell Physiology*, **12**, 633–640.

——— (1972) *Phytochemistry*, **11**, 2757–2762.

Mølgaard, P. & Ravn, H. (1988) *Phytochemistry*, **27**, 2411–2421.

Montague, M. J., Armstrong, T. A., & Jaworski, E. G. (1979) *Plant Physiology*, **63**, 341–345.

Montague, M. J., Koppenbrink J. W., & Jaworski, E. G. (1978) *Plant Physiology*, **62**, 430–433.

Morel, C., Villanueva, V. R., & Queiroz, O. (1980) *Planta*, **149**, 440–444.

Mudd, S. H. & Datko, A. H. (1990) *Plant Physiology*, **93**, 623–630.

Muhitch, M. J. & Fletcher, J. S. (1985) *Plant Physiology*, **78**, 25–28.

Mukhopadhyay, A., Choudhuri, M. M., Sen, K., & Ghosh, B. (1983) *Phytochemistry*, **22**, 1547–1551.

Murr, D. P. & Yang, S. F. (1975) *Phytochemistry*, **14**, 1291–1292.

Nagahama, K., Ogawa, T., Fujii, T., & Fukuda, H. (1992) *Journal of Fermentation and Bioengineering*, **73**, 1–5.

Nakajima, N., Nakagawa, N., & Imaseki, H. (1988) *Plant and Cell Physiology*, **29**, 1988.

Negrel, J. (1989) *Phytochemistry*, **28**, 477–481.

Negrel, J., Javelle, F., & Paynot, M. (1991) *Phytochemistry*, **30**, 1089–1092.

Negrel, J. & Martin, C. (1984) *Phytochemistry*, **23**, 279–280.

Negrel, J., Paynot, M., & Javelle, F. (1992) *Plant Physiology*, **98**, 1264–1269.

Negrel, J. & Smith, T. A. (1984a) *Phytochemistry*, **23**, 31–34.

———— (1984b) *Phytochemistry*, **23**, 739–741.

Negrel, J., Vallee, J., & Martin, C. (1984) *Phytochemistry*, **23**, 2747–2751.

Neilands, J. B. (1993) *Archives of Biochemistry and Biophysics*, **302**, 1–3.

Neilson, A. H. & Larsson, T. (1980) *Physiologia Plantarum*, **48**, 542–553.

Nieder, M., Yip, W., & Yang, S. F. (1986) *Plant Physiology*, **81**, 156–160.

Noh, E. W. & Minocha, S. C. (1994) *Transgenic Research*, **3**, 26–35.

Ogawa, T., Takahashi, M., Fujii, T., Tazaki, M., & Fukuda, H. (1990) *Journal of Fermentation and Bioengineering*, **69**, 287–291.

Ortiz de Montellano, P. R. & Mico, B. A. (1980) *Molecular Pharmacology*, **18**, 128–135.

Osswald, W. F., Schütz, W., & Elstner, E. F. (1989) *Journal of Plant Physiology*, **134**, 510–513.

Pakusch, A., Kneusel, R. E., & Matern, U. (1989) *Archives of Biochemistry and Biophysiology*, **271**, 488–494.

Palavan–Ünsal, N. (1987) *Plant and Cell Physiology*, **28**, 565–572.

Panagiotidis, C. A., Georgatsos, J. G., & Kyriakidis, D. A. (1982) *FEBS Letters*, **146**, 193–196.

Park, K. H. & Cho, Y. D. (1991) *Biochemical and Biophysical Research Communications*, **174**, 32–36.

Park, K. Y. & Lee, S. H. (1994) *Physiologia Plantarum*, **90**, 382–390.

Peck, S. C. & Kende, H. (1995) *Plant Molecular Biology*, **28**, 293–301.

Pedreño, M. A., Bouzayen, M., Pech, J., Marigo, G., & Latché, A. (1991) *Plant Physiology*, **97**, 1483–1486.

Peeters, K. M. U., Geuns, J. M. C., & Van Laere, A. J. (1993) *Journal of Experimental Botany*, **44**, 1709–1715.

Peiser, G. (1989) *Plant Physiology*, **90**, 21–24.

Peiser, G. D., Wang, R., Hoffman, N. E., Yang, S. F., Liu, H., & Walsh, C. T. (1984) *Proceedings of the National Academy of Sciences of the United States of America*, **81**, 3059–3063.

Peleman, J., Boerjan, W., Engler, G., Seurinck, J., Botterman, J., Alliotte, T., Van Montagu, M., & Inze, D. (1989) *Plant Cell*, **1**, 81–83.

Pelosi, L. A., Rother, A., & Edwards, J. M. (1986) *Phytochemistry*, **25**, 2315–2319.

Pérez-Amador, M. A., Carbonell, J. & Granell, A. (1995) *Plant Molecular Biology*, **28**, 997–1009.

Pfosser, M., Mengl, M., Königshofer, H., & Kandeler, R. (1992) *Journal of Plant Physiology*, **140**, 334–338.

Philosoph–Hadas, S., Meir, S., & Aharoni, N. (1985) *Physiologia Plantarum*, **63**, 431–437.

Phillips, R., Press, M. C., & Eason, A. (1987) *Journal of Experimental Botany*, **38**, 164–172.

Pirrung, M. C. (1985) *Biorganic Chemistry*, **13**, 219–226.

Pistocchi, R., Keller, F., Bagni, N., & Matile, P. (1988) *Plant Physiology*, **87**, 514–518.

Ponchet, M., Martin–Tanguy, J., Poupet, A., Marais, A., & Beck, D. (1982) *Journal of Chromotography*, **240**, 397–404.

Priebe, A. & Jäger, H. J. (1978) *Plant Science Letters*, **12**, 365–369.

Protacio, C. M. & Flores, H. E. (1992) *In Vitro Cellular Development and Biology–Plant*, **28P**, 81–86.

Rabe, E. & Lovatt, C. J. (1986) *Plant Physiology*, **81**, 774–779.

Rajam, M. V., Weinstein, L. H., & Galston, A. W. (1985) *Proceedings of the National Academy of Sciences of the United States of America*, **82**, 6874–6878.

———— (1986) *Plant Physiology*, **82**, 485–487.

Ramakrishna, S. & Adiga, P. R. (1974) *Phytochemistry*, **13**, 2161–2166.

———— (1975) *Phytochemistry*, **14**, 63–68.

Ramakrishna, S., Guarino, L., & Cohen, S. S. (1978) *Journal of Bacteriology*, **134**, 744–750.

Rastogi, R. & Davies, P. J. (1989) *Plant Physiology*, **94**, 1449–1455.

———— (1991) *Plant Physiology*, **95**, 41–45.

Ravanel, S., Droux, M., & Douce, R. (1995) *Archives of Biochemistry and Biophysics*, **316**, 572–584.

Reggiani, R., Aurisano, N., Mattana, M., & Bertani, A. (1993) *Journal of Plant Physiology*, **141**, 136–140.

Reggiani, R. & Bertani, A. (1989) *Journal of Plant Physiology*, **135**, 375–377.

Reggiani, R., Giussani, P., & Bertani, A. (1990) *Plant and Cell Physiology*, **31**, 489–494.

Reinhardt, D., Kende, H., & Boller, T. (1994) *Planta*, **195**, 142–146.

Rhodes, C. A., Pierce, D. A., Mettler, I. J., Mascarenhas, D., & Detmer, J. J. (1988) *Science*, **240**, 204–207.

Rhodes, D. & Hanson, A. D. (1993) *Annual Review of Plant Physiology and Plant Molecular Biology*, **44**, 357–384.

Riov, J., Dagan, E., Goren, R., & Yang, S. F. (1990) *Plant Physiology*, **92**, 48–53.

Riov, J. & Yang, S. F. (1982) *Plant Physiology*, **69**, 687–690.

Roberts, D. R., Walker, M. A., & Dumbroff, E. B. (1985) *Phytochemistry*, **24**, 1089–1090.

Roberts, D. R., Walker, M. A., Thompson, J. E., & Dumbroff, E. B. (1984) *Plant and Cell Physiology*, **25**, 315–322.

Robie, C. A. & Minocha, S. C. (1989) *Plant Science*, **65**, 45–54.

Robins, D. J. (1989) *Chemical Society Reviews*, **18**, 375–408.

Robins, D. J. & Sweeney, J. R. (1979) *Journal of the Chemical Society, Chemical Communications*, 120–121.

Robins, R. J., Parr, A. J., Bent, E. G., & Rhodes, M. J. C. (1991) *Planta*, **183**, 185–195.

Rodrigues–Pousada, R. A., De Rycke, R., Dedonder, A., Van Caeneghem, W., Engler, G., Van Montagu, M., & Van der Straeten, D. (1993) *Plant Cell*, **5**, 897–911.

Rombaldi, C., Lilièvre, J., Latché, A., Petitprez, M., Bouzayen, M., & Pech, J. (1994) *Planta*, **192**, 453–460.

Rosenthal, G. A. (1990) *Plant Physiology*, **94**, 1–3.

———— (1992) *Proceedings of the National Academy of Sciences of the United States of America*, **89**, 1780–1784.

Roustan, J., Latché, A., & Fallot, J. (1994) *Plant Science*, **103**, 223–229.

Rüdiger, W. & Thümmler, F. (1991) *Angewandte Chemie—International Edition in English*, **30**, 1216–1228.

Saftner, R. A. & Martin, M. N. (1993) *Physiologia Plantarum*, **87**, 535–543.

Sagee, O., Riov, J., & Goren, R. (1990) *Plant Physiology*, **91**, 54–60.

Samejima, K. B. & Yamanoha, B. (1982) *Archives of Biochemistry and Biophysiology*, **216**, 213–222.

Santanen, A. & Simola, L. K. (1992) *Journal of Plant Physiology*, **140**, 475–480.

——— (1994) *Physiologia Plantarum*, **90**, 125–129.

Sarjala, T. & Savonen, E. (1994) *Journal of Plant Physiology*, **144**, 720–725.

Sato, T. & Theologis, A. (1989) *Proceedings of the National Academy of Sciences of the United States of America*, **86**, 6621–6625.

Satoh, S. & Yang, S. F. (1989) *Archives of Biochemistry and Biophysiology*, **271**, 107–112.

Schröder, G. & Schröder, J. (1995) *European Journal of Biochemistry*, **228**, 74–78.

Schröder, J. (1989) *Nucleic Acids Research*, **17**, 460.

Secor, H. V., Izac, R. R., Hassam, S. B., & Frisch, A. F. (1993) *Journal of Labelled Compounds and Radiopharmaceuticals*, **34**, 421–426.

Sellstedt, A. & Atkins, C. A. (1991) *Journal of Experimental Botany*, **42**, 1493–1497.

Serafini–Fracassini, D. (1991) *Biochemistry and Physiology of Polyamines in Plants*, pp. 159–173, R. D. Slocum & H. E. Flores, Eds. Boca Raton, Fla.: CRC Press.

Serafini–Fracassini, D., Bagni, N., Cionini, P. G., & Bennici, A. (1980) *Planta*, **148**, 332–337.

Serafini–Fracassini, D., Del Duca, S., & D'Orazio, D. (1988) *Plant Physiology*, **87**, 757–761.

Shargool, P. D., Jain, J. C., & McKay, G. (1988) *Phytochemistry*, **27**, 1571–1574.

Shen, H. & Galston, A. W. (1985) *Plant Cell Regulation*, **3**, 353–363.

Shetty, K., Korus, R. A., & Crawford, D. L. (1989) *Applied Biochemistry and Biotechnology*, **20**, 825–843.

Shih, C. & Kao, C. H. (1996) *Plant Physiology*, **111**, 721–724.

Shiraishi, T., Owada, M. K., Tatsuka, M., Yamashita, T., Watanabe, K., & Kakunaga, T. (1989) *Cancer Research*, **49**, 2374–2378.

Signorini, M., Beninati, S., & Bergamini, C. M. (1991) *Journal of Plant Physiology*, **137**, 547–552.

Sindhu, R. K. & Cohen, S. S. (1984a) *Plant Physiology*, **74**, 645–649.

——— (1984b) *Plant Physiology*, **76**, 219–223.

Sitrit, Y., Riov, J., & Blumenfeld, A. (1986) *Plant Physiology*, **81**, 130–135.

——— (1988) *Plant Physiology*, **86**, 13–15.

Slocum, R. D., Bitonti, A. J., McCann, P. P., & Feirer, R. P. (1988) *Biochemical Journal*, **255**, 197–202.

Slocum, R. D. & Galston, A. W. (1985a) *Plant Physiology*, **79**, 336–343.

——— (1985b) *Plant and Cell Physiology*, **26**, 1519–1526.

Slocum, R. D. & Richardson, D. P. (1991) *Plant Physiology*, **96**, 262–268.

Slocum, R. D., Williamson, C. L., Poggenburg, C. A., & Lynes, M. A. (1990) *Plant Physiology*, **92**, 1205–1210.

Smith, J. J. & John, P. (1993) *Phytochemistry*, **32**, 1381–1386.

Smith, J. J., Ververides, P., & John, P. (1992) *Phytochemistry*, **31**, 1485–1493.

Smith, L. L. (1985) *Biochemical Society Transactions*, **13**, 332–334.

Smith, M. A. (1991) *Biochemistry and Physiology of Polyamines in Plants*, pp. 230–242, R. D. Slocum & H. E. Flores, Eds. Boca Raton, Fla.: CRC Press.

Smith, T. A. (1971a) *Biological Reviews*, **46**, 201–241.

——— (1971b) *Annals of the New York Academy of Science*, **171**, 988–1001.

——— (1981) *The Biochemistry of Plants*, pp. 249–268. New York: Academic Press.

——— (1982) *Plant Growth Substances*, pp. 483–490, P. F. Waring, Ed. London: Academic Press.

——— (1983) *Methods in Enzymology*, **94**, 176–180.

Smith, T. A. & Best, G. R. (1977) *Phytochemistry*, **16**, 841–843.

——— (1978) *Phytochemistry*, **17**, 1093–1098.

Smith, T. A., Best, G. R., Abbott, A. J., & Clements, E. D. (1978) *Planta*, **144**, 63–68.

Smith, T. A., Croker, S. J., & Loeffler, R. S. T. (1986) *Phytochemistry*, **25**, 683–689.

Smith, T. A., Negrel, J., & Bird, C. R. (1983) *Advances in Polyamine Research*, pp. 347–370, U. Bachrach, A. Kaye, & R. Chayen, Eds. New York: Raven Press.

Smith, T. A. & Wilshire, G. (1975) *Phytochemistry*, **14**, 2341–2346.

Songstad, D. D., Duncan, D. R., & Widholm, J. M. (1990) *Journal of Experimental Botany*, **41**, 289–294.

Spanu, P., Grosskopf, D. G., Felix, G., & Boller, T. (1994) *Plant Physiology*, **106**, 529–535.

Srivastava, S. K. & Mungre, S. M. (1986) *Phytochemistry*, **25**, 1563–1565.

Srivenugopal, K. S. & Adiga, P. R. (1980b) *Analytical Biochemistry*, **104**, 440–444.

Srivenugopal, K. S. & Adiga, P. R. (1980d) *Biochemical Journal*, **189**, 553–560.

——— (1980a) *FEBS Letters*, **112**, 260–264.

——— (1980c) *Biochemical Journal*, **190**, 461–464.

——— (1981) *Journal of Biological Chemistry*, **256**, 9532–9541.

Stammer, C. H. (1990) *Tetrahedron*, **46**, 2231–2254.

Stevenson, D. E., Akhtar, M., & Gani, D. (1990) *Biochemistry*, **29**, 7631–7647.

Strack, D., Eilert, O., Wray, V., Wolff, J., & Jaggy, H. (1990) *Phytochemistry*, **29**, 2893–2896.

Sun, L., Monneuse, M., Martin–Tanguy, J., & Tepfer, D. (1991) *Plant Science*, **80**, 145–156.

Sung, H., Liu, L., & Kao, C. H. (1994) *Plant and Cell Physiology*, **35**, 313–316.

Suresh, M. R. & Adiga, P. R. (1977) *European Journal of Biochemistry*, **79**, 511–518.

Sussex, I. M. (1989) *Cell*, **56**, 225–229.

Suttle, J. C. (1981) *Phytochemistry*, **20**, 1477–1480.

Suttle, J. C. & Kende, H. (1980) *Phytochemistry*, **19**, 1075–1079.

Suzuki, Y. & Hirasawa, E. (1980) *Plant Physiology*, **66**, 1091–1094.

Taibi, G. & Schiavo, M. R. (1993) *Journal of Chromotography*, **614**, 153–158.

Taylor, M. A., Burch, L. R., & Davies, H. V. (1993) *Journal of Plant Physiology*, **141**, 370–372.

Taylor, M. A., Mad Arif, S. A., Kumar, A., Davies, H. V., Scobie, L. A., Pearce, S. R., & Flavell, A. J. (1992) *Plant Molecular Biology*, **20**, 641–651.

Tiburcio, A. F. & Galston, A. W. (1986) *Phytochemistry*, **25**, 107–110.

Tiburcio, A. F., Kaur–Sawnhey, R., & Galston, A. W. (1986b) *Plant Physiology*, **82**, 375–378.

——— (1988) *Plant and Cell Physiology*, **29**, 1241–1249.

Tiburcio, A. F., Masdéu, M. A., Dumortier, F. M., & Galston, A. W. (1986a) *Plant Physiology*, **82**, 369–374.

Tonin, G. S., Wheeler, C. T., & Crozier, A. (1991) *Plant Cell and Environment*, **14**, 415–421.

Toppel, G., Witte, L., & Hartmann, T. (1988) *Phytochemistry*, **27**, 3757–3760.

Torrigiani, P., Altamura, M. M., Scaramagli, S., Capitani, F., Falasca, G., & Bagni, N. (1993) *Journal of Plant Physiology*, **142**, 81–87.

Torrigiani, P., Serafini–Fracassini, D., & Bagni, N. (1987) *Plant Physiology*, **84**, 148–152.

Torrigiani, P., Serafini–Fracassini, D., Biondi, S., & Bagni, N. (1986) *Journal of Plant Physiology*, **124**, 23–29.

Trebitsh, T., Goldschmidt, E. E., & Riov, J. (1993) *Proceedings of the National Academy of Sciences of the United States of America*, **90**, 9441–9445.

Tropea, J. E., Molyneux, R. J., Kaushal, G. P., Pan, Y. T., Mitchell, M., & Elbein, A. D. (1989) *Biochemistry*, **28**, 2027–2034.

Trull, M. C., Holaway, B. L., & Malmberg, R. L. (1992) *Canadian Journal of Botany*, **70**, 2339–2346.

Tsai, D., Arteca, R. N., Arteca, J. M., & Phillips, A. T. (1991) *Journal of Plant Physiology*, **137**, 301–306.

Uhlmann, A. & Ebel, J. (1993) *Plant Physiology*, **102**, 1147–1156.

Van Der Straeten, D., Rodrigues–Pousada, R. A., Villarroel, R., Hanley, S., Goodman, H. M., & Van Montagu, M. (1992) *Proceedings of the National Academy of Sciences of the United States of America*, **89**, 9969–9973.

Van Der Straeten, D., Van Viemeersch, L., Goodman, H. M., & Van Montagu, M. (1990) *Proceedings of the National Academy of Sciences of the United States of America*, **87**, 4859–4863.

Vaughn, K. C. & Duke, S. O. (1984) *Physiologia Plantarum*, **60**, 106–112.

Ververidis, P. & John, P. (1991) *Phytochemistry*, **30**, 725–727.

Vicente, C. & Legaz, M. E. (1982) *Physiologia Plantarum*, **55**, 335–339.

——— (1985) *Phytochemistry*, **24**, 217–219.

Villanueva, V. R., Adlakha, R. C., & Calvayrac, R. (1980a) *Phytochemistry*, **19**, 787–790.

——— (1980b) *Phytochemistry*, **19**, 962–964.

Villanueva, V. R., Adlakha, R. C., & Cantera–Soler, A. M. (1978) *Phytochemistry*, **17**, 1245–1249.

Vinkler, C. & Apelbawm, A. (1985) *Physiologia Plantarum*, **63**, 387–392.

Vioque, B. & Castellano, J. M. (1994) *Physiologia Plantarum*, **90**, 334–338.

Vogt, T., Aebershold, R., & Ellis, B. (1993) *Archives of Biochemistry and Biophysics*, **300**, 622–628.

Vollmer, J. J., Steiner, N. C., Larsen, G. Y., Muirhead, K. M., & Molyneux, R. J. (1987) *Journal of Chemical Education*, **64**, 1027–1030.

Vongs, A., Kakutani, T., Martienssen, R. A., & Richards, E. J. (1993) *Science*, **260**, 1926–1928.

Walbot, V. (1985) *Trends in Genetics*, **1**, 165–169.

Wallsgrove, R. M. & Mazelis, M. (1980) *FEBS Letters*, **116**, 189–192.

Walsh, C., Pascal, R. A., Johnston, M., Raines, R., Dikshit, D.,

Krantz, A., & Honma, M. (1981) *Biochemistry*, **20**, 7509–7519.

Walter, H. J. & Geuns, J. M. C. (1987) *Plant Physiology*, **83**, 232–234.

Walton, N. J., Peerless, A. C. J., Robins, R. J., Rhodes, M. C. J., Boswell, H. D., & Robins, D. J. (1994) *Planta*, **193**, 9–15.

Walton, N. J., Robins, R. J., & Peerless, A. C. J. (1990) *Planta*, **182**, 136–141.

Wang, S. Y., Wang, C. Y., & Wellburn, A. R. (1990) *Stress Responses in Plants*, pp. 147–173. New York: Wiley–Liss.

Wang, T. & Yang, S. F. (1987) *Planta*, **170**, 190–196.

Weinstein, L. H., Kaur–Sawhney, R., Rajam, M. V., Wettlaufer, S. H., & Galston, A. W. (1986) *Plant Physiology*, **82**, 641–645.

Weinstein, L. H., Osmeloski, J. F., Wettlaufer, S. H., & Galston, A. W. (1987) *Plant Science*, **51**, 311–316.

Werner, C., Hu, W., Lorenzi–Riatsch, A., & Hesse, M. (1995) *Phytochemistry*, **40**, 461–465.

White, J. A. & Kende, H. (1990) *Journal of Plant Physiology*, **136**, 646–652.

White, M. F., Vasquez, J., Yang, S. F., & Kirsch, J. F. (1994) *Proceedings of the National Academy of Sciences of the United States of America*, **91**, 12428–12432.

Wink, M. (1987) *Planta Medica*, **53**, 509–514.

Wink, M. & Hartmann, T. (1982) *Plant Physiology*, **70**, 74–77.

Woolhouse, H. W. (1984) *Canadian Journal of Botany*, **62**, 2934–2942.

Wright, L. C., Brady, C. J., & Hinde, R. W. (1981) *Phytochemistry*, **20**, 2641–2645.

Wyss–Benz, M., Streit, L., & Ebert, E. (1990) *Plant Physiology*, **92**, 924–930.

Yamaki, S. (1984) *Plant and Cell Physiology*, **25**, 151–166.

Yamanoha, B. & Cohen, S. S. (1985) *Plant Physiology*, **78**, 784–790.

Yanagisawa, H. & Suzuki, Y. (1981) *Plant Physiology*, **67**, 697–700.

——— (1982) *Phytochemistry*, **21**, 2201–2203.

Yang, C., Wilksch, W., & Wild, A. (1994) *Journal of Plant Physiology*, **143**, 389–395.

Yang, S. F. & Adams, D. O. (1980) *The Biochemistry of Plants*, pp. 163–175, E. Conn, Ed. New York: Academic Press.

Yip, W., Dong, J., Kenny, J. W., Thompson, G. A., & Yang, S. F. (1990) *Proceedings of the National Academy of Sciences of the United States of America*, **87**, 7930–7934.

Yip, W., Dong, J., & Yang, S. F. (1991) *Plant Physiology*, **95**, 251–257.

Young, N. D. & Galston, A. W. (1984) *Plant Physiology*, **76**, 331–335.

Yu, Y. & Yang, S. F. (1980) *Plant Physiology*, **66**, 281–285.

Yung, K. H., Yang, S. F., & Schlenk, F. (1982) *Biochemical and Biophysical Research Communications*, **104**, 771–777.

Zenk, M. H. (1978) *Recent Advances in Phytochemistry*, **12**, 139–176.

Zer, H., Peleg, I., & Chevion, M. (1994) *Physiologia Plantarum*, **92**, 437–447.

Zhang, Z., Schofield, C. J., Baldwin, J. E., Thomas, P., & John, P. (1995) *Biochemical Journal*, **307**, 77–85.

CHAPTER 21

Molecular Reactions at Cell Surfaces

Recent studies have considered the effects of metabolites and inhibitors on isolated intact cells, for example the penetrability of the surface and the relaying of substance, signals, or both to inner compartments. Problems of the nature and participation of the surface in growth, division, ingestion, virus infection, etc., have required data on surface membrane composition and alteration. Studies on tissues and organs have also revealed major components of the extracellular matrix, which provide connections and integrating elements among the embedded cells. Fibrous proteins, such as collagen, fibronectin, and the proteoglycans, as well as the cross-linked laminins, are found in the matrices of both vertebrate and invertebrate tissues.

In an intact animal, the medium bathing the tissues and cells may also contain other metabolizable polymers; in the mammal these include fibrinogen and other plasma proteins. Chemical change within the extracellular medium, as in the formation of a blood clot, is capable of changing the cell surface, occasionally with the participation of circulating polyamine. In some fluids that contain spermine, such as human and rat seminal fluids, the level of a specific polyamine-binding protein is androgen-determined, suggestive of a carefully monitored physiological role for both polyamine and binding protein. In other fluids like bile, the presence of a polyamine inhibits the crystallization of damaging bile stones that are composed of calcium hydroxyapatite (Crowther et al., 1993). The activation of alkaline phosphatase by polyamine also helps to dissolve crystals of calcium pyrophosphate (Shinozaki & Pritzker, 1995).

Relatively specific antibodies to polyamines bound to particular antigenic proteins may be generated in the mammal and some of these antibodies are reported to exist normally in mammalian sera (Bartos et al., 1980; Furuichi et al., 1980; Roch et al., 1978). It has been suggested that such antibodies are produced in response to polyamine complexes with proteins and that the antipolyamine antibodies may serve to scavenge or buffer the reactive circulating polyamine released by dying cells.

The cell surface is not merely a limiting membrane. It contains many specific receptors, as well as diverse mechanisms of controlled transport. In addition to the function of constraining both soluble and particulate cellular constituents, the surface of a given cell may consist of multiple functional layers. In the growing plant leaf cell, the outer rigid layers can contain the amine oxidases. In plant or red algae cells rich in outer polysaccharides, such as pectins and algal polymers, the polyamines themselves may bind tightly and cross-link uronic acid- and sulfate-containing structures (D'Orazi & Bagni, 1987; Scoccianti & Bagni, 1992). Bacterial and other cell surfaces can be toughened by some polyamines, thereby minimizing cell lysis.

The connections of the cell membrane to inner cell components might be expected to facilitate transmembrane signaling. The definition of a cell membrane therefore entails the description of some degree of depth in its connections to inner metabolic activity. This perception has led to the study of surface receptors as an approach to the study of the effects of pharmacologically reactive agents like insulin. Such a hormone at a specific surface receptor may initiate a cascade of metabolic events within the inner surface of the membrane.

A subgroup of excitatory amino acid receptors in neuronal cells of the central nervous system, known as N-methyl-D-aspartate (NMDA) receptors, have been found to contain polyamine recognition sites. These are sensitive to natural polyamines acting as agonists or to polyamine analogues acting as antagonists (Williams et al., 1990). Because the receptor controls the entrance of activating or potentially toxic cations, the effects of the polyamines are under active scrutiny for the possible mitigation or prevention of neurotrauma, stroke, etc. (Gilad & Gilad, 1992). This form of excitotoxicity is also reported to activate polyamine biosynthesis and the toxicity has been found to be minimized by difluoromethylornithine (DFMO) (Trout et al., 1993). Additionally, some natural toxins active at this receptor are known to contain unusual substituted polyamines (Fig. 21.1).

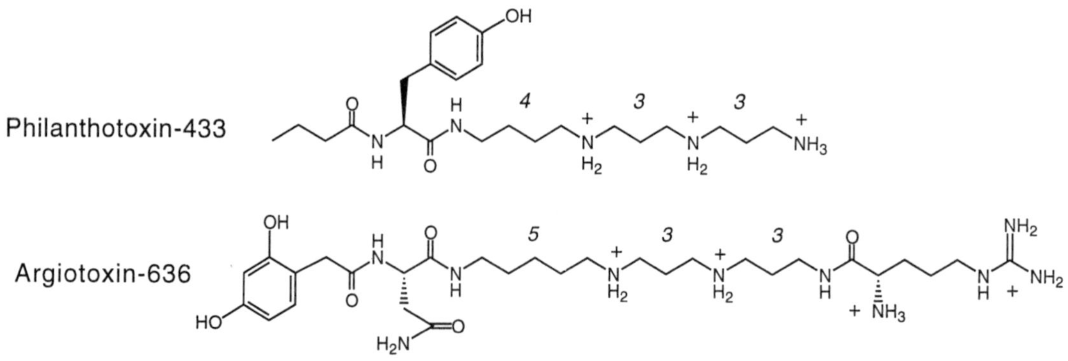

FIG. 21.1 Structures of some natural nerve toxins containing unusual polyamines; adapted from Choi et al. (1993).

In addition to opening ion channels, the arrival of an effector at the cell surface can elicit or trigger the action of protein kinases, or the activation of lipases. The resulting cascade of reactions may result in cell–cell interactions via adhesive properties of surface proteins such as the Ca^{2+}-dependent cadherins. The complex of the cadherins with cytoplasmic peripheral proteins known as catenins, links the cadherins to the internal actin filament network (Kemler, 1993). The membrane is thus linked to this internal filamentous network, in which the polymerization of actin may also be under the influence of the polyamines.

We shall begin with a transamidase enzyme known as *transglutaminase*, which does not react with free glutamine, but establishes inter- and intrachain cross-linkages of γ-glutamyl-containing proteins with other proteins containing ε-amino groups, that is, lysyl moieties. Transglutaminase can also use the polyamines as acceptors for the activated glutamyl carboxyls. The polyamines may then serve as modulating inhibitors of protein cross-linking, may themselves serve to cross-link proteins, or may merely be sopped up by activated proteins. The enzyme will be seen to be involved in problems central to blood clotting, wound healing, tissue development and organization, cancer and metastasis, aging, and apoptosis. The degree of participation of the polyamines in these biological activities is a significant problem.

Transglutaminase and Polyamine Participation

Prevention of blood loss in higher animals

Limited vascular damage results in the accumulation of platelets at the site of injury; the platelets aggregate and can plug the damaged site with the participation of bridging fibers of fibrinogen. The release of thromboplastic lipoproteins from cells in a damaged area also sets off a cascade of reactions leading to the proteolytic conversion of blood prothrombin to thrombin and of fibrinogen to form a fibrin clot (Davie et al., 1991). Thrombin is a multifunctional serine proteinase and cleaves a polypeptide (i.e., a fibrinopeptide) from fibrinogen that forms fibrin. Much of the thrombin is also sequestered in fibrin clots and is released by proteolysis (Fenton, 1988). The thrombin can also bind to endothelial cell surface proteins and some of these linkages may be covalent (Leroy–Viard et al., 1989).

The cleavage of four peptide bonds in fibrinogen to liberate four molecules of fibrinopeptide is a stepwise sequence, freeing some 3% of the substance and generating a soft fibrin clot containing 97% of the original fibrinogen mass. The single fibrin polymers are then cross-linked at newly unmasked sites by transglutaminase, generated from its precursor in plasma-(factor XIII), to form an insoluble aggregate in which the cross-links are γ-glutamyl-ε-lysine peptide bridges between the fibrin molecules (Lorand et al., 1966; Mosesson et al., 1995). A guinea pig liver enzyme has been shown to be capable of replacing the plasma enzyme in cross-linking fibrin.

The group of H. Waelsch showed that in the presence of Ca^{2+}, a liver enzyme could incorporate [^{14}C]-amines into various substrate proteins, including fibrinogen (Clarke et al., 1959). This enzyme, another transglutaminase, was found in many guinea pig tissues, of which liver possessed the highest activity. Among the active amines tested were putrescine, cadaverine, and spermine, as well as phenylethylamine, glycinamide, and histamine. A Ca^{2+}-activated production of ammonia was obtained during addition of the amine. The covalently bound amines were recoverable after acid hydrolysis of the protein product. These results were interpreted to signify a replacement of the amide group in an amino acid such as protein-bound glutamine by the amine, meaning a replacement of the carboxamide by the lysine isopeptide with liberation of ammonia in fibrin cross-linking or a replacement of ammonia in the amide by the amine in the incorporation of the latter.

$$R\text{—}CONH_2 + \text{cadaverine} \xrightarrow{Ca^{2+}}$$

$$NH_3 + R\text{—}CONHCH_2CH_2CH_2CH_2CH_2NH_2.$$

The cross-linking transglutaminase in blood is known as factor XIIIa and is produced by the action of the protease, thrombin, on the *pro*transglutaminase factor XIII. The latter is a tetramer comprising two types of subunits, a and b, of which the two a subunits of 83,000 molecular weight (MW) contain the catalytic domain. Cleavage from these a subunits of two polypeptides of 37 amino acids, each in the presence of Ca^{2+}, releases the two active a-derived subunits as the active enzyme a*₂, factor XIIIa. Factor XIIIa generates glutamyl-lysyl isopeptide bonds between fibrin monomers to form a strong insoluble fibrin clot. The production of NH_3 and its coupling to glutamate formation has been adapted in one clinical assay of factor XIII in the presence of thrombin and Ca^{2+} (Song et al., 1994).

Amines capable of inhibiting the cross-linking of fibrin monomers included the monodansyl derivative of cadaverine, the binding of which at a γ-glutamyl side chain provided a fluorescent label at an acceptor cross-linking site (Lorand et al., 1968). This reaction is one test for the activity of a transglutaminase. Reaction with a simple acceptor peptide, carbobenzoxyglutaminyl glycine, also resulted in the production of a readily separable fluorescent compound. The inhibitor was used early on in a search for individuals deficient in factor XIIIa.

The history of the various hypotheses concerning the mode of action of factor XIIIa was reviewed by A. Loewy (1968), who commented on the extraordinary dual role of thrombin in converting fibrinogen to a more exposed fibrin and factor XIII to the reactive enzyme, factor XIIIa. In subsequent studies the latter enzyme was found to further the cross-linking of the clot to tissue surface proteins such as fibronectin, increasing the fiber size and density of the fibrin clots and tying the clot to tissue.

Mechanism of Transglutaminase Reaction

Folk and Chung (1973) compared the properties of the guinea pig liver enzyme and factor XIIIa. Although asparaginyl peptides were inactive as substrates, the enzyme catalyzed the hydrolysis and aminolysis of *p*-nitrophenyl esters and some thioesters. The γ-carboxamide group of peptide-bound glutamine was described as a potential acyl donor in a Ca^{2+}-dependent acyl transfer to a primary amino group in an acyl acceptor. A cysteine sulfhydryl group, revealed by the addition of Ca^{2+}, was found essential to the activity. These workers detected the formation of a stable acyl–enzyme complex, a potentially reactive thiol ester, following incubation of *p*-nitrophenyl trimethylacetate with the enzyme. The slow release of *p*-nitrophenol in cleavage of the ester was dependent on Ca^{2+}. A peptide containing [¹⁴C]-trimethylacetate was cleaved by trypsin and chymotrypsin from the enzyme–substrate complex and was reported to contain a thioester of cysteine and trimethyl acetate (Folk, 1982). Thus, the immediate acyl donor to an amino group was suggested to be the protein-γ-glutaminyl thioester of the cysteinyl moiety in the transglutaminase.

$$\text{Protein–γ–glutamyl–}\overset{\displaystyle O}{\overset{\displaystyle \|}{C}}\text{–S–cyst-transglutaminase}$$

$$\downarrow RNH_2$$

$$\text{Protein–γ–glutamyl–}\overset{\displaystyle O}{\overset{\displaystyle \|}{C}}\text{–}\overset{\displaystyle H}{N}\text{R} + \text{HS-cysteinyl-transglutaminase}$$

Data on the cleavage of thioesters were extended to factor XIIIa (Curtis et al., 1974). Esters of thiocholine, of which the *trans*-cinnamoyl derivative was among the most active, were good inhibitors of the cross-linking of the fibrin clots by factor XIIIa. The acyl moiety was readily transferred to monodansylcadaverine to form readily isolated and characterized fluorescent amides.

Thioesters participate in the reactions of the $α_2$ macroglobulin and of the activated C3b and C4b components of complement with the ε-amino group of lysine of interacting proteins (Lorand, 1983). Human $α_2$-macroglobulin is a large glycoprotein of vertebrate plasma that forms inactive complexes with endoproteinases after an initial cleavage by the proteinase (Van Leuven, 1982). The ability to inhibit proteases is blocked by primary amines that cleave the internal protein thioester and expose a cysteinyl thiol. A similar inactivation of the C4b component of complement occurs with diamines, such as putrescine or 1,3-diaminopropane (DAP) (Dodds et al., 1996).

New classes of inhibitors capable of irreversible reaction with the thiol have been prepared to block the transglutaminase that may be involved in various diseases, such as acne, psoriasis, etc. The 3-halo-4,5-dihydroisoxazoles, modeled after the glutamine antagonist acivicin, are active inhibitors via this mechanism and have been shown to form a stable imino thioether enzyme adduct with the guinea pig liver enzyme (Auger et al., 1993). Freund et al. (1994) have recently described compounds that acetonylate the cysteine of factor XIIIa quite specifically.

The existence of the nonenzymatic inactivation by an amine of an existing protein containing a reactive thioester as in the $α_2$-macroglobulin or the activated complement component, indicates that the covalent binding of a diamine is not sufficient to demonstrate enzymatic activity. For example, it can be imagined that the increased rate of polyamine fixation by illumination of chloroplasts (Cohen et al., 1983, in Chap. 20, Fig. 20.8) relates to the increased formation of reactive thioesters in the organelles, and not to the activity of a transglutaminase. Additionally, the reaction mechanism requires the demonstration of an initially free cysteinyl thiol, the capability of generating γ-glutamyl-ε-lysyl isopeptides, and observations on the nature of the protein reactants and protein products. The fact that polyamines can be protein-acyl acceptors in in vitro reactions with the cross-linking transglutaminases does not prove that these substances are substrates in vivo and that their service as substrates is other than a random participation in the action of a browsing enzyme. Nevertheless, the fact that several transglutaminases can crosslink the glutamines of a test protein with the hypusine of EIF 5A, thereby generating γ-glutamyl-ω-hypusine may be quite significant (Beninati et al., 1995).

Folk and Chung (1973) also summarized properties among the protransglutaminases isolated from platelets and placentas. These zymogens appear to differ significantly from the plasma proenzyme, factor XIII. Nevertheless, the platelet and placental enzymes were immunologically cross-reactive with antibody to the catalytic subunit of factor XIII. These cellular proenzymes are thought to comprise only the two catalytic a subunits and not to contain the two noncatalytic b subunits found in the plasma factor XIII.

The human placental enzyme with a relatively simple subunit structure has been considered to be a potential product of biotechnology for use in wound healing (Takahashi et al., 1986). Various hemorrhagic disorders of fibrin processing have been detected and classified, and supplementation with factor XIII appears helpful in improving clot strength. The industrial production of the a_2 dimer characteristic of the placental proenzyme has been effected after isolation of the human gene for the a subunit (Ichinose et al., 1990) and its expression in yeast (Bishop et al., 1990). The structure of a homodimeric human factor XIII, comprising two a subunits and produced in a yeast, has recently been characterized by X-ray crystallography (Yee et al., 1994).

The thrombin-induced aggregation of platelets is inhibited by primary diamines, implicating transglutaminase in that aggregation (Ganguly & Bradford, 1982). The platelets also release the glycoprotein thrombospondin, which is a substrate for homopolymerization and amine acceptance and also forms complexes with fibrin fibrils (Lynch et al., 1987). Insoluble platelet structures are themselves rich in the distinctive isopeptides. Dimers of the platelet protransglutaminase are also activated by thrombin via removal of a 37-aminoacyl N-terminal peptide.

In the 1970s additional enzymes capable of generating the characteristic toughening γ-glutamyl-ε-lysyl isopeptide bonds were found in tissues such as hair follicles, skin, and rat prostate, but it had not been clearly established that the in vitro introduction of the polyamines into proteins was physiologically significant. Polyamines were found covalently bound to proteins in many fluids like human amniotic fluid, blood plasma, etc., but the origins of the amines and their roles were unclear. Studies on the incorporation of labeled amines in enzymatically active extracellular systems, such as clotting rat seminal plasma, did reveal numerous γ-glutamyl polyamines (Folk, 1980). In some instances [^{14}C]-putrescine fed to cell cultures led to the appearance of intracellular labeled spermidine and spermine in γ-glu-

taminyl linkage in proteins found later in the medium (Fan et al., 1983). The label of [^{14}C]-putrescine perfused into the rat portal vein was incorporated as protein-bound amine in various tissues. A rat liver homogenate incubated with [^{14}C]-putrescine produced protein-bound N^1-(γ-glutamyl)spermidine, N^1-(γ-glutamyl)spermine, and N^1,N^{12}-bis(γ-glutamyl)spermine, which were freed by exhaustive proteolytic degradation (Beninati et al., 1985). Tests on the incorporation of labeled polyamines into proteins in phytohemagglutinin-stimulated lymphocytes also revealed some γ-glutamyl polyamines.

In another type of experiment transglutaminase activity found in female rat tissues has been studied as a function of the stage of pregnancy. Although varying only slightly in the maternal liver, both the enzyme activity and putrescine contents show three similar spikes in very early, middle, and late pregnancy (Piacentini et al., 1986). Increased levels of diamine oxidase may well account for the decreases in the diamine and may contribute H_2O_2 for events in differentiation. However, it may be asked if the addition of putrescines to glutamyl moieties at these peaks of enzyme and diamine does not also participate in tissue molding.

Tying fibrin clots to cells

Circulating clots are a menace and it was found that tying these clots to tissue components could occur via the presence of soluble plasma components that were also components of cell surfaces and tissue stroma. Fibronectin is a high MW glycoprotein (450,000) present in and on cells and it exists in plasma as well (Mosesson & Amrani, 1980; Ruoslahti et al., 1982). The latter soluble form can be cross-linked by factor XIIIa to fibrin and to denatured collagen; these reactions are inhibited by spermidine or spermine.

Cell cultures may develop a pericellular matrix rich in filamentous fibronectin contained in a mixed population of polymers. Factor XIIIa can also bind fibronectin into this matrix in additional covalent isopeptide linkages (Barry & Mosher, 1989). Tissue enzyme, which is released from the cells after wounding, is also found in the matrix tied to fibronectin (Upchurch et al., 1991).

It is of interest that the fibronectin and fibrinogen of plasma can form insoluble noncovalently linked aggregates in the presence of spermine. A purified fibronectin, which polymerizes as thin filaments, becomes extensively polymerized as filamentous structures by the polyamines: spermine > spermidine > putrescine (Vuento et al., 1980). This noncovalent reaction is reversible on removal of the polyamine. The glycoprotein also forms a tight noncovalent complex in the absence of Ca^{2+} with erythrocyte or tissue transglutaminases. The complex of fibronectin with the former has a stoichiometry of 1 : 2 (Turner & Lorand, 1989). These complexes, derived from the intact plasma proteins or particular fragments, can also bind to collagen in ternary noncovalent complexes.

A 1 : 1 dimer (fibronectin C) has been detected in some pathophysiological conditions. Further, both fibrinogen and fibronectin have been observed to bind to rabbit hepatocyte surfaces in both covalent and noncovalent linkages. The former was apparently formed by transglutaminase, as determined by the inhibition by polyamines, as well as by insensitivity to heparin or hirudin. It was also found that the rabbit hepatocyte surface transglutaminase could serve as the binding site for both fibronectin and fibrinogen and can additionally covalently attach these proteins to the cell surface (Barsigian et al., 1991). In bovine aortic endothelial cell cultures, surface proteoglycans can be cross-linked to fibronectin in this matrix protein in reactions inhibited by monodansylcadaverine (Kinsella & Wight, 1990). A human milk protein, osteopontin, which becomes a component of bone matrices, can also be cross-linked to fibronectin by guinea pig liver transglutaminase, a reaction suggested to be important in generating bone matrices (Beninati et al., 1994).

The interaction between molecules of saliva and oral epithelial cells to form a salivary–mucosal pellicle also appears to involve cross-linking by a buccal cell transglutaminase (Bradway et al., 1992). The formation of a laminin–nidogen complex of some murine tumors has been shown to involve the action of a tissue transglutaminase (Aeschlimann et al., 1992). Such an enzyme can also attach fibronectin to the basal layer of endothelial cell layers, thereby anchoring the cells to the basement membrane (Martínez et al., 1994).

Erythrocyte Transglutaminase

The human erythrocyte represents a more readily available, apparently simple, model cell containing the enzyme. The Ca^{2+}-requiring enzyme was active before thrombin treatment; it was not present as a protransglutaminase. The penetration of Ca^{2+} into intact cells in the presence of an ionophore resulted in activation and the cross-linking of red cell membrane proteins (Lorand et al., 1976). It was asked if this phenomenon, with accompanying high levels of internal Ca^{2+}, did not produce the membrane deformability observed in sickle cell crisis (Siefring et al., 1978). An irreversible loss of the deformability of the membrane can be prevented by treatment with a small amine, such as histamine, or by other enzyme inhibitors. The sickle cell trait is marked by relatively high levels of intraerythrocytic spermine, and there is a question of whether the uptake of spermine in this system modulates or accentuates the membrane cross-linking.

The human erythrocytic enzyme was isolated and shown to cross-link several membrane-bound proteins, as well as to incorporate monodansylcadaverine or $[^{14}C]$-methylamine into several common soluble proteins (Brenner & Wold, 1978). In a later study with inside-out vesicles, a particular anion-transporting membrane protein was identified as a major substrate for the addition of dansylcadaverine in the presence of an active enzyme (Lorand et al., 1986). The further purification of the human erythrocyte transglutaminase to homogeneity and stability (Signorini et al., 1988) revealed

differences in amino acid content from the guinea pig liver enzyme, although numerous functional activities such as Michaelis constant (K_m), specificity for various proteins as substrates, requirements for Ca^{2+} and —SH, were earlier shown to be similar.

Two types of the enzyme may be found in some other mammalian cells, and the erythrocytic enzyme appears to be similar to a soluble transglutaminase increased in activity by treatment with retinoids and designated as a type 2 enzyme. Type 2 enzymes were subsequently found to be inhibited by GTP at low levels of Ca^{2+}. GTP apparently altered the conformation of the enzyme, protected it against cleavage by trypsin, and affected the binding of Ca^{2+} (Bergamini, 1988). Similar responses were found for other type 2 enzymes (Achyuthan & Greenberg, 1987). A periodate-oxidized GTP, containing a dialdehyde, binds irreversibly at a single site of the enzyme, which now is less sensitive to Ca^{2+} (Bergamini & Signorini, 1993).

Genetic tools for the study of the development of chicken erythrocyte transglutaminase were described. These include cDNAs derived from the poly(A)·RNA produced in chick embryos or erythroid cells of such embryos (Weraarchakul–Boonmark et al., 1992). The chicken enzyme with 697 amino acids and 78,621 MW proved to have a >60% identity with three related type 2 mammalian enzymes.

Possible roles for the erythrocyte transglutaminase in events such as aggregation, membrane deformation, and aging have compelled interest in other potentially similar tissue events. The aging lens and cataract involve an increasing insolubility, some of which arise from the formation of disulfide bonds. Nevertheless, reduction of these bonds does not dissociate human lens polymers, which also contain γ-glutamyl-ε-lysine isopeptides (Lorand et al., 1981). The lenses of various mammals contain a Ca^{2+}-dependent tissue enzyme that incorporates putrescine or dansylcadaverine into the subunits of β-crystallins present in a cortex homogenate. The β-crystallins become even better targets for the enzyme after oxidation by free radicals. Recombinant mutants of certain bovine lens crystallins have been used to determine the sequences at the lysine and glutamine sites that modify the reactivities of these residues (Groenen et al., 1994). Various types of oxidative stress in the development of cataracts and their prevention have been discussed by Spector (1995).

Transglutaminases of Scavenging Cells

The formation of a fibrin matrix can become both a route of macrophage migration and sites in which these cells can settle. Human monocytes are known to contain the dimeric a_2 factor XIII, which they synthesize. The proenzyme, expressed on the monocyte surface, may be activated by thrombin in a fibrin clot and contribute to cross-linking of the clot. Other vascular cells (e.g., vein endothelial cells) can also cross-link fibrinogen bound to the cells by adhesive proteins, the *integrins* (Clark & Brugge, 1995; Martínez et al., 1989).

In cultures of human blood monocytes, their content of factor XIII decreases as the cells mature into macrophages and an active cellular transglutaminase appears. The latter process is hastened by retinoic acid. Phagocytic activity is inhibited by monodansylthiacadaverine (Seiving et al., 1991). Among peritoneal macrophages, the cells highest in an internal transglutaminase have the greatest phagocytotoxic activity. The incorporation of putrescine into proteins in skin wounds during wound healing is well known (Bowness et al., 1987); it may be asked if the presence of the bound amines that compete with isopeptide formation affects the healing process or merely reflects the increased availability of polyamine and substrate proteins, such as fibronectin and partially degraded collagen.

The possible relation of the transglutaminase content of macrophages and the important functions of these cells have been examined in the context of several macrophage reactions: the inhibition of tumor cell growth, the release of toxic oxygen radicals, the response to cigarette smoke, etc. The Ca^{2+}-dependent fusion of mouse alveolar macrophages is induced by the $1\alpha,25$-dihydroxy vitamin D_3. The vitamin greatly increases transglutaminase content and cell fusion, and this effect is inhibited by the addition of amines. The induction by the vitamin is also inhibited by methylglyoxal-bis(guanylhydrazone) (MGBG) and this inhibition is reversed by exogenous spermidine (Tanaka et al., 1991). The relation of spermidine to the cross-linking enzyme in this system has not been clarified.

Receptor-mediated endocytosis has been suggested to be a transglutaminase-regulated function. The mobility of proteins within plasma membranes is clearly required, and it may be asked how the formation of relatively stable isopeptide bonds would facilitate such membrane activity. A membrane-bound transglutaminase participates in an ATP-dependent, Ca^{2+}-dependent endocytosis of the protein transferrin into the rat reticulocyte plasma membrane. [^{14}C]-polyamines are incorporated in the membrane during the endocytotic events. Monodansylcadaverine blocks this incorporation and decreases endocytosis (Katoh et al., 1994b).

Membranous Substrates

The association of fibrin and fibronectin, stromal components with tissue cell membranes, suggested the possible presence of a transglutaminase and its substrates in the membrane. The availability of antipolyamine antibody permitted a test of the presence of the polyamine on cell surfaces. These were found in a cytolytic test in the presence of complement and in tests of morphological change without complement. Antipolyamine antisera were cytotoxic to baby hamster kidney (BHK) cells, L cells, and mouse fibroblasts in the presence of complement, and L cells showed altered morphologies in the absence of complement (Quash et al., 1971). Antipoly-

amine antibodies covalently bound to latex spheres were used to detect the presence of polyamines on cell surfaces (Quash et al., 1978; Quemener et al., 1990) and to detect many polyamine-containing proteins in the plasma of mice bearing the Lewis lung carcinoma (Delcros et al., 1987).

Isopeptides of ε-(γ-glutamyl)lysine have been found in the membranes of cultured mouse fibroblast L cells (Birckbichler et al., 1973). In another approach to similar cells, the inner membranes of inside-out phagosomes were labeled at γ-glutamyl residues with dansylcadaverine by the guinea pig liver enzyme (Evans & Fink, 1977). It appeared that the inner (cytoplasmic surface) membrane is labeled more efficiently than the outer plasma membrane surface and some of the sites are in proteins bridging the membrane. Two human leukocyte antigens (HLAs) proved to be such transmembrane proteins with intracellular glutamine residues labeled by [^{3}H]-putrescine (Pober & Strominger, 1981).

Fibroblast transglutaminase was detected serologically as a membrane-associated enzyme. Inhibitors of the enzyme appeared to increase membrane fluidity and to decrease both isopeptide bonds and cholesterol content (Haugland et al., 1982). A transglutaminase was found to be associated specifically with rat liver plasma membranes (Slife et al., 1985). The incorporation of [^{14}C]-putrescine into exposed surface proteins of intact mouse neuroblastoma cells was demonstrated with the aid of the guinea pig liver enzyme (Chen, 1984).

The covalent cross-linking of cells would be expected to be significant in animal reactions such as wound healing or mucosal repair. In skin wounds, it can be supposed that collagen fragments known to be substrates will participate. However, the presence of an incorporated polyamine may inhibit effective cross-linking and indeed the topical application of putrescine has been found to decrease breaking strength in rat skin wounds and the incorporation of glycoprotein into the extracellular wound matrix (Dolynchuk et al., 1994). On the other hand, the presence of putrescine in gastrointestinal mucosa increases transglutaminase activity in the tissue and appears to facilitate healing (Wang et al., 1994a).

Diversity of Transglutaminase Functions

The discovery of the enzyme provided a possible insight into the problem of insoluble proteins and membranous components. In 1973 Folk and Chung described the enzymology in the formation of isopeptide bonds in the tough internal proteins in hair follicles. They detected two enzymes, one of which resembled the guinea pig liver transglutaminase, and a second that was an active (i.e., nonzymogen) Ca^{2+}-requiring enzyme of lower molecular weight. The pattern of sensitive substrates was quite different; these observations relate to keratinization.

Clotting reactions have been studied with the hemolymph or plasma of crustaceans, such as the lobster, shrimp, or horseshoe crab. One mode of minimizing blood loss in these organisms involves a clottable protein in the plasma that becomes cross-linked by the transglutaminase liberated from lysed hemocytes. The isolation and characterization of a partially purified transglutaminase from lobster muscle (*Homarus americanus*), which is capable of clotting *Homarus* plasma or of inserting dansylcadaverine into casein, were described by Myhrman and Brüner–Lorand (1970). The conversion of spiny lobster fibrinogens (*Panulirus interruptus*) into fibrin gels by the generation of ε-(γ-glutamyl)lysine bonds with extracts of lobster hemocytes was described by Fuller and Doolittle (1971). Similar results were observed in shrimp systems, the larger part of the clotting enzyme being localized in the smaller hyaline hemocytes (Martin et al., 1991). In the horseshoe crab, *Limulus polyphemus*, a relatively unstable enzyme was reportedly present in both hemocytes and amebocytes. Clots taken from the blood of *Limulus* also contained N^{ϵ}(γ-glutamyl)lysine cross-links (Wilson et al., 1992). The purified enzyme, which required Ca^{2+} and a free —SH group for activity, incorporated dansylcadaverine into dimethyl casein; this was uniquely inhibited by >0.5 M NaCl and was not inhibited by GTP (Tokunaga et al., 1993a). This group also obtained a cDNA, derived an amino acid sequence (764 residues), and demonstrated homology to various mammalian enzymes (Tokunaga et al., 1993b).

A detailed study was also developed on the clotting system in the Japanese species, *Tachypleus tridentatus*. Two proteins of the hemocytes, isolated from large granules of the hemocytes, were found to serve as crosslinkable substrates for the intracellular enzyme (Shigenaga et al., 1993).

Another clotting system included the coagulation of the seminal ejaculates of guinea pigs, rats, and mice (Williams–Ashman, 1989). The coagulable protein of guinea pig vesicular secretion was characterized and the clotting event caused by the enzyme of the anterior prostatic enzyme was shown to form γ-glutamyl-ε-lysine bridges. Clotting in rat semen was similar, with an incorporation of added polyamines into the coagulated protein. In this system, clotting and incorporation were both markedly enhanced by polyanions such as poly-L-glutamate (Williams–Ashman et al., 1977). The incorporated polyamines were present as both monosubstituted γ-glutamyl-spermidine and -spermine derivatives, as well as cross-bridging bisglutamyl derivatives of these substances (Folk et al., 1980). The transglutaminase, isolated from the rat coagulating gland, was purified and can serve as a cross-linker. The enzyme, now described as a type IV transglutaminse, was not inhibited by GTP and was immunologically not cross-reactive with the tissue enzymes or blood factor XIII (Seitz et al., 1991). The enzyme was both glycosylated and acylated and removal of these residues activated the protein, which was then coagulated and thereby partially inactivated. The initial reduced activity of the enzyme was suggested as a mechanism to prevent premature clotting until mixing of the components after copulation. Spermidine or spermine (10 mM) inhibited the coagulation and precipitation of the protein.

A cDNA characteristic of this transglutaminase was isolated (Ho et al., 1992). It was used to determine both amino acid sequence, revealing homology with factor XIIIa and tissue enzymes, and the controls on specific mRNA production from a single gene. Castration markedly reduced the production of the specific mRNA relating to this enzyme and this was restored by injection of androgen. The cDNA for a human prostate transglutaminase was also isolated and sequenced. Despite a lack of knowledge concerning the function of the human enzyme, it had a 51% identity with the type IV rat enzyme (Grant et al., 1994).

The rat ventral prostate was found to contain a relatively specific spermine-binding protein. This cytosolic acidic protein of about 30,000 MW was also under androgenic control and possibly functioned in the delivery of the newly synthesized broadly reactive spermine from cell to seminal fluid. The presence of an active transglutaminase, Ca^{2+}, and polyamine in a rat seminal ejaculate induced an examination of the rat sperm surface, which was found capable of covalently binding [^{14}C]-spermidine (Paonessa et al., 1984). Other competing amines reduced this binding.

The cross-linking of proteins and peptides by transglutaminase to form insoluble polymers containing isopeptide bonds has been observed in extracts of the postmortem brains of aged normal humans and patients with Alzheimer disease (Rasmussen et al., 1994). An amyloid protein polypeptide characteristic of this disease is also cross-linked by the guinea pig enzyme. Although it has been reported that the postmortem brains of Alzheimer patients are far lower in enzyme than those of control individuals, it has been asked if the cross-linking, inhibitable by transglutaminase inhibitors, contributes to the pathology in the humans (Miller & Johnson, 1995). Spermidine has recently been found to be significantly increased in the cortex of Alzheimer disease patients (Morrison & Kish, 1995).

An insoluble apolipoprotein, apolipoprotein B100, found in human chylomicrons, is a substrate for the tissue transglutaminase and contains cross-linking molecules of $N^{1},1N^{12}$-bis(γ-glutamyl)spermine, as well as the non-cross-linked N^{1}-(γ-glutamyl)spermine. The protein therefore appears to have many transglutaminase-reactive sites, although the possibility that the apolipoprotein contains a reactive thioester was not excluded (Cocuzzi et al., 1990). Lipoprotein (a), which contains both apolipoprotein B100 and another apolipoprotein (a) binds to the intimal surface of blood vessels in interaction with fibrin and cell-surface structures. This complex lipoprotein was found to be a substrate for tissue transglutaminase, and it has been asked if the atherogenic potential of deposition is in fact heightened by the enzymatic cross-linking (Borth et al., 1991).

Reactions with transglutaminase and a polyamine to

form an internal ε-(γ-glutamyl)polyamine linkage were found to increase the activity of the Ca^{2+}-dependent phospholipase A_2 up to threefold (Cordella–Miele et al., 1993). The enzyme, which hydrolyses the 2-acyl ester in 3-phosphoglycerides, was found in pancreas macrophages and in numerous snake venoms. This activation was suggested as the origin of some inflammatory diseases. The modified monomer can also be cross-linked to form a bis(polyamine) dimer. The incorporation of spermidine into γ-glutamyl linkage into the monomeric enzyme also increased the activity threefold, but did not lead to dimerization.

Empirical study has revealed amine-incorporating proteins in the nervous system, pancreas, testes, intestine, etc. Examples of such substrates include such important tissue proteins as actin and myosin and in plant chloroplasts, the large subunit of ribulose 1,5-biphosphate carboxylase/oxygenase. More recently it has been reported that several proteins of the human immunodeficiency virus (HIV) can incorporate spermidine at individual or cross-linking glutamyl sites, or can form isopeptide bonds with the aid of the guinea pig liver enzyme (Mariniello et al., 1993).

The effects of the modification of various proteins have been studied as exercises in protein chemistry, that is, the effects of cross-linking on the physical properties of polymer classes of β-lactoglobulin or the effects of amine incorporation on the molecular structure of this protein (Coussons et al., 1992). Protein emulsions have been gelated for possible use in food supplements or in tests of animal digestion. The potentialities of plant proteins as substrates for modification by transglutaminase, such as the covalent attachment of essential amino acids, have also been explored. The enzyme has been used to cross-link and coat other enzyme proteins, for example trypsin and amylase, on ion-exchanger supports that can then be used as immobilized and insoluble catalytic tools.

Studies in an Embryological System

The fertilization of sea urchin eggs and the development of the embryo is a much studied embryological system. A very early consequence of an initial fertilization is the production of a fertilization envelope that serves to block polyspermy (Heinecke & Shapiro, 1992). This follows an increase in intracellular Ca^{2+}, evoking the secretion of proteins that form the envelope. A cell surface-associated transglutaminase ties these proteins together into a soft envelope, and very soon thereafter the "soft" envelope is hardened by oxidation to produce o,o-dityrosine cross-links. Blocking transglutaminase with inhibitors limits the hardening process and subsequent cell divisions (Cariello et al., 1994).

Recognition of the increase of internal Ca^{2+} after fertilization induced an earlier study of the transglutaminase activity in several species of sea urchins (Cariello et al., 1984). At the 32-cell stage, the normal embryo was found to contain isopeptide. Incorporation of [³H]-

putrescine was demonstrated following fertilization, and this was inhibited by monodansylcadaverine, as shown in Figure 21.2. The inhibitory action of a noncompetitive inhibitor (DAPBT) of the enzyme suggested that the incorporation was enzymatic. Essentially all of the putrescine was incorporated as such into γ-glutamyl-putrescine in the proteins and was liberated by acid hydrolysis. The labeled γ-glutamyl-putrescine was also isolated. Several forms of the Ca^{2+}-dependent enzyme were described in sea urchin eggs. In a subsequent study, the incorporation of monodansylcadaverine in fertilized sea urchin eggs was used to demonstrate numerous proteins serving as substrates for the enzyme (Cariello et al., 1990). Increased constant rates of incorporation of labeled putrescine were observed at each doubling. It is of interest that this method of studying posttranslational modification of proteins during early stages of development (i.e., to the 64-cell stage) did not appear to provoke major pathology.

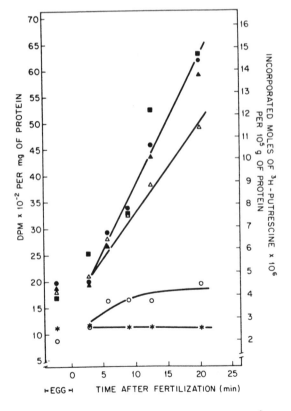

FIG. 21.2 Time course for the in vivo incorporation of [³H]-putrescine into proteins of *Arbacia punctulata* following fertilization; adapted from Cariello et al. (1984). (●) Control with the addition of either (○) 1 mM dansylcadaverine or (*) 0.05 mM 2-[3-(diallylamino)-propionyl]benzothiophene (DAPBT), (■) 0.1 mM MGBG, (▲) 0.1 mM emetine, or (△) 1 mM N^{α}-dimethylated dansylcadaverine.

Unexpected "Transglutaminases" of Plant, Microbial, and Animal Origin

Mention has been made (Chap. 20) of light-stimulated fixation of polyamine by isolated chloroplasts, the lack of a requirement for exogenous Ca^{2+}, and the curious sequestering of much noncovalently bound polyamine in an acid-insoluble fraction. A similar generation of noncovalently bound acid-precipitable polyamine has been reported more recently in parenchyma and greening cells of *Helianthus tuberosus* (Dinnella et al., 1992). Although a Ca^{2+} requirement has not been demonstrated (Falcone et al., 1993) and the enzyme has not been purified, the formation of isopeptides has recently been documented (Del Duca et al., 1995). The status of plant transglutaminases has been reviewed recently (Serafini–Fracassini et al., 1995).

It is relevant that a cyanobacterium, growing photosynthetically in a minimal medium at a high pH, accumulates high lethal concentrations of putrescine (see below) and incorporates large quantities of the diamine in an acid-insoluble form (Guarino & Cohen, 1979). Of the 10^6 molecules bound to structure in the cell after 150 min of incubation, some 2% were bound in two acid-insoluble molecules of putrescine per ribosome. The mechanism of this binding has not been studied.

The tough cyst coat proteins of the protozoan *Colpoda steinii* were found to contain significant amounts of covalently bound putrescine and spermidine (Tibbs, 1982). These amines were largely bound to core peptides rich in glutamic acid or glutamine. There was some intimation of a possible cross-linking by spermidine.

A Ca^{2+}-independent transglutaminase, which generates ε-(γ-glutamyl)lysine in casein and other proteins, has in fact been isolated from an actinomycete, *Streptoverticillium* (Nonaka et al., 1989). The enzyme has been purified to homogeneity, shown to possess a single cysteine residue, whose sulfhydryl is essential for the activity and its amino acid sequence of 331 amino acids has been established (Kanaji et al., 1993). The gene for this enzyme has been synthesized chemically and has been expressed in *Escherichia coli* (Takehana et al., 1994).

A transglutaminase was reported initially in female filariae *Brugia malayi* (Mehta et al., 1992). The enzyme, which is found in developing embryos in utero, generates a large number of isopeptide bonds important to the development of mature microfilariae within adult female worms. Monodansylcadaverine serves to inhibit the development of the embryo and points to the possible use of transglutaminase inhibitors in arresting filarial infections. In contrast to the *Streptoverticillium* enzyme whose amino acid sequence is quite different from the guinea pig liver enzyme, the filarial enzyme is antigenically cross-reactive with antibody to the guinea pig enzyme. The enzyme has been purified and found to be Ca^{2+} dependent (Singh & Mehta, 1994). Molting and development of *Onchocerca volvulus* also appears dependent on transglutaminase-catalyzed reactions (Lustigman et al., 1995).

A Ca^{2+}-dependent enzyme has been purified to apparent homogeneity from *Physarum polycephalum* as a dimer of 77 kDa (Klein et al., 1992). Activity is markedly increased (sixfold) in the differentiation of microplasmodia to spherules, whose extracts contain an actin which most actively binds applied monodansylcadaverine. Actin has been described as a preferred substrate for the incorporation of putrescine (also see Katoh et al., 1994a).

Cross-linking and γ-glutamyl amines

Schrode and Folk (1978) showed that the polyamines can serve as cross-linking agents between glutamine residues in peptides and proteins; that is, both primary amino groups can form amides to cross-link neighboring structures. Several mechanisms exist for cross-linking proteins other than that involving transglutaminase. Indeed the existence of a nonenzymatic cross-linking mechanism to generate the isopeptide bond has been noted, involving the cleavage of a thiol ester in the α_2-macroglobulin and a few other proteins. Also, a mechanism of cross-linking *o*-tyrosine moieties has been found in forming the fertilization envelope of sea urchin embryos. Additionally, several types of cross-linkages involving the ε-amino groups of lysine or hydroxylysine have been described. Lysyl oxidase produces the ε-aldehydo moiety that can form an aldol or Schiff base cross-link with lysine or histidine. Such structures have been demonstrated in *E. coli* envelopes and in chicken egg shell membranes, as well as in mammalian skin and bone. Lysine in proteins or a terminal amine also participates in the formation of aldosylamines that rearrange to a 1-deoxyfructosyl adduct. Such insolubilizing products have been found in erythrocyte membrane, lens crystallins, as well as in other isolated proteins such as hemoglobin. An additional oxidation product, N^ε-(carboxymethyl)lysine, accumulates with age in the human lens and is found in urine (Dunn et al., 1989). A fluorescent composite of lysine, arginine, and pentose, *pentosidine*, has also been described in collagen and other matrix proteins (Sell & Monnier, 1989).

Yet another reaction can account for the appearance of amines in proteins, particularly in cells such as leukocytes and monocytes active in oxidizing systems such as myeloperoxidase. This enzyme oxidizes chloride with H_2O_2 to form hypochlorous acid (HOCl) or hypochlorite ion (OCl⁻), which reacts with nitrogen to form mono- and dichloramine ($RNCl_2$) derivatives of primary amines. Polyamine reactants, like putrescine dichloramine, can react with imidazoles and phenols in proteins at physiological pH (Thomas et al., 1982). In this context it becomes important to estimate isopeptide content accurately and to define the bound γ-glutamyl polyamides to evaluate the role(s) of transglutaminase in the life of an organism.

Despite the resistance of the isopeptide to enzymatic release, it has been found that the compound is freed and appears seasonally in the blood plasma of the male winter flounder (Squires & Feltham, 1980). An enzyme present in animal tissues, γ-glutamylamine cyclotransferase, converts the isopeptide to free lysine and 5-oxo-L-proline (Fink & Folk, 1981; Harsfalvi et al., 1994). Other γ-substituted amines will be released as free amines, in the following reaction:

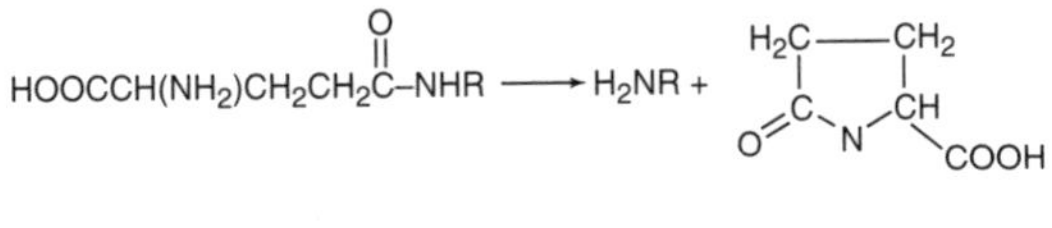

$$HOOCCH(NH_2)CH_2CH_2\overset{\overset{O}{\|}}{C}-NHR \longrightarrow H_2NR + $$

5-oxoproline

The cyclotransferase is a useful tool for the identification of γ-glutamyl polyamines. The presence of the isodipeptide has been demonstrated in the culture fluid of Chinese hamster ovary (CHO) cells (Fésüs & Tarcsa, 1989). The cyclotransferase, an enzyme also found in these cells, has been used to release lysine from the dipeptide.

The existence of the ε-(γ-glutamyl)lysine isopeptide in the proteins of various organisms and tissues was difficult to prove initially. As transglutaminase action became interesting in the food industry, newly discovered proteases were introduced to liberate the isopeptide (Sato et al., 1992). An isopeptidase, termed "destabilase," was isolated from the leech, Hirudo medicinalis; this enzyme cleaves dimeric polypeptides derived from "hard" fibrin (Baskova et al., 1990).

Many improvements have been made in the transglutaminase assay; numerous variations in acyl donor and amine acceptors and affinities and kinetic constants for these substrates were summarized by Lorand and Conrad (1984). Biotin-labeled derivatives of the substrates permitted the detection of cross-linked protein by avidin in microtiter plates. The use of a monoclonal antibody to detect the insertion of dansylcadaverine into protein substrates has markedly improved the identification of the most selective substrates in cellular extracts (Kvedar et al., 1992).

Some Cellular Systems

In cultures of synchronized CHO cells, a doubling of transglutaminase occurred very early after release from quiescence; this preceded a burst of ornithine decarboxylase (ODC) and was unaffected by cycloheximide (Scott & Russell, 1982). Multiple cycles of increase and decrease of transglutaminase activity insensitive to cycloheximide were found in a single cell cycle. The cystamine inhibition of the enzyme in cultured human lung cells decreased isopeptide formation and stimulated DNA synthesis and proliferation (Birckbichler et al., 1981). Also, an effective inhibitor in cells of the transglutaminase reaction, 1-(5-aminopentyl)-3-phenylthiourea, did not affect the proliferation of CHO cells (Lee et al., 1988). Conversely, DAP limitation of putrescine biosynthesis in these cells increased isopeptide

formation and inhibited proliferation (Patterson et al., 1982). Thus, the relation of the enzyme and cross-linking to cell division and proliferation in these systems remains poorly defined. It may be mentioned that DFMO is a substrate of transglutaminase but is far less active than putrescine (Delcros et al., 1984).

Nevertheless, it may be asked if the formation of adherent cells in cultures might not reflect a cross-linking activity of the enzyme in such cells. The enzyme was found in membrane components, cytosol, and nuclei in several rat tissues (Juprelle–Soret et al., 1988). Transfection of mouse fibroblasts by a plasmid containing a gene for the cytosolic enzyme increased the enzyme content of the cells. Under conditions of induction in such cells by sodium butyrate, the level of isopeptide and bis(γ-glutamyl)–polyamine conjugates was raised markedly, concomitant with increased adhesiveness of the cells (Gentile et al., 1992). Thus, the function of the enzyme may relate to phenomena of cell differentiation and tissue formation.

Metastasis

As summarized by McCormack et al. (1994), various studies have indicated a linkage of transglutaminase activity and cell motility and migration. These workers have described such a close relation in several mammalian cell lines and have demonstrated that the presence of polyamines facilitates migration. Thus, DFMO arrests cell motility, and addition of polyamine without cross-linking restores this activity. If transglutaminase affects adhesivity, it can be imagined that metastatic cancer cells will have a lower content of the enzyme, although metastasis will depend on many factors, of which the content of an enzyme substrate, like fibronectin, may be one. Although the content of the enzyme in normal tissues and in tumor cells does vary considerably, various transformed cells are low in the "tissue" type II transglutaminase (Birckbichler et al., 1977). In one study, differentiation with arrest of proliferation and increased adhesiveness of a human monocytic leukemia cell line caused by retinoic acid or o-tetradecanoylphorbol-13-acetate (TPA) was found to be associated with a 20–50-fold increase of the enzyme (Mehta & Lopez–Berestein, 1986).

In some metastatic rat sarcomas, a decrease in transglutaminase activity, a decrease in protein-conjugated polyamine, and a 20-fold increase in free putrescine marked the appearance of metastases (Hand et al., 1987). Highly metastatic cells of murine melanoma were quite low in tissue transglutaminase and high in spermidine. The cellular proteins were very high in N^1- and N^8-monoglutamyl spermidines, suggesting that the excess of spermidine served to block cross-protein linkages and thereby contributed to the ability of the cells to metastasize (Beninati et al., 1993a,b). It was reported that reduction of the spermidine content by DFMO and increase of the transglutaminase activity by treatment with theobromine markedly reduced the content of γ-glutamyl spermidines and decreased the invasiveness of the cells. It was shown that highly malignant tumor cells transfected to increase their tissue transglutaminase increase their adherence to other tumor cells and various other surfaces, for example, fibronectin-coated plastic. Such transfected cells were demonstrated to be less malignant and formed fewer primary tumors after injection into susceptible animals (Johnson et al., 1994). Despite the above, a recent study of human melanoma cell lines demonstrated a positive correlation between an increase in metastasis and an increase in the expression of tissue transglutaminase (Van Groningen et al., 1995). Takaku et al. (1995) did not accept this as a general rule in malignancy.

Apoptosis and isopeptide bonds

An intracellular elevation of Ca^{2+} can evoke enzymes and a rigidification of erythrocyte membranes. In a culture of mouse epidermal cells, an influx of Ca^{2+} and an increase of transglutaminase activity evokes a terminal differentiation of the cells marked by formation of a tough envelope containing cross-linked proteins (Hennings et al., 1981). Similar events characterize the apoptotic death of hepatocytes in a rat liver treated with lead nitrate (Fig. 21.3) or rat thymocytes induced by corticosteroids (Fésüs et al., 1987). The latter group, which has demonstrated the ubiquity of the tissue form of the

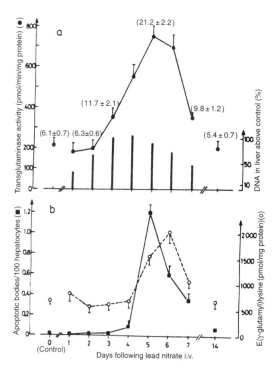

enzyme in all human organs (Thomázy & Fésüs, 1989), has pursued various aspects of the hypothesis that these events can be major components of programmed cell death.

In apoptotic hepatocytes isolated from rat livers after perfusion with collagenase, it was possible to detect both fragmented DNA and cornified membranes insoluble in urea, guanidine, and other agents (Fésüs et al., 1989). The insoluble cross-linked membranes contained both isolable isodipeptides and N^1,N^8-bis(γ-glutamyl)-spermidine. In fact the isodipeptides accumulate in the culture fluids of the hepatocytes or in the plasma of rats treated with lead nitrate. A protein similar to involucrin, the protein incorporated into the insoluble envelopes of terminally differentiated keratinocytes, has also been found in apoptotic hepatocytes (Tarcsa et al., 1992).

In metastatic cell lines derived from a hamster fibrosarcoma, both transglutaminase activity and spontaneous apoptosis were increased significantly during the mid S phase (El Alaoui et al., 1992). The amount of insoluble apoptotic envelopes rich in isopeptide decreased in cell lines of increasing metastatic potential, suggesting an inverse relationship between invasiveness and the cytosolic transglutaminase of the cells. In a human neuroblastoma cell line, transfection with the cDNA of a tissue transglutaminase, leading to an overexpression of this enzyme, markedly increased the cell death rate, with the development of internal insoluble envelopes in far less migratory cells (Melino et al., 1994).

In two human cancer cell lines in culture, retinoic acid increased "tissue" transglutaminase 6–12-fold with a similar rate of increase of apoptotic bodies in the culture. These retinoic acid-treated cultures had a significant increase of free putrescine and spermine and both N^1-(γ-glutamyl)spermine and N^1,N^{12}-bis(γ-glutamyl)-spermine were detected in the insoluble envelopes of the treated cells. Conversely, DFMO, which reduced the putrescine and spermidine content of the cells, reduced enzyme expression and the apoptotic index (Piacentini et al., 1991).

Psoriasis

This is a genetically determined epithelial disease characterized by unrestricted growth of the skin. It afflicts some 2–6% of world populations, with 6–8 million persons in the United States. The scope of the disease, its great nuisance value, exposed character, and resistance to therapy has presented psoriasis as a major problem to dermatologists. A possible etiologic role of polyamines began to be explored when the presence of these compounds was observed to accord with the marked increase of proliferating cells. One study of the distribution of free polyamines in human epidermis revealed a marked decrease in concentration of putrescine and spermine (nmol/mg DNA) from deep multiplying layers to the superficial layers. Spermine was surprisingly high in the epidermis and the ratio of spermine to spermidine

FIG. 21.3 (a) DNA content, transglutaminase activity, and quantity; (b) apoptotic index and protein-bound ε-(γ-glutamyl)-lysine concentration in rat liver following intravenous injection of lead nitrate. Adapted from Fésüs et al. (1987).

increased sharply in the nonmultiplying cells (El Baze et al., 1985).

In comparison of normal skin, psoriatic (scaly) skin, and abutting (uninvolved) skin, the psoriatic lesions had higher contents of putrescine, spermidine, and spermine whereas only the diamine was elevated in the uninvolved skin. Ameliorating chemotherapy reduced the polyamine levels in the lesions, and inhibitors of polyamine biosynthesis were suggested as possible topical chemotherapeutic agents. Urinary levels of polyamines were high in active disease and fell after clinical improvement. These results, as well as some with DFMO and MGBG, have alerted clinicians (McDonald, 1983). One group of dermatologists also obtained such improvements and correlative declines in urinary and epidermal polyamines with an oral retinoid (Lauharanta et al., 1981).

A relatively high level of arginase, a potential source of expanding ornithine production, was noted in this tissue. Additionally, a significant enhancement of this enzyme, as well as enzymes of polyamine biosynthesis, was detected in psoriatic lesions (Lauharanta & Käpyaho, 1983). Keratinocytes isolated from psoriatic skin appeared to be more sensitive to polyamine antimetabolites than were cells derived from normal epidermis (Käpyaho et al., 1984).

The covalent attachment of polyamines to proteins as γ-glutamyl derivatives was reexamined in cultured cells (Beninati et al., 1988). Retinoic acid increased proliferation of mouse epidermal cells and their content of free putrescine (Piacentini et al., 1988a,b). A concomitant increase in Ca^{2+} also increased transglutaminase activity, terminal differentiation, and the levels of protein-bound mono- and bis-γ-glutamyl derivatives of putrescine and spermidine. The latter compounds were particularly high in the insoluble cell envelopes found in the terminally differentiated cells. N^1,N^8-Bis(γ-glutamyl)-spermidine was then found to be present in unexpectedly high levels in the cell body envelopes of normal human skin and were fivefold higher in such envelopes from scales of psoriatic patients (Martinet et al., 1990). As shown in Table 21.1, the cross-linking polyamine markedly exceeded the content of ε-(γ-glutamyl)lysine in the psoriatic scales. These results paralleled the free spermidine levels of these tissues. Also, psoriatic skin was observed to have an increased activity (fivefold) of a membrane-associated transglutaminase (Esmann et al.,

TABLE 21.1 Bis(γ-glutamyl) spermidine and ε-(γ-glutamyl) lysine in envelopes of normal and psoriatic human skin.

Cross-linker	Cross-links/1000 amino acid residues	
	Normal skin	Psoriatic scales
ε-(γ-Glutamyl)lysine	4.16 ± 0.38	5.35 ± 1.10
Bis(γ-glutamyl)spermidine	1.66 ± 0.22	9.84 ± 1.40

Adapted from data of Martinet et al. (1990).

1989). In psoriasis the expression of the determining gene for this enzyme is found in early stages of skin differentiation whereas this enzyme appears in normal epidermis at relatively late stages (Nonomura et al., 1993). A similarly increased level of a tissue transglutaminase was reported by Bianchi et al. (1994). These results signal the potential importance of a specific inhibitor for the transglutaminases acting in psoriasis (Killackey et al., 1989).

Tissue or Type II Transglutaminases

Examination of certain tissues like skin revealed significant sequence differences among the transglutaminases. The cytosolic "tissue" enzymes were designated as type 2 enzymes; such enzymes are inhibited by GTP. The type 2 rabbit liver enzyme was purified to homogeneity and was shown to be a single polypeptide chain of about 80,000 MW with a single catalytic site per molecule (Abe et al., 1977). It is immunologically distinct from the rabbit plasma factor XIII and has some differences from the guinea pig enzyme in reactions of amine acceptance. The latter enzyme also consists of a single chain, and its sequence was deduced from cDNAs produced from poly(A+) RNA (Ikura et al., 1988). Several regions, including that surrounding the active site, were found to be strongly homologous to comparable regions in the a subunit of factor XIII. The enzyme was produced in *E. coli* infected with an expression plasmid (Ikura et al., 1990a). These workers observed a severalfold postnatal increase of the enzyme in guinea pig liver without an increase of mRNA (Ikura et al., 1990b). This group also isolated the gene and identified a promoter region. The human tissue enzyme, referred to currently as TGase 2, was implicated in many cellular processes, including apoptosis, and the gene was localized in the human chromosome 20 (Wang et al., 1994). It appears to be close to and related structurally to the gene for the soluble proenzyme TGase 3.

The type 2 enzyme is present in the bovine aortic endothelium, a tissue that turns over slowly and tends to maintain nonthrombogenic properties. The isolated enzyme of about 88,000 MW was cross-reactive with antibody to the rat liver enzyme and was mainly cytosolic, although up to 20% was found in a particulate fraction releasable by high salt concentrations (Korner et al., 1989). A human endothelial transglutaminase derived from cDNA showed >80% homology to the guinea pig liver enzyme (Gentile et al., 1991). Nakanishi et al. (1991) described a similar bovine enzyme structure derived from cDNA sequences and compared this sequence with those of various human, rat, and guinea pig transglutaminases, summarizing regions of apparent homology. The characteristic mRNA was found in many bovine tissues, and this mRNA was increased rapidly in cultures of vascular endothelial cells by treatment with 1 μM retinoic acid. The mRNA rose to particularly high levels as the cells became confluent, consistent with a relation of the enzyme to cell–cell connectedness (Nara et al., 1992).

As noted earlier, virus-transformed and chemically transformed cells are frequently low in transglutaminase. In such cells the existing enzyme is found mainly in a particulate form; the soluble tissue enzyme is thought to be downregulated in neoplastic cells. Nevertheless, a cell line derived from a benign tumor of rat adrenal produces a high level of soluble tissue transglutaminase in the presence of sodium butyrate (Byrd & Lichti, 1987). The low level of activity present in unstimulated cells is particulate and these workers postulated a reciprocal regulation of the two types of enzyme. A cell line derived from a human epidermal carcinoma grown in the presence of epidermal growth factor was also found to contain a relatively high level of the cytosolic tissue enzyme (Dadabay & Pike, 1989). This enzyme is inhibited by GTP, which can be used to elute the enzyme relatively specifically from affinity columns.

Epidermal Transglutaminase

The preparation and separation of epidermal tissues (e.g., snout, hair follicles, and layers) and the isolation of bound enzyme have often involved the use of proteases. The released active transglutaminases have frequently been degraded proteins (Martinet et al., 1988; Peterson & Wuepper, 1984). The insolubility of the original forms of these enzymes obviously creates difficulties in the clarification of the original structures, which may only be resolved by developing sequences through the analysis of the presumed mRNAs and cDNAs (Greenberg et al., 1991).

Despite the great interest in keratinization, skin structure, and development, even this limited layered tissue presented complexities limiting the interpretation of the biochemical data. However, the cultivation of keratinocytes permitted analysis of the cells and their contained transglutaminases at characteristic stages of growth. For example, an insoluble cross-linked protein envelope formed underneath the plasma membrane at the end of growth, and this formation could be activated by increasing internal Ca^{2+} (Rice et al., 1988). The process was transglutaminase generated and attached certain cytosolic proteins, involucrin, to the membrane, whose normal protein components were also cross-linked (Simon & Green, 1985). The recruitment of soluble monomeric involucrin and some other proteins (e.g., annexins, pancornulins) to the insoluble envelope has been studied. Involucrin, a protein of 585 amino acids containing several ("reiterated") glutamines has one glutamine that is highly preferred by the enzyme (Simon & Green, 1988). Green and Wang (1994) considered the reiteration of glutamine codons (CAG) in the context of codon reiteration in general and noted that glutamine is the most frequently reiterated amino acid. Proteins, of which involucrin is an example, containing highly reiterated amino acids proved to possess many special significant functions. Increased reiteration in CAG-trans-

lated proteins were detected in several genetic diseases like Huntington's (Green, 1993).

The major form of the keratinocyte transglutaminase was purified further and was shown by immunological tests to be different from one of two cytosolic transglutaminases, corresponding to the tissue type 2 transglutaminases. Keratinocytes contain both the cytosolic tissue type 2 enzyme and a distinct membrane-bound enzyme (type 1), a small fraction (5%) of which is found in the cytosol and is precipitable by antibody to the membrane-bound enzyme. Monoclonal antibody to the type 1 enzyme appeared at the interface of the stratum granulosum and the dead stratum corneum, at the layer in which involucrin had become serologically undetectable (Michel & Démarchez, 1988).

The type 1 transglutaminase of human keratinocytes has been purified from detergent extracts of the membrane fraction of a squamous carcinoma cell line. The 1000-fold purification revealed an immunologically distinctive 92 kDa protein (Thacher, 1989). Psoriatic epidermis appeared to contain the enzyme at an earlier stage of development than it appeared in normal epidermis. Indeed the mRNA for this enzyme proved to be markedly higher (3–7 times) in psoriatic skin than in normal skin (Schroeder et al., 1992).

The type 1 enzyme is downregulated by retinoic acid in contrast to the effect of this agent on type 2 tissue enzymes. The mRNA levels have been followed with cDNAs for the type 1 enzyme (Floyd & Jetten, 1989; Kim et al., 1991). cDNAs have also been prepared by several other groups who have used these materials to deduce the primary sequence of the enzymes of rats and humans and to study the regulation of gene expression in cultured cells. Phillips et al. (1992) have analyzed the structure of the human gene for the epidermal type 1 enzyme. The gene has been localized to human chromosome 14 and contains 15 exons and 14 introns. The studies have revealed a conservation of sequences within the family of transglutaminases (e.g., with factor XIII) and have led to suggestions of divergence by duplication and evolution from a common ancestor.

When cultured human epidermal cells were incubated with isotopic acetate, myristate, or palmitate, the type 1 enzyme contained labeled myristate or palmitate. Alkaline hydrolysis removed the fatty acid and solubilized the protein. Proteolysis by trypsin also solubilized the enzyme by removing a 10 kDa region containing the fatty acids (Rice et al., 1990). It appears that the acylation to a cluster of five cysteines provides the preconditions for membrane anchorage (Phillips et al., 1993). Chakravarty et al. (1990) also demonstrated a phosphorylation of serines in the membrane-bound type 1 enzyme after stimulation of cultured cells by phorbol ester.

Regulation by retinoic acid

Retinoic acid evoked a many-fold activation of the tissue (type 2) transglutaminase (Davies et al., 1984). Administration of retinoic acid invoked an early increase in the liver in putrescine specifically and transglutaminase activity but not in enzyme protein; a second increase in enzyme activity at 12 h required both transcription and protein synthesis (Piacentini et al., 1988a,b). Cells transformed by *ras* oncogenes do not produce the enzyme in response to retinoic acid (Kosa et al., 1993).

Retinoic acid was reported initially to reduce cross-linking of mouse epidermal basal cells and to inhibit the production of cross-linked envelopes (Yuspa et al., 1982). Studies with rabbit tracheal epithelial cells at terminal cell division (i.e., when they formed the cross-linked envelopes characteristic of a squamous phenotype) revealed a 20–30-fold increase of the type 1 membrane-bound enzyme (Jetten & Shirley, 1986). The expression of the activity of this enzyme and of the cross-linking of squamous cells were both inhibited by retinoic acid, further implicating the type 1 enzyme in epidermal cross-linking. These results also suggested two very different reciprocal effects of retinoic acid: an inhibition of the expression of the type 1 enzyme and an activation of expression of the type 2 enzyme. The inhibition of the production of the type 1 enzyme protein by retinoic acid was also obtained with cultured human keratinocytes (Michel et al., 1989).

The availability of specific cDNAs has permitted the estimation of mRNA in keratinocytes. Ross (1993) discussed the complexities of the conversion of the natural vitamin A and all-*trans*-retinoic acid and the delivery of the latter to a family of nuclear receptors regulating transcription at target genes.

Concluding remarks

Transglutaminases have become well-recognized participants in cellular and intracellular events. The initial studies in bloodclotting focused on protein–protein cross-linking. The discovery that the enzymes also catalyzed the formation of protein-glutaminyl polyamines that interfered with the protein–protein cross-linking was treated as an experimentally useful but nonphysiological aspect of the continuing studies. The existence of ε-glutaminyl polyamines in various cellular proteins was described, and the problem of the roles of the polyamines in modulating the formation of covalently linked proteins does exist. For example, cells containing glutaminyl-polyamines are more motile; the tying up of reactive protein-glutamines with polyamines may affect phenomena of metastasis. One may ask if these data can be extended to support the concept of functional and modulatory roles for the polyamines in the many biological effects of transglutaminase activity.

Matrix, Membranes, and Messages

Polyamine effects on intercellular reactions

Matrix-induced differentiation of cartilage and bone is accompanied by increases in cellular ODC. Parathyroid hormone, which stimulates synthesis of glycosaminoglycan, also increases ODC production. The addition of polyamines (10^{-7} M) to cultures of rabbit costal chondrocytes stimulates the synthesis of glycosaminoglycans and metachromasia in the cells, as well as the appearance of the differentiated phenotype (Takano et al., 1981). Proliferated cells active in polyamine production are contiguous to newly formed matrix particles in chondrogenesis and osteogenesis (Rath & Reddi, 1981). Cartilage calcification and mineralization is facilitated by the separation of collagen and proteoglycans and the activation of alkaline phosphatase in the extracellular space. Spermidine and spermine are found at high concentration in ossifying zones of preossifous cartilage. These polyamines do displace proteoglycan from a collagen-based affinity column and can activate alkaline phosphatase (Vittur et al., 1986).

Gamete fertilization is another system reportedly responsive to polyamine. Human seminal plasma (192 ultrafiltered samples) was recorded to contain 0–1.96 mM putrescine, 0.017–0.96 mM spermidine, and 0.13–20.8 mM spermine (Oefner et al., 1992). Although no correlation was found between polyamine concentration and sperm motility in individual human semen samples (Singer et al., 1989), a positive relationship was recorded between polyamine content of the seminal fluid and ram sperm motility (Melendrez et al., 1992).

It has been reported that the metabolism and motility of human sperm are affected by cAMP, whose levels are controlled by adenylate cyclase and cAMP phosphodiesterase. In sonicated bovine spermatozoa the former enzyme was found to be stimulated about threefold by millimolar spermine or spermidine (Casillas et al., 1980). The phosphodiesterase of sonicated human spermatozoa was reported to be inhibited about 60% by 3.8 mM spermine, a relatively high concentration. An enhancement (up to twofold) by spermine of the membrane-bound ATPase of human sperm was also found.

The formation of lactate from the seminal fructose, that is, fructolysis, is an activity of rat epididymal spermatozoa. This process is markedly stimulated (5–30-fold) in the presence of spermine (5 mM) and Ca^{2+} (1 mM). This effect does not occur in disrupted sperm and has been attributed to an effect on fructose uptake. Exogenous ATP further stimulates fructolysis and a conversion of lactate to CO_2 (Pulkkinen et al., 1978).

The polyamines have been implicated in the fusion of sperm acrosomal membranes and egg plasma membranes during fertilization, a reaction essential for release of the acrosomal contents and successful fertilization. The mammalian sperm contains an acrosomal endoproteinase as a proenzyme, proacrosin (Zaneveld & DeJonge, 1991). The conversion of the proenzyme to the active acrosin, possibly autocatalytic, is markedly inhibited by spermine (>10 mM) and less strongly by other polyamines (Parrish et al., 1979). Rubinstein and Breitbart (1991) found that [^{14}C]-spermine bound rapidly to ram sperm. The spermine was released in an ionic medium, and this was hastened by the polyanion, heparin.

The reduction of seminal viscosity arising from hyaluronic acid can be effected by hyaluronidase, an enzyme markedly activated by 0.1–1 mM spermine (Myaki et al., 1959). The reduction of viscosity in bull semen has been reported to be useful in accelerating the fertilizing efficacy of artificial insemination.

Human seminal fluid contains a specific prostatic acid phosphatase that is stabilized by 0.1–1 mM spermine (Jeffree, 1956). The enzyme is thought to be responsible, at least in part, for the cleavage of various organic phosphates like phosphorylcholine. In particular, this substance is suspected to be the major source of the phosphate subsequently precipitating as spermine phosphate (Lundquist, 1946). The presence of high levels of the specific enzyme protein in human serum may be diagnostic of prostatic cancer and possibly of metastasis to bone.

Charge, Lipids, and Amine Effects

Studies of the effects of the polyamines in reacting with acidic surface components have revealed binding and aggregating reactions of the amines with carboxyl- and sulfate-containing polysaccharides. It has been calculated that a carboxylate may be a stronger proton acceptor from a protonated amine, such as methylamine, than is a phosphate (Remko & Scheiner, 1991). Numerous studies have been carried out on binding of polyamines to liposomes or bilayer membranes of the phospholipids, such as phosphatidates, phosphatidyl ethanolamine, phosphatidyl serine, phosphatidyl choline, phosphatidyl inositol, and phosphatidyl glycerol. The heterogeneity of individual phospholipid preparations (i.e., their sources, fatty acid composition, degree of peroxidation, etc.) and differences in the methods of membrane formation result in differences in vesicle size and composition in various laboratory preparations. Nevertheless, it is generally agreed that spermidine and spermine aggregate liposomes comprising acidic phospholipids (i.e., of phosphatidates, phosphatidyl ethanolamine, and phosphatidyl serine), but will not aggregate the zwitterionic phosphatidyl choline (Schuber, 1989). Vesicles comprising both phosphatidyl choline and phosphatidyl inositol (4 : 1) do aggregate with both higher polyamines; a diamine is almost ineffective in increasing turbidity. The binding activities of the polyamines are in the order of the charge of the cations: spermine > spermidine > putrescine. However, the nature of the acidic phospholipid and its density in the liposomal vesicles does affect the interaction. Also, the 50% aggregation end point by spermine appears to be achieved cooperatively and depends on the concentration of spermine and the concentration of the acidic phospholipid on the external surface of the vesicle (Tadolini et al., 1985a). Ohki and Duax (1986) show that the degree of aggregation (Fig. 21.4) parallels the reduction of electrophoretic mobilities of the vesicles.

In studies of the binding of spermine with low concentrations (<50%) of phosphatidyl serine in the lipo-

somes after admixture with an unreactive phospholipid, complexes of spermine to phosphatidyl serine can attain a ratio of 1 : 1. This was interpreted to indicate a direct complexation; association of the polyamine at the liposome surface occurred with the acidic phospholipid. When the concentration of phosphatidyl serine exceeded 50% in the liposome membrane, complexes of spermine and phosphatidyl serine contained less than one molecule of polyamine per acidic phospholipid, implying a parallel configuration, with the extended cationic groups bridging several molecules of membrane phospholipid (Chung et al., 1985).

The polyamines evoke fusion of liposomal vesicles. The fusion (or separation) of membranes relates to both cell growth and cell division, among other aspects of membrane function. Model systems for the study of fusion have involved the preparation of two batches of liposomes, each of which contains relatively nonfluorescent substances such as dipicolinic acid and terbium III, which become highly fluorescent on mixing of their aqueous contents (Düzgünes et al., 1987). The loss of fluorescence can be a measure of the leakage of the chelation complex from the fused liposomes, or the leakage of contents from a preformed fluorescent liposome. Spermidine and spermine will aggregate liposomes of an acidic phospholipid like phosphatidyl serine but do not induce fusion; indeed such aggregates will dissociate on dilution (Schuber et al., 1983).

The apposition of membranes is required before fusion can occur. Liposomes composed of phosphatidate will fuse in the presence of Ca^{2+}, and spermine will both markedly reduce the threshold requirement of Ca^{2+} for fusion and increase the rate of this activity. Ca^{2+} is thought to effect a phase change from fluidity within the membrane to a crystalline alignment at the membrane boundaries. Liposomes containing phosphatidyl ethanolamine and acidic phospholipids, the former in high proportion, can be induced to fuse by spermine alone. Phase changes in contiguous microdomains of phosphatidyl ethanolamine have been suggested as the mechanism facilitating fusion in this instance.

The polyamines thus appear to participate in the movement of lipids within the membrane. Indeed it has been hypothesized that the ionic interactions of amines generally with the acidic phospholipids (e.g., of tissue cell membranes) may be responsible for the toxicity of those amines. Also, the topology of membrane proteins is thought to be affected by positively charged residues flanking or near the transmembrane boundaries. It has been suggested that positive charges reorder acidic lipid head groups and thereby permit the insertion of a protein into and through the membrane (Dalbey, 1990).

The reorganization of lipopolysaccharide-rich polypeptide micelles by spermine or some divalent cations, such as Ca^{2+}, has been demonstrated by Shilo (1971). His classical study of a toxin derived from the phytoflagellate *Prymnesium parvum* demonstrated the transformation of a cation-depleted low molecular nontoxic lipopolysaccharide by spermine or other multivalent

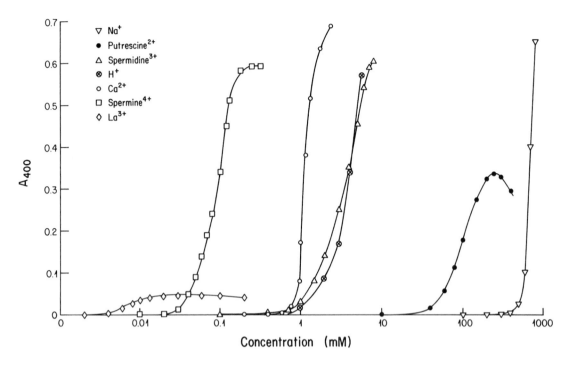

FIG. 21.4 Turbidity of small unilamellar phosphatidyl vesicles suspended in 0.1 M NaCl buffer solution as a function of concentration of metal cations or polyamines; adapted from Ohki and Duax (1986). ($\Diamond$) La^{3+}; ($\bigcirc$) Ca^{2+}; ($\square$) spermine^{4+}; ($\triangle$) spermidine^{3+}; ($\bullet$) putrescine^{2+}; ($\otimes$) H^+; ($\triangledown$) Na^+.

cations to a high molecular fish toxin. The toxin attacks gill surfaces of carp and the appearance of the algae and toxin in carp pools can be detected by the addition of spermine and bioassay of the lethality to fish. NH_3 may then be added to the pond, and this substance concentrated in the algae kills the algae selectively and minimizes the level of toxin.

The peroxidation of cellular and membrane lipid produces several aldehydic breakdown products. One of these, 4-hydroxy-nonenal, is toxic to the synthesis of DNA in cells in culture, as well as to the activity of enzymes, including ODC. Several groups have shown that peroxidation of membrane lipids results in the destruction of membrane structure and activities. One model for the study of a role of the polyamines in the inhibition of lipid peroxidation has been the effect of spermine on the production of malondialdehyde in a microsomal fraction exposed to the superoxide anion radical. Spermine did not inhibit the enzymatic generation of superoxide but did inhibit malondialdehyde production. In another model system, liposome vesicles containing acidic phospholipids were exposed to the Fenton reaction, produced by Fe^{2+} in the presence of H_2O_2. Spermine did not affect the production of malondialdehyde from vesicles of phosphatidyl choline. However, the oxidation of vesicles containing phosphatidyl choline and phosphatidyl inositol (1 : 1) were completely inhibited by spermine. A correlation was obtained between binding of the vesicle by spermine and protection against peroxidation. Tadolini et al. (1985b) asked if the

inhibition by spermine relates to a competition of binding with the iron of the Fenton reagent.

Reactions with Erythrocyte Membranes

Putrescine-binding sites on mouse or chick cell membranes have been detected by the use of antiputrescine antibody in observations of complement-aided lysis or the binding of microscopically visible antibody-coated latex spheres (Quash et al., 1978). The presence of external binding sites for putrescine in erythrocyte membranes has also been demonstrated by transglutaminase-catalyzed reactions.

A characterization of electrophoretic mobilities of the cells in defined media at various pH values (e.g., pH 7.4 in a glycine buffer) was used to demonstrate the binding of polyamines to cell surfaces (Chun et al., 1976). Normal human erythrocytes were found to have a greater mobility and surface charge density than sickled cells. The mobilities of the former were reduced significantly by the addition of spermine whereas the mobilities of sickle cells, already high in their content of spermine were affected only slightly.

Studies to determine the nature of the binding sites on the membranes showed that the neuraminidase removal of sialic acid from human erythrocytes actually increases the detectable binding sites for [^{14}C]-spermidine. Trypsin reduced the binding sites on untreated cells. A specific class of proteins containing bound spermidine was detected in the cell ghosts (Khan et al., 1989).

The binding of spermidine or spermine to erythrocytes or their membranes reduces the lateral diffusion of proteins within the membranes (Schindler et al., 1980) and increases the membrane's resistance to deformation, for example, by shear (Ballas et al., 1983). The effects of the polyamines apparently extend to inner skeletal proteins such as spectrin and other proteins like actin, which are no longer easily extracted from pretreated ghosts (Wyse & Butterfield, 1988). Spermine in particular is thought to increase protein cross-bridging to various components both within the membrane and to intracellular components, such as globin (Natta et al., 1980). Some effects of relatively high concentrations of the polyamines have been reported on some enzymes (e.g., adenylate cyclase) of erythrocyte membranes.

Receptors and Cascades

The association of an active hormonal agent with a target cell at a specific surface receptor can initiate a series of metabolic reactions that recognize a ligand at the cell surface, span the membrane, and exert its effect in the cytoplasm, often through the activation of a distinctive protein kinase. Eucaryotic cells frequently have more elaborate sequences of sensory response than do many procaryotes that possess "two-component pathways." In the latter systems the interaction of the stimulated receptor with a kinase produces an autophosphorylated kinase that then transfers a histidinyl phosphate to the γ-carboxyl of an aspartate on a regulatory protein. The latter binds and activates the next step in the sequence. Similar two-component pathways have recently been detected in some plant and yeast responses.

Cascades of kinases culminating in the increase of ODC have been described as a result of signaling by insulin, TPA, and forskolin (Shibley & Pennington, 1995). In the case of insulin, an early concern of mammalian biochemistry, the binding of the hormone to the insulin receptor activates its associated specific protein tyrosine kinase (PTK), which autophosphorylates (White & Khan, 1994). Subsequent events include changes in gene transcription and transport, as well as the activation of another protein kinase that phosphorylates serine and threonine hydroxyl groups in target proteins. In searches for functions of the polyamines it was observed that spermine and spermidine could mimic insulin action in effects of glucose and lipid metabolism in fat cells (Amatruda & Lockwood, 1974). Although it appears that the polyamines are bound to cell membranes at sites different from the sites of the insulin receptor, it is evident that the surface reactions of the polyamines can initiate significant cell responses.

In various mouse cells insulin increases the activity of casein kinase II (CKII), an enzyme stimulated by spermine and inhibited by heparin (Sommercorn et al., 1987). Nevertheless, the insulin receptor tyrosine kinase is not stimulated by polyamines, at least in the phosphorylation of calmodulin and other peptides, although this kinase is activated by some basic proteins (Sacks et al., 1989).

The protein kinases were discovered initially in studies of the interconvertibility of glycogen phosphorylase a, which is a phosphorylated tetramer, and phosphorylase b, which is a dephosphorylated dimer (Fischer & Krebs, 1989). A phosphorylase kinase was shown to be a cAMP-dependent protein kinase, which was also a kinase for glycogen synthase. The effects of phosphorylation are often canceled by the action of protein phosphatases that may be serine/threonine specific or tyrosine specific, and are themselves highly regulated (Cohen & Cohen, 1989). Details of the multiplicity of these enzymes and their responses to cations became available in the 1980s.

One liver phosphorylase kinase is regulated by Mg^{2+} and inhibited by spermine. Increased concentrations of Mg^{2+} reverse the inhibitory effect of spermine (Hashimoto et al., 1984). A particular glycogen synthase kinase 3 was inhibited about 50% by 0.7 mM spermine and even more actively by polylysine, and it was suggested that polycations that activate the phosphatase reactions regulate the phosphorylation state of glycogen synthase at certain serine sites (Hegazy et al., 1989). Several groups have also described glycogen synthase kinases designated as casein kinase I (CKI), which are cAMP and Ca^{2+} independent and are stimulated by spermine and spermidine. A third casein kinase, CK3, separable from CKI and CKII and stimulable by spermine, can phosphorylate and markedly activate phosphorylase kinase (Singh, 1989).

A classification has begun to achieve a modicum of order in describing these enzymes. Reviews by Taylor (1987) and Hanks et al. (1988) have stressed the conserved catalytic core within each of the members of this large family. This insight has culminated in a determination of the 3-dimensional structure of the catalytic subunit of the cAMP-dependent protein kinase, which phosphorylates phosphorylase kinase at serines or threonines in a particular consensus sequence (Knighton et al., 1991). It is known that numerous proteins are phosphorylated at multiple sites and that specific sequences of phosphorylation exist (i.e., hierarchical phosphorylation) for many of these (Roach, 1991).

Nevertheless, the breadth of the remaining problems can scarcely be described. For example, in a recent study of plant protein kinases found in extracts of the distantly related tobacco, cucumber, and *Arabidopsis*, the polyamines were found to increase the phosphorylation of many plasma membrane proteins significantly. Phosphorylation occurred at protein serines and threonines but was not found on tyrosine (Ye et al., 1994). Neither the nature of the enzymes, their specificities, nor the roles of their substrates have yet been addressed.

Inositol derivatives and second messengers

A discussion of the promotion of epidermal tumors in mice by phorbol esters (TPA; Chap. 16) included the role of protein kinase C and its activation in the cascade of reactions leading to a stimulated synthesis of ODC. Promotion by TPA evoked a rapid incorporation of ^{32}P into skin phospholipids. This section will discuss some details of the reactions culminating in the activation of protein kinase C and the various effects of the polyamines on these reactions.

In 1953 Hokins detected the rapid incorporation of ^{32}P into certain phospholipids, of which phosphatidyl inositols were major components, under conditions of agonist-stimulated transport into cells of various glands. The pursuit of a metabolic role of the inositol derivatives culminated in the discovery of details of their metabolism and the origin of a particular derivative that served to provide a major component, diacylglycerol (DAG), for the activation of protein kinase C.

The various components of the major precursor of DAG, phosphatidylinositol 4,5-bisphosphate presented in Figure 21.5, are produced in a maze of cellular reactions. These include the synthesis of saturated and unsaturated fatty acids, the synthesis and degradation of a triacylglycerol, phosphatidate, etc. The polyamines, including spermine in particular, have been reported to be involved in many of these reactions, such as fatty acid metabolism (Villanueva, 1983), and stimulation of sn-glycerol-3-phosphate acyl transferase and DAG acyl transferase. The effects of the polyamines in cells and crude homogenates have generally proved to be complex. For example, a brain lipase has been found to be activated by spermine in an extract. Triacylglycerol lipases containing sequences of basic amino acids are known to be inhibited by heparin and these inhibitions are in fact reversed by spermine and cationic polypeptides like polylysine. Effects such as the stimulation of an enzyme by spermine in an extract containing a heparinlike inhibitor have been observed in studies of the protein kinases and indeed have been used in early studies to distinguish the casein kinases. Such results appear to relate to the reaction of spermine with the true inhibitor heparin, but this conclusion has not been clearly proven. In a second instance, the arachidonate liberated by a microsomal lipase may be converted to a prostaglandin in microsomal preparations and this synthesis is inhibited by spermine. The inhibition does not occur with a purified prostaglandin synthetase but does occur when phosphatidyl choline is added to the system containing spermine (Igarashi et al., 1981). Obviously the effects of polyamines in reaction systems replete with a variety of complex ions present serious problems of interpretation. Most recently the many levels of phosphoinositide turnover in the rat brain are reported as being regulated by spermidine and spermine (Periyasamy et al., 1994). These authors have concluded that the polyamines play a role in the turnover cascade in the brain.

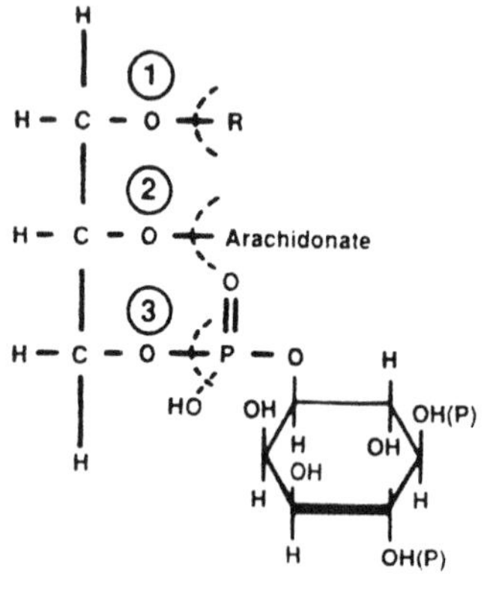

Spermidine and spermine do bind to acidic phospholipids, such as phosphatidyl inositol, and may affect various reactions, such as the fusion of the membrane containing such lipids. Suspensions of various inosisides (i.e., phosphatidylinositol (PI), phosphatidylinositol 4-phosphate (PI-P), and phosphatidylinositol 4,5-biphosphate (PI-P$_2$)) are readily precipitated by spermine and spermidine. When these inosisides were built into otherwise unreactive lysosomes at a concentration of 2%, these substances bound spermine in the order PI-P$_2$ > PI-P $\gg$ PI (Tadolini & Varani, 1986). PI-P$_2$ is thought to be situated largely in the inner monolayer of the cellular membrane. The binding to PI-P$_2$ in vesicles was analyzed in detail (Toner et al., 1988) and revealed a far lower ability to bind Ca^{2+} at pH 7 than Mg^{2+} or cationic spermine. It was calculated that even if the free concentration of spermine is on the order of 10^{-2} mM, 1% of the total cellular spermine, a significant fraction of PI-P$_2$ would have bound spermine.

Mg^{2+}-activated kinases of the phosphatidyl inositols were found to catalyze the reactions,

$$PI \xrightarrow[ATP]{①} PI-P \xrightarrow[ATP]{②} PI-P_2$$

The first of these is associated with plasma membranes; the second appears more cytosolic. More recently the phosphorylation of PI in plasma membranes of A431 cells by ATP to form PI-P was shown to be greatly stimulated (eightfold) by spermidine or spermine (Vogel & Hoppe, 1986). Similarly, a specific PI-P kinase was partially purified from rat brain cytosol and was found to be enhanced several-fold by spermine in the formation of PI-P$_2$ (Lundberg et al., 1986). This product can act as an inhibitor of this kinase. A stimulation (threefold) by spermine (0.2 mM) was also detected in rat liver plasma membranes (Lundberg et al., 1987). In their turn, Smith and Snyderman (1988) reported that spermine effects in the phosphorylations in human leukocyte membranes were determined largely by Mg^{2+} concentration, being stimulatory at low Mg^{2+}. These

FIG. 21.5 Structure of phosphatidylinositol; adapted from Majerus et al. (1986). Polyphosphoinositides contain additional phosphate monoesters (P). Breakdown is initiated by cleavage at bond 3; diacylglycerol may be cleaved by lipases at bond 1 and subsequently at bond 2.

workers reported spermine inhibition of phosphomono-esterase in the hydrolysis of PI-P$_2$. PI-P$_2$ particularly was shown to be a source of DAG released by phospholipase C, and the remainder of the molecule, inositol-1,4,5-triphosphate, was also a messenger that can mobilize Ca^{2+} stores.

The induced release of Ca^{2+} from microsomes by the triphosphate is inhibited by spermine reacting with the IP$_3$ receptor (Sayers & Michelangi, 1993). Additional activations and inhibitions by spermine in these sequences have been described by Singh et al. (1995).

Berridge and Irvine (1989) surveyed the metabolism of the inositol phosphates and their roles in cell signaling and described the large number of these compounds and paths to their syntheses and interconversions. Potter and Lampe (1995) discussed the roles of these compounds in the regulation of the oscillations of Ca^{2+} concentration. A delipidated plasma–membrane Ca^{2+}-transport ATPase can be reactivated by addition of mixtures containing 20% acidic lipids and 80% phosphatidyl choline; of these, PI-P$_2$ is most active in reactivating the enzyme and this effect is inhibited by spermine or neomycin (Missiaen et al., 1989). An agonist acting at a cell membrane activates a phosphatidylinositol-phosphodiesterase to liberate PI-P$_2$ + DAG (Majerus et al., 1986).

Several phospholipases can participate in the production of DAG for the activation of protein kinase C (PKC). These include phospholipase C, phospholipase D, phospholipase A$_2$, and phosphatidate phosphohydrolase (Nishizuka, 1992). Phospholipase C is specific for the phosphatidyl inositols and cleaves the three major forms (PI, PI-P, and PI-P$_2$) to produce DAG and several different inositol phosphates. Of these inositol (1,4,5) triphosphate is described as the "messenger" capable of mobilizing Ca^{2+}. Phosphatidyl inositol can be hydrolyzed by several mammalian phosphodiesterases. The hydrolysis by a rat brain Ca^{2+}-dependent soluble enzyme was subsequently shown to be inhibited by spermine and spermidine (Eichberg et al., 1981).

At least five serologically and genetically distinct phospholipase C types have been isolated and characterized (Rhee et al., 1989), of which several are both membrane associated and cytosolic. The appearance of inositol triphosphate in membranes after treatment with an agonist, presumably via the activation of phospholipase C, is enhanced by GTP and its analogues, suggesting the mediation of so-called G proteins. This specific stimulation was inhibited by spermine (0.25 mM) and

spermidine (2 mM) (Wojcikiewicz & Fain, 1988). On the other hand, the GTPase activity of purified GTP-binding proteins in phospholipid-containing vesicles is stimulated by spermine and spermidine (<0.1 mM). The latter effect simulates the effects of these amines on the release of histamine by agonist-treated rat mast cells (Bueb et al., 1992). These apparently contradictory observations led to the proposal that intracellular polyamines modulate signals mediated by G-protein coupled receptors. The phospholipase C of rat liver is reported to be markedly stimulated at 0.15 mM spermine and inhibited by this amine above 0.5 mM (Haber et al., 1991). A bovine heart enzyme was maximally stimulated at 0.1 mM spermine and inhibited at higher concentrations (McDonald & Mamrack, 1995). A complex fungal product, termed hispidospermidin, containing a trimethylspermidine side chain, was recently described as an inhibitor of the enzyme (Ohtsuka et al., 1994).

Phospholipase D is thought to convert phosphatidyl choline to phosphatidate, which is then cleaved by a phosphatidate monoesterase to DAG. Phospholipase D has been found in membrane fractions of a variety of mammalian tissues and may be activated by phorbol esters (Nishizuka, 1992). The enzyme will also exchange bases from phosphatidyl choline with serine to generate phosphatidyl serine.

Phospholipase A$_2$ liberates free fatty acids, like arachidonic acid, to generate lysophospholipids. It has been reported that typical unsaturated fatty acids liberated in this way enhance the activation of PKC. The structures of A$_2$ phospholipases are known mainly from the study of the enzymes present in snake or bee venom, which thereby lyse cells via the action of the product, lysophospholipids. The enzyme is activated several-fold by spermine and spermidine at relatively low concentrations in the cleavage of a phospholipid (e.g., a chemically defined phosphatidyl glycerol) and is inhibited at higher concentrations of the polyamines. Optimal activation by spermine occurs at a molar ratio of 1 : 1 spermine to phosphatidyl glycerol (Thuren et al., 1986).

DAG can also be generated by a Mg^{2+}-dependent phosphatidate phosphohydrolase. The enzyme is potentiated by polyamines (e.g., 0.5 mM spermine) and inhibited by still higher concentrations of spermine in the presence of Mg^{2+}. This activity can increase fourfold in a cytosolic aggregate produced by spermine (Jamdar et al., 1987).

Polyamines in activation and inhibition of PKC

The tumor-promoting activity of the phorbol ester TPA, which binds to the surface receptor containing PKC and evokes an increased synthesis of ODC, was discussed in Chapter 16. DAG markedly increases the affinity of PKC for Ca^{2+} whereas triacylglycerol, monoacyl glycerol, and free fatty acid are inactive. PKC consists of a single polypeptide of 77 kDa and comprises two domains. The hydrophobic domain enables the enzyme to bind to membranes. The hydrophilic catalytic domain may be removed by a protease to produce a fragment active in the absence of DAG, phosphatidyl serine, and Ca^{2+}. The enzyme phosphorylates serine- and threonine-hydroxyls exclusively. A nuclear histone is frequently used as a test substrate.

PKC has been characterized in plants as well as animals, and the plant enzyme cross-reacts with antisera to the animal PKC (Elliott et al., 1988). Several enzyme activities, distinguishable by Ca^{2+} requirements but similar serologically, are found in a rat liver homogenate, and a significant percentage of the activity (ca. 20%) is present in nuclei (Masmondi et al., 1989). The nuclear enzyme will phosphorylate various nuclear proteins, as well as histones; these include topoisomerase II and poly ADP-ribose polymerase, activating the former and inhibiting the latter (Bauer et al., 1992). In yeast nuclei the former is also activated by a CKII (Alghisi et al., 1994).

The various PKCs are currently described as a family of nine members, a homologous group similar in mechanism of activation, size, and structure (Azzi et al., 1992; Kikkawa et al., 1989). These reviews describe multiple genes and several isoenzymes and compare the various sections of these proteins, including the autophosphorylation sites. A conserved phorbol-binding region in the regulatory domain is cysteine- and histidine-rich and one of these enzymes has also been found to bind Zn^{2+} (Hubbard et al., 1991). The stimulation of secretion of insulin in pancreatic islets by TPA and the activation of PKC are inhibited by spermine (1.5 mM) and spermidine (5 mM) (Thams et al., 1986). Culture of islet cells in glucose-containing media stimulates the formation of insulin in mRNA but this is attenuated by an inhibition of polyamine biosynthesis (Welsh, 1990). Inhibitors such as DFMO lead to an accumulation of insulin-secretion granules (Sjöholm et al., 1991).

Qi et al. (1983) demonstrated that a phospholipid-sensitive Ca^{2+}-dependent protein kinase (presumably PKC) was markedly inhibited by spermine noncompetitively (50% at 0.84 mM). Mezzetti et al. (1988) showed that the separated catalytic domain of a brain PKC, incapable of binding TPA, was inhibited by spermine in the phosphorylation of histone H1; this particular inhibitory effect then is not merely an effect on phospholipids or other activating components. Spermidine and diamines did not inhibit this system.

Separated PKC will reassociate with erythrocyte membrane vesicles and in this form bind TPA with 1 : 1 stoichiometry (Moruzzi et al., 1987). This rapid reassociation of the enzyme with the membrane is markedly inhibited by 0.1 mM concentrations of spermine > spermidine ≫ putrescine. Similar results were obtained when the membranes were replaced by phospholipid vesicles containing different percentages of phosphatidyl serine. The level of inhibition by spermine was greatest at the highest concentration (50%) of that phospholipid.

The binding and insertion of PKC into unilamellar phospholipid vesicles has been found to depend on Ca^{2+}, whose presence permits an irreversible insertion of the enzyme. In the absence of Ca^{2+}, the binding and reversible association of PKC, estimated by the binding of [^{3}H]-TPA, was markedly decreased by the presence of 50 μM spermine. In the presence of Ca^{2+}, spermine was not inhibitory in the binding of TPA. The joint actions of the polyamine and Ca^{2+} affect the distribution of the enzyme in its states of association with the phospholipid-containing membrane (Moruzzi et al., 1993, 1995). The time-dependent inactivation of PKC by phosphatidyl serine in the absence of Ca^{2+} can be prevented by spermine (Monti et al., 1994).

It appears that the polyamines, spermine particularly, may act at many sites in affecting the activity of PKC. Spermine can bind to acidic phospholipids and alter their properties in membranes, and in so doing regulate the insertion of PKC itself into the membrane. This amine can alter reactions leading to the synthesis of PI-P$_2$ and to the formation of DAG. Spermine may also inhibit the catalytic action of the enzyme, independent of its activation by lipid cofactors. These possible effects have not been explored in spermine-depleted cells.

The role of sphingosine as an inhibitor of PKC was discussed in Chapter 16. The cation, sphingosine, may be present at inhibitory concentrations in membranes of normal cells as a result of the degradation of endogenous substrates and may neutralize activating acidic phospholipids (Slife et al., 1989). That sphingolipid metabolism and sphingosine in particular may modulate the activity of PKC in signal transduction has been suggested (Hannun & Bell, 1989; Merrill & Jones, 1990). The mechanisms by which a hydrophobic cation, such as sphingosine, can fulfill this role may include the inactivation or limitation of a structural framework and activators (e.g., phosphatidyl serine) in the membrane or on the regulatory domain, the removal of substrates (e.g., ATP or essential acidic groups on the protein), or the inactivation of PKC itself. The inhibition of PKC by sphingosine has not been compared in any of these respects with the inhibitions observed with spermine. Both sphingosine and spermine have been reported to activate CKII.

CKI and CKII

Families of protein kinases exist that are independent of cyclic nucleotides, phospholipids, and Ca^{2+}. Kinases capable of phosphorylating a dephosphorylated casein on serine- or threonine-hydroxyls are present in eucaryotic cytoplasm and nuclei containing numerous phosphorylatable protein and enzymic substrates of the casein kinases. Typically two fractions of such soluble kinase activity separable on various affinity columns are found in various extracts and one of these, CKI, was found to not be inhibited by heparin whereas CKII was inhibited by this polyanion (Mäenpää, 1977). One study of the nuclei of mouse neuroblastoma cells exemplified this distinction (Verma & Chen, 1986). The separated CKI proved to have a subunit of about 37,000 MW, and to be serologically reactive with antiserum to calf thymus CKI. The enzyme autophosphorylated and, when added to heat-treated nuclei, predominantly phosphorylated two low molecular nuclear proteins, which could not be phosphorylated by any other known protein kinase. These reactions were completely inhibited by 1 mM spermine.

An antiserum to CKI isolated from yeast reacted with the CKI of mouse ascites tumor cells and was used to detect a relatively high concentration of the enzyme in the ascites cell nucleoli (Grankowski & Issinger, 1990). CKI possesses functions in membranes and structures of cell mitosis, and in human erythrocytes is known to phosphorylate spectrin and the erythrocyte anion transporter. The assembly of CKI into membranes is regulated by PI-P$_2$, which is also inhibitory. A protein cross-reactive serologically with the erythrocytic CKI has been detected in the endoplasmic reticulum of mammalian cells, and in mitotic cells the protein is also clearly present in the spindle and centrosome (Brockman et al., 1992). This protein appears similar to that determined by a yeast gene that regulates various mitotic functions, for example, chromosomal segregation in that organism. CKII is also known to localize to the apparatus of nuclear division. In a recent study of yeast mutants, two homologous genes for CKI, which coded for products of higher molecular weight (62 kDa), were demonstrated. Loss of either gene did not affect growth. However, a loss of cell viability accompanied deletion of both genes (Robinson et al., 1992).

A cytosolic CKI from broccoli is a monomer of 36,037 Da and is cross-reactive with antisera to the mammalian CKI (Klimczak & Cashmore, 1993). The CKI utilized only ATP, unlike CKII which can also use GTP to phosphorylate. A purified CKII from the cytosol of the etiolated pea was stimulated by spermine and spermidine and inhibited by low concentrations of heparin. The enzyme was itself phosphorylated and lost activity in treatment with a phosphatase (Zhang et al., 1993).

CKII, reactive with both ATP and GTP, has often been found in an inhibited state in mammalian tissue extracts. The endogenous inhibitor has frequently proved to be a heparinlike glycosaminoglycan or mucopolysaccharide that is dissociable from the enzyme by spermine (Feige et al., 1980). A 1400-fold purified CKII has been isolated from bovine lung, an organ containing an endogenous inhibitor. The enzyme has been found to both autophosphorylate and phosphorylate casein, histones, protamines, phosphorylase b, phosphorylase kinase, and glycogen synthase (Cochet et al., 1983). The apparently homogenous protein required Mg^{2+}, was markedly activated by 2 mM spermine, and was inhibited by 10^{-5} M heparin. The purified tetrameric CKII contains both catalytic α_2 subunits and regulatory β_2 subunits, of which the regulatory subunits bind spermine preferentially, evoking a threefold stimulation of the enzyme (Filhol et al., 1991a; Leroy et al., 1994). A photoactivable derivative of spermine was used to identify the polyamine binding sites of the enzyme as both α and β subunits and to indicate the possible mechanism of polyamine activation of CKII (Leroy et al., 1995).

Extracts of mouse mammary gland have been found to contain both CKI and CKII and several endogenous phosphorylated substrates, characterized as caseins. The CKII of these extracts was inhibited by heparin and stimulated by polyamine and varied according to the developmental stage of the gland, increasing during pregnancy and achieving a high reactivity during lactation (Leiderman et al., 1985).

A study of a reticulocyte CKII as functions of concentration of polyamines, Mg^{2+}, and ionic strength, was occasioned by the observation that the red cell component 2,3-diphosphoglycerate was an inhibitor of the enzyme. The presence of polyamine not only reversed the effect of this inhibitor, but also diminished the dependence on Mg^{2+} (Hathaway & Traugh, 1984). The concentrations at which spermidine and spermine produced these effects were similar to those existing in red cells.

Both CKI and CKII have been isolated from the membrane and cytosolic fractions of the human erythrocyte (Wei & Tao, 1993). Essentially all of the major membrane proteins, with few exceptions, are substrates for protein kinases. The proteins and sites phosphorylated by CKI and CKII overlap in many respects, and phosphorylation tends to reduce protein interactions in the red cell cytoskeleton. The isolated CKII, which is larger than CKI, is markedly inhibited by heparin (90% at 1 µg/mL) and stimulated twofold by 1 mM spermine.

A complex CKII that preferentially phosphorylates acidic proteins, such as casein or the extensively phosphorylated phosvitin, has been isolated from the cytosol of chick duodenal mucosa (Mezzetti et al., 1985). The activity is also inhibited by endogenous heparinlike mucopolysaccharides, and this inhibition is reversed by spermine (Mezzetti et al., 1986). However, the phosphorylation of phosvitin by the purified enzyme was inhibited by spermine. This appears to be an effect of the combination of spermine with the phosphates of phosvitin because the phosphorylation of dephosphorylated phosvitin is stimulated by spermine (Ahmed et al., 1985). The latter group, who have studied stimulations of nuclear protein kinases by the polycation Co(NH$_3$)$^{3+}$ as well as by spermine, have concluded that in several instances the polyamines are acting nonspecifically in affecting charge. A suggestion is that the cations are primarily affecting the conformation of the substrate rather than the catalytic activity. A selective change of the circular dichroism of phosvitin has been detected following addition of spermine (Hara & Endo, 1982).

A rat heart ODC was shown to be phosphorylated on a single serine (0.8 mol/mol ODC) by a cytosolic rat liver CKII, and this reaction was greatly stimulated by 2 mM spermine or basic polypeptides (Meggio et al., 1984). The regulatory significance of the phosphorylation is unclear in this case. As discussed in Chapter 12, ODC may be phosphorylated at several sites and may be partially stabilized as a result.

The cationic aminoglycoside antibiotics are stimulatory, although less so than spermine, in the phosphorylation of casein (Ahmed et al., 1988). Basic polypeptides, such as polylysine, are often even more stimulatory than spermine (Pinna, 1990). A rabbit reticulocyte CKII may be autophosphorylated somewhat after isolation; addition to the β subunit is accentuated over that in the original isolation (Palen & Traugh, 1991). This in vitro reaction is inhibited by polylysine, which stimulates autophosphorylation of the α subunit nonstimulable by spermine. The activity of the catalytic α subunit, which can phosphorylate calmodulin alone, is downregulated toward certain proteins (e.g., calmodulin, ODC) by acidic stretches of the subunit. This reaction by the entire CKII can be activated by polybasic peptides that then appear to react primarily with the enzyme rather than with the substrate (Meggio et al., 1992). The production of relatively large amounts of subunits of CKII from human cDNA clones (Grankowski et al., 1991; Lozeman et al., 1990) suggests that direct tests of these interactions may be possible.

A nuclear CKII, frequently designated as NII, appears to concentrate in the nucleus during active proliferation. The nuclei of bovine adrenocortical cells will rapidly bind the enzyme in the presence of ATP, with spermine as a stimulant in an in vitro system (Filhol et al., 1990). Cells stimulated with ACTH and pretreated with DFMO do not accumulate the enzyme in their nuclei as do the stimulated cells cultivated in the absence of DFMO (Filhol et al., 1991b). The polyamine-stimulated CKII of estradiol-dependent rat tumors and kidneys was 3–5 times as active in the nuclei as in the cytoplasm. A single injection of the hormone evoked a significant increase of the enzyme (almost doubling in kidney nuclei) after only 30 min (Levashova & Morozova, 1988).

The dephosphorylation and activation of glycogen synthase in skeletal muscle initiates the stimulation of glycogen synthesis by insulin. Two types of protein phosphatases (1 and 2A) exist, and several oligomeric forms of each have been found to contain both catalytic (35 kDa) and regulatory subunits. Spermine and spermidine are good activators of specific dephosphorylation by these oligomeric phosphatases (Tung et al., 1985). Spermine (2 mM) was shown to maximally activate the dephosphorylation by protein phosphatase 2A of glycogen synthase, phenylalanine hydroxylase, and the several subunits of phosphorylase kinase. Although Mg^{2+} (5–10 mM) was also partially stimulatory for 2A, it did not substitute for spermine in stimulating protein phosphatase. Basic polypeptides also activate protein phosphatases and it has been supposed then that these cationic substances participate in vivo in the regulation of certain phosphorylating protein kinases and dephosphorylating protein phosphatases. In some reaction systems the polyamines are thought to react with the substrate rather than with the enzyme. On the other hand spermine inhibited dephosphorylation by the free catalytic subunit, and it has been proposed that the polyamine also reacts with the enzyme in these systems. It has also been observed that spermine inhibits myosin phosphatase and thereby increases Ca^{2+}-activated contractions, a role of polyamine suggested to be operative in smooth muscles (Sward et al., 1995).

In a third system, calmodulin, a calcium receptor, is known to regulate the phosphatase calcineurin and its effect is inhibited by spermine. In this case the inhibition correlated with the binding of spermine to a fluorescent form of calmodulin (Walters & Johnson, 1988). The proliferation of protein phosphatases and studies of their properties and regulation has been reviewed by Cohen (1989).

Endogenous nuclear proteins as well as casein are dephosphorylated by extracts of nuclei of HeLa cells and this hydrolysis is stimulated two- to fourfold by 0.1 mM spermine (Friedman, 1986). The polyamine is more effective than Mg^{2+} or Ca^{2+}.

Data on PTKs

These enzymes were first detected as products of viral oncogenes. New selective substrates were devised, such as polyglutamic tyrosine (4 : 1). Old casein kinase substrates, such as histones, are devoid of tyrosine whereas many former natural substrates, such as nonhistone nuclear proteins or calmodulin, contain serine/threonine and tyrosine hydroxyls. Esterified phosphates in the latter were selectively stable in N NaOH at 56°C. In a study of the phosphorylation of calmodulin, CKII phosphorylated serine/threonine mainly in the presence of polylysine and only one of three rat spleen PTKs phosphorylated tyrosine under such conditions. Ca^{2+} was inhibitory to both systems (Meggio et al., 1987).

A pig spleen PTK phosphorylated tyrosine in an angiotensin substrate and this was stimulated twofold by spermine and spermidine whereas the synthetic basic polypeptides activated some threefold. Several substrates like tubulin were similarly stimulated and it was concluded that the effects were on the enzyme. Heparin and other polyanions were active inhibitors (Sakai et al., 1988). Obviously the use of these criteria in the earlier distinction of CKI and CKII may have missed some PTK activity.

A cytosolic PTK has been isolated from human erythrocytes; spermine and polylysine were stimulatory (almost twofold) whereas heparin and other polyanions were slightly inhibitory. A similarly activable and inhibitable PTK has been isolated from particles of rat spleen. One PTK, a kinase kinase, can phosphorylate and activate ribosomal protein kinases that phosphorylate on serine/threonine. This is one example of crosstalking among kinases that may operate as a cascade (Erikson, 1991). Because the epidermal growth factor (EGF) and the v-src protein are both PTKs, it is supposed that cell proliferation may be induced as the result of a cascade begun by protein tyrosine phosphorylation. DFMO treatment reduced the phosphotyrosine content of some PTK substrates, and this was interpreted to suggest still another role for polyamines in growth inhibition (Oetken et al., 1992) (see following text).

Among the known inhibitors of PTK are compounds such as *erbstatin*, which approach the natural *trans*hydroxycinnamamide structure (Chang & Geahlen, 1992; Levitzki & Gazit, 1995). It would be of interest to know if the natural hydroxycinnamamides, like feruloylspermidine, have any inhibitory activity in these systems. Also, methylthioadenosine (MTA) is a specific reversible inhibitor of protein tyrosine phosphorylation stimulated in fibroblasts by the basic fibroblast growth factor (Maher, 1993).

Because several of the PTKs are associated with receptors, it was anticipated that membranes might also contain protein tyrosine phosphatases. These were found first in membranes, indeed in large excess over the PTK activity, are specific for phosphotyrosine, and act on many protein substrates (Fischer et al., 1991). A phosphotyrosine phosphatase from human placenta is inhibited by heparin and other polyanions. A homologous catalytic domain containing an essential cysteine has been found in all of these phosphatases. The problems of the interaction of kinases and phosphatases and their possible regulation by ligands in the control of cellular activities have barely begun to be addressed.

Tying Inner Structures to Cell Surface

The ubiquitous and highly conserved cytoskeletal protein, actin, is involved in processes of cell division, chromosome segregation, cytoplasmic streaming, and cellular motility. The movement of cells such as amebae or slime molds appears to be a function of the polymerization and depolymerization of actin filaments. In many changes of shape and motility, the actin cytoskeleton is associated with the inner side of the plasma membrane (Carroway & Carroway, 1989).

Analyses of the ATP-induced contraction of myosin B or actomyosin led to the isolation of globular (G) monomeric actin and the viscous filamentous (F) polymeric actin; F-actin mixed with myosin produced myosin B. Monomeric actin contains bound ATP that is rapidly hydrolyzed to ADP and inorganic phosphate (P_i) on polymerization to F-actin. In early studies of glycerinated muscle fibers, the ATPase activity of actomyosin was found to be inhibited by 5 mM spermine, which also relaxed contracted muscle preparations (DeMeis & DePaula, 1967).

The structures of monomeric actin, its clefts, and binding sites have been determined in crystallographic analyses of actin in 1:1 complexes with DNase 1 or the G-actin binding protein, profilin (Mannherz, 1992; Schutt et al., 1993). These determinations have assisted the numerous genetic studies in establishing the differences in sequence and family relations among the many known actins (Hennessey et al., 1993).

Oriol et al. (1977) crystallized monomeric G-actin at low ionic strength and demonstrated the reversible polymerization of G-actin by salts or 1 mM polyamines. Spermidine and spermine were most efficient and spermidine added to salt-induced F-actin increased its polymerization. F-actin induced by spermidine (2 mM) comprised single homogenous filaments; that induced by spermine (0.5 mM) had parallel arrangements of the protein, indicating some cross-linkages between filaments. F-actin organized in these forms by polyamines

was disaggregated rapidly by ATP (Oriol–Audit, 1982).

Oriol–Audit (1980) noted that cell cleavage furrows are related to the appearance of aligned actin microfilaments, a contractile ring, and that both this ring and polyamine-induced F-actin can be disaggregated by cytochalasins. At low concentrations spermine stabilizes microfilaments in the cortices of injected frog eggs and generates bundles at higher concentrations (Grant & Oriol–Audit, 1985). Further, Aimar and Grant (1992) have observed the induction of components of cleavage furrows in enucleate eggs by the injection of the polyamines, which by themselves are inadequate to organize cleavage.

MGBG can inhibit cell division in culture and leads to the accumulation of binucleate cells in phenomena marked by a disaggregation of actin filaments. The polyamines reverse this inhibition. Even more suggestive of a direct role of the polyamines are the effects of polyamine starvation developed in polyamine-auxotrophic cells. As Pohjanpelto et al. (1981) showed, polyamine depletion led to the disappearance of actin filaments and the arrest of division. Microtubules were also disaggregated, while cell shape was maintained through the unaffected intermediate filaments. Subsequently it was found that 1 mM spermine and spermidine are effective in polymerizing tubulin in the presence of GTP to form microtubules (Anderson et al., 1985). The disorganization of microtubules is demonstrated in chick fibroblasts treated with cyclohexylamine; this is preventable by spermidine (Caruso et al., 1994).

The polymerization of G-actin, which is rich in aspartic and glutamic acid residues, to F-actin has been observed with numerous cationic molecules. Macrocyclic polyamines are quite active in this respect (Oriol–Audit et al., 1985) and it may be asked if this may relate to the biological attributes of some plant alkaloids as well as to those of the synthetic macrocycles. Ordered complexes can exist between G-actin and positively charged liposomes, and monomeric G-actin can intercalate in a surface monolayer between lipid molecules rich in phosphatidyl choline (Grimard et al., 1993). Actin is also known to be a substrate for transglutaminase in the presence of putrescine and can be bound by this enzyme to liver membrane components (Katoh et al., 1994a).

The association of the actin cytoskeleton with buds and septa can be seen in budding and dividing yeast (Gabriel et al., 1992). After protoplasting, the asymmetric distribution around sites of active membrane is lost and sites of attachment are evenly distributed until after the regeneration of the cell wall. The completion of the wall permits the inception of budding, with actin appearing in the membrane initially at sites of bud formation and subsequently in a ring at the neck of the bud.

Monomeric G-actin is unavailable initially in cytoplasm as a result of its complex with one of at least four families of binding proteins. One of these may be profilin (Mannherz, 1992), a protein which can also combine with plasma membranes and phosphatidyl inositides. Indeed phosphatidyl inositides can dissociate profilin–actin complexes and possibly initiate actin assembly (Goldschmidt–Clermont et al., 1990). This protein may also regulate signal transduction by inhibiting phospholipase C and its hydrolysis of PI-P$_2$. In another related finding it was observed that another product of PI-P$_2$ hydrolysis, DAG, stimulates actin polymerization in isolated plasma membranes of the slime mold, *Dictyostelium* (Shariff & Luna, 1992). Stossel (1989, 1993) reviewed the mechanisms of actin polymerization and attachment to membrane in the development of cell crawling and indicated many of the proteins and phospholipids involved in making G-actin available for polymerization and in assisting assembly to form cables of F-actin, which can affect membrane movement.

The regulation of actin assembly and disassembly at crucial membrane sites is currently seen to be so complex that it is difficult to imagine how the possible roles of the polyamines can be tested or proven. For example, inhibitors of polyamine synthesis also inhibit the accumulation of mRNA for cytoskeletal elements including actin (Kamínska et al., 1992). Nevertheless, a potentially integrating set of experiments has emerged from a study to explore the role of ODC in carcinogenic transformation (Hölttä et al., 1993). This group infected rat cells with a temperature-sensitive strain of Rous sarcoma virus, which contains an oncogene, *v-src*, that encodes a PTK, a 60 kDa phosphoprotein, pp60$^{v\text{-}src}$. This protein is localized via myristoylation and membrane anchoring at the inner face of the plasma membrane, and the essential substrates for transformation as well as proteins important in signal transduction are also thought to exist at or near the membrane. Tyrosine phosphorylation in this virus infection was followed in the presence and absence of DFMO. It was observed that DFMO blocked transformation by v-src without inhibiting the synthesis of the v-src oncoprotein and the PTK. It did affect the PTK activity on some specific proteins, as well as the association of specific tyrosine-phosphorylated proteins with pp60$^{v\text{-}src}$ in a polyamine-dependent and transformation-dependent manner. Some inhibition was observed in the tyrosine phosphorylation of a kinase active on phosphatidyl inositol, which is thought to be involved in both signal transduction and the organization of actin filaments. Thus ODC, inhibited by DFMO, and polyamine availability are seen to be involved in the very complex sequence initiated by v-src at the cell membrane. Of some 40–50 proteins phosphorylated on tyrosine after this initiation, some of these appear to be both sensitive to DFMO and essential to transformation. The depolymerization of actin is an early change in transformation in this system and is inhibited by DFMO.

Eucaryotic Cells with Vacuoles

The existence of the vacuole in fungi and plant cells has complicated studies of transport of the polyamines and other molecules into these organisms. The vacuole of *Neurospora crassa* stores much of the cellular polyamine (Chap. 10). The existence of only a small metabolizable fraction of the total polyamine pointed to this almost inert polyamine compartment, but also stressed the existence of relatively inert bound polyamine within the cell that presumably associated with anionic polymers such as ribosomes. This binding and sequestration of polyamine, as well as further metabolism as in the further conversion of putrescine to spermidine and spermine, thereby establishes a difficult to interpret unidirectional flow.

Both intact vacuoles and right-side out vacuolar vesicles have been isolated from *Saccharomyces cerevisiae* (Ohsumi & Anraku, 1981). In the presence of ATP, vacuolar vesicles do accumulate spermidine and spermine, with uptake of the latter competitively inhibited by the former. [^{14}C]-Spermine, accumulated in the vesicles, was exchangeable with [^{12}C]-spermine or spermidine (Kakinuma et al., 1992). Release by a protonophore required the presence of a counterion, such as Ca^{2+}, suggesting the association of spermine with vesicular anions, for example, the significant fraction of polyphosphate remaining in the vesicles.

Polyamine transport systems have been described in *Aspergillus nidulans* in which the cell walls bind the amines rapidly; the amines are subsequently concentrated within the cells (Spathas et al., 1982). The fungus has an independent system for putrescine, noninhibitable by spermidine or spermine, but the spermidine system is inhibited by putrescine and spermine. In this system, the uptake of putrescine is accompanied by conversion to spermidine. Furthermore, putrescine can be used by *Aspergillus* as a nitrogen source.

Uptake of polyamine in wild-type *N. crassa* is slow in buffer but will, for example, concentrate 1 mM putrescine 16-fold in 10 min. Systems responsible for the uptake of putrescine, spermidine, and spermine are saturable (i.e., achieve a maximal rate) and nonsaturable (i.e., increase in rate) as a function of increasing polyamine concentration (Davis & Ristow, 1988). Both sets of systems are energy dependent; that is, they are extensively inhibited by cyanide and low temperature (4°C). Polyamine depletion in an ODC-deficient strain increases the rates of putrescine and spermidine uptake. Polyamine excess inhibits uptake, as does MGBG. Cations of the growth medium, including Ca^{2+}, are inhibitory.

Efflux or release is quite slow, except in cells containing an excess of polyamine. A diffusional exchange of putrescine and spermidine can occur after the internal spermidine-binding sites, as well as the putrescine pool, are saturated. Normal concentrations of polyamines are about 0.8 μmol putrescine per gram and 18 μmol spermidine per gram dry weight of mycelium (Davis & Ristow, 1989). Because *N. crassa* does not readily produce acetylated or otherwise conjugated amines, or utilize the polyamines as a nitrogen source, diffusional elimination is useful in removing potentially toxic excess.

Although the organism permits the equilibration of putrescine across the membrane during growth, a transport system for putrescine has been detected in a specific mutant. The system is normally inhibited by Ca^{2+} and this sensitivity has been lost in the mutant, resulting in a toxic concentration of putrescine (Davis et al.,

1991). Uptake of the diamine is accompanied by extrusion of K^+; a similar putrescine-K^+ antiporter has recently been described in *E. coli*. Much of the excess putrescine taken up in the mutant, to 80 mM levels, can be found in vacuoles. These also increase their polyphosphate content and reduce their content of monovalent cation, thereby serving to modulate the toxicity of the diamine (Davis & Ristow, 1991).

A rapid binding to cell wall anions or conversion to a toxic metabolite by an oxidase in cultured plant cells has been eliminated by the use of protoplasts in numerous studies of transport. Although the preparation of protoplasts induces arginine decarboxylase (ADC) and an accumulation of internal putrescine, a patch-clamp technique applied to protoplasts of *Arabidopsis* failed to detect plasma membrane channels specifically permeable to putrescine (Giromini et al., 1994).

When polyamine uptake was studied in carrot cells, most of the putrescine was found in the cytoplasm but spermidine was largely adsorbed on the cell walls (Pistocchi et al., 1987). Putrescine uptake could be inhibited by the presence of DAP, cadaverine, or spermidine. Efflux of the amines from the two compartments, the cells or walls, could be increased by an increase of exogenous amine or Ca^{2+}. Spermine, although bound mainly to cell walls in intact cell cultures, also enters protoplasts by saturable and nonsaturable systems. Ca^{2+} (1 mM) enhances uptake of spermine and other amines in a protoplast system.

$[^{14}C]$-Polyamines can absorb to protoplast membranes but this binding is minimized by a wash of the cells (or later the vacuoles) by 100 mM unlabeled polyamine. Labeled polyamines appear rapidly (to a maximum in 1–2 min) in the protoplasts and in the vacuoles. Vacuoles contain some 40% of the putrescine and 28% of the spermidine of the initially isolated protoplasts. After uptake of $[^{14}C]$-spermidine at a low concentration, some 10% of the total spermidine of the original protoplast and vacuole is radioactive. In all, 27% of the uptake of spermidine into the protoplast is localized in the vacuole.

Antognoni et al. (1993) also examined the competition between polyamine analogues and the natural polyamines in carrot protoplasts. In the absence of stimulatory Ca^{2+}, several amines (e.g., agmatine, cadaverine, spermidine, and MGBG), were significantly inhibitory to putrescine uptake whereas spermine and MGBG were inhibitory to spermidine uptake. In the presence of Ca^{2+}, which is thought to affect the cell surface primarily, spermidine was inhibitory to putrescine uptake and MGBG had no effect on this transport but did inhibit spermidine uptake. Curiously, spermine at concentrations >25 μM markedly enhanced uptake of both putrescine and spermidine. It was suggested that there is a specific transport system for spermine.

Transport in Animal Cells

The recognition of the ubiquity and essentiality of the polyamines in animal tissues, as well as the availability of the cells of these tissues by tissue culture methods, has evoked numerous studies of polyamine transport in these cells. The inhibition of growth of tumor cells by specific inhibitors of polyamine synthesis and the restoration of growth by exogenous polyamines have been a major stimulus to the study of polyamine uptake. The availability of tumor cells like Ehrlich ascites cells permitted the study of uptake of amino acids and of organic amines (Christensen & Liang, 1966). Several modes of uptake of $[^{14}C]$-benzylamine were evident: adsorption to the cell surface, removable in large part by washing; a saturable intracellular transport indicative of a specific process; and a slower nonsaturable uptake, suggestive of a mix of diffusion and internal binding. Christensen (1969) differentiated several of the modes of uptake.

Pinocytosis occurring in the uptake of putrescine by the myxamebae of *Dictyostelium discoideum* is more complex than that described as "mediated transport." Nevertheless, this organism employs pinocytosis with operating components of both a saturable system active at low (micromolar) putrescine concentrations and a nonsaturable system functioning at high (millimolar) concentrations (Turner et al., 1979).

The analysis of the mode of uptake of a metabolite of interest by a given cell in culture has usually involved:

1. a determination of the effect of cell density;
2. pH optimum of uptake;
3. the extent of conversion of the metabolite to other chemical species;
4. the time course of uptake at normal, 37°C, and low temperatures, 4°C;
5. the total uptake versus that in an acid extract;
6. the concentration dependence of the rate of uptake;
7. the effect of other media components like Na^+;
8. the effects of possible competitors of the uptake system;
9. the effects of inhibitors; and
10. the extent of uptake or efflux of a preloaded cell.

Experiments of these types relate to the specificity and energetic requirements of the process, the estimation of trapping during accumulation, and the possibility of regulation by the presence of internal metabolites. The existence of no fewer than five major systems of different substrate specificity has been revealed in the study of the uptake of neutral and basic amino acids in mammalian liver cells in addition to systems for imino acids (proline) and anionic amino acids (aspartate) (Cheeseman, 1991). Among these, a bidirectional system for cationic amino acids, like arginine, was found to be rate limiting in the conversion of arginine via arginase to ornithine and urea.

As summarized by Seiler and Dezeure (1990), polyamine uptake in mammalian cells is most generally specific, saturable, energy requiring, and carrier mediated. It can also be regulated in a process requiring RNA and protein synthesis. A review of the data on three membrane receptors of mammals has described marked similarities or homologies of these proteins to the numerous transporter proteins for amino acids and polyamines isolated from eubacteria and fungi (Reizer et al., 1993). These data suggest that the receptor proteins of mammalian cells may be described as modified transport proteins sharing a common origin with the transport proteins of these microbes.

Polyamine efflux from mammalian cells has seldom been studied as such. Hawel et al. (1994) have recently described a selective system of putrescine export in rat hepatoma cells. This is responsive to increased ODC activity within the cells evoked by exogenous ornithine or insulin and appears to indicate a homeostatic mechanism of reducing potentially toxic intracellular diamine.

Many cells have a single transporter for putrescine, spermidine, and spermine, with K_m values in the micromolar range. Both Na^+-dependent and -independent systems have been detected. Stressing the therapeutic significance of polyamine transport in cancer, reviewers pointed to the decreased virulence of a leukemia cell line lacking a transport system and the increased efficacy of DFMO in mice inoculated with such cells. As noted in Chapter 17, the depletion of putrescine and spermidine by DFMO markedly increased accumulation of MGBG and several polyamine analogues.

These data suggest the desirability of controlling polyamine transport in managing various pathologies and therapies. Several new inhibitors of putrescine uptake inhibit the uptake of the diamine more actively than that of spermidine (Minchin et al., 1991). These also indicate that the transporters of these different polyamines in melanoma cells are different. An attempt to inhibit the transport system for spermidine in Ehrlich ascites tumor cells was made with nitroimidazole–polyamine conjugates. A relatively nontoxic compound comprising a bis(dinitroimidazole) derivative of spermidine entered cells and inhibited uptake of the free triamine, suggesting the possibility of using these conjugates also to introduce potentially toxic substituents (Holley et al., 1992).

The uptake of putrescine in human platelets is saturable and energy dependent but appears complex. Studies with human erythrocytes, unable to synthesize the polyamines, demonstrated polyamine receptor sites at the cell surface and the uptake of putrescine, spermidine, and spermine largely (>95%) into an internal soluble compartment (Moulinoux et al., 1984). The uptake was minimal at 4°C, and this related mainly to binding to the cell stroma. At 37°C uptake of putrescine and spermidine was relatively rapid from serum whereas spermine entered slowly. The K_m values for the saturable putrescine and spermidine uptakes from plasma were 125 and 3.6 μM, respectively. Putrescine was lost by efflux at a far greater rate than was spermidine. Spermidine uptake was a Na^+-dependent process and was inhibited by putrescine and MGBG and much less by spermine.

In one strain of mouse cells, both Na^+ and spermidine were shown to enter in a 1:1 relationship (Khan et al., 1990a). The transport appeared to be ATP-independent because it was unaffected by 2-deoxyglucose, which depletes ATP. In examinations of many more strains of mammalian cells, spermidine uptake was affected by Na^+ concentration, although somewhat differently in each case (Khan et al., 1990b). The uptake was generally inhibited by ionophores and some polyamine analogues, the α,ω-dimethylaminoalkanes. Preloading of the cells with asparagine accelerated spermidine uptake in two strains, as well as of putrescine uptake in neuroblastoma cells (Rinehart & Chen, 1984). The latter effect was of interest because some amino acids, notably asparagine, had been found to induce ODC.

Exogenous putrescine markedly stimulates DNA, RNA, and protein synthesis in intestinal epithelial cells (IECs) (Ginty et al., 1989). Uptake of putrescine, spermidine, and ODC activities in these cells may be regulated via a Ca^{2+}/calmodulin-dependent mechanism (Groblewski et al., 1992; Scemama et al., 1993). Whereas the uptake of putrescine by normal rat IEC crypt cells is stimulated very slightly (1.3-fold) by 10 mM asparagine, human colon cancer cells are stimulated >300-fold in this uptake by asparagine (McCormack & Johnson, 1989, 1991). Thus, the nature of the asparagine effect on putrescine uptake appears to be greatly exaggerated in some human cancer cells.

Because asparagine activates a membrane Na^+/H^+ antiport, provokes an extrusion of H^+, and causes Na^+-dependent intracellular alkalinization, it has been asked if the induction of ODC produced by asparagine is not due to an increase of cellular pH (Fong & Law, 1988). Phorbol esters, which also induce ODC, evoke increases of cytoplasmic pH and have been found also to increase spermidine transport and concurrent activity of a Na^+-K^+ pump (Khan et al., 1992).

In studies of conA-activated lymphocytes, Kakinuma et al. (1988) found that a transport system common for the three natural polyamines had been activated, with affinities at the micromolar level in the order spermine > spermidine > putrescine. The inhibitory effects of carbonyl cyanide m-chlorophenylhydrazone and other ionophores suggested a major dependence of uptake on membrane potential. The depletion of intracellular polyamine markedly enhanced the activity of the transporter; this induction was repressed by exogenous spermidine or spermine. The rapidity of this regulation of transport may permit the maintenance of intracellular polyamine, despite a fall in ODC at different stages of the cell cycle in activated B lymphocytes (De Benedette et al., 1993).

Studies with a wide variety of animal cells indicated the common responses of transport mechanisms to intracellular polyamine concentrations. Nevertheless, some tissue cells and cancer cells appear to possess a single common mechanism whereas others, like the pig renal LLC-PK cell line, revealed several more discriminating systems. The latter cells contained both Na^+-dependent and Na^+-independent transporters localized in different cell areas (Van den Bosch et al., 1990). The former was more rapidly regulated, suggesting a possible role in excretion of urinary polyamines (Parys et al., 1990). Further, exogenous polyamines downregulated and depletion upregulated transporters seemingly unrelated to polyamine metabolism (Peng & Lever, 1993). The putrescine transporter of porcine aortic endothelial cells is similarly increased in activity in response to DFMO (Bogle et al., 1994).

Efforts were made on one hand to simplify the work on cells by use of membrane vesicles and on the other to extend the studies to larger physiologically significant anatomical units, such as tissues and organ systems. Studies on rat intestinal brush-border membrane vesicles demonstrated a major binding to membrane surface and brief periods of uptake for the three amines. The vesicles appeared to have two different Na^+-independent uptake systems for spermidine and spermine (Kobayashi et al., 1993).

Examinations of rat lens or entire digestive systems revealed properties potentially significant in the functions of these tissues. For example, the lens possessed energy-dependent transport. However, it was almost inactive in the de novo synthesis of polyamines but actively degraded transported spermine to spermidine and then to putrescine (Maekawa et al., 1989). The high levels of the polyamines in the lumens of various compartments of the rat gastrointestinal system, and the relation of these to transport, mucosal accumulation, and turnover, contributed to a systematic analysis of several organs and the digestive process (Osborne & Seidel, 1990).

As in the case of the bacterial transport systems, studies were carried out to identify and isolate the carrier molecules of animal cells. MGBG, the inhibitor of spermidine and spermine synthesis, competed with spermidine in transport. Some tumor cells resistant to MGBG were found to have a reduced drug uptake and this was exploited to develop mutant CHO cells and rat myoblasts with a considerable decrease (99%) in the uptake of the three polyamines. Several classes of deficiency and genetic controls were revealed (Heaton & Flintoff, 1988). Most recently spermidine conjugate photoaffinity labeling of plasma membranes of leukemic cells detected two proteins that may be related to carrier proteins (Felschow et al., 1995).

Polyamine transport proved to be essential in establishing the effects of exogenous polyamine and analogues in regulating ODC and spermidine/spermine N^1-acetyl transferase, as well as in demonstrating the toxicity of compounds such as the bis(ethyl) analogues (Byers & Pegg, 1989). The inability of some mutant cells to use polyamines to reverse the toxicity of DFMO permitted the use of this selection to isolate additional transport-negative mutants. The transfection of a transport-deficient CHO mutant with human DNA permitted the use of polyamine to restore growth to inhibited cells and to facilitate the isolation of transport-positive clones (Byers et al., 1989). The differences in transport of the individual polyamines among the individual clones indicated the existence of a multiplicity of polyamine carriers.

The availability of various mutant systems and inhibitors to eliminate biosynthetic steps has assisted the analysis of the regulation of transport. Depletion of polyamine content in ODC^- cells or in ODC^+ cells exposed to DFMO enhances polyamine transport in a process requiring protein synthesis. Conversely, high levels of exogenous amines, which markedly increase intracellular levels, cause a rapid drop in transport (Byers & Pegg, 1990). Bis(ethyl) polyamines acted like normal polyamines in this respect. However, the accumulation of MGBG in depleted cells did not cause a fall in transport of putrescine or spermine.

The regulation of transport has been studied in some detail with murine L1210 leukemia cells (Kramer et al., 1993). Depleted cells treated with inhibitors could be restocked with various polyamines whose uptake was then found to have increased. The accumulation of bis(ethyl)spermine, a compound in active exploration in cancer chemotherapy, decreased transport of spermidine or spermine unexpectedly rapidly (Fig. 21.6). Bis(ethyl)spermine eliminates spermidine and spermine by enhancing acetylation and degradation and by restricting the uptake of these normal amines. The analogue downregulates ODC also. Removal of the analogue in the presence of cycloheximide rapidly restored the polyamine transport, suggesting that downregulation was due to a change of state of the transporter protein rather than an arrest of its de novo synthesis.

An unexpected complexity has been found in the regulation of transport. As described in Chapter 13, a DFMO-resistant mutant of rat hepatoma cells was isolated and proved to be unable to degrade ODC; the cell accumulated very high levels (2000-fold increases) of the ODC protein. ODC did not decrease in the cells preferentially in the presence of cycloheximide, nor was the ODC activity regulated by exogenous spermidine. The mutant cells were killed and many were lysed following exposure to exogenous micromolar levels of spermidine; the triamine was accumulated to abnormally high levels and failed to downregulate the transport of spermidine and spermine. Protein synthesis was inhibited, a process unrelated to any of the possibly toxic metabolic fates of the amine (Mitchell et al., 1992a). Mg^{2+} and ATP concentrations were shown to be markedly decreased, with a concurrent inhibition of protein synthesis, following an accumulation of spermidine and spermine in mouse cells (He et al., 1993).

Downregulation of spermidine transport by exogenous spermidine (10–50 μM) was then found to be blocked by cycloheximide in HTC or CHO cells (Mitchell et al., 1992b). This result evoked the hypothesis that the rapid reversible inactivation of a spermidine carrier is effected by an unstable polyamine-induced protein. Recent studies of He et al. (1994), Mitchell et al. (1994), and Suzuki et al. (1994), indicated that antizyme, which regulates ODC activity and degradation and had been lost in the accumulating mutant, also serves to downregulate transport in animal cell systems in response to intracellular polyamine.

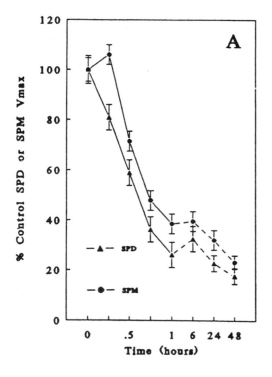

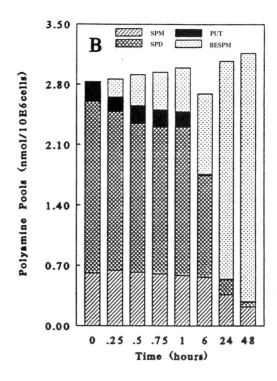

FIG. 21.6   Effects of treatment of L1210 cells with 10 μM bis(ethyl)spermine (BESPM) for 0–48 h on (**A**) spermidine (SPD) and spermine (SPM) transport and (**B**) polyamine pools; adapted from Kramer et al. (1993). PUT, putrescine.

Mitochondria, Spermine, and Ca²⁺

Mitochondria are the major cellular organelles concerned with O_2 consumption and the aerobic production of ATP. They possess their own characteristic genetic system of DNA, enzymes of RNA and protein synthesis, and several special enzymatic systems of mitochondrial synthesis and function. In *Neurospora*, for example, the latter include key steps of the biosynthesis of ornithine and its conversion to citrulline. This organelle in *Neurospora* also contains a system of arginine transport that both regulates ornithine biosynthesis and delivers arginine for protein synthesis (Yu & Weiss, 1992).

The properties of an ornithine-phosphate antiport in rat kidney mitochondria, and an ornithine/citrulline carrier in rat liver mitochondria have been described, as has an apparent defect in mitochondrial transport of ornithine in human hyperornithemia. The mitochondrial transport of the basic amino acids related to polyamine metabolism is well established in plants and animals. It is of interest that the transport of the protein precursor of the essential enzyme, ornithine transcarbamylase, into mitochondria is induced by polyamines, which are thought to facilitate the binding of the precursor to the organelle surface (González–Bosch et al., 1991). Spermine also promotes the translocation of CKII into this rat liver organelle (Bordin et al., 1994). On the other

hand, spermine causes a binding of CKI to the outer membrane of the organelle (Clari et al., 1994).

The organelle isolated from animal and plant tissues was also found to contain small amounts of polyamine, about 1–3% of the total cell content. Of this small amount, a significant fraction is adsorbed to the mitochondrial surface. Labeled polyamines adsorbed in a homogenate may be almost totally removed by chromatography of the rat heart organelles on heparin–Sepharose (Tadolini et al., 1985c). Residual putrescine is undetectable in the purified mitochondria and residual spermine and spermidine were estimated to remain at the levels of 2.66 and 0.36 nmol/mg protein, respectively. In another study in which rat liver mitochondria were isolated, putrescine was undetectable and the spermine and spermidine contents were 3.85 and 1.60 nmol/mg protein, respectively (Igarashi et al., 1989). The relatively large amount of spermine and its effects on pyruvate dehydrogenase in the mammalian organelle focused attention on this amine.

Plant mitochondria were reported to contain somewhat smaller amounts of each polyamine, including putrescine. These organelles were also reported to have low levels of ODC, ADC, and *S*-adenosylmethionine decarboxylase (AdoMetDC) (Torrigiani et al., 1986). With the exception of ADC, these levels are suspiciously low and raise questions of possible contamination. The problem was answered by demonstrating a release of spermine from liver mitochondria by an uncoupler of membrane potential and subsequent uptake by restoration of the membrane potential (Mancon et al., 1990). In *Helianthus* mitochondria, spermidine uptake is also a function of the membrane potential and is blocked by the ionophore, valinomycin. The accumulation of spermidine in the plant organelle was not affected by Ca^{2+} or inorganic phosphate.

That polyamines may have special functions in the organelle has been perceived as a result of many studies on the survival of mitochondrial structure and function. Spermine inhibits the swelling of the isolated organelle in hypoosmotic media (Tabor, 1960). The polyamines are adsorbed quite rapidly to the outer membranes of intact mitochondria. However, once the mitochondrial membranes are shocked in hypoosmotic media, the polyamines or MGBG are found to enter the matrix more readily and their inhibition of respiratory activity is enhanced (Diwan et al., 1988). Similar effects on the inhibition of respiration have been seen in studies that demonstrate a polyamine-induced collapse of inner and outer mitochondrial membranes as a result of an unregulated penetration of the polyamines.

The toxicity of MGBG, the inhibitor of an enzyme permitting the synthesis of spermidine and spermine, in animal cells was expressed in the loss of inner structure and the distension of mitochondria (Mikles–Robertson et al., 1979) with a concomitant depletion of ATP (Porter et al., 1979). The inhibitory effects of MGBG on mitochondrial oxidases and respiration are reversed by natural polyamines (Byczkowski & Porter, 1983). MGBG appears to affect Ca^{2+} cycling, which is also altered by spermine. More recently bis(ethyl)spermine was found to severely reduce the cellular content of mitochondrial DNA (Vertino et al., 1991).

The biosynthetically deficient mutants of *Saccharomyces cerevisiae* provided observations on the requirements for mitochondrial structure and function. A null mutant strain absolutely deficient in AdoMetDC and extremely depleted in spermidine and spermine was unable to grow even in the presence of putrescine (Balasundaram et al., 1991). The cells were deficient in budding during aerobic growth and had defects in the distribution of actin and chitin. Although the cells survived anaerobically and grew with spermine, they were in fact killed in air in the absence of spermidine or spermine, with destruction and loss of mitochondria (Balasundaram et al., 1993). The experiments indicated that the polyamines were somehow involved in overcoming oxygen toxicity. Spermine was also shown to inhibit glutathione release from mitochondria, as induced by Ca^{2+} and phosphate (Rigobello et al., 1993).

Mg^{2+} and spermine are known to act synergistically in determining the rate of respiration during the mitochondrial conversion of ADP to ATP. Specifically, low spermine concentrations increased O_2 consumption and the conversion of ADP to ATP by heat-aged rat liver mitochondria oxidizing β-hydroxybutyrate (Phillips & Chaffee, 1982). It was observed that spermine bound specifically to submitochondrial particles of beef heart

and increased the maximum velocity (V_{max}) of the complex oligomycin-sensitive ATPase of the particles. The effects occurred at the concentration of spermine found within functional mitochondria. The polyamine did not appear to affect the ATPase of isolated F_1-ATPase and was thought to bind at the site of interaction of membrane and F_1, the soluble coupling factor of the ATPase complex (Solaini & Tadolini, 1984). According to Igarashi et al. (1989), the F_1-ATPase of liver mitochondria lacks polyamine and phospholipid and is inhibited by spermine; this inhibition by spermine is much less in the presence of Mg^{2+}, which inhibits the binding of spermine to ATP. Polyamine-deficient cells were low in ATP, and it was suggested that spermine participates in maintaining a high ATP level.

In studies of pyruvate dehydrogenase and its regulation, the effects of polyamines were studied in permeabilized fat cells and mitochondria (Rutter et al., 1992). The enzyme is phosphorylated and inactivated by a kinase and dephosphorylated and reactivated by a phosphatase. The phosphatase requires Mg^{2+} or Mn^{2+} and the activity is stimulated by Ca^{2+} alone. The activity is further stimulated 20–30-fold by spermine at low Mg^{2+}, but this did not occur in the absence of Mg^{2+} or Ca^{2+} (Damuni et al., 1984). Spermine was also shown to increase the sensitivity of this enzyme to Mg^{2+} but not to Ca^{2+} (Thomas et al., 1986).

In addition to the stabilization of mitochondrial membranes, spermine was initially observed to be transported together with inorganic phosphate into the organelle matrix. The process was dependent on respiration and inhibited by Mg^{2+}. Both spermine and P_i were lost together by efflux after an initial preloading and treatment with a compound to decrease the transmembrane potential. Although spermine enhanced uptake of P_i, spermine uptake was subsequently found to be independent of P_i concentration and was related entirely to an increase of the electrical potential gradient (Toninello et al., 1988). These authors concluded that all polyamine transport in mitochondria was strictly electrophoretic via a channel mechanism (Toninello et al., 1992). Indeed this form of transport could occur bidirectionally as a function of the membrane potential and pH gradient (Siliprandi et al., 1992).

The significance of the sequestration of Ca^{2+} and its release to increase a low cytosolic concentration became a subject of active study in animal cell systems and yeasts (Gunter & Pfeiffer, 1990; Votyakova et al., 1990). The polyamines had been described as inhibitors of Ca^{2+} uptake but subsequent studies with rat liver organelles at low Ca^{2+} indicated an activation of Ca^{2+} uptake by the amines. The activation of uptake of Ca^{2+} (0.8 μM) by 350 μM spermine decreased the apparent K_m from 16 to 3.8 nmol Ca^{2+}/mg mitochondrial protein without a change of V_{max} (Nicchitta & Williamson, 1984). The stimulation of Ca^{2+} uptake was specific for mitochondria of various rat organs and spermine was far more effective than spermidine; Mg^{2+} was an antagonist (Lenzen et al., 1986). The polyamine effects were possibly of major significance in determining buffering of function by Ca^{2+} in developing rat brain (Jensen et al., 1989). A regulatory role of spermine on the activation of rat heart intramitochondrial Ca^{2+}-sensitive dehydrogenases (Fig. 21.7) was demonstrated by McCormack (1989).

The results, despite a lack of knowledge of the "free" spermine activity of liver cytosol (the total concentration of spermine in rat liver was 1 mM), have been thought to be physiologically relevant. More recent studies have demonstrated a more complex pattern of spermine effects on Ca^{2+} transport and have stressed a distinction between uptake and accumulation with spermine effecting an increase in mitochondrial Ca^{2+}-storage capacity (Lenzen et al., 1992). Spermine appears to limit Ca^{2+} binding to membrane phospholipids, and thereby facilitates diffusion at low Ca^{2+} concentrations (Rottenberg & Marbach, 1990).

Paraquat toxicity via polyamine transport

The herbicide, paraquat (1,1'-dimethyl-4,4'-bipyridylium) is quite toxic to humans in its commercial concentrate. The lungs of humans and experimental animals selectively accumulate the redox-cycling compound by a mechanism of polyamine transport (Smith, 1985; Wyatt et al., 1988). Specifically, rat lung slices accumulate diamines, spermidine, and spermine, as well as paraquat in an energy-dependent process that exhibits saturation kinetics. The accumulation occurred mainly in two types of cells, alveolar type I and type II, of the many types present in the lung. The accumulation of paraquat in human lung slices was markedly inhibited by 10 µM

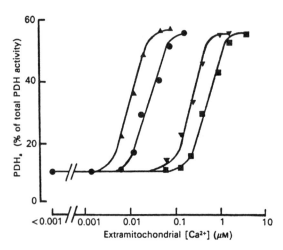

FIG. 21.7 Effects of spermine on the Ca^{2+}-dependent activation of pyruvate dehydrogenase (PDH) in coupled rat heart mitochondria incubated with or without Na^+ and Mg^{2+}; adapted from McCormack (1989). Additions to the incubation medium for 4 min include (●) none, (▲) 0.5 mM spermine, (■) 10 mM NaCl + 2 mM $MgCl_2$, or (▼) 10 mM NaCl + 2 mM $MgCl_2$ + 0.5 mM spermine.

putrescine, spermidine, or spermine. Optimal structural requirements for inhibitory compounds were found in α,ω-diaminoalkanes, of which 1,7-diaminoheptane was most active. N-(4-Aminobutyl)aziridine proved to be an even better inhibitor of the uptake of paraquat and putrescine (O'Sullivan et al., 1991). The possible exploitation of the polyamine uptake system for purposes of radioprotection and cancer chemotherapy was discussed (Cohen & Smith, 1990).

Paraquat was found to increase production of growth-inhibitory and lethal superoxide in E. coli. The toxicity observed in the presence of O_2 was markedly increased in mutants defective in polyamine synthesis, suggesting a possible role for the amines in eliminating oxygen-containing radicals. The presence of exogenous spermidine inhibited accumulation of paraquat, and endogenous spermidine conferred resistance to the agent (Minton et al., 1990).

In plants, paraquat normally accumulates in root cell vacuoles and is translocated to a growing shoot. Tobacco plants transgenic for the E. coli glutathione reductase proved to be resistant to the herbicide. Animal cells selected for resistance to paraquat were high in glutathione peroxidase, presumably active in detoxifying the H_2O_2 produced in metabolizing the agent (Krall et al., 1991).

The alveolar type II cells isolated from rat lungs, believed to accumulate the toxic herbicide, were shown to possess the expected polyamine–paraquat transport system. The study of competition with spermidine and putrescine uptake by several amines indicated the existence of both Na^+-independent transport, operative for spermine, and a Na^+-dependent system, which functioned for putrescine and paraquat particularly. Putrescine and paraquat were competitive in the latter system (Rannels et al., 1989).

Brief Summary

Chapter 21 has been concerned with four major subjects, connecting the molecular roles of the polyamines from tissue media and embedded cells to cellular membranes to activities in the inner cytosol. Among the latter reactions, mitochondria, despite their internal location, have been treated as specialized cells in a polyamine-containing medium. Reactions of transglutaminase function in many of these locales, clotting an external medium such as blood or plasma, tying cells or soluble proteins to embedded cell membrane, or constructing insoluble protein networks within cell membranes. The enzyme may cross-link proteins in its normal functions, but it can also add polyamines to suitable glutaminyl protein substrates to form amides under conditions of high polyamine levels. Such polyamine-containing amides have been found to be increased in several proteins found in particular biological systems or pathology.

Immunochemical methods and electron microscopy have been used to show that in several pathologies, polyamines can be bound noncovalently at specific sites at cell surfaces. Indeed the polyamines do bind at several types of anionic sites in membranes, and such associations tend to stabilize the membranes. Some of these anionic components, such as phosphorylated inositol derivatives, are very much involved in signal cascades, some of which are in fact modulated at defined enzymatic steps by the polyamines. The amines may penetrate the membranes of different types of cells by various mechanisms from simple diffusion to the involvement of elaborate protein carriers. The transport mechanisms may be regulated by intracellular polyamine. The polyamines may then accumulate in trapping environments involving pH, anion binding, or within specialized organelles, such as vacuoles or mitochondria. Efflux mechanisms are similarly of several kinds.

In the case of mitochondria, whose aerobic production of ATP often requires Ca^{2+}, the level of accumulation of Ca^{2+} and spermine has been used in animal cells both to buffer the cytosolic concentration of Ca^{2+} and to modulate enzyme activity within the organelle.

REFERENCES

Abe, T., Chung, S. I., DiAugustine, R. P., & Folk, J. E. (1977) *Biochemistry*, **16**, 5495.

Achyuthan, K. E. & Greenberg, C. S. (1987) *Journal of Biological Chemistry*, **262**, 1901–1906.

Aeschlimann, D., Paulsson, M., & Mann, K. (1992) *Journal of Biological Chemistry*, **267**, 11316–11321.

Ahmed, K., Goueli, S. A., & Williams–Ashman, H. G. (1985) *Biochemical Journal*, **232**, 767–771.

——— (1988) *Biochimica et Biophysica Acta*, **966**, 384–389.

Aimar, C. & Grant, N. (1992) *Biology of the Cell*, **76**, 23–31.

Alghisi, G., Roberts, E., Cardenas, M. E., & Gasser, S. M. (1994) *Cellular and Molecular Biology Research*, **40**, 563–571.

Amatruda, J. M. & Lockwood, D. H. (1974) *Biochimica et Biophysica Acta*, **372**, 266–273.

Anderson, P. J., Bardócz, S., Campos, R., & Brown, D. L. (1985) *Biochemical and Biophysical Research Communications*, **132**, 147–156.

Antognoni, F., Pistocchi, R., & Bagni, N. (1993) *Plant Physiology and Biochemistry*, **31**, 693–698.

Auger, M., McDermott, A. E., Robinson, V., Castelhano, A. L., Billedeau, R. J., Pliura, D. H., Krantz, A., & Griffin, R. G. (1993) *Biochemistry*, **32**, 3930–3934.

Azzi, A., Boscoboinik, D., & Hensey, C. (1992) *European Journal of Biochemistry*, **208**, 547–557.

Balasundaram, D., Tabor, C. W., & Tabor, H. (1991) *Proceedings of the National Academy of Sciences of the United States of America*, **88**, 5872–5876.

——— (1993) *Proceedings of the National Academy of Sciences of the United States of America*, **90**, 4693–4697.

Ballas, S. K., Mohandas, N., Marton, L. J., & Shohet, S. B. (1983) *Proceedings of the National Academy of Sciences of the United States of America*, **80**, 1942–1946.

Barry, E. L. R. & Mosher, D. F. (1989) *Journal of Biological Chemistry*, **264**, 4179–4185.

Barsigian, C., Stern, A. M., & Martínez, J. (1991) *Journal of Biological Chemistry*, **266**, 22501–22509.

Bartos, D., Bartos, F., Campbell, R. A., Grettie, D. P., & Smejtek, P. (1980) *Science*, **208**, 1178–1181.

Baskova, I. P., Timokhina, E. A., Nikonov, G. J., & Stepanov, V. M. (1990) *Biokhimiya*, **55**, 771–775.

Bauer, P. I., Farkas, G., Biday, L., Mikala, G., Meszaros, G., Kun, E., & Farrago, A. (1992) *Biochemical and Biophysical Research Communications*, **187**, 730–736.

Beninati, S., Abbruzzese, A., & Cardinali, M. (1993b) *International Journal of Cancer*, **53**, 792–797.

Beninati, S., Mantile, G., Passeggio, A., & Abbruzzese, A. (1993a) *Italian Journal of Biochemistry*, **42**, 66A–68A.

Beninati, S., Nicolini, L., Jakus, J., Passeggio, A., & Abbruzzese, A. (1995) *Biochemical Journal*, **305**, 725–728.

Beninati, S., Piacentini, M., Argento–Cerù, M. P., Russo–Caia, S., & Autuori, F. (1985) *Biochimica et Biophysica Acta*, **841**, 120–126.

Beninati, S., Piacentini, M., Cocuzzi, E. T., Autuori, F., & Folk, J. E. (1988) *Biochimica et Biophysica Acta*, **952**, 325–333.

Beninati, S., Senger, D. R., Cordella–Miele, E., Mukherjee, A. B., Chackalaparampil, I., Shanmugam, V., Singh, K., & Mukherjee, B. B. (1994) *Journal of Biochemistry*, **115**, 675–682.

Bergamini, C. M. (1988) *FEBS Letters*, **239**, 255–258.

Bergamini, C. M. & Signorini, M. (1993) *Biochemical Journal*, **291**, 37–39.

Berridge, M. J. & Irvine, R. F. (1989) *Nature*, **341**, 197–205.

Bianchi, L., Farrace, M. G., Nini, G., & Piacentini, M. (1994) *Journal of Investigative Dermatology*, **103**, 829–833.

Birckbichler, P. J., Dowben, R. M., Matacic, S., & Loewy, A. G. (1973) *Biochimica et Biophysica Acta*, **291**, 149–155.

Birckbichler, P. J., Orr, G. R., Conway, E., & Patterson, Jr., M. K. (1977) *Cancer Research*, **37**, 1340–1344.

Birckbichler, P. J., Orr, G. R., Patterson, Jr., M. K., Conway, E., & Carter, H. A. (1981) *Proceedings of the National Academy of Sciences of the United States of America*, **78**, 5005–5008.

Bishop, P. D., Lasser, G. W., Le Trong, I., Stenkamp, R. E., & Teller, D. C. (1990) *Journal of Biological Chemistry*, **265**, 13888–13889.

Bogle, R. G., Mann, G. E., Pearson, J. D., & Morgan, D. M. L. (1994) *American Journal of Physiology*, **266**, C776–C783.

Bordin, L., Cattapan, F., Clari, G., Toninello, A., Siliprandi, N., & Moret, V. (1994) *Biochimica et Biophysica Acta*, **1199**, 266–270.

Borth, W., Chang, V., Bishop, P., & Harpel, P. C. (1991) *Journal of Biological Chemistry*, **266**, 18149–18153.

Bowness, J. M., Henteleff, H., & Dolynchuk, K. N. (1987) *Connective Tissue Research*, **16**, 57–70.

Bradway, S. D., Bergey, E. J., Scannapieco, F. A., Ramasubbu, N., Zawacki, S., & Levine, M. J. (1992) *Biochemical Journal*, **284**, 557–564.

Brenner, S. C. & Wold, F. (1978) *Biochimica et Biophysica Acta*, **522**, 74–83.

Brockman, J. L., Gross, S. D., Sussman, M. R., & Anderson, R. A. (1992) *Proceedings of the National Academy of Sciences of the United States of America*, **89**, 945–948.

Bueb, J., Da Silva, A., Mousli, M., & Landry, Y. (1992) *Biochemical Journal*, **282**, 545–550.

Byczkowski, J. Z. & Porter, C. W. (1983) *General Pharmacology*, **14**, 615–621.

Byers, T. L. & Pegg, A. E. (1989) *American Journal of Physiology*, **257**, C545–C553.

——— (1990) *Journal of Cellular Physiology*, **143**, 460–467.

Byers, T. L., Wechter, R., Nuttall, M. E., & Pegg, A. E. (1989) *Biochemical Journal*, **263**, 745–752.

Byrd, J. C. & Lichti, U. (1987) *Journal of Biological Chemistry*, **262**, 11699–11705.

Cariello, L., Velasco, P. T., Wilson, J., Parameswaran, K. M., Karush, F., & Lorand, L. (1990) *Biochemistry*, **29**, 5103–5108.

Cariello, L., Wilson, J., & Lorand, L. (1984) *Biochemistry*, **23**, 6843–6850.

Cariello, L., Zanetti, L., & Lorand, L. (1994) *Biochemical and Biophysical Research Communications*, **205**, 565–569.

Carroway, K. L. & Carroway, C. A. (1989) *Biochimica et Biophysica Acta*, **988**, 147–171.

Caruso, A., Pellati, A., Bosi, G., Arena, N., & Stabellini, G. (1994) *European Journal of Histochemistry*, **38**, 245–252.

Casillas, E. T., Elder, C. M., & Hoskins, D. D. (1980) *Journal of Reproductive Fertility*, **59**, 297–302.

Chakravarty, R., Rong, X., & Rice, R. H. (1990) *Biochemical Journal*, **271**, 25–30.

Chang, C. & Geahlen, R. L. (1992) *Journal of Natural Products*, **55**, 1529–1560.

Cheeseman, C. I. (1991) *Progress in Biophysics & Molecular Biology*, **55**, 71–84.

Chen, K. Y. (1984) *Molecular and Cellular Biology*, **58**, 91–97.

Choi, S., Nakanishi, K., & Usherwood, P. N. R. (1993) *Tetrahedron*, **49**, 5777–5790.

Chun, P. W., Rennert, O. M., Saffen, E. E., & Taylor, W. J. (1976) *Biochemical and Biophysical Research Communications*, **69**, 1095–1101.

Chung, J., Kaloyanides, G., McDaniel, R., McLaughlin, A., & McLaughlin, S. (1985) *Biochemistry*, **24**, 442–452.

Clari, G., Toninello, A., Bordin, L., Cattapan, F., Piccinelli–Siliprandi, D., & Moret, V. (1994) *Biochemical and Biophysical Research Communications*, **205**, 389–395.

Clark, E. A. & Brugge, J. S. (1995) *Science*, **268**, 233–239.

Clarke, D. D., Mycek, M. J., Neidle, A., & Waelsch, H. (1959) *Archives of Biochemistry and Biophysics*, **79**, 338–354.

Cochet, C., Feige, J. J., & Chambaz, E. M. (1983) *Biochimica et Biophysica Acta*, **743**, 1–12.

Cocuzzi, E., Piacentini, M., Beninati, S., & Chung, S. I. (1990) *Biochemical Journal*, **265**, 707–713.

Cohen, G. M. & Smith, L. L. (1990) *Biochemical Society Transactions*, **18**, 743–745.

Cohen, P. (1989) *Annual Review of Biochemistry*, **58**, 453–508.

Cohen, P. & Cohen, P. T. W. (1989) *Journal of Biochemistry*, **264**, 21435–21438.

Cordella–Miele, E., Miele, L., Beninati, S., & Mukherjee, A. B. (1993) *Journal of Biochemistry*, **113**, 164–173.

Coussons, P. J., Price, N. C., Kelly, S. M., Smith, B., & Sawyer, L. (1992) *Biochemical Journal*, **283**, 803–806.

Crowther, R. C., Pritchard, C. M., Qiu, S. M., & Solloway, R. D. (1993) *Liver*, **13**, 141–145.

Curtis, C. G., Stenberg, P., Brown, K. L., Baron, A., Chen, K., Gray, A., Simpson, I., & Lorand, L. (1974) *Biochemistry*, **13**, 3257–3262.

Dadabay, C. Y. & Pike, L. J. (1989) *Biochemical Journal*, **264**, 679–685.

Dalbey, R. E. (1990) *Trends in Biochemical Sciences*, **15**, 253–257.

Damuni, Z., Humphreys, J. S., & Reed, L. J. (1984) *Biochemical and Biophysical Research Communications*, **124**, 95–99.

Davie, E. W., Fujikawa, K., & Kisiel, W. (1991) *Biochemistry*, **30**, 10363–10370.

Davies, P. J. A., Moore, Jr., W. T., & Murtaugh, M. P. (1984) *Bioessays*, **1**, 160–165.

Davis, R. H. & Ristow, J. L. (1988) *Archives of Biochemistry and Biophysics*, **267**, 479–489.

—— (1989) *Archives of Biochemistry and Biophysics*, **271**, 315–322.

—— (1991) *Archives of Biochemistry and Biophysics*, **285**, 306–311.

Davis, R. H., Ristow, J. L., Howard, A. D., & Barnett, G. R. (1991) *Archives of Biochemistry and Biophysics*, **285**, 297–305.

DeBenedette, M., Olson, J. W., & Snow, E. C. (1993) *Journal of Immunology*, **150**, 4218–4224.

Delcros, J., Roch, A., & Quash, G. (1984) *FEBS Letters*, **171**, 221–226.

Delcros, J., Roch, A., Thomas, V., El Alaoui, S., Moulinoux, J., & Quash, G. (1987) *Comptes Rendus de L'Academie des Sciences Paris*, **305**, 465–470.

Del Duca, S., Beninati, S., & Serafini–Fracassini, D. (1995) *Biochemical Journal*, **305**, 233–237.

Dinnella, C., Serafini–Fracassini, D., Grandi, B., & Del Duca, S. (1992) *Plant Physiology and Biochemistry*, **30**, 531–539.

Diwan, J. J., Yune, H. H., Bawa, R., Haley, T., & Manella, C. A. (1988) *Biochemical Pharmacology*, **37**, 957–961.

Dodds, A. W., Ren, X.-D., Willis, A. C., & Law, S. K. A. (1996) *Nature*, **379**, 177–179.

Dolynchuk, K. N., Bendor–Samuel, R., & Bowness, J. M. (1994) *Plastic and Reconstructive Surgery*, **93**, 567–573.

D'Orazi, D. & Bagni, N. (1987) *Biochemical and Biophysical Research Communications*, **148**, 1259–1263.

Dunn, J. A., Patrick, J. S., Thorpe, S. R., & Baynes, J. W. (1989) *Biochemistry*, **28**, 9464–9468.

Düzgünes, N., Hong, K., Baldwin, P. A., Bentz, J., Nir, S., & Papahadjopoulos, D. (1987) *Cell Fusion*, pp. 241–267, E. A. Sowers, Ed. New York: Plenum Press.

Eichberg, J., Zetusky, W. J., Bell, M. E., & Cavanagh, E. (1981) *Journal of Neurochemistry*, **36**, 1868–1871.

El Alaoui, S., Mian, S., Lawry, J., Quash, G., & Griffin, M. (1992) *FEBS Letters*, **311**, 174–178.

El Baze, P., Milano, G., Verrando, P., Renee, N., & Ortonne, J. P. (1985) *British Journal of Dermatology*, **112**, 393–396.

Elliott, D. C., Fournier, A., & Kokke, Y. S. (1988) *Phytochemistry*, **27**, 3725–3730.

Erikson, R. L. (1991) *Journal of Biological Chemistry*, **266**, 6007–6010.

Esmann, J., Voorhees, J. J., & Fisher, G. J. (1989) *Biochemical and Biophysical Research Communications*, **164**, 219–224.

Evans, R. M. & Fink, L. M. (1977) *Proceedings of the National Academy of Sciences of the United States of America*, **74**, 5341–5344.

Falcone, P., Serafini–Fracassini, D., & Del Duca, S. (1993) *Journal of Plant Physiology*, **142**, 265–273.

Fan, M., Chan, W., & Rennert, O. M. (1983) *Physiological Chemistry and Physics and Medical NMR*, **15**, 69–75.

Feige, J. J., Pirollet, F., Cochet, C., & Chambaz, E. M. (1980) *FEBS Letters*, **121**, 139–142.

Felschow, D. M., MacDiarmid, J., Bardos, T., Wu, R., Woster, P. M., & Porter, C. W. (1995) *Journal of Biological Chemistry*, **270**, 28705–28711.

Fenton, II, J. W. (1988) *Seminars in Thrombosis and Hemostasis*, **14**, 234–240.

Fésüs, L. & Tarcsa, E. (1989) *Biochemical Journal*, **263**, 843–848.

Fésüs, L., Thomázy, V., Autuori, F., Ceru, M. P., Tarcsa, E., & Piacentini, M. (1989) *FEBS Letters*, **245**, 150–154.

Fésüs, L., Thomázy, V., & Falus, A. (1987) *FEBS Letters*, **224**, 104–108.

Filhol, O., Cochet, C., & Chambaz, E. M. (1990) *Biochemistry*, **29**, 9928–9936.

Filhol, O., Cochet, C., Delagoutte, R., & Chambaz, E. M.

(1991a) *Biochemical and Biophysical Research Communications*, **180**, 945–952.

Filhol, O., Loue–Mackenbach, P., Cochet, C., & Chambaz, E. M. (1991b) *Biochemical and Biophysical Research Communications*, **180**, 623–630.

Fink, M. L. & Folk, J. E. (1981) *Molecular and Cellular Biochemistry*, **38**, 59–67.

Fischer, E. H., Charbonneau, H., & Tonks, N. K. (1991) *Science*, **253**, 401–406.

Fischer, E. H. & Krebs, E. G. (1989) *Biochimica et Biophysica Acta*, **1000**, 297–301.

Floyd, E. E. & Jetten, A. M. (1989) *Molecular and Cellular Biology*, **9**, 4846–4851.

Folk, J. E. (1980) *Annual Review of Biochemistry*, **49**, 517–531.

———— (1982) *Methods in Enzymology*, **87**, 36–42.

Folk, J. E. & Chung, S. I. (1973) *Advances in Enzymology*, **38**, 109–191.

Folk, J. E., Park, M. H., Chung, S. I., Schrode, J., Lester, E. P., & Cooper, H. L. (1980) *Journal of Biological Chemistry*, **255**, 3695–3700.

Fong, W. & Law, C. (1988) *Biochemical and Biophysical Research Communications*, **155**, 937–942.

Freund, K. F., Doshi, K. P., Gaul, S. L., Claremon, D. A., Remy, D. C., Baldwin, J. J., Pitzenberger, S. M., & Stern, A. M. (1994) *Biochemistry*, **33**, 10109–10119.

Friedman, D. L. (1986) *Biochemical and Biophysical Research Communications*, **134**, 1372–1378.

Fuller, G. M. & Doolittle, R. F. (1971) *Biochemistry*, **10**, 1311–1315.

Furuichi, K., Ezoe, H., Obara, T., & Oka, T. (1980) *Proceedings of the National Academy of Sciences of the United States of America*, **77**, 2904–2908.

Gabriel, M., Kopecká, M., & Svoboda, A. (1992) *Journal of General Microbiology*, **138**, 2229–2234.

Ganguly, P. & Bradford, H. R. (1982) *Biochimica et Biophysica Acta*, **714**, 192–199.

Gentile, V., Saydak, M., Chiocca, E. A., Akande, O., Birckbichler, P. J., Lee, K. N., Stein, J. P., & Davies, P. J. A. (1991) *Journal of Biological Chemistry*, **266**, 478–483.

Gentile, V., Thomázy, V., Piacentini, M., Fésüs, L., & Davies, P. J. A. (1992) *Journal of Cell Biology*, **119**, 463–474.

Gilad, G. M. & Gilad, V. H. (1992) *Biochemical Pharmacology*, **44**, 401–407.

Ginty, D. D., Osborne, D. L., & Seidel, E. R. (1989) *American Journal of Physiology*, **257**, G145–G150.

Giromini, L., Paina, A., Cerana, R., & Colombo, R. (1994) *Plant Physiology*, **105**, 921–926.

Goldschmidt–Clermont, P. J., Machesky, L. M., Baldassare, J. J., & Pollard, T. D. (1990) *Science*, **247**, 1575–1578.

González–Bosch, C., Jésus–Marcote, M., & Hernández–Yago, J. (1991) *Biochemical Journal*, **279**, 815–820.

Grankowski, N., Boldyreff, B., & Issinger, O. (1991) *European Journal of Biochemistry*, **198**, 25–30.

Grankowski, N. & Issinger, O. (1990) *Biochemical and Biophysical Research Communications*, **167**, 471–476.

Grant, F. J., Taylor, D. A., Sheppard, P. O., Mathewes, S. L., Lint, W., Vanaja, E., Bishop, P. D., & O'Hara, P. J. O. (1994) *Biochemical and Biophysical Research Communications*, **203**, 1117–1123.

Grant, N. J. & Oriol–Audit, C. (1985) *European Journal of Cell Biology*, **36**, 239–246.

Green, H. (1993) *Cell*, **74**, 955–956.

Green, H. & Wang, N. (1994) *Proceedings of the National Academy of Sciences of the United States of America*, **91**, 4298–4302.

Greenberg, C. S., Birckbichler, P. J., & Rice, R. H. (1991) *FASEB Journal*, **5**, 3071–3077.

Grimard, R., Tancrède, P., & Gicquaud, C. (1993) *Biochemical and Biophysical Research Communications*, **190**, 1017–1022.

Groblewski, G. E., Hargittai, P. T., & Seidel, E. R. (1992) *American Journal of Physiology*, **262**, C1356–C1363.

Groenen, P. J. T. A., Smulders, R. H. P. H., Peters, R. F. R., Grootjans, J. J., Van Den Ussel, P. R. L. A., Bloemendal, H., & De Jong, W. W. (1994) *European Journal of Biochemistry*, **220**, 795–799.

Guarino, L. A. & Cohen, S. S. (1979) *Proceedings of the National Academy of Sciences of the United States of America*, **76**, 3660–3664.

Gunter, T. E. & Pfeiffer, D. R. (1990) *American Journal of Physiology*, **258**, C755–C786.

Haber, M. T., Fukui, T., Lebowitz, M. S., & Lowenstein, J. M. (1991) *Archives of Biochemistry and Biophysics*, **288**, 243–249.

Hand, D., Elliott, B. M., & Griffin, M. (1987) *Biochimica et Biophysica Acta*, **930**, 432–437.

Hanks, S. K., Quinn, A. M., & Hunter, T. (1988) *Science*, **241**, 42–52.

Hannun, Y. A. & Bell, R. M. (1989) *Science*, **243**, 500–507.

Hara, T. & Endo, H. (1982) *Biochemistry*, **21**, 2632–2637.

Harsfalvi, J., Tarcsa, E., Udvardy, M., & Fésüs, L. (1994) *Fibrinolysis*, **8**, 378–381.

Hashimoto, E., Kobayashi, T., & Yamamura, H. (1984) *Biochemical and Biophysical Research Communications*, **121**, 271–276.

Hathaway, G. M. & Traugh, J. A. (1984) *Journal of Biological Chemistry*, **259**, 7011–7015.

Haugland, R. B., Lin, T., Dowben, R. M., & Birckbichler, P. J. (1982) *Biochemical Journal*, **37**, 191–193.

Hawel, III, L., Tjandrawinata, R. R., & Byus, C. V. (1994) *Biochimica et Biophysica Acta*, **1222**, 15–26.

He, Y., Kashiwagi, K., Fukuchi, J., Terao, K., Shirakata, A., & Igarashi, K. (1993) *European Journal of Biochemistry*, **217**, 89–96.

He, Y., Suzuki, T., Kashiwagi, K., & Igarashi, K. (1994) *Biochemical and Biophysical Research Communications*, **203**, 608–614.

Heaton, M. A. & Flintoff, W. F. (1988) *Journal of Cellular Physiology*, **136**, 133–139.

Hegazy, M. G., Schlender, K. K., Wilson, S. E., & Reimann, E. M. (1989) *Biochimica et Biophysica Acta*, **1011**, 198–204.

Heinecke, J. W. & Shapiro, B. M. (1992) *Journal of Biological Chemistry*, **267**, 7959–7962.

Hennessey, E. S., Drummond, D. R., & Sparrow, J. C. (1993) *Biochemical Journal*, **282**, 657–671.

Hennings, H., Steinert, P., & Buxman, M. M. (1981) *Biochemical and Biophysical Research Communications*, **102**, 739–745.

Ho, K., Quarmby, V. E., French, F. S., & Wilson, E. M. (1992) *Journal of Biological Chemistry*, **267**, 12660–12667.

Holley, J., Mather, A., Cullis, P., Symons, M. R., Wardman, P., Watt, R. A., & Cohen, G. M. (1992) *Biochemical Pharmacology*, **43**, 763–769.

Hölttä, E., Auvinen, M., & Andersson, L. C. (1993) *Journal of Cell Biology*, **122**, 903–914.

Hubbard, S. R., Bishop, W. R., Kirschmeier, P., George, S. J., Cramer, S. P., & Hendrickson, W. A. (1991) *Science*, **254**, 1776–1779.

Ichinose, A., Bottenus, R. E., & Davie, E. W. (1990) *Journal of Biological Chemistry*, **265**, 13411–13414.

Igarashi, K., Honma, R., Tokuno, H., Kitada, M., Kitagawa,

H., & Hirose, S. (1981) *Biochemical and Biophysical Research Communications*, **103**, 659–666.

Igarashi, K., Kashiwagi, K., Kobayashi, H., Ohnishi, R., Kakegawa, T., Nagasu, A., & Hirose, S. (1989) *Journal of Biochemistry*, **106**, 294–298.

Ikura, K., Nasu, T., Yokota, H., Tsuchiya, Y., Sasaki, R., & Chiba, H. (1988) *Biochemistry*, **27**, 2898–2905.

Ikura, K., Suto, N., & Sasaki, R. (1990b) *FEBS Letters*, **268**, 203–205.

Ikura, K., Tsuchiya, Y., Sasaki, R., & Chiba, H. (1990a) *European Journal of Biochemistry*, **187**, 705–711.

Jamdar, S. C., Osborne, L. J., Wells, G. N., & Cohen, G. M. (1987) *Biochimica et Biophysica Acta*, **917**, 381–387.

Jenson, J. R., Lynch, G., & Baudry, M. (1989) *Journal of Neurochemistry*, **53**, 1173–1181.

Jetten, A. M. & Shirley, J. E. (1986) *Journal of Biological Chemistry*, **261**, 15097–15101.

Johnson, T. S., Knight, C. R. L., El Alaoui, S., Mian, S., Rees, R. C., Gentile, V., Davies, P. J. A., & Griffin, M. (1994) *Oncogene*, **9**, 2935–2942.

Juprelle–Soret, M., Wattiaux–De Coninck, S., & Wattiaux, R. (1988) *Biochemical Journal*, **250**, 421–427.

Kakinuma, Y., Hoshino, K., & Igarashi, K. (1988) *European Journal of Biochemistry*, **176**, 409–414.

Kakinuma, Y., Masuda, N., & Igarashi, K. (1992) *Biochimica et Biophysica Acta*, **1107**, 126–130.

Kamínska, B., Kaczmarek, L., & Grzelakowska–Sztabert, B. (1992) *FEBS Letters*, **204**, 198–200.

Kanaji, T., Ozaki, H., Takao, T., Kawajiri, H., Ide, H., Motoki, M., & Shimonishi, Y. (1993) *Journal of Biological Chemistry*, **268**, 11565–11572.

Käpyaho, K., Kariniemi, A. L., Virtanen, I., & Jänne, J. (1984) *British Journal of Dermatology*, **111**, 403–411.

Katoh, S., Inoue, T., Kohno, H., & Ohkubo, Y. (1994a) *Biomedical Research*, **15**, 1–8.

Katoh, S., Midorikami, J., Takasu, S., & Ohkubo, Y. (1994b) *Biological and Pharmacological Bulletin*, **17**, 1003–1007.

Kemler, R. (1993) *Trends in Genetics*, **9**, 317–321.

Khan, N. A., Masson, I., Quemener, V., & Moulinoux, J. (1990a) *Cellular and Molecular Biology*, **36**, 345–348.

Khan, N. A., Quemener, V., & Moulinoux, J. (1989) *Biochimical Archives*, **5**, 321–329.

——— (1992) *Experimental Cell Research*, **199**, 378–382.

Khan, N. A., Quemener, V., Seiler, N., & Moulinoux, J. (1990b) *Pathobiology*, **58**, 172–178.

Kikkawa, U., Kishimoto, A., & Nishizuka, Y. (1989) *Annual Review of Biochemistry*, **58**, 31–44.

Killackey, J. J. F., Bonaventura, B. J., Castelhano, A. L., Billedeau, R. J., Farmer, W., Deyoung, L., Krantz, A., & Pliura, D. H. (1989) *Molecular Pharmacology*, **35**, 701–706.

Kim, H. C., Idler, W. W., Kim, I. G., Han, J. H., Chung, S. I., & Steinert, P. M. (1991) *Journal of Biological Chemistry*, **266**, 536–539.

Kinsella, M. G. & Wight, T. N. (1990) *Journal of Biological Chemistry*, **265**, 17891–17898.

Klein, J. D., Guzman, E., & Kuehn, G. D. (1992) *Journal of Bacteriology*, **174**, 2599–2605.

Klimczak, L. J. & Cashmore, A. R. (1993) *Biochemical Journal*, **293**, 283–288.

Knighton, D. R., Zheng, J., Ten Eyck, L. F., Ashford, V. A., Xuong, N., Taylor, S. S., & Sowadski, J. M. (1991) *Science*, **253**, 407–414.

Kobayashi, M., Iseki, K., Sugawara, M., & Miyazaki, K. (1993) *Biochimica et Biophysica Acta*, **1151**, 161–167.

Korner, G., Schneider, D. E., Purdon, M. A., & Bjornsson, T. D. (1989) *Biochemical Journal*, **262**, 633–641.

Kosa, K., Meyers, K., & De Luca, L. M. (1993) *Biochemical and Biophysical Research Communications*, **196**, 1025–1033.

Krall, J., Speranza, M. J., & Lynch, R. E. (1991) *Archives of Biochemistry and Physics*, **286**, 311–315.

Kramer, D. L., Miller, J. T., Bergeron, R. J., Khomutov, R., Khomutov, A., & Porter, C. W. (1993) *Journal of Cellular Physiology*, **155**, 399–407.

Kvedar, J. C., Pion, I. A., Bilodeau, E. B., Baden, H. P., & Greco, M. A. (1992) *Biochemistry*, **31**, 49–56.

Lauharanta, J. & Käpyaho, K. (1983) *Acta Dermato-Venerologica*, **63**, 277–282.

Lauharanta, J., Kousa, M., Käpyaho, K., Linnamaa, K., & Mustakallio, K. (1981) *British Journal of Dermatology*, **105**, 267–272.

Lee, K. N., Chung, S. I., Girard, J. E., & Fésüs, L. (1988) *Biochimica et Biophysica Acta*, **972**, 120–130.

Leiderman, L. J., Criss, W. E., & Oka, T. (1985) *Biochimica et Biophysica Acta*, **844**, 95–105.

Lenzen, S., Hickethier, R., & Panten, U. (1986) *Journal of Biological Chemistry*, **261**, 16478–16483.

Lenzen, S., Münster, W., & Rustenbeck, I. (1992) *Biochemical Journal*, **286**, 597–602.

Leroy, D., Schmid, N., Behr, J.-P., Filhol, O., Pares, S., Garin, J., Bougarit, J.-J., Chambaz, E. M. Y., & Cochet, C. (1995) *Journal of Biological Chemistry*, **270**, 17400–17406.

Leroy, D., Valero, E., Filhol, O., Heriché, J., Goldberg, Y., Chambaz, E. M., & Cochet, C. (1994) *Cellular and Molecular Biology Research*, **40**, 441–453.

Leroy–Viard, K., Jandrot–Perrus, A., Tobelem, G., & Guillin, M. C. (1989) *Experimental Cell Biology*, **181**, 1–10.

Levashova, Z. B. & Morozova, T. M. (1988) *Biokhimiya*, **53**, 251–255.

Levitzki, A. & Gazit, A. (1995) *Science*, **267**, 1782–1788.

Lorand, L. (1983) *Annals of the New York Academy of Science*, **421**, 10–27.

Lorand, L. & Conrad, I. M. (1984) *Molecular and Cellular Biochemistry*, **58**, 9–35.

Lorand, L., Hsu, L. K. H., Siefring, Jr., G. E., & Rafferty, N. S. (1981) *Proceedings of the National Academy of Sciences of the United States of America*, **78**, 1356–1360.

Lorand, L., Murthy, S. N. P., Velasco, P. T., & Karush, F. (1986) *Biochemical and Biophysical Research Communications*, **134**, 685–689.

Lorand, L., Weissmann, L., Epel, D. L., & Brüner–Lorand, J. (1976) *Proceedings of the National Academy of Sciences of the United States of America*, **73**, 4479–4481.

Lozeman, F. J., Litchfield, D. W., Piening, C., Takio, K., Walsh, K. A., & Krebs, E. G. (1990) *Biochemistry*, **29**, 8436–8447.

Lundberg, G. A., Jergil, B., & Sundler, R. (1986) *European Journal of Biochemistry*, **161**, 257–262.

Lundberg, G. A., Sundler, R., & Jergil B. (1987) *Biochimica et Biophysica Acta*, **922**, 1–7.

Lustigman, S., Brotman, B., Huima, T., Castelhano, A., Singh, R. N., Mehta, K., & Prince, A. M. (1995) *Antimicrobial Agents and Chemotherapy*, **39**, 1913–1919.

Lynch, G. W., Slayter, H. S., Miller, B. E., & McDonagh, J. (1987) *Journal of Biological Chemistry*, **262**, 1772–1776.

Maekawa, S., Hibasami, H., Uji, Y., & Nakashima, K. (1989) *Biochimica et Biophysica Acta*, **993**, 199–203.

Mäenpää, P. H. (1977) *Biochimica et Biophysica Acta*, **498**, 294–305.

Maher, P. A. (1993) *Journal of Biological Chemistry*, **268**, 4244–4249.

Majerus, P. W., Connolly, T. M., Deckmyn, H., Ross, T. S.,

Bross, T. E., Ishii, H., Bansal, V. S., & Wilson, D. B. (1986) *Science*, **234**, 1519–1526.

Mancon, M., Siliprandi, D., & Toninello, A. (1990) *Italian Journal of Biochemistry*, **39**, 278A–279A.

Mannherz, H. G. (1992) *Journal of Biological Chemistry*, **267**, 11661–11664.

Mariniello, L., Esposito, C., Di Pierro, P., Cozzolino, A., Pucci, P., & Porta, R. (1993) *European Journal of Biochemistry*, **215**, 99–104.

Martin, G. G., Hose, J. E., Omori, S., Chong, C., Hoodbhoy, T., & McKrell, N. (1991) *Comparative Biochemistry and Biophysiology B—Biochemistry & Molecular Biology*, **100B**, 517–522.

Martinet, N., Beninati, S., Nigra, T. P., & Folk, J. E. (1990) *Biochemical Journal*, **271**, 305–308.

Martinet, N., Kim, H. C., Girard, J. E., Nigra, T. P., Strong, D. H., Chung, S. I., & Folk, J. E. (1988) *Journal of Biological Chemistry*, **263**, 4236–4241.

Martínez, J., Chalupowicz, D. G., Roush, R. K., Sheth, A., & Barsigian, C. (1994) *Biochemistry*, **33**, 2538–2545.

Martínez, J., Rich, E., & Barsigian, C. (1989) *Journal of Biological Chemistry*, **264**, 20502–20508.

Masmoudi, A., Labourdette, G., Mersel, M., Huang, F. L., Huang, K., Vincendon, G., & Malviya, A. N. (1989) *Journal of Biological Chemistry*, **264**, 1172–1179.

McCormack, J. G. (1989) *Biochemical Journal*, **264**, 167–174.

McCormack, S. A. & Johnson, L. R. (1989) *American Journal of Physiology*, **256**, G868–G877.

——— (1991) *Experimental Cell Research*, **193**, 241–252.

McCormack, S. A., Wang, J., Viar, M. J., Tague, L., Davies, P. J. A., & Johnson, L. R. (1994) *American Journal of Physiology*, **267**, C706–C714.

McDonald, C. J. (1983) *Journal of Investigative Dermatology*, **81**, 385–387.

McDonald, L. J. & Mamrack, M. D. (1995) *Journal of Lipid Mediators*, **11**, 81–91.

Meggio, F., Boldyreff, B., Marin, O., Marchiori, F., Perich, J. W., Issinger, O., & Pinna, L. A. (1992) *European Journal of Biochemistry*, **205**, 939–945.

Meggio, F., Brunati, A. M., & Pinna, L. A. (1987) *FEBS Letters*, **215**, 241–246.

Meggio, F., Flamigni, F., Caldarera, C. M., Guarnieri, C., & Pinna, L. A. (1984) *Biochemical and Biophysical Research Communications*, **122**, 997–1004.

Mehta, K. & Lopez–Berestein, G. (1986) *Cancer Research*, **46**, 1388–1394.

Mehta, K., Rao, U. R., Vickery, A. C., & Fésüs, L. (1992) *Molecular and Biochemical Parasitology*, **53**, 1–16.

Melendrez, C. S., Ruttle, J. L., Hallford, D. M., Chaudhry, P. S., & Casillas, E. R. (1992) *Journal of Andrology*, **13**, 293–296.

Melino, G., Petruzzelli, M. A., Piredda, L., Candi, E., Gentile, V., Davies, P. J. A., & Piacentino, M. (1994) *Molecular and Cellular Biology*, **14**, 6584–6596.

Merrill, A. H. & Jones, D. D. (1990) *Biochimica et Biophysica Acta*, **1044**, 1–12.

Mezzetti, G., Monti, M. G., & Moruzzi, M. S. (1988) *Life Sciences*, **42**, 2293–2298.

Mezzetti, C., Moruzzi, M., Monti, M. G., Piccinini, G., & Barbirolo, B. (1985) *Molecular and Cellular Biochemistry*, **66**, 175–183.

Mezzetti, G., Moruzzi, M., Piccinini, G., Monti, M. G., & Barbiroli, B. (1986) *Molecular and Cellular Biochemistry*, **70**, 141–149.

Michel, S. & Démarchez, M. (1988) *Journal of Investigative Dermatology*, **90**, 472–474.

Michel, S., Reichert, U., Isnard, J. L., Shroot, B., & Schmidt, R. (1989) *FEBS Letters*, **258**, 35–38.

Mikles–Robertson, F., Feuerstein, B., Dave, C., & Porter, C. W. (1979) *Cancer Research*, **39**, 1919–1926.

Miller, M. L. & Johnson, G. V. W. (1995) *Journal of Neurochemistry*, **65**, 1760–1770.

Minchin, R. F., Raso, A., Martin, R. L., & Ilett, K. F. (1991) *European Journal of Biochemistry*, **200**, 457–462.

Minton, K. W., Tabor, H., & Tabor, C. W. (1990) *Proceedings of the National Academy of Sciences of the United States of America*, **87**, 2851–2855.

Missiaen, L., Wuytack, F., Raeymaekers, L., De Smedt, H., & Casteels, R. (1989) *Biochemical Journal*, **261**, 1055–1058.

Mitchell, J. L. A., Diveley, Jr., R. R., & Bareyal–Leyser, A. (1992a) *Biochemical and Biophysical Research Communications*, **186**, 81–88.

Mitchell, J. L. A., Diveley, Jr., R. R., Bareyal–Leyser, A., & Mitchell, J. L. (1992b) *Biochimica et Biophysica Acta*, **1136**, 136–142.

Mitchell, J. L. A., Judd, G. G., Bareyal–Leiser, A., & Ling, S. Y. (1994) *Biochemical Journal*, **299**, 19–22.

Miyaki, K., Mochida, I., Wada, T., & Kudo, T. (1979) *Chemical and Pharmacological Bulletin*, **7**, 123–126.

Monti, M. G., Marverti, G., Ghiaroni, S., Piccinini, H., Pernecco, L., & Moruzzi, M. S. (1994) *Experientia*, **50**, 953–957.

Morrison, L. D. & Kish, S. J. (1995) *Neuroscience Letters*, **197**, 5–8.

Moruzzi, M., Barbiroli, B., Monti, M. G., Tadolini, B., Hakim, G., & Mezzetti, G. (1987) *Biochemical Journal*, **247**, 175–180.

Moruzzi, M. S., Marverti, G., Piccinini, G., Frassineti, C., & Monti, M. G. (1993) *Molecular and Cellular Biochemistry*, **124**, 1–9.

——— (1995) *International Journal of Biochemistry and Cell Biology*, **27**, 783–788.

Mosesson, M. W. & Amrani, D. L. (1980) *Blood*, **56**, 145–158.

Mosesson, M. W., Siebenlist, K. R., Hainfeld, J. F., & Wall, J. S. (1995) *Journal of Structural Biology*, **115**, 88–101.

Moulinoux, J., Le Calve, M., Quemener, V., & Quash, G. (1984) *Biochimie*, **66**, 385–393.

Mylarman, R. & Brüner–Lorand, J. (1970) *Methods in Enzymology*, **19**, 765–770.

Nakanishi, K., Nara, K., Hagiwara, H., Aoyama, Y., Ueno, H., & Hirose, S. (1991) *European Journal of Biochemistry*, **202**, 15–21.

Nara, K., Aoyama, Y., Iwata, T., Hagiwara, H., & Hirose, S. (1992) *Biochemical and Biophysical Research Communationa*, **187**, 14–17.

Natta, C., Motyczka, A. A., & Kremzner, L. T. (1980) *Biochemical Medicine*, **23**, 144–149.

Nicchitta, C. V. & Williamson, J. R. (1984) *Journal of Biological Chemistry*, **259**, 12978–12983.

Nishizuka, Y. (1992) *Science*, **258**, 607–614.

Nonaka, M., Tanaka, H., Okiyama, A., Motoki, M., Ando, H., Umeda, K., & Matsuura, A. (1989) *Agicultural and Biological Chemistry*, **53**, 2619–2623.

Nonomura, K., Yamanishi, K., Hosokawa, Y., Doi, H., Hirano, J., Fukushima, S., & Yasuno, H. (1993) *British Journal of Dermatology*, **128**, 23–28.

Oefner, P. J., Wongyai, S., & Bonn, G. (1992) *Clinica Chimica Acta*, **205**, 11–18.

Oetken, C., Pessa–Morikawa, T., Autero, M., Andersson, L. C., & Mustelin, T. (1992) *Experimental Cell Research*, **202**, 370–375.

Ohki, S. & Duax, J. (1986) *Biochimica et Biophysica Acta*, **861**, 177–186.

Ohtsuka, T., Itezono, Y., Nakayama, N., Sakai, A., Shimma, N., & Yokose, K. (1994) *Journal of Antibiotics*, **47**, 6–15.

Ohsumi, Y. & Anraku, Y. (1981) *Journal of Biological Chemistry*, **256**, 2079–2082.

Oriol, C., Dubord, C., & Landon, F. (1977) *FEBS Letters*, **73**, 89–91.

Oriol–Audit, C. (1980) *Biochimie*, **62**, 713–714.

——— (1982) *Biochemical and Biophysical Research Communications*, **105**, 1096–1101.

Oriol–Audit, C., Hosseini, M. W., & Lehn, J. (1985) *European Journal of Biochemistry*, **151**, 557–559.

Osborne, D. L. & Seidel, E. R. (1990) *American Journal of Physiology*, **258**, G576–G584.

O'Sullivan, M. C., Golding, B. T., Smith, L. L., & Wyatt, I. (1991) *Biochemical Pharmacology*, **41**, 1839–1848.

Palen, E. & Traugh, J. A. (1991) *Biochemistry*, **30**, 5586–5590.

Paonessa, G., Metafora, S., Tajana, G., Abrescia, P., De Santis, A., Gentile, V., & Porta, R. (1984) *Science*, **226**, 852–855.

Parrish, R. F., Goodpasture, J. C., Zaneveld, L. J. D., & Polakoski, K. L. (1979) *Journal of Reproductive Fertility*, **57**, 239–243.

Parys, J. B., De Smedt, H., Van den Bosch, L., Geuns, J., & Borghgraef, R. (1990) *Journal of Cellular Physiology*, **144**, 365–375.

Patterson, Jr., M. K., Maxwell, M. D., Birckbichler, P. K., Conway, E., & Carter, H. A. (1982) *Biology International Reports*, **6**, 461–470.

Peng, H. & Lever, J. E. (1993) *Journal of Cellular Physiology*, **154**, 238–247.

Peiyasamy, S., Kothapalli, M. R., & Hoss, W. (1994) *Journal of Neurochemistry*, **63**, 1319–1327.

Peterson, L. L. & Wuepper, K. D. (1984) *Molecular and Cellular Biochemistry*, **58**, 99–111.

Phillips, J. E. & Chaffee, R. R. J. (1982) *Biochemical and Biophysical Research Communications*, **108**, 174–181.

Phillips, M. A., Qin, Q., Mehrpouyan, M., & Rice, R. H. (1993) *Biochemistry*, **32**, 11057–11063.

Phillips, M. A., Stewart, B. E., & Rice, R. H. (1992) *Journal of Biological Chemistry*, **267**, 2282–2286.

Piacentini, M., Fésüs, L., Farrace, M. G., Ghibelli, L., Piredda, L., & Melino, G. (1991) *European Journal of Cell Biology*, **54**, 246–254.

Piacentini, M., Fésüs, L., Sartori, C., & Cerù, M. P. (1988a) *Biochemical Journal*, **253**, 33–38.

Piacentini, M., Martinet, N., Beninati, S., & Folk, J. E. (1988b) *Journal of Biological Chemistry*, **263**, 3790–3794.

Piacentini, M., Sartori, C., Beninati, S., Bargagli, A. M., & Cerù–Argento, M. P. (1986) *Biochemical Journal*, **234**, 435–440.

Pinna, L. A. (1990) *Biochimica et Biophysica Acta*, **1054**, 267–284.

Pistocchi, R., Bagni, N., & Creus, J. A. (1987) *Plant Physiology*, **84**, 374–380.

Pober, J. S. & Strominger, J. L. (1981) *Nature*, **289**, 819–821.

Pohjanpelto, P., Virtanen, I., & Höltta, E. (1981) *Nature*, **293**, 475–477.

Porter, C. W., Mikles–Robertson, F., Kramer, D., & Dave, C. (1979) *Cancer Research*, **39**, 2414–2421.

Potter, B. V. L. & Lampe, D. (1995) *Angewandte Chemie International Edition English*, **34**, 1933–1972.

Pulkkinen, P., Piik, K., Koso, P., & Jänne, J. (1978) *Acta Endocrinologica*, **87**, 845–854.

Qi, D., Schatzman, R. C., Mazzei, G. J., Turner, R. S., Raynor, R. L., Liao, S., & Kuo, J. F. (1983) *Biochemical Journal*, **213**, 281–288.

Quash, G., Bonnefoy–Roch, A., Gazzolo, L., & Niveleau, A. (1978) *Advances in Polyamine Research*, vol. 2, pp. 85–92, R. A. Campbell, D. R. Morris, D. Bartos, G. D. Daves, & F. Bartos, Eds. New York: Raven Press.

Quash, G., Delain, E., & Huppert, J. (1971) *Experimental Cell Research*, **66**, 426–432.

Quemener, V., Moulinoux, J., Khan, N. A., & Seiler, N. (1990) *Biology of the Cell*, **70**, 133–139.

Rannels, D. E., Kameji, R., Pegg, A. E., & Rannels, S. R. (1989) *American Journal of Physiology*, **257**, L346–L353.

Rasmussen, L. K., Sørensen, E. S., Petersen, T. E., Gliemann, J., & Jensen, P. H. (1994) *FEBS Letters*, **338**, 161–166.

Rath, N. C. & Reddi, A. H. (1981) *Developmental Biology*, **82**, 211–216.

Reizer, J., Finley, K., Kakuda, D., MacLeod, C. L., Reizer, A., & Saier, Jr., M. H. (1993) *Protein Science*, **2**, 20–30.

Remko, M. & Scheiner, S. (1991) *Journal of Pharmaceutical Science*, **80**, 329–332.

Rhee, S. G., Suh, P., Ryu, S., & Lee, S. Y. (1989) *Science*, **244**, 546–550.

Rice, R. H., Chakravarty, R., Chen, J., O'Callahan, W., & Rubin, A. L. (1988) *Advances in Post-Translational Modifications of Proteins and Aging*, pp. 51–61, V. Zappia, P. Galletti, R. Porta, & F. Wold, Eds. New York: Plenum Press.

Rice, R. H., Rong, X., & Chakravarty, R. (1990) *Biochemical Journal*, **265**, 351–357.

Rigobello, M. P., Toninello, A., Siliprandi, D., & Bindoli, A. (1993) *Biochemical and Biophysical Research Communications*, **194**, 1276–1281.

Rinehart, Jr., C. A. & Chen, K. Y. (1984) *Journal of Biological Chemistry*, **259**, 4750–4756.

Roach, P. J. (1991) *Journal of Biological Chemistry*, **266**, 14139–14142.

Robinson, L. C., Hubbard, E. J. A., Graves, P. R., De Paoli–Roach, A. A., Roach, P. J., Kung, C., Haas, D. W., Hagedorn, C. H., Goebl, M., Culbertson, M. R., & Carlson, M. (1992) *Proceedings of the National Academy of Sciences of the United States of America*, **89**, 28–32.

Roch, A., Quash, G., & Huppert, J. (1978) *Comptes Rendus de l'Academie des Sciences, Paris*, **287D**, 1071–1074.

Ross, A. C. (1993) *FASEB Journal*, **7**, 317–327.

Rottenberg, H. & Marbach, M. (1990) *Biochimica et Biophysica Acta*, **1016**, 77–86.

Rubinstein, S. & Breitbart, H. (1991) *Biochemical Journal*, **278**, 25–28.

Ruoslahti, E., Pierschbacher, M., Hayman, E. G., & Engvall, E. (1982) *Trends in Biochemical Science*, **7**, 188–190.

Rutter, G. A., Diggle, T. A., & Denton, R. M. (1992) *Biochemical Journal*, **285**, 435–439.

Sacks, D. B., Fujita–Yamaguchi, Y., Gale, R. D., & McDonald, J. M. (1989) *Biochemical Journal*, **263**, 803–812.

Sakai, K., Sada, K., Tanaka, Y., Kobayashi, T., Nakamura, S., & Yamamura, H. (1988) *Biochemical and Biophysical Research Communications*, **154**, 883–889.

Sato, K., Tsukamasa, Y., Imai, C., Ohtsuki, K., Shimizu, Y., & Kawabata, M. (1992) *Journal of Agriculture and Food Chemicals*, **40**, 806–810.

Sayers, L. G. & Michelangeli, F. (1993) *Biochemical and Biophysical Research Communications*, **197**, 1203–1208.

Scemama, J., Grabié, V., & Seidel, E. R. (1993) *American Journal of Physiology*, **265**, G851–G856.

Schindler, M., Koppel, D. E., & Sheetz, M. P. (1980) *Proceedings of the National Academy of Sciences of the United States of America*, **77**, 1457–1461.

Schrode, J. & Folk, J. E. (1978) *Journal of Biological Chemistry*, **253**, 4837–4840.

Schroeder, W. T., Thacher, S. M., Stewart–Galetka, S., Annarella, M., Chema, D., Siciliano, M. J., Davies, P. J. A., Tang, H.-Y., Sowa, B. A., & Duvic, M. (1992) *Journal of Investigative Dermatology*, **99**, 27–34.

Schuber, F. (1989) *Biochemical Journal*, **260**, 1–10.

Schuber, F., Hong, K., Düzgünes, N., & Papahadjopoulos, D. (1983) *Biochemistry*, **22**, 6134–6139.

Schutt, C. E., Myslik, J. C., Rozycki, M. D., Goonesekere, N. C. W., & Lindberg, U. (1993) *Nature*, **365**, 810–816.

Scoccianti, V. & Bagni, N. (1992) *Plant Physiology and Biochemistry*, **30**, 135–138.

Scott, K. F. F. & Russell, D. H. (1982) *Journal of Cellular Physiology*, **111**, 111–116.

Seiler, N. & Dezeure, F. (1990) *International Journal of Biochemistry*, **22**, 211–218.

Seitz, J., Keppler, C., Hüntemann, S., Rausch, U., & Aumüller, G. (1991) *Biochimica et Biophysica Acta*, **1078**, 139–146.

Seiving, B., Ohlsson, K., Linder, C., & Stenberg, P. (1991) *European Journal of Hematology*, **46**, 263–271.

Sell, D. R. & Monnier, V. M. (1989) *Journal of Biological Chemistry*, **264**, 21597–21602.

Serafini–Fracassini, D., Del Duca, S., & Beninati, S. (1995) *Phytochemistry* **40**, 355–365.

Shariff, A. & Luna, E. J. (1992) *Science*, **256**, 245–247.

Shibley, Jr., I. A. & Pennington, S. N. (1995) *Biology of the Neonate*, **67**, 441–449.

Shigenaga, T., Takayenoki, Y., Kawasaki, S., Seki, N., Muta, T., Toh, Y., Ito, A., & Iwanaga, S. (1993) *Journal of Biochemistry*, **114**, 307–316.

Shilo, M. (1971) *Microbial Toxins, Volume VII*, pp. 67–103, S. Kadis, A. Ciegler, & S. J. Ajl, Eds. New York: Academic Press.

Shinozaki, T. & Pritzker, K. P. H. (1995) *Journal of Rheumatology*, **22**, 1907–1912.

Siefring, Jr., G. E., Apostol, A. B., Velasco, P. T., & Lorand, L. (1978) *Biochemistry*, **17**, 2598.

Signorini, M., Bortolotti, F., Poltronieri, L., & Bergamini, C. M. (1988) *Biological Chemistry Hoppe–Seyler*, **369**, 275–281.

Siliprandi, D., Toninello, A., & Via, L. D. (1992) *Biochimica et Biophysica Acta*, **1102**, 62–66.

Simon, M. & Green, H. (1985) *Cell*, **40**, 677–683.

—— (1988) *Journal of Biological Chemistry*, **263**, 18093–18098.

Singer, R., Sagiv, M., Levinsky, H., Maayan, R., Segenreich, E., & Allalouf, D. (1989) *International Journal of Fertility*, **34**, 224–230.

Singh, R. N. & Mehta, K. (1994) *European Journal of Biochemistry*, **225**, 625–634.

Singh, S. S., Chauhan, A., Brockerhoff, H., & Chauhan, V. P. S. (1995) *Life Sciences*, **57**, 684–694.

Singh, T. J. (1989) *FEBS Letters*, **243**, 289–292.

Sjöholm, A., Welsh, N., Hoftiezer, V., Blankston, P. W., & Hellerström, C. (1991) *Biochemical Journal*, **277**, 533–540.

Slife, C. W., Dorsett, M. D., Bouquett, G. T., Register, A., Taylor, E., & Conroy, S. (1985) *Archives of Biochemistry and Biophysics*, **241**, 329–336.

Slife, C. W., Wang, E., Hunter, R., Wang, S., Burgess, C., Liotta, D. C., & Merrill, Jr., A. (1989) *Journal of Biological Chemistry*, **264**, 10371–10377.

Smith, C. D. & Snyderman, R. (1988) *Biochemical Journal*, **256**, 125–130.

Smith, L. L. (1985) *Biochemical Society Transactions*, **13**, 332–334.

Solaini, G. & Tadolini, B. (1984) *Biochemical Journal*, **218**, 495–499.

Sommercorn, J., Mulligan, J. A., Lozeman, F. J., & Krebs, E. G. (1987) *Proceedings of the National Academy of Sciences of the United States of America*, **84**, 8834–8838.

Song, Y. C., Sheng, D., Taubenfeld, S. M., & Matsueda, G. R. (1994) *Analysis of Biochemistry*, **223**, 88–92.

Spathas, D. H., Pateman, J. A., & Clutterbuck, A. J. (1982) *Journal of General Microbiology*, **128**, 557–563.

Spector, A. (1995) *FASEB Journal*, **9**, 1173–1182.

Squires, E. J. & Feltham, L. A. W. (1980) *Biochemical Journal*, **185**, 761–766.

Stossel, T. P. (1989) *Journal of Biological Chemistry*, **264**, 18261–18264.

—— (1993) *Science*, **260**, 1086–1094.

Suzuki, T., He, Y., Kashiwagi, K., Murakami, Y., Hayashi, S., & Igarashi, K. (1994) *Proceedings of the National Academy of Sciences of the United States of America*, **91**, 8930–8934.

Sward, K., Pato, M. D., Nilsson, B.-O., Nordström, I., & Hellstrand, P. (1995) *American Journal of Physiology*, **269**, C563–C571.

Tadolini, B., Cabrini, L., Landi, L., Varani, E., & Pasquali, P. (1985b) *Biogenic Amines*, **3**, 97–106.

Tadolini, B., Cabrini, L., Piscinini, G., Davalli, P. P., & Sechi, A. M. (1985c) *Applied Biochemistry and Biotechnology*, **11**, 173–176.

Tadolini, B., Cabrini, L., Varani, E., & Sechi, A. M. (1985a) *Biogenic Amines*, **3**, 87–96.

Tadolini, B. & Varani, E. (1986) *Biochemical and Biophysical Research Communications*, **135**, 58–64.

Takahashi, N., Takahashi, Y., & Putnam, F. W. (1986) *Proceedings of the National Academy of Sciences of the United States of America*, **83**, 8019–8023.

Takaku, K. Futamura, M., Saitoh, S., & Takeuchi, Y. (1995) *Journal of Biochemistry*, **118**, 1268–1270.

Takano, T., Takigawa, M., & Suzuki, F. (1981) *Medical Biology*, **59**, 423–427.

Takehana, S., Washizu, K., Ando, K., Koikeda, S., Takeuchi, K., Matsui, H., Motoki, M., & Takagi, H. (1994) *Bioscience Biotechnology and Biochemistry*, **58**, 88–92.

Tanaka, H., Shinki, T., Takito, J., Jin, C. H., & Suda, T. (1991) *Experimental Cell Research*, **192**, 165–172.

Tarcsa, E., Kedei, M., Thomázy, V., & Fésüs, L. (1992) *Journal of Biological Chemistry*, **267**, 25648–25651.

Taylor, S. S. (1987) *Bioessays*, **7**, 24–29.

Thacher, S. M. (1989) *Journal of Investigative Dermatology*, **89**, 578–584.

Thams, P., Capito, K., & Hedeskov, C. J. (1986) *Biochemical Journal*, **237**, 131–138.

Thomas, A. P., Diggle, T. A., & Denton, R. M. (1986) *Biochemical Journal*, **238**, 83–81.

Thomas, E. L., Jefferson, M. M., & Grisham, M. B. (1982) *Biochemistry*, **24**, 6299–6308.

Thomázy, V. & Fésüs, L. (1989) *Cell and Tissue Research*, **255**, 215–224.

Thuren, T., Virtanen, J. A., & Kinnunen, P. K. J. (1986) *Journal of Membrane Biology*, **92**, 4–7.

Tibbs, J. (1982) *European Journal of Biochemistry*, **122**, 535–539.

Tokunaga, F., Muta, T., Iwanaga, S., Ichinose, A., Davie, E. W., Kuma, K., & Miyata, T. (1993b) *Journal of Biological Chemistry*, **268**, 262–268.

Tokunaga, F., Yamada, M., Miyata, T., Ding, Y., Hiranaga–Kawabata, M., Muta, T., Iwanaga, S., Ichinose, A., & Davie,

E. W. (1993a) *Journal of Biological Chemistry*, **268**, 252–261.

Toner, M., Vaio, G., McLaughlin, A., & Mclaughlin, S. (1988) *Biochemistry*, **27**, 7435–7443.

Toninello, A., Miotto, G., Siliprandi, D., Siliprandi, N., & Garlid, K. D. (1988) *Journal of Biological Chemistry*, **263**, 19407–19411.

Toninello, A., Via, L. D., Siliprandi, D., & Garlid, K. D. (1992) *Journal of Biological Chemistry*, **267**, 18393–18397.

Torrigiani, P., Serafini–Fracassini, D., Biondi, S., & Bagni, N. (1986) *Journal of Plant Physiology*, **124**, 23–29.

Trout, J. J., Koenig, H., Goldstone, A. D., Igbal, Z., Lu, C. Y., & Siddiqui, F. (1993) *Journal of Neurochemistry*, **60**, 352–355.

Tung, H. Y. L., Pelech, S., Fisher, M. J., Pogson, C. I., & Cohen, P. (1985) *European Journal of Biochemistry*, **149**, 305–313.

Turner, P. M. & Lorand, L. (1989) *Biochemistry*, **28**, 628–635.

Turner, R., North, M. J., & Harwood, J. M. (1979) *Biochemical Journal*, **180**, 119–127.

Upchurch, H. F., Conway, E., Patterson, Jr., M. K., & Maxwell, M. D. (1991) *Journal of Cellular Physiology*, **149**, 375–382.

Van den Bosch, L., De Smedt, H., Missiaen, L., Parys, J. B., & Borghgraef, R. (1990) *Biochemical Journal*, **265**, 609–612.

Van Groningen, J. J. M., Klink, S. L., Bloemers, H. P. J., & Swart, G. W. M. (1995) *International Journal of Cancer*, **60**, 383–387.

Van Leuven, F. (1982) *Trends in Biochemical Sciences*, **7**, 185–187.

Verma, R. & Chen, K. Y. (1986) *Journal of Biological Chemistry*, **261**, 2890–2896.

Vertino, P. M., Beerman, T. A., Kelly, E. J., Bergeron, R. J., & Porter, C. W. (1991) *Molecular Pharmacology*, **39**, 487–494.

Villanueva, V. R. (1983) *Advances in Polyamine Research*, pp. 297–306, V. Bachrach, A. Kaye, & R. Chayen, Eds. New York: Raven Press.

Vittur, F., Lunazzi, G., Moro, L., Stagni, N., de Bernard, B., Moretti, M., Stanta, G., Bacciottini, F., Orlandini, G., Reali, N., & Casti, A. (1986) *Biochimica et Biophysica Acta*, **881**, 38–45.

Vogel, S. & Hoppe, J. (1986) *European Journal of Biochemistry*, **154**, 253–257.

Votyakova, T. V., Bazhenova, E. N., & Zvjagilskaya, R. A. (1990) *FEBS Letters*, **261**, 139–141.

Vuento, M., Vartio, T., Saraste, M., von Bonsdorff, C., & Vaheri, A. (1980) *European Journal of Biochemistry*, **105**, 33–42.

Walters, J. D. & Johnson, J. D. (1988) *Biochimica et Biophysica Acta*, **957**, 138–142.

Wang, J., Viar, M. J., & Johnson, L. R. (1994a) *Proceedings of the Society for Experimental Biology and Medicine*, **205**, 20–28.

Wang, M., Kim, I., Steinert, P. M., & McBride, O. W. (1994b) *Genomics*, **23**, 721–722.

Wei, T. & Tao, M. (1993) *Archives of Biochemistry and Biophysics*, **307**, 206–216.

Welsh, N. (1990) *Biochemical Journal*, **271**, 393–397.

Weraarchakul–Boonmark, N., Jeong, J., Murthy, S. N. P., Engel, J. D., & Lorand, L. (1992) *Proceedings of the National Academy of Sciences of the United States of America*, **89**, 9804–9808.

White, M. F. & Kahn, C. R. (1994) *Journal of Biological Chemistry*, **269**, 1–4.

Williams, K., Dawson, V. L., Romano, C., Dichter, M. A., & Molinoff, P. B. (1990) *Neuron*, **5**, 199–208.

Williams–Ashman, H. G. (1989) *The Physiology of Amines*, pp. 3–20, U. Bachrach & Y. M. Heimer, Eds. Boca Raton, Fla: CRC Press.

Williams–Ashman, H. G., Wilson, J., Bell, R. E., & Lorand, L. (1977) *Biochemical and Biophysical Research Communications*, **79**, 1192–1198.

Wilson, J., Rickles, F. R., Armstrong, P. B., & Lorand, L. (1992) *Biochemical and Biophysical Research Communications*, **188**, 655–661.

Wojcikiewicz, R. J. H. & Fain, J. N. (1988) *Biochemical Journal*, **255**, 1015–1021.

Wyatt, I., Soames, A. R., Clay, M. F., & Smith, L. L. (1988) *Biochemical Pharmacology*, **37**, 1909–1918.

Wyse, J. W. & Butterfield, D. A. (1988) *Biochimica et Biophysica Acta*, **941**, 141–149.

Ye, X. S., Avdiushko, S. A., & Kuc, J. (1994) *Plant Science*, **94**, 109–118.

Yee, V. C., Pedersen, L. C., Trong, I. L., Bishop, P. D., Stenkamp, R. E., & Teller, D. C. (1994) *Proceedings of the National Academy of Sciences of the United States of America*, **91**, 7296–7300.

Yu, G. W. & Weiss, R. L. (1992) *Journal of Biological Chemistry*, **267**, 15491–15495.

Yuspa, S. H., Ben, T., & Steinert, P. (1982) *Journal of Biological Chemistry*, **257**, 9906–9908.

Zaneveld, L. J. D. & De Jonge, C. J. (1991) pp. 63–79, B. S. Dunbar & M. G. O'Rand, Eds. New York: Plenum Press.

Zhang, S., Jin, C., & Roux, S. J. (1993) *Plant Physiology*, **103**, 955–962.

CHAPTER 22

Molecular Effects on
Internal Cellular Polymers

With the exception of a few well-characterized reactions, such as the transglutaminases, the chemistries of external membrane-related structures are not clearly defined. In systems such as those of polyamine transport in *Escherichia coli* described in Chapter 7, the characterization and loci of the transport proteins are only initial studies of how these systems function at the molecular level. Some molecular interactions of the polyamines with internal cellular structures, such as some soluble proteins and nucleic acids, are becoming better known and will be the subject of this chapter and Chapter 23.

Observations with
Protein Enzymes

Early reviews pointed to numerous effects of the polyamines on catalytic activities. As the enzyme proteins were purified it appeared that in some instances the polyamines were affecting quaternary structure (i.e., associations and subunit interactions), for example, the sedimentation behavior of yeast uridine diphosphate galactose-4-epimerase and glycogen phosphorylase *b* in the presence of the polyamines (Darrow & Rodstrom, 1996; Wang et al., 1968). It was demonstrated that the partial activation and association of crystalline glycogen phosphorylase *b* by AMP was markedly enhanced by the polyamines, thereby demonstrating additional complexities.

The discovery of protein phosphorylation and the vast panoply of protein kinases and protein phosphatases began with studies on the interconversion of phosphorylases *b* and the phosphorylated *a*. Many of these reactions of phosphorylation and dephosphorylation have been found to be sensitive to the polyamines and polycations. New classes of these reactions with sensitivities to other cofactors, such as cAMP, Ca^{2+}, and calcitonin, have also been described (see Chap. 21).

The difficulty of understanding effects on structures embedded in complex membranes is underlined by two studies of acetylcholinesterase. One study examined effects on this enzyme, which is present in human erythrocyte ghosts; spermine in the range of 1–5 μM increased reactions at optimal substrate concentrations and inhibited the enzyme at 10–40 μM concentrations (Kossorotow et al., 1974). When similar studies were performed on the activity of the enzyme in insect and rat brains (Peter et al., 1979), low concentrations of spermine proved inhibitory and the modulation of the activities very much depended on substrate concentrations. Nevertheless, it was concluded that polyamines might participate in the regulation of the activity of this enzyme.

The mode of actions of the amine effects have been analyzed only occasionally to determine if the amines are affecting substrates, enzyme, products, impurities in the preparations, or some combinations of these. Effects on the phosphorylation of various nucleoside diphosphates by ATP in the presence of nucleoside diphosphokinase have been found to turn on the differences of the stability constants of spermidine, spermine, and Mg^{2+} with eight different ribonucleoside and deoxyribonucleoside diphosphates (Nakai & Glinsmann, 1977a). Titration of purified enzymes such as glucose oxidase (Voet & Andersen, 1984), yeast alcohol dehydrogenase (Dove & Tsai, 1976), and bovine brain α-aminobutyraldehyde dehydrogenase (Lee & Cho, 1992) with polyamines have demonstrated patterns of activation or inhibition of the activities as a result of effects on particular protein groups. In one study, the α-helical structure of a purified leaf proteinase was shown to be partially unraveled by 0.1 mM spermine (Balestreri et al., 1987). As noted earlier (Chap. 21), added polyamines often neutralize an inhibitory heparinlike component in a crude or even partially purified enzyme preparation. Additionally, the disruption of ribosomes by EDTA is known to release an activated proteinase, which is known to be inhibited by added polyamine (Levyant et al., 1979). In any case, these results have suggested the participation of the polyamines in the regulation of catalytic activities and cellular metabolism.

Protein Structure and Folding

By the early 1980s, the static structures of many small soluble proteins were defined in their primary sequences and details were known of local folding into β sheets and α helices. The degree of close packing of the latter secondary structures and the probable nature of the bonds conferring protein stability were also determined. The mechanisms of folding to form stable conformations were then studied on the premise that an amino acid sequence provides all the information necessary to gain a final tertiary structure. This was deduced initially from experiments on the denaturation and refolding of a very stable small ribonuclease, RNase A.

Nevertheless, the problems led to a search for substances facilitating the organization of tertiary structure (Gething & Sambrook, 1992). Two types of ubiquitous enzyme reactions were known to assist. A protein disulfide isomerase promoted the kind of protein disulfide that held a protein such as ribonuclease together, and a peptidyl prolyl *cis-trans*-isomerase affects the direction of a bend in the chain at a proline site. However, a new grouping of proteins, termed *chaperones*, was discovered; these helped to organize posttranslational folding and assembly, but they did not form part of the final structure. Chaperones exist in both procaryotic and eucaryotic cells and participate in protein maturation, in the rearrangement of cellular macromolecules during assembly and disassembly, and in targeting proteins for degradation. Antizyme may conceivably be classified as a chaperone in the specific degradation of ODC.

The view of a protein also grew as a dynamic breathing structure in which side chains and protein domains were continually moving. Indeed such movement may be essential to the acceptance of substrates and cofactors for enzymatic activity. A static structure of a protein, determined by the X-ray crystallography of the last half century, does not provide the dynamic picture obtained by newer experimental methods, such as nuclear magnetic resonance, isotope exchange, ligand binding, and catalytic activity. In the sense that a polyamine increases enzyme activity by reducing the motion that exposes an active site, it may be that the natural polyamines assume a function currently attributed to chaperones.

A fungal ribonuclease T1, which cleaves RNA specifically at guanyl residues, was found to bind tetraprotonated spermine at pH 6.0 at 25°C (Walz & Kitareewan, 1990). This binding of spermine is far tighter with the folded than the unfolded structure and in the former instance has a 38-fold higher association constant than does Mg^{2+}. Binding of spermine at a polyionic locus (three dicarboxylic amino acids) at the C-terminal end of the α helix occurs without loss of catalytic activity. The presence of spermine greatly slows unfolding of the native protein in 8 M urea. Reduction of the stabilizing disulfide bond results in the formation of inactive unfolded protein. The presence of 0.05 M spermine at 15°C stabilizes the activity and appears to increase the refolding; it may be asked if spermine or the less active spermidine fulfills this function in the cells producing RNase T1.

Ionic interactions of spermine were shown to increase the α helicity of synthetic peptides, and comparable reactive sites were identified in a range of native proteins (Tabet et al., 1993). Within somewhat disordered coiled-coil molecules of tropomyosin, spermine produced microcrystals of highly ordered tropomyosin (Xie et al., 1994). The Ca^{2+}-binding protein, parvalbumin, binds polyamines in its tryptophan-containing segment and appears to be stabilized thereby (Sudhakar et al., 1995).

Polyamines, Protein Isolation, and Characterization

McPherson et al. (1984) recognized that proteins of predominantly negative or positive character might be crystallized by appropriate electrostatic agents; for example, polyamines might be useful for the crystallization of acidic proteins. As described in Chapter 7, the recA protein can be crystallized directly from a lysate of *E. coli* by low concentrations of spermidine.

Ryanodine Receptor from Skeletal Muscle

The release of Ca^{2+} from the sarcoplasmic reticulum of skeletal muscle is triggered by a mechanism involving a receptor for the toxic alkaloid, ryanodine. Spermine stimulates the binding of ^{3}H-ryanodine to the receptor in the membrane some fivefold, with half-maximal stimulation at 3.5 mM. The effect is independent of the Ca^{2+} requirement for binding or the presence of phospholipids in the membrane. Spermidine and putrescine are essentially inactive in facilitating binding (Zarka & Shoshan–Barmatz, 1992). This specific effect of spermine facilitated a rapid purification of the high molecular weight (MW) group of soluble receptor polypeptides (~450 kDa, prelabeled with ^{3}H-ryanodine) by affinity chromatography on spermine-agarose, with specific elution in 2 mM spermine (Fig. 22.1). This phenomenon has contributed to the perception that polyamines modulate ion channel function (Uehara et al., 1996).

Activation of RNase A by Covalently Bound Spermine

The stable pancreatic RNase A was combined with spermine in the presence of the cross-linking agent, dimethyl suberimidate. A purified product, containing eight molecules of spermine per molecule of protein, was isolated in 5% yield. The polyspermine·RNase A had only a two- to fourfold increase of activity on single-stranded RNA but had a marked hydrolytic effect on

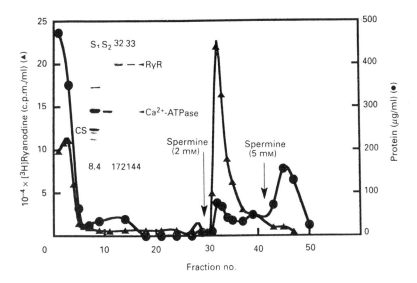

FIG. 22.1   Purification of ryanodine receptor on spermine-agarose column; adapted from Shoshan–Barmatz and Zarka (1992). Sarcoplasmic-reticulum membranes, labeled with (▲) [^{3}H]-ryanodine, were solubilized. (●) The preparation was fractionated on the column to reveal radioactivity and protein.

many double-stranded viral RNAs or a hybrid substrate, poly(dA)·poly(rI). The polyamine-RNase was 115 times as active as the original enzyme on poly(A)·poly(U) and 176 times as active toward poly(I)·poly(C) (Wang & Moore, 1977). Mere admixture of amine and enzyme in this ratio had essentially no effect on the enzyme activity.

The stimulation was thought to be a result of increasing the basic charge on this enzyme. Emulating the mechanism of RNase A, based on the activity of its histidine imidazoles, polyamine derivatives containing imidazoles were found to possess significant ability to cleave RNA (Shinozuka et al., 1994; Vlassov et al., 1995).

Nuclease Function and Polyamines

Polyamine activity became of interest with the recognition of the desirability of restricting nuclease activity in approaching the isolation of infectious intact viral RNA, studying the roles of mRNA or tRNA, or the participation of ribosomal RNA in the translation apparatus. The elimination of nucleic acids from protein-containing extracts with cationic precipitants such as streptomycin, led to the test of other precipitants such as spermine. Herbst and Doctor (1959) studied the effects of natural polyamines and their analogs on nucleates and various nuclease reactions in extracts of *Haemophilus parainfluenzae*. Gabbay and Shimshak (1968) demonstrated a spermidine and spermine inhibition of RNase-hydrolysis of poly(A) containing amine bound to adjacent phosphates. In general, agents that modify the secondary structure of RNA substrates markedly affect the degra-

dative patterns observed with various nucleases (Glitz et al., 1974).

On the other hand, Levy et al. (1974) described a homogeneous ribonuclease isolated from *Citrobacter* sp. that was stimulated by spermidine; this amine displaced inhibitory poly(G) and bound directly to the enzyme protein at two spermidine-binding sites. Schmukler et al. (1975) demonstrated a human plasma ribonuclease on which the polyamines had similar effects of activation and simultaneous removal of inhibitory poly(G). They linked these observations to the problem of the degradation of mRNA in eucaryotic cells. The existence of poly(A) as a 3′-terminating sequence of mRNA was known and Levy et al. (1975) found that poly(A) strongly inhibited RNA degradation by the *Citrobacter* nuclease and the human RNases; this inhibition was largely eliminated by spermidine and spermine. It was proposed that poly(A) at the end of mRNA prevents the degradation and that the polyamines regulate this effect. The synthesis of poly(A) by poly(A) polymerase on various RNA primers is inhibited almost completely by 0.4 mM spermine and somewhat less effectively by spermidine (Rose & Jacob, 1976). Karpetsky et al. (1980, 1981) produced a small poly(A)-terminated ribosomal RNA (5S) and described its altered properties and resistance to nucleases. This resistance was reversed by a mix of spermidine and Mg^{2+}.

Eventually the enormous multiplicity of ribonucleases, some 20 in *E. coli* alone, led to studies in which a considerable variety of polyamine effects was detected. These related to the degree of stimulation of hydrolysis, the nature of the linkages cleaved, base composition of the substrates, and the particular protein nuclease. The ready availability of the pancreatic RNase A and of the known synthetic substrates, the esters of 3'-pyrimidine nucleotides, facilitated concentration on these systems. With RNase A the polyamines did not change the Michaelis constant (K_m) for the substrates [e.g., 2',3'-cCMP or poly(C)], but did increase maximum velocity (V_{max}) significantly. In this system the polyamines apparently reacted primarily with the substrate (Kumagai et al., 1980).

A spermidine-dependent endonuclease has been discovered in mouse extracts (Nashimoto et al., 1991). The enzyme also requires an essential polynucleotide cofactor containing approximately 65 nucleotides, that is, partial tRNAs lacking 3' termini (Nashimoto, 1993).

Two Spermine-Binding Proteins

The rat ventral prostate was found to contain an acidic protein that bound spermine selectively: > thermine > spermidine > putrescine (Liang et al., 1978). The cytosolic protein, purified to homogeneity, was approximately 30,000 Da and noncovalently bound one molecule of spermine per molecule of protein (Mezzetti et al., 1979). The binding activity was largely lost after treatment with a phosphatase, and activity was regained by treatment with a protein kinase. The spermine-binding protein disappeared, with a half-life of 3.5 h, in the prostate of rats injected with cycloheximide. Intraperitoneal injection of an androgen rapidly restored the glycoprotein.

This spermine-binding protein could serve as a substrate for additional phosphorylation on a serine residue by cAMP-independent protein kinases. These reactions were stimulated by spermine (Fig. 22.2) and reduced in the prostate of castrated animals, suggesting androgen

control (Goueli et al., 1985). A cDNA for the binding protein was cloned, and its nucleotide sequence revealed a high content (33%) of dicarboxylic acids (Chang et al., 1987). The cDNA was used to demonstrate a marked decrease in prostatic mRNA after castration and restoration after androgen injection. Similar results were obtained with the mouse protein and an encoding cDNA that was described in terms of its exon composition (Mills et al., 1987).

A spermine-binding protein found in the duodenal mucosa of young chickens is heat sensitive, has a molecular weight of about 32,000, and its activity is insensitive to phosphatase (Mezzetti et al., 1981). This protein is quite specific for spermine and is produced rapidly in rachitic chicks after injection of 1,25-dihydroxycholecalciferol. Binding data, using a gel-filtration method, revealed a molar ratio of spermine to binding protein of 1 : 1, with a dissociation constant of about 4.1 µM. Another binding assay, using a cation-exchange resin, gave similar results, indicating a single class of binding sites on the protein (Mezzetti et al., 1983). Although the functions of the spermine-binding proteins are not clear, the rapid increase in the production of the proteins in response to a stimulation of growth suggests significant roles of bound spermine, possibly in transport of the tetramine. A spermine-binding protein has been found in rabbit serum and may reduce the concentration of spermine released by dying or dead cells.

Data on Eucaryotic DNA Polymerases

Studies with animal cells in tissue culture and with nuclei isolated from these cells demonstrated that the

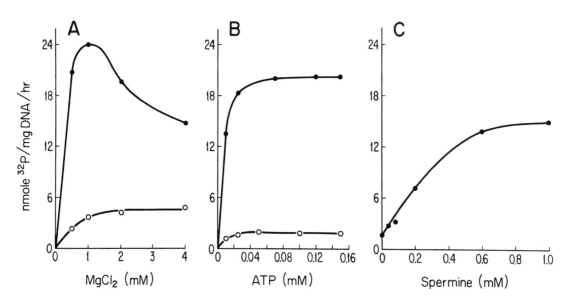

FIG. 22.2 Phosphorylation of spermine-binding protein by rat ventral-prostate chromatin-associated protein kinase activity; adapted from Goueli et al. (1985). (○) No spermine; (●) 1 mM spermine.

depletion of polyamines by treatment with inhibitors reduced rates of DNA synthesis. The inability to restore these rates to the normal levels by exogenous polyamines suggested some damage to the nuclear replication apparatus (Seyfried & Morris, 1979). Polyamine depletion by means of difluoromethylornithine (DFMO) causes cells to accumulate in the S phase. Depletion affected the rate of DNA elongation without affecting deoxynucleoside triphosphate pools (Oredsson et al., 1990). DNA repair was significantly slowed in depleted cells suffering DNA strand breaks as a result of X irradiation (Snyder, 1989).

The purification of DNA polymerase-α of calf thymus was described by Fisher and Korn (1977). They stated initially that the enzyme was stimulated two- to threefold by 1–4 μM spermidine. These workers studied the effect of spermidine on filling in gapped phage M13-DNA, recording stimulation at low concentrations of polyamine (1–2 mM) and increasingly inhibitory effects at higher concentrations (Fisher & Korn, 1979). The polyamine increased the K_m of the enzyme for DNA as well as the V_{max}. Their interpretations included a spermidine-induced dissociation of polymerase from its complex with DNA, that is, a decreased "processivity" induced by the amine. Nevertheless, they stated that spermidine did not affect the affinity of the polymerase for ssDNA.

DNA polymerase-α is recognized as the major DNA polymerase of eucaryotic cells; it is also active in DNA repair (Perrino & Loeb, 1990). Three families (A–C) of DNA polymerase have been recognized as possessing related amino acid sequences (Braithwaite & Ito, 1993) and DNA polymerase α (family B) is related to several enzymes, such as herpes simplex virus type I DNA polymerase for which some polyamine effects have been reported. The activity of the latter enzyme on gapped salmon sperm DNA is stimulated threefold and fourfold by spermidine and spermine (Wallace et al., 1981).

The search for increasingly active, larger complexes of DNA polymerase-α, comparable to the more efficient and active bacterial holoenzymes, detected a complex of DNA polymerase-α with primase. The latter synthesizes short polyribonucleotide primers that can be extended by the DNA polymerase. Polyamine depletion appears to block this process (Pohjanpelto & Hölttä, 1996). A dissociation of the primase from the complex and loss of DNA polymerase activity related to the elimination of some basic proteins, such as histones, which stimulated the activity of the complex on gapped DNA. Replacement of histone by 1 mM spermine effected a 10–15-fold increase of the activity (Hironaka et al., 1987). It is of interest that the yeast DNA telomerase, a ribonucleoprotein, is assayed in the presence of millimolar spermidine (Cohn & Blackburn, 1995).

As in the development of more effective bacterial replicative assemblies, the isolation of eucaryotic assemblies (So & Downey, 1992) has been aided by the use of small viral templates like simian virus 40 (SV40) DNA. The DNA polymerase α-primase complex found in essentially all animal cells and yeast is composed of four subunits of which two are associated with DNA primase. Numerous additional factors, replication proteins, and DNA helicases that unwind duplex DNA have been added to the postulated mechanism; and in vitro the entire assembly has appeared to bypass polyamine effects during synthesis. At present some nine different DNA helicases have been found in mammalian tissues. Thus, as in the study of the bacterial systems, effects on DNA synthesis by polyamines tested in highly structured cells and nuclei are not found with the complex composites of the isolated replicative components.

Retroviral DNA-Directed Synthesis

Retroviral reverse transcriptases were purified from preparations of several retroviruses and compared with DNA polymerase-α in their ability to replicate gapped thymus DNA. In systems in which α was stimulated two- to threefold by spermine, many of the viral DNA polymerases were stimulated 9–11-fold (Marcus et al., 1981). It had been noted that some trypanocides were quite active in inhibiting DNA polymerase-α and a retroviral reverse transcriptase, as well as the *Trypanosoma brucei brucei* DNA polymerases (Marcus et al., 1982). The inhibition of the reverse transcriptase could be overcome by the presence of spermine in the system. It was found that the DNA polymerases of the type C retroviruses (e.g., Rauscher leukemia virus) were particularly stimulated by spermine and spermidine and by some cationic trypanocides, such as antrycide and imidocarb. Indeed these trypanocides inhibited murine DNA polymerase-α and the *T. b. brucei* DNA polymerase, distinguishing the reverse transcriptase from the latter enzymes.

Additional Remarks on DNA Topoisomerases

Earlier chapters on these enzymes considered aspects of their discovery in all types of cells and roles in unraveling coils, thereby facilitating the formation of essential replication forks. Their differences and classification into type I and type II topoisomerases were described, as was the characterization of the multicomponent T4 topoisomerase. This enzyme increased the incidence of replicative forks in the synthesis of T4 viral DNA, which was in topological constraint in infected bacteria. Most cells, both bacterial and eucaryotic, appear to contain types I and II of the enzyme with which to control the level and direction of coiling of DNA. *E. coli* is now known to contain four topoisomerases (Luttinger, 1995). The coiling and uncoiling of the helical double-stranded structure is required in the functions of replication, transcription, and recombination.

A virus found in a culture of a thermophilic archaebacterium contained positively supercoiled DNA. This is an exception to almost all previously described duplex DNA, which is negatively supercoiled, that is, opposite to the direction of the double helix. An enzyme to effect this *reverse gyrase* activity was found in many thermophilic species of archaebacteria (Bouthier de la Tour et al., 1990). This reverse gyrase is a type I enzyme, operating on a single strand of duplex DNA, but is ATP-dependent, thereby unlike all other type I topoisomerases. Nadal et al. (1994) discussed the possible effects of spermidine in inhibition of reverse gyrase.

The initial discovery of the stimulation of DNA gyrase, a type II enzyme, by spermidine led to the demonstration that the increase of DNA concentration as a result of spermidine compaction was a major cause of the enhancement of enzymatic catenation evoked by the triamine (Krasnow & Cozzarelli, 1982). Although some workers reported that spermidine was not essential to the topoisomerase-catalyzed reaction, the addition of spermidine is frequently found in the assay conditions. Kosyavkin et al. (1990) reexamined the role of the polyamine as a function of the presence of other cations, Mg^{2+} and K^+, and of ionic strength, and described conditions of the parameters in which the *E. coli* DNA gyrase activity was maintained in the absence of spermidine. These authors interpreted their results to indicate that the ions, including spermidine, acted primarily to neutralize DNA charge and that spermidine was most effective at low Mg^{2+} concentrations and low ionic strength. The conclusion that spermidine induced compaction and that activation is a function of an increased DNA concentration is supported by the finding that a reverse topoisomerase derived from a thermophilic archaebacterium, *Sulfolobus acidocaldarius*, is activated by polyethylene glycol, which packs polymers by steric exclusion (Forterre et al., 1985). In a reaction run at 75°C, spermidine might well not effect compaction of DNA and it would be interesting to test the higher polyamines of the thermophiles, compounds that do support protein synthesis at temperatures at which spermidine is inactive.

Morris and colleagues observed the decline of synthesis of DNA (replication), RNA (transcription), and protein as function of polyamine depletion and suggested that the depletion of polyamine could cause a deficiency in topoisomerase action to account for all of these effects. They showed that spermidine-starved *E. coli* developed an abnormally high DNA content per cell and a decreased rate of DNA accumulation per starved cell, parameters consistent with a deceased rate of movement of DNA replication forks (Geiger & Morris, 1978). Marked reductions were also seen in growth rate, cell size, RNA, and protein content. It was suggested that a failure solely in the bacterial topoisomerase could produce this clinical picture. Nevertheless, the mechanism by which spermidine or other bacterial polyamines can control essential topoisomerase activities within the bacterial cell remains obscure.

According to Sinden and Pettijohn (1981), the bacterial chromosome in growing *E. coli* is constrained by torsional tension within 120 ± 30 domains per nucleoid. In eucaryotic cells the formation of nucleosomes by coiling DNA around histones imposes negative super-

coiling, which is subsequently relaxed enzymatically. Negative supercoiling imposed by bacterial gyrase requires wrapping of the DNA around the protein enzyme, a sequence also followed in generating nucleosomes. A one-sided cationic neutralization of DNA charge is now thought to account, at least in part, for the nucleosomic bending of DNA around histones (Crothers, 1994).

An assembly of DNA and histones into nucleosome-like particles on a DNA string occurs in the presence of the eucaryotic topoisomerase I, which then relaxes the superhelical turn imposed by wrapping DNA around histones (Germond et al., 1979). Chromatin itself contains this nicking–closing enzyme, which can be extracted by 150 mM sodium phosphate. On the other hand, a topoisomerase II appears to be essential for the condensation of mitotic chromosomes, and indeed is found to be a major constituent of the chromosomal scaffold and nuclear matrix (Adachi et al., 1991). This enzyme, which is also essential for the separation of chromatids during mitosis, is present in chromosomes as a phosphoprotein (Taagepera et al., 1993). Phosphorylation of the enzyme is effected in vivo by casein kinase II (CKII) and phosphorylation is essential to the activity (Cardenas & Gasser, 1993). Isolation of the enzyme by immunoprecipitation from a yeast extract has revealed a firm complex of the topoisomerase II and CKII, containing both catalytic activities (Bojanowski et al., 1993). Polyamines and polyanions are both important in the stabilization of the complex and spermine has been found to bind directly to an essential subunit of CKII. It was also found to bind to a subunit of CKII within the complex and to reduce the stability of the latter. Thus, in eucaryotic cells some of the effects of polyamine depletion may be mediated indirectly through the regulation of phosphorylation of topoisomerase II.

Among the eucaryotic topoisomerases, the type II eucaryotic enzyme requires Mg^{2+} and an energy source, such as ATP, in most systems to relax a supercoiled DNA; neither Mg^{2+} nor ATP are essential for type I enzymes (Osheroff, 1989). A variety of spermidine effects have been described, including stimulation of a type I topoisomerase found in vaccinia virus (Shuman et al., 1988).

Topoisomerase activity is very low in quiescent animal cells and very high in multiplying cells, including cancer cells. Specific inhibitors of the eucaryotic enzyme have been tested in the treatment of some cancers, compounds such as *m*-AMSA (see Chap. 19) and etoposide, which jam intermediate stages of DNA scission (Kohn et al., 1987). These workers have also reported that spermine and spermidine enhance the binding of a purified leukemic cell topoisomerase II to SV40 DNA and stimulate the relaxation of this supercoiled DNA (Pommier et al., 1989). Single-strand breaks of DNA in leukemic cells induced by *m*-AMSA were found to be increased by a polyamine deficiency produced by DFMO (Zwelling et al., 1985). Similarly a polyamine deficiency followed by etoposide increased cytotoxicity compared to the effects of either treatment alone (Dorr et al., 1986).

Eucaryotic Nucleus

The DNA-containing chromatin of nuclei is frequently described in terms of its condensation during the interphase. In contrast to the transcriptionally active relatively diffuse euchromatin, chromatin condensed in the interphase is designated as *heterochromatin*. The condensation of the chromosomes and all of the DNA normally takes place in preparation for mitotic separation and the control of these events is obviously of great interest. The condensed or less highly packed states of nuclear components may in themselves regulate important biological events, as seen by the effects of condensation of DNA by spermidine or polyethylene glycol on the activity of a DNA topoisomerase. Polyamines do condense structures within isolated rat liver cell nuclei. Major shrinkage and nucleolar condensation were obtained with 1 mM spermidine or spermine (Anderson & Norris, 1960). Eventually spermine was found in relatively large amounts in avian nucleated erythrocytes, compared to nonnucleated erythrocytes (Shimizu et al., 1965). Spermine promoted the development of young amphibian embryos when nuclei were transplanted into enucleated eggs.

After isolation of nuclei in presumably nonredistributive nonaqueous media (see Chap. 11), it was reported that 16–17% of the liver spermidine and spermine existed in the nuclei, which were only some 5% of the cell volume. This implied high concentrations of the polyamines in cell nuclei. The study of McCormick (1978), described in Chap. 11, Table 11.1, with cytochalasin-induced karyoplasts and cytoplasts derived from mouse fibroblasts pointed to about 55% of the total spermine and spermidine in the karyoplasts that contained only 9% of the ornithine decarboxylase (ODC), a largely cytosolic enzyme. In the nuclei of livers of aging rats (Symonds & Brosnan, 1977), protein and spermine contents of the nuclei increased with age, while the spermidine contents decreased.

Immunocytochemical and other cytochemical methods have demonstrated the presence of spermine and/or spermidine on the highly condensed chromatin of nucleated erythrocytes or in the metaphase and anaphase chromosomes of proliferating cells (Hougaard et al., 1987). These workers have suggested that the polyamines were normally important for chromatin condensation.

In a study of the effect of nucleases on nuclei isolated in a stabilizing buffer containing polyamine as the major cation, a nuclear Ca^{2+}-Mg^+-dependent endonuclease was found that, in contrast to pancreatic DNAase I, produced a stepladder of limit products (Hewish & Burgoyne, 1973). This implied a simple repeating substructure, now interpreted as a string of nucleosomes, tied together by linker DNA that was relatively unprotected

against endonuclease action. However, the condensed DNA of nuclei stabilized with polyamines is somewhat protected against DNAase II and micrococcal nuclease (Billett & Hall, 1979). A combination of Ca^{2+}, polyamine, and an inhibitor of endogenous proteinase in the isolation of mouse salivary gland nuclei permitted the isolation of relatively clean histone-containing mononucleosomes after treatment of the nuclei with micrococcal nuclease (Smith & Rowlatt, 1980).

In studies with the multinucleate slime mold *Physarum polycephalum*, the nuclei were labeled with isotopic metabolites in plasmodial cultures. Acidic soluble nuclear proteins were phosphorylated enzymatically in incubations of isolated nuclei and this was stimulated 5–15-fold by 1 mM polyamine and much less by Mg^{2+} (Atmar et al., 1978). A single unique nucleolar protein of 70,000 MW, phosphorylated in a polyamine-dependent reaction, was isolated and the essential protein kinase was purified (Daniels et al., 1981).

Thymic cells from normal or leukemic mice were separated in sucrose gradients and characterized in terms of size, DNA, and polyamine content (Heby et al., 1973). Nondividing G_0 cells were higher in spermine content than in a premitotic G_1 phase. All polyamines increase during the progression through the G and S phases and into mitosis, when the polyamine and DNA contents are twice that in G_1. Inhibition of polyamine synthesis in mammalian cells inhibited cytokinesis and resulted in many binucleate cells lacking well-organized actin filaments (Sunkara et al., 1979). Knuutila and Pohjanpelto (1983) showed that polyamine-dependent cells, depleted of polyamine, displayed many nuclear and chromosomal abnormalities, were often multinucleate, and lacked organized actin filaments and microtubules.

The incorporation of analogue polyamines, such as the bisethyl derivatives, into the nucleus is a major aspect of their mechanism of inhibition of tumor cell development. The effects of incorporation of 1,14-bis(ethylamino)-5,10-diazatetradecane (BE-4-4-4) into HeLa cells in culture was examined in nuclei in which histones had been removed. The DNA of nuclei of BE-4-4-4-treated cells was more sensitive than that of untreated cells to various nucleases and BE-4-4-4 condensed DNA and chromatin less effectively than did spermine. The incorporation of propidium iodide into DNA can be used to uncoil negatively supercoiled DNA and to estimate the degree of supercoiling in the DNA and the attachment of dehistonized DNA to the nuclear matrix. BE-4-4-4 incorporation reduced the unfolding pattern of the DNA and it was evident that the analogue affected the nuclear organization of that polymer.

Regulation of DNAases

Rat liver nuclei isolated in low salt buffer and extracted in 0.1 mM NaCl contained a soluble Mg^{2+}-dependent DNAase that also required spermine or spermidine (Nakano & Tsuboi, 1976). The activity was higher on ssDNA than on dsDNA. The spermine analogue sper- mindiol, N,N^1-bis(3-hydroxypropyl)-1,4-diaminobutane, also enhanced DNA hydrolysis by staphyloccal nuclease and spleen DNAase II (Yanagawa et al., 1976). Never- theless, many observations pointed to an action, by spermine particularly, against DNA cleavage by nucleases. For example, the course of micrococcal nuclease degradation of chromatin within isolated mouse nuclei was markedly affected by added poly- amine, which condensed various chromatin fractions (Rowlatt & Smith, 1981). The autodigestion of chroma- tin in yeast nuclei was prevented by 5 mM spermidine (Shalitin & Vishlizky, 1984).

The discovery and availability of specific restriction endonucleases enhanced the range of these observa- tions. These enzymes of bacterial origin were thought to eliminate foreign DNA without damaging host DNA. The well-known *E. coli* restriction nuclease, EcoRI, is localized in the peripheral periplasmic space, presum- ably to effect this role. Not only were the protections of DNA by 0.5–2 mM spermidine and spermine confirmed and extended to several widely used restriction nucleases, but many of the newer higher bacterial ho- mologues of the mammalian polyamines thermine, ho- mospermine, caldopentamine, caldohexamines, and the branched trisamines and tetrakisamine were even more markedly inhibitory (Kirino et al., 1990).

The most careful analysis of these effects required the use of DNA of known sequence and defined cleav- age sites, which might in fact be correlated with the genetic roles of the DNA sequences. Circular plasmids, whose physical properties are explorable after electro- phoresis in gels and whose genetic properties are even- tually testable in bacterial hosts, were substrates of choice for the study of the hydrolytic properties of par- ticular restriction nucleases. An early study of the ef- fects of the polyamines on EcoRI revealed a complex behavior in which the cleavage of plasmid and phage DNAs is enhanced at low concentrations of the poly- amines and inhibited by higher concentrations of these substances, as well as by two histonelike bacterial pro- teins (Pingoud et al., 1984). Pingoud (1985) discovered that a group of restriction endonucleases (type II) in- cluding EcoRI were relatively nonspecific in the ab- sence of spermidine. This polyamine markedly in- creased the specificity of cleavage by enhancing the cleavage rate at the specific site and slowing cleavage at other sites. Thus, spermidine increased the accuracy of cleavage by these nucleases.

A type II restriction endonuclease, NaeI, isolated from a *Nocardia* species, was found to be activated by 1 mM spermidine to cleave formerly resistant sites. With many of the 49 type II restriction enzymes, nuclease-resistant sites became cleavage sensitive on the addition of spermidine (Oller et al., 1991). It was asked if the variations of spermidine content as a func- tion of the bacterial cell cycle contributes to the regula- tion of gene expression by these nucleases. A similar transition in nucleosomal sites sensitive to and resistant to lysosomal DNase has been detected in *Drosophila*

DNA in response to the presence of spermidine (Gas- zner et al., 1993).

Chemical Cleavage of DNA

The activity of the lethal antibiotic, bleomycin, and its dependence on a terminal cation, as in bleomycin A_5, were discussed in Chapter 17. The cleavage of DNA by the various bleomycins begins with the elimination of a DNA base, and this aldehydic product is subject to subsequent β or δ eliminations. Recent studies on the more specific cleavage of RNA by bleomycin have been discussed by Hecht (1994).

The relative instability at low pH of the N-deoxyribo- syl purine bond in DNA leads to the formation of apuri- nic DNA, and the further cleavage of such nucleates is hastened by primary amines (Tamm et al., 1953). As described by Male et al. (1982), the order of effective- ness of the polyamines in such reactions was spermine > spermidine > putrescine > cadaverine; all apurinic sites in both free DNA and nucleohistones were hydrolyzed in the presence of these amines. In cleavage of apurinic sites of plasmid fragments by 1 mM spermidine or its acetyl derivatives, the efficiency of cleavage was in the order spermidine > N^8-acetylspermidine > N^1-acetylsper- midine (Haukanes et al., 1990).

The rate of depurination of a colicin DNA at 60°C and pH 4.3 is decreased some 10-fold by 10 mM poly- amine, with some twofold differences by these common amines (Andreev & Kaboev, 1982). The larger straight chain and branched polyamines found in thermophilic bacteria are far more effective than the triamines in pre- venting depurination at 70°C and pH 3.0 (Fig. 22.3). This may be one of the roles of these unusual amines

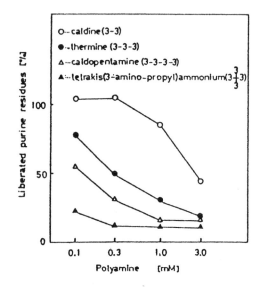

FIG. 22.3 Liberated purine residues after heating calf thymus DNA in solution at pH 3.0 with or without polyamines at 70°C for 3 h; adapted from Kobayashi et al. (1990).

in thermophiles. Simple diamines, but not monoamines, will cleave polyribonucleotides like poly(A) to small oligomers at 50°C and pH 8. This reaction, which does not occur with DNA, proceeds by the formation of 2′,3′-cyclic phosphates (Yoshinari et al., 1991).

The postulate of β elimination to produce a nick 3′ to the aldehydic site after spermine treatment of an abasic polydeoxyribonucleotide has been used to explain DNA nicks after the creation of apurinic or apyrimidinic sites (Lindahl & Andersson, 1972). This mechanism prevails in the cleavage of a model polynucleotide, but higher concentrations of spermine and longer periods of incubation also produce δ eliminations of an unsaturated sugar (Bailly et al., 1989). McHugh and Knowland (1995) have explored the use of diamines at neutrality to effect specific cleavage of abasic DNA.

An early enzymatic step in repair of a baseless DNA was found to involve endonucleases termed apurinic acid lyases, which catalyze strand scission on the 3′ side of the abasic site, and then eliminate an α,β-unsaturated aldose 5-phosphate. A synthetic molecule designed to evoke these reactions consists of 2-aminoadenine linked to a DNA intercalator, aminoacridine, by means of a polyamino chain. All of these components are essential to the activity, with the adenine residue affording abasic site recognition and the intercalator providing DNA binding. One very useful polyamine linker contains norspermidine, whose amine functions cleaved the abasic sites and bound DNA phosphate. In tests of the most efficient compound, an apurinic site in a plasmid of 4000 base pairs was recognized and cleaved at nanomolar concentrations of the synthetic nuclease (Fkyerat et al., 1993). Polyamine depletion is thought to increase sensitivity to some monofunctional alkylating agents by retarding DNA excision repair (Snyder & Bhatt, 1993). The application of a compound designed to cleave abasic nucleates under conditions of polyamine deficiency may prove useful in exacerbating the lethal action of such alkylating agents. Oxidative stress also affects DNA integrity and deoxyguanosine in DNA is particularly reactive with oxygen radicals. The presence of 0.1 mM spermine markedly reduces these effects (Muscari et al., 1995).

Polyamines and Chromatin

The current model of the nucleosome emerged relatively recently. Van Holde (1989) has presented a history of the evolution of ideas about the organization of chromatin until 1985. A nucleosome concept of chromatin as a sequence of partially protected beaded structure containing DNA linkers between the beads was formulated and substantiated mainly on biophysical, particularly crystallographic, evidence (Kornberg & Klug, 1981). Chromatin digested with micrococcal nuclease had been degraded to monomeric core particles and the purified monomers, lacking the noncore histone H1, were crystallized in various spermine-containing media. Hydrolysis by micrococcal nuclease has generally been retained as the method of fragmentation of the beaded

structures (Finch et al., 1977, 1981). Spermine was used earlier by many workers for the crystallization of tRNA (to be discussed in Chap. 23). The crystals of nucleosome cores were examined by X-ray diffraction and electron microscopy and led to the development of a detailed model for the structure of a nucleosome core at 7-Å resolution (Richmond et al., 1984). An earlier crystallization of chromatin subunits was effected as a cetyltrimethyl ammonium salt (Bakayev et al., 1975). The cation, cetyltrimethyl ammonium bromide, was introduced even earlier in the crystallization of tRNA.

The model of the nucleosome includes a repeating length of DNA wrapped around two copies, each of four different histones. The DNA, in a superhelix in a right-handed β form, is not bent uniformly and possesses some tight bends at contacts with particular histones, which protect against nuclease cleavage. The histone-DNA interactions occur on the inner surface of the superhelix and neutralize about 20% of the phosphate charge. Within the chromatin, the cores are linked in a string of beads to similar structures by short pieces of DNA carrying still another histone, H1, whose proximity to the core limits the action of a nuclease. H1 may be displaced by protamines or otherwise dissociated from the linker DNA. The loosening and even unfolding of the nucleosomal structure in increasing concentrations of NaCl may be detected in gel electrophoresis as the H1 proteins dissociate from the structure (Krajewski et al., 1993). Almost all eucaryotic chromatin is constructed in this way and the problems of gene transcription and replication as well as mitosis, which are mostly problems of nuclear function, are currently considered in terms of this complex structure.

The chromosomal preparations were found to contain polyamines. For example, metaphase HeLa cells were grown in the presence of radioactive polyamines, and the chromosomes were isolated from the frozen lyophilized cells by a nonaqueous system of centrifugation and layering (Goyns, 1979). The polyamine content of the slightly impure preparation of chromosome clusters was found to be 140 pmol spermine and 116 pmol spermidine/μg DNA, indicating a relative doubling of the spermine to spermidine ratio in these condensed chromosomes as compared to the ratio of the content of these polyamines in the intact cells.

That exogenous polyamines will prematurely condense chromosomes has long been known (Davidson & Anderson, 1960). The increased study of cells grown in culture revealed a variety of polyamine effects on the chromosomal state. Polyamine depletion in a polyamine-dependent Chinese hamster cell caused prolongation of the cell cycle and major chromosome aberrations: elongation and unpacking, fragmentation, gaps, and breaks (Pohjanpelto & Knuutila, 1982, 1984). An inhibition of chromosome condensation was perceived in HeLa cells treated with DFMO for 48 h, and this effect was reversed by exogenous putrescine (Sunkara et al., 1983). However, damaging events in an intact animal require severe depletion or, conversely, low levels of residual polyamine are structurally significant.

Chromatin damage by nucleases is more severe in DFMO- and methylglyoxal-bis(guanylhydrazone) (MGBG) depleted cells (Snyder, 1989b), as well as in cells treated with bisethylhomospermine, BE-4-4-4 (Basu et al., 1992). Polyamines or their derivatives binding to chromatin within the cells was demonstrated by a staining procedure that employed a spermine bisacridine, which enhances the intercalation and the binding pattern observed with the quinacrine technique (Van de Sande et al., 1979).

Various workers employed polyamines in the preparative media. The lability of the mononucleosomes liberated by micrococcal nuclease from nuclei of *Physarum* led to inclusion of 0.15 mM spermine and 0.5 mM spermidine during isolation stages (Stone et al., 1987). Similar results were obtained with the nucleosomes prepared from chicken erythrocyte nuclei (Colson & Houssier, 1989). The chromatin condensed by polyamine does not release polyamine readily by dialysis at low ionic strength, and indeed the continued presence of the polyamines affects some physical parameters, such as the sign of optical anisotropy of the preparations. In the isolation of chromatin from rice seedling nuclei, spermine specifically releases several chromosomal proteins (Van den Broeck et al., 1994).

A relaxed circular dsDNA, histones, and an embryo extract were mixed in attempts to produce beaded chains in vitro. The generation of nucleosomelike bodies was detected by the imposition of supercoiling on the reisolated DNA, an effect detectable in gel electrophoresis. The presence in the mix of a linker-bound histone (H1) and spermine stimulated nucleosome assembly (Bogdanova, 1984).

The thermal denaturation temperatures of DNA in mononucleosomes are significantly increased by spermidine and spermine but much less so by putrescine (Blankenship et al., 1987). The proportion of a major high-melting component induced by spermidine is reduced in the presence of monoacetylspermidine, pointing to a destabilizing effect of acetylation of nuclear polyamine. Effects on DNA supercoiling within the nucleosome were noted when only 30–40% of the DNA had been neutralized by the added natural polyamines, indicating a major histone participation (ca. 55%) in the neutralization (Morgan et al., 1987; Smirnov et al., 1988). In the latter study, exposed DNA sites in the nucleosomes were found every 10 base pairs; that is, DNA condensation in chromatin occurred with addition of two triamine molecules per turn of the DNA helix. Diamines and some synthetic triamines could not affect condensation of DNA and chromatin, indicating the importance of the specific charge distribution found in the natural polyamines. In studies of the compaction of chromatin and histone H1-depleted (i.e., linker-depleted) chromatin, spermine was found to react with the linker in the absence of H1.

The formation of condensed chromatin, found in interphase nuclei and compacted further in forming metaphase chromosomes, involves the folding of the beaded nucleosome filament (about 100 Å) into a helical solenoid of about six nucleosomes per turn in a filament of 300 Å. This compaction is effected by inorganic and organic cations, which act as counterions and compete among themselves (Widom, 1986). Both Na^+ and Mg^{2+} will fold the 100-Å filament, as will $Co(NH_3)_6^{3+}$, spermidine^{3+}, and spermine^{4+}. The counterion theory of Manning appears to apply, although a test of polyamine analogues for specificity other than charge has not been made.

The ready dissociability of the nucleosome at high osmotic strength, >0.5 M NaCl, is well known. Photoaffinity derivatives of the polyamines can be attached covalently to nucleosome components and located in the DNA and proteins. Among these derivatives were azidonitrobenzoyl (ANB)-spermine and the azidobenzamidino (ABA)-spermine and -spermidine (Clark et al., 1991; Morgan et al., 1989).

ANB-spermine

ABA-spermine

ANB-spermine bound noncovalently to mononucleosomes as if it were spermidine, that is, as a triamine. Covalent binding to DNA and histone occurred after illumination at 302 nm (close to the absorbance maximum of ANB-spermine). Seven preferred sites were found on the DNA near the center of each strand. ABA-spermine also bound to the nucleosome cores, shifting the temperature of denaturation (i.e., the melting temperatures, T_m) to higher values. It was bound covalently to DNA after illumination and was localized at sites different from those found with ANB-spermine. ABA-spermine proved to be a more potent inhibitor of *E. coli* ODC than ANB-spermidine. Unlike the latter, ABA-spermine can be taken up by yeast.

ANB-spermine has a clear preference for certain thymidine residues in a plasmid (Matthews, 1993; Xiao et al, 1991). A homopolymeric poly d(AT) is resistant to the binding agent whereas runs of alternating A and T, poly d(AT)·d(AT), bind well. A sequence TATATATA . . . placed in a yeast gene within the plasmid binds the compound, and it has been asked if this can produce effects on transcription of genes containing the regulatory TATA elements. The effects of ABA-spermine are thought to emulate the effects of spermine, particularly that of imposing a helical twist on DNA, more closely than those of ANB-spermine. In any case, these reactive derivatives appear to permit investigators to bypass some of the dissociation difficulties of ionic strength in identifying cellular sites of polyamine reactivity.

Metabolic Modifications of Chromatin

Despite the genetic stability of the chromosomal apparatus, it is a challenge to understand numerous metabolic alterations of chromosomal components within the context of the major nuclear functions of replication, transcription, repair, recombination, and mitosis. Components of chromatin can be modified by methylation; the polyamines and histones can be acetylated; the proteins can be phosphorylated and dephosphorylated; and a novel polymer of adenosine diphosphoribose, derived from the coenzyme NAD, can be attached to chromosomal proteins and removed.

The reactant for DNA methylation is S-adenosylmethionine (AdoMet), a precursor of synthesis of higher polyamines, and some interactions of these metabolites have been noted. Spermine (1 mM) is reported to be very inhibitory to a DNA methylase from rat liver (Cox, 1979). The methylation of chromosomal histone H4 in isolated nuclei of the rat brain was found to be sharply inhibited by 1.5 mM spermidine (Duerre et al., 1982). The inhibition by polyamines of methylation of chromatin DNA at the N^7 and O^6 positions of guanine and the N^3 position of adenine by a carcinogen, N-methyl N-nitrosourea, was discussed earlier. The pronounced effects of the polyamines on tRNA methylases will be described in Chapter 23.

Protein kinase reactions in nuclei resulted in the phosphorylation of chromosomal nonhistone proteins. Androgen control of these reactions in the prostate was pursued subsequently and polyamine effects on these reactions by chromatin-associated kinases were described by Ahmed et al. (1978). Spermidine and spermine were ineffective in in vitro tests in the presence of >150 mM NaCl, but were quite stimulatory at 1–2 mM concentrations in the absence of this added salt. The stimulations were maximal at the optimal Mg^{2+} concentration, 2–4 mM. The stimulations of phosphorylation were found mainly on acidic nonhistone proteins, that is, the nonnuclear dephosphophosvitin or on some endogenous chromosomal nonhistone proteins. The low rate of phosphorylation of lysine-rich histones was not affected by the polyamines. On the other hand, the dephosphorylation of a specific histone within a nucleosome preparation was markedly stimulated by spermine and spermidine (Kinohara et al., 1984), evoking speculation that this might be physiologically significant during the changes in nuclear polyamine content in the cellular growth cycle.

The posttranslational acetylation of NH_2-terminal lysines in the four core histones account for some 26 acetylation sites per core nucleosome (Loidl, 1988); the linker-associated H1 histone is not modified in this way. Several histone-specific transacetylases using acetyl coA were described in animals and plants, and an activity found in rabbit heart muscle was found to be en-

hanced by spermine (Moruzzi et al., 1974). The acetyl groups, which decrease the positive charge of these lysine-containing domains, may be removed by a deacetylase, thereby affecting the binding of the histone within the structure (Turner & O'Neill, 1995).

The acetylation and deacetylation of lysine in the various histones occurs on peripheral N-terminal fragments (tails) of the four core histones, fragments that are obviously accessible to the acetylase and deacetylase. The proteinase clostripain can selectively remove these peripheral acetylatable arms from the isolated mononucleosomes and produces a corelike particle that cannot be acetylated, having lost some 52 positively charged residues (Dumuis–Kervabon et al., 1986). As can be seen in Figure 22.4, the "native" nucleosome binds about 0.07–0.08 spermidines per nucleotide, a value which is increased to 0.12–0.13 spermidine molecules per nucleotide in the proteolyzed particles. The increased binding of spermidine was interpreted to signify a replacement of the basic histone contacts with the DNA by the polyamine. The removal of the amino terminal "tails" of the histones in the nucleosomes by proteolysis does not appear to affect the stability of the residual nucleosomal structure to salt, but does expose

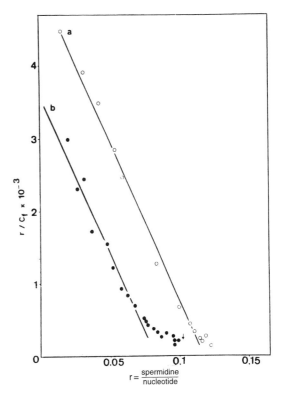

FIG. 22.4 Scatchard plots of the binding of [1,4-^{14}C]spermidine at 6.5°C to (**a**) clostripain-proteolyzed core particles and (**b**) to native core particles; adapted from Dumuis–Kervabon et al. (1986).

an early thermal melting component of the structure. The increase of acetylation of the native particle by polyamine was suggested to result from a competitive displacement of the histone arm from DNA to an enzymatically more vulnerable position (Parello, 1988). The structure of the clostripain-degraded nucleosome is being studied in greater detail (Banères et al., 1994).

Another reaction of chromatin proteins, specific for eucaryotic cells, is that of ADP ribosylation. The coenzyme NAD^+ is cleaved to form nicotinamide and generates the moiety of ADP-ribose, which forms a derivative of protein carboxyls. This may be represented as

Adenine
|
ribose-5'-pyrophosphoryl-OCH₂   O–C–protein

OH OH

Further addition of ADP-ribose can occur at the 2′ position of the bound ADP-ribose.

Kun et al. (1976) demonstrated the formation of (1 : 1) Schiff bases of spermine and spermidine with ADP-ribose and suggested a possible modulation of protein modification by this action of the polyamines. Actually, the presence of millmolar levels of polyamine increased nuclear ADP ribosylation several-fold, predominantly in histones in the presence of Mg^{2+} and in nonhistone proteins in the absence of that cation. In contrast to the lack of acetylation of histone H1, that protein in HeLa nuclei accepted poly(ADP-ribose) readily with added polyamine (Byrne et al., 1978). Significant poly(ADP) ribosylation of H1 in spermine-treated nuclei of the liver and hepatoma was observed by several groups, one of which also detected a concomitant stimulation of DNA synthesis (Tanigawa et al., 1980). Surprisingly, in this context ADP ribosylation was enhanced in mammalian fibroblasts by polyamine depletion (Wallace et al., 1984) and Palvimo et al. (1987) reported an increase of this modification in H1 in a depleted Chinese hamster ovary (CHO) cell.

Poly(ADP-ribose) polymerase is thought to act as a signal in DNA damage and repair. Substantial DNA damage and strand interruption evoke a great synthesis of the polymer and results in a depletion of both NAD and ATP. Satoh and Lindahl (1992) suggested that the enzyme binds to the DNA break and is itself autopolyribosylated, facilitating access of the site to enzymes of DNA repair. Additionally, automodification imitates a dissociation of histones from the DNA, assisting the development of access to the damaged DNA (Althaus et al., 1993). The present state of poly(ADP-ribose) metabolism was surveyed by Lautier et al. (1993). As summarized above, effects of the polyamines were found in many of the metabolic modifications affecting chromatin.

Nucleolar Activity

Staining of the nonmitotic nucleus revealed a more or less dense body, the nucleolus, within the structure; this membraneless body was subsequently shown to be the major site of DNA loops containing genes for ribosomal RNA (Spector, 1993). The nucleolus contains a DNA-dependent RNA polymerase, dubbed RNA polymerase I, or RPI, with the localized job of generating a preribosomal RNA. This is subsequently packaged within the organelle with ribosomal proteins synthesized in, and transferred from, the cytoplasm. The resulting ribonucleoprotein particle is processed to form large and small ribosomal subunits. The early recognition of this function and interest in the biosynthesis of RNA encouraged studies on the isolation of nucleoli, characterization of its components, mainly nucleates and enzymes, and their metabolic activity. A method for the isolation of nucleoli was developed in which stabilizing effects were obtained with spermine instead of a suspect Ca^{2+} (Busch et al., 1967).

Raina and Jänne (1970) sought effects of the polyamines on the synthesis of animal cell RNA in nuclei and nucleoli. They detected a stimulation of synthesis of nucleolar RNA by putrescine and spermidine, as well as an inhibition by spermidine of the degradation of newly synthesized RNA. Russell et al. (1972) used 5 mM spermidine in the isolation of rat liver nucleoli and found a higher content of DNA (2.5-fold) and of RNA polymerase (ninefold) in these structures. The nucleolar RNA polymerase was found to possess a short half-life (1.3 h) in the animals injected with cycloheximide, in contrast to the longer lived nucleoplasmic RNA polymerases II and III, which are active in the transcribing of other genes.

A stainable nucleolar protein, phosphorylated by a polyamine-mediated kinase, was isolated from Physarum and was found to stimulate (fivefold) the RNA synthesis by a nucleolar RNA polymerase I DNA complex (Kuehn et al., 1979). ODC bound by ³H-DFMO was demonstrated in the nucleoli and nucleoplasm of a polychaete oocyte, and DFMO suppressed the accumulation of ribosomal RNA and ribosomes in such cells (Emanuelsson & Heby, 1983).

Currently it is recognized that the nucleolus contains over 100 proteins and nucleic acids and may be characterized as possessing fibrillar and granular substructures. Significant variations are found in the eucaryotic organization of ribosomal rRNA genes and numerous RNA polymerase I transcription factors are found, as well as the presence of a topoisomerase I that eliminates torsional stress during transcription. Processing and splicing of preribosomal and other RNA is effected by nucleolar complexes of small ribonucleoproteins (snRNPs). Sorting out the components, functions, and mechanisms of this structure, which determines the synthesis of 80% of cellular RNA and of ribosomes, is obviously an active subject (Fischer et al., 1991; Hernandez–Verdun, 1991).

The nature of hormonal response of particular tissues

provides a continuing incentive for the study of poly-amine effects in generating hypertrophy and hyperplasia. For example, the administration of estradiol to ovari-ectomized rats stimulates both polyamine synthesis and nucleolar transcription. In such animals polyamines stimulate rRNA synthesis in the absence of injected estradiol. The polyamines plus ATP will produce this effect in isolated uterine nucleoli (Whelly, 1991). It was also observed that spermidine and spermine enhance the phosphorylation of uterine nucleolar proteins by $[\gamma^{32}P]$-ATP.

A stainable nucleolar phosphoprotein, nucleolin, is thought to be involved in regulating transcription by polymerase I and is cleaved readily by nucleolar proteinases. Nucleolin binds readily to spermine or various histones. The endogenous proteases degrade the 100–110 kDa protein to a 60 kDa polypeptide, the N-terminal half of the protein in which the major phosphorylation sites exist (Suzuki et al., 1993). Spermine apparently protects the 60 kDa phosphopeptide from further proteolytic degradation.

Transcription and Bacterial RNA Polymerases (RPs)

Three types of eucaryotic RPs have been distinguished, of which the nucleolar type I is mainly responsible for rRNA. Only a single RP is found in procaryotic cells and this enzyme will synthesize all bacterial transcripts. This single enzyme also possesses specific functions such as the transcription of rRNA genes to produce a functional rRNA. On the other hand, the apparently versatile T7 RP, operating in the presence of spermidine, will produce a nonfunctional "rRNA" (Lewicki et al., 1993). The E. coli RP comprises several subunits that dissociate reversibly at high ionic strength.

The genetic determinant of the E. coli RP is contained within a single operon concerned with replication, transcription, translation, and macromolecular synthesis (MMS) (Lupski & Godson, 1989). The gene for an RP component encodes a σ subunit that, within the composite RP, recognizes promoters and selects the position for the start of RNA synthesis. The gene for the DNA primase, which synthesizes the small RNA used to prime DNA synthesis, is also a component of the operon.

An in vitro DNA-dependent synthesis of RNA by the E. coli RP requires four nucleoside triphosphates and a divalent cation, Mg^{2+} or Mn^{2+}. In the hands of most workers, the synthesis slowed and essentially stopped after some 30–60 min; the inhibition and arrest of RNA is usually attributed to product inhibition by newly formed RNA chains. Exogenous RNA is similarly inhibitory and this inhibition may be prevented by the addition of spermidine. An early stimulation of synthesis or a reversal of complete inhibition may be accomplished by high ionic strength (e.g., 0.2 M KCl) and spermidine or Mg^{2+} (Fuchs et al., 1967).

Total RNA synthesis in a cell-free system is frequently followed by the incorporation of a labeled base in a nucleoside triphosphate. Initiations of new chains

are measured by the incorporation of a γ-^{32}P-labeled nucleoside triphosphate, which will be retained at the 5′ end of a new chain. Several groups have concluded that spermidine does not increase initiation and permits the continuing extension of the RNA chain. The conclusion that spermidine does not increase chain initiation has been challenged by Petersen et al. (1968).

Polyamines neither affect the pH optimum of the reaction nor stimulate synthesis on denatured DNA. No data are available on the effect of binding of the polyamines to an effective template, an emerging RNA, and the polymerase subunits during transcription. It should be noted that a useful current procedure for the isolation of the E. coli RP involves an early precipitation of the enzyme from a crude clarified extract by the cationic polyethyleneimine and selective elution from this readily sedimentable pellet (Burgess & Jendrisak, 1975). A similar procedure has served in the isolation of the RPII of wheat germ. Polyethyleneimine, synthesized by polymerization of aziridine, has been described as a polyamine and contains a mix of primary, secondary, and tertiary amines (Johnson & Klotz, 1974). It may be asked if the enzyme purified with this reagent is devoid of bound polyethylenimine.

Early studies with the enzymes isolated from Azotobacter vinlandii and Micrococcus lysodeikticus demonstrated the inhibitory effects of RNA and their release of inhibition by the polyamines (Fox & Weiss, 1963; Krakow, 1964). Both enzymes are capable of synthesizing an RNA from an RNA primer and these reactions are inhibited by polyamines. In the Azotobacter system it was also shown that ribonucleases would release pyrophosphate from nascent chains and thereby continue polymerase activity in the absence of accumulating inhibitory RNA (Krakow, 1966). With the Micrococcus system Fox et al. (1965) showed that the presence of RNA prior to the addition of DNA inhibited the formation of an enzyme–DNA complex and spermidine facilitated formation of the latter complex. Furthermore, the presence of spermidine affected strand selection considerably with the ds replicative form of ϕX174DNA (Gumport & Weiss, 1969).

In more recent studies of bacterial RPs, as in studies of the enzymes of Rickettsia prowazekii or Mycobacterium smegmatis, specific concentrations of spermidine were found to both stimulate and inhibit the activities. The stimulatory effect appeared to occur mainly in overcoming inhibition of elongation by RNA (Jain & Tyagi, 1987; also see Sarkar et al., 1995).

Phage-Induced DNA-Dependent RNA Polymerases

In infections of E. coli by bacteriophage T7, a protein synthesized very early proved to be a DNA-dependent RNA polymerase. A similar and even more stable enzyme is induced by the Salmonella phage, SP6. These are single subunit enzymes that are being widely used for the dissection of transcription and for the synthesis of specific RNA sequences (Melton et al., 1984). Isolation procedures involving polyethyleneimine and assays of these enzymes were described by Chamberlain et al. (1983), who noted that omission of spermidine imposed a 35% decrease in the activities of their T7 preparation. In the case of the SP6 polymerase, spermidine both stimulates synthesis twofold and reduces the rate of heat inactivation at 42°C (Butler & Chamberlain, 1982). The recent determination of the crystal structure of the T7 RNA polymerase indicated the similarities of structure and the existence of sequence homologies among many DNA- and RNA-dependent polymerases (McAllister & Raskin, 1993). There is a question of whether polyamine binds to any of these enzymes.

The fact that the nucleates are most frequently made with these enzymes in the presence of millimolar levels of spermidine or spermine raises the problems of whether these products also contain polyamines that may influence subsequent tests of these nucleates. Essentially all newly synthesized T7- or SP6-polymerase-generated RNAs are made in mixes containing 2–4 mM spermidine plus various concentrations (6–16 mM) of Mg^{2+} (Gurevich et al., 1991). Numerous procedures of protein denaturation do not remove polyamine from media containing nucleates (Shepherd & Hopkins, 1963). Recent chemical syntheses of ribozymes without the use of polyamines produced catalytically active products (Wincott et al., 1995).

The effects of linear and macrocyclic tetramines and hexamines on in vitro transcription by the T7 polymerase of ssDNA were reported recently (Frugier et al., 1994). All were found to be stimulatory at particular concentrations and a 12-fold stimulation on an ssDNA (twofold on dsDNA) was found with a macrocyclic hexamine (3 mM), which was highly positively charged at transcription pH (8.1). It was suggested that this amine both stabilized the ternary complex and activated the enzyme itself.

Eucaryotic DNA-Dependent RNA Polymerases

Three separate eucaryotic polymerases have been obtained, of which the type I RNA polymerase functions largely in the transcription of ribosomal genes (Jacob, 1995). Type II synthesizes mRNA, and type III transcribes genes for tRNA and 5S RNA. Nevertheless, many structural features and functions of procaryotic and eucaryotic RNA polymerases have been conserved (Eick et al., 1994). Both gene-specific factors and other regulatory cofactors modulate transcription by the type II polymerase. One of these, the nuclear activator PC4, is negatively regulated by CKII phosphorylation and activated by an in vitro dephosphorylation operating in the presence of spermidine (Ge et al., 1994). In addition to many reports of stimulation of activity by polyamines, the major foci of current work on transcription are the structure and topology of a site at which the enzyme can bind, separate the strands of dsDNA, and synthesize RNA on one or the other DNA strand. Obvious major problems include the choice of the transcription site

(i.e., the sequence of DNA) and the nature of additional transcription factors, many of which are isolable proteins (Conaway & Conaway, 1991; Latchman, 1990). Much effort has been expended in the analysis of a TATA-binding protein (TBP), which is a highly conserved eucaryotic transcription factor (Struhl, 1994). The mechanism of splitting strands and movement down the transcribed strand within a histone-containing complex supercoiled nucleosome (Felsenfeld, 1992) and the potential competition between replication and transcription (Georgiev et al., 1991) are subjects for active experimentation. The fact that the survival of casein mRNA in a lactating mammary gland is extended by spermidine (Oka & Perry, 1990) is an example of a regulatory mechanism bridging transcription and translation.

The mechanisms by which transcription factors, the DNA sequence-specific proteins, act are under active scrutiny. It was found that spermine and spermidine increase the rate of association of many of these proteins with their specific DNA sequences (Panagiotidis et al., 1995). This is also so for a herpes simplex virus gene regulator, ICP-4, effective in virus particles containing the DNA and these polyamines. Early studies with partially purified enzymes acting on definable templates, like SV40 DNA, demonstrated the hindrance of transcription by a superhelical configuration, potentially opened somewhat by divalent cations, notably Mn^{2+} (Mandel & Chambon, 1974). Resistance to the RNA polymerase on such a substrate was increased by spermine whereas the amine was stimulatory on less structured templates. DNA topoisomerase I acts as a cofactor in the initiation of transcription by RNA polymerase II.

In studies of the three separate polymerases on thymus DNA, spermine (1 mM) enhanced transcription three- to fourfold and a high concentration of $(NH_4)_2SO_4$ (30–100 mM) maintained the stimulation (Jacob & Rose, 1976). The polyamine increased the average length of the transcripts and the chain elongation rate (Jänne et al., 1975). Blair (1985) determined the extent of polyamine stimulation of the separated polymerases from several normal tissues and many more tumors of mice and found these effects to be similar among all tested; the responses of the various typed polymerases were unrelated to the growth rates of the tissues. Despite the numerous effects on RNA synthesis by an ionic milieu containing polyamines, by spermine particularly, as summarized by Blair and noted by Moruzzi et al. (1975), it has not been possible to define an in vivo role of the polyamines in nuclei. Among difficulties in extrapolating in vitro studies to functional cells or organelles, the problem of defining the ionic milieu in cells or cell compartments is just beginning to be addressed.

One line of investigation attempted to track the estrogen control of gene expression from receptors to nuclear DNA. The biosynthesis of polyamine is known to be stimulated in estrogen function and this function is disrupted by depletion of polyamine. In studies of the intracellular uterine receptors of estradiol and progesterone, binding to a thymus DNA-cellulose was increased significantly by spermidine. This increase was particularly large (22-fold) for polynucleotides containing poly(dG-dT) sequences (Thomas & Kiang, 1988a,b). Similar results were obtained in demonstrating a role of polyamine in decreasing the size of the intracellular receptor for vitamin D (Morishima et al., 1994) and the aryl hydrocarbon receptor (Thomas & Gallo, 1994), as well as increasing the binding of these receptors to DNA.

Mitochondrial DNA-Dependent RNA Polymerases

Gene expression from the separate genes contained in the mitochondria relates to the synthesis of respiratory proteins, two ribosomal RNAs, and a set of more than 20 tRNAs. The synthesis of RNA in isolated mitochondria and the discovery of DNA-dependent RNA polymerases in mitochondria from *Neurospora*, yeast, and animal tissues occurred in the early 1970s. The enzyme of the organelles of *Xenopus* ovaries was purified (Wu & Dawid, 1972), had a specific requirement for Mg^{2+}, and was stimulated about threefold by spermidine or spermine. The enzyme consisted of a single polypeptide chain of 46,000 MW and it aggregated at low ionic strength.

Mitochondrial transcription involves a core RNA polymerase acting on the organelle DNA plus a set of factors participating in initiation, specificity of strand selection, etc. (Shadel & Clayton, 1993). In yeast, the polymerase consists of a relatively nonspecific protein, apparently similar to the T7 polymerase, and is directed to promoter sequences by another specific protein.

Polyamines and Protein Synthesis

Zamecnik (1984) briefly reviewed early problems and studies in the clarification of protein synthesis in the 1950s. A cell-free system containing labeled amino acids, ATP, and a mix of sedimentable and nonsedimentable fractions could synthesize proteins. This opened the path to dissections of the activation of the amino acids, the role of a microsomal fraction containing ribosomes, and the roles of various other components. It became evident that amino acids were converted to their adenylates and were transferred to the ribosomes. The intermediary transferring device proved to be the tRNAs, each of which was somehow specific in bearing a single amino acid. The decade also revealed the existence of messenger RNAs that were constructed on DNA genes and transferred to ribosomes to instruct the course of protein synthesis. Subsequently the use of synthetic polynucleotides, such as poly(U), as mimics of mRNA, directed the synthesis of specific polypeptides like *polyphenylalanine*. This result initiated the detailed dissection of the genetic code via the study of codons in mRNA and anticodons in tRNA and nucleotide sequences in the various RNAs participating in protein synthesis. Cell-free systems of protein synthesis were obtained in procaryotic and eucaryotic cells and the path was taken to clarify the various steps and components and to maximize the rates of the individual reactions.

The apparently straightforward sequence,

$$\text{amino acids} + \text{ATP} \xrightarrow[\text{enzymes}]{\text{soluble}} \underset{\text{NH}_2 + \text{P·P}}{\text{R–C–CO–adenylate}} \xrightarrow{\text{tRNA}}$$

$$\underset{\text{NH}_2}{\text{R–C–CO–tRNA}} \xrightarrow[\text{ribosomes}]{\text{mRNA}} \text{polypeptides}$$

was readily amenable to an initial exploration of the roles of ionic strength and specific cations, such as Mg^{2+} and spermidine. These factors evidently affected the overall rates, the structural and functional states of the ribosomes, the conformation of nucleates, tRNA, and indeed the quality of the polypeptide product—the fidelity of translation.

Three major stages of the ribosomal process were described: the formation of an initiation complex, elongation, and termination (Mathews & Van Holde, 1991). Each of these involved many new cofactors, such as proteins and GTP. However, few of these dissections involved the further analysis of the roles of salts and cations. Nevertheless, the pursuit of the roles of the polyamines contributed to the investigations of protein synthesis. These substances became tools of structural studies on the various nucleates and cellular entities containing the nucleates, such as ribosomes and nucleosomes. Studies of the metabolism of spermidine revealed the existence of the new amino acid hypusine, which existed uniquely in a single protein that was apparently essential for protein synthesis in eucaryotic ribosomes. Finally, it appears that polyamine metabolism and essentiality in protein synthesis in thermophilic microorganisms may point to an evolutionary role of the polyamines in the diversity of procaryotic organisms and the origins of eucaryotic cells.

Ribosomal Subunits and Their Association

Some data on bacterial ribosomes, their subunits, and contents of RNA and proteins were summarized briefly in Chapter 7. In early studies ribosomes of guinea pig pancreas were also characterized with respect to size, RNA and Mg^{2+} content, and the presence of several enzymes, which were probably synthesized on these structures. Spermine was found to aggregate these eucaryotic ribosomes and displace some Mg^{2+} from the particles (Siekevitz & Palade, 1962).

The association of 30S and 50S ribosomal subunits of *E. coli* to form 70S monosomes and larger aggregates was aided by 10^{-2}–10^{-3} M Mg^{2+}. Goldberg (1966) quantitated the binding of Mg^{2+} to both ribosomes and ribosomal RNA, to the level of about one Mg atom to two RNA phosphates, that is, about 1600 Mg^{2+} ions per 70S ribosome. Eventually studies of the 70S particle progressed to the crystallization, X-ray analysis, and electron microscopy of these structures, as well as of the subunits, seeking contact points and spaces for the accommodation of components of protein synthesis (Yonath & Wittmann, 1988). Hansen et al. (1990) crystallized a complex of 70S ribosomes from a thermophilic eubacterium *Thermus thermophilus* that contained phenylalaninyl-tRNA[phe] and a short poly(U) chain in a mix containing 10 mM $MgCl_2$ and 2–10 mM spermidine. Crystallographic studies and computer graphic simulation detected the subunits, grooves rich in rRNA, and a tunnel thought likely to hold the mRNA and tRNA. Morphological and structural studies of archaebacterial ribosomes and their subunits are also available (Ramírez et al., 1993).

A role for spermidine was detected in the interaction of the two *E. coli* subunits to form a 70S monosome. These studies revealed a sharing of an associative function for divalent Mg^{2+} and the polyamines in ribosome organization. The replacement of more than 80% of the bound Mg^{2+} resulted in an inactivation of the particle, indicating the existence of some irreplaceable Mg^{2+} (Weiss & Morris, 1973). Excessive replacement of Mg^{2+} by polyamine destabilized the structures, as seen in the unfolding of the 30S subunit summarized in Table 22.1. These effects were reversible as measured by their participation in protein synthesis, a capability lost in the initial unfolding of the particle. Similar properties were observed for the 50S subunit, but a restoration of function could not be demonstrated despite a retention within the particle of the ribosomal proteins and two RNA species (Kimes & Morris, 1973). Such results pointed to the importance of including carefully balanced ions, including polyamines used in preparatory buffers.

The dissociation of the monosomes, 70S in eubacteria and archaebacteria and 80S in the eucaryotic cell, to their two subunits is an early step of protein synthesis and the maintenance of the *E. coli* monosome was examined as a function of temperature, sulfhydryl reagents, polyanions, and alkylating agents. In these studies the dissociated subunits were reassociated to actively synthetic monosomes with divalent cations or polyamines. Ribosomes were found to differ within the archaebacteria, as well as from those of eubacteria. Differential sensitivities to antibiotics clearly separated the

ribosomes of the sulfur-dependent extreme thermophiles from those of the methanogenic halophilic branch of archaebacteria, as well as from eubacterial and eucaryotic ribosomes (Cammarano et al., 1985).

Studies on in vitro protein synthesis detected a possible replacement of Mg^{2+} by the polyamines in the reaction mix in the synthesis of poly(U)-directed polyphenylalanine with poly(U) as a mRNA and in the synthesis of MS2 phage RNA-directed polypeptide (Martin & Ames, 1962). Such effects are presented in Figure 22.5 (Takeda, 1969). A similar replacement was found in facilitating the specific binding of phenylalanyl-tRNA in such syntheses (Takeda, 1968). Polyamine also increased the binding of poly(U) to 30S ribosomal subunits. This was also observed in rat liver systems of polyamine stimulation at suboptimal Mg^{2+} (Igarashi et al., 1973). The polyamines were also implicated in the formation of aminoacyl tRNA, the initiation and extension of peptide synthesis, and the fidelity of translation.

Jelenc (1980) developed a method of isolating highly active *E. coli* ribosomes in a salt- and polyamine-containing buffer that minimized the misincorporation of leucine in phenylalanine polymerization to in vivo levels. Additional improved systems approached the rate of synthesis found in in vivo systems described in the following. Studies with cell-free systems of protein synthesis derived from procaryotic eubacteria other than *E. coli* (e.g., spores of *Bacillus megaterium*, the pathogen; *B. thuringiensis*; and an extreme thermophile, *Thermus thermophilus*) demonstrated the Mg^{2+}-sparing action of spermidine and stimulation by a polyamine. However, protein synthesis in extracts of *B. thuringiensis* and *T. thermophilus* are absolutely dependent on polyamine. Further, *T. thermophilus* functioning at 50–80°C contained a sperminelike polyamine and synthesized protein best with spermine (Ohno–Iwashita et al., 1975). In *Pseudomonas* sp. strain Kim, spermidine was replaced by putrescine and S-(+)-2-hydroxyputrescine in stimulating ribosomal function (Rosano et al., 1983). In an *E. coli* cell-free system stimulation occurs in the order spermidine > hydroxyputrescine > putrescine. 1,3-Diaminopropane (DAP) can inhibit a spermidine-activated system competitively. Studies with sulfur-dependent thermophilic archaebacteria *Sulfolobus* species also demonstrated important roles of tetramines in activating in vitro protein synthesis: in stabilizing ribosomes at high temperatures, in forming initiation complexes, and in enhancing elongation. Friedman and Oshima (1989) compared the effects of polyamines in extracts of some *Sulfolobus* strains, all of which contained norspermidine, spermidine, thermine, and spermine, but which differed in their contents of these compounds. Thermine and spermine have similar activating effects in promoting the formation of 70S monosomes, as well as in the activation of poly(U)-directed polyphenylalanine synthesis.

Polyamine-Deficient Ribosomes

Maas and colleagues isolated polyamine-depletable mutants of *E. coli* and protein synthesis was examined sub-

TABLE 22.1   Properties of unfolded 30S ribosomal subunits of *E. coli*.

Property	Control	Polyamine replacing Mg^{2+}	
		Putrescine	Spermidine
s_{20} (S)	30.5	21.5 (25.5)	27.5 (29.5)
RNase sensitivity (%/min)	8.3	28.2 (18.0)	17.2 (16.2)
η_{sp}/c(mL/g)	8.0	13.8 (11.2)	9.4 (9.9)

Adapted from Weiss and Morris (1973).

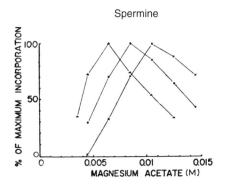
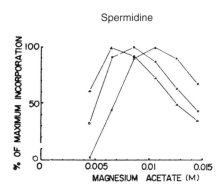

FIG. 22.5   The effect of polyamine on the optimal concentration of Mg^{2+} in cell-free polypeptide synthesis from MS2 phage RNA; adapted from Takeda (1969). The curves on the right in each section were determined without spermine or spermidine. Curves shifted to the left of these controls in each section were determined with 0.3 mM spermine and 0.35 mM spermine, or with 1 mM spermidine and 2 mM spermidine, respectively.

sequently in these cells and extracts. After a limited period of growth in the absence of polyamine to deplete polyamines, *E. coli* mutant MA261 blocked in the biosynthesis of putrescine and spermidine was markedly deficient in the ability of extracts to synthesize polyphenylalanine in response to poly(U) (Echandi & Algranati, 1975 a,b). The depleted bacteria were deficient in their content of 30S and 70S particles, both of which were restored by bacterial growth in the presence of putrescine. These effects could not be restored by Mg^{2+}. Abnormal ribosomal units smaller than 30S were detected in the depleted bacteria. These results suggested that polyamines participate in the assembly of 30S subunits.

It was found that the defective subunit alone contained a methyl-deficient 16S RNA, a deficiency centered mainly in adenines. In addition, the "30S" subunits were low in the large S1 protein (Igarashi et al., 1981) and the suggestion was made that S1, synthesized normally in polyamine deficiency, was integrated slowly into the methyl-deficient 16S RNA. S1, the largest *E. coli* ribosomal protein, is about as long as the 30S subunit, is weakly held in the structure, and is almost, but not quite, essential in translation (Subramanian, 1984). Spermidine was found to stimulate the rate of inclusion of ribosomal protein S1 in the in vitro assembly of subunits containing methyl-deficient RNA (Kakegawa et al., 1986).

A second mutation in the polyamine-deficient strain MA261 inhibited synthesis of the 30S subunit protein S1. The synthesis of this protein, as well as its assembly, were stimulated by the presence of putrescine in the growth medium (Kashiwagi et al., 1989).

Another polyamine-deficient mutant, dubbed BGA8, was deficient mainly in initiation of protein synthesis. Polyamine-deficient BGA8, which is sensitive to streptomycin when grown with polyamine, is relatively resistant when grown in its absence (Goldemberg & Algranati, 1981). The ribosomes of BGA8 bound streptomycin more readily when the cells were grown in

polyamine (Goldemberg et al., 1982). These results with BGA8 may have their explanation in part in the following observations. When *E. coli* strain MA261 was grown in the presence of polyamine, the polyamine preferentially stimulated the production of certain proteins, one of which was termed the PI protein (Mitsui et al., 1984). This protein, known from other studies as an oligopeptide-binding protein, the OAA protein, also proved to be a transport protein for aminoglycoside antibiotics, including streptomycin, and these cells thus became more sensitive to the latter antibiotic (Kashiwagi et al., 1992).

Additional Phenomena of Procaryotic Protein Synthesis

Takeda, Igarashi, and their colleagues explored the roles of the separate ribosomal subunits. Studies on initiation, requiring the binding of formyl methioninyl tRNA to 30S subunits, also suggested a selective stimulation of the binding of AUG-dependent mRNA by spermidine. Having detected such a stimulation with Qβ-phage RNA and MS2-phage RNA that coded for several proteins including an early RNA replicase, an A protein, and a coat protein, Watanabe et al. (1981) found that 1mM spermidine markedly stimulated the synthesis of the RNA replicases, but not of the coat protein. The explanation to this physiologically significant observation appears to turn on the true concentration of free spermidine within the cell, to be discussed in a later section.

The detection of a role for spermine in cell-free protein synthesis at a high temperature in thermophiles led to a study of other new polyamines at such temperatures. Kakegawa et al. (1988) tested these compounds in extracts of the eubacterium *T. thermophilus* at 60 and 70°C, as well as in *E. coli* systems. In *E. coli* systems an elevation of temperature from 37 to 47°C more than doubled the stimulatory effects of spermidine and some other triamines, as well as increasing the effects of spermine. However, at the higher temperature thermine and a series of pentamines were not more stimulatory than spermidine. Canavalmine, a 4,3,4 natural tetramine, was inactive in both *E. coli* and the thermophilic system. Subsequently it was found that the most active compound at high temperatures in the *Thermus* system is the branched quaternary polyamine, tetrakis(3-aminopropyl) ammonium (Uzawa et al., 1993a). In the *Thermus* systems at 60 and 70°C temperatures, which inactivate *E. coli* systems, the tetramines and pentamines were most stimulatory and Mg^{2+} sparing in polyphenylalanine synthesis.

MGBG has been found to be a noncompetitive inhibitor of the spermidine stimulation of protein synthesis in procaryotic and eucaryotic systems (Ohnishi et al., 1985). At low concentrations, MGBG also possesses an Mg^{2+}-sparing effect without stimulation and at higher concentrations is quite inhibitory. The cationic MGBG binds most strongly to ribosomal RNA and inhibits spermidine binding to this RNA.

Initially the duration and rates of polypeptide syntheses in cell-free systems were brief and very low, respectively, compared to the continuous rates observed in intact bacteria. Lucas–Lenard and Lipmann found that the compound *N*-acetylphenylalanine-tRNA (AcPhe-tRNA) was an excellent initiator of polyphenylalanine synthesis primed with poly(U). The addition of AcPhe-tRNA to washed polyU-primed ribosomes is stimulated by spermine (Kalpaxis & Drainas, 1992). Wagner et al. (1983) developed preformed ribosomal complexes with poly(U) and *N*-AcPhe-tRNA and incubated these initiation complexes with phenylalaninyl tRNA and the set of recently discovered elongation factors in mixes containing various ions, including 5 mM Mg^{2+}, 8 mM putrescine, and 1 mM spermidine plus ATP- and GTP-generating systems. Translation proceeded in a 10-s burst involving about 10% of the ribosomes at a rate of eight peptide bonds per ribosome per second, approaching that (10–20) in an intact bacterium. Furthermore, misreading in translation, the insertion of leucine into the polyphenylalanine chain, was quite low. A system of Mg^{2+}, NH_4^+, and buffers and reagents including polyamines that maximized the rate and accuracy of translation was developed by Bartetzko and Nierhaus (1988). The translation mix recently described by Dabrowski et al. (1995) includes 2 mM spermidine, 0.05 mM spermine, and 6 mM Mg^{2+}. *E. coli* or wheat germ systems of cell-free translation of various mRNAs operating continuously for 20 h were devised to produce many copies of viral coat protein per MS2 phage RNA (Spirin et al., 1988). The wheat germ system contained added spermidine whereas the *E. coli* system did not. More recently a system of continuous cell-free protein synthesis using the MS2 mRNA with components of the thermophilic eubacterium *T. thermophilus* functioned optimally at 65°C with the additions of 0.1 mM tetrakis(3-aminopropyl) ammonium and 1.0 mM spermine (Uzawa et al., 1993a,b).

Recent data on the initiation of mRNA translation in procaryotic ribosomes have been discussed by Gualerzi

and Pon (1990). Three initiation factors and GTP facilitate the addition of an mRNA to a complementary association with 16S RNA in a 30S ribosomal subunit. A ternary complex is formed containing formylmethionyl tRNA[fmet]. This complex is composed initially of a 30S ribosomal subunit, fmet-tRNA[fmet], GTP, and an mRNA in complementary association with 16S ribosomal RNA. It then combines with the 50S ribosomal subunit to form a 70S initiation complex containing a cleft between the particles within which the protein is formed. The monosome contains sites for the binding of aminoacyl-tRNAs (A site); the amino acid is transferred to a peptidyl tRNA by a peptidyl transferase (D site). The now freed nonacylated tRNA is released from an E site. Binding at these sites, initially facilitated by polyamine, is sensitive to the positioning of the 2′-OH (Potapov et al., 1995).

The nature of the ribosomal peptidyl transferase itself, whether protein, ribonucleoprotein, or RNA, is being studied actively (Noller, 1993). Spermine is reported to inhibit peptide bond formation during polyphenylalanine elongation; that is, it is believed to act as an inhibitor of ribosomal peptidyl transferase (Kalpaxis & Drainas, 1993). Exploring this system at low Mg^{2+} (6 mM), spermine has been shown to stimulate the initial activity of the peptidyl transferase by activating ribosomes, but to inhibit later phases of the early peptidyl transferase reaction, the puromycin reaction (Drainas & Kalpaxis, 1994).

Roles of Mg^{2+} and Polyamines

Despite the mutual sparing action of Mg^{2+} and the amines in protein synthesis, the *E. coli* ribosomal subunits have an essential requirement for a significant level of Mg^{2+}. On the other hand, functional ribosomes, on isolation, contain polyamine capable of neutralizing 20–25% of their RNA-phosphorus. Few data describe an inactivation of these structures resulting from loss of the polyamine. Bacteria blocked in polyamine synthesis do grow more slowly and produce incomplete ribosomes, but some mutant strains do grow at a slow, but significant, rate in the apparent absence of internal polyamine. Some *Pseudomonas* strains almost completely replace spermidine by putrescine plus 2-hydroxyputrescine. Nevertheless, among thermophilic organisms, their ribosomes have an almost essential requirement for thermine or spermine or even larger polyamines at the high temperatures in which the organisms live. The polyamines stimulate protein synthesis in extracts of both procaryotic and eucaryotic cells, and the requirement of spermidine for the synthesis of hypusine in the "initiation factor" eIF-5A accentuates the requirement for this polyamine in protein synthesis in all eucaryotic cells. There are therefore two sets of roles and functions to be clarified, those related to essentiality and those of mutual replacement. As indicated in previous chapters, studies with cells auxotrophic for the polyamines in media rich in Mg^{2+} have revealed essentiality for these substances in growth in some bacteria and in some eucaryotic cells like yeasts.

Webb (1949) demonstrated the requirement for Mg^{2+} in the growth of many bacteria in complex media, a requirement greater for Gram-positive organisms than for Gram-negative cells. He also described this requirement for many organisms, including *E. coli*, in chemically defined media, as well as their ability to accumulate and concentrate the cation. Subsequently two groups described a specific temperature-dependent transport system for Mg^{2+} in *E. coli* and demonstrated mediated exchange of internal Mg^{2+} with exogenous cation. Three genes for Mg^{2+} transport systems were since cloned from *Salmonella typhimurium*. Transport in bacteria and animal cells were reviewed by Romani and Scarpa (1992) and Flatman (1991).

Mg^{2+} is second only to K^+ as the most abundant cation in mammalian cells. Total cellular Mg^{2+} has been determined by atomic absorption microscopy. The ionized form of intracellular Mg^{2+} is a small fraction (<10%) of the total Mg, which is mainly bound to the range of anionic components, such as the structural nucleates and the metabolic nucleotides RNA and ATP, respectively. Free or ionized Mg^{2+} in animal cells is estimated as 0.2–1.0 mM by techniques of ^{31}P-NMR, ion-sensitive microelectrodes, or fluorescent indicators. Polyamines can interfere with the measurement of Mg^{2+} by a cation-selective electrode (Kent et al., 1974). Changes in the concentration of free Mg^{2+} may be expected to occur with changes in the concentrations of the polyamines and their ability to bind anions such as the nucleates and nucleotides and to displace bound Mg^{2+}.

The content of the divalent cation in *E. coli* growing exponentially in tryptone-glucose salts was reported as about 100 mM (Moncany & Kellenberger, 1981). Sophisticated microscopy of *E. coli* revealed individual cells containing up to 200 mmol Mg/kg uniformly distributed throughout the cells (Chang et al., 1986). Ca^{2+} was present at 1–2% of this value and was concentrated in the cell envelope. These divalent cations were more tightly bound in washing than were monovalent cations. Nevertheless, cellular Mg^{2+} could be severely depleted and growth of *E. coli* inhibited by treatment with a Mg^{2+}-ionophore in the presence of the sensitizer, polymyxin B (Alatossava et al., 1985). The addition of exogenous $MgCl_2$ to such cells could restore cellular Mg^{2+} and growth. Mg^{2+}-depleted *E. coli* were observed with electron microscopy to have lost ribosomes, without loss of viability. Restoration of Mg^{2+} over 5–6 h permits a resynthesis of rRNA and the reappearance of ribosomal subunits, as well as an apparently independent synthesis of the plasma membrane (Morgan et al., 1966).

The essentiality of Mg^{2+} was determined for the growth of *Neurospora crassa*, and Alberghina et al. (1971) noted the reduced content of ribosomes and mitochondria in depleted cells. Depletion or deficiency in the rat affected many tissues, leading to an increased water content, a loss of cellular K^+, and calcification. Swollen mitochondria and their calcification were evi-

dent fairly early (George & Heaton, 1975). These effects of Mg^{2+} depletion were duplicated in chick fibroblasts that also suffered a fall in DNA synthesis that was restorable by Mg^{2+} but not by Ca^{2+} (Sanui & Rubin, 1977). Also, a fall in Mg^{2+} content of 3T3 cells reduced protein synthesis (Terasaki & Rubin, 1985).

Thus, Mg^{2+} at the molecular level reacts with membrane anions, stabilizes intracellular polymers, regulates enzymes, and combines with many substrates and products of enzyme-catalyzed reactions (O'Rourke, 1993). Frey and Stuehr (1972) have published potentiometric data and stability constants as well as kinetic constants for the reaction of Mg^{2+} with a series of sugar phosphates and nucleoside mono-, di-, and triphosphates. Olson et al. (1987) have described a technique of equilibrium dialysis suitable for the determination of the binding of Mg^{2+} to proteins. The binding of the cation to 5S rRNA has been examined as a function of added monocations Na^+ and K^+, both of which appear to be able to affect ribosomal chemistry (Reid & Cowan, 1991).

The Mg^{2+} and polyamine contents of mycelia were estimated at various growth rates of *Aspergillus nidulans* in a chemostat (Bushell & Bull, 1974). The Mg^{2+} contents of the cells fell to 20% of that at maximal RNA contents per unit of mycelial weight whereas spermidine and spermine contents rose to maintain a constant molar ratio of polyamines plus Mg^{2+} to RNA. The Mg^{2+}-sparing effect of the polyamines in polyphenylalanine synthesis was explored in a cell-free barley system in which Mg^{2+} could also be replaced to some extent by Mn^{2+}, Ca^{2+}, and Ba^{2+} (Cohen & Zalik, 1978). Despite the apparent nonspecificity of these cation effects, spermine (0.03 mM) and spermidine (0.5 mM) were, respectively, 100 and 8 times more effective on a molar basis than the Mg^{2+} they replaced. In a search for the relation of charge to specificity, Fuchs (1976) studied the expression of phage T3 and T7 early and late enzymes as a function of Mg^{2+} and spermidine in an in vitro *E. coli* system starting from DNA. In such a system all the enzyme proteins were synthesized optimally at 3 mM spermidine; but a late enzyme, lysozyme, required substantially more Mg^{2+}. However, the requirement for the phosphoenolpyruvate energy source in the system, which buffered the free Mg^{2+} content, complicated the interpretation of the data.

Free Cations and Toxicity of Polyamine Excess

The interactions of the polyamines with nucleotides has underlined the difficulty of defining the content of free polyamine in various systems. Putrescine behaves like Mg^{2+} and binds one molecule of an adenine nucleotide per molecule at pH 7.5. However, two molecules of spermidine or spermine can bind a single nucleotide under these conditions (Nakai & Glinsmann, 1977a,b). These reactions appear to be almost entirely phosphate-dependent, independent of base or sugar. These workers supposed that the polyamines will compete with Mg^{2+}

for intracellular nucleotides as well as other metabolites, and indeed they have demonstrated effects on nucleoside diphosphate kinase attributable to reactions of the polyamine with the substrate. Various nucleotides and some sugar phosphates, 5-phosphoribosyl 1-pyrophosphate and 2,3-diphosphoglycerate, have been shown to form polyamine complexes and the spermine complexes of the two sugar phosphates mentioned have proved to be specific inhibitors of selected reactions (Yip & Balis, 1980). These workers have pointed to the competition of Mg^{2+} and polyamines in in vitro systems and in cells. It is possibly relevant that some nucleotides (e.g., adenosine cyclic 2′,3′-phosphate) can selfpolymerize 3′ → 5′ linkages on a poly(U) template in the presence of natural polyamines (Verlander et al., 1973).

The determinants of the levels of free Mg^{2+} and free polyamine within an intracellular milieu consist of both structural polymeric entities and soluble metabolites. The definition of the cellular levels of all of these competing and active cations presents an imposing problem. The scientists in the laboratory of Igarashi have investigated the Mg^{2+}-sparing effect of the polyamines in protein synthesis and their studies on the specificity of translation of multicistronic viral RNA indicate puzzling stimulations for the translation of early sequences, but not for late sequences. Mg^{2+}-sparing effects of spermidine have also been observed for apparently less complex enzyme reactions, as in the synthesis and degradation of guanosine 5′-diphosphate 3′-diphosphate (ppGpp) in E. coli in vitro systems as well as in a polyamine auxotroph (Igarashi et al., 1983).

These investigators posed the question of the concentration of free polyamine in various cells and calculated binding constants of spermidine and spermine for DNA, RNA, phospholipid, and ATP under "standard" conditions: 10 mM Tris-HCl, pH 7.5, low Mg^{2+} (2 mM), and 150 mM K^+ (Watanabe et al., 1991). Their results were estimated via Scatchard plots of data derived from column chromatographic determinations of binding and are given in Table 22.2. Binding constants for spermidine were lower than for spermine and were further reduced by increasing the KCl concentration. The contents of

the polymers and ATP were determined for both bovine lymphocytes and rat liver, and the contents of free spermidine and spermine were calculated for these cells and the liver. The levels of calculated free polyamine are seen to be quite low percentages of the total polyamine. However, the amount of spermine calculated to be bound to RNA in several in vitro systems (e.g., globin synthesis) was close to the amount found in the cells, and this and other correlations encouraged the view that a method could be found to determine the actual concentration of free polyamines in the cell and to correlate this with observed metabolic effects. Of great interest are the estimates of free polyamine in bovine lymphocytes on the order of 15% of the total spermidine and 5% of the total spermine and in rat liver only 7% of the total spermidine and 2% of the total spermine.

Similar estimates made for E. coli indicate that the free amines were about 4 and 39% of the total spermidine (6.88 mM) and putrescine (32.2 mM), respectively (Miyamoto et al., 1993). The E. coli data pointed to the existence of a significant pool of free putrescine and the presence of about a third of the ATP as a putrescine complex. Such a compound has never been isolated or characterized chemically.

In the bacteria the fraction of spermidine bound to RNA was about 90% of the bound amine whereas in the lymphocytes the fraction of bound spermidine and spermine on RNA were 67 and 68%, respectively. The estimate of spermidine bound to RNA in E. coli approached the concentration of RNA-bound spermidine found to stimulate the MS2 RNA-directed synthesis of the RNA replicase; that is, in a total of 1 mM spermidine added in the cell-free system, 3 mol spermidine were bound to 100 mol RNA-phosphorus. The binding constants indicated that under various conditions one to four and one to three molecules of spermine and spermidine, respectively, would be found on eucaryotic RNA, values close to those actually found in tRNA (Cohen, 1970).

It was assumed earlier that bound polyamines would decrease as Mg^{2+} was increased to 10 mM. The converse possibility was then explored that exogenous sper-

TABLE 22.2 Polyamine distribution in bovine lymphocytes.

| | Spermine | | | | | | Spermidine | | | |
| | 2 mM Mg^{2+}, 100 mM K^+ | | 2 mM Mg^{2+}, 150 mM K^+ | | 10 mM Mg^{2+}, 100 mM K^+ | | 2 mM Mg^{2+}, 100 mM K^+ | | 2 mM Mg^{2+}, 150 mM K^+ | |
	μM	N	μM	N	μM	N	μM	N	μM	N
Free	39.0		75.7		70.4		103		199	
DNA	421	1.16	287	0.79	378	1.04	276	0.76	167	0.46
RNA	945	3.39	1033	3.70	979	3.51	769	2.76	758	2.72
Phospholipid	59.0	0.36	37.3	0.23	13.6	0.08	35.0	0.22	38.0	0.23
ATP	106	3.36	137	4.34	129	4.10	137	4.35	158	5.02

Adapted from Watanabe et al. (1991). Free spermine and spermidine were set at the values at which total spermine and spermidine were calculated to be 1.57 and 1.32 mM, respectively. N represents the number of polyamine molecules bound to 100 mol of phosphate of macromolecules or ATP.

midine and spermine would be toxic as a function of decreasing Mg^{2+} (He et al., 1993). As Mg^{2+} concentration was reduced to 50 mM in media for mouse cells, the toxic effect of reduction of growth rate was demonstrated at low concentrations of exogenous spermidine (2 mM) and spermine (0.15 mM). This slowing of growth rate with decreased synthesis of DNA and protein correlated with a decrease of Mg^{2+} content in cells of normal polyamine content. Initially, the polyamines inhibited Mg^{2+} transport and did not affect ATP content. Subsequently, ATP content was also reduced and swollen mitochondria were found as the cells died. The uptake of N^1,N^{12}-bis(ethyl)spermine not only results in the cellular loss of spermidine and spermine, but also causes mitochondrial damage and a drop in ATP content (Fukuchi et al., 1992). Indeed, the synthesis of protein in mitochondria is almost selectively inhibited. The selective inhibition of ODC in tumor-bearing animals decreases DNA synthesis in the tumor but evokes an increase in nucleoside triphosphates. This suggests a possible increase of Mg^{2+} in such tissue (Westin et al., 1993).

Aspects of Eucaryotic Protein Synthesis

Useful model in vitro systems include extracts of rabbit reticulocytes, wheat germ, and yeast (Nygård & Nilsson, 1990). Protein synthesis in all of these is stimulated by polyamine. In the translation of the large RNA of plant viruses, the wheat germ system was frequently found to terminate proteins prematurely; this effect was occasionally mitigated by polyamine (Fritsch et al., 1977; Hunter et al., 1977).

The ribosomes of a liver microsomal fraction were the major functional entities in protein synthesis. Spermine was stimulatory in such a system at low Mg^{2+} (Hershko et al., 1961) and in a poly(U)-directed synthesis of polyphenylalanine by a system derived from an L-1210 mouse leukemia cell (Ochoa & Weinsten, 1965). The in vitro association of rat liver ribosomes to endoplasmic reticulum membranes was stimulated by a mix of spermine and Mg^{2+} (Khawaja, 1971). Although rat liver ribosomes contained far more spermine than spermidine, the 80S ribosomes of *Tetrahymena pyriformis* grown in a peptone broth were rich in putrescine and spermidine and contained less than 10% of the spermidine content as spermine (Weller et al., 1968).

A spermidine-stimulated, Mg^{2+}-sparing synthesis of polyphenylalanine in the presence of poly(U) was demonstrated readily in cell-free extracts of yeast (Bretthauer et al., 1963). A stimulation by spermidine of the transfer of phenylalanine from Phe-tRNA was also noted (Heredia & Halvorson, 1966). Effects of this kind were also observed with extracts of *Helianthus tuberosus* (Cocucci & Bagni, 1968).

Stimulatory effects of the polyamines on rabbit reticulocyte lysates that synthesized defined proteins in response to specific viral mRNAs were described by Atkins et al. (1975). The diamine DAP was slightly in-

hibitory at 1–5 mM. The stimulatory effects of spermidine or spermine were also demonstrated in reticulocyte lysates filtered through gels (Sephadex G-25) (Jackson et al., 1983).

Eucaryotic monosomes are larger (80S) than those of procaryotes and contain larger ribosomal subunits (40S and 60S) and larger rRNAs. In addition to the 18S, 28S, and 5S RNA, a new species of ribosomal RNA of 5.8S is known, and the larger eucaryotic ribosomes contain a greater number of proteins. Igarashi et al. (1980) described the stimulation of polypeptide synthesis and Mg^{2+}-sparing in fractionated extracts of rabbit reticulocytes or of wheat germ. The various effects included a stimulation of incorporation of initiating formylmethionine via fMet-tRNA. Although chain number was increased by spermidine, the length of the chains was not affected, suggestive of an effect of spermidine on initiation. Spermidine also stimulated the binding of fMet-tRNA in the presence of globin mRNA, but not in its absence. This result implied that spermidine stimulated the formation of the initiation complex of globin mRNA, 40S ribosomal subunits, and Met-tRNAs.

In a somewhat purified system, to which several isolated initiation factors were added, spermidine (0.4–0.6 mM) was again found to stimulate globin synthesis some six- to eightfold at optimal Mg^{2+} concentration. In the absence of added spermidine more α globin chains were made than β chains (Ogasawara et al., 1989). These authors concluded that the main effect of spermidine was in the formation of an initiation complex.

Marked differences between procaryotic and eucaryotic syntheses are found in the formation of initiation complexes, involving in the latter case an initiator tRNA for nonformylated methionine and some 10 eucaryotic initiation factors (IFs) instead of the three known for bacterial systems. One of these proteins, eIF-5A, contains the unique spermidine-derived amino acid, hypusine. Uniquely eucaryotic mRNA is capped at its 5′ beginning with 7-methylguanine nucleotide and terminates at the 3′ end with a long poly(A) tail. The biosyntheses and reactions of these features will obviously distinguish the processing of these messages. Thus, a preinitiation complex comprising the 40S subunit, eucaryotic IFs (eIFs), and Met-tRNA and GTP attaches the mRNA. The secondary structure of this mRNA has been opened in an ATP-driven helicase reaction involving another initiation factor that also identifies the AUG initiation codon. This 40+S preinitiation complex now associates with the 60S subunit with the aid of eIF-5 and GTP.

Some virus infections, such as by influenza virus, lead to the dephosphorylation of eIF-2. Phosphorylation of an essential initiation factor, eIF-2, inactivates the factor and thereby inhibits protein synthesis. This kinase reaction is inhibited by polyamines (Kuroda et al., 1982). The phosphorylation of many translational components controlling translation rates has been summarized by Hershey (1989). The regulation of the numerous translation factors of protein synthesis, with particular reference to initiation factors, has been reviewed by Rhoads (1993). The hypusine-containing eIF-5A, which is described by these reviewers as a protein of 17–18 kDa, is clearly different from eIF-5, a phosphoprotein with a monomeric mass of 58 kDa. The former was thought earlier to stimulate the synthesis of the first peptide bond whereas eIF-5 is considered to participate in the binding of the subunits. In a mouse liver system directed by endogenous or exogenous mRNA, spermidine or spermine stimulated both initiation and elongation and appeared to regulate differential translation of proteins coded by some multigenomic mRNAs such as tobacco mosaic virus (TMV) RNA (Giannakouros et al., 1990).

Newly purified wheat germ initiation factors were assayed in a wheat germ system in the presence of 0.1 mM spermine. The product of translation of mRNA for mouse interferon by a wheat germ system was quite active and accurate (Thang et al., 1975). Their observations indicated that the polyamines spermine or spermidine were essential and eliminated a lag in amino acid incorporation (Thang et al., 1976). Antispermine antibodies added to a wheat germ synthetic system sharply inhibited protein synthesis (Niveleau & Quash, 1979) (Fig. 22.6).

The RNA isolated from rat liver polysomes was used in a spermine-stimulated wheat germ system to produce a large polypeptide serologically characteristic of the rat liver fatty synthetase (Flick et al., 1978). Similarly, a poly(A)-containing RNA, the mRNA isolated from the French bean, was used to produce a bean seed protein in a spermidine-stimulated wheat germ system (Hall et al., 1978). In similar translations of rabbit globin mRNA and TMV RNA, the presence of spermidine permitted the synthesis of complete products, emphasizing a role in elongation (Abraham & Pihl, 1980). An excess

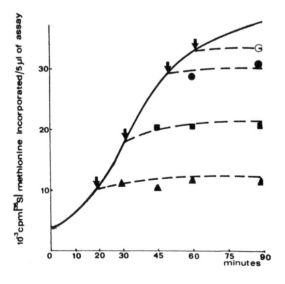

FIG. 22.6 Effect of antispermine antibodies on incorporation of [^{35}S]-methionine into protein in an in vitro wheat germ system active with animal cell RNA; adapted from Niveleau and Quash (1979).

of spermidine proved inhibitory in translation of some plant viral RNAs (Chroboczek, 1985).

Eucaryotic mRNA

Structural features of this RNA affecting the initiation of translation have been discussed by Kozak (1991). The existence of a 5′ cap of 7-methylguanine nucleotides has pointed to the use of 7-methylguanosine-5′-monophosphate as a potent inhibitor of initiation in a spermine-stimulated wheat germ system (Lucas et al., 1977). Many complexities of initiation, such as mRNA unwinding factors, reinitiation, and new specific mRNA binding factors, are being added in one or another synthetic system.

The existence on eucaryotic messages of a long 3′-poly(A), absent from rRNA and tRNA, has facilitated an isolation of most functional eucaryotic messages, which may be on the order of 1–2% of the total RNA. The early studies of protection against nucleases developed by C. Levy have been strengthened in recent experiments that have also been described an essential poly(A)-binding protein (Jackson & Standart, 1990). Although the presence of this segment does not determine initiation, it has been suggested to modulate the efficiency of translation. Polyadenylated RNAs have recently been found in procaryotic cells (Cohen, 1995; Manley, 1995).

Unfolding of Ribosomal Subunits and Ribosomal RNA

The existence of two classes of ribonucleates accounting for 80–90% of E. coli RNA was described by U. Littauer and H. Eisenberg in 1959. The hypochromicity of yeast ribosomal RNA was detected initially by heating detergent-treated ribosomes. Changes in the circular dichroism of ribosomes and of RNA were found after treatment with chelating agents or depletion of metals. Also, an acid extraction of ribosomal protein from 30S ribosomes permitted the isolation of a more native and open 16S RNA than did a method denaturing proteins with phenol. Additionally, the unfolding of ribosomes and RNA were effected by the intercalation of ethidium in double-stranded rRNA regions, an effect that can be inhibited partially by Mg^{2+}. Tight-binding sites for ethidium in 16S rRNA were more numerous than those present in the intact 30S ribosomal subunit, implying some cover-up by protein (Elson et al., 1979). In other tests of the compactness of structure, differential hydrogen exchange was hastened in a class of hydrogens in the subunits of eucaryotic ribosomes by divalent cations or polyamines (Moore & Spremulli, 1985). The penetrability of 30S ribosomal subunits to short complementary DNA sequences was found to be quite limited, although a very few regions are accessible for hybridization (Weller & Hill, 1992).

The conformation and thermal stability of 16S RNA in the presence of Mg^{2+} and K$^+$ were examined by Allen and Wong (1986), who were particularly concerned with conditions established as optimal for assembly into the 30S subunit. MgCl$_2$ (20 mM) facilitated pairing, stacking, and stability of the nucleotide bases in a compact secondary structure; the addition of 360 mM KCl loosened the structure and made the protein-binding sites more available.

Structure of Ribosomal RNA and Antibiotic Sensitivities

Among the various "footprinting" techniques used to establish the configuration of the RNA and its relation to the various proteins within the subunit and in the 70S monosome were the accessibility of the chain to nucleases and the modifiability of the phosphates by the alkylating agent, ethylnitrosourea (Baudin et al., 1989). Naked Mg^{2+}-containing 16S RNA proved to have a few interactions protecting only 7% of the phosphates, compared to an unfolded Mg^{2+}-free RNA; the addition of proteins in the subunit protected an additional 10%. The sensitivity of phosphate alkylation to ethylnitrosourea was also used to test the effect of the addition of IF-2 and initiator fmet-tRNA to the subunit. It was concluded that the 16S RNA thereby developed conformational changes within the subunit in response to components of initiation.

Various ribosomal inhibitors have been used to protect bases in the chain against reagents such as dimethylsulfate and kethoxal (Moazed & Noller, 1987). The inhibitory antibiotics tested include the aminoglycosides, such as streptomycin, and the spermidine-containing edeines. The latter structures contain five amino acids, of which four are unusual and the fifth, glycine, is bound in amide linkage to spermidine or its guanido derivative (Fig. 22.7).

The addition of streptomycin on 70S ribosomes produced a newly protected site, the specific adenine-containing nucleotides in the 16S RNA (positions 913–915) of the 1542 nucleotide chain, which were rendered resistant to methylation by dimethylsulfate. A mutation of *E. coli* resistant to streptomycin developed with a change of the 912 nucleotide from cytosine to uracil at

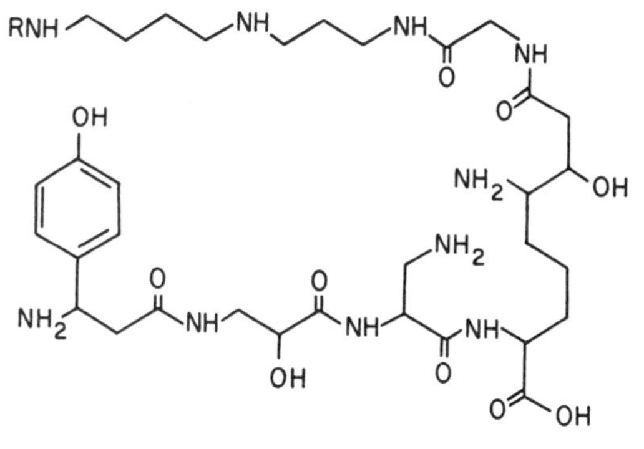

the fringe of the adenine site. These results were taken to suggest that this single region of the 16S RNA was a primary site of action of streptomycin. Thus, the cren-archaebacteria are insensitive to streptomycin and contain the C → U mutation expressed at the 912 site of their 16S rRNA (Amils et al., 1993).

The nature of the reaction of streptomycin, a strong cation, with the ribosome and the nature of its lethality to bacteria, which may well be separate problems, have a long history with some references to the polyamines. Mager et al. (1962) detected a partial protection by both polyamines and Mg^{2+} against the inhibition of protein synthesis by streptomycin and postulated a common site of action of antibiotic and polyamine.

The sensitivity to inhibition of protein synthesis was shown to be centered in the 30S ribosome that binds a molecule of ^{3}H-dihydrostreptomycin and results in the accumulation within the bacteria of aberrant initiation complexes containing mRNA and 70S monosomes. Because a relatively native 16S RNA of sensitive or resistant ribosomes can bind streptomycin tightly (Biswas & Gorini, 1972), it was thought that proteins may determine accessibility. Indeed, some 30S proteins were found to bind dihydrostreptomycin selectively. ^{3}H-Di-hydrostreptomycin can be cross-linked by phenyldiglyoxal to 16S RNA and three ribosomal proteins (Melançon et al., 1984). Cross-linking of the antibiotic with a bifunctional nitrogen mustard to 16S RNA revealed two cross-linked sites on the RNA, one of which included the 912 nucleotide (Gravel et al., 1987). Mutations in this site, the 915 region, confer resistance to streptomycin (Montandon et al., 1986) and reduce the binding of the antibiotic to the ribosome (Pinard et al., 1993). Still another site close to this major binding site appears to modulate the misreading effect of the antibiotic (Leclerc et al., 1991). In short, both RNA and proteins of the 30S subunit were strongly implicated as binding sites of the antibiotic. A region of 16S RNA (nucleotides 916–918) contiguous to the apparent streptomycin binding site (912–915) is believed to be associated with another domain (nucleotides 17–19) of the RNA, and this domain is postulated to be essential for the initiation of translation (Brink et al., 1993). Photoaffinity labeling with streptomycin derivatives has now established the binding site at the 16S rRNA-rich interface between ribosomal subunits close to two ribosomal proteins, S5 and L11. This result may explain most of the effects of the antibiotic on tRNA selection and protein synthesis (Abad & Amils, 1994).

Edeine is believed to block the binding site of tRNA, which appears to encompass four bases in various near contiguous folded positions of the 16S RNA chain. Spermidine was reported earlier to provide a competitive interference with edeine in globin chain initiation in a reticulocyte system (Konecki et al., 1975). Kozak and Shatkin (1978) also described an effect of edeine in mediating the formation of unusual complexes containing multiple 40S subunits bound to various mRNAs.

Detailed models of the 30S structure and the 16S RNA within it have been created from experiments on

FIG. 22.7 Structures of edeines A and B; adapted from Hettinger and Craig (1970). Edeine A, R = H; edeine B, R = CNHNH$_2$.

protected residues and sites and the experiments on cross-linking the RNA to protein (Brimacombe, 1995). None of these constructions refers explicitly to relevant data on the sites and roles of the polyamines. The studies have also been extended to the 23S RNA and the 50S subunits.

Polyamine-Dependent Ribosomal Systems

Studies with thermophilic bacteria led to the discovery of novel polyamine-dependent ribosomal systems. The thermoacidophilic archaebacterium *Caldariella acidophila* (or *Sulfolobus solfataricus*) possesses a system that synthesizes protein, with an optimum at 75–85°C dependent on 2–4 mM spermine in the presence of 18–20 mM Mg^{2+} (Cammarano et al., 1982). The systems of many other sulfur-dependent thermophilic archaebacteria, which also have spermine- or thermine-dependent synthesis of polyphenylalanine in response to poly(U), have easily dissociable ribosomes. Friedman (1986) found that spermine also protects the *Sulfolobus* system from thermal inactivation at 65°C. The systems are essentially inert to many antibiotics, including most aminoglycosides (Cammarano et al., 1985). An extract of the thermophilic eubacterium, *Thermatoga maritima*, with an optimum at 80°C, also had a requirement for spermine (0.5 mM) (Londei et al., 1988). At this temperature some other isolated ribosomal systems (e.g., *E. coli, B. stearothermophilus*) were somewhat active and still sensitive to streptomycin whereas the *Thermotoga* system was inert to both the inhibition of protein synthesis and misreading caused by the antibiotic in the sensitive systems. No reports have yet appeared on the possible role of spermine in limiting the inhibitory action of the antibiotic or on the sites at which the polyamine binds to the 30S subunit.

Uzawa et al. (1993b) constructed a continuous cell-free system from the eubacterium *T. thermophilus*, which operated in translating MS2 phage RNA for more than 5 h at 65°C. The system was inactive in the absence of added polyamine and was maximally active in the simultaneous presence of 1 mM spermine and 0.1 mM tetrakis(3-aminopropyl) ammonium.

Assembly of Ribosomal Subunits

Polyamine deficiency in growing bacteria provoked the formation of incomplete, functionally impaired subunits. Key proteins like S1 were missing in the structure, and the synthesis and integration of the protein to form a complete 30S subunit turned on the availability of polyamine. Further, deficiency in the cell restricted methylation of adenine in the 16S RNA and the methyl-deficient RNA was slower in the assembly. However, essential roles of polyamine in the assembly of functional ribosomal subunits was not detected very early

(Nomura & Held, 1974). In vitro assemblies frequently required incubations at high temperatures, suggestive of steps requiring as yet undetermined physiological processes.

Hosokawa et al. (1973) described a reconstitution in the presence of Mg^{2+} and polyamine of components of the *E. coli* 50S subunit to a slightly smaller 45–48S particle. However, this incomplete subunit was functionally inactive in protein synthesis, although the particle was more complete than that produced in Mg^{2+} alone. The partially functional assembly of the 50S subunit of the thermophile, *B. stearothermophilus*, was also obtained and was considered not to require polyamines. It may be noted that the crystallization of the 50S subunit of *B. stearothermophilus* was carried out with the inclusion of 10 mM spermine in the crystallization mix (Yonath et al., 1986), although the polyamine was not demonstrated to be an essential constituent (Berkovitch–Yellin et al., 1992). However, a complete functional reassembly was accomplished at high temperatures, 65 and 80°C, with the components of the larger ribosomal subunit of the thermophile *Sulfolobus solfataricus* in the presence of high concentrations of K^+ and Mg^{2+}, with an absolute requirement for thermine, spermine, or spermidine in the reconstitution mix (Londei et al., 1986) Figure 22.8.

The dissection of ribosomal subunits of *E. coli* revealed considerable complexity despite the presence of a single RNA (i.e., 16S in the 30S subunit) and two RNAs (i.e., 5S and 23S in the 50S subunit). The separation of these RNAs by gel filtration was developed and ribosomal RNA isolated from phenol-disrupted cells at low ionic strength contained large amounts of spermidine (see Chap. 23). The problem of reassembly of the fragments of the subunits to functionally competent

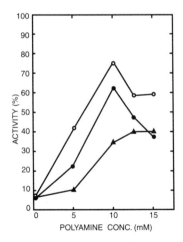

FIG. 22.8 Polyamine dependence of reconstitution of *Sulfolobus* large ribosomal subunits at 65°C for 45 min; adapted from Londei et al. (1986). (○) thermine; (●) spermine; (▲) spermidine.

structures assisted an understanding of the structures. Many studies were concerned with the analysis of the separated proteins and RNAs. The 30S ribosomal subunit and its 16S RNA received the greatest attention, but some data relating the cations to the other RNA molecules emerged.

The 5S ribosomal RNA of yeast was analyzed to estimate the primary sequence, the degree of base pairing, and the conformational changes in response to Mg^{2+}. Only two Mg^{2+} per molecule appear to produce a more ordered structure, detected by difference absorption spectroscopy and circular dichroism (Maruyama & Sugai, 1980). The 5S RNA of *E. coli* revealed an unexpected early melting transition, characterized by a decrease in T_m in response to cations, including Mg^{2+} and spermidine, for a part of the structure (Kao & Crothers, 1980). This was interpreted as a rapid conformational change possibly occurring cyclically during protein synthesis. Fox and Wong (1982) found that Mg^{2+} alone at 60°C was capable of reorganizing a correct folding pattern for the 5S RNA. The melting profiles of 5S rRNA from plant material were compared in the presence of Mg^{2+}, spermine^{4+}, or (permethyl)$_{10}$ spermine^{4+} (Barciszewski et al., 1988). The permethyl derivative had the weakest stabilizing effect, despite its tetravalent charge, and it was suggested that the purely electrostatic interactions of such a compound were inadequate to stabilize the structure. On the other hand, the tetracationic spermine was quite strongly stabilizing, more so than Mg^{2+}, a fact attributed to the hydrogen bonds afforded by all of the primary and secondary amines of spermine.

The suppression of mutations in the *Tetrahymena* ribozyme by spermidine has been attributed to the maintenance of a folded RNA structure (Hanna & Szostak, 1994).

Effects of Mg^{2+} and Polyamines on Polyribonucleotides

Polynucleotides formed well-defined complexes with metallic and organic cations. Conductimetric titration of the sodium salt of polyuridylate by Mg^{2+} revealed a replacement of Na^+ by Mg^{2+} until one equivalent of Mg^{2+} is bound strongly for each equivalent of negative charge (Felsenfeld & Huang, 1959). Calorimetric studies by Krakauer (1972) reported a heat of binding of Mg^{2+}, interpreted as arising in an increase of base stacking. This result and conclusion is consistent with the development of the hyperchromic effect induced on poly(U) by Mg^{2+} and the various polyamines, as seen in Figure 22.9.

The figure shows that the changing secondary structures induced by spermidine and spermine on poly(U) are markedly greater than that evoked by Mg^{2+} and the diamines. These effects are not obtained with poly(A) or (C). Although a fibrous complex of poly (U) and spermine or spermidine appeared double stranded in X-ray diffraction (Zimmerman, 1976), ethidium did not evoke a fluorescence characteristic of such a structure in poly(U)-spermine complexes in solution (Igarashi et al., 1977).

The binding constants of Mg^{2+} to nucleates in a neutral 5 mM sodium phosphate buffer were determined by Sander and Tso (1971) who reported that tRNA (ca. 1.7 $\times 10^4$ M^{-1}) > poly(A) = poly(A)·poly(U) = native DNA > denatured DNA = poly(U) = poly(C) (ca. 3×10^3 M^{-1}) ≫ monomers (ca. 2×10^2 M^{-1}). These results were interpreted as distinguishing the relative degree of folding of tRNA versus poly(A)·poly(U) vs. poly(U).

The formation of double-stranded structures, such as poly(A)·poly(U) were determined by following the decrease of optical density as a function of added cations. The completion of interaction required one equivalent

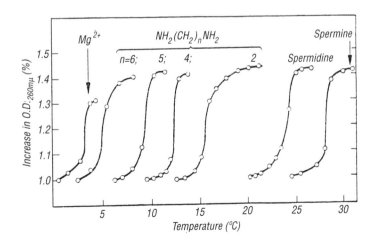

FIG. 22.9   Melting of poly(U) in the presence of Mg^{2+} and polyamines; adapted from Szer (1966). Polymer concentration was 4×10^{-5} M; concentration of divalent cations was 10^{-3} M. Concentration of polyamines was equivalent to polymer phosphate.

of divalent ion per mole of phosphate (Felsenfeld & Huang, 1960). The diamines were slightly less effective than Mg^{2+}, and increasing ionic strength decreased the strength of binding.

Polynucleotides of A and U can form both double-stranded and triple-stranded structures. Higuchi and Tsuboi (1966) demonstrated that spermine stabilized both the double- and three-stranded structures more than any of the others. The effect of spermine was markedly greater than that of Mg^{2+}. Spermine is also known to increase the T_m of polyinosinate : cytidylate; the resulting complex, which is capable of inducing interferon in rabbits, is known to have an increased resistance to pancreatic ribonuclease (Lampson et al., 1969). The formation of stable complexes of spermine with poly(A)·poly(U) and poly(I)·poly(C), as well as with tRNA, rRNA, and MS2 RNA, was demonstrated by Ikimura (1969). Viral RNAs containing polyamines were described in Chapter 19.

Derivatives of the polyamines, the diquaternary ammonium salts, have been used to study the T_m transitions, that is, the helix-coil transition of poly(A)·poly(U) and poly(I)·poly(C). When substituents on the spermine derivatives were large hydrophobic groups, the stabilization of poly(I)·poly(C) was less than that of poly(A)·poly(U). It was suggested that the helical structure of poly(I)·poly(C) was more dense and more sensitive to alkyl substituents than poly(A)·poly(U) (Glaser & Gabbay, 1968).

The early models of the reactions of polyamines with double-stranded DNA and RNA proposed electrostatic interactions of positively charged amino groups with phosphates and a bridge of the tetramethylene portion of spermidine or spermine across a narrow groove of a helical structure. Gabbay et al. (1976) proposed that diamines bind within the minor groove to phosphate oxygens by hydrogen bonding. Bolton and Kearns (1978) obtained evidence by proton magnetic resonance (^{1}H-NMR) that polyamines are simultaneously hydrogen bonded within an RNA complex to the 2′-OH and the oxygen of a 3′-phosphate, thereby displacing hydrogen-bonded water molecules. The elimination of resonances of the 2′-hydroxyl was readily obtained by the addition of spermine to poly(U), poly(A) and poly(C), and poly(I) poly(C), but slightly more difficult with E. coli mixed tRNA. On the other hand, Mg^{2+} was essentially ineffective. Obviously, such a mechanism would apply to RNA but not to DNA.

REFERENCES

Abad, J. P. & Amils, R. (1994) *Journal of Molecular Biology*, **235**, 1251–1260.

Abraham, A. K. & Pihl, A. (1980) *European Journal of Biochemistry*, **106**, 257–262.

Adachi, Y., Luke, M., & Laemmli, V. K. (1991) *Cell*, **64**, 137–148.

Ahmed, K., Wilson, M. J., Goueli, S. A., & Williams–Ashman, H. G. (1978) *Biochemical Journal*, **176**, 739–750.

Alberghina, F. A. M., Chimenti Signorini, R., Trezzi, F., & Viotti, A. (1971) *Journal of Submicroscopic Cytology and Pathology*, **3**, 9–18.

Alatossava, T., Jütte, H., Kuhn, A., & Kellenberger, E. (1985) *Journal of Bacteriology*, **162**, 413–419.

Allen, S. H. & Wong, K. (1986) *Archives of Biochemistry and Biophysics*, **249**, 137–147.

Althaus, F. R., Höfferer, L., Kleczkowka, H. E., Malanga, M., Naegeli, H., Panzeter, P., & Realini, C. (1993) *Environmental and Molecular Mutagenesis*, **22**, 278–282.

Amils, R., Cammarano, P., & Londei, P. (1993) *The Biochemistry of Archaebacteria*, pp. 393–438, M. Kates, D. J. Kushner, & A. T. Matheson, Eds. New York: Elsevier Science Publishers.

Andreev, O. A. & Kaboev, O. K. (1982) *Biochimica et Biophysica Acta*, **698**, 100–101.

Atkins, J. F., Lewis, J. B., Anderson, C. W., & Gesteland, R. F. (1975) *Journal of Biological Chemistry*, **250**, 5688–5695.

Atmar, V. J., Daniels, G. R., & Kuehn, G. D. (1978) *European Journal of Biochemistry*, **90**, 29–37.

Bailly, V., Derydt, M., & Verly, W. G. (1989) *Biochemical Journal*, **261**, 707–713.

Bakeyev, V. V., Melnickov, A. A., Osicka, V. D., & Varshavsky, A. J. (1975) *Nucleic Acids Research*, **2**, 1401–1419.

Balestreri, E., Cioni, P., Romagnoli, A., Bernini, S., Fissi, A., & Felicioli, R. (1987) *Archives of Biochemistry and Biophysics*, **255**, 460–463.

Banères, J. L., Essalouh, L., Jariel–Encontre, I., Mesnier, D., Garrod, S., & Parello, J. (1994) *Journal of Molecular Biology*, **243**, 48–59.

Barciszewski, J., Bratek–Wiewidorowska, M. D., Górnicki, P., Naskret–Barciszewska, M., Wiewidorowski, M., Zielenkiewicz, A., & Zienlenkiewicz, W. (1988) *Nucleic Acids Research*, **16**, 695–701.

Bartetzko, H. & Nierhaus, K. H. (1988) *Methods in Enzymology*, **164**, 650–658.

Basu, H. S., Sturkenboom, M. C. J. M., Delcros, J., Csokan, P. P., Szöllösi, J., Feuerstein, B. G., & Marton, L. J. (1992) *Biochemical Journal*, **282**, 723–727.

Baudin, F., Mougel, M., Romby, P., Eyermann, F., Ebel, J., Ehresmann, B., & Ehresmann, C. (1989) *Biochemistry*, **28**, 5847–5855.

Berkovitch–Yellin, Z., Bennett, W. S., & Yonath, A. (1992) *Critical Reviews in Biochemistry and Molecular Biology*, **27**, 403–444.

Billett, M. A. & Hall, T. J. (1979) *Nucleic Acids Research*, **6**, 2929–2945.

Biswas, D. K. & Gorini, L. (1972) *Proceedings of the National Academy of Sciences of the United States of America*, **69**, 2141–2144.

Blair, D. G. R. (1985) *International Journal of Biochemistry*, **17**, 23–30.

Blankenship, J. W., Morgan, J. E., & Matthews, H. R. (1987) *Molecular Biology Reports*, **12**, 21–26.

Bogdanova, E. S. (1984) *FEBS Letters*, **175**, 321–324.

Bojanowski, K., Filhol, O., Cochet, C., Chambaz, E. M., & Larsen, A. K. (1993) *Journal of Biological Chemistry*, **268**, 22920–22926.

Bolton, P. H. & Kearns, D. R. (1978) *Nucleic Acids Research*, **5**, 1315–1324.

Bouthier de la Tour, C., Portemer, C., Nadal, M., Stetter, K. O., Forterre, P., & Duguet, M. (1990) *Journal of Bacteriology*, **172**, 6803–6808.

Braithwaite, D. K. & Ito, J. (1993) *Nucleic Acids Research*, **21**, 787–802.

Brimacombe, R. (1995) *European Journal of Biochemistry*, **230**, 365–383.

Brink, M. F., Verbeet, M. P., & de Boer, H. A. (1993) *The EMBO Journal*, **12**, 3987–3996.

Burgess, R. R. & Jendrisak, J. J. (1975) *Biochemistry*, **14**, 4634–4638.

Bushell, M. E. & Bull, A. T. (1974) *Journal of General Microbiology*, **81**, 271–273.

Butler, E. T. & Chamberlin, M. J. (1982) *Journal of Biological Chemistry*, **257**, 5772–5778.

Byrne, R. H., Stone, P. R., & Kidwell, W. R. (1978) *Experimental Cell Research*, **115**, 277–283.

Cammarano, P., Teichner, A., Chinali, G., Londei, P., de Rosa, M., Gambacorta, A., & Nicolaus, B. (1982) *FEBS Letters*, **148**, 255–259.

Cammarano, P., Teichner, A., Londei, P., Acca, M., Nicolaus, B., Sanz, J. L., & Amils, R. (1985) *The EMBO Journal*, **4**, 811–816.

Cardenas, M. E. & Gasser, S. M. (1993) *Journal of Cell Science*, **104**, 219–225.

Chang, C., Saltzman, A. G., Hiipakka, R. A., Huang, I., & Liao, S. (1987) *Journal of Biological Chemistry*, **262**, 2826–2831.

Chang, C., Shuman, H., & Somlyo, A. P. (1986) *Journal of Bacteriology*, **167**, 935–939.

Chamberlin, M., Kingston, R., Gilman, M., Wiggs, J., & De-Vera, A. (1983) *Methods in Enzymology*, **101**, 540–568.

Chroboczek, J. (1985) *Plant Molecular Biology*, **4**, 23–30.

Clark, E., Swank, R. A., Morgan, J. E., Basu, H., & Matthews, H. R. (1991) *Biochemistry*, **30**, 4009–4020.

Cohen, A. S. & Zalik, S. (1978) *Phytochemistry*, **17**, 113–118.

Cohen, S. N. (1995) *Cell*, **80**, 829–832.

Cohen, S. S. (1970) *Annals of the New York Academy of Science*, **171**, 869–881.

Cohn, M. & Blackburn, E. H. (1995) *Science*, **269**, 396–400.

Colson, P. & Houssier, C. (1989) *FEBS Letters*, **257**, 141–144.

Conaway, J. W. & Conaway, R. C. (1991) *Journal of Biological Chemistry*, **266**, 17721–17724.

Cox, R. (1979) *Biochemical and Biophysical Research Communications*, **86**, 594–598.

Crothers, D. M. (1994) *Science*, **266**, 1819–1820.

Dabrowski, M., Spahn, C. M. T., & Nierhaus, K. H. (1995) *The EMBO Journal*, **14**, 4872–4882.

Daniels, G. R., Atmar, V. J., & Kuehn, G. D. (1981) *Biochemistry*, **20**, 2525–2532.

Dorr, R. T., Liddil, J. D., & Gerner, E. W. (1986) *Cancer Research*, **46**, 3891–3895.

Dove, M. J. & Tsai, C. S. (1976) *Canadian Journal of Biochemistry*, **54**, 432–437.

Drainas, D. & Kalpaxis, D. L. (1994) *Biochimica et Biophysica Acta*, **1208**, 55–64.

Duerre, J. A., Quick, D. P., Traynor, M. D., & Onisk, D. V. (1982) *Biochimica et Biophysica Acta*, **719**, 18–23.

Dumuis–Kervabon, A., Encontre, I., Etienne, G., Jaurequi–Adell, J., Mery, J., Mesnier, D., & Parello, J. (1986) *The EMBO Journal*, **5**, 1735–1742.

Echandi, G. & Algranati, I. D. (1975a) *Biochemical and Biophysical Research Communications*, **62**, 313–319.

——— (1975b) *Biochemical and Biophysical Research Communications*, **67**, 1185–1191.

Eick, D., Wedel, A., & Heumann, H. (1994) *Trends in Genetics*, **10**, 292–296.

Elson, D., Spitnik–Elson, P., Avital, S., & Abramowitz, R. (1979) *Nucleic Acids Research*, **7**, 465–480.

Emanuelsson, H. & Heby, O. (1983) *Cell Growth & Differentiation*, **12**, 277–285.

Felsenfeld, G. (1978) *Nature*, **271**, 115–122.

——— (1992) *Nature*, **355**, 219–224.

Finch, J. T., Brown, R. S., Rhodes, D., Richmond, T., Rushton, B., Lutter, L. C. & Klug, A. (1981) *Journal of Molecular Biology*, **145**, 757–769.

Finch, J. T., Lutter, L. C., Rhodes, D., Brown, R. S., Rushton, B., Levitt, M. & Klug, A. (1977) *Nature* **269**, 29–36.

Fischer, D., Weisenberger, D., & Scheer, V. (1991) *Chromosoma*, **101**, 133–140.

Fisher, P. A. & Korn, D. (1977) *Journal of Biological Chemistry*, **252**, 6528–6535.

——— (1979) *Journal of Biological Chemistry*, **254**, 11033–11039.

Fkyerat, A., Demeunynck, M., Constant, J., Michon, P., & Lhomme, J. (1993) *Journal of American Chemical Society*, **115**, 9952–9959.

Flatman, P. W. (1991) *Annual Reveiw of Physiology*, **53**, 259–271.

Flick, P. K., Chen, J., Alberts, A. W., & Vagelos, P. R. (1978) *Proceedings of the National Academy of Sciences of the United States of America*, **75**, 730–734.

Forterre, P., Mirambeau, G., Jaxel, C., Nadal, M., & Duguet, M. (1985) *The EMBO Journal*, **4**, 2123–2128.

Fox, J. W. & Wong, K. (1982) *Biochemisty*, **21**, 2096–2102.

Frey, C. M. & Stuehr, J. E. (1972) *Journal of the American Chemical Society*, **94**, 8898–8904.

Friedman, S. M. (1986) *Systematic and Applied Microbiology*, **7**, 325–329.

Friedman, S. M. & Oshima, T. (1989) *Journal of Biochemistry*, **105**, 1030–1033.

Fritsch, C., Mayo, M. A., & Hirth, L. (1977) *Virology*, **77**, 722–732.

Frugier, M., Florentz, C., Hosseini, M. W., Lehn, J., & Giegé, R. (1994) *Nucleic Acids Research*, **22**, 2784–2790.

Fuchs, E. (1976) *European Journal of Biochemistry*, **63**, 15–22.

Fukuchi, J., Kashiwagi, K., Kusama–Eguchi, K., Terao, K., Shirahata, A., & Igarashi, K. (1992) *European Journal of Biochemistry*, **209**, 689–696.

Gabbay, E. J., Adawadkor, P. D., & Wilson, W. D. (1976) *Biochemistry*, **15**, 146–151.

Gaszner, M., Haracska, L., & Udvardy, A. (1993) *Archives of Biochemistry and Biophysics*, **303**, 1–9.

Ge, H., Zhao, Y., Chait, B. T., & Roeder, R. G. (1994) *Proceedings of the National Academy of Sciences of the United States of America*, **91**, 12691–12695.

Geiger, L. E. & Morris, D. R. (1978) *Nature*, **272**, 730–732.

George, G. A. & Heaton, F. W. (1975) *Biochemical Journal*, **152**, 609–615.

Georgiev, G. P., Vassetzky, Jr., Y. S., Luchnik, A. N., Chernokhvostov, V. V., & Razin, S. V. (1991) *European Journal of Biochemistry*, **200**, 613–624.

Germond, J., Yaniv, J. R., Yaniv, M., & Brutlag, D. (1979) *Proceedings of the National Academy of Sciences of the United States of America*, **76**, 3779–3783.

Gething, M. & Sambrook, J. (1992) *Nature*, **355**, 33–45.

Giannakouros, T., Nikolakaki, H., & Georgatsos, J. G. (1990) *Molecular and Cell Biology*, **99**, 9–19.

Glitz, D. G., Eichler, D. C., & Angel, L. (1974) *Analytical Biochemistry*, **62**, 552–567.

Goldemberg, S. H. & Algranati, I. D. (1981) *European Journal of Biochemistry*, **117**, 251–255.

Goldemberg, S. H., Fernández–Velasco, J. G., & Algranati, I. D. (1982) *FEBS Letters*, **142**, 275–279.

Goueli, S. A., Davis, A. T., Hupakka, R. A., Liao, S., & Ahmed, K. (1985) *Biochemical Journal*, **230**, 293–302.

Goyns, M. H. (1979) *Experimental Cell Biology*, **122**, 377–380.

Gravel, M., Melançon, P., & Brakier–Gingras, L. (1987) *Biochemistry*, **26**, 6227–6232.

Gualerzi, C. O. & Pon, C. L. (1990) *Biochemistry*, **29**, 5881–5889.

Gurevich, V. V., Pokrovskaya, I. D., Obukhova, T. A., & Zozulya, S. A. (1991) *Analytical Biochemistry,* **195**, 207–213.

Hall, T. C., Ma, Y., Buchbinder, B. U., Pyne, J. W., Sun, S. M., & Bliss, F. A. (1978) *Proceedings of the National Academy of Sciences of the United States of America*, **75**, 3196–3200.

Hanna, M. & Szostak, J. W. (1994) *Nucleic Acids Research*, **22**, 5326–5331.

Hansen, H. A. S., Volkmann, N., Piefke, J., Glotz, C., Weinstein, S., Makowski, I., Meyer, S., Wittmann, H. G., & Yonath, A. (1990) *Biochimica et Biophysica Acta*, **1050**, 1–7.

Haukanes, B. I., Szajko, K., & Helland, D. E. (1990) *FEBS Letters*, **269**, 389–393.

He, Y., Kashiwagi, K., Fukuchi, J., Terao, K., Shirahata, A., & Igarashi, K. (1993) *European Journal of Biochemistry*, **217**, 89–96.

Heby, O., Sarna, G. P., Marton, L. J., Omine, M., Perry, S., & Russell, D. (1973) *Cancer Research*, **33**, 2959–2964.

Hecht, S. M. (1994) *Bioconjugate Chemistry*, **5**, 513–526.

Hernandez–Verdun, D. (1991) *Journal of Cell Science*, **99**, 465–471.

Hershey, J. W. B. (1989) *Journal of Biological Chemistry*, **264**, 20823–20826.

Hettinger, T. P. & Craig, L. C. (1970) *Biochemistry*, **9**, 1224–1232.

Hewish, D. R. & Burgoyne, L. A. (1973) *Biochemical and Biophysical Research Communications*, **52**, 504–510.

Hironaka, T., Itaya, A., Yoshihara, K., Minaga, T., & Kamiya, T. (1987) *Analytical Biochemistry*, **166**, 361–367.

Hosokawa, K., Kiho, Y., & Migita, L. K. (1973) *Journal of Biological Chemistry*, **248**, 4135–4143.

Hougaard, D. M., Bolund, L., Fujiwara, K., & Larsson, L. I. (1987) *European Journal of Cell Biology*, **44**, 151–155.

Hunter, A. R., Farrell, P. J., Jackson, R. J., & Hunt, T. (1977) *European Journal of Biochemistry*, **75**, 149–157.

Igarashi, K., Aoki, Y., & Hirose, S. (1977) *Journal of Biochemistry*, **81**, 1091–1096.

Igarashi, K., Hikami, K., Sugawara, K., & Hirose, S. (1973) *Biochimica et Biophysica Acta*, **299**, 325–330.

Igarashi, K., Kashiwagi, K., Kishida, K., Kakegawa, T. & Hirose, S. (1981) *European Journal of Biochemistry*, **114**, 127–131.

Igarashi, K., Kojima, M., Watanabe, Y., Maeda, K. & Hirose, S. (1980) *Biochemical and Biophysical Research Communications*, **105**, 164–167.

Igarashi, K., Mitsui, K., Kubota, M., Shirakuma, M., Ohnishi, R., & Hirose, S. (1983) *Biochimica et Biophysica Acta*, **755**, 326–331.

Jackson, R. J., Campbell, E. A., Herbert, P., & Hunt, T. (1983) *European Journal of Biochemistry*, **131**, 289–301.

Jackson, R. J. & Standart, N. (1990) *Cell*, **62**, 15–24.

Jacob, S. T. (1995) *Biochemical Journal*, **306**, 617–626.

Jacob, S. T. & Rose, K. M. (1976) *Biochimica et Biophysica Acta*, **425**, 125–128.

Jain, A. & Tyagi, A. K. (1987) *Molecular and Cell Biology*, **78**, 3–8.

Jänne, O., Bardin, C. W., & Jacob, S. T. (1975) *Biochemistry*, **14**, 3589–3597.

Jelenc, P. C. (1980) *Analytical Biochemistry*, **105**, 369–374.

Johnson, T. W. & Klotz, I. M. (1974) *Biopolymers*, **13**, 791–796.

Kakegawa, T., Hirose, S., Kashiwagi, K., & Igarashi, K. (1986) *European Journal of Biochemistry*, **158**, 265–269.

Kakegawa, T., Takamiya, K., Ogawa, T., Hayashi, Y., Hirose, S., Niitsu, M., Samejima, K., & Igarashi, K. (1988) *Archives of Biochemistry and Biophysics*, **261**, 250–256.

Kalpaxis, D. L. & Drainas, D. (1992) *Molecular and Cellular Biology*, **115**, 19–26.

——— (1993) *Archives of Biochemistry and Biophysics*, **300**, 629–634.

Kao, T. H. & Crothers, D. M. (1980) *Proceedings of the National Academy of Sciences of the United States of America*, **77**, 3360–3364.

Karpetsky, T. P., Shriver, K. K., & Levy, C. C. (1980) *Journal of Biological Chemistry*, **255**, 2713–2721.

——— (1981) *Biochemical Journal*, **193**, 325–337.

Kashiwagi, K., Miyaji, A., Ikeda, S., Tobe, T., Sasakawa, C., & Igarashi, K. (1992) *Journal of Bacteriology*, **174**, 4331–4337.

Kashiwagi, K., Sakai, Y., & Igarashi, K. (1989) *Archives of Biochemistry and Biophysics*, **268**, 379–387.

Kent, P., Bunce, S. C., Bailey, R. A., Aikens, D. A., & Hurwitz, C. (1974) *Analytical Biochemistry*, **62**, 75–80.

Khawaja, J. A. (1971) *Biochimica et Biophysica Acta*, **254**, 117–128.

Kimes, B. W. & Morris, D. R. (1973) *Biochemistry*, **12**, 442–449.

Kinohara, N., Usui, H., Yoshikawa, K., & Takeda, M. (1984) *Journal of Biochemistry*, **95**, 597–600.

Kirino, H., Kuwahara, R., Hamasaki, N., & Oshima, T. (1990) *Journal of Biochemistry*, **107**, 661–665.

Knuutila, S. & Pohjanpelto, P. (1983) *Experimental Cell Research*, **145**, 222–226.

Kobayashi, Y., Hamasaki, N., & Oshima, T. (1990) *International Symposia on Polyamines*, Kyoto, Japan, p. 10.

Konecki, D., Kramer, G., Pinphanichakarn, P., & Hardesty, B. (1975) *Archives of Biochemistry and Biophysics*, **169**, 192–198.

Kornberg, R. D. & Klug, A. (1981) *Scientific American*, **244**, 52–44.

Kossorotow, A., Wolf, H. U., & Seiler, N. (1974) *Biochemical Journal*, **144**, 21–27.

Kozak, M. (1991) *Journal of Biological Chemistry*, **266**, 19867–19870.

Kozak, M. & Shatkin, A. J. (1978) *Journal of Biological Chemistry*, **253**, 6568–6577.

Kozyavkin, S. A., Slesarev, A. I., Malkhosyan, S. R., & Panyutin, I. G. (1990) *European Journal of Biochemistry*, **191**, 105–113.

Krajewski, W. A., Panin, V. H., & Razin, S. V. (1993) *Biochemical and Biophysical Research Communications*, **1993**, 113–118.

Krakauer, H. (1972) *Biopolymers*, **11**, 811–828.

Krasnow, M. A. & Cozzarelli, N. R. (1982) *Journal of Biological Chemistry*, **257**, 2687–2693.

Kuehn, G. D., Affolter, H., Atmar, V. J., Seebeck, T., Gubler, U., & Braun, R. (1979) *Proceedings of the National Academy of Sciences of the United States of America*, **76**, 2541–2545.

Kumagai, H., Igarashi, K., Tsuji, I., Mori, C., & Hirose, S. (1980) *Chemical & Pharmaceutical Bulletin*, **28**, 1189–1195.

Kun, E., Chang, A. C. V., Sharma, M. L., Ferro, A. M., & Nitecki, D. (1976) *Proceedings of the National Academy of Sciences of the United States of America*, **73**, 3131–3135.

Kuroda, Y., Merrick, W. C., & Sharma, R. K. (1982) *Science*, **215**, 415–416.

Latchman, D. S. (1990) *Biochemical Journal*, **270**, 281–289.

Lautier, D., Lagueux, J., Thibodeau, J., Ménard, L., & Poirier, G. G. (1993) *Molecular and Cellular Biochemistry*, **122**, 171–193.

Leclerc, D., Melançon, P., & Brakier–Gingras, L. (1991) *Biochimie*, **73**, 1431–1438.

Lee, J. E. & Cho, Y. D. (1992) *Biochemical and Biophysical Research Communicatons*, **189**, 450–454.

Levy, C. C., Hieter, P. A., & LeGendre, S. M. (1974) *Journal of Biological Chemistry*, **249**, 6762–6769.

Levy, C. C., Schmukler, M., Frank, J. J., Karpetsky, T. P., Jewett, P. B., Hieter, P. A., LeGendre, S. M., & Dorr, R. G. (1975) *Nature*, **256**, 340–342.

Levyant, M. I., Bylinkina, V. S., & Orekhovich, V. N. (1979) *Biokhimiya*, **44**, 1454–1459.

Lewicki, B. T. U., Margus, T., Remme, J., & Nierhaus, K. H. (1993) *Journal of Molecular Biology*, **231**, 581–593.

Liang, T., Mezzetti, G., Chen, C., & Liao, S. (1978) *Biochimica et Biophysica Acta*, **542**, 430–441.

Lindahl, T. & Andersson, A. (1972) *Biochemistry*, **11**, 3618–3623.

Loidl, P. (1988) *FEBS Letters*, **227**, 91–95.

Londei, P., Altamura, S., Huber, R., Stetter, K. O., & Cammarano, P. (1988) *Journal of Bacteriology*, **170**, 4353–4360.

Londei, P., Teixido, J., Acca, M., Cammarano, P., & Amils, R. (1986) *Nucleic Acids Research*, **14**, 2269–2285.

Lucas, M. C., Jacobson, J. W., & Giles, N. H. (1977) *Journal of Bacteriology*, **130**, 1192–1198.

Lupski, J. R. & Godson, G. N. (1989) *Bioessays*, **10**, 152–157.

Luttinger, A. (1995) *Molecular Microbiology*, **15**, 601–606.

Male, R., Fosse, V. M., & Kleppe, K. (1982) *Nucleic Acids Research*, **10**, 6305–6318.

Mandel, J. & Chambon, P. (1974) *European Journal of Biochemistry*, **41**, 367–378.

Manley, J. L. (1995) *Proceedings of the National Academy of Sciences of the United States of America*, **92**, 1800–1801.

Marcus, S. L., Kopelman, R., Koll, B., & Bacchi, C. J. (1982) *Molecular and Biochemical Parasitology*, **5**, 231–243.

Marcus, S. L., Smith, S. W., & Bacchi, C. J. (1981) *Journal of Biological Chemistry*, **256**, 3460–3464.

Maruyama, S. & Sugai, S. (1980) *Journal of Biochemistry*, **88**, 151–158.

Mathews, C. K. & Van Holde, K. E. (1990) *Biochemistry*, pp. 954–988. Menlo Park, Calif.: Benjamin/Cummings Publishing Company, Inc.

Matthews, H. R. (1993) *Bioessays*, **15**, pp. 561–566.

McAllister, W. T. & Raskin, C. A. (1993) *Molecular Microbiology*, **10**, 1–6.

McHugh, P. J. & Knowland, J. (1995) *Nucleic Acids Research*, **23**, 1664–1670.

McPherson, A., Brayer, G. D., & Morrison, R. D. (1984) *Journal of Molecular Biology*, **189**, 305–327.

Melançon, P., Boileau, G., & Brakier–Gingras, L. (1984) *Biochemistry*, **23**, 6697–6703.

Melton, D. A., Krieg, P. A., Rebagliati, M. R., Maniatis, T., Zinn, K., & Green, M. R. (1984) *Nucleic Acids Research*, **12**, 7035–7056.

Mezzetti, G., Loor, R., & Liao, S. (1979) *Biochemical Journal*, **184**, 431–440.

Mezzetti, G., Monti, M. G., Moruzzi, M. S., Piccinimi, G., & Barbiroli, B. (1983) *Advances in Polyamine Research*, pp. 279–283, U. Bachrach, A. Kaye, & R. Chayen, Eds. New York: Raven Press.

Mezzetti, G., Moruzzi, M. S., & Barbiroli, B. (1981) *Biochemical and Biophysical Research Communications*, **102**, 287–294.

Mills, J. S., Needham, M., & Parker, M. G. (1987) *Nucleic Acids Research*, **15**, 7709–7724.

Mitsui, K., Igarashi, K., Kakegawa, T., & Hirose, S. (1984) *Biochimie*, **23**, 2679–2683.

Miyamoto, S., Kashiwagi, K., Ito, K., Watanabe, S., & Igarashi, K. (1993) *Archives of Biochemistry and Biophysics*, **300**, 63–68.

Moazed, D. & Noller, H. F. (1987) *Nature*, **327**, 389–394.

Moncany, M. L. J. & Kellenberger, E. (1981) *Experientia*, **37**, 845–847.

Montandon, P. E., Wagner, R., & Stutz, E. (1986) *The EMBO Journal*, **5**, 3705–3708.

Moore, M. N. & Spremulli, L. L. (1985) *Biochemistry*, **24**, 191–196.

Morgan, J. E., Blankenship, J. W., & Matthews, H. R. (1987) *Biochemistry*, **26**, 3643–3649.

Morgan, J. E., Calkins, C. C., & Matthews, H. R. (1989) *Biochemistry*, **28**, 5095–5106.

Morishima, Y., Inaba, M., Nishizawa, Y., Morii, H., Hasuma, T., Matsui–Yuasa, I., & Otani, S. (1994) *European Journal of Biochemistry*, **219**, 349–356.

Moruzzi, G., Barbiroli, B., Moruzzi, M. S., & Tadolini, B. (1975) *Biochemical Journal*, **146**, 697–703.

Moruzzi, G., Caldarera, C. M., & Casti, A. (1974) *Molecular and Cellular Biochemistry*, **3**, 153–161.

Muscari, C., Guarnieri, C., Stefanelli, C., Giaccari, A., & Calderera, C. M. (1995) *Molecular and Cellular Biochemistry*, **144**, 125–129.

Nadal, M., Couderc, E., Duguet, M., & Jaxel, C. (1994) *Journal of Biological Chemistry*, **269**, 5255–5263.

Nakai, C. & Glinsmann, W. (1977a) *Biochemical and Biophysical Research Communications*, **74**, 1419–1425.

——— (1977b) *Biochemistry*, **16**, 5636–5640.

Nakano, T. & Tsuboi, M. (1976) *Journal of Biochemistry*, **80**, 1435–1438.

Nashimoto, M. (1993) *Nucleic Acids Research*, **21**, 4696–4702.

Nashimoto, M., Sakai, M., & Nishi, S. (1991) *Biochemical and Biophysical Research Communications*, **178**, 1247–1252.

Niveleau, A. & Quash, G. A. (1979) *FEBS Letters*, **99**, 20–24.

Noller, H. F. (1993) *Journal of Bacteriology*, **175**, 5297–5300.

Nomura, M. & Held, W. A. (1974) *Ribosomes*, pp. 193–223. Cold Spring Harbor, N.Y.: Cold Spring Harbor Laboratory.

Nygård, O. & Nilsson, L. (1990) *European Journal of Biochemistry*, **191**, 1–17.

Ogasawara, T., Ito, K., & Igarashi, K. (1989) *Journal of Biochemistry*, **105**, 164–167.

Ohnishi, R., Nagami, R., Hirose, S., & Igarashi, K. (1985) *Archives of Biochemistry and Biophysics*, **242**, 263–268.

Ohno–Iwashita, Y., Oshima, T., & Imahori, K. (1975) *Archives of Biochemistry and Biophysics*, **171**, 490–499.

Oka, T. & Perry, J. W. (1990) *International Symposium on Polyamines*, Kyoto, Japan, p. 4.

Oller, A. R., Broek, W. V., Conrad, M., & Topal, M. D. (1991) *Biochemistry*, **30**, 2543–2549.

Olson, D. L., Deerfield, II, D. W., Berkowitz, P., Hiskey, R. G., & Pedersen, L. G. (1987) *Analytical Biochemistry*, **160**, 460–470.

Oredsson, S. M., Nicander, B., & Heby, O. (1990) *European Journal of Biochemistry*, **190**, 483–489.

O'Rourke, B. (1993) *Biochemical Pharmacology*, **46**, 1103–1112.

Osheroff, N. (1989) *Pharmacology & Therapeutics*, **41**, 223–241.

Palvimo, J., Pohjanpelto, P., Linnala–Kankkunen, A., & Maënpää, P. H. (1987) *Biochimica et Biophysica Acta*, **909**, 21–29.

Panagiotides, C. H., Artandi, S., Calame, K., & Silverstein, S. J. (1995) *Nucleic Acids Research*, **23**, 1800–1809.

Parello, J. (1988) *Journal of Cell Biology*, **107**, 1754 [Abstract].

Perrino, F. W. & Loeb, L. A. (1990) *Mutation Research*, **236**, 289–300.

Peter, H. W., Gies, A., Neumeier, M., Schädler, R., & Wegenir, I. (1979) *General Pharmacology*, **10**, 133–141.

Pinard, R., Payant, C., Melançon, P., & Brakier–Gingras, L. (1993) *The FASEB Journal*, **7**, 173–176.

Pingoud, A. (1985) *European Journal of Biochemistry*, **147**, 10–109.

Pingoud, A., Urbanke, C., Alves, J., Ehbrecht, H., Zabeau, M., & Gualerzi, C. (1984) *Biochemistry*, **23**, 5697–5703.

Pohjanpelto, P. & Hölttä, E. (1996) *The EMBO Journal*, **15**, 1193–1200.

Pohjanpelto, P. & Knuutila, S. (1982) *Experimental Cell Research*, **141**, 333–339.

———— (1984) *Cancer Research*, **44**, 4535–4539.

Pommier, Y., Kerrigan, D., & Kohn, K. W. (1989) *Biochemistry*, **28**, 995–1002.

Potapov, A. P., Triana–Alonso, F. J., & Nierhaus, K. H. (1995) *Journal of Biological Chemistry*, **270**, 17680–17684.

Raina, A. & Jänne, J. (1970) *Federation Proceedings*, **29**, 1568–1572.

Ramírez, C., Köpke, A. K. E., Yang, C., Boeckh, T., & Matheson, A. T. (1993) *The Biochemistry of Archaea (Archaebacteria)*, pp. 439–466, M. Kates, D. J. Kushner, A. T. Matheson, et al., Eds. New York: Elsevier Science Publishers B. V.

Reid, S. S. & Cowan, J. A. (1991) *Journal of the American Chemistry Society*, **113**, 673–675.

Rhoads, R. E. (1993) *Journal of Biological Chemistry*, **268**, 3017–3020.

Rich, A. (1977) *Accounts of Chemical Research*, **10**, 388–396.

Richmond, T. J., Finch, J. T., Rushton, B., Rhodes, D., & Klug, A. (1984) *Nature*, **311**, 532–537.

Romani, A. & Scarpa, A. (1992) *Archives of Biochemistry and Biophysics*, **298**, 1–12.

Rosano, C. L., Bunce, S. C., & Hurwitz, C. (1983) *Journal of Bacteriology*, **153**, 326–334.

Rose, K. M. & Jacob, S. T. (1976) *Archives of Biochemistry and Biophysics*, **175**, 748–753.

Rowlatt, C. & Smith, G. J. (1981) *Journal of Cell Science*, **48**, 171–179.

Russell, D. H., Levy, C. C., & Taylor, R. L. (1972) *Biochemical and Biophysical Research Communications*, **47**, 212–217.

Sander, C. & Tso, P. O. P. (1971) *Journal of Molecular Biology*, **55**, 1–21.

Sanui, H. & Rubin, H. (1977) *Journal of Cellular Physiology*, **92**, 23–31.

Sarkar, N., Shankar, S., & Tyagi, A. (1995) *Biochemistry and Molecular Biology International*, **35**, 1189–1198.

Satoh, M. S. & Lindahl, T. (1992) *Nature*, **356**, 356–358.

Scatchard, G. (1949) *Annals of the New York Academy of Science*, **51**, 660–672.

Schmukler, M., Jewett, P. B., & Levy, C. C. (1975) *Journal of Biological Chemistry*, **250**, 2206–2212.

Seyfried, C. E. & Morris, D. R. (1979) *Cancer Research*, **39**, 4861–4867.

Shadel, G. S. & Clayton, D. A. (1993) *Journal of Biological Chemistry*, **268**, 16083–16086.

Shalitin, C. & Vishlizky, A. (1984) *Biochimica et Biophysica Acta*, **782**, 328–330.

Shinozuka, K., Shimizu, K., Nakashima, Y., & Sawai, H. (1994) *Bioorganic & Medicinal Chemistry Letters*, **4**, 1979–1982.

Shoshan–Barmatz, V. & Zarka, A. (1992) *Biochemical Journal*, **285**, 61–64.

Shuman, S., Golder, M., & Moss, B. (1988) *Journal of Biological Chemistry*, **263**, 16401–16407.

Sinden, R. R. & Pettijohn, D. E. (1981) *Proceedings of the National Academy of Sciences of the United States of America*, **78**, 224–228.

Smirnov, I. V., Dimitrov, S. I., & Makarov, V. L. (1988) *Journal of Biomolecular Structure and Dynamics*, **5**, 1149–1161.

Smith, G. J. & Rowlatt, C. (1980) *Biochemical Journal*, **187**, 353–360.

Snyder, R. D. (1989a) *International Journal of Radiation Biology*, **55**, 773–782.

———— (1989b) *Biochemical Journal*, **260**, 697–704.

Snyder, R. D. & Bhatt, S. (1993) *Cancer Letters*, **72**, 83–90.

So, A. G. & Downey, K. M. (1992) *Critical Reviews in Biochemical and Molecular Biology*, **27**, 129–155.

Spector, D. L. (1993) *Annual Review of Cell Biology*, **9**, 265–315.

Spirin, A. S., Baranov, V. I., Ryabova, L. A., Ovodov, S. Y., & Alakhov, Y. B. (1988) *Science*, **242**, 1162–1164.

Stone, G. R., Baldwin, J. P., & Carpenter, B. G. (1987) *Biochimica et Biophysica Acta*, **908**, 34–45.

Struhl, K. (1994) *Science*, **263**, 1103–1104.

Subramanian, A. R. (1984) *Trends in Biochemical Sciences*, **9**, 491–494.

Sudhakar, K., Ericinska, M., & Vanderkooi, J. M. (1995) *European Journal of Biochemistry*, **230**, 498–502.

Sunkara, P. S., Chang, C. C., & Prakash, N. J. (1983) *Cell Biology International Reports*, **7**, 455–465.

Sunkara, P. S., Rao, P. N., Nishioka, K., & Brinkley, B. R. (1979) *Experimental Cell Research*, **119**, 63–68.

Suzuki, T., Suzuki, N., & Hosoya, T. (1993) *Biochemical Journal*, **289**, 109–115.

Symonds, G. W. & Brosnan, M. E. (1977) *Biochemical Society Transactions*, **5**, 1764–1766.

Szer, W. (1966) *Acta Biochimica Polonica*, **13**, 251–266.

Taagepera, S., Rao, P. N., Drake, F. H., & Gorbsky, G. J. (1993) *Proceedings of the National Academy of Sciences of the United States of America*, **90**, 8407–8411.

Tabet, M., Labroo, V., Sheppard, P., & Sasaki, T. (1993) *Journal of the American Chemical Society*, **115**, 3866–3868.

Takeda, Y. (1969) *Biochimica et Biophysica Acta*, **179**, 232–234.

Tanigawa, Y., Kawakami, K., Inai, Y., & Shimoyama, M. (1980) *Biochimica et Biophysica Acta*, **608**, 92–95.

Terasaki, M. & Rubin, H. (1985) *Proceedings of the National Academy of Sciences of the United States of America*, **82**, 7324–7326.

Thang, M. N., Dondon, L., & Mohier, E. (1976) *FEBS Letters*, **61**, 85–90.

Thang, M. N., Thang, D. C., De Maeyer, E., & Montagnier, L. (1975) *Proceedings of the National Academy of Sciences of the United States of America*, **72**, 3975–3977.

Thomas, T. & Gallo, M. A. (1994) *Toxicology Letters*, **74**, 35–49.

Thomas, T. & Kiang, D. T. (1988a) *Cancer Research*, **48**, 1217–1222.

———— (1988b) *Nucleic Acids Research*, **16**, 4705–4720.

Turner, B. M. & O'Neill, L. P. (1995) *Seminars in Cell Biology*, **6**, 229–236.

Uehara, A., Fill, M., Vélez, P., Yasukochi, M., & Imanaga, I. (1996) *Biophysical Journal*, **71**, 769–777.

Uzawa, T., Hamasaki, N., & Oshima, T. (1993a) *Journal of Biochemistry*, **114**, 478–486.

Uzawa, T., Yamagishi, A., Ueda, T., Chikazumi, N., Watanabe, K., & Oshima, T. (1993b) *Journal of Biochemistry*, **114**, 732–734.

Van den Broeck, D., Straeten, D. V. D., Montagu, M. V., & Caplan, A. (1994) *Plant Physiology*, **106**, 559–566.

Van de Sande, J. H., Lin, C. C., & Deugau, K. V. (1979) *Experimental Cell Biology*, **120**, 439–444.

Van Holde, K. E. (1989) *Chromatin*. New York: Springer–Verlag.

Verlander, M. S., Lohrmann, R., & Orgel, L. E. (1973) *Journal of Molecular Evolution*, **2**, 303–316.

Vlassov, V. V., Zuber, G., Felden, B., Behr, J.-P., & Giegé, R. (1995) *Nucleic Acids Research*, **23**, 3161–3167.

Voet, J. G. & Andersen, E. C. (1984) *Archives of Biochemistry and Biophysics*, **233**, 88–92.

Wagner, E. G. H., Jelenc, P. C., Ehrenberg, M., & Kurland, C. G. (1983) *European Journal of Biochemistry*, **122**, 193–197.

Wallace, H. M., Duff, P. M., Pearson, C. K., & Keir, H. M. (1981) *Biochimica et Biophysica Acta*, **652**, 354–357.

Wallace, H. M., Gordon, A. M., Keir, H. M., & Pearson, C. K. (1984) *Biochemical Journal*, **219**, 211–221.

Walz, Jr., F. G. & Kitareewan, S. (1990) *Journal of Biological Chemistry*, **265**, 7127–7137.

Wang, D. & Moore, S. (1977) *Biochemistry*, **16**, 2937–2941.

Watanabe, S., Kusama–Eguchi, K., Kobayashi, H., & Igarashi, K. (1991) *Journal of Biological Chemistry*, **266**, 20803–20809.

Watanabe, Y., Igarashi, K., & Hirose, S. (1981) *Biochimica et Biophysica Acta*, **656**, 134–139.

Weiss, R. L. & Morris, D. R. (1973) *Biochemistry*, **12**, 435–441.

Weller, J. W. & Hill, W. E. (1992) *Biochemistry*, **31**, 2748–2757.

Westin, T., Soussi, B., Idström, J., Lindnér, P., Edström, S., Lydén, E., Gustavsson, B., Hafström, L., & Lundholm, K. (1993) *British Journal of Cancer*, **68**, 662–667.

Whelly, S. M. (1991) *Journal of Steroid Biochemistry and Molecular Biology*, **39**, 161–167.

Wincott, F., DiRenzo, A., Shaffer, C., Grimm, S., Tracz, D., Workman, C., Sweedler, D., González, C., Scaringe, S., & Usman, N. (1995) *Nucleic Acids Reaearch*, **23**, 2677–2684.

Widom, J. (1986) *Journal of Molecular Biology*, **190**, 411–424.

Wu, G. & Dawid, I. B. (1972) *Biochemistry*, **11**, 3589.

Xiao, L., Swank, R. A., & Matthews, H. R. (1991) *Nucleic Acids Research*, **19**, 3701–3708.

Xie, X., Rao, S., Walian, P., Hatch, V., Phillips, Jr., G. N., & Cohen, C. (1994) *Journal of Molecular Biology*, **236**, 1212–1226.

Yanagawa, H., Ogawa, Y., & Egami, F. (1976) *Journal of Biochemistry*, **80**, 891–893.

Yip, L. C. & Balis, M. E. (1980) *Biochemistry*, **19**, 1849–1856.

Yonath, A., Saper, M. A., Makowski, I., Müssig, J., Piefke, J., Bartunik, H. D., Bartels, K. S., & Wittmann, H. G. (1986) *Journal of Molecular Biology*, **187**, 633–636.

Yonath, A. & Wittmann, H. G. (1988) *Methods in Enzymology*, **164**, 95–117.

Yoshinari, K., Yamazaki, K., & Komiyama, M. (1991) *Journal of the American Chemical Society*, **113**, 5899–5901.

Zamecnik, P. (1984) *Trends in Biochemical Science*, **9**, 464–466.

Zarka, A. & Shoshan–Barmatz, V. (1992) *Biochimica et Biophysica Acta*, **1108**, 13–20.

Zimmerman, S. B. (1976) *Journal of Molecular Biology*, **101**, 563–565.

Zwelling, L. A., Kerrigan, D., & Marton, L. J. (1985) *Cancer Research*, **45**, 1122–1126.

CHAPTER 23

Molecular Effects on Internal Cellular Polymers: Transfer RNA and DNA

Polyamines and Analysis of tRNA

The participation of tRNA in shuttling amino acids to ribosomes in the course of protein synthesis led to the isolation and fractionation of these transfer nucleates. The different tRNAs were relatively specific for single amino acids and were capable of incorporating all 20 amino acids into protein. *Escherichia coli* contains about 40 different tRNAs; several might serve a single amino acid. All have similar structures containing helical regions detectable by hyperchromicity in melting experiments and possess sequences of 70–80 nucleotides. The sequences of the separate isolated tRNAs could be drawn as composites of double-stranded and single-stranded regions in the familiar cloverleaf form.

Because the RNA content (mainly ribosomal) of *E. coli*, and indeed of mammalian liver as well, was found to parallel the spermidine content of the cells, it was asked if the apparent stoichiometric relation of polyamine to RNA-phosphorus pointed to some specific structural features of the ribonucleates with which spermidine might be specifically associated. At that time, isolated tRNAs were the sole nucleates whose size and known composition might provide clues to such a possible specific association. Also, the application of the dansyl reaction and fluorimetric scanning of dansyl amines on thin-layer chromatograms had increased the sensitivity of the analysis sufficiently to permit the detection of a single molecule of spermidine per molecule of tRNA. The RNAs were extracted from frozen cells lysed by phenol into an aqueous phase at low ionic strength at pH 5 and precipitated with ethanol. The RNAs were fractionated on Sephadex at low ionic strength to separate ribosomal RNA, a small fraction of 5S RNA, tRNA (40–60% of the total), and nucleate fragments. The ribosomal RNA from *E. coli* proved to

be relatively rich in spermidine at three to nine molecules per 77 nucleotides, depending on the conditions of RNA synthesis in the bacterium. The mixed tRNA obtained from relaxed cells attained a spermidine content of two molecules per 77 nucleotides (Cohen et al., 1969). The putrescine contents of these mixed tRNA samples were low, ca. 1/10 of the spermidine content. The tRNA of mouse fibroblasts in exponential and stationary phases of growth (Fig. 23.1) were devoid of putrescine and contained both spermidine and spermine, of which the latter was somewhat greater when isolated from stationary phase cells (Cohen, 1971b).

Similar methods of isolation were adopted for the isolation of tRNA from *Neurospora crassa* and from several plants (Bagni et al., 1973). In *N. crassa, Spinacia oleracea,* and *Pisum sativum* the isolated tRNA contained 1–2 mol spermidine/mol RNA whereas the plant rRNA was rich in both spermine and spermidine. In studies of the activation of dormant tuber tissue of *Helianthus tuberosus*, the polyamine content of the cells and the tRNA were both very low during dormancy and increased at the end of dormancy (Serafini–Fraccassini et al., 1984). The problem of whether spermidine exists on tRNA within the cell is not addressed by these data, because it can be supposed that the association of spermidine and tRNA occurs in the course of the steps of phenolic and aqueous extraction. Nevertheless, the binding of two to three molecules of polyamines to tRNA points to the existence of specific binding sites in the structure. The contents may only reflect characteristic association constants at these sites, which were in fact determined in later studies. Many more molecules (about 14–15) of spermidine can bind to tRNA at low ionic strength in vitro than that actually isolated from *E. coli* synthesizing excess spermidine. However, the cells convert excess spermidine to *N*-acetylspermidine

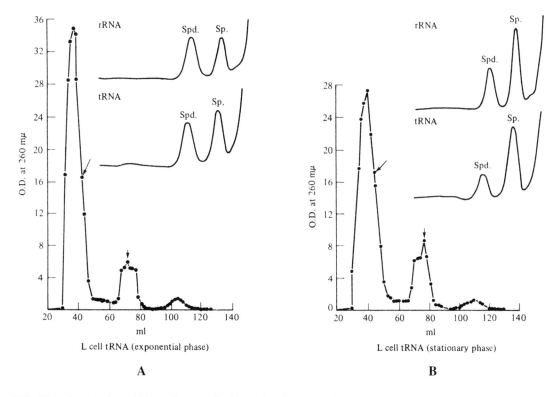

FIG. 23.1 Fractionation of RNA of mouse fibroblasts (L cells) grown in suspension cultures, with tracings of dansylated polyamines separated on TLC; adapted from Cohen (1970). sp, spermine; spd, spermidine; (**A**) exponential growth; (**B**) stationary phase.

and extrude this substance without accumulating the base on tRNA. Dialysis of tRNA containing 10 spermidines per molecule against Mg^{2+} produced a tRNA containing two spermidines per molecule.

Chemical Approaches to the Nature of Polyamine-Binding Sites in tRNA

Examination of the cloverleaf model revealed a limited number (three or four) of complementary base pairs in a single arm, the "dihydrouridine arm" or D stem (Fig. 23.2). Many types of data (e.g., sensitivity to heat, ribonuclease, and chemical reactivities) indicated early changes of tertiary structure in precisely this region of the structure. The question was asked if the tertiary structure of this region was not strengthened and maintained by cations, such as spermidine and Mg^{2+}. In *E. coli* many tRNA molecules contain a potentially fluorescent marker, 4-thiouridine at position 8, in proximity to this short, double-stranded region. Although position 8 in a cloverleaf representation is not itself in a double-stranded region, it will be seen that the 4-thiouridine or uridine at this position in tRNA is generally bonded in tRNAs to adenine at position 14 in another single-stranded region in such a way as to fold the dihydrouri-

dine loop back into the core of an L-shaped tertiary configuration. As a result of analysis of tRNA crystals, it was suggested that a further stabilization of this loop and fold is effected by hydrogen bonding of the 2′-hydroxyl of the sugar in the uridine at the 8 position to an unpaired adenine nitrogen (N1) at nucleotide 21. Some tRNAs lack the U8-A14 loop and some workers stressed the role of the G19-C56 interaction in locking the L.

The fluorescence of 4-thiouridine was established by proximity to nucleosides noncontiguous in the tRNA primary sequence; that is, fluorescence was low in a melted *E. coli* tRNA devoid of Mg^{2+} and spermidine and was increased more than threefold by Mg^{2+} (Fig. 23.3). Spermidine (10 mol/mol tRNA) doubled the fluorescence and this was increased further by the addition of Mg^{2+} (Pochon & Cohen, 1972). Each cation (i.e., Mg^{2+} and spermidine) cooperatively increased the development of the tertiary structure involving thiouridine at position 8. A stepwise titration to achieve maximal fluorescence was achieved by addition of two atoms of Mg^{2+}, eight molecules of spermidine, and finally two additional atoms of Mg^{2+}.

The ionization of 4-thiouridine to the anionic species shifts the absorption maximum to a lower wavelength (310 nm) than that of the neutral species (335 nm).

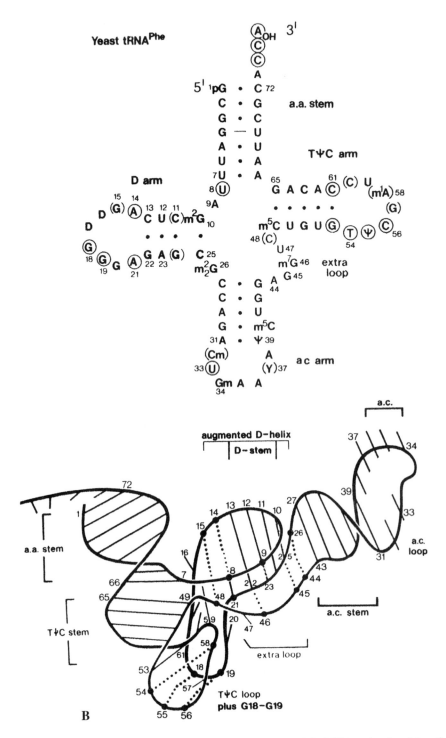

FIG. 23.2 Schematic diagram showing (**A**) the cloverleaf formula and (**B**) the chain folding and tertiary interactions between bases in yeast tRNA. Adapted from Jack et al. (1976). Long straight lines indicate base pairs in the double-helical stems. Shorter lines represent unpaired bases. Dotted lines represent base pairs outside the helices.

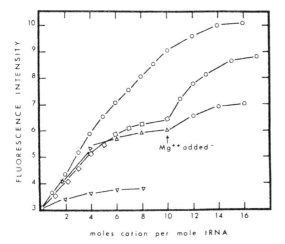

FIG. 23.3 Effect of cations on the intensity of fluorescence of thiouridine in *E. coli* tRNA; adapted from Pochon and Cohen (1972). Samples of tRNA (0.052 µmol) in 1.5 mL 0.01 M K acetate (pH 7.0) were titrated at 22°C with (○) MgCl$_2$, (□) spermidine, (△) spermine, or (▽) putrescine, and the emission at 530 nm was measured after excitation at 335 nm.

Whereas the pK of this shift for the free nucleoside is not affected by the cations, addition of the cations in tRNA does shift the pK of this ionization significantly (Sakai & Cohen, 1976). The melting temperatures of the hyperchromic shift of the absorption at 335 nm was increased significantly by 20 Mg^{2+} or 20 spermidines per tRNA some 20 and 27°C, respectively. These results were interpreted to indicate that the cations help to maintain the tertiary structure of tRNA, particularly in the region involving 4-thiouridine. Two *E. coli* tRNAs possessing three complementary base pairs in the region of the dihydrouridine (DHU) arm were demonstrated by ^{3}H-^{15}N-NMR to contain a base pair composed of 4-thiouracil 8-adenine 14 in the presence of 10 mM MgCl$_2$ (Griffey et al., 1986). The existence of an A14-U8 base pair in eucaryotic tRNA and effects of Mg^{2+} and spermidine on this particular structure and its environs were generally confirmed by NMR.

E. coli tRNA forms a photoproduct between 4-thiouridine at position 8 and cytidine at position 13 that is detected by the generation of fluorescence after reduction with NaBH$_4$. The rate of formation of the photoproduct is enhanced by spermidine more rapidly than by Mg^{2+} and spermidine is more active than Mg^{2+} in enhancing the fluorescence of the reduced photoproduct (Pochon & Cohen, 1972). The rates of the borohydride reduction and of the reactions of iodoacetamide and cyanogen bromide with the thiol were inhibited maximally by 10–15 Mg^{2+} or spermidine per tRNA.

Also, a bromoacetamido-spin label reacted specifically with thiouridine in *E. coli* tRNA and was hindered in its rotation after addition of spermidine (Sakai & Cohen, 1976). This suggested that spermidine closed the structure in the region of the spin-label substituent. A

similar methodology was used to explore an effect of polyamine and Mg^{2+} on a base specific to the anticodon arm, the fluorescent Y base, in yeast phenylalanine tRNA. The fluorescence of the Y base is increased by both Mg^{2+} and polyamine (spermidine or spermine) and the amino acid acceptance of preparations of this tRNA is increased in parallel fashion (Robison & Zimmerman, 1971a,b). Although spermidine is more active than spermine in producing this effect, the triamine is only 40% as active as Mg^{2+}. Fluorescent structures like 1-aminoanthracene or ethidium can be built into the anticodon arm and effects of the cations on the fluorescence of the region indicate that both the anticodon site and the dihydrouridine arm are indeed altered by spermidine and by Mg^{2+}, although somewhat differently. Although the chemical entities examined are within these specific regions, the effects might arise from a general tightening of the structure. Nevertheless, transitional effects demonstrated a stepwise initial loosening of a specific tertiary structure with heat or the stepwise tightening of such a tertiary structure in response to a stepwise addition of a cation (Heerschap et al., 1985).

The binding of Mg^{2+} to the anticodon site was reported to belong to a class of intermediate affinity sites, that is, equilibrium constants of 1–3 × 10^3 M^{-1}, in contrast to five other Mg^{2+} binding sites with equilibria constants of 2 × 10^4 to 5 × 10^6 M^{-1} (Labuda & Pörschke, 1982). Whereas the latter indicate a coordination of Mg^{2+} with two phosphate groups, the intermediate binding sites are described as the association of the cation with base atoms and a single phosphate.

Quantitation of tRNA Binding Sites for Polyamine

The finding of two spermidines for *E. coli* tRNA, the in vitro construction of tRNA containing 10+ spermidines, and the displacement of all but two by dialysis against buffer containing Mg^{2+} indicated the existence of several classes of binding sites. One method used to explore this quantitatively included the production of mixtures of tRNA (either yeast tRNA^{phe} or *E. coli* tRNA^{fmet}) and polyamine that were fractionated at low ionic strength on Sephadex to detect the formation of complexes and free polyamine (Fig. 23.4). In excess spermidine (40 mol/mol tRNA) complexes were formed containing 10–11 mol spermidine/mol with the remainder appearing as free spermidine (Sakai et al., 1975). With a mixture of spermidine and tRNA (2 : 1), no free spermidine was detected. Neither putrescine nor *S*-adenosylmethionine (AdoMet) formed a separable complex with yeast tRNA^{phe} under these conditions. Spermine complexes were formed and free spermine was not detected in a 12 : 1 mixture of spermine and tRNA; the spermine-tRNA complex precipitated in an 18–20 : 1 mixture. Obviously, this method can generate and isolate various cation complexes of the nucleates for structural and functional study.

This method was used to study the binding of the ultraviolet-absorbing planar triamine of ethidium, which

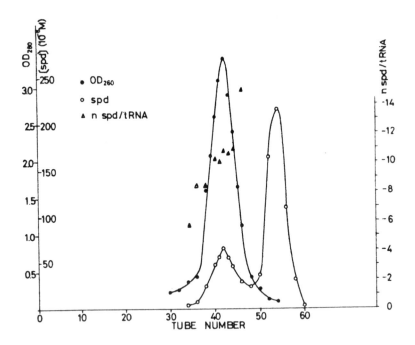

FIG. 23.4   The chromatographic separation on Sephadex G-100 of a mixture of spermidine and yeast RNA (40 : 1); adapted from Sakai et al. (1975).

is also fluorescent when bound to certain double-stranded regions of nucleic acid. The distribution of amino groups and carbon chains in spermidine can simulate the distribution of comparable groups and chains in the ethidium molecules, despite major differences in the pK values of the amino groups. Fractionation of a 40 : 1 mixture of ethidium and yeast tRNA^{phe} separated a complex of ethidium tRNA containing 10 molecules of the base, whose fluorescence was far lower and equivalent to that of three molecules of ethidium in a DNA complex. In a 1 : 1 mixture of ethidium and tRNA, all ethidium was bound in a fluorescent form. In a 2 : 1 mixture all the ethidium was bound, but a significant fraction (20–25%) was present in a nonfluorescent form, suggesting a relatively low affinity of the second fluorescent site. In a mixture of 1 : 1 : 1 spermidine : ethidium : tRNA, the resulting complex was similarly low in fluorescence, suggesting that in some molecules of tRNA a single molecule of spermidine occupied the primary fluorescent binding site for ethidium. Not only can spermidine displace fluorescent ethidium, but a large excess of ethidium can displace spermidine in a 10 : 1 complex and attain full fluorescence.

The fluorescence of ethidium was thought initially to be a function of its intercalation (see discussion on DNA below) between base pairs, and this was thought to occur in tRNA as well between two base pairs (U6-A67 and U7-A66) in the acceptor stem. When crystals of yeast tRNA^{phe} prepared in the presence of spermine (see below) were soaked in ethidium, the deeply stained

crystals contained nonintercalated ethidium bound in the hairpin turn established in the tertiary structure at the U8-A14 site (Liebman et al., 1977). Indeed, ethidium was stacked over U8. The idea that ethidium does lodge in this pocket of yeast tRNA^{phe} was supported in the interpretation of peak broadening studies of ^{31}P-NMR spectra of ethidium and tRNA (1.32 : 1) (Goldfield et al., 1983). Nevertheless, other studies of a single bound fluorescent ethidium in tRNA^{fmet} in solution preferred to site the compound in an intercalative mode in the amino acid acceptor stem (Ferguson & Yang, 1986).

Hexammine cobalt(III) chloride has been used to displace a spermine from spermine-containing crystals of yeast tRNA^{phe}. The Co(NH$_3$)$_6^{3+}$ molecule binds at the phosphates at nucleotides 22, 23, and 44 and to the bases of residues A44 and G45, which link the D helix to the anticodon stem (Hingerty et al., 1982).

Another approach to binding has involved equilibrium dialysis and the estimation of bound polyamine and the number of binding sites for a ligand per macromolecule via the use of the Scatchard plot (Scatchard, 1949). With a single class of independent noninteracting binding sites, the plot of v as the number of moles of ligand bound per mole of macromolecule and c as the concentration of free ligand gives a straight line with the slope equal to $-K_a$, where K_a is the association constant for the class, and the intercept on the abscissa is the total number of binding sites. However, when two or more classes of sites are present, the binding constants and number of binding sites require the solution of simultaneous equations based on the parameters of the Scatchard plots (Klotz & Hunston, 1971). These calculations have suggested the existence of two classes of binding sites in tRNA (assuming 80 nucleotides per molecule). The higher affinity sites correspond to three molecules of spermidine per yeast tRNA^{phe} ($K_a = 6.1 \times 10^4$ M^{-1}) within a total of 17–18 molecules of triamine per tRNA. The class of lower affinity shows $K_a = 5.4 \times 10^3$ M^{-1} (Sakai & Cohen, 1976).

In the presence of 9 mM Mg^{2+}, only a single class of about 13 spermidine-binding sites were found in yeast tRNA^{phe}, with an average K_a of 5.9×10^2 M^{-1}. Thus, a simple linear Scatchard plot is obtained under these conditions, implying a limited equivalence or overlap of the total Mg^{2+} and spermidine sites.

Crystallographic Analysis of tRNA

The direct visualization of the distribution of components of tRNA has been effected by X-ray crystallography, requiring the formation of highly ordered crystals stable in hours of X irradiation. Kim and Quigley (1979) summarized the major steps of determination of the crystal structure of tRNA: crystallization, collection of data, construction and interpretation of electron-density maps, and refinement of the structure. In 1968 several groups reported that large, relatively fragile single crystals of a single tRNA could be formed. The production of mixed crystals of yeast tRNA obtained from unfractionated preparations (Fresco et al., 1968) suggested that many tRNAs possessed similar conformations. Nevertheless, despite similarities in the folding of ribose-phosphate backbones, the crystals of different tRNAs, like yeast tRNA^{phe} and yeast tRNA^{asp}, are significantly different in stem positions and loop conformations (Moras et al., 1980).

Several groups found that spermidine facilitated crystallization in ethanol containing Mg^{2+} and buffer. The addition of spermine permitted the crystallization of yeast tRNA^{fmet} as hexagonal prisms containing 15% tRNA; spermine-containing crystals enabled a significantly higher degree of resolution, 6–7 Å, than did earlier crystals. It was also reported that crystallization of the numerous purified tRNAs required the presence of polyvalent cations.

Crystals of yeast tRNA^{phe} obtained with MgCl$_2$ and spermine were found to yield an X-ray diffraction pattern with a resolution of 2.3 Å (Kim et al., 1971). Some short helical regions were then found in several other crystals of tRNAs from yeast and *E. coli*. Systematic studies established the conditions involving both MgCl$_2$ and spermine for the formation of well-ordered crystals of yeast tRNA^{phe} (Ladner et al., 1972). Further, the conclusion was drawn that spermine, functioning as an elongated polycation, was essential not only in stabilizing individual molecules, but also in developing intermolecular contacts between the elongated molecules in the unit cell (Kim et al., 1973). Crystals prepared as

spermine derivatives soon permitted the perception that the initially indistinct tRNA structure approached an L shape in its tertiary structure. At 3 Å resolution, the electron density map not only traced most of the nucleotide chain, but also detected the presence of numerous cations in fixed positions (Robertus et al., 1974). It was noted that a base pair is formed between U8 and A14 in an alternative to a Watson–Crick pair (i.e., a reverse Hoogsteen type), as well as the possibility of an interaction of U8 with A21, as shown in Figure 22.2**B**. The refinements necessary to define the A21-U8-A14 positions and which indicate a stabilizing bond of the N1 of adenine 21 with the 2′-hydroxyl of uridine 8 were described by Jack et al. (1976).

Three strong binding sites were found for Mg^{2+}, one of which was at the U8 phosphate (Jack et al., 1977). In contrast to weak Mg-binding sites, the linkage of a strongly bound Mg^{2+} to a phosphate occurs with the loss of a molecule of water from the hexahydrated shell of Mg^{2+}, permitting the formation of a direct bond between the metallic cation and the oxygen of phosphate. Other workers reported the presence of four or five Mg atoms in crystals of yeast tRNA^{phe}. Quigley et al. (1978) presented a schematic diagram (Fig. 23.5**A**) of the positions of Mg atoms and two spermine molecules in yeast tRNA^{phe}. Holbrook et al. (1978) stated that it is difficult to place spermine in the model unambiguously. They tentatively reported that their analyses located one of the possible spermine molecules in the cleft of the double helix between the acceptor arm and the T stem and the other in the deep groove between the anticodon stem and the D arm. Nevertheless, Quigley et al. (1978) and Teeter et al. (1980) described one spermine as wrapped around P10 and one Mg^{2+} atom coordinating with P8, 9, 11, and 12 in the pocket containing P10 (Fig. 23.5**B**) in the beginning of the D stem. The other spermine, whose shape is described as something like a fish hook, was stated to sit in the major groove between the anticodon and D stem and appears to bring the strands together in that groove (Fig. 23.5**C**). Nevertheless, members of the latter group stated more recently that the precise location of spermine often requires a greater degree of resolution than that used earlier (Williams et al., 1991).

An L-shaped 3-dimensional form of yeast tRNA^{phe} can be constructed via computer by a modeling program based on the concept of satisfying constraint data obtained in the crystallographic analyses above (Major et al., 1993). A crucial element in this modeling was recognition and insertion of the U8-A14 interaction that led to the formation of the bend in the L structure.

Comparisons were made between properties of the tRNA crystals and samples taken before crystallization, mother liquors, or redissolved crystals. Despite many similarities the tRNA from redissolved crystals differed from uncrystallized tRNA in the melting temperature of the tRNA, the helicity of the structures, and their reactions with ethidium (Prinz et al., 1976). They suggested that the addition of spermine and Mg^{2+} in the crystallization imposed these new structural features.

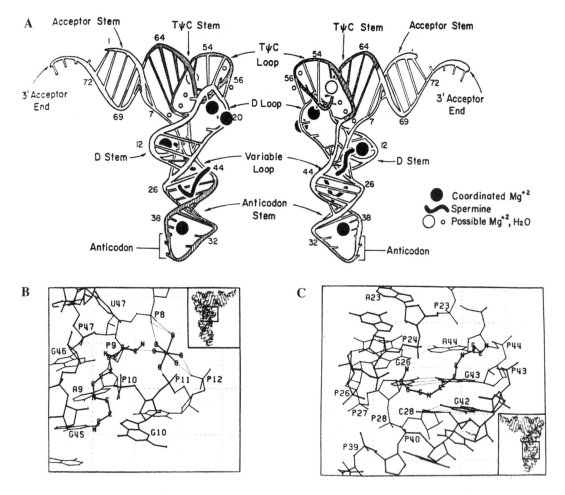

FIG. 23.5 (**A**) Schematic diagram showing two sides of yeast tRNA[Phe] in which the backbone is represented as a coiled tube; adapted from Quigley et al. (1978). (**B**) View of a region of the molecule including the P10 bend with spermine coiled around P10; adapted from Teeter et al. (1980). (**C**) View of the spermine molecule in the major groove between the anticodon and D stems; adapted from Teeter et al. (1980). (**B,C**) Insets show the locations of the diagrams relative to the whole molecule.

NMR and Binding to tRNA

The conclusions drawn from the crystallographic data were tested by many types of studies. In one of these, a distinction was made between secondary and tertiary structure on the basis of studies of thermal stability analyzed by differential ultraviolet absorption under a wide range of ionic conditions. Two melting steps were detected in yeast tRNA[phe]. At pH 7.0, maximal melting temperatures (T_m) of 76.5°C (Mg^{2+}), 74°C (spermine), and 71°C (spermidine) were attained at a 10–20-fold equivalence of added positive charge in the presence of 0.125 M Na^+ (Heerschap et al., 1985).

The NMR study of cation immobilization and base interactions in nucleates began in 1971. As described in Chapter 3, the widely used technique was applied to analysis and identification of amine structures and to

the determination of pK values, as well as to the formation and structures of complexes of polyamines and nucleotides. NMR has become useful in the estimation of free and bound Mg^{2+} in cells and tissues. The immobilization of [^{13}C]-putrescine in the membranes and ribosomes of *E. coli* was described in Chapter 7. NMR is currently useful in biochemistry in exploring the reactivities of various atomic nuclei—the ^1H of hydrogen, ^2H of deuterium, ^{13}C of carbon-13, ^{31}P of phosphorus-31, ^{25}Mg of magnesium, ^{14}N of nitrogen-14, and ^{15}N of nitrogen-15—all of which are relevant to the study of nucleate structures and complexes.

The earliest ^1H-NMR studies with yeast tRNA[phe] detected a loosening of hydrogen-bonded base pairs as a function of temperature increase at 35, 48, and finally at 59–61.5°C. These reversible changes of base pairs in the low field spectrum range were considered evidence for N—H ring protons participating in tertiary structure (Kearns et al., 1971). Subsequent studies undertook to determine the standard hydrogen-bonded Watson–Crick pairs, GC and AU, by determining the resonances of hydrogen-bonded protons of base pairs in tRNA. These data corroborated the existence of the double-stranded helical regions presented in the cloverleaf models (Kearns & Shulman, 1974).

Bolton and Kearns (1975) then demonstrated that in *E. coli* and yeast tRNAs the bonds between thiouracil 8 and adenine 14, and uracil 8 and adenine 14 were stabilized by Mg^{2+}. It may be supposed that actual base pairing will reduce tritium exchange in the purine of an apparently single-stranded region. This reduction of exchange was in fact realized for the A14 contained in the "single-stranded" loop of the cloverleaf of yeast $tRNA^{phe}$ (Gamble et al., 1976). Thus, a component (A14) of this single-stranded loop was bonded to a somewhat distant U8 in the tertiary structure.

Bolton and Kearns (1977) then demonstrated that in the presence of 0.17 M Na^+ both Mg^{2+} and spermine could stabilize several base-paired sites in the structures of various tRNAs at key temperature transitions. The cations did not entirely replace each other and a combination of low concentrations of Mg^{2+} and spermine formed an apparently native structure more effectively than either cation alone. For example, in 0.17 M NaCl at 44°C, the mixture of four spermines and four Mg^{2+} per *E. coli* tRNA evoked a 14.9-ppm peak at the resonance position characteristic of S^4U_8-A_{14}. The position of the S^4U_8-A_{14} peak at ca. 14.9 ppm was confirmed by studies with an undermodified *E. coli* tRNA. The bonding of U_8-A_{14} and other pairs in yeast $tRNA^{phe}$ were established by proton NMR with and without Mg^{2+}. Energy transfer between protons (nuclear Overhauser effects) assisted in the assignment of imino resonances (Roy & Redfield, 1983). Solvent exchange rates at particular base pairs were determined in the presence of the cations Na^+, Mg^{2+}, and spermidine as a function of increasing temperature. The melting observed between 32 and 39°C was accompanied by increases in the exchange rates. However, although five molecules of spermidine per tRNA stabilized the U_8-A_{14} base pair and was more effective than Na^+ in stabilizing the tertiary structure, it appeared that Mg^{2+} was somewhat more effective than spermidine (Trapp & Redfield, 1983). Spermine effects are considered to be relatively localized around S^4U_8-A_{14} in *E. coli* $tRNA^{phe}$ in contrast to Mg^{2+} that affects all stems (Hyde & Reid, 1985). Despite stabilization of the helical structure, spermidine and spermine increased solvent exchange for several types of imino protons; this increase appears to involve participation of the amines themselves in exchange with exposed bases (Heerschap et al., 1986).

Reference was made earlier to the NMR demonstration of the immobilization of [^{13}C]-putrescine on uptake into *E. coli* that was detected by the broadening of the resonance peak for ^{13}C (Frydman et al., 1984). An NMR study of [5,8-^{13}C$_2$]-spermidine mixes with *E. coli* mixed tRNA revealed that in contrast to the results with [^{13}C]-

putrescine, [^{13}C]-spermidine did in fact bind electrostatically to this nucleate (B. Frydman et al., 1990). The characteristic resonances with sharp peaks were found for the ^{13}C-5 (at 47.8 ppm) and for ^{13}C-8 (at 39.6 ppm) in free spermidine whereas these signals were much more broadened when bound to tRNA. The [^{13}C]-spermidine or [^{13}C]-spermine could be displaced by the homologous [^{12}C]-polyamine, resulting in a narrowing of the signal. In an experiment with [^{13}C]-spermidine, ~11 of the 12–14 bound spermidines could be easily displaced by increasing Mg^{2+}, whereas about three spermidines remained strongly bound to the tRNA. The line width for C-5, next to the secondary amine, was affected more than that of C-8, suggesting that the protonated secondary amine may be more tightly bound than the protonated primary amine.

The early hypothesis that the binding of the polyamines to nucleates is entirely a saltlike electrostatic interaction between proton-bearing amines and anionic phosphate was also challenged by more recent data. Indeed, a model of the spermidine complex with AMP developed with NMR data by Bunce and Kong (1978) and that of the spermine-contained crystal of yeast $tRNA^{phe}$ both introduced hydrogen bonds and depicted a primary amino group unreacted with anionic phosphate. Further, the study of the binding of [^{13}C]-spermidine to tRNA by B. Frydman et al. (1990) suggested a major participation of the secondary amine, possibly by hydrogen bonding. The subsequent study of the interaction of [^{15}N]-containing spermidine and spermine with tRNA permitted the determination of the participation of the primary and secondary amines and introduced another direct approach to the problem of the nature of the complex (Frydman et al., 1992). ^{15}N-NMR distinguished the reactions of a protonated primary amine, -$^{15}NH_3^+$, and a protonated secondary amine, -$^{15}NH_2^+$, in reactions of three molecules of isotopically enriched [1,4,9,12-$^{15}N_4$]-spermine and [1,4,8-$^{15}N_3$]-spermidine per *E. coli* mixed tRNA. The NMR spectra of the ^{15}N-spermidine and ^{15}N-spermine revealed (-NH_3^+) and (-NH_2^+) peaks at 11.7–11.8 and 24.9–25.0 ppm, respectively, downfield from the [$^{15}NH_4$]$_2SO_4$ resonance. Addition of tRNA to each polyamine caused slight shifts in both peaks and marked changes in peak breadths and heights, with significant differences in the NMR parameters (relaxation time) of the two nitrogens. The effects on the (-NH_2^+) peaks were clearly greater than those on the (-NH_3^+), as seen in Figure 23.6. The -$^{15}NH_3^+$ of -^{15}N-putrescine was unresponsive to the addition of tRNA.

The tumbling (correlation time) characteristics of each type of protonated amine was determined in the absence and presence of tRNA, as well as with the effect of temperature change. The data indicated that spermine binds more strongly to tRNA than does spermidine and more unequivocally that the secondary amines were major participants in the binding. ^{15}N-NMR also served to demonstrate a strong binding, via hydrogen bonds, of some bisethyl analogues to tRNA, in contrast to a weaker electrostatic binding of the parent unsubstituted polyamine analogue (Fernández et al., 1994).

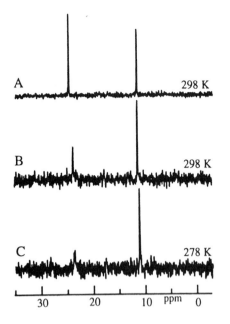

FIG. 23.6 (**A**) A ^{15}N-NMR spectrum of free [^{15}N$_4$]-spermine; (**B,C**) ^{15}N spectra of [^{15}N$_4$]-spermine in the presence of mixed tRNA of *E. coli*. Adapted from Frydman et al. (1992). The peaks of the nitrogens appear at 24.98 ppm (–NH$_2^+$) and 11.75 ppm (–NH$_3^+$) downfield from the (^{15}NH$_4$)$_2$SO$_4$ resonance.

Recently an NMR study of yeast tRNA^{phe} with spermine containing ^{13}C bonded to the primary and secondary amines (Frydman et al., 1996) has suggested a binding at the corner of the L-shaped structure where the D loop meets the TψC loop (see Fig. 23.2).

Modification of tRNA Bases

The presence of many unusual bases derived from the common purines and pyrimidines was detected in tRNA fairly early. Indeed, some bacteria like *E. coli* contain a thiolated uridine, 4-thiouridine, distinctively in the position of the eighth nucleotide from the 5′ end in the sequence; mammalian tRNAs contain an unmodified uridine at that position. Many puzzling methylations are known, and one methionine auxotroph of a relaxed strain of *E. coli* that synthesizes RNA in the absence of methionine is a source of undermethylated tRNA. However, the study of the various properties of completely unmodified tRNA awaits the production of DNA sequences that can be incorporated into a transcribable plasmid.

In one instance, the DNA sequence of yeast tRNA^{phe} was produced by a combination of chemical and enzymatic methods (Sampson & Uhlenbeck, 1988). In another, the DNA was that of a glycine tRNA gene (Samuelsson et al., 1988). The transcription was effected in each case with T7 polymerase in the presence of 6 mM MgCl$_2$ and 1 or 2 mM spermidine. In the case of the transcribed precursor tRNA^{gly}, the resulting RNA was trimmed to active tRNA with the tRNA-processing ribonuclease P-RNA. In both instances, the isolated transcripts of unmodified tRNA were functional. However, in the case of the yeast tRNA^{phe}, the unmodified RNA had a lower T_m in the presence of Mg^{2+} and a higher Michaelis constant (K_m) in aminoacylation than the modified structure.

Harrington et al. (1993) compared the properties of unmodified tRNA^{phe} derived from the *E. coli* gene with that of the normal modified *E. coli* tRNA^{phe}. The unmodified tRNA was denatured easily, and the absence of all modifications did affect the kinetic parameters of a series of reactions in the early steps of protein synthesis.

Methylation of tRNA

The hypothesis that tRNA modification might affect the quantity and quality of protein synthesis led to studies of the tRNAs in various healthy and pathological tissues and to a search for modifying enzymes. For example, the nucleoli of pea seedlings proved to contain a high concentration of RNA methylases (Birnstiel et al., 1963). High concentrations of NH$_4^+$ salts (0.3–0.4 M) significantly stimulated the activities of tRNA methylases that transfer methyl groups from AdoMet. Leboy (1971) also demonstrated such stimulations at concentrations of 0.1 mM spermine, 1 mM spermidine, 2–10 mM Mg^{2+}, and 10–30 mM putrescine by partially purified rat liver methylases on *E. coli* methyl-deficient and normal tRNA. These enzymes were almost inactive on a homologous rat liver tRNA.

Pegg (1971) also reported stimulations by the polyamines of the methylation of various bases in *E. coli* tRNA by rat tissue extracts and suggested that the effects were due to combination of the polyamine with the tRNA rather than the enzyme. Ethidium bromide and acriflavin inhibited methylation.

Young and Srinivasan (1971) also studied methylation by *E. coli* enzymes and the effects of polyamines and/or Mg^{2+}. Some 75–80% of the methylation stimulated by all of the cations was detectable in ribosyl thymine. Among the remaining newly methylated bases, polyamines facilitated the formation of 7-methylguanine and dimethylguanine whereas Mg^{2+} was much less active. The stimulatory effects of the polyamines on methylation also appeared to relate to the increase of T_m obtained in the newly methylated tRNAs.

Leboy (1971) detected differences in the effects of the polyamines in the methylation of different nucleotides within methyl-deficient *E. coli* tRNA by a rat liver soluble fraction, and detected inhibitors of methylase reactions in the extracts. Initially, three different tRNA methyltransferases were purified from rat liver and were found to be sensitive to the presence of cations (Leboy & Glick, 1976). The N^2-guanine and 1-adenine methyltransferases were optimally activated by 2 mM spermidine. The tRNA (adenine-1)-methyltransferase of rat liver was purified to homogeneity and was found to have a molecular weight (MW) of about 95,000 and a strict specificity for adenine in the GTψC loop of *E. coli* tRNA. With Mg-free tRNA the enzyme was optimally activated by 20–40 mM putrescine, which was about 40% as active at 3–4 mM spermidine. Complexes of Mg^{2+}-free tRNA + 4 mol bound spermidine/mol tRNA did not support the adenine methyltransferase activity. On the other hand, a spermidine–tRNA^{phe} complex (14–16 mol/mol) satisfied the ion requirement of guanine methylation by an N^2-guanine methyltransferase but not that of adenine methylation by an adenine methyltransferase. This implies that there is an effect of polyamine on the catalytic activity of the adenine methylase whereas the effect on guanine methylation appears to relate predominantly to the conformation of the tRNA.

Glick et al. (1978) also purified (6500-fold) two tRNA-(guanine) methyltransferases, one of which specifically methylates *E. coli* tRNA^{fmet} at N_1 and is maximally stimulated by 40 mM putrescine. The other guanine methyltransferase is reactive with several other *E. coli* tRNAs and introduces a methyl at the 2-amino group of guanine at nucleotide 10 in the sequence. This methylation is maximally stimulated by 0.1 mM spermidine. Low levels of spermine (0.02 mM) and putrescine (5 mM) permit a 90% level of stimulation whereas Mg^{2+} and NH_4^+ are less stimulatory.

Several *E. coli* tRNAs have been found to contain 3-(3-amino-3-carboxypropyl) uridine, in which the 3-amino-3-carboxypropyl moiety has been shown to derive from AdoMet via a Mg^{2+}-requiring enzyme found in *E. coli* (Nishimura et al., 1974).

In cells the regulation of levels of *S*-adenosylhomocysteine (AdoHcy) had significant effects on tRNA methylation. Arrest of polyamine biosynthesis in *Dictyostelium discoideum* also markedly decreased methylation of many bases in tRNA without effects on the incorporation of uridine into tRNA or on the net synthesis of protein or DNA (Mach et al., 1982).

Cadaverine (30 mM) also stimulates a guanine methyltransferase active on nucleotide 27 in tRNA^{fmet}. Cadaverine appears to destabilize a tertiary structure stabilized by Mg^{2+} at other sites, and it was suggested that polyamines may thereby open certain regions in tRNA^{fmet} for a complementary reaction with a ribosomal site. Such an opening may also make the previously stable tRNA region newly available for methylation, a phenomenon that was detected as an effect of cadaverine and attributed to a competition of cadaverine and Mg^{2+} (Wildenauer et al., 1974).

In yeast, as in *E. coli*, two m^1G methylating activities have been found. Both of the yeast enzymes are stimulated many-fold in their activity on undermethylated *E. coli* tRNA by spermine in the absence of Mg^{2+} (Smolar et al., 1975). Only one of these enzymes is active on normally methylated *E. coli* tRNA.

As described earlier, a polyamine deficiency in an *E. coli* auxotroph leads to the undermethylation of adenine in ribosomal 16S RNA. A purified rRNA-adenine (N^6)-methyltransferase stimulated by spermine and Mg^{2+} has been described in *E. coli* (Sipe et al., 1972). This 16S rRNA-adenine is in fact dimethylated at the adenine amino group, and an enzyme was isolated that methylates two adjacent adenines near the 3′ end of the 16S rRNA (Poldermans et al., 1979). Curiously, the enzyme that methylated adenine in both 23S and 30S ribosomal cores was stimulated by spermidine in the 23S core that contains 16S rRNA (Igarashi et al., 1980a).

Completing the Acceptor Stem of tRNA

The incorporation of two residues of cytidylate and one of adenylate from their triphosphates into the 3′-terminal sequence (CCA) of tRNA is effected by tRNA-nucleotidyltransferase. The enzyme, which provides an aminoacylatable terminus for tRNA, is essential for the growth of *E. coli*. Evans and Deutscher (1976) showed that polyamines markedly stimulate the completion from an RNA completely lacking CCA or from tRNA-C and tRNA-CC. Spermine was used in studies with yeast or liver mixed tRNA and a purified rabbit liver nucleotidyltransferase. This polyamine was ineffective in the complete absence of Mg^{2+}, but in the presence of 1 mM Mg^{2+} it was optimally stimulated by 5 mM spermine in the standard assay system. This concentration is possibly achieved uniquely in nuclei. Kinetic constants (i.e., K_m) were not affected. Spermine did not alter the number of CCA-accepting sites or the numbers of C and A at the acceptor stem. However, it did stimulate an incorporation of CMP into tRNA-C-C or tRNA-C-C-A. Pyrophosphorolysis, the reversal of the synthetic reaction, was inhibited by spermine. Further, spermine inhibited incorporation of nucleotides into some non-tRNA substrates such as rRNA. A tRNA lacking polyamines served as a substrate in the presence of Mg^{2+} and acceptance of nucleotides was stimulated by the addition of spermine. It was concluded that polyamines are not essential for a low rate of nucleotide addition, but spermine interaction with the tRNA was important for a high rate of such addition.

A plant tRNA nucleotidyltransferase was purified to homogeneity from seeds of *Lupinus lutiens*. This tissue contains a large pool of inactive tRNA molecules lacking a terminal -C-C-A sequence. The full activity of the enzyme on lupine seed tRNA is expressed with 1 mM Mg^{2+}, and this was not increased by addition of 3 mM spermine (Cudny et al., 1978). However, it is known that germination of some seeds may release much free polyamine from covalently bound compounds and it would be important to know if the substrate, lupine seed tRNA, was actually free of polyamine.

Turnip yellow mosaic virus (TYMV) contains a great deal of spermidine and spermine associated with its RNA. It was discovered by M. Beljanski in 1965 that this RNA would accept valine specifically in an *E. coli* extract and many plant viruses have been found to contain 3-termini that may apparently serve in this way as tRNAs, as amino acid acceptors. In the case of TYMV, this terminus can be extended with -CCA by an *E. coli* nucleotidyltransferase, and valine specifically is added by an *E. coli*-aminoacyl tRNA synthetase. Although this valine does not serve in protein synthesis, this termination sequence appears important for systemic infection. The 3′ terminus cannot be described in the usual cloverleaf form but the cleavage of the RNA to release the tRNA terminus can be effected by the catalytic RNAs of purified bacterial RNase P (Giegé et al., 1993; Mans et al., 1991).

It has been observed that 1–2 mM spermidine will serve with 10 mM Mg^{2+} to effect pre-tRNA cleavages (Guerrier–Takada et al., 1986). It may be asked if spermidine contained in viral RNA plays any role in the organization of the viral 3′ terminus.

Catalytic RNA in tRNA Processing and Splicing

The RNA component (M1RNA) of RNase P will act rapidly when the protein (the C5 protein) of the *E. coli* enzyme is replaced by low concentrations of spermidine. Although the interaction of the C5 protein and M1RNA have been studied (Talbot & Altman, 1994), a similar study has not yet been reported for an interaction of the M1RNA with spermidine or the possible competition of the polyamine with C5 for the RNA.

In a study of substrate recognition of a pre-tRNA synthesized by the T7 polymerase and processing to remove an intron by a yeast endonuclease and ligase, it was reported that U8 probably served as a recognition site for the endonuclease in the splicing reaction (Reyes & Abelson, 1988). Because spermidine and spermine markedly affect the tertiary structure at U8, it may be asked if U8 recognition by the endonuclease and the splicing sequence is affected by the polyamine.

Spermidine and Mg^{2+} have been included in selfsplicing reactions by both group I and group II intron ribozymes, although the reasons for the inclusion of the polyamine in the reaction system are most frequently not stated. However, in the system of the self-splicing class II mitochondrial intron mRNA operating at 45°C to excise an intron lariat, both Mg^{2+} at 10 mM and spermidine at 1–5 mM were reported to be essential (Peebles et al., 1986). It was stated subsequently that increased Mg^{2+}, up to 100 mM, abolished the "dependency" on spermidine.

In another instance a *Neurospora* mitochondrial plasmid transcript (164 nucleotides of VS RNA) underwent a self-cleavage, producing a cyclic phosphate terminus. The reaction requires a divalent cation such as Mg^{2+} and is stimulated about twofold by 2 mM spermidine; monovalent cations also stimulate (Collins & Olive, 1993). Neither of the latter can replace Mg^{2+}. It was suggested that spermidine and K^+ are structural cations facilitating folding of the RNA. A series of studies on the ribozyme derived from the self-splicing group I intron of *Tetrahymena thermophila* attempted to analyze the roles of Mg^{2+} and spermidine in its various activities. Doudna and Szostak (1989) tested the potentiality for an RNA-catalyzed synthesis of complementary strand RNA and

demonstrated that spermidine particularly facilitated such a reaction. In defining the minimal secondary structure necessary for phosphoester transfer of this ribozyme, Beaudry and Joyce (1990) eliminated some 300 nucleotides and retained a catalytic core of 114 nucleotides that were functional in 50 mM $MgCl_2$ at 50°C; the activity was significantly enhanced by 2 mM spermidine. The extent of tertiary folding of the ribozyme detected by reductive cleavage in FeII-EDTA was found to correlate with catalytic activity (Celander & Cech, 1991). This folding and formation of a catalytic core was shown to require a minimum of three Mg^{2+} ions and to be significantly increased by low concentrations of spermidine or higher concentrations of divalent cations. Thus, spermidine, as well as Mg^{2+}, were once again implicated in the 3-dimensional organization of an active structure. Although polyamines decrease the Mg^{2+} or Mn^{2+} requirement for group I ribozyme catalysis, they do not activate the system in the absence of Mg^{2+} or Mn^{2+} (Cech et al., 1992).

Three-dimensional ribozyme structures were studied by crystallization of T7 polymerase-synthesized RNAs and subsequent analysis by X-ray crystallography. "Hammerhead" ribozyme complexes with complementary DNA strands were crystallized in a variety of forms, of which several were generated in the presence of both Mg^{2+} and spermine (Pley et al., 1993). Similarly, several small ribozymes devoid of DNA were synthesized and subjected to the same approaches, with many being crystallized in the presence of spermine or spermidine (Scott et al., 1995; Takenaka et al., 1995;).

Synthesis of Aminoacyl-tRNA

The incorporation of amino acids into polypeptide chains begins with the enzymatic formation of a compound represented by a specific codon. The rate and quality of protein synthesis is determined in significant measure by the amount and activity of the specific aminoacyl-tRNA synthetase and the accuracy of its coupling of a single amino acid to its codon-specific tRNA. That polyamines contributed to in vitro protein synthesis was detected by Nathans and Lipmann in 1961, and the extension of this observation to cellular synthesis of proteins followed the isolation and exploitation of polyamine auxotrophs in the early 1970s. Doctor and Mudd (1963) found that the rate of extract acylation of phenylalanine in the mixed tRNAs of yeast and rat liver was markedly increased by spermine. Initially, the mechanism of the stimulation was discussed largely in terms of the two-stage theory of the reaction.

In the presence of an appropriate synthetase, an amino acid (aa) and ATP react to form an enzyme-bound aminoacyl-adenylate.

$$aa + ATP \rightleftharpoons aa - AMP + PP_i.$$

The activated amino acid is then transferred to form the ester of the 2′- or 3′-OH group of the tRNA.

$$aa\text{-}AMP + tRNA \rightleftharpoons aa - tRNA + AMP.$$

The requirement of specificity of each synthetase led to their isolation and characterization, resulting in the finding that the enzymes might be classified in two groups, depending on common partial sequences (Cramer & Freist, 1993). Several of these were crystallized and studied by X-ray crystallography. One crystalline complex of the *E. coli* glutaminyl-tRNA synthetase with tRNA^{gln} and ATP revealed the interactions of these reactants with the protein, as well as the conformations of these substances in the complex (Rould et al., 1989). The L-shaped phosphate backbone of the complexed tRNA^{gln} is very similar to that of uncomplexed yeast tRNA^{phe}, despite a major difference in the phase of the acceptor stem of the complexed tRNA. The nucleotide (G) at position 10 interacted with the synthetase. Thus, it might be asked if the L shape of the complexed tRNA, which is associated with the enzyme in step two, does contain the complement of cations, polyamine and Mg^{2+}, thought to strengthen, at least in part, the bond at SU8-A14.

The purification of the seryl-tRNA synthetase from yeast and of a yeast serine tRNA permitted the study of the formation of a seryladenylate-enzyme complex, as well as the isolation of the complex by gel filtration and its reactivity with serine tRNA (Bluestein et al., 1968). The formation of the complex was demonstrated with the addition of ATP and essential MgCl$_2$ in the ratio of 1 : 2 and in the presence of inorganic pyrophosphatase. Pyrophosphatase and the removal of pyrophosphate can drive the formation of the complex. Serine was transferred from the isolated complex to tRNA in the presence of Mg^{2+}. Omission of Mg^{2+} prevented this transfer, but spermidine could replace Mg^{2+} in this reaction. These experiments clearly separated the components and stages of a two-step reaction and demonstrated the requirement for cations in both, that for Mg^{2+} in the first step and an apparently less specific requirement for cations in the second. The formation of tightly bound spermine-tRNA complexes was demonstrated by Igarashi and Takeda (1970) who showed further that such complexes, presumably stabilizing tRNA, would stimulate formation of aminoacyl-tRNA in *E. coli* systems. The second step, the transfer reaction, was explored further with yeast tRNA^{phe} and the phenylalanine-adenylate complex containing the yeast phenylalanine-tRNA synthetase (Robison & Zimmerman, 1971a,b). It was found that the efficiency of the tRNA^{phe} as an acceptor correlated with the Mg^{2+} or polyamine-induced enhancement of fluorescence of the anticodon Y base. Thus, the transfer activity is a function of the tertiary structure of the tRNA and a well-ordered tertiary structure of tRNA is determined by two sets of cations, Mg^{2+} and polyamine, situated at different loci in the structure. The analysis of the requirements of the transfer reaction might have proceeded by the preparation and test of well-defined tRNAs containing the apparently essential numbers of Mg^{2+} atoms and polyamine and the tRNAs deficient in one or the other or both of the cations, but such studies have not yet been made.

Takeda et al. (1976) subsequently reported that puri-fied cation-tRNA complexes containing Mg^{2+} (50 : 1), spermine (10–12 : 1), or spermidine (10–12 : 1) can serve as isoleucine acceptors in a reaction mixture in which a highly purified enzyme, *E. coli* isoleucyl-tRNA synthetase, contained 4.6 molecules of Mg^{2+} per molecule of enzyme and small amounts of Mg^{2+} from other sources. In a study extended to the rat liver isoleucyl-tRNA synthetase, it was shown that [^{14}C]-spermine reacted with the rat liver tRNA^{leu} (5 mol/mol tRNA) in the presence of 20 mM KCl and 5 mM Mg^{2+} (Peng et al., 1990). These researchers reported that the spermine stimulation of rat liver isoleucine–tRNA appeared to correlate with an effect on the acceptor stem (Kusama–Eguchi et al., 1991).

The mechanism of the overall reaction became a matter of debate. Takeda and Igarashi (1969) demonstrated polyamine stimulation of formation of aminoacyl-tRNA by a partially purified *E. coli* extract and *E. coli* mixed tRNA. Further investigation suggested that the formation of amino acid adenylate did not occur when the system was supplemented with spermine. Spermine appeared to stimulate the overall reaction of formation of aminoacyl-tRNA without a separable initial reaction permitting ATP : PP$_i$ exchange. Pastuszyn and Loftfield (1972) proposed that the reaction was concerted and the three substrates bound to the enzyme reacted to produce the three final products.

Subsequent data indicated that the removal of Mg^{2+} from the enzyme and tRNA was frequently inadequate. Spermine only stimulated Mg^{2+}-containing tRNA in the overall aminoacylation (Santi & Webster, 1975). Spermine did not permit detectable aminoacylation or ATP–PP$_i$ exchange in Mg^{2+}-free reaction mixtures. Various aminoacyl-tRNA synthetases have different levels of ion requirements. The initial Mg^{2+} requirement in the overall reaction is attributed to the production of the essential metal chelate of ATP. Nevertheless, the *E. coli* arginine-tRNA synthetase is reported to not require or be stimulated by Mg^{2+} (Kim & Mehler, 1981). The kinetic studies of Mehler and others confirmed the existence of an aminoacyl adenylate as an intermediate. Lövgren et al. (1978) described the separate Mg^{2+} requirements for the individual steps and noted that the binding of spermine in the second step is far tighter than that of Mg^{2+}, and competition of spermine with Mg^{2+} can displace the latter. At a 1.0 mM concentration of *free* Mg^{2+} approaching that in a cell, which is virtually inactive in an in vitro system, 200 µM spermine stimulates an overall reaction about 50-fold. It has been concluded that the polyamines probably function in cells at this step of protein synthesis. Polyamine-sparing effects of Mg^{2+} dependency have now been documented for *E. coli* class I aminoacyl-tRNA synthetases specifically (Airas, 1996).

In a recent study of phenylalanyl-tRNA formation by extracts of *E. coli* and both archaebacterial and eubacterial thermophiles, it was found that the various polyamine homologues were stimulatory or inhibitory depending on the extract used (Uzawa et al., 1994, see Chap. 22). Thus, tris(3-aminopropyl) amine and tetrakis(3-aminopropyl) ammonium were inhibitory in a *Sulfolobus* extract whereas only the latter was inhibitory in a *Thermus* extract. The results do not permit a neat generalization.

Misacylation

Errors in the construction of proteins can occur as a result of changes in the structure and replication of genes and DNA sequences, errors in the transcription of DNA to produce aberrant RNA and in the splicing of exons, and in the translation of mRNA. Several types of errors may arise in the selection of amino acids to form aminoacyl-tRNA and in the selection of the latter to mRNA codons at the ribosome. Finally, a series of possible errors can occur in the readout of the message, that is, errors of processivity. Many of the problems of translational accuracy in bacteria have been discussed by Kurland (1992). Many of these phenomena of translation are affected by the polyamines.

This section will be concerned mainly with the fitting of an amino acid to a specific tRNA. The earliest studies, when pure tRNAs and synthetases had not yet been isolated, detected misincorporation in the course of overall synthesis. In a typical study, an amino acid other than phenylalanine, for example, leucine, was incorporated into protein in a polyuridylate-primed system. In an incorporation system derived from *Bacillus stearothermophilus*, leucine was incorporated as well as phenylalanine and this misincorporation was even extended to other amino acids in the presence of streptomycin (Friedman & Weinstein, 1964). Similar results were obtained with *E. coli* systems. A lack of fidelity in translation could be exacerbated by chemicals, changes of temperature, and changes in mRNA, as in the substitution of poly(U) by poly(T). As the various steps of translation were clarified and separated, the accuracy of amino acid selection was examined at each of these. In the early 1970s several groups detected "incorrect" aminoacylations catalyzed by purified fungal aminoacyl-tRNA synthetases with purified specific tRNAs. Thus, the recognition of a single tRNA and amino acid by a single aminoacyl synthetase was not absolute (Giegé et al., 1974). Nevertheless, the extraordinary specificity of these complex reactions involving a selectivity in a given cell among 20 amino acids, 40–60 tRNAs, and a large number of aminoacyl-tRNA synthetases prompted detailed studies of the recognition sites among the participants in both the charging and mischarging reactions.

The findings of a stimulation by polyamines of the aminoacylation in general and effects on the structures of tRNA compelled attention to the possible in vivo roles of the compounds. In general, in vitro misincorporation (1 in 100) might occur at rates greatly in excess of that detected in vivo (1 in 10,000). Jelenc and Kurland (1979) devised a salt medium containing both putrescine (8 mM) and spermidine (1 mM) and enzymes and substrates for the regeneration of ATP and GTP. This mix reduced the frequency of translation errors in an in vitro *E. coli* system to the in vivo level, as well

as markedly enhancing phenylalanine incorporation into the protein.

An analysis of the in vitro synthesis of phenylalanyl-tRNA by purified yeast components revealed that high concentrations of Mg^{2+} (15 mM) enhanced the charging reaction but also increased misacylation by valine to 20% of the correct acylation. On the other hand, 0.2 mM spermine + 1.0 mM Mg^{2+} gave high rates of correct aminoacylation and almost no misacylation, conditions thought to optimize the structure of yeast $tRNA^{phe}$ (Loftfield et al., 1981).

The search for the recognition sites of tRNA by the synthetases has led in two major and complementary directions: (a) the synthesis of minimal tRNA structures like minihelices, capable of aminoacylation; and (b) the isolation, crystallization, and X-ray crystallographic analysis of synthetase-tRNA complexes. Under (a) RNA sequences were synthesized from synthetic DNA templates by T7 RNA polymerase, and the amine was not demonstrated to have been eliminated from the purified products.

Under (b) two crystalline complexes of the tRNA with their corresponding aminoacyl-tRNA synthetases, *E. coli* glutaminyl-tRNA (class I) and yeast aspartyl-tRNA (class II), were described (Cavarelli et al., 1993; Rould et al., 1989). The synthetase classes and complexes containing L-shaped tRNAs are distinguished by the positions of the tRNAs that are associated with opposite sides of the proteins. Major contact areas of the tRNAs include their acceptor arms and anticodon stems. Additionally, in both studies the possibility that the now familiar G10 may contribute to tRNA discrimination was noted in a possible contact of the N_2 of G10 to a glutamic carboxyl in the glutaminyl system and in the formation of a tertiary interaction with G45 affecting the interactions of U11 and U12 with a protein hinge in the aspartyl system. Thus, both sets of analyses demonstrated contacts of L-shaped folded tRNAs and raised questions concerning the participation of the cations in generating specificity in the various systems. In a study of the sequence requirements of an *E. coli* $tRNA^{ser}$ by seryl-tRNA synthetase, some recognition sites were not base specific and a major element of recognition was the characteristic tertiary structure of $tRNA^{ser}$ (Asahara et al., 1994).

Fidelity of Translation at the Ribosome

In bacterial translation an mRNA is attached at a specific sequence to a complementary RNA sequence of a 30S ribosomal subunit primed with initiation factors and GTP (McCarthy & Brimacombe, 1994). The first aminoacyl-tRNA is charged with methionine and is subsequently formylated. This fMet-$tRNA^{fmet}$ is then added at the AUG codon of the mRNA to form the initiation complex, which releases an initiation factor; it can then acquire a 50S ribosomal subunit. Thus, *N*-formylmethionine will become the first N-terminal amino acid in a growing peptide chain.

Codons provided in the mRNA at the ribosomal site are the primary device for the selection of an anticodon-specific aminoacyl-tRNA. However, the loading mechanism for aminoacyl-tRNA at the 70S ribosome is more complicated than this, involving the initial formation in bacteria of a ternary complex composed of the acylated tRNA, an elongation factor *EF-Tu*, and a molecule of GTP. On deposition of the aminoacyl-tRNA complex at the ribosome, GTP is hydrolyzed if a correct tRNA has been loaded; the elongation factor is then released. Fidelity is increased in this sequence because the tRNA complex dissociates from the ribosome if the correct aminoacyl-tRNA has not been placed in the accepting site, the A site. An experimental system for the analysis of these complex steps may include the reactions shown in Figure 23.7 (Harrington et al., 1993).

Deacylases of initiating aminoacyl-tRNA are known (i.e., deacylases of met-tRNA^fmet) and have been found in the prostate (Mezzetti, 1981). A formyl methionyl-tRNA_f deacylase has been purified from *E. coli* (Igarashi et al., 1981). Both enzyme types are reported to be inhibited by polyamines.

In early studies Nathans and Lipmann (1961) examined the transfer of amino acids from aminoacyl-tRNA to protein in a relatively crude *E. coli* ribosomal system and detected protein factors active in the transfer of leucine via leucyl-tRNA to protein. They found that spermidine could replace Mg^{2+} in such a transfer, which released free reutilizable tRNA. These workers also detected the deacylation of charged tRNA in a puromycin-inhibited reaction: the use of the nucleoside-containing antibiotic puromycin as a terminal participant in peptidyl transfer on the ribosome.

Lucas–Lenard and Lipmann (1967) found that in the presence of initiating protein factors and an initiating *N*-acetylphenylalanine, an apparently optimal high Mg^{2+} concentration could be reduced significantly to below 8–12 mM. The apparent optimum can be shifted to 5 mM Mg^{2+} in the presence of polyamine (0.5 mM sper-

mine) (Rheinberger & Nierhaus, 1987; Takeda, 1969). Rheinberger and Nierhaus described conditions in which nearly 30–50% of the ribosomes are active in chain elongation and translocation and 80% of the ribosomes can bind tRNA.

In a yeast system programmed by poly(U), poly-amines stimulated the binding of phenylalanine-tRNA and *N*-acetylphenylalanyl-tRNA to the ribosomes, as well as the synthesis of polyphenylalanine (Wolska–Mitaszko et al., 1976). The binding of the former to the A site was best in the presence of an elongation factor and spermine. Spermine and Mg^{2+} greatly enhanced the nonenzymatic binding of the *N*-acetylphenylalanyl-tRNA to the ribosomes. This latter type of binding was inhibited by uncharged tRNA and could be reversed by spermine in an animal cell ribosomal system.

In the translation of natural mRNAs (i.e., plant virus RNA in cell-free ribosomal systems) short polypeptides were produced in Mg^{2+}-containing media, but a partial replacement of Mg^{2+} by spermidine facilitated elongation and the yield of full-length translation products. This effect was confirmed in a wheat germ system with tobacco mosaic virus (TMV)-RNA, but a test of the possible mechanism of aborted synthesis began a search for errors in amino acid insertion using various aminoacyl-tRNAs (Abraham et al., 1979). A wheat germ ribosomal system containing poly(U) inserted relatively large amounts of leucine from leucyl-tRNA, as well as phenylalanine from phenylalanyl-tRNA in the presence of 10 mM Mg^{2+}. However, the use of 2 mM Mg^{2+} + 1.6 mM spermidine reduced leucine incorporation almost 90%, despite a 20–30% enhancement of phenylalanine incorporation.

Misreading has been studied at the ribosome level in a wheat germ system. Spermidine reduced poly(U)-directed binding of leu-tRNA to the ribosomes; misreading caused by aminoglycoside antibiotics was not affected by the polyamine. Improvements by spermidine were also detected in the proofreading steps involved in

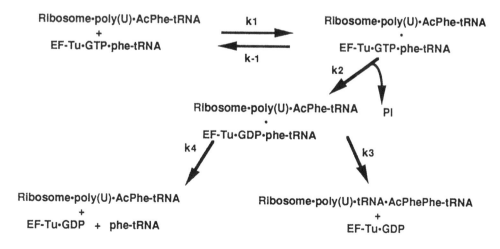

FIG. 23.7 Early steps of peptidyl synthesis on bacterial ribosomes, as formulated by Harrington et al. (1993).

the release of the noncognate aminoacyl-tRNA and GTP hydrolysis. In a subsequent analysis of protamine synthesis directed by protamine-mRNA in the wheat germ system, a spermidine-induced increase of fidelity of four different amino acids was found mainly in the binding of the various aminoacyl-tRNAs to ribosomes (Ito & Igarashi, 1986).

Discrimination at the *E. coli* ribosome was studied with ternary complexes comprising EF-Tu·GTP· and aminoacyl·tRNA (Kersten et al., 1981) in a reaction mixture of the Jelenc–Kurland type, that is, rich in polyamines. Tests were made of incorporation of several amino acids in tRNA^{phe} or its thymine-lacking precursor in systems coded by poly(U). Having minimized errors based on cation imbalances, it was possible to detect a higher level (10-fold) of incorporation of leucine in the tRNA lacking ribothymine. The tests revealed that this error was mainly a result of aberrant decoding at the initial accepting site of the ribosome.

In a poly(U)-programmed *E. coli* ribosomal system, the selection of the ternary complex containing phe-tRNA^{phe} occurs at about 50 times the rate of selection of the complex containing leu-tRNA^{leu}. Extensive rejection of leu-tRNA^{leu} after GTP hydrolysis markedly improves the accuracy of aminoacyl-tRNA selection. Mg^{2+} (6–12 mM) was found to increase the retention of leu-tRNA^{leu} at the initial binding site (Thompson et al., 1981). Unexpectedly, increased polyamine did not improve the proofreading steps, and it was concluded that optimal cation-containing assay systems have not yet been developed.

In a study of the binding of yeast tyr-tRNA and phe-tRNA as their ternary complexes with EF-Tu·GTP at ribosomal sites, Naranda and Kućan (1989) found that 3 spermine/tRNA stimulated the rate of binding of these aminoacyl-tRNAs to the poly(U$_{11}$A)-programmed A site greater than threefold and minimized nonspecific interactions. In a second part of this study, the amino groups of the *N*-aminoacyl-tRNAs were acetylated and these acetyl derivatives were able to enter the P site. Peptidic structures on tRNA in the P site transferred to the amino group of aminoacyl-tRNA in the A site. The tRNA in the P site was then released from the E site. The extended peptidyl-tRNA at the A site next moved to the now vacated P site. It was found that a 3 spermine-tRNA also markedly stimulated the binding of the acetyl-tRNAs to the P site. In the absence of spermine, binding to the P site was incomplete, and some low level of binding occurred even in the absence of mRNA, a situation conducive to error. The authors concluded that the stimulatory effect of the few molecules of spermine in the ribosomal polymerization was due essentially to its effect on the structural organization of tRNA. They suggested that the resulting accuracy in translation is the principal biological function of the polyamine.

In experiments with a reticulocyte lysate preparation, programmed by poly(U), the misincorporation of leucine instead of phenylalanine was tested with 33 different polyamine analogues (Snyder & Edwards, 1991).

Many of these, such as spermidine, reduced the initially optimal 5 mM Mg^{2+} concentration and enhanced polyphenylalanine synthesis. A large number of analogues in specific concentration ranges stimulated translation at suboptimal Mg^{2+} whereas high concentrations were inhibitory. A group of 10 substituted amines and a pentamine (4,3,3,4) enhanced misincorporation significantly.

Errors of Processivity in Translation

It may be anticipated that structural distortions of the fit of tRNA and mRNA on the ribosome will result in errors in the movements of tRNA and mRNA during translation, that is, in processivity. The presence of misincorporated aminoacyl-tRNAs at the ribosome can result in shifts in the translational reading frame as a result of "slipping" and "hopping" (Kurland, 1992). In hopping, a gap is produced when a dissociated peptidyl-tRNA lands on an out of phase codon. In "drop-off" a truncated polypeptide is generated when a peptidyl-tRNA hydrolase releases the tRNA from the incomplete polypeptide contained in the prematurely released peptidyl-tRNA. Normally "stop" codons determine the termination of translation, but a "false termination" can occur when a sense codon is misread as a stop codon.

Mutations of DNA leading to the production of mRNA containing sequences signaling stop during translation were detected in phages; these phages were described as *amber* mutants. The termination codons were identified as -UAG, -UAA, and -UGA. However, such mutants were viable on selected bacteria containing an *amber* suppressor gene; this provides a tRNA whose anticodon corresponds in some degree to the new mutant codon and whose structure permitted aminoacylation to form the needed aminoacyl-tRNA. In most instances the suppression (i.e., provision of the amino acid) is not very efficient.

The discovery that the translation of plant virus RNAs like TYMV and TMV resulted in products of varying length suggested that the above errors of processivity might be involved. Polyamines apparently diminished these errors and increased "read-through." The addition to the translation system of a yeast *amber* suppressor tRNA, which corrected a UAG termination codon, was tested in a reticulocyte lysate and showed that spermidine and spermine stimulated the reading of termination codons in both TYMV and TMV RNA (Morch & Benicourt, 1980). In the TYMV system the polyamine effect is obtained without addition of a heterologous suppressor tRNA, implying that the "normal" lysate contains a specific RNA that can respond to polyamines and read UAG codons in this way. Similar examples were observed in extracts of procaryotes and other eucaryotes such as plants.

The Tabors (1982) showed that polyamine deficiency decreases the suppression of *amber* mutants of T7 in *E. coli* carrying *amber* suppression. In this instance, the *amber* mutation in the phage related specifically to the aborted synthesis of the T7 gene/protein, corrected in the polyamine-supplemented suppressor bacterium. These workers considered the conclusion of Morch and Benicourt to be relevant. More recently, this group also detected an increase in +1 ribosomal frameshifting in yeast under conditions of spermidine deficiency and putrescine overproduction (Balasundaram et al., 1994). They suppose that this type of frameshift may arise during a translational pause occurring when a "rare" tRNA is temporarily unavailable. Overproduction of a particular rare tRNA^{arg} (CCU) inhibited frameshifting in this system. They suggest that elevated levels of putrescine in the absence of spermidine increase +1 ribosomal frameshifting.

Recent evidence indicates that +1 frameshifting determined by high internal concentrations of cellular spermidine is involved in controlling the cellular expression of antizyme (Matsufuji et al., 1995; Rom & Kahana, 1994). In the antizyme system, the mechanism is thought to relate to the reading of a fourth base after a termination codon in an initial open reading frame of antizyme mRNA, thereby introducing a +1 frameshift in translation (Hayashi & Murakami, 1995). Although either the structure of the mRNA or of tRNA at the frameshift site may be affected by spermidine (Gesteland & Atkins, 1996; Farabaugh, 1996), the effect has most frequently been attributed to mRNA interactions on the ribosome. It may be noted that frameshifting in response to viral RNA, seen often in retroviral systems, is mainly of the −1 form (Brierley, 1995).

Early Studies of DNA–Polyamine Complexes

A bihelical structure of DNA and its possible mode of replication were inferred initially from X-ray crystallographic studies. The indication that polyamines might be involved in the organization of DNA structure was turned up serendipitously with the discoveries of the transfer of polyamines with T-even phage DNA. Studies in the reactions of polyamines with "native" and frequently partially degraded DNA and RNA followed, as well as studies with the newly characterized enzymatically synthesized polyribonucleotides (Cohen, 1971a). In the physiological studies the correlations of polyamine content and synthesis with growth suggested a possible relation to RNA content and synthesis whereas polyamine production in a phage-infected cell appeared independent of the synthesis and accumulation of DNA and rRNA. As described above, the parallel accumulation of spermidine and RNA also led to the study of the effects of the polyamines on the then newly characterized tRNA structure.

The existence of exploitable DNA-determined biological systems (e.g., phage, transformation, mutagenesis) suggested several instructive studies of the effects of the polyamines and other cationic structures on DNA stability and form. Interest in DNA as a genetic element stimulated improvements in the isolation of the nucleate

and analysis of base composition and physical organization as detected by electron microscopy, sedimentation, viscosity, ultraviolet absorption, circular dichroism, light scattering, etc. Physical properties of DNA were determined in stabilizing media and buffers (e.g., the effects of ionic strength and pH) and were then explored as functions of major environmental change in temperature and extremes of pH.

The hyperchromic effect is the increase of ultraviolet absorption on treatment of DNA with acid or heat, as well as the sensitivity of this change to cation concentration, and was detected rather early (Thomas, 1993). This measure of DNA denaturation is currently ascribed to a transition from helix to random coil and to the loss of both base pairing and base stacking in the double-stranded nucleate. The readily characterizable phenomenon, which on denaturation of DNA provided an approximately 30% increase in ultraviolet absorption at about 260 nm, was sparing of DNA and easily analyzed with a common laboratory instrument. A similar separation of paired bases can be detected in infrared spectra after treatment with formamide (Kyogoku et al., 1961).

The polyamines were found to have both stabilizing and lethal effects, depending on the microorganism tested, and this suggested a competition of the polyamines for acidic cell components. Razin and Rozansky (1959) demonstrated adsorption of spermine to *Staphylococcus aureus* and the formation of precipitates of phospholipid, RNA, and DNA by the polyamine; these precipitates were dissociable in salt solutions. Cantoni (1960), in studies of mixed tRNA, detected a selective precipitation of tRNA^{val} and of tRNA^{pro} by spermine at pH 5.6. Huang and Felsenfeld (1960) observed the marked precipitation of spermine complexes of poly(A), poly(inosinic), poly(C), and poly(U); among duplexes, poly(adenylic + uridylic) was much more completely precipitated in the pH range (5.1–6.8) than was poly-(inosinic + cytidylic).

Precipitation is visually interesting but offers limited instruction. H. Tabor (1961, 1962) demonstrated that spermine stabilized a *B. subtilis* transforming DNA to heat, an effect subsequently correlated with an increase of T_m of the DNA. Similarly, a partially purified T2 phage DNA capable of infecting *E. coli* protoplasts was shown to be labile to freezing and thawing but retained its activity in the presence of spermidine and diamines (Fraser & Mahler, 1958). The DNA of λ phage was readily broken by the hydrodynamic shearing of stirring, and these effects of shear were prevented by the presence of spermine (Kaiser et al., 1963).

Thermal denaturation of DNA was found to be markedly affected by polyamines and was studied initially with a series of diamines (Mahler et al., 1961). The presence of Mg^{2+} in the salt buffer tested had only a small (ca. 1°C) effect on the T_m, the temperature of the midpoint of the transition. The most effective diamine, cadaverine, had a much greater effect (5.2–5.3°C) on the T_m values of thymus DNA and T2 DNA. Denaturation at slightly acidic pH was also markedly inhibited by diamines (Mehrotra & Mahler, 1964). A competition between various di- and triamines and Na⁺ in the stabilization of DNA was demonstrated in the work of Horacek and Cernohorsky (1968). Thus, there are differences among diamines of different lengths, despite identities of charge (Mahler & Mehrotra, 1963).

Subsequently, Stevens (1967) demonstrated even more striking increases of T_m in low ionic strength media by spermidine, spermine, and analogues of the latter, containing central units of two to six methylenes. However, the effects are quite sensitive to ionic strength and to the buffer used. A study of the differential thermal denaturation curves of DNA, a plot of the first derivative of hyperchromicity with respect to temperature, in the presence of known amounts of ligand per DNA-P, permitted a determination of the association constants of many of the natural amines under standard conditions (Morgan et al., 1986).

Initially the increase in T_m effected by spermine appeared to be a function of the adenine-thymine content of the DNA (Mandel, 1962). M. Tsuboi (1964) concluded that the effect was primarily an effect of temperature on the equilibrium constants of the formation of the various spermine-DNA complexes and was not directly related to the base compositions of the DNA. Hirschman et al. (1967) observed rapid changes of the pK values of the amino groups of spermine as a function of temperature and were unable to detect a variation in the binding sites for spermine of various DNAs of different base compositions using equilibrium dialysis. These data then suggested that polyamine binding probably arose primarily in electrostatic interaction with the chains of phosphate anions in DNA. Further, when two DNAs of very different base compositions (i.e., high AT vs. high GC), were used to compete for spermine or Ca²⁺ in the presence of Na⁺, Li⁺, K⁺, and Cs⁺ on different sides of a membrane, there was no preferential binding on either DNA. Similarly when sonicated DNA (T7 or calf thymus) of about 10⁶ MW was equilibrated with Mg²⁺, putrescine, spermidine, or spermine over a range of salt concentrations, binding of these cations to DNA was dependent on ionic strength and temperature in a manner suggestive mainly of an electrostatic interaction (Braunlin et al., 1982).

As described in Chapter 19, a bacteriophage ϕW-14 DNA was found that contains putrescinyl-thymine substituting for about 50% of the normal thymine. The DNA has an abnormally high T_m and the hypothesis was offered that the presence of the basic groups reduced electrostatic repulsion between the phosphate moieties and contributed to the efficiency of DNA packaging. Synthesis of deoxyoligonucleotides (decamers and dodecamers of thymine) containing putrescinylthymine that partially replaces the thymines then permitted the production of double-stranded molecules annealed to poly(dA). Unexpectedly the T_m values of these molecules were not increased by the presence of the unusual base; indeed increased amounts from two to four molecules in a dodecamer decreased the T_m (Takeda et al., 1987).

Crystals of Polyamine Phosphates and Early Models of DNA Complexes

Several groups undertook the X-ray crystallography of polyamine salts, which were readily crystallized. The phosphates appeared to be particularly relevant to the interactions of the amines and nucleates. Spermine phosphate was recrystallized from water as thin flakes with the composition of a diphosphate $2(HPO_4)^{2-}$ and hexahydrate. Both primary and secondary amines were protonated in tetravalent spermine (Iitaka & Huse, 1965). Sheets of spermine molecules were separated by a sheet of phosphate ions and water molecules, and these were held together mainly by hydrogen bonds between spermine-amino groups and phosphate-oxygen. Spermine molecules in the crystal existed as extended zigzag chains of 16.2-Å length, which were centrosymmetric and planar except for a slight deviation of the terminal primary amines. Curiously, the molecules of spermine within a crystalline tetrahydrochloride were not zigzag planar and contained two gauche C—C bonds in the diaminobutane moiety (Giglio et al., 1966). The crystal packing of the tetrahydrochloride appeared to be determined by interactions between charged amino and imino groups and chloride ions.

Spermidine phosphate trihydrate was crystallized as a dimer of spermidine, containing three monoprotonated anions $(HPO_4)^{2-}$ and six molecules of H_2O. This also crystallized from water as flat plates, and the crystals comprised layers of spermidine and phosphate water. Within the former, spermidine molecules existed in parallel in extended zigzag chains (Huse & Iitaka, 1969).

sym-Homospermidine triphosphate monohydrate, isolated as needles, was examined more recently (Ramaswamay & Murthy, 1991). The composition was that of a monoanionic phosphate, distinguishing the substance and probably its crystalline structure from that of spermidine phosphate trihydrate. The amine also appeared to exist in an extended all-trans conformation.

Putrescine diphosphate crystallized without water of hydration. Each protonated amino group was hydrogen bonded to three $H_2PO_4^-$ monoanions. The diamine was centrosymmetric in the trans conformation about the central C—C bond and gauche about the remaining C—C bonds (Woo et al., 1979). The observed structure cannot bridge the groove of a DNA helix, although this would be possible in an all-trans conformation, as is found in the putrescine portion of spermine. Apparently no more than two hydrogen bonds can be formed, possibly explaining the small effect of putrescine in increasing the T_m of DNA. Putrescine can span the deep major groove of an RNA complex.

Tsuboi noted that the distance between successive phosphate anions in the DNA helix (7.3 Å) might be compared with a phosphate to phosphate distance (7.96 Å) in the spermine phosphate crystals. The distance across a shallow minor groove of B DNA is about 13 Å, compared to the separation of 11.6 Å between phosphates in sheets in the crystal. He proposed an oblique bridge of spermine across the minor groove of DNA as a clamp to bind and stabilize the DNA duplex.

Liquori et al. (1967) completed a crystallographic X-ray study of spermine·4HCl and proposed the familiar models of spermine and spermidine binding to contiguous phosphates with aminopropane moieties on a single DNA chain and crossing the minor groove of the duplex with the putrescinyl moiety, as in Figure 23.8. A study of X-ray fiber photographs of the DNA-spermine complex at high humidity did reveal parameters of the B form of DNA (Suwalsky et al., 1969). However, the analyses also suggested spermine cross-links between DNA molecules.

Condensed DNA, Compaction, and Nonspecific Cationic Interaction

DNA is generally found in a compact state in the genomes of viruses, bacteria, and chromatin. As described in Chapter 19, the increased concentration of DNA created in polyamine-induced collapse facilitates the activity of topoisomerases, and these observations have also been extended to effects of spermidine analogues (Srivenugopal et al., 1987). DNA-spermidine complexes are significantly resistant to DNase I and other nucleases but remain transcriptionally active in the presence of *E. coli* RNA polymerase (Baeza et al., 1987).

DNA in solution can form ordered aggregates in a DNA-rich phase by the addition of neutral polymers plus salt. Condensation of DNA effected by flexible anionic polymers like polyglutamate generates more complex micellar scaffolds (Ghirlando et al., 1992). In a study of sized DNA molecules (50 mM), the liquid-crystalline phase created by 0.1 M NaCl and 0.25 M ammonium acetate was characterized by polarizing microscopy, electron microscopy, and X-ray diffraction as possessing a columnar longitudinal order and a hexagonal lateral order reminiscent of the packing of TMV in similarly precipitated tactoids (Livolant et al., 1989). This degree of order encouraged efforts in DNA crystallization, which had been observed first with herring sperm DNA in ethanolic solution (Giannoni et al., 1969). Crystals were subsequently obtained with sonicated calf thymus DNA, as well as with the DNAs of *E. coli* and phage T7; in each case the DNA molecules were believed to fold, bend, or loop (Maniatis et al., 1974).

Obviously, the neutralization of repulsive negatively charged structures by spermidine to produce compact toroids differs from both the phase separation effected by neutral and anionic polymers (Gosule & Schellman, 1978) and the course of viral packaging described in Chapter 19. Sonicated *E. coli* DNA plus spermidine also formed toroidal particles (Skuridin et al., 1978).

Studies with trivalent metal ion complexes, for example, $Co(NH_3)_6^{3+}$, were consistent with hypotheses that the self-association of DNA and its condensation and collapse merely requires the neutralization of 80–90% of the DNA phosphate (Wilson & Bloomfield, 1979).

However, in tests of condensation with spermine, the addition of concentrations lower than that required for collapse (Sp/P < 0.10) first produced a stiffening of the DNA molecules as detected by analysis of birefringence in an electric field, followed at Sp/P = 0.20–0.30 by an apparent bending of the molecules (Marquet et al., 1985). Toroids proved to be the most stable complexes of spermine with DNA of >700 base pairs (Marquet et al., 1987). Permethylated spermidine produced a high proportion of rods, suggesting a participation of both the primary and secondary amines of natural spermidine and spermine in the binding process (Bloomfield, 1991).

Initially, the phenomena of precipitation of the nucleates by the polyamines were discussed simply as consequences of charge neutralization. The concept of counterion condensation had been stated by Manning (1978) and developed to suggest that there was relatively little chemical specificity in the interaction. An electrostatic binding of Mg^{2+} to DNA phosphate was demonstrated, illustrative of a postulated point-charge interaction. The models of Tsuboi, Liquori, and Suiwalsky suggested that the protonated amines of extended polyamines might bind to phosphate within the minor groove of DNA. Nevertheless, the crystals of spermine-tRNA did not fit this model and a limited number of spermine molecules were participating in an organization of helical regions of the structure.

Despite the evident participation of charge neutralization in cation-induced collapse of DNA, the effects of several spermidine analogues in varying helical separation and cross-linkage within the toroids implied marked effects of polycation structure on the arrangement of the DNA strands (Schellman & Parthasarathy, 1984). Indeed, the effects of various polyamines and spermidine analogues on T_m, indicated a significant degree of structural specificity as a function of salt concentration (Thomas & Bloomfield, 1984), challenging the hypothesis of a mere point-charge neutralization by these compounds. Vertino et al. (1987) demonstrated a range of different effects on the T_m of calf thymus DNA as a function of the number and location of protonated amines, the aliphatic chain length, and the nature of amine substituents. Finally, Plum and Bloomfield (1988) compared the trivalent cations, spermidine and hexammine cobalt(III), and observed similar qualitative effects in collapsing DNA; nevertheless, they demonstrated large differences (fivefold) in the requirement for these counterions in causing collapse, despite indistinguishable binding constants in a range of Na^+ concentrations. They concluded that the differences in the effects of these ions must be assigned to specific ion effects. Studies of polyamine interactions with tRNA and DNA affirmed the conclusion that these effects were due to both charge and the specific structures of the amines tested.

Spermine is inhibitory to *E. coli*, but little has been done to pinpoint the effects at specific areas of metabolism. The fact that putrescine is frequently at high concentration (10–30 mM) in this bacterium specifically, and plays a role in tail assembly and DNA packaging

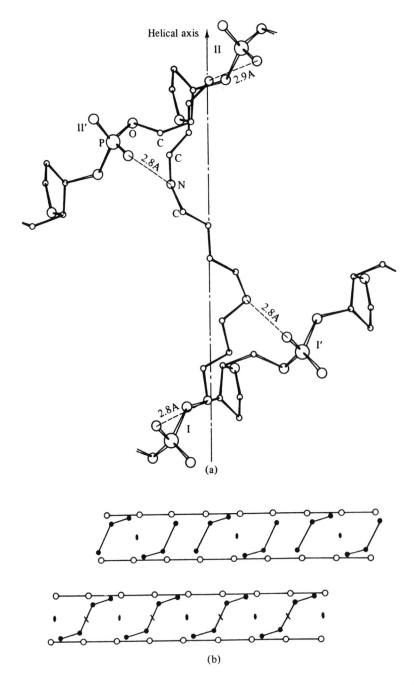

Helical axis

II

II'

2.9Å

O

C

C

2.8Å

N

C

2.8Å

I'

2.8Å

I

(a)

(b)

FIG. 23.8 (a) Detailed scheme of the complex between DNA and spermine projected on a cylindrical surface. (b) Models showing spermidine and spermine molecules in the narrow grooves of DNA. Adapted from Liquori et al. (1967).

of the λ phage, elicited a study of the interaction of *E. coli* DNA and this diamine (Takeda et al., 1983). Putrescine and DNA were tested at a ratio of 1 diamine to 1 P. Increased ratios in DNA solutions (6 putrescines per P) were required to approach a maximal T_m (ΔT_m of ca. 12°C). An approximate association constant for putrescine in reaction with DNA was significantly less than that for spermine. The X-ray diffraction pattern of the putrescine-DNA fiber indicated the retention of the B form at different humidities, in contrast to the change to the A form of uncomplexed DNA at 75% relative humidity (see below). It was suggested that the properties of reversible binding and stabilization of the B form contributed to the selection of putrescine in the life cycle of λ within *E. coli*.

Flexibility of DNA Structure

Initially mechanical models of the double helix created an image of rigid DNA. Direct evidence of a flexible structure was seen in the earliest electron micrographs of DNA, and more recently in timed views of the tube-like motion of single viral DNA molecules (Perkins et al., 1994). That the internal structure of DNA might change was perceived in the X-ray structures of DNA at various degrees of hydration.

Two forms of right-handed helical DNA were detected initially in the X-ray diffraction studies of DNA fibers: high humidity (ca. 92%) evoked the B configuration of the double helix whereas an A configuration was found at a humidity of about 75%. B DNA has been considered to be the major form in biological systems. The geometries of the DNA ladders were different with base pairs astride and perpendicular to the helix axis in B; the base pairs were pulled away from the axis and significantly tilted in the A form. The A DNA possessed a minor groove that was more shallow and wider than that of B DNA and contained a deeper major groove with internally oriented phosphates. All double helical RNA was subsequently shown to be of the A form.

Changes in salt content in solutions of a double-stranded poly(dGdC)·poly(dGdC) produced a reversible conformational transition (Pohl & Jovin, 1972). Minyat et al. (1978) showed, mainly by study of circular dichroism (CD) spectra, that an increased salt content or the addition of ethanol, that is, a decrease of water activity, evoked a transition from the B to the A form. Most diamines (putrescine, cadaverine, and 1,6-hexanediamine) maintained the B form, but spermidine and spermine stabilized the A configuration and indeed evoked a shift from the B to the A form. The CD spectrum of the B DNA contains a negative band at 246 nm and a positive band at ~280 nm; as shown in Figure 23.9, spermine reduces the negative band and magnifies a peak at 270 nm. At a ratio of one spermine to five base pairs, this peak approaches the molar CD value of the A form at 270 nm ($\Delta\varepsilon$ ~10) and can occur without aggregation. Spermidine acts similarly with smaller changes, and it was suggested that the diaminopropane moiety of spermidine and spermine was the effective

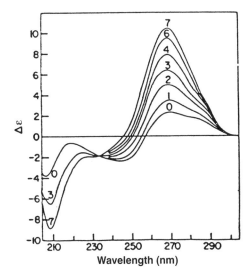

FIG. 23.9 Circular dichroism spectra of *Micrococcus luteus* DNA in 60% (w/v) water/ethanol solutions in the presence of different spermine concentrations ($\times 10^{-6}$ M); adapted from Minyat et al. (1978). DNA concentration = 5×10^{-5} M (as P); NaCl concentration = 3×10^{-4} M.

linking unit to phosphates in the deep "major" groove (Zhurkin et al., 1980). Indeed, diaminopropane and propylenediamine stabilized the A form.

The chemical synthesis and crystallization of deoxyoligonucleotides of defined sequence permitted the extension of the X-ray methodology to the analysis of configurational diversity. Spermine was effective in crystallizing these synthetic polynucleotides, and an early octamer was crystallized in an A form with a symmetrical bend of the helix axis (Shakked et al., 1981). It was seen that a high C-high G sequence in the B form in low salt, the tetramer d(CCGG), was converted in that solution at low temperature to the A form by spermine whereas the A tetramer d(GGCC) was not affected by spermine in similar conditions. Both tetramers crystallized in the A form in the presence of spermine (Majumder & Brahmachari, 1989). Mixes of spermine and Mg^{2+} were recently used to crystallize two different hexanucleotides in the A form (Mooers et al., 1995). Thus, spermine plays a significant role in developing the structure of the nucleates during crystallization and is sensitive to sequence in this role.

A deoxyoligonucleotide with a sequence common in natural DNA, d(GTGTACAC), crystallized as tetragonal bipyramids in the presence of $MgCl_2$ and spermine (Jain et al., 1989). The resolution at 2 Å permitted a localization of the single molecule of spermine in the deep groove, as shown in Figure 23.10; the amino groups of spermine were described as interacting with bases of one strand and crossing the floor of the cleft and interacting with bases on the other strand. Additionally, methylene groups of spermine formed hydrophobic

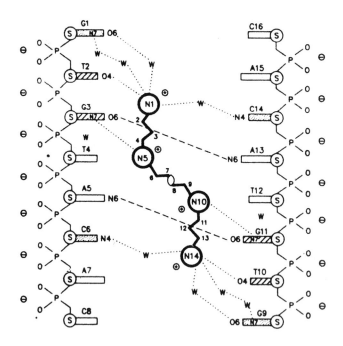

FIG. 23.10 Schematic drawing showing the hydrogen-bonding interactions between spermine and the octomer duplex; adapted from Jain et al. (1989).

bonds with methyl groups of thymines. A different crystalline A form of this deoxyoligonucleotide was also discovered that did not contain spermine (Jain et al., 1991). In solution the octamer was entirely B and was A only in the solid state; indeed, the positions of the bases of the B conformation in solution would not accommodate a molecule of spermine.

The A form prevails in RNA–DNA hybrids and it has been asked if this can relate to steps in nucleic acid syntheses. Following the synthesis of RNA–DNA chimeras, a DNA oligonucleotide d(GGGTATACGC) was annealed to the chimeric sequence r(GGG)d(TATA-CCC); that is, it contained an Okazaki fragment capable of initiating DNA replication (Egli et al., 1992). The complex, crystallized in the presence of 7.8 mM $MgCl_2$ and 75 mM spermine tetrachloride, was shown to be entirely in the A form without a structural A–B transition. Although some duplex chimeras were not in the A form, the insertion of a single 2'-hydroxyl into a B-DNA decamer compelled its conversion into a crystalline A form (Ban et al., 1994a).

Extending this study, a chimeric deoxynucleotide decameric duplex containing ribonucleotide base pairs at each terminus was crystallized with spermine and the structure was determined at 1.9-Å resolution (Ban et al., 1994b). The entire duplex was driven into an A-DNA conformation and a severe bend was imposed by spermine at the fourth base pair. The well-ordered spermine molecule was trapped within the bend and bound entirely to phosphates.

Crystallization of B-DNA

The form assumed in solutions and crystals of DNA is a function of both sequence and the environment of the nucleates (Shakked et al., 1981). Nevertheless, Dickerson et al. (1994) consider base sequence to be primary in determining several important parameters of B-DNA, for example, helix axis, bending, and minor groove width. The organization of water molecules within the structure also appears to be important to the adoption of particular forms. A dodecameric self-annealing deoxyoligonucleotide, d(CGCGAATTCGCG), was crystallized as B-DNA in the presence of spermine and Mg^{2+} per double-stranded molecule (Drew & Dickerson, 1981). The positions of electron density detected a monolayer of water molecules within the major groove, as well as an ordered stabilizing hydration of the minor groove. The single spermine, detected as a snakelike electron density revealed in multiple cycles of refinement, appeared to span the upper end of the major groove (Fig. 23.11). The protonated primary amines were described as forming salt links with two phosphates, and the protonated secondary amines were linked via a hydrogen bond to a guanine O-6 and via a bridging H_2O to a cytosine N-4 a base pair away.

According to Timsit and Moras (1992), of the A, B, and Z forms described, the B forms exhibit the most variety of packing in the crystals and specific interactions of parts of the DNA. Packing families appeared to relate to defined ratios of the cations spermine and Mg^{2+}

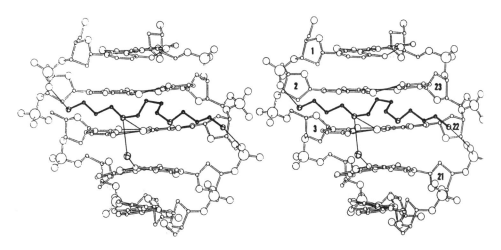

FIG. 23.11 Geometry of spermine binding at the upper end of the major groove of a B-DNA dodecamer; adapted from Drew and Dickerson (1981).

to specific base pairs, but the data are not yet sufficient to be generalized. Among five "native" orthorhombic B forms, the cation/base pair ratios used in the crystallizations were 1–2 for Mg^{2+} and 0.05–0.20 for spermine. One dodecamer, named *tet*, is described as requiring spermine for crystallization (Timsit & Moras, 1995). Resolutions of 1.9–2.6 Å are listed for most of the B forms and in this range many solvent molecules can be located in the map.

The effects of spermine in crystallizing B-DNAs were recently extended to the crystallization of DNA complexes with proteins, such as human transcription factors (Ghosh et al., 1995; Müller et al., 1995). The structures of these complexes were determined at resolutions of 2.6–2.8 Å and the precise location of spermine often requires a greater degree of resolution. As described by Williams et al. (1991), the position of the "putative" spermine in the B-DNA octamer described above (Drew & Dickerson, 1981) was only tentatively defined. The paucity of detailed data on the disposition of spermine within nucleate structures is understandable therefore, and awaits the production of crystals displaying greater degrees of order as has been found in some A and Z forms and only occasionally in a B-DNA. Although many carefully synthesized unusual deoxyoligonucleotides were crystallized in the B form with spermine and were analyzed, including sequences with mismatches, unpaired bases, nicks, and bending properties, the various degrees of resolution were not adequate to define the sites of spermine unequivocally in the structures.

Williams et al. (1990) described the structure of anthracycline-hexameric DNA complexes, crystallized in the presence of spermine, at near 1.5-Å resolution. These drugs intercalate at the d(CG) dinucleotide steps of the complex and contain two spermine molecules in the major groove. As shown in Figure 23.12, the three anthracyclines, each hydroxylated differently, modify

the internal structure of the groove sufficiently to alter the configuration and interactions of the spermine molecules. New crystalline forms of these spermine complexes were described recently (Dautant et al., 1995). Another B-DNA duplex, present with the operator of the *gal* operon, was crystallized with spermine to produce a novel opening of the duplex, suggested to relate to transcription of the operon (Tari & Secco, 1995).

Formation and Structure of Z-DNA Containing Polyamines

A DNA hexamer d(CGCGCGCG) was synthesized and crystallized as rectangular plates in a spermine-$MgCl_2$ mix containing 5% isopropanol (Wang et al., 1979). The highly ordered crystals permitted a resolution of 0.9 Å and were found to contain antiparallel double helices organized in a left-handed sense, a Z form. A brief comparison of A-, B-, and Z-DNA forms is presented in Table 23.1.

Some years earlier it had been observed that a right-handed B-DNA form of poly(dG-dC) could be converted by high concentrations of NaCl or $MgCl_2$ to an apparently left-handed form, as shown by CD. The high degree of order obtained in the Z-DNA crystals and the unusual zigzag structure constructed of 12 instead of 10 base pairs per turn, posed many chemical and biological questions. The crystallographic analysis demonstrated the existence of tilted dinucleotide repeating units in which guanosine is in the all syn conformation alternating with an anti conformation of cytidine. This permits an organization, unlike that of anti configurations in B-DNA, with the retention of a Z configuration after the introduction of a 2′-hydroxyl.

The formation of anti Z-DNA antibodies permitted the detection of the presence of Z-DNA or the potential of its formation in the DNA of many organisms, such as native calf thymus DNA (Hasan et al., 1995). Not all

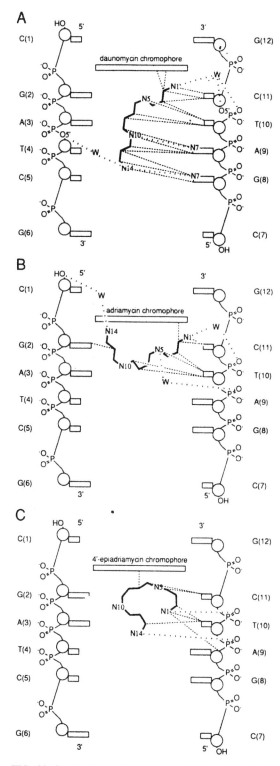

inversions of CD spectra are due to the formation of Z-DNA, as shown in tests of mitomycin-modified DNAs by antibody to Z-DNA (Tomász et al., 1983). Methylation of cytosine in a GC sequence of DNA facilitates transition to the Z form and it has been asked if such a structure can regulate transcription.

The existence of a nucleosome apparently stabilizes the B-DNA contained within it, while the presence of the Z form prevents the formation of native nucleosomes (Garner & Felsenfeld, 1987). Nevertheless, it was reported that an *E. coli* RNA polymerase and a wheat germ RNA polymerase can transcribe both B and Z conformations, although the Z-DNA template is significantly less efficient than the B form. On the other hand, an *E. coli* DNA polymerase is almost totally inactive on a Z form. Sequences of potential Z-DNA in *E. coli* were found to be hot spots of spontaneous deletion (McLean & Wells, 1988). Relatively specific antibodies to brominated [r(BG)] in a mix of A- and Z-RNA, in which bromoguanosine (BG) is in the syn configuration were used to detect the presence of Z-RNA in animal cells (Hardin et al., 1988). When poly(dAdC)·poly-(dGdT) was injected into rabbits with or without spermidine or spermine, only the presence of combined polyamine elicited the production of antibody to Z-DNA (Gunnia et al., 1991). The question was asked if this relates to the high level of antibody to Z-DNA in the sera of lupus patients.

The conversion of a B-DNA containing 5-methylcytosine (poly(dG-me^{5}dC)·poly(dG-me^{5}dC)) to Z-DNA is also effected by spermine or $[Co(NH_3)_6]^{3+}$ at far lower concentrations of these ions than of Mg^{2+} (Behe & Felsenfeld, 1981). A tetrameric d(CGCG) was crystallized in the Z form with and without spermine (Crawford et al., 1980). These were in slightly different conformations and a single spermine was found in an asymmetric unit containing three chains of the tetramer. However, a slightly disordered structure in the crystals did not permit conclusions on the disposition of the cations.

An analysis of the Z forms of the hexameric d(CGCGCG) crystallized with Mg^{2+} and Na^+ alone or with spermine alone permitted a determination of the Z structure at 1.0-Å atomic resolution. The two spermine molecules were found in slightly different positions, both of which were located on the convex surface along the edge of the deep groove, as postulated earlier by Basu et al. (1988). The molecules formed both direct and water-bridged hydrogen bonds to phosphates, guanosine N-7, and O-6 positions. The spermine-containing structure was drawn as an elongated pincushion in which spermines do not enter the grooves of the DNA (Fig. 23.13) (Gessner et al., 1989). A spermine-containing Z form of dCGCGCG, lacking any inorganic polyvalent cations, differed from the original Wang et al. Z-form hexamer crystallized with both Mg^{2+} and spermine (Egli et al., 1991). The pure spermine form was more compact and possessed a compressed helical axis and a narrower minor groove. The three spermines per duplex were bound to the convex surface in a zigzag form with bends at

FIG. 23.12 Schematic diagrams of the spermine molecules bound to DNA-anthracycline complexes; adapted from Williams et al. (1990). (**A**) A daunomycin complex; (**B**) an adriamycin complex; (**C**) an epiadriamycin complex.

TABLE 23.1 Comparison of three DNA forms.

	A-DNA	B-DNA	Z-DNA
Shape	Wide, stubby	Slim, elongated	Slimmer, more elongated
Grooves			
Minor	Shallow	Narrow, equal depth	Deep
Major	Deep	Wide	Flattened
Direction of helix	Right handed	Right handed	Left handed

either end. At low temperature (−110°C) to restrict thermal motion, the electron density map of the pure spermine form revealed a spermine within the narrow minor groove, in addition to the "intrahelix" spermines seen at room temperature (Bancroft et al., 1994).

Trivalent spermidine was less effective than spermine in inducing transitions to apparent Z forms in synthetic and natural DNAs as tested by CD spectroscopy. In comparing the effects of spermidine and its analogues in promoting the transition from B to Z, an immunoassay with an antibody to Z gave results comparable to the CD scans (Thomas & Messner, 1988).

A comparison of spermine and two natural pentamines, the 3,3,3,3 and 4,4,4,4 compounds, indicated that the pentamines aggregated DNA at lower polyamine/base-pair ratios. The 3,3,3,3 compound caldopentamine was similar to spermine in its considerable effects in increasing the T_m of DNA and less effective than spermine in causing the CD transition of poly(dG-me^{5}dC) (Basu & Marton, 1987). Compounds that facilitated aggregation and the B → Z transition were not good inhibitors of cell growth.

Z-DNA hexamer d[(CG)$_3$]$_2$ complexes containing different polyamines and Mg^{2+} were crystallized and the crystallographic data obtained at a resolution of 1.0–1.1 Å. The positions of the various polyamines and the differences in the DNA structures were described by Tomita et al. (1989). A crystal structure containing one molecule of spermidine linking two hexamers within a dimer was described and also contains three Mg^{2+} ions, an Na$^+$ ion, and 97 H$_2$O molecules. In this study it was concluded that the polyamine is not always extended, and the spermidine in the structure resembles a fishhook.

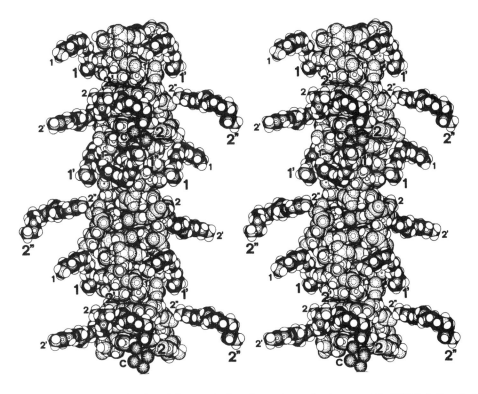

FIG. 23.13 Stereo van der Waals drawing of the organization of spermine molecules around the Z-DNA helix; adapted from Gessner et al. (1989).

CD has made it feasible to follow the induction of the B → Z transition by a relatively simple method. Ivanov and Minyat (1981) demonstrated a slower transition of B → Z for poly(dG-dC) than of B → A and also detected transitions of Z → A and Z → B (Fig. 23.14). Similar studies (Chen et al., 1984) revealed that as little as one spermine per 40–50 nucleotides in 0.1 M Na$^+$ could effect a transition of poly(dG-me^{5}dC) from B → Z. Spermine was far more effective in inducing the transition than putrescine or either the N^1- or N^8-acetylspermidines. A 23 base-pair sequence of (dG-dC)$_n$·(dG-dC)$_n$ was inserted into a large plasmid of B-DNA and the transition of this small block from B-Z within the large B structure was effected by spermidine or spermine, as detected by CD or immunoassay. Putrescine or the acetyl-spermidines could not evoke this transition (Thomas et al., 1991).

Several studies, including the rates of the various transitions, have pointed to the possible existence of intermediate conformations between the B- and Z-DNAs. A CD study indicated the successive conversion of B → A by spermine and the subsequent transformation of the A → Z by the addition of EDTA (Russell et al., 1983). Saenger and Heinemann (1989) suggested that a rotation of the guanosine sugar into the syn orientation might occur readily in an A form of poly[d(GC)]·poly[d(GC)] and thereby initiate an easy similar transformation to the Z form, in short that B → Z is really B → A → Z.

Despite the many studies on the possibility of generating Z-DNA, Wahl and Sundaralingam (1995) assert that Z-DNA has not been found in vivo. A recent review (Herbert & Rich, 1996) has been more sanguine of the possible biological significance of Z-DNA.

NMR Studies of DNA in Solution

The problem of the representation of the molecules in solution is most frequently approached by NMR. De-

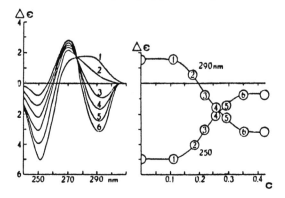

FIG. 23.14 The circular dichroism (CD) spectra (left) and a change of the CD magnitude (right) for an aqueous solution of poly(dG-dC) in the presence of spermine; adapted from Ivanov and Minyat (1981). C is the spermine content per nucleotide pair.

spite a low natural abundance of ^{13}C, NMR with this isotope permits the exploration of signals from base and sugar carbons during many types of physical and chemical changes. Zanatta et al. (1987) described the NMR spectra and ^{13}C chemical shifts as a function of temperature in a hexamer such as [d(CG)$_3$]$_2$. [^{13}C]-NMR is more sensitive to hydrogen bonding and steric effects than ^{1}H- or proton NMR.

Proton NMR has been used to study the equilibrium between the syn and anti conformations of guanosine in poly(dG-dC) in salt solutions, that is, the prevalence of anti at low salt and the switch to syn in high salt (Patel et al., 1982a). The interaction between sugar H-1′ protons and purine H-8 and pyrimidine H-6 protons can be examined by application of proton nuclear Overhauser effects, which measure the change in intensity of a given spin as a function of a nearby coupled spin. Such analyses begin to dissect the overall effects surveyed in CD spectra. ^{1}H-NMR techniques were also used to study duplex opening via proton exchange with water melting transition, etc. ^{31}P-NMR enabled an examination of phosphodiester backbones (Patel et al., 1982b).

The rates of exchange of the imino proton of thymine in deuterated water was found to indicate base-pair opening reactions of a DNA duplex. The exchange was sensitive to temperature, Na$^+$, and spermidine and was used to monitor both the dissociation of the chains of the helix and of a limited number of individual base pairs via an increase in solvent accessibility. Spermidine particularly diminishes dissociation in poly[dA]-poly-[dT] (Plum & Bloomfield, 1990), but increases proton exchange (solvent accessibility) in the base pairs.

Several groups have followed the interaction of spermidine and spermine with DNA phosphate by the displacement of bound ^{23}Na. Bound ^{23}Na is detected in excess linewidths in the spectra and, as spermine N to P approaches 1.0, these excess linewidths are almost completely eliminated (Burton et al., 1981). Spermidine, norspermidine, and diamines do not displace Na$^+$ as effectively.

Some NMR studies failed to detect a restriction of tumbling of presumably bound polyamine on DNA. This result was reexamined in a study of the self-diffusion of completely N-methylated spermidine added to DNA (Andreasson et al., 1993). The diffusion coefficient of the compound, determined by a ^{1}H-NMR method, in which all of the methyl protons give a single resonance, decreased with increasing DNA length. It fell to that approaching the diffusion coefficient of the DNA itself at low concentration ratios of the amine to DNA phosphate. This was interpreted as demonstrating that the polyamine interacts with DNA in a highly localized manner, which is considered to be representative of the natural unmethylated polyamines as well. On the other hand Symons (1995) asked if polyammonium cations, capable of acting as drugs at specific DNA sites, cannot be made to diffuse along DNA and react at the desired site.

Molecular Modeling and Dynamics

The interaction of spermine and nucleates has been simulated on theoretical grounds using the known geometry of the interacting structures to estimate a minimal, and hence preferred, energy of the complex. One study of this type concluded that such a complex could be formed in several types of bonds to the bases in the minor groove of an AT sequence in a B-DNA (Zakrzewska & Pullman, 1986). In an A form, stable complexes would involve spermine along the major groove between the two phosphate chains. In these calculations the geometry of the DNA was presumed to remain constant.

These workers also made calculations on the likely positions of spermine in yeast tRNA[phe], whose geometry was known. They found the best fit in the deep pocket around P10, with hydrogen bonds from the amino hydrogens of spermine to the anionic oxygens of P9, P10, P11, P12, P46, and P47. The position of the second spermine was also calculated and found to form a second-best fit within the major groove of the anticodon stem.

The theoretical work of Feuerstein et al. (1986) on various DNA forms was begun with the explicit intent of exploring the possible metabolic and medical effects of polyamine inhibitors and analogues. These workers calculated conformational energies and docked spermine into various B-DNA structures in computer graphics programs before and after energy minimization. Energy minima with interactions at phosphates were detected at specific almost central sites in the major groove of heteropolymeric $d(GC)_5 \cdot d(GC)_5$ and $d(AT)_5 \cdot d(AT)_5$. In the former, electrostatic interactions evoked a significant bend, not dissimilar to that seen in yeast tRNA[phe]. The presence of spermine decreased the width of the major groove and increased the width of the minor groove. Some similar effects were seen in interactions with the AT heteropolymer. In both instances a preference was found for spermine bridging the major groove. Specific effects of sequence on bending by spermine were detected by electrophoresis and CD measurements (Bordin et al., 1992).

The methodology employed in the calculation of the energy minima of spermine conformers and the various spermine DNA complexes has been described (Feuerstein et al., 1990, 1991). A review (Haworth et al., 1991) of the studies of the molecular mechanics has stressed reasons to be careful about the initial interpretations of the Feuerstein model, and also commented on the difficulty of including the effects of solvent in the earlier calculations.

Basu et al. (1988) also calculated the docking of spermine within Z-DNA and Z-RNA and compared these results with the effects of spermine in inhibiting the binding of area-specific anti Z antibodies (Basu et al., 1988). Three such antibodies were available, two of which recognized free or substituted cytosines on the major convex surface and a third that recognized phosphates of the backbone. The Z configurations were generated by bromine-containing syn-nucleotides. The binding of the third antibody was specifically inhibited by spermine and spermidine, neither of which altered the test Z-DNA and Z-RNAs. This result suggested a binding of the polyamines near the phosphate backbone on the major convex surface of the helix, as suggested subsequently by the crystallographic data.

Despite large areas of agreement, some experimental data appeared to challenge some of the theoretical results. For example, a low ratio of spermine to P, added to poly(dGdC), stiffens the molecules in contrast to the binding leading to bends and toroid formation evoked in poly(dAdT) (Marquet & Houssier, 1988). Thus, DNA condensation into toroidal particles was thought to depend on spermine-induced binding of AT-rich regions. However, the synthesis of polyaminobenzenediazonium salts enabled a test of the DNA-binding sites of sperminelike molecules (Schmid & Behr, 1991). The structure given below can apparently bind to DNA as does spermine, and on illumination with 320–400 nm UV light induces a single strand break.

The site of the attachment and break indicated that the spermine derivative was bound in the minor groove of the B-DNA studies. Further, the molecule appeared to react with GC regions, despite a two- to threefold preference for AT-rich regions.

Structural Distortions in DNA–Drug Interactions

Lindahl (1993) summarized the many ways the primary structure of DNA can be damaged in a cellular environment: hydrolysis, oxidation, radiation, etc. The presence of the 2'-OH in the ribose-phosphodiesters of RNA facilitates RNA hydrolysis and is suggested to minimize the possibilities of RNA survival in an early RNA world, as well as assisting in the selection of a DNA-based genetic system. In its turn, however, the N-glycosyl bond of deoxyribosyl purines is far more labile than the ribosyl bond; the resulting depurination and repair of DNA is estimated at 2×10^3 to 10×10^3 DNA purine bases per day in a human cell. As noted earlier, apurinic DNA is hydrolyzed ever more rapidly in the presence of polyamine, spermine being the most active and millimolar spermidine increasing the rate of hydrolysis about

10-fold. On the other hand, putrescine and cadaverine reduced the X-irradiation effects on DNA, that is reduction of T_m, or ultraviolet-induced thymine dimerization. The effects of polyamine excess or depletion on these types of normal DNA degradation in vivo have not been reported.

Many mismatched or modified DNAs have been crystallized, often with the aid of spermine, and the effects on structure have been dissected by X-ray diffraction methods (Hunter, 1992). The addition of the carcinogen *N*-2-acetylaminofluorene to a specific guanine in one defined sequence has proven to evoke a mutation in a mutation hot spot, and the crystallization of the original sequence (with spermine) has permitted the demonstration that this guanine is somewhat exposed in the specific B structure (Timsit et al., 1989).

Early observations of the inhibitory properties of proflavine in phage multiplication and the mutagenic effects of ethidium on yeast stimulated interest in the interactions for these and other heterocyclic "drugs" with DNA. L. Lerman (1963) discovered intercalation, which is the insertion of the drug between successive base pairs of a DNA structure, and the biological and chemical significance of intercalation was underlined by his work on insertion of actinomycin D between GC base pairs. Subsequently spermine was shown to displace actinomycin D from complexes with DNA (D'Orazi et al., 1979).

Proflavine inactivates penetrable phages in the dark and this effect is maximally prevented by the presence of cadaverine among the aliphatic diamines (Fraser & Mahler, 1961). Cadaverine was also found to maximize the increase in T_m of isolated T2 DNA. This led to a study of the effects of a series of steroidal diamines.

Two compounds of the type included irehdiamine whose -3β,20α- substituents were primary amines, and malouetine whose similar substituents were trimethyl quaternary amines. After an initial stabilization of the DNA to heat, the former depressed the T_m but did not cause a decrease of T_m at higher ratios of sterol to DNA (Mahler et al., 1966). Both compounds were good inhibitors of multiplication of DNA phages selectively, with the diprimary amine being the most active (Mahler & Baylor, 1967). Both compounds were mutagens and rapidly turned off DNA synthesis. The diprimary amine is believed to be capable at a high ratio of transforming the two strands of a ds-DNA into linked random coils, in contrast to malouetine, which stabilizes the DNA at all concentrations. Studies with proton magnetic resonance indicated a restriction of rotation of the steroid nucleus in the DNA structure but complete intercalation was ruled out. Nevertheless, a partial insertion of the malouetine steroid between unstacked base pairs was demonstrated (Gourevitch & Lainé, 1984). More recently new even more tightly binding steroidal polyamines, including dimeric tetramines, have been synthesized (Hsieh et al., 1994).

The synthesis of compounds (e.g., derivatives of aliphatic amines) whose reactions with polynucleotides might be detected by spectral shifts, the "reporter" molecules, was undertaken. Some of these intercalated their aromatic rings completely into DNA, with an increase in length of the helices and an increase in viscosity (Gabbay et al., 1972; Long & Barton, 1990). Bifunctional intercalators, such as the diacridines, linked by polymethylene bridges or even by spermidine or spermine through amino groups on the aromatic structure were constructed and proved to be even more tightly bound in DNA (Fico & Canellakis, 1977).

The classic intercalator ethidium bromide is a trypanocidal phenanthradinium triamine in which a quaternary amine within the central aromatic ring is separated from the aromatic primary amines in the attached rings by bond distances comparable to those relating primary amines to the secondary amine of spermidine. Binding to DNA and RNA evokes a spectrophotometric shift to longer wavelengths, which is a metachromatic effect, and enabled Waring (1965) to study a relatively tight saturable binding. This type of binding also increases the fluorescence of ethidium some 20–25-fold and this has been used to estimate both DNA and RNA. Nucleases have been detected by their ability to decrease this ethidium-induced fluorescence of the substrates. The enhanced fluorescence with DNA at least is a consequence of intercalation (Le Pecq & Paoletti, 1967). Intercalation within circular duplex DNA (e.g., some viral DNAs and mitochondrial DNAs) markedly affects superhelical turns and the density of the DNA; linear or nicked molecules containing more dye molecules have changed densities (Radloff et al., 1967). Ethidium analogues containing azides in place of primary amines are photoaffinity reagents that can label DNA irreversibly and prevent the conversion of B-DNA to Z-DNA. Further, when Z-DNA binds ethidium it is converted to B-DNA. As in the study of bisacridines, bis derivatives of ethidium like bis(methidium)spermine have been synthesized and proved to bind very tightly to DNA, intercalating between four base pairs with the two planar phenanthridinium structures (Dervan & Becker, 1978; Kuhlmann et al., 1978).

Various polyamines can displace fluorescent ethidium from calf thymus DNA in the order pentamine (4,4,4,4) > spermine > spermidine ≫ putrescine. The simplicity of the test facilitates the categorization of DNA-binding agents. Basu et al. (1990) compared many polyamine analogues in their efficacy in establishing a B–Z transition of poly(dG-me^{5}dC), their effects on T_m and aggregation of DNA, and their ability to displace ethidium from DNA. Spermine proved to be the most effective natural tetramine in these activities. An additional detailed comparison of polyamine effects on the ethidium-bound poly(dA-dT) and poly(dG-dC) complexes was described by Delcros et al. (1993). Not all ethidium is displaced from these polymers by high concentrations of spermine, and the former hold significantly more than the latter.

In any case it is clear that both the conformation of DNA and its capacity to bind drugs, as a function of nucleotide sequence, are markedly affected by the number and distribution of charges on interacting polyamines.

On the Cutting Edge

The characterization and manipulation of DNA is currently both a major activity and perspective of future work. As indicated in the above sections, the use of cations, including the polyamines, for purposes of isolation, trimming, and modification of DNA is both embedded in the experimental literature and is being exploited increasingly in current work. The selectivity of the precipitation of the nucleate by spermine was explored by Hoopes and McClure (1981), and a straightforward precipitation of a purified DNA by spermine can be applied to tissue DNA samples treated with proteinase and RNase (Ross et al., 1991). A rapid isolation of DNA from complex samples can result from the capture of the DNA and associated structures on intercalating methidium tied to Sepharose via an intermediate spermine (Harding et al., 1992). The preparation of libraries of large fragments (a megabase) of mammalian DNA to be cloned in yeast (YACS) for purposes of generating transgenic mice may employ spermine or spermidine in various stages of isolation, visualization, fragmentation, and transfection (Montoliu et al., 1995). New polycationic liposomal reagents were constructed to improve the efficiency of transfection endeavoring to optimize in vitro gene delivery (Remy et al., 1995). Interestingly, spermidine and spermine derivatives of 3-carboxycholesterol proved to be among the most effective of these new reagents (Guy–Caffey et al., 1995).

The reisolation of DNA from agarose gels frequently carries over a product inhibitory to some restriction nucleases; such inhibitions, presumably by extracted polyanions, can often be eliminated by a preincubation with spermidine (Bouché, 1981). The amplification of some DNA samples, particularly those obtained from plants by the polymerase chain reaction (PCR), has been described as difficult or inconsistent. Empirically spermidine has been found to enhance and improve the PCR amplification of some plant genes (Wan & Wilkins, 1993). On the other hand, the presence of polyamines in the "natural" samples is thought to introduce some variability in PCR yields (Ahokas & Erkkilä, 1993).

The increasing knowledge of nucleic acid structure and its flexibility is being exploited to approach problems of the control of function. Site-specific reagents, which can affect a very few base pairs, are simply not specific enough and it is understood that sequences of perhaps as many as a dozen bases are necessary to complement a unique genetic sequence in place within a genome. The formation of triple-stranded helices was discovered in the late 1950s and such DNA-containing structures are being explored currently as a participant in genetic recombination and as possible approaches to the inhibition of functions of DNA or of cutting DNA

at specific sites (Frank–Kamenetskii & Mirkin, 1995). The former is proposed in the control of site-specific transcription or of site-specific modification (Takasugi et al., 1991). Antisense oligonucleotides can block the multiplication of viruses but the feasibility of such an effort turns on the penetrability of the inhibitory oligonucleotide at the sites of multiplication. In any case the formation of triplex DNA is markedly favored by spermine or spermidine and can induce a dismutation of poly[d(TC)]·poly[(GA)] to form a triplex plus the extrusion of a free strand (Hampel et al., 1991). A comparison of the stabilizing effects of various polyamines and their analogues on the formation of triplex DNA has been presented (Musso & Van Dyke, 1995; Thomas & Thomas, 1993). Further, a spermine-linked oligonucleotide will form a triplex whereas the oligonucleotide lacking the unappended polyamine does not do so under otherwise comparable conditions (Tung et al., 1993).

The ability to modify a genetically active sequence specifically is among the interests of an increasing number of chemists. A. Eschenmoser has approached such chemical problems as the biological choice of ribose or deoxyribose in the construction of the nucleates. J. M. Lehn and others have explored the syntheses and self-assembly of alternative architectures for deoxyribonucleohelicates. Wittung et al. have replaced the sugar phosphates by a backbone of N-(2-aminoethyl)glycine units to form hybridizable mimics of RNA and DNA. Approaching the problem of antiviral agents, McConnaughie et al. have designed nonintercalative ligands for preferentially binding to A-form duplex RNA. The design of polyamine-carbodiimides for the chemical ligation of oligodeoxynucleotides provides new reagents for syntheses (von Kiedrowski & Dörwald, 1988). The problem of clipping a DNA in a specific site has become the cutting edge for several groups who will then find it expedient to turn their reagents over to the many biological and biochemical workers who are looking forward to their successes in the control of gene function.

REFERENCES

Abraham, A. K., Olsnes, S., & Pihl, A. (1979) *FEBS Letters*, **101**, 93–96.

Ahokas, H. & Erkkilä, M. J. (1993) *PCR Methods and Applications*, **3**, 65–68.

Airas, R. K. (1996) *European Journal of Biochemistry*, **240**, 223–231.

Andreasson, B., Nordenskiöld, L., Braunlin, W., Schultz, J., & Stilbs, P. (1993) *Biochemistry*, **32**, 961–967.

Asahara, H., Himeno, H., Tamura, K., Nameki, N., Hasegawa, T., & Shimizu, M. (1994) *Journal of Molecular Biology*, **236**, 738–748.

Baeza, I., Gariglio, P., Rangel, L. M., Chavez, P., Cervantes, L., Arguello, C., Wong, C., & Montanez, C. (1987) *Biochemistry*, **26**, 6387–6392.

Bagni, N., Stabellini, G., & Fracassini, D. S. (1973) *Physiologia Plantarum*, **29**, 218–222.

Balasundaram, D., Dinman, J. D., Tabor, C. W., & Tabor, H. (1994) *Journal of Bacteriology*, **176**, 7126–7128.

Ban, C., Ramakrishnan, B., & Sundaralingam, M. (1994a) *Journal of Molecular Biology*, **236**, 275–285.

———— (1994b) *Nucleic Acids Research*, **22**, 5466–5476.

Bancroft, D., Williams, L. D., Rich, A., & Egli, M. (1994) *Biochemistry*, **33**, 1073–1086.

Basu, H. S., Feuerstein, B. G., Zarling, D. A., Shafer, R. H., & Marton, L. J. (1988) *Journal of Biomolecular Structure and Dynamics*, **6**, 299–309.

Basu, H. S. & Marton, L. J. (1987) *Biochemical Journal*, **244**, 243–246.

Basu, H. S., Schwietert, H. C. A., Feuerstein, B. G., & Marton, L. J. (1990) *Biochemical Journal*, **269**, 329–334.

Beaudry, A. A. & Joyce, G. F. (1990) *Biochemistry*, **29**, 6534–6539.

Behe, M. & Felsenfeld, G. (1981) *Proceedings of the National Academy of Sciences of the United States of America*, **78**, 1619–1623.

Bloomfield, V. A. (1991) *Biopolymers*, **31**, 1471–1481.

Bolton, P. H. & Kearns, D. R. (1975) *Nature*, **255**, 347–349.

Bordin, F., Cacchione, S., Savino, M., & Tufillaro, A. (1992) *Biochemistry International*, **27**, 891–901.

Bouché, J. P. (1981) *Analytical Biochemistry*, **115**, 42–45.

Braunlin, W. H., Strick, R. J., & Record, Jr., M. T. (1982) *Biopolymers*, **21**, 1301–1314.

Brierley, I. (1995) *Biochemical Journal*, **76**, 1885–1892.

Bunce, S. & Kong, E. S. W. (1978) *Biophysical Chemistry*, **8**, 357–368.

Burton, D. R., Forsén, S., & Reimarsson, P. (1981) *Nucleic Acids Research*, **9**, 1219–1228.

Cavarelli, J., Rees, B., Ruff, M., Thierry, J., & Moras, D. (1993) *Nature*, **362**, 181–184.

Cech, T. R., Herschlag, D., Piccirilli, J. A., & Pyle, A. M. (1992) *Journal of Biological Chemistry*, **267**, 17479–17482.

Celander, D. W. & Cech, T. R. (1991) *Science*, **251**, 401–407.

Chen, H. H., Behe, M. J., & Rau, D. C. (1984) *Nucleic Acids Research*, **12**, 2381–2389.

Cohen, S. S. (1970) *Annals of the New York Academy of Science*, **171**, 869–881.

———— (1971a) *Introduction to the Polyamines*, pp. 129–134. Englewood Cliffs, N.J.: Prentice–Hall, Inc.

———— (1971b) *Introduction to the Polyamines*, pp. 139–147. Englewood Cliffs, N.J.: Prentice–Hall, Inc.

Cohen, S. S., Morgan, S., & Streibel, E. (1969) *Proceedings of the National Academy of Sciences of the United States of America*, **64**, 669–676.

Collins, R. A. & Olive, J. E. (1993) *Biochemistry*, **32**, 2795–2799.

Cramer, F. & Freist, W. (1993) *Angewandte Chemie–International Edition in English*, **32**, 190–200.

Crawford, J. L., Kolpak, F. J., Wang, A. H., Quigley, F. J., Van Boom, J. H., Van der Marel, G., & Rich, A. (1980) *Proceedings of the National Academy of Sciences of the United States of America*, **77**, 4016–4020.

Cudny, H., Pietrzak, M., & Kaczkowski, J. (1978) *Planta*, **142**, 29–36.

Dautant, A., d'Estaintot, B. L., Gallois, B., Brown, T., & Hunter, W. N. (1995) *Nucleic Acids Research*, **23**, 1710–1716.

Delcros, J., Sturkenboom, M. C. J. M., Basu, H. S., Shafer, R. H., Szöllosi, J., Feuerstein, B. G., & Marton, L. J. (1993) *Biochemical Journal*, **291**, 269–274.

Dervan, P. B. & Becker, M. M. (1978) *Journal of the American Chemical Society*, **100**, 1968–1970.

Dickerson, R. E., Goodsell, D. S., & Neidle, S. (1994) *Proceedings of the National Academy of Sciences of the United States of America*, **91**, 3579–3583.

D'Orazi, D., Serafini–Fracassini, D., & Bagni, N. (1979) *Biochemical and Biophysical Research Communications*, **90**, 362–367.

Doudna, J. A. & Szostak, J. W. (1989) *Nature*, **339**, 519–522.

Drew, H. R. & Dickerson, R. E. (1981) *Journal of Molecular Biology*, **151**, 535–556.

Egli, M., Usman, N., Zhang, S., & Rich, A. (1992) *Proceedings of the National Academy of Sciences of the United States of America*, **89**, 534–538.

Egli, M., Williams, L. D., Gao, Q., & Rich, A. (1991) *Biochemistry*, **30**, 11388–11402.

Evans, J. A. & Deutscher, M. P. (1976) *Journal of Biological Chemistry*, **251**, 6646–6652.

Farabaugh, P. J. (1996) *Microbiological Reviews*, **60**, 103–134.

Ferguson, B. Q. & Yang, D. C. H. (1986) *Biochemistry*, **25**, 5298–5304.

Fernández, C. D., Frydman, B., & Samejima, K. (1994) *Cellular and Molecular Biology*, **40**, 933–944.

Feuerstein, B. G., Pattabiraman, N., & Marton, L. J. (1986) *Proceedings of the National Academy of Sciences of the United States of America*, **83**, 5948–5952.

———— (1990) *Nucleic Acids Research*, **18**, 1271–1282.

Feuerstein, B. G., Williams, L. D., Basu, H. S., & Marton, L. J. (1991) *Journal of Cellular Biochemistry*, **46**, 37–47.

Fico, R. M. & Canellakis, E. S. (1977) *Biochemical Pharmacology*, **26**, 269–273.

Frank–Kamenetskii, M. D. & Mirkin, S. M. (1995) *Annual Review of Biochemistry*, **64**, 65–95.

Frydman, B., Frydman, R. B., De Los Santos, C., Garrido, D. A., Goldemberg, S. H., & Algranati, I. D. (1984) *Biochimica et Biophysica Acta*, **805**, 337–344.

Frydman, B., de los Santos, C., & Frydman, R. B. (1990) *Journal of Biological Chemistry*, **265**, 20874–20878.

Frydman, B., Westler, W. M., & Samejima, K. (1996) *Journal of Organic Chemistry*, **61**, 2588–2589.

Frydman, L., Rossomondo, P. C., Frydman, V., Fernandez, C. O., Frydman, B., & Samejima, K. (1992) *Proceedings of the National Academy of Sciences of the United States of America*, **89**, 9186–9190.

Frydman, L., Rossomando, P., Frydman, V., Frydman, B., & Samejima, K. (1990) *International Symposium on Polyamines*, Kyoto, Japan, p. 14.

Gabbay, E. J., DeStefano, R., & Sanford, K. (1972) *Biochemical and Biophysical Research Communications*, **46**, 155–161.

Gamble, R. C., Schoemaker, H. J. P., Jekowsky, E., & Schimmel, P. R. (1976) *Biochemistry*, **15**, 2791–2799.

Garner, M. M. & Felsenfeld, G. (1987) *Journal of Molecular Biology*, **196**, 581–590.

Gessner, R. V., Frederick, C. A., Quigley, G. J., Rich, A., & Wang, A. H. (1989) *Journal of Biological Chemistry*, **264**, 7921–7935.

Gesteland, R. F., & Atkins, J. F. (1996) *Annual Review of Biochemistry*, **65**, 741–768.

Ghirlando, R., Wachtel, E. J., Arad, T., & Minsky, A. (1992) *Biochemistry*, **31**, 7110–7119.

Ghosh, G., Van Duyne, G., Ghosh, S., & Sigler, P. B. (1995) *Nature*, **373**, 303–310.

Giegé, R., Florentz, C., & Dreker, T. W. (1993) *Biochimie*, **75**, 569–582.

Giegé, R., Kern, D., Ebel, J., Grosjean, G., De Henau, S., & Chantrenne, H. (1974) *European Journal of Biochemistry*, **45**, 351–362.

Glick, J. M., Averyhart, V. M., & Leboy, P. S. (1978) *Biochimica et Biophysica Acta*, **518**, 158–171.

Goldfield, E. M., Luxon, B. A., Bowie, V., & Gorenstein, D. G. (1983) *Biochemistry*, **22**, 3336–3344.

Gosule, L. C. & Schellman, J. A. (1978) *Journal of Molecular Biology*, **121**, 311–326.

Gourévitch, M. & Lainé, F. K. (1984) *Comptes Rendus de l'Academie de Science Paris*, **III**, 275–278.

Griffey, R. H., Davis, D. R., Yamaizumi, Z., Nishimura, S., Hawkins, B. L., & Poulter, C. D. (1986) *Journal of Biological Chemistry*, **261**, 12074–12078.

Guerrier–Takada, C., Haydock, K., Allen, L., & Altman, S. (1986) *Biochemistry*, **25**, 1509–1515.

Gunnia, U. B., Thomas, T., & Thomas, T. J. (1991) *Immunological Investigations*, **20**, 337–350.

Guy–Caffey, J. K., Bodepudi, V., Bishop, J. S., Jayaraman, K., & Chaudhary, N. (1995) *Journal of Biological Chemistry*, **270**, 31391–31396.

Hampel, K. J., Crosson, P., & Lee, J. S. (1991) *Biochemistry*, **30**, 4455–4459.

Hardin, C. C., Zarling, D. A., Wolk, S. K., Ross, W. S., & Tinoco, Jr., I. (1988) *Biochemistry*, **27**, 4169–4177.

Harding, J. D., Bebee, R. L., & Gebeyehu, G. (1992) *Methods in Enzymology*, **216**, 29–39.

Harrington, K. M., Mazarenko, I. A., Dix, D. B., Thompson, R. C., & Uhlenbeck, O. C. (1993) *Biochemistry*, **32**, 7617–7622.

Hasan, R., Moinuddin, K., & Alam, R. A. (1995) *FEBS Letters*, **368**, 27–30.

Haworth, I. S., Rodger, A., & Richards, W. G. (1991) *Proceedings of the Royal Society of London*, **244B**, 107–116.

Hayashi, S. & Murakami, Y. (1995) *Biochemical Journal*, **306**, 1–10.

Heerschap, A., Walters, J. A. L. I., & Hilbers, C. W. (1985) *Biophysical Chemistry*, **22**, 205–215.

——— (1986) *Nucleic Acids Research*, **14**, 983–998.

Herbert, A. & Rich, A. (1996) *Journal of Biological Chemistry*, **271**, 11595–11598.

Hingerty, B. E., Brown, R. S., & Klug, A. (1982) *Biochimica et Biophysica Acta*, **697**, 78–82.

Holbrook, S. R., Sussman, J. L., Warrant, R. W., & Kim, S. (1978) *Journal of Molecular Biology*, **123**, 631–660.

Hoopes, B. C. & McClure, W. R. (1981) *Nucleic Acids Research*, **9**, 5493–5504.

Hsieh, H., Müller, J. G., & Burrows, C. J. (1994) *Journal of the American Chemical Society*, **116**, 12077–12078.

Hunter, W. N. (1992) *Methods in Enzymology*, **211**, 221–231.

Hyde, E. I. & Reid, B. R. (1985) *Biochemistry*, **24**, 4315–4325.

Igarashi, K., Kishida, K., & Hirose, S. (1980a) *Biochemical and Biophysical Research Communications*, **96**, 678–684.

Igarashi, K., Kojima, M., Watanabe, Y., Maeda, K., & Hirose, S. (1980b) *Biochemical and Biophysical Research Communications*, **97**, 480–486.

Igarashi, K., Matsunaka, M., Kashiwagi, K., Mitsui, K., & Hirose, S. (1981) *Biochimica et Biophysica Acta*, **656**, 240–245.

Igarashi, K. & Takeda, Y. (1970) *Biochimica BioPhysica Acta*, **213**, 240–244.

Ito, K. & Igarashi, K. (1986) *European Journal of Biochemistry*, **156**, 505–510.

Ivanov, V. I. & Minyat, E. E. (1981) *Nucleic Acids Research*, **9**, 4782–4798.

Jack, A., Ladner, J. E., & Klug, A. (1976) *Journal of Molecular Biology*, **108**, 619–649.

Jack, A., Ladner, J. E., Rhodes, D., Brown, R. S., & Klug, A. (1977) *Journal of Molecular Biology*, **111**, 315–328.

Jain, S., Zon, G., & Sundaralingam, M. (1989) *Biochemistry*, **28**, 2360–2364.

——— (1991) *Biochemistry*, **30**, 3567–3576.

Jelenc, P. C. & Kurland, C. G. (1979) *Proceedings of the National Academy of Sciences of the United States of America*, **76**, 3174–3178.

Kearns, D. R., Patel, D. J., & Shulman, R. G. (1971) *Nature*, **229**, 338–339.

Kearns, D. R. & Shulman, R. G. (1974) *Accounts of Chemical Research*, **7**, 33–39.

Kersten, H., Albani, M., Männlein, E., Praisler, R., Wurmbach, P., & Nierhaus, K. H. (1981) *European Journal of Biochemistry*, **114**, 451–456.

Kim, J. P. & Mehler, A. H. (1981) *Archives of Biochemistry and Physics*, **209**, 465–470.

Kim, S. & Quigley, G. J. (1979) *Methods in Enzymology*, **59**, 3–21.

Kim, S. H., Quigley, G., Suddath, F. L., McPherson, A., Sneden, D., Kim, J. J., Weinzierl, J., & Rich, A. (1973) *Journal of Molecular Biology*, **75**, 429–432.

Kim, S. H., Quigley, G., Suddath, F. L., & Rich, A. (1971) *Proceedings of the National Academy of Sciences of the United States of America*, **68**, 841–845.

Klotz, I. M. & Hunston, D. L. (1971) *Biochemistry*, **10**, 3065–3069.

Kuhlmann, K. F., Charbeneau, H. J., & Mosher, C. W. (1978) *Nucleic Acids Research*, **5**, 2629–2641.

Kurland, C. G. (1992) *Annual Review of Genetics*, **26**, 29–50.

Kusama–Eguchi, K., Watanabe, S., Irisawa, M., Watanabe, K., & Igarashi, K. (1991) *Biochemical and Biophysical Research Communications*, **177**, 745–750.

Labuda, D. & Pörschke, D. (1982) *Biochemistry*, **21**, 49–53.

Ladner, J. E., Finch, J. T., Klug, A., & Clark, B. F. C. (1972) *Journal of Molecular Biology*, **72**, 99–101.

Leboy, P. S. (1971) *FEBS Letters*, **16**, 117–120.

Leboy, P. S. & Glick, J. M. (1976) *Biochimica et Biophysica Acta*, **435**, 30–38.

Liebman, M., Rubin, J., & Sundaralingam, M. (1977) *Proceedings of the National Academy of Sciences of the United States of America*, **74**, 4821–4825.

Lindahl, T. (1993) *Nature*, **362**, 709–715.

Liquori, A. M., Constantino, L., Crescenzi, V., Elia, V., Giglio, E., Puliti, R., De Santis–Savino, M., & Vitagliano, V. (1967) *Journal of Molecular Biology*, **24**, 113–122.

Livolant, F., Levelut, A. M., Doucet, J., & Benoit, J. P. (1989) *Nature*, **339**, 724–726.

Loftfield, R. B., Eigner, E. A., & Pastuszyn, A. (1981) *Journal of Biological Chemistry*, **256**, 6729–6735.

Long, E. C. & Barton, J. K. (1990) *Accounts of Chemical Research*, **23**, 271–273.

Lövgren, T. N. E., Petersson, A., & Loftfield, T. B. (1978) *Journal of Biological Chemistry*, **253**, 6702–6710.

Mach, M., Kersten, H., & Kersten, W. (1982) *Biochemical Journal*, **202**, 153–162.

Major, F., Gautheret, D., & Cedergren, R. (1993) *Proceedings of the National Academy of Sciences of the United States of America*, **90**, 9408–9412.

Majumder, M. & Brahmachari, S. K. (1989) *Biochemistry International*, **18**, 455–465.

Maniatis, T., Venable, J. H., & Lerman, L. S. (1974) *Journal of Molecular Biology*, **84**, 37–64.

Manning, G. S. (1978) *Quarterly Reviews of Biophysics*, **11**, 179–246.

Marquet, R. & Houssier, C. (1988) *Journal of Biomolecular Structure and Dynamics*, **6**, 235–246.

Marquet, R., Houssier, C., & Fredericq, E. (1985) *Biochimica et Biophysica Acta*, **825**, 365–374.

Marquet, R., Wyart, A., & Houssier, C. (1987) *Biochimica et Biophysica Acta*, **909**, 165–172.

Matsufuji, S., Matsufuji, T., Miyazaki, Y., Murakami, Y., Atkins, J. F., Gesteland, R. F., & Hayashi, S. (1995) *Cell*, **80**, 51–60.

McCarthy, J. E. G. & Brimacombe, R. (1994) *Trends in Genetics*, **10**, 402–407.

McLean, M. J. & Wells, R. D. (1988) *Biochimica et Biophysica Acta*, **950**, 243–254.

Mezzetti, G. (1981) *Italian Journal of Biochemistry*, **30**, 90–98.

Minyat, E., Ivanov, V. I., Kritzyn, A. M., Minchenkova, L. E., & Schyolkina, A. K. (1978) *Journal of Molecular Biology*, **128**, 397–409.

Montoliu, L., Bock, C.-T., Schütz, G., & Zentgraf, H. (1995) *Journal of Molecular Biology*, **246**, 486–492.

Mooers, B. H. M., Schroth, G. P., Baxter, W. W., & Ho, P. S. (1995) *Journal of Molecular Biology*, **249**, 772–794.

Moras, D., Comarmond, M. B., Fischer, J., Weiss, R., Thierry, J. C., Ebel, J. P., & Giegé, R. (1980) *Nature*, **288**, 669–674.

Morch, M. & Benicourt, C. (1980) *European Journal of Biochemistry*, **105**, 445–451.

Morgan, J. E., Blankenship, J. W., & Matthews, H. R. (1986) *Archives of Biochemistry and Biophysics*, **246**, 225–232.

Müller, C. W., Rey, F. A., Sodeoka, M., Verdine, G. L., & Harrison, S. C. (1995) *Nature*, **373**, 311–317.

Musso, M. & Van Dyke, M. W. (1995) *Nucleic Acids Research*, **23**, 2320–2327.

Naranda, T. & Kućan, Z. (1989) *European Journal of Biochemistry*, **182**, 291–297.

Nishimura, S., Taya, Y., Kuchino, Y., & Ohashi, Z. (1974) *Biochemical and Biophysical Research Communications*, **57**, 702–708.

Pande, S., Vimaladithan, A., Zhao, H., & Farabaugh, P. J. (1995) *Molecular and Cellular Biology*, **15**, 298–304.

Pastuszyn, A. & Loftfield, R. B. (1972) *Biochemical and Biophysical Research Communications*, **47**, 775–783.

Patel, D. J., Kozlowski, S. A., Nordheim, A., & Rich, A. (1982a) *Proceedings of the National Academy of Sciences of the United States of America*, **79**, 1413–1417.

Patel, D. J., Pardi, A., & Itakura, K. (1982b) *Science*, **216**, 581–590.

Peebles, C. L., Perlman, P. S., Mecklenburg, K. L., Petrillo, M. L., Tabor, J. H., Jarrell, K. A., & Cheng, H. L. (1986) *Cell*, **44**, 213–223.

Pegg, A. E. (1971) *Biochimica et Biophysica Acta*, **232**, 630–642.

Peng, Z., Kusama–Eguchi, K., Watanabe, S., Ito, K., Watanabe, K., Nomoto, Y., & Igarashi, K. (1990) *Archives of Biochemistry and Biophysics*, **279**, 138–145.

Perkins, T. T., Smith, D. E., & Chu, S. (1994) *Science*, **264**, 819–822.

Pley, H. W., Lindes, D. S., De Luca–Flaherty, C., & McKay, D. B. (1993) *Journal of Biological Chemistry*, **268**, 19656–19658.

Plum, G. E. & Bloomfield, V. A. (1988) *Biopolymers*, **27**, 1045–1051.

——— (1990) *Biochemistry*, **29**, 5934–5940.

Pochon, F. & Cohen, S. S. (1972) *Biochemical and Biophysical Research Communications*, **47**, 720–726.

Pohl, F. M. & Jovin, T. M. (1972) *Journal of Molecular Biology*, **67**, 375–396.

Poldermans, B., Roza, L., & Van Knippenberg, P. H. (1979) *Journal of Biological Chemistry*, **254**, 9094–9100.

Prinz, H., Furgac, N., & Cramer, F. (1976) *Biochimica et Biophysica Acta*, **447**, 110–115.

Quigley, G. J., Teeter, M. M., & Rich, A. (1978) *Proceedings of the National Academy of Sciences of the United States of America*, **75**, 64–68.

Ramaswamy, S. & Murthy, M. R. N. (1991) *Indian Journal of Biochemistry and Biophysics*, **28**, 504–512.

Remy, J., Kichler, A., Mordvinov, V., Schuber, F., & Behr, J. P. (1995) *Proceedings of the National Academy of Sciences of the United States of America*, **92**, 1744–1748.

Reyes, V. M. & Abelson, J. (1988) *Cell*, **55**, 719–730.

Rheinberger, H. & Nierhaus, K. H. (1987) *Journal of Biomolecular Structure and Dynamics*, **5**, 435–446.

Robertus, J. D., Ladner, J. E., Finch, J. T., Rhodes, D., Brown, R. S., Clark, B. F. C., & Klug, A. (1974) *Nature*, **250**, 546–551.

Robison, B. & Zimmerman, T. P. (1971a) *Journal of Biological Chemistry*, **246**, 110–117.

——— (1971b) *Journal of Biological Chemistry*, **246**, 4661–4670.

Rom, E. & Kahana, C. (1994) *Proceedings of the National Academy of Sciences of the United States of America*, **91**, 3959–3963.

Ross, J. A., Nelson, G. B., & Holden, K. L. (1991) *Nucleic Acids Research*, **19**, 6053.

Rould, M. A., Perona, J. J., Söll, D., & Steitz, T. A. (1989) *Science*, **246**, 1135–1142.

Roy, S. & Redfield, A. G. (1983) *Biochemistry*, **22**, 1386–1390.

Russell, W. C., Precious, B., Martin, S. R., & Bayley, P. M. (1983) *EMBO Journal*, **2**, 1647–1653.

Saenger, W. & Heinemann, U. (1989) *FEBS Letters*, **257**, 223–227.

Sakai, T. T. & Cohen, S. S. (1976) *Progress in Nucleic Acid Research and Molecular Biology*, **17**, 15–42.

Sakai, T. T., Torget, R., I, J., Freda, C. E., & Cohen, S. S. (1975) *Nucleic Acids Research*, **2**, 1005–1022.

Sampson, J. R. & Uhlenbeck, O. C. (1988) *Proceedings of the National Academy of Sciences of the United States of America*, **85**, 1033–1037.

Samuelsson, T., Bofen, T., Johansen, T., & Lustig, F. (1988) *Journal of Biological Chemistry*, **263**, 13692–13699.

Santi, D. V. & Webster, Jr., R. W. (1975) *Journal of Biological Chemistry*, **250**, 3874–3877.

Scatchard, G. (1949) *Annals of the New York Academy of Science*, **51**, 660–672.

Schellman, J. A. & Parthasarathy, N. (1984) *Journal of Molecular Biology*, **175**, 313–329.

Schmid, N. & Behr, J. (1991) *Biochemistry*, **30**, 4357–4361.

Scott, W. G., Finch, J. T., & Klug, A. (1995) *Cell*, **81**, 991–1002.

Serafini–Fracassini, D., Torrigiani, P., & Branca, C. (1984) *Physiologia Plantarum*, **60**, 351–357.

Shakked, Z., Rabinovich, D., Cruse, W. B. T., Egert, E., Kennard, O., Sala, G., Salisbury, S. A., & Viswamitra, M. A. (1981) *Proceedings of the Royal Society of London*, **B213**, 479–487.

Sipe, J. E., Anderson, Jr., W. M., Remy, C. N., & Love, S. H. (1972) *Journal of Bacteriology*, **110**, 81–91.

Skuridin, S. G., Kadykov, V. A., Shaskov, V. S., Evdokimov, Y. M., & Varshavskii, Y. M. (1978) *Molekulyarnaya Biologiya*, **12**, 413–420.

Smolar, N., Hellman, U., & Svensson, I. (1975) *Nucleic Acids Research*, **2**, 993–1004.

Snyder, R. D. & Edwards, M. L. (1991) *Biochemical and Biophysical Research Communications*, **176**, 1383–1392.

Srivenugopal, K. S., Wemmer, D. E., & Morris, D. R. (1987) *Nucleic Acids Research*, **15**, 2563–2580.

Symons, M. C. R. (1995) *Free Radicals Research*, **22**, 1–9.

Tabor, H. & Tabor, C. W. (1982) *Proceedings of the National Academy of Sciences of the United States of America*, **79**, 7087–7091.

Takasugi, M., Guendouz, A., Chassignol, M., Decout, J. L., Lhomme, J., Thuong, N. T., & Hélène, C. (1991) *Proceedings of the National Academy of Sciences of the United States of America*, **88**, 5602–5606.

Takeda, T., Ikeda, K., Mizuno, Y., & Ueda, T. (1987) *Chemical and Pharmaceutical Bulletin*, **35**, 3558–3567.

Takeda, Y., Nara, H., Iwahashi, K., Mitsui, Y., & Iitaka, Y. (1983) *Journal of Biochemistry*, **94**, 275–282.

Takeda, Y., Ohnishi, T., & Ogiso, Y. (1976) *Journal of Biological Chemistry*, **80**, 463–469.

Takenaka, A., Matsumoto, O., Chen, Y., Hasegawa, S., Chatake, T., Tsunoda, M., Ohta, T., Komatsu, Y., Koizumi, M., & Ohtsuka, E. (1995) *Journal of Biochemistry*, **117**, 850–855.

Talbot, S. J., & Altman, S. (1994) *Biochemistry*, **33**, 1399–1405.

Tari, L. W. & Secco, A. S. (1995) *Nucleic Acids Research*, **23**, 2065–2073.

Teeter, M. M., Quigley, G. J., & Rich, A. (1980) *Nucleic Acid–Metal Ion Interactions*, pp. 146–177, T. G. Spiro, Ed. New York: Wiley.

Thomas, R. (1993) *Gene*, **135**, 77–79.

Thomas, T. & Thomas, T. J. (1993) *Biochemistry*, **32**, 14068–14074.

Thomas, T. J. & Bloomfield, V. A. (1984) *Biopolymers*, **23**, 1295–1306.

Thomas, T. J., Gunnia, U. B., & Thomas, T. (1991) *Journal of Biological Chemistry*, **266**, 6137–6141.

Thomas, T. J. & Messner, R. P. (1988) *Journal of Molecular Biology*, **201**, 463–467.

Thompson, R. C., Dix, D. B., Gerson, R. B., & Karim, A. M. (1981) *Journal of Biological Chemistry*, **256**, 6676–6681.

Timsit, Y. & Moras, D. (1992) *Methods in Enzymology*, **211**, 409–429.

—— (1995) *Journal of Molecular Biology*, **251**, 629–647.

Timsit, Y., Westhof, E., Fuchs, R. P. P., & Moras, D. (1989) *Nature*, **341**, 459–462.

Tomász, M., Barton, J. K., Magliozzo, C. C., Tucker, D., Lafer, E. M., & Stollar, B. D. (1983) *Proceedings of the National Academy of Sciences of the United States of America*, **80**, 2874–2878.

Tomita, K., Hakoshima, T., Inubushi, K., Kunisawa, S., Ohishi, H., van der Marel, G. A., van Boom, J. H., Wang, A. H., & Rich, A. (1989) *Journal of Molecular Graphics*, **7**, 71–75.

Tropp, J. S. & Redfield, A. G. (1983) *Nucleic Acids Research*, **11**, 2121–2134.

Tung, C., Breslauer, K. J., & Stein, S. (1993) *Nucleic Acids Research*, **21**, 5489–5494.

Vertino, P. M., Bergeron, R. J., Cavanaugh, Jr., P. F., & Porter, C. W. (1987) *Biopolymers*, **26**, 691–703.

von Kiedrowski, G. & Dörwald, F. Z. (1988) *Leibig's Annals of Chemistry*, 787–794.

Wahl, M. C. & Sundaralingam, M. (1995) *Current Opinions in Structural Biology*, **5**, 282–295.

Wan, C. & Wilkins, T. A. (1993) *PCR Methods and Amplifications*, **3**, 208–210.

Wang, A. H., Quigley, G. J., Kolpak, F. J., Crawford, J. L., van Boom, J. H., van der Marel, G., & Rich, A. (1979) *Nature*, **282**, 680–686.

Wildenauer, D., Gross, H. J., & Riesner, D. (1974) *Nucleic Acids Research*, **1**, 1165–1182.

Williams, L. D., Frederick, C. A., Gessner, R. V., & Rich, A. (1991) *Molecular Conformation and Biological Conformations*, pp. 295–309, P. Balaram & S. Ramaseshan, Eds. Bangalore: Indian Academy of Sciences.

Williams, L. D., Frederick, C. A., Ughetto, G., & Rich, A. (1990) *Nucleic Acids Research*, **18**, 5533–5541.

Wilson, R. W. & Bloomfield, V. A. (1979) *Biochemistry*, **18**, 2192–2196.

Wolska–Mitaszko, B., Jakubowicz, T., & Gasior, E. (1976) *Acta Microbiologica Polonica*, **25**, 187–197.

Woo, N., Seeman, N., & Rich, A. (1979) *Biopolymers*, **18**, 539–552.

Young, D. V. & Srinivasan, P. R. (1971) *Biochimica et Biophysica Acta*, **238**, 447–463.

Zakrzewska, K. & Pullman, B. (1986) *Biopolymers*, **25**, 375–392.

Zanatta, N., Borer, P. N., & Levy, G. (1987) *Recent Advances in Organic NMR Spectroscopy*, pp. 89–110, J. B. Lambert & R. Rittner, Eds. Landisville, N.J.: Norell Press.

Zhurkin, V. B., Lysov, Y. P., & Ivanov, V. I. (1980) *Biopolymers*, **19**, 1415–1434.

Figure and Table Credits

Chapter 2

Figure 2.2a   With kind permission from J.-P. Behr.

Figure 2.2b   Reprinted from A. Garcia, R. Giege, and J.-P. Behr, *Nucleic Acids Research*, 18:89–95, copyright 1990, with kind permission of Oxford University Press.

Figure 2.3   Adapted with kind permission from R. J. Bergeron, K. A. McGovern, M. A. Channing, and P. S. Barton, *Journal of Organic Chemistry*, 45:1589–1592, copyright 1980 American Chemical Society.

Chapter 3

Figure 3.1   Reprinted with kind permission from M. M. Kimberly and J. H. Goldstein, *Analytical Chemistry*, 53:789–793, copyright 1981 American Chemical Society.

Figure 3.2   Reprinted with kind permission from D. L. Van Rheenen, *Nature*, 193:170–171, copyright 1962 Macmillan Magazines Limited.

Figure 3.3   Reprinted from S. Kanda, M. Takahashi, and S. Nagase, *Analytical Biochemistry*, 180:307–310, copyright 1989, with kind permission from S. Kanda and Academic Press, Inc.

Figure 3.4   Reprinted from H. Tabor, C. W. Tabor, and F. Ineverre, *Analytical Biochemistry*, 55:457–467, copyright 1973, with kind permission from H. Tabor and Academic Press, Inc.

Figure 3.5   Reprinted from N. Seiler and B. Knödgen, *Journal of Chromatography B—Biomedical Applications*, 339:45–57, copyright 1985, with kind permission from Elsevier Science-NL, Sara Burgerhartstraat 25, 1055 KV Amersterdam, The Netherlands.

Figure 3.6   Reprinted from N. Seiler, B. Knödgen, and F. Eisenbeiss, *Journal of Chromatography B—Biomedical Applications*, 145:29–39, copyright 1978, with kind permission from Elsevier Science-NL, Sara Burgerhartstraat 25, 1055 KV Amersterdam, The Netherlands.

Figure 3.7   Reprinted from R. K. Sindhu and S. S. Cohen, in U. Bachrach, A. Kaye, and R. Chayen (eds.), *Advances in Polyamine Research*, vol. 4, pp. 371–380, copyright 1983, with kind permission from S. S. Cohen and Lippincott-Raven Publishers.

Figure 3.8   Reprinted from S. Kochhar, P. K. Mehta, and P. Christen, *Analytical Biochemistry*, 179:182–185, coyright 1989, with kind permission from S. Kochhar, P. Christen, and Academic Press, Inc.

Figure 3.9   Reprinted from T. Walle, in D. H. Russell (ed.), *Polyamines in Normal and Neoplastic Growth*, pp. 355–365, copyright 1973, with kind permission from Lippincott-Raven Publishers.

Figure 3.10   With kind permission from S. S. Cohen.

Table 3.3   Reprinted from U. Bachrach, in R. A. Campbell et al. (eds.), *Advances in Polyamine Research* 2:5–11, copyright 1978, with kind permission from U. Bachrach and Lippincott-Raven Publishers.

Chapter 4

Figure 4.1   Adapted with kind permission from M. W. Hosseini and J. Lehn, *Journal of the American Chemical Society*, 109:7047–7058, copyright 1987 American Chemical Society.

Diagram 1   Adapted with kind permission from J. E. Richman and T. J. Atkins, *Journal of the American Chemical Society*, 96:2268–2270, copyright 1974 American Chemical Society.

Diagram 2   Reproduced by kind permission of the Royal Society of Chemistry from D. Parker, *Chemical Society Reviews*, 19:271–291, copyright 1990.

Figure 4.2   Adapted with kind permission from H. Jahansouz, Z. Jiang, R. H. Himes, M. P. Mertes, and K. B. Mertes, *Journal of the American Chemical Society*, 111:1409–1413, copyright 1989 American Chemical Society.

Chapter 5

Figure 5.2   Reprinted from C. R. Woese, in M. E. A. Kates (ed.), *The Biochemistry of Archaea*, copyright 1993, pp. vii–xxix, with kind permission from Elsevier Science-NL, Sara Burgerhartstraat 25, 1055 KV Amersterdam, The Netherlands.

Figure 5.3   Reprinted from M. De Rosa, A. Gambacorta, M. Cartini-Farina, and V. Zappia, in J. M. Gaugas (ed.), *Polyamines in Biomedical Research*, pp. 255–272, copyright 1980, with kind permission from John Wiley and Sons, Ltd.

Figure 5.5   Reprinted from R. Shapira, A. Altman, Y. Henis, and I. Chet, *Journal of General Microbiology*, 135:1361–1367, copyright 1989, with kind permission from R. Shapira and the Society for General Microbiology.

Figure 5.6   Reprinted from N. Seiler, *Journal of Chromatography*, 379:157–176, copyright 1986, with kind permission from N. Seiler and Elsevier Science-NL, Sara Burgerhartstraat 25, 1055 KV Amersterdam, The Netherlands.

Table 5.1   Reprinted by kind permission of S. Bardocz and the publisher from S. Bardocz, G. Grant, D. S. Brown, A. Ralph, and A. Pusztai, *Journal of Nutritional Biochemistry*, 4:66–71, copyright 1993 by Elsevier Science Inc.

Chapter 6

Figure 6.2   Reprinted with kind permission from T. P. Singer, in B. Mondovi (ed.), *Structure and Functions of Amine Oxidases*, pp. 219–229, copyright CRC Press, Boca Raton, Florida, copyright 1985.

Figure 6.3   Reprinted with kind permission from D. M. Dooley, M. A. McGuirl, D. E. Brown, P. N. Turowski, W. S. McIntire, and P. F. Knowles, *Nature*, 349:262–264, copyright 1991 Macmillan Magazines Limited.

Figure 6.5   Reprinted from D. W. Lundgren, H. A. Lloyd, and J. J. Hankins, *Biochemical and Biophysical Research Communications*, 97:667–672, copyright 1980, with kind permission from Academic Press, Inc.

Figure 6.6   With kind permission from T. Nakajima.

Table 6.1   Reprinted from A. A. Coleman, C. H. Scaman, Y. J. Kang, and M. M. Palcic, *Journal of Biological Chemistry*, 266:6795–6800, copyright 1991, with kind permission from M. Palcic and the American Society for Biochemistry and Molecular Biology.

Chapter 7

Figure 7.1   Reprinted from B. Elliott and I. A. Michaelson, *Proceedings of the Society for Experimental Biology and Medicine*, 131:105–108, copyright 1969, with kind permission from Blackwell Science Publishers, Inc.

Figure 7.2   Reprinted from B. Frydman, T. B. Frydman, C. De Los Santos, D. A. Garrido, S. H. Goldemberg, and I. D. Agranati, *Biochimica et Biophysica Acta*, 805:337–344, copyright 1984, with kind permission from Elsevier Science-NL, Sara Burgerhartstraat 25, 1055 KV Amersterdam, The Netherlands.

Drawing	Reprinted from K. S. Moore, S. Wehrli, H. Roder, M. Rogers, J. N. Forrest, D. McCrimmon, and M. Zasloff, *Proceedings of the National Academy of Sciences of the United States of America*, 90:1354–1358, copyright 1993, with kind permission from M. Zasloff and the National Academy of Sciences, U.S.A.
Figure 7.3	Reprinted from G. F. Munro, K. Hercules, J. Morgan, and W. Sauerbier, *Journal of Biological Chemistry*, 247:1271–1280, copyright 1972, with kind permission from the American Society for Biochemistry and Molecular Biology.
Figure 7.4	Reprinted from C. W. Tabor and H. Tabor, *Journal of Biological Chemistry*, 241:3714–3723, copyright 1966, with kind permission from the American Society for Biochemistry and Molecular Biology.
Figure 7.5	Reprinted with kind permission from I. Flink and D. E. Pettijohn, *Nature*, 253:62–63, copyright 1975 Macmillan Magazines Limited.
Figure 7.6	Reprinted with kind permission from R. Schekman, A. Weiner, and A. Kornberg, *Science*, 186:987–993, copyright 1974 American Association for the Advancement of Science.
Figure 7.7	Reprinted from A. Raina, M. Jansen, and S. S. Cohen, *Journal of Bacteriology*, 94:1684–1696, copyright 1967, with kind permission from S. S. Cohen and the American Society for Microbiology.
Figure 7.8	Reprinted from S. Ramakrishna, L. Guarino, and S. S. Cohen, *Journal of Bacteriology*, 134:744–750, copyright 1978, with kind permission from S. S. Cohen and the American Society for Microbiology.
Figure 7.9	Reprinted from K. Igarashi, K. Mitsui, M. Kubota, M. Shirakuma, R. Ohnishi, and S. Hirose, *Biochimica et Biophysica Acta*, 755:326–331, copyright 1983, with kind permission from K. Igarashi and Elsevier Science-NL, Sara Burgerhartstraat 25, 1055 KV Amersterdam, The Netherlands.
Figure 7.10	With kind permission from S. S. Cohen.
Table 7.1	Reprinted from C. Chang, H. Shuman, and A. P. Somlyo, *Journal of Bacteriology*, 167:935–939, copyright 1986, with kind permission from A. P. Somylo and the American Society for Micobiology.
Table 7.2	Reprinted from C. Hurwitz and C. L. Rosano, *Journal of Biological Chemistry*, 242:3719–3722, copyright 1967, with kind permission from the American Society for Biochemistry and Molecular Biology.
Table 7.3	Reprinted from A. Raina, M. Jansen, and S. S. Cohen, *Journal of Bacteriology*, 94:1684–1696, copyright 1967, with kind permission from S. S. Cohen and the American Society for Microbiology.
Table 7.4	Reprinted from R. Michaels and T. T. Tchen, *Journal of Bacteriology*, 95:1966–1967, copyright 1968, with kind permission from the American Society for Microbiology.

Chapter 8

Figure 8.2	Reprinted from D. R. Morris and A. B. Pardee, *Biochemical and Biophysical Research Communications*, 20:697–702, copyright 1965, with kind permission from Academic Press, Inc.
Figure 8.3	Reprinted from W. H. Wu and D. R. Morris, *Journal of Biological Chemistry*, 248:1687–1695, copyright 1973, with kind permission from the American Society for Biochemistry and Molecular Biology.
Figure 8.5	Reprinted from J. R. Fozard and J. Koch-Weser, *Trends in Pharmacological Sciences*, March, pp. 77–110, copyright 1982, with kind permission from Elsevier Trends Journals.
Figure 8.7	Reprinted from G. N. Cohen and I. Saint-Girons, in F. C. Neidhardt (ed.), *Escherichia coli and Salmonella typhimurium: Cellular and Molecular Biology*, 1:429–444, copyright 1987, with kind permission from G. N. Cohen and the ASM Press.
Table 8.1	Reprinted from D. R. Morris and E. A. Boeker, *Methods in Enzymology*, 94:125–134, copyright 1983, with kind permission from Academic Press, Inc.
Table 8.2	Reprinted from D. R. Morris and E. A. Boeker, *Methods in Enzymology*, 94:125–134, copyright 1983, with kind permission from Academic Press, Inc.

| Table 8.4 | Reprinted from D. R. Morris and C. M. Jorstad, *Journal of Bacteriology*, 101:731–737, copyright 1970, with kind permission from the American Society for Microbiology. |
| Table 8.5 | Reprinted from C. Satischandran, G. D. Markham, R. C. Moore, and S. M. Boyle, *Journal of Bacteriology*, 172:4748, copyright 1990, with kind permission from the American Society for Microbiology. |

Chapter 9

Figure 9.6	Adapted with kind permission from G. Orr, D. Danz, G. Pontoni, P. Prabhakaran, S. Gould, and J. K. Coward, *Journal of the American Chemical Society*, 110:5791–5799, copyright 1988 American Chemical Society.
Figure 9.7	Reprinted from R. W. Myers and R. H. Abeles, *Journal of Biological Chemistry*, 265: 16913–16921, copyright 1990, with kind permission from R. H. Abeles and the American Society for Biochemistry and Molecular Biology.
Table 9.1	Reprinted from K. Kashiwagi and K. Igarashi, *Journal of Bacteriology*, 170:3131–3135, copyright 1988, with kind permission from the American Society for Microbiology.
Table 9.2	Reprinted from C. Su and S. S. Cohen, in D. H. Russell (ed.), *Polyamines in Normal and Neoplastic Growth*, pp. 299–306, copyright 1973, with kind permission from Lippincott-Raven Publishers.
Table 9.4	Reprinted from G. Cacciapuoti, M. Porcelli, M. Carteni-Farina, A. Gambacorta, and V. Zappia, *European Journal of Biochemistry*, 161:263–271, copyright 1986, with kind permission from G. Cacciapuoti and Springer-Verlag New York, Inc.
Table 9.5	Adapted with kind permission from T. M. Chu, M. F. Mallette, and R. O. Mumma, *Biochemistry*, 7:1399–1406, copyright 1968 American Chemical Society.
Table 9.6	With kind permission from S. Yamamoto.
Drawing	Reprinted from H. Elo, I. Mutikainen, L. Alhonen-Hongisto, R. Laine, J. Jänne, and P. Lumme, *Zeitschrift für Naturforschung*, 41C:851–855, copyright 1986, with kind permission from Verlag der Zeitschrift der Naturforschung.

Chapter 10

Figure 10.1	Reprinted from J. Webster, *Introduction to Fungi*, p. 177, copyright 1970, with kind permission from Cambridge University Press.
Figure 10.2	Reprinted from J. Webster, *Introduction to Fungi*, p. 177, copyright 1970, with kind permission from Cambridge University Press.
Figure 10.3	Reprinted from R. H. Davis, *Microbiological Reviews*, 50:280–313, copyright 1986, with kind permission from R. H. Davis and the American Society for Microbiology.
Figure 10.4	Reprinted from P. A. Whitney and D. R. Morris, *Journal of Bacteriology*, 134:214–220, copyright 1978, with kind permission from D. R. Morris and the American Society for Microbiology.
Figure 10.5	Reprinted from W. A. Fonzi, *Biochemical and Biophysical Research Communications*, 162: 1409–1416, copyright 1989, with kind permission from Academic Press, Inc.
Figure 10.6	Reprinted from G. Hardin, *Biology: Its Human Implications*, copyright 1949 by Garrett Hardin. Used with kind permission from W. H. Freeman and Co.
Table 10.1	Reprinted from P. A. Whitney and D. R. Morris, *Journal of Bacteriology*, 134:214–220, copyright 1978, with kind permission from D. R. Morris and the American Society for Microbiology.
Table 10.2	Reprinted from C. L. Cramer and R. H. Davis, *Journal of Biological Chemistry*, 259:5152–5157, copyright 1984, with kind permission from the American Society for Biochemistry and Molecular Biology.

Chapter 11

| Figure 11.1 | Reprinted from R. G. Ham, *Biochemical and Biophysical Research Communications*, 14: 34–38, copyright 1964, with kind permission from Academic Press, Inc. |

Figure 11.2   Reprinted from S. Kusunoki and I. Yasumasu, *Biochemical and Biophysical Research Communications*, 68:881–885, copyright 1976, with kind permission from Academic Press, Inc.

Figure 11.3   Reprinted from V. Rosander, I. Holm, B. Grahn, H. Løvtrup-Rein, M.-O. Mattson, and O. Heby, *Biochimica et Biophysica Acta*, 1264:121–128, copyright 1995, with kind permission from Elsevier Science-NL, Sara Burgerhartstraat 25, 1055 KV Amersterdam, The Netherlands.

Figure 11.4   Reprinted from F. McCormick, *Biochemical Journal*, 174:427–434, copyright 1978, with kind permission from Portland Press Limited.

Figure 11.5   Reprinted from S. M. Oredsson, S. Anehus, and O. Heby, Fig. 1, *Molecular and Cellular Biochemistry*, 64(1984):163–172, with kind permission from Academic Publishers.

Figure 11.6   Reprinted from S. Nagarajan, B. Ganem, and A. E. Pegg, *Biochemical Journal*, 254:373–378, copyright 1988, with kind permission from A. E. Pegg and Portland Press Limited.

Figure 11.7   Reprinted from U. Bachrach, M. Mensche, J. Faber, H. Desser, and N. Seiler, in C. M. Calderara, V. Zappia, and U. Bachrach (eds.), *Advances in Polyamine Research*, 3:259–274, copyright 1981, with kind permission from U. Bachrach and Lippincott-Raven Publishers.

Figure 11.8   Reprinted from R. H. Fillingame and D. R. Morris, in D. H. Russell (ed.), *Polyamines in Normal and Neoplastic Growth*, pp. 249–260, copyright 1973, with kind permission from Lippincott-Raven Publishers.

Figure 11.9   Reprinted from S. W. Snyder, M. J. Egorin and L. A. Geelhaar, *Cancer Research*, 48:3613–3616, copyright 1988, with kind permission from M. J. Egorin and the American Association for Cancer Research, Inc.

Figure 11.11   Reprinted from E. Hölttä, L. Sistonen, and K. Alitalo, *Journal of Biological Chemistry*, 263:4500–4507, copyright 1988, with kind permission from E. Holtta and the American Society for Biochemistry and Molecular Biology.

Table 11.1   Reprinted from F. McCormick, *Journal of Cellular Physiology*, 93:285–292, copyright 1977, with kind permission from Wiley-Liss, Inc., a subsidiary of John Wiley and Sons, Inc.

Chapter 12

Figure 12.4   With kind permission from T. Noto.

Figure 12.7   Reprinted from N. Seiler, F. N. Bolkenius, and S. Sarhan, *International Journal of Biochemistry*, 13:1205–1214, copyright 1981, with kind permission from Elsevier Science Ltd., The Boulevard, Langford Lane, Kidlington OX5 1GB, UK.

Figure 12.9   Reprinted from N. Seiler and F. N. Bolkenius, *Neurochemical Research*, 10:529–544, copyright 1985, with kind permission from Plenum Publishing Corp.

Figure 12.10   Reprinted from S. H. Snyder and D. H. Russell, *Federation Proceedings*, 29:1575–1582, copyright 1970, with kind permission from S. Snyder and FASEB.

Figure 12.11   Reprinted from A. Raina, J. Jänne, P. Hannonen, E. Hölttä, and J. Ahonen, in D. H. Russell (ed.), *Polyamines in Normal and Neoplastic Growth*, pp. 167–180, copyright 1973, with kind permission from J. Janne and Lippincott-Raven Publishers.

Chapter 13

Figure 13.1   Reprinted from C. Guarnieri, A. Lugaresi, C. Muscari, F. Flamigni, and C. M. Caldarera, *Italian Journal of Biochemistry*, 31:444–445, copyright 1982, with kind permission from C. Guarnieri and Il Pensiero Scientifico Editore.

Figure 13.2   Reprinted from J. E. Kay and V. J. Lindsay, *Biochemical Journal*, 132:791–796, copyright 1973, with kind permission from J. Kay and Portland Press Limited.

Figure 13.3   Reprinted from Y. Murakami and S. Hayashi, *Biochemical Journal*, 226:893–896, copyright 1985, with kind permission from Portland Press Limited.

Figure 13.4   Reprinted from O. Heby and L. Persson, *Trends in Biochemical Science*, 15:153–158, copyright 1990, with kind permission from O. Heby and Elsevier Trends Journals.

Figure 13.5   Reprinted from Y. Murakami, K. Tanaka, S. Matsufuji, Y. Miyazaki, and S. Hayashi, *Biochemical Journal*, 283:661–664, copyright 1992, with kind permission from Portland Press Limited.

Figure 13.7   Reprinted from R. Poulin, L. Lu, B. Ackermann, P. Bey, and A. E. Pegg, *Journal of Biological Chemistry*, 267:150–158, copyright 1992, with kind permission from the American Society for Biochemistry and Molecular Biology.

Figure 13.8   Reprinted from P. Bey, C. Danzin, and M. Jung, in P. P. McCann, A. E. Pegg, and A. Sjoerdsma (eds.), *Inhibition of Polyamine Metabolism*, pp. 1–31, copyright 1987, with kind permission from P. Bey and Academic Press, Inc.

Table 13.1   With kind permission from A. E. Pegg.

Table 13.2   Reprinted from F. Solano, R. Peñafiel, J. A. Solano, and J. A. Lozano, *International Journal of Biochemistry*, 20:463–470, copyright 1988, with kind permission from J. Solano and Pergamon-Elsevier Science Ltd., The Boulevard, Langford Lane, Kidlington OX5 1GB, UK.

Table 13.3   Reprinted from Y. Murakami, S. Matsufuji, M. Nishiyama, and S. Hayashi, *Biochemical Journal*, 259:839–845, copyright 1989, with kind permission from Portland Press Limited.

Chapter 14

Figure 14.3   Reprinted from P. Mamont, C. Danzin, J. Wagner, M. Siat, A. Joder-Ohlenbusch, and N. Claverie, *European Journal of Biochemistry*, 123:499–504, copyright 1982, with kind permission from P. Mamont and Pergamon-Elsevier Science Ltd., The Boulevard, Langford Lane, Kidlington OX5 1GB, UK.

Figure 14.4   Reprinted from A. Shirahata and A. Pegg, *Journal of Biological Chemistry*, 260:9583–9588, copyright 1985, with kind permission from the American Society for Biochemistry and Molecular Biology.

Figure 14.5   Adapted with kind permission from A. Pegg, K. Tang, and J. Coward, *Biochemistry*, 21:5082–5089, copyright 1982 American Chemical Society.

Table 14.1   Reprinted from J. Wagner, N. Claverie, and C. Danzin, *Analytical Biochemistry*, 140:108–116, copyright 1984, with kind permission from J. Wagner and Academic Press, Inc.

Table 14.2   Reprinted from A. Raina, T. Eloranta, and O. Kajander, *Biochemical Society Transactions*, 4:968–971, copyright 1976, with kind permission from O. Kajander and Portland Press Limited.

Table 14.3   Reprinted from J. Baillon, P. Mamont, J. Wagner, F. Gerhart, and P. Lux, *European Journal of Biochemistry*, 176:237–242, copyright 1988, with kind permission from P. Mamont and Pergamon-Elsevier Science Ltd., The Boulevard, Langford Lane, Kidlington OX5 1GB, UK.

Table 14.4   Reprinted from P. Kajander, L. Kauppinen, R. Pajula, K. Karkola, and T. Eloranta, *Biochemical Journal*, 259:79–86, copyright 1989, with kind permission from O. Kajander and Portland Press Limited.

Chapter 15

Figure 15.1   Reprinted from M. Murray and P. Correa, in J. M. Gaugas (ed.), *Polyamines in Biomedical Research*, pp. 221–235, copyright 1980, with kind permission from John Wiley and Sons, Ltd.

Figure 15.2   Reprinted from C. Leaf, J. Weshnok, and S. Tannenbaum, *Biochemical, Biophysical Research Communications*, 163:1032–1037, with kind permission from Academic Press, Inc.

Figure 15.3   Adapted with kind permission from M. Marletta, *Journal of Medicinal Chemistry*, 37:1899–1907, copyright 1994 American Chemical Society.

Chapter 16

Figure 16.2   Reprinted from A. Merrill, D. C. Liotta, and R. E. Riley, in R. M. Bell (ed.), *Handbook of Lipid Research*, Vol. 8: *Lipid Second Messengers*, p. 208, copyright 1996, with kind permission from A. Merrill and Plenum Publishing Corp.

Figure 16.3	Reprinted from T. O'Brien, L. Dzubow, A. Dlugosz, S. Gilmour, K. O'Donnell, and O. Hietala, in T. E. A. Slaga (ed.), *Skin Carcinogenesis: Mechanisms and Human Relevance*, pp. 213–231, copyright 1989, with kind permission from Wiley-Liss, Inc., a subsidiary of John Wiley and Sons, Inc.
Figure 16.4	Reprinted from N. Seiler and B. Knödgen, *Biochemical Medicine*, 21:168–181, copyright 1979, with kind permission from Academic Press, Inc.
Figure 16.6	Reprinted from N. Seiler, in D. R. Morris and L. J. Marton (eds.), *Polyamines in Biology and Medicine*, pp. 169–180, copyright 1981, by kind courtesy of Marcel Dekker, Inc.
Figure 16.7	Reprinted with kind permission from S. Dredar, J. Blankenship, P. Marchaut, V. Manneh, and D. Fries, *Journal of Medicinal Chemistry*, 32:984–989, copyright 1989 American Chemical Society.
Figure 16.9	With kind permission from R. Casero.
Table 16.4	Reprinted from F. Della Ragione and A. Pegg, *Biochemical Journal*, 213:701–706, copyright 1983, with kind permission from Portland Press Limited.

Chapter 17

Figure 17.1	With kind permission from T. Takita.
Figure 17.2	With kind permission from V. Quemener.
Figure 17.3	Reprinted from V. Quemener, J.-P. Moulinoux, C. Bergeron, F. Darcel, B. Cipolla, A. Denais, R. Havonis, C. Martin, and N. Seiler, Fig. 1, *Polyamines in the Gastrointestinal Tract*, ed. R. H. Dowling, U. R. Fölsch, and C. Löser, copyright 1992, with kind permission from Kluwer Academic Publishers.

Chapter 18

Figure 18.1	Reprinted by kind permission of C. Bacchi and the publisher from C. J. Bacchi, H. C. Nathan, S. H. Hutner, D. S. Duch, and C. A. Nichol, *Biochemical Pharmacology*, 30:883–886, copyright 1981 by Elsevier Science Inc.
Figure 18.4	Reprinted from A. H. Fairlamb and A. Cerami, *Annual Review of Microbiology*, 46:695–729, copyright 1992, with kind permission from A. Fairlamb and Annual Reviews Inc.
Figure 18.5	Reprinted from H. Shim and A. H. Fairlamb, *Journal of General Microbiology*, 134:807–817, copyright 1988, with kind permission from A. Fairlamb and the Society for General Microbiology.
Figure 18.6	With kind permission from A. Fairlamb.
Table 18.1	Reprinted from P. P. McCann, C. J. Bacchi, H. C. Nathan, and H. Sjoerdsma, in T. P. Senger and R. N. Ondarza (eds.), *Mechanisms of Drug Action*, pp. 159–173, copyright 1983, with kind permission from Academic Press, Inc.
Table 18.2	Reprinted from N. Yarlett and C. J. Bacchi, *Molecular and Biochemical Parasitology*, 27: 1–10, copyright 1988, with kind permission of N. Yarlett and Elsevier Science-NL, Sara Burgerhartstraat 25, 1055 KV Amersterdam, The Netherlands.
Table 18.3	Reprinted from A. H. Fairlamb and A. Cerami, *Annual Review of Microbiology*, 46:695–729, copyright 1992, with kind permission from A. Fairlamb and Annual Reviews Inc.
Table 18.4	Reprinted from N. Yarlett, *Parasitology Today*, 4:357–360, copyright 1988, with kind permission from N. Yarlett and Elsevier Trends Journals.

Chapter 19

Figure 19.1	Reprinted from A. S. Dion and S. S. Cohen, *Journal of Virology*, 9:419–422, copyright 1972, with kind permission from S. S. Cohen and the American Society for Microbiology.
Figure 19.2	Reprinted from E. W. Hafner, C. W. Tabor, and H. Tabor, *Journal of Biological Chemistry*, 254:12419–12426, copyright 1979, with kind permission from the American Society for Biochemistry and Molecular Biology.

Figure 19.3	Reprinted from M. A. Krasnow and N. R. Cozzarelli, *Journal of Biological Chemistry*, 257: 2687–2693, copyright 1982, with kind permission from M. Krasnow and the American Society for Biochemistry and Molecular Biology.
Figure 19.4	Reprinted from C. Christianssen and R. L. Baldwin, *Journal of Molecular Biology*, 115: 441–454, copyright 1977, with kind permission from Academic Press (London), Ltd.
Figure 19.5	Reprinted from A. Osland and K. Kleppe, *Nucleic Acids Research*, 4:685–694, copyright 1977, with kind permission of A. Osland and Oxford University Press.
Figure 19.6	Reprinted from L. W. Black, *Annual Review of Microbiology*, 43:267–292, copyright 1989, with kind permsission from L. W. Black and Annual Reviews Inc.
Figure 19.7	Reprinted from B. Leipold, *Journal of Virology*, 21:445–450, copyright 1977, with kind permission from the American Society for Microbiology.
Figure 19.8	Reprinted by kind permission of P. Pohjenpelto and the publisher from P. Pohjanpelto, A. Sekki, V. Hukkanen, and C. von Bonsdorff, *Life Sciences*, 42:2011–2018, copyright 1988 Elsevier Science Inc.
Figure 19.9	Reprinted from H. C. Isom, *Journal of General Virology*, 42:265–278, copyright 1979, with kind permission from H. Isom and the Society for General Microbiology.
Figure 19.10	Reprinted from R. Virudachalam, K. Sitaraman, K. L. Heuss, P. Argos, and J. L. Markley, *Virology*, 130:360–371, copyright 1983, with kind permission from Academic Press, Inc.
Figure 19.11	With kind permission from J. M. Kaper.
Table 19.1	Reprinted from B. N. Ames and D. T. Dubin, *Journal of Biological Chemistry*, 235:769–775, copyright 1960, with kind permission from the American Society for Biochemistry and Molecular Biology.
Table 19.2	Reprinted from I. Fukuma and S. S. Cohen, *Journal of Virology*, 16:222–227, copyright 1975, with kind permission from S. S. Cohen and the American Society for Microbiology.
Table 19.3	Reprinted from G. Shay Fout, K. C. Medappa, J. E. Mapoles, and R. R. Rueckert, *Journal of Biological Chemistry*, 259:3639–3643, copyright 1984, with kind permission from R. Rueckert and the American Society for Biochemistry and Molecular Biology.
Table 19.4	Reprinted from R. Torget, L. Lapi, and S. S. Cohen, *Biochemical and Biophysical Research Communications*, 87:1132–1139, copyright 1979, with kind permission from S. S. Cohen and Academic Press, Inc.
Table 19.5	Reprinted from Balint and S. S. Cohen, *Virology*, 144:194–203, copyright 1985, with kind permission from S. S. Cohen and Academic Press, Inc.

Chapter 20

Figure 20.1	Reprinted from O. J. Crocomo and L. C. Basso, *Phytochemistry*, 13:2659–2665, copyright 1974, with kind permission from O. J. Crocomo and Elsevier Science Ltd., The Boulevard, Langford Lane, Kidlington OX5 1GB, UK.
Figure 20.2	Reprinted from V. R. Villanueva, R. C. Adlakha, and R. Calvayrac, *Phytochemistry*, 19: 962–964, copyright 1980, with kind permission from V. R. Villanueva and Elsevier Science Ltd., The Boulevard, Langford Lane, Kidlington OX5 1GB, UK.
Figure 20.3	With kind permission from U. Bachrach.
Figure 20.4	Reprinted from K. S. Srivenugopal and P. R. Adiga, *FEBS Letters*, 112:260–264, copyright 1981, with kind permission from P. R. Adiga and the American Society for Biochemistry and Molecular Biology.
Figure 20.5	Reprinted from D. P. Murr and S. F. Yang, *Phytochemistry*, 14:1291–1292, copyright 1975, with kind permission from D. P. Murr and Elsevier Science Ltd., The Boulevard, Langford Lane, Kidlington OX5 1GB, UK.
Figure 20.6	Reprinted from R. L. Malmberg, J. McIndoo, A. H. Hiatt, and B. A. Lowe, *Cold Spring Harbor Symposia*, 40:475–482, copyright 1985, with kind permission from R. Malmberg and the Cold Spring Harbor Laboratory Press.
Figure 20.7	Reprinted from T. A. Smith, *The Biochemistry of Plants*, pp. 249–268, copyright 1981, with kind permission from T. A. Smith and Academic Press, Inc.

Figure 20.8	With kind permission from S. S. Cohen.
Figure 20.9	Reprinted from C. R. Bird and T. A. Smith, *Annals of Botany*, 53:483–488, copyright 1984, with kind permission from T. A. Smith and Academic Press Ltd.
Figure 20.11	With kind permission from J. Berlin.
Figure 20.12	Reprinted from K. H. Yung, S. F. Yang, and F. Schlenk, *Biochemical and Biophysical Research Communications*, 104:771–777, copyright 1982, with kind permission from F. Schlenk and Academic Press, Inc.
Figure 20.13	Reprinted from S. S. Cohen, R. Balint, R. K. Sindhu, and D. Marcu, *Medical Biology*, 59: 394–402, copyright 1981, with kind permission from S. S. Cohen and The Finnish Medical Society Duodecim.
Table 20.1	Reprinted from T. A. Smith, in P. F. Waring, *Plant Growth Substances 1982*, pp. 483–490, copyright 1982, with kind permission from T. A. Smith and Academic Press, Inc.
Table 20.2	Reprinted from T. A. Smith and G. R. Best, *Phytochemistry*, 16:841–843, copyright 1977, with kind permission from T. A. Smith and Elsevier Science Ltd., The Boulevard, Langford Lane, Kidlington OX5 1GB, UK.

Chapter 21

Figure 21.1	Reprinted from S. Choi, K. Nakanishi, and P. N. R. Usherwood, *Tetrahedron*, 49:5777–5790, copyright 1993, with kind permission from S.-K. Choi and Elsevier Science Ltd., The Boulevard, Langford Lane, Kidlington OX5 1GB, UK.
Figure 21.2	Reprinted with kind permission from L. Cariello, J. Wilson, and L. Lorand, *Biochemistry*, 23:6843–6850, copyright 1984 American Chemical Society.
Figure 21.3	Reprinted from L. Fésüs, V. Thomázy, and A. Falus, *FEBS Letters*, 224:104–108, copyright 1987, with kind permission of Elsevier Science-NL, Sara Burgerhartstraat 25, 1055 KV Amersterdam, The Netherlands.
Figure 21.4	Reprinted from S. Ohki and J. Duax, *Biochimica et Biophysica Acta*, 861:177–186, copyright 1986, with kind permission of S. Ohki and Elsevier Science-NL, Sara Burgerhartstraat 25, 1055 KV Amersterdam, The Netherlands.
Figure 21.5	Reprinted with kind permission from P. W. Majerus, T. M. Connolly, H. Deckmyn, T. S. Ross, T. E. Bross, H. Ishii, V. S. Bansal, and D. B. Wilson, *Science*, 234:1519–1526, copyright 1986 American Association for the Advancement of Science.
Figure 21.6	Reprinted from D. L. Kramer, J. T. Miller, R. J. Bergeron, R. Khomutov, A. Khomutov, and C. W. Porter, *Journal of Cellular Physiology*, 155:399–407, copyright 1993, with kind permission from C. W. Porter, D. L. Kramer and Wiley-Liss, Inc., a subsidiary of John Wiley and Sons, Inc.
Figure 21.7	Reprinted from J. G. McCormick, *Biochemical Journal*, 264:167–174, copyright 1989, with kind permission from Portland Press Limited.

Chapter 22

Figure 22.1	Reprinted from V. Shoshan-Barmatz and A. Zarka, *Biochemical Journal*, 285:61–64, copyright 1992, with kind permission from V. Shoshan-Barmatz and Portland Press Limited.
Figure 22.2	Reprinted from S. A. Goueli, A. T. Davis, R. A. Hupakka, S. Liao, and K. Ahmed, *Biochemical Journal*, 230:293–302, copyright 1985, with kind permission from K. Ahmed and Portland Press Limited.
Figure 22.3	With kind permission from T. Oshima.
Figure 22.4	Reprinted from A. Dumuis-Kervabon, I. Encontre, G. Etienne, J. Jaurequi-Adell, J. Mery, D. Mesnier, and J. Parello, *EMBO Journal*, 5:1735–1742, copyright 1986, with kind permission from Oxford University Press.
Figure 22.5	Reprinted from Y. Takeda, *Biochimica et Biophysica Acta*, 179:232–234, copyright 1969, with kind permission of Y. Takeda and Elsevier Science-NL, Sara Burgerhartstraat 25, 1055 KV Amersterdam, The Netherlands.
Figure 22.6	Reprinted from A. Niveleau and G. A. Quash, *FEBS Letters*, 99:20–24, copyright 1979, with kind permission of Y. Takeda and Elsevier Science-NL, Sara Burgerhartstraat 25, 1055 KV Amersterdam, The Netherlands.

Figure 22.7	Reprinted with kind permission from T. P. Hettinger and L. C. Craig, *Biochemistry*, 9: 1224–1232, copyright 1970 American Chemical Society.
Figure 22.8	Reprinted from P. Londei, J. Teixido, M. Acca, P. Cammarano, and R. Amils, *Nucleic Acids Research*, 14:2269–2285, copyright 1986, with kind permission from Oxford University Press.
Figure 22.9	Reprinted from W. Szer, *Acta Biochimica Polonica*, 13:251–266, copyright 1966, with kind permission from Acta Biochimica Polonica.
Table 22.2	Reprinted with kind permission from R. L. Weiss and D. R. Morris, *Biochemistry*, 12: 435–441, copyright 1973 American Chemical Society.
Table 22.3	Reprinted from S. Watanabe, K. Kusama-Eguchi, H. Kobayashi, and K. Igarashi, *Journal of Biological Chemistry*, 266:20803–20809, copyright 1991, with kind permission from the American Society for Biochemistry and Molecular Biology.

Chapter 23

Scheme 1	Reprinted with kind permission from K. M. Harrington, I. A. Mazarenko, D. B. Dix, R. C. Thompson, and O. C. Uhlenbeck, *Biochemistry*, 32:7617–7622, copyright 1993 American Chemical Society.
Figure 23.1	Reprinted from S. S. Cohen, *Annals of the New York Academy of Science*, 171:869–881, copyright 1970, with kind permission of the New York Academy of Science.
Figure 23.2	Reprinted from A. Jack, J. E. Ladner, and A. Klug, *Journal of Molecular Biology*, 108: 619–649, copyright 1976, with kind permission from Academic Press (London), Ltd.
Figure 23.3	Reprinted from F. Pochon and S. S. Cohen, *Biochemical and Biophysical Research Communications*, 47:720–726, copyright 1972, with kind permission from Academic Press, Inc.
Figure 23.4	Reprinted from T. T. Sakai, R. Torget, J. I, C. E. Freda, and S. S. Cohen, *Nucleic Acids Research*, 2:1005–1022, copyright 1975, with kind permission of Oxford University Press.
Figure 23.5A	With kind permission from A. Rich.
Figure 23.5B	Reprinted from M. M. Teeter, G. J. Quigley, and A. Rich, in T. B. Spiro (ed.), *Nucleic Acid–Metal Ion Interactions*, pp. 146–177, copyright 1980, by kind permission of A. Rich and John Wiley and Sons, Inc.
Figure 23.5C	Reprinted from M. M. Teeter, G. J. Quigley, and A. Rich, in T. B. Spiro (ed.), *Nucleic Acid–Metal Ion Interactions*, pp. 146–177, copyright 1980, by kind permission of A. Rich and John Wiley and Sons, Inc.
Figure 23.6	With kind permission from B. Frydman.
Figure 23.7	Reprinted from A. M. Liquori, L. Constantino, V. Crescenzi, V. Elia, E. Giglio, R. Puliti, M. De Santis-Savino, and V. Vitagliano, *Journal of Molecular Biology*, 24:113–122, copyright 1967, with kind permission from Academic Press (London), Ltd.
Figure 23.8	Reprinted from E. Minyat, V. I. Ivanov, A. M. Kritzyn, L. E. Minchenkova, and A. K. Schyolkina, *Journal of Molecular Biology*, 128:397–409, copyright 1978, with kind permission from E. Minyat and Academic Press (London), Ltd.
Figure 23.9	Reprinted with kind permission from S. Jain, B. Zon, and M. Sundaralingam, *Biochemistry*, 28:2360–2364, copyright 1989 American Chemical Society.
Figure 23.10	Reprinted from H. R. Drew and R. E. Dickerson, *Journal of Molecular Biology*, 151:535–556, copyright 1981, with kind permission from Academic Press (London), Ltd.
Figure 23.11	Reprinted from L. D. Williams, C. A. Frederick, G. Ughetto, and A. Rich, *Nucleic Acids Research*, 18:5533–5541, copyright 1990, with kind permission of A. Rich and Oxford University Press.
Figure 23.12	Reprinted from R. V. Gessner, C. A. Frederick, G. J. Quigley, A. Rich, and A. H. Wang, *Journal of Biological Chemistry*, 264:7921–7935, copyright 1989, with kind permission from A. Rich and the American Society for Biochemistry and Molecular Biology.
Figure 23.13	Reprinted from V. I. Ivanov and E. E. Minyat, *Nucleic Acids Research*, 9:4782–4798, copyright 1981, with kind permission of E. Minyat and Oxford University Press.

Index

precursor expression and processing, 407
protoplasts, 404, 405
synthesis and turnover, 406, 407
Arginine deiminase
depletion of arginine by *Mycoplasma*, 134
pathway of energy production, 133, 134
Ascorbic acid (ascorbate)
antimutagen, 291, 292
role in ACC oxidase reaction, 429
Asparagine
induction of ODC, 196
root component and transport, 396
store of ammonia, 403
stimulation of putrescine uptake, 468
Azetidine-2-carboxylic acid
biosynthesis from methionine and AdoMet, 412
discovery, 411
presence in siderophores, 411
proline analogue, 411, 412

Bacteria
Acidiphilium facilis, 100 (*see also* Acidity of medium)
Acinetobacter tartarogenes
homospermidine synthase, 161, 162
presence of diaminopropane, 58
Aerobacter (Enterobacter) aerogenes
cleavage of AdoMet and MTA, 158, 159
growth in Mg^{2+}-limited chemostat, 103, 104
putrescine transaminase, 128
Agrobacterium tumefaciens
β-aspartic semialdehyde as aminopropyl source for spermine and thermospermine, 160
Anacystis nidulans
absence of conjugated polyamines, 61
content of putrescine and spermidine, 61
degradation of spermidine to diaminopropane, 61, 112, 113
formation of covalently bound putrescine, 117
toxicity of putrescine, 61, 112, 116, 117
transport, 95, 116, 117
Archaebacteria, 96, 147, 148
Azotobacter vinlandii
biosynthesis of spermine, 144
RNA polymerase and polyamine effects, 493
Bacillus megaterium
accumulation of ODC, RNA and spermidine in sporulation, 143
increased spermidine content by exogenous supply, 143
Bacillus stearothermophilus
crystallization of ribosomal subunit, 504, 505
spermine content, 60, 61
spermine content in ribosomes, 106
Bacillus subtilis, 97, 108, 109, 289
Bacterium cadaveris, isolation and size of LDC, 126
Citrobacter, presence of glutathionyl spermidine, 115

Clostridia
conversion of ornithine to 2,4-diaminopenta-noate, 134
presence of spermine, 61
Clostridium perfringens, gene for prohistidine decarboxylase, 152
Corynebacteria, penetrability of AdoMet, 147
Cyanobacteria. See Bacteria: *Anacystis nidulans*
Erwinia uredovora, spermidine in ice nucleation, 97
Escherichia coli, 94–119, 122–133, 142–161
absence of spermine, 57
adaptation to salt, 65, 66, 187
antimutagenicity of spermine, 289
biosynthesis of spermidine, 144–145
diamines plus spermidine as sole polyamines, 142
free and bound polyamine, 500
impenetrability of AdoMet, 147
MAO-containing Cu and TOPA, 78, 81
methionine increases RNA and spermidine content, 147, 148
polyamines facilitate synchronization of division, 149
presence of polyamines, 33
putrescine and spermidine content, 142, 143
sulfate starvation blocks cell division, 144
Halobacteria
absence of usual polyamines, 66
presence of agmatine, 66
Haemophilus influenzae, genes for ODC and transport of polyamines, 118
Haemophilus parainfluenzae, nutritional requirement for polyamines, 53
Halococcus acetinofaciens, presence of higher polyamines, 59
Klebsiella pneumoniae
conversion of 5-methylthioribose to methionine, 159, 160
inhibition by 5-trifluoromethyl ribose, 159
Lactobacillus casei, 96, 111
Lactobacillus sp.30a
histidine decarboxylase as pyruvoyl enzyme, 151
isolation and characterization of ODC and gene, 126
Micrococcus rubens, putrescine oxidase, 82, 83
Mycobacteria, 70, 99 (*see also* Bacteria: cell wall)
Mycobacterium smegmatis
non-Mg^{2+}-requirement for ADC, 127
PLP requirement for ADC, 128
regulation of ODC by RNA, 132
RNA polymerase and polyamine effects, 493
Mycoplasma, 94, 97, 118
arginine deiminase and energy production, 133, 134
genes for putrescine and spermidine transport, 118
LDC and its inhibition, 139
synthesis of cadaverine and aminopropylcadaverine, 126

RETURN
TO ➡

CHEMISTRY LIBRARY
100 Hildebrand Hall • 642-3753

LOAN PERIOD 1	2	3
4	5 1 MONTH	6

ALL BOOKS MAY BE RECALLED AFTER 7 DAYS
Renewable by telephone

DUE AS STAMPED BELOW

DEC 1 9 2002		
May 24 2003		
AUG 1 5 2003		
DEC 1 8 2003		

FORM NO. DD5

UNIVERSITY OF CALIFORNIA, BERKELEY
BERKELEY, CA 94720-6000